Livestock Diseases and Management

This book series discusses the various infectious diseases affecting the livestock, principle of the disease control, and specific disease management. It discusses the existing strategies to control infectious disease includes animal management programs, vaccination, targeted antimicrobial use, and food hygiene.

Despite public health and veterinary public health improvement within the last century, animal populations remain vulnerable to health threats caused by infectious diseases.

It reviews the current understanding of the zoonotic, emerging, and transboundary animal infections in relation to their transmission, epidemiology, clinical and pathological effects, diagnosis and treatment.

It also examines the recent advancements in the veterinary diagnostics including the existing capabilities, constraints, opportunities, and future potentials. In addition, it elaborates on the conventional and recombinant vaccines that are used in the veterinary medicines and the molecular approaches that have led to the development of new vaccines in recent years. A volume focusing on the various water- and food-borne diseases and its impact on the domestic animals is also a part of this series.

The book series examines the emergence of antimicrobial resistance in livestock, ongoing global surveillance, and monitoring program, its impact on the animal-human interface and strategies for combating resistance.

Minakshi Prasad • Rajesh Kumar •
Mayukh Ghosh • Shafiq M. Syed •
Soumendu Chakravarti
Editors

Nanotechnology Theranostics in Livestock Diseases and Management

Springer

Editors
Minakshi Prasad
Department of Animal Biotechnology
LLR University of Veterinary and Animal,
ICAR-National Research Centre on Equines
Hisar, Haryana, India

Rajesh Kumar
Department of Veterinary Physiology and Biochemistry
LLR University of Veterinary and Animal
Hisar, Haryana, India

Mayukh Ghosh
Department of Veterinary Physiology and Biochemistry
Banaras Hindu University
Mirzapur, Uttar Pradesh, India

Shafiq M. Syed
Biomedical Sciences and Pharmacy
University of Newcastle Australia
Callaghan, NSW, Australia

Soumendu Chakravarti
Pirbright Institute
Pirbright, UK

ISSN 2662-4346 ISSN 2662-4354 (electronic)
Livestock Diseases and Management
ISBN 978-981-16-1609-9 ISBN 978-981-16-1610-5 (eBook)
https://doi.org/10.1007/978-981-16-1610-5

This Springer imprint is published by the registered company Springer Nature Singapore Pte Ltd.
The registered company address is: 152 Beach Road, #21-01/04 Gateway East, Singapore 189721, Singapore

If disposing of this product, please recycle the paper.

Preface

Welcome to the world of nanotechnology theranostics in livestock diseases and management. Over the past few decades, the field of nanotechnology has witnessed tremendous advancements, revolutionizing various sectors, including healthcare and agriculture. Nanotechnology offers unprecedented opportunities to develop novel diagnostic and therapeutic approaches, and its application in the livestock industry holds great promise for improving animal health, welfare, and productivity.

This book is a comprehensive guide that explores the potential of nanotechnology in addressing the challenges faced in livestock diseases and management. It aims to provide researchers, veterinarians, and professionals in the field with a deeper understanding of the principles, methodologies, and applications of nanotechnology theranostics.

The term "theranostics" represents the convergence of therapeutics and diagnostics, emphasizing the integrated approach of using nanotechnology-based tools for both treatment and diagnosis. In the context of livestock diseases and management, nanotechnology theranostics offer multifunctional platforms that can simultaneously detect, treat, and monitor various conditions. These advanced technologies can revolutionize disease prevention, enhance treatment efficacy, minimize antimicrobial resistance, and improve overall animal health outcomes.

This book is divided into several sections, each focusing on a specific aspect of nanotechnology theranostics in livestock diseases and management. It begins with an introduction to the fundamental concepts of nanotechnology, including the synthesis and characterization of nanomaterials, as well as their interactions with biological systems. Building upon this foundation, subsequent chapters delve into the applications of nanotechnology in livestock disease diagnosis, such as the development of nanosensors and point-of-care devices for rapid and accurate detection.

Furthermore, the book explores the therapeutic applications of nanotechnology in livestock, encompassing targeted drug delivery systems, antimicrobial nanomaterials, and nanovaccines. These nanoscale interventions have the potential to revolutionize treatment strategies, enabling precise and effective delivery of therapeutics while minimizing side effects and improving animal welfare.

In addition to diagnostics and therapeutics, the book also highlights the role of nanotechnology in disease monitoring and management. It explores the use of nanoscale biosensors for continuous real-time monitoring of livestock health parameters, as well as the potential of nanomaterials in bioimaging techniques to visualize disease progression and assess treatment outcomes.

Throughout this book, we have assembled contributions from leading experts in the field who share their knowledge, expertise, and cutting-edge research findings. Their collective insights provide a comprehensive overview of the current state-of-the-art in nanotechnology theranostics in livestock diseases and management.

As editors, our goal is to bridge the gap between nanotechnology and livestock health, promoting interdisciplinary collaboration and inspiring further research in this exciting field. We hope that this book serves as a valuable resource for scientists, veterinarians, policymakers, and anyone interested in the application of nanotechnology for the improvement of livestock health and management.

We would like to express our gratitude to all the contributors for their outstanding work and dedication. We also extend our appreciation to the editorial team for their support in bringing this book to fruition. Finally, we thank the readers for their interest in this subject matter and their commitment to advancing the field of nanotechnology theranostics in livestock diseases and management.

Together, let us embark on this journey of exploration and innovation, uncovering the immense potential of nanotechnology in transforming the livestock industry for a healthier and more sustainable future.

Hisar, Haryana, India — Minakshi Prasad
Hisar, Haryana, India — Rajesh Kumar
Mirzapur, Uttar Pradesh, India — Mayukh Ghosh
Callaghan, NSW, Australia — Shafiq M. Syed
Pirbright, UK — Soumendu Chakravarti

Contents

Editors and Contributors

About the Editors

Minakshi Prasad is a Professor in the Department of Animal Biotechnology, LLR University of Veterinary and Animal Sciences (LUVAS), India. Her research interest is in Bioinformatics and Biotechnology with specialization in viruses; molecular characterization, diagnosis and vaccine design, study of genome analysis and diversity among viruses. She has been conferred with various prestigious awards notably, International Lifetime Achievement Award 2019 by PISRF(I) in Thailand; International Research Ratna Award, Malaysia. She has more than 20 years of teaching and research experience in Microbiology and Biotechnology, Molecular Diagnosis, Forensics, Vaccinology, Comparative Genomics, and Bioinformatics. She has also published more than 70 research articles in the peer-reviewed international journal and authored or co-authored books and book chapters. She is a member of many international scientific societies and organizations importantly, the National Academy of Agricultural Sciences, India; the National Academy of Veterinary Sciences, International Academy of Biosciences, UK.

Rajesh Kumar is working as Scientist in the Department of Veterinary Physiology and Biochemistry, Lala Lajpat Rai University of Veterinary and Animal Sciences, Hisar, Haryana, India. Earlier, he was engaged as an Assistant Professor, Kerala Veterinary and Animal Science University, Kerala, India. His research interests are in Reproductive Physiology, Environmental Physiology, Proteomics and Metabolomics, Nanotechnology, Antioxidant and Redox Biology. Dr. Kumar has to his credit a patent for the development of diagnosis method of anoestrous animals which have genetic linkage. He has been awarded with several prestigious fellowships and worked as a referee for several prestigious International journals. He has authored or co-authored over 40 research articles in peer-reviewed international journals, several books, and book chapters. He is a member of different national and international scientific societies and organizations.

Mayukh Ghosh is an Assistant Professor in the Department of Veterinary Physiology and Biochemistry, RGSC, Banaras Hindu University, UP, India. His research interests are in Proteomics and Metabolomics, Nanotechnology, Reproduction, Molecular Parasitology, Antioxidant and Redox Biology. He has authored or co-authored more than 30 research articles in the peer-reviewed international journals and several books and book chapters. He is a member of several national and international scientific societies and organizations.

Shafiq M. Syed is currently a Lecturer at the University of Newcastle. His research work is focused on defining the molecular pathways involved in the pathogenesis of gynaecological cancers, for which he has developed several models including cancer patient-derived xenograft and organoid models, and genetically modified mouse models of cancer progression and metastasis. He has published more than 30 research articles and book chapters. He is a member of the Hunter Cancer Research Alliance (HCRA) and Australian Society of Reproductive Biology.

Soumendu Chakravarti is a Scientist (Sr. Scale) in ICAR-Indian Veterinary Research Institute, Izatnagar, India and has more than 12 years of research and teaching experience in Molecular Virology. He has been awarded ICAR International Fellowship and is currently on deputation as a Visiting Scientist, The Pirbright Institute, UK. He is currently working on a project to identify host-restriction factors mediated by Interferon-stimulated genes (ISGs) in virus infections and factors governing species-specific host restriction against viral diseases at the Pirbright Institute. He has published more than 50 research and review articles in peer-reviewed journals. He is a reviewer of several peer-reviewed journals and is Life member of several scientific societies such as Society of General Microbiology (UK), Indian Science Congress Association, Indian Society for Veterinary Immunology and Biotechnology, Veterinary Council of India.

Contributors

Akshita Department of Genetics and Plant Breeding, Institute of Agricultural Science, Rajiv Gandhi South Campus, Banaras Hindu University, Mirzapur, Uttar Pradesh, India
Department of Plant Biotechnology, Institute of Agricultural Sciences, RGSC, Banaras Hindu University, Varanasi, Uttar Pradesh, India
Department of Plant Biotechnology, Institute of Agricultural Sciences, Rajiv Gandhi South Campus, Banaras Hindu University, Mirzapur, Uttar Pradesh, India

Neeraj Arora Department of Veterinary Surgery and Radiology, College of Veterinary Sciences, Lala Lajpat Rai University of Veterinary and Animal Sciences, Hisar, Haryana, India

Somesh Banerjee Department of Veterinary Microbiology, Immunology Section, Lala Lajpat Rai University of Veterinary and Animal Sciences, Hisar, Haryana, India

Subhasis Batabyal Department of Veterinary Biochemistry, Faculty of Veterinary and Animal Sciences, West Bengal University of Animal and Fishery Sciences, Kolkata, West Bengal, India

Arpita Bharali Department of Animal Biotechnology, College of Veterinary Science, Assam Agricultural University, Khanapara, Guwahati, Assam, India

Suman Biswas Department of Avian Sciences, Faculty of Veterinary and Animal Sciences, West Bengal University of Animal and Fishery Sciences, Kolkata, West Bengal, India

Swaraj Biswas Department of Veterinary Biochemistry, Faculty of Veterinary and Animal Sciences, West Bengal University of Animal and Fishery Sciences, Kolkata, West Bengal, India

Lukumoni Buragohain Department of Animal Biotechnology, College of Veterinary Science, Assam Agricultural University, Khanapara, Guwahati, Assam, India

Soumendu Chakravarti The Pirbright Institute, Woking, UK

Alonkrita Chowdhury Department of Plant Biotechnology, Institute of Agricultural Sciences, RGSC, Banaras Hindu University, Mirzapur, Uttar Pradesh, India
Department of Genetics and Plant Breeding, Institute of Agricultural Science, Rajiv Gandhi South Campus, Banaras Hindu University, Mirzapur, Uttar Pradesh, India
Department of Plant Biotechnology, Institute of Agricultural Sciences, RGSC, Banaras Hindu University, Varanasi, Uttar Pradesh, India

Dhruv N. Desai Department of Pathobiology, School of Veterinary Medicine, University of Pennsylvania, Philadelphia, PA, USA

Mayukh Ghosh Department of Veterinary Physiology and Biochemistry, Faculty of Veterinary and Animal Sciences, Institute of Agricultural Sciences, Rajiv Gandhi South Campus, Banaras Hindu University, Mirzapur, Uttar Pradesh, India
Department of Veterinary Physiology and Biochemistry, RGSC, Banaras Hindu University, Varanasi, Uttar Pradesh, India

Kajal Kumar Jadav Animal Biotechnology Centre, Nanaji Deshmukh Veterinary Sciences University, Jabalpur, Madhya Pradesh, India

Shobha Jawre School of Wildlife Forensic and Health, Nanaji Deshmukh Veterinary Sciences University, Jabalpur, Madhya Pradesh, India

Shyam Sundar Kesh Department of Veterinary Clinical Complex, Faculty of Veterinary and Animal Sciences, West Bengal University of Animal and Fishery Sciences, Kolkata, West Bengal, India

Alla Yaswanth Naveen Kumar Department of Genetics and Plant Breeding, Institute of Agricultural Science, Rajiv Gandhi South Campus, Banaras Hindu University, Mirzapur, Uttar Pradesh, India
Department of Veterinary Physiology and Biochemistry, Faculty of Veterinary and Animal Sciences, Rajiv Gandhi South Campus, Banaras Hindu University, Mirzapur, Uttar Pradesh, India

Anil Kumar Department of Agricultural Meteorology, Chaudhary Charan Singh Haryana Agricultural University, Hisar, Haryana, India

Ankit Kumar Haryana Pashu Vigyan Kendra, Lala Lajpat Rai University of Veterinary and Animal Sciences (LUVAS), Hisar, Haryana, India

Komal Kumari Department of Veterinary Physiology and Biochemistry, Faculty of Veterinary and Animal Sciences, Institute of Agricultural Sciences, Rajiv Gandhi South Campus, Banaras Hindu University, Mirzapur, Uttar Pradesh, India

Nitin Kumar Centre for Medical Biotechnology, Maharshi Dayanand University, Rohtak, Haryana, India

Rajesh Kumar Department of Veterinary Physiology and Biochemistry, College of Veterinary Sciences, Lala Lajpat Rai University of Veterinary and Animal Sciences, Hisar, Haryana, India
Department of Genetics and Plant Breeding, Mirzapur, Uttar Pradesh, India

Sandeep Kumar Department of Veterinary Surgery and Radiology, Lala Lajpat Rai University of Veterinary and Animal Sciences, Hisar, Haryana, India

Subodh Kumar Centre for Medical Biotechnology, Maharshi Dayanand University, Rohtak, Haryana, India

Sushil Kumar Department of Biosciences and Bioengineering, Indian Institute of Technology Guwahati, Guwahati, Assam, India

Tarun Kumar Veterinary Medicine, Lala Lajpat Rai University of Veterinary and Animal Sciences (LUVAS), Hisar, Haryana, India

Anu Kurian Department of Genetics and Plant Breeding, Institute of Agricultural Science, Rajiv Gandhi South Campus, Banaras Hindu University, Mirzapur, Uttar Pradesh, India

Neeti Lakhani Department of Animal Nutrition, College of Veterinary and Animal Sciences, Guru Angad Dev Veterinary and Animal Sciences University, Rampuraphul, Punjab, India

Preeti Lakhani Department of Veterinary Physiology and Biochemistry, Lala Lajpat Rai University of Veterinary and Animal Sciences, Hisar, Haryana, India
Department of Genetics and Plant Breeding, Mirzapur, Uttar Pradesh, India

Apratim Maity Department of Veterinary Biochemistry, Faculty of Veterinary and Animal Sciences, West Bengal University of Animal and Fishery Sciences, Kolkata, West Bengal, India

H. N. Malik Odisha University of Agriculture and Technology, Odisha, India

Dodla Mamatha Department of Genetics and Plant Breeding, Institute of Agricultural Science, Rajiv Gandhi South Campus, Banaras Hindu University, Mirzapur, Uttar Pradesh, India
Department of Plant Biotechnology, Institute of Agricultural Sciences, RGSC, Banaras Hindu University, Varanasi, Uttar Pradesh, India

Manjeet Chahar Centre for Medical Biotechnology, Maharshi Dayanand University, Rohtak, Haryana, India

Kratika Maurya Department of Genetics and Plant Breeding, Institute of Agricultural Science, Rajiv Gandhi South Campus, Banaras Hindu University, Mirzapur, Uttar Pradesh, India
Department of Plant Biotechnology, Institute of Agricultural Sciences, RGSC, Banaras Hindu University, Varanasi, Uttar Pradesh, India
Department of Plant Biotechnology, Institute of Agricultural Sciences, Rajiv Gandhi South Campus, Banaras Hindu University, Mirzapur, Uttar Pradesh, India

Saumya Maurya Department of Genetics and Plant Breeding, Institute of Agricultural Science, Rajiv Gandhi South Campus, Banaras Hindu University, Mirzapur, Uttar Pradesh, India

Anamika Mazumdar Department of Animal Biotechnology, College of Veterinary Science, Assam Agricultural University, Khanapara, Guwahati, Assam, India

Bhabesh Mili Department of Veterinary Physiology and Biochemistry, College ofVeterinary Sciences and Animal Husbandry, CAU (I), Jalukie, Peren, Nagaland, India

Hari Mohan Centre for Medical Biotechnology, Maharshi Dayanand University, Rohtak, Haryana, India

S. K. Mohapatra Odisha University of Agriculture and Technology, Odisha, India

Nikhil Mudgil LalaLajpat Rai University of Veterinary and Animal Sciences (LUVAS), Hisar, Haryana, India

Nidhi Centre for Medical Biotechnology, Maharshi Dayanand University, Rohtak, Haryana, India

Harshad Sudhir Patki Department of Veterinary Anatomy and Histology, College of Veterinary and Animal Sciences, Kerala Veterinary and Animal Science University, Pookode, Wayanad, Kerala, India

Shamik Polley Department of Veterinary Biochemistry, Faculty of Veterinary and Animal Sciences, West Bengal University of Animal and Fishery Sciences, Kolkata, West Bengal, India

Minakshi Prasad Department of Animal Biotechnology, Lala Lajpat Rai University of Veterinary and Animal Sciences, Hisar, Haryana, India
National Research Center on Equines, Hisar, Haryana, India

Ravina Centre for Medical Biotechnology, Maharshi Dayanand University, Rohtak, Haryana, India

Harsh Kumar Rawat Centre for Medical Biotechnology, Maharshi Dayanand University, Rohtak, Haryana, India

Manoj Kumar Rose Department of Veterinary Physiology and Biochemistry, COVS, Lala Lajpat Rai University of Veterinary and Animal Sciences, Hisar, Haryana, India
Department of Animal Nutrition, College of Veterinary and Animal Sciences, Guru Angad Dev Veterinary and Animal Sciences University, Rampuraphul, Punjab, India

Sandeep Saharan Department of Veterinary Surgery and Radiology, Lala Lajpat Rai University of Veterinary and Animal Sciences, Hisar, Haryana, India

Tarun Saini Centre for Medical Biotechnology, Maharshi Dayanand University, Rohtak, Haryana, India

Krishna Kant Sharma Department of Microbiology, Maharshi Dayanand University, Rohtak, Haryana, India
Centre for Medical Biotechnology, Maharshi Dayanand University, Rohtak, Haryana, India

Neelesh Sindhu Veterinary Medicine, Lala Lajpat Rai University of Veterinary and Animal Sciences (LUVAS), Hisar, Haryana, India
Department of Veterinary Clinical Complex, Lala Lajpat Rai University of Veterinary and Animal Sciences, Hisar, Haryana, India

Sonia Sindhu Department of Veterinary Physiology and Biochemistry, Lala Lajpat Rai University of Veterinary and Animal, Hisar, Haryana, India
Department of Animal Nutrition, College of Veterinary and Animal Sciences, Guru Angad Dev Veterinary and Animal Sciences University, Rampuraphul, Punjab, India

Subodh Soni Centre for Medical Biotechnology, Maharshi Dayanand University, Rohtak, Haryana, India

Manjeet Subodh Soni Centre for Medical Biotechnology, Maharshi Dayanand University, Rohtak, Haryana, India

Sourabh Sulabh Department of Animal Science, Kazi Nazrul University, Asansol, West Bengal, India

S. Suvetha Department of Genetics and Plant Breeding, Institute of Agricultural Science, Rajiv Gandhi South Campus, Banaras Hindu University, Mirzapur, Uttar Pradesh, India

Kumari Swati Department of Genetics and Plant Breeding, Institute of Agricultural Science, Rajiv Gandhi South Campus, Banaras Hindu University, Mirzapur, Uttar Pradesh, India

Shafiq M. Syed Biomedical Sciences and Pharmacy, University of Newcastle Australia, Callaghan, NSW, Australia

Deepak Kumar Tiwari Department of Veterinary Surgery and Radiology, College of Veterinary Sciences, Lala Lajpat Rai University of Veterinary and Animal Sciences, Hisar, Haryana, India

Rajat Varshney Department of Veterinary Microbiology, Faculty of Veterinary and Animal Sciences, Rajiv Gandhi South Campus, Banaras Hindu University, Mirzapur, Uttar Pradesh, India

Ritu Varshney Biological Engineering, Indian Institute of Technology, Gandhinagar, Gujarat, India

Annu Yadav Veterinary Medicine, Lala Lajpat Rai University of Veterinary and Animal Sciences (LUVAS), Hisar, Haryana, India

Neelam Yadav Department of Biotechnology, Deenbandhu Chotu Ram University of Science and Technology, Sonepat, Haryana, India

Pavan Kumar Yadav Department of Veterinary Physiology and Biochemistry, Faculty of Veterinary and Animal Sciences, Rajiv Gandhi South Campus, Banaras Hindu University, Mirzapur, Uttar Pradesh, India

Pooja Yadav University Institute of Engineering and Technology, Maharshi Dayanand University, Rohtak, Haryana, India

Chapter 1
Introduction and Scope of Nanotechnology in Veterinary Science

Mayukh Ghosh, Alonkrita Chowdhury, Rajesh Kumar, Shafiq M. Syed, Soumendu Chakravarti, and Minakshi Prasad

Abstract Nanotechnology offers several potential applications in veterinary science, enhancing various aspects of animal healthcare, diagnostics, and treatment. As drug delivery systems, nanoparticles can be designed to encapsulate drugs and target specific tissues or cells, allowing for controlled release and improved drug efficacy. Nanoscale sensors and imaging techniques can enable early detection and accurate diagnosis of diseases in animals. Nanosensors can detect biomarkers associated with specific diseases, providing rapid and sensitive diagnostic tools. Nanoparticles can also be used as contrast agents in imaging techniques such as magnetic resonance imaging (MRI) or computed tomography (CT) scans, enhancing the visualization of internal organs and structures. Nanomaterials can promote wound healing and tissue regeneration in animals. Nanofiber scaffolds and dressings can create a favorable environment for cell growth and tissue regeneration. Nanoparticles can also be used to deliver growth factors and therapeutic agents directly to the injured tissues,

M. Ghosh (✉)
Department of Veterinary Physiology and Biochemistry, Banaras Hindu University, Mirzapur, Uttar Pradesh, India

A. Chowdhury
Department of Plant Biotechnology, Institute of Agricultural Sciences, RGSC, Banaras Hindu University, Varanasi, Uttar Pradesh, India

R. Kumar
Department of Veterinary Physiology and Biochemistry, Lala Lajpat Rai University of Veterinary and Animal Sciences, Hisar, Haryana, India

S. M. Syed
Biomedical Sciences and Pharmacy, University of Newcastle Australia, Callaghan, NSW, Australia

S. Chakravarti
The Pirbright Institute, Woking, UK

M. Prasad
Department of Animal Biotechnology, Lala Lajpat Rai University of Veterinary and Animal Sciences, Hisar, Haryana, India

LLR University of Veterinary and Animal, ICAR-National Research Centre on Equines, *Nanotechnology Theranostics in Livestock Diseases and Management*, Livestock Diseases and Management,
https://doi.org/10.1007/978-981-16-1610-5_1

enhancing the healing process. Nanoparticles with antimicrobial properties can combat drug-resistant bacteria and reduce the need for antibiotics in veterinary medicine. These nanoparticles can disrupt the bacterial cell membrane or inhibit their growth, providing an alternative approach to controlling infections. Nanotechnology can contribute to improving animal nutrition and ensuring food safety. Nanoencapsulation of nutrients can enhance their bioavailability and stability, allowing for efficient nutrient delivery to animals. Nanosensors can be employed to detect contaminants and toxins in animal feed, ensuring the safety and quality of the food supply. Nanoparticles can be used as carriers for vaccines, improving their stability and delivery. Nano-formulated vaccines can enhance immune responses and provide prolonged protection, reducing the frequency of vaccinations and improving animal health. It is important to note that while nanotechnology shows promising potential in veterinary science, many of these applications are still in the research and development stage.

Keywords Nanodiagnostics · Nanosensors · Nano-drug delivery system · Nano-vaccines · Animal health · Food safety

1.1 Introduction

Nanotechnology, a multidisciplinary field that deals with the manipulation and control of matter at the nanoscale, has the potential to revolutionize various industries, including veterinary science. By harnessing the unique properties exhibited by materials at the nanoscale, researchers and veterinarians can develop novel approaches for disease diagnosis, drug delivery, tissue engineering, and improved animal health management (Minakshi et al. 2020; Prasad et al. 2021).

The scope of nanotechnology in veterinary science includes but not limited to disease diagnosis, drug and vaccine delivery, antimicrobial applications, wound healing and tissue engineering, fracture management, parasite and vector control, food safety, imaging, and monitoring. Nanotechnology offers innovative tools for early detection and accurate diagnosis of diseases in animals. Nanosensors can be designed to detect specific biomarkers, pathogens, or toxins, enabling rapid and sensitive diagnostic tests. These nanosensors can be integrated into portable devices, providing veterinarians with on-site diagnostic capabilities. Nanoparticles (NPs) can be used as carriers for targeted drug and vaccine delivery in veterinary medicine. These nanoparticles can encapsulate drugs and release them in a controlled manner, improving therapeutic outcomes and minimizing side effects. Additionally, nanocarriers can enhance the stability and bioavailability of drugs, allowing for reduced dosages and extended drug release profiles. Antibiotic resistance is a significant concern in veterinary medicine. Nanoparticles with antimicrobial properties can combat bacterial infections, offering an alternative to conventional antibiotics. These nanoparticles can disrupt the bacterial cell membrane or inhibit essential cellular processes, making them effective against antibiotic-resistant strains (Minakshi et al. 2019, 2020; Prasad et al. 2021). Nanotechnology-based approaches can accelerate wound healing in animals. Nanofiber scaffolds and hydrogels can be used to create three-dimensional structures that mimic the extracellular matrix (ECM), promoting cell growth and tissue regeneration (Seyyed Tabaei et al. 2020). Furthermore, nanoparticles can be loaded with growth factors or stem cells to enhance tissue repair and regeneration (Wang et al. 2019; Gainza et al. 2014). Nanoparticles can be utilized for controlling and eliminating parasites in animals. Nanoemulsions and nanosprays can deliver anti-parasitic agents to target parasites, improving efficacy and reducing the need for frequent administration (Sun et al. 2019). Additionally, nanoparticles can be used for developing novel vaccines against parasitic diseases (Ottino et al. 2022). Nanotechnology can contribute to enhancing food safety in animal husbandry. Nanosensors can detect contaminants, such as heavy metals or mycotoxins, in animal feed or water sources (Kumar et al. 2020). Nanomaterials can also be used to develop antimicrobial coatings for food storage containers, reducing the risk of foodborne illnesses (Kuswandi 2017). Nanoparticles can enhance imaging techniques, such as magnetic resonance imaging (MRI) and computed tomography (CT), in veterinary diagnostics (Aslan et al. 2020; Naseri et al. 2017). These nanoparticles can be designed to specifically target tissues or cells of interest, providing detailed imaging and monitoring capabilities.

The scope of nanotechnology in veterinary science is continuously expanding as researchers explore new applications and technologies. While there are still challenges to overcome, such as ensuring the safety and regulatory aspects of nanomaterials, nanotechnology holds immense potential for improving animal health, disease management, and overall veterinary care.

1.2 Scope of Nanotechnology in Veterinary Disease Diagnosis

The scope of nanotechnology in veterinary disease diagnosis is significant and holds great promise for improving the accuracy, speed, and efficiency of diagnosing diseases in animals. Some specific areas where nanotechnology can contribute to disease diagnosis in veterinary medicine have been delineated hereunder.

Nanosensors can be developed to detect specific biomarkers associated with various diseases in animals. These sensors can be designed to detect proteins, nucleic acids, or metabolites indicative of a particular disease condition. By utilizing nanomaterials, such as carbon nanotubes or nanoparticles, these biosensors can provide highly sensitive and specific detection, enabling early and accurate diagnosis. A novel gold nanoparticles (AuNPs) biosensor has been customized using thiol-linked oligonucleotides that detect the conserved 3D gene of foot-and-mouth disease virus (FMDV). The sensitivity of the biosensor was reported to be 10 copy number RNA standards in RT-PCR and 1 copy number RNA standard in real-time rRT-PCR yielding 94.5% efficiency and 100% specificity (Hamdy et al. 2018).

Nanotechnology-based diagnostic devices can be created for point-of-care testing in veterinary clinics or field settings. These portable devices can integrate nanosensors and microfluidics to analyze small samples of blood, urine, or other body fluids. They can provide rapid and on-site diagnosis, reducing the need for sending samples to a laboratory and enabling immediate treatment decisions. Black carbon nanoparticles have been used for point-of-care detection of mastitis pathogens on a paper-based microarray format in less than 3 h (Mujawar et al. 2013).

Nanoparticles can be engineered as imaging agents for various imaging modalities used in veterinary diagnostics, such as magnetic resonance imaging (MRI), computed tomography (CT), or ultrasound. These nanoparticles can enhance contrast, improve signal-to-noise ratios, and target specific tissues or cells of interest. This allows for better visualization of anatomical structures, identification of abnormalities, and tracking of disease progression in real time. For instance, gadolinium-based contrast agents are most frequently used for MRI, but their toxicity induces severe side effects raising alarm from the Food and Drug Administration (FDA). Iron oxide nanoparticles (IONPs) have emerged as a promising alternative and offer good biocompatibility and magnetic properties. In addition, their ease of surface functionalization with diverse targeting ligands such as aptamers, antibodies, peptides, and sugars facilitates the scope of molecular MRI. Evidently, IONPs have found significant interest in cancer imaging recently (Avasthi et al. 2020; Jeon et al. 2021).

Nanotechnology can aid in the rapid and accurate detection of pathogens causing infectious diseases in animals. Nanoparticles or nanomaterial-based assays can be developed to detect viral, bacterial, or fungal pathogens in clinical samples. These assays can provide sensitive, specific, and rapid identification of pathogens, enabling timely intervention and disease control measures. Gold NPs (AuNPs), silver NPs (Ag NPs), magnetic nanoparticles (MNPs), graphene and carbon nanotubes (CNTs), quantum dots (QDs), etc., have been frequently used in various pathogen detection modalities. The nanoscale size, unique optical, electrical, and magnetic properties, ample surface modification properties, etc., have rendered these materials suitable for rapid and enhanced pathogen detection with multiplexing options. Several biosensors and point-of-care devices have been customized using these novel materials to detect various pathogens of veterinary importance (Minakshi et al. 2020; Prasad et al. 2021).

Nanotechnology offers not only animal-side diagnosis but also advanced tools for molecular diagnostics in veterinary medicine. Nanoparticles can be functionalized with probes, such as DNA or RNA sequences, to detect specific genetic variations or mutations associated with inherited diseases or cancer in animals. This enables the identification of genetic markers and personalized treatment strategies for affected animals. Customization of fluorescence-based biosensors for cancer diagnostics frequently employs QDs, polymer dots (PDs), upconversion nanoparticles (UCNPs), and AuNPs for their enhanced optical properties. Antibodies against cancer antigens or complementary nucleic acid probes against cancer genes can be functionalized with these NPs for early and sensitive detection of cancer biomarkers (Chinen et al. 2015; Brar et al. 2021).

Microfluidics, integrated with nanotechnology, allows for miniaturized diagnostic platforms that can handle small sample volumes efficiently. These devices can perform multiple diagnostic tests simultaneously, utilizing nanosensors or nanoparticles for various detection methods. Microfluidic devices offer automation, portability, and cost-effectiveness, making them suitable for veterinary diagnostics. Microfluidic devices can automate and streamline sample preparation steps, such as sample enrichment and DNA extraction, which are crucial for pathogen detection. For bacterial detection, microfluidics can provide controlled environments for bacterial culture and growth, allowing rapid detection. These systems can create optimal growth conditions, monitor bacterial growth in real time, and provide a means for continuous monitoring of bacterial populations. Microfluidic immunoassays, such as enzyme-linked immunosorbent assays (ELISAs) or immunomagnetic separation, can be implemented for pathogen detection. Microfluidic channels and chambers can be functionalized with antibodies or capture probes specific to the target pathogen, facilitating the capture, concentration, and detection of the pathogen from complex samples. Microfluidic devices can be utilized for polymerase chain reaction (PCR) or loop-mediated isothermal amplification (LAMP) to amplify the nucleic acids from the target organism. These amplification techniques allow for highly sensitive and specific detection of the pathogen. Integration of amplification and detection steps on a single microfluidic chip enables rapid and on-site analysis. Microfluidics offers the opportunity to develop lab-on-a-chip systems for complete pathogen detection workflows. These systems can integrate sample preparation, amplification, and detection modules onto a single chip, reducing the time, cost, and complexity of the analysis (Verma et al. 2022).

Nanotechnology-based approaches can enable non-invasive diagnostic methods in veterinary medicine. For example, NPs can be engineered to cross the blood-brain barrier and act as contrast agents for imaging brain tumors or neurological conditions (Faiz et al. 2022; Zottel et al. 2019). Non-invasive techniques reduce the need for invasive procedures, minimize patient stress, and enhance animal welfare.

The scope of nanotechnology in veterinary disease diagnosis is continuously evolving, driven by ongoing research and technological advancements. By leveraging nanotechnology, veterinary professionals can achieve more accurate and timely diagnoses, leading to improved animal health outcomes and better disease management.

1.3 Nanotechnology in Drug Delivery

Nanotechnology has the potential to revolutionize drug delivery in veterinary science by enhancing the efficacy, safety, and targeted delivery of therapeutic agents. The applications of nanotechnology in drug delivery for veterinary medicine are described hereunder.

NPs can be designed to deliver drugs specifically to the target tissues or cells in animals. Functionalized nanoparticles can be engineered to recognize and bind to specific receptors or markers present on diseased cells, enabling targeted delivery of drugs to the affected site. This approach minimizes off-target effects, reduces systemic toxicity, and enhances therapeutic efficacy. Nanoparticles can be used to encapsulate drugs and provide controlled-release profiles. These nanoparticles can release drugs in a sustained and controlled manner over an extended period, improving drug stability and reducing the need for frequent dosing. Controlled-release systems ensure a consistent drug concentration at the target site, optimizing therapeutic outcomes. For instance, liposomes provide a good platform for in vivo delivery of several anti-tumor drugs, such as doxorubicin and paclitaxel, as well as nucleic acids for cancer therapy (Yao et al. 2020).

NPs also offer enhanced bioavailability of drugs in veterinary medicine. By encapsulating poorly soluble drugs or unstable compounds within nanoparticles, their solubility and stability can be enhanced. This allows for improved drug absorption and distribution in the animal's body, resulting in better therapeutic response (Gunasekaran et al. 2014). For instance, ellagic acid, a natural molecule that has potent anticancer potential, suffers from high hydrophobicity and poor gastrointestinal tract (GIT) absorption. Encapsulation of ellagic acid in polycaprolactone nanoparticles has found to increase its oral bioavailability and activity (Mady and Shaker 2017).

Sensitive drugs or biologics can also be protected from degradation or premature metabolism through NP implementation. They can shield the active ingredients from enzymatic degradation in the gastrointestinal tract or rapid clearance in the bloodstream. This protective effect enhances the stability and bioactivity of drugs, ensuring their efficacy upon administration. Nanoparticles can enhance the oral bioavailability of peptide drugs by protecting them from proteolytic enzymes (Li et al. 2022).

Nanotechnology enables the development of combination therapies by co-encapsulating multiple drugs or therapeutic agents within nanoparticles. This approach allows for synergistic effects, where different drugs work together to enhance therapeutic outcomes. Combination therapies can be particularly beneficial for treating complex or multi-factorial diseases in animals (Gurunathan et al. 2018).

NPs can serve as carriers for vaccines in veterinary medicine. These nanoparticles can enhance the delivery and presentation of vaccine antigens to the immune system, improving the immune response and the duration of immunity. Nanoparticle-based vaccines can elicit stronger and longer-lasting immune responses, reducing the frequency of vaccinations in animals (Hajiaghapour Asr et al. 2023).

NPs can be utilized for targeted and localized drug delivery in veterinary cancer treatment. They can selectively accumulate in tumor tissues through passive or active targeting mechanisms, delivering anticancer drugs directly to the tumor cells while minimizing toxicity to healthy tissues. This approach improves the effectiveness of chemotherapy while reducing adverse effects (Zabielska-Koczywąs and Lechowski 2017).

Nanotechnology enables the development of specialized nanomedicines tailored for veterinary use (Hua et al. 2018). These nanomedicines can be formulated as injectable nanoparticles, oral formulations, topical gels, or implantable devices, depending on the specific therapeutic requirements. Veterinary nanomedicines offer improved treatment options for a wide range of diseases and conditions in animals (Patra et al. 2018).

Nanotechnology-based drug delivery systems in veterinary medicine are still in the early stages of development, but they hold immense potential for improving treatment outcomes, reducing side effects, and enhancing animal health and welfare. Continued research and development in this field can lead to the translation of these innovations into practical and effective veterinary therapeutics.

1.4 Nanotechnology in Antibacterial Applications

Nanotechnology offers promising solutions for combating bacterial infections in veterinary science by providing effective antibacterial applications.

Some NPs, such as AgNPs, ZnONPs, or CuNPs, possess inherent antimicrobial properties (Prasad et al. 2021). Hence, they can directly interact with bacterial cells, disrupting their membranes or interfering with essential cellular processes. They exhibit broad-spectrum activity against bacteria, including antibiotic-resistant strains, making them valuable tools in veterinary antibacterial applications (Minakshi et al. 2019). It also facilitates the development of targeted and controlled-release antibacterial drug delivery systems including liposomes, nanoemulsions, solid lipid nanoparticles (SLNs), polymeric NPs, and metal NPs (MNPs). These antibiotics or antimicrobial agent-encapsulated NPs enable the controlled release of drugs at the infection site. Such targeted delivery enhances the concentration of the drug at the site of infection, minimizing systemic exposure and reducing the development of resistance (Zong et al. 2022).

Advancement in nanotechnology has also aided in the development of antimicrobial and antiviral surface coatings with cationic polymers, metal coatings, and antifouling micro- or nanostructures. These coatings provide a physical barrier and release antimicrobial agents to prevent bacterial colonization and infection. Hence, clinical equipment, implants, or other biomedical devices functionalized with antimicrobial surface coatings can inhibit the growth of bacteria, preventing biofilm formation and reducing the risk of bacterial contamination (Erkoc 2021).

Nanotherapeutic applications also include the implementation of NPs in wound treatment. Colonization of bacteria or other microorganisms in the wound site can slow down the natural wound healing processes, causing deterioration of the wound. However, wound dressings based on antimicrobial NPs, for instance, metal or metal oxide nanoparticles, such as AgNPs, AuNPs, and ZnONPs can promote the healing process through interaction with the bacterial microenvironment eliciting a therapeutic response (Jiang and Loo 2020; Naskar and Kim 2020). Besides, several nanotechnology-based delivery systems including nanofibers, hydrogels, hydrocolloids, and nanohybrids have been evolved for wound management. Nanofiber dressings can create a protective barrier over wounds while allowing oxygen and moisture exchange. These dressings can be functionalized with antimicrobial nanoparticles or loaded with antimicrobial agents to provide sustained antibacterial activity, reducing the risk of wound infections (Ataide et al. 2022).

Furthermore, nanoparticle-functionalized fabrics or nanotextiles have been developed that are more advantageous in comparison to conventional fabrics in terms of water repellency, elasticity or tensile strength, fire retardancy, etc. Moreover, it can facilitate UV absorption, drug delivery, microbial resistance, etc. Fabrics or textiles treated with antimicrobial nanoparticles namely, AgNP, AuNP, TiO_2, ZnO, Se, SiO_2, and CuO, can inhibit bacterial growth and contamination. Hence, these textiles can be utilized in veterinary clinics, animal bedding, or protective clothing for veterinarians and animal handlers, minimizing the spread of infectious diseases (Aguda and Lateef 2022).

NP-aided novel vaccines can be engineered to display bacterial antigens, enhancing their immune recognition and immune response. These nanoscale vaccines can stimulate a stronger and more targeted immune response, improving protection against bacterial pathogens. For instance, antigen-conjugated AuNPs have been found to elicit an antibody response against *Pseudomonas aeruginosa*, *Listeria monocytogenes*, *Streptococcus pneumoniae*, and *Yersinia pestis* in the subject. Besides, several polymeric NPs, such as poly(lactic-co-glycolic acid) (PLGA) and polypropylene sulfide (PPS), are used in inducing effective immune response against *Chlamydia trachomatis*, *Staphylococcus aureus*, *Mycobacterium tuberculosis*, etc. (Chien-Wei Lin et al. 2018).

In veterinary medicine, rapid and sensitive detection of bacterial pathogens can also be achieved through nanotechnology-based approaches. Nanosensors or nanoprobes can be designed to specifically recognize and bind to bacterial antigens or genetic material, facilitating the identification of bacterial infections. For instance, *S. aureus*, *Escherichia coli*, *Cryptosporidium parvum*, *S. typhimurium*, and such other bacteria can be detected through magnetic NP (AuNPs, AgNPs, gold magnetite nanocomposites)-employed techniques, namely plasmon absorbance, fluorescence, surface-enhanced Raman (SER) scattering, LSPR signal sensing, sandwich immunoassay (electrochemical immunosensor), and dark-field microscopy imaging. These nanosensors can provide a quick and accurate diagnosis, allowing for timely treatment interventions (Hajipour et al. 2021). In addition, the employment of NPs in antibacterial applications shows promise in overcoming antibiotic resistance, improving treatment outcomes, and reducing the spread of infectious diseases in animals (Lee et al. 2019). However, it is essential to address safety considerations and potential environmental impacts associated with the use of nanomaterials in veterinary settings. Further research and development are needed to optimize the effectiveness and practicality of these nanotechnology-based antibacterial solutions in veterinary medicine.

1.5 Nanotechnology in Wound Healing and Tissue Engineering

Wound healing is a biological process comprising three stages, namely inflammatory, proliferative, and remodeling phases. The latter two phases are involved with the production of collagen and other extracellular matrices due to fibroblast activation as well as neo-angiogenesis, leading to the regeneration of tissue morphology and physiology. However, the delineated process may get interrupted due to bacterial colonization or biofilm formation, resulting in the spread of infection in the deeper layer and subsequent chronicization. *P. aeruginosa*, *E. coli*, *S. aureus*, etc., are some of the mostly detected pathogens causing severe wounds (Mihai et al. 2019).

Advancements in wound healing and tissue engineering processes can be achieved with the aid of nanotechnology. Exploiting the unique properties exhibited by nanoscale materials, several innovative strategies have been developed to stimulate tissue regeneration, promote wound healing, and improve overall tissue engineering approaches.

Nanofiber scaffolds and three-dimensional structures can be fabricated using nanotechnology with precise control over their physical and chemical properties. These scaffolds mimic the extracellular matrix (ECM) and provide a supportive framework for cell attachment, proliferation, and differentiation. Nanoscale features, such as nanotopography and nanoscale porosity, can be incorporated into the scaffolds to guide cell behavior and tissue regeneration. For example, berberine-functionalized chitosan/polyethylene oxide nanoscaffolds (CS/PEO/BBR) have shown rapid healing of the cutaneous wound caused by Leishmania major in BALB/c mice through enhanced epithelialization, leading to the speedy regeneration of damaged tissues (Seyyed Tabaei et al. 2020).

NP-aided controlled and targeted delivery of growth factors, cytokines, or other therapeutic agents to the wound or tissue site can promote wound healing and tissue regeneration. Implementation of NPs in such delivery systems can prevent the degradation of bioactive molecules enhancing their stability and provide sustained release kinetics. Several nano-drug delivery systems based on liposomes, polymeric or inorganic NPs, nanohydrogel, nanofibers, etc., have been formulated for the effective and sustainable delivery of drugs. For example, antimicrobial peptide LL_{37}-functionalized PLGA nanoparticles (PLGA-LL_{37} NPs) exhibit rapid healing of wounds caused by *E. coli* through enhanced tissue regeneration, higher collagen accumulation, re-epithelialization, and neo-angiogenesis. Epidermal growth factor (EGF)-loaded solid lipid nanoparticles (SLNs) and nanostructured lipid carriers (NLCs) have been found to accelerate wound healing with restoration of the inflammatory phase and re-epithelialization (Wang et al. 2019; Gainza et al. 2014).

Nanomaterials can also be modified to enhance cell adhesion and proliferation. Surface modifications with nanoscale features, such as nanotopography or nanoroughness, can promote cell attachment and spreading, facilitating tissue integration and regeneration. Additionally, nanomaterials can be functionalized with specific ligands or peptides that promote cell adhesion, migration, and proliferation (Chou et al. 2017).

The construction of biomimetic structures that closely resemble the natural tissue environment is also advantageous in wound healing and tissue engineering. Nanoparticles or nanofibers can be functionalized with ECM components, growth factors, or bioactive molecules to mimic the biological cues present during tissue development and repair. These biomimetic approaches provide a favorable micro-environment for cell behavior and tissue regeneration. Several dermal fibroblasts have been cultured as skin substitutes for wound healing. Examples of some nanofibrous scaffolds as dermal substitutes for tissue engineering include polycaprolactone (PCL)/gelatin, polyurethane (PU)/polyvinylidene fluoride (PVDF), poly(L-lactic acid)-co-poly(ecaprolactone) (PLACL)/gelatin, etc. (Jayarama Reddy et al. 2013).

Stem cell-based therapy can also be performed for the regeneration of tissues in the wound site. Mesenchymal stem cells derived from bone marrow exhibit rapid wound closure with proliferation of cells, re-epithelialization, and neo-angiogenesis. Nanoparticles can be used to deliver genetic material or reprogramming factors to stem cells, inducing their differentiation into specific cell lineages for tissue engineering purposes. Nanoscale cues can also guide the differentiation of stem cells toward desired tissue types. Furthermore, several polymeric nanofiber scaffolds (NFSs) have been constructed for better attachment of stem cells along with enhanced survivability and proliferation, leading to advanced tissue organization (Jayarama Reddy et al. 2013; Ma et al. 2011).

Microbial contamination during the wound healing process is a matter of concern as it can delay the healing process, making it more severe. Hence, utilizing the antimicrobial properties of nanoparticles, several nano-formulations have been evolved that facilitate anti-infectious wound treatment. NPs with inherent antimicrobial properties can be incorporated into wound dressings or scaffolds to inhibit bacterial growth and prevent infections. Metal NPs such as AgNPs, AuNPs, and CuNPs have gained attention as antimicrobial agents owing to their antibacterial properties and low toxicity. While chitosan and its derivatives are exploited as organic antibacterial compounds, metal or metal oxide-based nanomaterials such as AgNP, CuNP, AuNP, TiO_2NP, ZnONP, MgONP, CuONP, and Fe_2O_3NP can destroy biofilms through collapsing enzyme activity or the respiratory chain in bacteria (Jiang et al. 2022).

Monitoring the physiological and morphological changes of tissues at the wound site is essential for successful wound treatment. In this regard, nanotechnology-aided imaging techniques involving quantum dots or superparamagnetic NPs enable non-invasive imaging and monitoring of tissue-engineered constructs. These NPs can be employed as contrast agents for techniques like magnetic resonance imaging (MRI) or fluorescence imaging to track the distribution and integration of tissue-engineered constructs in vivo. Several nanosensors have also been constructed that provide real-time imaging of wound healing. For instance, AgNP-functionalized nanofiber-based biosensor system has been developed that facilitates enhanced sensitivity for the detection of biomarkers in the wound site due to its large surface area (Jiang et al. 2022; Sattar et al. 2019).

Therefore, nanotechnology offers tremendous potential for enhancing wound healing, tissue regeneration, and tissue engineering in veterinary medicine. Continued research and development in this field can lead to the translation of these nanotechnology-based approaches into practical and effective veterinary treatments, promoting improved healing outcomes and overall animal well-being.

1.6 Nanotechnology in Parasite Control

Nanotechnology offers promising solutions for parasite control in veterinary science by providing innovative approaches to combat and eliminate parasites that affect animals. Ways of utilizing nanotechnology in parasite control have been summarized hereunder.

The employment of nanoparticles as carriers for delivering anti-parasitic drugs to specific sites in the animal's body is the foremost approach in parasite management. It can be achieved through encapsulating the drugs within nanomaterials followed by the subsequent release in a controlled manner, ensuring the targeted delivery of the therapeutic agents. This approach enhances the efficacy of the drugs by improving their bioavailability, sustained release, and intracellular penetration capability while minimizing their systemic toxicity to the animal. Sustained release of anti-parasitic drugs over an extended period ensures a continuous and effective concentration of the drug against the parasites. Hence, it eliminates the need for frequent dosing and improves compliance with enhanced treatment outcomes. Several anti-parasitic drug delivery systems have been developed based on nanomaterials such as liposomes, polymeric NPs, SLNs, nanosuspensions, etc. (Sun et al. 2019). For example, polyethylene glycol (PEG)-functionalized liposome has been constructed as an effective delivery system for the treatment of echinococcosis. This PEGylated-liposomal formulation encapsulating albendazole facilitates the controlled release of drug at the target site for a prolonged period (Panwar et al. 2010).

Furthermore, such nano-formulations can improve the stability of anti-parasitic drugs, which may be prone to degradation or breakdown. By encapsulating the drugs within nanoparticles, their stability can be enhanced, protecting them from degradation in the harsh environment of the gastrointestinal tract or other physiological conditions. This ensures that a sufficient amount of the drug reaches the parasites and maintains its potency. For instance, anti-parasitic drugs ivermectin and praziquantel are vulnerable to enzymatic degradation, leading to its inactivation. Hence, liposome- or SLN-mediated nano-carriers enable effective transport of respective drugs to target tissues as well as enhance bioavailability (Xie et al. 2010; Gamboa et al. 2016). Arteether (ART), an antimalarial drug, is poorly water-soluble due to higher lipophilicity and thus less stable in the gastric environment, resulting in poor absorption upon oral administration. However, ART-encapsulated SLN

(ART-SLN) has shown oral bioavailability of the drug by avoiding gastric acid degradation (Dwivedi et al. 2014).

Nanoparticles can also be functionalized or engineered to specifically target parasites. By modifying the surface of the nanoparticles, they can recognize and bind to parasite-specific molecules or structures, allowing for precise parasite targeting. This targeted approach ensures that the drugs are delivered directly to the parasites, increasing their effectiveness and minimizing the impact on non-target organisms. For instance, cysteine-functionalized AuNPs offer precise targeting of *Plasmodium falciparum* through the binding of cysteine-rich proteins on its surface. Hence, it can be conjugated with antimalarial agents such as ciprofloxacin for the prevention as well as treatment of malaria (Varela-Aramburu et al. 2020).

Inhibiting the transmission of parasites via disease-carrying vectors, such as ticks, fleas, or mosquitoes, with the aid of nanomaterials is another major aspect of parasite control. Nanoparticles can be incorporated into topical treatments or insecticides, providing enhanced efficacy against these vectors. They can disrupt the life cycle of the vectors, inhibit their reproduction, or act as insecticidal agents, reducing the risk of parasite transmission to animals. Several AuNP formulations are found to possess a toxic effect on arthropod vectors, such as mosquitoes (Benelli 2018). Single-chain polymer nanoparticles (SCNPs) hold the potential to target ookinetes of *Plasmodium berghei* present in the mosquito midgut and deliver antimalarial drug atovaquone to inhibit ookinete maturation. Hence, treating the mosquito vector of *P. berghei* with atovaquone-loaded SCNPs can result in the prevention of malaria by disrupting the parasitic life cycle (Hamelmann et al. 2022).

The development of novel vaccines using nanotechnology-based approaches holds good promise against parasitic diseases. Nanoparticles can serve as delivery systems for parasite antigens, enhancing their presentation to the immune system and improving immune responses. These nanoparticle-based vaccines can elicit stronger and longer-lasting immune protection against parasites, reducing the severity and prevalence of parasitic infections in animals. *Leishmania braziliensis* antigen-loaded poly(D,L-lactic) acid (PLA) polymer-based nano-formulation has been developed that trigger total anti-Leishmania IgG antibodies in the liver and spleen. Therefore, it can reduce parasite load in vaccinated animals along with minimum histological damage (Ottino et al. 2022).

Furthermore, the development of sensitive and specific diagnostic tools can facilitate the detection of parasites, ensuring effective treatment. Nanosensors or nanoprobes can be designed to recognize parasite-specific molecules or genetic material, enabling rapid and accurate detection of parasites in clinical samples. These diagnostic tools facilitate early detection, timely treatment, and effective management of parasitic infections in animals.

The application of nanotechnology in parasite control for veterinary science holds great promise for improving treatment outcomes, reducing the prevalence of parasitic diseases, and enhancing animal health and welfare. However, it is important to consider safety aspects and potential environmental impacts associated with the use of nanomaterials in veterinary settings. Continued research and development in this field can lead to practical and effective nanotechnology-based solutions for parasite control in veterinary medicine.

1.7 Nanotechnology in Food Safety

Nanotechnology has the potential to play a significant role in ensuring food safety by offering innovative approaches for the detection, monitoring, and prevention of foodborne hazards. Nanosensors and nanoprobes can be designed for the detection and identification of foodborne pathogens quickly and accurately. These nanoscale devices can recognize specific pathogens or their biomarkers, providing rapid and sensitive detection methods.

Nanotechnology-based sensors enable real-time monitoring of food samples, reducing the time required for pathogen detection and facilitating timely interventions to prevent foodborne illnesses. Therefore, single-wall surface carbon nanotubes (SWCNTs) are employed for the detection of *Salmonella*, whereas multi-walled carbon nanotube (MWCNTs) electrodes functionalized with indium tin oxide (ITO) are used for the detection of *Salmonella enteritidis* and *Salmonella typhimurium* in milk or other food substances. The presence of *E. coli* O157:H7 in milk powder can also be captured via an AuNP-conjugated DNA sensor. Carbon quantum dots (CDs) can detect *S. typhimurium* as well as aflatoxin in different food materials (Kumar et al. 2020).

Nanosensors can be employed to monitor the quality and freshness of food products. These sensors can detect indicators of spoilage, such as microbial metabolites or volatile compounds, allowing for early identification of food putrefaction and potential health risks. The freshness of fish can be determined by tin dioxide-zinc oxide (SnO_2-ZnO) nanocomposite owing to its gas-sensing properties. It allows the detection of trimethylamine (TMA) produced during the decay of fish or shellfish (Zhang and Zhang 2008). Vitamins and other antioxidant compounds present in food products get easily degraded, deteriorating their quality. Hence, several nanosensors, such as MWCNTs, SWCNTs, nickel oxide NPs, and *N*-(3,4-dihydroxyphenethyl)-3,5-dinitrobenzamide-functionalized MWCNT, have been developed for sensing vitamins, viz. folic acid, ascorbic acid, essential amino acids such as tryptophan, etc., in different food samples (Kumar et al. 2017). Therefore, nanotechnology-based monitoring systems can assist in maintaining the quality and shelf life of food products, reducing food waste, and ensuring consumer safety. Nanotechnology offers the potential to develop intelligent and active packaging systems for food. Nanomaterials can be incorporated into packaging films or coatings to provide enhanced barrier properties, preventing the transfer of gases, moisture, and contaminants between food and the external environment, thereby extending the shelf life of food products and reducing the risk of foodborne pathogens. Furthermore, such nano-packaging allows bio-degradation, eliminating the risk of pollution. Active nano-packaging possesses antimicrobial agents and oxygen-scavenging reagents that prevent food contamination, whereas a smart packaging system is employed with nanosensors and nanodevices for determining the freshness of foods as well as monitoring disintegrity (Kuswandi 2017).

Detection and monitoring of chemical contaminants in food, such as pesticides, toxins, or heavy metals, can also be carried out through nanotechnology-based approaches. Thus, various nanosensors have been designed to selectively detect and quantify these contaminants, offering sensitive and specific detection methods. Metal NPs, such as AuNPs, AgNPs, and CuNPs, employed colorimetric assays, viz. surface plasmon resonance, surface-enhanced Raman scattering, etc., have been established for the quantification of heavy metals in food materials (Nam et al. 2022). β-Lactoglobulin (β-Lg), a major allergen of milk, can be detected through SER employing Au-Ag nanourchins functionalized with the hybridization of β-Lg capturing aptamer and Raman reporter molecule (6-Carboxyl-X-Rhodamine: ROX)-labeled complementary DNA (cDNA) (Duan et al. 2022). Similarly, antibiotics, pesticides, and such other chemical impurities can also be detected through various NP-incorporated modalities. Nanomaterials can also be employed in filtration or purification systems for removing contaminants from food or water sources.

In addition, nanotechnology can contribute to ensuring the authenticity and traceability of food products. Nanoscale markers or tags can be incorporated into food packaging or labeling, enabling accurate tracking of food products throughout the supply chain. These markers can provide information about the origin, production processes, and quality of the food, assisting in preventing food fraud and ensuring consumer confidence (Aggarwal and Idrishi 2023).

Nanotechnology-based encapsulation systems can be used for the controlled release of food additives, such as antioxidants or flavor enhancers. Nanoparticles can encapsulate these additives, protecting them from degradation or volatilization. Controlled-release systems ensure the gradual release of additives during food processing or storage, enhancing their effectiveness and improving the overall quality and safety of the food (Moradi et al. 2022). However, the incorporation of such nanosized particles in food can exert major adverse effects through improper metabolization, leading to their accumulation in several organs. Therefore, it is important to note that the safety and regulatory aspects of nanotechnology in food safety are crucial considerations. The potential risks associated with the use of nanomaterials in food applications should be thoroughly evaluated to ensure consumer safety. Regulations and guidelines governing the use of nanotechnology in the food industry are continuously evolving to address these concerns (Medina-Reyes et al. 2020).

Overall, nanotechnology offers exciting opportunities to improve food safety by enabling rapid and sensitive detection methods, enhancing packaging, monitoring food quality, and controlling contamination. Continued research and development in this field can lead to practical and effective nanotechnology-based solutions that ensure safer and healthier food for consumers.

1.8 Nanotechnology in Imaging and Monitoring

Nanotechnology plays a significant role in advancing imaging and monitoring techniques in veterinary science. By leveraging the unique properties of nanoscale materials, many innovative approaches have been developed for imaging, diagnostics, and monitoring of various conditions in animals.

NPs have a huge implementation as contrast agents in various imaging modalities, such as magnetic resonance imaging (MRI), computed tomography (CT), or ultrasound. These NPs enhance the visibility of specific tissues or structures, allowing for improved imaging and the detection of abnormalities in animals. Functionalization of NPs allows specific targeting of cells or tissues, providing selective contrast enhancement and better diagnostic accuracy. For instance, AuNPs are widely employed as contrast agents in CT scanning in addition to bismuth, manganese, tungsten, and such other NPs. Functionalization of AuNPs with ligands, viz. antibodies, folic acids, or peptides, allows targeting of specific organs/tissues that can facilitate early detection of malignant tumors. MRI is mostly assisted with gadolinium NPs, Fe_2O_3NPs, and such other heavy metal NPs as they show better sensitivity in the presence of a magnetic field. Gd-Au nanostructures have also been formulated for CT/MRI dual imaging (Aslan et al. 2020; Naseri et al. 2017). NPs can be engineered to carry specific molecules, such as antibodies or peptides, that target molecular markers associated with diseases or conditions. These targeted nanoparticles can accumulate at the site of interest, enabling molecular imaging to visualize specific cellular or molecular events. Molecular imaging techniques provide valuable insights into the underlying mechanisms of diseases in animals and aid in treatment planning and monitoring (Woźniak et al. 2022).

Nanosensors and nanoprobes can be designed to detect specific biomarkers or molecules in animals. These nanoscale devices can provide real-time monitoring of physiological parameters, such as glucose levels, oxygen saturation, or pH, in vivo. Hence, several biosensors based on QDs, graphene oxide (GO), carbon nanotubes, silica nanoparticles (SiNPs), and metal and metal oxide nanoparticles, viz. AuNPs, AgNPs, ZnONPs, IONPs, CuNPs, etc., have been constructed (Mokhtarzadeh et al. 2017). Localized surface plasmon resonance (LSPR)-based nanosensors provide rapid and quantitative detection of biomarkers (Hong et al. 2012). Nanoprobes can also be used to monitor the distribution and efficacy of therapeutic agents or track the progression of diseases over time. These tools offer non-invasive or minimally invasive monitoring options for animals, reducing stress and discomfort.

Nanotechnology-based approaches can improve the accuracy and sensitivity of the diagnostic sector. Nanoparticles can carry specific biomarkers or antibodies that bind to disease-associated molecules, enabling early detection and diagnosis of diseases or infections. Nanoscale diagnostic tools provide rapid and precise results, allowing for timely treatment interventions and improved prognosis for animals. For instance, AuNP with peptide nucleic acid (PNA) enables label-free detection of viral RNA in Newcastle disease owing to the agglomerative property of PNA (Joshi et al. 2013). Antibody-conjugated MNPs offer precise detection of biomarkers such as carcinoembryonic antigen via enzyme-linked immunosorbent assay (ELISA) (Tang et al. 2021).

Nanotechnology facilitates theranostics, which combines diagnostics and therapeutics in a single platform. Theranostic nanoparticles can serve as both imaging agents and drug carriers. They can be used to visualize and monitor diseases while delivering therapeutic agents directly to the affected site. This integrated approach enables targeted therapy, personalized medicine, and real-time monitoring of treatment response in animals. For instance, L-3,4-dihydroxyphenylalanine (L-DOPA)-functionalized MnONPs act as contrast agents in MRI as well as it carries DOPA across the blood-brain barrier, providing a therapeutic application for patients suffering from Parkinson's disease (McDonagh et al. 2016).

Additionally, nanotechnology enables the development of wearable sensors for monitoring animal health parameters. Nanomaterials can be incorporated into wearable devices, such as smart collars or patches, to track vital signs, activity levels, or other physiological parameters. These sensors provide continuous monitoring of animals' health, enabling early detection of abnormalities or changes in their condition. For instance, carbon nanotubes and graphene-based wearable salivary electrochemical sensors have been developed that analyze oral conditions for confirming gastro-phageal reflux diseases as well as monitor lactate variations in saliva determining the health conditions in animals. Several body-sensing technologies have been evolved for the detection of stress levels in animals by sensing metabolites and electrolytes with the help of band-aid like radio-frequency identification (RFID) sensor patches or temporary tattoo-based sensors. Hence, different wearable nanobiosensors have emerged as breath analyzer, saliva analyzer, toxin detector, body temperature detector, stress detector, and so on (Neethirajan 2017).

Hence, the use of nanotechnology in imaging and monitoring in veterinary science holds great promise for improving diagnostics, treatment monitoring, and overall animal care. However, it is important to consider safety aspects and potential regulatory requirements associated with the use of nanomaterials in veterinary applications. Continued research and development in this field will contribute to the translation of these nanotechnology-based approaches into practical tools that enhance veterinary imaging, diagnostics, and monitoring capabilities.

1.9 Conclusion

In conclusion, nanotechnology offers a vast scope of applications in veterinary science, revolutionizing various aspects of animal healthcare. The unique properties and capabilities of nanomaterials enable the development of innovative approaches in disease diagnosis, drug delivery, wound healing, parasite control, food safety, imaging, and monitoring. By harnessing nanotechnology, veterinary professionals can enhance the accuracy, efficiency, and effectiveness of veterinary diagnostics, treatments, and preventive measures. Nanotechnology-based approaches enable targeted drug delivery, providing precise and controlled release of therapeutic agents, reducing side effects, and improving treatment outcomes. They also contribute to the development of advanced imaging techniques, facilitating the detection and monitoring of diseases, and enabling real-time assessment of treatment responses. Furthermore, nanotechnology plays a crucial role in wound healing and tissue engineering, promoting tissue regeneration and enhancing the integration of biomaterials with host tissues. In the field of parasite control, nanotechnology provides innovative solutions for the targeted delivery of anti-parasitic drugs, vector control, and the development of novel vaccines. Additionally, nanotechnology enhances food safety by enabling rapid and sensitive detection of pathogens, monitoring food quality, and improving packaging to prevent contamination and spoilage. Nanotechnology also contributes to the development of wearable sensors and biosensors, facilitating continuous monitoring of animal health parameters. However, it is essential to consider the safety, ethical considerations, and regulatory aspects associated with the use of nanomaterials in veterinary science. Continued research, standardization, and collaboration among scientists, veterinarians, and regulatory bodies are crucial to ensure the safe and responsible implementation of nanotechnology in veterinary practices. Overall, the scope of nanotechnology in veterinary science is wide ranging and promising. As research and development in this field progress, nanotechnology-based solutions will continue to advance the field of veterinary medicine, improving animal health, well-being, and the overall practice of veterinary care.

References

Aggarwal D, Idrishi R (2023) Nanotechnology applications for food traceability. In: Nanotechnology applications for food safety and quality monitoring. Academic, pp 457–472. https://doi.org/10.1016/B978-0-323-85791-8.00011-2

Aguda ON, Lateef A (2022) Recent advances in functionalization of nanotextiles: a strategy to combat harmful microorganisms and emerging pathogens in the 21st century. Heliyon 8(6): e09761. https://doi.org/10.1016/j.heliyon.2022.e09761

Aslan N, Ceylan B, Koç MM, Findik F (2020) Metallic nanoparticles as X-ray computed tomography (CT) contrast agents: a review. J Mol Struct 1219:128599. https://doi.org/10.1016/j.molstruc.2020.128599

Ataide JA, Zanchetta B, Santos ÉM, Fava AL, Alves TF, Cefali LC, Chaud MV, Souto EB, Mazzola PG (2022) Nanotechnology-based dressings for wound management. Pharmaceuticals 15(10):1286. https://doi.org/10.3390/ph15101286

Avasthi A, Caro C, Pozo-Torres E, Leal MP, García-Martín ML (2020) Magnetic nanoparticles as MRI contrast agents. In: Surface-modified nanobiomaterials for electrochemical and biomedicine applications. Springer, pp 49–91. https://doi.org/10.1007/s41061-020-00302-w

Benelli G (2018) Gold nanoparticles—against parasites and insect vectors. Acta Trop 178:73–80. https://doi.org/10.1016/j.actatropica.2017.10.021

Brar B, Ranjan K, Palria A, Kumar R, Ghosh M, Sihag S, Minakshi P (2021) Nanotechnology in colorectal cancer for precision diagnosis and therapy. Front Nanotechnol 3:699266. https://doi.org/10.3389/fnano.2021.699266

Chien-Wei Lin L, Chattopadhyay S, Lin C, Hu MJ (2018) Advances and opportunities in nanoparticle- and nanomaterial-based vaccines against bacterial infections. Adv Healthc Mater 7(13):1701395. https://doi.org/10.1002/adhm.201701395

Chinen AB, Guan CM, Ferrer JR, Barnaby SN, Merkel TJ, Mirkin CA (2015) Nanoparticle probes for the detection of cancer biomarkers, cells, and tissues by fluorescence. Chem Rev 115(19): 10530–10574. https://doi.org/10.1021/acs.chemrev.5b00321

Chou SF, Gunaseelan S, Kiellani MHH, Thottempudi VVK, Neuenschwander P, Nie H (2017) A review of injectable and implantable biomaterials for treatment and repair of soft tissues in wound healing. J Nanotechnol. https://doi.org/10.1155/2017/6341710

Duan N, Yao T, Li C, Wang Z, Wu S (2022) Surface-enhanced Raman spectroscopy relying on bimetallic Au–Ag nanourchins for the detection of the food allergen β-lactoglobulin. Talanta 245:123445. https://doi.org/10.1016/j.talanta.2022.123445

Dwivedi P, Khatik R, Khandelwal K, Taneja I, Raju KS, Wahajuddin, Paliwal SK, Dwivedi AK, Mishra PR (2014) Pharmacokinetics study of arteether loaded solid lipid nanoparticles: an improved oral bioavailability in rats. Int J Pharm 466(1–2):321–327. https://doi.org/10.1016/j.ijpharm.2014.03.036

Erkoc P (2021) Nanotechnology-based antimicrobial and antiviral surface coating strategies. Prosthesis 3(1):25–52. https://doi.org/10.3390/prosthesis3010005

Faiz K, Lam FC, Chen J, Kasper EM, Salehi F (2022) The emerging applications of nanotechnology in neuroimaging: a comprehensive review. Front Bioeng Biotechnol 10:855195. https://doi.org/10.3389/fbioe.2022.855195

Gainza G, Pastor M, Aguirre JJ, Villullas S, Pedraz JL, Hernandez RM, Igartua M (2014) A novel strategy for the treatment of chronic wounds based on the topical administration of rhEGF-loaded lipid nanoparticles: in vitro bioactivity and in vivo effectiveness in healing-impaired db/db mice. J Control Release 185:51–61. https://doi.org/10.1016/j.jconrel.2014.04.032

Gamboa GV, Palma SD, Lifschitz A, Ballent M, Lanusse C, Passirani C, Benoit JP, Allemandi DA (2016) Ivermectin-loaded lipid nanocapsules: toward the development of a new antiparasitic delivery system for veterinary applications. Parasitol Res 115(5):1945–1953. https://doi.org/10.1007/s00436-016-4937-1

Gunasekaran T, Haile T, Nigusse T, Dhanaraju MD (2014) Nanotechnology: an effective tool for enhancing bioavailability and bioactivity of phytomedicine. Asian Pac J Trop Biomed 4(Suppl 1):S1–S7. https://doi.org/10.12980/APJTB.4.2014C980

Gurunathan S, Kang MH, Qasim M, Kim JH (2018) Nanoparticle-mediated combination therapy: two-in-one approach for cancer. Int J Mol Sci 19(10):3264. https://doi.org/10.3390/ijms19103264

Hajiaghapour Asr M, Dayani F, Saedi Segherloo F, Kamedi A, Neill AO, MacLoughlin R, Doroudian M (2023) Lipid nanoparticles as promising carriers for mRNA vaccines for viral lung infections. Pharmaceutics 15(4):1127. https://doi.org/10.3390/pharmaceutics15041127

Hajipour MJ, Saei AA, Walker ED, Conley B, Omidi Y, Lee B, Mahmoudi M (2021) Nanotechnology for targeted detection and removal of bacteria: opportunities and challenges. Adv Sci 8(21):2100556. https://doi.org/10.1002/advs.202100556

Hamdy ME, Del Carlo M, Hussein HA, Salah TA, El-Deeb AH, Emara MM, Compagnone D (2018) Development of gold nanoparticles biosensor for ultrasensitive diagnosis of foot and mouth disease virus. J Nanobiotechnol 16:1–12. https://doi.org/10.1186/s12951-018-0374-x

Hamelmann NM, Paats JWD, Avalos-Padilla Y, Lantero E, Spanos L, Siden-Kiamos I, Paulusse JM (2022) Single-chain polymer nanoparticles targeting the ookinete stage of malaria parasites. ACS Infect Dis 9:56. https://doi.org/10.1021/acsinfecdis.2c00336

Hong Y, Huh YM, Yoon DS, Yang J (2012) Nanobiosensors based on localized surface plasmon resonance for biomarker detection. J Nanomater 2012:759830. https://doi.org/10.1155/2012/759830

Hua S, de Matos MB, Metselaar JM, Storm G (2018) Current trends and challenges in the clinical translation of nanoparticulate nanomedicines: pathways for translational development and commercialization. Front Pharmacol 9:790. https://doi.org/10.3389/fphar.2018.00790

Jayarama Reddy V, Radhakrishnan S, Ravichandran R, Mukherjee S, Balamurugan R, Sundarrajan S, Ramakrishna S (2013) Nanofibrous structured biomimetic strategies for skin tissue regeneration. Wound Repair Regen 21(1):1–16. https://doi.org/10.1111/j.1524-475X.2012.00861.x

Jeon M, Halbert MV, Stephen ZR, Zhang M (2021) Iron oxide nanoparticles as T1 contrast agents for magnetic resonance imaging: fundamentals, challenges, applications, and prospectives. Adv Mater 33(23):1906539. https://doi.org/10.1002/adma.201906539

Jiang L, Loo SCJ (2020) Intelligent nanoparticle-based dressings for bacterial wound infections. ACS Appl Bio Mater 4(5):3849–3862. https://doi.org/10.1021/acsabm.0c01168

Jiang T, Li Q, Qiu J, Chen J, Du S, Xu X, Chen T (2022) Nanobiotechnology: applications in chronic wound healing. Int J Nanomedicine 17:3125–3145. https://doi.org/10.2147/IJN.S372211

Joshi VG, Chindera K, Singh AK, Sahoo AP, Dighe VD, Thakuria D, Tiwari AK, Kumar S (2013) Rapid label-free visual assay for the detection and quantification of viral RNA using peptide nucleic acid (PNA) and gold nanoparticles (AuNPs). Anal Chim Acta 795:1–7. https://doi.org/10.1016/j.aca.2013.06.037

Kumar V, Guleria P, Mehta SK (2017) Nanosensors for food quality and safety assessment. Environ Chem Lett 15:165–177. https://doi.org/10.1007/s10311-017-0616-4

Kumar H, Kuča K, Bhatia SK, Saini K, Kaushal A, Verma R, Bhalla TC, Kumar D (2020) Applications of nanotechnology in sensor-based detection of foodborne pathogens. Sensors 20(7):1966. https://doi.org/10.3390/s20071966

Kuswandi B (2017) Environmental friendly food nano-packaging. Environ Chem Lett 15(2):205–221. https://doi.org/10.1007/s10311-017-0613-7

Lee N, Ko W, Hsueh P (2019) Nanoparticles in the treatment of infections caused by multidrug-resistant organisms. Front Pharmacol 10:1153. https://doi.org/10.3389/fphar.2019.01153

Li Y, Zhang W, Zhao R, Zhang X (2022) Advances in oral peptide drug nanoparticles for diabetes mellitus treatment. Bioact Mater 15:392–408. https://doi.org/10.1016/j.bioactmat.2022.02.025

Ma K, Liao S, He L, Lu J, Ramakrishna S, Chan CK (2011) Effects of nanofiber/stem cell composite on wound healing in acute full-thickness skin wounds. Tissue Eng Part A 17(9–10):1413–1424. https://doi.org/10.1089/ten.TEA.2010.0373

Mady F, Shaker M (2017) Enhanced anticancer activity and oral bioavailability of ellagic acid through encapsulation in biodegradable polymeric nanoparticles. Int J Nanomedicine 12:7405–7417. https://doi.org/10.2147/IJN.S147740

McDonagh BH, Singh G, Hak S, Bandyopadhyay S, Augestad IL, Peddis D, Sandvig I, Sandvig A, Glomm WR (2016) L-DOPA-coated manganese oxide nanoparticles as dual MRI contrast agents and drug-delivery vehicles. Small (Weinheim an der Bergstrasse, Germany) 12(3):301–306. https://doi.org/10.1002/smll.201502545

Medina-Reyes EI, Rodríguez-Ibarra C, Déciga-Alcaraz A, Díaz-Urbina D, Chirino YI, Pedraza-Chaverri J (2020) Food additives containing nanoparticles induce gastrotoxicity, hepatotoxicity and alterations in animal behavior: the unknown role of oxidative stress. Food Chem Toxicol 146:111814. https://doi.org/10.1016/j.fct.2020.111814

Mihai MM, Dima MB, Dima B, Holban AM (2019) Nanomaterials for wound healing and infection control. Materials (Basel, Switzerland) 12(13):2176. https://doi.org/10.3390/ma12132176

Minakshi P, Ghosh M, Brar B, Kumar R, Lambe UP, Ranjan K, Manoj J, Prasad G (2019) Nano-antimicrobials: a new paradigm for combating mycobacterial resistance. Curr Pharm Des 25(13):1554–1579. https://doi.org/10.2174/1381612825666190620094041

Minakshi P, Ghosh M, Kumar R, Brar B, Lambe UP, Banerjee S, Ranjan K, Kumar B, Goel P, Malik YS, Prasad G (2020) An insight into nanomedicinal approaches to combat viral zoonoses. Curr Top Med Chem 20(11):915–962. https://doi.org/10.2174/1568026620666200325114400

Mokhtarzadeh A, Eivazzadeh-Keihan R, Pashazadeh P, Hejazi M, Gharaatifar N, Hasanzadeh M, Baradaran B, de la Guardia M (2017) Nanomaterial-based biosensors for detection of pathogenic virus. TrAC Trends Anal Chem 97:445–457. https://doi.org/10.1016/j.trac.2017.10.005

Moradi M, Razavi R, Omer AK, Farhangfar A, McClements DJ (2022) Interactions between nanoparticle-based food additives and other food ingredients: a review of current knowledge. Trends Food Sci Technol 120:75–87. https://doi.org/10.1016/j.tifs.2022.01.012

Mujawar LH, Moers A, Norde W, van Amerongen A (2013) Rapid mastitis detection assay on porous nitrocellulose membrane slides. Anal Bioanal Chem 405:7469–7476. https://doi.org/10.1007/s00216-013-7192-7

Nam NN, Do HD, Trinh KT, Lee NY (2022) Recent progress in nanotechnology-based approaches for food monitoring. Nanomaterials 12(23):4116. https://doi.org/10.3390/nano12234116

Naseri N, Ajorlou E, Asghari F, Pilehvar-Soltanahmadi Y (2017) An update on nanoparticle-based contrast agents in medical imaging. Artif Cells Nanomed Biotechnol. https://doi.org/10.1080/21691401.2017.1379014

Naskar A, Kim KS (2020) Recent advances in nanomaterial-based wound-healing therapeutics. Pharmaceutics 12(6):499. https://doi.org/10.3390/pharmaceutics12060499

Neethirajan S (2017) Recent advances in wearable sensors for animal health management. Sens Biosens Res 12:15–29. https://doi.org/10.1016/j.sbsr.2016.11.004

Ottino J, Leite JC, Alves O, González MA, de Carvalho TF, Garcia GM, Batista MA, Silveira P, Cardoso MS, Bueno LL, Fujiwara RT, Santos RL, Paes PR, Assis O, Galdino AS, Angel M, Dutra WO, Mosqueira VC, Giunchetti RC (2022) Nanoformulations with Leishmania braziliensis antigens triggered controlled parasite burden in vaccinated Golden hamster (Mesocricetus auratus) against visceral Leishmaniasis. Vaccine 10(11):1848. https://doi.org/10.3390/vaccines10111848

Panwar P, Pandey B, Lakhera PC, Singh KP (2010) Preparation, characterization, and in vitro release study of albendazole-encapsulated nanosize liposomes. Int J Nanomedicine 5:101–108. https://doi.org/10.2147/ijn.s8030

Patra JK, Das G, Fraceto LF et al (2018) Nano based drug delivery systems: recent developments and future prospects. J Nanobiotechnol 16:71. https://doi.org/10.1186/s12951-018-0392-8

Prasad M, Ghosh M, Kumar R, Brar B, Surjith KP, Lambe UP, Ranjan K, Banerjee S, Prasad G, Kumar Khurana S, Kharb P (2021) The importance of nanomedicine in prophylactic and theranostic intervention of bacterial zoonoses and reverse zoonoses in the era of microbial resistance. J Nanosci Nanotechnol 21(6):3404–3452. https://doi.org/10.1166/jnn.2021.18999

Sattar H, Sarwar I, Ulamin R, Muhammad J, Mushtaq M, Kazmi R, Akram M, Ashraf M, Shafi U (2019) Smart wound hydration monitoring using biosensors and fuzzy inference system. Wirel Commun Mob Comput 2019:8059629. https://doi.org/10.1155/2019/8059629

Seyyed Tabaei SJ, Rahimi M, Akbaribazm M, Ziai SA, Sadri M, Shahrokhi SR, Rezaei MS (2020) Chitosan-based nano-scaffolds as antileishmanial wound dressing in BALB/c mice treatment: characterization and design of tissue regeneration. Iran J Basic Med Sci 23(6):788–799. https://doi.org/10.22038/ijbms.2020.41361.9770

Sun Y, Chen D, Pan Y, Qu W, Hao H, Wang X, Liu Z, Xie S (2019) Nanoparticles for antiparasitic drug delivery. Drug Deliv 26(1):1206–1221. https://doi.org/10.1080/10717544.2019.1692968

Tang J, Yang H, Gao X, Zeng X, Wang F (2021) Directional immobilization of antibody onto magnetic nanoparticles by Fc-binding protein-assisted photo-conjugation for high sensitivity detection of antigen. Anal Chim Acta 1184:339054. https://doi.org/10.1016/j.aca.2021.339054

Varela-Aramburu S, Ghosh C, Goerdeler F, Priegue P, Moscovitz O, Seeberger PH (2020) Targeting and inhibiting Plasmodium falciparum using ultra-small gold nanoparticles. ACS Appl Mater Interfaces 12(39):43380–43387. https://doi.org/10.1021/acsami.0c09075

Verma N, Kulkarni R, Pandya A (2022) Microfluidic tools for veterinary and zoonotic disease diagnostics. Prog Mol Biol Transl Sci 187(1):281–293. https://doi.org/10.1016/bs.pmbts.2021.07.026

Wang W, Lu K, Yu C et al (2019) Nano-drug delivery systems in wound treatment and skin regeneration. J Nanobiotechnol 17:82. https://doi.org/10.1186/s12951-019-0514-y

Woźniak M, Płoska A, Siekierzycka A, Dobrucki LW, Kalinowski L, Dobrucki IT (2022) Molecular imaging and nanotechnology—emerging tools in diagnostics and therapy. Int J Mol Sci 23(5):2658. https://doi.org/10.3390/ijms23052658

Xie S, Pan B, Wang M, Zhu L, Wang F, Dong Z, Zhou W (2010) Formulation, characterization and pharmacokinetics of praziquantel-loaded hydrogenated castor oil solid lipid nanoparticles. Nanomedicine 5(5):693–701. https://doi.org/10.2217/nnm.10.42

Yao Y, Zhou Y, Liu L, Xu Y, Chen Q, Wang Y, Wu S, Deng Y, Zhang J, Shao A (2020) Nanoparticle-based drug delivery in cancer therapy and its role in overcoming drug resistance. Front Mol Biosci 7:193. https://doi.org/10.3389/fmolb.2020.00193

Zabielska-Koczywąs K, Lechowski R (2017) The use of liposomes and nanoparticles as drug delivery systems to improve cancer treatment in dogs and cats. Molecules (Basel, Switzerland) 22(12):2167. https://doi.org/10.3390/molecules22122167

Zhang WH, Zhang WD (2008) Fabrication of SnO_2–ZnO nanocomposite sensor for selective sensing of trimethylamine and the freshness of fishes. Sensors Actuators B Chem 134(2): 403–408. https://doi.org/10.1016/j.snb.2008.05.015

Zong T, Silveira AP, Morais JA, Sampaio MC, Muehlmann LA, Zhang J, Jiang C, Liu S (2022) Recent advances in antimicrobial nano-drug delivery systems. Nanomaterials 12(11):1855. https://doi.org/10.3390/nano12111855

Zottel A, Videtič Paska A, Jovčevska I (2019) Nanotechnology meets oncology: nanomaterials in brain cancer research, diagnosis and therapy. Materials (Basel, Switzerland) 12(10):1588. https://doi.org/10.3390/ma12101588

Chapter 2
General Methods for Generation, Characterization, and Functionalization of Different Types of Nanomaterials

Alla Yaswanth Naveen Kumar, Anu Kurian, Pavan Kumar Yadav, Ravindra Prasad, and Mayukh Ghosh

Abstract Nanotechnology has proved its worth in diverse scientific arenas, which include but are not limited to nano-vaccines and adjuvants facilitating prophylaxis; nanosensors, biomedical devices, and imaging modalities assisting in state-of-the-art diagnostics; antimicrobials; targeted nano-delivery systems; photothermal therapy enabling cutting-edge therapeutics or theranostics; wound and fracture management; and cosmetics. Myriads of engineered nano-delivery systems possessing several novel attributes provide unique benefits, such as enhanced and specific penetration, crossing the physiological barriers, facilitating targeted drug delivery, and preventing rapid degradation and clearance of the encapsulated drugs, precisely delivering gene or si-RNAs or biologicals. Further, nano-formulations can exert synergistic action or even act itself as drugs to overcome tricky situations like drug resistance or tolerance. The nanoscale materials of 1–10 nm size at least in one dimension are having narrow size distribution, a high surface area-to-volume ratio, enormous scope of surface functionalization, and unique optical, electrical, and magnetic properties which render them suitable for multi-arrayed biomedical applications. Nanomaterials can be synthesized either by top-down or bottom-up approach. In top-down approach, the bulk material is broken down into sub-micronic particles by applying chemical, physical, and biological energy. In bottom-up approach, the nanomaterials are synthesized from their atomic-level counterparts using various chemical or biological methods. Characterization of the synthesized nanomaterials is essential to verify their physicochemical attributes, which may be performed using diverse spectroscopic, X-ray-based, magnetometric,

A. Y. N. Kumar · A. Kurian · R. Prasad
Department of Genetics and Plant Breeding, Institute of Agricultural Science, Rajiv Gandhi South Campus, Banaras Hindu University, Mirzapur, Uttar Pradesh, India

P. K. Yadav · M. Ghosh (✉)
Department of Veterinary Physiology and Biochemistry, Faculty of Veterinary and Animal Sciences, Rajiv Gandhi South Campus, Banaras Hindu University, Mirzapur, Uttar Pradesh, India

LLR University of Veterinary and Animal, ICAR-National Research Centre on Equines, *Nanotechnology Theranostics in Livestock Diseases and Management*, Livestock Diseases and Management,
https://doi.org/10.1007/978-981-16-1610-5_2

or microscopic techniques, such as X-ray diffraction, FTIR, NMR, UV-visible spectroscopy, DLS, mass spectrometry, superconducting quantum interference device magnetometry, TEM, SEM, and measurement of zeta potential. Functionalization modifies the surface chemistry of nanoparticles by the addition of new properties or features such as ligand molecules, polymers, and biomolecules through covalent or non-covalent interactions. Functionalization enhances the biocompatibility, bioimaging potential, drug targeting and sustained release, intracellular delivery, and theranostic applications of the nanomaterials.

Keywords Nanomaterials · Green synthesis · Nanotheranostics · Photothermal therapy · Bioimaging · Nano-vehicles · Antimicrobials

2.1 Introduction

Nanotechnology has achieved immense attention and growth over time. *Nanomaterials* (NMs) are the fundamental components of nanotechnology that have emerged as an amazing class of materials, depicting a wide array of applications and substantially improved performances. *Nanomaterials* are defined as materials having a size in the range of *1–100 nm*, at least in any one of their dimensions. Despite ample diversity, they usually comprise carbon, metal, metal oxides, or organic matters (Ealia and Saravanakumar 2017). Nanomaterials (NMs) are prime scientific materials that already have enormous potential in diverse research arenas, especially from the current perspective of biomedical and pharmaceutical applications. Nobel laureate Richard P. Feynmann during his well-known lecture in 1959 expressed "nanotechnology" in precise terms i.e., "*there's plenty of room at the bottom*" (Khan et al. 2017). Since then, continuous progress has taken place in the field of nanotechnology, with persistent revolutionary developments by the addition of novel NMs to the repository as well as finding newer applications.

Nanomaterials surprisingly display distinctive chemical, physical, and biological properties at the nanoscale range in juxtaposition with their respective bulk counterparts. These novel attributes are oftentimes underlined by a comparatively enhanced surface area-to-volume ratio, improvised functionalization, stability, or reactivity in a chemical reaction and amplified mechanical strength. The predominant attributes that strongly govern the physicochemical properties of NMs are size, composition, and shape. In fact, one-billionth of a meter is a nanometer; thus, it does not drive a surprise that nanoparticles (NPs) of varied elements show diverse properties in contrast to its larger counterparts. For instance, when an element of a given volume is composed of NPs, the surface area of a particle escalates. In general, as the particle size reaches to the nanoscale range, a greater segment of particles is found on the surface. This in turn leads nanoparticles to become more chemically reactive to themselves and to their contiguous milieu (Kolahalam et al. 2019). Of note, it has been proven that the various biological functions and mechanisms that are occurring inside an organism are in turn having a strong relation with the size of the elements (Gatoo et al. 2014). The surface of NPs can be purposefully adorned with a wide array of minuscule entities such as functional groups, small molecules, metal ions, or polymeric substances. Based upon the surface property, the customized NMs can be exploited in divergent fields, such as catalysis, functional coatings, sensors, nanoelectronics, adsorbents, and biomedical and agricultural applications. Nanoparticles have unrolled various applications in developing dynamic catalysts by utilizing the enhanced surface area-to-volume ratio and aspect ratio. For instance, materials that are considered indigent catalysts, such as gold, are established to have prominent catalytic properties in their NP form.

Besides the size, the shape of NMs also has deterministic role in exhibiting their properties. Nanomaterials are of diverse shapes such as nanorods, nanoparticles, nanotubes, and nanosheets, which is characterized based on their dimensionality. The striking feature is that the modulation of nanomaterial properties is possible by regulating their size during their syntheses, accomplished using distinct methods. Even the impact of NMs on health and environment is influenced by their size and shape. Nanoparticles of certain materials are thought to pose greater risks of toxicity than larger-sized particles of the same element. Above all, the biodistribution of the nano-elements and the accumulation of a specific type of particles in a precise location in the body, in turn, are regulated by the particle size and surface characteristics (Prasad et al. 2021). Dominant distribution sites and precise target organs for NPs are frequently unspecified; it often appears liver and spleen as the major target organs. For instance, cadmium selenide quantum dots (CdSe QDs) persist in tissue up to 8 months (Shin et al. 2015) and cause severe damage inducing cell death as a result of membrane blebbing, chromatin condensation, and free radical formation, which are characteristic symptoms of apoptosis. Another effect of cytotoxicity leading to the loss of function arises as a consequence of QD interaction with intracellular components (Hardman 2006). Nanoparticles of diameter less than 6 nm are not able to be secreted by kidneys, resulting into their accumulation in liver and spleen imparting serious side effects. Pharmacokinetics (distribution of drug or NP in vivo), which is dependent on NP size and surface chemistry, has prime role in determining cytotoxicity.

Magnetic properties of nanoscale materials have also drawn a lot of attention in research zone, prominently in the biomedical applications. It is the uneven electronic distribution present in NPs that brings this amazing change in magnetic property at nanoscale level. Iron oxide NPs have claimed a pivotal position in bioimaging, toward the diagnosis of atherosclerosis, various types of cancer, as drug vehicles, etc. Further, several of these magnetic NMs have been depicted as potential "theranostic" agents, and some are undergoing through different phases of clinical trials for 'bench to bedside' transition (Gul et al. 2019). Superparamagnetic iron oxide nanoparticles (SPION) falling in the range of 50–100 nm are well exploited toward molecular imaging and therapeutic efficacy assessment. The wide biomedical applications of SPION include, but are not limited to, magnetic resonance imaging (MRI), magnetic particle imaging (MPI), magnetic fluid hypothermia (MFH), separation of biomolecules, targeted drug and gene delivery into the tissues, etc. (Dadfar et al. 2020).

Metallic nanoparticles have excellent thermal conductivity than those of fluids in the solid form. The reason underlies the heat transfer that occurs at the surface of the particles. Obviously, it is preferable to use particles with large total surface area. This paves the way for nanofluids. Nanofluids are produced by dispersing the nanometric-scale solid particles into liquids like water, ethylene glycol, or oils (Minea 2019). Nanomaterials have relevant influence on electrical and optical properties, depending on the nature of the element. Electrical conductivity of the nanofluid depends on the capacity of the charged nanoparticles in the fluid to transport the charges toward opposite electrodes once an electric potential is applied (Chereches and Minea 2019). Metal NPs (silver and gold) possessing unique optical properties have been widely employed for the diagnosis and treatment of diverse types of cancer. These metal NPs are extensively used in cell imaging, detection of DNA hybridization, proteins interaction, and photothermal therapy due to their extremely strong absorption and light dispersion in the plasmon resonance. The tumor detection and treatment can be further improved by regulating the size and shape of the NPs or conjugation of these NPs with specific metal ions, ligands, and markers (Ficai et al. 2015).

Another class of NM called quantum dots (QDs), which are crystals of a fluorescent semiconductor material ranging 2–10 nm, are bestowed with excellent opto-electronic properties based on their size and shape that have been extensively exploited for intracellular tracking, biomedical imaging and diagnosis, efficient nano-carriers for targeted drug delivery and cancer therapy, thus currently become the hotspot in the arena of nano-drug research (Wagner et al. 2019). QD nano-carrier system for drugs boosts up the diagnosis and treatment by early detection, monitoring, localized treatment of specific disease sites, improving the stability of drugs, and enhancing targeted absorption (Zhao and Zhu 2016). The diverse attributes of NPs render them fit into the multifaceted urge of current biomedical applications.

2.2 Types of Nanoparticles

NMs can be broadly categorized based on their dimensionality, morphology, state, and chemical composition as below:

2.2.1 Carbon-Based NMs

Carbon is the principal constituent of this type of NMs and can be of following types

2.2.1.1 Carbon Nanotubes (CNTs)

Carbon nanotubes (CNTs) are the one-dimensional allotropes of carbon. Depending on their structure, CNTs are of three different types: single-walled carbon nanotubes (SWCNTs), double-walled carbon nanotubes (DWCNTs), and multi-walled carbon nanotubes (MWCNTs), consisting of one, two, and several graphene sheets respectively. Further, based on the chiral angles, SWCNTs can be metallic or semiconducting that differ markedly in terms of applications (Li et al. 2012). CNTs show up extraordinary strength, current capacity, large flexibility, rapid electron transfer, high aspect ratio, and metallic or semiconducting characteristics.

Carbon nanotubes (CNTs) and graphene have attracted enormous attention and are in the spotlight of twenty-first century, depicting miscellaneous applications in the electronics, biomedical field, tissue engineering, neuroengineering, gene therapy, and biosensor technology. Single-walled and multi-walled carbon nanotubes vary in their arrangements of graphene cylinders. SWCNTs contain a single layer of graphene, while MWCNTs are composed of multiple layers (Al Moustafa et al. 2016). CNTs are sorted out as drug delivery vehicles due to their ability to cross cell membranes with relative ease and are provided with appreciable multiple attachment sites for drug targeting (Kakran and Li 2012). Such NMs can be customized to control and assist the release of the drug during transportation as well as at precise distant locations. Since the distribution and subsequent clearance of the drug from the body can be interfered and altered through nano-intervention, an increase in drug therapeutic efficacy and reduction in side effects can also be achieved through these NM-assisted targeted and controlled drug delivery systems.

2.2.1.2 Fullerenes (C_{60})

Fullerenes (C_{60}) are an incredible allotrope of carbon discovered in 1985, exhibiting cage-like structures having sp^2 hybridized carbon atoms linked via covalent bond with icosahedral symmetry (Minakshi et al. 2020). The prominent electron affinities of fullerenes have shown great potential for eliminating reactive oxygen species (ROS). Surface-modified fullerenes possess excellent potential for drug delivery and are used as nano-vesicles developed for targeted delivery and sustained release of the encapsulated drugs. The water-soluble cationic fullerene, tetra(piperazino)fullerene epoxide (TPFE) is successfully used for the targeted delivery of DNA and siRNA, specifically to lungs.

2.2.1.3 Carbon-Based QDs

Novel form of NMs comprised discrete and quasi-spherical nanoscale particles of carbon that are less than 10 nm in size (Baig et al. 2021). Carbon QDs offer enhanced chemiluminescence property that have been well exploited to develop highly sensitive and selective sensors thus attained significant priority in clinical imaging and nanomedicine. These carbon QDs are water-soluble, non-toxic, and can be adorned with diverse surface functional groups as well as bioactive molecules to be employed in multi-arrayed biomedical applications. Graphene QDs are another class of zero-dimensional, quasi-spherical carbon NP-based QDs that are composed of single or multiple crystalline graphene layers and possess similar attributes in terms of size and photoluminescence properties to be suitable for similar biomedical relevance (Liu et al. 2020).

2.2.1.4 Carbon Nanohorns (CNHs)/Nanocones

They express conical nanocages or nanocone structures consisting of sp^2-hybridized carbon sheets. Carbon nanocones are of three types: dahlia-like CNHs, bud-like CNHs, and seed-like CNHs. Carbon nanohorns possess some crucial advantages over CNTs, for instance, the ability for toxic-metal-catalyst-free synthesis and large-scale production at room temperature, ample specific surface area for functionalization, high purity and excellent chemical stability, outstanding catalytic properties, enhanced porosity and admirable conductivity, good absorption capabilities and low toxicity, etc. All these attributes render them suitable for diverse applications including but not limited to biosensing, biofuel cell, as electrodes, supercapacitor, biomedical devices, etc. (Karfa et al. 2019).

2.2.1.5 Nanodiamonds (NDs)

Nanodiamonds are nanoscopic truncated semi-octahedral sp^3-hybridized carbon-based materials of 5–50 nm diameter, usually synthesized by detonation of graphite or other energy-intensive processes. NDs have unique characteristics in terms of magnetic and optical properties that render them suitable for semiconductors, coating and abrasive, biosensing, drug delivery, imaging, etc. Biocompatibility is another important feature of these NMs to be used for wound healing treatments (Medina-Cruz et al. 2020). In addition, surface functionalization of nanodiamonds with diverse functional groups render them fit for targeted delivery of drugs, genes, tissue labeling, etc. (Qin et al. 2021).

2.2.2 *Metallic and Metal Oxide NMs*

Commonly explored metals for the synthesis of NPs include aluminum (Al), cadmium (Cd), copper (Cu), cobalt (Co), gold (Au), lead (pb), iron (Fe), silver (Ag), zinc (Zn), etc. (Ealia and Saravanakumar, 2017). Upon the synthesis of their particular NPs, they attain good desorption and adsorption ability, magnetic, optical, electronic and catalytic properties as well as high surface area and have potential applications in cellular and magnetic imaging, photodynamic therapy, catalysis, insulation, biosensors, opto-electronics, antimicrobials, antibacterial coatings, cosmetics and skin protectants, molecular sieves and fillers, and as vehicles for the targeted delivery of therapeutics into tissues, by conjugation of various ligands, antibodies, and functional groups on the surface of the metallic NPs (Mody et al. 2010). However, several metallic NPs have been reported to be toxic in nature. Encapsulation of metal or metal oxide NPs into the core with a shell of organic or inorganic NMs or incorporation of the reactive metallic NPs into host films or doped metallic NPs is the potential resolution derived to evade the impending health hazard of these metallic NMs (Yaqoob et al. 2020).

Metal oxide NPs are currently considered as an irreplaceable component in nanotechnology because of their superparamagnetic behavior, selective activity, non-linear optical properties, and improved ductility at high temperatures. Size of these particles has a strong relationship with their optical properties. Of note, they produce a highly reactive surface, which can be modified accordingly by attaching a polymer chain, doping metal ions, or coupling agents. Iron oxides, especially magnetite, have achieved special interest to material scientists since they contribute in applications stretching from magnetic resonance imaging (MRI) to various magnetic storage devices and gas sensors used for health care and environmental monitoring systems (Chavali and Nikolova 2019).

2.2.3 Metal-Organic Framework (MOF)

MOFs are synthesized from inorganic constituents such as metal ions and are interlinked by organic linker molecules, creating organic-inorganic crystalline materials with pores that are in nanometer range. MOFs exhibit high porosity since most atoms are present on their internal surface, intermediate chemical stability (predominantly under strong basic conditions), but improved mechanical and thermal stability. MOFs have high surface area; to be noted their internal surface area is extremely high exceeding 6000 m^2/g, thus they significantly show selective absorption, tunable record porosity with 90% free volume, crystallinity, and robust metal-ligand interaction contributing toward their structural diversity. MOFs are having excellent potential future application promisingly in biomedical arena as a new contrasting material for magnetic resonance imaging (MRI) and to be used as a theranostic agent (Peller et al. 2018).

2.2.4 Organic NPs

Organic NPs have gained major attraction in life science and cancer therapies since they are biodegradable polymers, offering enhanced stability in biological fluids, and also they address the issues of cytotoxicity. In addition, they are applicable in different fields of immunoassays, nucleic acid delivery, and cell adhesion. Polymeric NPs have achieved considerable attention over recent years due to their potential for targeted drug delivery, efficacy in protecting the encapsulated bioactive molecules from premature release and degradation by the biological activity persisting in the localized cellular environment, and improving their therapeutic index and bioavailability. These polymeric NPs are having significant role in developing and delivering novel vaccines, antigens, and proteins. Different prominent polymer-based NMs include, but are not limited to, polylactic acid (PLA), polylactic glycolic acid (PLGA), polyglutamic acid (PGA), chitosan, etc. (Han et al. 2018). Surface functionalization of organic NPs has encountered certain limitations such as premature erosion and curtailed circulation time that has now been surmounted by using several specific approaches. For example, hydrophilic polyethylene glycol (PEG)-coated NPs can extend blood circulation time of the loaded bioactive cargo and improve tumor targeting (Feracci et al. 2012).

2.2.5 *Nanocomposites*

Nanocomposites are solid materials having polyphase domains, and at least one of them has nanoscale structures. Various types of nanocomposites include ceramic matrix nanocomposites (CMNC), metal matrix nanocomposites (MMNC), polymer matrix nanocomposites (PMNC), and magnetic nanocomposites. They present high aspect ratio and adequately improve the performance and properties of polymer (Kolahalam et al. 2019). In material processing, employment of nanocomposites is speculated to have a significant impact on electro-ceramic, sensor, optical, and structural ceramic materials (Komarneni 1992).

2.2.6 *Ceramic NMs*

These inorganic NMs are composed of ceramics that can be synthesized from non-metallic and metallic compounds exhibiting chemical immobility and increased heat resistance property. Ceramic NPs are synthesized by successive heating of the bulk material at elevated temperature and fast cooling and primarily comprise oxides, carbides, phosphates, and carbonates of metals and metalloids such as Ca, Ti, and Si. They exist in diverse forms such as amorphous, crystalline, porous, dense, or hollow materials. Their supportive properties like high heat resistance and chemical inertness render them suitable to be explored in the biomedical field as imaging agents, excellent delivery vehicle of drugs, genes, and proteins, and so on (Thomas et al. 2015). A size of less than 50 nm and good shielding ability to protect the encapsulated cargo from fluctuation in external temperature or pH facilitate ceramic NPs to be employed as good drug delivery vehicles. Magnetism, specific optical and dielectric properties, and chemical inertness of these NMs also assist them to be used in orthopedic applications and photodynamic therapy of cancers. However, non-biodegradability is one of the inherent limitations of these NMs similar to the metal-based carriers exerting undesirable effects (Carvalho et al. 2019).

2.2.7 Lipid NPs

Lipid NPs have emerged as the most promising non-viral gene delivery system. They are extensively used in the effective protection and transfer of mRNA into the cell and are currently in the spotlight since now they are a key component of COVID-19 mRNA vaccines. Incorporating nucleic acids with positively charged lipids enhances their stability and precisely delivers them to target cells, as they protect them from nuclease degradation. These lipid NPs get adsorbed by the cell surface and undergo subsequent endocytosis, and thus the nucleic acids gain entry into the cells (Tenchov et al. 2021). As the barrier in the development and use of genetic medicine is progressively getting surmounted, the application of lipid NPs has been spiked, as they possess high nucleic acid encapsulation capacity and express improved penetration into tissues to deliver drugs (Hou et al. 2021). Several types of lipid NPs have currently been employed for the delivery of drugs that include, but are not limited to, diverse types of liposomes, cationic lipid nanoparticles, solid lipid nanoparticles (SLN) and nanostructured lipid carriers (NLC), non-lamellar lipid nanoparticles, ethosomes, etc. (Tenchov et al. 2021).

2.2.8 Mesoporous Silica NPs (MSN)

These are currently one of the most extensively studied inorganic NPs possessing unique properties such as high drug-loading capacity attributed by large surface area as well as high and tunable pore volume, easy chemical functionalization to facilitate stimuli-responsive drug delivery and ligand-mediated enhanced positive targeting, low toxicity, and good overall biocompatibility. Evidently, it has been effectively used to deliver anticancerous agents such as drugs and siRNA immobilized on the external surface of MSN, thus a promising type of NM for cancer therapy. To be noted, MSN with large pores make encapsulation of genes possible thus protecting against nuclease degradation (Wang and Huang 2014; Doan et al. 2018).

2.2.9 Protein Nanoparticles

Protein nanoparticles can be synthesized either by crosslinking the native protein molecules harboring pre-existing functional groups or by crosslinking the protein molecules followed by purposeful modification of their surface functional groups (Hasan et al. 2020). These nanoparticles can be manufactured using various proteins such as albumin, gelatin, whey protein, sericin, zein, soy protein, ferritin, and lipoprotein employing various techniques such as electrospray, desolvation, emulsion, and complex coacervation techniques. Protein NPs have several superiorities over other NPs, as they are abundant naturally, biocompatible, biodegradable, cost-effective, easily metabolizable, and provide surface modifications, even the loading up of multiple drugs. Toxicity is not an issue in the case of protein NPs as they are degraded by the action of enzymes already present inside the host body. For instance, ferritin is a major iron storage protein having a hollow interior cavity; thus, ferritin NPs have emerged as a potential protein NP of biomedical importance that can be functionalized by incorporating various metals and ligands on their surface with high affinity and encapsulation of probes or therapeutics in their interior cavity. Ferritins contain high number of hydrogen bonds and salt bridges between their subunits that confer it with a unique property to withstand high temperatures and also make it stable in distinct denaturants such as guanidium chloride, sodium hydroxide, and urea. An incredible feature of ferritin is that it can be dissociated in an acidic environment (a pH range of 2.5) and restored to normal when pH backs to physiological conditions. This property has rendered ferritin a mighty NP that is well exploited in disease diagnosis, cancer treatment, vaccine development, bioassays, nano-devices, and drug delivery vehicles (Wang et al. 2017).

2.3 Methods for the Synthesis of Nanoparticles

The attributes of nanomaterials are integral aspects of their synthesis methods. The reproducibility of the sub-micronic particles and, in turn, their applicability in the biomedical arena are enormously dependent on stringent customization protocols. Several synthesis methods, such as hydrothermal synthesis, biological synthesis, and template synthesis, are currently in operation; however, they fall under the two basic approaches: the top-down approach and the bottom-up approach (Fig. 2.1) (Rane et al. 2018).

2.3.1 Top-Down Approach

The principle of this approach is to break down the bulk materials into their nano-sized particles by using different techniques or methods (Fig. 2.1). The most commonly employed methods in top-down approach are mechanical milling, etching, laser ablation, sputtering, electro-explosion, etc. (Baig et al. 2021) that have been elaborated hereunder.

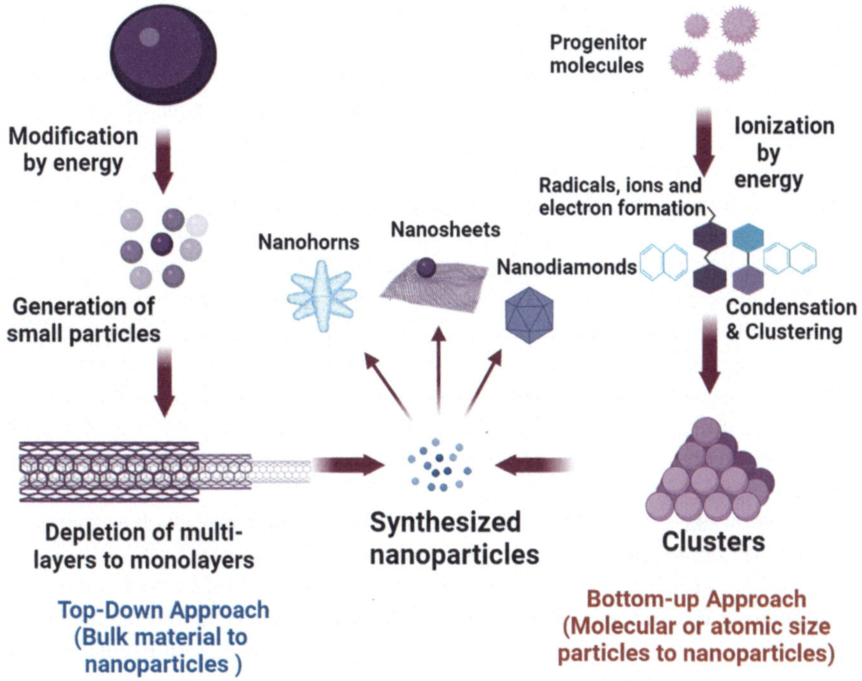

Fig. 2.1 General approaches for nanoparticle synthesis

2.3.1.1 Mechanical Milling

The basic principle of this method is "impact and attrition." The ball mill is a cylindrical device that rotates in horizontal axis to grind the bulk materials. The materials to be grinded in ball mill are usually ores, chemicals, ceramic raw materials, paints, etc. Initially ball mill is filled with half-grounded material and grinding medium. Size reduction is carried out by impact as balls are dropped from the top of a shell which creates internal cascading effect reducing larger materials to fine powder followed by mixing and defect formation (Takacs 2002). The primary aim of this method is the production of complex oxide dispersion-strengthened (ODS) alloys for high-temperature structural applications. This method helps in successfully producing fine and uniform dispersion of oxide particles, such as aluminum trioxide (Al_2O_3), yttrium oxide (Y_2O_3), thorium dioxide (ThO_2) in aluminum/nickel/magnesium/copper-based nanoalloys which are not achievable through conventional powder metallurgy methods (Yadav et al. 2012). The mechanical milling can be used for the manufacturing of nanoalloy, nanocomposites, and nano-quasi-crystalline materials in considerably high amount (Yadav et al. 2012).

2.3.1.2 Sputtering

Sputtering is a process through which the surface of substrates is bombarded with high-energy ionic gas or gaseous plasma leading to the ejection of NPs from the surface of the bulk material. The basic concept of sputtering processes is the bombardment of ions on the solid surface, resulting in the interaction between the surface and the ions. The incident ions on the surface get neutralized and reflected. Due to the impact of ions, secondary electrons are emitted from the surface of the parent material. The target material gets structurally rearranged because of ion impact. Collision on the surface is a continuous process, so there is an incessant ejection of electrons from the surface. This phenomenon is known as sputtering. The various types of sputtering are DC-diode sputtering, DC-triode sputtering, RF sputtering, magnetron, and reactive sputtering. These processes are effective for producing a thin film of NPs. Magnetron sputtering is used for the production of tungsten-diselenide (WSe_2)-layered nano-films on silicon oxide (SiO_2) and carbon paper substrates. This process is also employed in optical applications by depositing a thin layer on glass. Several advantages are inherent to sputtering, such as sputtered nanomaterial composition remains the same as target material with only fewer impurities, but it is very cost effective as compared to electron beam lithography. The sputtering is also used in industries such as micro-electronics, semiconductors, and optical coatings (Wender et al. 2011; Asanithi et al. 2012) (https://nanografi.com/blog/sputtering-process-in-nanotechnology/).

2.3.1.3 Laser Ablation

It is also called photo-ablation. In this process the powerful laser beam is focused at the surface of the target material, leading to the vaporization of the light-absorbing surface materials into a laser plume due to high-energy laser irradiation under non-equilibrium vapor/plasma conditions, followed by nucleation and agglomeration of the vaporized species, resulting in the generation of NMs. As there is no use of hazardous agents or chemicals, it is also called green technique. Generally, NMs with high purity can be synthesized by this method, which depends on the purity of the target precursor and the background media. Further, the size, shape, and distribution of the NPs are crucially controlled by several factors, such as laser pulse parameters, laser focusing parameters, and the ablation medium. The size of the NPs can be regulated based on the number of laser pulse and the photon energy (wavelength) of the laser applied. Generally, a laser of higher wavelength produces larger particles and vice versa. Laser ablation time and laser fluence also critically regulate the NP size distribution and yield. Usually, laser ablation under low-pressure background gas is favorable to generate smaller particles. The synthesis of NPs using pulsed laser ablation can be carried out in the solid as well as liquid phases. Pulsed laser ablation in liquids is extremely useful for producing monodispersed colloidal NP solutions without using a surfactant or ligand. Further, in the liquid phase, the surfactant-free synthesis of NPs provides ease of functionalization if required. Diverse types of NPs can be fabricated using this method, which include, but are not limited to, metallic, semiconducting, magnetic, and organic NPs (Naser et al. 2019).

2.3.1.4 Lithography

Diverse nano-architectures are developed through lithography by using a focused beam of light/electrons at the surface of the bulk material. Basically this process is useful for nano/micro-patterning. Basically nanolithography can be broadly categorized into photolithography, scanning lithography, soft lithography, and other miscellaneous lithographic techniques.

Photolithography is the most commonly used technique in the mass production of micro-electronics and semiconductor devices. It is characterized by both high production throughput and small-sized features of the patterns. The photolithography includes optical lithography and quantum optical lithography (QOL). Optical lithography is one of the most significant and established sets of techniques in the nanolithography field. Several techniques in optical lithography use short wavelengths to change the solubility of particular molecules. Major optical lithographic techniques include multi-photon lithography, X-ray lithography, light coupling nanolithography (LCM), and extreme ultraviolet lithography (EUVL). Quantum optical lithography (QOL) is capable of unlimited diffraction and able to fabricate NMs at 1 nm resolution by using a red laser diode of 650 nm wavelength. Geometrical figures and patterns which are complex in structures like nano-patterned graphene of 20 nm can be drawn by employing high-resolution QQL method (Pavel et al. 2019a, b).

Scanning lithography includes techniques such as electron beam lithography, scanning probe lithography, proton beam lithography, and charged particle lithography. Scanning lithography employs a directed scanning probe, electron beam, high-energy proton beam, or beam of energetic lightweight ions either to create nano-patterns by directly writing on the substrate or by etching away the sensitive coating from the surface of the thin-film-covered substrate (Soh et al. 2001; Altissimo 2010).

Soft lithography is another important technique to create micro- or nano-patterns by using elastomeric stamp or mold of chemical compounds like polydimethylsiloxane (PDMS). PDMS is bestowed with extensive elasticity, ample optical transparency, hydrophobicity, biocompatibility, and permeability to be preferred for engineering the mold or stamp to create the nano-patterns. Soft lithography is extremely useful for the fabrication of micro- or nano-patterned chips used in microfluidic devices. However, the technique has to rely on photolithography or electron beam lithography to customize the elastomeric mold to be used for soft lithographic printing (Whitesides et al. 2001). Other miscellaneous lithographic techniques include, but are not limited to, nanoimprint lithography, nanosphere lithography, neutral particle lithography, plasmonic lithography, and magnetolithography (Xie et al. 2011; Venugopal and Kim 2013; Bardea and Yoffe 2017; Sreenivasan 2017; Baig et al. 2021).

2.3.1.5 The Arc-Discharge Method

This method is used for producing carbon-based NPs such as CNTs, carbon nanohorns (CNHs), grapheme films, and fullerenes, along with metal and metal oxide NPs. In arc-discharge synthesis system, usually two graphite rods are adjusted in a chamber: one acts as the cathode, and another acts as the anode; to exert pressure helium (inert gas) is filled in it. The DC current is passed through the system, resulting in the vaporization of carbon rod. After vaporization, condensation, nucleation, and nanoparticle synthesis take place, respectively. Different types of carbon-based NMs are collected at different positions in the system based on mechanisms. The conditions present during the synthesis can affect the NP production and properties. For instance, the graphene sheets which are prepared under hydrogen arc-discharge exfoliation are more effective and thermally stable than the sheets prepared under argon arc discharge. The presence of moisture or oxygen inhibits the synthesis of fullerenes (Baig et al. 2021).

2.3.2 *Bottom-Up Approach*

The principle of this approach is to synthesize NPs by combining smaller-size particles such as molecules or atomic-sized entities (Fig. 2.1). The most conveniently employed methods in this approach are as follows: chemical vapor deposition (CVD), solvo-thermal and hydrothermal method, sol-gel method, soft and hard templating method, reverse micelle method, etc.

2.3.2.1 Chemical Vapor Deposition (CVD)

It is one of the best methods for the synthesis of pure and crystal-quality NPs. It is mainly used for the production of two-dimensional NPs, such as a thin film of NMs. In this method the substrate is placed in an oven and heated to a high temperature, and simultaneously carbon-containing gas (hydrocarbon) is introduced slowly into the chamber as a precursor. As the temperature rises the carbon atoms are slowly released due to decomposition, and again the released carbon atoms recombine to form NMs on the heated substrate by means of some chemical reactions. The process of CVD can be performed in different formats, depending upon the means of initiation of the chemical reaction for NP generation. Based upon the operating condition, CVD can be classified as atmospheric pressure CVD (APCVD), sub-atmospheric CVD (SACVD), low-pressure CVD (LPCVD), and ultra-high vacuum CVD (UHVCVD). The most advanced CVD formats among the aforesaid are LPCVD and UHVCVD. Depending upon the physical characteristics of vapor, CVD formats can also be classified as aerosol-assisted CVD (AACVD) and direct liquid injection CVD (DLICVD). Further, CVD formats can be classified as hot wall CVD and cold wall CVD, based on substrate heating methods. Catalyst also plays a crucial role in the synthesis of material type and morphology. For instance, copper is used as a catalyst for the synthesis of monolayer graphene, and nickel and cobalt are used as catalysts for the synthesis of multilayered graphene (https://doi.org/10.1039/d0ma00807a). The commercially important NMs prepared by CVD technique include, but are not limited to, polycrystalline silicon, silicon dioxide, silicon nitride, carbide-derived carbon, CNTs, graphene, nanodiamond, fullerenes, nanoparticle-embedded thin films, metal, and metal-organic frameworks that are used in semi-conductors, solar cell devices, opto-electronics, and membrane coatings (Stassen et al. 2015; Manawi et al. 2018). Even the properties of substrate, such as conductivity and hydrophobicity, are improved due to the deposition of NMs on the substrate.

2.3.2.2 Solvo-Thermal and Hydrothermal Methods

The process involves the direct synthesis of NMs from the solution of precursor substrate mixed in an aqueous or non-aqueous solvent under high temperature and pressure in a closed system. The high temperature (usually between 100 and 1000 °C) and pressure (generally between 1 and 10,000 atm) facilitate the dissolution of the substrate mixture and the interaction of the precursors to synthesize NMs of desired properties. Commonly, ethanol, methanol, ammonia, carbon dioxide, hydrochloric acid, hydrofluoric acid, etc., are used as solvents. When water is taken as solvent, "solvo-thermal" processes are termed "hydrothermal." The substrate is dissolved in the solvent and kept in an autoclave. The autoclave is heated to a temperature above the boiling point of the solvent. As the temperature and pressure increase, crystallization of the dissolved material in the solvent takes place (Nair and Dhoble 2021). Basically, increasing temperature and pressure alter the fundamental properties of the solvent, such as the dielectric constant, ionic product, density, viscosity, thermal conductivity, and heat capacity, which promote the precipitation and crystallization of the dissolved precursor moieties to form respective NMs. Diverse geometries of NMs, such as thin films, bulk powders, and nanocrystals of different morphologies like sphere, rod, or wire, can be achievable through this method by modulating solvent supersaturation, precursor chemical concentration, and kinetic control. Quantum dots and metallic and metal oxide NPs, such as magnetite, zinc oxide, and aluminum oxide, are produced by this method, depicting prominent biomedical applications such as in biosensors, nano-optical circuits, nanomagnetics, for chemotherapy, cell separation, scaffold-based tissue engineering, and organ-tissue imaging (Ortega and Reguera 2019).

2.3.2.3 Sol-Gel Method

Sol-gel method is a wet chemical technique in which the molecules of precursor substrate remain dissolved as chemical solution or colloidal particles in water or alcohol that subsequently converted into an integrated network as gel upon heating and continuous stirring by hydrolysis/alcoholysis. Despite customizing various types of nanostructures, the method is trendy for the synthesis of high-quality metal oxide NMs and diverse microcrystalline porous nano-architectures (Schubert 2015; Bokov et al. 2021). Metal alkoxides and metal chlorides are commonly used as precursors for the synthesis of NMs in this technique. This process involves multiple steps such as sol formation, condensation, aging, drying, and calcination. The metal oxides are initially hydrolyzed in water or alcohol for sol formation. After sol formation, condensation takes place, which results in an increase in the viscosity of the solvent, and porous structures are formed. In condensation or poly-condensation, oxo (M-O-M) or hydroxyl (M-OH-M) bridges are formed. These bridges yield the formation of metal-oxo or metal-hydroxy poly-layers. In aging the poly-condensation continues with the changes taking place in porosity, structure, and properties. The porosity decreases and the distance between colloidal particles increases during aging. The drying proceeds next to removing water and organic solvents from the gel. Finally calcination takes place, removing the volatile substances by heating to high temperature, and NMs are formed (Baig et al. 2021). Sol-gel synthesized NMs can be used in opto-electronics, surface engineering, coating films, orthopedic implants, biosensors, bioactive glasses, fiber-optic applications, and separation technologies (Bokov et al. 2021). Sol-gel method is also used for the generation of superparamagnetic NMs having diverse biomedical applications (Gupta and Kumar 2008).

2.3.2.4 Template Synthesis Method

Templated synthesis employs a nanoporous template to facilitate facile synthesis of ordered mesoporous nanostructures of desired morphology and attributes from diverse materials, such as polymers, metals, carbon, and semiconductors. Templated synthesis generally involves template customization followed by template-guided synthesis of the target NMs and subsequently removal of the template if required (Liu et al. 2013; Xie et al. 2016). Templated synthesis can be classified as hard and soft templating methods, depending upon the physical properties of the template. Hard templating, also known as "nanocasting," employs rigid materials as a template with stable desired nanoscale features that determine the size and morphological attributes of the customized NMs. Colloidal particles, nanoporous membranes, and patterned solid surfaces, such as polymeric microspheres, plastic foam, ion exchange resin, carbon fiber, and porous metal oxide structures, can be used as hard templates for NM synthesis. Further, either these templates can be used as a purely physical structure-based templating agent or the physical template can also participate in the chemical transformation during NM synthesis, acting as a chemical template.

Generally, different methods such as sol-gel coating, layer-by-layer assembly, adsorption-reduction process, graft-polymerization, electrodeposition, and chemical vapor deposition are used for the filling of the precursor substrate molecules into the mesopores of the nano-templates for customizing the nano-architectures (Liu et al. 2013; Savic et al. 2018). Contrary to hard templating, the soft templating method employs non-rigid flexible nanostructures as a template, such as macro- and micro-emulsions, micelles or vesicles, polymers, and nanostructured biological molecular aggregates. Inorganic precursor species are deposited at the surface or the internal cavities of these porous, soft templates by precipitation, electrochemical deposition, and other suitable methods to generate desired nanoscale materials (Xie et al. 2016). Soft templating is advantageous due to its low experimental nature, straightforward action, synthesis of materials with various morphological attributes, etc. Block copolymers, flexible organic compounds, and cationic, anionic, and non-ionic surfactants are commonly used as soft templates for guided NM synthesis. The interaction between precursor and soft template occurs through hydrogen bonding, electrostatic forces, and van der Waals forces. Soft templates of three-dimensional specifically arranged, liquid crystalline micelles are used for the synthesis of 3D mesoporous structures; for instance, alkyl-tri-methyl-ammonium is used to produce mesoporous solids such as cubic (MCM-48), lamellar (MCM-50), and hexagonal (MCM-41) ordered mesoporous silica. Usually two methods such as cooperative self-assembly and true liquid crystal templating are adopted for synthesizing the ordered mesoporous materials. The structure of 3D mesoporous material is affected by several factors, such as precursor and surfactants concentration, precursor and surfactant ratio, environmental conditions, and surfactant structure. Different types of nanomaterials such as single-crystal nanorods, mesoporous N-doped graphene, porous aluminas, and mesoporous polymeric carbonaceous nanospheres are produced by soft templating method. The mesoporous templates have been used to synthesize nanowires, nanorods, 3D nanostructured materials, and diverse other porous nano-architectures that have several biomedical applications. Hallow mesoporous silica nanomaterials (HMSNs) exhibited tremendous potential in cancer imaging and therapy. HMSNs are also effective in molecular imagining, optical imaging, nanomedicine, and positron emission tomography (PET) (Shi et al. 2013).

2.3.2.5 Reverse Micelle Method

The reverse micelle technique can synthesize NPs with a well-defined shape, size, and composition without involving too much specialized or expensive equipment as required in different physical synthesis methods. Normal micelles are formed by an oil-in-water emulsion, in which hydrophilic tails are pointed toward core and holding oil droplets in it. In reverse micelles the hydrophilic heads are pointing toward the core, which contains water in it. Reverse micelles remain dispersed in the organic solvent medium stabilized by a surfactant, and the aqueous core of the reverse micelles acts as nanoreactors for the synthesis of NMs. The size of the synthesized NPs can be controlled by water concentration, for instance, larger-sized particles are generated with an increase in water content. The size of NMs can also be controlled by adjusting water-to-surfactant ratio. Nanomaterials produced by this method are homogeneous and monodispersed in nature, devoid of the aggregation issue (Baig et al. 2021). Reverse micelles are used for the synthesis of magnetic NPs, QDs, metallic and bimetallic NPs of various morphological features, electrochemical nanosensors, nanocomposites, nanoelectrodes, etc., that depicted versed biomedical applications (Chaurasiya and Hebbar 2017; Arsene et al. 2021).

2.3.2.6 Biological Synthesis of Nanomaterials

The existence of metals in the ecology of several unicellular and multicellular organisms has been well established. Some of these metals such as zinc, cobalt, copper, and nickel possess nutritive value to be used in cellular processes, while several other metals are nonessential and toxic to the organisms. Evidently, the existence of several genes providing resistance to the metal ion toxicity is also identified in various unicellular as well as multicellular organisms. The resistance mechanism helps to generate metallic or metal oxide NPs naturally by converting metal ions into non-toxic or less toxic form; such NPs have diverse biomedical applications (Fig. 2.2). The resistance mechanisms can be broadly classified as permeability restriction through barrier function, intra- and extracellular sequestration of the metal ions, outflow of the metal ions through active transport-coupled efflux pumps, enzymatic detoxification, and reducing the susceptibility of cellular targets to toxic metal ions (Bruins et al. 2000) (Table 2.1). The biological synthesis

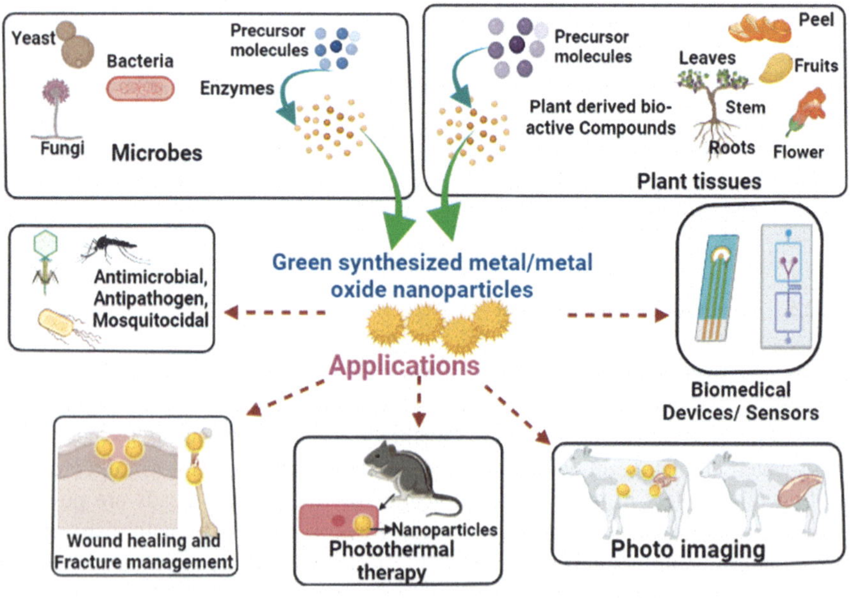

Fig. 2.2 The biological synthesis of nanoparticles and their common biomedical applications

Table 2.1 Biological synthesis of nanoparticles using various enzymes/components

Species	Type of nanoparticles synthesized	Enzymes required	Key components
Bacterial synthesis			
Pseudomonas stutzeri	Silver nanoparticles	Nitrate reductase	Intracellular synthesis and synthesized nanoparticles are triangular and hexagonal in shape
Morganella sp.	Silver nanoparticles	Silver resistance by silver-binding proteins	Crystalline silver nanoparticles are synthesized extracellularly
Lactobacillus	Silver and gold nanoparticles	Nitrate reductase	Sub-micronic silver and gold nanoparticles are synthesized
Leuconostoc lactis	Silver nanoparticles	Nitrate reductase	Round-shaped and stabilized silver nanoparticles are synthesized
Cyanobacteria *Oscillatoria limnetica* aqueous extracts	Silver nanoparticles	Nitrate reductase	Free amino acids and sulfur-containing amino acids of bacterial extract serve as capping agents
Rhodopseudomonas capsulata	Gold nanoparticles	NADH-dependent reductase	Required specific pH to synthesize nanoparticles, at pH 4 spherical-shaped nanoparticles are synthesized and at pH 7 nanoplates are synthesized
Pseudomonas aeruginosa	Gold nanoparticles	Reduction of gold ion by bacterial cell supernatant	Extracellular synthesis of gold nanoparticles of size 15–30 nm
Cupriavidus metallidurans	Gold nanoparticles	Energy-dependent reductive precipitation of gold complex involving oxido-reductase	Include various processes such as efflux, methylation, and reduction
Desulfovibrio alaskensis	Platinum oxide nanoparticles	Chloride reductase	Requires temp 30 °C and pH 7, and the time required is 2 h
Acinetobacter calcoaceticus	Platinum oxide nanoparticles	Hydrogenase	Cuboidal-shaped nanoparticles synthesized and requires temperature of 30 °C and pH 7; production time is 24 h
Cyanobacteria *Plectonema boryanum*	Amorphous platinum oxide nanoparticles	Reductase, organic sulfur, and organic phosphorus	Synthesize spherical platinum oxide nanoparticles of size 30 nm

(continued)

Table 2.1 (continued)

Species	Type of nanoparticles synthesized	Enzymes required	Key components
Aquaspirillum magnetotacticum	Iron oxide nanoparticles	Oxido-reductase	Octahedral prism-shaped nanoparticles of size 40–50 nm are synthesized
Magnetospirillum magnetotacticum	Iron oxide nanoparticles	Ferredoxin oxido-reductase	Cubo-octahedron-shaped nanoparticles of size 47 nm
Morganella morganii	Copper nanoparticles	Metal ion reductase	In the processes copper ions change their color from blue to pickle green
Morganella psychrotolerans	Copper nanoparticles	Metal ion reductase	pH plays an important role in determining the size of nanoparticles
Cyanobacterium *Nostoc* sp.	Zinc oxide nanoparticles	Oxido-reductase	Synthesize star-shaped nanoparticles of size 50–80 nm
Lactobacillus johnsonii, *Lactobacillus paracasei*, and *Lactobacillus sporogenes*	Zinc oxide nanoparticles	Oxido-reductase	Spherical- and hexagonal-shaped nanoparticles are synthesized
Fungal synthesis			
Epicoccum nigrum	Silver nanoparticles	Nitrate reductase	Required alkaline pH to synthesize nanoparticles
Colleotrichum sp. ALF2-6	Gold nanoparticles	Nitrate reductase	Required alkaline pH and temperature of 50 °C
Aspergillus fumigates	Zinc oxide nanoparticles	Oxido-reductase	The synthesized ZnO NPs exhibit high monodispersity and without agglomeration
Fusarium oxysporum	Platinum oxide nanoparticles	Oxido-reductase	Fungal extract is shaken with Pt ions at 200 rpm for 96 h at room temperature
Plant-based synthesis			
Skimmia laureola leaf extract	Silver nanoparticles	Flavonoids and terpenoids are responsible for capping and the stabilization of nanoparticles	Synthesize spherical nanoparticles of size 38 nm in size
Prosopis farcta leaf extract	Silver nanoparticles	Chemical reduction by plant-origin reducing and capping agents	Requires room temperature, and synthesized nanoparticles are 10.8 nm in size
Hydrocotyle rotundifolia	Silver nanoparticles	Chemical reduction by plant-origin phytochemicals	Ultra-small size nanoparticles of size 7.38 nm are synthesized

(continued)

Table 2.1 (continued)

Species	Type of nanoparticles synthesized	Enzymes required	Key components
Anthemis atropatana stem extract	Silver nanoparticles	Chemical reduction by plant-origin reducing and capping agents	Exhibit antimicrobial activity against some pathogens such as *P. aeruginosa* and *S. aureus*
Lemon extract	Silver nanoparticles	Chemical reduction by plant-origin reducing agent	Shows antimicrobial action against some microbes such as *P. aeruginosa*, *E. coli* and *S. aureus*
Azadirachta indica gum extract	Silver nanoparticles	Heteropolysaccharide present in gum acts as reducing agent	Synthesize spherical shape and doesn't exhibit any aggregations
Allium cepa bulb extract	Silver nanoparticles	Chemical reduction by plant-origin reducing and stabilizing agents	Synthesizes nanoparticles of size 10–23 nm and also acts as both reducing and capping agents
The *magnolia* leaf extract	Copper nanoparticles	Chemical reduction by plant-origin reducing agents	The particle size depends upon various factors such as temperature, leaf broth concentration, $CuSO_4{\cdot}5H_2O$ concentration
Cissus vitiginea leaf extract	Copper nanoparticles	Reduction by plant-origin reducing agent	Synthesized nanoparticles were used to treat urinary tract infections caused by microbes such as *Enterococcus*, *E. coli*, *Klebsiella* sp.
Azadirachta indica leaf extract	Copper oxide nanoparticles	Phenolic compounds and plant-derived flavonoids act as capping and stabilizing agent	The nanoparticles reduced the breast tumor by inducing apoptosis through the activation of TNF-α and caspases signaling
Cochlospermum gossypium gum extract	Platinum oxide nanoparticles	Hydroxyl groups present in the gum matrix mediated the reduction of metal cations	Requires 120 °C temperature, pressure 15 psi, and pH 8
Laurus nobilis aqueous leaf extract	Zinc oxide nanoparticles	Plant-derived biomolecules such as α-pinene, β-pinene, myrcene, limonene, linalool, methyl chavicol, neral, a-terpineol, geranyl acetate, eugenol, and	Antibacterial activity of the nanoparticle was greater against gram-positive (*S. aureus*) bacteria than gram-negative (*P. aeruginosa*), also

(continued)

Table 2.1 (continued)

Species	Type of nanoparticles synthesized	Enzymes required	Key components
		chavicol form complexing agents with zinc acetate	inhibited biofilm production
Orange peel extract	Zinc oxide nanoparticles	Reduction by orange peel extract-origin reducing agents	The size and morphology of synthesized nanoparticles depend on annealing temperature and pH
Ziziphus zizyphus aqueous leaf extract	Gold nanoparticles	Chemical reduction by plant-origin reducing agent	Synthesized nanoparticles exhibit antimicrobial activity, which depends on factors such as method of synthesis, size, shape, and concentration of nanoparticles
Salvia officinalis, *Lippia citriodora*, *Pelargonium graveolens*, and *Punica granatum*	Gold nanoparticles	Reduction by plant-origin reducing and capping agents	The plant extracts of these species acts as both reducing and capping agents

of metallic and metal oxide NPs is basically the outcome of these resistance mechanisms, primarily involving the processes of "bioreduction" and "biosorption."

Bioreduction is the process by which the metal ions are reduced by enzymes of an organism and also capped by biological components (Table 2.1). It facilitates harmless recovery of the synthesized NPs. Bioreduction is carried out by several microbial enzymes such as hydrogenases (Hyd-1, Hyd-2, and Hyd-3) produced by *Escherichia coli* reducing palladium (Pd) to form catalytic palladium (Deplanche et al. 2010) and nitrogen reductase produced by *Bacillus licheniformis* reducing silver ions to silver NPs (Kalishwaralal et al. 2008). It has also been found that plant extract components such as flavonoids, sugars, terpenoids, polyphenols, amides, phenolic acids, polyphenols, and alkaloids facilitate the synthesis of metallic NPs by bioreducing metallic salts (Iravani 2011). For instance, the terpenoids derived from geranium (*Pelargonium graveolens*) leaves have the capability to act as both reducing and capping agents. This terpenoids help in the synthesis of gold NPs by bioreducing chloroaurate ions (Shankar et al. 2003). Capping agents help in controlling overgrowth and agglomeration (cluster formation) of the NPs. Similarly neem (*Azadirachta indica*) leaf extract used in the biosynthesis of gold, silver, and bimetallic NPs by reducing the respective metallic salts (Singh et al. 2010).

Biosorption is a nutrient-independent mechanism for the biological synthesis of NPs. In this mechanism the metal ions which are present in aqueous microenvironment or soil will bind on the cell wall of the whole organism, such as bacteria, algae, or fungi, and biopolymers derived from the plants or other organisms (Muthu et al. 2017). In cell wall-based mechanism the metal ions bind to the exposed functional groups mostly carboxyl and polysaccharides such as chitosan and alginate on the extracellular space. The cell wall with high mucilaginous polysaccharides and more complex structure has better metal biosorption capacity. The biopolymers which contain several functional groups, such as hydroxyl, sulfhydryl, sulfonate, thioether, carboxyl, carbonyl, imine, amine, amide, imidazole, phosphate, and phosphodiester groups, also serve as metal binders (Muthu et al. 2017). The biosorption mechanism includes different steps, such as complexation, ion exchange, physical adsorption, chelation, and precipitation (Karman et al. 2015). Complexation is the process in which metal ions bind to the exposed functional groups on cell surface, and complex formation takes place simultaneously. Ion exchange involves the exchange of ion between metal ions and cell wall polysaccharide ions such as Na^+, K^+, Ca^{2+}, Mg^{2+}, etc. Physical adsorption is mediated by Vander Waals forces generated between the metal ions present in the medium and the cell wall of microbes. In precipitation, the metal ions are removed by microbial defense mechanism (Muthu et al. 2017). Biosorption mechanism takes place even in dead biomasses, for instance, brown algae *Fucus vesiculosus* synthesize gold NPs from their metal ions (Mata et al. 2009).

Bacterial Synthesis of Nanoparticles

Several bacterial species have the ability to synthesize various kinds of NPs with multiple inherent advantages that include, but are not limited to, their low energy requirement, freedom from using toxic agents and chemicals, ease of scaling, ease of manipulation, etc. (Table 2.1 and Fig. 2.2). The bacterial synthesis of NPs can be both intracellular and extracellular. However, intracellular synthesis is a complex process, and extracellular synthesis is often preferred due to the ease of extraction of the synthesized NPs.

Silver nanoparticles (AgNPs) of different sizes and morphologies are one of the most prominent metallic NMs that have been generated through bacterial synthesis and have numerous applications in the biomedical field. The size and shape of the NMs play important role in determining their functions. AgNPs can be synthesized intracellularly as well as extracellularly by using various bacteria. For, instance, *Pseudomonas stutzeri* intracellularly synthesized crystalline AgNPs of 200 nm size having triangular and hexagonal shapes (Klaus et al. 1999). Whereas *Morganella* sp. extracellularly synthesize crystalline AgNPs of size 15–20 nm (Parikh et al. 2008). *Lactobacillus* intracellularly synthesizes sub-micronic AgNPs, gold nanoparticles (AuNPs), and bimetallic Au/Ag crystals by reducing auric chloride and silver nitrate salts (Nair and Pradeep 2002). The rate of synthesis of NPs is affected by factors such as pH and bacterial chemical environment. Rate of AgNPs reduction increases with increase in pH; it also increases the recovery rate. The bacterial components not only synthesize NPs but also serve as capping agents to prevent NP overgrowth and aggregation (Sintubin et al. 2009). The *Leuconostoc lactis* exopolysaccharides reduce the silver nitrate solution to synthesize and stabilize round AgNPs of size 35 nm (Saravanan et al. 2017). The temperature also influences NP synthesis; increase in temperature enhances the rate of synthesis. More evidently under abiotic stress condition only cyanobacteria can reduce silver nitrate into AgNPs, but not by any chemical reaction. Cyanobacteria *Oscillatoria limnetica* aqueous extract reduces the silver nitrate to quasi-spherical AgNPs of 3.30–17.97 nm size. In this processes, sulfur or phosphorous functional groups help in metal binding, whereas free amino acids and sulfur-containing amino acids of bacterial extract serve as capping agents (Hamouda et al. 2019). In addition, there are also several other microbes that can synthesize AgNPs such as *Klebsiella pneumonia*, *Escherichia coli*, and *Enterobacter* (Shahverdi et al. 2007).

Gold nanoparticles (AuNPs) are another type of metallic NPs having therapeutic importance that can be synthesized by various bacteria with different sizes and morphologies. The bacteria *Rhodopseudomonas capsulata* bioreduces auric chloride ($HAuCl_4$) into AuNPs of different sizes and shapes. The size and shape of the synthesized NPs depend upon their pH. Spherical AuNPs of 10–20 nm size have been synthesized at pH 7, while at pH 4 nanoplates of 50–400 nm are synthesized (He et al. 2007). The bacteria *Pseudomonas aeruginosa* extracellularly synthesizes AuNPs of 15–30 nm size (Husseiny et al. 2007). The size of the synthesized NPs also depends on the type of bacterial strain. The bacteria *Cupriavidus metallidurans*, which is a gold resident, has the capability to biomineralize and accumulate AuNPs through a combination of efflux, reduction, and methylation processes resulting from the gold detoxification mechanisms of the bacteria (Reith et al. 2009). The bacteria *Halomonas salina* has the ability of biogenically synthesize of AuNPs by reducing auric nitrate (Sathiyanarayanan et al. 2014). The membrane-bound components of *Escherichia coli* K12 have synthesized AuNPs by reducing $HAuCl_4$ solution (Srivastava et al. 2013). There are various bacterial species such as *Caldicellulosiruptor changbaiensis*, *Bacillus cereus*, *Desulfovibrio desulfuricans*, *Shewanella loihica*, *Rhodopseudomonas capsulata*, *Micrococcus yunnanensis*, *Mycobacterium* sp., and actinomycete *Thermomonospora* sp., *Acinetobacter*, *E. coli* DH5α, *Klebsiella pneumonia*, *Pseudomonas putida*, and *Pseudomonas fluorescens*, which have the ability to synthesize AuNPs of different sizes and morphologies (Ahmad et al. 2003; Thakkar et al. 2010; Ovais et al. 2018; Singh et al. 2018). These metallic NPs have depicted potential antimicrobial, anticancer, and antioxidant activities along with their versatile use in photothermal therapy, drug delivery, biosensors, etc. (Mikhailova 2021).

Platinum nanoparticles (PtNPs) of therapeutic importance can be synthesized biologically by several bacteria (Riddin et al. 2010). The anaerobic bacteria *Desulfovibrio alaskensis* is a sulfate-reducing bacteria that synthesizes AuNPs by reducing $PtCl_4$ salt extracellularly. It requires the temperature of 30 °C at pH 7 and the production time of 2 h (Capeness et al. 2015), whereas another bacteria *Acinetobacter calcoaceticus* synthesizes cuboidal-shaped PtNPs of 2–3.5 nm size at pH 7 and temperature of 30 °C with a production time of 24 h (Gaidhani et al. 2014). Cyanobacteria *Plectonema boryanum* can synthesize amorphous spherical PtNPs of size 30 nm (Lengke et al. 2006). Several other species of bacteria are also used for the synthesis of PtNPs, such as *Psychrobacter faecalis*, *Vibrio fischeri*,

Jeotgalicoccus coquinae, *Sporosarcina psychrophila*, *Kocuria rosea*, *Pseudomonas kunmingensis*, and *Pseudomonas putida* (Bloch et al. 2021).

Microaerophilic bacteria *Aquaspirillum magnetotacticum* can synthesize iron oxide (Fe_2O_3) nanoparticles of octahedral prism shape and 40–50 nm size, and the NPs remain as a chain inside the organism (Mann et al. 1984). The magnetotactic bacteria *Magnetospirillum magnetotacticum* has also been reported to synthesize handle-shaped cluster of iron oxide (Fe_3O_4) nanoparticles and individual NPs of cubo-octahedron shape and 47 nm in size (Philipse and Maas 2002). Spherical iron oxide nanoparticles (IONPs) of diameter 19–30 nm have been synthesized extracellularly from the *Proteus vulgaris* ATCC-29905 (Majeed et al. 2020). Similarly, several other bacteria such as *Actinobacter* sp., *Shewanella oneidensis*, *Paenibacillus polymyxa*, *Geobacter sulfurreducens*, *Klebsiella oxytoca*, *Lactobacillus fermentum*, *Proteus mirabilis*, *Escherichia coli*, *Pseudomonas aeruginosa*, *Desulfotomaculum acetoxidans*, *Pseudomonas stutzeri*, *Bacillus subtilis*, *Streptococcus suis*, *Thiobacillus thioparus*, and *Alcaligenes faecalis* have been reported to synthesize IONPs of diverse size and morphology either by extracellular or by intracellular method (Nadeem et al. 2022). The bacterial origin of these sub-micronic metal oxide NPs has depicted significant antimicrobial activity and anticancer activity, along with several other clinical application such as antioxidant therapy, drug delivery, and cancer treatment. However, the potential cytotoxic effect and environmental pollution through eutrophication associated with these metal-based NMs need to be introspected properly (Alphandéry 2014; Nadeem et al. 2022).

The copper nanoparticles (CuNPs) are synthesized by incubating cell supernatant of *Morganella morganii* with copper sulfate ($CuSO_4$) solution. In this process copper ions will change their color from light blue to pickle green by bioreduction. The CuNPs synthesized by this method are having spherical shape and size 13.5 nm (Lalitha et al. 2020). The bacteria *Morganella psychrotolerans* can synthesize the CuNPs from $CuSO_4$ solution having a size of 4–60 nm. The pH of the medium plays a crucial role in determining the size and shape of the NPs (Pantidos et al. 2018). The metal-reducing bacteria *Shewanella oneidensis* intracellularly synthesize CuNPs of size 20–40 nm (Kimber et al. 2018). Anaerobic bacteria *Thermoanaerobacter* sp. X513 has been reported to synthesize CuNPs extracellularly by bioreduction of copper salts in an aqueous solution (Jang et al. 2015). Several other microbial species such as *Phormidium cyanobacterium*, *Serratia* sp., and *E. coli* can also synthesize copper oxide nanoparticles (CuO NPs) of different morphologies. These green-synthesized metallic NPs also depicted antibacterial activity, antioxidant and anticancer potential, and efficient photocatalytic activity for diverse applications (Akintelu et al. 2020).

Bacterial synthesis of zinc oxide nanoparticles (ZnO NPs) has been reported in several bacterial species by extracellular as well as intracellular methods. Star-shaped ZnO NPs of 50–80 nm diameters have been synthesized by bacterial cell extract of the cyanobacterium *Nostoc* sp. The ZnO NPs exhibited antibacterial action against different bacteria such as *E. coli*, *P. aeruginosa*, and *S. aureus*. These NPs are also found to be less toxic to normal human lung fibroblast cells as compared to cancerous lung alveolar epithelial cells (Ebadi et al. 2019). Flower-shaped ZnO NPs of 191 nm and 291.1 nm size have been reported to be synthesized by cell biomass and culture supernatant of zinc-tolerant bacteria *Lactobacillus plantarum* TA4 respectively (Yusof et al. 2020). Similarly, spherical- and hexagonal-shaped ZnO NPs are synthesized by several species of *Lactobacillus* such as *L. johnsonii*, *L. paracasei*, and *L. sporogens* (Yusof et al. 2019). Several bacteria of different species such as *Pseudomonas aeruginosa*, *Rhodococcus pyridinivorans*, *Sphingobacterium thalpophilum*, *Staphylococcus aureus*, *Escherichia coli*, and *Streptomyces* sp. can synthesize ZnO NPs of different sizes, mostly below 100 nm, and shapes like triangular, spherical, etc. (Yusof et al. 2019). The biomedical and industrial applications of ZnO NPs include, but are not limited to, anticancer potential, antibacterial activity, anti-inflammatory activity, treatment of diabetes, bioimaging, as a dietary supplement in animal feed, as antimicrobial agents in wound management, and in food packaging (Jiang et al. 2018; Yusof et al. 2019, 2020).

Titanium oxide nanoparticles (TiO_2 NPs) are another type of metal oxide NPs that have been synthesized by several bacterial species. TiO_2 NPs are potentially used in photodynamic therapy (PDT). The bacteria *Bacillus subtilis* extracellularly synthesize TiO_2 NPs by reducing the $TiO(OH)_2$ solution. The shapes of TiO_2 NPs synthesized by this process are spherical and oval in shape and 66–77 nm in size (Kirthi et al. 2011). The other *Bacillus* species such as *B. amyloliquefaciens* and *B. mycoides* synthesized spherical (22.1–97.2 nm) and polydispersed (40–60 nm) TiO_2NPs respectively (Nadeem et al. 2018). The other bacterial species *Lactobacillus* sp. and *Planomicrobium* sp. have also been reported to synthesize spherical TiO_2 NPs of size below 100 nm (Nadeem et al. 2018). Spherical TiO_2 NPs of 20 nm average diameter have been reported to be synthesized by *Staphylococcus aureus* that depicted antibacterial property and inhibited biofilm production (Landage et al. 2020). Chemoorganotrophic proteobacteria *Halomonas elongata* IBRC-M 10214 has been reported to synthesize spherical-shaped TiO_2 and ZnO NPs of 104.63 ± 27.75 and 18.11 ± 8.93 nm average diameters, respectively, that depicted antimicrobial activity against multi-drug-resistant (MDR) strains of *E. coli* and *Staphylococcus aureus* (Taran et al. 2017).

Fungal Synthesis of Nanoparticles

Fungi-mediated biosynthesis of metallic NPs takes place as their common response to toxic stress from the metal ions, which is alleviated through the bioreduction of metal ions to generate harmless protein metal nanoconjugates by the action of fungal-origin bioreducing and capping agents (Table 2.1 and Fig. 2.2). Biogenic synthesis of silver NPs having therapeutic implications has been reported by various fungi. By adjusting various factors such as temperature, pH, time, and fungus used can produce NPs of desired morphology. The temperature plays a crucial role in determining the size, shape, and speed of the NP synthesis. For instance, the optimum rate of the synthesis of AgNPs using *Trichoderma harzianum* can be achieved at 40 °C. Interestingly no synthesis of AgNPs took place at temperature of 80 °C or 10 °C using *Rhizopus stolonifer* due to the inactivation of proteins, enzymes, and other molecules. Another factor, the pH, determines the nitrate reductase enzyme activity. At higher pH there was more competition between protons and metal ions for establishing bonds in negatively charged regions. So at alkaline pH, the synthesis rate of NPs was higher. The fungus *Epicoccum nigrum* was incubated in $AgNO_3$ solution at alkaline pH to synthesize AgNPs. The synthesis of gold NPs using the fungus *Colleotrichum* sp. ALF2-6 at alkaline pH and at a temperature of 50 °C was faster than at lower pH. The $AgNO_3$ concentration can also influence the amount of NPs synthesized. For instance, an increase in the quantity of NP has been observed employing *Fusarium oxysporum* when the concentration of $AgNO_3$ increased from 0.1 to 1.5 mM. Further increase in concentration effected no additional change in synthesis. Another factor, culture medium, can also exert effect on NM synthesis. The fungus *Sclerotinia sclerotiorum* in potato dextrose can yield the highest production of NPs than any other medium. The amount of biomass used for the synthesis can also influence the properties of NPs. The fungus *Penicillium oxalicum* can produce high amount of NPs when more amount of biomass is used (Guilger-Casagrande and de Lima 2019).

The synthesis of gold nanoparticles (AuNPs) using fungi, although common, is relatively less exploited than bacterial synthesis, despite the former having several advantages over the latter. The NPs produced by the fungi depicted greater size distribution and morphological attributes. Nanoparticle synthesis by the fungi mostly takes place through extracellular means, and the synthesis rate is faster (usually <1 h) in cell-free filtrate than bacterial synthesis (commonly >24 h). The fungi-mediated NP synthesis rate may be enhanced in the presence of light. Further, the scalability of NP synthesis to industrial processes is more likely achievable through fungal means than bacterial processes. Almost 30 fungal species including *Trichoderma viride*, *Cylindrocladium floridanum*, *Verticillium* sp., *Fusarium* sp., *Penicillium* sp., and *Aspergillus* sp. have been reported to synthesize AuNPs. The biosynthetic process is multi-factorial, which depends on temperature, pH, metal ion concentration, fungal species, inoculum age, etc. (Kitching et al. 2014; Omar et al. 2019). One important consideration regarding the biomedical application of green-synthesized NPs through fungal or bacterial means is that the NPs may contain fungal- or bacterial-origin capping agents that can elicit an immune response in the target subject, mimicking fungal or bacterial infection as potential side effects (Kitching et al. 2014). Several other metal and metal oxide NPs are also being reported to be synthesized by fungal means. For instance, spherical-shaped platinum oxide nanoparticles can be produced using *Fusarium oxysporum* fungal extract when Pt ions are shaken at 200 rpm for 96 h at room temperature (Fahmy et al. 2020). Zinc oxide nanoparticles (ZnO NPs) can also be synthesized by using different fungi, such as *Aspergillus fumigatus*-synthesized ZnO NPs of 3.8 nm in size, by bioreducing Zn ions. The synthesized ZnO NPs exhibited high monodispersity without agglomeration. *Fusarium* sp. biomass has been reported to act as capping agents in the synthesis of ZnO NPs (Velmurugan et al. 2010; Yusof et al. 2019).

Plant-Based Synthesis of Nanoparticles

Nanoparticle synthesis using plant extract is a conventional, economic, and ecofriendly method. Plant extracts mitigate the use of chemicals and other toxic agents for NP synthesis. Different parts of the plant such as leaves, roots, fruits, rhizome, and flowers are used for the synthesis of nanoparticles (Fig. 2.2). Metal and metal oxide NPs are the principal output of plant extract-based synthesis. The basic mechanism for NP synthesis is bioreduction, in which metal ions or metal oxides are reduced to synthesize their respective NPs by using various plant-origin enzymes such as hydrogenases, reductases, and nitrases (Table 2.1). Some plant components act as reducing agents as well as capping agents, which synthesize and stabilize the NPs. For instance, plant-derived polyphenols act as capping agents in the synthesis of AgNPs. The different types of metal and metal oxide NPs synthesized by the plant system include, but are not limited to, AgNPs, AuNPs, CuNPs, ZnO NPs, and platinum oxide NPs (Vanlalveni et al. 2021).

Silver nanoparticles (AgNPs) of various morphological attributes have been synthesized by using various plant extracts. The plant species *Skimmia laureola* leaf extract has been employed in the synthesis of spherical-shaped AgNPs of about 38 nm in size. These NPs are tested against some microbes such as *E. coli*, *K. pneumoniae*, *P. aeruginosa*, *P. vulgaris*, *S. aureus*, and depicted antimicrobial activity. Another plant species *Prosopis farcta* leaf extract assisted in the biosynthesis of 10.8-nm-sized particles at room temperature. The NPs synthesized by this method are tested against bacteria such as *Staphylococcus aureus* (PTCC 1431), *Bacillus subtilis* (PTCC 1420), and *E. coli* (PTCC 1399), *Pseudomonas aeruginosa* (PTCC 1074) by using disk diffusion method and showed considerable antimicrobial activity. The leaf extract of *Mukia maderaspatana* was used for the synthesis of AgNPs of size 58–458 nm. The antimicrobial activity of synthesized NPs was tested against pathogens such as *Bacillus subtilis*, *K. pneumonia*, *S. typhi*, and *S. aureus*. The bioreduction of $AgNO_3$ to AgNPs has been carried out by using the leaf extract of some other plant species, such as *Salvinia molesta*, *Sesbania grandiflora*, *Indoneesiella echioides* and *Phlomis*. The ultra-small size AgNPs of diameter 7.38 nm was prepared by employing the leaf extract of *Hydrocotyle rotundifolia*. The leaf extract of *Amaranthus gangeticus* was used to produce AgNPs of globular shape. This globular-shaped AgNPs exhibited inhibitory action against gram-positive and gram-negative bacteria as well as some fungi.

Various plant seed extracts were also used for the biosynthesis of AgNPs (Vanlalveni et al. 2021). The seeds of *Sinapis arvensis* were used for the synthesis of AgNPs, which showed strong inhibitory action against fungus *N. parvum*. The AgNPs prepared by *Vigna radiate* exhibited more inhibitory action against gram-negative bacteria than gram-positive bacteria as the gram-positive bacteria possess thicker peptidoglycan layer in cell wall than gram-negative bacteria. The AgNPs synthesized by employing the seed extract of *Phoenix sylvestris* depicted strong biocidal activity against some skin bacteria such as *P. acnes* and *S. epidermidis* advocating for cosmetics application (Qidwai et al. 2018). The seed extract of various other plant species such as *Tectona grandis*, *Persea americana*, *Salvia hispanica*, *Trigonella foenum-graecum*, and *Pimpinella anisum* have also been used for the synthesis of AgNPs.

The flower extracts of some plant species are also being employed for the synthesis of various NPs (Vanlalveni et al. 2021). The flower extract of *Tagetes erecta* was used in the bioreduction of $AgNO_3$ to AgNPs, and these NPs exhibited antibacterial action against some bacteria such as *S. aureus*, *B. cereus*, *E. coli*, *P. aeruginosa*, *C. glabrata*, *C. albicans*, and *C. neoformans*. The flower extract of *Tecoma stans* was used for the synthesis of spherical-shaped AgNPs of 50–60 nm size. These NPs exhibited stronger antibacterial action against *S. aureus* than *E. coli*; similarly the flower extract of *Moringa oleifera* was employed to synthesize ultra-small AgNPs that exhibited antibacterial action against *S. aureus*. The polydispersed AgNPs synthesized by the flower extract of *Syzygium aromaticum* also acted as capping agents. *Pseudomonas* sp. depicted susceptibility to these AgNPs.

The biosynthesis of NPs by using root extract has depicted immense significance. The root extract of *Alpinia calcarata* was used as both bioreducing and stabilizing agents to synthesize AgNPs that were spherical in shape and induced cell disruption of bacteria (Vanlalveni et al. 2021). The root-bark extract of *Annona muricata* Linn was also used for the synthesis of AgNPs that depicted dose-dependent antimicrobial action against pathogenic bacteria such as *E. coli*, *B. subtilis*, *S. aureus*, *K. pneumonia*, and *Pseudomonas aeruginosa*. The AgNPs and AuNPs synthesized by employing *Arctium lappa* depicted different morphological attributes and anti-microbial potential. The AgNPs predominantly depicted spherical morphology with 21.3 nm of average diameter, while AuNPs showed versatility in shapes such as spherical, geometrical, triangular, and hexagonal with an average size of 24.7 nm. Such difference in morphological attributes of these two metallic NPs was underlined by the differential reduction potential as well as the NP-type specific capping agents of the plant source.

The fruit extracts of some plant species are also being employed in NP synthesis (Vanlalveni et al. 2021). For instance, lemon extract was used for the biosynthesis of AgNPs; antimicrobial action of these NPs was also tested against some pathogenic microbes such as *P. aeruginosa*, *E. coli*, and *S. aureus* yielding promising results. The fruit extracts of *Adansonia digitata* and *Momordica charantia* were employed for spherical AgNPs synthesis. The gum extracts were widely employed as bioreducing and stabilizing agents in various NP syntheses. The gum extract of *Azadirachta indica* was used for the synthesis of AgNPs with spherical shape that did not exhibit any aggregations. Another plant species *Buchanania lanzan* gum extract was also used for biosynthesizing AgNPs of size 14.74–19.86 nm. This AgNPs showed antimicrobial activity against *E. coli* and *Schizophyllum commune*.

Stem extract is another plant-origin component which act as bioreducing agent and has been widely used for the synthesis of diverse NPs. The aqueous stem extract of waste grass, *Moringa oleifera* and *Swertia paniculate*, has been reported to mediate the biosynthesis of AgNPs. The waste grass extract-synthesized AgNPs were reported to be smaller in size and, evidently, showed greater antimicrobial activity by disrupting the bacterial cell as compared to the green-synthesized AgNPs using *Moringa oleifera* and *Swertia paniculate* extracts. Similarly stem extract of

C. pulcherrima was employed in the green synthesis of AgNPs that depicted stronger biocidal activity against bacteria than fungi. Further, the AgNPs depicted synergistic effect when used along with conventional antibiotics like chloramphenicol and amphotericin B. The stem extract of Anthemisa tropatana was also utilized for the green synthesis of AgNPs that exhibited antimicrobial activity against *P. aeruginosa* and *S. aureus* (Vanlalveni et al. 2021).

Recently bark extracts of various plants were also extensively used for the green synthesis of AgNPs. The bark extract acts as bioreducing as well as stabilizing agent (Vanlalveni et al. 2021). The green-synthesized AgNPs using bark extracts of *Afzelia quanzensis*, *Syzygium alternifolium*, and *Cochlospermum religiosum* depicted considerable antimicrobial activity. *Ficus benghalensis* and *Azadirachta indica* bark extracts were used in the green synthesis of AgNPs of size 90.13 nm, which showed antimicrobial activity against pathogens such as *E. coli*, *P. aeruginosa*, *V. cholera* and *B. subtilis*. Usually, most of the green-synthesized AgNPs has been depicted to be more effective against Gram-negative bacteria than the Gram-positive bacteria. However, AgNPs synthesized by *Handroanthus impetiginosus* bark extract were reported to show higher antibacterial against Gram-positive *S. aureus* as compared to Gram-negative *E. coli* (Tormena et al. 2020).

Rhizome extract has also been used for the biosynthesis of size-selective AgNPs; spherical is the most commonly found shape, although other morphology is also available. For instance, rhizome extracts of *Bergenia ciliate* and *Dryopteris crassirhizoma* facilitated the synthesis of spherical AgNPs that depicted bactericidal activity against bacterial strains like *B. cereus* and *P. aeruginosa*. Most uniquely, AgNPs synthesized by using *D. crassirhizoma* rhizome extract as the reducing as well as capping agent under green LED light depicted the strongest antimicrobial activity as compared to the AgNPs synthesized under red and blue lights. Further, AgNPs synthesized in white LED light and sunlight depicted smallest zone of inhibition against *B. cereus* and *P. aeruginosa* (Lee et al. 2016). *Coptis chinensis* rhizome extract was employed for the synthesis of AgNPs of 15 nm size. The synthesized AgNPs as well as the chitosan-modified AgNPs both depicted bactericidal activity against *E. coli* and *B. subtilis*. Interestingly, the chitosan-modified AgNPs exhibited more antibacterial action than the free AgNPs. Similarly, ginger rhizome extract-synthesized AgNPs depicted more prominent antimicrobial activity against various aquatic pathogens than chemically synthesized AgNPs (Vanlalveni et al. 2021).

For the synthesis of nanoparticles with the desired shape plant peel extract is used as a reducing agent. The peel extract of *Citrus maxima* is used as both reducing and capping agents to synthesize AgNPs ranging from 4 to 11 nm in size. The citrus-mediated AgNPs exhibit antibacterial property against bacterial strains such as *E. coli* and *S. aureus*. The AgNPs mediated by *Citrus limetta* peel extract exhibit greater antifungal action. The bulb/peel extract was used for the biosynthesis of nanoparticles. The bulb extract of *Allium cepa* is used as both reducing and capping agents for the synthesis of AgNPs of size 10–23 nm. The onion-mediated AgNPs exhibit antimicrobial against various pathogens such as *B. subtilis*, *B. cereus*, *B. licheniformis*, *S. aureus*, *S. mutans*, *E. coli*, *K. pneumoniae*, *S. typhimurium*,

P. aeruginosa, *P. vulgaris*, *S. marcescens*, and *C. albicans*. The spherical AgNPs of size 10–25 nm are synthesized by reducing 20 mM $AgNO_3$ with 0.5 mL *Crocus haussknechtii* Bois extract at a temperature of 75 °C and at pH 7. The AgNPs synthesized by employing aqueous extract of *Allium sativum* have antimicrobial activity against pathogenic bacteria such as *P. aeruginosa* and *B. licheniformis*. These nanoparticles are used as antimicrobial drugs due to their cell-permeable ability. The antimicrobial activity is higher at a temperature of 95 °C and leaf broth concentration of 15% and at these conditions (Vanlalveni et al. 2021).

Copper nanoparticles (CuNPs) have been synthesized by using plant leaf extract as a reducing agent, and they also depicted significant bactericidal activity. The *Magnolia kobus* leaf extract was treated with an aqueous solution of $CuSO_4 \cdot 5H_2O$ for the synthesis of stable CuNPs. The formation of CuNPs was monitored using UV-vis spectroscopy. The particle size depends on various factors, such as temperature, leaf broth concentration, and $CuSO_4 \cdot 5H_2O$ concentration. The synthesized NP size decreased with increase in leaf broth concentration up to 15% and increased at 20% leaf broth concentration (Lee et al. 2013). The leaf extract of *Cissus vitiginea* was utilized for the synthesis of CuNPs. This CuNPs were used to treat urinary tract infections caused by microbes such as *Enterococcus*, *E. coli*, *Klebsiella* sp. (Wu et al. 2020).

Platinum nanoparticles (PtNPs) were synthesized by using various plant extracts. The leaf broth of *Azadirachta indica* was incubated with Pt ions for 1 h at 100 °C to synthesize PtNPs. The terpenoids present in the leaf broth act as reducing as well as capping agents. *Nigella sativa* (black cumin) seed extract was used to synthesize PtNPs; in these processes black cumin seed extract and Pt ions were stirred together for 2 days at 200 rpm and at 75 °C temperature. The green synthesis of PtNPs was obtained by employing the leaf extract of *Diospyros kaki* (Persimmon). Using 10% leaf broth concentration at 95 °C for 2–3 h, more than 90% of Pt ions are bioreduced to PtNPs. The gum extract of *Cochlospermum gossypium* was mixed with Pt metal ions at 120 °C and pressure 15 psi (autoclave) at pH 8 to synthesize PtNPs. Spherical PtNPs were obtained by bioreducing Pt ions with the leaf extract of *Mentha piperita* (Peppermint) at 60 °C for 90 min (Fahmy et al. 2020).

Zinc oxide NPs were synthesized by using various plant species and their extracts. Some of the plant species used to synthesize ZnO NPs include *Cassia fistula*, *Trifolium pratense*, and *Ocimum basilicum*. The aqueous leaf extract of *Laurus nobilis* was mixed with two different zinc salts (zinc acetate and zinc nitrate) to synthesize ZnO NPs. The process of the synthesis of the ZnO NPs includes different steps such as mixture preparation and grinding the mixture to powder, centrifugation, stirring of the mixture, and precipitation. The synthesized NPs from zinc acetate and zinc nitrate were 21.49 nm and 25.26 nm in size and exhibited bullet and flower-like structures respectively (Fakhari et al. 2019). Orange peel extract was also used to synthesize ZnO NPs. The orange peel extract acts as reducing agent and reduces zinc acetate to ZnO NPs. The size and morphology of the synthesized NPs depend on annealing temperature and pH. The synthesized NPs exhibited strong antibacterial action against some bacteria such as *E. coli* and *S. aureus* (Thi et al. 2020). The root extract of *Rubus fairholmianus* acted as a reducing agent to synthesize ZnO NPs. It reduced zinc nitrate solution to form ZnO NPs, due to phytochemicals present in *Rubus fairholmianus* root extract. The synthesized NPs exhibited antimicrobial property against pathogenic *Staphylococcus aureus* (Rajendran et al. 2021).

The synthesis of AuNPs was also reported from several plant extracts. The aqueous leaf extract of *Ziziphus zizyphus* acted as a reducing agent due to the presence of phytochemicals. The leaf extract reduced Au^{3+} ions to Au^0 to synthesize AuNPs. The antimicrobial activity of the AuNPs was dependent upon parameters such as the method of synthesis, size, shape, and concentration of NPs (Aljabali et al. 2018). The plant extracts of different other species such as *Salvia officinalis*, *Lippia citriodora*, *Pelargonium graveolens* and *Punica granatum* were also used to synthesize AuNPs. The plant extracts of these species served as both reducing and capping agents. The AuNPs are formed by the reduction of Au^+ ions to Au atoms and followed by aggregation of the Au atoms to generate the NPs. The biomedical applications include, but are not limited to, drug delivery, gene delivery, pathogen detection, fluorescent labeling, and tissue engineering (Zeiri et al. 2014).

2.4 Characterization of Nanoparticles

The arena of nanomaterials is currently growing through an exponential phase as it has gained wide attraction due to their potential applications in divergent aspects of nanomedicine, drug and vaccine delivery, agriculture, environmental science, and engineering. Thus, the characterization of NPs has utmost importance, as these unique properties at the nanoscale level that are not expressed in their bulk counterparts make them an enthralling topic of research. Nanoparticles are being synthesized by a wide number of techniques, which consist of chemical, mechanical, and other pathways, as we have discussed in the previous section. Thus, NPs are currently produced in larger amounts than they were a decade ago, demanding the customization of unambiguous and reliable protocols for the characterization of NPs to surmount the inherent struggle for analyzing nanomaterials because of their comparatively low quantity and nano-size in contrast to their bulk particles. Herein we mark out the distinct characterization methods of NPs that are extensively used for studying various properties of nanomaterials (Table 2.2).

2.4.1 X-Ray-Based Techniques

2.4.1.1 X-Ray Diffraction (XRD)

This is one of the cornerstone techniques used for the characterization of NPs. X-ray is having wavelength in the atomic-scale range, and being a non-destructive technique, it provides credible information regarding the structural properties of NPs.

Table 2.2 Techniques for the characterization of nanoparticles

Techniques	Properties of NM explained	Advantages	Limitations
1. X-ray-based techniques			
XRD	Size, composition, crystalline structure	Required only minimal samples; widely available	Detection limit of approximately 1% by volume
XAS	NP chemical and elemental composition, oxidation state, interatomic distance, Debye-Waller factor	Operated at high temperature; good spatial resolution; structural characterization of NP alloys possible	Especially powerful source of X-rays needed
XPS	Oxidation state, ligand binding properties, structural composition of particles, growth kinetics	Can probe down to core electrons; differentiate between oxidation states of molecules; suitable for elucidating uppermost layers	Long time for the generation of spectra; limited to elements having atomic number ≤ 3
XMCD	Magnetic and electronic structure; site symmetry; magnetic moments	Direct determination of spin polarization; size, direction, anisotropies of magnetic NPs that are not perfectly crystallized can be determined	Element specific
2. Spectroscopy techniques			
FTIR	Surface composition, ligand binding, nature, and strength of bonds	Good spectral quality; increased data collection speed; data reproducibility; relative ease and maintenance of use	Sampling chamber produces limitation due to its small size; mounted pieces sometimes disturb IR beam; certain materials completely absorb IR, making it impossible to get credible result
NMR	Intramolecular ligand dynamics; size, structure, and composition change under the influence of ligands; electronic structure of materials and bonding environment	View molecular dynamics in liquid or solid state; robust and reproducible method; produce rich structural information; simplify and speed up the acquisition of data	Insufficiently concentrated samples show low sensitivity to technique; magnetic field drift
UV-visible spectroscopy	Optical properties, shape, size, agglomeration state	Quick analysis; easy to use; most useful for organometallic species	Cannot quantify the extent of metal NP functionalization
PL spectroscopy	Optical properties by analyzing the electronic structure of NMs	Free of photoblinking and photobleaching, making advantageous for metals; non-destructive and direct technique;	Disappearance of luminescence signals at certain times; many optical centers having multiple excited states create problem

(continued)

Table 2.2 (continued)

Techniques	Properties of NM explained	Advantages	Limitations
		require low sample temperature	
DLS	Size of magnetic NP and those in colloidal suspension; detection of agglomerates; diffusion coefficient of NPs	Low cost and robust method; reduce the screening time in comparison with conventional technique; higher sensitivity; accurate detection limit	Occasional multiple scattering effect; cannot precisely characterize polydispersed sample
NTA	NP size and its distribution; hydrodynamic size	Superior to DLS as it can characterize ploy- and mono-disperse samples; reproducibility; verified accuracy	As particle size increases it leads to more scattering of light
LEIS	Chemical composition and thickness of self-assembled monolayers	Excellent surface sensitivity; fast and direct analytical technique; giving credible information on the composition of outer atomic surface	
Mossbauer spectroscopy	Chemical, electronic, and magnetic properties of NPs, oxidation state, symmetry and spin state	Excellent technique for quantitative determination of valence state, esp. iron-containing particles	Effect can be detected only in those isotopes with low-lying excited states; resolution depends on the lifetime of produced excited states; bulk technique
MS	Size; elemental composition; chemical state; charge and size of metallic core NPs	Sensitive and highly compatible technique; can attain real-time information; quantify bioconjugation of target molecule	Unable to feel optical and geometrical isomers apart
ICP-MS	Elemental analysis of metallic NPs; analyze metallic impurities in NPs	High selectivity and robustness; interference-free detection	
ICP-OES	Characterize functional groups; density of ligand binding on engineered NPs	Reproducible and sensitive; detect variations at trace-level concentration and identify multiple elements; analyze liquid sample	
3. Magnetic property-based techniques			
SQUID	Magnetization saturation and remanence; blocking temperature	Magnetic imaging; detect changes in magnetic	Noise limitation

(continued)

Table 2.2 (continued)

Techniques	Properties of NM explained	Advantages	Limitations
		properties even in small spin system	
VSM	Magnetic moment; magnetization saturation and remenescence	Changes in the range of 10^{-5} and 10^{-6} emu can be verified; highly accurate	Suffer from temperature dependence; cannot be used on fragile samples
FMR	Shape, size distribution, surface composition, magnetic anisotropism	Wide range of measuring frequencies; information on crystallographic imperfection; easier and simple manipulation of data	Care must be given for the applied magnetic field as spin precession depends on it
SPMR	Magnetic flux density	Ideal approach for the detection of cancer and the treatment monitoring; determining antibodies	
4. Optical and electron microscopy techniques			
TEM	Size, distribution, morphology of NPs; chemical nature	Detect the location of NPs within cells; determine subcellular-level variation in the case of NP-based treatment	Large number of particles cannot be analyzed; not often produce wrong images
Liquid TEM	Can detect NP growth in real time; study superlattice formation; help study the growth mechanism of NP	Used in analyzing biological analytes; overcome the limitation of conventional TEM	Certain times using SiN grids causes scattering of electron beam
Cryo-TEM	Study complex growth mechanism; information on aggregation process	Determine structure with high resolution; provide images during growth time; free energy of QDs determined	Very low signal-to-noise ratio
SEM	Surface chemistry; structure; size; height of NP	High resolution esp. spatial resolution at sub-nanometer level and interactions between them	Large size; expensive
STEM	Variation in the KE of electrons; analyze chemical state and atomic-level structure; crystal structure	Using EELS or EDX produces elemental maps showing features down atomic level; imaging distinct material interface	Sensitive to vibration and electromagnetic field
AFM	Size and shape in 3D mode	High resolution and magnification; generate 3D images and thus can determine height; lower scanning time	Thermal drift becomes a hurdle during precise and high-resolution imaging, particularly in studies at small scales

(continued)

Table 2.2 (continued)

Techniques	Properties of NM explained	Advantages	Limitations
5. Miscellaneous techniques			
Zeta potential	Surface charge in colloidal NPs; physical stability	Allows simultaneous size and surface charge measurements	
BET	Surface area	Simple, robust, accurate method	Occasionally the resulting agglomeration induces certain error
TGA	Mass and composition of stabilizers	Analyze properties of polymers, residual metal content of sample; analyze surface-coated materials for NPs	Variation lower than 1 nanogram cannot be detected in conventional TGA
RMM-MEMS	Detection and measurement of sub-micron and sub-visible information on particles and their mass and size buoyancy, density	Helps in characterization of nanobubble; differentiate positive and negative buoyant particle in NP sample	
DSC	Difference in melting temperature and analysis of crystallinity; specific heat capacity detected	Characterize based on thermal properties; automated	

XRD reveals information on crystallinity, grain size, lattice variables, and regarding the nature of phase of NPs. The XRD method is often carried out in the powder form of samples; it provides certain merits due to as they are volume averaged values represented statistically (Whitfield and Mitchell 2004). Deduction of composition of particles using XRD is achieved by collating the intensity and position of peaks obtained, with the database of International Centre for Diffraction Data (ICDD). Peak broadening can lead to significant complications in the XRD analysis of NPs; thereby, it doesn't fit particles having a size less than 3 nm and of amorphous materials (Mourdikoudis et al. 2018).

2.4.1.2 X-Ray Absorption Spectroscopy (XAS)

This technique provides the measurement of chemical and elemental composition as well as the crystallinity of NPs on the basis of X-ray absorption attributes, which is assessed as a function of energy. XAS comprises both X-ray absorption near edge structures (XANES) within 50–100 eV above the edge and extended X-ray absorption fine structure (EXAFS) having spectral features near 500 eV or above (Holbrook et al. 2015). The EXAFS and XANES analyses help to ascertain the electronic properties, atomic structure, Debye-Waller factors, and oxidation state of NPs. Synchrotrons, an especially powerful source of X-rays, are necessary for acquiring XAS spectra, and that adds a limitation as the technique is not readily available. The EXAFS technique can be operated at a temperature higher than 200 °C and aids in the determination of both kinetic and structural characters in supported metal catalysts for the reactions taking place over a few seconds. The structural analysis of NPs of size less than 10 nm can be effectively performed using EXAFS as it exhibits good spatial resolution and information regarding inter atomic distances and co-ordination numbers. The structural characterization of NP alloys can also be performed by EXAFS. XANES is significant for determining vacant orbitals, oxidation states, site uniformity of the absorbing atom, and electronic configuration. Of note, superiority of EXAFS over radial electron density distribution (REDD) is the selectivity, and REDD provides precise data of interatomic distances. Ideally along with EXAFS, Fourier-transform infrared spectroscopy (FTIR), XANES, and the inclusion of REDD are recommended to derive the knowledge on physical, chemical, and structural features of NMs (Mourdikoudis et al. 2018).

2.4.1.3 X-Ray Photoelectron Spectroscopy (XPS)

The method is having an alternative acronym electron spectroscopy for chemical analysis (ESCA), and it is an extensively exploited technique for the analysis of the surface chemistry of NMs. The fundamental physical principle underlying the mentioned technique is photoelectric effect (Mourdikoudis et al. 2018). XPS predominantly facilitates the analysis of composition of the mesoporous materials on the stand of electronic state of the particle. Illumination of mesoporous particle with a specific energy (eV) of X-ray beam by maintaining high vacuum, ionization, and emission of electrons from the inner shell is assessed by electron energy analyzer and creates a photoelectron spectrum ranging between 0 and 1400 eV. Each element present in the mesoporous material has a distinct binding energy that results in a characteristic peak of the spectrum. It also provides knowledge about the possible changes, the functional group may traverse during chemical conversions; for instance, differences in the oxidation states during catalysis. The data acquired from the spectrum helps in deducting the kind of molecular species or elements present. XPS operating under ultra-high vacuum condition aids in analyzing surface functionalization, dielectric properties, and ligand-exchange interactions of NMs. XPS is also an excellent technique to determine the internal heterostructures of NPs. Near-ambient pressure (NAP)-XPS technique promotes the in situ study of the steady-state conditions prevailing at solid-gas interface, which are having significant relevance in the domains of electrochemistry, environmental science, catalysis, and so on. Since X-rays can penetrate only up to a depth of <20 nm in the sample, the technique is mainly suited for elucidating the superficial layers (Asefa and Dubovoy 2017).

2.4.1.4 XMCD (X-Ray Magnetic Circular Dichroism)

This is a state-of-the-art characterization technique that provides in-depth information about the magnetic and electronic structure of nanoparticles. Dichroism is the ability of material to rotate the plane polarized light after absorption, and the symmetry of material is lost during the event. Circular dichroism detects the variation in right and left polarized light by the substance. XMCD is providing a difference in x-ray absorption spectra (XAS) obtained by right and left circularly polarized light. This technique is well suited for the analysis of ferrimagnets and ferromagnets, and assists in the quantitative study of the direction and size of spin as well as orbital moments. One of the striking features of this technique is that it is successful for the direct determination of spontaneous spin polarization of gold (Au) nanoparticles; the development of this spin polarization is a result of interaction between protecting polymer and nanoparticles. X-ray magnetic circular dichroism is very sensitive to the site symmetry of absorbing ions and the amplitude and orientation of magnetic momentum. This novel approach thus brings into sight the data on size, direction, and anisotropies (sizes in various directions) of the atomic-level magnetic moment for even the particles that have not perfectly undergone crystallization (Stöhr et al. 1998).

2.4.2 Spectroscopy-Based and Allied Techniques

2.4.2.1 Fourier-Transform Infrared Spectroscopy (FTIR)

This is an analytical technique based on the quantification of the electromagnetic radiation absorption, having a wavelength within the range of 4000–400 cm^{-1}, i.e., the mid-infrared region. When the molecule absorbs infrared radiation, it turns into infrared active due to the modifications exhibited in the dipole moment. Precise interpretation of spectrum yields knowledge on, the specific functional groups if present, the relative nature and strength of bonds and details regarding molecular structures and interactions, as FTIR is sensitive toward the changes in the molecular structure. FTIR technique is employed for analyzing polymeric, organic, and inorganic materials, and deciphers surface composition and ligand binding. A characteristic change in the pattern of absorption bands is an obvious indication of change in the composition of material. FTIR spectroscopy can effectively display the distinct chemical bonding present in a particle (Mourdikoudis et al. 2018).

2.4.2.2 Nuclear Magnetic Resonance (NMR) Spectroscopy

It is an exemplary technique that provides credible knowledge on the interactions and nature of NMs. It's a non-destructive, robust, and reproducible method that can be effectively employed for the quantitative and qualitative characterization; it provides deep insight on shape, chemical structure, size, diffusion properties, intermolecular interactions, and intramolecular ligand dynamics. The pivotal principle underlying NMR is that atomic nuclei express certain magnetic properties which can be used potentially to extract the information on chemical and molecular environments. Ideally, NMR is widely exploited to study co-ordination or interaction between the surface of antiferromagnetic or diamagnetic NPs and the ligands. However, it is not advisable for the characterization of ferro- or ferri-magnetic particles, as the large-scale magnetization shown by these materials leads to disparity in the local magnetic field, resulting in notable broadening of signal peaks making the data inutile. Of note, NMR is useful for determining both the development and final structure of noble metallic NMs. Capping ligands are one of the major elements studied by NMR, and the obtained information includes atomic composition, bonding environment, electronic structure, in general, various features of the particle core. NMR is also utilized for the direct evaluation of the adsorbed gas diffusing onto the metallic NPs. Diffusion-ordered NMR spectroscopy (DOSY-NMR) put forward the possibility to differentiate bound ligands and in situ free ligands; at the same time the relative distribution of the ligands and its influence on the size and shape of NPs can be measured (Mourdikoudis et al. 2018). NMR measurements boost up the knowledge on transport and release of drugs especially using AuNPs. Metal oxide NPs can be effectively characterized with ease using solution NMR spectroscopy, and SS (solid state) NMR spectroscopy is a potential tool to analyze the robust behavior of solid catalysts and the vigorous chemical reactions occurring on their surface. This provides credible insight into the bonding nature of ligand particles (Agarwal et al. 2018).

2.4.2.3 UV-Visible Spectroscopy

Ultraviolet (UV) visible (vis) spectroscopy is an extensively used characterization technique to analyze the optical properties of NPs. Nanoparticles possess optical properties which in turn are sensitive to shape, agglomeration state, size, refractive index adjacent to NP surface, and concentration that makes uv-vis spectroscopy a credible tool for their characterization. This spectroscopy technique underlies the principle of Beer-Lambert law, which states that the quantity of incident light absorbed by the solution of a particular wavelength is in turn related to the path length and the concentration of absorbing species. It evaluates the intensity of reflected light from a sample and cross-checks with the intensity of reflected light from the reference material.

UV-visible spectroscopy is an excellent technique for the characterization of microgels loaded with noble metal nanoparticles. Time-course UV-visible is exploited to analyze the kinetics of microgel-related swelling and deswelling. Certain metallic NPs made of gold, copper, and silver show strong interaction with specific wavelengths of light as a result of localized surface plasmon resonance signals present in the visible spectrum and exhibit unique optical properties. At certain times, from UV-visible spectroscopy spectrum data, information regarding the molar concentration and size can be obtained, for instance, in AuNPs. An ability to visually identify molecules such as thiamine by the change in color of the solution (e.g., if it would be gold NPs color changes from red to greenish gray) adds in another application of uv-vis spectroscopy. It is also a robust approach to monitor the effect of various dyes on distinct NPs by analyzing variations in the visible spectrum of dyes (Mourdikoudis et al. 2018).

2.4.2.4 Photoluminescence (PL) Spectroscopy

This is another approach to study and characterize nanoscale materials. This technique is preferably suited for fluorescent NMs such as metal nanoclusters and quantum dots; it analyzes the light emitted from molecules or atoms that have received the photon. This spectroscopic characterization method is a non-destructive and direct technique that provides insight into the optical properties by analyzing the electronic structure of NMs. The variation in electronic properties of the nanoscale material from its respective bulk counterparts is due to the differences in the electron density states (Singha et al. 2005). At relatively low sample temperatures, the associated characteristic PL spectrum displays certain spectral peaks, related even to the extremely low impurities contained in the sample material, which otherwise would have influenced the quality and performance of NMs (Abraham et al. 2020). Photoluminescence of metals adds an advantage, making it free from photoblinking and photobleaching; thus, it serves as a better option for optical labeling.

2.4.2.5 Dynamic Light Scattering (DLS)

DLS is also referred to as photon correlation spectroscopy, which is one of the widely used techniques to calculate the size of magnetic nanoparticles (MNPs) and also of those in colloidal suspensions where they are in a continuous Brownian motion. In DLS technique, the MNPs are exposed to an electromagnetic wave (beam of light), and the resultant scattering brings changes in the intensity and path of light; as these particles are in a random motion, these variations exhibited can be utilized to quantify diffusion coefficients of the NPs. For MNPs possessing spherical shape, the hydrodynamic radius can be calculated by Stokes-Einstein equation. Care must be taken so that the concentration of NPs is low to avoid multiple scattering effects. The time taken for measurement is relatively short, and the process is mostly automated, which makes the technique less labor-intensive. One of the important features of DLS is since the intensity of the scattered light and the sixth power of the radius of the particle are directly proportional, it makes the DLS technique especially sensitive to the existence of even small aggregates (Lim et al. 2013). DLS is also used for measuring the formation of aggregates as a consequence of virus-antibody binding. This innovative approach effectively reduced the screening time in comparison with the conventional methods as ELISA assays. Being a low-cost and robust method, along with its higher sensitivity and accurate detection limits, made DLS a potential technique for the characterization of the colloidal stability of magnetic NPs in suspension (Mourdikoudis et al. 2018).

2.4.2.6 Nanoparticle Tracking Analysis (NTA)

It is a relatively newly developed tool for size characterization that determines the size of the particles by measuring the diffusion coefficient. As NTA uses identical physical properties as that of DLS, it is a significant method to verify the results obtained through DLS. Furthermore, in comparison with DLS, NTA maintains lower concentration detection limit. NTA assesses the diffusion coefficient by analyzing the vigorous movement of each particle, recorded and visualized using a high-resolution camera having a microscope (Zhou et al. 2015). Molecules are visible as a result of the light scattering when we illuminate the sample by an integrated laser. From the obtained diffusion coefficient, we are able to determine the hydrodynamic radius through Stokes-Einstein equation. The reproducibility and the verified accuracy of NTA made it a suitable technique to quantify the size population of bimodal samples. Recently, it has gained significance in the characterization of nanoparticles of those developed as therapeutic protein products, for instance, interferon-β-1a and IgG. The particle sizing property of NTA is superior to that of DLS, as NTA calculates the accurate size value for both mono- and polydispersed samples (Mourdikoudis et al. 2018).

2.4.2.7 Low-Energy Ion Scattering Spectroscopy (LEIS)

One of the critical surface properties of NPs that requires to be quantified is the chemical composition and the thickness of SAM (self-assembled monolayers), which aids in the assembly and ordering of SAM. LEIS, a modern analytical technique with excellent surface sensitivity, assists in the rapid analysis of the elemental composition of self-assembled monolayers. It is a fast and direct technique in which the samples are initially exposed to some low-energy noble gas ions; the subsequent scattering and the resultant reduction in energy of these ions, displayed via spectrum, give the relevant information regarding the composition of the outer atomic surface (Bruener et al. 2019). High-sensitivity-low-energy ion scattering (*HS-LEIS*) provides improved sensitivity, approximately 10 nm from the top atomic layer, that leads to surface characterization and collects distinct and accurate information with the least amount of surface damage. LEIS technique has been successfully exploited in the characterization of NPs in divergent fields, such as thickness determination, catalysis, and behavior of the NPs in ionic liquids. The thickness values obtained using LEIS were in agreement with the data gained using atomic force microscope (AFM), sputter depth profiling and x-ray reflection techniques (Mourdikoudis et al. 2018).

2.4.2.8 Mossbauer Spectroscopy

This technique is also referred to as nuclear gamma resonance spectroscopy. It is a versatile spectroscopic technique for the determination of structural, chemical, electronic, and magnetic characteristics of NPs, underlying the principle of the Mossbauer Effect, proposed by R.L. Mossbauer. The technique utilizes two parameters—the Mossbauer Effect and the Doppler shift—to study the relatively fine transitions of the nucleus from the ground state to a higher energy level. The nuclear energy level varies due to the interaction of radioactive nucleus with its immediately surrounded solid in which it is immersed; this is the fundamental concept utilized in Mossbauer spectroscopy, and the extensively studied element is iron as its isotopes are showing the Mossbauer Effect (recoilless resonant absorption of the nucleus). In Mossbauer spectroscopy, the solid sample is irradiated by gamma rays, and the detector measures the intensity of the transmitted light and then represents it in the form of a spectrum, in which the intensity of the gamma ray is represented as a function of velocity. The peaks in the spectrum yields information regarding the chemical environment of the sample. This technique can be used to analyze the spin state, oxidation state, and symmetry, along with the magnetic ordering of iron (Fe) atoms in the given sample, and subsequently determine the magnetic phase. Furthermore, the Mossbauer spectrums plotted as a function of temperature for a magnetic material can also be utilized to measure superparamagnetism and calculate magnetic anisotropic energy (Mourdikoudis et al. 2018). In this spectroscopic technique isomer shift is the chief parameter that shows up due to the nuclear energy shift contributed by coulombic interactions occurring between the electron-dense site of the nucleus and the nucleus itself. This parameter can be exploited to calculate the valency, oxidation state, and electron shielding effect. Secondly quadrupole splitting is an important parameter which helps to determine the arrangement of ligands, site symmetry, and spin state. Here the excited states of nucleus are cleaved into two degenerate doublets given that the magnetic interactions are negligible. There are two major limitations for Mossbauer spectroscopy—firstly, the relative occurrence of recoil-free events is strongly related to the energy of gamma ray, thus the effect can be detected only in those isotopes with low-lying excited states. Next, the resolution depends on the lifetime of the produced excited states. These both factors together reduce the number of isotopes that can be used under this spectroscopy (http://www.rsc.org/Membership/Networking/InterestGroups/MossbauerSpect/Intropart1.asp).

2.4.2.9 Mass Spectrometry (MS)

The technique has currently emerged as a strong tool for the characterization of size, composition, chemical state, and structure of NPs. Under the MS technique the sample undergoes ionization and further sorting on the basis of the mass-to-charge ratio in electric and magnetic fields. Recent development in MS has paved the way for precise characterization of the charge and size of the metallic core of ultra-small NPs (1–3 nm range) and also for quantifying the number of ligands present. MS also provides the additional advantage of monitoring the evolution of these properties at the time of their synthesis. Further, MS technique can be used to quantify the bioconjugation of target molecules. Being a sensitive and highly compatible technique, it can be integrated with various separation modalities to attain real-time information, providing novel insights into the characters of NPs, which in turn leads to an increase in their potentiality.

Inductively Coupled Plasma MS (**ICP-MS**), with its high selectivity, robustness, and sensitivity and possessing a wide dynamic range and simple protocols, is well exploited for the elemental analysis of metallic NPs. ICP serves as an excellent ionization source in the mass spectrometry technique, which is capable of ionizing more than 90% elements. Of note, ICP-MS can analyze the metallic impurities present in the non-metallic NPs. Coupling MS with matrix-assisted laser desorption/ionization (MALDI) and electrospray ionization (ESI) can yield information on the ligands surrounding the NPs. Integration of ICP-MS with size-exclusion chromatography provides data on the elemental and size characterization of AuNPs. *Field Flow Fractionation* (*FFF*) technique coupled with ICP-MS helps to analyze the size and multi-elemental composition of natural colloids.

A major breakthrough in mass spectroscopy is the introduction of single particle operation mode ICP-MS (**spICP-MS**), a high throughput, low-cost method, yielding information on the size distribution and concentration of NPs; of note, the basic principle of spICP-MS is that NP suspension used for characterization must be highly diluted so that it can be analyzed individually. On the basis of mass-to-charge ratio the NP constituents generate a distinct pulse of ions, and the frequency of these signals can be correlated with particle number concentration in solution. These, in turn, help to calculate the average size and distribution of NP (Mozhayeva and Engelhard 2019). Of the many uses of spICP-MS, one of the significant properties is that it is employed to analyze and characterize TiO_2 and Au NPs that are being used in water purification systems. Molecular chemical information of NPs can be extracted using a surface analysis technique known as secondary ion mass spectroscopy (**SIMS**), where primary ions are employed to sputter negatively and positively charged secondary ions. Under SIMS technique, a beam of primary ions possessing high energy (Cs^+, O, Ar^+) is directed onto the analytes in order to obtain a secondary mass spectrum which will be recorded and analyzed further. Time-of-flight secondary ion mass spectroscopy (**ToF-SIMS**) is another analytical variant with high surface, molecular, and chemical sensitivities and is used for the characterization of NPs in drug delivery formulations. Nanoparticle-based SIMS is having paramount importance since these NPs can be architectured accordingly to measure enzyme activity directly. ToF-SIMS also gives information about the molecular confirmation and orientation in addition to the functional groups present. Currently ToF-SIMS has been employed for their potential in characterizing functional groups or surface coating of nanomaterials (Mourdikoudis et al. 2018).

2.4.2.10 Inductively Coupled Plasma-Optical Emission Spectrometry (ICP-OES)

This is a highly sensitive and reproducible technique that can effectively characterize the functional groups or ligands present on the engineered NPs, which are having relevant importance in nanomedicine (Mourdikoudis et al. 2018). In this technique, external plasma energy (high electron density and temperature) is passed to the sample, which makes the components in the sample excited to a higher energy state, and when it returns from higher to lower energy states, there is the production of emission rays. The detector system measures the wavelength and intensity of emitted light and produces the spectrum. As each element produces different intensity peaks, the determination of the elements can be carried out with ease. ICP-OES is able to detect variations even at trace-level concentration and has the capability to simultaneously identify multiple elements. Thus, it provides elemental information about the conjugated surface groups on AuNPs, which have attracted great attention in the field of biomedicine, chiefly as effective drug delivery agents (Elzey et al. 2012). ICP-OES also measures the density of ligand packing, and recently it has gained wide popularity as a leading technology for the analysis of mainly liquid samples.

2.4.3 Magnetometry-Based and Allied Techniques

2.4.3.1 Superconducting Quantum Interference Device Magnetometry (SQUID)

This is an important tool developed for characterizing the magnetic properties of nanoparticles. As a particle undergoes a decrease in size, it advances from multi-domain to single domain and eventually to superparamagnetic status, and these changes are having quite high significance at nano level. The SQUID device comprises two superconductors that are set apart by a thin layer that acts as an insulator, forming two parallel Josephson junctions, thus utilizing two major features of superconductors—the Josephson effect and flux quantization. When the analyte is transferred to the superconducting device, its relative up and down movement generates varying magnetic flux in the coil, which is then converted to voltage using a magnetic flux-to-voltage converter and subsequently amplified. This technique is capable of detecting even low magnetic moments with high resolution. Analysis using SQUID yields data on blocking temperature, magnetization saturation, and magnetization remanence. Recent innovation of nano-SQUID is a promising tool for nanoscale spectroscopy and also for magnetic imaging. Being a sensitive technique nano-SQUID provides an additional advantage for direct measurements of changes in magnetic properties even in small spin systems and can also be utilized to analyze magnetic properties, especially of iron oxide nanoparticles (Mourdikoudis et al. 2018).

2.4.3.2 Vibrating Sample Magnetometry (VSM)

This is an adaptable technique in order to characterize the magnetic moment of sample when it is exposed to a magnetizing field. Added advantage is that changes even within the range of 10^{-5} to 10^{-6} emu can be verified using VSM. It underlies the principle of Faraday's law of magnetic induction, developed by Foner in the 1950s. Its relative ease and accuracy made the technique popular. In VSM, initially we mount the sample onto a sample holder placed across a horizontally aligned electromagnetic pole. Further, it is exposed to a constant vibration at a constant frequency. The change in the magnetic flux will be in direct proportion to the magnetic moment of the sample; this results in AC voltage formation, i.e., detected by certain coils, and further, it is led into the amplifier, which yields data on the sample's magnetic moment as a function of the imposed magnetic field, time, and temperature. It also provides information on magnetization remanence and magnetization saturation, as given by SQUID (Adeyeye and Shimon 2015). VSM technique has been exploited to obtain first-order reversal curve (a broad quantitative technique that describes general magnetic properties of a system) values, such as composition and effective magnetic particle size (Ruta et al. 2017).

2.4.3.3 Ferromagnetic Resonance (FMR)

It is a robust technique that aids in the quantitative analysis of shape, size, and particle distribution, surface composition, and magnetic anisotropism of magnetic particles. The basic experimental setup of FMR includes placing the sample in a magnetic field to which a DC (direct current) is applied perpendicularly (Callori et al. 2020). The FMR can be utilized in two different ways: (1) exposing the sample placed in a resonant cavity to a microwave and magnetic field and (2) electrical measurements on the basis of stripe lines and coplanar waveguides. The latter is an innovative approach significantly used in the study of thin magnetic films (Thomson 2014). A signal is generated in FMR as a result of spin precession, and it is in direct proportion to the absorbed microwave power. To be noted, during resonance they exhibit a sharp increase in absorption. The spin precession is in turn influenced by the power of the magnetic field applied and the relative orientation of the sample (Indeck et al. 2016). Generated FMR spectra can be interpreted to determine the average shape and size of catalyst NPs made of ferromagnetic particles, which are widely used in synthesizing carbon nanotubes. FMR is also a sensitive technique that provides precise information on crystallographic imperfections (Mourdikoudis et al. 2018).

2.4.3.4 Superparamagnetic Relaxometry (SPMR)

This is an emerging technology that is becoming popular for the characterization of magnetic NPs. It uses magnetic sensors and superparamagnetic features of Fe_3O_4 NPs in combination. In this, magnetic moments of the NPs are made to align by a pulse of magnetizing field for a few seconds (1–2 s) with an amplitude of some Millitesla (1–2 mT). After stopping the flow of the field, the decay of the net magnetic moment of the samples can be precisely recorded. Taking the help of SQUID, it is able to predict the magnetic flux density of the sample from its net magnetic moment; SQUID-dependent relaxometry in comparison with MRI is offering a better accuracy in NP quantification and used more in therapeutic monitoring. Using incubation measurements SPMR's potentiality is exploited for analyzing different cancer lines and determining antibodies. Thus, it is a significant technique for the diagnosis and monitoring of the treatment of cancer (Mourdikoudis et al. 2018).

2.4.4 *Microscopy Techniques*

2.4.4.1 Transmission Electron Microscopy (TEM)

This is an indispensable technique for the characterization of size, morphology, distribution, and chemical nature of NPs. TEM operates under the principle of optical microscopy, utilizing the interaction between a thin sample and an electron beam of uniform density; the interaction is influenced by the elemental composition, size, and density of the sample. When the electron beam is incident on the sample, only a fraction of the electron gets transmitted; the rest gets inelastically or elastically scattered. A final image is acquired using information from the transmitted electrons, using which we obtain information about the physical properties and the homogeneity of NPs (Kumar et al. 2019). A large number of particles cannot be analyzed using the TEM technique, and projection limitations, depth sensitivity, black-and-white images, and potential artifacts are the major cons of the TEM analysis. Incorporating uv-vis spectroscopy and DLS along with TEM overcomes the limitation to measure a large quantity of particles. Interparticle distance is also an important parameter of NP characterization; hence TEM is also exploited for the characterization of aggregate formation, which has significance in biomedicine. TEM has been utilized to determine the location of NPs within the cells. For instance, TEM has detected the presence of apoptosis-related vacuoles in cancerous cells after a treatment with iron oxide NPs. Thus, it shows that TEM is an excellent technique to analyze the variations occurring at subcellular level as a consequence of NP-based treatment; in addition it yields information on the degree of penetration of NPs into tissues, especially in sunscreen application which involves the use of TiO_2NPs.

High-resolution transmission electron microscopy (HRTEM) uses both scattered and transmitted electrons to produce the image via phase contrast imaging. Having very high resolution, it provides information on atoms present in crystalline structures thus bringing credible information about the NP structure. HRTEM has emerged as a promising technology to study the internal structural features and provide information on their characteristics, such as structural transitions and effects imposed by ligands in the final structure, and to recognize structural defects from single-crystal images (Herrera-Ramirez et al. 2019).

Liquid transmission electron microscopy overcomes the limitations of conventional TEM and helps to observe processes that are unable to be assessed by normal TEM techniques. For instance, biological analytes are degraded within a few seconds in conventional TEM as a result of high vacuum. Therefore, liquid TEM is an effective technique to conserve the liquid state of biological specimen inside TEM vacuum and aids in situ observation of biological processes. TEM liquid cell is fabricated using silicon nitride (SiN), an electron-transparent membrane that incorporates the liquid sample and preserves the microscope vacuum. Electrodes are attached to produce an electric potential in order to observe the particle movement or for the induction of electrochemical reactions for imaging (Pu et al. 2020). The drawback of SiN membrane grids is that it promotes the scattering of electron beam, resulting in lower image resolution. Therefore, the alternative to SiN is imaging in low-vapor pressure ionic liquids (ILs). It has the advantage of low scattering and improved image resolution. A recent development has emphasized the usage of two thin graphene sheets enclosing the liquid sample, which helps in achieving atomic-level resolution. Unlike other techniques which give information only about the final structure, liquid TEM allows to observe the NP trajectory during its growth. The assembly of NPs and the superlattice formation can be monitored by liquid TEM. Thus, these are now emerging as major domains of research in nano-sciences because these novel NMs possess different properties in comparison with their bulk materials and individual NPs (Mourdikoudis et al. 2018).

Cryo-TEM (cryo-electron microscopy) is a variant of transmission electron microscopy that aids in the visualization of samples in their frozen native form, which is contained in vitreous condition at cryogenic temperature. In this technique liquid nitrogen (N_2) is used to freeze the analytes. The entire process can be described in the following different stages: initially, the liquid sample is placed onto a copper grid that is coated with carbon, and the excess liquid is removed, followed by the formation of a thin film using plunge freezing and then exposing the grid to liquid nitrogen, finally storing the analytes in a box with liquid nitrogen. Cryo-TEM helps in determining the structure of molecules with very high resolution. It provides the images of NPs at the time of their growth in their native condition, characterizes the colloidal drug delivery system, and also provides a technique that precisely helps to obtain sub-nanomolar resolution. Cryo-TEM is exploited to study the aggregation process of complex NPs; in addition it also quantifies the involvement of thermodynamic forces in the NP assembly. For instance, free energy of quantum dots (QDs) was determined using cryo-TEM images (Mourdikoudis et al. 2018).

2.4.4.2 Scanning Electron Microscopy (SEM)

This is an extensively applied technique for the characterization of nanostructures. In SEM, a beam of electrons ejected from a source is made to incident on a sample by using a set of lenses, which subsequently generates various signals that are detected and visualized on a screen (Amidon et al. 2009). This yields information on the surface chemistry of NPs, their texture, and size, and achieves spatial resolution at sub-nanometer level. Thus, it is exploited in various domains, such as biomedicine, nanofabrication, and material science. The development of high-resolution SEM (HR-SEM) has improved the resolution and extended it to study the specific spatial arrangements of NMs and all possible interactions occurring between them. And the sample preparation in HR-SEM is relatively easy in comparison with SEM alone. In the diagnosis point of view HR-SEM is a valuable tool as it helps in the imaging of AuNPs in tissues or cells. The SEM technique is able to give precise information about the height of NPs. It is now developed as a potential tool to determine the lateral dimensions of nanostructures (Mourdikoudis et al. 2018).

2.4.4.3 Scanning Transmission Electron Microscopy (STEM)

In STEM, microscope lenses are adjusted in such a manner to create a highly focused electron beam on the sample surface. This focusing of the electron beam across the sample generates various signals which are then collected point to point to form an image. Taking use of *EELS* (electron energy loss spectroscopy) or *EDX* (energy-dispersive X-ray spectroscopy) spectroscopy along with STEM gives the advantage of obtaining elemental maps that incorporate features down to the atomic level. In EELS we introduce the sample to a beam of electrons with known kinetic energy (KE). Subsequently some electrons undergo inelastic scattering, lose energy, and get deflected from their path. The energy loss is then quantified using an electron spectrometer. The variations in kinetic energy of electrons can be determined to analyze the chemical state and atomic-level structure of elements (Ponce et al. 2012).

High-angle annular dark field imaging (HAADF) is one of the methods for mapping the particles in STEM. The images are generated by collecting scattered electrons using an annular dark field detector. These are known as Z-contrast images and are too sensitive toward the variations shown in the atomic number of elements. HAADF-STEM is an indispensable tool for determining the changes at atomic level during structural variations and is successfully established for the imaging of distinct material interfaces. The recent innovation of aberration correctors for objective lens produced a radical advancement in the resolution limits of microscopes used in HAADF-STEM (Mourdikoudis et al. 2018).

2.4.4.4 Atomic Force Microscopy (AFM)

AFM was developed by Heinrich Rohrer and Gerard Binning in 1986. It is a microscopic analysis which produces three-dimensional (3-D) images of surfaces at a high resolution and magnification. The technique calculates the interacting forces present between the sample and the probe, which is attached to the tip of a cantilever and is made of either SiN (silicon nitride) or silicon. The deflection shown by cantilever due to repulsive or attractive forces between sample and tip is quantified by a laser beam. The interacting forces are then calculated by combining the data obtained from laser difference and cantilever stiffness. AFM scans under three distinct modes—contact, noncontact, tapping mode or oscillating mode—on the basis of the degree of proximity between sample and probe. Oscillating mode is commonly used for characterizing NPs; to be noted, it is highly sensitive toward the free amplitude of an oscillating tip. As it generates 3-D image, the height of NPs can also be calculated. Furthermore, AFM doesn't need any modification or coating on the surface before imaging; this feature makes it a useful technique to characterize the surface properties of NPs with a relatively lower scanning time (Mourdikoudis et al. 2018).

2.4.5 Additional Characterization Techniques

2.4.5.1 Zeta Potential

The measurement of zeta potential (ZP) is an excellent technique for the estimation of the surface charge of NPs present in colloidal dispersions, which further adds information on their physical stability too. A layer termed the stern layer will be formed over the surface of a charged particle, which will comprise the opposite charge of the NP surface due to attraction toward the opposite charge (Shnoudeh et al. 2019). Further, when a particle undergoes diffusion in solution, there is the formation of an outer layer with loosely associated ions, resulting in the formation of an electrical double layer. The electric potential of the so-formed double layer is known as zeta potential, and ZP is determined by measuring the velocity of charged particles in the sample solution containing two electrodes and exposed to an external electric field. The measured velocity of the particle will be in direct proportion to the degree of zeta potential measured via Doppler anemometry (Watters ct al. 2012), which also produces information on direction and speed. The values of ZP generally ranges between +100 and −100 mV and can be interpreted to calculate the colloidal stability. To be noted, the magnitude of charge on the surface of NP depends on the pH of the solution. If the magnitude of ZP is higher [±20–30 mV], it results in increased stability due to the exhibition of high electrostatic repulsion. It is observed that a zeta potential value less than or near to −28 mV will result in the agglomeration of NPs. The factors influencing ZP include the composition of the solvent, additives, and concentration of the sample. Thus, the mentioned technique which measures the surface charge is significant to determine the fate of NPs in physiological conditions (Mourdikoudis et al. 2018).

2.4.5.2 Brunauer-Emmett-Teller (BET) Technique

Research in the arena of nanotechnology has expanded out to various fields of energy, information technology, medicine, and so on. Many potential applications of NPs arise from their relatively small size; thus, the characterization of surface area is important. BET is a robust, simple, and accurate method for determining the surface area of NMs. The name of the technique was originated from the surnames of its three developers—Stephen Brunauer, Paul Emmett, and Edward Teller. BET technique underlies the principle of physical adsorption of gas molecules onto the solid surface; and BET theory has acquired its principle from the extension of Langmuir Theory, developed by Irving Langmuir (1916) (Mourdikoudis et al. 2018).

Adsorption is the adhesion of gas molecules or atoms to a surface, and the quantity of atoms adsorbed chiefly relies on the exposed surface area, as well as the solidity of the interaction between solid and gas, temperature, and gas pressure. Surface area determination via BET technique basically uses nitrogen gas at a temperature of 77 K due to accessibility of the gas in high purity and stability, strong interaction with majority of solids. Since there is a frail interaction among solid and gaseous phases, it is advised to cool the surface using liquid nitrogen to attain an appreciable amount of adsorption. The known quantity of gaseous nitrogen is then released into the sample. Accurate and credible pressure transducers evaluate the pressure variations arising due to adsorption process. Ensuing the formation of adsorption layers, the sample is withdrawn from the nitrogen atmosphere, allowed to be heated to result in the release of adsorbed nitrogen from the material, and subsequently quantified. The data achieved is expressed as BET isotherm in which the amount of gas adsorbed is plotted against relative pressure. From the displayed information in BET isotherms, the determination of the surface area of materials can be carried out. Occasionally the resulting agglomeration of smaller NMs induces certain errors (Ambroz et al. 2018).

2.4.5.3 Thermal Gravimetric Analysis (TGA)

This is a simple and direct experimental technique for the characterization of mass and composition of stabilizers. Currently NPs are a salient constituent in nanotherapeutics. Certain NPs, for instance, metallic NPs, comprise a surface modification that assists in the ability and dispersion. Surface coating influences the behavior of NMs in the biological environment and, thus, is a critical factor to be characterized. TGA measures the variation in sample mass as a function of time/temperature in a controlled and inert environment. Generally, TGA analysis is performed at an optimum temperature of 1000 °C. Initially, the NM sample undergoes heating; subsequently the constituents of sample with distinct degradation temperatures are subjected to decomposition and vaporization, which contributes to a change in mass. Resulting change in mass is recorded using a thermogravimetric analyzer, which displays it as a thermogravimetric curve taking into account the mass of the starting sample, quantity, and type of organic ligands present in NP. Derivative thermogravimetric curve (DTG), which is a derivative plot of TGA curve, displays the rate of variation in mass and graphically expresses the mass loss versus temperature curve, as peaks or curves.

TGA has been significantly exploited to analyze the properties of polymers, absorbed moisture content, the decomposition temperature of polymers, and the residual metal content of samples. TGA combined with FTIR can be used to evaluate the out-turn of different dopants. TGA can also be integrated with MS (mass spectrometry), GC (gas chromatography), and IR (infrared spectroscopy) efficiently to examine the residual components or the involved gaseous components. These excellent capabilities of TGA make it a valuable tool for analyzing the surface-coated materials of NPs (Dongargaonkar and Clogston 2017). Thus, it facilitates the prediction of the thermal stability of samples and also the residual metallic contents in the sample. The innovative micro-thermogravimetric analysis technique exploits the identical thermal decomposition principle of TGA; the only variation arises is that the mass of the sample taken must be in the order of one microgram. This helps to overcome the limits of conventional TGA technique by allowing the detection of the variations lower than 1 ng (Mourdikoudis et al. 2018).

2.4.5.4 Resonant Mass Measurement Micro-Electro Mechanical System (RMM-MEMS)

This is a recently emerged technique that aids in the detection and measurement of sub-visible as well as sub-micron particles and analyzing their properties, such as mass and size. The characterization of nanobubbles, which is having significant applications in various domains such as biomedicine, agriculture, and environmental science, can also be effectively performed using resonant mass measurement (RMM). Recently, nanobubbles have been in the spotlight for researchers because of their excellent properties to act as a contrast agent for ultra-sound diagnosis and also as a potential drug delivery vehicle because of their increased efficacy to accumulate in tumor-containing tissues. These signify the importance of RMM technique and its gaining popularity in NP characterization. This technology makes use of a MEMS (micro-electro mechanical system) sensor containing a microfluidic channel within a resonating cantilever in vacuum; fluids with particles possessing a size range from 50 nm to 5 μm bring alterations in the resonating frequency of the cantilever while passing through the channel, and these variations are used in the analysis of buoyant mass (Mourdikoudis et al. 2018). A resonant positive frequency shift is obtained when the fluid density is higher than the particle density, and a negative shift is achieved when the particle density is higher. Of note, RMM is a credible technique that provides accurate measurements of buoyancy, size, density, concentration of sample, volume, stability, and viscosity, as well as having a unique capability for effectively differentiating the negative and positive buoyant particles within the given nano-formulation. It is also performed to verify the changes occurring in the concentration and diameter of NPs within the solutions (Hernandez et al. 2019).

2.4.5.5 Differential Scanning Calorimetry (DSC)

The technique was developed during the 1960s. It is a sensitive technique to characterize NPs on the basis of their thermal properties. The particle will be exposed to a controlled temperature during the period of analysis, and DSC is able to detect both exothermic and endothermic transitions. Enthalpy can also be calculated as a function of temperature. The sample placed in a calorimeter is heated to a definite temperature. The variation in temperature between the sample and the surrounding medium (inside calorimeter) measures the heat capacity, consumption, and the release of heat from the sample. In DSC technique, there will be a sample pan with NMs as well as a reference pan under identical conditions; the difference in the signal produced by both sample and reference will be analyzed to detect the thermal variations (Koshy et al. 2017). Nowadays, the entire procedure of DSC has been automated. Using the DSC technique, the detection of thermal conductivity and specific heat capacity of NMs is possible (Mourdikoudis et al. 2018). It can be utilized to demonstrate the difference in melting temperature, which is influenced by the size of the particle, and its additional applications include the analysis of crystallinity.

2.5 Functionalization of Nanoparticles

Functionalization is the processes through which the surface chemistry of particle changes by the addition of new properties or features. Nanoparticles are ultrafine particles composed of different organic or inorganic materials. Due to unique properties that nanoparticles possess they are used for various purposes in biomedical field, such as gene and drug delivery, regenerative medicines, vaccines, tissue engineering, cancer therapy, biomolecule detection, and theranostics. While using nanoparticles in the biomedical application, two important parameters are taken into consideration. They are toxicity and uptake efficiency. The cytotoxic effect on the cells limits their use in clinical practices. These effects can be reduced by modifying physicochemical properties such as size, shape, surface charge, and surface composition. The modification of the physicochemical properties of nanoparticles includes functionalization. Surface functionalization improves the biocompatibility and uptake of nanoparticles. Generally molecules such as oligosaccharides, peptides, proteins, ligands, and antibodies are employed to modify the surface of the nanoparticles. Ligands are mainly used for the surface modification of silver nanoparticles, which reduces toxicity and provides stability in biological fluids (Sanità et al. 2020).

2.5.1 Functionalization Using Ligand Molecules

The surface of NPs or the cavities of the porous NMs are oftentimes adorned with a variety of ligands through covalent or non-covalent conjugation. The conjugated ligands facilitate cell and biomolecule sensing, disease diagnosis, and targeted delivery. Nanoparticles modified with ligands have the ability to selectively and specifically identify their cognate cell surface, proteins, and other biomacromolecules. Nanoparticles electrostatically complexed with green fluorescent protein (GFP) have been used for discriminating healthy, cancerous, and metastatic cells in both mouse and humans. During sensing, surface charge along with hydrophobicity and aromatic assembling plays a crucial role. Generally, positively charged ligand-functionalized NPs exhibit higher cell internalization than negatively and neutrally charged particles (Mout et al. 2012). However, iron oxide NPs functionalized with negatively charged carboxylate were observed to be easily taken up by cells through direct diffusion of cell membrane and pinocytosis. Functionalized NP-assisted targeted delivery of covalently attached drugs to the tumor cells depicts that positive charge functionalization promotes drug uptake into proliferating cells, while negatively charged particles diffuse rapidly to facilitate delivering drugs quickly and deep into tissues, whereas non-specific biological interactions are prevented by neutral functional groups. Further, NP-assisted intracellular delivery often gets obstructed due to endosomal entrapment, limiting the potential of cytosolic drug delivery that can be surmounted through the functionalization of NPs with signaling peptides. For instance, the functionalization of AuNPs with nuclear localizing signal (NLS) peptide has enabled them to avert endosomes and enter the cancer cell nucleus, inducing DNA damage. Frequently the payloads are surface functionalized, but in mesoporous silica nanoparticles (MSNs), the payload is attached to the internal pores of the NPs, thus providing an alternative means of nano-delivery. Functionalized MSNs with amine have been used for gene delivery in in vivo as well as in vitro. Further, capping of the pores of MSNs with appropriate ligands can assist to avoid premature and non-specific leakage of the cargo (Mout et al. 2012).

2.5.2 Functionalization Using Polymers

The NPs functionalized with polymers at the surface will impart the properties of macromolecules. Polyethylene glycol (PEG)-functionalized NPs enable passive targeting to the tumor tissue by EPR effect. The coating of NPs with PEG or other polymers increases circulation time, prevents blood serum protein adsorption, and increases particle penetration ability into tumor tissues. There is clear visualization of accumulating AuNPs which are PEGylated and coated with human transferrin in solid tumor cells, which is not possible with unmodified NPs. Functionalization of NPs can facilitate controlled drug release in the target location. For instance, poly(*N*-isopropylacrylamide)-functionalized gold nanocages are sensitive to temperature. At higher temperatures the pores of nanocages are exposed due to the collapse of polymer chains, so the loaded anticancer drug is released. The temperature can be increased by exposure to infrared radiation. When temperature decreases the polymer get to its original structure. These nanocages were used to control the release of drug dosage delivery by changing power density and time of irradiation (Mout et al. 2012).

2.5.3 Functionalization Using Biomolecules

Nanoparticles functionalized with biomolecules facilitate efficient delivery of biomacromolecules with minimum toxicity to cells. The AuNPs functionalized with DNA are capable of initiating gene silencing through antisense mechanism. The DNA-AuNPs are efficient in cell internalization compared to normal DNA molecules. The siRNA can be attached with AuNPs covalently by ethylene glycol spacer and alkylthiol linkage. The polyvalent siRNA-functionalized AuNPs act as delivery agents to enable gene regulation in in vivo conditions. Further, an additional benefit of using siRNA-functionalized AuNPs is that they generate less innate immune response as compared to different commercially available transfection agents. The AuNPs can be functionalized with oligo-nucleotides through gold thiol bond formation. The oligonucleotide-functionalized AuNPs help in molecular diagnostics, which include intracellular molecule detection, such as ATP and nucleic acids. Further, visualization and quantification of mRNA in living cells have been carried out by using novel oligonucleotide-AuNP probes that have been customized by fluorophore-labeled complementary sequence hybridization. These probes are non-toxic, resistant to enzymatic degradation, and also do not require any additional agents or microinjections for entering into the cells (Mout et al. 2012).

Some other biomolecules such as proteins and antibodies are frequently employed for the functionalization of NPs. The NPs functionalized with antibodies are used as a tool for imaging and photothermal thermal therapy due to their precise delivery mediated by specific interaction with the target ligands. For instance, Her-2 receptor gets overexpressed in both ovarian and breast tumor cells. The NPs functionalized with Herceptin, the cognate antibody of the Her-2 receptor, facilitate targeting these tumor cells. The light scattering by gold core NPs functionalized with anti-epidermal growth factor receptor (anti-EGFR) has been used for cancer cell imaging. Also the magnetic nanoparticles (MNPs) coated with proteins are also used as agents for imaging. Specific tumor tissues are localized by targeting protein moiety. So through functionalization of NPs, MNP probes are developed, which help in detailed visualization of critical biological events by ultra-sensitive diagnostic imaging and are also used for therapeutic purposes (Mout et al. 2012).

2.5.4 Methods of Nanoparticle Functionalization

Functionalization can be done by deposition of particular compounds on the NP surface or in the interior cavities of porous NMs. These depositions can be done through various methods, such as covalent binding, non-covalent binding, and surface deposition. Functionalization improves some characteristics of NPs, such as physicochemical properties, specific binding, and stability. Usually, functionalization is more efficient when the surface area is greater. Some of the basic methods of functionalization are discussed hereunder (Thiruppathi et al. 2016).

2.5.4.1 Non-covalent Binding

Surface functionalization of NPs can be done by the attachment of secondary ligands through non-covalent interactions. Biomolecules are attached to the surface of NPs through an affinity-based receptor ligand system. The non-covalent binding takes place through various approaches such as π–π stacking, electrostatic interactions, and biomolecule entrapment in biomolecule films, such as polymer, DNA, and phospholipids. Non-covalent attachment of t-RNA molecules to chitosan NPs by electrostatic interactions and hydrogen bonding through G-C and A-U base pairs are some examples. Cadmium telluride (CdTe) quantum dots (QDs) attached non-covalently to the lysozyme binding DNA through electrostatic interactions are used in luminescent sensing of lysozyme (Thiruppathi et al. 2016).

2.5.4.2 Covalent Binding

Covalent binding method for the functionalization of NPs involves sequential steps. This method results in the production of nanostructures with various functions focused on diagnosis and therapy, sometimes using the theranostic approach (Mout et al. 2012). Here, functionalization is done by direct binding of ligand on the surface of NPs through covalent conjugation. These bindings consist of linkage reactions, and in terms of stability, they prefer unspecific physisorption. For the release of small interfering RNA (SiRNA) molecules, covalent linkers used to form conjugates of SiRNA and poly (D,L-lactic-*co*-glycolic acid) (PLGA). So electrostatic attractions are not noticed. Covalent binding involves various approaches such as chemisorptions through thiol derivatives, adapter molecules such as streptavidin and biotin, and bi-functional or mediator linkers (Thiruppathi et al. 2016).

2.5.4.3 Amorphous Coating

This method involves the coating of amorphous materials such as silica and other polymers on the surface of NPs. Surface coating attachment of targeting ligand is possible on the surface. In rats ZnO NPs are introduced by coating with amorphous silica, which alters biological response by influencing the protein corona, zeta potential, and agglomerate size. Nanoparticles can be coated with both organic and inorganic layer compounds. Gold (Au) and silica (Si) are extensively used for coating, which gives high stability and conductivity to the environment for subsequent functionalization. This coating may change the physical or magnetic properties of NPs, which is helpful in optical imaging (Thiruppathi et al. 2016).

2.5.4.4 Surface Epitaxial Growth

This approach involves the deposition of organic or inorganic layer on the surface of NPs. Through surface deposition bi-functional composite NPs can be obtained. Through this approach Au-Fe_3O_4 nanoparticles can be synthesized by the surface growth of iron oxide nanoparticles (IONPs) on gold seeds by the decomposition of $Fe(CO)_5$ on the surface of AuNPs through oxidation. Microwave-assisted gold nanotriangles can be synthesized by combining with IONPs. Synthesized nanotriangles provide plasmonic as well as magnetic properties to the particle. This approach helps in the development of multifunctional nanoprobes which can facilitate diagnostic imaging in the medical field, disease detection such as cancer, and sensing (Thiruppathi et al. 2016).

Overall, functionalization has potentiated nano-intervention in several ways, for instance, by enhancing biocompatibility, bioimaging potential, drug targeting and sustained release, intracellular delivery, theranostic applications, reducing toxicity, facilitating multimodal imaging, and potentiating vaccines through functionalized nano-adjuvants and delivery systems. Despite prominent advantages, the issues of reproducibility, low yield and difficulty of scaling up, biocompatibility, biodistribution and clearance, and regulatory issues for clinical transition are some of the major limitations that still exist in the path of commonplace biomedical applications of functionalized nanoconjugates (Thiruppathi et al. 2016).

In conclusion, the methods for generation, characterization, and functionalization of different types of nanomaterials and their applications are evolving very rapidly in a need-based manner; evidently the knowledge in this arena needs to be updated accordingly. In the limited purview of this chapter, the fundamental aspects are being focused on a few prominent examples to provide an outline of the contents. Further, specialized nano-interventions in veterinary perspectives will be delineated in the subsequent chapters.

References

Abraham J, Jose B, Jose A, Thomas S (2020) Characterization of green nanoparticles from plants. In: Phytonanotechnology. Elsevier, pp 21–39. https://doi.org/10.1016/b978-0-12-822348-2.00002-4

Adeyeye AO, Shimon G (2015) Growth and characterization of magnetic thin film and nanostructures. In: Handbook of surface science. Elsevier, pp 1–41. https://doi.org/10.1016/b978-0-444-62634-9.00001-1

Agarwal N, Nair MS, Mazumder A, Poluri KM (2018) Characterization of nanomaterials using nuclear magnetic resonance spectroscopy. In: Characterization of nanomaterials. Elsevier, pp 61–102. https://doi.org/10.1016/b978-0-08-101973-3.00003-1

Ahmad A, Senapati S, Khan MI, Kumar R, Sastry M (2003) Extracellular biosynthesis of monodisperse gold nanoparticles by a novel extremophilic actinomycete, Thermomonospora sp. Langmuir 19(8):3550–3553. https://doi.org/10.1021/la0267721

Akintelu SA, Folorunso AS, Folorunso FA, Oyebamiji AK (2020) Green synthesis of copper oxide nanoparticles for biomedical application and environmental remediation. Heliyon 6(7):e04508. https://doi.org/10.1016/j.heliyon.2020.e04508

Al Moustafa A-E, Mfoumou E, Roman D, Nerguizian V, Alazzam A, Stiharu I, Yasmeen A (2016) Impact of single-walled carbon nanotubes on the embryo: a brief review. Int J Nanomedicine 11: 349. https://doi.org/10.2147/ijn.s96361

Aljabali A, Akkam Y, Al Zoubi M, Al-Batayneh K, Al-Trad B, Abo Alrob O et al (2018) Synthesis of gold nanoparticles using leaf extract of Ziziphus zizyphus and their antimicrobial activity. Nanomaterials 8(3):174. https://doi.org/10.3390/nano8030174

Alphandéry E (2014) Applications of magnetosomes synthesized by magnetotactic bacteria in medicine. Front Bioeng Biotechnol 2:5. https://doi.org/10.3389/fbioe.2014.00005

Altissimo M (2010) E-beam lithography for micro-/nanofabrication. Biomicrofluidics 4(2):026503. https://doi.org/10.1063/1.3437589

Ambroz F, Macdonald TJ, Martis V, Parkin IP (2018) Evaluation of the BET theory for the characterization of meso and microporous MOFs. Small Methods 2:1800173. https://doi.org/10.1002/smtd.201800173

Amidon GE, Secreast PJ, Mudie DM (2009) Particle, powder, and compact characterization. In: Qiu Y, Chen Y, Zhang GGZ, Liu L, Porter WR (eds) Developing solid oral dosage forms: pharmaceutical theory and practice. Academic, New York, pp 163–186

Arsene M-L, Răut I, Călin M, Jecu M-L, Doni M, Gurban A-M (2021) Versatility of reverse micelles: from biomimetic models to nano (bio)sensor design. Processes 9(2):345. https://doi.org/10.3390/pr9020345

Asanithi P, Chaiyakun S, Limsuwan P (2012) Growth of silver nanoparticles by DC magnetron sputtering. J Nanomater 2012:1–8. https://doi.org/10.1155/2012/963609

Asefa T, Dubovoy V (2017) Ordered mesoporous/nanoporous inorganic materials via self-assembly. Comprehens Supramol Chem II:157–192. https://doi.org/10.1016/b978-0-12-409547-2.12649-6

Baig N, Kammakakam I, Falath W (2021) Nanomaterials: a review of synthesis methods, properties, recent progress, and challenges. Mater Adv 2(6):1821–1871. https://doi.org/10.1039/d0ma00807a

Bardea A, Yoffe A (2017) Magneto-lithography, a simple and inexpensive method for high throughput, surface patterning. IEEE Trans Nanotechnol 16(3):439–444. https://doi.org/10.1109/tnano.2017.2672925

Bloch K, Pardesi K, Satriano C, Ghosh S (2021) Bacteriogenic platinum nanoparticles for application in nanomedicine. Front Chem 9:624344. https://doi.org/10.3389/fchem.2021.624344

Bokov D, Turki Jalil A, Chupradit S, Suksatan W, Javed Ansari M, Shewael IH, Valiev GH, Kianfar E (2021) Nanomaterial by sol-gel method: synthesis and application. Adv Mater Sci Eng 2021:5102014, 21 pages. https://doi.org/10.1155/2021/5102014

Bruener P, Grehl T, Brongersma HH (2019) Low-energy ion scattering (LEIS). In: Hodoroaba V-D, Unger WES, Shard AG (eds) Characterization of nanoparticles: measurement processes for nanoparticles. Elsevier, pp 419–439. https://doi.org/10.1016/B978-0-12-814182-3.00022-5

Bruins MR, Kapil S, Oehme FW (2000) Microbial resistance to metals in the environment. Ecotoxicol Environ Saf 45(3):198–207. https://doi.org/10.1006/eesa.1999.1860

Callori SJ, Saerbeck T, Cortie DL, Lin K-W (2020) Using polarized neutron reflectometry to resolve effects of light elements and ion exposure on magnetization. Solid State Phys 71:73–116. https://doi.org/10.1016/bs.ssp.2020.09.002

Capeness MJ, Edmundson MC, Horsfall LE (2015) Nickel and platinum group metal nanoparticle production by Desulfovibrio alaskensis G20. New Biotechnol 32(6):727–731. https://doi.org/10.1016/j.nbt.2015.02.002

Carvalho A, Fernandes AR, Baptista PV (2019) Nanoparticles as delivery systems in cancer therapy. In: Applications of targeted nano drugs and delivery systems. Elsevier, pp 257–295. https://doi.org/10.1016/b978-0-12-814029-1.00010-7

Chaurasiya RS, Hebbar HU (2017) Reverse micelles for nanoparticle synthesis and biomolecule separation. Nanosci Food Agric 4:181–211. https://doi.org/10.1007/978-3-319-53112-0_5

Chavali MS, Nikolova MP (2019) Metal oxide nanoparticles and their applications in nanotechnology. SN Appl Sci 1(6):607. https://doi.org/10.1007/s42452-019-0592-3

Chereches EL, Minea AA (2019) Electrical conductivity of new nanoparticle enhanced fluids: an experimental study. Nanomaterials 9(9):1228. https://doi.org/10.3390/nano9091228

Dadfar SM, Camozzi D, Darguzyte M, Roemhild K, Varvarà P, Metselaar J et al (2020) Size-isolation of superparamagnetic iron oxide nanoparticles improves MRI, MPI and hyperthermia performance. J Nanobiotechnol 18(1):22. https://doi.org/10.1186/s12951-020-0580-1

Deplanche K, Caldelari I, Mikheenko IP, Sargent F, Macaskie LE (2010) Involvement of hydrogenases in the formation of highly catalytic Pd(0) nanoparticles by bioreduction of Pd(II) using Escherichia coli mutant strains. Microbiology 156(9):2630–2640. https://doi.org/10.1099/mic.0.036681-0

Doan TLH, Mai NXD, Matsumoto K, Tamanoi F (2018) Tumor targeting and tumor growth inhibition capability of mesoporous silica nanoparticles in mouse models. Enzymes. https://doi.org/10.1016/bs.enz.2018.09.002

Dongargaonkar AA, Clogston JD (2017) Quantitation of surface coating on nanoparticles using thermogravimetric analysis. In: Characterization of nanoparticles intended for drug delivery. Springer, pp 57–63. https://doi.org/10.1007/978-1-4939-7352-1_6

Ealia SAM, Saravanakumar MP (2017) A review on the classification, characterisation, synthesis of nanoparticles and their application. IOP Conf Ser Mater Sci Eng 263:032019. https://doi.org/10.1088/1757-899x/263/3/032019

Ealia SAM, Saravanakumar MP. A review on the classification, characterisation, synthesis of nanoparticles and their application. In: IOP Conference Series: Materials Science and Engineering. IOP Publishing; 2017. p. 32019

Ebadi M, Zolfaghari MR, Aghaei SS, Zargar M, Shafiei M, Zahiri HS, Noghabi KA (2019) A bio-inspired strategy for the synthesis of zinc oxide nanoparticles (ZnO NPs) using the cell extract of cyanobacterium Nostoc sp. EA03: from biological function to toxicity evaluation. RSC Adv 9(41):23508–23525. https://doi.org/10.1039/c9ra03962g

Elzey S, Tsai D-H, Rabb SA, Yu LL, Winchester MR, Hackley VA (2012) Quantification of ligand packing density on gold nanoparticles using ICP-OES. Anal Bioanal Chem 403(1):145–149. https://doi.org/10.1007/s00216-012-5830-0

Fahmy SA, Preis E, Bakowsky U, Azzazy HME-S (2020) Platinum nanoparticles: green synthesis and biomedical applications. Molecules 25(21):4981. https://doi.org/10.3390/molecules25214981

Fakhari S, Jamzad M, Kabiri Fard H (2019) Green synthesis of zinc oxide nanoparticles: a comparison. Green Chem Lett Rev 12(1):19–24. https://doi.org/10.1080/17518253.2018.1547925

Feracci H, Gutierrez BS, Hempel W, Gil IS (2012) Organic nanoparticles. Front Nanosci 4:197–230. https://doi.org/10.1016/b978-0-12-415769-9.00008-x

Ficai D, Ficai A, Andronescu E (2015) Advances in cancer treatment: role of nanoparticles. In: Soloneski S, Larramendy ML (eds) Nanomaterials—toxicity and risk assessment. IntechOpen. https://doi.org/10.5772/60665

Gaidhani SV, Yeshvekar RK, Shedbalkar UU, Bellare JH, Chopade BA (2014) Bio-reduction of hexachloroplatinic acid to platinum nanoparticles employing Acinetobacter calcoaceticus. Process Biochem 49(12):2313–2319. https://doi.org/10.1016/j.procbio.2014.10.002

Gatoo MA, Naseem S, Arfat MY, Mahmood Dar A, Qasim K, Zubair S (2014) Physicochemical properties of nanomaterials: implication in associated toxic manifestations. Biomed Res Int 2014:498420. https://doi.org/10.1155/2014/498420

Guilger-Casagrande M, de Lima R (2019) Synthesis of silver nanoparticles mediated by fungi: a review. Front Bioeng Biotechnol 7:287. https://doi.org/10.3389/fbioe.2019.00287

Gul S, Khan SB, Rehman IU, Khan MA, Khan MI (2019) A comprehensive review of magnetic nanomaterials modern day theranostics. Front Mater 6:179. https://doi.org/10.3389/fmats.2019.00179

Gupta R, Kumar A (2008) Bioactive materials for biomedical applications using sol–gel technology. Biomed Mater 3(3):034005. https://doi.org/10.1088/1748-6041/3/3/034005

Hamouda RA, Hussein MH, Abo-elmagd RA, Bawazir SS (2019) Synthesis and biological characterization of silver nanoparticles derived from the cyanobacterium Oscillatoria limnetica. Sci Rep 9(1):13071. https://doi.org/10.1038/s41598-019-49444-y

Han J, Zhao D, Li D, Wang X, Jin Z, Zhao K (2018) Polymer-based nanomaterials and applications for vaccines and drugs. Polymers 10(1):31. https://doi.org/10.3390/polym10010031

Hardman R (2006) A toxicologic review of quantum dots: toxicity depends on physicochemical and environmental factors. Environ Health Perspect 114(2):165–172. https://doi.org/10.1289/ehp.8284

Hasan MM, Chisty AH, Rahman MM, Khan MN (2020) Bioprotein based IPN nanoparticles as potential vehicles for anticancer drug delivery: fabrication technology. In: Jana S, Jana S (eds) Interpenetrating polymer network: biomedical applications. Springer, Singapore. https://doi.org/10.1007/978-981-15-0283-5_7

He S, Guo Z, Zhang Y, Zhang S, Wang J, Gu N (2007) Biosynthesis of gold nanoparticles using the bacteria Rhodopseudomonas capsulata. Mater Lett 61(18):3984–3987. https://doi.org/10.1016/j.matlet.2007.01.018

Hernandez C, Abenojar EC, Hadley J, de Leon AC, Coyne R, Perera R et al (2019) Sink or float? Characterization of shell-stabilized bulk nanobubbles using a resonant mass measurement technique. Nanoscale. https://doi.org/10.1039/c8nr08763f

Herrera-Ramirez JM, Perez-Bustamante R, Aguilar-Elguezabal A (2019) An overview of the synthesis, characterization, and applications of carbon nanotubes. In: Carbon-based nanofillers and their rubber nanocomposites. Elsevier, pp 47–75. https://doi.org/10.1016/b978-0-12-813248-7.00002-x

Holbrook RD, Galyean AA, Gorham JM, Herzing A, Pettibone J (2015) Overview of nanomaterial characterization and metrology. Front Nanosci 8:47–87. https://doi.org/10.1016/b978-0-08-099948-7.00002-6

Hou X, Zaks T, Langer R, Dong Y (2021) Lipid nanoparticles for mRNA delivery. Nat Rev Mater 6:1078. https://doi.org/10.1038/s41578-021-00358-0

Husseiny MI, El-Aziz MA, Badr Y, Mahmoud MA (2007) Biosynthesis of gold nanoparticles using Pseudomonas aeruginosa. Spectrochim Acta A Mol Biomol Spectrosc 67(3–4):1003–1006. https://doi.org/10.1016/j.saa.2006.09.028

Indeck RS, Muller MW, Avazpour L (2016) Magnetic recording measurements. In: Reference module in materials science and materials engineering. Elsevier. https://doi.org/10.1016/b978-0-12-803581-8.01756-2

Iravani S (2011) Green synthesis of metal nanoparticles using plants. Green Chem 13(10):2638. https://doi.org/10.1039/c1gc15386b

Jang GG, Jacobs CB, Gresback RG, Ivanov IN, Meyer HM III, Kidder M et al (2015) Size tunable elemental copper nanoparticles: extracellular synthesis by thermoanaerobic bacteria and capping molecules. J Mater Chem C 3(3):644–650. https://doi.org/10.1039/c4tc02356k
Jiang J, Pi J, Cai J (2018) The advancing of zinc oxide nanoparticles for biomedical applications. Bioinorg Chem Appl 2018:1062562. https://doi.org/10.1155/2018/1062562
Kakran M, Li L (2012) Carbon nanomaterials for drug delivery. Key Eng Mater 508:76–80. https://doi.org/10.4028/www.scientific.net/kem.508.76
Kalishwaralal K, Deepak V, Ramkumarpandian S, Nellaiah H, Sangiliyandi G (2008) Extracellular biosynthesis of silver nanoparticles by the culture supernatant of Bacillus licheniformis. Mater Lett 62(29):4411–4413. https://doi.org/10.1016/j.matlet.2008.06.051
Karfa P, De S, Majhi KC, Madhuri R, Sharma PK (2019) Functionalization of carbon nanostructures. In: Andrews DL, Lipson RH, Nann T (eds) Comprehensive nanoscience and nanotechnology, 2nd edn. Academic, pp 123–144. https://doi.org/10.1016/B978-0-12-803581-8.11225-1
Karman SB, Diah SZM, Gebeshuber IC (2015) Raw materials synthesis from heavy metal industry effluents with bioremediation and phytomining: a biomimetic resource management approach. Adv Mater Sci Eng 2015:1–21. https://doi.org/10.1155/2015/185071
Khan I, Saeed K, Khan I (2017) Nanoparticles: properties, applications and toxicities. Arab J Chem 12:908. https://doi.org/10.1016/j.arabjc.2017.05.011
Kimber RL, Lewis EA, Parmeggiani F, Smith K, Bagshaw H, Starborg T et al (2018) Biosynthesis and characterization of copper nanoparticles using Shewanella oneidensis: application for click chemistry. Small 14(10):1703145. https://doi.org/10.1002/smll.201703145
Kirthi AV, Rahuman AA, Rajakumar G, Marimuthu S, Santhoshkumar T, Jayaseelan C et al (2011) Biosynthesis of titanium dioxide nanoparticles using bacterium Bacillus subtilis. Mater Lett 65(17–18):2745–2747. https://doi.org/10.1016/j.matlet.2011.05.077
Kitching M, Ramani M, Marsili E (2014) Fungal biosynthesis of gold nanoparticles: mechanism and scale up. Microb Biotechnol 8(6):904–917. https://doi.org/10.1111/1751-7915.12151
Klaus T, Joerger R, Olsson E, Granqvist C-G (1999) Silver-based crystalline nanoparticles, microbially fabricated. Proc Natl Acad Sci U S A 96(24):13611–13614. https://doi.org/10.1073/pnas.96.24.13611
Kolahalam LA, Kasi Viswanath IV, Diwakar BS, Govindh B, Reddy V, Murthy YLN (2019) Review on nanomaterials: synthesis and applications. Mater Today Proc. https://doi.org/10.1016/j.matpr.2019.07.371
Komarneni S (1992) Feature article. Nanocomposites. J Mater Chem 2(12):1219. https://doi.org/10.1039/jm9920201219
Koshy O, Subramanian L, Thomas S (2017) Differential scanning calorimetry in nanoscience and nanotechnology. In: Thermal and rheological measurement techniques for nanomaterials characterization. Elsevier, pp 109–122. https://doi.org/10.1016/b978-0-323-46139-9.00005-0
Kumar PS, Pavithra KG, Naushad M (2019) Characterization techniques for nanomaterials. In: Nanomaterials for solar cell applications. Elsevier, pp 97–124. https://doi.org/10.1016/b978-0-12-813337-8.00004-7
Lalitha K, Kalaimurgan D, Nithya K, Venkatesan S, Shivakumar MS (2020) Antibacterial, antifungal and mosquitocidal efficacy of copper nanoparticles synthesized from entomopathogenic nematode: insect–host relationship of bacteria in secondary metabolites of Morganella morganii sp. (PMA1). Arab J Sci Eng 45(6):4489–4501. https://doi.org/10.1007/s13369-020-04487-6
Landage KS, Arbade GK, Khanna P, Bhongale CJ (2020) Biological approach to synthesize TiO_2 nanoparticles using Staphylococcus aureus for antibacterial and antibiofilm applications. J Microbiol Exp 8(1):36–43. https://doi.org/10.15406/jmen.2020.08.00283
Lee H-J, Song JY, Kim BS (2013) Biological synthesis of copper nanoparticles using Magnolia kobus leaf extract and their antibacterial activity. J Chem Technol Biotechnol 88:1971–1977. https://doi.org/10.1002/jctb.4052

Lee J-H, Lim J-M, Velmurugan P, Park Y-J, Park Y-J, Bang K-S, Oh B-T (2016) Photobiologic-mediated fabrication of silver nanoparticles with antibacterial activity. J Photochem Photobiol B Biol 162:93–99. https://doi.org/10.1016/j.jphotobiol.2016.06.029

Lengke MF, Fleet ME, Southam G (2006) Synthesis of platinum nanoparticles by reaction of filamentous cyanobacteria with platinum(IV)–chloride complex. Langmuir 22(17):7318–7323. https://doi.org/10.1021/la060873s

Li J, He Y, Han Y, Liu K, Wang J, Li Q, Fan S, Jiang K (2012) Direct identification of metallic and semiconducting single-walled carbon nanotubes in scanning electron microscopy. Nano Lett 12(8):4095–4101. https://doi.org/10.1021/nl301561f

Lim J, Yeap S, Che H, Low S (2013) Characterization of magnetic nanoparticle by dynamic light scattering. Nanoscale Res Lett 8(1):381. https://doi.org/10.1186/1556-276x-8-381

Liu Y, Goebl J, Yin Y (2013) Templated synthesis of nanostructured materials. Chem Soc Rev 42(7):2610–2653. https://doi.org/10.1039/c2cs35369e

Liu J, Li R, Yang B (2020) Carbon dots: a new type of carbon-based nanomaterial with wide applications. ACS Central Sci 6(12):2179–2195. https://doi.org/10.1021/acscentsci.0c01306

Majeed S, Danish M, Mohamad Ibrahim MN, Sekeri SH, Ansari MT, Nanda A, Ahmad G (2020) Bacteria mediated synthesis of iron oxide nanoparticles and their antibacterial, antioxidant, cytocompatibility properties. J Clust Sci 32(4):1083–1094. https://doi.org/10.1007/s10876-020-01876-7

Manawi Y, Ihsanullah, Samara A, Al-Ansari T, Atieh M (2018) A review of carbon nanomaterials' synthesis via the chemical vapor deposition (CVD) method. Materials 11(5):822. https://doi.org/10.3390/ma11050822

Mann S, Frankel RB, Blakemore RP (1984) Structure, morphology and crystal growth of bacterial magnetite. Nature 310(5976):405–407. https://doi.org/10.1038/310405a0

Mata YN, Torres E, Blázquez ML, Ballester A, González F, Muñoz JA (2009) Gold(III) biosorption and bioreduction with the brown alga Fucus vesiculosus. J Hazard Mater 166(2–3):612–618. https://doi.org/10.1016/j.jhazmat.2008.11.064

Medina-Cruz D, Saleh B, Vernet-Crua A, Ajo A, Roy AK, Webster TJ (2020) Drug-delivery nanocarriers for skin wound-healing applications. In: Wound healing, tissue repair, and regeneration in diabetes. Elsevier, pp 439–488. https://doi.org/10.1016/b978-0-12-816413-6.00022-8

Mikhailova EO (2021) Gold nanoparticles: biosynthesis and potential of biomedical application. J Func Biomater 12(4):70. https://doi.org/10.3390/jfb12040070

Minakshi P, Ghosh M, Brar B, Ranjan K, Patki HS, Kumar R (2020) Separation techniques with nanomaterials. In: Hussain CM (ed) Handbook of nanomaterials in analytical chemistry. Elsevier, Amsterdam, pp 99–158. https://doi.org/10.1016/b978-0-12-816699-4.00006-2

Minea AA (2019) A review on electrical conductivity of nanoparticle-enhanced fluids. Nanomaterials 9(11):1592. https://doi.org/10.3390/nano9111592

Mody V, Siwale R, Singh A, Mody H (2010) Introduction to metallic nanoparticles. J Pharm Bioallied Sci 2(4):282. https://doi.org/10.4103/0975-7406.72127

Mourdikoudis S, Pallares RM, Thanh NTK (2018) Characterization techniques for nanoparticles: comparison and complementarity upon studying nanoparticle properties. Nanoscale 10(27): 12871–12934. https://doi.org/10.1039/c8nr02278j

Mout R, Moyano DF, Rana S, Rotello VM (2012) Surface functionalization of nanoparticles for nanomedicine. Chem Soc Rev 41(7):2539. https://doi.org/10.1039/c2cs15294k

Mozhayeva D, Engelhard C (2019) A critical review of single particle inductively coupled plasma mass spectrometry—a step towards the ideal method for nanomaterials characterization. J Anal At Spectrom 35:1740. https://doi.org/10.1039/c9ja00206e

Muthu M, Wu H-F, Gopal J, Sivanesan I, Chun S (2017) Exploiting microbial polysaccharides for biosorption of trace elements in aqueous environments—scope for expansion via nanomaterial intervention. Polymers 9(12):721. https://doi.org/10.3390/polym9120721

Nadeem M, Tungmunnithum D, Hano C, Abbasi BH, Hashmi SS, Ahmad W, Zahir A (2018) The current trends in the green syntheses of titanium oxide nanoparticles and their applications. Green Chem Lett Rev 11(4):492–502. https://doi.org/10.1080/17518253.2018.1538430

Nadeem M, Khan R, Shah N, Bangash IR, Abbasi BH, Hano C, Liu C, Ullah S, Hashmi SS, Nadhman A, Celli J (2022) A review of microbial mediated iron nanoparticles (IONPs) and its biomedical applications. Nanomaterials 12:130. https://doi.org/10.3390/nano12010130

Nair GB, Dhoble SJ (2021) Phosphor-converted LEDs. In: The fundamentals and applications of light-emitting diodes. Woodhead Publishing, pp 87–126. https://doi.org/10.1016/b978-0-12-819605-2.00004-5

Nair B, Pradeep T (2002) Coalescence of nanoclusters and formation of submicron crystallites assisted by Lactobacillus strains. Cryst Growth Des 2(4):293–298. https://doi.org/10.1021/cg0255164

Naser H, Alghoul MA, Hossain MK, Asim N, Abdullah MF, Ali MS et al (2019) The role of laser ablation technique parameters in synthesis of nanoparticles from different target types. J Nanopart Res 21(11):249. https://doi.org/10.1007/s11051-019-4690-3

Omar RA, Afreen S, Talreja N, Chauhan D, Ashfaq M (2019) Impact of nanomaterials in plant systems. In: Prasad R (ed) Plant nanobionics, Advances in the understanding of nanomaterials research and applications, vol 1. Springer International Publishing, Cham. https://doi.org/10.1007/978-3-030-16383-9_5

Ortega G, Reguera E (2019) Biomedical applications of magnetite nanoparticles. Mater Biomed Eng 2019:391–428. https://doi.org/10.1016/b978-0-12-816913-1.00013-1

Ovais M, Khalil A, Ayaz M, Ahmad I, Nethi S, Mukherjee S (2018) Biosynthesis of metal nanoparticles via microbial enzymes: a mechanistic approach. Int J Mol Sci 19(12):4100. https://doi.org/10.3390/ijms19124100

Pantidos N, Edmundson MC, Horsfall L (2018) Room temperature bioproduction, isolation and anti-microbial properties of stable elemental copper nanoparticles. New Biotechnol 40:275–281. https://doi.org/10.1016/j.nbt.2017.10.002

Parikh RY, Singh S, Prasad BLV, Patole MS, Sastry M, Shouche YS (2008) Extracellular synthesis of crystalline silver nanoparticles and molecular evidence of silver resistance from Morganella sp.: towards understanding biochemical synthesis mechanism. Chembiochem 9(9):1415–1422. https://doi.org/10.1002/cbic.200700592

Pavel E, Prodan G, Marinescu V, Trusca R (2019a) Recent advances in 3- to 10-nm quantum optical lithography. J Micro/Nanolithogr MEMS MOEMS 18(2):020501. https://doi.org/10.1117/1.JMM.18.2.020501

Pavel E, Marinescu V, Lungulescu M (2019b) Graphene nanopatterning by quantum optical lithography. Optik 203:163532. https://doi.org/10.1016/j.ijleo.2019.163532

Peller M, Böll K, Zimpel A, Wuttke S (2018) Metal–organic framework nanoparticles for magnetic resonance imaging. Inorg Chem Front 5(8):1760–1779. https://doi.org/10.1039/C8QI00149A

Philipse AP, Maas D (2002) Magnetic colloids from magnetotactic bacteria: chain formation and colloidal stability. Langmuir 18(25):9977–9984. https://doi.org/10.1021/la0205811

Ponce A, Mejía-Rosales S, José-Yacamán M (2012) Scanning transmission electron microscopy methods for the analysis of nanoparticles. Methods Mol Biol 906:453–471. https://doi.org/10.1007/978-1-61779-953-2_37

Prasad M, Kumar R, Buragohain L, Kumari A, Ghosh M (2021) Organoid technology: a reliable developmental biology tool for organ-specific nanotoxicity evaluation. Front Cell Dev Biol 9:696668. https://doi.org/10.3389/fcell.2021.696668

Pu S, Gong C, Robertson AW (2020) Liquid cell transmission electron microscopy and its applications. R Soc Open Sci 7(1):191204. https://doi.org/10.1098/rsos.191204

Qidwai A, Kumar R, Dikshit A (2018) Green synthesis of silver nanoparticles by seed of Phoenix sylvestris L. and their role in the management of cosmetics embarrassment. Green Chem Lett Rev 11(2):176–188. https://doi.org/10.1080/17518253.2018.1445301

Qin J-X, Yang X-G, Lv C-F, Li Y-Z, Liu K-K, Zang J-H et al (2021) Nanodiamonds: synthesis, properties, and applications in nanomedicine. Mater Des 210:110091. https://doi.org/10.1016/j.matdes.2021.110091

Rajendran NK, George BP, Houreld NN, Abrahamse H (2021) Synthesis of zinc oxide nanoparticles using Rubus fairholmianus root extract and their activity against pathogenic bacteria. Molecules 26(10):3029. https://doi.org/10.3390/molecules26103029

Rane AV, Kanny K, Abitha VK, Thomas S (2018) Methods for synthesis of nanoparticles and fabrication of nanocomposites. In: Synthesis of inorganic nanomaterials. Elsevier, pp 121–139. https://doi.org/10.1016/b978-0-08-101975-7.00005-1

Reith F, Etschmann B, Grosse C, Moors H, Benotmane MA, Monsieurs P et al (2009) Mechanisms of gold biomineralization in the bacterium Cupriavidus metallidurans. Proc Natl Acad Sci U S A 106(42):17757–17762. https://doi.org/10.1073/pnas.0904583106

Riddin T, Gericke M, Whiteley CG (2010) Biological synthesis of platinum nanoparticles: effect of initial metal concentration. Enzym Microb Technol 46(6):501–505. https://doi.org/10.1016/j.enzmictec.2010.02.006

Ruta S, Hovorka O, Huang P-W, Wang K, Ju G, Chantrell R (2017) First order reversal curves and intrinsic parameter determination for magnetic materials; limitations of hysteron-based approaches in correlated systems. Sci Rep 7(1):45218. https://doi.org/10.1038/srep45218

Sanità G, Carrese B, Lamberti A (2020) Nanoparticle surface functionalization: how to improve biocompatibility and cellular internalization. Front Mol Biosci 7:587012. https://doi.org/10.3389/fmolb.2020.587012

Saravanan C, Rajesh R, Kaviarasan T, Muthukumar K, Kavitake D, Shetty PH (2017) Synthesis of silver nanoparticles using bacterial exopolysaccharide and its application for degradation of azo-dyes. Biotechnol Rep 15:33–40. https://doi.org/10.1016/j.btre.2017.02.006

Sathiyanarayanan G, Vignesh V, Saibaba G, Vinothkanna A, Dineshkumar K, Viswanathan MB, Selvin J (2014) Synthesis of carbohydrate polymer encrusted gold nanoparticles using bacterial exopolysaccharide: a novel and greener approach. RSC Adv 4(43):22817–22827. https://doi.org/10.1039/c4ra01428f

Savic S, Vojisavljevic K, Počuča-Nešić M, Zivojevic K, Mladenovic M, Knezevic N (2018) Hard template synthesis of nanomaterials based on mesoporous silica. Metall Mater Eng 24(4):225–241. https://doi.org/10.30544/400

Schubert U (2015) Chemistry and fundamentals of the sol–gel process. In: Levy D, Zayat M (eds) The sol-gel handbook. https://doi.org/10.1002/9783527670819.ch01

Shahverdi AR, Minaeian S, Shahverdi HR, Jamalifar H, Nohi A-A (2007) Rapid synthesis of silver nanoparticles using culture supernatants of Enterobacteria: a novel biological approach. Process Biochem 42(5):919–923. https://doi.org/10.1016/j.procbio.2007.02.005

Shankar SS, Ahmad A, Pasricha R, Sastry M (2003) Bioreduction of chloroaurate ions by geranium leaves and its endophytic fungus yields gold nanoparticles of different shapes. J Mater Chem 13(7):1822. https://doi.org/10.1039/b303808b

Shi S, Chen F, Cai W (2013) Biomedical applications of functionalized hollow mesoporous silica nanoparticles: focusing on molecular imaging. Nanomedicine (Lond) 8(12):2027–2039. https://doi.org/10.2217/nnm.13.177

Shin S, Song I, Um S (2015) Role of physicochemical properties in nanoparticle toxicity. Nanomaterials 5(3):1351–1365. https://doi.org/10.3390/nano5031351

Shnoudeh AJ, Hamad I, Abdo RW, Qadumii L, Jaber AY, Surchi HS, Alkelany SZ (2019) Synthesis, characterization, and applications of metal nanoparticles. In: Biomaterials and bionanotechnology. Academic, pp 527–612. https://doi.org/10.1016/b978-0-12-814427-5.00015-9

Singh AK, Talat M, Singh DP, Srivastava ON (2010) Biosynthesis of gold and silver nanoparticles by natural precursor clove and their functionalization with amine group. J Nanopart Res 12(5):1667–1675. https://doi.org/10.1007/s11051-009-9835-3

Singh J, Dutta T, Kim K-H, Rawat M, Samddar P, Kumar P (2018) "Green" synthesis of metals and their oxide nanoparticles: applications for environmental remediation. J Nanobiotechnol 16(1):84. https://doi.org/10.1186/s12951-018-0408-4

Singha A, Dhar P, Roy A (2005) A nondestructive tool for nanomaterials: Raman and photoluminescence spectroscopy. Am J Phys 73(3):224–233. https://doi.org/10.1119/1.1819933

Sintubin L, De Windt W, Dick J, Mast J, van der Ha D, Verstraete W, Boon N (2009) Lactic acid bacteria as reducing and capping agent for the fast and efficient production of silver nanoparticles. Appl Microbiol Biotechnol 84(4):741–749. https://doi.org/10.1007/s00253-009-2032-6

Soh HT, Guarini KW, Quate CF (2001) Introduction to scanning probe lithography. In: Scanning probe lithography. Springer, pp 1–22. https://doi.org/10.1007/978-1-4757-3331-0_1

Sreenivasan SV (2017) Nanoimprint lithography steppers for volume fabrication of leading-edge semiconductor integrated circuits. Microsyst Nanoeng 3(1):17075. https://doi.org/10.1038/micronano.2017.75

Srivastava SK, Yamada R, Ogino C, Kondo A (2013) Biogenic synthesis and characterization of gold nanoparticles by Escherichia coli K12 and its heterogeneous catalysis in degradation of 4-nitrophenol. Nanoscale Res Lett 8(1):70. https://doi.org/10.1186/1556-276x-8-70

Stassen I, Styles M, Grenci G, Gorp HV, Vanderlinden W, Feyter SD et al (2015) Chemical vapour deposition of zeolitic imidazolate framework thin films. Nat Mater 15(3):304–310. https://doi.org/10.1038/nmat4509

Stöhr J, Padmore HA, Anders S, Stammler T, Scheinfein MR (1998) Principles of X-ray magnetic dichroism spectromicroscopy. Surf Rev Lett 05(06):1297–1308. https://doi.org/10.1142/s0218625x98001638

Takacs L (2002) Self-sustaining reactions induced by ball milling. Prog Mater Sci 47(4):355–414. https://doi.org/10.1016/s0079-6425(01)00002-0

Taran M, Rad M, Alavi M (2017) Biosynthesis of TiO_2 and ZnO nanoparticles by Halomonas elongata IBRC-M 10214 in different conditions of medium. Bioimpacts 8(2):81–89. https://doi.org/10.15171/bi.2018.10

Tenchov R, Bird R, Curtze AE, Zhou Q (2021) Lipid nanoparticles—from liposomes to mRNA vaccine delivery, a landscape of research diversity and advancement. ACS Nano 15:16982. https://doi.org/10.1021/acsnano.1c04996

Thakkar KN, Mhatre SS, Parikh RY (2010) Biological synthesis of metallic nanoparticles. Nanomedicine 6(2):257–262. https://doi.org/10.1016/j.nano.2009.07.002

Thi TUD, Nguyen TT, Thi YD, Ta Thi KH, Phan BT, Pham KN (2020) Green synthesis of ZnO nanoparticles using orange fruit peel extract for antibacterial activities. RSC Adv 10(40):23899–23907. https://doi.org/10.1039/d0ra04926c

Thiruppathi R, Mishra S, Ganapathy M, Padmanabhan P, Gulyás B (2016) Nanoparticle functionalization and its potentials for molecular imaging. Adv Sci 4(3):1600279. https://doi.org/10.1002/advs.201600279

Thomas SC, Sharma H, Mishra PK, Talegaonkar S (2015) Ceramic nanoparticles: fabrication methods and applications in drug delivery. Curr Pharm Des 21(42):6165–6188. https://doi.org/10.2174/1381612821666151027153246

Thomson T (2014) Magnetic properties of metallic thin films metallic films for electronic. In: Barmak K, Barmak K, Coffey K (eds) Optical and magnetic applications. Woodhead Publishing, Cambridge, pp 454–546. https://doi.org/10.1533/9780857096296.2.454

Tormena IRP, Rosa EV, Oliveira Mota BF, Chaker JA, Fagg CW, Freire DO et al (2020) Evaluation of the antimicrobial activity of silver nanoparticles obtained by microwave-assisted green synthesis using Handroanthus impetiginosus (Mart. ex DC.) Mattos underbark extract. RSC Adv 10(35):20676–20681. https://doi.org/10.1039/d0ra03240a

Vanlalveni C, Lallianrawna S, Biswas A, Selvaraj M, Changmai B, Rokhum SL (2021) Green synthesis of silver nanoparticles using plant extracts and their antimicrobial activities: a review of recent literature. RSC Adv 11(5):2804–2837. https://doi.org/10.1039/d0ra09941d

Velmurugan P, Shim J, You Y, Choi S, Kamala-Kannan S, Lee K-J et al (2010) Removal of zinc by live, dead, and dried biomass of Fusarium spp. isolated from the abandoned-metal mine in South

Korea and its perspective of producing nanocrystals. J Hazard Mater 182(1–3):317–324. https://doi.org/10.1016/j.jhazmat.2010.06.032
Venugopal G, Kim S (2013) Nanolithography. In: Advances in micro/nano electromechanical systems and fabrication technologies. IntechOpen. https://doi.org/10.5772/55527
Wagner AM, Knipe JM, Orive G, Peppas NA (2019) Quantum dots in biomedical applications. Acta Biomater 94:44–63. https://doi.org/10.1016/j.actbio.2019.05.022
Wang Y, Huang L (2014) Composite nanoparticles for gene delivery. Adv Genet 88:111–137. https://doi.org/10.1016/b978-0-12-800148-6.00005-5
Wang Z, Gao H, Zhang Y, Liu G, Niu G, Chen X (2017) Functional ferritin nanoparticles for biomedical applications. Front Chem Sci Eng 11(4):633–646. https://doi.org/10.1007/s11705-017-1620-8
Watters RJ, Kester M, Tran MA, Loughran TP, Liu X (2012) Development and use of ceramide nanoliposomes in cancer. Methods Enzymol 508:89–108. https://doi.org/10.1016/b978-0-12-391860-4.00005-7
Wender H, Gonçalves RV, Feil AF, Migowski P, Poletto FS, Pohlmann AR et al (2011) Sputtering onto liquids: from thin films to nanoparticles. J Phys Chem C 115(33):16362–16367. https://doi.org/10.1021/jp205390d
Whitesides GM, Ostuni E, Takayama S, Jiang X, Ingber DE (2001) Soft lithography in biology and biochemistry. Annu Rev Biomed Eng 3(1):335–373. https://doi.org/10.1146/annurev.bioeng.3.1.335
Whitfield P, Mitchell L (2004) X-ray diffraction analysis of nanoparticles: recent developments, potential problems and some solutions. Int J Nanosci 03(06):757–763. https://doi.org/10.1142/s0219581x04002620
Wu S, Rajeshkumar S, Madasamy M, Mahendran V (2020) Green synthesis of copper nanoparticles using Cissus vitiginea and its antioxidant and antibacterial activity against urinary tract infection pathogens. Artif Cells Nanomed Biotechnol 48(1):1153–1158. https://doi.org/10.1080/21691401.2020.1817053
Xie Z, Yu W, Wang T, Zhang H, Fu Y, Liu H et al (2011) Plasmonic nanolithography: a review. Plasmonics 6(3):565–580. https://doi.org/10.1007/s11468-011-9237-0
Xie Y, Kocaefe D, Chen C, Kocaefe Y (2016) Review of research on template methods in preparation of nanomaterials. J Nanomater 2016:1–10. https://doi.org/10.1155/2016/2302595
Yadav TP, Yadav RM, Singh DP (2012) Mechanical milling: a top down approach for the synthesis of nanomaterials and nanocomposites. Nanosci Nanotechnol 2:22. https://doi.org/10.5923/j.nn.20120203.01
Yaqoob AA, Ahmad H, Parveen T, Ahmad A, Oves M, Ismail IMI et al (2020) Recent advances in metal decorated nanomaterials and their various biological applications: a review. Front Chem 8:341. https://doi.org/10.3389/fchem.2020.00341
Yusof HM, Mohamad R, Zaidan UH, Abdul Rahman NA (2019) Microbial synthesis of zinc oxide nanoparticles and their potential application as an antimicrobial agent and a feed supplement in animal industry: a review. J Anim Sci Biotechnol 10(1):57. https://doi.org/10.1186/s40104-019-0368-z
Yusof HM, Abdul Rahman N, Mohamad R, Zaidan UH, Samsudin AA (2020) Biosynthesis of zinc oxide nanoparticles by cell-biomass and supernatant of Lactobacillus plantarum TA4 and its antibacterial and biocompatibility properties. Sci Rep 10(1):19996. https://doi.org/10.1038/s41598-020-76402-w
Zeiri Y, Elia P, Zach R, Hazan S, Kolusheva S, Porat Z (2014) Green synthesis of gold nanoparticles using plant extracts as reducing agents. Int J Nanomedicine 9:4007. https://doi.org/10.2147/ijn.s57343

Zhao M-X, Zhu B-J (2016) The research and applications of quantum dots as nano-carriers for targeted drug delivery and cancer therapy. Nanoscale Res Lett 11(1):207. https://doi.org/10.1186/s11671-016-1394-9

Zhou C, Krueger AB, Barnard JG, Qi W, Carpenter JF (2015) Characterization of nanoparticle tracking analysis for quantification and sizing of submicron particles of therapeutic proteins. J Pharm Sci 104(8):2441–2450. https://doi.org/10.1002/jps.24510

Chapter 3
Nanosensors: Types and Customisation

Sushil Kumar, Nidhi, Subodh Soni, Krishna Kant Sharma, Rajesh Kumar, and Hari Mohan

Abstract The unique properties of nanoparticles are recently exploited in various areas specially for sensing purposes. Nanosensors with electrical, optical and magnetic transducers are used for highly specific and sensitive detection of various analytes. Innovative use of nanosensors in different fields has expanded the worldwide market for the devices. Nanosensors can offer advantages in various applications such as disease diagnosis, environmental pollutants monitoring, food safety testing and pharmaceutical testing. The multidisciplinary domains of nanosensing technology have opened various new prospects for fabrication, characterisation of nanoparticles, molecular chemistry and signal analysis. The possibility of developing nanosensors in portable forms and measuring any analyte in real time with a comparatively higher sensitivity than conventional methods has been responsible for increased attention towards them. The current chapter focuses on the principles behind nanosensing technology, types of nanosensors, sensing techniques and customisation of nanosensors according to molecules of interest to be analysed.

Keywords Nanotechnology · Nanosensors · Transducers · Analyte · Immunosensor · Sensing technique

---

Sushil Kumar
Department of Biosciences and Bioengineering, Indian Institute of Technology Guwahati, Guwahati, Assam, India

Nidhi · S. Soni · H. Mohan (✉)
Centre for Medical Biotechnology, Maharshi Dayanand University, Rohtak, Haryana, India
e-mail: harimohan.cmbt@mdurohtak.ac.in

K. K. Sharma
Department of Microbiology, Maharshi Dayanand University, Rohtak, Haryana, India

R. Kumar
Department of Veterinary Physiology and Biochemistry, Lala Lajpat Rai University of Veterinary and Animal Sciences, Hisar, Haryana, India

LLR University of Veterinary and Animal, ICAR-National Research Centre on Equines, *Nanotechnology Theranostics in Livestock Diseases and Management*, Livestock Diseases and Management,
https://doi.org/10.1007/978-981-16-1610-5_3

3.1 Introduction

Nanotechnology has become one of the emerging fields in providing highly sensitive alternative for monitoring various physical and chemical changes. This discipline has inherent potential to attract much scientific research attention and revolutionise important areas like medicine, diagnostics and therapeutics (Sharma et al. 2021; Munawar et al. 2019; Vo-Dinh 2005). The extraordinary knowledge-based technical developments have contributed to the emergence of nanotechnology. Nanotechnology is a multidisciplinary field; hence, it needs collaborative research with experts from physics, chemistry, biotechnology, engineering and medicine to solve the complexity of the field and lead to comprehensive applications. The prefix 'nano' is used to denote one-billionth of any unit as per scientific reference. Hence, nanotechnology basically deals with properties and functions of nanoscale objects to construct and design devices and systems. The shape and sizes of the objects are controlled on nanometre scale for various applications (Mousavi et al. 2018). A wide range of nanomaterials have emerged in recent years for diverse applications. Nanoscale materials are those with smallest dimensions in the range of up to 100 nm. Various biological materials like proteins, bacteria and viruses are considered in the same range. Also, due to the specific properties and functions of various other materials they have high tendency to be designed as non-biological nanomaterials.

Many of the modern gadgets used by population to interact with physical environment perform with the help of sensors. Sensors are devices which convert variable quantities such as pressure, movement, force and electric current into specific signals to detect them precisely. These devices detect physical, chemical and mechanical changes/stimuli and convert them into measurable signals for further processing (Saleh Ahammad et al. 2009). The conversion of one form of energy into another is done with the help of transducers. Nanosensors are a category of sensing devices having sensing dimensions up to 100 nm, i.e., in nanoscale. These devices monitor physical and chemical phenomena in various regions which are normally laborious to reach, like measuring nanoparticles in environment and industry, detecting biomolecules in cells like disease-related metabolites, proteins, pathogens etc. The nanoparticles in these nanosensors are conjugated to various ligands, which attract particular marker of interest for detection and impart specificity to the nanosensors. Various nanostructured materials used in nanosensor assemblies include carbon nanotubes, thin films, nanoscale wires and polymer nanomaterials. Biological and non-biological nanomaterials prepared via biological methods have also been proven very effective in various applications. With research advances, nanotechnology is moving towards fabricating more flexible and versatile nanosensors with high sensitivity and specificity. Various sensing approaches are being assimilated to develop such sensors to increase sensitivity, compact instrument size and expand analyte detection. Also preprocessing of samples, signal amplification and transduction approaches are employed for high sensitivity analyte recognition.

Nanosensors have the ability to detect metabolites, pathogens, cells and other molecules, which led them to extensive applications in clinical, industrial, agricultural and environmental research developments (Kamal 2018). The existing techniques for detection and diagnostic applications have one limitation or the other, like some are difficult to use and sometimes they do not provide specific, quick and stringent results. For instance, conventional techniques for pathogen detection like cell culture and colony count are time consuming, while more specific techniques like ELISA and RT-PCR are costly. Other techniques need extensive sample preparation procedures from biological samples. Hence, nanosensors are preferred as point-of-care detection methods which are easy to handle, give uniform results in different environmental conditions and provide cheap, sensitive alternative for the detection of multiple analytes.

3.2 Principle of Nanosensing

Nanosensing technology employs nanomaterials like nanoparticles, nanowires, nanotubes, nanofilms, nanocoatings, nanocomposites, quantum dots and biological nanomaterials for the development of efficient and excellent nanosensors. These nanomaterials have various fascinating properties like exceptional magnetic properties, biocompatibility, enhanced electrical conductivity, high reactivity and large surface area-to-volume ratio, which provide unmatched sensing potential (Petros and Desimone 2010). For example, high surface area-to-volume ratio of these materials contributes to the detection of analyte in low sample volume offered by immobilised nanoparticles. These materials hence provide robust sensing systems which are uniquely fabricated to be used in smaller volumes as compared to traditional diagnostic methods. These sensors consist of two central elements: a recognition element and a transduction element. Various categories of nanosensors are divided on the basis of these two components. The sensor receives input from the sample, which is detected by recognition elements, for instance, by interaction between the receptor as a recognition element and its target analyte from the sample and is converted into a computable signal like electric current. The recognition element in case of nanosensors essentially comprises at least one dimension up to 100 nm. The fabrication and optimisation of nanosensors are done according to the marker of interest to be detected, for example, fabrication of sensor for protein detection is usually on the basis of protein conformation. Most of the nanosensors have a layer of biological receptors such as molecular biotic element (like antibody, enzyme, nucleic acid and protein) or living biotic component such as cells and tissues, which act as a diagnostic element (Colombo et al. 2009). These biological receptors act as a substrate for the attachment of analyte to the sensing element. The interaction between the receptor and the analyte is used to generate the signal to be computed and analysed. The sensing nanoparticles are synthesised by two steps: nucleation and expansion. Metallic nanoparticles are often used in connection with microorganisms.

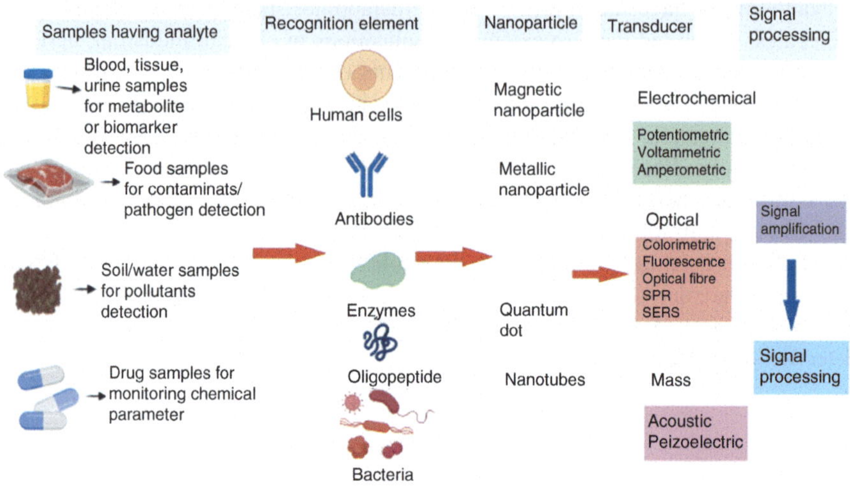

Fig. 3.1 Schematic illustration of nanosensing process

Using this principle, nanosensor devices are used to diagnose various biochemicals in tissues, measure nanoscale particles in industries and administer physical and chemical elements in areas difficult to reach. A schematic illustration of nanosensing process has been given in Fig. 3.1.

3.3 Classification of Nanosensors

Nanosensors can be divided into various subtypes on the basis of categorisation, like structure, receptor molecules, energy sources and applications (Fig. 3.2). Various transduction systems can be used for sensing technology, which form the basis for three major categories of nanosensors according to their structure. These three common methods are based on optical detection, electrochemical detection and mass-based detection, i.e. vibrational/mechanical detection by different transducer usages. Other methods like magnetic, voltaic and combinations of various methods can also be used (Vo-Dinh 2005).

On the basis of energy source nanosensors can be classified into active and passive nanosensors, wherein active ones use energy sources like thermistors and passive ones don't need any energy sources like thermocouple.

Four categories of nanosensors are divided on the basis of application. Chemical nanosensors consist of electronics and capacitive cantilevers. These are used in detecting various chemicals in samples, for example, detecting drugs from environmental samples. Biosensors mainly contain some biological detecting material in combination with physicochemicals to detect the analyte under study. These can be used for cancer detection, specific DNA detection, etc. Deployable nanosensor is

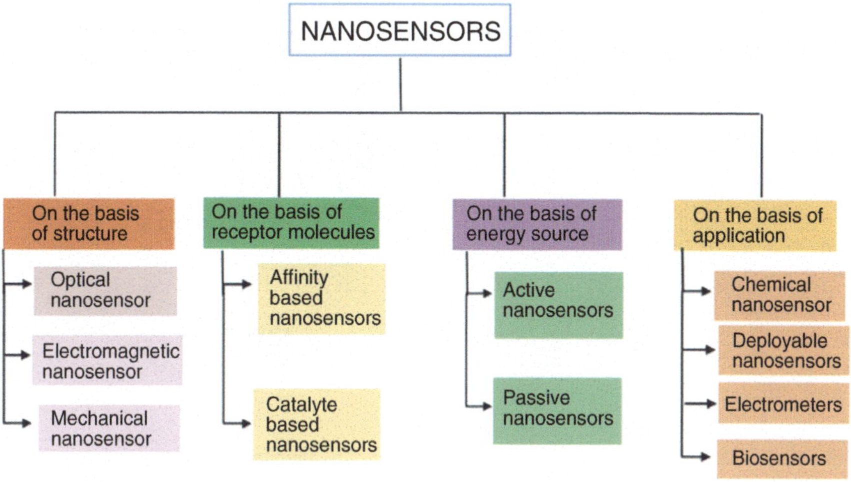

Fig. 3.2 Types of nanosensors

another category used in national security, military and other such applications. Electrometers are also a category of nanosensors which are used to couple charge to a mechanical element using a torsional mechanical resonator.

Various kinds of receptor molecules used for nanosensor fabrication also form the basis of categorisation. Affinity-based nanosensor is the subtype that includes receptors such as nucleic acids, antibodies and hormones which have affinity to irreversibly bind analytes of interest in a noncatalytic manner. Another subtype is catalytic-based sensors which contain cells or enzyme-based receptors to catalytically bind an analyte (Devreese 2007).

3.4 Major Types of Nanosensors

3.4.1 Optical Nanosensors

The first reported optical nanosensor was used to measure pH, which was fluorescein based and trapped within polyacrylamide nanoparticles (Sasaki et al. 1996). For fluorescent detection, the nanoparticles either use free dye or use labelled nanoparticles which contain attached reporter molecules. The fluorescent detection is based on the interaction of dyes with molecules of interest followed by the emission of a quantum of light. Other than fluorescent analysis, optical nanosensors are used for several other analyses like visible radiation, absorption, refraction, Raman scattering and dispersion. Nanoparticle components of optical nanosensors act as sources of signal production, which reflect the analyte concentration. The changes due to molecular binding of the analyte with the recognition component can be analysed by referring to a change in the excitation spectra of the nanomaterials

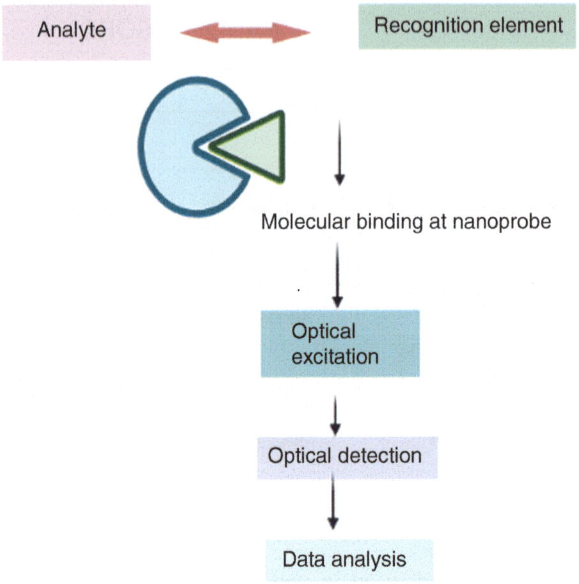

Fig. 3.3 Optical nanosensing

used. Sometimes changes on the surface of nanoparticles in physiological conditions create issues for optical detection, like harsh conditions such as high salt concentration, which can be resolved using stabilising agents and combinatorial approaches (Korzeniowska et al. 2013). The signals produced by the nanosensors can be amplified by using a combination of nanoparticles with distinctive optical properties, and these approaches have been proven useful in the early diagnosis of diseases (Afzal et al. 2013). A schematic representation of optical nanosensing process has been depicted in Fig. 3.3. Although optical sensors have proven safe and non-toxic, but in vivo testing needs to be explored with more detailed data in support. Major issues related to optical sensing technology include achieving reproducibility, controlling shape/size and detecting background fluorescence from tissues. The latter one can be avoided by upconverting nanoparticles by exciting such particles with low-energy infrared photons and emitting signals in visible range (Hao et al. 2013).

Typical optical nanosensor examples include localised surface plasmon resonance (LSPR) nanosensors, which contain plasmon system oscillating around nanoparticles. These nanosensors offer label-free detection with a 1 nm limit of detection. One such nanosensor is developed for the diagnosis of Alzheimer's disease. Specific silver nanoparticles are used to detect Alzheimer's biomarker, i.e. amyloid-derived diffusible ligands. The LSPR signals are produced by secondary proteins bound to the analyte-primary protein complex in a sandwich assay manner. The signals were used for qualitative analysis via optically coupled spectroscope fibres using UV-visible excitation (Munawar et al. 2019).

Another important property other than LSPR which is more commonly used in optical nanosensing method is absorption followed by scattering using metal

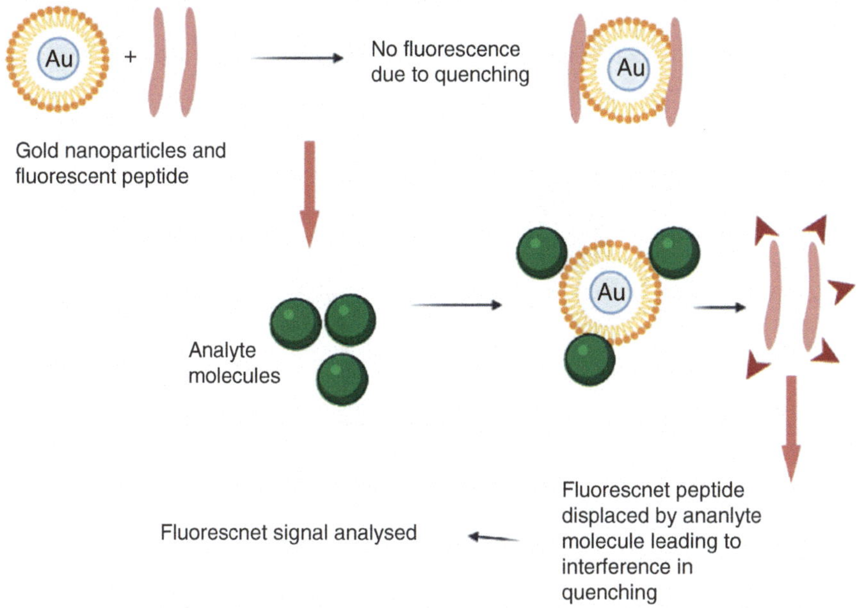

Fig. 3.4 Gold nanoparticle-based optical nanosensor

nanoparticles that offer robust absorption. Differences of charges between fluorescent molecule and targeted molecule form the basis of these detectors. Normally metal NPs like AuNPs react with fluorescent compounds, leading to quenching of the visible reaction, but in the presence of the target compound, disruption of the signal takes place, leading to the release of fluorescent compounds (Fig. 3.4) (Schmid 2008). This principle is employed for the quantification of various proteins (You et al. 2007).

3.4.2 Mechanical Nanosensors

These nanosensors are devices which use the principle of mechanics to measure the physical and chemical properties of molecules in a nanoscale region. The measurement of nanoscale mechanical properties by these particular nanosensors is the major advantage over the other two kinds, i.e. optical and electromagnetic nanosensors (Chi et al. 2013). Common examples of these nanosensors are nanomechanical cantilever sensors, which use an atomic force microscope and CNT-based (molecular-scale tubes of graphite carbon) fluidic shear stress sensors. The physical properties measured by these sensors include strain, temperature, pressure, nanoscale mass, force, cantilever beam bending and elastic modulus measurement. Cantilever sensors measure elastic modulus and unbinding force of protein-ligand interactions and also detect the presence of specific molecule by beam bending due to adsorption (Liu

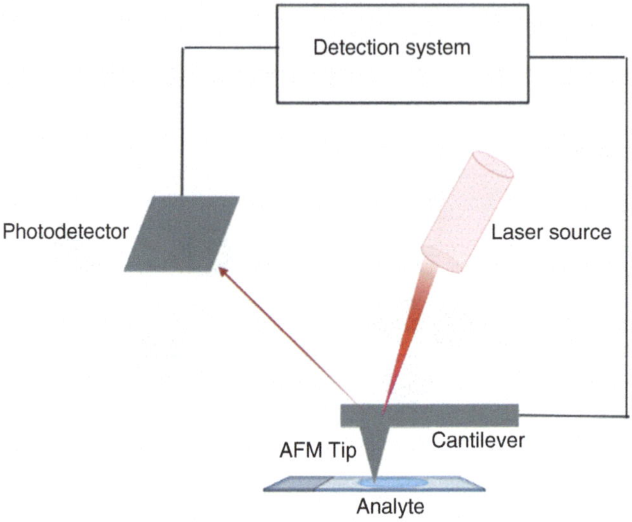

Fig. 3.5 Nanomechanical cantilever sensor

et al. 2010). They detect molecular binding on the surface of the cantilever through its nanomechanical motion with rapid signal transduction and real-time monitoring of biomolecular interactions in gaseous, liquid as well as vacuum mediums (Fig. 3.5). Most of the microcantilevers are composed of silicon, silicon oxide and silicon nitride, which are microfabricated using micromachining techniques employing polymers or metals with separate power and wavelength lasers (Zhang and Xu 2004). These sensors were used for detecting specific DNA molecules related to particular diseases (Tschulena 1988). Complementary-oligo-coated cantilever system was immobilised on gold covered side of the cantilever and binding of analyte DNA to the complementary surface induced changes in surface stress and bending which can be measured for detection. Also explosives can be detected using these sensors by employing the usage of low vapour pressure property of the explosives such as dinitrotoluene, hexohydro1,3,5 trinitroazine and trinitrotoluene (Senesac and Thundat 2008). The CNT tubes-based sensors measure nanoscale mass by calculating resonant frequency and changes in strain, pressure and temperature (Sinha et al. 2006). The strain measurement in different locations can be effectively done by employing the isotopic nature of CNT tubes. CNT sensors can be used for detecting sodium or measuring the pH of blood, meat freshness evaluation (Funazaki et al. 1995), environmental contaminants and detecting bioweapons (Mahdavian et al. 2009). Mechanical nanosensors are also useful in microelectronic devices as important components, such as nanodevices and nanoscale assemblies.

3.4.3 *Electromagnetic Nanosensors*

These nanosensors employ the use of electromagnetic communication, which includes transmission and reception of electromagnetic radiation from nanomaterial components. These sensors monitor either electric current or magnetism for detection (Abdel-Karim et al. 2020). Electric current measurement is employed in gold nanoparticle-based nanosensor to detect hydrogen sulphide gas. The H_2S forms sulphide shell around AuNPs, which inhibit charge transfer among nanoparticles, and this hopping of electrons can be determined by using current and voltage across electrodes (Akyildiz and Jornet 2010). Various magnetic nanosensors have also been developed to detect biomolecules like proteins, viruses, and enzymes using magnetic nanoparticles like iron oxide. The stable nanoassemblies formed after the interaction of an analyte with magnetic nanoparticles lead to a decrease in spin-spin relaxation time of surrounding water molecules, which is detected using magnetic resonance techniques (Hassan 2017).

3.4.4 *Biosensors*

Biosensors are devices which contain integrated receptor-transducer complex to convert biological response into electrical signals. The biorecognition element in conjugation with nanoparticles recognises the analyte, and the electronic component detects the physiological change or the analyte's presence by converting into measurable signals (Sharma et al. 2021). The instrumentation of the biosensor essentially comprises of samples having analyte, biorecognition element/receptor, transducer, electronics and display system. Samples can be cells, tissues or any suspension having analyte molecules such as glucose, ammonia and proteins. The bioreceptor component contains any biological element which possesses strong affinity to bind the target substrate, i.e. an analyte molecule. The bioreceptors usually used are enzymes, cells, aptamers, antibodies and nucleic acids. When these receptors interact with suitable target molecules then there is a generation of signals in the form of pH, charge, light, mass change or any other product formation (Turner 2013). These changes or biorecognition events are converted into measurable signals with the help of transducers. The electronic system processes and prepares the signal for display, and the amplified signal in the form of digital values is displayed on the display unit. Based on the primary component that is the bioreceptor, the biosensors can be classified into various subtypes such as enzymatic biosensors, immunosensors and aptamer-based biosensors. A gold coated tungsten (W-Au) electrode-based enzymatic biosensor has been developed for the real-time monitoring of glucose changes in the brain. Another such nanosensor has been developed to detect urea using potentiometric transduction (Cordeiro et al. 2018). An antibody-based biosensor has enabled label-free detection of Aflatoxin B1 with a linear range of 0.1–2 nM. Another important type of biosensors is whole-cell-based biosensors, which exhibit the special property of self-replicating and producing biorecognition elements such as antibodies. This has removed the need for the purification of the bioreceptor for immobilisation (Berepiki et al. 2020). A label-free whole-cell optical biosensor developed using *Escherichia coli* was found to have a detection limit of 3 ng/mL for detecting insecticide pyrethroid (Riangrungroj et al. 2019). Nanoparticles such as QDs, CNTs and graphene can act as both bioreceptors and transducers. These can also be used in combination with other bioreceptors for better sensitivity.

3.5 Customisation of Nanosensors

A prominent advantage offered by nanosensors over the other diagnostic devices is that they can be customised according to the application or the molecule of interest for detection. For customisation, one needs to have minute information regarding components of the nanosensors. Various components of the nanosensors can be selected and optimised according to the molecule of interest to be detected. First component is nanoparticles, which can be metallic or non-metallic (Abdel-Karim et al. 2020). Au, Ag, Pt, Cu etc. are some of the metal nanoparticles which have unique physical and chemical properties important for sensing technology. Gold nanoparticles act as effective electrocatalysts, which can be fabricated for electrochemical nanosensor formation with high sensitivity (Chen and Huang 2014). These nanosensors have been proven effective for protein (haem proteins, redox enzymes) (Vidotti et al. 2011) and antibiotic (ciprofloxacin, chloramphenicol) (Yuan et al. 2018) detection. Other than metal NPs, metal oxide nanoparticles have ultrahigh surface area that facilitate low-cost detection with high efficiency and sensitivity in response to changes in physical properties and chemical environment (Maduraiveeran and Jin 2017). TiO_2 nanostructures have been developed to detect biomolecules such as cytochrome C, myoglobin and haemoglobin for medical and pharmaceutical applications. Polymeric nanomaterials such as nanocomposites are used for detecting food and environmental pollutants i.e. polyaniline nanofibres have applications in analysis of toxic gases such as toluene, acetone, benzene etc. Other than these components, bionanomaterials have been synthesised to employ catalytic properties of biomolecules in combination with nanoscale materials. Biogenic magnetic nanoparticles have been synthesised to remove azo dyes from waste water.

After customised nanoparticles, there are sensing techniques which vary for different analytes. Different sensing techniques or detection systems offer a range of limits of detection, such as femto, pico, atto and zepto scales. Other than optical and mechanical sensing, another important technique is electrochemical detection. The nanosensors employing this sensing type are called chemical nanosensors. The principle behind the working of these sensors is the chemical reaction taking place between the restrained ligand and the analyte, which leads to a measurable signal from the transducer (Munawar et al. 2019). The electrochemical approaches used for quantifying the analyte using an electrochemical signal include potentiometry, voltametry, electrochemical impedance and amperometry. When this sensing technique is conjugated with immunoassays, then the resulting sensor is termed as immunosensor. ELISA-based electrochemical sensors are considered to be the most common type of amperometric immunosensor. In these sensors the redox species generated by the labelled enzyme due to the formation of Ag-Ab complex is converted into current, which is measured to indicate the results (Koryta 1976). These immunosensors have diverse applications in detecting pathogens, biomarkers, toxins and antibodies with high specificity and sensitivity. Three basic processes need to be considered while customising nanosensor for specific application:

1. Sample preparation is the first step which needs to be considered. The sample could be homogeneous or complex, depending on the source and the situation. It can be challenging when the sample suspension has impurities that may interfere with the results. The sample may contain diverse analyte molecules, such as antibiotics, hormones, enzymes, DNA/RNA, metals and pH, which are specific functional groups/molecules/organisms, should be precisely detected by the sensor. The sensors target the analyte molecules. So, the nanosensors can be customised according to the target analyte in the sample.
2. Recognition process is the second step, which needs customised recognition elements. These elements can be antibodies, aptamers or enzymes, which have high affinity for the target analyte. These recognition elements bind specifically to the analytes due to their selective nature and quantify them.
3. Signal transduction is the last step in the sensing process, which needs customisation. The signal transduction methods compute the recognition process and convert it into measurable signals. The signals are further computed and processed to produce the data. Suitable transduction methods are employed as per the analyte or application, such as optical, mechanical, electronic and pyroelectric (Sharma et al. 2021).

After selecting the appropriate nanomaterials, the process of nanofabrication is performed. Nanofabrication can be achieved in two ways, i.e. bottom-up to sculpt the base material to a desired shape and top-down by placing nanomolecules one by one to form the nanostructure. Different immobilisation techniques to immobilise the recognition elements on the surface of transducer can be employed for sensor preparation (Fig. 3.6). These techniques are customised to preserve the activity, structure and function of the recognition element used. Physical and chemical immobilisation techniques are commonly employed. The choice of the technique depends on the biorecognition element chosen, characteristics of the analyte, the physico-chemical environment and the transducer (Naresh and Lee 2021).

1. Physical immobilisation is also referred to as reversible immobilisation. This technique attaches the recognition element on the surface of the transducer by processes such as physical adsorption, physical entrapment and microencapsulation.
2. Chemical immobilisation or irreversible immobilisation attaches the recognition element's functional groups and the transducer via strong chemical bonding.

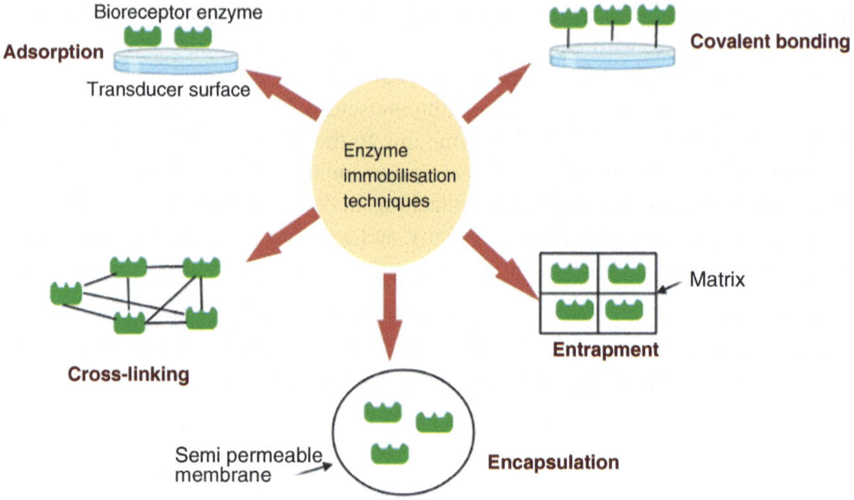

Fig. 3.6 Immobilisation techniques for enzymatic receptors

Various examples of customised nanosensors according to different applications can help in better understanding the concept. Customised nanosensors can be designed by transitioning the speculative properties of different nanosensor components into practical output.

3.6 Nanosensors for Microorganism Detection

Accurate diagnosis of pathogenic microbes is very important for correct assessment and treatment. The traditional diagnostic approaches have low sensitivity and are time-consuming. To improve the situation, nanosensors have been developed for high-speed diagnosis. For in-situ pathogen biorecognition, nanoparticle-based sensors have been developed (Bosch et al. 2010). Silver nitride cantilever sensor has been used to diagnose *Salmonella enterica* in even smaller amount such as 25 microbes in the sample. Electrical and optical transduction methods have proven efficient in detecting even single virus particle as utilised for fluorescently labelled influenza detection.

3.7 Nanosensors for Cancer Diagnosis

Quantum dots bioconjugates are able to detect cancer cells in the organisms by identifying specific cancer cell-specific proteins. Magnetic microparticle probes can detect prostate cancer by using conjugated antibodies specific to prostate-specific antigen. Label-free biomarker detection using nanosensor is quite possible to detect BRAF mutation for diagnosing virulent melanoma (Huber et al. 2013). Double-stranded hairpin-structured oligonucleotide was used on the surface of cantilever biosensor to detect transcription factor NF-κB for cancer diagnosis (Huber et al. 2006). For viral cancer diagnosis, virus detection is employed. Surface immobilised ligands are used, which attach the virus and lead to a change in the resonance frequency of the cantilever surface. This change is associated with a change in the mass of the surface cantilever, which can be converted into computable signals (Braun et al. 2009). These nanomechanical cantilever sensors have been proven very efficient for cancer diagnosis. Some customised nanosensors for cancer diagnosis have been mentioned in Table 3.1.

3.8 Nanosensors for Pharmaceutical Industry

Biosensors can be employed to monitor the therapeutic efficacy of drugs by recognising the bioanalyte products with fast response times and high resolution (Shafiee 2019). Acute rejections after organ transplantation can be detected using nanosensors for real-time monitoring of patient's condition. Bioanalytes which are detrimental for cells and are produced by excess drug uptake can be analysed using nanosensors (Cash et al. 2015). Using microfluidic technology a DNA aptamer-based electrochemical sensor has been developed. This nanosensor has been assembled on an organ-on-a-chip to detect small molecules such as various antibiotics. This nanosensor showed remarkable results in the real-time monitoring of doxorubicin and kanamycin for up to 4 h (Ferguson et al. 2013). The fouling of the sensor by blood was prevented using continuous-flow diffusion filter. Clenbuterol, a β-adrenergic agonist, can be detected by differential pulse voltammetry (DPV) using sulphonated grapheme sheets-carbon nanotube-based nanosensor (Zhai et al. 2015). Also the persistence of drugs in water and other environmental resources can be monitored using nanosensors.

3.9 Nanosensors in Agroecosystems

Nanosensors can be employed to accelerate agricultural productivity and for improving agricultural sustainability. Real-time monitoring of temperature, soil moisture content, soil nutrients and soil microenvironment can be done using nanosensors. Even nanosensors have been customised to detect residual agrochemicals in soil and water such as pesticides, heavy metals, fertilisers, toxins and plant pathogens (Table 3.1). These applications of nanosensors can help in predicting information needed prior and would help in mitigating economic loss. Metal-engineered nanoparticles have been proven effective in monitoring different molecules which affect plant growth-promoting bacteria such as rhizobacteria (Mishra and Kumar

Table 3.1 Some customised nanosensors and their applications

Nanomaterial	Analyte	Transducer	Application	References
SiNPs	Ascorbic acid, glutathione	Fluorescent	To identify intracellular reducing chemicals in breast cancer cells	Li et al. (2021)
Cd-S:Eu nanocrystals	Exosomes	Aptasensor	To diagnose breast cancer cells secreted exosomes from serum sample	Fang et al. (2019)
CdS:Mn carbon dots	miR141	Electrochemiluminescence	cDNA bound to CDs detect miR142 abnormally expressed in various cancer types	Israeli et al. (2009)
Polymer dots	Human epididymis protein 4	Ratiometric assay	Detects HE4 protein, which serves as a biomarker for ovarian cancer	Fang et al. (2019)
PANI (polyaniline) and PDMS (polydimethylsiloxane)	Creatinine	Enzymatic assay with triboelectric output	Detects creatinine levels which correspond to kidney disorders and thyroid malfunction	Luo et al. (2021)
N,P-doped carbon quantum dots (N, P-CQDs)	New coccine	Fluorescent	Detects synthetic food colour, new coccine, in food samples	Liu et al. (2022)
Graphene nanosheets	Saxitoxin	Electrochemical	Detects saxitoxins from shellfish samples and lake water samples	Nikoleli et al. (2018)

2009). PLP-conjugated lysozyme-cocooned gold nanoclusters have been proven efficient in detecting zinc ions in agroecosystems, even at nanomolar concentrations (Bothra et al. 2017). Optical nanosensors designed using silver nanodendrites and fabricated by laser-assisted photochemical method of immobilisation to detect pesticide dimethoate in soil have been found to be a rapid, real-time and non-destructive method in detecting residual pesticides in the environment.

3.10 Future Prospects

Although nanosensing technology is going to face major advancements in the upcoming years, the insights of nanoscale phenomena would need more exploration to overcome the present limitations. Existing nanosensors are working quite fine, but more insights are needed to enhance their performance. New nanosensors would be developed using novel mechanisms. More studies are required to understand the in vivo applications and to prepare appropriate biocompatible nanosensors. Nanosensors could provide foundation for technology to monitor neurotransmitters and other biomolecules in human body. This could prove really crucial for long-term neurochemical monitoring inside tissues. Smart delivery systems can be created using nanosensors by coupling them with GPS systems and robotics to detect as well as treat specific areas in vivo prior to the onset of symptoms. By integrating nanosensors with microarray technology, a wide range of nanosensors can be developed to achieve more specific diagnosis using small sample size and automated measurements. Nanosensors can be developed for customised applications, such as clinical diagnosis like measuring pathogenic entities, molecular markers, immunological markers and biochemical parameters; food quality control like monitoring toxins in food, microbes, genetically modified food products and quality assessment; and environmental monitoring for soil, water, air, fauna and flora assessment. Nanosensors could also prove a blessing for assessing therapeutic agents and then optimising their dosage. Some major limitations associated with this wonderful technique include cytotoxicity issues; fabrication scalability and signal dilution that affect their widespread implementation. But by increasing research to overcome these limitations, nanosensors can provide significant benefits as compared to other techniques. In the long term, the successful development of nanosensors can be adapted for the real-time monitoring of phenomena and molecules, along with providing analyte flexibility.

References

Abdel-Karim R, Reda Y, Abdel-Fattah A (2020) Review—Nanostructured materials-based nanosensors. J Electrochem Soc 167(3):037554. https://doi.org/10.1149/1945-7111/ab67aa

Afzal A, Iqbal N, Mujahid A, Schirhagl R (2013) Advanced vapor recognition materials for selective and fast responsive surface acoustic wave sensors: a review. Anal Chim Acta 787: 36–49. https://doi.org/10.1016/j.aca.2013.05.005

Akyildiz IF, Jornet JM (2010) Electromagnetic wireless nanosensor networks. Nano Commun Netw 1(1):3–19. https://doi.org/10.1016/j.nancom.2010.04.001

Berepiki A, Kent R, Machado LFM, Dixon N (2020) Development of high-performance whole cell biosensors aided by statistical modelling. ACS Synth Biol 9(3):576–589. https://doi.org/10.1021/acssynbio.9b00448

Bosch J, Heister K, Hofmann T, Rainer U (2010) Nanosized iron oxide colloids strongly enhance microbial iron reduction. Appl Environ Microbiol 76(1):184–189. https://doi.org/10.1128/AEM.00417-09

Bothra S, Babu LT, Paira P, Kumar SKA (2017) A biomimetic approach to conjugate vitamin B6 cofactor with the lysozyme cocooned fluorescent AuNCs and its application in turn-on sensing of zinc (II) in environmental and biological samples. Anal Bioanal Chem 410(1):201–210. https://doi.org/10.1007/s00216-017-0710-2

Braun T, Ghatkesar MK, Backmann N, Grange W, Boulanger P, Letellier L, Lang H, Bietsch A, Gerber C, Hegner M (2009) Membrane protein–ligand interactions using microcantilever array sensors. Nat Nanotechnol 4(3):179–185. https://doi.org/10.1038/NNANO.2008.398

Cash KJ, Li C, Xia J, Wang LV, Clark HA, Al CET (2015) Optical drug monitoring: photoacoustic imaging of nanosensors to monitor therapeutic lithium in vivo. ACS Nano 9(2):1692–1698. https://doi.org/10.1021/nn5064858

Chen HH, Huang JF (2014) EDTA assisted highly selective detection of As^{3+} on Au nanoparticle modified glassy carbon electrodes: facile in situ electrochemical characterization of Au nanoparticles. Anal Chem 86(24):12406–12413. https://doi.org/10.1021/ac504044w

Chi R, Ching M, Lim T (2013) Encyclopedia of microfluidics and nanofluidics. Springer, pp 1–5. https://doi.org/10.1007/978-3-642-27758-0

Colombo M, Ronchi S, Monti D, Corsi F, Trabucchi E, Prosperi D (2009) Femtomolar detection of autoantibodies by magnetic relaxation nanosensors. Anal Biochem 392(1):96–102. https://doi.org/10.1016/j.ab.2009.05.034

Cordeiro CA, Sias A, Koster T, Westerink BHC, Cremers TIFH (2018) In vivo "real-time" monitoring of glucose in the brain with an amperometric enzyme-based biosensor based on gold coated tungsten (W-Au) microelectrodes. Sensors Actuators B Chem 263:605. https://doi.org/10.1016/j.snb.2018.02.116

Devreese J (2007) Importance of nanosensors: Feynman's vision and the birth of nanotechnology. MRS Bull 32(9):718–725. https://doi.org/10.1557/mrs2007.14

Fang D, Zhang S, Dai H, Lin Y (2019) An ultrasensitive ratiometric electrochemiluminescence immunosensor combining photothermal amplification for ovarian cancer marker detection. Biosens Bioelectron 146:111768. https://doi.org/10.1016/j.bios.2019.111768

Ferguson BS, Hoggarth DA, Maliniak D, Ploense K, White RJ, Woodward N, Hsieh K, Bonham AJ, Eisenstein M, Kippin TE, Plaxco KW, Soh HT (2013) Real-time, aptamer-based tracking of circulating therapeutic agents in living animals. Sci Transl Med 5(213):213ra165. https://doi.org/10.1126/scitranslmed.3007095

Funazaki N, Hemmi A, Ito S, Asano Y, Yano Y, Miura N, Yamazoe N (1995) Application of semiconductor gas sensor to quality control of meat freshness in food industry. Sensors Actuators B Chem 25(1–3):797–800. https://doi.org/10.1016/0925-4005(95)85177-1

Hao S, Chen G, Yang C (2013) Sensing using rare-earth-doped upconversion nano-particles. Theranostics 3(5):331–345. https://doi.org/10.7150/thno.5305

Hassan TAA (2017) Development of nanosensors in nuclear technology. AIP Conf Proc 1799. https://doi.org/10.1063/1.4972925

Huber F, Hegner M, Gerber C, Peter H (2006) Label free analysis of transcription factors using microcantilever arrays. Biosens Bioelectron 21:1599–1605. https://doi.org/10.1016/j.bios.2005.07.018

Huber F, Lang HP, Backmann N, Rimoldi D, Gerber C (2013) Direct detection of a BRAF mutation in total RNA from melanoma cells using cantilever arrays. Nat Nanotechnol 8(2):125–129. https://doi.org/10.1038/nnano.2012.263

Israeli E, Agmon-Levin N, Blank M, Shoenfeld Y (2009) Adjuvants and autoimmunity. Lupus 18(13):1217–1225. https://doi.org/10.1177/0961203309345724

Kamal I (2018) Prospects of some applications of engineered nanomaterials: a review. Open Access J Biomed Eng Biosci 2(5):245–252. https://doi.org/10.32474/oajbeb.2018.02.000149

Koryta J (1976) Fundamentals and applications of ISE's. Anal Chem 48(4):407A–412A. https://doi.org/10.1021/ac60368a765

Korzeniowska B, Nooney R, Wencel D, McDonagh C (2013) Silica nanoparticles for cell imaging and intracellular sensing. Nanotechnology 24(44):442002. https://doi.org/10.1088/0957-4484/24/44/442002

Li C, Xu Y, Liu S, Zhang Y, Yin W, Dai Z (2021) Cancer cell identification by facile imaging of intracellular reductive substances with fluorescent nanosensor. Talanta 234:122650. https://doi.org/10.1016/j.talanta.2021.122650

Liu Y, Wang W, Shu W (2010) Nanomechanical cantilever sensors. Nanosensors 69–96. https://doi.org/10.1201/b10450-4

Liu L, Mi Z, Huo X, Yuan L, Bao Y, Liu Z, Feng F (2022) A label-free fluorescence nanosensor based on nitrogen and phosphorus co-doped carbon quantum dots for ultra-sensitive detection of new coccine in food samples. Food Chem 368:130829. https://doi.org/10.1016/j.foodchem.2021.130829

Luo Y, Zhao T, Dai Y, Li Q, Fu H (2021) Flexible nanosensors for non-invasive creatinine detection based on triboelectric nanogenerator and enzymatic reaction. Sensors Actuators A Phys 320:112585. https://doi.org/10.1016/j.sna.2021.112585

Maduraiveeran G, Jin W (2017) Nanomaterials based electrochemical sensor and biosensor platforms for environmental applications. Trends Environ Anal Chem 13:10–23. https://doi.org/10.1016/j.teac.2017.02.001

Mahdavian L, Monajjemi M, Mangkorntong N (2009) Sensor response to alcohol and chemical mechanism of carbon nanotube gas sensors. Fuller Nanotub Carbon Nanostruct 17(5):484–495. https://doi.org/10.1080/15363830903130044

Mishra VK, Kumar A (2009) Impact of metal nanoparticles on the plant growth promoting rhizobacteria. Digest J Nanomater Biostruct 4(3):587–592

Mousavi SM, Hashemi SA, Zarei M, Amani AM, Babapoor A (2018) Nanosensors for chemical and biological and medical applications. Med Chem 08(08):205–217. https://doi.org/10.4172/2161-0444.1000515

Munawar A, Ong Y, Schirhagl R, Tahir MA, Khan WS, Bajwa SZ (2019) Nanosensors for diagnosis with optical, electric and mechanical transducers. RSC Adv 9(12):6793–6803. https://doi.org/10.1039/c8ra10144b

Naresh V, Lee N (2021) A review on biosensors and recent development of nanostructured materials-enabled biosensors. Sensors 21(4):1109

Nikoleli GP, Nikolelis D, Siontorou CG, Karapetis S (2018) Lipid membrane nanosensors for the opportunities, and the challenges. Sensors 18(1):284. https://doi.org/10.3390/s18010284

Petros RA, Desimone JM (2010) Strategies in the design of nanoparticles for therapeutic applications. Nat Rev Drug Discov 9(8):615–627. https://doi.org/10.1038/nrd2591

Riangrungroj P, Bever CS, Hammock BD, Polizzi KM (2019) A label-free optical whole-cell *Escherichia coli* biosensor for the detection of pyrethroid insecticide exposure. Sci Rep 9:1–9. https://doi.org/10.1038/s41598-019-48907-6

Saleh Ahammad AJ, Lee JJ, Rahman MA (2009) Electrochemical sensors based on carbon nanotubes. Sensors 9(4):2289–2319. https://doi.org/10.3390/s90402289

Sasaki K, Shi ZY, Kopelman R, Masuhara H (1996) Three-dimensional pH microprobing with an optically-manipulated fluorescent particle. Chem Lett (2):141–142. https://doi.org/10.1246/cl.1996.141

Schmid G (2008) The relevance of shape and size of Au55 clusters. Chem Soc Rev 37(9): 1909–1930. http://www.ncbi.nlm.nih.gov/pubmed/18762839

Senesac L, Thundat TG (2008) Nanosensors for trace explosive detection: the need for explosive trace detectors. Mater Today 11(3):28

Shafiee A (2019) Nanosensors for therapeutic drug monitoring: implications for transplantation. Nanomedicine (Lond) 14(20):2735–2747

Sharma P, Pandey V, Mukut M, Sharma M, Patra A, Singh B (2021) A review on biosensors and nanosensors application in agroecosystems. Nanoscale Res Lett 16:136. https://doi.org/10.1186/s11671-021-03593-0

Sinha N, Ma J, Yeow JTW (2006) Carbon nanotube-based sensors. J Nanosci Nanotechnol 6(3): 573–590. https://doi.org/10.1166/jnn.2006.121

Tschulena G (1988) Sensors for process control. Phys Scr 1988(T23):293–298. https://doi.org/10.1088/0031-8949/1988/T23/054

Turner APF (2013) Biosensors: sense and sensibility. Chem Soc Rev. https://doi.org/10.1039/c3cs35528d

Vidotti M, Carvalhal RF, Mendes RK, Ferreira DCM, Kubota LT (2011) Biosensors based on gold nanostructures. J Braz Chem Soc 22(1):3–20. https://doi.org/10.1590/S0103-50532011000100002

Vo-Dinh T (2005) Optical nanosensors for detecting proteins and biomarkers in individual living cells. Methods in Mol Biol (Clifton, NJ) 300:383–401. https://doi.org/10.1385/1-59259-858-7:383

You CC, Miranda OR, Gider B, Ghosh PS, Kim IB, Erdogan B, Krovi SA, Bunz UHF, Rotello VM (2007) Detection and identification of proteins using nanoparticle-fluorescent polymer "chemical nose" sensors. Nat Nanotechnol 2(5):318–323. https://doi.org/10.1038/nnano.2007.99

Yuan Y, Zhang F, Wang H, Gao L, Wang Z (2018) A sensor based on Au nanoparticles/carbon nitride/graphene composites for the detection of chloramphenicol and ciprofloxacin. ECS J Solid State Sci Technol 7(12):M201–M208. https://doi.org/10.1149/2.0111812jss

Zhai H, Liu Z, Chen Z, Liang Z, Su Z, Wang S (2015) A sensitive electrochemical sensor with sulfonated graphene sheets/oxygen-functionalized multi-walled carbon nanotubes modified electrode for the detection of clenbuterol. Sensors Actuators B Chem 210:483–490. https://doi.org/10.1016/j.snb.2014.12.121

Zhang XR, Xu X (2004) Development of a biosensor based on laser-fabricated polymer microcantilevers. Appl Phys Lett 85(12):2423–2425. https://doi.org/10.1063/1.1791731

Chapter 4
Nanotechnology in Bacterial Livestock Disease Diagnosis

Dodla Mamatha, Saumya Maurya, Kumari Swati, Alla Yaswanth Naveen Kumar, Rajat Varshney, Pavan Kumar Yadav, Minakshi Prasad, and Mayukh Ghosh

Abstract Trending nanotechnology has gained popularity in various research including health care, agriculture, and veterinary science. It has become an attractive and feasible option for rapid diagnosis, subsequent directed therapy, antibiotic stewardship, and control of diseases. The proficiency of these nanomaterials is due to their unique physicochemical properties and biocompatibility. Various conventional diagnostic tools viz. isolation, biochemical, serological, and molecular characterization of pathogens are though commonplace, but often pose intricacy where nanomodalities can enhance the diagnostic efficacy. Moreover, nanosensors offer rapid and animal-side diagnosis to widen their clinical utility. Therefore, the current discussion is focused toward various nanomaterials which have already ushered ample promise with escalated sensitivity and specificity for the diagnosis of livestock bacterial diseases.

Keywords Biosensors · Aptamers · Immunochromatography assay · Lateral flow assay · ELISA · CNTs · Quantum dots · Magnetic nanoparticles · Gold nanoparticles · Nanocomposites · Liposomes

D. Mamatha · S. Maurya · K. Swati · A. Y. N. Kumar
Department of Genetics and Plant Breeding, Institute of Agricultural Science, Rajiv Gandhi South Campus, Banaras Hindu University, Mirzapur, Uttar Pradesh, India

R. Varshney
Department of Veterinary Microbiology, Faculty of Veterinary and Animal Sciences, Rajiv Gandhi South Campus, Banaras Hindu University, Mirzapur, Uttar Pradesh, India

P. K. Yadav · M. Ghosh (✉)
Department of Veterinary Physiology and Biochemistry, Faculty of Veterinary and Animal Sciences, Rajiv Gandhi South Campus, Banaras Hindu University, Mirzapur, Uttar Pradesh, India

M. Prasad
Department of Animal Biotechnology, Lala Lajpat Rai University of Veterinary and Animal Sciences, Hisar, Haryana, India

LLR University of Veterinary and Animal, ICAR-National Research Centre on Equines, *Nanotechnology Theranostics in Livestock Diseases and Management*, Livestock Diseases and Management,
https://doi.org/10.1007/978-981-16-1610-5_4

4.1 Introduction

The microorganisms implicated in emerging, re-emerging, food, water, and air-borne zoonotic diseases pose major constrain in the public health predicament and socio-economic development of the country. The various pathogenic bacteria, including *Staphylococcus*, *Streptococcus*, *Bacillus*, *Clostridium*, *Erysipelothrix*, *Rhodococcus*, *Nocardia*, *Corynebacterium*, *Mycobacterium*, *Listeria*, *Escherichia*, *Salmonella*, *Yersinia*, *Klebsiella*, *Proteus*, *Campylobacter*, *Vibrio*, *Pseudomonas*, *Pasteurella*, *Mannheimia*, *Burkholderia*, *Moraxella*, *Taylorella*, *Haemophilus*, *Acinetobacter baumannii*, *Mycoplasma*, *Leptospira*, *Rickettsia*, and *Chlamydia*, pose enormous loss in terms of health and productivity.

Early disease diagnosis is needed for directed therapy and to coordinate antibiotic stewardship for combating various microbial diseases (Agrawal et al. 2020; Varshney et al. 2020). It also facilitates disease surveillance in animal population which holds utmost importance for global trade of animals and animal products. In addition, it is beneficial to design "test and segregation" or "test and slaughter" policies, especially in the absence of reliable therapy and vaccine. Delayed or improper diagnosis and undiagnosed cases foil the attempt of efficient disease control as well as containment as it can facilitate the spread of zoonotic infections among the animal handler, in-contact persons, veterinarians, and person implicated in industrial processing of animal-derived products, viz. milk, meats, hide, wools, etc. The other complications may be disease spill over to and from wild animals, emergence of resistant pathogen, and hindrance to the eradication policies.

Till now various methods including bacterial cultivation and serological tests, nucleotide-based detection such as polymerase chain reaction (PCR), DNA micro-array, loop-mediated isothermal amplification (LAMP) and polymerase spiral reaction (PSR), and biosensors-based on electrochemical, impedimetric, amperometric, optical, colorimetric, fluorescence and microfluidic principles have been described and employed for detection of pathogenic bacteria (Váradi et al. 2017).

In-vitro cultivation of microorganism can provide ample diagnostic authenticity, but some of the pathogens like *Rickettsia* except *Bartonella* (formerly *Rochalimaea*) *quintana* and *Chlamydia* may not grow in ordinary media. Even, some of the bacteria are yet to be cultured successfully in in-vitro media. For instance, *Mycoplasma* sp. requires additional supplementation of sterols for growth in culture media. Another pathogen, *Mycobacterium* sp. takes prolonged time to grow in in-vitro media. So, intra-dermal tests like Mantoux and Stormont test based on hypersensitivity are frequently carried out to diagnose *Mycobacterial* infection. However, animals showing energy or over-activity can pose limitations to intradermal tests. Some bacteria such as *Salmonella* require pre-enrichment and/or enrichment media for recovery of the organism from clinical samples before subsequent plating and characterization. Maintenance of the *Leptospira* sp. for microscopic agglutination test in the laboratory is quite cumbersome. Moreover, some zoonotic organisms such as *Brucella* and *Mycobacterium* can spill out to pose a threat to the media handlers and veterinarians. In biochemical-based diagnosis, in-vitro pathogen cultivation is a prerequisite in most cases. Additionally, the observed strain-wise variations in biochemical tests limit its diagnostic value (Minakshi et al. 2019; Prasad et al. 2021).

In case of acute and per-acute cases antigen detection is preferable to antibody detection as the subject gets very little time to generate antibody against such pathogens. Further, the generation of antibody during the earlier exposure and its sustenance can yield false-positive or false-negative outcome to detect ongoing infection. Individual subjects varying in genetic make-up, nutritional status, the exact physiochemical properties (affinity, avidity) of the generated antibodies, immune complex clearance capacity of the guarding cells, and level and time of antigen exposure can lead to complexities in selecting apposite test, in interpreting the results, and in standardizing serological tests. Despite serological tests exhibiting low cost and highly efficient performance, it is facing some challenges such as low-stability, cross-reactions, non-adsorption, and false-positive results. Therefore, serological diagnosis can raise intricacy in disease diagnosis (Fierz 1999; Huang et al. 2021). Complications in molecular diagnostics come in terms of recourse intensive, requiring skilled personnel and often lack portability to be applicable at the farm-level diagnosis.

The aforementioned limitations of conventional diagnostics can be overcome by using various peptide functionalized polymeric nanoparticles, graphene-based polymer composites, clustered regularly interspaced short palindromic repeats (CRISPR)-based probes, and nanomaterials. Trending nanoparticles (NPs) such as carbon NPs (CNPs), metallic NPs, and metal oxide NPs and different kinds of nanocomposites have acquired significant attention as a building block for diverse types of biosensors and bacterial-responsive probes intended for pathogen detection (Table 4.1). These particles have favorable characteristics such as enhanced surface area, manageable morphology, distinctive and controllable optical properties, and remarkable mechanical, electromagnetic, and chemical properties which empower them to improve the test performance. Over the last few decades, CNPs has been preferred over other class of NPs due to the ease of designing, imaging, and sensing applications. The other classes of NPs such as metallic and metal oxide NPs possess unique properties that include Surface Plasmon Resonance (SPR), making them efficient particles for developing colorimetric, fluorescent, and non-linear optical sensors for detecting bacterial pathogens (Alafeef et al. 2020). Functionalization of NPs with recognition elements such as antibody, phage. or aptamer may widen its diagnostic applicability. Nanoparticles magnificently improve the diagnostic sensitivity and reduce the operation time by lowering the steps of pre-enrichment and isolation of bacteria, lastly combining the functions of detection, capture, and elimination of pathogenic bacteria into a single matrix (Shen et al. 2021; Mi et al. 2022).

Table 4.1 Nanotechnology in bacterial livestock disease diagnosis

Detection system	Nano-component	Bacterial species	Description	Reference
Carbon-based NMs in diagnostics				
Fluorescence aptasensor	Multi-walled carbon nanotube (MWCNTs)-based fluorescence aptasensor	*B. anthracis* (rPAD4)	The immobilized aptamer was labeled with Gel Green, and a sensitivity of 20 and 62.5 ng/mL purified and unpurified rPAD4 protein	Karimi and Dabbagh (2019)
Aptamer	Single-walled carbon nanotube	*E. coli*	Identify *E. coli* $\approx 2.7 \times 10^5$ CFU/mL	So et al. (2008)
Fluorescence spectroscopy and microscopy	Amphiphilic carbon dots (CDs)	*E. coli*, *S. typhimurium*, *P. aeruginosa*	Immediately bind to bacterial cells, intensity depend on bacterial species, and detection through both fluorescence spectroscopy and microscopy	Ritenberg et al. (2016)
Silica-based NPs in diagnostics				
Antibody	Dye-doped silica NPs	*E. coli*, *Salmonella*, *S. aureus*	Multiplexing available, fast detection	Santra et al. (2001)
Europium complex and organic-dye-doped NP	Silica NP	*B. anthracis*	Rapid and ultrasensitive, superior to terbium-based sensors	Ai et al. (2009)
Quantum dots (QDs) in diagnostics				
QD conjugated to streptavidin	CdSe/ZnS core/shell QDs, semiconductor quantum dots (QDs)	*E. coli* O157:H7	Identifies the pathogen rapidly and accurately, more sensitive than conventional dyes	Valizadeh et al. (2012)
Magnetic beads coated with anti-*E. coli* O157 antibodies	QDs	*E. coli* O157:H7	The modality has 100 times lower detection limit as compared to FITC-based method	Zhu et al. (2011)
Antibody-based fluorescence spectroscopy	QDs	*E. coli* O157:H7 and *S. typhimurium*	Detects bacteria within 2 h with availability of multiplexing	Yang and Li (2006)
Antibody	QDs	*S. aureus*	Detection limit 10^3 CFU/mL but in food it was not satisfactory	Cui et al. (2021)

Toxin antibody	QDs	Cholera toxin, ricin, Shiga-like toxin 1, and S. enterotoxin B	Multiplexing with four colors of fluoro-reagents	Pöhlmann and Elßner (2020)
Metallic and metal oxide NMs in diagnostics				
Quartz crystal microbalance (QCM) biosensor with single-stranded modified AuNPs probes	Gold NPs	*B. anthracis Ba813*	Depicted a sensitivity of 3.5×10^2 CFU/mL	Hao et al. (2011)
Colorimetric assay	Gold NPs	*B. anthracis*	Yielded detection limit 10 pg/mL with no cross-reactivity	Deng et al. (2013)
Sensitive aptamer-based biosensor	Gold NPs	*E. coli* O157:H7	As low as 10 CFU of *E. coli* O157:H7 were detected. The generated signals (red bands on the test zone of a lateral flow strip) can be unambiguously read out by the naked eye	Wu et al. (2021)
An oligonucleotide-modified AuNP-based colorimetric assay	Gold NPs	*Brucella* spp. BCSP31	LOD of 10^3 CFU/mL in bovine (urine and semen samples) and 10^4 CFU/mL in milk with no cross-reactivity, detection in 30 min	Pal et al. (2017)
Nanobiosensor based on gold nanoparticles (AuNPs) and oligonucleotide probe	Oligonucleotide probe IS711 gene functionalized with AuNPs	*Brucella*	Achieved LOD is 1.09 pg/μL, highly specific, low cost, rapid, direct, and visual detection	Sattarahmady et al. (2015)
Surface-enhanced Raman scattering (SERS)	Silver nanorod arrays	Against bacteria	Used for detection	Chu et al. (2008)
Antibody	Gold NP	*Salmonella* spp.	Specificity is 99.1% and 100% sensitivity	Fang et al. (2009)
Bio-barcoded electrochemical biosensor	Gold NPs, MNPs, and nanoparticle tracers (NTs, such as PbS and CdS)	*B. anthracis* and *S. enteritidis*	Detection limit 0.2 pg/mL for *B. anthracis* and 50 pg/mL for *S. enteritidis*	Zhang et al. (2010)
DNA probe	Gold and MNPs	*S. enteritidis*	Detect *S. enteritidis* at a concentration as low as 1 ng/mL	Amini et al. (2017)

(continued)

Table 4.1 (continued)

Detection system	Nano-component	Bacterial species	Description	Reference
Antibody	AuNP-based strip	*S. aureus*	Fast detection within 10 min and very low cost	Chen et al. (2022)
3D-printed chip	Gold NP	*S. aureus* (MRSA)	Easy-to-use colorimetric pathogen detection, very low cost, and rapid	Chudobova et al. (2015)
Immuno chromatographic detection	Gold NP-based	*S. aureus*	100% sensitivity and 94.7–100% specificity	Huang (2006)
Immunomagnetic separation	Gold NP/MNP	*S. aureus*	Detection limit 1.5×10^5 CFU in milk	Sung et al. (2013)
DNA based	Unmodified gold NP	*L. monocytogenes*	Detects 1.304 fg/μL of DNA and compared to PCR 70-fold highly sensitive	Liu et al. (2015)
Conjugated with aptamer	Fe_3O_4 NP	*L. monocytogenes*	Yielded detection limit 5.4×10^3 CFU/mL	Wang et al. (2020)
Lateral flow test strip	Ferro MNPs	*C. jejuni*	Limit of detection for pure culture was 100 or 1 CFU/mL and in spiked stool sample $10–10^2$ CFU/mL	Poonlapdecha et al. (2018)
LipL32-coupled immunochromatographic test	Gold NP	*Leptospira*	Sensitivity was 10^3/mL with no cross-reactivity	Chirathaworn et al. (2014)
Antibody	Gold NP	*Leptospira*	With general bacteria sensitivity of detection was 10 leptospires/mL with no cross-reactivity	Othman et al. (2022)
Antibody	Streptavidin-coated magnetic NP	*E. coli* O157:H7	Detects 1.6×10 to 7.2×10^7 CFU/mL in 15 min	Varshney et al. (2005)
Antibody	Immunomagnetic beads	*E. coli* O157:H7	Sensitive and economically beneficial	Tu et al. (2009)
Antimicrobial peptide functionalized NPs	Detection by magnetic NP and urease catalyzed signal	*E. coli* O157:H7	Have low sensitive detection of 12–233 CFU/mL bacteria within 30 min	Qiao et al. (2017)

Paper-based peptide probe	Magnetic NP	*E. coli* O157:H7	Limits of detection of 30–300 CFU/mL in spiked complex food matrices	Suaifan et al. (2017a)
Diagnosis by multiplex loop-mediated isothermal amplification (m-LAMP) lateral flow dipstick	Gold NPs	*Leptospira*	Detect DNA at concentrations as low as 3.95×10^{-1} genomic equivalent/mL	Najian et al. (2016)
Antibody	Europium NP	*B. anthracis*	Compared to traditional method, it is 100-folds more sensitive	Tang et al. (2009)
Antibody	Magnetic NPs	*E. coli* and *S. typhimurium*	Detection limit 10^4–10^5 CFU/mL within 30 min	Ravindranath et al. (2009)
10 primers used lateral flow biosensor	Gold NPs	*V. cholerae*	Rapid, visual and sensitive detection	Yu et al. (2011)
Cell wall adherence by D-Ala-D-Ala units	MNPs bound with vancomycin	*S. saprophyticus* and *S. aureus*	Detection limit 7×10^4 CFU/mL	Lin et al. (2005)
Antibody-based electrochemical impedance spectroscopy (EIS)	Gold NPs	*E. coli* 0157:H7	LOD is 10 cells/0.173 cm^2	Lin et al. (2019)
Surface plasmon resonance (SPR) (antibody)	Gold NPs	*E. coli* K12; *L. fermentum*	10^4 and 10^3 CFU/mL	Baccar et al. (2010)
Biosensor-mediated detection based on *S. aureus* proteases proteolytic activity on a peptide substrate	MNP and gold surface	*S. aureus*	Detection limit of 7.5 and 100 CFU/mL bacteria in food products and environmental sample, rapid and low-cost paper-based biosensor	Suaifan et al. (2017b)
Antibody	Magnetic NPs (MNPs)	*E. coli* and *S. typhimurium*	Detection limit 10^4–10^5 CFU/mL within 30 min	Quintela et al. (2022)
Metal organic framework (MOF)				
Electrochemical impedimetric immunosensor for label-free detection	Immunosensor using *O*-carboxymethyl chitosan surface-modified Fe_3O_4 NP	*C. jejuni*	Long stability, reusable immunosensors, fast response, wide linear range and acceptable reproducibility	Huang et al. (2010)
Nanocomposites				
Biosensors based detection		*Vibrio cholerae*	Limit of detection of 1.0×10^{-21} M	Rahman et al. (2017)

(continued)

Table 4.1 (continued)

Detection system	Nano-component	Bacterial species	Description	Reference
	Vibrio cholerae DNA with polystyreneco-acrylic acid composite nanospheres			
Immunoassay	Anti-*L. monocytogenes* poly-styrene coated NP	*L. monocytogenes*	Compared to traditional method, it's more sensitive	Shi et al. (2015)
Lipid NPs				
Lateral flow assays (LFAs)	Combination of capture aptamer and GM1 cell receptor-labeled liposomes	Cholera toxin	Achieved limit of detection of 2 ng/mL (by the 3σ rule)/10 ng/mL (visual), detection in ~15 min	Frohnmeyer et al. (2019)
Bacterial toxins	Liposomes	Tetanus, botulinum, and cholera toxins	Tetanus, botulinum toxins specifically bind to gangliosides G1b series while cholera toxin binds selectively to GM1	Singh et al. (2000)
Cholera toxin	Ganglioside liposome immunoassay	Cholera toxin	Limit of detection of 0.1–30 pg/mL in 20 min	Ahn-Yoon et al. (2003)
Yersinia pestis F1 capsular antigen	Lateral flow with blue carboxylate-modified nanoparticles (CMLNPs)	*Yersinia pestis*	Rapid test, for wildlife in field condition	Hsu et al. (2018)

4.2 Multifaceted Nano-Interventions in Bacterial Diagnostics

4.2.1 Nanoparticles in Sample Preparation for Diagnosis

Reduction of detection time and planning of therapy in the separation of bacteria is carried out by quick processing and precise techniques. Enrichment step is necessary before the separation of target bacteria if the concentration of bacteria is below the detection limit. A minimum of 10^3–10^4 and 10^4–10^5 CFU/mL is required for detection in methods such as PCR and Enzyme-linked immunosorbent assay (ELISA); pre-enrichment step is unavoidable before analysis. Large amount of bacteria separation within small sample size is done in an ideal enrichment step. The most common method is immune magnetic separation (IMS) in which magnetic materials coated with target-specific antibodies are used to capture bacteria. Still, because antibody has a high cost of production and low stability of antibodies in unfavorable environment, the use of IMS is restricted. The limitations of traditional methods have been eliminated with the help of advanced NP designs. The shape and surface of Gold nanoparticles (AuNPs) have changed to confirm the interaction of the NP with gram-positive and gram-negative bacteria. The attached AuNPs to the bacterial cell walls exhibited direct correlation between the number of the adhered NP and the potential of the bacterial strains using electron microscopy observation. NPs are altered for the separation of pathogenic bacteria from the blood such as MNP modified with zinc-bis(dipicolylamine) (ZnDPA) ligands and vancomycin targeting phosphatidylserines on gram-negative and gram-positive bacteria surface and peptidoglycan on gram-positive bacteria is prepared. Vancomycin-attached MNPs have the ability for the enrichment of *S. aureus* from the complete blood matrix within 2.5 h. Separation of the gram-negative bacteria such as *E. coli* and endotoxins has been achieved by using the ZnDPA and bis-Zn-DPA modified polyethylene glycol coated (PEG) MNP and minimal reaction with red blood cells could be observed. Recently, identification of the pathogen and confirmation of bacterial blood stream infections are carried out by a core-shell-structured NP-based new capture platform made up of mesoporous titanium dioxide TiO_2-coated magnetic NPs functionalized with targeting aptamers (designated as $AptFe_3O_4@mTiO_2$) (Şen Karaman et al. 2021).

Magnetic detection encloses the measurement of stray magnetic fields generated into the particles by an external field, for example, by means of Giant magneto-resistance (GMR) and diagnostic magneto-resistance (DMR). This technique is based on several steps. On the sensor surface, the probe is immobilized and binds the target. Consequently, a biotinylated reporter probe is attached to the sequestered targets. Finally, captured targets are labeled by streptavidin-coated magnetic particles and are easily detected (Ranzoni et al. 2013). Koets et al. described the detection of double-stranded bacterial DNA at a concentration of 4–250 pM by employing magneto-resistance sensor within 3 min of assay time (Koets et al. 2009). Magnetoresistive immunosensor integrated with a microfluidic network has also been employed for the detection of *E. coli* O157:H7 strains in food and clinical samples with a limit of detection (LOD) of 10–1000 CFU/mL (Mujika et al. 2009).

4.2.2 Electrochemical Biosensors

Biosensors are tools which can detect chemical constituent in biological reactions and chemical reactions with the help of producing signals proportional to the chemical constituent concentrations. These are of various kinds depending on the signal that is produced which includes electricity, heat, and light. Electrochemical biosensors are a principal branch of biosensors containing the longest history, broad application range, and good strength. It is a type of detecting tool which converts the biochemical reactions to electrical signals. On the surface of electrodes, these biosensors affix biological elements such as antibodies and aptamers by biological modification. The target molecules in the solution are recognized by the fixed biological sensitive elements and generate a specific binding reaction. A quantitative or semi-quantitative detection is achieved by converting the binding reaction into an electrical signal through the electrode. These biosensors can be classified into amperometric, impedance, potential, and capacitance types. The rapid and sensitive detection of food-borne pathogens has been carried out by using impedimetric biosensors. For instance, specific immobilized antibody on screen-printed gold electrode has been employed to capture *S. typhimurium* resulting in impedance change between the gold electrodes to facilitate bacterial detection within 20 min with LOD 1×10^3 CFU/mL (Farka et al. 2016). Impedimetric biosensor employing anti-*E. coli* antibody immobilized on modified screen-printed electrode has also been used for quantitative detection of *E. coli* in drinking water with LOD as low as 3×10 CFU/mL (Cimafonte et al. 2020). In another experiment, magnetic NPs coated with anti-*Listeria* monoclonal antibodies have been employed to capture and concentrate the *Listeria* cells from the target sample. Subsequently, AuNPs functionalized with urease enzyme and the polyclonal antibodies have been targeted to interact with the *Listeria* cells to form a sandwich complex of the bacteria captured between magnetic and AuNPs. Following resuspension of the complex in the presence of urea, liberation of ammonium ions and carbonate ions takes place from urea due to urease action of the functionalized AuNPs resulting in impedance change to be measured by electrode/biosensor. The limit of detection of the sensor for *L. monocytogenes* has been found to be 1.6×10^3 CFU/mL (Wang et al. 2017). Currently, reduction in the detection time and LOD has been done by using amperometric biosensors based on cyclic voltammetry combined with immune technology. So, they are broadly used in rapid and sensitive detection of multi-target molecules and food-borne pathogenic bacteria (Huang et al. 2021). For instance, extraction of bacteria from complex samples followed by determination of the concentration of extracted bacteria has been carried out by carbohydrate ligands-coated magnetic nanoparticles strip and cyclic voltammetry. The LOD of this method has found to be 10^2 CFU/mL for *E. coli* O157:H7, *S. enteritidis*, and

L. monocytogenes and detection within 30 min (Matta et al. 2018). Core-shell magnetic beads surface-functionalized with polymeric nanocomposites containing glucose oxidase (GOx)–polydopamine, AuNPs, and antibodies specifically identify and bind with *E. coli* O157:H7 via antibody recognition to form a complex. This bacteria-nanocomposite complex produces a redox reaction by resuspension in glucose solution catalyzed by GOx. *E. coli* concentration and the current change caused by the reaction have found to maintain a linear relationship as measured by Prussian blue (PB)-modified screen-printed microelectrodes. LOD of the method for the target bacteria has found to be 10^2 CFU/mL by cyclic voltammetry (Xu et al. 2016).

Electrochemical impedance spectroscopy (EIS) has been reported as the rapid and simple method of quantitative detection of *Brucella melitensis*. Immobilization of *B. melitensis* antibody on the surface of gold nanoparticle-modified screen-printed carbon electrodes (GNP-SPCEs) used for the construction of label-free immunosensors. The *B. melitensis* antigen reaction on the surface of GNP-SPCEs with antibody was identified by cyclic voltammetry and EIS. Analysis of the impedance components of the system using general equivalent design of an electrochemical cell. The results observed that the change in electron transfer resistance varied prominently due to the binding of *B. melitensis* cells. These biosensors are capable to detect as low as 1×10^4 and 4×10^5 CFU/mL of *B. melitensis* in pure culture and milk samples, respectively, in less than 1.5 h (Wu et al. 2013).

The most common cause of gastroenteritis in developed countries is *Campylobacter jejuni*. Rapid detection of these bacteria was carried out by a novel electrochemical impedimetric immunosensor based on *O*-Carboxymethyl chitosan surface-modified Fe_3O_4 NP (denoted as OCMCS-Fe_3O_4). On the surface of OCMCS-Fe_3O_4 NPs, anti-Fla A monoclonal antibodies 2D12 (denoted as 2D12McAbs) are immobilized. Estimate the relative change in impedance before and after 2D12McAbs-*C. jejuni* interaction with the technique of electrochemical impedance spectroscopy accomplished the detection. Under optimum conditions the comparative difference in impedance is proportional to the *C. jejuni* concentration between the range 1×10^3 and 1×10^7 CFU/mL. The advantages of the OCMCS-Fe_3O_4 NPs-based immunosensor are simplicity of use, rapid response, broad linear range, reproducibility, and long stability (Huang et al. 2010).

4.2.3 *Nanomaterials in Aptamer-Based Detection*

Aptamer means a bioaffinity ligand which can specifically bind to target. They are divided into peptide aptamer and oligonucleotide aptamer. The peptide aptamer consists of a protein reagent and its target and focus would be only on protein. Oligonucleotide aptamer means a single-stranded DNA or RNA sequence selected using systematic evolution of ligands by exponential enrichment (SELEX) techniques also known as chemical antibodies because of their enhanced attraction toward its targets. Usually, aptamers always refer to oligonucleotide one.

Aptamer-based biosensors for the detection of bacteria include optical, electrochemical, or surface-enhanced Raman scattering (SERS) signal. Usually sandwich structures are used to capture and detect bacteria in aptamer-based colorimetric biosensors. A gold nanoparticle-based (AuNP) Enzyme-linked antibody-aptamer sandwich (Nano-ELAAS) method is used to detect *S. typhimurium* by Wu et al. (2014). Capture probes are prepared with the help of magnetic microparticles altered with *S. typhimurium*-specific aptamers (MMP-aptamers) and enriching the *S. typhimurium* from samples, then MMP aptamers are used to sandwich the *S. typhimurium* and detection of antibodies. Subsequently, this nanoprobe with a reporter antibody and horseradish peroxidase (HRP) interacted with sandwich complex (MMP-aptamer/*S. typhimurium*/detector antibody), which produces a color change that could be observed with unaided eye. Sensitivity of this technique is much more compared to colorimetric ELAAS and chemiluminescent ELAAS. The same structure is used to detect *L. monocytogenes* and LOD ranging from 10 to 10^6 CFU/mL. It is detected by aptamer-modified magnetic nanoparticles (Apt-MNPs) and then identified by immunoglobulin Y-coated silver nanoclusters. Aptamer-based fluorescent biosensors are also based on sandwich structure which is used in colorimetric biosensors. Identification and capturing of *S. aureus* using vancomycin-stabilized fluorescent gold nanoclusters (AuNCs@Van) and aptamer-coated magnetic beads have been described by Cheng et al. (2016). Improved sensitivity and specificity of the detection is due to dual recognition strategy of vancomycin and aptamer. It can be used to detect about 70 CFU/mL of *S. aureus* in complex samples. A similar dual identification technique is designed with AuNCs@Van and aptamer-altered AuNPs (Apt-AuNPs) in which AuNPs act as an energy donor and Apt-AuNPs act as an energy acceptor. Fluorescence resonance energy transfer (FRET) mechanism is capable to detect *S. aureus* with sensitivity and selectivity within 30 min (Li et al. 2021).

E. coli O157:H7 detection has been carried out by using sensitive aptamer-based biosensor. In this assay, two aptamers specific for the outer membrane of these bacteria are used. One of the aptamers is used for magnetic bead enrichment while the other is used as a signal reporter, which is amplified by isothermal strand displacement amplification (SDA) and later detected by a lateral flow biosensor. The generated signals (red bands on the test zone of a lateral flow strip) can be unambiguously read out by the naked eye (Wu et al. 2015).

4.2.4 Nanoparticles in Immunochromatographic Technique

Immunochromatography assay (ICA), namely lateral flow test, is a common detection tool designed to diagnose the presence or absence of the target analyte. It is a union of chromatography (components of a sample separated based on changes in their movement through sorbent) and immunochemical reaction. The test strip is widely used in immunochromatographic technique (Fig. 4.1). This essay is employed to detect *S. typhimurium* and *S. enteritidis* in a single strip. The technique

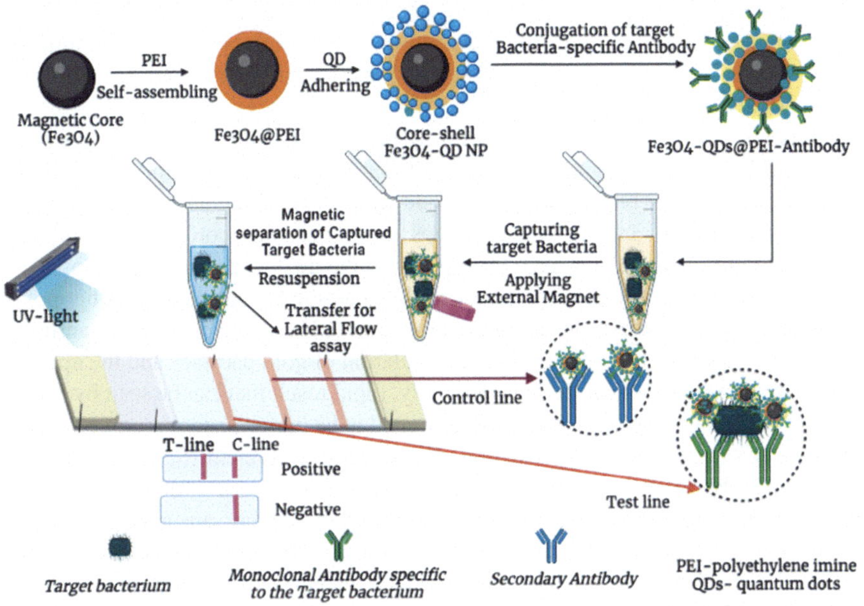

Fig. 4.1 Immunochromatography-based lateral flow test for bacterial detection employing core-shell magnetic nanoparticles

has been designed in the form of sandwich, on a nitrocellulose membrane 2 specific anti-*S. typhimurium* and anti-*S. enteritidis* antibodies are immobilized at separate test lines, whereas the other specific antibody to *Salmonella* is attached with gold NPs. This assay can detect *S. typhimurium* and *S. enteritidis* immediately in a culture medium at a range as low as 10^4 and 10^6 CFU/mL, respectively. It has some advantages such as simplicity and fast diagnosis of these specific bacterial contaminants in chicken and can also be useful for mass detection in chicken farms and other livestock products (Moongkarndi et al. 2011). *S. aureus* has also been detected using immunochromatographic assay based on gold NPs. This assay is assembled by using anti-protein A IgG with two different specificities in the form of sandwich. One anti-protein A IgG is impaired in known detection zone on a porous nitrocellulose membrane, while the other one is linked with AuNPs. The cell wall protein of *S. aureus* is Staphylococcus protein A which has the capability to react with various host components, probably denoting a role as a virulence factor in *S. aureus* infections. Sensitivity in the immunochromatographic test of 306 *S. aureus* strains was 100% and the specificity of 44 non-*S. aureus* strains was 96–100%. This test can detect less number of *S. aureus* in processed foods with total analytical time of only 25 h when compared to conventional methods which take 5–6 days to complete (Huang 2007).

Immunochromatographic strip has been developed for the rapid diagnosis of *E. coli* O157:H7 using murine monoclonal antibody to *E. coli* O157:H7 lipopolysaccharides attached with 40 nm colloidal gold particles by citrate method. The sensitivity of this method detected using tenfolds diluted *E. coli* O157:H7 with a level of 1.8 × 10^7 to 1.8 CFU/mL in enriched raw beef. A minimum of 1.8 × 10^5 CFU/mL of *E. coli* was detected without enrichment and 1.8 CFU/mL after enrichment (Jung et al. 2005).

Epidemic and pandemic cholera is mostly related with *Vibrio cholera* serogroups O1 and O139. Rapid diagnosis of cholera to prevent spread of outbreaks, a direct one-step lateral flow biosensor has been developed for simultaneous detection of both *V. cholera* O1 and O139 serogroups using alkaline peptone water culture. Colloidal gold nanoparticles have been used for the dual detection in the biosensor functionalized by serogroup-specific monoclonal antibodies raised against lipopoly-saccharides. It is based on immunochromatographic principle where antigen-antibody reaction would result in the accumulation of gold particles and the appearance of red line on the strip. It is user-friendly, rapid result that can be seen by naked eye and cold chain free storage that is well suited to be performed in resource-limited setting (Yu et al. 2011). The detection of surface array protein (Sap), which is an important biomarker for *B. anthracis* has been detected using immunochromatographic test. Mouse monoclonal antibodies are generated using the recombinant sap. These antibodies are attached with colloidal gold for arrangement of ICT. The assay completes the sample testing within 2 min compared to 4–5 h in ELISA (Puranik et al. 2020).

4.2.5 Nanoparticles in ELISA

ELISA can provide a useful measurement of antigen-antibody concentration. Specific antibodies are employed to detect the presence and quantity of antigen in a target sample or vice versa. Lack of multiplexing to detect multiple targets at one go and cross-reactivity are certain important limitations of conventional ELISA. Nanointervention to modify the conventional ELISA has erased some of these limitations. For instance, a nanocoaxial-based electrochemical sensor comprised of vertically oriented nanoscale coaxial electrodes in array format has provided the facility of multiplexing along with point-of-care detection besides comparable sensitivity to conventional ELISA. The modality has employed the working principle of electrochemical ELISA and differential pulse voltammetry or square wave voltammetry. This modality has efficiently detected cholera toxin and the LOD was found to be 2 ng/mL (Archibald et al. 2015).

The detection of *E. coli* O157:H7 using functional nanoparticle-enhanced ELISA (FNP-ELISA) has found to be twofold more sensitive to immunomagnetic separation ELISA (IMS-ELISA) and fourfold more sensitive to Conventional ELISA (C-ELISA). The FNP-ELISA has been customized on a two-step process: first, separation of the target bacteria from sample using mouse monoclonal anti-O157: H7 antibody functionalized immunomagnetic NPs (IMMNPs) followed by bacterial detection using Beacon gold NPs (B-GNPs) attached with polyclonal anti-O157:H7 antibody and biotin single-stranded DNA (B-DNA). The addition of streptavidin-horseradish peroxidase (Strep-HRP) interacts with biotinylated counterpart and a change in color can be visible as the chromogenic substrate 3,3′,5,5′--tetramethylbenzidine (TMB-H_2O_2) is added. The LOD is 6.8×10^2 to 6.8×10^3 CFU/mL in food samples (Shen et al. 2014).

4.3 Diverse Nanomaterials in Bacterial Disease Detection

4.3.1 Carbon-Based NMs in Diagnostics

Based on the number of graphene sheets, carbon nanotubes (CNTs) are divided into single-walled carbon nanotubes (SWCNTs) and multi-walled carbon nanotubes (MWCNTs). MWCNTs have certain advantages over SWCNTs such as they are cheaper, stiffer, and easier to produce on a large scale. MWCNTs-based fluorescence aptasensor has been used for the diagnosis of anthrax based on recognition of recombinant protective antigen domain 4 (rPAD4), the most important key factor in anthrax detection. The limit of detection of the aptasensor was found to be 20 ng/mL for purified and 62.5 ng/mL for unpurified rPAD4 protein (Manhas et al. 2021). Along with CNTs, multilayer biosensors and microfluidic chip-based LAMP assays have also been used for the detection of *E. coli* O157:H7 and its toxic gene. The target *E. coli* O157:H7 in complex samples was selectively captured by the anti-*E. coli* O157:H7 antibody functionalized CNT multilayer biosensor. As the anti-*E. coli* O157:H7 antibody interaction gets dismantled, the captured bacteria can be released. The DNA concentration of the captured bacteria was consequently analyzed with microfluidic chip-based LAMP. The proposed sensing platform was capable of detecting bacterial concentration as low as 1 CFU/mL after systematic increase of capturing and detecting conditions. The noticeable advantage of this method includes high specificity, low cost, good reproducibility, and ability to regenerate, making it an efficient platform for detecting *E. coli* O157:H7 in clinical diagnosis as well as in food samples (Li et al. 2017).

4.3.2 Silica-Based NPs in Diagnostics

Fluorescent dyes including organic and inorganic have acquired attention because of the ease of diagnosis under fluorescent microscope, when used in designing diagnostic tools and protocols for bio-applications. Usually traditional organic fluorophore-based probes carry some intrinsic flaws such as low fluorescent intensity, low photostability, and high cytotoxicity. In bioimaging, conventional dyes such as fluorescein, Cy3, Cy5, fluorescein isothiocyanate, and Rhodamine B are broadly used, but these dyes have some limitations such as photobleaching and chemical instability. These limitations can be overcome by advanced NP-based fluorescent probes which exhibit high sensitivity and high photostability. Extensive use of fluorescent dye-doped silica NPs has been evidenced for bioimaging and biosensing applications such as cellular detection, cell surface receptor targeting, and detection of bacteria. Silica NPs are more suitable for biological applications because of their biocompatibility and low cytotoxicity, even though the other NPs and quantum dots have high fluorointensity (Ananda et al. 2021).

Detection of anthrax spore biomarker dipicolinic acid (DPA) has been done by many lanthanide ions-based probes. For instance, determination of the DPA in *Bacillus subtilis* spores has been carried out by a novel radiometric fluorescent nanoprobe based on europium (Eu)-doped silicon nanoparticles Eu@SiNPs. In addition, detection of micro DPA in *Bacillus subtilis* spores in real environmental samples has also been carried out by using a new radiometric fluorescent method. LOD was found to be as low as 2.38×10^4 spore/mL (Na et al. 2020). The silica NPs have also been used for the sensitive and multiplexed detection of bacterial pathogens using multi-colored FRET (fluorescence resonance energy transfer) within 30 min. Artificially prepared NPs emit distinct colors upon excitation with a single wavelength by varying the ratio of three tandem dyes co-encapsulated into the NPs. Detection of *E. coli*, *S. typhimurium*, and *S. aureus* is achieved when these NPs have been conjugated with monoclonal antibodies specific for these pathogenic bacteria (Wang et al. 2007). Rapid detection of pathogenic *E. coli* O157:H7 has been carried out by fluorescent NP-based indirect immunofluorescence assay employing either fluorescence microscopy or flow cytometry. Microemulsion (w/o) is used for the preparation of dye-doped silica NPs with the combination of 3-amino propyl tri ethoxy silane (APTES), fluorescein isothiocyanate (FITC), and polymerization reaction with carboxy ethyl silane triol sodium salt (CEOS). On the surface of these NPs, protein A is immobilized by covalent binding to the carboxy linkers. Recognition of *E. coli* O157:H7 has been carried out by using rabbit anti-*E. coli* O157:H7 antibody as primary antibody. Fluorescent signals are generated by antibody-binding protein A labeled with FITC-doped silica NPs. *E. coli* in buffer and bacterial mixture has been efficiently detected by this method (Chen et al. 2015).

4.3.3 Quantum Dots (QDs) in Diagnostics

Simultaneous detection of *E. coli* and *S. typhimurium* can be carried out by using quantum dots (QDs) as fluorescence labels in immunoassays. Different emission peaks can be measured using various-sized QDs when excited with a single wavelength of light. Anti-*E. coli* and anti-*Salmonella* antibodies are conjugated by highly fluorescent semiconductor QDs with different emission wavelengths (525 and 705 nm). Specific antibody-coated magnetic beads have been used for capturing the target bacteria from complex samples. Bead cell QD complex is formed by interaction between bead cell complex and antibody-coated QD. QD antibody conjugates can uniformly and absolutely attach to the surface of the bacterial cells as elucidated by fluorescent microscopic images of QD labeled *E. coli* and *Salmonella* cells, showing that the conjugated QD molecules still maintain their effective fluorescence while binding with specific target bacteria in a complex mixture recognized by active cognate antibody molecules. Quantitative detection of *E. coli* O157:H7 and *S. typhimurium* has been carried out by this method with the limit of detection of 10^4 CFU/mL and it can be completed within 2 h (Yang and Li 2006).

Staphylococcus aureus is one of the major food-borne pathogenic bacteria. Encouraging food safety by the development of rapid detection methods for *S. aureus*, an immunoassay has been established for the quantitative detection of *S. aureus* using QDs acting as fluorescent labels. Separation of *S. aureus* cells using conjugation of biotin-labeled anti-*S. aureus* antibody with streptavidin-coated magnetic nanobeads (180 nm diameter) followed by mixing of the fluorescence labels with the separated *S. aureus* and, finally, observing the fluorescence intensity of the bead cell QD at a wavelength of 620 nm has enabled measurement of pathogenic load in a complex sample. The limit of detection has found to be 10^3 CFU/mL (Yaohua et al. 2014). Mannose-coated cadmium sulfide (CdS) QDs have been synthesized under one pot facile aqueous condition as the presence of CdS under reducing conditions facilitates self-assembly of thiolated mannose. At a wavelength of 550 nm the nanoparticles having ~15 nm diameter produce an intense, broad luminescence emission. These mannose-CdS QDs have been used to induce luminescent aggregates of *E. coli* by which bacteria in cell suspension as few as 10^4 *E. coli* per mL can be detected. *E. coli* cell surface FimH mannose-specific lectins have facilitated the aggregation process (Mukhopadhyay et al. 2009). Even, multiplexed detection of cholera toxin, ricin, Shiga-like toxin 1, and Staphylococcal enterotoxin has been performed simultaneously in single well of a microtiter plate by sandwich fluoroimmuno assay using bioinorganic conjugates comprised of highly luminescent semiconductor nanocrystals (CdSe-ZnS core–Shell QDs) and cognate antibodies. A mixture of all four QD-antibody reagents has been used for the simultaneous detection of four toxins from a single sample (Goldman et al. 2004).

4.3.4 Metallic and Metal Oxide NMs in Diagnostics

Metal NPs and metal oxide NPs have drawn greater attention because of their unique optical, magnetic, and electrical properties. Selective detection of bacteria using metal-based NPs comparatively has more efficiency. Various metallic and metal oxide NPs used for preparation of diagnostic tools to detect bacteria include but not limited to gold, silver, magnetic oxide, europium NPs, etc. (Table 4.1).

4.3.4.1 Gold NPs in Bacterial Detection

Detection of bacterial growth is tedious. Conventional method used for detection such as ELISA and Polymerase chain reaction (PCR) have notable drawback of long processing time. It takes about 6–24 h for bacterial growth followed by 1–3 days for morphological and biochemical analysis. AuNPs are excellent platforms for rapid biosensing due to their distinct features such as surface plasmon band localization in the visible spectrum, various functionalization capabilities, along with ease synthesis, etc. Bacterial detection based on sensitivity of the surface plasmon resonance of the AuNPs at aggregation state produces visible color change (Mocan et al. 2017). Based upon this principle, a calorimetric method has been developed for the detection *E. coli* employing cysteine gold nanoparticles (CAuNPs) of 2–20 nm size range depicting limit of detection about 100 CFU/mL. Electrostatic adhesion between the positively charged cysteine in CAuNPs and the negatively charged lipopolysaccharides of the bacterial outer membrane has demonstrated the binding affinity of *E. coli* O157:H7 to H2 receptor present in CAuNPs. The cross-linking of CAuNPs in presence of known number of bacteria has yielded a clear red-to-blue shift in sample color that can be easily observed by the unaided eye (Raj et al. 2015). Similarly, a novel aptasensor has been developed for the calorimetric detection of *Salmonella typhimurium* based on color change. AuNPs are functionalized with aptamers which recognize *S. typhimurium* and act as a stabilizer for the AuNPs. Bacterial binding with the aptamers at NP surface results in aggregation of the AuNPs producing a visible color change. The limit of detection of this method has found to be as low as 56 CFU/mL (Ma et al. 2017).

Bimetallic Au@Ag core-shell NPs have been constructed and employed an immunomagnetic separation-polymerase chain reaction (IMS-PCR) method for the detection of *Campylobacter jejuni*. The size range of the NPs was 6.8–27.7 nm and detection limit was 10 CFU/mL. In this method, after capturing the target bacteria by immunomagnetic separation, gold-tagged immunoprobes have been added to create a signal on solid surface followed by amplification of the signal, resulting in highly sensitive detection of bacteria through naked eye (Cao et al. 2011).

Surface-enhanced Raman spectroscopy (SERS)-based method has also been employed for the detection of *E. coli* and *S. aureus* in which polyethylenimine (PEI)-modified Au-coated magnetic microspheres (Fe_3O_4@Au@PET) and concentrated Au@Ag NPs are used for capturing the bacteria and signal generation, respectively. The method works on the principle of three-step process, i.e., capture-enrichment-enhancement. Here, the capture and enrichment properties of PEI-conjugated gold-coated magnetic microsphere have been employed as electrostatic adhesion between positively charged PEI and bacterial cells with negative charge facilitates rapid and efficient bacterial capture and enrichment followed by SERS enhancement by Au@Ag NPs resulting in sensitive bacterial detection. LOD was found to be 1000 CFU/mL with total assay time of 10 min (Wang et al. 2016).

AuNPs functionalized with oligonucleotide probe targeting IS711 gene have facilitated rapid, direct, and visual detection of *Brucella* spp. with LOD of 1.09 pg/μL (Sattarahmady et al. 2015). Similarly, hybridization assay employing AuNPs functionalized with oligonucleotide probe targeting the gene of BCSP31 outer membrane protein has found to specifically detect *Brucella* spp. within 30 min in a colorimetric assay with LOD of 10^3 CFU/mL in bovine urine and semen samples, and 10^4 CFU/mL in milk with no cross-reactivity (Pal et al. 2017). A similar competitive lateral flow assay based on AuNPs probe has been designed to detect Cholera toxin. The principle of this assay is the exploitation of the competition of cholera toxin in solution with immobilized cholera toxin for binding with aptamer-coated AuNPs. The assay produced visual color change within 10 min (Frohnmeyer et al. 2019). AuNPs functionalized with oligonucleotide probes specific to Ba813 and pag of *B. anthracis* have been employed in a Quartz Crystal Microbalance (QCM) biosensor yielding a sensitivity of 3.5×10^2 CFU/mL for detecting the organism (Hao et al. 2011). Longer single-stranded DNA specific for *B. anthracis* has been generated by asymmetric polymerase chain reaction (As-PCR) followed by wrapping them around AuNPs which has found to detect the organisms colorimetrically without any visual aids (Deng et al. 2013). AuNPs coated with anti-*Salmonella* antibody have been successfully employed to detect the organisms from stool samples through a colony print test with a specificity of 99.1%. The method does not require any sophisticated equipment and results can be obtained within 2 h after the screening test (Fang et al. 2009). A highly sensitive bio-barcode DNA assay has been customized employing AuNPs coated with DNA probe specific for insertion element (*Iel*) gene of *Salmonella enteritidis* and fluorescein-labeled barcode DNA. The biosensor transducer also contained magnetic nanoparticles (MNPs) coated with the second target-specific DNA probe. First the target bacteria are captured from a complex sample in a sandwich format (MNPs-second DNA probe/Target DNA/first DNA probe-AuNPs-barcode DNA) followed by release of the labeled barcode and measuring the fluorescence intensity. The limit of detection of this assay has found to be as low as 2.15×10^{-16} mol (1 ng/mL) (Zhang et al. 2009). Even multiplexing in the bio-barcode DNA assay using AuNPs labeled with probes specific for the protective antigen A (*pag*A) gene of *Bacillus anthracis* and *Iel* gene of *Salmonella enteritidis* has also been successfully tried for simultaneous detection of multiple pathogens. Here instead of using fluorescent label, nanoparticle tracers (NT, such as PbS and CdS) have been used and the signal generated by the NT^{2+} ions has been detected by electrochemical biosensor. LOD of this multiplex bio-barcoded DNA sensor has found to be 0.5 ng/mL and 50 pg/mL for the *Iel* gene of *S. enteritidis* using CdS NTs and *pag*A gene of *B. anthracis* using PbS NTs, respectively (Zhang et al. 2010).

An immunochromatographic assay has been developed for the detection of *S. aureus* employing AuNPs coated with anti-Staphylococcal protein A IgG in the sandwich format. On a porous nitrocellulose membrane, first anti-Staphylococcal protein A IgG has been immobilized in a defined detection zone while the other anti-Staphylococcal protein A IgG has been conjugated with AuNPs. The sample mixture containing the bacteria has then passed along the porous membrane by capillary action yielding red color. The test depicted 100% and 94.7–100% sensitivity for 130 *S. aureus* strains and 36 non-*S. aureus* strains, respectively. This method is quick and easy to perform and has a long shelf-life at room temperature (Huang 2006). Similarly, detection of *S. aureus* has been carried out by an immunochromatographic test strip using AuNPs coated with anti-*S. aureus* monoclonal antibody employed as a double-antibody sandwich format. For this test strip, LOD and rate of detection have found to be 10^3 CFU/mL and 98.7%, respectively. This test strip method can be used for rapid detection of *S. aureus* in food samples and the result can be visually identified by unaided eye. The immunochromatographic assay-based test strip has extended several advantages such as simple, fast, low-cost, favorable sensitivity, and specificity (Niu et al. 2014). For rapid detection of Methicillin-resistant *Staphylococcus aureus* (MRSA), a 3D-printed chip has been customized which is suitable for bacterial cultivation, DNA isolation, PCR, and detection of amplified gene using AuNP probes as an indicator of MRSA binding. Presence of MRSA in the samples can be confirmed based on a specific interaction between *mec*A gene and the AuNP probes. Later, the non-cross-linking aggregation of DNA-functionalized AuNPs can be exploited for calorimetric detection which is rapid, less-expensive, and easy for MRSA detection (Chudobova et al. 2015).

Listeria monocytogenes is another bacterial pathogen of zoonotic importance and has been detected by unmodified AuNPs of 13 nm average diameter. In the presence of PCR target products of *L. monocytogenes*, species-specific probes targeting the *lmo0202hly* gene fragment caused irreversible aggregation of the AuNPs resulting in a red to purple colorimetric change. Visual detection limit of 1.304 fg/μL of genomic DNA from *L. monocytogenes* has been achieved under unaided eye by this method without requiring any exclusive instrumentation, which is also nearly 70-fold higher sensitive than the conventional PCR and gel electrophoresis-based detection (Liu et al. 2015).

Immunochromatographic strip employing AuNPs, coupled with antibody to LipL32, a protein commonly found in pathogenic *Leptospira*, has yielded sensitive detection of the pathogen at the level of 10^3/mL. The method has produced no positive result when tested with non-pathogenic *Leptospira*, or other bacterial genera such as *S. aureus*, *Streptococcus* group B, *Acinetobacter baumannii*, *E. coli*, *Klebsiella pneumoniae*, *Salmonella typhi*, and *Enterococcus faecalis* (Chirathaworn et al. 2014).

4.3.4.2 Silver NPs in Bacterial Detection

Silver nanorod array has been used as a surface-enhanced Raman scattering (SERS) active substrate to produce differences in spectral signal generation when bound to bacteria of different Gram types, species, strains, and their mixtures (*E. coli* O157: H7, *E. coli* DH 5α, *Staphylococcus aureus*, *S. epidermidis*, and *Salmonella typhimurium*). Bacterial detection even at single cell level as well as difference between viable and non-viable cells with minimal sample processing has been achieved with this modality (Chu et al. 2008). Lectin sensitized anisotropic AgNPs with two bands of surface plasmon resonance has produced substantial changes in optical spectra when interacted with different bacteria such as *E. coli*, *S. aureus*, and *Bacillus subtilis* which facilitated their detection. The spectral deviation was found to be correlated with the bacterial titer (Gasparyan and Bazukyan 2013). *E. coli* bound to citrate-capped AgNPs of around 9.4 nm diameter can be detected by anodic particle coulometry even up to single bacteria level. Here the inherent NP binding property of bacteria empowered by their surface property and strong electrochemical activity of the AgNPs are exploited for the sensitive detection (Sepunaru et al. 2015).

4.3.4.3 Magnetic Nanoparticles (MNPs)

In earlier discussion, the role of magnetic NPs (MNPs) in immunomagnetic separation of bacteria from complex samples has already been depicted which facilitates pathogen detection (Fig. 4.2). Here, the further applications of MNPs in bacterial detection have been elaborated with suitable examples. A conductometric immunosensor employing anti-LPS antibody functionalized MNPs has been developed for the capture and detection of gram-negative bacteria on graphite ink electrode. Sensitive detection of *E. coli* or *Serratia marcescens* has been carried out using the biosensor from cultures having the concentration of 1–10^3 CFU/mL. Direct detection of *Pseudomonas aeruginosa* and *Acinetobacter baumannii* strains at 10–10^3 CFU/mL concentration has also been performed with the modality (El Ichi et al. 2014). A biosensing platform employing a magnetostrictive particle (MSP) coated with layers of silica functionalized with anti-*Campylobacter jejuni* antibodies has been used for the detection of *C. jejuni* in water released from the chicken factories yielding a sensitive limit of detection of 10^2 CFU/mL (Zhang et al. 2015). Similarly, bifunctional MNPs coupled with fluorescent probe (a conjugate of vancomycin and fluorescein) [FePt@Van] have been employed for efficient capture of bacteria from blood stream infections through multivalent interactions and rapid detection within 2 h yielding sensitivity of 10 CFU/mL (Gao et al. 2006). As the antibiotics, vancomycin and daptomycin bind with the cell wall of Gram-positive bacteria, so their *trans*-cyclooctene (TCO) derivatives capture the bacteria, as well as bind with tetrazine-modified MNPs facilitating optical and magnetic detection with

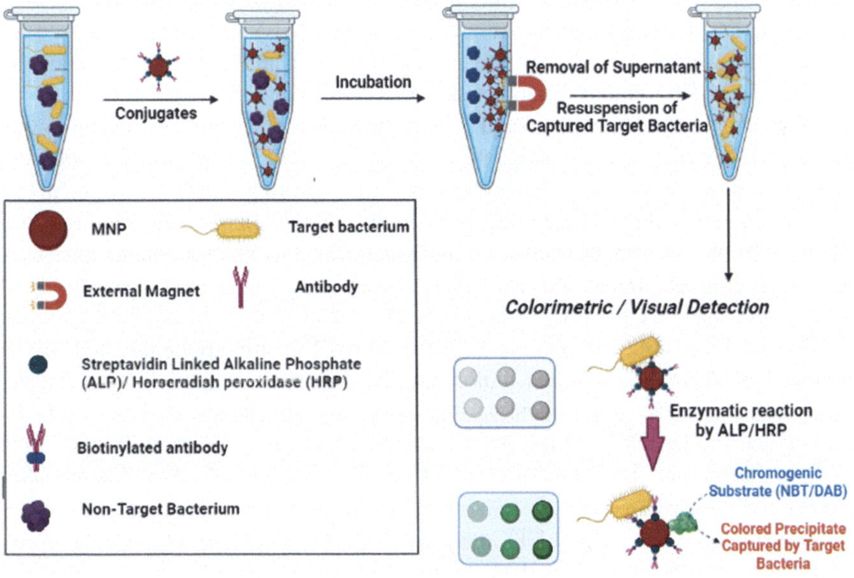

Fig. 4.2 Magnetic nanoparticle-assisted immunomagnetic separation followed by colorimetric detection of target bacteria from complex samples

extreme sensitivity as evidenced for *S. aureus*, *S. pneumonia*, *S. epidermis*, and *E. faecalis* (Chung et al. 2011). Application of MNPs for immunomagnetic separation of *Yersinia pestis* from environmental samples has facilitated fast sorting and bacterial detection, even at high or low bacterial concentration (Zahavy et al. 2012). Simultaneous detection of *E. coli*, *S. aureus*, and *Salmonella* species based on immunomagnetic bead (IMB) has been developed and tested in four animal-based foods: beef, ham, egg, and ricotta cheese. It detected all three pathogens with LOD 10 CFU/g without requiring any additional platforms (Kim et al. 2018).

Antibody functionalized magnetic beads have been used to capture *E. coli* O157: H7 from samples. Impedance spectroscopy has been used to detect pure *E. coli* cells and magnetic bead captured *E. coli* immobilized on nanoporous alumina membrane. Capturing of *E. coli* by magnetic bead has enhanced the sensitivity of detection achieving the LOD of 10 CFU/mL (Chan et al. 2013). Detection of *E. coli* O157:H7 has also been carried out in ground beef sample by immunomagnetic separation with MNP-antibody conjugates (MNCs). On the streptavidin-coated MNPs, the biotin-labeled polyclonal goat anti-*E. coli* antibodies were immobilized for capturing the bacteria from food matrix by using a magnet followed by plating onto sorbitol MacConkey agar. The capture effectiveness was found to be ranged from 1.6×10 to 7.2×10^7 CFU/mL with 15 min reaction time without any enrichment (Varshney et al. 2005). Streptavidin-coated MNPs (SAMNPs) have also been employed as visible labels in a multiplex PCR and microarray-based simultaneous detection of food-borne pathogens *E. coli* O157:H7, *Salmonella enterica*, *Vibrio cholera*, and *Campylobacter jejuni* (Li et al. 2013). Similarly, SAMNPs have also been employed for target enrichment in SERS-based detection of *Mycoplasma mycoides* causing contagious bovine pleuropneumonia (Strelau et al. 2011). Multifunctional magnetic-plasmonic Fe_3O_4-Au core-shell nanoparticles (Au-MNPs) have been employed for rapid concentration of bacteria under magnetic field and sensitive detection by generation of SERS signal (Zhang et al. 2012; Mocan et al. 2017).

As amino-modified silica-coated MNPs have found to bind trace amount of genomic DNA of various pathogenic bacteria from milk samples under optimum condition, these MNPs have been employed for capturing the DNA fragments of *Salmonella enteritidis* (Gram-negative) and *Listeria monocytogenes* (Gram-positive) from milk samples followed by PCR (termed as immunomagnetic separation polymerase chain reaction [IMS-PCR]) based detection achieving sensitivity of 8 CFU/mL and 13 CFU/mL, respectively (Bai et al. 2013). In another example, *Salmonella* from milk samples were first captured by antibody-conjugated MNPs followed by their dissolution in buffer and binding with antibody-coated Titania (TiO_2) nanocrystals. Subsequently, the absorbance of unbound Titania (TiO_2) nanocrystals under UV light has been observed which stood inversely proportional to the bacterial titer. The LOD of 100 CFU/mL has been achieved by this modality (Joo et al. 2012).

A lateral flow test strip (LFT) assay has been developed employing antibody-conjugated ferromagnetic NPs (FMNs) for rapid detection of *Campylobacter jejuni* from poultry samples. In this method, first the bacterial cells were captured using the FMNs functionalized with *C. jejuni*-specific polyclonal antibodies followed by amplification of the DNA released from lysis of the captured cells by PCR using the hipO gene-specific primers to detect the amplicons with the lateral flow strip assay. The LOD was found to be 1 CFU/mL for pure culture and 10–10^2 CFU/mL for target cells spiked in poultry sample (Poonlapdecha et al. 2018). A nanoparticle cluster (NPC)-based biosensor has been developed for rapid and sensitive detection of food-borne pathogens. *Listeria monocytogenes* has been captured by using the cell wall-binding antibiotic vancomycin and Fe_3O_4 NPC-modified aptamer as the signal amplification probe in a sandwich format. NPC showed collective effect-enhanced catalytic activity to induce the color reaction of chromogenic substrate for visual detection of the bacteria within a linear range of 5.4×10^3–10^8 CFU/mL and visual LOD of 5.4×10^3 CFU/mL (Zhang et al. 2016). A visual biosensor has been developed for rapid detection of *E. coli* O157:H7 based on the ability of the *E. coli* proteases to change the colorimetric response of a surface-modified, MNP-specific peptide probe which gradually intensified the golden color upon proteolysis and the intensity of color change was found to be proportional to the *E. coli* concentration. The sensor has found to be highly sensitivity with LOD of 12 CFU/mL in broth sample and 30–300 CFU/mL in spiked complex food matrices (Suaifan et al. 2017a). The most prevalent causal agents in hospital-acquired infections and food-borne diseases are *S. aureus*. Circulation of these *S. aureus* toxins and virulence proteases in host blood vessels leads to life-threatening consequences. The bacterial culturing is standard method which takes many days and nucleic acid-based method is costly and requires skilled personnel. To overcome these limitations, a paper-based biosensor has been employed for visual detection of *S. aureus*. In this method, a specific peptide substrate has been sandwiched between magnetic nanobeads and gold surface on a paper support. The *S. aureus* captured by the magnetic beads exerted proteolytic activity on the peptide substrate resulting in color change which could be detected by unaided eye. LOD was found to be 7, 40, and 100 CFU/mL for *S. aureus* in broth culture, inoculated food produces, and environmental samples, respectively (Suaifan et al. 2017b).

Antimicrobial peptide (AMP) functionalized MNPs (MNPs-AMP) and urease catalyzed signal amplification system have been employed for rapid and sensitive visual detection of *E. coli* O157:H7 from spiked apple juice and ground beef samples. The MNP-AMP efficiently captured *E. coli* O157:H7 through electrostatic and hydrophobic interactions. As the binding sites for urease were blocked on the MNP-AMP surfaces, so, a high amount of urease was accumulated in the supernatant after magnetic separation of the bacteria resulting in catalytic hydrolysis of urea to produce ammonium that induced the color change of the pH indicator. LOD of this assay was found to be 12 CFU/mL with 30 min detection time (Qiao et al. 2017). Sensitive detection of anthrax protective antigen (PA) has been carried out by a europium nanoparticle-based immunoassay (ENIA). The ENIA shows the linear dose-dependent pattern within the detection range of 0.01–100 ng/mL, and it is relatively 100-fold more sensitive than ELISA (Tang et al. 2009). A biosensor has been customized employing electrically active magnetic (EAM) NPs as concentrator of DNA targets as well as electrochemical transducers for the detection of *Bacillus anthracis* protective antigen A (*pag*A) gene. The detection system consisted of two sets of target gene-specific DNA probes: the biotinylated capture probe and the detector probe coated with the EAM NPs. The targets were captured on streptavidin-modified screen-printed carbon electrodes through the biotinylated capture probes and the detection was carried out electrochemically by estimating the oxidation-reduction signal of the EAM NPs. The biosensor was capable to detect target DNA concentrations as low as 0.01 ng/μL (Pal and Alocilja 2010). Vancomycin-modified MNPs as affinity probes have been used for capturing *Staphylococcus saprophyticus* and *Staphylococcus aureus* from urine samples followed by detection of the micro-organisms based on mass spectral fingerprinting through Matrix-assisted laser desorption/ionization mass spectrometry (MALDI-MS). The LOD achieved for both *S. saprophyticus* and *S. aureus* in urine samples was 7×10^4 CFU/mL approximately (Lin et al. 2005).

4.3.5 Nanocomposites

Different nanocomposites have been used for the detection of various pathogenic microorganisms from biological samples. For instance, polystyrene-co-acrylic acid (PSA) latex nanospheres-gold nanoparticles composite (PSA-AuNPs) DNA matrix has been employed in an electrochemical biosensor for the detection of pathogenic *Vibrio cholerae* nucleic acid (Rahman et al. 2017). Rabbit anti-*Listeria monocytogenes*-conjugated carboxyl-modified fluorescent polystyrene nanoparticles have been successfully employed for the detection of *L. monocytogenes* in biofilms. This NP-based immunoassay has found to be more sensitive than the conventional immunoassay (Yang et al. 2007).

4.3.6 Liposome NPs in Bacterial Detection

In the field of nanodevices, drug delivery and gene delivery through liposomes are broadly used as carriers of molecules and also as a mimic for cell membranes. Due to their outstanding carrier properties for encapsulation of signal marker compounds across broad spectrum of sensing techniques, liposomes find special application in signal amplification. Also, the capability to modify the liposome surface structure to perform recognition functions with a broad range of types of analytes is a prominent aspect of their use in bioanalysis.

Lateral flow assay (LFA) has been developed for the detection of cholera toxin (CT), using combination of capture aptamer and GM1 cell receptor labeled liposomes. This assay is developed with very good sensitivity so that insertion of one or more antibodies into the assay is not required. These artificially prepared liposomes exhibit the oligosaccharide tail of ganglioside GM1 on their surface as signaling elements and the tail is easily united into the liposome lipid bilayer. The cell membrane receptor for the CT is GM1, so the interaction between these two is generally high. It allows the LOD of 2 ng/mL (as per 3 sigma rule) and 10 ng/mL by visual detection in ~15 min (Frohnmeyer et al. 2019). Similarly, ganglioside incorporated liposomes are developed to detect the cholera toxin (CT). Liposomes are combined with ganglioside GM1 by spontaneous insertion into the phospholipid bilayer. The strong relation between GM1 and CT shows the CT identification and signal generation. In a sandwich immunoassay, CT is detected as a stained band on the nitrocellulose membrane strip as immobilized antibodies capture the CT which are bound to GM1 liposomes. The intensity of the band is measured using computer software and densitometry. LOD of this modality was found to be 0.1–30 pg/mL (Ahn-Yoon et al. 2003).

Liposomes containing ganglioside mimic cells that are captured by bacterial toxins. These can be used as sensitive probes for detecting these toxins. Ganglioside bearing liposomes are used for the detection of cholera, botulinum, and tetanus toxins. Tetanus and botulinum toxins specifically bind gangliosides of the G1b series, namely GT1b, GD1b, and GQ1b, present in neuronal membranes and cholera toxin binds GM1 selectively on the surface of enterocytes and different eukaryotic cell types. The fluorophore-labeled lipids are inserted in the bilayer of liposomes to convey signal generation ability to these liposomes. The sandwich fluoro immunoassay for the detection of cholera, botulinum, and tetanus toxins based on fluorescent liposomes containing both a marker (rhodamine) and a receptor (GT1b or GM1) in the bilayer. The LOD was found to be as low as 1 nM of each toxin (Singh et al. 2000).

4.4 Conclusion

The chapter outlines the prominent bacterial diseases of livestock including the zoonotic ones and highlights the significance of nanodiagnostics toward improved detection of these diseases to customize better controlling strategies. The employed functionalization and improvisations in the nanomaterials science as reported in the aforementioned discussion for efficient sample enrichment and sensitive detection of the bacterial pathogens have provided viable solutions and techniques to the researchers and the livestock industry to minimize the devastating impacts of these diseases. However, it is important to consider the advantages and disadvantages of using nanomaterials in the perspective of short- and long-term toxicity elicited by these nanoscale materials. Considering the success of nanotechnology in biomedical science and the emerging market, the future prospect looks bright enough to overcome the limitations progressively which may pave the way for designing new and effective combat strategies against these bacterial pathogens.

References

Agrawal A, Varshney R, Gattani A, Kirthika P, Khan MH, Singh R, Kodape S, Patel SK, Singh P (2020) Gold nanoparticle based immunochromatographic biosensor for rapid diagnosis of Mycobacterium avium subspecies paratuberculosis infection using recombinant protein. J Microbiol Methods 177:106024. https://doi.org/10.1016/j.mimet.2020.106024

Ahn-Yoon S, DeCory TR, Baeumner AJ, Durst RA (2003) Ganglioside-liposome immunoassay for the ultrasensitive detection of cholera toxin. Anal Chem 75(10):2256–2261. https://doi.org/10.1021/ac026428t

Ai K, Zhang B, Lu L (2009) Europium-based fluorescence nanoparticle sensor for rapid and ultrasensitive detection of an anthrax biomarker. Angew Chem 121(2):310–314. https://doi.org/10.1002/ange.200804231

Alafeef M, Moitra P, Pan D (2020) Nano-enabled sensing approaches for pathogenic bacterial detection. Biosens Bioelectron 165:112276. https://doi.org/10.1016/j.bios.2020.112276

Amini B, Kamali M, Salouti M, Yaghmaei P (2017) Fluorescence bio-barcode DNA assay based on gold and magnetic nanoparticles for detection of Exotoxin A gene sequence. Biosens Bioelectron 92:679–686. https://doi.org/10.1016/j.bios.2016.10.030

Ananda P, Tillekaratne A, Hettiarachchi CM, Lalichchandran N (2021) Sensitive detection of E. coli using bioconjugated fluorescent silica nanoparticles. Appl Surf Sci Adv. https://doi.org/10.1016/j.apsadv.2021.100159

Archibald MM, Rizal B, Connolly T, Burns MJ, Naughton MJ, Chiles TC (2015) A nanocoaxial-based electrochemical sensor for the detection of cholera toxin. Biosens Bioelectron 74:406–410. https://doi.org/10.1016/j.bios.2015.06.069

Baccar H, Mejri MB, Hafaiedh I, Ktari T, Aouni M, Abdelghani A (2010) Surface plasmon resonance immunosensor for bacteria detection. Talanta 82(2):810–814. https://doi.org/10.1016/j.talanta.2010.05.060

Bai Y, Song M, Cui Y, Shi C, Wang D, Paoli GC, Shi X (2013) A rapid method for the detection of foodborne pathogens by extraction of a trace amount of DNA from raw milk based on amino-modified silica-coated magnetic nanoparticles and polymerase chain reaction. Anal Chim Acta 787:93–101

Cao C, Gontard LC, Thuy Tram LL, Wolff A, Bang DD (2011) Dual enlargement of gold nanoparticles: from mechanism to scanometric detection of pathogenic bacteria. Small 7(12):1701–1708. https://doi.org/10.1002/smll.201100294

Chan KY, Ye WW, Zhang Y, Xiao LD, Leung PH, Li Y, Yang M (2013) Ultrasensitive detection of E. coli O157:H7 with biofunctional magnetic bead concentration via nanoporous membrane based electrochemical immunosensor. Biosens Bioelectron 41:532–537. https://doi.org/10.1016/j.bios.2012.09.016

Chen ZZ, Cai L, Chen MY, Lin Y, Pang DW, Tang HW (2015) Indirect immunofluorescence detection of E. coli O157:H7 with fluorescent silica nanoparticles. Biosens Bioelectron 66:95–102. https://doi.org/10.1016/j.bios.2014.11.007

Chen W, Lai Q, Zhang Y, Liu Z (2022) Recent advances in aptasensors for rapid and sensitive detection of Staphylococcus aureus. Front Bioeng Biotechnol 10:889431. https://doi.org/10.3389/fbioe.2022.889431

Cheng D, Yu M, Fu F, Han W, Li G, Xie J, Song Y, Swihart MT, Song E (2016) Dual recognition strategy for specific and sensitive detection of bacteria using aptamer-coated magnetic beads and antibiotic-capped gold nanoclusters. Anal Chem 88(1):820–825. https://doi.org/10.1021/acs.analchem.5b03320

Chirathaworn C, Janwitthayanan W, Sereemaspun A, Lertpocasombat K, Rungpanich U, Ekpo P, Suwancharoen D (2014) Development of an immunochromatographic test with anti-LipL32-coupled gold nanoparticles for Leptospira detection. New Microbiol 37(2):201–207

Chu H, Huang Y, Zhao Y (2008) Silver nanorod arrays as a surface-enhanced Raman scattering substrate for foodborne pathogenic bacteria detection. Appl Spectrosc 62(8):922–931. https://doi.org/10.1366/000370208785284330

Chudobova D, Cihalova K, Skalickova S, Zitka J, Rodrigo MA, Milosavljevic V, Hynek D, Kopel P, Vesely R, Adam V, Kizek R (2015) 3D-printed chip for detection of methicillin-resistant Staphylococcus aureus labeled with gold nanoparticles. Electrophoresis 36(3): 457–466. https://doi.org/10.1002/elps.201400321

Chung HJ, Reiner T, Budin G, Min C, Liong M, Issadore D, Lee H, Weissleder R (2011) Ubiquitous detection of Gram-positive bacteria with bioorthogonal magnetofluorescent nanoparticles. ACS Nano 5(11):8834–8841

Cimafonte M, Fulgione A, Gaglione R, Papaianni M, Capparelli R, Arciello A, Bolletti Censi S, Borriello G, Velotta R, Della Ventura B (2020) Screen printed based impedimetric immunosensor for rapid detection of Escherichia coli in drinking water. Sensors (Basel) 20(1):274. https://doi.org/10.3390/s20010274

Cui J, Zhou M, Li Y, Liang Z, Li Y, Yu L, Liu Y, Liang Y, Chen L, Yang C (2021) A new optical fiber probe-based quantum dots immunofluorescence biosensors in the detection of Staphylococcus aureus. Front Cell Infect Microbiol 11:665241. https://doi.org/10.3389/fcimb.2021.665241

Deng H, Zhang X, Kumar A, Zou G, Zhang X, Liang X-J et al (2013) Long genomic DNA amplicons adsorption onto unmodified gold nanoparticles for colorimetric detection of Bacillus anthracis. Chem Commum 49:51–53. https://doi.org/10.1039/C2CC37037A

El Ichi S, Leon F, Vossier L, Marchandin H, Errachid A, Coste J, Jaffrezic-Renault N, Fournier-Wirth C (2014) Microconductometric immunosensor for label-free and sensitive detection of Gram-negative bacteria. Biosens Bioelectron 54:378–384. https://doi.org/10.1016/j.bios.2013.11.016

Fang SB, Tseng WY, Lee HC, Tsai CK, Huang JT, Hou SY (2009) Identification of Salmonella using colony-print and detection with antibody-coated gold nanoparticles. J Microbiol Methods 77(2):225–228. https://doi.org/10.1016/j.mimet.2009.02.008

Farka Z, Juřík T, Pastucha M, Kovář D, Lacina K, Skládal P (2016) Rapid immunosensing of Salmonella typhimurium using electrochemical impedance spectroscopy: the effect of sample treatment. Electroanalysis 28:1803–1809. https://doi.org/10.1002/elan.201600093

Fierz W (1999) Basic problems of serological laboratory diagnosis. Mol Biotechnol 13(2):89–111. https://doi.org/10.1385/MB:13:2:89

Frohnmeyer E, Tuschel N, Sitz T, Hermann C, Dahl GT, Schulz F, Baeumner AJ, Fischer M (2019) Aptamer lateral flow assays for rapid and sensitive detection of cholera toxin. Analyst 144(5): 1840–1849. https://doi.org/10.1039/c8an01616j

Gao J, Li L, Ho P, Mak GC, Gu H, Xu B (2006) Combining fluorescent probes and biofunctional magnetic nanoparticles for rapid detection of bacteria in human blood. Adv Mater 18(23): 3145–3148. https://doi.org/10.1002/adma.200601058

Gasparyan VK, Bazukyan IL (2013) Lectin sensitized anisotropic silver nanoparticles for detection of some bacteria. Anal Chim Acta 766:83–87. https://doi.org/10.1016/j.aca.2012.12.015

Goldman ER, Clapp AR, Anderson GP, Uyeda HT, Mauro JM, Medintz IL, Mattoussi H (2004) Multiplexed toxin analysis using four colors of quantum dot fluororeagents. Anal Chem 76(3): 684–688. https://doi.org/10.1021/ac035083r

Hao R-Z, Song H-B, Zuo G-M, Yang RF, Wei H-P, Wang D-B et al (2011) DNA probe functionalized QCM biosensor based on gold nanoparticle amplification for Bacillus anthracis detection. Biosens Bioelectron 26:3398–3404. https://doi.org/10.1016/j.bios.2011.01.010

Hsu HL, Chuang CC, Liang CC, Chiao DJ, Wu HL, Wu YP, Lin FP, Shyu RH (2018) Rapid and sensitive detection of Yersinia pestis by lateral-flow assay in simulated clinical samples. BMC Infect Dis 18(1):402. https://doi.org/10.1186/s12879-018-3315-2

Huang SH (2006) Gold nanoparticle-based immunochromatographic test for identification of Staphylococcus aureus from clinical specimens. Clin Chim Acta 373(1–2):139–143. https://doi.org/10.1016/j.cca.2006.05.026

Huang S-H (2007) Gold nanoparticle-based immunochromatographic assay for the detection of Staphylococcus aureus. Sens Actuators B Chem 127:335–340. https://doi.org/10.1016/j.snb.2007.04.027

Huang J, Yang G, Meng W, Wu L, Zhu A, Jiao X (2010) An electrochemical impedimetric immunosensor for label-free detection of Campylobacter jejuni in diarrhea patients' stool based on O-carboxymethylchitosan surface modified Fe_3O_4 nanoparticles. Biosens Bioelectron 25(5):1204–1211. https://doi.org/10.1016/j.bios.2009.10.036

Huang F, Zhang Y, Lin J, Liu Y (2021) Biosensors coupled with signal amplification technology for the detection of pathogenic bacteria: a review. Biosensors 11(6):190. https://doi.org/10.3390/bios11060190

Joo J, Yim C, Kwon D, Lee J, Shin HH, Cha HJ, Jeon S (2012) A facile and sensitive detection of pathogenic bacteria using magnetic nanoparticles and optical nanocrystal probes. Analyst 137(16):3609–3612

Jung BY, Jung SC, Kweon CH (2005) Development of a rapid immunochromatographic strip for detection of Escherichia coli O157. J Food Prot 68(10):2140–2143. https://doi.org/10.4315/0362-028x-68.10.2140

Karimi F, Dabbagh S (2019) Gel green fluorescence ssDNA aptasensor based on carbon nanotubes for detection of anthrax protective antigen. Int J Biol Macromol 140:842–850. https://doi.org/10.1016/j.ijbiomac.2019.08.219

Kim JH, Yoo JG, Ham JS, Oh MH (2018) Direct detection of Escherichia coli, Staphylococcus aureus, and Salmonella spp. in animal-derived foods using a magnetic bead-based immunoassay. Korean J Food Sci Anim Resour 38(4):727–736. https://doi.org/10.5851/kosfa.2018.e11

Koets M, van der Wijk T, van Eemeren JT, van Amerongen A, Prins MW (2009) Rapid DNA multi-analyte immunoassay on a magneto-resistance biosensor. Biosens Bioelectron 24(7): 1893–1898. https://doi.org/10.1016/j.bios.2008.09.023

Li S, Liu H, Deng Y, Lin L, He N (2013) Development of a magnetic nanoparticles microarray for simultaneous and simple detection of foodborne pathogens. J Biomed Nanotechnol 9(7): 1254–1260

Li T, Zhu F, Guo W, Gu H, Zhao J, Yan M, Liu S (2017) Selective capture and rapid identification of E. coli O157:H7 by carbon nanotube multilayer biosensors and microfluidic chip-based LAMP. RSC Adv 7:30446–30452. https://doi.org/10.1039/C7RA04583B

Li D, Liu L, Huang Q et al (2021) Recent advances on aptamer-based biosensors for detection of pathogenic bacteria. World J Microbiol Biotechnol 37:45. https://doi.org/10.1007/s11274-021-03002-9

Lin YS, Tsai PJ, Weng MF, Chen YC (2005) Affinity capture using vancomycin-bound magnetic nanoparticles for the MALDI-MS analysis of bacteria. Anal Chem 77(6):1753–1760. https://doi.org/10.1021/ac048990k

Lin D, Pillai RG, Lee WE, Jemere AB (2019) An impedimetric biosensor for E. coli O157:H7 based on the use of self-assembled gold nanoparticles and protein G. Mikrochim Acta 186(3):169. https://doi.org/10.1007/s00604-019-3282-3

Liu ZM, Zhu JC, Yang CY, Li XH (2015) Visual detection of Listeria monocytogenes using unmodified gold nanoparticles based on a novel marker. Anal Methods 7:8159–8164. https://doi.org/10.1039/C5AY01367D

Ma X, Song L, Zhou N, Xia Y, Wang Z (2017) A novel aptasensor for the colorimetric detection of S. typhimurium based on gold nanoparticles. Int J Food Microbiol 245:1–5. https://doi.org/10.1016/j.ijfoodmicro.2016.12.024

Manhas PK, Quintela IA, Wu VCH (2021) Enhanced detection of major pathogens and toxins in poultry and livestock with zoonotic risks using nanomaterials-based diagnostics. Front Vet Sci 8:673718. https://doi.org/10.3389/fvets.2021.673718

Matta LL, Harrison J, Deol GS, Alocilja EC (2018) Carbohydrate-functionalized nanobiosensor for rapid extraction of pathogenic bacteria directly from complex liquids with quick detection using cyclic voltammetry. IEEE Trans Nanotechnol 17:1006–1013. https://doi.org/10.1109/TNANO.2018.2841320

Mi F, Hu C, Wang Y, Wang L, Peng F, Geng P, Guan M (2022) Recent advancements in microfluidic chip biosensor detection of foodborne pathogenic bacteria: a review. Anal Bioanal Chem 414(9):2883–2902. https://doi.org/10.1007/s00216-021-03872-w

Minakshi P, Ghosh M, Brar B, Kumar R, Lambe UP, Ranjan K, Manoj J, Prasad G (2019) Nano-antimicrobials: a new paradigm for combating mycobacterial resistance. Curr Pharm Des 25(13):1554–1579. https://doi.org/10.2174/1381612825666190620094041

Mocan T, Matea CT, Pop T, Mosteanu O, Buzoianu AD, Puia C, Iancu C, Mocan L (2017) Development of nanoparticle-based optical sensors for pathogenic bacterial detection. J Nanobiotechnol 15(1):25. https://doi.org/10.1186/s12951-017-0260-y

Moongkarndi P, Rodpai E, Kanarat S (2011) Evaluation of an immunochromatographic assay for rapid detection of Salmonella enterica serovars Typhimurium and Enteritidis. J Vet Diagn Investig 23(4):797–801. https://doi.org/10.1177/1040638711408063

Mujika M, Arana S, Castaño E, Tijero M, Vilares R, Ruano-López JM, Cruz A, Sainz L, Berganza J (2009) Magnetoresistive immunosensor for the detection of Escherichia coli O157:H7 including a microfluidic network. Biosens Bioelectron 24(5):1253–1258. https://doi.org/10.1016/j.bios.2008.07.024

Mukhopadhyay B, Martins MB, Karamanska R, Russell DA, Field RA (2009) Bacterial detection using carbohydrate-functionalised CdS quantum dots: a model study exploiting E. coli recognition of mannosides. Tetrahedron Lett 50(8):886–889. https://doi.org/10.1016/j.tetlet.2008.12.029

Na M, Zhang S, Liu J, Ma S, Han Y, Wang Y, He Y, Chen H, Chen X (2020) Determination of pathogenic bacteria-Bacillus anthrax spores in environmental samples by ratiometric fluorescence and test paper based on dual-emission fluorescent silicon nanoparticles. J Hazard Mater 386:121956. https://doi.org/10.1016/j.jhazmat.2019.121956

Najian AN, Syafirah EEN, Ismail N, Mohamed M, Yean CY (2016) Development of multiplex loop mediated isothermal amplification (m-LAMP) label-based gold nanoparticles lateral flow dipstick biosensor for detection of pathogenic Leptospira. Anal Chim Acta 903:142–148

Niu K, Zheng X, Huang C, Xul K, Zhi Y, Shen H, Jia N (2014) A colloidal gold nanoparticle-based immunochromatographic test strip for rapid and convenient detection of Staphylococcus aureus. J Nanosci Nanotechnol 14(7):5151–5156. https://doi.org/10.1166/jnn.2014.8703

Othman S, Lee PY, Lam JY, Philip N, Azhari NN, Affendy NB, Masri SN, Neela VK, Mohd-Taib FS, Chee HY (2022) A versatile isothermal amplification assay for the detection of leptospires from various sample types. PeerJ 10:e12850. https://doi.org/10.7717/peerj.12850

Pal S, Alocilja EC (2010) Electrically active magnetic nanoparticles as novel concentrator and electrochemical redox transducer in Bacillus anthracis DNA detection. Biosens Bioelectron 26(4):1624–1630. https://doi.org/10.1016/j.bios.2010.08.035

Pal D, Boby N, Kumar S, Kaur G, Ali SA, Reboud J, Shrivastava S, Gupta PK, Cooper JM, Chaudhuri P (2017) Visual detection of Brucella in bovine biological samples using DNA-activated gold nanoparticles. PLoS One 12(7):e0180919. https://doi.org/10.1371/journal.pone.0180919

Pöhlmann C, Elßner T (2020) Multiplex immunoassay techniques for on-site detection of security sensitive toxins. Toxins 12(11):727. https://doi.org/10.3390/toxins12110727

Poonlapdecha W, Seetang-Nun Y, Wonglumsom W, Tuitemwong K, Erickson LE, Hansen RR, Tuitemwong P (2018) Antibody-conjugated ferromagnetic nanoparticles with lateral flow test strip assay for rapid detection of Campylobacter jejuni in poultry samples. Int J Food Microbiol 286:6–14. https://doi.org/10.1016/j.ijfoodmicro.2018.07.009

Prasad M, Ghosh M, Kumar R, Brar B, Surjith KP, Lambe UP, Ranjan K, Banerjee S, Prasad G, Kumar Khurana S, Kharb P (2021) The importance of nanomedicine in prophylactic and theranostic intervention of bacterial zoonoses and reverse zoonoses in the era of microbial resistance. J Nanosci Nanotechnol 21(6):3404–3452. https://doi.org/10.1166/jnn.2021.18999

Puranik N, Pal V, Tripathi NK, Goel AK (2020) Development of a rapid immunochromatographic assay for detection of surface array protein (Sap), a potent biomarker of Bacillus anthracis. Biologia 75:613–617. https://doi.org/10.2478/s11756-019-00379-9

Qiao Z, Lei C, Fu Y, Li Y (2017) Rapid and sensitive detection of E. coli O157:H7 based on antimicrobial peptide functionalized magnetic nanoparticles and urease-catalyzed signal amplification. Anal Methods 9:5204–5210. https://doi.org/10.1039/C7AY01643C

Quintela IA, Vasse T, Lin CS, Wu VCH (2022) Advances, applications, and limitations of portable and rapid detection technologies for routinely encountered foodborne pathogens. Front Microbiol 13:1054782. https://doi.org/10.3389/fmicb.2022.1054782

Rahman M, Heng LY, Futra D, Ling TL (2017) Ultrasensitive biosensor for the detection of Vibrio cholerae DNA with polystyrene-co-acrylic acid composite nanospheres. Nanoscale Res Lett 12(1):474. https://doi.org/10.1186/s11671-017-2236-0

Raj V, Vijayan AN, Joseph K (2015) Cysteine capped gold nanoparticles for naked eye detection of E. coli bacteria in UTI patients. Sens BioSens Res 5:33–36. https://doi.org/10.1016/j.sbsr.2015.05.004

Ranzoni A, Sidjabat H, Cooper MA (2013) Nanoparticle sample preparation and mass spectrometry for rapid diagnosis of microbial infections. Microbiol Aust 34:170–174. https://doi.org/10.1071/MA13057

Ravindranath SP, Mauer LJ, Deb-Roy C, Irudayaraj J (2009) Biofunctionalized magnetic nanoparticle integrated mid-infrared pathogen sensor for food matrixes. Anal Chem 81(8):2840–2846

Ritenberg M, Nandi S, Kolusheva S, Dandela R, Meijler MM, Jelinek R (2016) Imaging Pseudomonas aeruginosa biofilm extracellular polymer scaffolds with amphiphilic carbon dots. ACS Chem Biol 11(5):1265–1270. https://doi.org/10.1021/acschembio.5b01000

Santra S, Wang K, Tapec R, Tan W (2001) Development of novel dye-doped silica nanoparticles for biomarker application. J Biomed Opt 6(2):160–166. https://doi.org/10.1117/1.1353590

Sattarahmady N, Tondro GH, Gholchin M, Heli H (2015) Gold nanoparticles biosensor of Brucella spp. genomic DNA: visual and spectrophotometric detections. Biochem Eng J 97:1–7. https://doi.org/10.1016/j.bej.2015.01.010

Şen Karaman D, Pamukçu A, Karakaplan MB, Kocaoglu O, Rosenholm JM (2021) Recent advances in the use of mesoporous silica nanoparticles for the diagnosis of bacterial infections. Int J Nanomedicine 16:6575–6591. https://doi.org/10.2147/IJN.S273062

Sepunaru L, Tschulik K, Batchelor-McAuley C, Gavish R, Compton RG (2015) Electrochemical detection of single E. coli bacteria labeled with silver nanoparticles. Biomater Sci 3(6):816–820

Shen Z, Hou N, Jin M, Qiu Z, Wang J, Zhang B, Wang X, Wang J, Zhou D, Li J (2014) A novel enzyme-linked immunosorbent assay for detection of Escherichia coli O157:H7 using immunomagnetic and beacon gold nanoparticles. Gut Pathog 6:14. https://doi.org/10.1186/1757-4749-6-14

Shen Y, Zhang Y, Gao ZF, Ye Y, Wu Q, Chen H-Y, Xu J-J (2021) Recent advances in nanotechnology for simultaneous detection of multiple pathogenic bacteria. Nano Today 38:101121. https://doi.org/10.1016/j.nantod.2021.101121

Shi L, Wu F, Wen Y, Zhao F, Xiang J, Ma L (2015) A novel method to detect Listeria monocytogenes via superparamagnetic lateral flow immunoassay. Anal Bioanal Chem 407(2):529–535. https://doi.org/10.1007/s00216-014-8276-8

Singh AK, Harrison SH, Schoeniger JS (2000) Gangliosides as receptors for biological toxins: development of sensitive fluoroimmunoassays using ganglioside-bearing liposomes. Anal Chem 72(24):6019–6024. https://doi.org/10.1021/ac000846l

So H-M, Park D-W, Jeon E-K, Kim Y-H, Kim BS, Lee C-K et al (2008) Detection and titer estimation of Escherichia coli using aptamer-functionalized single-walled carbon-nanotube field-effect transistors. Small 4(2):197–201. https://doi.org/10.1002/smll.200700664

Strelau KK, Brinker A, Schnee C, Weber K, Möller R, Popp J (2011) Detection of PCR products amplified from DNA of epizootic pathogens using magnetic nanoparticles and SERS. J Raman Spectrosc 42(3):243–250

Suaifan GA, Alhogail S, Zourob M (2017a) Paper-based magnetic nanoparticle-peptide probe for rapid and quantitative colorimetric detection of Escherichia coli O157:H7. Biosens Bioelectron 92:702–708. https://doi.org/10.1016/j.bios.2016.10.023

Suaifan GA, Alhogail S, Zourob M (2017b) Rapid and low-cost biosensor for the detection of Staphylococcus aureus. Biosens Bioelectron 90:230–237. https://doi.org/10.1016/j.bios.2016.11.047

Sung YJ, Suk HJ, Sung HY, Li T, Poo H, Kim MG (2013) Novel antibody/gold nanoparticle/magnetic nanoparticle nanocomposites for immunomagnetic separation and rapid colorimetric detection of Staphylococcus aureus in milk. Biosens Bioelectron 43:432–439. https://doi.org/10.1016/j.bios.2012.12.052

Tang S, Moayeri M, Chen Z, Harma H, Zhao J, Hu H, Purcell RH, Leppla SH, Hewlett IK (2009) Detection of anthrax toxin by an ultrasensitive immunoassay using europium nanoparticles. Clin Vaccine Immunol 16(3):408–413. https://doi.org/10.1128/CVI.00412-08

Tu SI, Reed S, Gehring A, He Y, Paoli G (2009) Capture of Escherichia coli O157:H7 using immunomagnetic beads of different size and antibody conjugating chemistry. Sensors (Basel, Switzerland) 9(2):717–730. https://doi.org/10.3390/s90200717

Valizadeh A, Mikaeili H, Samiei M, Farkhani SM, Zarghami N, Kouhi M, Akbarzadeh A, Davaran S (2012) Quantum dots: synthesis, bioapplications, and toxicity. Nanoscale Res Lett 7(1):480. https://doi.org/10.1186/1556-276X-7-480

Váradi L, Luo JL, Hibbs DE, Perry JD, Anderson RJ, Orenga S, Groundwater PW (2017) Methods for the detection and identification of pathogenic bacteria: past, present, and future. Chem Soc Rev 46(16):4818–4832. https://doi.org/10.1039/c6cs00693k

Varshney M, Yang L, Su XL, Li Y (2005) Magnetic nanoparticle-antibody conjugates for the separation of Escherichia coli O157:H7 in ground beef. J Food Prot 68(9):1804–1811. https://doi.org/10.4315/0362-028x-68.9.1804

Varshney R, Varshney R, Chaturvedi VK, Rawat M, Saminathan M, Singh V, Singh R, Sahoo M, Gupta PK (2020) Development of novel iron-regulated Pasteurella multocida B: 2 bacterin and refinement of vaccine quality in terms of minimum variation in particle size and distribution vis-a-vis critical level of iron in media. Microb Pathog 147:104375. https://doi.org/10.1016/j.micpath.2020.104375

Wang L, Zhao W, O'Donoghue MB, Tan W (2007) Fluorescent nanoparticles for multiplexed bacteria monitoring. Bioconjug Chem 18(2):297–301. https://doi.org/10.1021/bc060255n

Wang C, Wang J, Li M, Qu X, Zhang K, Rong Z, Xiao R, Wang S (2016) A rapid SERS method for label-free bacteria detection using polyethylenimine-modified Au-coated magnetic microspheres and Au@Ag nanoparticles. Analyst 141(22):6226–6238. https://doi.org/10.1039/c6an01105e

Wang D, Chen Q, Huo H, Bai S, Cai G, Lai W, Lin J (2017) Efficient separation and quantitative detection of Listeria monocytogenes based on screen-printed interdigitated electrode, urease and magnetic nanoparticles. Food Control 73:555–561. https://doi.org/10.1016/j.foodcont.2016.09.003

Wang L, Huang X, Wang C, Tian X, Chang X, Ren Y, Yu S (2020) Applications of surface functionalized Fe_3O_4 NPs-based detection methods in food safety. Food Chem 342:128343. https://doi.org/10.1016/j.foodchem.2020.128343

Wu H, Zuo Y, Cui C, Yang W, Ma H, Wang X (2013) Rapid quantitative detection of Brucella melitensis by a label-free impedance immunosensor based on a gold nanoparticle-modified screen-printed carbon electrode. Sensors (Basel, Switzerland) 13(7):8551–8563. https://doi.org/10.3390/s130708551

Wu W, Li J, Pan D, Li J, Song S, Rong M, Li Z, Gao J, Lu J (2014) Gold nanoparticle-based enzyme-linked antibody-aptamer sandwich assay for detection of Salmonella typhimurium. ACS Appl Mater Interfaces 6(19):16974–16981. https://doi.org/10.1021/am5045828

Wu W, Zhao S, Mao Y, Fang Z, Lu X, Zeng L (2015) A sensitive lateral flow biosensor for Escherichia coli O157:H7 detection based on aptamer mediated strand displacement amplification. Anal Chim Acta 861:62–68. https://doi.org/10.1016/j.aca.2014.12.041

Xu M, Wang R, Li Y et al (2016) Analyst 141:5441–5449. https://doi.org/10.1039/C6AN00873A

Yang L, Li Y (2006) Simultaneous detection of Escherichia coli O157:H7 and Salmonella typhimurium using quantum dots as fluorescence labels. Analyst 131(3):394–401. https://doi.org/10.1039/b510888h

Yang H, Qu L, Lin Y, Jiang X, Sun Y-P (2007) Detection of Listeria monocytogenes in biofilms using immunonanoparticles. J Biomed Nanotechnol 3(2):131–138. https://doi.org/10.1166/jbn.2007.019

Yaohua H, Chengcheng W, Bing B, Mintong L, Wang R, Li Y (2014) Detection of Staphylococcus aureus using quantum dots as fluorescence labels. Int J Agric Biol Eng 7(1):77–83

Yu CY, Ang GY, Chua AL, Tan EH, Lee SY, Falero-Diaz G, Otero O, Rodríguez I, Reyes F, Acosta A, Sarmiento ME, Ghosh S, Ramamurthy T, Yean Yean C, Lalitha P, Ravichandran M (2011) Dry-reagent gold nanoparticle-based lateral flow biosensor for the simultaneous detection of Vibrio cholerae serogroups O1 and O139. J Microbiol Methods 86(3):277–282. https://doi.org/10.1016/j.mimet.2011.04.020

Zahavy E, Ber R, Gur D, Abramovich H, Freeman E, Maoz S, Yitzhaki S (2012) Application of nanoparticles for the detection and sorting of pathogenic bacteria by flow-cytometry. In: Nanobiotechnology for biomedical and diagnostic research. Springer

Zhang D, Carr DJ, Alocilja EC (2009) Fluorescent bio-barcode DNA assay for the detection of Salmonella enterica serovar Enteritidis. Biosens Bioelectron 24(5):1377–1381. https://doi.org/10.1016/j.bios.2008.07.081

Zhang D, Huarng MC, Alocilja EC (2010) A multiplex nanoparticle-based bio-barcoded DNA sensor for the simultaneous detection of multiple pathogens. Biosens Bioelectron 26(4):1736–1742. https://doi.org/10.1016/j.bios.2010.08.012

Zhang L, Xu J, Mi L, Gong H, Jiang S, Yu Q (2012) Multifunctional magnetic–plasmonic nanoparticles for fast concentration and sensitive detection of bacteria using SERS. Biosens Bioelectron 31(1):130–136

Zhang L, Wang O, Cheng ZY, Huang TS (2015) Silica-coated magnetostrictive biosensors for real-time detection of Campylobacter jejuni in washing water. In: Meeting abstracts. The Electrochemical Society, Pennington, p 1832

Zhang L, Huang R, Liu W, Liu H, Zhou X, Xing D (2016) Rapid and visual detection of Listeria monocytogenes based on nanoparticle cluster catalyzed signal amplification. Biosens Bioelectron 86:1–7. https://doi.org/10.1016/j.bios.2016.05.100

Zhu P, Shelton DR, Li S, Adams DL, Karns JS, Amstutz P, Tang CM (2011) Detection of E. coli O157:H7 by immunomagnetic separation coupled with fluorescence immunoassay. Biosens Bioelectron 30(1):337–341. https://doi.org/10.1016/j.bios.2011.09.029

Chapter 5
Nanointerventions for Detections of Viral Livestock Diseases

Lukumoni Buragohain, Rajesh Kumar, Alla Yaswanth Naveen Kumar, and Mayukh Ghosh

Abstract Timely and rapid diagnosis always favours the outcome of all diseases including the viral ones and thus stands key for successful control and corrective measures. Most pathogenic viruses are contagious and fast acting in subjects with respect to any other diseases; hence early and rapid diagnosis with high sensitivity and specificity is more crucial for viral diseases. Various diagnostic tests had been developed in the past primarily based upon detecting viral nucleic acids, antigens or their cognate antibody in the host-body fluid through serological tests like ELISA, histochemistry, fluorescence antibody test or nucleic acid amplification techniques such as conventional polymerase chain reaction (PCR), real-time PCR (q-PCR), loop-mediated isothermal amplification (LAMP) and nested PCR. However, sensitivity, specificity, cost, throughput or robustness are some of the major aspects that urge further improvements and most importantly additional focus needs to be paid on developing point-of-care or animal-side diagnostics that will certainly benefit the terminal farmers the most. In order to address such issues of conventional diagnostic techniques nanoparticle-based diagnostics have yielded favourable outcomes. Various nanoparticles due to their high surface area, modified functionalization, enhanced porosity, distinct opto-electronics and magnetic properties have proven

L. Buragohain
Department of Animal Biotechnology, College of Veterinary Science, Assam Agricultural University, Guwahati, Assam, India

R. Kumar
Department of Veterinary Physiology and Biochemistry, Lala Lajpat Rai University of Veterinary and Animal Sciences, Hisar, Haryana, India

A. Y. N. Kumar
Department of Genetics and Plant Breeding, Institute of Agricultural Science, Rajiv Gandhi South Campus, Banaras Hindu University, Mirzapur, Uttar Pradesh, India

M. Ghosh (✉)
Department of Veterinary Physiology and Biochemistry, Faculty of Veterinary and Animal Sciences, Rajiv Gandhi South Campus, Banaras Hindu University, Mirzapur, Uttar Pradesh, India

LLR University of Veterinary and Animal, ICAR-National Research Centre on Equines, *Nanotechnology Theranostics in Livestock Diseases and Management*, Livestock Diseases and Management,
https://doi.org/10.1007/978-981-16-1610-5_5

advantageous in diagnostics; some of them impart simultaneous diagnostic as well as therapeutic values to bear theranostic capability. Exploitation of these unique physico-chemical properties specific to the type of nanoparticles has been employed to develop diagnostic tests that are more efficient and convenient than the traditional methods. Recently, nanoconjugated biochemicals are used oftentimes in sample preparation or detection systems that have potentiated the diagnostic efficiency of conventional pathogen detection modalities. Further, various nanoparticle-based biosensors, lateral flow modalities have extended the opportunity of rapid and animal-side diagnostics depicting marked benefits over the conventional systems. Handful of such nanodiagnostics is already in the pipeline for commercialization and several more will certainly become reality imminently. The enormous input directed towards developing state-of-the-art nanodiagnostic platforms will certainly have a lusty impression in the landscape of viral livestock disease diagnosis and for controlling the transmission of the contagious pathogens through rapid detection followed by isolation measures. This will be beneficial for not only animal health but also human subjects by limiting the transmission of zoonotic viral diseases and providing food safety.

Keywords Viral diseases · Livestock · zoonotic diseases · Nanodiagnostics · Sensors

5.1 Introduction

Despite being suitable for all diseases, the golden words "Prevention is better than cure" are even more rational for viral diseases as the option for therapeutic intervention is certainly very limited against viruses in comparison to other infectious aberrations. Viral diseases in livestock can be well prevented by vaccination and good management practices. But a number of viral diseases in animals cannot be shielded by vaccines. Although, some of the antiviral therapeutics are available to control viral diseases, in the case of animals the option is very limited. Therefore, one of the best ways to control viral disease in livestock or animals is the early diagnosis by using rapid, sensitive, specific, cost-effective and point-of-care diagnostic assay. The rapid diagnostic tool is very important in emerging and re-emerging viral diseases in order to inhibit their hasty transmission and enforce effective control. Various viral diseases have emerged or re-emerged in both animal and human subjects in recent times like SARS, wild poliovirus type1, Hepatitis, Influenza A, Ebola virus etc. In these cases vaccination is not yet well-developed; therefore the only effective way to contain such diseases refers rapid diagnosis, segregation and stringent measures to stop the spreading of the pathogens. For instance, to date there is no vaccine or effective treatment available for the prevention and control of African Swine Fever (ASF). However, ASF can be controlled by early diagnosis so that infected animals can be identified and culled to forbid the spread of the disease. The diagnostic test or assay designed for diagnosis of animal viral diseases must be simple, potable and easy to handle with field applicability as the availability of sophisticated laboratory or skilled personnel is very limited at the unorganized terminal farming locations. The importance of animal virus diagnosis is yet impending despite the immense significance not only for conservation and well-being perspective of the animals but also for human health aspect as most of the viral diseases in humans are zoonotic in nature. In this regard, point-of-care or pen-side diagnostics are high appreciable to be used in field condition for screening of viral diseases in animals. However, laboratory diagnostic assays cannot be neglected because they are essential for final confirmation as well as to diagnose the rare, new, emerging and re-emerging viral diseases in a particular geographical location or in a new host species.

There are different serological and molecular diagnostics platforms available for the detection of viral diseases of livestock which include but are not limited to enzyme-linked immunosorbent assay (ELISA), indirect fluorescence antibody test (i-FAT), Dot-blot, histochemical techniques, antibody arrays, lateral flow assay, polymerase chain reaction (PCR), real-time PCR (q-PCR), nested PCR, loop-mediated isothermal amplification (LAMP), microarrays, etc. But the introduction of nanotechnology has empowered the field of diagnostics to another dimension. The use of nanotechnology makes diagnostics assays more robust, rapid, highly sensitive, accurate, selective and specific. Even by using nanotechnology, the available diagnostics assays could be further refined to increase the speed, sensitivity and specificity. Nanoparticles (NPs), such as carbon nanotubes (CNTs), graphene, quantum dots, silica, gold, silver, magnetic, zinc and many other metal and metal oxide NPs, and nanocomposites are reported to be used in the development of diagnostic tools for detection of diverse viral diseases of both human and animals (Ehtesabi 2020; Kobayashi et al. 2021; Manhas et al. 2021). Different diagnostics platforms have utilized NPs for detection of several livestock and poultry viruses such as Japanese encephalitis virus, avian influenza virus, Newcastle disease virus (NDV), porcine circoviruses, pseudorabies virus, porcine parvovirus and hepatitis E virus, with good sensitivity and specificity (Kobayashi et al. 2021; Manhas et al. 2021; Chowdhury et al. 2019; Belza et al. 2021; Lai et al. 2017). In this chapter, an overview of different NP-based diagnostic tools has been delineated that have been used to detect animal viruses including certain viruses of zoonotic importance.

5.2 Nanoparticle in Diagnostics

Various metallic and non-metallic, carbon-based and engineered NPs have been employed in effective viral diagnosis of different livestock diseases exploiting the principle of immunological methods, biosensors or nucleic acid-based complementary binding or amplification techniques in effective ways (Fig. 5.1 and Table 5.1). The intelligent nanomaterial (NMs) have found utility in almost each and every stage of diagnostics which includes sample preparation to processing, identification and imaging of results to impart effective diagnosis, even beyond in theranostic applications.

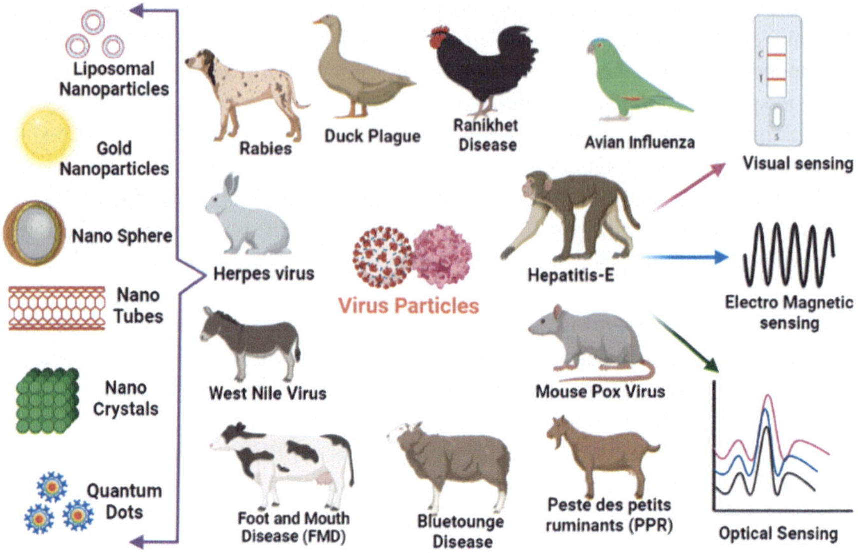

Fig. 5.1 Nanoparticle-assisted diagnosis of different viral livestock diseases exploiting various sensing principles

5.2.1 *Carbon-Based NMs in Viral Diagnostics*

Detection of viruses in complex clinical materials can be done by using carbon-based NMs with simplicity and rapidity. Different types and forms of carbon NMs like carbon nanotubes, carbon nanofibers, graphene or graphene oxide, fullerenes, nanodiamonds and carbon dots are most commonly used in diagnostic modalities (Ehtesabi 2020; Belza et al. 2021) (Table 5.1). Many researchers have widely reported the use of carbon-based biosensor assays for detection of viruses (Ehtesabi 2020). Besides human viruses, carbon NP-based diagnostics tools are also employed for the detection of various animal viruses.

5.2.1.1 Carbon Nanoparticles in Virus Detection

Chin et al. (2017) have developed a rapid and cost-effective immunosensor strip-based assay for the detection of Japanese encephalitis virus (JEV) in serum samples. In this assay, the screen-printed carbon electrode (SPCE) was modified with carbon nanoparticles (CNPs) synthesized from starch nanoparticles. The incorporation of CNPs has tremendously improved the immunosensor assay by enhancing the electron transfer kinetics and current intensity. The JEV detection range of the assay was between 5 and 20 ng/mL and limit of detection (LOD) was 2.0 ng/mL. Another advantage of this assay was the time factor; it takes only 20 min to get the result which is far less compared to many other diagnostic assays. Moreover, the

Table 5.1 Nanotechnology in viral livestock disease diagnosis

Detection system	Nano-component	Viral species	Description	Reference
Carbon nanotubes in virus detection				
Electrochemical biosensor strip	Carbon NP	Japanese encephalitis	Linear detection ranges between 1 and 20 ng/mL with LOD of 0.36 ng/mL, result within 10 min	Chin et al. (2017)
Geno-sensor	Gold nano-structured screen-printed carbon electrodes	SARS virus	DOL 2.5 pmol/L with linearly in the range between 2.5 and 50 pmol/L	Martínez-Paredes et al. (2009)
Graphene NPs in virus detection				
Electrochemical biosensor	Graphene oxide polymer nanocomposite	Dengue virus	0.12 PFU/mL detection limit	Navakul et al. (2017)
Graphene-based biosensor	Graphene nano-materials	Zika virus	Detection at concentrations as low as 0.45 nM without any cross-reactivity	Afsahi et al. (2018)
Screen-printed electrochemical sensor	Graphene oxide	Influenza A virus	Linearity 10 ng/mL to 10 μg/mL with a limit of detection of 10 ng/mL	Kinnamon et al. (2018)
Quantum dots (QDs) in viral diagnostics				
Microarray	Quantum dots, Nanowires and carbon nanotubes	Congo-Crimean haemorrhagic fever, Rift Valley fever, Marburg Ebola and Hanta virus	Oligo-nucleotide probes on differential microarray platform	Valdivia-Granda et al. (2005)
Biochip platform using a QDs-conjugated RNA oligonucleotide	Quantum dots (QDs)	SARS-CoV nucleocapsid (N) protein	DOL of 0.05 μg/mL, sensitive and rapid	Roh (2012)
Optical QDs-based RNA aptamer chip	Quantum dots (QDs)	SARS-CoV N protein	SARS N protein was detected with lower concentration of 0.1 pg/mL	Roh and Jo (2011)

(continued)

Table 5.1 (continued)

Detection system	Nano-component	Viral species	Description	Reference
Surface plasmon resonance-mediated fluorescence-based detection	Semiconductor quantum dot (QD) nanocrystals	Zika virus	LODs of alloyed Au/AgNP Qdot 646-MB (1.7 copies/mL) > CS Au/AgNP-Qdot646-MB (LOD 2.4 copies/mL) > AuNP-Qdot646-MB (LOD 2.9 copies/mL) > AgNP-Qdot646-MB (LOD 7.6 copies/mL)	Adegoke et al. (2017)
Optical biosensor	Molybdenum disulphide (MoS_2) quantum dots	Avian influenza H5N1	LOD 7.35 and 80.92 pg/mL QD and MoSQD	Ahmed and Neethirajan (2018)
Fluorescent	Quantum dots QDs	H5N1	Rapid detecting, detecting range of $2^{-1.2}$ to 2^6 HAU 20/μL	Xu et al. (2016)
Fluorescence	CdZnTeS Quantum dots	H1N1	Detect extremely low concentrations	Adegoke et al. (2016)
Gold NPs in viral detection				
Nanoparticles-assisted PCR of 3D gene	AuNPs	Encephalomyocarditis	LOD 1.2×10^1 copies/μL without cross-reaction	Yuan et al. (2016)
Nano-PCR using 3D gene of FMDV	Gold nanoparticles	Foot and mouth disease virus strain, A, and SAT2	LOD was 10 copy number, no cross-reactivity with SVDV	Jain et al. (2018)
Plasmonic bioassay	Gold NPs+ PEG	Avian influenza	Visually detect within 30 min	Marín et al. (2013)
Immunomagnetic separation coupled with enzyme-induced metallization	Gold NP	Avian influenza	Colorimetric Detection, LOD detection was 0.6 amol/50 μL	Zhou et al. (2014)
Electrochemical impedance biosensor	Gold NP	Japanese encephalitis virus	LOD was 167 PFU/mL with a linear range of test between 500 and 5×10^5 PFU/mL	Geng et al. (2016)

Label-free ampere metric immune sensor	Au NP	Japanese B encephalitis	LOD 6×10^{-9} lg PFU/mL	Yuan et al. (2005)
Potentiometric immune-sensor	Gold NPs	Japanese B encephalitis	LOD 3.5×10^{-8} lg PFU/mL	Zhang et al. (2006)
RT-PCR and real-time quantitative RT-PCR assay	Gold NPs	Japanese encephalitis virus	Increased the amplification yield and shortened time, with LOD (1–10,000 copies) of the JEV RNA genomes	Huang et al. (2008)
A surface-enhanced Raman scattering (SERS) immunoassay	Gold nanoparticles	West Nile virus	LOD 2 ng/mL	Neng et al. (2010)
Colorimetric detection of DNA sequences	Gold NPs	SARS	Visible detection within 5 min, and DOL	Li and Rothberg (2004)
Colorimetric detection	Gold NP	Rift Valley fever	High accuracy and specificity, faster within 30 min with LOD 10 RNA copies/reaction	Zaher et al. (2018)
Multi-target programmable nanoarray	Gold NP	All Ebola virus subtypes	Visual detection LOD 400 amols (24×10^6 molecules)	Balcioglu et al. (2017)
Monoplex platform	AuNP	Ebola virus	100% sensitivity and 98% specificity compared to standard (ELISA)	Brangel et al. (2018)
Field-effect transistor (FET) biosensor	Gold NP	Ebola	Faster (within a few seconds), DOL as low as 1 ng/mL of EGP	Chen et al. (2017)
Multiplexed detection by surface-enhanced Raman scattering	Gold NP	Hepatitis A and hepatitis B virus, human immunodeficiency virus, Ebola virus, variola virus and *Bacillus anthracis*	LOD 20 fM	Cao et al. (2002)
Anti-NS1 affibody functionalized AuNPs	AuNPs	Dengue virus (DENV)	Enhanced ELISA resulted in signal amplification in tune of 14.2-fold	Bang et al. (2018)

(continued)

Table 5.1 (continued)

Detection system	Nano-component	Viral species	Description	Reference
Raman Fingerprinting	Gold NP	Dengue virus (DENV) and West Nile virus (WNV)	LOD 10 plaque-forming units (PFU)/mL of DENV-2 and WNV	Paul et al. (2015)
Immunochromatographic detection	Gold NP	Four dengue virus serotypes and Zika viral infections	Detects all four dengue serotypes, with sensitivity/specificity ranged from 0.76 to 1.00. Sensitivity/specificity for the ZIKV rapid test was 0.81/0.86, respectively	Bosch et al. (2017)
Label-free optical detection	Gold nanoparticles	Dengue virus (DENV)	LOD was 0.12 μM at RNA: DNA ratio of 1:0.01	Rahman et al. (2014)
Optical caustic plasmonic light scattering sensor	Gold NP	Dengue	LOD 10 μg/mL of anti-protein E antibody in PBS	García et al. (2017)
All-optical fibre sensor based on localized surface plasmon resonance (LSPR) with antibody	AuNPs	Dengue virus	LOD was 0.074 μg/mL = 1.54 nM	Camara et al. (2013)
Surface plasmo-resonance (SPR) and Ig-based detection on biochip	Gold NP	Dengue virus	Need only 1 μL sample for test with 83–93% sensitivity and 100% specificity	Jahanshahi et al. (2014)
SPR-based immunosensor	Gold NP	Dengue virus	Sensitive comparable to MAC-ELISA	Kumbhat et al. (2010)
Quartz crystal microbalance DNA sensing system	Oligo-nucleotide functionalized AuNP probes	Dengue virus	Linear detection range from $2 \times 10^{(0)}$ to 2×10^{6} PFU/mL, LOD2 PFU/mL	Chen et al. (2009)
Paper-based plasmonic biosensor	Gold nanorods	Zika virus	The bioplasmonic paper-based device, excellent sensitivity and selectivity	Jiang et al. (2017)

Surface-enhanced Raman spectroscopy-based sandwich immunoassays	Gold NP	Multiplexed detection of Zika and Dengue viruses	The LOD was 0.72 ng/mL for ZIKV NS1 and 7.67 ng/mL for DENV NS1	Sánchez-Purrà et al. (2017)
Plasmonic-gold nanotechnology platform	Nanostructured plasmonic gold	Multiplexed detection of Zika and Dengue viruses	For early acute-phase ZIKV infections, detected with 47% of specificity. In the convalescent phase, sensitivity and specificity are greater than 90% and 98%, respectively. LOD 1 μL and results in ~2 h	Zhang et al. (2017a)
Electrochemical impedance spectroscopy	Gold NPs	Zika virus	Linearity 10 pM to 1 nM and LOD 10 pM, high sensitivity	Kaushik et al. (2018)
An optofluidic nanoplasmonic biosensor	Gold film	Vesicular stomatitis virus, pseudo-typed Ebola and viruses (vaccinia virus)	Direct detection of a broad range of live viral pathogens	Yanik et al. (2010)
Immunosensor	Gold NP	Newcastle disease virus	5 ~ 10 times improved detection with detection limit 5 pg	Luo et al. (2018)
Magnetic nanozyme-linked immunosorbent assay	Gold NPs	Influenza A	Rapid ELISA. Sensitivity 5×10^{-15} g/mL	Oh et al. (2018)
Nano-ELISA	Gold nanoparticles	H9N2 avian influenza	Sensitivity of detection 100%	Gholami et al. (2016)
Biosensor based on luminescence resonance energy transfer LRET	Gold NP	Avian influenza viruses	LOD 10 pmol, linearity range 10 pm–10 nm	Ye et al. (2014)
Biosensor	Gold NP	14 Influenza virus strains	Single-step, portable colorimetric procedure	Zheng et al. (2017)
Surface plasmon resonance	Gold nanoparticles	Influenza virus H3N2	RNA aptamer specific to bind with hemagglutinin of human influenza virus 120 pM	Misono and Kumar (2005)
Wave guide biosensor	Gold nanoparticle	H3N2	LOD 8×10^3 PFU/mL	Gopinath et al. (2013)

(continued)

Table 5.1 (continued)

Detection system	Nano-component	Viral species	Description	Reference
Microfluidic biosensor	Gold NP	H1N1	Rapid detection	Tseng et al. (2016)
Nano RT-PCR	Sugar chain immobilized GNPs	Influenza	Increase sensitivity	Suda et al. (2015)
Silver NPs in viral detection				
Multiplexed rapid lateral flow diagnostics	Multi-coloured silver nanoparticles	Yellow fever, Ebola and Dengue	LODs of 150 ng/mL for each disease	Yen et al. (2015)
Optical sensing probe	Silver NP	Japanese encephalitis	LOD of 12.8 ng/mL with 1 h assay time	Lim et al. (2017)
Surface acoustic wave (SAW) sensors	Silicon dioxide	Ebola virus	LOD 1.9×10^4 PFU/mL	Baca et al. (2015)
Colorimetric detection	Triangular silver NP	Dengue virus	Detection of different dengue serotype RNA with high sensitivity and specificity	Vinayagam et al. (2018)
Electrochemical sensor	Silver nanoparticle	Influenza virus	Single virus detection change in the current magnitude direct proportion to virus concentration	Sepunaru et al. (2016)
Surface-enhanced Raman scattering	Silver and gold nanoparticle	H1N1, H3N2 and H2N2	LOD 10^6 PFU/mL	Sivashanmugan et al. (2013)
Magnetic NPs in viral detection				
Antibody-based assay	Magnetic nanoparticles	Herpes simplex virus-1	LOD 5 viral particles in 10 μL	Perez et al. (2003)
Magnetic capture based detection of viral DNA	Gold-paramagnetic nanoparticle	West Nile Virus	LOD was 10 pM for target DNA	Zhang et al. (2011)
Antibody-based surface-enhanced Raman scattering (SERS) detection	Multiplex detection Au and paramagnetic NP	West Nile virus and Rift Valley fever virus	LOD 5 fg/mL in PBS and 25 pg/mL for PBS spiked with foetal bovine serum	Neng et al. (2013)

Engineered magnetic nanosensors	Magnetic relaxation sensitive nanoparticles	Zika virus	Examine and quantify host-Zika interactions	Shelby et al. (2017)
Immunoassay by lab-on-a-chip device	Magnetic nanoparticle	Hendra virus	Rapid, on field application with limit of detection of nearly 0.48 ng/mL	Petkovic et al. (2016)
Immunosensor	Magnetic and luminescent functional particles	Hendra virus	Fast and portable biosensor using at least two tags	Lisi et al. (2012)
Immunoassay	Magnetic nanoparticles and phycoerythrin	Hendra virus	Test completed within 1 h	Gao et al. (2015)
Breath sensor designed with electronic detection system	Magnetic nanoparticles	Influenza A as well as H1N1	Rapid, can use exhaled breath for virus detection	Shen et al. (2012)
Other metallic and metal oxide NPs in viral detection				
Electrochemical immune sensors	Titanium, platinum and silicon dioxide nanomaterials	Japanese encephalitis virus	LOD was 10 ng/mL a linear range of test between 25 ng/mL to 1 µg/mL	Tran et al. (2012)
Enzyme-linked immunosorbent assay (ELISA)	Iron oxide NP	West Nile virus or human herpes virus	Shows better linearity, higher stability as compared to ELISA	Zhang et al. (2016)
Platinum nanoparticles incorporated volumetric bar-chart chip (PtNPs-V-Chip)	Platinum nanoparticles	Ebola virus DNA	Can detect as low as 16 pM Ebola Virus DNA	Wang et al. (2016)
Electrochemical nanoporous membrane-based biosensor	Alumina nanoporous membrane and platinum wire electrodes	Dengue virus	LOD 9.55×10^{-12} M and linearity over 6 order of magnitude	Rai et al. (2012)
Electrochemical genosensor	ZnO/Pt-Pd NP	Dengue virus	The sensor LOD as 4.3×10^{-5} M with linearity range of 1×10^{-6} M to 100×10^{-6} M	Singhal et al. (2017)
Electrochemical biosensor	A printed gold electrode on disposable screen-(SPGE) + silicon nanowires (SiNWs) + GNP	Dengue virus	Good storage stability, LOD 1.63×10^{-12} M and linearity in range of 1×10^{-11} to 1×10^{-7} M DNA concentration	Rashid et al. (2015)

(continued)

Table 5.1 (continued)

Detection system	Nano-component	Viral species	Description	Reference
Nanoparticle-based LDI-MS immunoassay	Antibody-modified metal nanoparticles	Enterovirus 71, Japanese encephalitis virus, Zika virus	Multiple diagnosis of viral infections	Chu et al. (2019)
Motion-based immunological detection	Platinum (Pt)-nanomotors	Zika virus	LOD 1 particle/μL	Draz et al. (2018)
Microfluidic electrochemical sensor	ZnO nanorods	H1N1, H5N1 and H7N9	LOD 1 pg/mL–10 ng/mL for different strain	Han et al. (2016)
Organic NPs in viral diagnostics				
Diamond electrode-based immune sensor	Diamond NP	Influenza virus	Rapid, within 5 min LOD 1 fg/mL and 5–10 viruses/sample in saliva	Nidzworski et al. (2017)
Lateral flow assay	Carbon nanostrings	Influenza A virus	Within 15 min, LOD containing 2.8×10^3 TCID 50/mL	Wiriyachaiporn et al. (2013)
Air sampling electrochemical detection	Polystyrene particles	H1N1	Cyclone-based air sample device	Kim et al. (2018)
Nano-composites in viral diagnostics				
Biosensor	Molybdenum disulphide nanosheets (MoS_2 NSs)	Chikungunya virus	Chikungunya virus DNA with limit of detection in linear way between linear range of 0.1 nM to 100 μM	Singhal et al. (2018)
Electrical detection with antibody	Silicon nanowire	Avian adenovirus group III, influenza A, and avian paramyxovirus virus	Faster, showed discrete conductance for influenza A, reliably at single virus level	Patolsky et al. (2004)
Impedimetric DNA biosensor	Nanoporous alumina membrane	Dengue virus	Pore resistance (Rp) linearly increases in response towards the increasing concentration of the target DNA in the range of 1×10^{-12} to 1×10^{-6} M	Deng and Toh (2013)

Electrochemical impedance spectroscopy based detection	Nanoporous alumina membrane	Dengue serotype 2 (Denv2)	Linear range of detection was 1–900 plaque forming unit per mL (Pfu/mL)	Nguyen et al. (2012)
Electrochemical impedance spectroscopy	Nanoporous alumina membrane	Dengue virus	Faster detection with detection limit of 0.230 PFU/mL and 0.710 PFU/mL respectively for dengue 2 and dengue 3 viruses	Peh and Li (2013)
Silicon nanowire biosensor for detection of RT-PCR product	Silicon nanowire	Dengue serotype 2	LOD below 10 fM and faster within 30 min	Zhang et al. (2010a)
Electrical detection DNA oligomer	Silicon nanowire	Dengue virus	LOD as low as 2.0 fM, high sensitivity of 45.0 μA M	Nuzaihan et al. (2016)
Silicon nanowire (SiNW) biosensing platform	Silicon nanowire	Dengue virus	Serotype-specific identification, multiplexing	Huang et al. (2013)
Optical DNA biosensor	Porous silica nanospheres (PSiNs)	Dengue virus	Linear range of detection between 1.0×10^{-16} and 1.0×10^{-10} M with LOD at 0.2 aM	Ariffin et al. (2018)
Field-effect transistor	Poly-silicon nanowire	H5N7	Sensitivity in fM range	Lin et al. (2009)
Label free electric diagnosis	Si_3N_4 nano pores	Influenza A, H1N1, H3N2 and B	Single virus detection	Arima et al. (2018)
Lipid NPs in viral diagnostics				
Biosensor coupled to the isothermal nucleic acid sequence based amplification (NASBA) technique	Liposomes	Dengue virus	Rapid, serotype specific within 15 min. With detection in the range of 100–1000 PFU/mL	Baeumner et al. (2002)
Microfluidic biosensor with fluorescence detection	Liposomes	Dengue virus	Rapid serotype-specific detection with LOD 0.125 nM and 50 pM for intact and lysed liposome	Zaytseva et al. (2005)

(continued)

Table 5.1 (continued)

Detection system	Nano-component	Viral species	Description	Reference
Surface Plasmon resonance	Bovine brain lipid neomembranes	Influenza A virus H1N1	Effective	Critchley and Dimmock (2004)
Miscellaneous NMs in viral diagnostics				
Nucleoprotein capture based detection	Nanotrap particles	Rift Valley fever virus	Enhanced detection	Shafagati et al. (2015)
Optical interferometry based detection	Nanofluidic channels	HIV and Sindbis viruses	Ability of detecting and recognizing different species of human viruses from a mixture	Mitra et al. (2010)
POC	Virus-like particles (VLPs)	Nipah virus	ELISA revealed that the NiV M protein produced in *P. pastoris* reacted strongly with positive NiV sera demonstrating its potential as a diagnostic reagent	Joseph et al. (2016)
Immunodiagnostic	Virus-like particles and magnetic microspheres	Alpha virus	Improves sensitivity by up to 2-logs, faster	Ricks et al. (2018)
Antibody based	VLPs	Hantavirus	Generation of virus-reactive MAb against hantavirus Gc glycoprotein	Zvirbliene et al. (2014)
Microarray technique	Nanogold-labelled oligonucleotide probes along with silver	Hepatitis E	Identified by naked eyes with only 2 min development time	Liu et al. (2006)

immunosensor strip was found to be stable for 6 months at 4 °C or at dry atmosphere. Similarly, Lai et al. (2017) have designed another sensitive and rapid electrochemical biosensor strip for detection of JEV by using CNPs modified SPCE. In this experiment, the CNPs were synthesized from chitosan NPs. The designed diagnostic model is more sensitive and rapid as its detection range of JEV was 1–20 ng/mL with limit of detection (LOD) of 0.36 ng/mL and the results could be obtained within 10 min. Recently, an immunoassay was developed with improvised carbon dot-based silica spheres to detect the nucleocapsid protein of severe acute respiratory syndrome coronavirus 2 (SARS-CoV-2) in clinical samples (Xu et al. 2021). The sensitivity of the test was 100 pg/mL when observed under ultraviolet light, but the sensitivity was 10 pg/mL when observed under a fluorescence microscope.

5.2.1.2 Graphene NPs in Virus Detection

A hybrid optoelectronic sensor was designed by Ahmed et al. (2018) for the detection of fowl adenovirus (FAdVs). It consists of gold nanobundles (AuNBs) and graphene quantum dots (GQDs). The target FAdVs and the electric signal of the sensor showed detection ability up to 10 Plaque forming unit (PFU)/mL in buffer and 50 PFU/mL in blood media. LOD of the sensor was 8.75 PFU/mL which is several times more sensitive than ELISA. Anik et al. (2017) reported an electrochemical biosensor for detection of influenza virus A (H9N2). The electrode used was screen-printed with graphene-gold (Au) NPs for detecting neuraminidase (N) activity of influenza virus A. The detection range of the sensor was between 10^{-8} and 10^{-1} U/mL with LOD value of 10^{-8} U/mL. Avian influenza virus H7 (AIV H7) was detected and quantified by using a special electrochemical sensor. The sensor was designed with electrode coated with gold nanoparticle-graphene-chitosan nanocomposite and electrochemical signal material used was silver nanoparticle-graphene-chitosan nanocomposite. The sensor exhibited detection range of 1.6×10^{-3} to 16 ng/mL with detection limit of 1.6 pg/mL. Afsahi et al. (2018) demonstrated a rapid, sensitive and specific graphene-based biosensor for the detection of Zika virus. It can detect Zika viral antigen at a very low concentration (as low as 450 pM) under optimum condition.

5.2.2 *Quantum Dots (QDs) in Viral Diagnostics*

Quantum dots (QDs) are solid nanodimension semiconductor crystals with special electronic and optical properties (Singh et al. 2020). The QDs have size-dependent emission wavelength, narrow emission peak and broad excitation range which have wide application including the development of robust and versatile diagnostic systems (Singh et al. 2020) (Table 5.1).

Roh and Jo (2011) have developed a chip-based system for the detection of SARS-CoV. The detection system used a glass chip immobilized with nucleocapsid

(N) protein of SARS-CoV. A specific QDs-conjugated with RNA aptamer was designed which hybridized with the immobilized N protein. Once the QDs-conjugated RNA aptamer hybridized specifically with N protein, there are changes in optical signal, and the variations in optical signal are the basis for the detection of SARS-CoVN protein. This optical QDs-based RNA aptamer chip could detect N protein of SARS-CoV as low as 0.1 pg/mL. Avian influenza virus (H5N1) is being detected by using a dual mode optical sensor system (Ahmed and Neethirajan 2018). The detection system was constructed using Molybdenum disulphide quantum dots (MoS_2QDs) which are based on fluorescence and chirality. Compared to ELISA, it was found to be more sensitive in detecting avian influenza virus. In another experiment, hydrogel-based fluorescent aptasensor was constructed to detect H5N1 avian influenza virus (Xu et al. 2016). Quantum dots were used in the system as fluorescence reporters. Swelling of the hydrogel indicates the presence of the target (virus) in a sample which results from the binding of aptamer and target. Quantitation is done based on the changes in fluorescent emission peak. The lowest detection limit of the system was 0.4 hemagglutination units (HAU) and it can detect $2^{-1.2}$ to 2^6 HAU/20μL. The sensor was cheap, simple and specific for detection of avian influenza virus H5N1, and it was a rapid technique as result can be obtained within 30 min. Adegoke et al. (2016) have demonstrated a very sensitive nanodiagnostic system using near-infrared alloyed CdZnSeTeS quantum dots conjugated with molecular beacon for the detection of H1N1 influenza virus RNA. The LOD of the diagnostic system was about ~1 copy/mL which is sensitive than many other molecular diagnostic assays. An electrochemical biosensor was designed with graphene quantum dots and gold-embedded polyaniline nanowires for the detection of hepatitis E virus (Chowdhury et al. 2019). The sensitivity of the sensor was similar to that of qPCR.

5.2.3 Metallic and Metal Oxide NMs in Viral Diagnostics

Several metallic and metal oxide NPs such as gold, silver, magnetic oxide and zinc oxide are utilized to design unique diagnostic tools for the detection of different viruses (Table 5.1). The metallic and metal oxide NPs depicted great value as diagnostic tools as they are readily available and their synthesis protocols are well-documented.

5.2.3.1 Gold NPs in Viral Detection

Nanoparticle-assisted PCR (nanoPCR) assay is an advanced form of PCR which is more sensitive and specific compared to conventional PCR (Wang et al. 2015; Cui et al. 2014). In nanoPCR assay, the PCR buffer contains nanoparticles, particularly gold NPs that form colloidal nanofluids which enhance the thermal conductivity, thereby decreasing nonspecific amplification (Li et al. 2005; Shen et al. 2009). The application of nanoparticle-assisted PCR assay has been reported in many viral diseases of animals which include porcine parvovirus (Cui et al. 2014), pseudorabies virus (Ma et al. 2013), canine coronaviruses (Qin et al. 2021), mink enteritis virus (Wang et al. 2015), duck tembusu virus (Wanzhe et al. 2016), encephalomyocarditis virus (Yuan et al. 2016), Japanese encephalitis virus (Huang et al. 2008) and several others.

Zhang et al. (2010b) developed an improved chemiluminescence immunoassay using gold NPs to detect porcine circovirus type 2 (PCV2) with high sensitivity and reliability. The detection limit of the assay was 1.71×10^3 copies/mL, and it was further reduced to 2.67×10^2 copies/mL when the assay was modified with hydroxylamine-amplified gold NPs. The detection of PCV2 by surface-enhanced Raman scattering (SERS)-based immunoassay using multi-branched gold NPs was reported by Luo et al. (2013). The detection limit of the assay was 8×10^2 copies/mL and linear detection of PCV2 ranges from 8×10^2 to 8×10^6 copies/mL. A rapid gold NP-based dot blot assay was standardized to detect antibodies against Foot-and-mouth disease (FMD) vaccine (Jain et al. 2018). It can detect antibodies in serum up to 10^{-4} dilution. Newcastle disease virus (NDV) was detected through an immunosensor which was developed by using gold nanosphere coated excessively tilted fibre grating (Ex-TFG). It has also been observed that the limit for NDV detection was 5 ~ 10 times higher when Ex-TFG was used alone without the coat of gold nanospheres signifying the impact of gold NPs to improve the detection sensitivity (Luo et al. 2018).

Marín et al. (2013) have demonstrated a plasmonic bioassay for the detection of human influenza virus X31 (H3N2). In this assay, gold NPs are functionalized with thiolated trivalent α2,6-thio-linked sialic acid derivative (glyconanoparticles). In presence of influenza virus, there is aggregation of the glyconanoparticles due to the binding of haemagglutinin of the virus with the trivalent α2,6-thio-linked sialic acid ligand resulting in colour changes in the solution within half an hour. Moreover, this assay has the power to discriminate between human (H3N2) and avian influenza virus (H5N1). Gholami et al. (2016) have optimized a nano-ELISA for the detection of H9N2 avian influenza antibody in chicken serum samples. In this ELISA, the secondary antibody was conjugated with gold NPs to increase the sensitivity up to 100%. Yang et al. (2020a) constructed a colloidal gold-based immunochromatographic test strip to detect low pathogenic H7N9 avian influenza virus with high sensitivity, specificity and repeatability.

A sensitive electrochemical-based immunosensor was designed by Geng et al. (2016) to detect JEV. The sensor was designed with a screen-printed carbon electrode fabricated with gold NPs. The detection limit of the sensor was 167 PFU/mL and the range of detection is between 500 and 5×10^5 PFU/mL. A gold NP-based colorimetric detection assay was developed by Zaher et al. (2018) for the detection of Rift Valley fever virus (RVFV) within 30 min. In this assay pre-amplification step is not necessary and detection can be done by the naked eye by observing the colour change between positive and negative samples. The LOD value of this assay is 10 RNA copies/reaction which is very sensitive than many other molecular tests. A stable affibody-functionalized gold NP was designed to detect NS1 antigen of dengue virus by ELISA (Bang et al. 2018). The use of this gold NPs-based affibody could increase the signal of ELISA by 14.2-fold. Chen et al. (2009) developed a biosensing method to detect Dengue virus by combining circulating-flow quartz crystal microbalance (QCM) and oligonucleotide-functionalized gold nanoparticles as probe. The linear detection range of the method is from 2×10 to 2×10^6 PFU/mL and the detection limit is 2 PFU/mL. An immune-sensing chip was developed by Kaushik et al. (2018) for the detection of Zika virus. The sensor was designed by using a functionalized interdigitated micro-electrode of gold array. It is very selective in detecting protein of Zika virus having the detection range of 10 pM to 1 nM and LOD of 10 pM.

5.2.3.2 Silver NPs in Viral Detection

A multiplexed lateral flow assay was designed by Yen et al. (2015) for simultaneous detection of Dengue, Yellow fever and Ebola viruses. The assay used multi-coloured silver NPs for the detection of multiple pathogens. Lim et al. (2017) developed an optical sensing probe based on silver NPs for the detection of Japanese encephalitis virus. The principle behind the detection system is that in the presence of JEV antigen there is a change in the light absorbance property of silver NPs. The detection limit of this system is 12.8 ng/mL and the result could be obtained in 1 h. Single influenza virus particle was detected by tagging with silver NPs based on nano-electrochemical technique (Sepunaru et al. 2016). In this technique, influenza virus present in a solution is detected based on the magnitude and frequency of current spikes produced on the application of optimum potential.

5.2.3.3 Magnetic NPs in Viral Detection

Virus can be detected by using magnetic NPs conjugated with virus-surface-specific antibodies. By using this principle, a method was developed by Perez et al. (2003) to detect Herpes simplex virus-1 and adenovirus-5 at very low concentration (5 viral particles/10 μL). Neng et al. (2013) reported a surface-enhanced Raman scattering (SERS)-based immunoassay for simultaneous detection of Rift Valley fever virus and West Nile virus. The detection system uses Raman reporter-coated gold and paramagnetic NPs conjugated with polyclonal antibodies specific for Rift Valley fever virus and West Nile virus antigen. Both the viruses are distinguished by using two different Raman reporter dyes that produce two Raman spectra specific for antigen of Rift Valley fever virus and West Nile virus. By using magnetic NPs, a chip-based immunoassay detection system was developed to detect Hendra virus-specific antibodies (Petkovic et al. 2016). The system could detect antibodies against Hendra virus in horse serum within 15 min. The LOD value of the detection system was about 0.48 ng/mL. Another magnetic NP-based technique was also demonstrated for the detection of Hendra virus-specific IgG antibody (Gao et al. 2015). This technique took advantage of magnetic NPs and fluorescence label

R-phycoerythrin. The technique could also estimate Hendra virus IgG antibody titre in two dilutions (1:20 and 1:1000) and the whole process of IgG detection could be completed within 60 min. The resonance light scattering (RLS) technique based on magnetic silica NPs was reported for the detection of avian influenza virus with a detection range of 0.5–50 ng/mL and detection limit of 0.15 ng/mL (Zou et al. 2012).

5.2.3.4 Other Metallic and Metal Oxide NPs in Viral Detection

Besides gold, silver and magnetic NPs, several other metallic and metal oxide NPs like iron oxide, aluminium, zinc oxide and platinum are used for the detection of several animal and human viruses with high specificity and sensitivity (Table 5.1).

For the detection of Dengue virus cDNA, ultrasensitive DNA biosensor was designed with nanoporous alumina membrane and platinum wire electrode (Rai et al. 2012). It was found to be very specific and sensitive having limit of detection value of 9.55×10^{-12} M. Singhal et al. (2017) reported an electrochemical biosensor composed of Zinc oxide/platinum-palladium fabricated fluorine doped tin oxide (FTO) glass plate for the detection of dengue virus genomic sequence. The sensor exhibited LOD value 4.3×10^{-5} M and limit of quantification (LOQ) value 9.5×10^{-5} M with detection range from 1×10^{-6} to 100×10^{-6} M. Another electrochemical biosensor based on graphene quantum dots (GQDs) and apoferritin-encapsulated Cu (Cu-apoferritin) NPs were employed for the detection of avian leukosis virus (Wang et al. 2013). The detection range of the biosensor was between $10^{2.08}$ and $10^{4.50}$ $TCID_{50}$/mL and the LOD value was found to be 115 $TCID_{50}$/mL. Wang et al. (2016) used platinum NPs in a volumetric bar-chart chip for the detection of Ebola virus nucleic acid with 16 pM LOD. A rapid fluorescent immunochromatographic strip test (FICT) was developed to detect hemagglutinin antigen of avian influenza H7 subtype virus and the test was found to be more sensitive compared to other rapid tests that use gold NPs (Yeo et al. 2017). Zinc oxide (ZnO) nanorods-based nano-flow immunosensor was reported to be used for the simultaneous detection of multiple viruses like H1N1, H5N1 and H7N9 influenza viruses (Han et al. 2016). The detection limit of the sensor for each virus was 1 pg/mL.

5.2.4 *Nanocomposites in Viral Diagnostics*

Different types of nanocomposites such as nanospheres, nanotubes, nanowires, nanopores and nanosheets are being used for the detection of viruses with high confidence (Table 5.1).

For the detection of chikungunya virus nucleic acid, two-dimensional molybdenum disulphide nanosheets (MoS_2 NSs) based biosensor was employed by Singhal et al. (2018). In this biosensor system, the screen-printed gold electrodes (SPGEs) were coated with MoS_2 NSs and detection was done by using electrochemical voltammetric technique. The sensor exhibited linear detection range of chikungunya virus nucleic acid from 0.1 nM to 100 μM. Deng and Toh (2013) showed the utilization of nanoporous alumina membrane-based biosensor for the detection of dengue virus nucleic acid with linear detection range from 1×10^{-12} to 1×10^{-6} M concentration of the target nucleic acid. Besides this, many other alumina membrane-based sensor platforms were also used for the detection of dengue virus (Nguyen et al. 2012; Peh and Li 2013) with high specificity and sensitivity. Similarly, silicon nanowire or silica nanospheres based biosensors were also reported to be used for the detection of dengue virus (Zhang et al. 2010a; Huang et al. 2013; Nuzaihan et al. 2016).

A silicon nanowire field-effect transistor was designed for the detection of pathogenic H5H7 avian influenza virus nucleic acid (Lin et al. 2009). The detection method was simple, low-cost and ultrasensitive with a detection range of fM to pM level for avian influenza virus nucleic acid. Based on solid-state nanopores (Si_3N_4), the influenza virus A (H1N1 and H3N2) and B could be detected at the single-virus level and the process could be standardized for the development of point-of-care testing of flu viruses (Arima et al. 2018).

5.2.5 Miscellaneous Other NMs in Viral Diagnostics

In addition to the above-mentioned NMs, there are also certain unique NPs which have great potential in the field of diagnostics, and virus-like particles (VLPs) are one such NP that have been extensively used in the development of ELISA or other diagnostics. Besides VLPs, there are also other NPs such as metal-organic framework nanoparticles, protein nanoparticles, solid-lipid NPs and polymeric NPs that can be explored for the development of novel and promising viral diagnostics tools to use in field condition.

A diamond-based biosensor was utilized for the detection of M1 protein of Influenza virus (Nidzworski et al. 2017). The electrode of the biosensor consists of diamond functionalized with polyclonal anti-M1 antibodies to detect M1 protein of influenza virus. The detection limit of the sensor in saliva buffer was recorded to be as low as 1 fg/mL. Shafagati et al. (2015) screened multiple nanotrap particles for the detection of Rift Valley fever virus and observed that a special nanotrap particle (NT45) has significantly improved the detection limit of Rift Valley fever virus nucleoprotein. Moreover, it was observed that the use of nanotrap particle could protect the nucleoprotein from degradation for a long period of time at 37 °C. Joseph et al. (2016) produced VLPs (200 nm) of Nipah virus by expressing matrix protein in *Pichia pastoris*. The expressed matrix protein reacted strongly with Nipah virus positive serum, thereby exhibiting its potential to be used as diagnostic reagent for the detection of Nipah virus antibodies in clinical samples. Detection of hepatitis E virus (HEV) nucleic acid was demonstrated by employing nano-gold labelled oligonucleotide probes in microarray technique (Liu et al. 2006). Virus-like particles (VLPs) were generated by expressing recombinant capsid protein of porcine circovirus type 3 (PCV3) in *Escherichia coli*. The capsid protein-based VLPs were used as coating antigen in indirect ELISA for the detection of PCV3-specific antibodies in swine serum samples and it was observed that VLPs of PCV3 are promising tool for the detection of PCV3-specific antibodies. A VLP-based ELISA was developed for the detection of swine vesicular disease virus (SVDV)-specific antibodies in pig serum samples (Ko et al. 2005). The VLPs of SVDV were generated by expressing P1 and 3CD proteins of SVDV using baculovirus expression system. The specificity of the VLP-based ELISA was found to be 99.9%.

5.3 Multifaceted Nanointervention in Viral Diagnostics

Viruses are always evolving and new strains are continuously emerging due to higher rate of mutation, thus looming as one of the major threats to various flora and fauna on the earth. The application of nanotechnology has left no step untouched in various diagnostic platforms: from sample preparation to detection module development due to the unique physico-chemical properties of engineered NPs providing advantages over conventional materials. This section will illuminate the multifaceted applications of various NPs in allied techniques empowering viral disease diagnosis.

5.3.1 Nanoparticles in Sample Preparations for Diagnosis

Sample preparation is one of the critical steps to achieve efficient diagnostic outcome. Various NPs have been effectively employed in sample preparations. They enhanced the diagnostic ability by increasing sensitivity and specificity, enhancing rapidity and making the diagnosis point-of-care and cost-effective. For instance, magnetic nanoparticles (MNPs) have been successfully used in the extraction of nucleic acids (DNA and RNA), proteins and enzymes from various biological samples and also have been commercialized. They are effective in enrichment of targeted samples facilitating effective diagnosis. Silica-magnetite NPs, carboxyl-coated MNPs, functionalized MNPs, silica-coated MNPs and amino-functionalized methacrylate-based NPs have been effectively used in DNA and RNA enrichment and isolation. Similarly poly(amino ester) with carboxyl groups (PC)-coated magnetic nanoparticles (pcMNPs) has been effectively used in viral RNA extraction (Zhao et al. 2020). Because of ultra-fast diffusion times, with low detection limit and high-resolution separation, various carbon-based NPs, silica NPs and other polymeric NPs have been utilized in chromatography which makes the process very fast even at point-of-care diagnosis that otherwise conventionally need separate set with a time-consuming process. Affinity and magnetic-based filtration techniques using various NPs like Poly(*N*-isopropylacrylamide-co-acrylic acid) NPs, iron oxide@silica core-shell magnetic NPs, silica-coated magnetite Fe_3O_4@SiO_2 @$GMANHNH_2$ and Boronic acid functionalized SiNPs have been employed to enrich and isolate various protein, peptides, glycoproteins, phosphopeptides and various enzymes from biological samples which may have diverse diagnostics applications including in different viral diseases. Even NPs have been used in isolation of viruses and bacteria as such in effective ways where they selectively isolate such biologicals. FACS employing NP-conjugated antibodies using γ-Fe_2O_3 MNPs has been successfully used to isolate various kinds of cells including viruses. Even the application of SiO_2/Fe_2O_3 composite NPs along with gold NPs has been demonstrated for performing both extraction of DNA and qPCR in one go (Minakshi et al. 2020). Therefore, certainly NPs have advocated that either they are more effective or they have one or more positive attributes in terms of rapidness, lower LOD, selectivity, cost or time in order to separate targeted cells or biological materials as diverse as protein, peptide, glycoprotein and nucleic acid which is not only helpful in effective diagnosis but also in the understanding of patho-physiology of diseases. Recently, microfluidics-based techniques using diverse NPs have also been successfully used in virus diagnosis.

5.3.2 Nanoparticles in Immunochromatographic Technique

Immunochromatography-based lateral flow assays principally employ two components: one antigen-detection component which is performed on a nitrocellulose or similar membrane and another component includes using fluorescent or alike materials to enhance the sensitivity of the detection. Such system has been successfully employed for enhancing the sensitivity of diagnosis in various cases like in avian influenza where influenza related antibody is conjugated with fluorescent beads referred as fluorescent immunochromatographic test (Sakurai et al. 2015). Similarly, the application of superparamagnetic nanoparticles for H1N1 lateral-flow immunoassay depicted lower detection limit, lower cross-reactivity with H3N2 influenza A with additional feature of quantification of the amount of antigens captured by using magnetic reader that rendered it superior with respect to colloidal gold-based strips (Sun et al. 2013). Limit of detection was further reduced to 31 ng/mL when europium NPs were used along with monoclonal antibody (MAb) in fluorescent immunochromatographic test. This system has extended advantage over presently used RT-PCR test for avian influenza detection in many ways with point-of-care application. Porcine circovirus type 2 (PCV2) causing post-weaning multi-systemic wasting syndrome (PMWS) has also been detected using the hydroxylamine amplified AuNPs (III)-based chemiluminescence immunoassay with LOD of 2.67×10^2 copies/mL (Zhang et al. 2010b). Immunochromatographic assays using viral protein antigens and AuNPs have been successfully developed Newcastle disease (ND) diagnosis with better LOD and no cross-reactivity (Yang et al. 2020b). Furthermore, fluoroimmunoassay using magnetic separation has been employed for ND diagnosis with higher sensitivity than without NP-based enrichment method (Wang et al. 2010). Similarly swine fever virus detection sensor has been developed using quartz crystal microbalance (QCM), gold and protein p12. The biosensor works on changes in resonance frequency which in turn is dependent on mass change at the surface (Abad et al. 1998).

5.3.3 Nanoparticles in Electrochemical Sensors

Diverse types of sensors have been recently employed successfully and also commercialized for various diagnostics applications ranging from various metabolic to infectious disorders by either exploiting changes in physico-chemical properties of biological samples or alteration in some chemical properties due to analyte interaction. Frequently, the sensors exploit classical immune reaction or aptamer-based approach to develop reliable, sensitive, low-cost and rapid point-of-care detection system. Graphene NPs possessing honeycomb-like two-dimensional lattice structure have been used successfully in various diagnostic sensors because of advantages of higher electrical conductivity, surface-to-volume ratio, electron transfer rate and better thermal stability than the conventional immune sensor where horseradish peroxidase is employed having lower stability. Gold nanoparticle-graphene nanocomposites (AuNPs-G) based immune electrochemical sensor has yielded sensitivity of 1.6 pg/mL with excellent selectivity for avian influenza H7 against non-target H7N9, H7N2, H7N7, H1N1, H2N3, H3N2, H4N6, H6N8, H8N4, H9N2, H10N3, H11N9, H12N5, H13N5, H14N5, H15N9, H16N3, ILTV, infectious bronchitis virus (IBV) and NDV (Newcastle disease virus) (Huang et al. 2016). The Au/Ag-coated tilted fibre Bragg grating (TFBG)-surface plasmon resonance (SPR) optical biosensor has depicted nearly 100 times sensitivity with LOD as low as 25 pg/mL for detecting NDV (Luo et al. 2018). Similarly, other properties such as electric current changes, magnetic properties, voltammetry, conductometry, amperometry and optical properties have been explored in the sensor development. Silver NP as a reagent has been used by Teengam et al. (2017) for the development of colorimetric paper-based analytical device with a detection limit of 1.53 nM to detect DNA-associated viral infection like MERS-CoV. Similarly, electrochemiluminescence-based device based on voltammetric response has been developed using AuNPs in carbon electrode for the detection of Human CoV (HCoV) and MERS-CoV. Good linear response was obtained at the concentration range from 0.001 to 100 ng/mL and 0.01 to 10,000 ng/mL for MERS-CoV and HCoV, respectively. The assay time was 20 min with LOD of 0.4 and 1.0 pg/mL for HCoV and MERS-CoV, respectively (Layqah and Eissa 2019). Electrochemical impedance spectroscopy-based label-free electrochemical immunosensors have been developed using thermally decomposed reduced grapheme oxide (TrGO) nanosheets to diagnose influenza virus (H1N1) that depicted an LOD of 26 and 33 plaque-forming units (PFU) in phosphate-buffered saline and diluted saliva, respectively (Joshi et al. 2020). Similarly, electrochemical biosensor using functionalized TiO_2 nanotube has been found to be effective in rapid detection of corona virus within 30 s in a concentration range of 14–1400 nM (Vadlamani et al. 2020). A self-sacrificial biosensor has been developed using Fe_3O_4 magnetic NPs

(MNPs) for detecting avian influenza virus (H5N1) that works by electrochemical conversion of the MNPs to electroactive Prussian blue analogue generating detection signal. In this process strong acidic condition is generated through high-potential step due to the splitting of H_2O releasing Fe^{3+} from the MNPs followed by reduction of coexisting potassium ferricyanide $[K_3Fe(CN)_6]$ and Fe^{3+} to potassium ferrocyanide $[K_4Fe(CN)_6]$ and Fe^{2+}, respectively under low-potential step that react to produce Prussian blue analogue. The LOD achieved in this technique was 0.0022 hemagglutination units (Zhang et al. 2017b). A boron-doped diamond electrode electrochemically surface functionalized with 4-aminobenzoic acid self-assembled monolayer (SAM), along with polyclonal anti-M1 antibodies, has been used for the detection of influenza viruses. The sensor detected as low as 1 fg/mL or 5–10 viruses with a rapidity of test within 5 min (Nidzworski et al. 2017).

5.3.4 Nanotechnology in Sequencing Method

Nucleic acid sequencing is a common method for the detection of viruses and their strains. It is the first step to knowing any mutation or strain-specific sequences of viruses for developing aptamer-based diagnostic system. Earlier sequencing technique based on Sanger's method was usually used that was cumbersome, time-consuming and low-throughput system. With the advent of nanopore-based high throughput sequencing, the landscape of sequencing-based diagnostics has changed a lot. It is reliable, cheaper and faster. Here single DNA molecule is passed through protein nanopore chamber and the changes in electric current on passage of nucleic acid through nanopore are measured. The method has increased sensitivity and performs real-time sequencing even of longer RNA/DNA samples. Because of longer reads, portability and lower cost, it has been employed for virus detection in clinical samples. For instance, the third-generation nanopore sequencing using MinION technology has enabled rapid and efficient detection of porcine kobuvirus from the faeces of Belgian diarrheic piglets (Theuns et al. 2018). Detection of influenza A, Peste des petits ruminants (PPR), Bovine viral diarrhoea (BVD), Zika virus and Ebola virus has been achieved in various species using this state-of-the-art sequencing modality (Schuele et al. 2020; Xu et al. 2018; Torsson et al. 2020; King et al. 2021).

5.3.5 Nanomaterials in Nucleic Acidamplification-Based Diagnosis

Nanomaterials are also being incorporated in PCR-based diagnostics to enhance the sensitivity of detection (Manhas et al. 2021). For instance, human influenza A virus has been successfully diagnosed by using AuNPs-RT loop-mediated isothermal amplification (AuNPs-RT-LAMP) with 100% specificity and 98.6% sensitivity (Nikbakht et al. 2014). Gold NPs-based RT-PCR and real-time quantitative RT-PCR assay have also been developed for Japanese encephalitis virus (JEV). Similarly highly sensitive, as well as specific, nanomaterials-assisted PCR (nano-PCR) using graphene oxide (GO) and gold nanoparticles (AuNPs) has been employed to detect FMDV yielding sensitivity up to 1000 times than conventional methods (Kim et al. 2020a). A novel AuNPs biosensor has been developed based on the same principle that detected FMDV with sensitivity of 10 copy number RNA standards in RT-PCR while as low as 1 copy number RNA standard in real-time rRT-PCR (Hamdy et al. 2018). Application of using NPs like gold and magnetic nanoparticles, titanium oxide, quantum dots and carbon-based materials in PCR/qPCR/LAMP-based diagnostics has been established to be highly beneficial due to enhanced consistency, sensitivity, shorter thermal cycling time and accuracy which has also been demonstrated by Sang et al. (2015) using CdTe quantum dot (QD) in hotstart PCR (Sang et al. 2015).

5.3.6 Nanomaterials in Aptamer-Based Detection

Aptamers are target-specific 10–100 nucleotides long (nt/DNA or RNA) single-stranded artificial nucleotides with a remarkable target-specific binding ability that are used for the diagnosis of various virus-borne diseases. These aptamers are extensively tried for developing sensor-based diagnostic tools called aptasensors. Depending upon the detection mechanism and physical properties, various types of aptasensors have been used for disease diagnosis such as surface plasmon resonance (SPR) aptasensors, colorimetric aptasensors, chemiluminescence aptasensors, fluorescence aptasensors, surface-enhanced Raman scattering aptasensors, interferometry aptasensors, enzyme-linked aptamer assays (ELAA) and lateral flow based aptamer. They generate specific signals when recognizing the target and are cost-effective, simple to use and portable in nature. Various NPs have been used to develop such sensors. AuNPs-based SPR chips have been developed for the detection of H5N1 influenza virus. Aptamer-coated MB along with concanavalin A (ConA), glucose oxidase (GOx) and gold nanoparticles (AuNPs-GOx-ConA) complexes has been used to develop electrochemical impedance spectroscopy (EIS)-based sensor for the detection of H5N1 influenza virus with an LOD of 8×10^{-4} HAU (Ghanbari et al. 2017). Glucose is used for enhancing the ionic strength through the generation of gluconic acid and AuNPs help in proper recording of signal with lower noise in aptamer-based influenza virus (IV) detection (Kim et al. 2020b). Similarly, for the detection of red-spotted grouper nervous necrosis virus (RGNNV) in fishes, anaptamer-based lateral flow using AuNPs-DNA conjugates has been developed with an LOD of RGNNV-CP protein as low as 5 ng/mL and the test is completed within 5 min (Liu et al. 2020). Similarly, AuNP-based sensor using aptamer of conserved 3D gene of FMDV has been developed that showed no cross-reactivity with swine vesicular disease virus (SVDV) and yielded 94.5% efficiency

(Hamdy et al. 2018). Sensor using peptide nucleic acid (PNA) probes and AuNPs is effective to diagnose as low as 5–10 ng of viral RNA from biological samples indicating enormous sensitivity of the system (Joshi et al. 2013). Electrophoretic mobility shift assay (EMSA) and surface plasmon resonance (SPR) analysis using RNA aptamer have been successfully employed for the detection of Porcine circovirus type 2 (PCV2) (Yoon et al. 2010). Nowadays, instead of using single output system, aptamer-based assay has been made smarter by using dual electrode-based sensors. This helps in self-calibration of the result, hence further increasing the efficiency of diagnostics. In avian influenza viruses (AIVs), electrochemical aptasensor using tungsten rods surface-modified with a 3D nanostructured porous silica film (3DNRE) in both electrode has been developed where one electrode uses target-specific binding and other binds with control aptamer. This enhances the result output, stabilizes the signal and avoids the necessity of further processing such as purification and washing of sample as the control aptamer will help in countering various noises like change in electrolyte ionic strength, cell debris and pH (Lee et al. 2020). The aptamer used as control has the same nucleotide length of target but is made of sequences which does not match to subject or target or any other possible agent, hence acting in baseline correction. The application of epoxy-phenanthroline-Fe(III)-NH_2-ssDNA as probe layer was used to develop aptamer-based RNA detecting sensor for avian influenza which show linearity of detection in range of 3.0×10^3 to 3.0×10^5 copies/μL (Malecka et al. 2019). Further, various aptamer-based diagnostics have also been tried for Bovine herpes virus 1, FMD virus, prion disease, Bovine Parainfluenza, Pseudorabies virus, Porcine circo virus type 2, etc. where addition of NPs has shown many positive attributes in comparison to conventional techniques (Devi et al. 2021).

5.3.7 Nanoparticles in ELISA or Lateral Flow

It is an antigen-antibody interaction-based method where colour is developed due to the reaction of secondary antibody-conjugated enzyme and chromogenic substrate. Nowadays various lateral flow or ELISA-based virus diagnostic kits have been developed. In these kits, gold NPs are frequently conjugated with the antibody due to enhanced sensitivity and stability, lowering of detection limit, enhanced signal generation and cost effectiveness (Billingsley et al. 2017; Liu et al. 2020). For instance, the application of AuNPs in ELISA-based detection of respiratory syncytial virus (RSV) has increased the sensitivity and decreased the assay time. Gold NPs have been exploited as carriers of the signalling anti-RSV-horseradish peroxidase antibody to amplify the signal (Zhan et al. 2014). Similarly, to overcome the stability issue of horseradish peroxidase in conventional ELISA, the application of gold-platinum and silica dioxide nanospheres has shown encouraging results to eliminate the stability issue (Manhas et al. 2021). Detection of influenza A virus by ELISA using immunomagnetic beads and biotinylated silica NPs is another such example where the nano-conjugation has been exploited for amplification of the colorimetric signal. Without depicting any cross-reactivity with the porcine respiratory virus, the modality achieved LOD as low as 3×10^3 PFU/mL for H1N1 strain and 4×10^4 PFU/mL for the H3N2 strain while LOD of 1.2×10^7 PFU/mL was achieved for H1N1 using the conventional ELISA (Farre et al. 2022).

Various other techniques have also successfully incorporated NPs to detect viruses of interest; for instance, multi-branched surface-enhanced Raman scattering (SERS) along with AuNPs identified Porcine circovirus type 2 cap protein with LOD of 8×10^2 copies/mL and for influenza virus the LOD was 1 pg/μL (Lin et al. 2014). Similarly, microfluidics along with NPs have also been employed for the effective detection of viruses in very low concentration. Influenza virus H5N1 gene detection system was developed using a microfluidic chip integrated with reduced graphene oxide transistor that achieved LOD as low as 5 pM (Chan et al. 2017). Magnetic NPs are frequently used in microfluidic sensors for virus detection; several human as well as animal viruses such as Dengue virus, Ebola virus, Influenza virus and Zika virus have been detected by this modality employing diverse sensing systems like electrochemical, fluorescence, magnetoresistance, optomagnetic signalling, surface-enhanced Raman scattering, etc. (Wu et al. 2020). The NPs facilitated to enhance specificity and sensitivity, reduce assay time and cost, therefore improving the performance.

5.3.8 Nanoparticles in Microscopy Systems

Study of virosphere, i.e. virus and its surrounding atmosphere, and understanding their interaction, is a key requirement for the development of robust and effective diagnostics and control strategy. Electron microscopy (EM) has been employed to explore the aspect along with facilitating effective and reliable virus diagnostic. High-power EM with nanoscale descriptive potential has further shown enhancement of resolution capacity due to the advent of lanthanide-doped nanoparticles (Eu, Er, Ho, Tb, Sm, Dy, Nd, Tm and Yb) because of their cathodoluminescence property reducing the signal-to-noise ratio providing better quality. Furthermore nowadays, the resolution limit has been extended to infinity with the advent of super-resolution for microscopy which helps in obtaining more discrete spectra to elucidate the 2D and 3D structures, proficiently employing advanced image-processing systems. Various NPs applications like semiconductor quantum dots (QDots), polymer dots (PDots), carbon-based nanodots (CDots), upconversion nanocrystals (UCNPs), fluorescent nanodiamonds (FNDs) and nonfluorescent surface-enhanced Raman scattering (SERS) have enhanced the capacity of microscopy-based diagnosis and structural tracking to a single molecule resolution. Further application of scanning electron microscope (SEM) has also been acknowledged for accurate quantification of viruses in addition to diagnosis (Richert-Pöggeler et al. 2019). Along with transmission electron microscopy (TEM), it provides the advantage of detecting viruses from diverse kinds of biological materials like cell culture, fluid or tissues. Diagnostic electron microscopy has been successfully used in the detection of viruses like Vaccinia virus, TMV, rotavirus, Acanthamoeba polyphaga mimivirus, Culex theileriflavivirus, Yellow fever virus, alphavirus, etc. where the application of NPs with distinct colours can further enhance diagnostic versatility by exploiting bioimaging through multicolour electron microscopy in the future (Prigozhin et al. 2019).

In conclusion, the entire arsenal of NMs and the full potential of nanotechnology have been directed towards diverse aspects of disease management including diagnostics and viral diseases are not any exception. Myriads of engineered NMs possessing numerous novel attributes extend unique benefits in virus detection in terms of early detection, enhanced sensitivity and specificity, lowering of signal-to-noise ratio, reduced assay time and inputs, possibility of point-of-care diagnosis, etc., clearly providing the advantage over the conventional diagnostics. Although extensive effort has been extended for the diagnosis of viral diseases, oftentimes the major focus is human as target subject or the zoonotic diseases. More emphasis needs to be directed towards controlling the viral diseases of animals also from economics and public health perspective. Crucial challenges in terms of potential nanotoxicity, reproducibility of synthesis, rigorous regulatory issues for nanointerventions in biomedical applications, etc. are resulting in sluggish bench-to-beside transition of the nano-formulations, but the future prospect looks bright enough as the obstacles are getting cleared progressively through nanotechnology hightail.

References

Abad JM, Pariente F, Hernandez L, Lorenzo E (1998) A quartz crystal microbalance assay for detection of antibodies against the recombinant African swine fever virus attachment protein p12 in swine serum. Anal Chim Acta 368:183–189. https://doi.org/10.1016/S0003-2670(98)00205-0

Adegoke O, Kato T, Park EY (2016) An ultrasensitive alloyed near-infrared quinternary quantum dot-molecular beacon nanodiagnostic bioprobe for influenza virus RNA. Biosens Bioelectron 80:483–490. https://doi.org/10.1016/j.bios.2016.02.020

Adegoke O, Morita M, Kato T, Ito M, Suzuki T, Park EY (2017) Localized surface plasmon resonance-mediated fluorescence signals in plasmonic nanoparticle-quantum dot hybrids for ultrasensitive Zika virus RNA detection via hairpin hybridization assays. Biosens Bioelectron 94:513–522. https://doi.org/10.1016/j.bios.2017.03.046

Afsahi S, Lerner MB, Goldstein JM, Lee J, Tang X, Bagarozzi DA Jr, Pan D, Locascio L, Walker A, Barron F, Goldsmith BR (2018) Novel graphene-based biosensor for early detection of Zika virus infection. Biosens Bioelectron 100:85–88. https://doi.org/10.1016/j.bios.2017.08.051

Ahmed SR, Mogus J, Chand R, Nagy E, Neethirajan S (2018) Optoelectronic fowl adenovirus detection based on local electric field enhancement on graphene quantum dots and gold nanobundle hybrid. Biosensors & bioelectronics, 103, 45–53. https://doi.org/10.1016/j.bios.2017.12.028.

Ahmed SR, Neethirajan S (2018) Chiral MoS_2 quantum dots: dual mode detection approaches for avian influenza viruses. Glob Chall 2(4):1700071. https://doi.org/10.1002/gch2.201700071

Anik Ü, Tepeli Y, Sayhi M, Nsiri J, Diouani MF (2017) Towards the electrochemical diagnostic of influenza virus: development of a graphene-Au hybrid nanocomposite modified influenza virus biosensor based on neuraminidase activity. Analyst 143(1):150–156. https://doi.org/10.1039/c7an01537b

Ariffin EY, Tan LL, AbdKarim NH, YookHeng L (2018) Optical DNA biosensor based on square-planar ethyl piperidine substituted Nickel(II) Salphen complex for dengue virus detection. Sensors (Basel) 18(4):E1173. https://doi.org/10.3390/s18041173

Arima A, Tsutsui M, Harlisa IH, Yoshida T, Tanaka M, Yokota K, Tonomura W, Taniguchi M, Okochi M, Washio T, Kawai T (2018) Selective detections of single-viruses using solid-state nanopores. Sci Rep 8(1):16305. https://doi.org/10.1038/s41598-018-34665-4

Baca JT, Severns V, Lovato D, Branch DW, Larson RS (2015) Rapid detection of Ebola virus with a reagent-free, point-of-care biosensor. Sensors (Basel) 15(4):8605–8614. https://doi.org/10.3390/s150408605

Baeumner AJ, Schlesinger NA, Slutzki NS, Romano J, Lee EM, Montagna RA (2002) Biosensor for dengue virus detection: sensitive, rapid, and serotype specific. Anal Chem 74(6):1442–1448. https://doi.org/10.1021/ac015675e

Balcioglu M, Rana M, Hizir MS, Robertson NM, Haque K, Yigit MV (2017) Rapid visual screening and programmable subtype classification of Ebola virus biomarkers. Adv Healthc Mater 6(2):1600739. https://doi.org/10.1002/adhm.201600739

Bang J, Park H, Choi WI, Sung D, Lee JH, Lee KY, Kim S (2018) Sensitive detection of dengue virus NS1 by highly stable affibody-functionalized gold nanoparticles. New J Chem 42(15): 12607–12614. https://doi.org/10.1039/C8NJ02244E

Belza J, Opletalová A, Poláková K (2021) Carbon dots for virus detection and therapy. Mikrochim Acta 188(12):430. https://doi.org/10.1007/s00604-021-05076-6

Billingsley MM, Riley RS, Day ES (2017) Antibody-nanoparticle conjugates to enhance the sensitivity of ELISA-based detection methods. PLoS One 12(5):e0177592. https://doi.org/10.1371/journal.pone.0177592

Bosch I, de Puig H, Hiley M, Carré-Camps M, PerdomoCelis F et al (2017) Rapid antigen tests for dengue virus serotypes and Zika virus in patient serum. Sci Transl Med 9(409):eaan1589. https://doi.org/10.1126/scitranslmed.aan1589

Brangel P, Sobarzo A, Parolo C, Miller BS, Howes PD, Gelkop S, Lutwama JJ, Dye JM, McKendry RA, Lobel L, Stevens MM (2018) A serological point-of-care test for the detection of IgG antibodies against Ebola virus in human survivors. ACS Nano 12(1):63–73. https://doi.org/10.1021/acsnano.7b07021

Camara AR, Gouvêa PM, Dias AC, Braga AM, Dutra RF, de Araujo RE, Carvalho IC (2013) Dengue immunoassay with an LSPR fiber optic sensor. Opt Express 21(22):27023–27031. https://doi.org/10.1364/OE.21.027023

Cao YC, Jin R, Mirkin CA (2002) Nanoparticles with Raman spectroscopic fingerprints for DNA and RNA detection. Science 297(5586):1536–1540. https://doi.org/10.1126/science.297.5586.1536

Chan C, Shi J, Fan Y, Yang M (2017) A microfluidic flow-through chip integrated with reduced graphene oxide transistor for influenza virus gene detection. Sens Actuators B Chem 251:927–933. https://doi.org/10.1016/j.snb.2017.05.147

Chen SH, Chuang YC, Lu YC, Lin HC, Yang YL, Lin CS (2009) A method of layer-by-layer gold nanoparticle hybridization in a quartz crystal microbalance DNA sensing system used to detect dengue virus. Nanotechnology 20(21):215501. https://doi.org/10.1088/0957-4484/20/21/215501

Chen Y, Ren R, Pu H, Guo X, Chang J, Zhou G, Mao S, Kron M, Chen J (2017) Field effect transistor biosensor for rapid detection of Ebola antigen. Sci Rep 7(1):10974. https://doi.org/10.1038/s41598-017-11387-7

Chin SF, Lim LS, Pang SC, Sum MSH, Perera D (2017) Carbon nanoparticle modified screen printed carbon electrode as a disposable electrochemical immunosensor strip for the detection of Japanese encephalitis virus. Mikrochim Acta 184:491–497. https://doi.org/10.1007/s00604-016-2029-7

Chowdhury AD, Takemura K, Li TC, Suzuki T, Park EY (2019) Electrical pulse-induced electrochemical biosensor for hepatitis E virus detection. Nat Commun 10:3737. https://doi.org/10.1038/s41467-019-11644-5

Chu HW, Lai CS, Ko JY, Harroun SG, Chuang CI, Wang RYL, Unnikrishnan B, Huang CC (2019) Nanoparticle-based LDI-MS immunoassay for the multiple diagnosis of viral infections. ACS Sens 4(6):1543–1551. https://doi.org/10.1021/acssensors.9b00054

Critchley P, Dimmock NJ (2004) Binding of an influenza A virus to a neomembrane measured by surface plasmon resonance. Bioorg Med Chem 12(10):2773–2780. https://doi.org/10.1016/j.bmc.2004.02.042

Cui Y, Wang Z, Ma X, Liu J, Cui S (2014) A sensitive and specific nanoparticle-assisted PCR assay for rapid detection of porcine parvovirus. Lett Appl Microbiol 58(2):163–167. https://doi.org/10.1111/lam.12171

Deng J, Toh CS (2013) Impedimetric DNA biosensor based on a nanoporous alumina membrane for the detection of the specific oligonucleotide sequence of dengue virus. Sensors (Basel) 13(6):7774–7785. https://doi.org/10.3390/s130607774

Devi S, Sharma N, Ahmed T, Huma ZI, Kour S, Sahoo B, Singh AK, Macesic N, Lee SJ, Gupta MK (2021) Aptamer-based diagnostic and therapeutic approaches in animals: current potential and challenges. Saudi J Biol Sci 28(9):5081–5093. https://doi.org/10.1016/j.sjbs.2021.05.031

Draz MS, Lakshminaraasimulu NK, Krishnakumar S, Battalapalli D, Vasan A, Kanakasabapathy MK, Sreeram A, Kallakuri S, Thirumalaraju P, Li Y, Hua S, Yu XG, Kuritzkes DR, Shafiee H (2018) Motion-based immunological detection of Zika virus using Pt-nanomotors and a cellphone. ACS Nano 12(6):5709–5718. https://doi.org/10.1021/acsnano.8b01515

Ehtesabi H (2020) Application of carbon nanomaterials in human virus detection. J Sci Adv Mater Devices 5:436. https://doi.org/10.1016/j.jsamd.2020.09.005

Farre C, Viezzi S, Wright A, Robin P, Lejal N, Manzano M, Vidic J, Chaix C (2022) Specific and sensitive detection of Influenza A virus using a biotin-coated nanoparticle enhanced immunomagnetic assay. Anal Bioanal Chem 414:265–276. https://doi.org/10.1007/s00216-020-03081-x

Gao Y, Pallister J, Lapierre F, Crameri G, Wang LF, Zhu Y (2015) A rapid assay for Hendra virus IgG antibody detection and its titre estimation using magnetic nanoparticles and phycoerythrin. J Virol Methods 222:170–177. https://doi.org/10.1016/j.jviromet.2015.05.008

García AA, Franco LS, Pirez-Gomez MA, Pech-Pacheco JL, Mendez-Galvan JF, Machain-Williams C, TalaveraAguilar L, Espinosa-Carrillo JH, Duarte-Villaseñor MM, BeOrtiz C, Espinosa-de Los Monteros LE, Castillo-Pacheco A, Garcina-Rejon JE (2017) Feasibility study of an optical caustic plasmonic light scattering sensor for human serum anti-dengue protein E antibody detection. Diagnostics (Basel) 7(3):E47. https://doi.org/10.3390/diagnostics7030047

Geng X, Zhang F, Gao Q, Lei Y (2016) Sensitive impedimetric immunoassay of Japanese encephalitis virus based on enzyme biocatalyzed precipitation on a gold nanoparticle-modified screen-printed carbon electrode. Anal Sci 32(10):1105–1109. https://doi.org/10.2116/analsci.32.1105

Ghanbari K, Roushani M, Azadbakht A (2017) Ultra-sensitive aptasensor based on a GQD nanocomposite for detection of hepatitis C virus core antigen. Anal Biochem 534:64–69. https://doi.org/10.1016/j.ab.2017.07.016

Gholami H, Emami T, Golchinfar F, Madani R, Gheshlaghchaei SI (2016) Development of a Nano-ELISA system for the rapid and sensitive detection of H9N2 avian influenza. Arch Razi Inst 71(1):29–34

Gopinath SCB, Awazu K, Fujimaki M, Shimizu K, Shima T (2013) Observations of immuno-gold conjugates on influenza viruses using waveguide-mode sensors. PLoS One 8(7):e6912.1. https://doi.org/10.1371/journal.pone.0069121

Hamdy ME, Del Carlo M, Hussein HA, Salah TA, El-Deeb AH, Emara MM, Pezzoni G, Compagnone D (2018) Development of gold nanoparticles biosensor for ultrasensitive diagnosis of foot and mouth disease virus. J Nanobiotechnol 16(1):48. https://doi.org/10.1186/s12951-018-0374-x

Han JH, Lee D, Chew CHC, Kim T, Pak JJ (2016) A multi-virus detectable microfluidic electrochemical immunosensor for simultaneous detection of H1N1, H5N1, and H7N9 virus using

ZnO nanorods for sensitivity enhancement. Sens Actuators B Chem 228:36–42. https://doi.org/10.1016/j.snb.2015.07.068

Huang SH, Yang TC, Tsai MH, Tsai IS, Lu HC, Chuang PH, Wan L, Lin YJ, Lai CH, Lin CW (2008) Gold nanoparticle-based RT-PCR and real-time quantitative RT-PCR assays for detection of Japanese encephalitis virus. Nanotechnology 19(40):405101. https://doi.org/10.1088/0957-4484/19/40/405101

Huang MJ, Xie H, Wan Q, Zhang L, Ning Y, Zhang GJ (2013) Serotype-specific identification of Dengue virus by silicon nanowire array biosensor. J Nanosci Nanotechnol 13(6):3810–3817. https://doi.org/10.1166/jnn.2013.7220

Huang J, Xie Z, Xie Z, Luo S, Xie L, Huang L, Fan Q, Zhang Y, Wang S, Zeng T (2016) Silver nanoparticles coated graphene electrochemical sensor for the ultrasensitive analysis of avian influenza virus H7. Anal Chim Acta 913:121–127. https://doi.org/10.1016/j.aca.2016.01.050

Jahanshahi P, Zalnezhad E, Sekaran SD, Adikan FR (2014) Rapid immunoglobulin M-based dengue diagnostic test using surface plasmon resonance biosensor. Sci Rep 4:3851. https://doi.org/10.1038/srep03851

Jain B, Lambe U, Tewari A, Kadian SK, Prasad M (2018) Development of a rapid test for detection of foot-and-mouth disease virus specific antibodies using gold nanoparticles. Virusdisease 29(2):192–198. https://doi.org/10.1007/s13337-018-0450-8

Jiang Q, Chandar YJ, Cao S, Kharasch ED, Singamaneni S, Morrissey JJ (2017) Rapid, point-of-care, paper-based plasmonic biosensor for Zika virus diagnosis. Adv Biosyst 1:e1700096. https://doi.org/10.1002/adbi.201700096

Joseph NM, Ho KL, Tey BT, Tan CS, Shafee N, Tan WS (2016) Production of the virus-like particles of Nipah virus matrix protein in Pichiapastoris as diagnostic reagents. Biotechnol Prog 32(4):1038–1045. https://doi.org/10.1002/btpr.2279

Joshi VG, Chindera K, Singh AK, Sahoo AP, Dighe VD, Thakuria D, Tiwari AK, Kumar S (2013) Rapid label-free visual assay for the detection and quantification of viral RNA using peptide nucleic acid (PNA) and gold nanoparticles (AuNPs). Anal Chim Acta 795:1–7. https://doi.org/10.1016/j.aca.2013.06.037

Joshi SR, Sharma A, Kim GH, Jang J (2020) Low cost synthesis of reduced graphene oxide using biopolymer for influenza virus sensor. Mater Sci Eng C Mater Biol Appl 108:110465. https://doi.org/10.1016/j.msec.2019.110465

Kaushik A, Yndart A, Kumar S, Jayant RD, Vashist A, Brown AN, Li CZ, Nair M (2018) A sensitive electrochemical immunosensor for label-free detection of Zika-virus protein. Sci Rep 8(1):9700. https://doi.org/10.1038/s41598-018-28035-3

Kim HU, Min J, Park G, Shin D, Sung G, Kim T, Lee MH (2018) Electrochemical detection of airborne influenza virus using air sampling system. Aerosol Air Qual Res 18:2721–2727. https://doi.org/10.4209/aaqr.2018.06.0221

Kim JW, Kim M, Lee KK, Chung KH, Lee CS (2020a) Effects of graphene oxide-gold nanoparticles nanocomposite on highly sensitive foot-and-mouth disease virus detection. Nanomaterials (Basel, Switzerland) 10(10):1921. https://doi.org/10.3390/nano10101921

Kim SM, Kim J, Noh S, Sohn H, Lee T (2020b) Recent development of aptasensor for influenza virus detection. Biochip J 14(4):327–339. https://doi.org/10.1007/s13206-020-4401-2

King J, Pohlmann A, Dziadek K, Beer M, Wernike K (2021) Cattle connection: molecular epidemiology of BVDV outbreaks via rapid nanopore whole-genome sequencing of clinical samples. BMC Vet Res 17(1):242. https://doi.org/10.1186/s12917-021-02945-3

Kinnamon DS, Krishnan S, Brosler S, Sun E, Prasad S (2018) Screen printed graphene oxide textile biosensor for applications in inexpensive and wearable point-of-exposure detection of influenza for at-risk populations. J Electrochem Soc 165(8):B3084–B3090. https://doi.org/10.1149/2.0131808jes

Ko YJ, Choi KS, Nah JJ, Paton DJ, Oem JK, Wilsden G, Kang SY, Jo NI, Lee JH, Kim JH, Lee HW, Park JM (2005) Noninfectious virus-like particle antigen for detection of swine vesicular disease virus antibodies in pigs by enzyme-linked immunosorbent assay. Clin Diagn Lab Immunol 12(8):922–929. https://doi.org/10.1128/CDLI.12.8.922-929.2005

Kobayashi RKT, Dhakal S, Nakazato G (2021) Editorial: The use of nanoparticles in the diagnosis and therapy of infectious disease in animals. Front Vet Sci 8:829540. https://doi.org/10.3389/fvets.2021.829540

Kumbhat S, Sharma K, Gehlot R, Solanki A, Joshi V (2010) Surface plasmon resonance based immunosensor for serological diagnosis of dengue virus infection. J Pharm Biomed Anal 52(2): 255–259. https://doi.org/10.1016/j.jpba.2010.01.001

Lai HC, Chin SF, Pang SC, Henry Sum MS, Perera D (2017) Carbon nanoparticles based electrochemical biosensor strip for detection of Japanese encephalitis virus. J Nanomater 2017:3615707. https://doi.org/10.1155/2017/3615707

Layqah LA, Eissa S (2019) An electrochemical immunosensor for the corona virus associated with the Middle East respiratory syndrome using an array of gold nanoparticle-modified carbon electrodes. Mikrochim Acta 186(4):224. https://doi.org/10.1007/s00604-019-3345-5

Lee I, Kim SE, Lee J, Woo DH, Lee S, Pyo H, Song CS, Lee J (2020) A self-calibrating electrochemical aptasensing platform: correcting external interference errors for the reliable and stable detection of avian influenza viruses. Biosens Bioelectron 152:112010. https://doi.org/10.1016/j.bios.2020.112010

Li H, Rothberg L (2004) Colorimetric detection of DNA sequences based on electrostatic interactions with unmodified gold nanoparticles. Proc Natl Acad Sci U S A 101(39):14036–14039. https://doi.org/10.1073/pnas.0406115101

Li H, Huang J, Lv J, An H, Zhang X, Zhang Z, Fan C, Hu J (2005) Nanoparticle PCR: nanogold-assisted PCR with enhanced specificity. Angew Chem Int Ed 44:5100–5103. https://doi.org/10.1002/anie.200500403

Lim LS, Chin SF, Pang SC, Sum MSH, Perera D (2017) A novel silver nanoparticles-based sensing probe for the detection of Japanese encephalitis virus antigen. Sains Malays 46(12):2447–2454. https://doi.org/10.17576/jsm-2017-4612-21

Lin CH, Hung CH, Hsiao CY, Lin HC, Ko FH, Yang YS (2009) Poly-silicon nanowire field-effect transistor for ultrasensitive and label-free detection of pathogenic avian influenza DNA. Biosens Bioelectron 24(10):3019–3024. https://doi.org/10.1016/j.bios.2009.03.014

Lin YJ, Wu CY, Li T, Hsiao PW, Chang DK (2014) A rapid and sensitive early diagnosis of influenza virus subtype via surface enhanced Raman scattering. J Biosens Bioelectron 5:150. https://doi.org/10.4172/2155-6210.1000150

Lisi F, Falcaro P, Buso D, Hill AJ, Barr JA, Crameri G, Nguyen TL, Wang LF, Mulvaney P (2012) Rapid detection of Hendra virus using magnetic particles and quantum dots. Adv Healthc Mater 1(5):631–634. https://doi.org/10.1002/adhm.201200072

Liu HH, Cao X, Yang Y, Liu MG, Wang YF (2006) Array-based nano-amplification technique was applied in detection of hepatitis E virus. J Biochem Mol Biol 39(3):247–252

Liu J, Qin Q, Zhang X, Li C, Yu Y, Huang X, Mukama O, Zeng L, Wang S (2020) Development of a novel lateral flow biosensor combined with aptamer-based isolation: application for rapid detection of grouper nervous necrosis virus. Front Microbiol 11:886. https://doi.org/10.3389/fmicb.2020.00886

Luo Z, Li W, Lu D, Chen K, He Q, Han H, Zou M (2013) A SERS-based immunoassay for porcine circovirus type 2 using multi-branched gold nanoparticles. Microchim Acta 180(15–16): 1501–1507. https://doi.org/10.1007/s00604-013-1032-5

Luo B, Xu Y, Wu S, Zhao M, Jiang P, Shi S, Zhang Z, Wang Y, Wang L, Liu Y (2018) A novel immunosensor based on excessively tilted fiber grating coated with gold nanospheres improves the detection limit of Newcastle disease virus. Biosens Bioelectron 100:169–175. https://doi.org/10.1016/j.bios.2017.08.064

Ma X, Cui Y, Qiu Z, Zhang B, Cui S (2013) A nanoparticle-assisted PCR assay to improve the sensitivity for rapid detection and differentiation of wild-type pseudorabies virus and gene-deleted vaccine strains. J Virol Methods 193(2):374–378. https://doi.org/10.1016/j.jviromet.2013.07.018

Malecka K, Świętoń E, Verwilst P, Stachyra A, Sirko A, Dehaen W, Radecki J, Radecka H (2019) Ultrasensitive electrochemical genosensor for direct detection of specific RNA sequences derived from avian influenza viruses present in biological samples. Acta Biochim Pol 66(3): 299–304. https://doi.org/10.18388/abp.2019_2774

Manhas PK, Quintela IA, Wu VCH (2021) Enhanced detection of major pathogens and toxins in poultry and livestock with zoonotic risks using nanomaterials-based diagnostics. Front Vet Sci 8:673718. https://doi.org/10.3389/fvets.2021.673718

Marín MJ, Rashid A, Rejzek M, Fairhurst SA, Wharton SA, Martin SR, McCauley JW, Wileman T, Field RA, Russell DA (2013) Glyconanoparticles for the plasmonic detection and discrimination between human and avian influenza virus. Org Biomol Chem 11(41):7101–7107. https://doi.org/10.1039/c3ob41703d

Martínez-Paredes G, González-García MB, Costa-García A (2009) Genosensor for SARS virus detection based on gold nanostructured screen. Electroanalysis 21(3–5):379–385. https://doi.org/10.1002/elan.20080439

Minakshi P, Ghosh M, Brar B, Ranjan K, Patki HS, Kumar R (2020) Separation techniques with nanomaterials. In: Hussain CM (ed) Handbook of nanomaterials in analytical chemistry. Elsevier, Amsterdam, pp 99–158. https://doi.org/10.1016/b978-0-12-816699-4.00006-2

Misono TS, Kumar PKR (2005) Selection of RNA aptamers against human influenza virus hemagglutinin using surface plasmon resonance. Anal Biochem 342(2):312–317. https://doi.org/10.1016/j.ab.2005.04.013

Mitra A, Deutsch B, Ignatovich F, Dykes C, Novotny L (2010) Nano-optofluidic detection of single viruses and nanoparticles. ACS Nano 4(3):1305–1312. https://doi.org/10.1021/nn901889v

Navakul K, Warakulwit C, Yenchitsomanus PT, Panya A, Lieberzeit PA, Sangma C (2017) A novel method for dengue virus detection and antibody screening using a graphene-polymer based electrochemical biosensor. Nanomedicine (Lond) 13(2):549–557. https://doi.org/10.1016/j.nano.2016.08.009

Neng J, Harpster MH, Zhang H, Mecham JO, Wilson WC, Johnson PA (2010) A versatile SERS-based immunoassay for immunoglobulin detection using antigen-coated gold nanoparticles and malachite green-conjugated protein A/G. Biosens Bioelectron 26(3):1009–1015. https://doi.org/10.1016/j.bios.2010.08.015

Neng J, Harpster MH, Wilson WC, Johnson PA (2013) Surface enhanced Raman scattering (SERS) detection of multiple viral antigens using magnetic capture of SERS-active nanoparticles. Biosens Bioelectron 41:316–321. https://doi.org/10.1016/j.bios.2012.08.048

Nguyen BT, Peh AE, Chee CY, Fink K, Chow VT, Ng MM, Toh CS (2012) Electrochemical impedance spectroscopy characterization of nanoporous alumina dengue virus biosensor. Bioelectrochemistry 88:15–21. https://doi.org/10.1016/j.bioelechem.2012.04.006

Nidzworski D, Siuzdak K, Niedziałkowski P, Bogdanowicz R, Sobaszek M, Ryl J, Weiher P, Sawczak M, Wnuk E, Goddard WA III, Jaramillo-Botero A, Ossowski T (2017) A rapid-response ultrasensitive biosensor for influenza virus detection using antibody modified boron-doped diamond. Sci Rep 7(1):15707. https://doi.org/10.1038/s41598-017-15806-7

Nikbakht H, Gill P, Tabarraeic A, Niazia A (2014) Nanomolecular detection of human influenza virus type A using reverse transcription loop-mediated isothermal amplification assisted with rod-shaped gold nanoparticles. RSC Adv 4:13575. https://doi.org/10.1039/C3RA47398H

Nuzaihan MNM, Hashim U, MdArshad MK, Kasjoo SR, Rahman SF, Ruslinda AR, Fathil MF, Adzhri R, Shahimin MM (2016) Electrical detection of dengue virus (DENV) DNA oligomer using silicon nanowire biosensor with novel molecular gate control. Biosens Bioelectron 83: 106–114. https://doi.org/10.1016/j.bios.2016.04.033

Oh S, Kim J, Tran VT, Lee DK, Ahmed SR, Hong JC, Lee J, Park EY, Lee J (2018) Magnetic nanozyme-linked immunosorbent assay for ultrasensitive influenza A virus detection. ACS Appl Mater Interfaces 10(15):12534–12543. https://doi.org/10.1021/acsami.8b02735

Patolsky F, Zheng G, Hayden O, Lakadamyali M, Zhuang X, Lieber CM (2004) Electrical detection of single viruses. Proc Natl Acad Sci U S A 101(39):14017–14022. https://doi.org/10.1073/pnas.0406159101

Paul AM, Fan Z, Sinha SS, Shi Y, Le L, Bai F, Ray PC (2015) Bio-conjugated gold nanoparticle based SERS probe for ultrasensitive identification of mosquito-borne viruses using Raman fingerprinting. J Phys Chem C Nanomater Interfaces 119(41):23669–23775. https://doi.org/10.1021/acs.jpcc.5b07387

Peh AE, Li SF (2013) Dengue virus detection using impedance measured across nanoporous alumina membrane. Biosens Bioelectron 42:391–396. https://doi.org/10.1016/j.bios.2012.10.054

Perez JM, Simeone FJ, Saeki Y, Josephson L, Weissleder R (2003) Viral-induced self-assembly of magnetic nanoparticles allows the detection of viral particles in biological media. J Am Chem Soc 125(34):10192–10193. https://doi.org/10.1021/ja036409g

Petkovic K, Metcalfe G, Chen H, Gao Y, Best M, Lester D, Zhu Y (2016) Rapid detection of Hendra virus antibodies: an integrated device with nanoparticle assay and chaotic micromixing. Lab Chip 17(1):169–177. https://doi.org/10.1039/C6LC01263A

Prigozhin MB, Maurer PC, Courtis AM, Liu N, Wisser MD, Siefe C, Tian B, Chan E, Song G, Fischer S, Aloni S, Ogletree DF, Barnard ES, Joubert LM, Rao J, Alivisatos AP, Macfarlane RM, Cohen BE, Cui Y, Dionne JA et al (2019) Bright sub-20-nm cathodoluminescent nanoprobes for electron microscopy. Nat Nanotechnol 14(5):420–425. https://doi.org/10.1038/s41565-019-0395-0

Qin T, Wang J, Cui SJ (2021) Development of a nanoparticle-assisted PCR assay to distinguish canine coronaviruses I and II. J Vet Diagn Investig 33(1):104–107. https://doi.org/10.1177/1040638720974114

Rahman SA, Saadun R, Azmi NE, Ariffin N, Abdullah J, Yusof NA, Sidek H, Hajian R (2014) Label-free dengue detection utilizing PNA/DNA hybridization based on the aggregation process of unmodified gold nanoparticles. J Nanomater 2014:839286

Rai V, Hapuarachchi HC, Ng LC, Soh SH, Leo YS, Toh CS (2012) Ultrasensitive cDNA detection of dengue virus RNA using electrochemical nanoporous membrane-based biosensor. PLoS One 7(8):e42346. https://doi.org/10.1371/journal.pone.0042346

Rashid JIA, Yusof NA, Abdullah J, Hashim U, Hajian R (2015) A novel disposable biosensor based on SiNWs/AuNPs modified screen printed electrode for dengue virus DNA oligomer detection. IEEE Sensors J 15:4420–4427. https://doi.org/10.1109/JSEN.2015.2417911

Richert-Pöggeler KR, Franzke K, Hipp K, Kleespies RG (2019) Electron microscopy methods for virus diagnosis and high resolution analysis of viruses. Front Microbiol 9:3255. https://doi.org/10.3389/fmicb.2018.03255

Ricks KM, Shoemaker CJ, Dupuy LC Jr, Flusin O, Voorhees MA, Fulmer AN, Six CM, Badger CV, Schmaljohn CS, Schoepp RJ (2018) Virus-like particles and magnetic provide a flexible and sustainable multiplexed alphavirus immunodiagnostic platform. bioRxiv 2018:335315. https://doi.org/10.1101/335315

Roh C (2012) A facile inhibitor screening of SARS coronavirus N protein using nanoparticle-based RNA oligonucleotide. Int J Nanomedicine 7:2173–2179. https://doi.org/10.2147/IJN.S31379

Roh C, Jo SK (2011) Quantitative and sensitive detection of SARS coronavirus nucleocapsid protein using quantum dots-conjugated RNA aptamer on chip. J Chem Technol Biotechnol 86:1475–1479. https://doi.org/10.1002/jctb.2721

Sakurai A, Takayama K, Nomura N, Kajiwara N, Okamatsu M, Yamamoto N, Tamura T, Yamada J, Hashimoto M, Sakoda Y, Suda Y, Kobayashi Y, Kida H, Shibasaki F (2015) Fluorescent immunochromatography for rapid and sensitive typing of seasonal influenza viruses. PLoS One 10(2):e0116715. https://doi.org/10.1371/journal.pone.0116715

Sánchez-Purrà M, Carré-Camps M, de Puig H, Bosch I, Gehrke L, Hamad-Schifferli K (2017) Surface-enhanced Raman spectroscopy-based sandwich immunoassays for multiplexed detection of Zika and dengue viral biomarkers. ACS Infect Dis 3(10):767–776. https://doi.org/10.1021/acsinfecdis.7b00110

Sang FM, Yang Y, Yuan L, Ren J, Zhang Z (2015) Development of a high-throughput real time PCR based on a hot-start alternative for Pfu mediated by quantum dots. J Nanoscale 7(38):15852–15862

Schuele L, Cassidy H, Lizarazo E, Strutzberg-Minder K, Schuetze S, Loebert S, Lambrecht C, Harlizius J, Friedrich AW, Peter S, Niesters H, Rossen J, Couto N (2020) Assessment of viral

targeted sequence capture using nanopore sequencing directly from clinical samples. Viruses 12(12):1358. https://doi.org/10.3390/v12121358
Sepunaru L, Plowman BJ, Sokolov SV, Young NP, Compton RG (2016) Rapid electrochemical detection of single influenza viruses tagged with silver nanoparticles. Chem Sci (Camb) 7(6): 3892–3899. https://doi.org/10.1039/C6SC00412A
Shafagati N, Lundberg L, Baer A, Patanarut A, Fite K, Lepene B, Kehn-Hall K (2015) The use of Nanotrap particles in the enhanced detection of Rift Valley fever virus nucleoprotein. PLoS One 10(5):e0128215. https://doi.org/10.1371/journal.pone.0128215
Shelby T, Banerjee T, Zegar I, Santra S (2017) Highly sensitive, engineered magnetic nanosensors to investigate the ambiguous activity of Zika virus and binding receptors. Sci Rep 7(1):7377. https://doi.org/10.1038/s41598-017-07620-y
Shen C, Yang W, Ji Q, Maki H, Dong A, Zhang Z (2009) NanoPCR observation: different levels of DNA replication fidelity in nanoparticle-enhanced polymerase chain reactions. Nanotechnology 20(45):455103. https://doi.org/10.1088/0957-4484/20/45/455103
Shen F, Wang J, Xu Z, Wu Y, Chen Q, Li X, Jie X, Li L, Yao M, Guo X, Zhu T (2012) Rapid flu diagnosis using silicon nanowire sensor. Nano Lett 12(7):3722–3730. https://doi.org/10.1021/nl301516z
Singh S, Dhawan A, Karhana S, Bhat M, Dinda AK (2020) Quantum dots: an emerging tool for point-of-care testing. Micromachines 11(12):1058. https://doi.org/10.3390/mi11121058
Singhal C, Pundir CS, Narang J (2017) A genosensor for detection of consensus DNA sequence of Dengue virus using ZnO/Pt-Pd nanocomposites. Biosens Bioelectron 97:75–82. https://doi.org/10.1016/j.bios.2017.05.047
Singhal C, Khanuja M, Chaudhary N, Pundir CS, Narang J (2018) Detection of chikungunya virus DNA using two-dimensional MoS_2 nanosheets based disposable biosensor. Sci Rep 8(1):7734. https://doi.org/10.1038/s41598-018-25824-8
Sivashanmugan K, Liao JD, You JW, Wu CL (2013) Focused-ionbeam-fabricated Au/Ag multi-layered nanorod array as SERS-active substrate for virus strain detection. Sens Actuators B Chem 181:361–367. https://doi.org/10.1016/j.snb.2013.01.035
Suda Y, Nagatomo M, Yokoyama R, Ohzono M, Aoyama K, Zhang X, Nakajima K, Murakami N, Shinoda T, Hirota T, Yanagihara S, Nishi JI (2015) Highly sensitive detection of influenza virus in saliva by real-time PCR method using sugar chain immobilized gold nanoparticles; application to clinical studies. Biotechnol Rep (Amst) 7:64–71. https://doi.org/10.1016/j.btre.2015.05.004
Sun J, Lei X, Wang W, Liu Y, Liang P, Bao H, Wang Q, Guo Y, Yang J, Yan Z (2013) Development and evaluation of a paramagnetic nanoparticle based immunochromatographic strip for specific detection of 2009 H1N1 influenza virus. J Nanosci Nanotechnol 13:1684–1690. https://doi.org/10.1166/jnn.2013.7127
Teengam P, Siangproh W, Tuantranont A, Vilaivan T, Chailapakul O, Henry CS (2017) Multiplex paper-based colorimetric DNA sensor using pyrrolidinyl peptide nucleic acid-induced AgNPs aggregation for detecting MERS-CoV, MTB, and HPV oligonucleotides. Anal Chem 89(10): 5428–5435. https://doi.org/10.1021/acs.analchem.7b00255
Theuns S, Vanmechelen B, Bernaert Q, Deboutte W, Vandenhole M, Beller L, Matthijnssens J, Maes P, Nauwynck HJ (2018) Nanopore sequencing as a revolutionary diagnostic tool for porcine viral enteric disease complexes identifies porcine kobuvirus as an important enteric virus. Sci Rep 8(1):9830. https://doi.org/10.1038/s41598-018-28180-9
Torsson E, Kgotlele T, Misinzo G, Johansson Wensman J, Berg M, Karlsson Lindsjö O (2020) Field-adapted full genome sequencing of Peste-Des-Petits-ruminants virus using nanopore sequencing. Front Vet Sci 7:542724. https://doi.org/10.3389/fvets.2020.542724
Tran QH, Nguyen THH, Mai AT, Nguyen TT, Vu QK, Phan TN (2012) Development of electrochemical immunosensors based on different serum antibody immobilization methods for detection of Japanese encephalitis virus. Adv Nat Sci Nanosci Nanotechnol 3:3015012. https://doi.org/10.1088/2043-6262/3/1/015012

Tseng YT, Wang CH, Chang CP, Lee GB (2016) Integrated microfluidic system for rapid detection of influenza H1N1 virus using a sandwich-based aptamer assay. Biosens Bioelectron 82:105–111. https://doi.org/10.1016/j.bios.2016.03.073

Vadlamani BS, Uppal T, Verma SC, Misra M (2020) Functionalized TiO_2 nanotube-based electrochemical biosensor for rapid detection of SARS-CoV-2. Sensors (Basel, Switzerland) 20(20): 5871. https://doi.org/10.3390/s20205871

Valdivia-Granda W, Keating CD, Kann M, Beresford R, Kelley SO (2005) Detection of encephalic and hemorrhagic viruses: integration of micro-and nano-fabrication with computational tools. In: Proceedings of the international conference on MEMS, NANO and smart systems, ICMENS, pp 408–414. [1540865]

Vinayagam S, Rajaiah P, Mukherjee A, Natarajan C (2018) DNA triangular silver nanoparticles nanoprobe for the detection of dengue virus distinguishing serotype. Spectrochim Acta A Mol Biomol Spectrosc 202:346–351. https://doi.org/10.1016/j.saa.2018.05.047

Wang G, Xie P, Xiao C, Yuan P, Su X (2010) Magnetic fluorescent composite nanoparticles for the fluoroimmunoassays of Newcastle disease virus and avian virus arthritis virus. J Fluoresc 20: 499–506. https://doi.org/10.1007/s10895-009-0573-9

Wang X, Chen L, Su X, Ai S (2013) Electrochemical immunosensor with graphene quantum dots and apoferritin-encapsulated Cu nanoparticles double-assisted signal amplification for detection of avian leukosis virus subgroup. J Biosens Bioelectron 47:171–177. https://doi.org/10.1016/j.bios.2013.03.021

Wang J, Cheng Y, Zhang M, Zhao H, Lin P, Yi L, Tong M, Cheng S (2015) Development of a nanoparticle-assisted PCR (nanoPCR) assay for detection of mink enteritis virus (MEV) and genetic characterization of the NS1 gene in four Chinese MEV strains. BMC Vet Res 11:1. https://doi.org/10.1186/s12917-014-0312-6

Wang Y, Zhu G, Qi W, Li Y, Song Y (2016) A versatile quantitation platform based on platinum nanoparticles incorporated volumetric bar-chart chip for highly sensitive assays. Biosens Bioelectron 85:777–784. https://doi.org/10.1016/j.bios.2016.05.090

Wanzhe Y, Jianuan L, Peng L, Jiguo S, Ligong C, Juxiang L (2016) Development of a nanoparticle-assisted PCR assay for detection of duck tembusu virus. Lett Appl Microbiol 62(1): 63–67. https://doi.org/10.1111/lam.12509

Wiriyachaiporn N, Sirikett H, Dharakul T (2013) Rapid influenza A antigen detection using carbon nanostrings as label for lateral flow immunochromatographic assay. In: Proceedings of the IEEE 13th international conference on nanotechnology (IEEE-NANO), Beijing, China, August 5–8

Wu K, Saha R, Su D, Krishna VD, Liu J, Cheeran MC-J, Wang JP (2020) Magnetic-nanosensor-based virus and pathogen detection strategies before and during COVID-19. ACS Appl Nano Mater 3:9560–9580. https://doi.org/10.1021/acsanm.0c02048

Xu L, Wang R, Kelso LC, Ying Y, Li Y (2016) A target-responsive and size-dependent hydrogel aptasensor embedded with QD fluorescent reporters for rapid detection of avian influenza virus H5N1. Sens Actuators B Chem 234:98–108. https://doi.org/10.1016/j.snb.2016.04.156

Xu Y, Lewandowski K, Lumley S, Pullan S, Vipond R, Carroll M, Foster D, Matthews PC, Peto T, Crook D (2018) Detection of viral pathogens with multiplex nanopore MinION sequencing: be careful with cross-talk. Front Microbiol 9:2225. https://doi.org/10.3389/fmicb.2018.02225

Xu LD, Zhu J, Ding SN (2021) Immunoassay of SARS-CoV-2 nucleocapsid proteins using novel red emission-enhanced carbon dot-based silica spheres. Analyst 146(16):5055–5060. https://doi.org/10.1039/d1an01010g

Yang F, Xiao Y, Chen B, Wang L, Liu F, Yao H, Wu N, Wu H (2020a) Development of a colloidal gold-based immunochromatographic strip test using two monoclonal antibodies to detect H7N9 avian influenza virus. Virus Genes 56(3):396–400. https://doi.org/10.1007/s11262-020-01742-8

Yang F, Li Y, Jin X, Xu Q, Cheng F, Wang X (2020b) Immunosensor-based rapid quantitative detection of Newcastle disease virus antibodies using innovative gold immunochromatographic assay. J Appl Microbiol 129:1751–1757. https://doi.org/10.1111/jam.14688

Yanik AA, Huang M, Kamohara O, Artar A, Geisbert TW, Connor JH, Altug H (2010) An optofluidic nanoplasmonic biosensor for direct detection of live viruses from biological media. Nano Lett 10(12):4962–4969. https://doi.org/10.1021/nl103025u

Ye WW, Tsang MK, Liu X, Yang M, Hao J (2014) Upconversion luminescence resonance energy transfer (LRET)-based biosensor for rapid and ultrasensitive detection of avian influenza virus H7 subtype. Small 10(12):2390–2397. https://doi.org/10.1002/smll.201303766

Yen CW, de Puig H, Tam JO, Gómez-Márquez J, Bosch I, Hamad-Schifferli K, Gehrke L (2015) Multicolored silver nanoparticles for multiplexed disease diagnostics: distinguishing dengue, yellow fever, and Ebola viruses. Lab Chip 15(7):1638–1641. https://doi.org/10.1039/c5lc00055f

Yeo SJ, Bao DT, Seo GE, Bui CT, Kim DTH, Anh NTV, Tien TTT, Linh NTP, Sohn HJ, Chong CK, Shin HJ, Park H (2017) Improvement of a rapid diagnostic application of monoclonal antibodies against avian influenza H7 subtype virus using Europium nanoparticles. Sci Rep 7(1):7933. https://doi.org/10.1038/s41598-017-08328-9

Yoon S, Lee G, Han D, Song JY, Kang KS, Lee YS (2010) Neutralization of infectivity of porcine circovirus type 2 (PCV2) by capsid-binding 2′F-RNA aptamers. Antivir Res 88(1):19–24. https://doi.org/10.1016/j.antiviral.2010.06.012

Yuan R, Zhang L, Li Q, Chai Y, Cao S (2005) A label-free amperometric immunosenor based on multi-layer assembly of polymerized o-phenylenediamine and gold nanoparticles for determination of Japanese B encephalitis vaccine. Anal Chim Acta 531(1):1–5. https://doi.org/10.1016/j.aca.2004.10.072

Yuan W, Li Y, Wang J, Wang J, Sun J (2016) A nanoparticles assisted PCR assay for the detection of encephalomyocarditis virus. Vet Arhiv 86(1):1–8

Zaher MR, Ahmed HA, Hamada KEZ, Tammam RH (2018) Colorimetric detection of unamplified Rift Valley fever virus genetic material using unmodified gold nanoparticles. Appl Biochem Biotechnol 184(3):898–908. https://doi.org/10.1007/s12010-017-2592-3

Zaytseva NV, Montagna RA, Baeumner AJ (2005) Microfluidic biosensor for the serotype-specific detection of dengue virus RNA. Anal Chem 77(23):7520–7527. https://doi.org/10.1021/ac0509206

Zhan L, Wu WB, Yang XX, Huang CZ (2014) Gold nanoparticle-based enhanced ELISA for respiratory syncytial virus. New J Chem 38(7):2935–2940. https://doi.org/10.1039/C4NJ00253A

Zhang L, Yuan R, Chai Y, Chen S, Wang N, Zhu Q (2006) Layerby-layer self-assembly of films of nano-Au and Co(bpy)33+ for the determination of Japanese B encephalitis vaccine. Biochem Eng J 28(3):231–236. https://doi.org/10.1016/j.bej.2005.11.014

Zhang GJ, Zhang L, Huang MJ, Luo ZHH, Tay GKI, Lim EJA, Kang TG, Chen Y (2010a) Silicon nanowire biosensor for highly sensitive and rapid detection of dengue virus. Sens Actuators B Chem 146(1):138–144. https://doi.org/10.1016/j.snb.2010.02.021

Zhang H, Li W, Sheng Z, Han H, He Q (2010b) Ultrasensitive detection of porcine circovirus type 2 using gold(III) enhanced chemiluminescence immunoassay. Analyst 135(7):1680–1685. https://doi.org/10.1039/c0an00025f

Zhang H, Harpster MH, Park HJ, Johnson PA, Wilson WC (2011) Surface-enhanced Raman scattering detection of DNA derived from the West Nile virus genome using magnetic capture of Raman-active gold nanoparticles. Anal Chem 83(1):254–260. https://doi.org/10.1021/ac1023843

Zhang L, Tong S, Zhou J, Bao G (2016) Accurate quantification of disease markers in human serum using iron oxide nanoparticle linked immunosorbent assay. Theranostics 6(9):1353–1361. https://doi.org/10.7150/thno.16093

Zhang B, Pinsky BA, Ananta JS, Zhao S, Arulkumar S, Wan H, Sahoo MK, Abeynayake J, Waggoner JJ, Hopes C, Tang M, Dai H (2017a) Diagnosis of Zika virus infection on a nanotechnology platform. Nat Med 23(5):548–550. https://doi.org/10.1038/nm.4302

Zhang Q, Li L, Qiao Z, Lei C, Fu Y, Xie Q, Yao S, Li Y, Ying Y (2017b) Electrochemical conversion of Fe_3O_4 magnetic nanoparticles to electroactive Prussian blue analogues for self-sacrificial label biosensing of avian influenza virus H5N1. Anal Chem 89(22):12145–12151. https://doi.org/10.1021/acs.analchem.7b02784

Zhao Z, Cui H, Song W, Ru X, Zhou W, Yu X (2020) A simple magnetic nanoparticles-based viral RNA extraction method for efficient detection of SARS-CoV-2. bioRxiv. 2020.02.22.961268. https://doi.org/10.1101/2020.02.22.961268

Zheng L, Wei J, Lv X, Bi Y, Wu P, Zhang Z, Wang P, Liu R, Jiang J, Cong H, Liang J, Chen W, Cao H, Liu W, Gao GF, Du Y, Jiang X, Li X (2017) Detection and differentiation of influenza viruses with glycan-functionalized gold nanoparticles. Biosens Bioelectron 91:46–52. https://doi.org/10.1016/j.bios.2016.12.037

Zhou CH, Zhao JY, Pang DW, Zhang ZL (2014) Enzyme-induced metallization as a signal amplification strategy for highly sensitive colorimetric detection of avian influenza virus particles. Anal Chem 86(5):2752–2759. https://doi.org/10.1021/ac404177c

Zou X, Huang H, Gao Y, Su X (2012) Detection of avian influenza virus based on magnetic silica nanoparticles resonance light scattering system. Analyst 137(3):648–653. https://doi.org/10.1039/c1an16041a

Zvirbliene A, Kucinskaite-Kodze I, Razanskiene A, PetraityteBurneikiene R, Klempa B, Ulrich RG, Gedvilaite A (2014) The use of chimeric virus-like particles harbouring a segment of hantavirus Gc glycoprotein to generate a broadly-reactive hantavirus-specific monoclonal antibody. Viruses 6(2):640–660. https://doi.org/10.3390/v6020640

Chapter 6
Nanotechnology in Protozoan Livestock Disease Diagnosis

Rajesh Kumar, Alonkrita Chowdhury, Dodla Mamatha, Manoj Kumar Rose, and Mayukh Ghosh

Abstract Livestock plays a major role in increasing food security, rural livelihood, and developing countries' economies. It also significantly impacts the "One Health" perspective. However, livestock faces severe threats due to various microorganisms such as bacteria, viruses, and protozoan pathogens. Babesiosis, theileriosis, anaplasmosis, trypanosomosis, coccidiosis, leucocytozoonosis, toxoplasmosis, sarcocystis, trichomonosis, etc. are some of the examples of protozoan parasitic diseases of animals. Diagnosis of these diseases through the conventional methods such as microscopy, serological tests, conventional PCR, and ELISA is time-consuming, resource-intensive, and often results in low sensitivity and specificity. Nanotechnology proposes various ways in overcoming these challenges. It is very helpful in diagnosing diseases rapidly, cost-effectively and also with enhanced sensitivity. Various nanoparticles (NPs) have already been employed in the development of sensitive biosensors and bioimaging systems for animal-side diagnosis. The prominent sensor types include but not limited to lateral flow biosensors, electrochemical biosensors, genosensors, aptasensors, and various immunosensors which have lower detection limit compared to conventional methods. PCR-based assays using NPs increase sensitivity of detection compared to conventional PCR. Similarly, the nano-ELISA showed more sensitivity and specificity as compared to the conventional ELISA. The application of NPs in diagnostic methods such as immune-based assays and nucleic acid amplification-based assays are more advantageous in terms of cost, speed, and sensitivity. Here the current chapter discusses the application of NPs in

R. Kumar · M. K. Rose
Department of Veterinary Physiology and Biochemistry, COVS, Lala Lajpat Rai University of Veterinary and Animal Sciences, Hisar, Haryana, India

A. Chowdhury · D. Mamatha
Department of Plant Biotechnology, Institute of Agricultural Sciences, RGSC, Banaras Hindu University, Varanasi, U.P., India

M. Ghosh (✉)
Department of Veterinary Physiology and Biochemistry, RGSC, Banaras Hindu University, Mirzapur, U.P., India

LLR University of Veterinary and Animal, ICAR-National Research Centre on Equines, *Nanotechnology Theranostics in Livestock Diseases and Management*, Livestock Diseases and Management,
https://doi.org/10.1007/978-981-16-1610-5_6

various state-of-the art modalities for the diagnosis of protozoan diseases of livestock along with their prominent recompense.

Keywords Protozoan diseases · Nano-PCR · Genosensors · Electrochemical sensors · Nano-ELISA

6.1 Introduction

Protozoan pathogens are one of the important groups of etiological agents affecting various livestock species, pets as well as the poultry birds. The most important ones include Bovine babesiosis of cattle inflicted by *Babesia bovis* and *Babesia bigemina*; Ovine babesiosis of sheep and goats by *Babesia ovis* and *Babesia motasi*; Tropical theileriosis and East coast fever of cattle caused by *Theileria annulata* and *Theileria parva*, respectively; Ovine theileriosis by *Theileria lestoquardi* in sheep and goats; Equine piroplasmosis by *Theileria equi* and *Babesia caballi* in horses, donkeys, and mules; Trypanosomiasis by *Trypanosoma brucei*, *T. congolese*, and *T. vivax* not only affecting the cattle but also pigs, camels, goats, and sheep; Neosporosis in cattle by *Neospora caninum*; Coccidiosis by *Eimeria* spp. affecting the poultry birds along with minor infection in cattle, sheep, goat, horse, and rabbit; Cryptosporidiosis in neonatal farm ruminants by *Cryptosporidium* sp.; Sarcocystosis or Sarcocystis in farm animals; Trichomoniasis in cattle and cat by *Tritrichomonas foetus* while in chicken and wild birds by *Trichomonas gallinae*; Amoebiasis in bovines, horses, and dogs by *Acanthamoeba* spp. while in horses and sheep by *Balamuthia mandrillaris*; *Besnoitia besnoiti* infection in cattle. Some other zoonotic protozoan diseases include Toxoplasmosis caused by *Toxoplasma gondii* (cats act as reservoir host), Cryptosporidiosis by *Cryptosporidium parvum* and *C. hominis* (calves act as reservoir host), Visceral leishmaniasis by *Leishmania donovani*, *L. infantum* in the Old World, whereas by *L. chagasi* in the New World (dogs and wild canines act as reservoir host), etc. (Schnittger and Florin-Christensen 2018). Early and precise diagnosis of these protozoan diseases can significantly reduce the morbidity and mortality, thus optimize the production output from the livestock. Conventional diagnosis of these diseases includes blood smear preparation and microscopic examination for the hemo-protozoan diseases; immunodiagnostic assays such as complement fixation test (CFT), immunodiffusion (ID), indirect hemagglutination (IHA), latexagglutination, indirect fluorescent antibody test (IFAT), radio-immunoassay (RIA) and the enzyme-linked immune-sorbent assay (ELISA), etc. and nucleic acid-based diagnosis by conventional polymerase chain reaction (PCR) as well as its variants like real-time PCR (qPCR), nested PCR, hot-start PCR, loop-mediated isothermal amplification (LAMP)-PCR, etc. The most commonly employed blood smear examination often fails to detect sub-clinical infections and the infected animals may serve as carrier of the disease predisposing the entire herd. Conventional immunodiagnostics are usually very specific but often require longer detection time. The cost-intensiveness and sensitivity also often pose limitations. Further, the animals once infected with certain protozoan infection earlier but does not carry live infection currently can also give positive reaction in immune-diagnosis due to pre-existing circulating antibodies. Nucleic acid-based diagnostics are usually very sensitive but require sophisticated instrumentation which may not be available to the terminal livestock owners. Therefore, the quest for specific as well as sensitive, rapid, cost-effective, and user-friendly diagnostics suitable for field application still remains as the need of the hour to control the protozoan diseases. This is not only helpful for the animals but also to secure "one health" perspective encompassing the human beings as several protozoan diseases are zoonotic in nature and animals act as reservoir hosts. Nanotechnology has achieved significant penetration in biomedical science recently with several successful interventions in nano-diagnostics, particularly for designing biosensors and point-of-care devices. The diagnostic modalities usually offer precise diagnosis with adequate sensitivity, require less time and resources, most importantly the point-of-care detection which render these nano-diagnostics suitable for field-level diagnosis, thus removing majority of the limitations of conventional diagnostics (Prasad et al. 2021a; Minakshi et al. 2020a). A step ahead, nanotechnology may revolutionize the field of biomedical imaging with several nanoparticles-based targeting and contrast agents enhancing the capacity of different diagnostic imaging modalities. For instance, improved image generation can be achieved in positron-emission tomography (PET) and magnetic resonance imaging (MRI) instrumentation by employing radio-labeled iron oxide nanoparticles (IONPs). Applications of gadolinium-based contrast agents in MRI enhance the resolution, use of super-paramagnetic iron oxide NPs enhance ultrastructural penetration, soft tissue contrast; while in computed tomography (CT), use of iodine-based micelles and liposome, barium NPs, AuNPs, BiNPs has found to be beneficial to enhance penetration, contrast, and resolution (Baetke et al. 2015). Similarly, application of lanthanide-doped (Eu, Er, Ho, Tb, Sm, Dy, Nd, Tm, and Yb) nanomaterials as cathodoluminescence agent has empowered the high-resolution electron microscopy imagining and other microscopic techniques (Prasad et al. 2021b). So, in near future, nanotechnology may play a significant role to upgrade the capacity of the conventional diagnostics as well as to propel the arena of state-of-the art diagnostics imaging modalities.

6.2 Application of Nanomaterials in Disease Diagnosis

6.2.1 Nanomaterials in Immunodiagnostic Assays

Nanomaterials have been found to possess beneficial impacts on disease diagnosis through the advancement of conventional diagnostic modalities including ELISA, PCR, qPCR-based assays. NP-aided immune assay facilitates label-free detection of biomolecules. For instance, the employment of AuNPs and AgNPs in immune-based diagnostic kit development provides efficient antigen/antibody-based diagnosis of the human T-lymphotropic virus, hepatitis B virus, and H1N1 virus. Lateral flow, Raman scattering, fluorescence-based assays, microfluidic, immunoaffinity assay, and conventional ELISA are some other immune-based diagnostic platforms where nanomaterials are being exploited (Lambe et al. 2016). These modalities coupled with diverse nanomaterials facilitate in diagnosis of various kinds of diseases along with respective pathogens such as mycotoxin assay by MNP, AuNP for the diagnosis of HIV-1, enterobacteriaceae, Hepatitis C, and TB. Similarly, the detection of influenza viruses (H1N1, H5N1, H7N9) can be facilitated by the exploitation of AuNPs, ZnONPs, AgNPs, and other NPs in immune-chromatographic assay-based rapid diagnostic kits, AuNPs immunosensor, surface plasmon resonance, electrochemical biosensor, and so on. The application of NPs in conventional ELISA is advantageous in terms of stability, cost, time, and diagnosis efficiency. AuNP functionalized ELISA-based immune sensor enables improved sensitivity in the diagnosis of influenza (Dalal et al. 2020). Owing to the peroxidase-like activity along with enzyme mimicking properties of hemin and MnO_2, hemin-MnO_2 nanocomposite-based ELISA assay has been constructed for the determination of human immunoglobulin G. It is way more advantageous in terms of cost, stability, sensitivity, and specificity (Linyan et al. 2017). Similarly, NPs aided diagnostic assays, including ELISA, PCR, and immuno-PCR offer enhanced sensitivity, simplicity, and specificity as well. Regarding ELISA and immuno-PCR assays, NPs facilitate the loading of multiple biomolecules on its surface, stability as well as lowering the diagnostic reaction time. For instance, the incorporation of gold nanospheres and nanorods in the conventional ELISA has improved the detection sensitivity of osteopontin in saliva to 0.03 ng/mL and 0.02 ng/mL, respectively, against 0.14 ng/mL by conventional ELISA for the diagnosis of oral cancer (Chakraborty et al. 2018).

6.2.2 *Nanomaterials in Nucleic Acid Amplification-Based Diagnostic Assays*

NPs also possess a massive role in the nucleic acid-based molecular diagnostic field. For instance, the detection of nucleic acid or protein can be achieved by enhancing light absorption through the accumulation of AuNPs (Mieszawska et al. 2013). Application of NPs in DNA- and RNA-based diagnostic assay includes isolation, purification, and confirmation of nucleic acid in the target samples. Magnetic nanoparticles (MNPs) have been frequently employed for DNA isolation to facilitate enhanced diagnostics. MNPs assist in the separation of target nucleic acid from complex mixtures with ample rapidity and in a cost-effective manner along with chances of lesser organic contaminants during target capturing. MNPs provide magnetic force-based aggregation and dispersion, affinity toward specific targets by employing suitable functionalization along with a high surface-to-volume ratio facilitating high-quality purification and capturing of the target(s) (Nguyen et al. 2022). MNPs can be used for the simultaneous isolation of DNA and RNA from cells which can be further processed for direct detection as performed by gold-plated

MNP networks modified with probe DNA (Tavallaie et al. 2018). The efficiency of nucleic acid visualization and extraction by using MNPs has been made rapid enough now a day that the entire process can be completed within 15 min even without requiring any electric instrumentation (Kang et al. 2021). Similarly, specific target RNAs have also been extracted and documented efficiently by using MNPs. The application of Poly-NH2-MNPs not only facilitated the isolation of RNA from nasopharyngeal swab samples at a very low cost but also the captured RNA was found to be compatible with qPCR-based assay (Chacón-Torres et al. 2020). Application of other nanomaterials such as nanoparticle-graphene oxide complex and amino-modified magnetic nanoparticles (AMNPs), AuNPs, chitosan gold nanoparticles, etc. has also been found to improve nucleic acid capturing from complex samples by enhanced adsorption efficiency, and the captured nucleic acids are found to be compatible with PCR-based diagnosis (Xu et al. 2021; Seyyedi et al. 2020; Tammam et al. 2017; Ghahari et al. 2018). The application of NPs in the nucleic acid-based diagnostic platforms is not only confined to the purification of cell extract obtained from lysis buffer treatment, but it can also act as a lysis agent for efficient nucleic acid extraction from the cells. Silver nanoparticles (AgNPs) are mostly employed for the lysis of microbial cells for nucleic acid extraction. Such extract can be stored for a substantial time without deteriorating its quality facilitating field-level diagnosis (Goswami et al. 2017). Analogously, nanowires can also be effectively employed in the isolation of microbial genomes compatible with LAMP and PCR by rupturing cells enabling a simplified and rapid identification of nucleic acid in a single-step (Yasui et al. 2019). Faster detection of bacteria through microchip PCR-based assay is also performed employing surface functionalized AuNPs with poly-(quaternary ammonium) for rupturing cells (Wan and Yeow 2012). Hence, several NP-aided approaches are prevailing for cell lysis followed by extraction and identification of nucleic acid which are way more advantageous than conventional methods. Implementations of NPs in conventional nucleic acids amplification-based methods such as LAMP, PCR, immuno-PCR, and q-PCR have enhanced the rapidity, cost-effectiveness, and overall success of these modalities. AuNPs are found to possess lower toxicity and better biocompatibility when employed in PCR with significant improvement in PCR efficacy as well as specificity of nano-PCR assay. A remarkable heat transfer property of AuNPs facilitates the upregulation of sensitivity of an end-point PCR up to 10,000-fold to conventional qPCR assay, however, the effect must be carefully observed prior inclusion of AuNPs in PCR assays as non-specific nucleic acid binding at the nanoparticle surface may reduce the specificity of diagnosis (Li et al. 2005; Haber et al. 2008). Such attributes of NPs are significant in the diagnosis of diseases where inoculum is less either due to early pathogenesis or sub-clinical cases. Thiol-linked oligonucleotides functionalized AuNP biosensors have been synthesized for the detection of a global veterinary-concerned organism, foot and mouth disease virus (FMDV). Such nanomaterials-assisted PCR or nano-PCR avails ultrasensitive diagnosis by means of precise recognition of FMDV genome without any cross-reactivity with other correlated family members such as swine vesicular disease virus (SVDV) (Hamdy et al. 2018). Similarly, respiratory syncytial virus of bovines can also be diagnosed through amplification using AuNP-assisted PCR with advanced sensitivity in comparison to the conventional method (Liu et al. 2019). However, destabilization of double-stranded DNA by using AuNPs has also arisen as a negative aspect in some cases, though it can be overcome via a thorough evaluation of the concerned nanomaterial for specific conditions (Haber et al. 2008). Nanomaterials also possess implementations in multiplexing of PCR. For example, three chicken diseases, viz. fowl adenovirus (FAdV), infectious bursal disease virus (IBDV), and chicken anemia virus (CAV) can be simultaneously diagnosed through triplex nanoparticle-assisted PCR (nano-PCR) assay using primers specific for respective viral genome. This is especially simple, sensitive as well as highly specific assay with sensitivity up to 1000-fold in comparison to conventional multiplex PCR (Luan et al. 2022). Besides, qPCR employing plasmonic NPs can successfully detect SARS-CoV-2 RNA obtained from human saliva and nasal specimens through amplification with 100% sensitivity and specificity (Blumenfeld et al. 2022). Immuno-PCR provides a rapid diagnosis of several diseases such as hepatitis B virus (HBV), human thyroid stimulating hormone (hTSH), human chorionic gonadotropin (hCG), HIV-1 virion, and so on by exploiting specific antibody-based probes for amplification as well as quantification of gene. In general, PCR amplification can be optimized up to 100–1000 copies of DNA; however, the issue of non-specific amplification can be overcome by immune PCR with an enhanced sensitivity of 10^5-fold against ELISA. AuNP-aided sandwich immuno-complex has been formulated with dually functionalized AuNP with oligonucleotides and pathogen-specific monoclonal antibody L13 (mAb L13) bound with pathogen-specific polyclonal antibody (pAb) arrested antigen. Subsequently, the release of signal DNA (S-DNA), complementary with the AuNP captured DNA was performed by heating that act as a pathogenic marker and treated with PCR amplification to assist pathogenic identification (Chen et al. 2009). Hence, the employment of nanomaterials in thermocycler-based platforms advances the opportunities in the diagnosis of pathogenic organisms in comparison to conventional modalities. Moreover, the implementation of nanomaterials can also provide visualization of amplified PCR products through changes in color excluding the need for further equipment. For example, MNP-aided PCR facilitates the visualization of desired, amplified genomes due to the aggregation of NPs. Such applications may expedite diagnosis through PCR enabling the development of rapid, simplified, cost-effective diagnostic point-of-care tests (POCTs). For example, nanoparticles-DNA aggregation (STat aggregation of tagged DNA, STAT-DNA)-based method has been developed for the visual detection of the hepatitis C virus (HCV) genome. Here, biotin and digoxigenin-labeled primer were used to obtain targeted amplicons through PCR which were subsequently treated with streptavidin particles and anti-DXN nanoparticles resulting in amplicon-nanoparticle aggregations. Such aggregated complexes can be visualized using the naked eye (Kim et al. 2022). As outcomes of positive impacts imparted by nanomaterials in almost all steps of thermocycler-based diagnostic modalities, various associated kits as well as chemicals are being commercialized nowadays (Minakshi et al. 2020b).

6.2.3 Nanotechnology in Point-of-Care Devices

The incorporation of nanomaterials for biotechnological tool development employing the concepts of biomedical engineering has yielded a number of point-of-care diagnostic devices in recent times. These devices are mostly user-friendly and may be used single or multiple times with less recourse involvement yet provide real-time data. Nano-sensors working on various physicochemical properties of the nanomaterials remain at the core of these smart devices to identify the target molecule(s) which can be indicative of any disease signatures, either notable or even recordable by coupling with smartphone-like modality. Regardless of the type of nanomaterial used or underlying law of operation, such devices often provide facilities to differentiate the target analytes from the closely related biomolecules or capture them from complex samples, depict ample sensitivity with a low limit of detection; provide better analyte-binding stability in time course and temperature range; and show compatibility at a dynamic range of analyte concentrations. Various types of nano-sensors have been developed such as capacitive sensors, cantilever sensors, chemiluminescent, colorimetric, disposable lateral flow, electrochemical, fluorescence, optic system-based sensors, etc. which can generate detectable signals exploiting a change in pH, temperature, redox potential, color, antigen–antibody binding, aptamer binding resulting into a change in electric strength or any other physical or chemical parameters. The customization of such sensors considers the properties of targeted analyte(s), the nature of the detection reaction, sample types, and expected concentration of the target analyte(s), etc. (Akgönüllü and Denizli 2022). Various nanoparticles-based diagnostic devices have been approved by the U.S. Food and Drug Administration (US FDA) like "Verigene" for pathogen detection using DNA barcodes where oligonucleotides are conjugated with AuNP to provide high sensitivity for several proteins (10^{-18} M) and nucleic acid (10^{-19} M); T2Candida panel for the detection of sepsis-causing fungal pathogens from blood. Similarly, SPIONs and magnetic NP-based contrast agents like "Feridex" have also been approved by US FDA for MRI applications (Papafilippou et al. 2021). Similarly, magnetic nanosensor-based point-of-care devices for the diagnosis of coronavirus disease (Wu et al. 2020), electrochemical sensor using AuNPs for SARS-CoV, and MERS-CoV pathogens (Bardhan et al. 2021) are some of the examples of US FDA approved diagnostics and pathogen detection systems under commercialization stages (Pradhan et al. 2021; Adam and Gopinath 2022). This indicates that nano-sensors can be used in various disease diagnoses including protozoan disease of veterinary importance.

6.3 Conventional Diagnostic Techniques for Protozoan Disease of Veterinary Importance

Protozoan diseases in livestock hold significant impacts on global agriculture as well as animal welfare. Such parasitic infections not only compromise the health and productivity of livestock but also possesses the risk of zoonotic transmission to humans. Traditional diagnostic methods, such as microscopic examinations, serological tests, etc. are time-consuming, and labor-intensive, along with limited sensitivity and specificity. The tentative diagnosis that is used to perform by the veterinarians is based on disease pattern or history, epidemiology, and symptoms, followed by the confirmatory tests. Protozoa-causing diseases in veterinary subjects can be detected through several diagnostic modalities such as microscopic observation from fecal and blood samples, antigen–antibody-based tests, thermocycler-based amplification techniques, and so on. Generally, in most veterinary hospitals, microscopic observations of relevant samples are carried out for diagnosis of Babesiosis, Theileriosis, Trypanosomiasis, Giardiasis, Trichomoniasis, Balantidiasis, etc. Immunodiagnostic assays through antigen or antibody detection such as complement fixation test (CFT), immunodiffusion (ID), indirect hemagglutination (IHA), the latex agglutination (LA), the indirect fluorescent antibody test (IFAT), the radio-immunoassay (RAI), the enzyme-linked immune-sorbent assay (ELISA), blotting, etc. are also performed for epidemiological studies or due to hospital referrals. However, the presence of pre-existing antibodies even after recovery often gives false results. Moreover, such conventional techniques also possess many drawbacks such as a higher limit of detection (LOD), i.e., they need a larger quantity of analytes for assessment, high-priced, poor stability, and need for specific instrumentation. However, several nucleic acid-based diagnostic modalities such as blotting, PCR, Q-PCR, and LAMP have been advanced resulting in higher sensitivity as well as lower LOD. Detection of *Trypanosoma* and *Eimeria* species can be accomplished by Multi-locus Enzyme Electrophoresis (MEE) (Atuhaire et al. 2021; Morris and Gasser 2006). This is used for characterizing organisms; however, it can detect genetic variations also through the relative mobility of organisms from a complex sample under the electrophoresis-based principle. Polymerase chain reaction (PCR)-based methods including its variants such as random amplified polymorphic DNA (RAPD), PCR-restriction fragment length polymorphism (PCR-RFLP), amplification fragment length polymorphism (AFLP), single-strand conformation polymorphism (SSCP), and loop-mediated isothermal amplification (LAMP) are being performed nowadays for the detection and visualizations of the nucleic acid of target organism through amplification. Various protozoa including *Giardia* sp., *Entamoeba* sp., *Babesia* sp., *Theileria* sp., etc. have been identified from their respective samples through such methods (Satbige et al. 2018; Maharana et al. 2016). In comparison to conventional PCR, RT-PCR is found to be more effective in the diagnosis of early infection of *Babesia* sp. while for the detection of *Theileria* and *Babesia*, reverse line blot (RLB) offers enhanced sensitivity (Kim et al. 2007; Teal et al. 2012; Gubbels et al. 1999). Thermocycler-based diagnostic techniques are not only advantageous in terms of enhanced sensitivity but it also eliminates the drawbacks of antibody-based assay including late antibody generation and false positive assessment. However, these methods also need concern regarding specialization, sensitivity for sub-clinical disease diagnosis, real-time-based assay, cost-effectiveness, etc. Such limitations for currently available diagnostic methods can be overcome through implementations of nano-technological interventions which are found to be beneficial in bacterial and viral disease diagnosis. Moreover, the employment of NPs in diagnosis can also be beneficial regarding the development of more accurate, rapid, and cost-effective techniques for the detection and monitoring of animal protozoan diseases.

6.4 Nano-diagnostics for Protozoan Diseases

Zoonotic disorders caused by protozoans often consequences in hindering yield from livestock. For example, productivity of cattle, poultry, sheep, goats, etc. gets reduced due to coccidiosis by *Eimeria* sp. causing diarrhea and weight loss (Chapman et al. 2010). Giardiasis, caused by *Giardia* sp., affects pet animals, and livestock, causing malabsorption and diarrhea subsequently (Thompson 2004). Toxoplasmosis, caused by *Toxoplasma gondii*, affects a wide range of animals, and can be transmitted to humans also, leading to reproductive failure and neurological disorders in infected hosts (Dubey 2010). Cryptosporidiosis caused by *Cryptosporidium* sp. affects several animals resulting in malnutrition, dehydration, and diarrhea (Ryan and Hijjawi 2015). Hence, animals harbor a group of protozoa that can infect multiple organisms including mankind through intimacy with companion animals as well as livestock. Nearly 62% of households have at least one pet as companion animal in USA while in India, more than 70% of people are dependent upon livestock and tend to rear in vicinity enhancing the chances of zoonotic transmission to human (https://dahd.nic.in; Department of Animal Husbandry and Dairying 2023; Esch and Petersen 2013). Hence, advancement in diagnostic modalities for both animals and humans is required for timely detection of pathogen followed by proper treatment. Several conventional methods such as, microscopic examination, leishmanin skin test (LST) or Montenegro test, agglutination test, rK39 dipstick test, etc. are there for the diagnosis of leishmaniasis in both human and animal subjects such as dogs, rodents, cats, etc. However, such techniques also have limitations. Microscopic examination lacks sensitivity. Range of detection of leishmanin skin test (LST) or Montenegro test is limited to cutaneous leishmaniasis (CL) only even though it is a simple, cost-effective, sensitive, and specific method. In addition to the requirement of high intensity of infection, rk39 dipstick test and FAST agglutination screening test (FAST) generate false positive results. False positivity can also be obtained from direct agglutination test. Latex agglutination test is a complex method with poor sensitivity. In conventional PCR, dead parasitic genome can also be amplified resulting in false positivity. Furthermore, it is an expensive procedure along with the requirement of expertise, as well as additional instrumentation for visualization.

Similarly, quantitative real-time PCR is also expensive with the need of expertise personel and develops complex interpretation. Rapidity and improved visualization are required in enzyme-linked immunosorbent assay (ELISA), whereas in loop-mediated isothermal amplification (LAMP), limited temperature range, higher specificity and sensitivity, cost-effective instrumentation is needed (Gedda et al. 2021).

Research in the area of protozoan diseases always remains behind the study of viral or bacterial disorders reflecting the same in the field of diagnostic tools development. For instance, a group working on bacterial and viral vaccine development is merely concerned with protozoan diseases. Same scenario can be found in diagnostic research arenas also. However, application of nano-technological tools in all kinds of diagnostic platforms has been found to enhance the efficiency and utility of such modalities (Prasad et al. 2021b; Minakshi et al. 2019, 2020b). Employment of NPs in protozoan disease diagnosis is also advantageous regarding enhanced sensitivity, specificity, cost-effectiveness, time-management, higher stability, along with elimination of the need of instrumentation. Such advancement also facilitates evolution of point-of-care diagnostic method with ease of operation.

6.4.1 Biosensors for Protozoan Disease Diagnosis

The advancement of conventional diagnostic modalities through the implementation of nanomaterials helps in getting over the limitations of existing methods. Such as the development of biosensors employing metallic nanostructures along with fluorescent dyes, QDs, dye-loaded silica, as well as polystyrene functionalized with super-paramagnetic NPs-based probes for capturing nucleic acid and proteins results in real-time bioimaging with advanced sensitivity and specificity. Nano-structured nickel oxide-based sensitive and stable DNA biosensor system has been developed for the diagnosis of Visceral Leishmaniasis (Kala-azar) (Mohan et al. 2011). Several other sensors such as leishmania-specific probes or AuNP-aided lateral flow biosensors, and electrochemical genosensor containing magnetic cobalt–zinc ferrite QDs have been evolved. AuNP-employed plasmonic detection of PCR amplified genetic biomarker has LOD 212.5–6.64 ng/μL. However, such AuNP-based applications for targeting kinetoplastid minicircle DNA further lowered the LOD to 11.5 ng/μL. Further advancement of electrochemical genosensor employing AuNPs on the gold electrode provides LOD as low as 1 ZM. Another novel DNA-based biosensor using immobilized thiolated sequences of the *Leishmania* sp. genome on the surface of AuNPs on the gold electrode provides rapid as well as accurate ultrasensitive diagnosis of *Leishmania* infections (Mobed et al. 2020). Similarly, other diagnostic modalities such as lateral flow immunoassay, surface plasmon resonance immunosensor, and electrochemical immunosensor have been developed for specific detection of leishmania-specific marker RNA employing AuNPs that aid in lowering the detection limits (Gedda et al. 2021). An electrical impedance-based device was developed for the real-time detection of parasitic stage-wise development of *Cryptosporidium* sp. with alternating zoite and meront predominances to quantify sample infectivity (Dibao-Dina et al. 2015). Such constructs can provide confirmation of endoparasitic growth at a faster rate than existing tools. Nanoparticle-based biosensors not only increase the sensitivity and specificity of diagnosis but also reduce the limit of detection relying upon the inherent properties NPs. For instance, functionalized MNPs can help in the aggregation and capturing of target analytes from complex samples, higher surface area to volume ratio provides ease of functionalization with more reactivity potential, thus reducing the detection time, favorable optical properties of NPs facilitating signal amplification and pathogen detection. Evidently, immunomagnetic separation (IMS) is now considered a standard method for the diagnosis of *Cryptosporidium* spp., and *Giardia lamblia* and the separation of the oocysts/cysts from various aquatic samples [USEPA (2012) http://www.epa.gov/safewater]. Various sensors have been tried to identify protozoa from aquatic samples which serve as reservoirs and source of transmission to both humans and animals. For instance, lead–zirconium–titanium oxide (PZT) based piezoelectric ceramics films and glass layered immunosensor have successfully detected *G. lamblia* from water samples with a LOD of 1–10 cysts/mL within 15 min of detection time. Antigen-coated AuNP-based piezoelectric immunosensor has detected *T. gondii* from infected rabbit serum at antibody titer as low as 1:5500 (Wang et al. 2004). Similarly, magnetic-fluorescent nanoparticle (Fe_3O_4/CdTe) and magnetic-fluorescent CdTe@Ni quantum dots (mQDs) based genosensors have detected *T. gondii* DNA with a LOD of 8.339×10^{-9} M and 2.70×10^{-9} mol/L, respectively (Nemati et al. 2023). Gold nanoparticles electrodeposited carbon electrode-based microfluidic immunosensor has also been used for the detection of anti-*T. cruzi* antibodies with a LOD of 3.065 ng/mL (Pereira et al. 2011). Plasmodium infection is endemic among diverse human populations. Various biosensors using NPs have already been customized to diagnose human plasmodium infection. Hemozoin, a magnetic nanoparticle by-product of malarial parasites has been detected by a magneto-optic detection system with a detection limit of 8.1 ng/mL for β-hematin that mimics hemozoin (Armani et al. 2019). Similar kind of biosensors can also be implemented for protozoan disease diagnosis in animals. Additionally, electrochemical nanosensor using CuO NPs coated gold electrode has detected *P. berghei* in infected mice serum and *P. falciparum* in human blood serum samples (sensitivity 3.60–4.8 mM and 0.65–1.35 mM, respectively) (Obisesan et al. 2019). Zinc oxide nanoparticles-based microfluidic laser-induced fluorescence immunosensor has quantitatively detected anti-*T. gondii* specific IgG antibodies from serum samples at a concentration range of 0–200 U/mL within 31 min of detection time (Medawar Aguilar et al. 2019). Recently, surface-enhanced Raman spectroscopy (SERS)-based methods employing NPs are being applied for the detection of various pathogens through ultrasensitive optical detection and spectroscopy. For instance, rhodamine B isothiocyanate and malachite green isothiocyanate conjugated AuNPs were used for the rapid detection of *Cryptosporidium parvum* and *Giardia lamblia*. Such a system is faster and can be operated at room temperature with significant reproducibility. Such systems may also be implemented in protozoa detection (Rule and Vikesland 2009).

6.4.2 Nanoparticles in Nucleic Acid-Based Diagnostics of Protozoan Diseases

Diagnosis of protozoan diseases can be effectively carried out through PCR-based assay. An intestinal protozoan parasite, *Cryptosporidium* sp. causing diarrhea in animals, including humans can be visually detected through nano-PCR with sensitivity up to 10 times in comparison to conventional PCR. Hence, AuNP-employed nano-PCR facilitates faster, specific, and accurate detection of the oocytes in fecal samples of goats, camels, bovines, and chickens infected with *Cryptosporidium* sp. (Yin et al. 2021; Yao et al. 2022). PCR multiplexing can be successfully performed for the diagnosis of variants of three main diarrhea-causing intestinal parasites, namely *Giardia* sp., *Cryptosporidium* sp., and *Entamoeba* sp. However, the efficacy of such diagnostic methods can be advantageous through the employment of NPs as discussed earlier. For instance, streptavidin-functionalized AuNP-assisted multiplex recombinase polymerase amplification (RPA) assay based on lateral flow has been established to detect the DNA of intestinal protozoa eliminating the requirement for thermal cycling equipment. Hence, the exploitation of NP in such diagnostic techniques facilitates cost-effective performance along with high sensitivity and specificity (Crannell et al. 2016). Similarly, NP-aided aggregation of DNA in samples can also be achieved for the identification of pathogens. The most notorious malaria-causing protozoan species, *Plasmodium falciparum* can be effectively detected by advanced LAMP technique assisted with streptavidin-coated gold nanoparticles with LOD of 1 parasite/μL (Suwannin et al. 2021). In comparison to conventional PCR, PCR using ZnO nanoflower provide a diagnosis of *Babesia canis vogeli* and *Hepatozoon canis* in dogs with better results (Upadhyay et al. 2020).

6.4.3 Nanomaterials in Immunodiagnostics of Protozoan Diseases

Most homoiothermic animals including all farm animals and humans act as hosts for *Toxoplasma gondii*. These protozoan parasites possess a worldwide distribution showing diverse signs and symptoms based on their hosts and genetic variances. Serological diagnosis of toxoplasmosis can be accomplished through AuNP-conjugated nano-ELISA kits with enhanced sensitivity and specificity in contrast to conventional ELISA (Khodadadi et al. 2020; Hegazy et al. 2015; Al Hamshary et al. 2020). Nano-gold beads based-ELISA kit constructed for the detection of *Giardia* sp. is more sensitive and specific than the usual sandwich ELISA technique (Moharm et al. 2014). Similarly, a novel nano magnetic beads dot ELISA immunoassay has been developed for the detection of *Giardia lamblia* enabling better sensitivity and specificity along with easy interpretation in a faster and simpler way (Shaban Aly et al. 2018). Luciferase-linked Antibody Capture Assay (LACA) is another technique involved in the diagnosis of *Toxoplasma gondii*, where recombinant nano-luciferase is fused with *T. gondii* antigen resulting in the development of a diagnostic platform with comparatively higher sensitivity and similar specificity to conventional PCR (Duong et al. 2022). Similarly, AuNPs employed RPA-based cryptosporidium assay (RPAC) has also been evolved for the detection of *Cryptosporidium parvum* oocysts parasite. This technique is devoid of any instrumentation besides a micropipette, pipette tips, and a single temperature heater, hence, can be exploited as a point-of-care device in the future (Crannell et al. 2014). Recently, an immune-chromatographic assay kit has been constructed employing fluorescent silica NPs attached with anti-Acanthamoeba antibodies for the diagnosis of Acanthamoeba keratitis (Toriyama et al. 2015). In comparison to conventional ELISA, AuNPs functionalized poly antibodies (pAb)-aided sandwich ELISA has been found to possess enhanced sensitivity and specificity up to 92% and 94%, respectively, for the diagnosis of *Giardia* sp. (Chen et al. 2020). Hence, such NP-aided diagnostic modalities may ensure better aggregation and improved Ag–Ab reaction leading to enhancement in sensitivity and specificity of diagnosis. Another exploitation of AuNP includes AuNP-assisted RPA-LFD assay for the detection of *Babesia* sp. Similarly, a lateral flow-based kit has also been developed employing AuNPs to diagnose *Theileria equi* infection with 94% diagnostic sensitivity as well as specificity of up to 91% [Agrinnovate (2021); http://www.agrinnovateindia.co.in]. Recently, NanoCHIP® molecular microarray system has been commercialized for the diagnosis of intestinal protozoa *Giardia duodenalis, Cryptosporidium* spp., *Entamoeba histolytica, Entamoeba dispar, Dientamoeba fragilis,* and *Blastocystis hominis* along with a simultaneous diagnosis of other organisms, hence facilitating time management (Ken Dror et al. 2016). Similarly, another real-time-based platform, Open Array™ provides simultaneous detection of multiple organisms including *Giardia lamblia* and *Cryptosporidium* sp. through NP-assisted multiplexing is also valuable for future diagnostics purposes.

6.5 Conclusion

Conventional diagnostic modalities such as microscopic observations, immunodiagnostic assays, and thermocycler-based amplification techniques are commonly used to detect protozoan diseases. However, it possesses several drawbacks. For instance, the method of diagnosis using microscopic examination lacks sensitivity and often fails to detect sub-clinical infections. Several existing immunodiagnostic assays including CFT, immunodiffusion, hemagglutination, latex agglutination test, IFAT, ELISA, blotting, etc. are unlikely to be performed effectively as they are prone to provide false positivity along with longer detection time. Furthermore, such traditional diagnostic techniques are not pocket-friendly along with their limitation in terms of stability, sensitivity, and requiring sophisticated instrumentation. Whereas, thermocycler-based diagnostic techniques are comparatively advantageous in terms of sensitivity, proper assessment, etc. but are resourse-intensive. Implementation of nanomaterials can enhance the specificity and sensitivity of such conventional diagnostic techniques along with enabling sub-clinical disease diagnosis, real-time assessment, and cost-effectiveness as well. Proper isolation, purification, and confirmation of nucleic acid in the target samples can be obtained at a single go by employing NPs in nucleic acid amplification-based assays. For instance, nanoparticle-graphene oxide complex, amino-modified magnetic nanoparticles (AMNPs), AuNPs, Chitosan gold nanoparticles, etc. facilitate nucleic acid capturing from complex samples by enhanced adsorption efficiency. High-quality purification and capturing of the target(s) can be achieved using magnetic force-based aggression and dispersion of MNPs. Moreover, AgNP-aided lysis of microbial cells can also carry out efficient nucleic acid extraction. Hence, applications of NMs in several thermo-cycler-based diagnoses such as LAMP, PCR, Immuno-PCR, and qPCR have enhanced the rapidity, cost-effectiveness, and overall success of such modalities. As a result, several NP functionalized probe-based PCR has been performed successfully resulting in the commercialization of several diagnostic kits as well as chemicals. For instance, nano-gold beads based-ELISA kit has been industrialized for the diagnosis of *Giardia* sp. with better sensitivity and specificity. In addition, conventional laboratory-based systems are cumbersome to perform in veterinary diagnosis, whereas advanced biosensors-assisted point-of-care diagnosis provides field-level applications. Hence, several DNA biosensors, lateral flow biosensors, electrochemical immunosensors, genosensors, etc. have been evolved for the detection of a wide range of gastro-intestinal protozoan parasites, such as *Cryptosporidium* sp., and *Giardia* sp., *Leishmania* sp., *Entamoeba* sp., etc. For the detection of *Cryptosporidium parvum* and *Giardia lamblia*, NPs employed surface-enhanced Raman spectroscopy (SERS) has been established to ensure ultrasensitive optical detection of pathogens with rapidity. However, every diagnostic modality has its own boundaries and negative sides no matter how advanced technologies are implemented. Similarly, the impact of nanotoxicity on subjects should also be considered. Hence, further introspection and efforts are required for the development of better and optimal detection approaches for protozoa parasites.

Acknowledgments The support from National Agricultural Science Fund (NASF)-ICAR is duly acknowledged.

References

Adam T, Gopinath SC (2022) Nanosensors: recent perspectives on attainments and future promise of downstream applications. Process Biochem 117:153–173

Agrinnovate (2021). http://www.agrinnovateindia.co.in

Akgönüllü S, Denizli A (2022) Nano-sensors and nano-devices for biological disaster monitoring (virus/disease epidemics/animal plagues detections). In: Nanotechnology-based smart remote sensing networks for disaster prevention. Elsevier, Amsterdam, pp 43–57

Al Hamshary AS, Bayoumi IR, Aly NS, Omar RE, Mohammed DA, Marei YM, Rashed GA (2020) Evaluation of nano-based-ELISA for serodiagnosis of human toxoplasmosis. Benha J Appl Sci 5(2 part (1)):161–167

Armani A, Chen D, McBirney S, Kaypaghian K, Huber H, Ameri H (2019) Portable optical diagnostics for early malaria detection. In: Frontiers in biological detection: from nanosensors to systems XI, vol 10895. SPIE, pp 29–33

Atuhaire DK, Muleya W, Mbao V, Niyongabo J, Nyabongo L, Nsanganiyumwami D, Musoke AJ (2021) Molecular characterization and population genetics of *Theileria parva* in Burundi's unvaccinated cattle: towards the introduction of East Coast fever vaccine. PLoS One 16(5): e0251500

Baetke SC, Lammers TGGM, Kiessling F (2015) Applications of nanoparticles for diagnosis and therapy of cancer. Br J Radiol 88(1054):20150207

Bardhan NM, Jansen P, Belcher AM (2021) Graphene, carbon nanotube and plasmonic nanosensors for detection of viral pathogens: opportunities for rapid testing in pandemics like COVID-19. Front Nanotechnol 3:733126

Blumenfeld NR, Bolene MAE, Jaspan M, Ayers AG, Zarrandikoetxea S, Freudman J, Sia SK (2022) Multiplexed reverse-transcriptase quantitative polymerase chain reaction using plasmonic nanoparticles for point-of-care COVID-19 diagnosis. Nat Nanotechnol 17(9): 984–992

Chacón-Torres JC, Reinoso C, Navas-León DG, Briceño S, González G (2020) Optimized and scalable synthesis of magnetic nanoparticles for RNA extraction in response to developing countries' needs in the detection and control of SARS-CoV-2. Sci Rep 10(1):1–10

Chakraborty D, Viveka TS, Arvind K, Shyamsundar V, Kanchan M, Alex SA, Mukherjee A (2018) A facile gold nanoparticle-based ELISA system for detection of osteopontin in saliva: towards oral cancer diagnostics. Clin Chim Acta 477:166–172

Chapman HD, Jeffers TK, Williams RB (2010) Forty years of monensin for the control of coccidiosis in poultry. Poult Sci 89(9):1788–1801

Chen L, Wei H, Guo Y, Cui Z, Zhang Z, Zhang XE (2009) Gold nanoparticle enhanced immuno-PCR for ultrasensitive detection of Hantaan virus nucleocapsid protein. J Immunol Methods 346(1–2):64–70

Chen CY, Lee CH, Lee HM, Yin WY, Chin WL, Lee MH, Chen JH (2020) Check point to get adequate weight loss within 6-months after laparoscopic sleeve gastrectomy for morbid obesity in Asian population. Sci Rep 10(1):1–9

Crannell ZA, Castellanos-Gonzalez A, Irani A, Rohrman B, White AC, Richards-Kortum R (2014) Nucleic acid test to diagnose cryptosporidiosis: lab assessment in animal and patient specimens. Anal Chem 86(5):2565–2571

Crannell Z, Castellanos-Gonzalez A, Nair G, Mejia R, White AC, Richards-Kortum R (2016) Multiplexed recombinase polymerase amplification assay to detect intestinal protozoa. Anal Chem 88(3):1610–1616

Dalal A, Mohan H, Prasad M, Pundir CS (2020) Detection methods for influenza A H1N1 virus with special reference to biosensors: a review. Biosci Rep 40(2):BSR20193852

Department of Animal Husbandry and Dairying (2023) https://dahd.nic.in

Dibao-Dina A, Follet J, Ibrahim M, Vlandas A, Senez V (2015) Electrical impedance sensor for quantitative monitoring of infection processes on HCT-8 cells by the waterborne parasite cryptosporidium. Biosens Bioelectron 66:69–76

Dubey JP (2010) *Toxoplasma gondii* infections in chickens (*Gallus domesticus*): prevalence, clinical disease, diagnosis and public health significance. Zoonoses Public Health 57(1):60–73

Duong HD, Taniguchi Y, Takashima Y, Sekiguchi S, Aye KM, Ahmadi P, Yoshida A (2022) Diagnostic value of recombinant nanoluciferase fused *Toxoplasma gondii* antigens in luciferase-linked antibody capture assay (LACA) for toxoplasma infection in pigs. J Vet Med Sci 84(7): 905–913

Esch KJ, Petersen CA (2013) Transmission and epidemiology of zoonotic protozoal diseases of companion animals. Clin Microbiol Rev 26(1):58–85

Gedda MR, Madhukar P, Shukla A, Mudavath SL, Srivastava ON, Singh OP, Sundar S (2021) Nanodiagnostics in leishmaniasis: a new frontiers for early elimination. Wiley Interdiscip Rev Nanomed Nanobiotechnol 13(2):e1675

Ghahari S, Ghahari S, Nematzadeh GA (2018) Magnetic nano fluids for isolation of genomic DNA and total RNA from various prokaryote and eukaryote cells. J Chromatogr B 1102:125–134
Goswami G, Boruah H, Gautom T, Hazarika DJ, Barooah M, Boro RC (2017) Chemically synthesized silver nanoparticles as cell lysis agent for bacterial genomic DNA isolation. Adv Nat Sci Nanosci Nanotechnol 8(4):045015
Gubbels JM, de Vos AP, van der Weide M, Viseras J, Schouls LM, de Vries E, Jongejan F (1999) Simultaneous detection of bovine Theileria and Babesia species by reverse line blot hybridization. J Clin Microbiol 37(6):1782–1789
Haber AL, Griffiths KR, Jamting ÅK, Emslie KR (2008) Addition of gold nanoparticles to real-time PCR: effect on PCR profile and SYBR Green I fluorescence. Anal Bioanal Chem 392:887–896
Hamdy ME, Del Carlo M, Hussein HA, Salah TA, El-Deeb AH, Emara MM, Compagnone D (2018) Development of gold nanoparticles biosensor for ultrasensitive diagnosis of foot and mouth disease virus. J Nanobiotechnol 16:1–12
Hegazy S, Farid A, Rabae I, El-Amir A (2015) Novel IMB-ELISA assay for rapid diagnosis of human toxoplasmosis using SAG1 antigen. Jpn J Infect Dis 68(6):474–480
Kang J, Li Y, Zhao Y, Wang Y, Ma C, Shi C (2021) Nucleic acid extraction without electrical equipment via magnetic nanoparticles in Pasteur pipettes for pathogen detection. Anal Biochem 635:114445
Ken Dror S, Pavlotzky E, Barak M (2016) Evaluation of the NanoCHIP® gastrointestinal panel (GIP) test for simultaneous detection of parasitic and bacterial enteric pathogens in fecal specimens. PLoS One 11(7):e0159440
Khodadadi A, Madani R, Atyabi N (2020) Development of nano-ELISA method for serological diagnosis of toxoplasmosis in mice. Arch Razi Inst 75(4):419
Kim C, Iseki H, Herbas MS, Yokoyama N, Suzuki H, Xuan X, Fujisaki K, Igarashi I (2007) Development of TaqMan-based real-time PCR assays for diagnostic detection of *Babesia bovis* and *Babesia bigemina*. Am J Trop Med Hyg 77(5):837–841
Kim SK, Oh YH, Ko DH, Sung H, Oh HB, Hwang SH (2022) Nanoparticle-based visual detection of amplified DNA for diagnosis of hepatitis C virus. Biosensors 12(9):744
Lambe U, Prasad M, Brar B, Guray M, Ikbal, Ranjan K et al (2016) Nanodiagnostics: a new frontier for veterinary and medical sciences. J Exp Biol Agric Sci 4:307–320. https://doi.org/10.18006/2016.4(3S).307.320
Li M, Lin YC, Wu CC, Liu HS (2005) Enhancing the efficiency of a PCR using gold nanoparticles. Nucleic Acids Res 33(21):e184. https://doi.org/10.1093/nar/gni183
Linyan G, Qian P, Yang M (2017) Determination of immunoglobulin G by a hemin–manganese (iv) oxide-labeled enzyme-linked immunosorbent assay. Anal Lett 50(11):1803–1811
Liu Z, Li J, Liu Z, Li J, Li Z, Wang C, Guo L (2019) Development of a nanoparticle-assisted PCR assay for detection of bovine respiratory syncytial virus. BMC Vet Res 15(1):1–6
Luan Q, Jiang Z, Wang D, Wang S, Yin Y, Wang J (2022) A sensitive triple nanoparticle-assisted PCR assay for detection of fowl adenovirus, infectious bursal disease virus and chicken anemia virus. J Virol Methods 303:114499
Maharana BR, Tewari AK, Saravanan BC, Sudhakar NR (2016) Important hemoprotozoan diseases of livestock: challenges in current diagnostics and therapeutics: an update. Vet World 9(5):487
Medawar-Aguilar V, Jofre CF, Fernández-Baldo MA, Alonso A, Angel S, Raba J, Messina GA (2019) Serological diagnosis of Toxoplasmosis disease using a fluorescent immunosensor with chitosan-ZnO-nanoparticles. Anal Biochem 564:116–122
Mieszawska AJ, Willem JM, Fayad ZA, Cormode DP (2013) Multifunctional gold nanoparticles for diagnosis and therapy of disease. Mol Pharm 10(3):831–847
Minakshi P, Ghosh M, Brar B, Kumar R, Lambe UP, Ranjan K, Prasad G (2019) Nano-antimicrobials: a new paradigm for combating mycobacterial resistance. Curr Pharm Des 25(13):1554–1579

Minakshi P, Ghosh M, Brar B, Ranjan K, Patki HS, Kumar R (2020a) Separation techniques with nanomaterials. In: Handbook of nanomaterials in analytical chemistry. Elsevier, Amsterdam, pp 99–158

Minakshi P, Ghosh M, Kumar R, Brar B, Lambe UP, Banerjee S, Prasad G (2020b) An insight into nanomedicinal approaches to combat viral zoonoses. Curr Top Med Chem 20(11):915–962

Mobed A, Mehri P, Hasanzadeh M, Mokhtarzadeh A (2020) Binding of Leishmania spp. with gold nanoparticles supported polyethylene glycol and its application for the sensitive detection of infectious photogenes in human plasma samples: a novel biosensor. J Mol Recognit 33(7):e2839

Mohan S, Srivastava P, Maheshwari SN, Sundar S, Prakash R (2011) Nano-structured nickel oxide based DNA biosensor for detection of visceral leishmaniasis (Kala-azar). Analyst 136(13): 2845–2851

Moharm IM, El Enain G, Bahaa El Deen W, Mahgoub A, Hendawy M, Aly I (2014) Evaluation of nano-gold beads based-ELISA for detection of *Giardia lamblia* antigen in stool samples of infected patients. MEJSR 21(12):2264–2273

Morris GM, Gasser RB (2006) Biotechnological advances in the diagnosis of avian coccidiosis and the analysis of genetic variation in Eimeria. Biotechnol Adv 24(6):590–603

Nemati S, Shalileh F, Mirjalali H, Omidfar K (2023) Toward waterborne protozoa detection using sensing technologies. Front Microbiol 14:363

Nguyen LT, Thi Le NH, Thi Ta HK, Dang Nguyen K (2022) Isolation of DNA from *Arthrospira platensis* and whole blood using magnetic nanoparticles (Fe_3O_4@ OA and Fe_3O_4@ OA@ SiO_2). J Anal Sci Technol 13(1):1–11

Obisesan OR, Adekunle AS, Oyekunle JA, Sabu T, Nkambule TT, Mamba BB (2019) Development of electrochemical nanosensor for the detection of malaria parasite in clinical samples. Front Chem 7:89

Papafilippou L, Claxton A, Dark P, Kostarelos K, Hadjidemetriou M (2021) Nanotools for sepsis diagnosis and treatment. Adv Healthc Mater 10(1):2001378

Pereira SV, Bertolino FA, Fernández-Baldo MA, Messina GA, Salinas E, Sanz MI, Raba J (2011) A microfluidic device based on a screen-printed carbon electrode with electrodeposited gold nanoparticles for the detection of IgG anti-*Trypanosoma cruzi* antibodies. Analyst 136(22): 4745–4751

Pradhan A, Lahare P, Sinha P, Singh N, Gupta B, Kuca K, Krejcar O (2021) Biosensors as nano-analytical tools for COVID-19 detection. Sensors 21(23):7823

Prasad M, Ghosh M, Kumar R, Brar B, Surjith KP, Lambe UP, Kharb P (2021a) The importance of nanomedicine in prophylactic and theranostic intervention of bacterial zoonoses and reverse zoonoses in the era of microbial resistance. J Nanosci Nanotechnol 21(6):3404–3452

Prasad M, Ghosh M, Patki HS, Kumar S, Brar B, Sindhu N, Kumar R (2021b) Imaging techniques in veterinary disease diagnosis. In: Advances in animal disease diagnosis. CRC Press, Boca Raton, pp 103–145

Rule KL, Vikesland PJ (2009) Surface-enhanced resonance Raman spectroscopy for the rapid detection of *Cryptosporidium parvum* and *Giardia lamblia*. Environ Sci Technol 43(4): 1147–1152

Ryan U, Hijjawi N (2015) New developments in Cryptosporidium research. Int J Parasitol 45(6): 367–373

Satbige AS, Rajendran C, Patil NA, Sandeep H (2018) Advances in diagnosis of important protozoan diseases: old and new approaches. Int J Curr Microbiol Appl Sci 7(7):3177–3189

Schnittger L, Florin-Christensen M (2018) Introduction into parasitic protozoa. In: Florin-Christensen M, Schnittger L (eds) Parasitic protozoa of farm animals and pets. Springer, Cham

Seyyedi N, Farjadian F, Farhadi A, Rafiei Dehbidi G, Ranjbaran R, Zare F, Behzad-Behbahani A (2020) High yield gold nanoparticle-based DNA isolation method for human papillomaviruses genotypes from cervical cancer tissue samples. IET Nanobiotechnol 14(7):555–562

Shaban Aly N, Bayoumi I, Selem R, Kardoush M, Rashed G, Moharam A (2018) A novel nano magnetic beads dot ELISA immunoassay and its application on the detection of *Giardia lamblia* coproantigen. Iran J Parasitol 13(4):532–540

Suwannin P, Polpanich D, Leelayoova S, Mungthin M, Tangboriboonrat P, Elaissari A, Tangchaikeeree T (2021) Heat-enhancing aggregation of gold nanoparticles combined with loop-mediated isothermal amplification (HAG-LAMP) for *Plasmodium falciparum* detection. J Pharm Biomed Anal 203:114178

Tammam SN, Khalil MA, Gawad EA, Althani A, Zaghloul H, Azzazy HM (2017) Chitosan gold nanoparticles for detection of amplified nucleic acids isolated from sputum. Carbohydr Polym 164:57–63

Tavallaie R, McCarroll J, Le Grand M, Ariotti N, Schuhmann W, Bakker E, Gooding JJ (2018) Nucleic acid hybridization on an electrically reconfigurable network of gold-coated magnetic nanoparticles enables microRNA detection in blood. Nat Nanotechnol 13(11):1066–1071

Teal AE, Habura A, Ennis J, Keithly JS, Madison-Antenucci S (2012) A new real-time PCR assay for improved detection of the parasite *Babesia microti*. J Clin Microbiol 50(3):903–908

Thompson RA (2004) The zoonotic significance and molecular epidemiology of giardia and giardiasis. Vet Parasitol 126(1–2):15–35

Toriyama K, Suzuki T, Inoue T, Eguchi H, Hoshi S, Inoue Y, Ohashi Y (2015) Development of an immunochromatographic assay kit using fluorescent silica nanoparticles for rapid diagnosis of Acanthamoeba keratitis. J Clin Microbiol 53(1):273–277

Upadhyay A, Yang H, Zaman B, Zhang L, Wu Y, Wang J, Han Q (2020) ZnO nanoflower-based NanoPCR as an efficient diagnostic tool for quick diagnosis of canine vector-borne pathogens. Pathogens 9(2):122

USEPA (2012). http://www.epa.gov/safewater

Wan W, Yeow JT (2012) Integration of nanoparticle cell lysis and microchip PCR for one-step rapid detection of bacteria. Biomed Microdevices 14:337–346

Wang H, Lei C, Li J, Wu Z, Shen G, Yu R (2004) A piezoelectric immunoagglutination assay for *Toxoplasma gondii* antibodies using gold nanoparticles. Biosens Bioelectron 19(7):701–709

Wu K, Saha R, Su D, Krishna VD, Liu J, Cheeran MCJ, Wang JP (2020) Magnetic-nanosensor-based virus and pathogen detection strategies before and during COVID-19. ACS Appl Nano Mater 3:9560–9580

Xu J, Chen D, Yang Y, Gong H, Gao W, Xiao H (2021) A one step method for isolation of genomic DNA using multi-amino modified magnetic nanoparticles. RSC Adv 11(6):3324–3332

Yao Q, Yang X, Wang Y, Wang J, Huang S, Song J, Zhao G (2022) Development and preliminary evaluation of a nanoparticle-assisted PCR assay for the detection of *Cryptosporidium parvum* in calves. Animals 12(15):1953

Yasui T, Yanagida T, Shimada T, Otsuka K, Takeuchi M, Nagashima K, Baba Y (2019) Engineering nanowire-mediated cell lysis for microbial cell identification. ACS Nano 13(2):2262–2273

Yin YL, Wang Y, Lai P, Yao Q, Li Y, Zhang LX, Zhao GH (2021) Establishment and preliminary application of nanoparticle-assisted PCR assay for detection of *Cryptosporidium* spp. Parasitol Res 120:1837–1844

Chapter 7
Nanointerventions for the Detection of Fungal Livestock Diseases

Ravina, Subodh Soni, Manjeet Chahar, Minakshi Prasad, and Hari Mohan

Abstract The field of nanotechnology offers potential solutions for early detection and intervention of fungal infections in veterinary medicine. By integrating biological and non-biological systems to create intelligent therapeutic delivery devices, it is possible to detect and treat diseases at the nanoscale before macro symptoms manifest. Fungal infections can cause systemic diseases that can affect multiple body organs, and antifungal drugs are commonly used to treat such infections. Despite advancements in vaccine development and biosecurity protocols, the persistence of animal diseases poses a threat to animal health, food safety, the economy, and the environment. Mitigating the risks associated with animal diseases is crucial for ensuring animal health, profitability of farming operations, and sustainability of animal agriculture. The current chapter will enlighten diverse nanointerventions towards sensitive and precise detection of fungal diseases of livestock.

Keywords Nanotechnology · Fungal disease · Sensor · Mycosis · Mycosensor

7.1 Introduction

Nanotechnology is the application of science and engineering at the nanoscale to create and manipulate materials and devices. It involves integrating biological and non-biological systems to develop new products and technologies, focusing on the material's size. Nanotechnology spans various industries and fields and is not limited to a specific sector (Yallapu et al. 2010). In contemporary veterinary science, the detection of disease in an animal through whole-organism symptoms may take an extended period, potentially resulting in the spread of the infection and requiring the

Ravina · S. Soni · M. Chahar · H. Mohan (✉)
Centre for Medical Biotechnology, Maharshi Dayanand University, Rohtak, Haryana, India
e-mail: harimohan.cmbt@mdurohtak.ac.in

M. Prasad (✉)
National Research Center on Equines, Hisar, Haryana, India

LLR University of Veterinary and Animal, ICAR-National Research Centre on Equines, *Nanotechnology Theranostics in Livestock Diseases and Management*, Livestock Diseases and Management,
https://doi.org/10.1007/978-981-16-1610-5_7

culling of the entire herd. Nanotechnology offers the potential for early diagnosis and intervention, as it operates at the same scale as viral particles or infectious agents along with the possibility of utilizing "intelligent" therapeutic delivery devices that can detect diseases prior to the manifestation of macro symptoms. For example, an implanted miniature device that regularly analyzes body fluid samples, such as saliva, could identify illness and trigger a customized therapeutic delivery system before the appearance of fever or other symptoms. These intelligent treatment delivery systems could come up with a range of products, including drugs, nutrients, probiotics, nutraceuticals, and implantable cell bioreactors, in an animal-specific manner (Scott 2007). In biology, fungi is classified as living organisms with a distinct life cycle. The dispersal of fungal spores or fragments, which can give rise to a fungal colony or grow into hyphae under favorable conditions, contributes to their widespread presence in the environment. These fungal parts can be easily spread through wind or other media, causing infections in livestock as well as canines.

Fungal infections in veterinary medicine are treated with antifungal drugs. These infections can be either localized, affecting only a specific body surface, or systemic, involving the spread of the fungal organism throughout multiple body organs. Systemic fungal diseases are characterized by the introduction and subsequent dissemination of the fungal organism in the host. On the basis of diverse species, fungi used to colonize on specific organs including the lungs, eyes, and lymph nodes. These fungi are commonly found in soil and tend to occur in specific geographic regions (Jungerman & Schwartzman 1972).

The persistence of animal diseases, despite advancements in vaccine development and the implementation of strict biosecurity protocols, poses a persistent threat to animal health, food safety, the economy, and the environment (Wilkinson et al. 2011). The risk of devastating losses in animal populations due to infectious diseases has increased with the enhancement of travel, higher livestock production, and variations in climatic conditions (Gilroyed et al. 2016; Natural Resources Conservation Service 2015). The mitigation of the risks posed by animal diseases is of utmost importance, as it concerns the health of the animals, the profitability of farming operations, and the sustainability of animal agriculture.

7.2 Fungal Infections

The opportunistic saprobic infection known as mucormycosis is the result of fungi categorized under the order Mucorales formerly classified as zygomycetes. The most commonly identified species within this order are genera such as *Rhizopus*, *Mucor*, *Rhizomucor*, *Lichtheimia* (formerly *Absidia*), *Apophysomyces*, *Cunninghamella*, and *Saksenaea*.

The natural habitat of fungi from the order Mucorales is soil, and they can be commonly isolated from decaying organic matter. These fungi are often found in environmental sources such as dust, food, and air, both indoors and outdoors

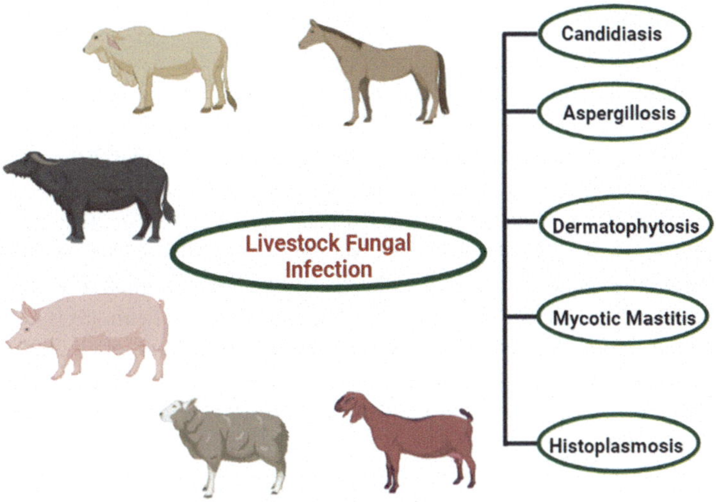

Fig. 7.1 Different fungal infections in livestock

(Hoffmann et al. 2013). In terms of epidemiology, mucormycosis is a fungal infection that can affect both domestic and wild animals, as well as mammals and non-mammals. The route of entry, site of localization, and lesion formation vary (Jensen et al. 1996).

The opportunistic pathogens belonging to the order Mucorales are prevalent in indoor environments and the domestic setting of animals. However, the fungal infections become prevalent only when the normal balance between the host and the pathogen is disrupted.

The spread of mucormycotic infections in the rumens and lymph nodes of cattle is significant, as the disruption of the balance between the animal and the pathogen results in a favorable environment for fungal growth. Usually, the rumen is anaerobic, but the ruminal wall constitutes an interface between aerobes and anaerobes, allowing the colonization of microaerophilic bacteria (Jensen 1994) that are disrupted with antibiotics, enhancing the potentiality for mucormycotic ruminitis. Such condition often arises as a result of severe antibiotic therapy in cattle (Jensen et al. 1989). However, the infection of intestinal lymph nodes can occur due to prolonged exposure to Mucorales fungi through contaminated food (Ortega et al. 2010).

Mucormycosis in ruminants can also extend to affect the respiratory apparatus and other anatomical regions like the gastrointestinal tract, as well as the entire body (Jensen et al. 1994). The Mucorales fungi are known to cause systemic spread via the hematogenous route, which is regularly evidenced by their tendency to angioinvade. In pregnant cows, the fungal infection can sometime spread to the placenta, although *Aspergillus fumigatus* is the most commonly observed cause of bovine mycotic placentitis and abortion (Jensen et al. 1991; Al-Doory 1980). Thus, different fungi are reported to be responsible for infections as represented in Fig. 7.1.

7.3 Candidiasis

A phylogenetic classification approach is being applied to re-categorize the genus *Candida*, which encompasses more than 200 species. Among them 15 species have been isolated from humans and animals (Moran et al. 2012; Yapar 2014). The most commonly encountered pathogens include *Candida albicans, Candida glabrata, Candida parapsilosis, Candida tropicalis,* and *Candida krusei* (Bassetti et al. 2013; Diekema et al. 2012; Lewis 2009; Maubon et al. 2014; Pfaller and Diekema 2007). These species are considered to be commensal and facultatively pathogenic as they are often found in the microbiota of healthy humans and animals (Odds 1988; Al-Yasiri et al. 2017; Pacynska 2013; Różański et al. 2013; Brito et al. 2009). Other *Candida* sp. capable of causing infections have been isolated from environmental sources; however, *C. albicans* and *C. glabrata* appear to be associated with warm-blooded hosts only.

The zoonotic potential of *Candida* sp. is considered to be limited, as infections are typically caused by the strains that have already colonized the host as a commensal rather than through vertical or longitudinal transfer. Although *Candida albicans* is considered the most virulent among the *Candida* sp., other species might predominate in certain animals depending on the site of infection. For example, mastitis in cattle is caused by *Candida albicans, Candida catenulata, Candida guilliermondii, Candida kefyr, Candida krusei, Candida maltosa, Candida rugosa,* and other fungi. Abortions are caused by *Candida parapsilosis* and *Candida tropicalis*, while otitis externa, gastroenteritis, and disseminated candidiasis are caused by different *Candida* sp., including *Candida albicans* and *Candida glabrata* (Costa et al. 1993; McDonald 2009).

7.4 Diagnosis

Implementing advanced diagnostic techniques can significantly enhance the early detection of diseases and improve the prognosis of patients. Most diseases can only be diagnosed based on the observable symptoms, which decreases the likelihood of a positive response to therapy wheras the probability of curing a disease increases with early diagnosis. In contemporary healthcare, biomedical laboratory diagnostics play a crucial role. Nanotechnology has paved the way for new opportunities in the field of diagnostic techniques, leading to more subtle, effective, and economical solutions. Researches and applications of nanotechnology in medicine and veterinary science have gained significant progress. Nanomaterials are effective in preventing, diagnosing, and treating diseases in animals and humans because of their distinctive physical and chemical characteristics and their ability to be well tolerated by living organisms. Rapid diagnostic tools are crucial in poultry and livestock industries, where pathogens and toxins persist. While there are several auspicious diagnostic techniques based on nanomaterial for animal pathogens; many have yet to achieve a

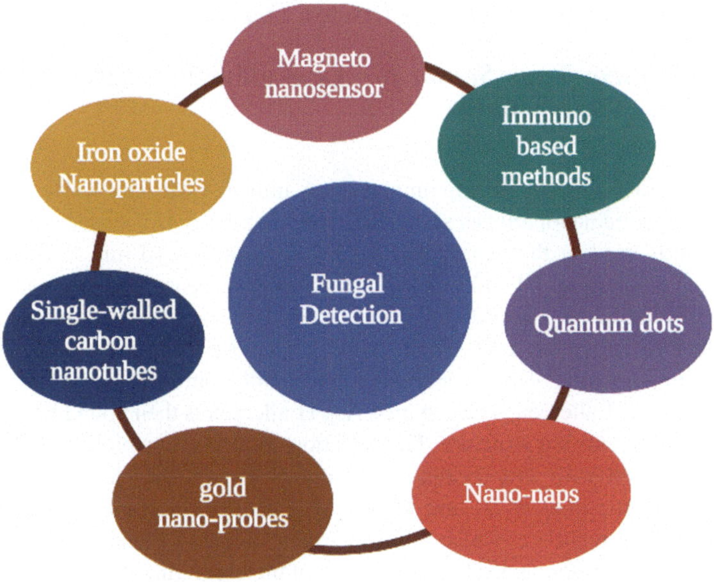

Fig. 7.2 Nanotechnology in detection of various fungal pathogens

balance between sensitivity, specificity, repeatability, and cost-effectiveness. Figure 7.2 represents all the mentioned ways of detection of various fungal pathogens with the help of nanotechnology.

7.5 Fungal Nanotechnology and Biomedicine

The utilization of fluorescent nano-systems such as quantum dots, gold/metal nanoparticles, and carbon-based nanoparticles is being investigated for the early detection of cancer and other diseases by the assessment of biomarkers. The high surface area of nanoparticles provides a significant advantage for the detection of biomarkers. Research is ongoing to develop nanotechnology-based diagnostic modalities that are quite specific, simple with proper identification of various biological molecules including bacteria, viruses, DNA, proteins, and antibodies. This technology holds significant potential to enhance diagnostic capabilities, increasing the diagnosis efficiency and fidelity of samples containing body-fluid. An enterprise in Illinois called Northbrook has created a method for the optical detection of genetic compositions of biological specimens through single target detectors and gold nanospheres. The analyte, tagged with quantum nanocrystals, can be directly analyzed by optical methods such as confocal fluorescence microscopy when they exhibit fluorescence.

7.6 Nano Sensor for Fungal Disease Detection

7.6.1 Magneto-Nano Sensor for Aspergillus fumigatus *Allergen Asp f1*

The magneto-nano sensor biochip immunoassay utilizes GMR spin-valve sensors to quantify magnetic nanoparticles selectively bound to the sensor surface. The prototype biochips used in the experiment have dimensions of 10 mm × 12 mm and consist of an array of 64 individually programmable sensors, each made up of 48 spin-valve strips that are 90 mm × 0.75 mm in size and protected by a biochemically stable ultrathin layer of silicon oxide passivation. The detection procedure involves immobilizing capture antibodies covalently onto the sensor surface to capture the Asp f1 allergen. The captured Asp f1 allergen is then bound by detection antibodies in a sandwich structure. The detection antibodies are biotinylated beforehand and labeled with streptavidin-coated magnetic nanoparticles, which have a diameter of approximately 50 nm. Finally, the magnetic nanoparticles produce a magnetic field in real-time, which influences the oscillating external magnetic field applied during measurement. The perturbation of the oscillating magnetic field leads to alteration in the resistance of the spin-valve, as a result of the Giant Magnetoresistance (GMR) effect. This change in resistance is used to quantify the number of magnetic nanoparticle labels bound to the sensor surface, thereby providing a specific biological signal for measurement. The magneto-nano sensor biochip is expected to have a high potential for the ultra-sensitive detection of several allergic illnesses and fungal pathogen-caused infections. Its detection capability surpasses that of the commercially available ELISA kit (Kim and Wang 2012).

7.6.2 Gold Nanoparticle-Based Sensor

A highly sensitive electrochemical nano-biosensor has been developed for detecting invasive aspergillosis in a miniature experimental setup. The biosensor was created by self-assembling glip probes onto a gold electrode using 1,6-hexanedithiol and chitosan-coated gold nanoparticles. Detection of the glip target gene relied on hybridization and toluidine blue as the indicator molecule. Characterization of the probe attachment was done using cyclic voltammetry, UV-visible spectroscopy, and impedance spectroscopy. The biosensor demonstrated a limit of detection $0.32 \pm 0.01 \times 10^{-14}$ and a dynamic range of 1×10^{-14} to 1×10^{-2} M (Bhatnagar et al. 2018).

7.6.3 Using Gold Nanoprobe

The fungal pathogen *Paracoccidioides brasiliensis* is the causative agent of paracoccidioidomycosis, a disease primarily observed in Central and South American countries. The study aimed to create a new diagnostic method for *P. brasiliensis* using gold nanoprobes, as identifying it can be difficult due to its similarity to other fungal infections. Gold nanoparticles were made by mixing gold chloride and sodium citrate, producing a red solution that absorbed light around 520–530 nm. The process followed with the addition of a unique *P. brasiliensis* genetic sequence to the nanoparticles develops a wine-red color of the solution. The hydrogen bond interaction between the DNA and its complementary sequence allowed the strands to stay together, forming a double-helical structure and preserving the colloid stability. In contrast, the absence of complementary sequences caused the nanoprobes to aggregate, leading to a change in light absorption (Martins et al. 2012).

7.6.4 Single-Walled Carbon Nanotubes

A biosensor based on a field-effect transistor (FET) architecture has been developed to detect *Candida albicans* with improved diagnostic accuracy. Single-walled carbon nanotubes (SWCNTs) were used as a conductive channel, and monoclonal antibodies specific to *C. albicans* were added to create binding sites for fungal antigens. A blocking agent (0.5%) was used to prevent non-specific binding, and the FET devices were able to detect as little as 50 cfu/mL of *C. albicans* in 1 h. The device remained stable for over 10 days. Specificity was checked using non-specific potential organisms *Cryptococcus albidus* and *Saccharomyces cerevisiae* with *C. albicans*. The results indicated that the carbon nanotube-based FET biosensors are effective tools for detecting pathogenic yeasts like *C. albicans* at low concentrations, displaying high selectivity.

A new method for the rapid and accurate detection of pathogenic DNA has been developed, which is feasible to enhance early diagnosis and provides infectious disease treatment. The approach combines a surface-enhanced Raman scattering (SERS) sensor with a patterned gold nanowire (Au NW)-on-film and an exonuclease III-assisted target DNA recycling reaction. Exonuclease III selectively digests the complementary probe DNA, decreasing its concentration, while target DNAs are recycled for detection using the SERS technique. The Au NW-on-film sensor and the target recycling reaction have a sensing limit of 100 femtomolar, allowing for the simultaneous identification of multiple pathogen DNAs. Tests using pathogen's multiple genomic DNA demonstrate the diagnostic capability of this sensor (Yoo et al. 2011).

7.6.5 *Iron Oxide Nanoparticles*

In biomedical research, iron oxide NPs are readily incorporated as magnetic resonance (MR) contrast agents. In MR, signal variations are depicted on high intracellular levels using iron oxide NPs. The impact of endosomal localization on the degradation and magnetic resonance (MR) contrast of iron oxide nanoparticles (NPs) was analyzed by scientists. The results demonstrate that the rate of decaying and decreased MR contrast of NPs is possibly determined by the stability of their coating. It was found that lipid-coated NPs exhibited the highest stability and intracellular clustering, which significantly improved their MR properties and intracellular persistence. However, the rapid degradation of NPs with unstable coatings leads to the complete loss of MR contrast and negative impacts on cell functionality. These findings are of significant importance in the field of biomedical research, as they highlight the importance of selecting NPs with stable coatings to ensure their effectiveness in functional in vivo studies (Soenen et al. 2010). The study showed that the stability of the coating is the main factor determining the degradation rate and reduced MR contrast of NPs. Lipid-coated NPs were found to be the most stable and effective in terms of MR properties and intracellular persistence due to their sturdy nature and intracellular clustering.

7.7 Nanoformulated Napthalocyanines (Nanonaps)

Nanonaps were previously discovered by Zhang et al. (2014), and they possess unique properties including (1) strong absorption in the near-infrared region (NIR) in solution, with a value over 1000, almost 100 times compared with gold nanorods at similar concentrations; (2) the absence of toxicity from heavy metals is a result of the organic composition of nanonaps; (3) nanonaps can perform multispectral imaging, possess broad range NIR region spectral tuning; and (4) the absence of spectral shifts was observed, indicating the multi-color stability of nanonaps at extreme optical densities. The formation of nanonaps was accomplished by utilizing copolymers made from Pluronic triblock consisting of two blocks which are biocompatible made using polyethylene and were kinetically frozen.

The hydrophobic nature of napthalocyanine allows for the removal of excess surfactants through membranes while retaining the dye within the micelles at low temperatures. In this in vivo study, dual-color nanonaps and photoacoustic imaging (PAI) were used to noninvasively delineate two different SLNs and lymphatic systems. Additionally, the depth of imaging penetration in biological tissues was approximately 10 mm accomplished through the use of highly absorbing nanonaps in vivo. Therefore, it is anticipated that nanonaps could be beneficial as PA lymph node markers for breast cancer staging that require minimal invasion or no invasion.

The use of dual-color photoacoustic lymphangiography, utilizing all-organic nano-formulated napthalocyanine (referred to as "nanonaps"), has been demonstrated for the non-invasive, non-ionizing, and high-resolution identification of separate lymphatic drainage systems in vivo. Nanonaps frozen micelles, consisting of two different napthalocyanine dyes with near-infrared absorption at 707 nm or 860 nm, were self-assembled to facilitate this imaging method. According to the result, deeply buried rat lymph nodes beneath a 10 mm thick layer of chicken breast was successfully visualised, indicating the potential of this multispectral photoacoustic imaging technique with nanonaps for the detailed mapping of lymphatic drainage systems, which is important for cancer staging in clinical practice (Lee et al. 2015).

7.7.1 Quantum Dots

A type of semiconductor nanoparticles known as quantum dots show quantum confinement effects' inception. Quantum dots (QDs) are nanoscale semiconductor nanoparticles that exhibit quantum confinement effects, resulting in discrete electronic energy states and unique optical properties. Their tunable band gap energy enables size-dependent absorption and enhances photoluminescent efficiency. The optical properties of QDs, including peak maximum, depend on their nanocrystal size. Surface shells of ligands on colloidal quantum dots (CQDs) determine their solubility and affect their physical and photophysical properties. These extraordinary properties make QDs highly versatile and valuable luminophores in the field of nanotechnology, significantly impacting many modern biology fields. CQDs are commonly used as contrast agents and labels in both in vivo and in vitro bioimaging and biosensing applications (Jeong et al. (2016); Kairdolf et al. 2013; Arya et al. 2005; Le Trequesser et al. 2013). A single quantum dot (QD) typically consists of thousands of atoms group II–VI elements (e.g., CdSe, CdTe, CdS, ZnSe, ZnS, or ZnTe), group I–III–VI2 elements (e.g., AgInS2 or CuInS2), group III–V elements (InAs or InP), group IV elements (e.g., C, Si, or Ge), and group IV–VI elements (e.g., PBSe, PbTe, or PbS) (Martynenko et al. 2017).

A research examines the efficiency of a novel imaging technique that employs nanotechnology for visualizing mammalian reproductive cells in diverse conditions such as in vitro, in situ, and ex vivo. The validation of the proposed nano-based labeling method was carried out using conventional and innovative imaging methods. This approach has the potential to track reproductive cells noninvasively in their physiological environments and measure the molecular and cellular events that affect gamete quality. The study also explores the possibility of utilizing QD-BRET nanoparticles to image spermatozoa in the female reproductive tract, cultured ovarian follicles, and in vitro matured oocytes. However, further experimentation is required to optimize nanoparticle uptake by ovarian follicular cells and subsequent bioimaging. Moreover, the research conducted targeted bioimaging using QD-BRET labeled with anti-porcine plasminogen antibodies (Feugang et al. 2015).

7.7.2 *Immuno-Based Methods*

Nanoparticle-based immunoassays utilize the interactions between antibodies and antigens to improve diagnostic procedures. Nanoparticle-conjugated antibodies are successful biorecognition agents as they are durable and compatible with living systems, and possess the ability to detect low levels of target antigens. These immunoassays have demonstrated superior performance compared to traditional immunoassays and have been widely employed in various diagnostic applications (Vrublevskaya et al. 2017). The identification of particular antibodies in urine samples, animal blood, and serum, can act as a dependable indicator of pathogen infection and exposure. The detection of specific antibodies in animal sera, blood, and urine can serve as a reliable indicator of pathogen exposure and infection (Wu et al. 2016). Nanomaterials can be labeled with different antigens to pick out these antibodies at different stages of infections, providing a valuable tool to detect various infectious agents (Li et al. 2011; Wu et al. 2011; Sayed et al. 2019). The utilization of nanomaterials in diagnostic assays has become a commonly employed approach for the diagnosis of new and recurring infectious diseases in veterinary medicine. Functionalized nanomaterials can be employed in diagnostic assays for the rapid and precise detection of emerging and re-emerging veterinary infectious agents by conjugating either antigens or antibodies, enabling the detection of either antigens or antibodies.

7.8 Future Perspective

A number of fungal infections are reported in livestock. Nanotechnology-based therapies are present against fungi but the diagnostic area is in developing phase. There is a lack of research in diagnostics of fungal infections using nano-interventions. The nano-interventions can help in saving time and money of farmers. Also, livestock plays an important role in the economy of a country. Fungal infections cause harm in many ways to people. Early diagnosis helps in fast treatment which saves the resources. The use of nanotechnology is based on nanomaterials such as metal nanoparticles, inclusions in coating materials, composites synthesized using green chemistry or composites combined with polymers, which enhances the performance of diagnostic method. Improved performance of diagnostic methods also reduces the side effects, especially in case of long-term therapy. Continued advances in the field of nanotechnology have facilitated the search and development of new and more effective diagnostic alternatives for detecting invasive fungal pathogens.

References

Al-Doory Y (1980) Laboratory medical mycology. Lea & Febiger, Philadelphia, p 219

Al-Yasiri MH, Normand AC, Piarroux R, Ranque S, Mauffrey JF (2017) Gut yeast communities in *Larus michahellis* from various breeding colonies. Med Mycol 55(4):436–444

Arya H, Kaul Z, Wadhwa R, Taira K, Hirano T, Kaul SC (2005) Quantum dots in bio-imaging: revolution by the small. Biochem Biophys Res Commun 329(4):1173–1177

Bassetti M, Merelli M, Righi E, Diaz-Martin A, Rosello EM, Luzzati R, Tumbarello M (2013) Epidemiology, species distribution, antifungal susceptibility, and outcome of candidemia across five sites in Italy and Spain. J Clin Microbiol 51(12):4167–4172

Bhatnagar I, Mahato K, Ealla KKR, Asthana A, Chandra P (2018) Chitosan stabilized gold nanoparticle mediated self-assembled gliP nanobiosensor for diagnosis of invasive aspergillosis. Int J Biol Macromol 110:449–456

Brito EH, Fontenelle RO, Brilhante RS, Cordeiro RA, Monteiro AJ, Sidrim JJ, Rocha MF (2009) The anatomical distribution and antimicrobial susceptibility of yeast species isolated from healthy dogs. Vet J 182(2):320–326

Costa EOD, Gandra CR, Pires MF, Coutinho SD, Castilho W, Teixeira CM (1993) Survey of bovine mycotic mastitis in dairy herds in the state of São Paulo, Brazil. Mycopathologia 124: 13–17

Diekema D, Arbefeville S, Boyken L, Kroeger J, Pfaller M (2012) The changing epidemiology of healthcare-associated candidemia over three decades. Diagn Microbiol Infect Dis 73(1):45–48

Feugang JM, Youngblood RC, Greene JM, Willard ST, Ryan PL (2015) Self-illuminating quantum dots for non-invasive bioluminescence imaging of mammalian gametes. J Nanobiotechnol 13: 1–16

Gilroyed BH, Conrad C, Hao X, McAllister TA, Stanford K, Reuter T (2016) Composting for biocontained cattle mortality disposal and associated greenhouse gas and leachate emissions. J Environ Qual 45(2):646–656

Hoffmann K, Pawłowska J, Walther G, Wrzosek M, De Hoog GS, Benny GL, Voigt K (2013) The family structure of the Mucorales: a synoptic revision based on comprehensive multigene-genealogies. Persoonia 30(1):57–76

Jensen HE (1994) Systemic bovine aspergillosis and zygomycosis in Denmark with reference to pathogenesis, pathology, and diagnosis. APMIS Suppl 42:1–48

Jensen HE, Basse A, Aalbcek B (1989) Mycosis in the stomach compartments of cattle. Acta Vet Scand 30:409–423

Jensen HE, Krogh HV, Schønheyder H (1991) Bovine mycotic abortion—a comparative study of diagnostic methods. J Vet Med Ser B 38(1–10):33–40

Jensen HE, Olsen SN, Aalbaek B (1994) Gastrointestinal aspergillosis and zygomycosis of cattle. Vet Pathol 31(1):28–36

Jensen HE, De Los Monteros AE, Carrasco L (1996) Caprine mastitis due to aspergillosis and zygomycosis: a pathological and immunohistochemical study. J Comp Pathol 114(2):183–191

Jeong K, Kim Y, Kang CS, Cho HJ, Lee YD, Kwon IC, Kim S (2016) Nanoprobes for optical bioimaging. Opt Mater Express 6(4):1262–1279

Jungerman PF, Schwartzman RM (1972) Veterinary medical mycology. Lea & Febiger, Philadelphia

Kairdolf BA, Smith AM, Stokes TH, Wang MD, Young AN, Nie S (2013) Semiconductor quantum dots for bioimaging and biodiagnostic applications. Annu Rev Anal Chem 6:143–162

Kim D, Wang SX (2012) A magneto-nanosensor immunoassay for sensitive detection of *Aspergillus fumigatus* allergen Asp f 1. IEEE Trans Magn 48(11):3266–3268

Le Trequesser Q, Seznec H, Delville MH (2013) Functionalized nanomaterials: their use as contrast agents in bioimaging: mono-and multimodal approaches. Nanotechnol Rev 2(2):125–169

Lee C, Kim J, Zhang Y, Jeon M, Liu C, Song L, Kim C (2015) Dual-color photoacoustic lymph node imaging using nanoformulated naphthalocyanines. Biomaterials 73:142–148

Lewis RE (2009) Overview of the changing epidemiology of candidemia. Curr Med Res Opin 25(7):1732–1740

Li H, Ding X, Peng Z, Deng L, Wang D, Chen H, He Q (2011) Aptamer selection for the detection of K88. Can J Microbiol 57(6):453

Martins JF, Castilho ML, Cardoso MA, Carreiro AP, Martin AA, Raniero L (2012) Identification of *Paracoccidioides brasiliensis* by gold nanoprobes. In: Biomedical vibrational spectroscopy V: advances in research and industry, vol 8219. SPIE, pp 144–149

Martynenko IV, Litvin AP, Purcell-Milton F, Baranov AV, Fedorov AV, Gun'Ko YK (2017) Application of semiconductor quantum dots in bioimaging and biosensing. J Mater Chem B 5(33):6701–6727

Maubon D, Garnaud C, Calandra T, Sanglard D, Cornet M (2014) Resistance of Candida spp. to antifungal drugs in the ICU: where are we now? Intensive Care Med 40:1241–1255

McDonald (2009) Mastitis in cow. Dairy Cattle Production. Department of Animal Science, Montreal, pp 1–12

Moran G, Coleman D, Sullivan D (2012) An introduction to the medically important *Candida* species. In: Calderone RA, Clancy CJ (eds) Candida and candidiasis. ASM Press, Washington, DC

Natural Resources Conservation Service (2015) New emergency animal mortality management (368) conservation practice standard and highly pathogenic avian influenza-webinar [WWW Document]. http://www.conservationwebinars.net/webinars/emergency-animal-mortality-standard-and-hpai. Accessed 15 Jan 19

Odds FC (1988) Candida and candidosis: a review and bibliography. Bailliere Tindall, London

Ortega J, Uzal FA, Walker R, Kinde H, Diab SS, Shahriar F, Read DH (2010) Zygomycotic lymphadenitis in slaughtered feedlot cattle. Vet Pathol 47(1):108–115

Pacynska JA (2013) Yeast-like fungi and yeasts isolated from healthy breeding horses. Pol J Vet Sci 16(1):69

Pfaller MA, Diekema D (2007) Epidemiology of invasive candidiasis: a persistent public health problem. Clin Microbiol Rev 20(1):133–163

Różański P, Ślaska B, Różańska D (2013) Prevalence of yeasts in English full blood mares. Mycopathologia 175:339–344

Sayed RH, Abousenna MS, Mohamoud D, Saad MA (2019) Development of a lateral flow kit for detection of IgG and IgM antibodies against rift valley fever virus in sheep. Indian J Vet Sci Biotechnol 15(2):63–68

Scott NR (2007) Nanoscience in veterinary medicine. Vet Res Commun 31:139–144

Soenen SJ, Himmelreich U, Nuytten N, Pisanic TR, Ferrari A, De Cuyper M (2010) Intracellular nanoparticle coating stability determines nanoparticle diagnostics efficacy and cell functionality. Small 6(19):2136–2145

Vrublevskaya VV, Afanasyev VN, Grinevich AA, Skarga YY, Gladyshev PP, Ibragimova SA, Morenkov OS (2017) A sensitive and specific lateral flow assay for rapid detection of antibodies against glycoprotein B of Aujeszky's disease virus. J Virol Methods 249:175–180

Wilkinson K, Grant WP, Green LE, Hunter S, Jeger MJ, Lowe P, Waage J (2011) Infectious diseases of animals and plants: an interdisciplinary approach. Philos Trans R Soc B Biol Sci 366(1573):1933–1942

Wu S, Duan N, Zhu C, Ma X, Wang M, Wang Z (2011) Magnetic nanobead-based immunoassay for the simultaneous detection of aflatoxin B1 and ochratoxin A using upconversion nanoparticles as multicolor labels. Biosens Bioelectron 30(1):35–42

Wu L, Yin W, Tang K, Shao K, Li Q, Wang P, Han H (2016) Highly sensitive enzyme-free immunosorbent assay for porcine circovirus type 2 antibody using Au-Pt/SiO_2 nanocomposites as labels. Biosens Bioelectron 82:177–184

Yallapu MM, Jaggi M, Chauhan SC (2010) Scope of nanotechnology in ovarian cancer therapeutics. J Ovarian Res 3:1–10

Yapar N (2014) Epidemiology and risk factors for invasive candidiasis. Ther Clin Risk Manag 10: 95–105

Yoo SM, Kang T, Kang H, Lee H, Kang M, Lee SY, Kim B (2011) Combining a nanowire SERRS sensor and a target recycling reaction for ultrasensitive and multiplex identification of pathogenic fungi. Small 7(23):3371–3376
Zhang Y, Jeon M, Rich LJ, Hong H, Geng J, Zhang Y, Lovell JF (2014) Non-invasive multimodal functional imaging of the intestine with frozen micellar naphthalocyanines. Nat Nanotechnol 9(8):631–638

Chapter 8
Nanotechnology in Metabolic Disease Diagnosis of Livestock Animals

Swaraj Biswas, Suman Biswas, Subhasis Batabyal, Apratim Maity, Shamik Polley, Shyam Sundar Kesh, and S. Suvetha

Abstract Aberration in metabolic homeostasis resulting into under-utilization, over-production, deficiency or excessive storage of certain metabolites may lead to the onset of metabolic disorders. The most prevalent metabolic diseases that occur in livestock are ketosis, parturient paresis, diabetes mellitus, ovine pregnancy toxaemia, hypoglycaemia in baby pigs, hyperlipidaemia, cardiovascular diseases and genetically linked metabolic storage diseases. Deranged metabolic pathways often resulting from deficient enzymes or hormones, nutritional deficiencies, and genetic inheritance are the most common predisposing factors although acquired metabolic disorders are not uncommon in animals. Metabolic disorders such as diabetes mellitus can affect individual subjects or it can be manifested as herd disease like underfeeding ketosis affecting a group of animals. Despite having severe impact over optimal animal health and production, most of the metabolic disorders remain unnoticed, particularly in subclinical cases. Poor awareness and management, low productivity, cost-effectiveness concern and lack of adequate diagnostic facilities are some of the major factors obstructing the stewardship of such metabolic disorders. Conventional biochemical diagnostic methods or diagnostic imaging modalities often lack adequate sensitivity, rapidity, or resolution for early and precise detection

S. Biswas (✉) · S. Batabyal · A. Maity · S. Polley
Department of Veterinary Biochemistry, Faculty of Veterinary and Animal Sciences, West Bengal University of Animal and Fishery Sciences, Kolkata, West Bengal, India

S. Biswas
Department of Avian Sciences, Faculty of Veterinary and Animal Sciences, West Bengal University of Animal and Fishery Sciences, Kolkata, West Bengal, India

S. S. Kesh
Department of Veterinary Clinical Complex (Veterinary Biochemistry), Faculty of Veterinary and Animal Sciences, West Bengal University of Animal and Fishery Sciences, Kolkata, West Bengal, India

S. Suvetha
Department of Genetics and Plant Breeding, Institute of Agricultural Science, Rajiv Gandhi South Campus, Banaras Hindu University, Mirzapur, Uttar Pradesh, India

LLR University of Veterinary and Animal, ICAR-National Research Centre on Equines, *Nanotechnology Theranostics in Livestock Diseases and Management*, Livestock Diseases and Management,
https://doi.org/10.1007/978-981-16-1610-5_8

of the aberration. In addition, lack of portability of the usual diagnostic modalities is another major issue for animal disease diagnosis. The progress in nanotechnology has led to the development of diverse rapid, sensitive, and accurate diagnostic techniques to overcome the issues. Several nanomaterial-based sensors have been developed which even facilitate portable animal-side disease diagnosis. Advancement in nano-science has also paved the way for customizing novel contrast agents to improve the resolution of different bioimaging modalities such as computed tomography (CT), magnetic resonance imaging (MRI), ultrasonography, radiographic imaging, etc. Nanomedicine has already shown ample potential in disease diagnosis including the metabolic disorders, although it is yet to full-bloom, but definitely holds a bright prospect in near future.

Keywords Livestock metabolic disorders · Ketosis · Parturient paresis · Diabetes mellitus · Hyperlipidaemia · Cardiovascular diseases · Genetic storage diseases · Nanoparticles · Biosensors

8.1 Introduction

In context of biological sciences, nanotechnology encompasses the materials, devices and systems whose structure and function are relevant for small length scales or nanoscale (1–100 nm) (Whitesides 2003). The nanotechnology was first used in the year of 1974 for manipulation of materials. In veterinary science, the term nanotechnology is applied in the form of nanobiotechnology, nanomedicine and nanotheranostics. The first term nanobiotechnology is used to describe the applications of nanotechnology in wider portion of biological science including veterinary science (Troncarelli et al. 2013). Likewise, nanomedicine is defined as applications of nanotechnological tools for efficient and faster management of animal diseases. Similarly, nanotheranostics are the advanced formulations which hold both diagnostic and therapeutic character in one time application. These applications are helpful for targeted drug delivery, understanding of different biological and disease processes, treatment of infected or malignant cells, specific hormone and enzyme therapy, etc. (El-Sayed and Kamel 2020). But the action of nanomaterials depends on the body pH, temperature and presence of certain chemicals. Nowadays, smart delivery system can be formulated which is self-regulatory (Manuja et al. 2012).

Moreover, recently developed nano-biomaterials having structural similarity with the different body receptors, DNA and proteins and also having bio-degradable and lipid-based carrier properties. Their use prevents long-term toxicity and also facilitates interaction with cellular membrane and tissues at molecular level. They are easily engulfed and degraded by circulating macrophages (Dobrovolskaia et al. 2016).

In veterinary science, nanotechnology has provided considerable respite by eliminating the threat from various dreadful intracellular pathogens and blood parasites through early diagnosis, effective therapeutics or even preventive measures (Troncarelli et al. 2013). In current scenario, maximization of animal production is the prime focus in livestock farming. Health status of animals exclusively reflects the productivity. But different types of emerging and re-emerging diseases are continuously affecting the animal health every year. Along with these diseases, metabolic diseases are also economically important and having direct relation with farmers profit and livelihood. The rapid detection of these metabolic diseases is very important because they are silent killer of the high yielders. But lack of adequate veterinary facilities in terms of infrastructure, resources, modern laboratory facilities and cost-effective methods are the major concern to address these metabolic diseases. This gap can be bridged by adept nanointerventions such as introduction of high-sensitivity biosensors, state-of the art imaging modalities, point-of-care diagnostics, etc.

8.2 Nanoparticles in Metabolic Disease Diagnosis

8.2.1 Ketosis

Bovine ketosis is an important metabolic disease during peri- and post-parturient periods of high producing animals. Subclinical ketosis may rise 40% and above in dairy herds. The subclinical from of ketosis is mostly measured by the presence of β-hydroxy butyric acid (BHBA) in milk and blood. Previously human diabetic ketoacidosis methods were employed for the detection of β-hydroxy butyrate (BHB) in blood. But this method was unscientific due to different blood grouping system in cattle and antigen expression (Neethirajan 2017). Therefore, the sensors that have been developed based on microfluidic systems are capable of accurately measuring BHBA in animal samples due to high sensitivity as well as specificity and remain as very few options to serve the purpose (Neethirajan et al. 2016). Plasma BHBA concentrations in normal cattle are less than 1000 μmol/L, while those in subclinical ketosis are larger than 1400 μmol/L and exhibit diurnal variation (Radostits et al. 2007). For the first 30 days in milk, every 0.1 mmoL/L increase in BHB at preclinical ketosis is related with a reduction in milk output of 0.5 kg/day (Han van der Kolk et al. 2017). Different durable and portable instruments can be used for the quick and effective detection of subclinical ketosis. Electrochemical sensor based on BHB was developed in the year of 2013 that covalently immobilize the NAD+ on single-walled carbon nanotubes (SWCNT) on the surface of the screen-printed electrode (Khorsand et al. 2013). This sensor-based method can provide paper-based result of BHB up to 0.009 mmol/L. By immobilizing the 3-hydroxybutyrate dehydrogenase enzyme, an electrochemical biosensor has also been used to detect β-hydroxy butyrate (Veerapandian et al. 2016). Ferrocenemethanol and thionine were used to create a screen-printed biosensor. Microfluidic biosensor has also been employed to detect BHB which is having high sensitivity and specificity, provides results within 1 min with detection limit 0.05 nM (Weng et al. 2015). Researchers studied how BHB behaved electrochemically during its oxidation in the presence of nicotinamide adenine dinucleotide and β-hydroxybutyrate dehydrogenase, and they suggested a chronoamperometric approach to measure BHB with a detection limit of 16 M (Feng and Liu 2015).

8.2.2 Parturient Paresis

Parturient paresis or hypocalcaemia is a metabolic disease of dairy cows characterized by decline in plasma calcium concentration soon after calving. In normal physiological state or during pregnancy period, if blood calcium concentration fluctuates between 8.5 and 10 mg/dL, cows may suffer from subclinical hypocalcaemia. Parturient paresis may start to develop if calcium concentration falls below 8.5 mg/dL and typically reaches below 5.5 mg/dL. Hypocalcaemia in subclinical stages may induce displacement of abomasum, uterine prolapse, retained placenta, postpartum paralysis, endometritis, ketosis, etc. In clinical hypocalcaemia, animals lie on the groud with flaccid limb paralysis; loss of appetite and coma or even death may occur (Oetzel 2011). So, in order to treat postpartum hypocalcaemia or subclinical hypocalcaemia during pregnancy, various rapid blood calcium detection methods have been developed. Parturient paresis could also be developed in pregnant and lactating ewes and does. With the use of nanotechnology-based biosensors, it is easy to estimate blood calcium in a more accurate, reliable and affordable way.

An octadecyltrichlorosilane-cellulose paper-based microfluidic field-effect biosensor coupled with single-walled carbon nanotubes and DNAzyme was developed for rapid and inexpensive detection of calcium in blood (Wang et al. 2022). Octadecyltrichlorosilane was used to increase the mechanical strength and hydrophobicity of cellulose paper. Semiconducting SWCNTs and DNAzyme were used in order to produce ultra-sensitive detection, minimize detection time with supreme accuracy. This biosensor could detect calcium concentration ranged between 25 μM and 5 mM, with a lower detection limit of 10.7 μM. This dual-field-effect biosensor was employed to minimize the interference of impurities and superior detection of hypocalcaemia in blood. A new electrochemical biosensor based on nitrogen-doped graphene modified glass carbon electrodes, gold nanoparticles (Au NPs) and DNAzyme was developed by Yu et al. (2022). Nitrogen-doped graphene and gold nanoparticles were used for better electrical conductivity, electrocatalytic activity, sensitivity and selectivity. This instrument depicted detection range of 5–50 μM and 50–400 μM, with lowest detection level up to 3.8 μM. This biosensor was equally sensitive and accurate to detect serum calcium level in comparison to atomic absorption spectrometry.

8.2.3 *Diabetes Mellitus*

Diabetes mellitus (DM) is a chronic disorder due to impaired metabolism of carbohydrate linked with faulty or complete deficiency of insulin. Most of times small animals like dogs and cats are affected, particularly middle-aged dogs and cats older that 7 years. Dogs are prone to develop insulin dependent diabetes mellitus (type I, IDDM), while cats are mostly affected by diabetes mellitus type II similar to human type. Poorly controlled IDDM of dogs may lead to ketoacidosis similar to human type I diabetes mellitus (Qiu et al. 2022). Some dog breeds are commonly affected like Dachshunds, Cairn Terriers, Miniature Poodles, Australian terrier, Beagles, etc. and cat breeds are Norwegian Forest Cat, Tonkinese, and Burmese (Greco 2018; Dotsenko et al. 2022). Persistent hyperglycaemia with concurrent glycosuria is the major clinical sign in DM. Fasting blood glucose level in dogs and cats ranges between 75 and 120 mg/dL. Conventional diagnostic methods for DM include estimation of fasting blood glucose level, glucose tolerance test and measurement of glycated haemoglobin (HbA1c) level in blood. Estimation of blood fructosamine is helpful in determining DM in cats from stress-induced hyperglycaemia. All of this procedure depends on age of the animal, testing time, food habits, stress condition and various other physiological factors. Also, measurement of HbA1c needs sophisticated instruments and blood processing time. Biosensors built on cutting-edge nanotechnologies have been designed to enable quick and invasive or non-invasive diabetes testing in order to get around the drawbacks of these conventional diagnostic techniques. Most of the biosensors were tested using human sample but it seems to be also useful in diagnosing the diabetic biomarker of small animals.

Acetone level in breath could be an important non-invasive biomarker for detection of DM. Normal concentration in human breath ranges between 0.3 and 0.9 ppm but it exceeds >1.8 ppm in diabetic people (Deng et al. 2004). Mostly WO_3 and SnO_2 based chemiresistive nano-biosensors were used for the detection of acetone in human breath. Normal acetone level in animal breath particularly in dogs and cats is still not measured as biomarker for DM. In future days, this metal oxide-based chemiresistive nano-biosensors can be utilized for the detection of DM in animals (Righettoni et al. 2015; Gong et al. 2020). Invasive biomarkers of DM, i.e., blood glucose, insulin and HbA1c were mostly detected using electrochemical and optical detection methods.

8.2.3.1 Electrochemical Method

Electrochemical biosensors provide an appealing point-of-care strategy in clinical diagnosis of DM. This electrochemical biosensor combined with nanostructural materials creates a highly sensitive and specific, low-cost diagnostic tool for DM. Sensor based on gold nano-flowers coated screen-printed carbon electrode could be used for quantitative estimation of HbA1c. HbA1c could be detected in the range of 2–20% (Wang et al. 2019). Using the similar mechanism, nanocomposites combination of reduced graphene oxide and gold was used for development for another biosensor. HbA1c specific aptamers were linked to the electrode for efficient binding of HbA1c (Jaberi et al. 2019). Multi-wall carbon nanotube-based electrochemical biosensor was also used for dual detection of HbA1c and total Hb (Moon et al. 2017). HbA1c and Hb could be detected in a finger prick blood. Like HbA1c, glucose and insulin were also detected particularly using metal or graphene nanoparticles. Among metals, gold and silver nanomaterial-based biosensors are the most commonplace in use (Ge et al. 2019).

8.2.3.2 Optical Method

Diabetic biomarker detection using optical sensors based upon colorimetry, fluorescence, chemiluminescence, surface plasmon resonance and surface-enhanced Raman scattering, etc. are the most important ones. The principle of fluorescent nano-biosensor is to label the specific target biomolecule, and standard fluorophores are incorporated in nanoparticle matrices. Most commonly used fluorescent sources are small organic fluorescent molecule, quantum dots, gold nanoclusters, single-walled carbon nanotubes, upconversion nanoparticles, etc. A fluorescent microfluidic system was developed by Li and co-workers for estimation of blood HbA1c (Li et al. 2016). Here, a fluorescent aptamer specific to HbA1c was used for accurate detection of glycated haemoglobin. Similarly, glucose oxidase enzyme coated with nanoparticles, coumarin 6 fluorescent dye and an oxygen sensitive probe were used to develop a biosensor for the detection of glucose (Gao et al. 2016). Enzyme-mimicking nanosensor was developed using AuNPs quantum dots-encapsulated mesoporous silica microsphere. AuNPs mimics as glucose oxidase that could oxidize glucose and produce H_2O_2 which ultimately quenched the photoluminescence signal of quantum dots and detected the plasma glucose level. Zubkovs et al. developed a biosensor based on single-walled carbon nanotubes wrapped with glucose oxidase. SWCNTs could sense fluorescent signal in near-infrared range and detect glucose in blood (Zubkovs et al. 2017). Carboxyl-modified graphene oxide nanoenzyme has the ability to mimic the horseradish peroxidise enzyme activity. So, this strategy was utilized to oxidize chromogenic substrates in order to produce coloured products and used for colorimetric detection of glucose in biofluids (Song et al. 2010). Similarly, graphene combined with Fe_3O_4 nanoparticle and MoS_2 and graphene oxide (MoS_2/GO) nanohybrid were replaced in order to improve peroxidase-like activity, sensitivity and reliability of glucose detection (Wang et al. 2017; Peng and Weng 2017). A further Janus nanoparticle was established with both glucose oxidase and peroxidase-like activity. Using this property, sensor has been customized for glucose determination in plasma (Lu et al. 2015). Now another multianalyte detection system has been developed by Ahmad et al. for the detection of glucose, cholesterol and urea. Integrated field-effect transistors biosensor based on ZnO nanorods could efficiently detect glucose, cholesterol and urea in mice and dog blood (Ahmad et al. 2015).

8.2.4 *Hyperlipidaemia*

Hyperlipidaemia is defined as the condition in which total cholesterol, triglyceride or low-density lipoprotein (LDL) levels become more than 90th percentile on comparing to the population. Abnormal lipid metabolism which includes the disorder in the synthesis and degradation mechanism carried out in liver or the transport of lipid molecules is responsible for the hyperlipidaemic condition. It is classified into two, namely familial (primary) and acquired (secondary) hyperlipidaemia. The primary case is due to genetic disorders, whereas the secondary case is predominantly because of the unhealthy diet, uncontrolled diabetes, hypothyroidism and certain medications such as glucocorticoids. The lipid level, particularly of total cholesterol and triglyceride when rises above the normal limit (178 ± 38 mg/dL and 20–112 mg/dL respectively, in dog), it is said to be hyperlipidaemia. It is the major cause of coronary heart disease and atherosclerosis. Thus, it has to be diagnosed at early stage (Kaneko et al. 2009; Hill and Bordoni 2022). However, it has no specific symptom to characterize. Most of the time hyperlipidaemia becomes the secondary cause for symptomatic diseases such as coronary artery disease, peripheral arterial disease and acute pancreatitis. It is usually determined by monitoring the serum lipid profile and C-reactive protein. The assistance of nanoparticles in quantitative estimation of lipid profile has been demonstrated in multiple literatures. Wang et al. (2014) found that the lipoprotein profile can be examined using gold nanoparticles conjugated with small, dense and low-density lipoprotein (sdLDL). As a result, it improved the peak efficiency. Further the early detection of atherosclerotic diseases was made feasible by obtaining the lipoprotein profile via microchip-based capillary electrophoresis assay. Similarly, electrochemical cholesterol biosensor was developed by Alagappan et al. (2020). In this biosensor, cholesterol oxidase (ChOx) enzyme is immobilized on gold nanoparticles functionalised by multi-walled CNT and the electrode was modified using a nanocomposite, polypyrrole (PPy). As the cholesterol level increases, the current in cyclic voltammeter decreases. The AuNPs-f-MWCNT-PPy-ChOx/GCE showed a detection limit and sensitivity at 0.1×10^{-3} M and 10.12 μA mM^{-1} cm^{-2}, respectively.

8.2.5 *Cardiovascular Diseases*

Cardiovascular diseases (CVDs) are the major cause of increasing mortality rate in most of the developing nations. It causes around 32% of the death occurring in the world and in that 80% is due to stroke and heart attack among human population (World Health Organization 2021). The most common causes of CVD are cholesterol metabolic disorder and atherosclerosis which ultimately result in coronary artery diseases (CAD). Conventional diagnostic methods for CAD are electrocardiography and imaging techniques. These methods can be enhanced for early diagnosis with the help of nanotechnology. The detection biomarkers of CAD, namely myoglobin (Myo), C-reactive protein (CRP), and cardiac troponins (cTns) which are released in blood stream under abnormal heart condition can be detected precisely by binding it to nanoparticles that helps in minimizing the nonspecific adsorption in mass spectrometry. Biomarker cardiac troponins (cTns) bound by GPRu-Au, Au-CNN were sensed by antibody and quantitatively detected using electrochemiluminescence method at 10 fg/mL–10 ng/mL range (Ye et al. 2019). Similarly, Ji et al. used upconverting nanoparticles (UCNPs) to enhance the recognition of myoglobin, and this was effectively detected by antibodies embedded in lateral flow assay (Ji et al. 2019). Other than improving the detection of biomarkers by biosensing methods such as electrochemistry, chemiluminescence, electrochemiluminescence, photo-electrochemistry, electrochemical impedance spectroscopy, lateral flow assay and surface-enhanced Raman spectroscopy; nanoparticles were also used to enhance the contrast in imaging techniques such as MRI, optical imaging, nuclear scintigraph and PAI (Photo Acoustic Imaging). P-selectin on activated ECs (endothelial cells) and platelets were targeted by the self-assembled nanoparticle complex of low-molecular-weight fucoidan and thermolysin-hydrolysed protamine peptide (Gd-TPP_{1880}/$LMWF_{8775}$ CNPs) in order to visualize the inflammatory endothelial cells under MRI by targeting the P-selectin, penetrating the target cell and loading Gd-DTPA (Hu et al. 2022).

8.2.6 Genetic Storage Diseases

There are number of genetic disorders inherited from parent to offsprings that interfere in the metabolism. Such metabolic disorders are called genetic storage metabolic diseases. Familial hypercholesterolemia, Gaucher disease, Hunter syndrome, Krabbe disease, Maple syrup urine disease, Metachromatic leukodystrophy, mitochondrial encephalopathy, lactic acidosis, stroke-like episodes (MELAS), Niemann–Pick, Phenylketonuria (PKU), Porphyria, Tay–Sachs disease and Wilson's disease are some of the most common metabolic disorders. Along with human beings, canines are also susceptible to several aforesaid genetic storage metabolic diseases. For instance, several dog breeds such as Beagle crosses, German Shorthaired Pointers, Japanese Spaniels, Portuguese Water Dogs, etc. and cat breeds like Siamese, Korat, domestic cats are susceptible to GM1 and GM2 gangliosidosis. Similarly, German Shepherd and Poodle dogs as well as Siamese and domestic shorthair cats are prone to Niemann–Pick Disease (sphingomyelinosis), an autosomal recessive hereditary disorder due to lack of sphingomyelinase enzyme. Australian Silky Terriers and Dalmatian dogs are susceptible to glucocerebrosidosis while ceroid lipofuscinosis may occur in dog breeds like English Setters, Cocker Spaniels, Dachshunds, Chihuahuas, Salukis, Border Collies, etc. and also in domestic cats. Glycogenosis occurs in Silky Terrier dogs and in some cats. Mucopolysaccharidosis occurs in several dog breeds that include but not limited to Rottweilers, Labrador Retrievers, German Shepherd, Rat Terriers, Wirehaired Dachshunds, etc. (MSD Manual 2022) (https://www.msdvetmanual.com/metabolic-disorders/metabolic-disorders-introduction/metabolic-storage-disorders-and-inborn-errors-of-metabolism-in-animals). Genetic storage diseases are not only restricted to monogastric animals but several large as well as small ruminants suffer from such kind of metabolic disorders. For instance, alpha-mannosidosis has been reported in large animals like Angus, Simmental, Galloway and Holstein cattle while beta-mannosidosis is found in Saler cattle as well as Nubian and Nubian-cross goats. Generalized glycogenosis is reported in Holstein, Shorthorn and Brahman cattle as well as in Suffolk sheep and pigs. Globoid cell leukodystrophy (Krabbe disease), a lethal demyelinating disease characterized by the formation of giant, multinucleated globoid cells occurs in polled Dorset sheep. Maple syrup urine disease due to impaired branched chain amino acid metabolism occurs in Hereford and polled Shorthorn cattle. Goitre, Ruminal Parakeratosis, osteogenesis imperfect, photosensitization, Ehlers–Danlos syndrome, hemochromatosis, porcine stress syndrome, etc. are the other metabolic diseases that also occur in sheep, cattle and pigs (MSD Manual 2022) (https://www.msdvetmanual.com/metabolic-disorders/metabolic-disorders-introduction/metabolic-storage-disorders-and-inborn-errors-of-metabolism-in-animals). Most of such disorders make the system deficit for certain enzymes which leads the individual to suffer. Few disorders can be diagnosed by the screening test carried out in newborns, whereas others could be detected only when the individual shows symptoms. When the symptom appears, it can be diagnosed by screening the DNA (Mayo Clinic 2017) (https://www.mayoclinic.org/diseases-conditions/inherited-metabolic-disorders/symptoms-causes/syc-20352590). Many of these have specific biomarkers which help to diagnose the defective condition. In case of phenylketonuria, the initial test has a high risk of false positive. This can be normalized, and PKU can be diagnosed accurately by using an electrochemical nano-biosensor in which the carbon electrode is coated with gold nanoparticles attached to the reduced graphene sheet. Over this gold nanoparticles, a specific alkanethiol ssDNA probes (complementary oligo of the mutated PHA gene, i.e., phenylalanine hydroxylase) are fastened along with the inter chelating electrochemical label, oracet blue. This method is found to have detection limit of 21.3 fM and a great specificity (Seifati et al. 2018).

8.2.7 *Nanoparticle-Assisted Molecular Imaging in Metabolic Disease Diagnosis*

Various molecular imaging modalities such as optical imaging, surface-enhanced Raman scattering (SERS) ultrasonography, photoacoustic (PA) imaging, magnetic resonance imaging (MRI), computed tomography (CT), positron emission tomography (PET), single-photon emission computerized tomography (SPECT), nuclear scintigraphy, radiographic imaging, etc. along with multimodal imaging have already proved their utility in diagnosis of several metabolic disorders that include but not limited to cardiovascular diseases, diabetes and insulin resistance, abdominal obesity, various organ damage like fatty liver and fibrosis, ischemia, inflammation, aberration in brain macrostructure, renal insufficiency, occlusion and clearance (Han et al. 2010; Kim and Lerman 2018; Prasad et al. 2021; Hu et al. 2022). Despite the proven efficiency of these imaging modalities, they often suffer from poor resolution and contrast, high signal to noise ratio, lack of portability for point-of-care diagnosis, etc. Incorporation of nanomaterials in these imaging platforms has assisted to overcome the aforesaid difficulties to a great extent. For instance, electrochemiluminescence immunosensor based on the enhanced optical properties of metal organic framework (MOF)-CdTe quantum dot nanocomposites, sensitive detection of a biomarker of myocardial infarction, cardiac troponin-I antigen has been carried out with a very low detection limit of 0.46 fg/mL (Yang et al. 2018). An aptasensor comprised of nanodiamonds (NDs) and hydrogen-substituted graphdiyne (HsGDY) has efficiently sensed myoglobin and cardiac troponin I with ultra-low detection limits of 9.04 and 6.29 fg/mL, respectively (Wang et al. 2021). Even considering the portability of the imaging modality, a smartphone-based point-of-care diagnostics has been customized targeting the myocardial infarction related miRNAs in serum with sensitivity as low as 1 fM when seen with naked eyes. Here, a chemiluminescence (CL) imager has been developed using a hemin-functionalized MOF that is having an inherent peroxidase-like catalytic activity as well as enhanced surface area facilitating enhanced binding of G-quadruplex (G4) DNAzymes produced by miRNA-triggered rolling circle amplification reactions, thus displaying 100-fold intense catalytic activity in the luminol/H_2O_2 CL system at the MOF-bound interface as compared to in-solution (Mi et al. 2020). Ferumoxtran-10, a dextran-coated iron oxide nanoparticle has been employed as MRI contrast agent for monitoring pancreatitis and islet inflammation as these NPs are readily taken up by macrophages due to smaller size and conducive surface properties. Several NP-based contrast agents have been employed for early diagnosis of type-I diabetes. Magnetic NPs such as SPION-based exendin-4 have been used as contrast agents for MRI in beta cell imaging. Nanoparticle-based implantable glucose sensor has also been employed for continuous and real-time monitoring of blood glucose to detect diabetes (Veiseh et al. 2015). Even, carbon nanotubes and graphene-polypyrrole nanocomposites have been employed for electrochemical detection of insulin (Lemmerman et al. 2020). Moreover, application of angiotensin II-functionalized Ag_2S nanodots in infrared fluorescence imaging of ischemic myocardial tissues, Gd-DTPA contrast agent loaded into a self-assembled complex nanoparticles comprised of depolymerized low-molecular-weight fucoidan (LMWF8775) and a thermolysin-hydrolysed protamine peptide (TPP1880) for MRI of inflammatory endothelial cells, anti-CD36 decorated semiconducting polymer NPs for photoacoustic imaging of inflammatory cells are certain examples of NP-assisted molecular imaging to detect CVD. Folate conjugated two-dimensional (2D) Pd@Au nanoprobe has facilitated multimodal imaging of advanced atherosclerotic plaques using SPECT, CT and PA imaging (Hu et al. 2022). Strawberry-like Fe_3O_4-Au hybrid nanoparticles have been employed as dual-modality contrast agent for MR and CT imaging to detect progressive liver disease such as fatty liver, cirrhosis, hepatocellular carcinoma, etc. The contrast agent has facilitated 34.61-fold contrast enhancement in MRI, and the highest Hounsfield Unit (HU) obtained in CT was 174 at 30 min post-injection of the NPs (Zhao et al. 2015). The aforesaid instances vividly suggest that the impact of nanotechnology in the domain of diagnostic imaging is certainly going to escalate in upcoming days.

8.3 Conclusion and Future Prospect

The key to improved animal productivity is an early and precise identification of pathogen, management and treatment of complications. Numerous nanotechnology-based detection methods or biosensors have recently been customized for the diagnosis of metabolic diseases of animals. The objective of the chapter is to examine the literature on the use of nanotechnology-based disease diagnosis methods in metabolic disorders of animals. Although nanomaterials has several benefits, but there are still a number of issues that need to be resolved. The important concerns for future progress are the effective, quick, accurate and affordable detection of biomarkers.

Most of the metabolic disorders are either get unnoticed or least cared in case of livestock species. Howbeit, the monogastric animals such as dog, cat, bear, etc. have considerable similarity with human beings. Therefore, they hold similar kind of potential risk towards various metabolic disorders. In the context of progressively gained attention towards pet management, nano-diagnostics of metabolic disorders in pets can yield significant opportunity in upcoming days. In addition, conventional diagnostic imaging for animals is getting common place progressively. The nanotechnology can also fit into the niche area to enhance the potency of different diagnostic imaging modalities such as MRI, ultrasound imaging, positron emission tomography (PET) and computed tomography (CT). Although the necessity is being perceived late, the scope of nanotechnology in diagnosis of metabolic disorders in animals seems to be bright enough imminently.

References

Ahmad R, Tripathy N, Park JH, Hahn YB (2015) A comprehensive biosensor integrated with a ZnO nanorod FET array for selective detection of glucose, cholesterol and urea. Chem Commun 51(60):11968–11971. https://doi.org/10.1039/C5CC03656A

Alagappan M, Immanuel S, Sivasubramanian R, Kandaswamy A (2020) Development of cholesterol biosensor using Au nanoparticles decorated f-MWCNT covered with polypyrrole network. Arab J Chem 13(1):2001–2010. https://doi.org/10.1016/j.arabjc.2018.02.018

Deng C, Zhang J, Yu X, Zhang W, Zhang X (2004) Determination of acetone in human breath by gas chromatography–mass spectrometry and solid-phase microextraction with on-fiber derivatization. J Chromatogr B 810(2):269–275. https://doi.org/10.1016/j.jchromb.2004.08.013

Dobrovolskaia MA, Shurin M, Shvedova AA (2016) Current understanding of interactions between nanoparticles and the immune system. Toxicol Appl Pharmacol 299:78–89. https://doi.org/10.1016/j.taap.2015.12.022

Dotsenko R, Vashchyk Y, Zakhariev A, Zemlianskyi A, Seliukova N, Gliebova K, Kostyshyn LM (2022) Prevalence of internal diseases of dogs and cats: a retrospective analysis (1994–2014). ScienceRise Biol Sci 2(31):8–12. https://doi.org/10.15587/2519-8025.2022.260493

El-Sayed A, Kamel M (2020) Advanced applications of nanotechnology in veterinary medicine. Environ Sci Pollut Res 27(16):19073–19086. https://doi.org/10.1007/s11356-018-3913-y

Feng B, Liu YN (2015) The fabrication of screen printed electrode mixed ferrocenemethanol and thionin for β-hydroxybutyrate biosensor. Int J Electrochem Sci 10:947–955

Gao SW, Peng HS, Wang XH, You FT, Teng F, Wang HX (2016) Preparation of photoluminescent enzymatic nanosensors for glucose sensing. Sens Actuators B Chem 222:638–644. https://doi.org/10.1016/j.snb.2015.08.092

Ge Y, Lakshmipriya T, Gopinath SC, Anbu P, Chen Y, Hariri F, Li L (2019) Glucose oxidase complexed gold-graphene nanocomposite on a dielectric surface for glucose detection: a strategy for gestational diabetes mellitus. Int J Nanomed 14:7851. https://doi.org/10.2147/IJN.S222238

Gong H, Zhao C, Niu G, Zhang W, Wang F (2020) Construction of 1D/2D α-Fe_2O_3/SnO_2 hybrid nanoarrays for sub-ppm acetone detection. Research 2020:2196063. https://doi.org/10.34133/2020/2196063

Greco DS (2018) Chapter 37: Diabetes mellitus in animals: diagnosis and treatment of diabetes mellitus in dogs and cats. In: Bagchi D, Nair S (eds) Nutritional and therapeutic interventions for diabetes and metabolic syndrome, 2nd edn. Academic Press, pp 507–517. https://doi.org/10.1016/B978-0-12-812019-4.00037-4

Han W, Chuang KH, Chang YT, Olivo M, Velan SS, Bhakoo K, Townsend D, Radda GK (2010) Imaging metabolic syndrome. EMBO Mol Med 2(6):196–210. https://doi.org/10.1002/emmm.201000074

Hill MF, Bordoni B (2022) Hyperlipidemia. In: StatPearls. StatPearls Publishing, Treasure Island (FL)

Hu Q, Fang Z, Ge J, Li H (2022) Nanotechnology for cardiovascular diseases. Innovation 3(2):100214

Jaberi SYS, Ghaffarinejad A, Omidinia E (2019) An electrochemical paper based nano-genosensor modified with reduced graphene oxide-gold nanostructure for determination of glycated hemoglobin in blood. Anal Chim Acta 1078:42–52. https://doi.org/10.1016/j.aca.2019.06.018

Ji T, Xu X, Wang X et al (2019) Point of care up conversion nanoparticles-based lateral flow assay quantifying myoglobin in clinical human blood samples. Sens Actuator B Chem 282:309–316

Kaneko JJ, Harvey JW, Bruss ML (2009) Clinical biochemistry of domestic animals, 6th edn. Lloyd-Luke, London

Khorsand F, Darziani Azizi M, Naeemy A, Larijani B, Omidfar K (2013) An electrochemical biosensor for 3-hydroxybutyrate detection based on screen-printed electrode modified by coenzyme functionalized carbon nanotubes. Mol Biol Rep 40(3):2327–2334. https://doi.org/10.1007/s11033-012-2314-4

Kim SR, Lerman LO (2018) Diagnostic imaging in the management of patients with metabolic syndrome. Transl Res 194:1–18. https://doi.org/10.1016/j.trsl.2017.10.009

Lemmerman LR, Das D, Higuita-Castro N, Mirmira RG, Gallego-Perez D (2020) Nanomedicine-based strategies for diabetes: diagnostics, monitoring, and treatment. Trends Endocrinol Metab 31(6):448–458. https://doi.org/10.1016/j.tem.2020.02.001

Li J, Chang KW, Wang CH, Yang CH, Shiesh SC, Lee GB (2016) On-chip, aptamer-based sandwich assay for detection of glycated hemoglobins via magnetic beads. Biosens Bioelectron 79:887–893. https://doi.org/10.1016/j.bios.2016.01.029

Lu C, Liu X, Li Y, Yu F, Tang L, Hu Y, Ying Y (2015) Multifunctional janus hematite–silica nanoparticles: mimicking peroxidase-like activity and sensitive colorimetric detection of glucose. ACS Appl Mater Interfaces 7(28):15395–15402. https://doi.org/10.1021/acsami.5b03423

Manuja A, Kumar B, Singh RK (2012) Nanotechnology developments: opportunities for animal health and production. Nanotechnol Dev 2:4

Mayo Clinic (2017). https://www.mayoclinic.org/diseases-conditions/inherited-metabolic-disorders/symptoms-causes/syc-20352590

Mi L, Sun Y, Shi L, Li T (2020) Hemin-bridged MOF interface with double amplification of G-quadruplex payload and DNAzyme catalysis: ultrasensitive lasting chemiluminescence MicroRNA imaging. ACS Appl Mater Interfaces 12(7):7879–7887. https://doi.org/10.1021/acsami.9b18053

Moon JM, Kim DM, Kim MH, Han JY, Jung DK, Shim YB (2017) A disposable amperometric dual-sensor for the detection of hemoglobin and glycated hemoglobin in a finger prick blood sample. Biosens Bioelectron 91:128–135. https://doi.org/10.1016/j.bios.2016.12.038

MSD Manual (2022). https://www.msdvetmanual.com/metabolic-disorders/metabolic-disorders-introduction/metabolic-storage-disorders-and-inborn-errors-of-metabolism-in-animals

Neethirajan S (2017) Recent advances in wearable sensors for animal health management. Sens Bio-Sens Res 12:15–29. https://doi.org/10.1016/j.sbsr.2016.11.004

Neethirajan S, Weng X, Chen L (2016) U.S. Patent No. 9,316,591. Washington, DC: U.S. Patent and Trademark Office

Oetzel GR (2011) Diseases of dairy animals | non-infectious diseases: milk fever. In: Fuquay JW (ed) Encyclopedia of dairy sciences, 2nd edn. Academic Press, pp 239–245. https://doi.org/10.1016/B978-0-12-374407-4.00145-X

Peng J, Weng J (2017) Enhanced peroxidase-like activity of MoS_2/graphene oxide hybrid with light irradiation for glucose detection. Biosens Bioelectron 89:652–658. https://doi.org/10.1016/j.bios.2015.12.034

Prasad M, Ghosh M, Suman, Patki HS, Kumar S, Brar B et al (2021) Imaging techniques in veterinary disease diagnosis. In: Advances in animal disease diagnosis. CRC Press, Boca Raton, pp 103–145. https://doi.org/10.1201/9781003080282-8

Qiu LN, Cai SV, Chan D, Hess RS (2022) Seasonality and geography of diabetes mellitus in United States of America dogs. PLoS One 17(8):e0272297. https://doi.org/10.1371/journal.pone.0272297

Radostits OM, Gay CC, Hinchcliff KW, Constable PD (2007) Veterinary medicine: a textbook of the diseases of cattle, horses, sheep, pigs and goats, 10th edn. WB Saunders Co. Ltd, Philadelphia, pp 673–762

Righettoni M, Amann A, Pratsinis SE (2015) Breath analysis by nanostructured metal oxides as chemo-resistive gas sensors. Mater Today 18(3):163–171. https://doi.org/10.1016/j.mattod.2014.08.017

Seifati SM, Nasirizadeh N, Azimzadeh M (2018) Nano-biosensor based on reduced graphene oxide and gold nanoparticles, for detection of phenylketonuria-associated DNA mutation. IET Nanobiotechnol 12(4):417–422

Song Y, Qu K, Zhao C, Ren J, Qu X (2010) Graphene oxide: intrinsic peroxidase catalytic activity and its application to glucose detection. Adv Mater 22(19):2206–2210. https://doi.org/10.1002/adma.200903783

Troncarelli MZ, Brandão HM, Gern JC, Guimarães AS, Langoni H (2013) Nanotechnology and antimicrobials in veterinary medicine Badajoz, Spain. FORMATEX. http://www.formatex.info/microbiology4/vol1/543-556.pdf

van der Kolk JH, Gross JJ, Gerber V, Bruckmaier RM (2017) Disturbed bovine mitochondrial lipid metabolism: a review. Vet Q 37(1):262–273. https://doi.org/10.1080/01652176.2017.1354561

Veerapandian M, Hunter R, Neethirajan S (2016) Lipoxygenase-modified Ru-bpy/graphene oxide: electrochemical biosensor for on-farm monitoring of non-esterified fatty acid. Biosens Bioelectron 78:253–258. https://doi.org/10.1016/j.bios.2015.11.058

Veiseh O, Tang B, Whitehead K et al (2015) Managing diabetes with nanomedicine: challenges and opportunities. Nat Rev Drug Discov 14:45–57. https://doi.org/10.1038/nrd4477

Wang H, Zhang W, Wan J, Liu W, Yu B, Jin Q, Guan M (2014) Microchip-based human serum atherogenic lipoprotein profile analysis. Anal Biochem 467:75–83

Wang Q, Zhang X, Huang L, Zhang Z, Dong S (2017) One-pot synthesis of Fe_3O_4 nanoparticle loaded 3D porous graphene nanocomposites with enhanced nanozyme activity for glucose detection. ACS Appl Mater Interfaces 9(8):7465–7471. https://doi.org/10.1021/acsami.6b16034

Wang X, Su J, Zeng D, Liu G, Liu L, Xu Y, Wang C, Liu X, Wang L, Mi X (2019) Gold nanoflowers (Au NFs) modified screen-printed carbon electrode electrochemical biosensor for label-free and quantitative detection of glycated hemoglobin. Talanta 201:119–125. https://doi.org/10.1016/j.talanta.2019.03.100

Wang C, Li J, Kang M, Huang X, Liu Y, Zhou N, Zhang Z (2021) Nanodiamonds and hydrogen-substituted graphdiyne heteronanostructure for the sensitive impedimetric aptasensing of myocardial infarction and cardiac troponin I. Anal Chim Acta 1141:110–119. https://doi.org/10.1016/j.aca.2020.10.044

Wang H, Chen R, Zhang F, Yu Z, Wang Y, Tang Z, Yang L, Tang X, Xiong B (2022) Superhydrophobic paper-based microfluidic field-effect transistor biosensor functionalized with semiconducting single-walled carbon nanotube and DNAzyme for hypocalcemia diagnosis. Int J Mol Sci 23(14):7799. https://doi.org/10.3390/ijms23147799

Weng X, Chen L, Neethirajan S, Duffield T (2015) Development of quantum dots-based biosensor towards on-farm detection of subclinical ketosis. Biosens Bioelectron 72:140–147. https://doi.org/10.1016/j.bios.2015.05.008

Whitesides GM (2003) The 'right' size in nanobiotechnology. Nat Biotechnol 21(10):1161–1165. https://doi.org/10.1038/nbt872

World Health Organization (2021) Cardiovascular diseases (CVDs). World Health Organization, Geneva

Yang X, Yu YQ, Peng LZ, Lei YM, Chai YQ, Yuan R, Zhuo Y (2018) Strong electrochemiluminescence from MOF accelerator enriched quantum dots for enhanced sensing of trace cTnI. Anal Chem 90(6):3995–4002. https://doi.org/10.1021/acs.analchem.7b05137

Ye J, Zhu L, Yan M et al (2019) Dual-wavelength ratiometric electrochemiluminescence immunosensor for cardiac troponin I detection. Anal Chem 91:1524–1531

Yu Z, Wang H, Zhao Y, Zhang F, Tang X, Xiong B (2022) Electrochemical biosensor using nitrogen-doped graphene/Au nanoparticles/DNAzyme for Ca^{2+} determination. Biosensors 12(5):331. https://doi.org/10.3390/bios12050331

Zhao HY, Liu S, He J, Pan CC, Li H, Zhou ZY, Ding Y, Huo D, Hu Y (2015) Synthesis and application of strawberry-like Fe_3O_4-Au nanoparticles as CT-MR dual-modality contrast agents in accurate detection of the progressive liver disease. Biomaterials 51:194–207. https://doi.org/10.1016/j.biomaterials.2015.02.019

Zubkovs V, Schuergers N, Lambert B, Ahunbay E, Boghossian AA (2017) Mediator less, reversible optical nanosensor enabled through enzymatic pocket doping. Small 13(42):1701654. https://doi.org/10.1002/smll.201701654

Chapter 9
Nanotechnology in Cancer Diagnostics

S. Suvetha, Alonkrita Chowdhury, Kumari Swati, Dodla Mamatha, Saumya Maurya, Pavan Kumar Yadav, and Mayukh Ghosh

Abstract Early and precise detection of cancer decreases its fatality rate to a great extent. Conventionally, diverse screening methods are there to detect any abnormality or phenotypical changes in tissues or cell masses. For instance, MRI for sensing cancerous growth in CNS; CT scan for diagnosing lung cancer, colorectal cancer, and others with an accuracy of 93.5%. Chest X-ray is performed for symptomatic lung cancer with 77–80% sensitivity. For lung, colon, stomach, or rectum cancer, endoscopy is carried out. However, all these traditional methods have their limitations such as prostate cancer is very much difficult to be diagnosed by CT scan technology because of its interior soft tissue characterization. Incorporating nanotechnology in this discipline enhances their diagnostic efficiency. It also provides the opportunity for therapeutic intervention. Consequently, diverse NPs have been evolved for imaging, drug delivery, biosensing, and photothermal therapy such as MNPs, AuNPs, QDs, CNPs, SNPs, polymer-based NPs (liposome, dendrimers, core-shell NPs), and others. Due to unique optical and electronic properties, AuNPs are exploited as biosensors, contrast agents, and therapeutics. It is employed as contrasting agent in MRI, CT scan, PFTI, XRFI, and other imaging modalities. QDs are also applied in diagnostic methods for sensing malignancy in tissues of lung, breast, and other organs. Correspondingly, polymeric NPs, CNPs, SNPs are also incorporated for better tomography. However, retention of these NPs inside cells for longer time can lead to cytotoxicity, ROS generation, fibrosis, allergy, or organ failure. Therapeutic applications of these NPs can also develop toxic effects in biological system. So, a regulated and balanced nanointervention in cancer diagnosis

S. Suvetha · A. Chowdhury · K. Swati · D. Mamatha · S. Maurya
Department of Genetics and Plant Breeding, Institute of Agricultural Science, Rajiv Gandhi South Campus, Banaras Hindu University, Mirzapur, Uttar Pradesh, India

P. K. Yadav · M. Ghosh (✉)
Department of Veterinary Physiology and Biochemistry, Faculty of Veterinary and Animal Sciences, Rajiv Gandhi South Campus, Banaras Hindu University, Mirzapur, Uttar Pradesh, India

LLR University of Veterinary and Animal, ICAR-National Research Centre on Equines, *Nanotechnology Theranostics in Livestock Diseases and Management*, Livestock Diseases and Management,
https://doi.org/10.1007/978-981-16-1610-5_9

is highly desirable for early and precise detection besides surmounting the limitations of conventional diagnostics as well as avoiding nanotoxicology concern.

Keywords Magnetic resonance imaging · Computerized tomography · Mammography · Position emission tomography · Ultrasonography · Biopsy · Magnetic NPs · Gold NPs · Quantum dots · Core-shell NPs

9.1 Introduction

Cancer is found to be the major cause of death all around the world, which has accounted 9,958,133 deaths and 19,292,789 new cases in 2020 (https://gco.iarc.fr/today/data/factsheets/cancers/39-All-cancers-fact-sheet.pdf). In this, almost 33.3% occurrence is due to the use of tobacco and improper diet whereas the most important reason for high mortality rate is delayed diagnosis. If cancer is diagnosed at an early stage, it can be effectively treated and even cured. The diagnostic methods that are conventionally used to detect cancerous growth are laboratory tests, imaging, and biopsy. Among these methods, laboratory test is the basic method carried out to determine the level of chemical compounds present in body fluids and tissues. Blood test, tissue examination, and tumor marker tests are frequently used to diagnose cancer. The complete blood count (CBC) is used to measure the size, number, and maturity of different blood cells. This measure helps to diagnose the occurrence of leukemia and lymphoma. Similarly, urine analysis is carried out to diagnose multiple myeloma and prostate cancer. Recent approaches which are developed based on circulating cell-free DNA (biomarker) are the CancerSeek that detects mutation in the DNA and the GRAIL that helps in cancer diagnosis by detecting the methylation occurred. CancerSeek helps in diagnosing the cancer at an early stage with almost 99% specificity (7 out of 812 healthy controls were reported false), whereas the sensitivity is 73% for stage II cancer and 43% for stage I. It is used to detect around 8 types of cancers, namely breast, lung, colorectal, ovarian, liver, stomach, pancreatic, and esophageal cancer. Similarly, GRAIL detects 50 different types of cancers with >99% specificity; 69% sensitivity for stage II and 39% for stage I cancer (Lokshin et al. 2021). As the abnormalities in the body fluids might also be from other reasons, there is a high probability of false reports. In such case, imaging technique is used to confirm and complete the diagnosis of cancer. By imaging internal body parts, abnormal growth of tissues can be detected. Magnetic resonance imaging (MRI) and computerized tomography (CT) scan are the most commonly used diagnostic imaging techniques. Radiographic techniques such as mammography, position emission tomography (PET), and ultrasonography are also used. MRI is efficient at locating the abnormal growth, especially brain tumor and spinal cord tumor (https://www.cancer.org/cancer/brain-spinal-cord-tumors-adults/detection-diagnosis-staging/how-diagnosed.html). Whereas CT scan is used to diagnose lung cancer, colon cancer, rectum cancer, and others with the accuracy of 93.5% which is 76.09% for MRI (Bradley et al. 2019). In the case of chest X-ray, it has 77–80% sensitivity for symptomatic lung cancer (Feng et al. 2017). However, these screening methods cannot be used to detect all types of cancer. For example, prostate cancer is very much difficult to be diagnosed by CT scan technology because of its interior soft tissue characterization. Similarly, in ovarian cancer, a tumor or mass of cells can be detected by X-ray but whether it is cancerous or not, that cannot be confirmed. In such case, an age-old technology, biopsy is used to confirm the cancerous mass. In this procedure, a piece of tissue or a sample of cells is removed from a suspicious area and examined for abnormality in size and organization. It is the most effective technique for cancer diagnosis. There are various types of biopsies for different cancer. For example, a needle biopsy is done for collecting tissues from thyroids, lymph nodes, and breasts. More than one million breast biopsy is performed each year in the United States (effectivehealthcare.ahrq.gov/products/breast-biopsy-update/clinician). Endoscopy is done for lung, colon, stomach, or rectum cancer. Bone marrow biopsy is particularly done in case of blood cancer, leukemia, lymphoma, and multiple myeloma. Biopsy is reported to have 88–100% diagnostic accuracy (wiki.cancer.org.au). However, there are some major difficulties to collect a sample from an inaccessible or undetectable anatomical location, which leads to a rise in the false report. For example, TRUS biopsy is reported to yield up to 30% false-negative reports in prostate cancer due to unknown tumor location.

A notable scientific advancement in nano-diagnostic procedures is found to help in early diagnosis of diseases and treatment. This techniques involve employment of nanoparticles (NPs) in diagnosing the cancer. Recent studies have shown various successful applications of NPs in this regard. Nanoparticles easily penetrate the cell or any biological barrier to reach at the tumor site and facilitate to find out the abnormalities at the cellular or sub-cellular level. Hence, cancer can be detected at the earliest stage when only one or a few cells are affected. In case of any radiological techniques, subsidiary cells around the cancerous part get damaged. The radiation used to diagnose also has the risk of inducing oncogenes in neighboring cells. However, in nanodiagnostics, these chances are meager due to the ability of NPs to specifically target the cancerous cells. Moreover, the accessibility of NPs to transport across any biological barriers inside body, i.e., dense stomatal tissue of pancreas or the blood–brain barriers (BBB) makes it highly effective to diagnose the disease. Nanotechnology-based bioassay can be used in place of TRUS biopsy in detecting prostate cancer. It helps to detect the stages of prostate cancer using different biomarkers released at various stages of cancer. Nanoparticles are used as molecular imaging agent, as they help in increasing the intensity of signal and imaging the few cells with greater depth. It also helps in providing stable and strong signal over a long period of time. The most common existing screening methods of cancer and the different NPs that are used in cancer diagnosis are discussed in this chapter with appropriate examples.

9.2 Existing Screening Methods for Cancer Diagnosis

There are methods to diagnose different types of cancers such as laboratory test, radiological test, histological, cytological, hematological, and molecular diagnosis.

Clinical laboratory tests are recommended to detect the abnormalities in the body fluids that help in cancer diagnosis. Blood and urine tests are commonly performed to measure the level of cancer-specific chemical substances present in it. Body fluid-based analysis such as CBC is helpful to detect leukemia, multiple myeloma, and other types of cancer. Blood and tissue samples are obtained in order to perform tumor marker test. Tumor markers are either the protein or other substances produced by cancer cells in higher amounts as compared to other healthy cells. It helps to diagnose cancer, determine its stage, and to estimate prognosis. Tumor markers are of two types: one that circulates freely in body fluids (circulating tumor marker) and tumor tissue markers. In case of tumor tissue marker, the tissue mass from the tumor is obtained by biopsy. For example, PSA or prostate-specific antigen test is used to detect prostate cancer; detection of alpha-fetoprotein (AFP) in blood, JAK2 gene mutation in blood and bone marrow, urine leukemia, and neuroblastoma, respectively. However, in some conditions, non-cancerous cells can also elevate the tumor marker level. Hence, tumor marker tests solely cannot confirm the cancer. Techniques such as biopsy and imaging are also performed to diagnose cancer. Moreover, tumor marker tests often fail to detect cancer at its early stage making it more complex for the treatment. A few liquid biopsy-based assays, such as PapSEEK, CancerSEEK, and UroSEEK, are useful in this regard as they provide early detection of multiple tumor markers. PapSEEK and UroSEEK detect ovarian or endometrial cancer and bladder cancer, respectively, whereas CancerSEEK can determine 8 different types of cancer (www.cancer.gov). In these clinical laboratory tests, the presence of carcinoma in the body is determined, whereas the exact location of the malignant cells cannot be detected.

Various radiological diagnostic methods help in imaging and detecting the location of tumor and in real-time monitoring of the cancerous mass. There are several imaging techniques such as X-ray, computed tomography (CT), magnetic resonance imaging (MRI), ultrasonography, elastography, and radionuclide imaging that includes scintigraphy, positron emission tomography (PET), single photon emission computed tomography (SPECT), etc.

In X-ray radiography, a beam of X-ray is passed through the body. The denser part of the body prevents more rays than the lighter part. Thus, denser part such as bone or thick tissue appears as white or light gray, whereas the thinner area such as soft tissue appears as shades of gray. Lung, as an air-filled organ looks black on the detector plate. A tumor is generally made of thick tissues. Therefore, they are showed up as light gray color in contrasting organs, e.g., chest X-ray is performed to diagnose lung cancer. To enhance the picture quality of soft tissues, some iodine-based dyes or contrast media are used along with X-rays. These special types of X-rays are called contrast studies. For example, barium enema is used for diagnosing aberration at the gastrointestinal tract (mainly colon) and detects colorectal cancer.

Here, the barium-sulfate-containing liquid is injected to the colon which coats its lining and silhouette the soft tissues of the colon (www.mayoclinic.org/tests-procedures/barium-enema). Similarly, an intravenous pyelogram (IVP) is used to diagnose aberration at the urinary tract and detect kidney cancer and bladder cancer. It is accomplished with the help of a contrasting dye (iodine contrast solution) (www.mayoclinic.org/tests-procedures/intravenous-pyelogram). Another radiographic test, mammogram (a breast X-ray) is especially done for breast cancer diagnosis. According to the U.S. Food and Drug Administration, 40 million mammograms are performed each year in the U.S. (www.hf.org/breasthealth). Finally, X-ray is fast-performing, easy to get, and less costly than other techniques. In addition, X-ray is good at finding the bone issues but poor to diagnose the aberrations in soft tissues and organs. In that case, MRI and CT scans provide better images. Instead of using a broad beam of X-rays from one angle, in computed tomography, pencil-thin beams of X-rays are directed from different angles to create a 3-D image of the part. Therefore, from the CT scan image, the shape, size, and location of the tumor can be determined. It is also very useful in staging cancer. Sometimes, a CT scan is also done for image-guided biopsy tests which are called CT-guided biopsy. An abdominal CT scan is done for diagnosing bladder cancer, colorectal cancer, kidney cancer, ovarian cancer, and stomach cancer. Lung cancer is also detected by a CT scan. However, prostate cancer, uterine cancer, and certain liver cancers are hard to detect in CT scans but can be easily detected through MRI. Unlike radiation in X-ray and CT scan, MRI uses powerful magnetic pulse and produces detailed image of cancerous mass. It can determine whether the tumor is malignant or benign; if it is metastasized or not, and also the stage of cancer. MRI creates a higher quality picture of soft tissues than other imaging techniques. Therefore, it is an important imaging tool to detect tumors in soft tissue. Standard MRI cannot detect fluid that is moving, such as blood inside blood vessels. Hence, a contrast dye (gadolinium) is injected into the bloodstream to avoid “flow voids.” For example, brain or spinal cord tumors are best viewed in MRI with contrast dye. Breast MRI is performed to diagnose breast cancer and metastatic axillary lymph nodes near the breast area. It is also used to detect colorectal cancer. MRI urogram is a special type of MRI, used to detect cancer in the upper part of the urinary system, i.e., kidney, ureters, and bladder. Liver cancer, prostate cancer, and many other cancers can also be diagnosed by the magnetic resonance imaging but it cannot detect ovarian cancer because of its deeper position and small size. In such cases, a CT or USG is recommended.

Ultrasonography is an imaging technique in which high-frequency sound waves are directed that will pass through the body and bounce off organs or tissues making echoes. These echoes are converted to real-time pictures by a transducer. Ultrasonography is great at picturing soft tissues that are not captured well on X-rays. It can also differentiate between a liquid-filled cyst and solid tumors as the echoes differ. Some special ultrasonography machines, known as Doppler flow machines, can detect the speed and direction of blood flow. This helps to detect tumors as the direction of blood flow differs in tumors from normal tissues. Ultrasounds are also used to guide a needle in a biopsy, i.e., ultrasonography-guided needle biopsy.

Cancer in lymph nodes, breasts, and liver are mostly diagnosed by this process. USG is also performed to diagnose liver cancer, breast cancer, and colon cancer. Ultrasound cannot create images as detailed as an X-ray or CT scan but provide information about the location and size of a tumor. However, whether the tumor is malignant or not, it cannot be detected by ultrasound alone.

Briefly, the imaging tests can detect irregular masses of tissue; but they alone fail to determine the cancerous nature of a mass. In such case, cytological method of diagnosis is carried out, most commonly biopsy. From biopsy results, it can be determined whether the cells are cancerous. If it is cancerous, its origin, type of cancer, and cancer grade, i.e., the aggressiveness of the cancer can also be obtained. Needle biopsy, endoscopy, skin biopsy, bone marrow aspiration and biopsy, and surgical biopsy are some conventionally used biopsy methods. The most common test to diagnose cancer is needle biopsy test. It is a procedure to draw a few masses of tissues and blood samples for laboratory testing. Fine needle aspiration (FNA), core needle biopsy (CNB), and vacuum-assisted biopsy (VAB) are some common types of needle biopsy. In fine needle aspiration, a small amount of tissue and fluid is removed from the suspicious area and used to evaluate metastasis to diagnose breast cancer, thyroid cancer, and others. However, in core needle biopsy, a larger amount of tissue sample is obtained using a long, hollow tube. Core needle biopsy is used in more than half of all breast biopsies done because of its 90–99% accuracy (Dillon et al. 2006). Vacuum assisted biopsy uses a greater number of tissues to examine under a microscope. A vacuum power-assisted tool is used in this regard. Vacuum-assisted biopsy for breast cancer diagnosis is also called a minimal invasive breast biopsy (MIBB). CNB and VAB are the most commonly used techniques for the detection of breast cancer. There is an interdisciplinary biopsy test, namely image guided biopsy, in which a needle biopsy and different imaging technology such as ultrasound, fluoroscopy, CT scan, X-ray, and MRI are used. Core needle biopsy using MRI, USG, or mammography techniques have been employed to diagnose ovarian and breast cancer. Similarly, colorectal cancer is determined using needle biopsy with the aid of CT. In the case of soft tissue mass diagnosis, the accuracy of core needle biopsy is more than that of fine needle aspiration. However, if the determination of malignancy, diagnosis perfection, appropriate guiding, and treatment is considered, then an open biopsy is more accurate than CNB and FNA. Endoscopic biopsy is another type of biopsy test that is commonly used for cancer diagnosis. Endoscopy is a technique frequently used to visualize the gastrointestinal architecture of the body using an endoscope, which is a long thin tube with a camera and light, inserted through any natural orifice. Therefore, biopsy, performed with the help of the endoscopy method, is considered as an endoscopic biopsy. Cystoscopy, bronchoscopy, and colonoscopy are some endoscopic techniques used for biopsy to diagnose bladder, lung, and colon cancer, respectively. Laparoscopy is similar to endoscopy where it needs a small incision on the abdomen or pelvis to insert the laparoscope. Therefore, a laparoscopic biopsy is usually done for the tissues that are deep and difficult to collect by endoscopic biopsy. It is generally used to diagnose gynecological cancers, such as cervical cancer, ovarian cancer, endometrial cancer as well as colon and kidney cancer (www.cancer.org). A skin biopsy is performed to diagnose melanoma. It is done by removing cells from the skin surface in order to assess under a microscope. There are different skin biopsy methods such as shave biopsy, punch biopsy, incisional biopsy, and excisional biopsy. Shave biopsy is performed to detect an intraepidermal carcinoma or basal cell carcinoma. Whereas punch biopsy is used to diagnose skin cancer, vulvar and cervical cancer (www.cancer.gov). Incisional or excisional biopsy is carried out to diagnose melanoma. In excisional biopsy, entire mass of the tumor is removed and in the case of incisional biopsy, a part of the tumor is removed. Bone marrow aspiration and biopsy are used to diagnose leukemia, lymphoma, and multiple myeloma. As bone marrow tissues are the place for erythropoiesis, therefore analyzing bone marrow sample helps us to detect abnormalities in the blood. Sometimes, breast lumps are removed through surgery for diagnosing breast cancer. This is called a surgical biopsy. Lymph nodes are also removed similarly, in case of lymphoma diagnosis. Transurethral resection of bladder tumor (TURBT) is a surgical procedure used for the early detection and treatment of bladder cancer. All over, an open surgical biopsy is 100% accurate. However, regarding malignancy determination, FNA and CNB have 75.4% and 80.7% accuracy, respectively. Besides these techniques, nanomodalities can also be used to improve the diagnostic efficacy of cancer which can eliminate the stumbling blocks in conventional diagnostics techniques.

9.3 Different Nanoparticles in Cancer Nanodiagnosis

The fatality rate of cancer increases due to late diagnosis of its phenotypical manifestation. Early detection of malignant tumors preferably before the metastasis stage enhances the chances of therapeutic intervention. There are various laboratory methods, radiological and cytological techniques for cancer diagnosis. However, nanoparticles are recently being employed to enhance the efficiency of aforesaid methods with early and precise detection. Particles show distinctive physicochemical features when their size is reduced to <100 nm. Therefore, using this unique features, nanodevices are developed which help in monitoring biological incidents with unprecedented efficiency and sensitivity. The NPs can be functionalized with small molecules, polyethylene glycol (PEG), antibodies, fluorescent probes, targeting peptides, aptamers, and other moieties. These ligands interact with the biological system and recognize molecular level changes which enhance the specificity and sensitivity of an assay. Hence, NPs are being exploited in imaging, drug delivery, biosensing, and photothermal therapy (Chen et al. 2013a). Different types of NPs frequently used in cancer diagnosis are magnetic NPs (MNPs), gold NPs (AuNPs), quantum dots (QDs), carbon-based NPs, silicon-based NPs, polymer-based NPs (liposome, dendrimers, core-shell NPs), and others (Table 9.1). Most commonly used NPs are metallic nanoparticles made of pure metals (e.g., gold, platinum, silver, titanium, zinc, cerium, iron, and thallium) or their derivatives (e.g., oxides, hydroxides, sulfides, phosphates, fluorides, and chlorides). Core-shell NPs is one of the popular type in recent time which consists of a core of inorganic metal or

Table 9.1 Application of different nanoparticles in cancer diagnostic modalities

Nanoparticles	Diagnostic platforms	Application	References
Zinc oxide NPs (ZnONPs)	Laser-induced fluorescence (LIF)	Polyvinyl alcohol-coated zinc oxide nanoparticles (ZnONPs-PVA) are integrated into a microfluidic immunosensor to detect epithelial cell adhesion molecule (EpCAM) in epithelial carcinomas of lung, colorectal, breast, prostate, head, neck, and hepatic origin	Fernández-Baldo et al. (2016)
Iron oxide NPs (IONPs)	MRI	Fe_3O_4-SiO_2-HPG NPs conjugated with folate (FA) as a targeting agent is employed for breast cancer diagnosis	Heydari Sheikh Hossein et al. (2020)
	MRI	Block copolymer poly(ethylene oxide)-*block*-poly(-γ-methacryloxypropyltrimethoxysilane) (PEO-*b*-PγMPS) coated iron oxide nanoparticles (IONPs) conjugated with HER2 antibody applied for breast cancer diagnosis	Chen et al. (2013b)
	MRI	PEG-coated iron oxide nanoparticles employed as contrast agents for long lasting molecular imaging in cancer diagnosis	Lazaro-Carrillo et al. (2020)
Gold NPs (AuNPs)	X-ray fluorescence imaging (e.g., XFCT)	In prostate cancer diagnosis, PEGylated GNPs coupled with anti-prostate-specific membrane antigen (PSMA) antibodies are implemented	Hara et al. (2021)
	ELISA technique	In bladder cancer, early detection is provided by anti-survivin antibody-conjugated GNPs	Jazayeri et al. (2020)
Quantum dots (QDs)	Confocal laser scanning microscopy-based imaging modality	Poly(aspartate)–Na–*graft*–poly(ethylene glycol)–dodecylamine (PASP–Na–*g*–PEG–DDA)-coated QDs modified with anti-vascular endothelial growth factor (anti-VEGF) antibody is employed to detect liver cancer owing to their targeting property	Sun et al. (2014)
	Confocal fluorescence microscopy	Bovine serum albumin (BSA)-coated CdTe/ZnS quantum dots (BSA–QDs), conjugated with folic acid (FA), used for diagnosis of embryonic kidney cells, epithelial carcinoma cells, and so on	Meng et al. (2011)
Liposomes	MRI	Neural cell adhesion molecule (NCAM)-coated liposomes loaded with doxorubicin and a lipophilic gadolinium derivative facilitate simultaneous targeted drug delivery as well as MRI-based	Grange et al. (2010)

		detection in a severe combined immunodeficient mouse model of NCAM-positive Kaposi's sarcoma	
	Confocal laser scanning microscopy (CLSM)	D-alpha-tocopheryl polyethylene glycol 1000 succinate mono-ester (TPGS) coated theranostic liposomes containing docetaxel and quantum dots for therapy as well as imaging of MCF-7 breast cancer cells having folate receptor overexpression	Muthu et al. (2012)
	Fluorescence microscopic imaging	Transferrin conjugated TPGS-coated liposomes containing docetaxel and quantum dots for therapy as well as imaging of brain cancer	Sonali et al. (2016)
Dendrimers	Electrochemiluminescence (ECL) biosensing	Poly-amidoamine (PAMAM) dendrimers functionalized to Fe_3O_4 nanoparticles for the detection of liver cancer targeting alpha-fetoprotein (AFP) and glypican-3 (GPC-3). Detection limit for AFP and GPC-3 was 6 pg/mL and 0.03 pg/mL. The NPs enhanced ECL emission to achieve better sensitivity	Chikhaliwala and Chandra (2020)
	Fluorescence-based imaging	Paclitaxel-encapsulated PAMAM dendrimer functionalized with tumor-penetrating cyclic peptide for tumor targeting and NIR carbocyanine fluorescent dye for in vivo imaging facilitated theranostic intervention of liver cancer	Cao et al. (2018)
	Near-infrared-II (NIR-II, 1000–1700 nm) imaging	PEGylated dendrimer encapsulated silver sulfide quantum dots for in-vivo imaging of A549 cancer cells (adenocarcinomic human alveolar basal epithelial cells)	Awasthi et al. (2020)
Polymeric NPs	T1-weighted MRI (NPs as T1 contrast agent)	Hyperbranched poly(glycerol) conjugated with gadolinium chelates and folic acid ligands for the imaging of tumor-bearing mice	Cao et al. (2017)
	MRI	Synthetic polymer, polyvinyl alcohol (PVA)–diethylenetriamine pentaacetate (DTPA)–gadolinium (Gd) conjugate NPs as contrast agent for the magnetic resonance imaging	Xu et al. (2019)
	MRI	Poly-*N*-(2-hydroxypropyl) methacrylamide (HPMA)-DTPA/dodecane tetraacetic acid (DOTA)-Gd conjugates NPs as contrast agent for the magnetic resonance imaging	Xu et al. (2018)

(continued)

Table 9.1 (continued)

Nanoparticles	Diagnostic platforms	Application	References
Core-shell NPs	MRI and CT	Gold-coated iron oxide nanoparticles capped with polyvinylpyrrolidone (Fe_3O_4-Au Core-Shell NPs) is used in in-vivo imaging modalities and focused photothermal therapy (PTT) against glioblastoma	Caro et al. (2021)
	MRI	NaDyF4–NaGdF4 core shell nanoparticles (NPs) composed of paramagnetic Dy3+ and Gd3+ are employed for detecting prostate cancer cells	Dash et al. (2021)
Silicon NPs (SNPs)	Fluorescence imaging	Dual-template molecularly imprinted polymer (MIP) coupled with fluorescent silicon nanoparticles (Si NPs) for focused imaging and therapy in HER2 positive breast cancer cells	Wang et al. (2019)
	Fluorescence microscopy	Dual-labeled iRGD-modified multifunctional porous silicon nanoparticles (PSi NPs), prepared from dibenzocyclooctyl (DBCO) modified Psi, are used in prostate cancer diagnosis	Wang et al. (2015)
	Visual lateral flow immunoassay (LFIA)	Silica-coated silver-assembled silica nanoparticles (SiO_2–Ag–SiO_2 NPs) for detection of PSA in prostate cancer diagnosis	Kim et al. (2021)
Carbon NPs	Near-infrared-II (NIR-II, 1000–1700 nm) imaging	Hollow carbon-ammonia borane nanoparticles (HC-AB NPs) are developed by loading ammonia borane (AB) into hollow carbon nanoparticles (HCs) for NIR-II photoacoustic (PA) imaging and NIR-II hydrogenothermal therapy	Jia et al. (2021)

metal oxides covered with a shell made up of organic or inorganic material or metal oxide (Sharma et al. 2015).

Metal NPs are advantageous as theranostic agents over other nanoparticles because of their anti-cancerous properties. Therapeutic moieties can be conjugated with them and thus they can serve as carriers of diverse therapeutic agents. They can also be attached to different imaging agents, such as fluorescent dye, radioisotopes, and optical imaging agents for multimodal imaging. Hence, metal NPs can be employed as therapeutic agents, cancer diagnostic tools, and also as a monitor. For example, a des-carboxy-prothrombin (DCP) immunosensor has been developed using zinc oxide NPs (ZnONPs) for accurate assessment of DCP which is an abnormal form of prothrombin produced by hepatic cells in case of liver cancer. In case of breast cancer, there is an elevated level of choline and related chemicals. Thus, ZnO NPs interfaced electrochemical biosensor has been developed for the detection of choline in cancerous cells (Anjum et al. 2021). Whereas, iron oxide nanoparticles (IONPs) are used as contrasting agents in MRI technology. In liver cancer diagnosis, IONPs enter healthy liver Kupffer cells but they are excluded from cancerous cells, thereby creating a contrasting image of dark healthy cells with bright cancerous tissues. Similarly, these IONPs are also being used in the diagnosis of spleen, liver, and lymph node cancers. Super paramagnetism is an exclusive feature of IONPs. Recently, superparamagnetic iron oxide nanoparticles (SPIONs) and ultrasmall superparamagnetic iron oxide nanoparticles (USPIONs) are being used as contrasting agents for MRI in cancer nano diagnostics. As an example, Ferumoxtran, an USPIONs, is employed for imaging in lymph node metastasis (Martinkova et al. 2018). C595 monoclonal antibody-conjugated SPIONs are used to detect mucin 1 (MUC1) using MRI for ovarian cancer diagnosis. Such functionalized MNPs facilitate improved imaging as well as therapeutic intervention in cancer by precise targeting (Fig. 9.1). Gold NPs (AuNPs) can be exploited in all aspects of cancer, i.e., imaging, sensing, and detection due to their unique optical and electronic properties. Hence, AuNPs are remarkable tools to be used as biosensors, contrast agents, and therapeutics. Being a powerful contrasting agent, it is applied to almost all types of imaging techniques, such as computed tomography (CT), magnetic resonance (MRI), positron emission tomography (PET), fluorescence imaging (FI), photoacoustic imaging (PAI), and X-ray fluorescence imaging (XRFI). Gold NPs function as multimodal agents because of their ability to perform two or more aforesaid diagnostic methods at a time. For example, AuNP coated electrochemical immunosensor is employed for the detection of a biomarker, GM2 activator protein (GM2AP), in case of lung cancer. Zr-labeled gold nanoparticle antibody conjugates are used in PET imaging for pancreatic tumor detection. AuNPs, conjugated with concanavalin A on an interdigitated electrode has been employed to detect prostate-specific membrane antigens (PSMA) for prostate cancer detection. Similarly, gold NPs coupled with PSA-specific aptamer are used on a mini-electrode to detect prostate-specific antigen (PSA) (Medici et al. 2021).

Nanoparticles such as quantum dots (QDs) also play a major role in cancer diagnosis. These are semiconductor nanocrystals along with typical optical properties upon light excitation. High brightness, resistance to photobleaching, and tunable

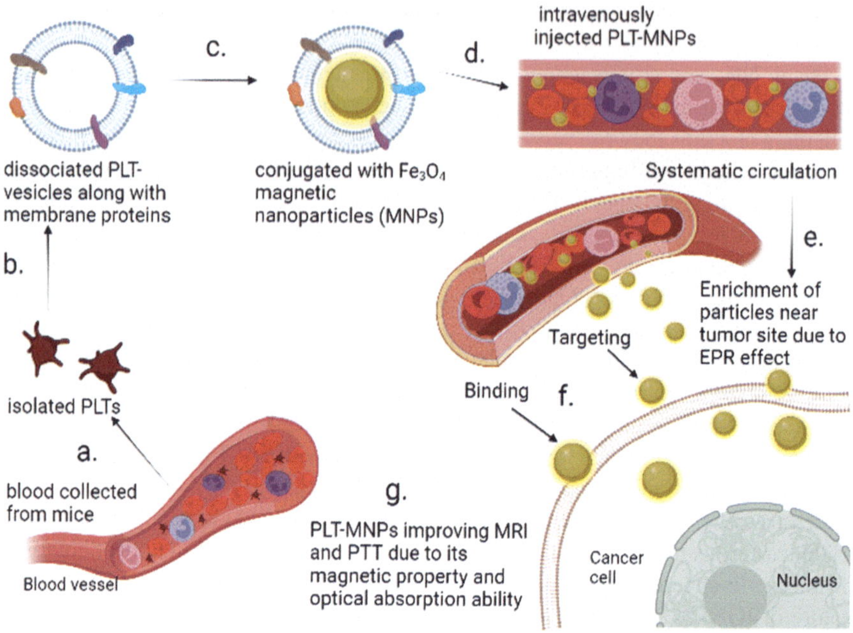

Fig. 9.1 Enhanced imaging of malignant cells by targeting with MNPs: (**a**) Platelets (PLTs) isolated from mice blood sample, (**b**) membrane vesicles along with membrane proteins are separated from PLTs, (**c**) these are coated onto MNP (Fe_3O_4), (**d**) PLT membrane-coated MNPs (PLT-MNPs) are intravenously injected back into the donor mice, (**e**) after systematic circulation, NPs gather around tumor site, (**f**) target and closely bind to the malignant cells. (**g**) Now, due to optical absorption ability, PLT-MNPs enhance the resonance of magnetic resonance imaging (MRI) of tumor in vivo

wavelength, i.e., ability to fluorescence at different wavelengths are some of its excellent optical properties. Therefore, due to its optical advantages, unique size, and surface effects, QDs show great potential in biomedical applications (Fang et al. 2017). In cases of invasive breast cancer, human epidermal growth factor receptor 2 (HER2) is overexpressed in 25–30% case. It is located in the extracellular domain (ECD) of cancerous cells. Hence, several QDs-based methods have been developed, such as an electrochemical immunosensing strategy based on QDs as an electrochemical label for the detection of HER2-ECD in breast cancer (Freitas et al. 2020). In prostate cancer, QDs probes conjugated with prostate-specific membrane antigen (PSMA) are used to obtain sensitive and multicolor imaging of cancer cells. CA125, a tumor marker for ovarian cancer is detected with the help of QDs. QDs-based multiplexed DNA sensor is reported to show a great potential for the diagnosis of breast cancer, ovarian cancer, and papillary thyroid carcinoma. In the case of lung cancer, quantum dots-based immunofluorescence histochemistry (QDs-IHC) is much more sensitive for quantification of protein markers than traditional IHC; such as detection of caveolin-1 and proliferating cell nuclear antigen (PCNA) in lung cancer is more efficient by QDs-IHC than the conventional one. QDs combined with magnetic NPs are also used to perceive metastases in lung cancer (Baetke et al. 2015).

Using polymer chemistry, macromolecules are designed that combine multiple bio-active elements into a single system. Thus, polymeric NPs are developed that can self-assemble to form amphiphilic macromolecules under appropriate conditions with an apposite monomer distribution. In contrast to inorganic NPs, polymeric NPs are more advantageous in terms of stability, biocompatibility, biodegradability, tailorability, and low-cost. There are various types of polymeric NPs such as liposomes, dendrimers, core-shell NPs, and others. Liposomes are artificially formed spherical structures of phospholipid bilayer enclosing an aqueous core. Liposomes are considered as the most effective drug delivery systems into cells due to their smaller size, ability to incorporate various substances, and phospholipid bilayer in nature along with fewer side-effects. Here, imaging agents such as iron oxide NPs, quantum dots, and gold nanoparticles are incorporated inside the liposome or linked covalently with the bilayer, and anticancer drugs are also attached similarly to provide both therapy and diagnosis simultaneously. This type of theranostic liposomes is called as liposome-nanoparticles hybrid. It is further conjugated with a molecular probe for targeting. This facilitates advanced prognosis and monitoring of disease. Dendrimers are synthetic polymeric NPs composed of radially emerging highly branched monomer subunits. Radiolabeled dendrimers are used to detect tumor cells. Gadolinium (^{153}Gd) is the best-known magnetic resonance (MR) contrasting agent. Gadolinium conjugated dendrimers are employed for selective targeting and imaging of tumors in the kidney, liver, and other organs. Nowadays, core/shell nanoparticles are being used in the field of bio-imaging for cell labeling, controlled drug delivery, and tissue engineering practices. Nanosphere and nano-capsules are two such NPs differing in their morphology. Nanocapsules are comprised of an oily core where the drugs remain dissolved and surrounded by a polymeric shell. Whereas, nanospheres are structured as a continuous polymeric network having the drug retained inside (reservoir system) or adsorbed on the surface (matrix system). For example, silica core-gold shell nanoparticles conjugated with anti HER2 are used for both imaging and therapy of SKBr3 breast cancer cells.

Carbon nanoparticles (CNPs) are also a popular nanomaterial being widely used for gene and drug delivery, biosensor, and molecular imaging. Carbon nanotubes (CNTs) are a type of CNP formed by rolling carbon sheets into a tubular shape. CNTs are diversely applied in the biomedical field owing to their unique physical and chemical properties such as high aspect ratio, ultralight weight, high mechanical strength, high electrical conductivity, and high thermal conductivity. CNTs have been reported as a promising tool for the diagnosis of carcinoma at its early stages. Carcinoembryonic antigen (CEA), alpha-fetoprotein (AFP), carbohydrate antigen 19–9 (CA19–9), carcinoma antigen 125 (CA125), prostate-specific antigen (PSA), and human chorionic gonadotropin (hCGs) are some important tumor markers which can be sensed by CNT-based techniques for early detection of colorectal cancer, hepatocellular carcinoma, pancreatic cancer, epithelial ovarian tumor, prostate cancers, and others (Ji et al. 2010).

Silicon nanoparticles (SNPs) are also used in diagnostics. These are highly porous, biocompatible, and easily functionalizable nanoparticles. Other metal-based NPs such as gold, silver, titanium oxide, cadmium selenide, etc. often get stuck in internal organs and cannot be excreted. Thus, it affects the organism due to prolonged toxicity. However, due to bio-degradable properties, silicon benefits the organism without imparting significant toxicity. SNPs are broadly classified as xerogels and mesoporous silica nanoparticles (MSN). Various silica-based NPs have been developed for the detection of cancer that combines imaging techniques such as MRI, fluorescence imaging, Raman spectroscopy, and therapeutic agents. These are also functionalized by conjugating tumor-specific ligands, such as antibodies, proteins, and synthetic peptides for pinpointing the tumor cells. For example, various silica NPs such as Nile Red doped ORMOSIL NPs, FITC doped mesoporous NPs, and others are employed for the detection of breast cancer. These are dye-doped SNPs for targeted imaging in cancer (Arap et al. 2013).

9.4 Challenges of Nanoparticles in Cancer Diagnostics and Nanotoxicology

Even though nanotechnology-based cancer diagnostic methods are showing high progress, only few techniques are found to be commercialized till now. There is a need to deal with the challenges to increase the translation of nanotechnology into clinical trials. Reliability is the first challenge of nano-based cancer diagnosis. Many factors affect the diagnostic property of NPs that includes but not limited to nonspecific binding of probes, accumulation, and undesirable detection conditions. The next notable challenge is the mass production of nanoprobes adhering to the stringent customization protocol. The basic diagnostic techniques (histology, electron microscopy, magnetic resonance imaging, computed topography) amalgamated with nanoparticles also have some disadvantages. For instance, histology has some limitations as it is a time taking and laborious process, chances of error during manual slide preparation, evaluation of NP distribution in limited number of tissue sections, etc. Limitations in electron microscopy include inability to evaluate large tissue sections, a greater number of nanoparticles are needed, all nanoparticles not suitable for sample preparation, etc. Computed topography includes no protection from harmful ionizing radiation, not provides detail information of nanoparticles association with the cell, it needs insertion of some contrasting agents for clear demarcation of tissues, and it alters the physiochemical properties. Difficulties in magnetic resonance imaging include it needs a greater number of contrast agents, poor sensitivity, not suitable for subjects with pre-inserted metallic devices, takes more time to get an image, and long processing time. In vivo optical imaging includes less image clarity compared to CT and MRI, it is unable to examine single nanoparticles but have more intensity for fluorescence (Arms et al. 2018).

Further, some of these NPs are found to have toxic impact during systemic administration. Changes made in the structural and physiochemical properties of nanomaterials in order to facilitate molecular interaction, improper biodistribution, and elimination are the potential reasons underlying such toxicological implications. Most of the particles are delivered into the biological system by one of the following ways: oral, inhalation, dermal, intravenous, subcutaneous, and intraperitoneal. They retain in the cells for a period of time before it gets processed. This retention in the cells and organ may lead to cytotoxicity, ROS generation, fibrosis, allergy, and organ failure. This is due to the modified surface area to volume ratio of nanoparticles in order to increase interaction and sensitivity. While using liposome as a carrier with nanoparticles like Abelcet, Caelyx, DaunoXome, DepoDur, and AmBlsome, the formulations have caused adverse effects such as anaphylaxis, skin related rashes on feet and legs, damage to liver (toxic hepatitis), kidneys malfunction, liver failure, respectively. Use of iron oxide nano-formulations like Feraheme and Feridex shows harmful effects like constipation, swelling of tissues, severe allergic reactions, and hypotension. Polymeric nanoparticles like Copaxone and Macugen show damage symptoms such as death of skin cells and discomfort in chest, inflammation of tissues in eye, respectively (Prasad et al. 2021). Sometimes larger size of nanoparticles prevents their penetration depth to 1–2 cell layers from blood vessels, and the drugs slowly released by nanoparticles. Thus, in-depth introspection should be carried out to avoid accumulation of such inert particles which leads to the toxic effect to the biological components. Therefore, risk assessment is mandatory to understand the implication of the diagnostic methods using nanomaterials (Chenthamara et al. 2019).

9.5 Conclusion and Future Prospects

Materials, at nanoscale size, exhibit distinct physical and biochemical properties than the same at bulk volume. Based on this unique property, novel imaging agents and biosensors have been developed. Nanoparticles such as MNPs, QDs, liposomes, or dendrimers can be used in imaging modalities for greater stability and longer circulation half-life, stronger binding to target epitopes, and higher contrast compared with existing solutions. Moreover, owing to its significant size, that is even smaller than a cell size, it can deliver a large cargo of contrast agent or fluorescent probe onto the surface or interior of the cell, without disrupting its function.

Nanotechnology offers advancement in diagnostics by increasing sensitivity and spatial resolution, improving stability and safety of imaging agents, and enabling molecular targeting, multiplexed analysis, real-time imaging, and monitoring of disease progression and therapeutic intervention. For progression in nanodiagnostics, only technological advancements are not sufficient, but, key regulatory policies, interdisciplinary collaboration, and training are also needed. Over the last 10 years, much development has been accomplished in joining these disparate fields. However, more integration and cross-training are looked-for. New developments are allowing functionality to evolve multimodal imaging agents as well as multiplex and quantitative sensors. Improvements in nanoparticles as contrast agents can lead to non-invasive detection and characterization of unique biochemical signatures that differentiate tissues even before phenotypical variations become apparent.

Furthermore, nanomaterials used in nanodiagnostics also exhibit toxic effects. Among all the NPs, QDs are one of the most commonly employed material in diagnostics that are often composed of heavy metal ions (such as Cd^{2+}) leading to lung and kidney damage. Therefore, concerns about the safety of quantum dots may hinder their progress to clinical application. Further studies, hence, needed for reducing its toxic effects for safer application in future. Finally, owing to the remarkable capabilities of nanomaterials in detecting precancerous and neoplastic lesions, and the biological processes that lead to cellular transformation, a significant increase in its application to cancer diagnostics is expected in upcoming years.

References

Anjum S, Hashim M, Malik SA, Khan M, Lorenzo JM, Abbasi BH, Hano C (2021) Recent advances in zinc oxide nanoparticles (ZnO NPs) for cancer diagnosis, target drug delivery, and treatment. Cancers 13(18):4570

Arap W, Pasqualini R, Montalti M, Petrizza L, Prodi L, Rampazzo E, Marchiò S (2013) Luminescent silica nanoparticles for cancer diagnosis. Curr Med Chem 20(17):2195–2211

Arms L, Smith DW, Flynn J, Palmer W, Martin A, Woldu A, Hua S (2018) Advantages and limitations of current techniques for analyzing the biodistribution of nanoparticles. Front Pharmacol 9:802

Awasthi P, An X, Xiang J, Kalva N, Shen Y, Li C (2020) Facile synthesis of noncytotoxic PEGylated dendrimer encapsulated silver sulfide quantum dots for NIR-II biological imaging. Nanoscale 12(9):5678–5684

Baetke SC, Lammers TGGM, Kiessling F (2015) Applications of nanoparticles for diagnosis and therapy of cancer. Br J Radiol 88(1054):20150207

Bradley SH, Abraham S, Callister ME, Grice A, Hamilton WT, Lopez RR, Neal RD (2019) Sensitivity of chest X-ray for detecting lung cancer in people presenting with symptoms: a systematic review. Br J Gen Pract 69(689):e827–e835

Cao Y, Liu M, Zu G, Kuang Y, Tong X, Xiong D, Pei R (2017) Hyperbranched poly(glycerol) as a T1 contrast agent for tumor-targeted magnetic resonance imaging in vivo. Polym Chem 8(6):1104–1113

Cao J, Ge R, Zhang M, Xia J, Han S, Lu W, Sun Y (2018) A triple modality BSA-coated dendritic nanoplatform for NIR imaging, enhanced tumor penetration and anticancer therapy. Nanoscale 10(19):9021–9037

Caro C, Gámez F, Quaresma P, Páez-Muñoz JM, Domínguez A, Pearson JR, García-Martín ML (2021) Fe_3O_4-Au core-shell nanoparticles as a multimodal platform for in vivo imaging and focused photothermal therapy. Pharmaceutics 13(3):416

Chen H, Zhen Z, Todd T, Chu PK, Xie J (2013a) Nanoparticles for improving cancer diagnosis. Mater Sci Eng R Rep 74(3):35–69

Chen H, Wang L, Yu Q, Qian W, Tiwari D, Yi H, Mao H (2013b) Anti-HER2 antibody and ScFvEGFR-conjugated antifouling magnetic iron oxide nanoparticles for targeting and magnetic resonance imaging of breast cancer. Int J Nanomed 8:3781–3794

Chenthamara D, Subramaniam S, Ramakrishnan SG, Krishnaswamy S, Essa MM, Lin FH, Qoronfleh MW (2019) Therapeutic efficacy of nanoparticles and routes of administration. Biomater Res 23(1):1–29

Chikhaliwala P, Chandra S (2020) Poly-amidoamine dendrimers@ Fe_3O_4 based electrochemiluminescent nanomaterials for biosensing of liver cancer biomarkers. Electroanalysis 32(11):2404–2414

Dash A, Blasiak B, Tomanek B, Latta P, van Veggel FC (2021) Target-specific magnetic resonance imaging of human prostate adenocarcinoma using NaDyF4–NaGdF4 core–shell nanoparticles. ACS Appl Mater Interfaces 13(21):24345–24355

Dillon MF, Quinn CM, McDermott EW, O'Doherty A, O'Higgins N, Hill ADK (2006) Diagnostic accuracy of core biopsy for ductal carcinoma in situ and its implications for surgical practice. J Clin Pathol 59(7):740–743

Fang M, Chen M, Liu L, Li Y (2017) Applications of quantum dots in cancer detection and diagnosis: a review. J Biomed Nanotechnol 13(1):1–16

Feng M, Yang X, Ma Q, He Y (2017) Retrospective analysis for the false positive diagnosis of PET-CT scan in lung cancer patients. Medicine 96(42):e7415

Fernández-Baldo MA, Ortega FG, Pereira SV, Bertolino FA, Serrano MJ, Lorente JA, Messina GA (2016) Nanostructured platform integrated into a microfluidic immunosensor coupled to laser-induced fluorescence for the epithelial cancer biomarker determination. Microchem J 128:18–25

Freitas M, Neves MM, Nouws HP, Delerue-Matos C (2020) Quantum dots as nanolabels for breast cancer biomarker HER2-ECD analysis in human serum. Talanta 208:120430

Grange C, Geninatti-Crich S, Esposito G, Alberti D, Tei L, Bussolati B, Camussi G (2010) Combined delivery and magnetic resonance imaging of neural cell adhesion molecule-targeted doxorubicin-containing liposomes in experimentally induced Kaposi's sarcoma. Cancer Res 70(6):2180–2190

Hara D, Tao W, Totiger TM, Pourmand A, Dogan N, Ford JC, Pollack A (2021) Prostate cancer targeted X-ray fluorescence imaging via gold nanoparticles functionalized with prostate-specific membrane antigen (PSMA). Int J Radiat Oncol Biol Phys 111(1):220–232

Hossein HSH, Jabbari I, Zarepour A, Zarrabi A, Ashrafizadeh M, Taherian A, Makvandi P (2020) Functionalization of magnetic nanoparticles by folate as potential MRI contrast agent for breast cancer diagnostics. Molecules 25(18):4053

Jazayeri MH, Aghaie T, Nedaeinia R, Manian M, Nickho H (2020) Rapid noninvasive detection of bladder cancer using survivin antibody-conjugated gold nanoparticles (GNPs) based on localized surface plasmon resonance (LSPR). Cancer Immunol Immunother 69(9):1833–1840

Ji SR, Liu C, Zhang B, Yang F, Xu J, Long J, Yu XJ (2010) Carbon nanotubes in cancer diagnosis and therapy. Biochim Biophys Acta 1806(1):29–35

Jia Z, Dai R, Zheng Z, Qin Y, Duan A, Peng X, Zhang R (2021) Hollow carbon-based nanosystem for photoacoustic imaging-guided hydrogenothermal therapy in the second near-infrared window. RSC Adv 11(20):12022–12029

Kim HM, Kim J, Bock S, An J, Choi YS, Pham XH, Jun BH (2021) Silver-assembled silica nanoparticles in lateral flow immunoassay for visual inspection of prostate-specific antigen. Sensors 21(12):4099

Lazaro-Carrillo A, Filice M, Guillén MJ, Amaro R, Viñambres M, Tabero A, Marciello M (2020) Tailor-made PEG coated iron oxide nanoparticles as contrast agents for long lasting magnetic resonance molecular imaging of solid cancers. Mater Sci Eng C 107:110262

Lokshin A, Bast RC, Rodland K (2021) Circulating cancer biomarkers. Cancers 13(4):802

Martinkova P, Brtnicky M, Kynicky J, Pohanka M (2018) Iron oxide nanoparticles: innovative tool in cancer diagnosis and therapy. Adv Healthc Mater 7(5):1700932

Medici S, Peana M, Coradduzza D, Zoroddu MA (2021) Gold nanoparticles and cancer: detection, diagnosis and therapy. In: Seminars in cancer biology, vol 76. Academic Press, pp 27–37

Meng H, Chen JY, Mi L, Wang PN, Ge MY, Yue Y, Dai N (2011) Conjugates of folic acids with BSA-coated quantum dots for cancer cell targeting and imaging by single-photon and two-photon excitation. J Biol Inorg Chem 16(1):117–123

Muthu MS, Kulkarni SA, Raju A, Feng SS (2012) Theranostic liposomes of TPGS coating for targeted co-delivery of docetaxel and quantum dots. Biomaterials 33(12):3494–3501

Prasad M, Kumar R, Buragohain L, Kumari A, Ghosh M (2021) Organoid technology: a reliable developmental biology tool for organ-specific nanotoxicity evaluation. Front Cell Dev Biol 9: 696668

Sharma H, Mishra PK, Talegaonkar S, Vaidya B (2015) Metal nanoparticles: a theranostic nanotool against cancer. Drug Discov Today 20(9):1143–1151

Sonali, Singh RP, Singh N, Sharma G, Vijayakumar MR, Koch B, Muthu MS (2016) Transferrin liposomes of docetaxel for brain-targeted cancer applications: formulation and brain theranostics. Drug Deliv 23(4):1261–1271

Sun X, Li Y, Huang H, Yang B, Wang Y (2014) Synthesis and application of a targeting diagnosis system via quantum dots coated by amphiphilic polymer for the detection of liver cancer cells. Luminescence 29(7):831–836

Wang CF, Sarparanta MP, Mäkilä EM, Hyvönen ML, Laakkonen PM, Salonen JJ, Santos HA (2015) Multifunctional porous silicon nanoparticles for cancer theranostics. Biomaterials 48: 108–118

Wang HY, Cao PP, He ZY, He XW, Li WY, Li YH, Zhang YK (2019) Targeted imaging and targeted therapy of breast cancer cells via fluorescent double template-imprinted polymer coated silicon nanoparticles by an epitope approach. Nanoscale 11(36):17018–17030

Xu W, Long H, Xu X, Fu G, Pu L, Ding L (2018) Poly(HPMA)-DTPA/DOTA-Gd conjugates for magnetic resonance imaging. New J Chem 42(24):19344–19348

Xu W, Lin Z, Li G, Long H, Du M, Fu G, Pu L (2019) Linear PVA–DTPA–Gd conjugate for magnetic resonance imaging. RSC Adv 9(63):37052–37056

Chapter 10
Nanosensors in Veterinary Disease Detection

Nidhi, Sushil Kumar, Neelam Yadav, Rajesh Kumar, and Hari Mohan

Abstract The health of livestock and poultry is a significant concern due to their major contribution to agricultural output. Infectious diseases pose a major threat to this output as they account for economic losses and risk for zoonotic transmissions. Hence, rapid detection methods for diagnosing veterinary pathogens are highly desirable. Early disease intervention for veterinary infections requires on-site pathogen detection. Conventional diagnostic methodologies lack the ability to provide point-of-care detection and also possess numerous other limitations. However, nanosensors offer a great advantage in providing the best application for pathogen detection for animal health management. This technology is an emerging market for producing accurate health status of animals and can offer numerous advantages for contributing to precision livestock farming. This review discusses the scope of different nanosensors for animals and the uses of these nanosensors as diagnostic techniques for detecting various infectious diseases in livestock. It also enlists the advantages of nanosensors over conventional diagnostic methods and the major drawbacks associated with conventional methods in the domain of disease detection. The chapter provides an insight into all the recent advancements in the field of nanosensors regarding their applications for veterinary disease detection and contribution to enhanced animal welfare.

Nidhi · H. Mohan (✉)
Centre for Medical Biotechnology, Maharshi Dayanand University, Rohtak, Haryana, India
e-mail: harimohan.cmbt@mdurohtak.ac.in

S. Kumar
Department of Biosciences and Bioengineering, Indian Institute of Technology Guwahati, Guwahati, Assam, India

N. Yadav
Department of Biotechnology, Deenbandhu Chotu Ram University of Science and Technology, Sonepat, Haryana, India

R. Kumar
Department of Veterinary Physiology and Biochemistry, Lala Lajpat Rai University of Veterinary and Animal Sciences, Hisar, Haryana, India

LLR University of Veterinary and Animal, ICAR-National Research Centre on Equines, *Nanotechnology Theranostics in Livestock Diseases and Management*, Livestock Diseases and Management,
https://doi.org/10.1007/978-981-16-1610-5_10

Keywords Nanosensors · Veterinary diseases · On-site detection · Livestock · Immunosensor · Sensing technique · Brucellosis · Influenza

10.1 Introduction

Global population growth is causing a major increase in food demands which put livestock welfare under pressure of intensifying the production operations. This intensification makes the livestock more vulnerable to infections which can also affect human faction. Infectious animal diseases caused by microbes cause deaths among humans, livestock, and wildlife. This results in altered food supply, economic losses, and threats to global biodiversity. Rapid selective and sensitive assays for diagnosis of these pathogens can lead to efficient management. Currently, various veterinary pathogens like *E. coli*, *Salmonella*, avian influenza, *Mycoplasma*, *Clostridium*, and haemorrhagic diseases pose significant risks. Early pathogen detection is majorly crucial for better intervention and management. The current research in the field is mainly directed towards early identification and controlling the infectious disease spread in time.

Conventional techniques for pathogen detection from animals require microbe propagation and culturing for the isolation of culture to detect the agent. This method tends to have more drawbacks besides high sensitivity like time-consuming and labor-intensive nature. Other methods like RT-PCR take less time but introduce the need to isolate the genetic material and hence can't be applied for point-of-care detection (Shakuntala n.d.). Serological analysis, microscopic examination, and other highly sensitive techniques need sophisticated equipment as well as trained personnel. For on-time monitoring, methods easy to handle without compromising the specificity and sensitivity need to be explored.

Nanosensors are analytical devices that can be the most suitable alternative to conventional methods. These devices do not need much sample manipulation and can help achieve rapid pathogen detection. The simplicity, easy handling, and real-time analysis potential suggest them to be the perfect candidates for veterinary disease detection. The nanosensing technology provides a promising method for improving the productivity of livestock by monitoring the disease-causing microorganisms.

Early detection of veterinary diseases would provide a left shift to the epidemiological curve as it would enable rapid response, managing disease spread, and providing suitable treatment in time leading to significant social and economic benefits.

This review focuses on the nanosensors that can be applied to animal pathogen diagnostics. It discusses the conventional diagnosis methods along with their drawbacks and the major veterinary pathogens that pose a threat to livestock as well as mankind. The various biomarkers that can be used as markers for nanosensing techniques to detect the pathogen and some examples of successful prototypes of the nanosensor devices in this field are also reviewed in detail.

10.2 Common Veterinary Pathogens

Veterinary diseases can be of various types based on the type of pathogen responsible for the infection. Viral, bacterial, and fungal animal pathogens have been responsible for causing most of the veterinary diseases worldwide (Fig. 10.1). Zoonotic bacteria possess the most risk among all the pathogens as diseases by zoonotic pathogens can be transmitted from animals to humans. Zoonotic bacteria include *Staphylococcus*, *Campylobacter*, and *Bartonella* sp. (Shin and Park 2018). These bacteria are responsible for causing fever, diarrhoea, and other illnesses in animals as well as humans. Undercooked meals, unpasteurized dairy products, and aerosols from animals can be some of the modes of transmission of the diseases to humans. Some pathogens spread among animals via the reproductive tract include *Brucella*, *Leptospira*, and bovine herpes viruses. These STI pathogens may result in acute diseases as well as long-term complications resulting in adverse outcomes affecting livestock welfare (Yang et al. 2018). Many veterinary diseases lead to huge economic repercussions like swine fever, infectious bursal disease, foot and mouth disease, Newcastle disease, Marek's disease, swine influenza, chicken infectious influenza, bluetongue disease, haemorrhagic septicaemia, and many more (Swanepoel et al. 2021). Viral veterinary diseases have always constituted a large proportion of infectious diseases which usually emerge and re-emerge leading to catastrophic effects in humans, wild animals, and domestic animals. The most commonly observed veterinary viruses are foot and mouth disease virus, bovine

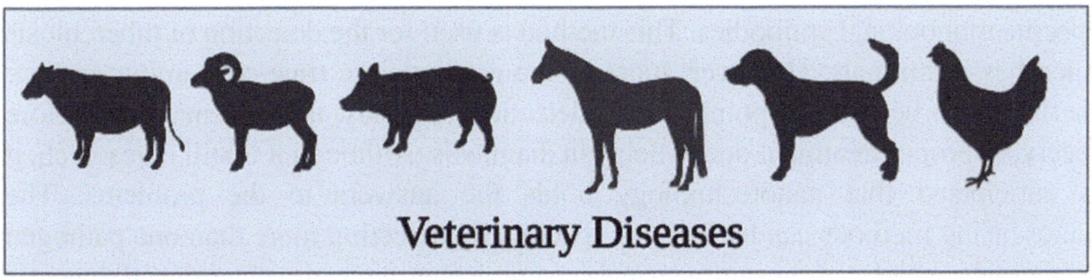

Bacterial diseases: Anthrax, Hemorrhagic septicemia, Tuberculosis, Tetanus, Salmonellosis, Brucellosis, etc.

Parasitic diseases: Liver fluke infection, Hook worm infection, Coccidiosis, Trypanosomiasis, Babesiosis, Haemonchosis, Taeniasis, etc.

Fungal disease: Dermatophytosis, Aspergillosis, Candidiasis, etc.

Viral diseases: Foot and mouth disease, New castle disease, Avian influenza, Bovine viral diarrhoea, Blue tongue disease, African horse sickness, Rabies, Pests des Petits Ruminants (PPR), etc.

Fig. 10.1 Common veterinary diseases

respirovirus, pestivirus, influenza A virus, rabies virus, and bluetongue virus. These diseases also pose a threat to the conservation of certain wildlife species. Other than clinical and zoonotic significance, these diseases have economic implications as well.

10.3 Conventional Diagnostic Methods

The most necessary step to ensure proper preventive and control measures for infectious veterinary diseases is establishing rapid diagnostic methods. For ages, various methods have been employed to detect infectious diseases in animals. Culturing the pathogens and then screening them have been the oldest method which has been a time-consuming affair taking 2–8 days to give results. Immunoassays have been conventionally used for detecting antibodies against the pathogen of interest using ELISA (Enzyme-linked immunosorbent assay), immunodiffusion, virus neutralization, and other such techniques. These assays are based on the principle of interaction between polyclonal antibodies against pathogens and then detected by some cytopathic or haemolytic effect (Shakuntala n.d.). By use of gene cloning and other gene expression methods, the immunoassays have been made more advanced such as employment of flow cytometry, immunohistochemistry, and other techniques in conjugation. Molecular methods such as PCR (Polymerase chain reaction) are more specific and sensitive, but they need extensive machinery and hence can't be used for on-site monitoring. Detection of gamma interferon is also one of the popular methods that need host fresh blood as it employs the use of host-specific monoclonal antibodies. This method is used for the detection of tuberculosis microbes in animals. However, most of the methods are time-consuming and are inefficient to be used as point-of-care detection methods, thus, animals die before receiving proper treatment due to delay in diagnosis. Although it is still in research, it is anticipated that nanotechnology holds the answers to the problems. The nanosensing methods can be used for specifically detecting more than one pathogen in seconds which promises to provide a good alternative in obtaining diagnostic tools with better speed and accuracy (Emerich and Thanos 2003).

10.4 Nanosensors

Nanosensors are devices that have electrical, optical, and magnetic transducers, and they can be used for highly sensitive and specific detection of a wide range of analytes. Disease diagnosis is one of the various applications of nanosensors. The principle behind nanosensing technology is the interaction between some biorecognition elements to determine an analyte from a sample. Various sensing nanoparticles conjugated with the transducers are available. The analyte from the sample can be any biomarker like a monoclonal antibody, RNA, DNA, lectin, enzyme, tissue, or cells. The interaction between the bioreceptor and the analyte is converted into a signal by the transducer which is used for quantitative or qualitative identification of pathogens. Nanosensors can diagnose the disease with high specificity even before clinical symptoms appear in the animal.

10.4.1 Principle of Nanosensing Technology

Nanosensors basically recognize biomarker molecules that are specific for pathogens via binding to bioreceptors. The bioreceptor element performs highly sensitive and specific interaction with the molecule of interest without interfering with other molecules in the sample, and this interaction is then converted to measurable signals with the help of a transducer (Fig. 10.2) (Yoo and Lee 2015). This principle can be employed for both quantitative and qualitative analyses of analyte molecules using the signals produced by the transducer. Two major issues which if targeted nicely can make nanosensors the most convenient and effective diagnostic systems are detecting biomarkers from complex media and appropriate bioassays for the detection. As pathogens spread rapidly inside the animal body, it increases the need to develop highly sensitive methods to detect the pathogen before clinical symptoms appear. Nanosensors can offer appropriate techniques for real-time analysis of veterinary diseases (Zhu et al. 2015).

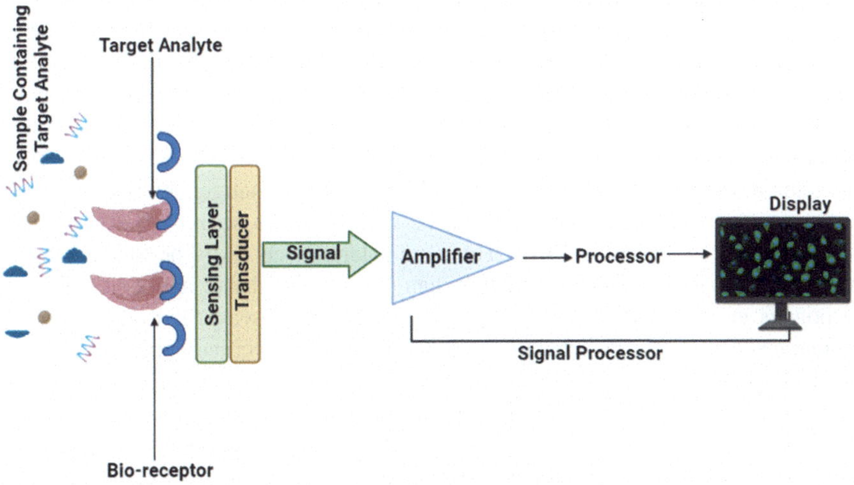

Fig. 10.2 Scheme representing components of nanosensor used for diagnosis purposes: the sensing layer generally contains nanoparticles conjugated with the bioreceptor molecules specific for the analyte. The electrodes and nanoparticles represent the electrical interface

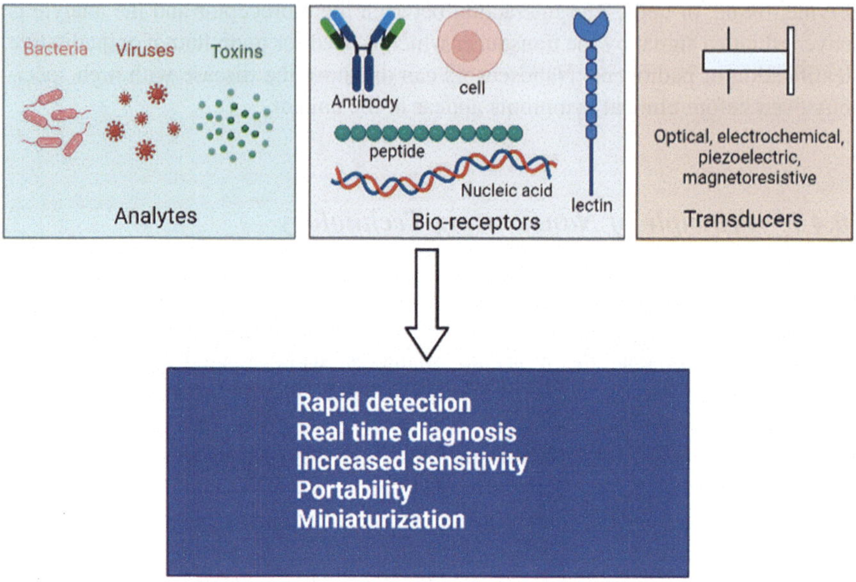

Fig. 10.3 Components involved in nanosensor-based diagnosis and the advantages of such diagnosis over conventional diagnostic methods

The ductile property and combined strength of nanomaterials have made nanomaterials quite an impactful alternative for diagnostics. Nanoparticles such as AuNPs, Ag NPs, and quantum dots offer unique properties such as forming stable conjugates and thereby offer unique functions for diagnostic applications (Sau et al. 2001). As nanoparticles in themselves are inert in nature, their functionalization needs integration and encapsulation with certain molecules such as biomolecules, ligands, or polymers (Mout et al. 2012). Antibodies act as excellent biorecognition elements in conjugation with nanomaterials. By utilizing antigen-antibody interaction property, veterinary pathogens can be detected from serum, urine, or blood samples (Draz and Shafiee 2018). The nanomaterials can easily be functionalized with conjugation to antigens and then can be used to detect antibodies against pathogens in veterinary practices. Not only veterinary diseases but other sorts of biomarkers can also be detected from animal samples in veterinary practices such as the widespread use of antibiotics on animals poses a serious threat to humans consuming animal products, hence, nanosensors can be employed to detect antibiotics in animal meat samples (Zbor et al. 2016). Components involved in nanosensor-based diagnosis and the advantages of such diagnosis over conventional diagnostic methods have been presented in Fig. 10.3.

10.5 Nanosensor Applications in Veterinary Practices

10.5.1 Anthrax Detection

Anthrax is a fatal disease for ruminants as well as humans. It is an infectious bacterial disease that spreads through water and feed and is caused by the bacterium *Bacillus anthracis* (Park et al. 2019). It forms highly resistant spores which make its transmission more efficient and fast, contributing to it being considered a biowarfare agent. These spores get attached to animal bodies and also cause zoonotic transmission to humans causing serious infection with about 30–60% fatality rate (Garofolo et al. 2011). The conventional diagnostic methods for anthrax include PCR, ELISA, and bacterial culture-based detection, but there is a need for more rapid diagnostic methods to prevent the animals and human population from the risk of getting infected by controlling the transmission at early stages. The rAPD4 (recombinant protective antigen domain 4) can be detected using a nanosensing technique by fluorescence aptasensor. This fluorescence aptasensor was developed by coupling multi-walled carbon nanotubes with an aptasensor which not only reduces the cost but also increases the sensitivity of the assay by labeling this aptasensor with gel green this technique offered sensitivity of 20 ng/mL for purified antigen and 62.5 ng/mL for unpurified one (Seleem et al. 2010). Other than this aptasensor, a quartz crystal microbalance biosensor has been developed using AuNPs probes specific for ba813 and pag of the bacterium (Hao et al. 2011). AuNPs-based detection methods for anthrax diagnosis have been proven to be highly selective showing no cross reactivity (Deng et al. 2013).

10.5.2 Avian Influenza

Avian influenza viruses have been a serious global threat to not only the poultry industry but also the human population. The huge economic losses due to the pandemics caused by the emergence of avian influenza point to concerned need to control the disease (Allen et al. 2017). Usually, aquatic birds are considered the main reservoir for the virus, but it also spreads rapidly in domestic poultry. Sixteen hemagglutinin and 9 neuraminidase subtypes have been categorized based on antigenic specificities of the birds-infecting influenza A virus. Out of all the subtypes, H5, H7, and H9 are considered highly pathogenic avian influenza viruses (HPAIVs). The influenza viruses belong to the orthomyxoviridae family and the AIV genome contains negative stranded RNA. Various serological tests are conventionally employed for the diagnosis of AIVs. HA tests, HI tests, isolation in cell culture/embryonated eggs, and NT tests are some of the methods used for the detection of AIVs. But serological analyses are time-consuming and labour intensive which require sample collection and a specific amount of virus particles tending to make the whole process costly. Other molecular methods like PCR and RT-PCR

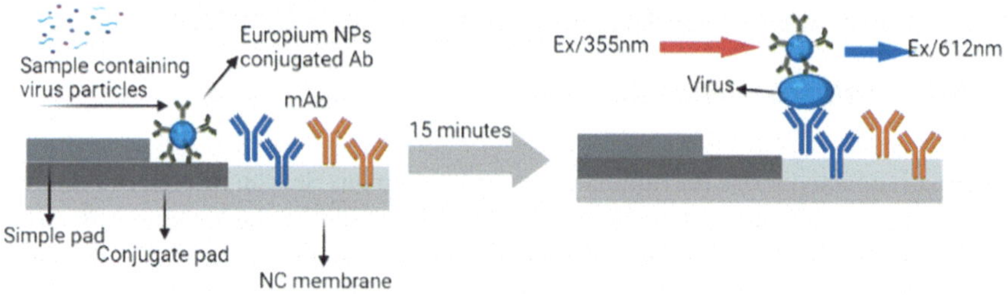

Fig. 10.4 Schematic diagram of rapid diagnostic biosensor for avian influenza: the diagnostic system employs Europium conjugated H7 subtype-specific antibody and 6D7 mAb. When the sample containing lysis buffer and the virus particles is dropped on the strip, the virus gets sandwiched between the two antibodies. The resulting complex can get excited with 355 nm light and emits 612 nm light which can be detected for a positive test reaction (Yeo et al. 2017)

require the isolation of genetic materials and specific laboratory machinery. On-site detection is needed for early monitoring to contain the disease which is quite rare. Highly sensitive point-of-care detection methods are needed to reduce the economic losses due to the spread of the disease. Monoclonal antibody (MAb)-modified europium nanoparticles were used in a fluorescent immunochromatographic test for specific detection of hemagglutinin. This method was found to be highly specific and sensitive with a detection limit of 30 ng/mL (Fig. 10.4) (Yeo et al. 2017).

Another immunochromatographic strip test was developed for colloidal Au nanoparticles; these test strips were found to have 98.6% sensitivity and 71.4% specificity as compared to PCR methods (Luo et al. 2017). Another nanosensor which was an antibody-based method and utilized magnetic silica nanoparticles with a resonance light scattering system showed no cross-reactivity against other influenza strains and had a sensitivity range of 0.5–50 ng/mL for AIVs (Zou 2012). Specific anti-M1 antibodies were used in conjugation with gold nanoparticles-based carbon nanotubes to develop a sensitive plasmon-assisted fluoro-immunoassay to detect the influenza virus (Vidic et al. 2017). A uniform fluorescence quenching-based assay has been developed using two probes: HA-specific antibody-modified gold nanoparticles and glycan-conjugated luminescent quantum dots (QDs). This sensor brings the two probes together when exposed to strain-specific hemagglutinin protein due to the specific binding of HA with the two probes (Li et al. 2012). The non-radioactive energy transfer from QDs to AuNPs diminishes the fluorescence intensity of QDs and this assay is a highly sensitive tool for influenza virus discrimination (Neethirajan 2017). The quantum dots-based chip biosensors for detecting avian influenza are found to be more sensitive than QD-based immunoassays (Chen et al. 2010). The portable diagnostic methods developed by integrating nanoparticles with suitable biorecognition elements have been suitable point-of-care detection methods without altering the sensitivity or specificity.

10.5.3 Newcastle Disease

It is a viral disease caused by Newcastle disease virus (NDV) leading to respiratory manifestations, lesions, nervous implications, and cessation of egg production. The NDV also possesses the zoonotic risk of causing conjunctivitis to those in contact with infected poultry or wild avian species. The conventional diagnostic method of virus isolation and characterization usually takes 2–12 days to give results. An immunosensor with tilted fibre grating (Ex-TFG) coated with staphylococcal protein A-Au nanospheres has successfully been developed to detect NDV antibodies within 25 min with high sensitivity (Luo et al. 2017). For avoiding cross-reactivity against other antibodies quantitative antigen-based Au-immunochromatographic technique incorporated with viral protein antigens/AuNPs was developed which showed sensitivity up to 2^2 titres (Nodoushan et al. 2019). A nanosensor based on cadmium telluride quantum dots conjugated with NDV antibodies and mercaptosuccinic acid coupled with iron oxide nanoparticles was developed to be able to detect even low antigen amounts up to 1.5 ng/mL, and the method was also found to be highly specific. Fluoroassays coupled with nanoparticles have also been developed to rapidly detect NDV.

10.5.4 Chlamydiosis

Chlamydiae causes various diseases in animals and human hosts. They not only infect not only livestock but also wildlife and exotic species. The bacteria can persist inside the host body as asymptomatic and can lead to life-threatening illnesses. The bacteria possess extreme zoonotic risks as the small-sized bacteria render difficulty to analyse using light microscopy. The microscopic and histological findings are not specifically indicative of Chlamydia and can give false positive results in the case of no infection at all (Borel et al. 2018). The characterization studies related to these bacteria only rely on molecular findings, and the nucleic acid amplification tests are considered effective for the diseases. In spite of high sensitivity and specificity of the amplification assays, the tests are not as gold standards due to the need for genome isolation and expertise while handling. Various Chlamydia family bacteria causing veterinary diseases include *C. psittaci* in horses, *C. abortus* in ruminants, *C. avium* in avian species, and *C. felis* in cats. A biosensor to diagnose *Chlamydia* sp. using quartz crystal microbalance has been successfully developed which analyses bacteria in association with an antibody-functionalized electrode. The crystals assembled with anti-chlamydia antibodies in conjugation with anti-mouse IgGAb revealed the long-term stability of the biosensor (Ben-Dov et al. 1997). Another rapid and multiplexed diagnostic optical sensor developed using plasmonic microarray and gold nanohole arrays successfully demonstrated the analysis of Chlamydia without the need for DNA extraction from the sample (Soler et al. 2017). DNA-based LSPR sensors developed using AuNPs are more portable and durable, eliminating the need for DNA amplification which is efficient for in-field sensing (Cary 2020). These nanosensors can revolutionize the field of veterinary pathogen detection and offer a sensitive method for the same.

10.5.5 Foot and Mouth Disease

Another highly contagious and infectious veterinary pathogen is the foot and mouth disease virus (FMDV) belonging to the *Aphthovirus* genus of the Picornaviridae family which infects cattle, sheep, pigs, and many other wild species. This disease affects a large number of animals worldwide and shows high morbidity. The clinical symptoms of this include the formation of painful blisters on mouth tissues like the tongue and lips, and it also affects the skin tissues where the epidermis layer is thin such as between the toes. Anorexia, reluctance to move, depression, lameness, and excessive salvation are the supplementary symptoms. Classical diagnostic methods for this single-stranded positively charged RNA virus are ELISA and RT-PCR (Jamal and Belsham 2013). Due to the complexity and time-consuming nature of existing tests, various sensors have been developed to ease the diagnosis. The lateral flow device developed for this virus detection improved the sample preservation during transport which is a dry, non-hazardous transport system of samples for further testing (Fowler et al. 2014). Viral imprinted polymer-based biosensors for rapid and selective detection of foot and mouth disease offered 50-fold lower LOD in comparison to ELISA and PCR. The nanosensor used oxidized *O*-aminophenol film imprinted with FMDV on a gold screen-printed electrode and the results were demonstrated using cyclic voltametry. This method offered the possibility of on-site detection and monitoring of viruses in the field with no prior sample preparation (Hussein et al. 2019). Another nanosensor with superior analytical and clinical performance in detecting FMDV has been developed using gold nanoparticles which can accelerate the FMDV diagnosis in farms and endemic areas to offer better surveillance with 95.4% efficiency and 100% specificity (Hamdy et al. 2018). This biosensor was developed using thiol-linked oligonucleotide to recognize the conserved 3D gene of FMDV. These nanoparticle-based sensors offer enhanced specificity and efficiency for FMDV detection and can be extensively used for routine diagnostic purposes.

10.5.6 Brucellosis

Brucellosis is a bacterial disease caused by *Brucella* sp. bacteria which are facultative intracellular bacteria affecting humans and ruminants. Arthralgia, osteomyelitis, and endocarditis include fever, chill, weight loss, and arthralgia and the chronic affects are osteomyelitis, and endocarditis. This disease contains zoonotic potential as it can infect humans through contaminated milk and contact with animal faeces.

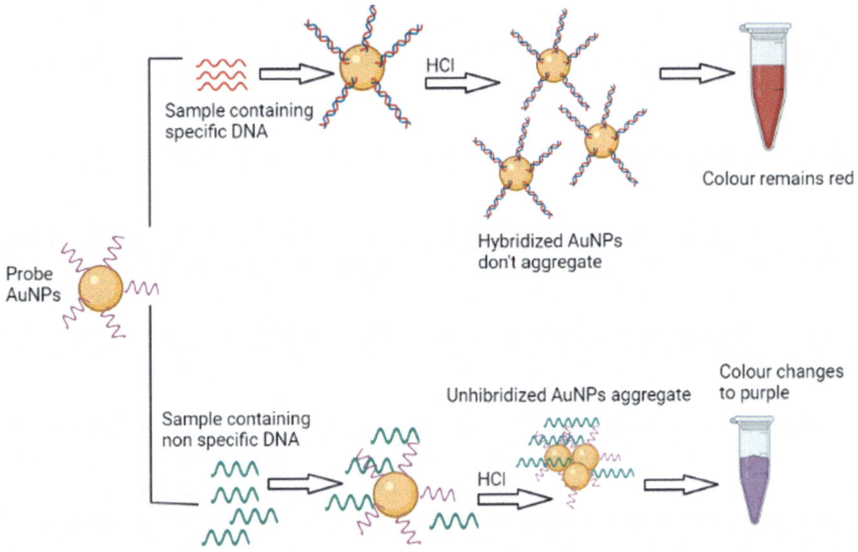

Fig. 10.5 Schematic representation of biosensor for diagnosis of *Brucella* from a blood sample. The sample that is positive for *Brucella* contains a specific conserved gene against which the probe is conjugated to AuNPs. For a positive reaction, the gene gets hybridized with a probe molecule which prevents aggregation of AuNPs and hence the colour of the sample remains red. But for a negative reaction, there is no binding between the probe and DNA, and hence the unhybridized AuNPs get aggregated leading to a change in colour. The purple colour represents a negative reaction and hence dictates the absence of *Brucella* in the sample

Milk ring tests, serum tube agglutination tests and culture on selective media are common tests for this disease. But to increase sensitivity various nanomaterials-based assays and sensors have been developed. Oligonucleotide-modified AuNPs-based assay was able to detect *Brucella* spp. within 30 min with no cross-reactivity (Fig. 10.5) (Pal et al. 2017).

DNA-activated gold nanoparticles-based sensor using DNA derived from the pathogen demonstrated the possibility of using the biosensor for in-field analysis using milk urine or semen samples. This method showed the detection limit of 10^3 cfu/mL within 30 min and can be implemented readily in a field with the advantages of cost and convenience (Pal et al. 2017). Another rapid culture-free detection method was developed using ionic self-assembled multilayer film coupled with optical fibres. This nanomaterials-based biosensor could detect a low number of bacteria in clinical samples (Mccutcheon et al. 2019). Magnetic and silica nanoparticles have significant advantages over other nanoparticles such as cheap production and simple synthesis and hence are employed for optical biosensor formation to detect *Brucella* spp. This method has various advantages including short detection time, high specificity and sensitivity, and ease of application without needing any specific dedicated equipment (Ahangari et al. 2022). Multi-cross displacement amplification in conjugation with nanoparticles-based lateral flow biosensor was employed to detect these bacteria and promised to be a rapid, simple, and reliable method for screening *Brucella* strains (Li et al. 2019).

10.5.7 Clostridium Enteric Disease

Clostridium enteric diseases are caused by *Clostridium perfringens* which is a gram-negative fermentative soil bacterium that forms highly resistant spores and exotoxins. The various toxins produced by this bacterium include alpha, beta, epsilon toxins, enterotoxin E, and NetB toxin. The NetB toxin is responsible for necrotic enteritis in chickens. *C. perfringens* strain can infect birds, goats, horses, pigs, cattle, sheep, and dogs. The diseases caused by the bacteria are responsible for huge economic losses and high mortality among animals. Some strains of the bacteria are even capable of infecting humans by their enterotoxins produced during sporulation. The food-borne diseases caused by these bacteria in humans via animal products are highly dangerous. The main challenge for diagnostic techniques to detect the disease-causing pathogen in this case is rapid strain identification along with differentiation from non-pathogenic strains. Various cell-based nanosensors have been proven effective in detecting the presence of pathogens or active toxins in clinical as well as food samples (Banerjee and Bhunia 2010). The toxins produced by the bacteria can also be detected using an immunosensor which comprises specific antibodies integrated into the nanosensor. The immunosensor contained carbon nanotubes assembled to analytically detect nM ranges of concentrations of the toxin with high sensitivity as compared to ELISA (Chen et al. 2016). DNA-based biosensors can be employed for the detection of *C. perfringens* from extracted DNA samples. This biosensor was developed using a DNA probe immobilized on CeO_2/chitosan modified electrode surface. The range of detection for this nanosensor is 10^{-14} to 10^{-7} mol/L (Qian et al. 2018). For rapid diagnosis, the lateral flow biosensor is developed in combination with loop-mediated isothermal amplification technique which gives high relative accuracy (Sridapan et al. 2021). It is a low-cost qualitative method which can be used for routine diagnosis.

10.5.8 Other Deadly Veterinary Diseases

There are various limitations in conventional diagnosis techniques for many veterinary diseases such as low sensitivity, specificity, time consumption, and the need for sample treatment and dedicated machinery. Nanosensing techniques in the field of diagnosis of these diseases can overcome many of the limitations. Mycoplasmas are the smallest bacterial cells which are insensible to antibiotics due to lack of cell wall. *M. bovis* causes pneumonia, mastitis, or arthritis in cattle. The disease caused by mycoplasma needs high cost for treatment, but many biosensors have been developed to detect bacterial proteins NAGs and haptoglobin from various samples

(Pemberton et al. 2001). Electrochemical biosensors using screen-printed carbon electrodes and biotin-labeled aptamer assay have proven effective in detecting *Mycoplasma* (Vidic et al. 2017). *Salmonella* bacteria, haemorrhagic bacteria and *Campylobacter* also need focus to develop such nanosensors for rapid diagnosis.

10.6 Conclusion

This review compiled some advanced biotechnological diagnostic approaches for detecting veterinary diseases that affect livestock and poultry. These pathogens lead to significant economic losses by increasing veterinary costs, decreasing food production, and possessing global threat of zoonosis. Conventional diagnosis techniques are usually time-consuming and need sophisticated equipment handled with expertise. The nanosensing techniques can overcome these limitations by providing rapid, accurate, and affordable biosensor devices. Most of the biosensors are developed for food and dairy product samples from animals which need modification to detect the pathogens from clinical animal samples (Funazaki et al. 1995). For certain zoonotic pathogens, biosensors have been developed to detect the disease in human samples such as for brucellosis which needs special emphasis to develop similar assays for veterinary samples. Nanomaterials like AuNPs, AgNPs, iron oxide NPs, silica NPs, quantum dots, and polymeric nanoparticles have proven to be valuable for detecting pathogens and toxins in livestock and poultry due to their biocompatibility and cheap production. Modifications and functionalization of these nanoparticles in conjugation with bioreceptor components contribute to rapid detection and improve the range of detection. As most of the diagnostic methods are operated only in laboratories, nanosensors offer great possibilities as analytical devices which can be operated in the field. There is a strong need to research nanosensors for point-of-care detection methods, and these can be developed as portable, multitargeting miniature devices for detecting veterinary diseases. To harmonize and validate these devices for on-site diagnosis, there is a strong need for inter-laboratory testing and modifications to adapt these sensors to adapt for in-field conditions.

References

Ahangari A, Mahmoodi P, Mohammadzadeh A (2022) Biosensors functionalized with nanoparticles for rapid detection of Brucella. Microchem J 181:107697

Allen T, Murray KA, Zambrana-torrelio C, Morse SS, Rondinini C, Di Marco M, Breit N, Olival KJ, Daszak P (2017) Zoonotic diseases. Nat Commun 8:1124. https://doi.org/10.1038/s41467-017-00923-8

Banerjee P, Bhunia AK (2010) Cell-based biosensor for rapid screening of pathogens and toxins. Biosens Bioelectron 26(1):99–106. https://doi.org/10.1016/j.bios.2010.05.020

Ben-Dov I, Willner I, Zisman E (1997) Piezoelectric immunosensors for urine specimens of Chlamydia trachomatis employing quartz crystal microbalance microgravimetric analyses. Anal Chem 69(17):3506–3512

Borel N, Polkinghorne A, Pospischil A (2018) A review on Chlamydial diseases in animals: still a challenge for pathologists? Vet Pathol 55(3):374–390. https://doi.org/10.1177/0300985817751218

Cary RM (2020) Sensing of small molecules, biomarkers, and pathogens using unique plasmonic assay platforms. Doctoral dissertation, University of Cincinnati

Chen L, Sheng Z, Zhang A, Guo X, Li J (2010) Quantum-dots-based fluoroimmunoassay for the rapid and sensitive detection of avian influenza virus subtype H5N1. Luminescence 25:419–423. https://doi.org/10.1002/bio.1167

Chen H, Huang J, Fam DWH, Tok AY (2016) Horizontally aligned carbon nanotube based biosensors for protein detection. Bioengineering 3:23. https://doi.org/10.3390/bioengineering3040023

Deng H, Zhang X, Kumar A, Zou G (2013) Long genomic DNA amplicons adsorption onto unmodified gold nanoparticles for colorimetric detection of Bacillus anthracis. Chem Commun 49:51–53. https://doi.org/10.1039/c2cc37037a

Draz MS, Shafiee H (2018) Applications of gold nanoparticles in virus detection. Theranostics 8(7):1985–2017. https://doi.org/10.7150/thno.23856

Emerich DF, Thanos CG (2003) Nanotechnology and medicine. Expert Opin Biol Ther 3(4):655–663

Fowler V, Bashiruddin JB, Belsham GJ, Stenfeldt C, Bøtner A, Knowles NJ, Bankowski B, Parida S, Barnett P (2014) Characteristics of a foot-and-mouth disease virus with a partial VP1 G-H loop deletion in experimentally infected cattle. Vet Microbiol 169(1–2):58–66. https://doi.org/10.1016/j.vetmic.2013.12.008

Funazaki N, Hemmi A, Ito S, Asano Y, Yano Y, Miura N, Yamazoe N (1995) Application of semiconductor gas sensor to quality control of meat freshness in food industry. Sensors Actuators B Chem 25(1–3):797–800. https://doi.org/10.1016/0925-4005(95)85177-1

Garofolo G, Serrecchia L, Corrò M, Fasanella A (2011) Anthrax phylogenetic structure in Northern Italy. BMC Res Notes 4:273

Hamdy ME, Del Carlo M, Hussein HA, Salah TA, El Deeb AH, Emara MM, Pezzoni G, Compagnone D (2018) Development of gold nanoparticles biosensor for ultrasensitive diagnosis of foot and mouth disease virus. J Nanobiotechnol 16(1):48. https://doi.org/10.1186/s12951-018-0374-x

Hao R, Song H, Zuo G, Yang R, Wei H, Wang D, Cui Z, Zhang Z, Cheng Z, Zhang X (2011) DNA probe functionalized QCM biosensor based on gold nanoparticle amplification for Bacillus anthracis detection. Biosens Bioelectron 26(8):3398–3404. https://doi.org/10.1016/j.bios.2011.01.010

Hussein HA, Hassan RYA, Mohamed R, Nashar E, Khalil SA, Salem SA, El-sherbiny IM (2019) Designing and fabrication of new VIP biosensor for the rapid and selective detection of foot-and-mouth disease virus (FMDV). Biosens Bioelectron 141:111467. https://doi.org/10.1016/j.bios.2019.111467

Jamal SM, Belsham GJ (2013) Foot-and-mouth disease: past, present and future. Vet Res 44(1):116

Li X, Lu D, Sheng Z, Chen K, Guo X, Jin M, Han H (2012) A fast and sensitive immunoassay of avian influenza virus based on label-free quantum dot probe and lateral flow test strip. Talanta 100:1–6. https://doi.org/10.1016/j.talanta.2012.08.041

Li S, Liu Y, Wang Y, Wang M, Liu C, Wang Y, Wang Y (2019) Rapid detection of Brucella spp. and elimination of carryover using multiple cross displacement amplification coupled with nanoparticles-based lateral flow biosensor. Front Cell Infect Microbiol 9:78. https://doi.org/10.3389/fcimb.2019.00078

Luo B, Xu Y, Wu S, Jiang P, Shi S, Wang Y, Wang L, Liu Y, Luo B, Xu Y, Wu S, Zhao M, Shi S, Zhang Z, Wang Y, Wang L, Liu Y (2017) A novel immunosensor based on excessively tilted fiber grating coated with gold nanospheres improves the detection limit of Newcastle disease virus. Biosens Bioelectron. https://doi.org/10.1016/j.bios.2017.08.064

Mccutcheon K, Bandara AB, Zuo Z, Heflin JR, Inzana TJ (2019) The application of a nanomaterial optical fiber biosensor assay for identification of Brucella nomenspecies. Biosensors 9(2):64

Mout R et al (2012) Surface functionalization of nanoparticles for nanomedicine. Chem Soc Rev 41(7):2539–2544. https://doi.org/10.1039/c2cs15294k

Neethirajan S (2017) Recent advances in wearable sensors for animal health management. Sens Biosens Res 12:15–29. https://doi.org/10.1016/j.sbsr.2016.11.004

Nodoushan SM, Nasirizadeh N, Kachuei R, Ali A, Fooladi I (2019) Electrochemical detection of a flatoxin B1: an aptasensor prepared using graphene oxide and gold nanowires. Anal Methods 11:6033–6042. https://doi.org/10.1039/c9ay01673b

Pal D, Boby N, Kumar S, Kaur G, Ali SA, Reboud J, Shrivastava S, Gupta PK, Cooper JM, Chaudhuri P (2017) Visual detection of Brucella in bovine biological samples using DNA-activated gold nanoparticles. University of Glasgow. https://doi.org/10.5525/gla.researchdata.406

Park BN, Haufiku B, Katjiuanjo P, Likando A, Shikesho R, Nyarko K, Miller LA, Agolory S, Vieira AR, Salzer JS, Bower WA, Campbell L, Kolton CB, Marston C, Gary J, Bollweg BC, Zaki SR, Hoffmaster A, Walke H (2019) Anthrax epizootic in wildlife, Bwabwata National Park, Namibia, 2017. Emerg Infect Dis 25(5):947–950. https://doi.org/10.3201/eid2505.180867

Pemberton RM, Hart JP, Mottram TT (2001) An assay for the enzyme N-acetyl-β-d-glucosaminidase (NAGase) based on electrochemical detection using screen-printed carbon electrodes (SPCEs). Analyst. https://doi.org/10.1039/b104874k

Qian X, Qu Q, Li L, Ran X, Zuo L, Huang R, Wang Q (2018) Ultrasensitive electrochemical detection of Clostridium perfringens DNA based morphology-dependent DNA adsorption properties of CeO_2 nanorods in dairy products. Sensors 18(6):1878. https://doi.org/10.3390/s18061878

Sau TK, Pal A, Jana NR, Wang ZL, Pal T (2001) Size controlled synthesis of gold nanoparticles using photochemically prepared seed particles. J Nanoparticle Res 3:257–261

Seleem MN, Boyle SM, Sriranganathan N (2010) Brucellosis: a re-emerging zoonosis. Vet Microbiol 140:392–398. https://doi.org/10.1016/j.vetmic.2009.06.021

Shakuntala I (n.d.) Advanced techniques in diagnosis of livestock diseases and zoonotic pathogens

Shin B, Park W (2018) Zoonotic diseases and phytochemical medicines for microbial infections in veterinary science: current state and future perspective. Front Vet Sci 5:166. https://doi.org/10.3389/fvets.2018.00166

Soler M, Belushkin A, Cavallini A, Kebbi-beghdadi C, Greub G (2017) Multiplexed nanoplasmonic biosensor for one-step simultaneous detection of Chlamydia trachomatis and Neisseria gonorrhoeae in urine. Biosens Bioelectron 94:560–567. https://doi.org/10.1016/j.bios.2017.03.047

Sridapan T, Tangkawsakul W, Janvilisri T, Kiatpathomchai W, Dangtip S, Ngamwongsatit N, Nacapricha D et al (2021) Rapid detection of Clostridium perfringens in food by loop-mediated isothermal amplification combined with a lateral flow biosensor. PLoS One 16:e0245144. https://doi.org/10.1371/journal.pone.0245144

Swanepoel H, Crafford J, Quan M (2021) A scoping review of viral diseases in African ungulates. Vet Sci 8(2):17

Vidic J, Manzano M, Chang CM, Renault NJ (2017) Advanced biosensors for detection of pathogens related to livestock and poultry. Vet Res 48(1):11. https://doi.org/10.1186/s13567-017-0418-5

Yang Z, Xu G, Reboud J, Ali SA, Kaur G, Mcgiven J, Gupta P, Chaudhuri P, Cooper JM (2018) Rapid veterinary diagnosis of bovine reproductive infectious diseases from semen using paper-origami DNA microfluidics. ACS Sens. https://doi.org/10.1021/acssensors.7b00825

Yeo S, Bao DT, Seo G, Bui CT, Thi D, Kim H, Thi N (2017) Improvement of a rapid diagnostic application of monoclonal antibodies against avian influenza H7 subtype virus using Europium nanoparticles. Sci Rep 206:1–11. https://doi.org/10.1038/s41598-017-08328-9

Yoo SM, Lee SY (2015) Optical biosensors for the detection of pathogenic microorganisms. Trends Biotechnol 34:7. https://doi.org/10.1016/j.tibtech.2015.09.012

Zbor R, Jakubec P, Urbanová V, Medr Z (2016) Advanced sensing of antibiotics with magnetic gold nanocomposite: electrochemical detection of chloramphenicol. Chemistry. https://doi.org/10.1002/chem.201602434

Zhu C, Yang G, Li H, Du D, Lin Y (2015) Electrochemical sensors and biosensors based on nanomaterials and nanostructures. Anal Chem 87:230–249

Zou X (2012) Detection of avian influenza virus based on magnetic silica nanoparticles resonance light scattering system. Analyst 137:648–653. https://doi.org/10.1039/c1an16041a

Chapter 11
Nanotechnology in Diagnostic Imaging

Alonkrita Chowdhury, Kratika Maurya, Akshita, and Mayukh Ghosh

Abstract Early and precise diagnosis is pivotal to prevent disease progression or spillover or to determine the suitable therapeutic regimen. Based upon the type of disease and its phenotypic manifestations, several diagnostic methods are employed, and in this context, diagnostic imaging techniques have found niche applications. Different popular diagnostic imaging modalities such as optical imaging, radiographic techniques, magnetic resonance imaging (MRI), computed tomography (CT), and ultrasonography and their nano-enhancements for improved diagnosis have constituted the core of the current discussion. For instance, gold nanoparticles (AuNPs) and silver nanoparticles (AgNPs) based dark-field microscopy, multiphoton plasmon resonance microscopy, optical coherence microscopy, hyper-Rayleigh scattering spectroscopy, and Raman spectroscopy have been evolved for biomedical imaging. In ultrasound imaging, mesoporous silica nanoparticles (MSNs), polystyrene particles, etc. are employed as contrasting agents. Using fluorescent mesoporous silica nanoparticles as a contrast agent, targeted endoscopic detection is performed in colorectal cancer. There are also two other imaging techniques, i.e., thermal imaging and nuclear imaging based on infrared radiation and ionizing electromagnet radiation, respectively. Iron oxide NPs or magnetic NPs (MNPs), Au core/hollow mesoporous silica shell nanoparticles, etc. are exploited in thermal imaging, whereas AuNPs are utilized as X-ray contrast agents in radiographic imaging to enhance the resolution of the images. Three-dimensional cross-sectional image with high spatial and temporal resolution is developed in computed tomography technique. AuNPs, iodine-based NPs are employed in this technique to obtain images with enhanced contrast and better resolution. Iodine-based NPs offer

A. Chowdhury · K. Maurya · Akshita
Department of Genetics and Plant Breeding, Institute of Agricultural Science, Rajiv Gandhi South Campus, Banaras Hindu University, Mirzapur, Uttar Pradesh, India

M. Ghosh (✉)
Department of Veterinary Physiology & Biochemistry, Faculty of Veterinary and Animal Sciences, Rajiv Gandhi South Campus, Banaras Hindu University, Mirzapur, Uttar Pradesh, India

LLR University of Veterinary and Animal, ICAR-National Research Centre on Equines, *Nanotechnology Theranostics in Livestock Diseases and Management*, Livestock Diseases and Management,
https://doi.org/10.1007/978-981-16-1610-5_11

better tumor penetration ability in addition to their biodegradable and clearance properties, making them an appropriate contrast agent. In magnetic resonance imaging, superparamagnetic iron oxide nanoparticles (SPIONS), gadolinium-based dendrimers are efficiently being exploited as contrast agent. Positron emission tomography (PET) and single-photon emission computerized tomography (SPECT) are mostly used nuclear imaging modalities in the clinical diagnosis. In terms of sensitivity and resolution, PET is more gainful in comparison to SPECT. A variety of liposomes have been evolved in addition to other nanomaterials that act as a contrasting agent in PET imaging. Functionalized gold NPs aided SPECT is employed in lymphatic mapping and early-stage cervical cancer detection. On the basis of electrical properties like conductivity, permittivity, and impedance, electrical impedance tomography (EIT) technique has been evolved. It can provide images of a specific region in the human body based on the electrical conductivity of the tissues. So diagnostic imaging aided by nanotechnology has already proved to be efficient to gain ample attention and application imminently.

Keywords Disease diagnosis · Microscopy · Raman spectroscopy · Radiographic imaging · Magnetic resonance imaging · Computed tomography · Ultrasonography · Thermal imaging · Nanoparticles

11.1 Introduction

In disease management and control, probably half the battle can be conquered by early and precise diagnosis. Traditionally, hemato-biochemical laboratory methods, serological diagnosis, nucleic acid-based detection, isolation and culture of causative agents, imaging techniques, biopsy, etc. are in practice for disease diagnosis. Among the aforementioned methods, diagnostic imaging holds a pivotal position in the detection of certain complex diseases like cancer. For instance, according to a 2020 report, the imaging sector is dominated by more than 65% of the cancer diagnostics market share (https://www.gminsights.com/industry-analysis/cancer-diagnostics-market). However, biomedical imaging faces limitations in light penetration depth and signal attenuation in tissues. Therefore, technological advancements in medical imaging systems have been urged. In this regard, nanotechnology plays a pivotal role. It facilitates acquiring higher-resolution images and greater efficacy in the detection of various aberrations due to their physical, pharmacokinetic, and nanoscopic properties. Nanotechnology is an emerging field in every branch of science. It is remarkably the most researched area in biomedical science, especially in oncology. In 1954, a polymer-drug conjugate was designed by Jatzkewitz which was probably the beginning of the nano-oncology (Aslan et al. 2013). In 1964, liposomes were discovered by Bangham and Horne (Bayda et al. 2019; Aslan et al. 2013). Subsequently, different nanoparticles (NPs) such as quantum dots (Louis Brus 1985), carbon nanotubes (CNTs) (Sumio Iijima 1991), gold nanoshells (Naomi Halas 2003), and fluorescent carbon dots (Xu et al. 2004) were constructed. These are widely used in biomedical sensing tools due to their unique optical properties and biocompatibility. In comparison to traditional techniques, optical imaging using nanomaterials is much more sensitive. It provides higher-resolution images and greater efficacy. NPs such as semiconductors, quantum dots, and iron oxide nanocrystals possess exclusive optical, magnetic, or structural properties which encourage their application in molecular imaging. Conventional screening methods are often time-consuming, thus limiting early detection. For instance, traditional imaging modalities can detect tumors only subsequent to phenotypic manifestation, i.e., after cell proliferation and metastasis. Moreover, it fails to determine the characteristics of the tumor whether it is malignant or benign. For this, a biopsy is needed. Contrastingly, NPs aided imaging technology provides early detection of cancer. Different nanomaterials are also employed as contrasting agents to enhance the performance of traditional screening methods, such as magnetic resonance imaging (MRI), computed tomography (CT), positron emission tomography (PET), and ultrasonography. Additionally, novel imaging modalities such as photoacoustic tomography (PAT), Raman spectroscopic imaging, and multimodal imaging are also being functionalized as current nanoscale imaging platforms (https://www.cancer.gov/nano/cancer-nanotechnology/detection-diagnosis). Labeling of a highly specific tumor using NP-conjugated anti-tumor drugs and biomolecules (such as peptides, antibodies, or other chemicals) emphasizes the detection and screening of cancer cells. MRI aided with immune superparamagnetic iron oxide nanoparticles (SPIONs) offers more anatomical information and detailed visualization than other methods. For instance, metastases in lung cancer can be determined by MRI-SPION imaging. The dendritic multivalence property and the enhanced penetration and retention (EPR) effect of the dendritic molecules make it favorable for effective accumulation in the tumor site. As a result, dendrimer-based PET shows extremely sensitive and specific imaging. Recently, nanoprobe biotechnology also plays a key role in the discovery of novel disease-specific biomarkers. Hence, different optical imaging methods using fluorescent probes, phosphorescent probes, photoacoustic (PA) tomography, etc. have been developed in the fields of fluorescence, phosphorescence, and photoacoustic imaging (Zhang et al. 2022). Therefore, the amalgamation of these NPs with conventional screening technologies provides a great opportunity to obtain improved resolution and better imaging of specific aberrations for early detection of diseases with enhanced sensitivity (Table 11.1).

11.2 Nanotechnology Advancements in Microscopy

Microscopy deals with the study of objects under a microscope that is beyond the visibility range of the naked eye. Different microbes, histopathological lesions, abnormal configuration of tumor cells, presence of biomarkers, or unwanted cells in an organ due to metastasis can be detected by means of microscopy technique. Light, electromagnetic waves, sound waves, particle waves, etc. are directed to illuminate the target object as a primary wave, and subsequently, the reflecting secondary wave is captured to detect the targeted object. Based on the energy source

Table 11.1 Application of NP-based diverse optical imaging modalities in disease diagnosis

Imaging techniques	NPs used	Applications	References
Topically applied surface-enhanced resonance Raman ratiometric spectroscopy (TAS3RS or tasers)	Folate receptor (FR)-targeted surface-enhanced resonance Raman scattering (SERRS) nanoparticles (NPs) (FR-SERRS-NPs)	1. Detection of tumor lesions in a murine model of ovarian adenocarcinoma regardless of their size or localization 2. Intraoperative detection of microscopic residual tumors and reduction in recurrence rates in ovarian cancer	Oseledchyk et al. (2017)
Surface plasmon resonance (SPR) absorption spectroscopy	Monoclonal anti-epidermal growth factor receptor (anti-EGFR) antibodies conjugated gold NPs	Due to specific and homogenous binding to the surface of malignant oral epithelial cells (HOC 313 clone 8 and HSC 3), it provides a sharper SPR absorption band in contrast to non-malignant cells. Therefore, it acts as molecular biosensor for the diagnosis and investigation of oral epithelial living cancer cells	El-Sayed et al. (2005)
In vitro dark-field scattering imaging	Polyethylene glycol-transferrin-conjugated nanoparticles (PEG-Tf-NPs)	Used as a contrast agent for cancer diagnostics due to enhancement of scattering-based optical contrast	Li et al. (2009)
UV–Vis spectroscopy and scanning electron microscopy (SEM)	AuNPs functionalized with thiol-modified oligonucleotide	For specific and sensitive detection of paratuberculosis or Johne's disease-causing *Mycobacterium avium* subsp. paratuberculosis (MAP) DNA in ruminants	Ganareal et al. (2018)
Dual-modal ultrasound (US)/magnetic resonance (MR) imaging	Targeted gold-nanoshelled poly(lactic-co-glycolic acid) (PLGA) magnetic hybrid nanoparticles carrying anti-human epidermal growth factor receptor 2 (Her2) antibodies (Her2-GPH NPs)	1. It prominently enhances the US/MR molecular imaging effect of targeted cells by cell-specific binding 2. It also serves in PTT as effective photoabsorbers to specifically induce SKBR3 cell death upon near-infrared laser irradiation	Dong et al. (2019)
Photoacoustic imaging	Colloidal gold nanoparticles (AuNPs)	To serve as an exogenous contrast agent for the	Tian et al. (2016)

(continued)

Table 11.1 (continued)

Imaging techniques	NPs used	Applications	References
	functionalized with polyethylene glycol (PEG) and (Arginine(R)–Glycine(G)–Aspartic (D))4 (RGD) peptides	detection and imaging of single live cancer cells	
T2-weighted magnetic resonance (MR) and X-ray computed tomography (CT) dual-modality imaging	Lectin conjugated Fe_2O_3@Au core@shell nanoparticle (lectin–Fe_2O_3@Au NP)	Employed as molecular probes for MR/CT imaging of tumors. It also satisfies tumor MR imaging targeting ability and strong CT contrast enhancement effects in the subcutaneous tumor model	He et al. (2014)
Computed tomography	Glycol-chitosan-coated gold NPs, conjugated with fibrin-targeting peptides (fib-GC-AuNPs)	Employed as an imaging agent for direct imaging of cerebral thromboembolism and carotid thrombosis	Kim et al. (2015)
Computed tomography	Dextran-coated cerium oxide nanoparticles (Dex-CeNP)	Used as a CT contrast agent for imaging gastrointestinal tract (GIT) with inflammatory bowel disease (IBD)	Naha et al. (2020)
Computed tomography imaging	Cyclo (Arg-Gly-Asp-D-Phe-Lys) peptide (RGD)-modified PEGylated dendrimer-entrapped gold nanoparticles (PEGylated Au DENPs-RGD)	It can be used as targeted nanoprobes with good biocompatibility for targeted CT imaging and diagnosis of integrin-positive tumors. For instance, a mice model with MDA-MB-435 xenograft tumors showed the accumulation of nanoprobes in the targeted area enabling the tumor to be detected by CT imaging	Li et al. (2015)
Magnetic resonance imaging	Cetuximab (C225) anti-human IgG functionalized ultra-small superparamagnetic iron oxide (USPIO) particles (C225-USPIO)	1. Due to good targeting and specificity, it is exploited in focusing the epidermal growth factor receptor (EGFR) of A549 lung cancer cells leading to a significant reduction in the MRI T2 value 2. It is also employed as MRI contrast agent in non-invasive detection of	Zhang et al. (2016)

(continued)

Table 11.1 (continued)

Imaging techniques	NPs used	Applications	References
		early-stage nasopharyngeal carcinoma with EGFR overexpression	
Magnetic resonance imaging	Superparamagnetic iron oxide nanoparticles (SPIONs)	It acts as a tissue-specific MR imaging contrast agent in the imaging of hepatocellular carcinoma (HCC). As SPIONs are taken up by Kupffer cells that are lacking in cancerous liver, hepatic metastases do not take up SPIONs; results in hyperintense appearance relative to the surrounding healthy tissues. Hence, SPIONs are effectively employed for the detection of metastatic liver cancers	Zhang et al. (2016)
Magnetic resonance imaging	Polysorbate 80 coated temozolomide-loaded PLGA-based superparamagnetic nanoparticles (P80-TMZ/SPIO-NPs)	Used as a good MRI contrast agent for malignant brain glioma Due to high drug loading and encapsulation efficiency, as well as good sustained drug release capability, it is also characterized as drug carriers. It also possesses antiproliferative effect for C6 glioma cells. It holds a good promise to multifunctional theranostic carrier of brain cancer	Ling et al. (2012)
Single-photon emission computerized tomography and computed tomography (SPECT/CT) dual-modal imaging	Annexin V and radionuclide Tc-99m conjugated gold NPs (^{99m}Tc-GNPs-Annexin V)	By means of specific targeting of the apoptotic macrophages, accurate localization and precise diagnosis of vulnerable atherosclerotic plaques was accomplished	Li et al. (2016)

applied for imaging the entity, multiple types of microscopes have been evolved. Utilizing visible light, compound microscope, dark-field microscope, phase-contrast microscope, fluorescent and confocal microscopes have been developed. Raman microscope and two-photon microscope employ the near-infrared region of light;

electron beams of extremely short wavelength are used in transmission electron microscopy (TEM), scanning electron microscopy (SEM), scanning transmission electron microscopy (STEM), and cryogenic electron microscopy (Cryo-EM); and of normal wavelength are used in atomic force microscope (AFM). X-ray microscope is developed as per the difference in absorbing soft X-ray (Prasad et al. 2021).

Particles at the nanoscale range exhibit unique optical, electronic, and magnetic properties which are absent in bulk scale. In comparison to other NPs (e.g., semiconductor and magnetic NPs), metallic NPs show more advantages for bioimaging as they scatter light intensely, and are much brighter than chemical fluorophores. Moreover, they are devoid of photo-bleaching and can be easily detected at the single particle limit. It also exhibits enhanced scattering at wavelengths of their localized plasmon resonance. Due to optical excitation, electrons in metallic NPs show oscillation which in turn results in its localized surface plasmon resonance (LSPR) properties. Using this property of Au nano-rods and AgNPs in NP-based dark-field microscope, high-contrast multiplex imaging of pancreatic cancer cells is obtained. In dark-field microscopy, multiphoton plasmon resonance microscopy, optical coherence microscopy, hyper-Rayleigh scattering spectroscopy, and Raman spectroscopy, AuNPs and AgNPs are mostly employed as contrast agents for biomedical imaging (Hu et al. 2009). Another optical microscopy used in bioimaging is fluorescence microscopy. Instead of scattering, reflection, or absorption, it exploits the ability of fluorochromes to emit light after being excited with light of a certain wavelength. It is more useful over phase-contrast and differential interference-contrast microscopy for live cell imaging owing to its high spatial and temporal resolution.

Förster resonance energy transfer (FRET) is a phenomenon that occurs when an excited donor chromophore transfers energy to an acceptor chromophore present within a few nanometers via nonradiative dipole-dipole coupling. Nanoparticle probes are often designed using FRET for cancer diagnosis. These FRET probes provide real-time spatial measurements between the donor and acceptor fluorophores and thus are employed to detect cancer biomarkers and cells. Hence, it allows the design of more sensitive bioassays for cancer biomarkers and cell detection. The most widely used nanoparticle probes employed in diagnostics are quantum dots (QDs), polymer dots (PDs), upconversion nanoparticles (UCNPs), gold nanoparticles (AuNPs), silicon nanoparticles (SNPs), etc. For example, luminescent terbium chelates (LTC)–QD FRET pair is used to detect PSA cancer biomarkers (Chinen et al. 2015). Amine-labeled Sgc8 aptamer conjugated carboxyl-modified fluorescent silica NPs (FSNPs) are used to detect leukemia cells with high sensitivity and specificity. This ability to detect leukemia cells using Sgc8 aptamer is evaluated using fluorescence microscopy (Tan et al. 2016).

Multiphoton microscopy is much more advantageous than conventional one-photon microscopy; thus, it can be employed as novel noncontact biomedical diagnostic tool for three-dimensional (3D) resolved fluorescence imaging. In this imaging modality, higher fluorescence intensity and good photostability are provided by gold NPs as compared to traditional fluorescent dye. In lymphoma cell line Karpas-299, ACT-1 antibody gets attached to the membrane receptor CD25.

Therefore, ACT-1 antibody conjugated gold NPs specifically couple with the Karpas-299 cell membrane making it fluorescent upon laser excitation (Qu et al. 2008).

Raman microscopy is a fast, label-free, non-invasive chemical analysis technique based on the inelastic scattering of photons (Raman scattering) discovered in 1928 by the Indian physicist Sir C.V. Raman. It provides detailed information about chemical structure, phase, polymorphy, and crystallinity owing to the interaction of light with the chemical bonds of a substance and high-resolution imaging at relatively low-cost compared to other well-established medical imaging techniques (e.g., ultrasound and magnetic resonance imaging (MRI)). However, it has its limitations in terms of producing weak signal, a much stronger fluorescence background, and a partial penetration depth. In contrast, an advanced Raman-based technique, surface-enhanced Raman spectroscopy (SERS), is developed that in conjugation with NPs provides targeting and imaging tumors and multiple cell-surface biomarkers of cancer. Au and Ag nanoshells are mostly used nanomaterials in SERS. For instance, in human breast cancer, gold nano popcorn (GNPOP) conjugated single-wall carbon nanotube (SWCNT) based SERS assay shows highly selective binding to SKBRk3 cells (overexpressing HER2); but MDA-MB cells (HER2-negative breast cancer cell line) lack this binding. Hence, aggregation of SER probes in SKBRk3 cells provides significant enhancement of Raman signal intensity. In colorectal cancer, maghemite (Fe_2O_3, γ-Fe_2O_3) NPs precoated with dimercaptosuccinic acid and surface functionalized with anti-carcinoembryonic antigen (anti-CEA) have been successfully used to target CEA expressing colorectal cancer (CRC) cells. AgNP-based SERS are employed to detect nasopharyngeal cancer via biochemical analysis of blood plasma samples (Ravanshad et al. 2018). SERS is more advantageous over fluorescence-based NPs in different aspects like multiplex detection capability due to spectral fingerprinting, invincibility to photobleaching, and less cytotoxicity. SERS is also employed to detect multiple pathogenic antigens from West Nile virus, Rift Valley fever virus, and Yersinia pestis in fetal bovine serum by means of silica-encapsulated nanotags and magnetic nanoparticles conjugated with polyclonal antibodies. Here, antigens get bound to the polyclonal antibodies conjugated on magnetic nanoparticles and nanotags resulting in the formation of immune complexes. Consequently, discrete spectra corresponding to each captured antigen are obtained by the silica-encapsulated nanotags (Neng et al. 2018).

In contrast to optical microscopy, electron microscopy allows the visualization of optically invisible objects. Instead of using visible light as the energy source, it produces high-resolution images employing electron beams of extremely shorter wavelengths. Localized nanoparticles in malignant tumors are detected using electron microscopes. Thus, NPs are exploited by EM to distinguish cancerous tissues from healthy cells. For example, intravenously injected AuNPs get accumulated only in the areas of mice brain tissues that contain tumor cells. These NPs are then readily identified using an electron microscope to obtain the exact location of the tumor cells without hampering surrounding healthy cells (Kempen et al. 2015).

X-ray microscopy allows visualizing the fine structures of the natural dynamic state of living cells. Its aptitude for imaging both internal and external morphologies with minimal sample preparation, high spatial resolution, and high penetration power provides some extra edge over conventional microscopy techniques in the field of diagnosis. However, small blood vessels are difficult to identify under an X-ray microscope. Hence, vascular angiogenesis, which is critical for the early diagnosis of cancer as the tumors become malignant after vascularization, remains untraceable. Incidentally, NPs can be employed to enhance the contrast level. For example, AuNPs injected with heparin act as contrasting agents producing very high spatial resolution image in angiography (Zhou et al. 2012).

11.3 Ultrasonography Technique

Not only electromagnetic waves but sound waves are also utilized to visualize the morphological appearance of any tissue or organ. Hence, based on ultrasound, one such technique has evolved named ultrasonography. On the basis of the principle of the Doppler effect, pulses of sound waves with frequencies beyond the audible range (</>20 Hz–20 kHz) are directed to the soft tissues through a probe, and the reflected echo is then converted to electrical energy using a transducer which is then captured by a biometer for real-time imaging of internal organs and soft tissues. Piezoelectric effect, i.e., generation of electrical potential due to distortion of the lattice structure of piezoelectric material upon mechanical stress, is applied in a transducer. Any deviation in the amplitude of the reflected wave from that of healthy tissues depicts abnormality. However, sometimes deviation occurs due to the inability of waves to travel through hard tissues or other hindrances. To overcome this difficulty, 3D ultrasound imaging (Van Sloun et al. 2019) and ultrasound localization microscopy (Van Sloun et al. 2019) have been developed for more specific spatial information of the area and higher-resolution imaging. As this technique is free from any ionizing radiation, the risk of radical damage is absent in sonography. Recent advances in nanotechnology have led to the advancement of ultrasonography for diagnostic and therapeutic applications. For instance, there are three ways in which ultrasound provides the information: first, B-Mode or Brightness mode, then M (motion) Mode, and, lastly, the Doppler processing. Solid NPs (e.g., Silica nanospheres and Polystyrene particles) are exploited to enhance the image brightness, i.e., mean gray scale level. For example, solid nanoparticle suspension in saline can be intravenously injected into mice to enhance the mean gray scale level of mouse liver. Thus, NPs are acting as contrast-enhancing agents for ultrasonic imaging. This B-mode enhancing property is associated with the size and concentration of the nanoparticles (Liu et al. 2006). For the same purpose, NPs can be functionalized with antibodies. For example, mesoporous silica nanoparticles (MSNs) functionalized with the monoclonal antibody Herceptin, can be used as an effective ultrasound contrast agent (UCA) for targeting human epidermal growth factor receptor 2 positive (HER2+) cancer cells in breast cancer (Milgroom et al. 2014). By means of folate receptor-mediated perfluorooctylbromide nanoparticles (FR-mediated PFOB NPs), enhanced target-specific US imaging of FR-overexpressing tumors can be obtained. It remains for longer periods in the circulatory system, provides a relatively stable echogenic response without increasing the background signal, and exhibits lower acoustic attenuation in comparison to commercial UCAs (Li et al. 2018).

In combination of the high-contrast and spectroscopic-based specificity of optical imaging with the high spatial resolution of ultrasound imaging, a new biomedical imaging modality has been developed, i.e., photoacoustic imaging. It employs the laser-generated ultrasound; hence, it can be regarded as an ultrasound image that relies on the optical properties, specifically optical absorption of a tissue, instead of its mechanical and elastic properties. As a consequence, it offers greater specificity than conventional ultrasound imaging, with greater penetration depth and visualization of anatomical structures such as the microvasculature. Functional information regarding blood oxygenation, blood flow, and temperature can also be obtained through this technique (Beard 2011). Intravenous administration of gold NPs as a novel class of probes provides monitoring of the early phase of kidney injury via photoacoustic imaging (Pan et al. 2018).

Despite being employed as a diagnostic agent, the ultrasound method is also applied for therapeutic purposes. For instance, Magnetic Resonance Guided Focused Ultrasound (HIFU), which is an amalgam of MRI and high-intensity ultrasound, is a non-invasive method to be used as an alternative form of surgery. As in cancer, HIFU can be used to cause necrosis in tumor cells by absorbing high energy (Huisman and van den Bosch 2011).

Ultrasonography can also be used for other important purposes like drug release which can be achieved through the grafting of a copolymer on the surface of mesoporous silica nanoparticles. The copolymer presents an open conformation at 4 °C, allowing the NPs to be loaded with cargo, and at 37 °C, it collapses closing the pore entrances which subsequently allows the nanoparticles to carry the drugs at physiological temperature without premature release. Upon ultrasound irradiation, the hydrophobicity of the copolymer changes resulting in alternation in conformation. Consequently, the cargo gets released in the targeted area (Paris et al. 2015).

11.4 Endoscopy

To visualize the internal morphology of any hollow tube-like body parts, such as the gastrointestinal (GI) tract, the light-based imaging technique or endoscopy is put forth. This method of scanning tissue-organ relies on the principle of total internal reflection. Thereby, a long thin tube, namely an endoscope, made with optical fiber for illumination and imaging is exploited. According to conventional endoscopy procedure, optical signals coming from the illuminated target area get captured via the optical lens on the charge-coupled device (CCD) photosensitive surface which in turn is converted to an electrical signal and subsequently displayed on the monitor (https://www.xzykd.com/news/principle-of-medical-endoscope.html). Nowadays, advancements have been made in endoscopic procedures using nanotechnology. For instance, in colorectal adenocarcinoma, biodegradable near-infrared fluorescent silica NPs (NIR FSNs) provide early detection of colonic dysplasia (precursor lesion to colorectal adenocarcinoma) via fluorescence-assisted white-light colonoscopy. As in APC1311/+ pig model, intravenous administration of FSNs (with a dose of $\approx$5 fmol/g) highlighted the adenomas and enabled real-time, combined Near Infra-Red/White Light (NIRF/WL) endoscopy surveillance. FSNs have a shorter half-life ($t_{1/2}$ of 2.7 weeks), thus, are biodegradable and well-tolerated, and provide depiction and identification of adenomas as small as 0.5 mm^2 with a high tumor-to-background ratio (Rogalla et al. 2019).

Conventional colonoscopy technique enables the detection of polyps and carcinomas by only morphology. Consequently, smaller polyps, such as flat or sessile serrated polyps, as well as nascent neoplasms of recurrent disease, get overlooked due to a lack of detection sensitivity in the traditional diagnostic method. Therefore, an urge for early detection of colorectal adenomatous polyps and their precursors arises before their malignancy transformation. As a result, fluorescent mesoporous silica nanoparticles (MSNs) are utilized as contrast agents for targeted endoscopic detection in colorectal cancer. Glycosylation of normally occurring mucins that protect the GI mucosa and expression of abnormal glycoproteins on colorectal polyps and tumors signifies the malignant transformation of healthy colorectal mucosa. An elevated expression of α-L-fucose on luminal surfaces of glycoproteins implies colorectal dysplastic adenomas and early cancer. Hence, α-L-fucose targeting lectin Ulex Europaeus Agglutinin-1 (UEA1) has been developed that binds to colorectal adenocarcinomas, but not the healthy epithelium. Later on, UEA1-labeled fluorescent/PEGylated MSNs were constructed to serve as α-L-fucose targeted endoscopic contrast agents for more sensitive detection of early-stage colorectal polyps and nascent colorectal cancers (Chen et al. 2017). Functionalization of MSNs with near-infrared (NIR) fluorescent contrast agent, ATTO647N, provides relative transparency of most tissues at NIR wavelengths, maximizes MSN surface area, and lowers systematic toxicities. That is, the cell-specific targeting moieties of MSNs diminish the chance of collateral damage to healthy cells. The increased surface area of MSN facilitates subsequent conjugation of drugs for therapeutic intervention and provides targets to biomolecular ligands for highly targeted particle delivery. Thus, tri-functionalized mesoporous silica nanoparticles have been developed as theranostic compounds for imaging, targeting, and therapy in a single particle (Cheng et al. 2010). In recent studies, endoscopy is being employed to combine with Raman spectroscopy for more sensitive detection of dysplastic lesions. For instance, ^{64}Cu labeled SERS (surface-enhanced Raman scattering) nanoparticles are exploited as tumor-targeting contrast agents in colonoscopy for early detection of colon cancer (Zavaleta et al. 2011).

11.5 Thermal Imaging

Thermal imaging is a recently introduced imaging-based diagnostic method. The principle behind it depicts the conversion of infrared (IR) radiation (heat) into visible images to capture the spatial distribution of temperature differences using thermal camera. Conventionally, it is employed for the tracking of animals in wildlife for their welfare; but in recent studies it is being utilized to detect any abnormalities in the body. For example, irregularity in breast tissue temperature distribution specifies the presence of embedded tumors. However, thermal imaging lacks the ability to determine the location of an embedded tumor precisely. Therefore, NPs can be employed to endure this limitation. For example, polyethylene glycol (PEG) coated iron oxide MNP enhances the thermographic potential in thermal scanning. These MNPs when placed in a magnetic field cause vibration, resulting in an increase in the temperature of the tumor tissues. As a result, it allows selective aiming of the breast tumor tissue (Kirubha and Rajput 2019). Perfluorohexane (PFH)-filled and polydopamine (PDA)-coated Au core/hollow mesoporous silica shell nanoparticles (Au@mSiO_2-PFH-PDA, or ASPP) are also employed as contrast agents in thermal imaging. Intra-tumoral injection of ASPP NPs enhances the temperature of the tumor site. Thus, it acts as a thermal imaging agent for tumor diagnosis in vivo (Cai et al. 2019).

11.6 Radiographic Imaging

Utilizing ionizing electromagnet radiation, such as X-ray, a new branch of diagnostic technique, referred to as radiography, has been evolved to visualize objects. It is probably the most commonly used diagnostic imaging method for medical purposes. In thermal imaging, infrared radiation of 780–1000 nm wavelength is applied. Therefore, owing to its low penetration power, it can only portray surface abnormality based on temperature distribution. Contrastingly, radiography provides deeper penetration of Rontgen rays (0.01–10 nm), thus allowing visualization of inner structure. In recent times, several advancements have been introduced in this technology. For instance, iodinated molecules and barium sulfate suspensions are conventionally used as X-ray contrast agents. But it has limitations regarding improper information, safety concerns, etc. Accordingly, NPs offer a lot of desirable features such as high contrasting ability, the potential for extensive circulation times, tunable physicochemical properties, and low toxicity (Zavaleta et al. 2011). Biocompatible heavy metal nanoparticles such as AuNPs are competent to target the tumor by the EPR effect, and they offer enhanced penetration in radiography due to a very high X-ray absorption coefficient. Hence, the AuNP-based radiography technique is widely used in biomedical applications. For example, biocompatible gum Arabic stabilized gold nanocrystals (GA-AuNPs) get accumulated in the tumor site and, hence, these are exploited as X-ray contrast agents in small and large animals (Chanda et al. 2014).

11.7 Computed Tomography

CT scan is a diagnostic method using X-ray images taken from different angles around the object, and subsequently, it develops a computed three-dimensional cross-sectional image with high spatial and temporal resolution. It relies on the principle that the density of the tissue passed by the X-ray beam can be measured from the calculation of the attenuation coefficient, i.e., the fraction of a beam of X-rays or gamma rays that is absorbed or scattered per unit thickness of the absorber. Contrast agents are employed to enhance the contrast of computed tomography (CT) imaging, thus, improving diagnostic accuracy. Traditionally, iodine-based contrast agents are applied for this purpose. However, due to low molecular weight, these agents are rapidly excreted from the kidney or moved to extravascular tissues through the capillary vessels, leading to nonspecific enhancement of contrast images for tissues. Hence, NPs are put forth to act as contrast-enhancing agent (Table 11.1). Gold nanoparticles are mostly used as contrast agents in CT scans. They possess properties, like high X-ray absorption and longer vascular retention time, which makes them potent for this purpose. In a study, it is showed that the novel porous gold NPs (PGNPs) enhance CT scan images with higher efficacy as compared to solid gold NPs (SGNPs) (Aziz et al. 2017). PEGylated AUNPs conjugated with cancer-specific antibodies, such as HER2 (Au-PEG-HER2ab), are designed as a new contrast agent to specifically scan breast tumors via CT using an enhanced permeability and retention (EPR) effect (Nakagawa et al. 2016). Polydi(carboxylatophenoxy)phosphazene (PCPP) encapsulated AuNPs are employed to produce strong computed tomography contrast leading to the disintegration of the nanosphere into biodegradable PCPP shell and AuNPs. Consequently, these smaller-sized AuNPs get excreted from the body (Cheheltani et al. 2016). In atherosclerosis, the risk of acute ischemic events depends on the composition and stage of atherosclerotic plaques. Rupture-prone plaques are prone to contain high-density macrophages that lead to the formation of a thrombus causing myocardial ischemia. Accumulation of NPs on macrophage-rich tissues results in a significant enhancement of the plaque contrast. Consequently, iodinated NPs and gold NPs are employed as CT contrast agents. CT can provide information regarding plaque composition due to less attenuation of X-ray via lipid-rich tissues as presented in high-risk plaques. Gold NPs also provide a theranostic platform for atherosclerosis and cardiovascular diseases as a photothermal therapy (PTT) agent (Shilo et al. 2012; Wu et al. 2021). A better tumor penetration ability of iodine-based polymer iodine NPs, in addition to their biodegradable and clearance properties makes it an appropriate contrast agent. It gets efficiently accumulated in tumors and provides high-contrast vascular tumor imaging. In comparison to AuNPs, iodine-based NPs offer better tumor uptake and clearance from the liver owing to their small size and long blood half-life (40 h). A sentinel lymph node tracer consisting of iodine and docetaxel was developed for early detection of carcinoma metastasis. It facilitates simultaneous sentinel lymph node CT and locoregional chemotherapy of the draining lymphatic system (Siddique and Chow 2020). Another appealing contrast agent in CT is bismuth-based NPs. It is considered the best biocompatible material among other heavy elements; however, its ionic form can be toxic at 0.2 M (Luo et al. 2012). By co-precipitating bismuth citrate and sodium sulfide in the presence of 3-mercaptopropionic acid, bismuth sulfide nanoparticles were developed which acted as a surfactant, followed by coating these nanoparticles with PVP resulting in plate-shaped nanoparticles. They were used in cytotoxicity assay where no changes were observed in the viability of macrophages or hepatocytes below 20 mM bismuth concentration, but devaluation in the viability of macrophages was seen above 20 mM concentration. Functionalization of iron oxide nanoparticles with a bismuth oxide shell was formed by co-precipitating bismuth nitrate and iron sulfate (Andrés-Vergés et al. 2012). These nanoparticles allow combined imaging through CT and MRI because of their iron oxide core.

11.8 Magnetic Resonance Imaging

With the advances in multimodal imaging techniques, it has become easier to improve the imaging system. One such technique is the non-intrusive imaging technique named magnetic resonance imaging which provides multiparametric and complete information (Yousaf et al. 2018). Magnetic resonance imaging employs powerful magnets which produce a strong magnetic field that forces protons in the body to align with that field. When a radiofrequency current is pulsed through the patient, the protons are stimulated, and spin out of equilibrium, straining against the pull of the magnetic field, then a computer is used to generate a detailed image of the internal structures of the body. In conventional MRI, it has been observed that the contrast of the image produced is not detailed enough to obtain a better image of internal organs. Hence, contrast agents have shown great potential in improving the quality and resolution of the image. In 2015, approximately 17 million contrast agents were used in the United States for MRI examinations. Among them contrast agents based on gadolinium (Gd) are being used efficiently for MRI diagnosis. It has been observed that collagen becomes imbalanced in cancer or infected cells and the formation of excess collagen in the liver condition is seen in cases of alcohol and drug abuse. To detect excess collagen, gadolinium nanoparticle-based contrast agents have been used (Siddique and Chow 2020). Dendrimer-based gadolinium is also exploited as contrast agent that showed greater contrast on MRI angiograms. Gadolinium dendrimers when bound to a specific target can prominently improve cellular uptake. Research has shown that Arg-Gly-Asp-Phe-Cys peptide complex has been utilized as a targeting molecule when combined with multimodal gadolinium dendrimers contrast agents and Au nanoparticles as carriers (Siddique and Chow 2020). However, gadolinium accumulation in bone and brain tissue is observed which can be the result of the use of gadolinium-based contrast agents (Ramalho et al. 2016; Layne et al. 2018). Hence, an alternative to gadolinium, superparamagnetic iron oxide nanoparticles (SPIONS) are efficiently being used as a contrast agent (Table 11.1). In one report, SPIONS such as Fere Track have been used as tracking agents of stem cells. Another report concluded that it was successful in tracking the stem cells in mice bearing Glioma (Kim et al. 2016). SPIONS when combined with antibodies can be exploited as a carrier of drug in cancer therapy (Dulińska-Litewka et al. 2019). Drug delivery through nanodiamonds is also possible. These are non-toxic substances. A study showed that on the surface of a non-fluorescent nanodiamond, a chemotherapeutic agent Doxorubicin (DOX) is absorbed passively, which makes a complex called nanodiamond drug complex (NDX). This NDX is then transferred into cancer cells maintaining competency with ease and persistent DOX release (Huang et al. 2007; Chow et al. 2011). Fluorescent nanodiamonds can also be used for drug delivery that is made from crushed high-pressure high-temperature (HPHT) bulk diamonds or in some cases chemical vapor deposition grown nanodiamonds. Hence the FNDs are used for image-monitored drug delivery (Slegerova et al. 2015; Wu et al. 2015; Reineck et al. 2018). Biopolymers and proteins can be coated on the FND to assure colloidal stability and also associated with DOX molecules for the treatment of cancer cell lines (Alkahtani et al. 2018). A further kind of nanomaterial is one of carbon's allotrope Graphene. It has physiochemical, surface engineering, and biological properties such as hydrodynamic size, low toxicity, and biocompatibility and hence can be used as a biomedical imaging efficiently. Examples of nanomaterials based on Graphene are Graphene dye conjugates, Graphene Antibody conjugates, Graphene Nanoparticles composites, and Graphene quantum dots that can be used for molecular imaging as in vitro and in vivo agents (Garg et al. 2015). Nanodots are found to be potent imaging agents to detect cancer as they enhance the visualization of tumor cells in mice (Nurunnabi et al. 2014). Curcumin-conjugated MNPs coated with polyethylene glycol-polylactic acid block copolymer and polyvinylpyrrolidone have been employed in MRI to detect amyloid plaques of mice Alzheimer's disease (Cheng et al. 2015).

11.9 Nuclear Imaging

Molecular imaging is the area of radiology that provides visualization, characterization, and comprehension of cellular as well as molecular processes; i.e., it offers both anatomical and functional information. For instance, it allows early detection of tumors and the occurrence of metastasis. These imaging modalities also act as a prevailing means in the real-time monitoring of treatments. Recently, a wide range of transmission methods, such as radionuclide-based positron emission tomography (PET), single-photon emission computerized tomography (SPECT), and ionizing radiation-based CT, are being implemented solely or in combination with other techniques such as MRI, fluorescence-based optical imaging, ultrasound imaging, and photoacoustic imaging. Among the currently available imaging modalities, SPECT and PET are mostly used in the clinical diagnosis of diseases. PET is more gainful, compared to SPECT, in terms of sensitivity and resolution. It employs radiolabeled contrast agents or radiotracers, holding radioactive compounds such as ^{11}C, ^{13}N, ^{15}O, and ^{18}F, that emit short-lived positron as the radiotracer decays. As a result, images of the target tissues are generated to study the dynamics of ongoing metabolic processes. For example, radioactive fluorine (^{18}F) labeled deoxyglucose (^{18}FDG) is exploited as a tracer for imaging tumors in cancer diagnosis. However, due to its limitations showing high accumulation in infectious or inflammatory tissues, it results in an elevation of false-negative and false-positive cases. This drawback of ^{18}FDG led to the demand for designing new radiotracers for clinical diagnosis and imaging. Subsequently, antibodies, peptides, or ligand-conjugated radioactive tracers are constructed to target overexpressed receptors or antigens on diseased cells. However, this shows poor penetration or low bioavailability, incompetent binding of antibodies, peptides, or ligands, triggering low yield of the radiotracer. Consequently, the application emerges as potential delivery agent for PET imaging agents. This NP-based platform provides the contrast of the disease site relative to surrounding healthy tissues advancing the sensitivity as well as specificity of PET. The most applied nanomaterials include liposomal carriers, magnetic IONPs, quantum dots (QDs), polymeric nanoparticles, silica nanoparticles, dendrimers, carbon nanotubes, gold NPs, etc. For instance, liposomes get accumulated in higher quantities at tumor site and act as a contrasting agent in PET imaging. Hence, a variety of liposomes formulated with different radionuclides, such as ^{111}In, ^{99m}Tc, ^{67}Ga, ^{64}Cu, and ^{18}F have evolved (Srivatsan and Chen 2014). In brain tumors, ^{124}I-labeled gold nanostar probe employed PET is applied for intracranial tumor detection (Siddique and Chow 2020).

SPECT, analogous to PET, is a hybrid technology where radiolabeled pharmaceutical or metabolites are employed for scanning the internal organs and detecting them on a computer as two- or three-dimensional images. It differs from PET as it emits single-photon instead of two coincident 511-keV annihilation photons, hence also reducing the cost (Prasad et al. 2021). Instead of applying X-ray as in CT, it employs gamma-ray for imaging. This gamma-ray is converted into light energy in gamma-camera crystal and subsequently transformed into an electric pulse which is recorded, followed by digitization and reconstruction into tomographic images (Teleanu et al. 2019). Advancements in nanotechnology platforms also influenced SPECT. Gold NPs are commonly utilized as a contrasting agent in SPECT imaging. For instance, Technetium (^{99m}Tc)-radiolabeled AuNPs functionalized with mannose is used as a SPECT contrast agent for lymphatic mapping. In early-stage cervical cancer, ^{99m}Tc conjugated SPECT-MRI can be exploited in sentinel lymph nodes (SLNs) mapping for the assessment of SLN metastases. IONPs-adapted SPECT/MRI-based agents can also be developed for tumor imaging. In contradiction to PET, SPECT shows significantly lower molecular sensitivity. However, due to its low cost, the longer half-life of the radionuclides used, and the engagement of dual-labeled compounds, this technique is widely applied in clinical practice and preclinical research (Teleanu et al. 2019).

11.10 Electrical Impedence Tomography

As ultrasonography and thermal imaging techniques, another non-invasive diagnosis method has been evolved based on electrical energy, named electrical impedance tomography (EIT). It relies on electrical properties like conductivity, permittivity, and impedance. As per the electrical conductivity of biological tissue, images of a specific region in the body are reconstructed using a surface electrode in EIT technique. The electrodes are placed on the body surface, through which low-frequency electrical current is passed. The voltage generated is measured via another electrode and electrical conductivity difference is determined in different parts of the body. Subsequently, images are obtained. Conductivity and permittivity of a tumor significantly vary from that of healthy tissue. Therefore, exploiting these criteria, tumor cells are identified via EIT. NPs are employed to enhance the tumor characteristics and determine the infiltration vector for easier detection. In different medical imaging technologies, sensitizers are used that enhance the signal strength and improve diagnostic accuracy. Hence, nanoscale sensitizers have been developed to obtain more clear distinction of electrical characteristics between healthy and aberrant tissues. They also improve the quality of EIT imaging providing more successful application in clinical diagnosis. For example, AuNPs can act as sensitizers. In EIT, image reconstruction depends on the potential distribution of the sensitizer material. AuNPs get uniformly distributed in tissue and deeply influence the electrical properties of both normal and cancerous tissues. Therefore, upon electromagnetic field stimulation, the final conductivity can be obtained with diagnostic accuracy. Metal NPs or composites can also be employed in electro-analysis. For example, carbon nanotubes (CNTs) show extraordinary mechanical strength, high electrical conductivity, and chemical stability that have led them to their application as advanced electrodes in electrochemical systems. It provides many advantages including increased surface area, added porosity, improved adsorption properties, higher mass transport, and enhanced catalytic activity. In addition, it also allows high integration to the structure of electrochemical electrodes. Correspondingly, AuNPs are also a promising candidate for electrode surface modifications owing to their large surface area, chemical stability, and biocompatibility. Hence, a novel composite electrode with surface deposition of Gold NPs-Multiwalled carbon nanotube (AuNPs/MWCNT) has been designed for micro-electrical impedance tomography. This conjugation offers improvement in the effective surface area, increase in double-layer capacitance of electrodes, and enhancement in electron transfer on the electrode–analyte interface. It also exhibits lower frequency dependent electrical properties in impedance measurement (Rezanejad Gatabi et al. 2020). Due to its affordability and lack of any harmful effects, nowadays, EIT is gaining popularity. It also allows monitoring for a long time and characterization of tissues with fast and easy collection of data (Brown 2003).

11.11 Conclusion

In vivo diagnostic imaging is one of the most prominent areas focusing the nanointervention. Within a few decades, nanotechnology and nanoscience have gained the fundamental importance in biomedical applications such as diagnostic biosensors, drug delivery systems, and imaging probes. Nanomaterials combined with imaging are being developed for high-quality assays and improved tumor biopsies. An ideal nanoparticle imaging probe for clinical use should be biodegradable or rapidly excreted and has low toxicity. Here nanoparticles-aided different optical imaging modalities regarding disease diagnosis have been discussed. To obtain vivid information about the location and properties of the anomalies, molecular imaging requires a multimodal approach with multiple imaging moieties combined in a single nanoplatform. It offers the greatest potential for clinical use by enabling compatibility in multiple imaging modalities (e.g., MR/CT, MR/PET, optical imaging, and others) and holds fascinating prospects. For instance, in the future, depending upon the identification of genetic or epigenetic markers with in vivo and ex vivo diagnostics, NP-based imaging agents can provide classification of tumor subtypes in highly heterogeneous tumors, leading to personalized therapies.

However, the accumulation of biopersistant nanomaterials, such as some metallic particles, within organs can cause chronic toxicities. It can lead to increased systemic exposure to toxic components of nanomaterial-based imaging agents, as a result of delayed clearance and increased systemic circulation relative to conventional imaging agents. Nanoparticles may also form micron-scale aggregates upon injection into the circulation, leading to microcirculation compromise, particularly in the capillary beds of the lungs. Consequently, inflammation and granuloma formation may occur in lung tissue. Finally, the judicial and appropriate application of nanotechnology can revolutionize the field of medical imaging. Nanotheranostic property, the ability to enhance early detection, and the ability to have multiple contrast agents on a single platform can be a game-changer in future medical imaging.

References

Alkahtani MH, Alghannam F, Jiang L, Almethen A, Rampersaud AA, Brick R et al (2018) Fluorescent nanodiamonds: past, present, and future. Nanophotonics 7(8):1423–1453

Andrés-Vergés M, del Puerto Morales M, Veintemillas-Verdaguer S, Palomares FJ, Serna CJ (2012) Core/shell magnetite/bismuth oxide nanocrystals with tunable size, colloidal, and magnetic properties. Chem Mater 24(2):319–324

Aslan B, Ozpolat B, Sood AK, Lopez-Berestein G (2013) Nanotechnology in cancer therapy. J Drug Target 21(10):904–913

Aziz F, Ihsan A, Nazir A, Ahmad I, Bajwa SZ, Rehman A, Khan WS (2017) Novel route synthesis of porous and solid gold nanoparticles for investigating their comparative performance as contrast agent in computed tomography scan and effect on liver and kidney function. Int J Nanomedicine 12:1555

Bayda S, Adeel M, Tuccinardi T, Cordani M, Rizzolio F (2019) The history of nanoscience and nanotechnology: from chemical–physical applications to nanomedicine. Molecules 25(1):112
Beard P (2011) Biomedical photoacoustic imaging. Interface Focus 1(4):602–631
Brown BH (2003) Electrical impedance tomography (EIT): a review. J Med Eng Technol 27(3): 97–108
Cai C, Li X, Wang Y, Liu M, Shi X, Xia J, Shen M (2019) Polydopamine-coated gold core/hollow mesoporous silica shell particles as a nanoplatform for multimode imaging and photothermal therapy of tumors. Chem Eng J 362:842–850
Chanda N, Upendran A, Boote EJ, Zambre A, Axiak S, Selting K, Kannan R (2014) Gold nanoparticle based X-ray contrast agent for tumor imaging in mice and dog: a potential nanoplatform for computer tomography theranostics. J Biomed Nanotechnol 10(3):383–392
Cheheltani R, Ezzibdeh RM, Chhour P, Pulaparthi K, Kim J, Jurcova M, Cormode DP (2016) Tunable, biodegradable gold nanoparticles as contrast agents for computed tomography and photoacoustic imaging. Biomaterials 102:87–97
Chen NT, Souris JS, Cheng SH, Chu CH, Wang YC, Konda V, Lo LW (2017) Lectin-functionalized mesoporous silica nanoparticles for endoscopic detection of premalignant colonic lesions. Nanomedicine 13(6):1941–1952
Cheng SH, Lee CH, Chen MC, Souris JS, Tseng FG, Yang CS, Lo LW (2010) Tri-functionalization of mesoporous silica nanoparticles for comprehensive cancer theranostics—the trio of imaging, targeting and therapy. J Mater Chem 20(29):6149–6157
Cheng KK, Chan PS, Fan S, Kwan SM, Yeung KL, Wang YXJ et al (2015) Curcumin-conjugated magnetic nanoparticles for detecting amyloid plaques in Alzheimer's disease mice using magnetic resonance imaging (MRI). Biomaterials 44:155–172
Chinen AB, Guan CM, Ferrer JR, Barnaby SN, Merkel TJ, Mirkin CA (2015) Nanoparticle probes for the detection of cancer biomarkers, cells, and tissues by fluorescence. Chem Rev 115(19): 10530–10574. https://doi.org/10.1021/acs.chemrev.5b00321
Chow EK, Zhang XQ, Chen M, Lam R, Robinson E, Huang H, Ho D (2011) Nanodiamond therapeutic delivery agents mediate enhanced chemoresistant tumor treatment. Sci Transl Med 3(73):73ra21
Dong Q, Yang H, Wan C, Zheng D, Zhou Z, Xie S, Li F (2019) Her2-functionalized gold-nanoshelled magnetic hybrid nanoparticles: a theranostic agent for dual-modal imaging and photothermal therapy of breast cancer. Nanoscale Res Lett 14(1):1–16
Dulińska-Litewka J, Łazarczyk A, Hałubiec P, Szafrański O, Karnas K, Karewicz A (2019) Superparamagnetic iron oxide nanoparticles—current and prospective medical applications. Materials 12(4):617
El-Sayed IH, Huang X, El-Sayed MA (2005) Surface plasmon resonance scattering and absorption of anti-EGFR antibody conjugated gold nanoparticles in cancer diagnostics: applications in oral cancer. Nano Lett 5(5):829–834
Ganareal TACS, Balbin MM, Monserate JJ, Salazar JR, Mingala CN (2018) Gold nanoparticle-based probes for the colorimetric detection of Mycobacterium avium subspecies paratuberculosis DNA. Biochem Biophys Res Commun 496(3):988–997
Garg B, Sung CH, Ling YC (2015) Graphene-based nanomaterials as molecular imaging agents. Wiley Interdiscip Rev Nanomed Nanobiotechnol 7(6):737–758
He X, Liu F, Liu L, Duan T, Zhang H, Wang Z (2014) Lectin-conjugated Fe_2O_3@Au core@shell nanoparticles as dual mode contrast agents for in vivo detection of tumor. Mol Pharm 11(3): 738–745
Hu R, Yong KT, Roy I, Ding H, He S, Prasad PN (2009) Metallic nanostructures as localized plasmon resonance enhanced scattering probes for multiplex dark-field targeted imaging of cancer cells. J Phys Chem C 113(7):2676–2684
Huang H, Pierstorff E, Osawa E, Ho D (2007) Active nanodiamond hydrogels for chemotherapeutic delivery. Nano Lett 7(11):3305–3314
Huisman M, van den Bosch MAAJ (2011) MR-guided high-intensity focused ultrasound for noninvasive cancer treatment. Cancer Imaging 11(1A):S161

Kempen PJ, Kircher MF, de la Zerda A, Zavaleta CL, Jokerst JV, Mellinghoff IK, Sinclair R (2015) A correlative optical microscopy and scanning electron microscopy approach to locating nanoparticles in brain tumors. Micron 68:70–76
Kim JY, Ryu JH, Schellingerhout D, Sun IC, Lee SK, Jeon S, Kim DE (2015) Direct imaging of cerebral thromboemboli using computed tomography and fibrin-targeted gold nanoparticles. Theranostics 5(10):1098
Kim SJ, Lewis B, Steiner MS, Bissa UV, Dose C, Frank JA (2016) Superparamagnetic iron oxide nanoparticles for direct labeling of stem cells and in vivo MRI tracking. Contrast Media Mol Imaging 11(1):55–64
Kirubha SA, Rajput A (2019) Enhancement of thermal imaging by iron oxide nanoparticle—preliminary study. Biocatal Agric Biotechnol 17:352–360
Layne KA, Dargan PI, Archer JR, Wood DM (2018) Gadolinium deposition and the potential for toxicological sequelae—a literature review of issues surrounding gadolinium-based contrast agents. Br J Clin Pharmacol 84(11):2522–2534
Li JL, Wang L, Liu XY, Zhang ZP, Guo HC, Liu WM, Tang SH (2009) In vitro cancer cell imaging and therapy using transferrin-conjugated gold nanoparticles. Cancer Lett 274(2):319–326
Li K, Zhang Z, Zheng L, Liu H, Wei W, Li Z, Zhang G (2015) Arg-Gly-Asp-D-Phe-Lys peptide-modified PEGylated dendrimer-entrapped gold nanoparticles for targeted computed tomography imaging of breast carcinoma. Nanomedicine 10(14):2185–2197
Li X, Wang C, Tan H, Cheng L, Liu G, Yang Y, Shi H (2016) Gold nanoparticles-based SPECT/CT imaging probe targeting for vulnerable atherosclerosis plaques. Biomaterials 108:71–80
Li X, Sui Z, Li X, Xu W, Guo Q, Sun J, Jing F (2018) Perfluorooctylbromide nanoparticles for ultrasound imaging and drug delivery. Int J Nanomedicine 13:3053
Ling Y, Wei K, Zou F, Zhong S (2012) Temozolomide loaded PLGA-based superparamagnetic nanoparticles for magnetic resonance imaging and treatment of malignant glioma. Int J Pharm 430(1–2):266–275
Liu J, Levine AL, Mattoon JS, Yamaguchi M, Lee RJ, Pan X, Rosol TJ (2006) Nanoparticles as image enhancing agents for ultrasonography. Phys Med Biol 51(9):2179
Luo Y, Wang C, Qiao Y, Hossain M, Ma L, Su M (2012) In vitro cytotoxicity of surface modified bismuth nanoparticles. J Mater Sci Mater Med 23(10):2563–2573
Milgroom A, Intrator M, Madhavan K et al (2014) Mesoporous silica nanoparticles as a breast-cancer targeting ultrasound contrast agent. Colloids Surf B Biointerfaces 116:652–657. https://doi.org/10.1016/j.colsurfb.2013.10.038
Naha PC, Hsu JC, Kim J, Shah S, Bouché M, Si-Mohamed S, Cormode DP (2020) Dextran-coated cerium oxide nanoparticles: a computed tomography contrast agent for imaging the gastrointestinal tract and inflammatory bowel disease. ACS Nano 14(8):10187–10197
Nakagawa T, Gonda K, Kamei T, Cong L, Hamada Y, Kitamura N, Ohuchi N (2016) X-ray computed tomography imaging of a tumor with high sensitivity using gold nanoparticles conjugated to a cancer-specific antibody via polyethylene glycol chains on their surface. Sci Technol Adv Mater 17(1):387–397
Neng J, Li Y, Driscoll AJ, Wilson WC, Johnson PA (2018) Detection of multiple pathogens in serum using silica-encapsulated nanotags in a surface-enhanced Raman scattering-based immunoassay. J Agric Food Chem 66(22):5707–5712
Nurunnabi M, Khatun Z, Reeck GR, Lee DY, Lee YK (2014) Photoluminescent graphene nanoparticles for cancer phototherapy and imaging. ACS Appl Mater Interfaces 6(15):12413–12421
Oseledchyk A, Andreou C, Wall MA, Kircher MF (2017) Folate-targeted surface-enhanced resonance Raman scattering nanoprobe ratiometry for detection of microscopic ovarian cancer. ACS Nano 11(2):1488–1497
Pan W, Peng W, Ning F, Zhang Y, Zhang Y, Wang Y, Zhang X (2018) Non-invasive detection of the early phase of kidney injury by photoacoustic/computed tomography imaging. Nanotechnology 29(26):265101

Paris JL, Cabañas MV, Manzano M, Vallet-Regí M (2015) Polymer-grafted mesoporous silica nanoparticles as ultrasound-responsive drug carriers. ACS Nano 9(11):11023–11033

Prasad M, Ghosh M, Patki HS, Kumar S, Brar B, Sindhu N, Kumar R (2021) Imaging techniques in veterinary disease diagnosis. In: Advances in animal disease diagnosis. CRC Press, pp 103–145

Qu X, Wang J, Zhang Z, Koop N, Rahmanzadeh R, Hüttmann G (2008) Imaging of cancer cells by multiphoton microscopy using gold nanoparticles and fluorescent dyes. J Biomed Opt 13(3): 031217

Ramalho J, Semelka RC, Ramalho M, Nunes RH, AlObaidy M, Castillo M (2016) Gadolinium-based contrast agent accumulation and toxicity: an update. Am J Neuroradiol 37(7):1192–1198

Ravanshad R, Karimi Zadeh A, Amani AM, Mousavi SM, Hashemi SA, Savar Dashtaki A, Zare B (2018) Application of nanoparticles in cancer detection by Raman scattering based techniques. Nano Rev Exp 9(1):1373551

Reineck P, Lau DW, Wilson ER, Nunn N, Shenderova OA, Gibson BC (2018) Visible to near-IR fluorescence from single-digit detonation nanodiamonds: excitation wavelength and pH dependence. Sci Rep 8(1):1–7

Rezanejad Gatabi Z, Mohammadpour R, Rezanejad Gatabi J, Mirhoseini M, Sasanpour P (2020) A novel composite gold/gold nanoparticles/carbon nanotube electrode for frequency-stable micro-electrical impedance tomography. J Mater Sci Mater Electron 31(13):10803–10810

Rogalla S, Flisikowski K, Gorpas D, Mayer AT, Flisikowska T, Mandella MJ, Harmsen S (2019) Biodegradable fluorescent nanoparticles for endoscopic detection of colorectal carcinogenesis. Adv Funct Mater 29(51):1904992

Shilo M, Reuveni T, Motiei M, Popovtzer R (2012) Nanoparticles as computed tomography contrast agents: current status and future perspectives. Nanomedicine 7(2):257–269

Siddique S, Chow JC (2020) Application of nanomaterials in biomedical imaging and cancer therapy. Nanomaterials 10(9):1700

Slegerova J, Hajek M, Rehor I, Sedlak F, Stursa J, Hruby M, Cigler P (2015) Designing the nanobiointerface of fluorescent nanodiamonds: highly selective targeting of glioma cancer cells. Nanoscale 7(2):415–420

Srivatsan A, Chen X (2014) Recent advances in nanoparticle-based nuclear imaging of cancers. Adv Cancer Res 124:83–129

Tan J, Yang N, Hu Z, Su J, Zhong J, Yang Y, Zhao Y (2016) Aptamer-functionalized fluorescent silica nanoparticles for highly sensitive detection of leukemia cells. Nanoscale Res Lett 11(1): 1–8

Teleanu DM, Chircov C, Grumezescu AM, Volceanov A, Teleanu RI (2019) Contrast agents delivery: an up-to-date review of nanodiagnostics in neuroimaging. Nanomaterials 9(4):542

Tian C, Qian W, Shao X, Xie Z, Cheng X, Liu S, Wang X (2016) Plasmonic nanoparticles with quantitatively controlled bioconjugation for photoacoustic imaging of live cancer cells. Adv Sci 3(12):1600237

Van Sloun RJ, Cohen R, Eldar YC (2019) Deep learning in ultrasound imaging. Proc IEEE 108(1): 11–29

Wu Y, Ermakova A, Liu W, Pramanik G, Vu TM, Kurz A, Weil T (2015) Programmable biopolymers for advancing biomedical applications of fluorescent nanodiamonds. Adv Funct Mater 25(42):6576–6585

Wu Y, Vazquez Prada KX, Liu Y, Whittaker AK, Zhang R, Ta HT (2021) Recent advances in the development of theranostic nanoparticles for cardiovascular diseases. Nanotheranostics 5(4): 499

Xu X, Ray R, Gu Y, Ploehn HJ, Gearheart L, Raker K, Scrivens W (2004) A electrophoretic analysis and purification of fluorescent single-walled carbon nanotube fragments. J Am Chem Soc 126:12736–12737. https://doi.org/10.1021/ja040082h

Yousaf T, Dervenoulas G, Politis M (2018) Advances in MRI methodology. Int Rev Neurobiol 141: 31–76

Zavaleta CL, Hartman KB, Miao Z, James ML, Kempen P, Thakor AS, Gambhir SS (2011) Preclinical evaluation of Raman nanoparticle biodistribution for their potential use in clinical endoscopy imaging. Small 7(15):2232–2240

Zhang C, Yan Y, Zou Q, Chen J, Li C (2016) Superparamagnetic iron oxide nanoparticles for MR imaging of pancreatic cancer: potential for early diagnosis through targeted strategies. Asia Pac J Clin Oncol 12(1):13–21

Zhang Q, Hou K, Chen H, Zeng N, Wu Y (2022) Nanotech probes: a revolution in cancer diagnosis. Front Oncol 12:933125

Zhou Y, Kong Y, Kundu S, Cirillo JD, Liang H (2012) Antibacterial activities of gold and silver nanoparticles against Escherichia coli and bacillus Calmette-Guérin. J Nanobiotechnol 10(1): 1–9

Chapter 12
Overview of Nanovaccines and Nanoadjuvants

Subodh Soni, Pooja Chugh, Krishna Kant Sharma, and Hari Mohan

Abstract Immunization is still the most efficient intervention to prevent the spread of diseases. Despite the tremendous scientific advancements in vaccine development, there is no effective vaccine against cancer, AIDS, malaria, etc. Furthermore, most of the existing vaccines are suffering from the limitations of incomplete stimulation of immune system, stability at room temperature, in vivo toxicity, multiple boosters requirement, etc. Advancements in technology are helping in the development of effective vaccines for the prevention of disease; still, a lot of people remain unvaccinated throughout the world. World Health Organization (WHO) has considered mass immunization as the main objective in their immunization schedule 2030. Nanovaccines may be an essential intervention for mass immunization in low-income countries because of their stable nature and storage ability at room temperature. Nanomaterial's high surface area to volume ratio has enhanced its immunogenicity, biocompatibility, slow release, and safety. Although vaccines are practical tools for immune response generation, some fail to generate an effective immune response alone; at that point, adjuvants come into play. Nanoadjuvants are tiny particles of the nm range, providing stability, increasing vaccine efficacy, and generating a robust immune response. They also include the types of damage-associated molecular patterns (DAMPs) and pathogen-associated molecular patterns (PAMPs) and also include aluminum, emulsion, saponin-based adjuvants, and toll-like receptor (TLR) adjuvants, respectively. This chapter discusses the nanovaccines and nanoadjuvants types, their mode of action, and their applications in various diseases.

S. Soni · K. K. Sharma · H. Mohan (✉)
Centre for Medical Biotechnology, Maharshi Dayanand University, Rohtak, Haryana, India
e-mail: harimohan.cmbt@mdurohtak.ac.in

P. Chugh
Department of Microbiology, Maharshi Dayanand University, Rohtak, Haryana, India

LLR University of Veterinary and Animal, ICAR-National Research Centre on Equines, *Nanotechnology Theranostics in Livestock Diseases and Management*, Livestock Diseases and Management,
https://doi.org/10.1007/978-981-16-1610-5_12

Keywords Nanovaccines · Nanoadjuvants · DAMPs · PAMPs · Nanomaterials · Immunity

12.1 Introduction

Globally diseases cause a negative impact on health and wealth of life forms. To control the spread of diseases, novel treatments and prevention strategies need to be evolved regularly. For mass immunization, nanotechnology provides the choice of efficient nanocarriers, routes of administration according to the needs of the patient's convenience, improved stability of products, and efficient nano-adjuvants (Cordeiro et al. 2021). Nanovaccine is a nanosized formula that mimics the size of a targeted pathogen. It provides immunization that can prevent the spreading of diseases because of its ability to use the body's immune system to kill pathogens (Azharuddin et al. 2022). Antigens are either encapsulated (liposome, micelle, dendrimer, nanosphere core, etc.) or adsorbed on the nanomaterial surface. Encapsulation protects the antigen from degradation under physiological conditions and an encapsulated antigen's endocytosis is carried out by antigen-presenting cells (APCs). The endosomal-lysosomal complex processes it, and the MHC molecules present the processed antigen on APCs. Antigens are loaded on the cell surface of nanomaterials by chemical bonding and physical attachment. Immunogenicity of nanovaccine depends on size, charge, nature, and hydrophobicity of antigen and nanomaterial. Nanovaccines have several advantages, including improved blood flow stability, increased immune system stimulation, no need for booster doses, devoid of requiring cold chain, and the potential to produce active targeting (Gheibi Hayat and Darroudi 2019). To develop proper immunity before the onset of diseases, prophylactic nanovaccines are given, and after the development of diseases, therapeutic vaccinations are given so that the immune system fights the infections (Bhardwaj et al. 2020). In the endosome, acidic pH derives transformable nanovaccine change into a bigger size for an escape out into cytoplasm and enhance antigen cross-presentation (Gong et al. 2020).

Vaccines are the most potent method of providing humoral and cell-mediated immunity, and the live attenuated vaccines are the earliest vaccines used to cure smallpox (Plotkin 2014). Other types of vaccines include only a tiny part of a pathogen, like mRNA and subunit vaccines, but they often fail to generate an effective immune response in many cases. Some nanoparticles as adjuvants can bind with a vaccine and intensifies immune response, and these adjuvants can also differ in their mode of stimulation of immune response. Adjuvants include damage-associated molecular proteins (DAMPs) (Ong et al. 2021), pathogen-associated molecular patterns (PAMPs), and also the aluminum adjuvants like alums, aluminum phosphates, and aluminum hydroxides (Dandashli et al. 2002). Some emulsion adjuvants like AS03 and MF59 generate the immune response by antigen-presenting cell (APC) activation. These emulsions are formed by combining either water-in-oil or oil-in-water formulation (O'Hagan et al. 2020). Saponins-based adjuvants are isolated from isoprenoids, and they are also DAMPs type. These possess anticancerous, antiviral, anti-inflammatory, anti-microbial, and anti-fungal activities (Biswas and Dwivedi 2019). Pathogen-associated molecular patterns (PAMPs) are composed of microbial products that recognize specific toll-like receptors (TLRs) and activate the dendritic cells. These adjuvants can contain microbial molecules like flagellin, a DNA containing CpG oligodeoxynucleotides, and are classified as TLR5 and TLR9 ligand. Some adjuvants contain a mixture of both DAMPs and PAMPs and are highly immunogenic such as TLRs agonists like monophosphoryl lipid A (MPL), macrophage-activating lipopeptide-2 (MALP-2), and VLA-15 (Narayanan and Park 2015). In this chapter, we will discuss the nanovaccines, types of nanovaccines, their mode of action, the necessity of nanovaccines, and the nanoadjuvants, their types, working, and roles in various diseases.

12.2 Why Do We Need Nanovaccine?

Nanovaccines are required to solve the problem of poor immune response, multiple-dose requirements, and poor stability of vaccines (Bhardwaj et al. 2020). Nanosize range, high antigen loading, enhanced immunogenicity, more retention period in lymph nodes, biocompatible, biodegradable, and many other factors make nanovaccine more efficient than conventional vaccine (Azharuddin et al. 2022). In a therapeutic nanovaccine tumor cell lysate is used as an antigen and personalized for the patient. Nanovaccines' stability at room temperature helps supply chain and storage in low-income countries. To stop the metastasis and prevent tumor regrowth after surgery, these therapeutic nanovaccines are used (Kim et al. 2019). Single nanovaccine can be used for multiple diseases by conjugating multiple epitopes; for example, nanovaccines for diseases like BRDC (bovine respiratory disease complex) include numerous pathogens that protect against up to 5–8 different infections. By the use of this method, we can reduce the handling of vaccines and the number of vaccines required, and also can ensure that before exposure the animals have been vaccinated against various diseases (Maina et al. 2020). Generally, orally administrated vaccines are degraded by the acidic pH of the stomach, but the use of nanomaterials protects the antigens and adjuvants from such kinds of chemical and enzymatic degradation (Azharuddin et al. 2022).

12.3 Different Types of Nanomaterials That Can be Used in Nanovaccine

Nanomaterials are nanosized and mimic the size of targeted pathogens. Biolosome, buckyballs, carbon nanotubes, chitosan, liposome, micelle, polymers, quantum dot nanocrystals, self-assembled membrane, etc. can be used in the formation of nanovaccines (Fig. 12.1).

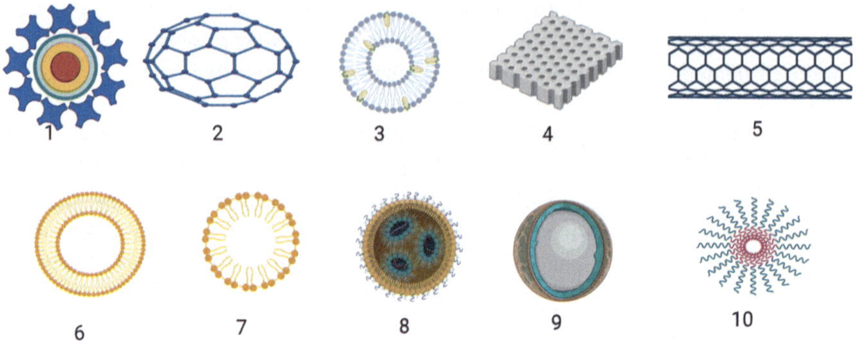

Fig. 12.1 Different types of nanomaterials that can be used in nanovaccine: (1) Quantum dot nanocrystals, (2) Buckyball, (3) Biolosome, (4) Self-assembled membrane, (5) Carbon nanotubes, (6) Liposome, (7) Micelle, (8) Lipid NPs, (9) Nanosphere, (10) Polymeric micelle

12.3.1 Carbon-Based Nanomaterials

Carbon nanotubes, graphene oxides, and quantum dots are examples of some carbon-based nanomaterials. Due to the presence of electrical, mechanical, optical, thermal, and structural features, these are very useful in disease diagnosis and treatments (Tao et al. 2019). Multiwalled carbon nanotubes (MWCNT) have a diameter of 0.4–2 nm, while single-walled carbon nanotubes (SWCNT) have a diameter of 1 nm and due to their biocompatibility, they are more suitable antigen carriers (Nunes De Carvalho Lima et al. 2021). MWCNTs can be an effective delivery vehicle for PA-based vaccine formulations against anthrax because they improve rPAD4 stability and self-adjuvant property, and graphene oxide-OVA-C_PG formulated nanovaccine can be used against B16-OVA-melanoma tumors (Zhang et al. 2022).

12.3.2 Natural Polymers

Dextran, chitosan, alginate, silver, iron, and gold oxides are natural nanomaterials. They are formed by natural phenomena like volcanic eruptions, radioactive decay, rock weathering, water streams, etc. IUPAC dextran is a branched poly-D-glucoside of microbiological origin and contains C_1–C_6 glycosidic linkage (Annu and Ahmed 2021). Dextran NPs containing OVA and LPS as antigens are engulfed by bone marrow dendritic cells (DC), activating T_h and T_c cells proliferation. Dextran nanovaccine also elicited a more prominent delayed-type hypersensitivity response and more potent cytotoxic $CD8^+$ T cells activation than soluble OVA and LPS administration (Shen et al. 2013). In insects and crustaceans, natural chitosan polymer is an amino polysaccharide made of deacetylating chitin (Jamiołkowska and Kopacki 2019). An inactivated influenza virus chitosan nanovaccine induced the proliferation of antigen-specific IFNγ secreting T-helper/memory and γδ T cells and an increased TH1 (proinflammatory) (IFNγ, IL-6, and IL-2) and TH2 (anti-inflammatory) (IL-10 and IL-13) cytokines mRNA expression in the tracheobronchial lymph nodes (Renu et al. 2020). In the brown algae, natural polymer alginate is present, which is a colloidal heteropolysaccharide and an anionic, hydrophilic, chain-forming, and contains randomly ordered linear unbranched chains of -L-guluronate (G block) and -D-mannuronate (M block) residues (Fernando et al. 2020). Alginate-chitosan-LPS-selenium NP composite is an oral vaccine that extends protection against *Vibrio cholerae* by increasing the release of anti-inflammatory cytokines and decreasing the release of proinflammatory cytokines (Bagheri-Josheghani and Bakhshi 2022).

12.3.3 Synthetic Polymers

Synthetic nanomaterials are formulated for a specific purpose and have particular sizes and chemical compositions. Some synthetic nanomaterials are dendrimers, poloxamer, polyetherimide, polymethylmethacrylate, polylactide-co-glycolide (PLGA) copolymer, polyvinylpyrrolidone, polyphenylene sulfide, etc. (Rosana et al. 2020). Dendrimers are homogenous and monodispersed structures and contain the asymmetric core, an inner shell, and an outer shell (Abbasi et al. 2014). Dendrimer modified with guanidinobenzoic acid (DGBA)-Ovalbumin (OVA)-unmethylated cytosine guanine dinucleotide (CpG) formulated nanovaccine can induce an immune response against B16—OVA melanoma (Xu et al. 2019). The use of polylactic-co-glycolic acid with monophosphoryl lipid adjuvants prevents diseases like visceral leishmaniasis by increasing the production of IL12 and IFN gamma and T cell proliferation (Athanasiou et al. 2017). Polycationic polyethyleneimine (PEI)-OVA-Cyclic phosphate guanine (CPG) formulated nanovaccine has been used for cancer immunotherapy. Cationic PEI increases antigen uptake by DC and induces OVA antigen cross-presentation and CPG get recognized by the toll-like receptor 9 (Guan et al. 2018) (Table 12.1).

12.3.4 Liposome and Micelle

Liposomes are tiny vesicles formed from the phospholipid bilayer. MANα1-2MAN loaded in liposome increases the immune response of bovine herpes simplex virus-1 DNA vaccine (Kornuta et al. 2021). Micelles are spherical amphiphilic structures consisting of a hydrophobic core and hydrophilic shell. An antigen can be loaded into a hydrophobic core (Silindir-Gunay and Ozer 2018). For cancer immunotherapy, imiquimod $R8_{37}$ and $Zn_{1.15}Fe_{1.85}O_4$ are encapsulated in the core of pluronic F_{127} copolymer self-assembled micelle (Jiexin et al. 2021).

Table 12.1 Nanovaccines for cancer immunotherapy

Nanovaccine	Nanomaterial	Antigen	Targeted pathway	Model tumor cells	References
SeMac (self-adjuvanted molecular activator) nanovaccine	Mal PEG_{500}-b-$PC7A_{45}$	C_7 monomer	STING (stimulator of interferon gene)	Colon carcinoma 26 (CT26) and B16F10 tumor cells	Luo et al. (2021)
OVA@FG nanovaccine	Fe-polyphenol	OVA	NOX2 pathway	B16 OVA cells	Chong et al. (2022)
Photosensitizer indocyanine Green nanovaccine	ICG	OVA	ROS	EG7 tumor cells	Cao et al. (2018)
OMPN (Ovalbumin MnO_2 polydopamine) nanovaccine	MnO_2	OVA	Dendritic cell	B16 OVA cells and DC2.4 cells	Xiao et al. (2021)
Proton driven transformable nanovaccine	Naphthalene conjugated D Peptide	OVA	NLRP3 inflammatory pathway	B16F10 OVA and HPV E6/E7 tumor cells	Gong et al. (2020)

12.4 Mode of Action of Nanovaccines

Nanovaccines elicit both cell-mediated and humoral immunity, and the different size of NPs activates a different type of immunity (Fig. 12.2). Nanosized particles (NPs) with 200–250 nm diameter, are endocytosis by APCs and activate cell-mediated immunity, whereas NPs of 0.5–5 μm are phagocytosed by immune cells and trigger a humoral immune response (Kim et al. 2019). Dendritic cell interconnects cell-mediated and humoral immunity. Intravenously administrated nanovaccine enters the local lymph node and in the lymph node, it is endocytosed by APCs (mostly dendritic cells). The anovaccines antigen (Ag) is processed by the endocytotic-lysosomal complex and then APCs represent the antigen on their surface by the MHCII molecules. This Ag MHCII is encountered by T_H cells. T_H cells then release cytokines and costimulatory molecules, start the proliferation of T cells, and activate B cells in the germinal center of the lymph node. Activated B cells differentiate into memory cells and plasma cells. A plasma cell's half-life is 14 days and in most cases, IgM is produced by the plasma cells and then converted into other antibodies depending on the type of antigen, antigen entry pathway in the body, and the size and chemical nature of the antigen. So in this way, plasma cells secrete antigen-specific antibodies. The generation of immune response is influenced by the size and chemical nature of NPs, antigens, and adjuvants (Kim et al. 2019). The success rate of a nanovaccine also depends on the types of nanocarrier used and the route of administration (Chenthamara et al. 2019). Vaccines with nanoadjuvant increase the antibody titers and facilitate slow and prolonged presence of antigen (Wagner et al. 2019). Sometimes there is no need for an adjuvant because secretion of the antigen from nanovaccine occurs slowly and in a controlled manner (Afshari et al. 2020).

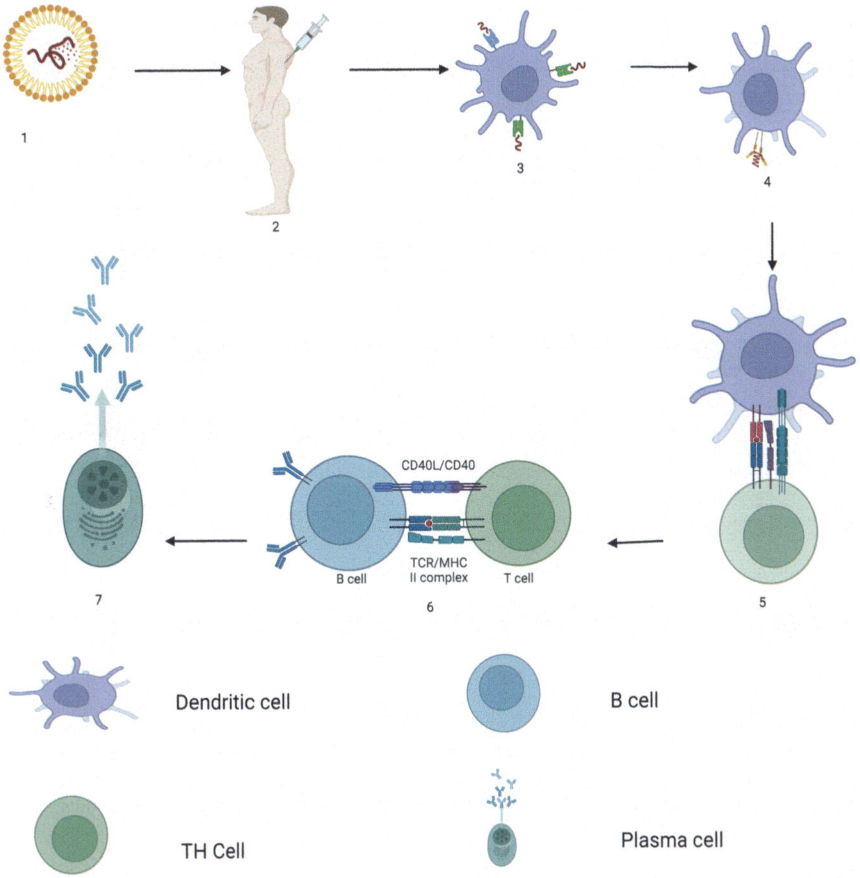

Fig. 12.2 Mode of action of nanovaccine. (1) Micelle NP core contains peptide antigen and adjuvant, (2) Intravenous administration of nanovaccine, (3) Uptake of antigen by dendritic cell, (4) Antigen representation on MHCII of dendritic cell, (5) Dendritic cell activate T_H cell, (6) T_H cell activate B cell, (7) Plasma cell (Differentiated B cell) secrete antigen-specific antibodies

12.5 Nanoadjuvants

Vaccines play immensely for eliciting an innate immune response in mammals to provide adaptive immunity that helps in fighting foreign antigens, they are used for eliciting both humoral- and cell-mediated immunity although these responses can be weak or strong. Vaccines are considered an effective tool for the prevention of diseases; generally inactivated and attenuated vaccines are used for immune response generation as they are cost-effective, easy to develop, and possess the tendency to generate effective immune responses. Due to the advancement in technology, a lot of new methods are developed for purifying vaccines and making them highly safer to use, but as we know a coin has two sides, these highly safe and purified vaccines are found to be less immunogenic when used solitarily (McKee et al. 2010). Scientists used adjuvants in combination with vaccines to overcome the

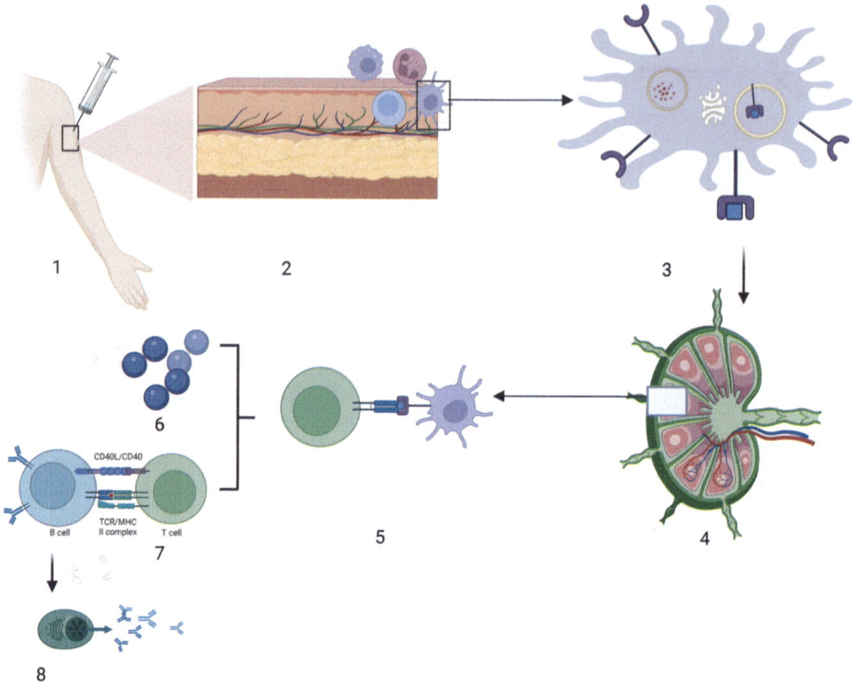

Fig. 12.3 Mode of action of nanoadjuvant. (1) Intravenously administrated nanovaccine with nanoadjuvants, (2) Immune cells secrete out from blood vessels, (3) Antigen processed and present in dendritic cells, (4) Dendritic cells move to local lymph node, (5) Dendritic cell activate T_H cell, (6) T_H cells secrete cytokines, (7) T_H cell activate B cell, (8) Plasma cell secretes antibodies

problem of low immunogenicity of purified vaccines, such combination increases vaccine efficacy, triggers an adaptive immune response, and superior antigen presentation which in turn provide interminable protection in opposition to pathogens (Fig. 12.3). There are three different types of adjuvants that rely on stimulation of immune response—(1) Damage-associated molecular patterns (DAMPs) include alum, emulsion, and saponins, (2) Pathogen-associated molecular patterns (PAMPs) include lipopolysaccharides (LPS), TLR ligands, mycobacterial compounds and liposomes, (3) Particulate adjuvants include ISCOMs and nanoparticles (Ong et al. 2021). The release of these adjuvants is triggered by modified vaccines for an effective immune response generation (Tizard 2021).

12.5.1 Damage-Associated Molecular Patterns (DAMPs)

DAMPs include adjuvants like saponin, aluminum, and emulsion, and these are produced in response to damaging tissues. Generally, these DAMPs cause chemical-mediated irritations or possess direct toxic effects.

DAMPs also differ in their nature of actions like

1. Aluminum-based adjuvants release uric acid, adenosine, and DNA by acting on dying cells.

2. Emulsions contain a short chain of detergent molecules that are responsible for cell lysis.
3. Saponins have amphipathic soaps-like nature which disrupts the cell membrane.

Despite the different mode of action, all these adjuvants release DAMPs which activate the inflammatory response by binding to the antigen-presenting cells (APC) (Ong et al. 2021; Tizard 2021).

12.5.1.1 Aluminum-Based Adjuvant

Aluminum adjuvants are well established and lie among the earliest adjuvants. They are used in veterinary and human prophylactic vaccines because they are non-pyrogenic and safer to use in nature (Hamaguchi 1926; Liang et al. 2020; Shi et al. 2019). In old times, alum adjuvants (aluminum potassium sulfate) were prepared through precipitation but reproducible results were not found, subsequently it was replaced with an adsorption process in which aluminum gels were prepared (Gupta 1998). The size of adjuvants plays a significant role in immunogenicity. Sometimes it shows negative effects on the immune system, as the nanoscale concentration of adjuvants is highly immunogenic and produces a long-lasting and robust immune response. Adjuvants of small size easily accumulate in tumors and lymph nodes. Nanoparticles like aluminum, gold, and iron are more effective than microparticles (Bowen et al. 2018). The two most common vaccine adjuvants are aluminum phosphate adjuvants (AP) and aluminum hydroxide adjuvants (AH). For the preparation of aluminum phosphate adjuvants, either phosphate solution and aluminum salts are mixed and sodium hydroxide is used for precipitation, or trisodium phosphate basic solution is mixed with $AlK(SO_4)_2$ or $AlCl_3$ salts of aluminum. The aluminum-based adjuvants can be non-crystalline or crystalline substances. Aluminum phosphate adjuvants are non-crystalline substances that alter the crystallization process because of the presence of phosphate (Klein et al. 2000). However, some other aluminum adjuvants have low or high crystallization, in these, the speed of dilution and adsorptive capacities are altered by the degree of crystallinity which depends on manufacturing conditions. Crystalline aluminum adjuvants are aluminum hydroxides like aluminum oxyhydroxide (AlO(OH)) which is formed by the addition of sodium hydroxide in $AlK(SO_4)_2$ or $AlCl_3$ salts of aluminum, and the highly crystalline solutions possess a low adsorptive capacity and lower surface are vice versa with low crystalline solutions (Dandashli et al. 2002; Yau et al. 2006).

Theories described that the alum has little effect on cytotoxic T-cell-mediated response but is the most potent stimulator of Th2 cells, amplifying antibody-mediated reactions. Nanoparticles have been given more preference in biomedical applications as they manifest unique properties, so in a study researchers compared the effect of alum with nano-alum (Berd et al. 2004; Jain 2008). For this, they characterized nano-alum properties and observed their histological analysis, tumor size, and immune cytotoxicity. With a scanning electron microscope, size of the nano-alum was measured, and the observed diameter of nanoparticles was in the range of 20–30 nm. Then D-Hank solution was mixed with fresh nano-alum adjuvants, and this mixture was instantly sonicated preceding the injection. For the animal study, a mature tumor model H22 murine hepatoma was used and this tumor size is further inspected for the consequences of immune treatments (Luo et al. 2006). In tumor tissue, histological studies of necrosis were observed in the nano-alum + tumor cell vaccine group, which indicated an active anti-tumor immune reaction, and lymphocytes infiltrated tumor tissue in treatment groups. Anti-tumor cytotoxicity evaluation is done by randomly picking animals from all the groups, LDH assay was performed to check lymphocytes cytotoxicity effect on H22, and nearly 68% of the cell was killed in nano-alum + tumor cancer vaccine group. The group treated with nano-alum and a tumor cell vaccine showed the smallest average tumor size, indicating a significant enhancement of the anti-tumor response, hence concluded that nano-alum intensify anti-tumor response and is more effective adjuvant than the conventional one (Sun et al. 2010).

12.5.1.2 Emulsion-Based Adjuvants

The emulsion can be defined as the mixture of liquids that are surfactant stabilized and immiscible. Generally, there are two types of emulsion: one is oil-in-water and the other water-in-oil. They differ in the disperse phase, and water serves as disperse phase in oil-in-water emulsions. Emulsion adjuvants were developed as an advancement method in the adjuvants system, although aluminum-based adjuvants (alum) were first developed; adjuvants are very effective and safer to use but at some points, they show reduced potency that led to the development of these emulsion adjuvants. MF59, AS04 and AS03, AS01, and CpG1018 are examples of some emulsion adjuvants (O'Hagan et al. 2020). All emulsion adjuvants comprised of squalene found in both plants and animals as a metabolic intermediate. Shark liver oil is a major source of squalene (Fox et al. 2019; O'Hagan et al. 2020), although squalene solitarily doesn't behave as adjuvants, stabilizer molecules like surfactant are combined to make effective emulsion solution that generates an effective immune response (Fig. 12.4), AS03 and MF59 are widely used adjuvants for clinical purposes as observed in the pandemic flu vaccine and Novartis vaccines, respectively (Fox and Haensler 2013). High-pressure homogenization, microfluidizers, and emulsification (Lodaya et al. 2019) are used for squalene-based emulsion preparation.

MF59 is composed of squalene (250 nm), a metabolizable oil and is categorized under the oil-in-water emulsion; it was firstly approved for adults aged 65 and above in 1997 in the influenza vaccine and positive effects are observed, vaccines developed with MF59 adjuvant were highly safe for use in all age-groups (Podda 2001; Podda and Del Giudice 2003). However, a study revealed enhanced efficacy of the influenza vaccine, low dosage provides outermost protection and improved immunogenicity due to MF59 adjuvant in young children and infants (O'Hagan et al. 2013; Stassijns et al. 2016). Scientists believe that MF59 is the most potent adjuvant for enhancing subunit vaccine immunogenicity. Elevated levels of IgG1 and IgG2a

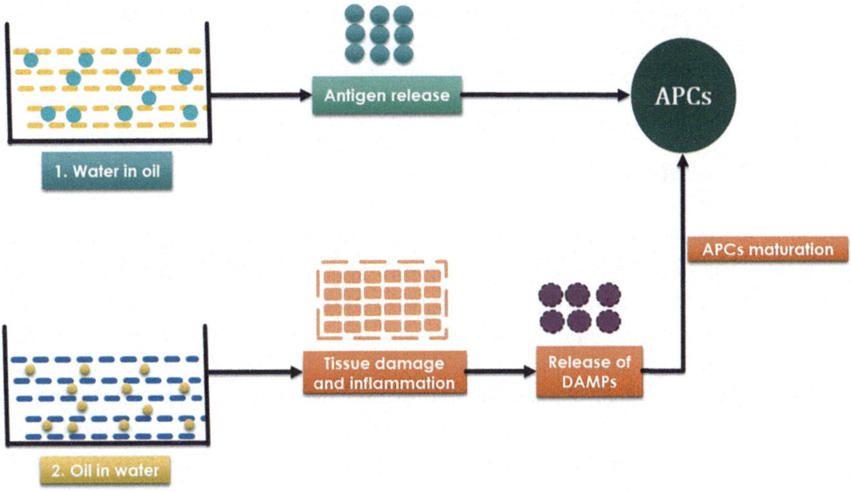

Fig. 12.4 The action of emulsion adjuvants. (1) Water-in-oil emulsion adjuvants facilitate the release of antigens that produce antigen-producing cells (APCs) and generate the immune response. (2) Oil-in-water emulsion first led to the tissue-damaging reaction and inflammation occurs at that site, then there is the release of damage-associated molecular patterns (DAMPs) that will give signals for APCs maturation, and eventually, antigen-presenting cells will be formed and immune response will be generated

antibodies were observed in mouse models immunized with MF59 conjugated adjuvant with MERS-CoV RBD protein (Zhang et al. 2016). Moreover, a study was performed to compare MF59-adjuvanted CTEP vaccine therapeutic effects with Alum/CpG in atherosclerosis rabbit models, and anti-atherosclerosis property was shown by MF59-conjugated CETP vaccine. Still, the immune response doesn't provide direct protection (Aghebati et al. 2016).

Adjuvant system 03 (AS03) was developed for producing an effective influenza vaccine, AS03 is a type of oil-in-water emulsion system composed of squalene and DL-α-tocopherol, TLR-independent, studies revealed it elevates the immune cell stimulation, APCs recruitment to lymph nodes (Morel et al. 2011), stress signals, complement activation, rapid antigen delivery, and also an expression of immune response gene (Caproni et al. 2012; Garçon et al. 2012; Leroux-Roels 2009).

However, scientists studied the AS03 adjuvanted influenza vaccine's long-term evolution of cellular and humoral response effects in BALB/c mice. Although short-term effects were already measured (Yam et al. 2015), for long-term study, mice models were provided with high and low booster doses and their serological analysis was done by multiplex ELISA, ELISpot, and flow cytometry for cytokines IFNγ, TNFα, IL-2, and IL-5. Results showed that in comparison with the unadjuvanted group, AS03 adjuvanted higher dose generated highest titers while low doses showed intermediate results, antibody titers were detected through endpoint ELISA assay, (HA) hemagglutinin content. Antibody-secreting cells (ASCs) and memory antibody-secreting cells that are influenza-specific were counted using

ELISpot, and their spot size was measured in the spleen. The highest number of plasma cells and memory cells specific for influenza were observed in AS03 adjuvanted animals (Yam et al. 2017).

12.5.1.3 Saponin-Based Adjuvants

Saponins are derived from the Latin word *Sapo*, posing foam-generating properties like soaps. Saponins are of two types: steroid saponins and triterpenoid saponins based on structural classification, a sapogenin aglycone derived from isoprenoid, ester or ether linkage is responsible for attachment with sugar chains, and these are responsible for amphiphilic properties (Vincken et al. 2007). In both plants and animals, saponins protect against herbivores and pathogens attack, and they also have anticancerous, antiviral, anti-inflammatory, anti-microbial, and anti-fungal properties, hence used in a variety of drugs (Biswas and Dwivedi 2019; Papadopoulou et al. 1999).

Quil A® is a triterpenoid saponin, which is widely used as an adjuvant in veterinary sciences. It is extracted from *Quillaja saponaria*, Quil A® elicit a robust humoral and cellular response, although it possess some side effects like systemic toxicity, local reactions, and hemolytic activity (Sivakumar et al. 2011). A saponin from *Q. brasiliensis* is lesser toxic than Quil A® and generates strong immune response, possess similar biological and structural activities like *Q. saponaria* (Cibulski et al. 2017), and generate vigorous delayed-type hypersensitivity reaction, intensifies Th-2 and Th-1 response. QB-80 and QB-90 saponins adjuvants from *Q. brasiliensis* elevates IFN-γ production by $CD4^+$ and $CD8^+$ T cells and IL-2 and IFN-γ cytokines associated with Th-1 response (Cibulski et al. 2016; Yendo et al. 2016). Another saponin-based adjuvant QS-21 is remarkably best, protecting against intracellular pathogens and cancer (Bigaeva et al. 2016; Sun et al. 2009). Saponin binds with cholesterol with high affinity but in most conditions fails to generate the immune response that's why (ISCOMs) immune-stimulatory complexes (Morein et al. 1984) are used to reduce the hemolytic effect and provide stability, formed by combining saponin Quil A®, phospholipid, and cholesterol, and the antigen. These are 40 nm self-assembled cage-like structures which also act as antigen delivery system (Chauhan et al. 2017; O'Hagan and Fox 2015). Another vaccine adjuvant ISCOMATRIX™ formed without the involvement of antigen (Morelli et al. 2012).

However, saponins-based adjuvants are used to produce herpes zoster virus, malaria, and the SARS-CoV-2 vaccine (NVX-CoV2373). These saponin-based adjuvants are believed to activate $CD8^+$ T cell, antigen cross-presentation by dendritic cells, and $CD4^+$ T-cell mediated immunity; still, scientists are unable to understand the exact mechanism of dendritic cell cross-presentation, some subsets like CD8α + DCs cross-present without exogenous adjuvants, although vacuolar pathway and cytosolic pathway are two major cross-presentation pathways in dendritic cells (Kotsias et al. 2019; Nierkens et al. 2013).

Moreover, to check the toxicity of saponins adjuvants, QB-90 ISCOMs are used with influenza antigen IQB-90-Flu in the mouse model and observed no damage to the membrane in hemolysis assay. A transmission electron microscope is used for the study, in mouse models after two-dose administration no swelling, hair loss, lethality was observed that confirms QB-90 ISCOMs are safer to use. Mice were also administrated with IQB-90-Flu subcutaneously and intravenously to check humoral response, robust IgM and IgG levels were found in mice with adjuvanted vaccine injected subcutaneously and intravenously respectively. The immune response induced was of Th1 type as confirmed through IgG2a/IgG1 ratio in intranasal and subcutaneous. Hemagglutinin inhibition(HIA) test was performed for determining influenza-specific antibodies production, and higher levels were observed in adjuvanted IQB90-Flu vaccine (Rivera-Patron et al. 2021).

12.5.2 Pathogen-Associated Molecular Patterns (PAMPs) Adjuvants

Pathogen-associated molecular patterns are composed of microbial products designed to target pattern recognition receptors (Behzadi and Behzadi 2016; Kawai and Akira 2010). These molecules are toll-like receptors (TLRs), TLRs are transmembrane biomolecules found in a variety of hosts such as mammals, arthropods like insects providing both innate and adaptive immunity to host (Luo et al. 2019; Wang et al. 2020). Pattern recognition receptors are divided into five families based on the protein domain homology, retinoic acid-inducible gene-I-like receptors (RLRs), c-type lecithin receptors (CLRs), (TLRs) toll-like receptors, and NOD-like receptors composed of (LRR) leucine-rich repeats (Brubaker et al. 2015). TLRs stimulate the production of cytokines like IL-12, and IL-1 by activating the (DCs) dendritic cells that intensify the Th-1 and Th2-mediated immune response (Narayanan and Park 2015). However, toll-like receptors protect from H1N1 and swine influenza and also enhance cytokines production like IL-23 and IL-12 by activation of TLR7, TLR8, TLR9, TLR3, and TLR4 in humans as well as mouse dendritic cells DCs, generating (Th1) T-helper-1 response (Kasturi et al. 2011; Kwissa et al. 2012).

Moreover, *Chlamydophila abortus* in their outer membrane possesses a 90 kDa protein POMP91B, belongs to protein group (POMPs) polymorphic outer membrane protein, has exposed N-terminal extracellular region made up of Beta-barrel, is believed to be highly immunogenic even in small amounts as confirmed through serum analysis in sheep (Buendía et al. 2001; Longbottom et al. 1998; Vretou et al. 2003). Experimental analyses were done with POMP91B for immune response generation, and cell proliferation studies that revealed low IFN-γ production and paucity in cell proliferation, which confirms reduced Th-1 response. At the same time, POMP91B generates robust IL-4 production and demonstrates strong Th-2 response (Le Moigne et al. 2009).

Studies revealed TLRs agonists which possess powerful immunostimulatory properties used as adjuvants, the TLR9 agonist is CpG-1018 while the TLR4 agonist

Table 12.2 Nanovaccine for cattle diseases

Nanoparticles	Pathogen	Antigen	Adjuvant	References
Multiwalled carbon tubes	*Bacillus anthracis*	rPA (Protective Antigen) domain 4	Freund's complete	Karimi et al. (2018)
Polyanhydride NP	*Bacillus anthracis*	Protective antigen	Cyclic di-GMP	Kelly et al. (2021)
PLGA (Polylactic-Co-Glycolic acid) NP	*Pasteurella multocida*	Inactivated Ag 4	Squalene	Kole et al. (2019)
PLGA NP	*Brucella abortus*	LPS (Lipopolysaccharide)	–	Afshari et al. (2020)
Mucoadhesive tamarind seed Polymer (Xyloglucan) NP	*Brucella*	LPS	Quillaja saponin	Vyas et al. (2020)
Chitosan NP	*Salmonella*	Outer membrane and flagellin protein	–	Renu et al. (2018)

is monophosphoryl lipid A (MPL), macrophage-activating lipopeptide-2 (MALP-2) serves as a TLR2 agonist, tripalmitoyl-*S*-glycerylcysteine (Pam3Cys) and dipalmitoyl-*S*-glycerylcysteine (Pam2Cys), VLA-15 targets *Borrelia* outer surface protein A of spirochete Lyme disease (Turk et al. 2019) (Table 12.2).

12.6 Challenges and Future Perspective

Nanotechnology is a promising technology for developing a safe and efficient vaccine. Different nanovaccines have already shown promising results in preclinical and clinical studies (Gheibi Hayat and Darroudi 2019). A general overview of nanovaccine production has been given in Fig. 12.5. There are some challenges such as the need for more methods for scaling up nanovaccine production, the knowledge of the toxicity of nanomaterials, in vivo aggregation, and crossing the blood–brain barrier. However, diligent research and rigorous analyses of pre-clinical and clinical data in nanovaccine domain can certainly overcome the hurdles. In future, it will be deciphered which nanomaterials and immunogens are suitable for targeted biological pathways in a disease-specific manner. The efficiency of a nanovaccine depends on age, health of patients, and the route of administration. Such analyses may help to develop a personalized nanovaccine with a preferred administration route. Single nanovaccine carrying multiple epitopes may generate broad-spectrum immunogenicity to protect against multiple pathogens (Azharuddin et al. 2022). Although, specific vaccines are available for particular diseases like influenza, malaria, and respiratory disease but still, some of them are less effective. Adjuvants are often used to enhance immune response, particularly for sub-unit vaccines. Adjuvants are classified as (DAMPs) damage-associated molecular proteins (Ong et al. 2021) and (PAMPs) pathogen-associated molecular proteins (Narayanan and Park 2015),

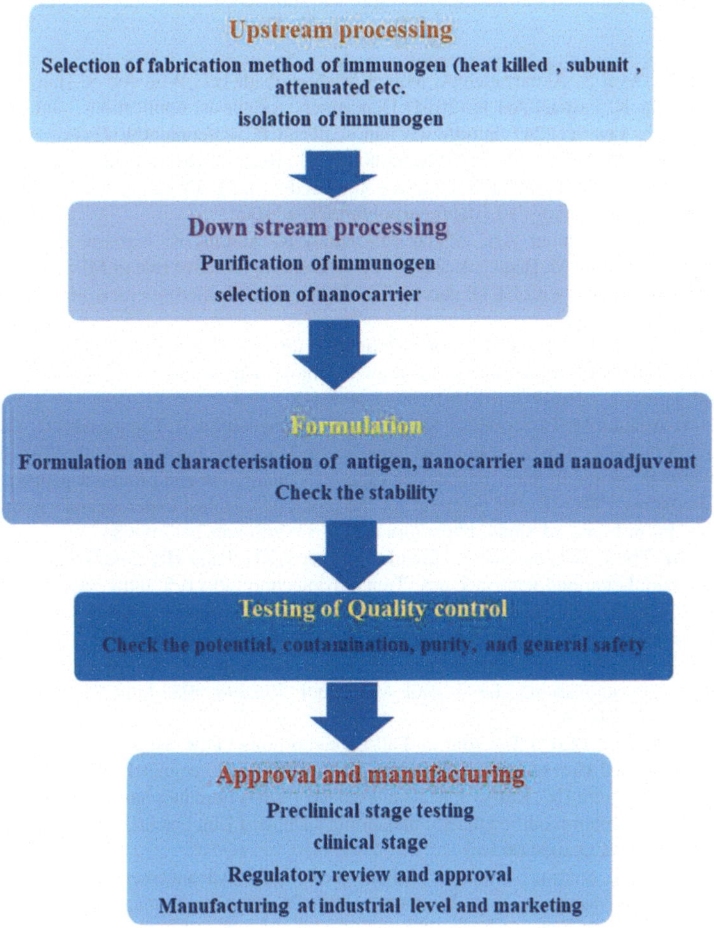

Fig. 12.5 Flowchart showing steps involved in the development of nanovaccine

some are made from mixing both and act as an agonist to traditional adjuvants, DAMPs include saponins adjuvants, aluminum adjuvants (Dandashli et al. 2002), emulsion formed from mixing oil-in-water or vice versa (O'Hagan et al. 2020). PAMPs include toll-like receptors present on microorganism membranes that interact with specific toll-like ligand TLRs example TLR4, TLR8, and TLR8 (Narayanan and Park 2015), regardless of type and mode of action they aim to intensify the cellular and humoral immunity.

Although we have found a way to generate effective nanovaccines, they are limited to certain diseases; even nanoadjuvants can work effectively with a few systems; sometime carrier system is required. So, further research is advocated in this arena to exploit full potential of nanomaterials in vaccine formulations with wider coverage against several neglected pathogens.

References

Abbasi E, Fekri Aval S, Akbarzadeh A, Milani M, Nasrabadi HT, Woo Joo S, Hanifehpour Y, Nejati-Koshki K, Pashaei-Asl R (2014) Dendrimers: synthesis, applications, and properties. Nanoscale Res Lett 9(1):247. http://www.nanoscalereslett.com/content/9/1/247

Afshari H, Maleki M, Salouti M (2020) Immunological effects of two new nanovaccines against Brucella based on OPS and LPS antigens conjugated with PLGA nanoparticles. Eur Polym J 139:110021. https://doi.org/10.1016/j.eurpolymj.2020.110021

Aghebati T, Mohammadpour AH, Afshar M, Jaafari R, Abnous K, Nazemi S, Issazadeh S, Hashemzadeh S, Zare M, Badiee A (2016) A novel atheroprotective role of MF59-like adjuvant when co-administered with CETP vaccine in rabbit model of atherosclerosis. Iran J Basic Med Sci 19(12):1345–1352

Annu, Ahmed S (2021) Advanced green materials: an overview. In: Advanced green materials: fabrication, characterization and applications of biopolymers and biocomposites, pp 1–13. https://doi.org/10.1016/B978-0-12-819988-6.00001-X

Athanasiou E, Agallou M, Tastsoglou S, Kammona O, Hatzigeorgiou A, Kiparissides C, Karagouni E (2017) A poly(lactic-co-glycolic) acid nanovaccine based on chimeric peptides from different Leishmania infantum proteins induces dendritic cells maturation and promotes peptide-specific IFNγ-producing CD8+ T cells essential for the protection against experimental visceral leishmaniasis. Front Immunol 8:684. https://doi.org/10.3389/fimmu.2017.00684

Azharuddin M, Zhu GH, Sengupta A, Hinkula J, Slater NKH, Patra HK (2022) Nano toolbox in immune modulation and nanovaccines. Trends Biotechnol 40:1195. https://doi.org/10.1016/j.tibtech.2022.03.011

Bagheri-Josheghani S, Bakhshi B (2022) Formulation of selenium nanoparticles encapsulated by alginate-chitosan for controlled delivery of Vibrio cholerae LPS: a novel delivery system candidate for nanovaccine. Int J Biol Macromol 208:494–508. https://doi.org/10.1016/J.IJBIOMAC.2022.03.087

Behzadi E, Behzadi P (2016) The role of Toll-Like receptors (TLRs) in urinary tract infections (UTIs). Cent Eur J Urol 69(4):404–410. https://doi.org/10.5173/ceju.2016.871

Berd D, Sato T, Maguire HC, Kairys J, Mastrangelo MJ (2004) Immunopharmacologic analysis of an autologous, hapten-modified human melanoma vaccine. J Clin Oncol 22(3):403–415. https://doi.org/10.1200/JCO.2004.06.043

Bhardwaj P, Bhatia E, Sharma S, Ahamad N, Banerjee R (2020) Advancements in prophylactic and therapeutic nanovaccines. Acta Biomater 108:1–21. https://doi.org/10.1016/j.actbio.2020.03.020

Bigaeva E, van Doorn E, Liu H, Hak E (2016) Meta-analysis on randomized controlled trials of vaccines with QS-21 or ISCOMATRIX adjuvant: safety and tolerability. PLoS One 11(5):e0154757. https://doi.org/10.1371/journal.pone.0154757

Biswas T, Dwivedi UN (2019) Plant triterpenoid saponins: biosynthesis, in vitro production, and pharmacological relevance. Protoplasma 256(6):1463–1486. https://doi.org/10.1007/s00709-019-01411-0

Bowen WS, Svrivastava AK, Batra L, Barsoumian H, Shirwan H (2018) Current challenges for cancer vaccine adjuvant development. Expert Rev Vaccines 17(3):207–215. https://doi.org/10.1080/14760584.2018.1434000

Brubaker SW, Bonham KS, Zanoni I, Kagan JC (2015) Innate immune pattern recognition: a cell biological perspective. Annu Rev Immunol 33:257. https://doi.org/10.1146/annurev-immunol-032414-112240

Buendía AJ, Cuello F, Del Rio L, Gallego MC, Caro MR, Salinas J (2001) Field evaluation of a new commercially available ELISA based on a recombinant antigen for diagnosing Chlamydophila abortus (Chlamydia psittaci serotype 1) infection. Vet Microbiol 78(3):229–239. https://doi.org/10.1016/S0378-1135(00)00298-4

Cao FQ, Yan MM, Liu YJ, Liu LX, Lu L, Wang H, Zhang C, Sun HF, Kong DL, Ma GL (2018) Photosensitizer-induced self-assembly of antigens as nanovaccines for cancer immunotherapy. Biomater Sci 6(3):473–477. https://doi.org/10.1039/c7bm01082f

Caproni E, Tritto E, Cortese M, Muzzi A, Mosca F, Monaci E, Baudner B, Seubert A, De Gregorio E (2012) MF59 and Pam3CSK4 boost adaptive responses to influenza subunit vaccine through an IFN type I-independent mechanism of action. J Immunol 188(7):3088–3098. https://doi.org/10.4049/jimmunol.1101764

Chauhan N, Tiwari S, Iype T, Jain U (2017) An overview of adjuvants utilized in prophylactic vaccine formulation as immunomodulators. Expert Rev Vaccines 16(5):491–502. https://doi.org/10.1080/14760584.2017.1306440

Chenthamara D, Subramaniam S, Ramakrishnan SG, Krishnaswamy S, Essa MM, Lin F, Qoronfleh MW (2019) Therapeutic efficacy of nanoparticles and routes of administration. Biomater Res 23:20

Chong G, Su R, Gu J, Yang Y, Zhang T, Zang J, Zhao Y, Zheng X, Liu Y, Ruan S, He R, Yin W, Li Y, Dong H, Li Y (2022) Catalytic nanovaccine for cancer immunotherapy: a NADPH oxidase-inspired Fe-polyphenol network nanovaccine for enhanced antigen cross-presentation. Chem Eng J 435:134993. https://doi.org/10.1016/j.cej.2022.134993

Cibulski SP, Silveira F, Mourglia-Ettlin G, Teixeira TF, dos Santos HF, Yendo AC, de Costa F, Fett-Neto AG, Gosmann G, Roehe PM (2016) Quillaja brasiliensis saponins induce robust humoral and cellular responses in a bovine viral diarrhea virus vaccine in mice. Comp Immunol Microbiol Infect Dis 45:1–8. https://doi.org/10.1016/j.cimid.2016.01.004

Cibulski S, Rivera-Patron M, Suárez N, Pirez M, Rossi S, Yendo AC, de Costa F, Gosmann G, Fett-Neto A, Roehe PM, Silveira F (2017) Leaf saponins of Quillaja brasiliensis enhance long-term specific immune responses and promote dose-sparing effect in BVDV experimental vaccines. Vaccine 36(1):55–65. https://doi.org/10.1016/j.vaccine.2017.11.030

Cordeiro AS, Patil-Sen Y, Shivkumar M, Patel R, Khedr A, Elsawy MA (2021) Nanovaccine delivery approaches and advanced delivery systems for the prevention of viral infections: from development to clinical application. Pharmaceutics 13(12):2091. https://doi.org/10.3390/pharmaceutics13122091

Dandashli EA, Zhao Q, Yitta S, Morefield GL, White JL, Hem SL (2002) Effect of thermal treatment during the preparation of aluminum hydroxide adjuvant on the protein adsorption capacity during aging. Pharm Dev Technol 7(4):401–406. https://doi.org/10.1081/PDT-120015042

Fernando IPS, Lee WW, Han EJ, Ahn G (2020) Alginate-based nanomaterials: fabrication techniques, properties, and applications. Chem Eng J 391:123823. https://doi.org/10.1016/J.CEJ.2019.123823

Fox CB, Haensler J (2013) An update on safety and immunogenicity of vaccines containing emulsion-based adjuvants. Expert Rev Vaccines 12(7):747–758. https://doi.org/10.1586/14760584.2013.811188

Fox CB, Van Hoeven N, Granger B, Lin S, Guderian JA, Hartwig A, Marlenee N, Bowen RA, Soultanov V, Carter D (2019) Vaccine adjuvant activity of emulsified oils from species of the Pinaceae family. Phytomedicine 64:152927. https://doi.org/10.1016/j.phymed.2019.152927

Garçon N, Vaughn DW, Didierlaurent AM (2012) Development and evaluation of AS03, an Adjuvant System containing α-tocopherol and squalene in an oil-in-water emulsion. Expert Rev Vaccines 11(3):349–366. https://doi.org/10.1586/erv.11.192

Gheibi Hayat SM, Darroudi M (2019) Nanovaccine: a novel approach in immunization. J Cell Physiol 234(8):12530–12536. https://doi.org/10.1002/jcp.28120

Gong N, Zhang Y, Teng X, Wang Y, Huo S, Qing G, Ni Q, Li X, Wang J, Ye X, Zhang T, Chen S, Wang Y, Yu J, Wang PC, Gan Y, Zhang J, Mitchell MJ, Li J, Liang XJ (2020) Proton-driven transformable nanovaccine for cancer immunotherapy. Nat Nanotechnol 15(12):1053–1064. https://doi.org/10.1038/s41565-020-00782-3

Guan X, Chen J, Hu Y, Lin L, Sun P, Tian H, Chen X (2018) Highly enhanced cancer immunotherapy by combining nanovaccine with hyaluronidase. Biomaterials 171:198–206. https://doi.org/10.1016/J.BIOMATERIALS.2018.04.039

Gupta RK (1998) Aluminum compounds as vaccine adjuvants. Adv Drug Deliv Rev 32(3): 155–172. https://doi.org/10.1016/S0169-409X(98)00008-8

Hamaguchi H (1926) Immunological notes on human prolactin. Acta Obstet Gynaecol Jpn 22(9): 995–1004

Jain KK (2008) Nanomedicine: application of nanobiotechnology in medical practice. Med Princ Pract 17(2):89–101. https://doi.org/10.1159/000112961

Jamiołkowska A, Kopacki M (2019) Natural compounds against plant pests and pathogens. In: Natural remedies for pest, disease and weed control. Elsevier, pp 55–63. https://doi.org/10.1016/B978-0-12-819304-4.00005-1

Jiexin L, Ren H, Sun Y, Liu G, Yang X, Qiu Q, Ding Y, Lovell JF, Zhang Y (2021) Magnetic metal micelles for enhanced delivery of self-immolating CD8+ T-cell epitopes for cancer immunotherapy. Chem Mater 33:167–209. https://doi.org/10.1007/978-88-7642-577-6_6

Karimi F, Alizadeh S, Alizadeh H (2018) Immunogenicity of multiwalled carbon nanotubes functionalized with recombinant protective antigen domain 4 toward development of a nanovaccine against anthrax. J Drug Deliv Sci Technol 47:322–329. https://doi.org/10.1016/j.jddst.2018.07.020

Kasturi SP, Skountzou I, Albrecht RA, Koutsonanos D, Hua T, Nakaya HI, Ravindran R, Stewart S, Alam M, Kwissa M, Villinger F, Murthy N, Steel J, Jacob J, Hogan RJ, García-Sastre A, Compans R, Pulendran B (2011) Programming the magnitude and persistence of antibody responses with innate immunity. Nature 470(7335):543–550. https://doi.org/10.1038/nature09737

Kawai T, Akira S (2010) The role of pattern-recognition receptors in innate immunity: update on toll-like receptors. Nat Immunol 11(5):373–384. https://doi.org/10.1038/ni.1863

Kelly SM, Larsen KR, Darling R, Petersen AC, Bellaire BH, Wannemuehler MJ, Narasimhan B (2021) Single-dose combination nanovaccine induces both rapid and durable humoral immunity and toxin neutralizing antibody responses against Bacillus anthracis. Vaccine 39(29): 3862–3870. https://doi.org/10.1016/J.VACCINE.2021.05.077

Kim GC, Kye CY, Yun HC (2019) The role of nanovaccine in cross-presentation of antigen-presenting cells for the activation of CD8+ T cell responses. Pharmaceutics 11(11):612. https://doi.org/10.3390/pharmaceutics11110612

Klein J, Ushio M, Burrell LS, Wenslow B, Hem SL (2000) Analysis of aluminum hydroxyphosphate vaccine adjuvants by 27Al MAS NMR. J Pharm Sci 89(3):311–321. https://doi.org/10.1002/(SICI)1520-6017(200003)89:3<311::AID-JPS3>3.0.CO;2-8

Kole S, Qadiri SSN, Shin SM, Kim WS, Lee J, Jung SJ (2019) PLGA encapsulated inactivated-viral vaccine: formulation and evaluation of its protective efficacy against viral haemorrhagic septicaemia virus (VHSV) infection in olive flounder (Paralichthys olivaceus) vaccinated by mucosal delivery routes. Vaccine 37(7):973–983. https://doi.org/10.1016/j.vaccine.2018.12.063

Kornuta CA, Bidart JE, Soria I, Gammella M, Quattrocchi V, Pappalardo JS, Salmaso S, Torchilin VP, Cheuquepán Valenzuela F, Hecker YP, Moore DP, Zamorano PI, Langellotti CA (2021) MANα1-2MAN decorated liposomes enhance the immunogenicity induced by a DNA vaccine against BoHV-1. Transbound Emerg Dis 68(2):587–597. https://doi.org/10.1111/tbed.13718

Kotsias F, Cebrian I, Alloatti A (2019) Antigen processing and presentation. Int Rev Cell Mol Biol 348:69–121. https://doi.org/10.1016/bs.ircmb.2019.07.005

Kwissa M, Nakaya HI, Oluoch H, Pulendran B (2012) Distinct TLR adjuvants differentially stimulate systemic and local innate immune responses in nonhuman primates. Blood 119(9): 2044–2055. https://doi.org/10.1182/blood-2011-10-388579

Le Moigne V, Robreau G, Mahana W (2009) Immune response to Chlamydophila abortus POMP91B protein in the context of different Pathogen Associated Molecular Patterns

(PAMP); role of antigen in the orientation of immune response. Toxins 1(2):59–73. https://doi.org/10.3390/toxins1020059

Leroux-Roels G (2009) Prepandemic H5N1 influenza vaccine adjuvanted with AS03: a review of the pre-clinical and clinical data. Expert Opin Biol Ther 9(8):1057–1071. https://doi.org/10.1517/14712590903066695

Liang Z, Zhu H, Wang X, Jing B, Li Z, Xia X, Sun H, Yang Y, Zhang W, Shi L, Zeng H, Sun B (2020) Adjuvants for coronavirus vaccines. Front Immunol 11:589833. https://doi.org/10.3389/fimmu.2020.589833

Lodaya RN, Kanitkar AP, Friedrich K, Henson D, Yamagata R, Nuti S, Mallett CP, Bertholet S, Amiji MM, O'Hagan DT (2019) Formulation design, optimization and in vivo evaluations of an α-tocopherol-containing self-emulsified adjuvant system using inactivated influenza vaccine. J Control Release 316:12–21. https://doi.org/10.1016/j.jconrel.2019.10.042

Longbottom D, Findlay J, Vretou E, Dunbar SM (1998) Immunoelectron microscopic localisation of the OMP90 family on the outer membrane surface of Chlamydia psittaci. FEMS Microbiol Lett 164(1):111–117. https://doi.org/10.1016/S0378-1097(98)00187-6

Luo Y, Wen YJ, Ding ZY, Fu CH, Wu Y, Liu JY, Li Q, He QM, Zhao X, Jiang Y, Li J, Deng HX, Kang B, Mao YQ, Wei YQ (2006) Immunotherapy of tumors with protein vaccine based on chicken homologous tie-2. Clin Cancer Res 12(6):1813–1819. https://doi.org/10.1158/1078-0432.CCR-05-1990

Luo L, Lucas RM, Liu L, Stow JL (2019) Signalling, sorting and scaffolding adaptors for Toll-like receptors. J Cell Sci 133(5):jcs239194. https://doi.org/10.1242/jcs.239194

Luo Z, He T, Liu P, Yi Z, Zhu S, Liang X, Kang E, Gong C, Liu X (2021) Self-adjuvanted molecular activator (SeaMac) nanovaccines promote cancer immunotherapy. Adv Healthc Mater 10(7):e2002080. https://doi.org/10.1002/adhm.202002080

Maina TW, Grego EA, Boggiatto PM, Sacco RE, Narasimhan B, Mcgill JL (2020) Applications of nanovaccines for disease prevention in cattle. Front Bioeng Biotechnol 8:608050. https://doi.org/10.3389/fbioe.2020.608050

McKee AS, MacLeod MKL, Kappler JW, Marrack P (2010) Immune mechanisms of protection: can adjuvants rise to the challenge? BMC Biol 8:1–10. https://doi.org/10.1186/1741-7007-8-34

Morein B, Sundquist B, Höglund S, Dalsgaard K, Osterhaus A (1984) Iscom, a novel structure for antigenic presentation of membrane proteins from enveloped viruses. Nature 308(5958):457–460. https://doi.org/10.1038/308457a0

Morel S, Didierlaurent A, Bourguignon P, Delhaye S, Baras B, Jacob V, Planty C, Elouahabi A, Harvengt P, Carlsen H, Kielland A, Chomez P, Garçon N, Van Mechelen M (2011) Adjuvant system AS03 containing α-tocopherol modulates innate immune response and leads to improved adaptive immunity. Vaccine 29(13):2461–2473. https://doi.org/10.1016/j.vaccine.2011.01.011

Morelli AB, Becher D, Koernig S, Silva A, Drane D, Maraskovsky E (2012) ISCOMATRIX: a novel adjuvant for use in prophylactic and therapeutic vaccines against infectious diseases. J Med Microbiol 61(PART7):935–943. https://doi.org/10.1099/jmm.0.040857-0

Narayanan KB, Park HH (2015) Toll/interleukin-1 receptor (TIR) domain-mediated cellular signaling pathways. Apoptosis 20(2):196–209. https://doi.org/10.1007/s10495-014-1073-1

Nierkens S, Tel J, Janssen E, Adema GJ (2013) Antigen cross-presentation by dendritic cell subsets: one general or all sergeants? Trends Immunol 34(8):361–370. https://doi.org/10.1016/j.it.2013.02.007

Nunes De Carvalho Lima E, Sobhie Diaz R, Justo JF, Roberto J, Piqueira C (2021) Advances and perspectives in the use of carbon nanotubes in vaccine development. Int J Nanomedicine 16:5411–5435. https://doi.org/10.2147/IJN.S314308

O'Hagan DT, Fox CB (2015) New generation adjuvants—from empiricism to rational design. Vaccine 33(S2):B14–B20. https://doi.org/10.1016/j.vaccine.2015.01.088

O'Hagan DT, Ott GS, Van Nest G, Rappuoli R, Del Giudice G (2013) The history of MF59® adjuvant: a phoenix that arose from the ashes. Expert Rev Vaccines 12(1):13–30. https://doi.org/10.1586/erv.12.140

O'Hagan DT, Lodaya RN, Lofano G (2020) The continued advance of vaccine adjuvants—'we can work it out'. Semin Immunol 50:101426. https://doi.org/10.1016/j.smim.2020.101426

Ong GH, Lian BSX, Kawasaki T, Kawai T (2021) Exploration of pattern recognition receptor agonists as candidate adjuvants. Front Cell Infect Microbiol 11:1–17. https://doi.org/10.3389/fcimb.2021.745016

Papadopoulou K, Melton RE, Leggett M, Daniels MJ, Osbourn AE (1999) Compromised disease resistance in saponin-deficient plants. Proc Natl Acad Sci U S A 96(22):12923–12928. https://doi.org/10.1073/pnas.96.22.12923

Plotkin S (2014) History of vaccination. Proc Natl Acad Sci U S A 111(34):12283–12287. https://doi.org/10.1073/pnas.1400472111

Podda A (2001) The adjuvanted influenza vaccines with novel adjuvants: experience with the MF59-adjuvanted vaccine. Vaccine 19(17–19):2673–2680. https://doi.org/10.1016/S0264-410X(00)00499-0

Podda A, Del Giudice G (2003) MF59-adjuvanted vaccines: increased immunogenicity with an optimal safety profile. Expert Rev Vaccines 2(2):197–204. https://doi.org/10.1586/14760584.2.2.197

Renu S, Markazi AD, Dhakal S, Lakshmanappa YS, Gourapura SR, Shanmugasundaram R, Senapati S, Narasimhan B, Selvaraj RK, Renukaradhya GJ (2018) Surface engineered polyanhydride-based oral Salmonella subunit nanovaccine for poultry. Int J Nanomedicine 13:8195–8215. https://doi.org/10.2147/IJN.S185588

Renu S, Feliciano-Ruiz N, Ghimire S, Han Y, Schrock J, Dhakal S, Patil V, Krakowka S, Renukaradhya GJ (2020) Poly(I:C) augments inactivated influenza virus-chitosan nanovaccine induced cell mediated immune response in pigs vaccinated intranasally. Vet Microbiol 242: 108611. https://doi.org/10.1016/J.VETMIC.2020.108611

Rivera-Patron M, Moreno M, Baz M, Roehe PM, Cibulski SP, Silveira F (2021) Iscom-like nanoparticles formulated with Quillaja brasiliensis saponins are promising adjuvants for seasonal influenza vaccines. Vaccine 9(11):1–18. https://doi.org/10.3390/vaccines9111350

Rosana S-V, Mercedes P, África G-F (2020) Polymeric nanostructure vaccines: applications and challenges. Exp Opin Drug Deliv 17(7):1007–1023. https://doi.org/10.1080/17425247.2020.1776259

Shen L, Higuchi T, Tubbe I, Voltz N, Krummen M (2013) A trifunctional dextran-based nanovaccine targets and activates murine dendritic cells, and induces potent cellular and humoral immune responses in vivo. PLoS One 8(12):80904. https://doi.org/10.1371/journal.pone.0080904

Shi S, Zhu H, Xia X, Liang Z, Ma X, Sun B (2019) Vaccine adjuvants: understanding the structure and mechanism of adjuvanticity. Vaccine 37(24):3167–3178. https://doi.org/10.1016/j.vaccine.2019.04.055

Silindir-Gunay M, Ozer AY (2018) Liposomes and micelles as nanocarriers for diagnostic and imaging purposes. In: Design of nanostructures for theranostics applications. Elsevier, pp 305–340. https://doi.org/10.1016/B978-0-12-813669-0.00008-7

Sivakumar SM, Safhi MM, Kannadasan M, Sukumaran N (2011) Vaccine adjuvants—current status and prospects on controlled release adjuvancity. Saudi Pharm J 19(4):197–206. https://doi.org/10.1016/j.jsps.2011.06.003

Stassijns J, Bollaerts K, Baay M, Verstraeten T (2016) A systematic review and meta-analysis on the safety of newly adjuvanted vaccines among children. Vaccine 34(6):714–722. https://doi.org/10.1016/j.vaccine.2015.12.024

Sun HX, Xie Y, Ye YP (2009) Advances in saponin-based adjuvants. Vaccine 27(12):1787–1796. https://doi.org/10.1016/j.vaccine.2009.01.091

Sun Z, Wang W, Wang R, Duan J, Hu Y, Ma J, Zhou J, Xie S, Lu X, Zhu Z, Chen S, Zhao Y, Xu H, Wang C, Yang XD (2010) Aluminum nanoparticles enhance anticancer immune response induced by tumor cell vaccine. Cancer Nanotechnol 1(1–6):63–69. https://doi.org/10.1007/s12645-010-0001-5

Tao W, Ni D, Liu G, Huang P, Mou X, Yang K, Maiti D, Tong X (2019) Carbon-based nanomaterials for biomedical applications: a recent study. Front Pharmacol 9:1401. https://doi.org/10.3389/fphar.2018.01401

Tizard IR (2021) Adjuvants and adjuvacanticity. Vaccines for Veterinarians 2021:75–86.e1. https://doi.org/10.1016/b978-0-323-68299-2.00016-2

Turk SP, Lumbard K, Liepshutz K, Williams C, Hu L, Dardick K, Wormser GP, Norville J, Scavarda C, McKenna D, Follmann D, Marques A (2019) Post-treatment Lyme disease symptoms score: developing a new tool for research. PLoS One 14(11):1–17. https://doi.org/10.1371/journal.pone.0225012

Vincken JP, Heng L, de Groot A, Gruppen H (2007) Saponins, classification and occurrence in the plant kingdom. Phytochemistry 68(3):275–297. https://doi.org/10.1016/j.phytochem.2006.10.008

Vretou E, Giannikopoulou P, Longbottom D, Psarrou E (2003) Antigenic organization of the N-terminal part of the polymorphic outer membrane proteins 90, 91A, and 91B of Chlamydophila abortus. Infect Immun 71(6):3240–3250. https://doi.org/10.1128/IAI.71.6.3240-3250.2003

Vyas S, Dhoble S, Ghodake V, Patravale V (2020) Xyloglucan based mucosal nanovaccine for immunological protection against brucellosis developed by supercritical fluid technology. Int J Pharm X 2:100053. https://doi.org/10.1016/J.IJPX.2020.100053

Wagner DA, Kelly SM, Petersen AC, Peroutka-Bigus N, Darling RJ, Bellaire BH, Wannemuehler MJ, Narasimhan B (2019) Single-dose combination nanovaccine induces both rapid and long-lived protection against pneumonic plague. Acta Biomater 100:326–337. https://doi.org/10.1016/j.actbio.2019.10.016

Wang Y, Zhang S, Li H, Wang H, Zhang T, Hutchinson MR, Yin H, Wang X (2020) Small-molecule modulators of Toll-like receptors. Acc Chem Res 53(5):1046–1055. https://doi.org/10.1021/acs.accounts.9b00631

Xiao B, Li D, Xu H, Zhou X, Xu X, Qian Y, Yu F, Hu H, Zhou Z, Liu X, Gao J, Slater NKH, Shen Y, Tang J (2021) An MRI-trackable therapeutic nanovaccine preventing cancer liver metastasis. Biomaterials 274:120893. https://doi.org/10.1016/j.biomaterials.2021.120893

Xu J, Wang H, Xu L, Chao Y, Wang C, Han X, Dong Z, Chang H, Peng R, Cheng Y, Liu Z (2019) Nanovaccine based on a protein-delivering dendrimer for effective antigen cross-presentation and cancer immunotherapy. Biomaterials 207:1–9. https://doi.org/10.1016/j.biomaterials.2019.03.037

Yam KK, Gupta J, Winter K, Allen E, Brewer A (2015) AS03-adjuvanted, very-low-dose influenza vaccines induce distinctive immune responses compared to unadjuvanted high-dose vaccines in BALB/c mice. Front Immunol 6:207. https://doi.org/10.3389/fimmu.2015.00207

Yam KK, Brewer A, Bleau V, Beaulieu É, Mallett CP, Ward BJ (2017) Low hemagglutinin antigen dose influenza vaccines adjuvanted with AS03 alter the long-term immune responses in BALB/c mice. Hum Vaccin Immunother 13:561–571

Yau KP, Schulze DG, Johnston CT, Hem SL (2006) Aluminum hydroxide adjuvant produced under constant reactant concentration. J Pharm Sci 95(8):1822–1833. https://doi.org/10.1002/jps.20692

Yendo ACA, de Costa F, Cibulski SP, Teixeira TF, Colling LC, Mastrogiovanni M, Soulé S, Roehe PM, Gosmann G, Ferreira FA, Fett-Neto AG (2016) A rabies vaccine adjuvanted with saponins from leaves of the soap tree (Quillaja brasiliensis) induces specific immune responses and protects against lethal challenge. Vaccine 34(20):2305–2311. https://doi.org/10.1016/j.vaccine.2016.03.070

Zhang N, Channappanavar R, Ma C, Wang L, Tang J, Garron T, Tao X, Tasneem S, Lu L, Tseng CTK, Zhou Y, Perlman S, Jiang S, Du L (2016) Identification of an ideal adjuvant for receptor-binding domain-based subunit vaccines against Middle East respiratory syndrome coronavirus. Cell Mol Immunol 13(2):180–190. https://doi.org/10.1038/cmi.2015.03

Zhang L, Xu L, Wang Y, Liu J, Tan G, Huang F, He N, Lu Z (2022) A novel therapeutic vaccine based on graphene oxide nanocomposite for tumor immunotherapy. Chin Chem Lett 33:4089. https://doi.org/10.1016/J.CCLET.2022.01.071

Chapter 13
Nanotechnology in Vaccine Development to Combat Bacterial Livestock Diseases

Rajat Varshney, Dhruv N. Desai, Sourabh Sulabh, Ritu Varshney, and Mayukh Ghosh

Abstract Prophylaxis is undoubtedly the best strategy for disease management, particularly to combat acute and per acute fatal diseases but effective commercialized vaccines are unavailable for several economically important livestock diseases. And if available, the conventional inactivated vaccines often suffer from the limitations such as high antigenic biomass requirement, denaturation of conformational epitopes, and residual virulence, whereas live attenuated vaccines carry the drawbacks of reversion to virulence and cold-chain maintenance. Although new-generation subunit vaccines have eliminated the concern of safety but often require adjuvantation and multiple doses to elicit satisfactory level of protective immune response. Classical adjuvants such as alum and Freund's complete and incomplete adjuvants (CFA and IFA) also have some limitations such as toxicity, hyperreactivity, eliciting strong humoral but poor cellular immunity, and others. At this juncture, incorporation of suitable nanoparticles (NPs) in vaccine formulations has extended multi-arrayed benefits. As vaccine delivery system, nano-encapsulation of antigens has extended several advantages in terms of reducing antigenic biomass requirement, stabilization of labile protein or nucleic acid cargo, targeted delivery and sustained release of antigen(s), improved pharmacokinetic

R. Varshney
Department of Veterinary Microbiology, Faculty of Veterinary and Animal Sciences, RGSC, Banaras Hindu University, Mirzapur, Uttar Pradesh, India

D. N. Desai
Department of Pathobiology, School of Veterinary Medicine, University of Pennsylvania, Philadelphia, PA, USA

S. Sulabh (✉)
Department of Animal Science, Kazi Nazrul University, Asansol, West Bengal, India

R. Varshney
Biological Engineering, Indian Institute of Technology, Gandhinagar, Gujarat, India

M. Ghosh
Department of Veterinary Physiology and Biochemistry, Faculty of Veterinary and Animal Sciences, RGSC, Banaras Hindu University, Mirzapur, Uttar Pradesh, India

LLR University of Veterinary and Animal, ICAR-National Research Centre on Equines, *Nanotechnology Theranostics in Livestock Diseases and Management*, Livestock Diseases and Management,
https://doi.org/10.1007/978-981-16-1610-5_13

properties, biodegradability and biocompatibility, etc. Polymeric NPs, liposomes, and dendrimers have already been employed in nanovaccine formulation as delivery vehicle against various bacterial infections including *Salmonella, Brucella, Mycobacterium, E. coli,* etc. with promising outcome. Various NPs have also been used as nano-adjuvants to potentiate different multi-subunit/fusion antigens such as aluminum hydroxide NPs with *Bacillus anthracis* and *Mycobacterium tuberculosis* antigen EsxV, oil in water adjuvants MF59 and AS03, and liposome-based adjuvants AS01E and CAF01 with *Mycobacterial* fusion antigens. These nano-adjuvants often elicit strong humoral as well as cell-mediated immune response which is one of the notable advantages of using NPs in vaccine formulations. Incorporation of NPs can also increase the room temperature stability of vaccines to surmount the necessity of cold-chain maintenance. Although nanovaccines have yet to full-bloom but already ushered enough promise toward a bright prospect in near future.

Keywords Prophylaxis · Nanovaccines · Nano-adjuvants · Vaccine delivery system · Humoral immunity · Cell-mediated immunity · Microbial diseases · Animals

13.1 Introduction

The vaccine development and vaccination technology in the modern world has been one of the most instrumental achievements in disease management (Hilleman 2000; Ada 2001). Vaccination is considered the better possible solution to combat acute and per acute fatal diseases rather than directed therapy (Varshney et al. 2020). It has not only resulted in the eradication of deadly diseases such as smallpox, but it has also significantly reduced deaths and disability induction by preventing diseases at various stages of life (Andre 2003).

The requirement of high antigenic biomass, denaturation of conformational epitopes, residual virulence, etc. are primary problems in chemically inactivated vaccine production (Jain et al. 2016). Conversely, the major obstacles in the use of live attenuated vaccine are reversion to virulence and cold-chain maintenance (Solanki et al. 2021). The safety concern can be resolved by new-generation subunit vaccines, but the inclusion of adjuvant in subunit vaccines and other vaccines is necessary to elicit ample protective immune response (Singh et al. 2023; Panda 2012).

Adjuvants (from Latin "adjuvare," to help) are substances that, when mixed with an antigen and injected with it, enhance the immunogenicity of that antigen (Kindt et al. 2007). Adjuvants are often used to boost the immune response in case of low immunogenicity or low antigenic biomass (Kindt et al. 2007). The concept of adjuvant was proposed at the beginning of the twentieth century when traditional vaccines failed to produce an effective immune response with diphtheria and tetanus purified toxoids (Shi et al. 2019). The components of adjuvants have gone through a process of changes from natural ingredients to artificial synthetic compounds, and the discovery of the adjuvant effect of aluminum salts in 1926 is a milestone (Shi et al. 2019).

With advancement in immunology and molecular biology, it has been possible to develop newer and more efficacious vaccines. In current scenario, nanotechnology-based formulations offer numerous advantages over conventional vaccine formulations. The nanocarriers such as nano-aluminum salts, nano-polymeric particles, liposomes, lipid nanoparticles (NPs), and virosomes not only protect vaccine antigen from degradation by improving their stability but ensure targeted delivery and efficient antigenic presentation (Panda 2012). Nanotechnology has been defined as "the design, characterization, production, and application of structures, devices, and systems by controlled manipulation of size and shape at the nano-meter scale that produce structures, devices, and systems with at least one novel or superior characteristic and property." (Bawa 2007).

Nanotechnology plays an imperative role in the delivery of drugs and vaccines, where systems like nanomaterials in the form of particles, emulsion, liposomes, micelles, dendrimers, etc. usually give higher therapeutic indices in comparison to traditional adjuvants (Bawarski et al. 2008; Xiang et al. 2006; Peek et al. 2008). These small carrier systems have the tendency and efficacy to enter APCs and deliver the optimum amount of antigen (Look et al. 2010). Further, adjuvant activity vis a vis immune responses against specific diseases can be escalated by using these carrier systems in two alternate ways. The first is by co-entrapping an immunomodulator and the antigen, and the second is by surface-engineering the antigen or carrier particles to activate the innate as well as the adaptive immune system. The nanotechnology particulate system entraps candidate vaccines along with nano-adjuvant or nano-immunomodulator with dendritic cell activator, supporting an opportunity for vaccine development. It is especially useful to combat intracellular infections where both T cell and B cell-mediated response are required for protective immunity. Thus, immunizations with suitable nano-systems provide a practical solution. Nanoparticle delivery systems can also be used for genetic immunization and for delivery of antigens along with immunomodulators and toll-like receptors (TLRs). With the increase in the success of vaccination against several infectious diseases, focus is being turned toward modern vaccine strategies for chronic noninfectious diseases like allergy and cancer, along with other inflammatory diseases and the problem of contraception. In the present scenario and with more research, nanotechnology-based vaccine delivery systems can be a game changer in developing new and efficient vaccine strategies.

13.2 Aluminum Hydroxide NPs

Aluminum hydroxide is the most common vaccine adjuvant and has been used in various human and animal vaccines. The safety profile of aluminum hydroxide is very high, but it has also been reported that it can induce only weak to moderate antigen-specific antibody responses from the immune system (Li et al. 2014). The actual size of the primary fibers of aluminum hydroxide is in nanometers. However, when they are dispersed in an aqueous solution, aluminum hydroxide forms aggregated particulates of 1–20 μm in size (Romero Mendez et al. 2007; Hem and Hogenesch 2007). Another type of aluminum salt, the aluminum phosphate, is also used as an adjuvant, the size of primary particle of which is around 20 nm but it also shows similar properties of aggregation to that of aluminum hydroxide when mixed in aqueous solution (Romero Mendez et al. 2007; Hem and Hogenesch 2007). NPs around or less than 200 nm as vaccine or antigen carriers have more potent adjuvant activity than larger microparticles (Li et al. 2014).

Till now, the efficacy of aluminum nano-adjuvants has been tested in few studies. In one such study, researchers used ovalbumin (OVA) and *Bacillus anthracis* as protective antigens to induce inflammatory responses by adsorbing them on small aluminum hydroxide NPs, which showed a stronger OVA-specific antibody immune response in comparison to OVA-adsorbed large aluminum hydroxide, whose size was bigger (microparticles) (Li et al. 2014). The untoward complications just after immunization were also less in aluminum hydroxide NPs in comparison to aluminum hydroxide microparticles. The potent adjuvant activity of the aluminum hydroxide NPs was ascribed by enhanced uptake of the adsorbed antigens by antigen-presenting cells. The overall efficiency has found to be dependent on the particle size of the aluminum hydroxide adjuvant (Li et al. 2014). Later the same concept was also used for the administration of *Mycobacterium tuberculosis* antigen EsxV in mice through subcutaneous injection and the efficacy of the aluminum hydroxide NPs were found to be significantly better in inducing release of IFN-γ and activating Th-1 response from the immune system when compared to the antigen alone (Amini et al. 2016).

The nano-aluminum hydroxide as a potential adjuvant for delivery of antigens through the pulmonary route has also been evaluated. Alum microparticles were converted into alum NPs (aluminum oxide) of about 30 nm, which were used as pulmonary vaccine adjuvant delivery systems for antigens and elicited a strong immune response (in the form of IgG and IgA in saliva, nasal, respiratory, and vaginal fluids), showing prospects of its use in future vaccines as a tool for simplified delivery systems (Wang et al. 2020). The aforesaid instances are explanatory regarding the potential of alum-based nano-adjuvant and pave way for the development of an efficient adjuvant as well as delivery vehicle.

13.3 Oil in Water Emulsions

Emulsions can be defined as the mixture of two immiscible liquid phases whose stability is maintained by using mechanical shear and surfactant (Kale and Deore 2017). The most commonly used oils in water adjuvants are MF59 (Novartis) and AS03 (GlaxoSmithKline). Both of these are squalene-based emulsions, and the droplet size of both of them is around 160 nm. MF59 and AS03 have already been licensed for human (Gasparini et al. 2010; Lee and Nguyen 2015). MF59 and AS03 can be considered nano-adjuvants as they are less than 200 nm in size. Though both are squalene based, MF59 is composed of squalene and two surfactants, Tween

80 and Span 85, whereas AS03 is composed of squalene, polysorbate 80, and α-tocopherol (vitamin E) (Wilkins et al. 2017). MF59 can recruit local monocytes, macrophages, and dendritic cells for antigen uptake and can activate CD4 T cells (O'Hagan et al. 2012; Calabro et al. 2011). On the other hand, AS03 has been shown to have nonspecific activation of the immune system in mice (Morel et al. 2011). MF-59 and aluminum hydroxide-adjuvanted vaccine has been tested in mice to combat *Staphylococcus aureus* infection. The vaccine induced persistent T cell response as well as antigen-specific humoral response and afforded protection in mice up to 4-month post-challenge (Monaci et al. 2015). The study of oil in water emulsion based adjuvant show potential for development of effective subunit vaccine; further investigation into the property and side effects should be considered before wider application.

13.4 Nano-Liposomes

Liposomes are composed of a double layer of phospholipids, forming a sphere when present within water as a result of self-assembly under the driving force of hydrophobicity (Wang et al. 2019). Liposomes are made of various kinds of amphiphilic phospholipids, which can also be combined with other lipids like cholesterol to increase membrane stability. Another way to increase the stabilization of liposomes is to use PEGylation. PEGylation prevents liposomes from coming together so that fusion or coagulation can be avoided. Also, PEGylation prevents liposomes from non-specific binding with plasma proteins but allows binding with other specific plasma proteins, increasing the NP circulation time in vivo (Schottler et al. 2016).

Freeze-drying of liposomes in the presence of a liposome cryoprotectant like sugar can be an alternative strategy to increase the stability of the liposomal structure (Barenholz 2012). Among different types of nanocarriers, liposomes have been frequently used as a major drug delivery system or vaccine adjuvant delivery system. It has been reported that according to the size of liposomes, different types of immune reactions can be observed when they are used as a delivery system or adjuvant. Subcutaneous injection or oral inoculation of smaller liposomes (less than 150 nm) influences the development of Th-2 type of response, whereas larger liposome particles (more than 200 nm) promote more IFN-γ and Th-1 type of response (Mann et al. 2009). Phospholipids are the major constituent of liposomes which are known to play important mediators of innate and adaptive immune system (Tretiakova and Vodovozova 2022). Liposomes have the inbuilt capacity to activate the innate immune system but can also effectively present the encapsulated antigen to APCs. According to their physiochemical properties, liposomes can stimulate both naive T-lymphocytes into CD4+ or CD8+ cells, as well as B-lymphocytes to induce the humoral immune response (Tretiakova and Vodovozova 2022).

The particles smaller than 20 nm may be distributed and dispersed into the tissues and then enter into the systemic circulation of the body, resulting in a minute chance of lymphatic uptake (Wang et al. 2019). However, bigger particles of more than

100 nm have better chances of getting trapped in tissues, thus creating depot effects at the site of injection so as to be easily identified and picked up by APCs. Subsequently, the mature APCs will present the requisite epitopes and migrate to draining lymph nodes for further adaptive immune responses (Wang et al. 2019). Monophosphoryl lipid A, which is a detoxified part of an endotoxin lipid A along with a TLR4 ligand, has been identified as a safe and efficacious lipid-based adjuvant which has been used in the development of emerging vaccines to combat diseases such as malaria, leishmania, and some other infections, as well as several cancers (Alving et al. 2012; Vandepapelière et al. 2008).

GlaxoSmithKline (GSK) developed a subunit vaccine [M72 (recombinant fusion antigen) + AS01E (liposome-based adjuvant system)] that induced a strong T cell-mediated response against *M. tuberculosis* (Minakshi et al. 2019). Another subunit vaccine H56/CAF01 [fusion antigens (Ag85B, ESAT-6, and Rv2660) + cationic liposome CAF01 (an immunomodulatory glycolipid adjuvant)], also conferred protection against tuberculosis by inducing localized CD4 T cell response in lung parenchyma (Minakshi et al. 2019). Liposome (CAF01)-based vaccine carrying recombinant *Mycobacterium avium paratuberculosis* proteins induced age-independent humoral response and age-dependent IFNγ response in claves. Liposome-based vaccine carrying *Salmonella enteritidis* fimbrial protein, SEF14 conferred high IgG, IgA in intestinal mucus as well as serum and limit the infection of *S. enteritidis* in chickens (Schwendener 2014).

13.5 Immune Stimulating Complexes (ISCOM)

ISCOM is an important lipid-based adjuvant or vaccine delivery system. Like the liposomes, the lipid particles are made up of cholesterol and phospholipids but also contain some protein antigens and saponin Quil A (Skene and Sutton 2006). They are 40–50 nm large and are cage-like structure (Sun et al. 2009). The antigens carried by the ISCOM are presented to both Major Histocompatibility Complex I and II (MHC I and MHC II). ICOSAMATRIX is also a type of ISCOM that does not contain any antigen (Lee and Nguyen 2015). The higher affinity between saponin and cholesterol, which are part of ISCOM and ICOSAMATRIX, is responsible for their higher stability (García and Sanctis 2014). They can elicit a balanced Th1 and Th2 response from the immune system, maintaining a long-lasting antibody titre and effective cell-mediated immunity. The disadvantage of using these carrier molecules is the presence of the saponin, which can cause toxicity. Therefore, saponin-rich fraction of *Asparagus adscendens* Roxb. fruit (AA-SRF) can help to minimize the toxicity as an alternative source to standard saponin Quil A (Singh et al. 2023). Both ISCOM and ICOSAMATRIX have been experimentally used for veterinary purposes. A single immunization with *Mycoplasma gallisepticum* immunostimulating complexes (ISCOMS) has conferred protective immunity in chicken (Sundquist et al. 1996)

13.6 Toll-Like Receptors (TLRs)-Based Adjuvants and Nanovaccines

TLRs, which are pattern recognition receptors, are crucial for microbial detection. TLR-based adjuvants can augment immune response to vaccine by multiple ways such as enhancement of phagocytosis, antigenic processing and presentation via up-regulation of MHC and co-stimulatory molecules, cross-priming/presentation, T-cell survival and memory response, etc. Principally, boosting of innate immune response through TLR-agonists evidently strengthen the adaptive immune response too, therefore potentiating the efficacy of nano-vaccines containing TLR -based adjuvants. Many adjuvants are built on the PAMPs that bind to TLRs. TLRs on cell surfaces, such as TLR1, TLR2, TLR4, TLR5, TLR6, and TLR11, are capable of detecting extracellular antigens, whereas TLRs found inside of cells, such as TLR3, TLR7, TLR8, TLR9, and TLR10, are capable of detecting internal intruders. TLR 13 has been found, although it's not present in humans.

The activation of transcription factors like AP-1, NF-κB, and interferon regulatory factors begins with the stimulation of TLRs by the relevant PAMPs (IRFs). The creation of interferons (IFNs), pro-inflammatory cytokines, and effector cytokines that control the adaptive immune response are only a few of the cellular responses triggered by TLR signaling. NF-κB, MAP kinase, and IRF3 are three important transcription factors that are activated by all TLRs except TLR 3. In addition to the activation, enzyme caspase-1, NF-κB and MAPK activate the genes for the three important proteins IL-1, IL-6, and TNF-alpha. Another transcription factor, IRF3, which is activated by the TRIF complex, turns on the gene for the antiviral cytokine Interferon beta. TLR ligands include synthetic substances that cause professional APC to mature and become activated as well as induce the release of inflammatory cytokines and chemokines.

TLR-7 and TLR-7/8 are ligands for the small molecule nucleoside analogues imiquimod and resiquimod, respectively. Imiquimod is approved to treat HPV and basal cell carcinoma in addition to treating leishmaniasis (BCC) (Vidal and Alomar 2008; Hanna et al. 2016). Although the precise mechanism of action of imiquimod is unknown, it is believed that its activity as a TLR-7 agonist mimics a microbial antigen and induces the expression of various cytokines that includes IL-1, IL-6, IL-12, IFN-α and TNF-α, which stimulate or augment both the innate immune system and the cell-mediated immune response as well as enhances migration of Langerhans' cells from the dermis to local lymph nodes. Although inclusion of TLR-based adjuvants in nano-vaccines are mostly targeted towards viral diseases and cancer but the same strategy can also be applied to combat bacterial diseases.

13.7 Polymeric NPs-Based Vaccine

Various polymers, namely, poly lactic-co-glycolic acid (PLGA), chitosan, 1,8-bis-(p-carboxyphenoxy)-3,6-dioxaoctane (CPTEG), 1,6-bis-(p-carboxyphenoxy) hexane (CPH), and sebacic acid polymers, etc. have been extensively used to formulate nanovaccines against diverse pathogens. Polymeric NP-based vaccine facilitates stabilization of labile protein antigens and efficacious adjuvanticity.

Nanovaccine containing oligopolysaccharide (OPS) antigen of *Brucella melitensis* and PLGA NPs has significantly enhanced IgG and IgM antibody level, opsonophagocytosis effect and post-challenged protection in mice (Maleki et al. 2019). Balb/c mice immunized with the subunit nanovaccine containing quillaja saponin adjuvanted antigenic acellular lipopolysaccharides from *B. abortus* (S19) and mucoadhesive tamarind seed polymer (xyloglucan)-based NPs via intranasal route elicited prolonged cell-mediated and humoral (increase both Ig A and Ig G) immunity. Moreover, a mixed Th1/Th2 type of immunity with polarized Th1 response was observed suggesting the potential of polymeric NPs as efficient vaccine delivery system (Vyas et al. 2020).

Intranasal immunization of White Leghorn chickens with nanovaccine comprised of chitosan-DNA NPs carrying a gene for the major structural protein FlaA, conferred significantly increased levels of intestinal Ig A and serum anti-*Campylobacter jejuni* IgG response. The chitosan acted as natural mucus absorption enhancer that facilitated transgenic DNA expression in chicken nasopharynx (Huang et al. 2010).

An alum-adjuvanted pneumococcal surface protein A (PspA)-based polyanhydride [comprised of 1,8-bis-(p-carboxyphenoxy)-3,6-dioxaoctane (CPTEG), 1,6-bis-(p-carboxyphenoxy) hexane (CPH), and sebacic acid polymers] nanovaccine has induced protective immunity against *Streptococcus pneumoniae* in mice. The nanovaccine has significantly reduced the requirement of antigenic biomass and precluded the cold-chain maintenance via improving thermostability (Wagner-Muñiz et al. 2018).

The nano-vaccine comprised of cyclic di-GMP, polyanhydride [comprised of CPTEG and CPH] NPs has induced high titer of neutralizing antibody against the protective antigen of *Bacillus anthracis* and long-lasting humoral immunity in mice (Kelly et al. 2021).

Nanovaccine comprised of chitosan (CS) or ascorbate chitosan (AsCS) NPs and outer membrane or flagellar antigen extracted from *E. coli* serogroups O1 and O78 has been employed for immunization in specific pathogens free (SPF) 3-weeks old broiler chickens. The inclusion of CS NPs, particularly to encapsulate the antigens in vaccine formulation has improved vaccine efficacy to control colibacillosis (Mohammed et al. 2020).

The oral administered nano-vaccine containing outer membrane proteins (OMPs) and flagellin (F) antigen of *Salmonella* and CS NPs up-surged the expression of TLRs, and Th1 and Th2 cytokines in chickens. Consequently, up-regulation of mucosal IgA and systemic IgY antibodies along with reduction intestinal *Salmonella* load have also been observed (Renu et al. 2020). The similar use of CS NPs along with OMPs and F antigen in nano-vaccine has curtailed the *Salmonella enterica serovar Enteritidis* infection in broilers (Han et al. 2020).

13.8 Dendrimers

The word dendrimer has been derived from the Greek word "dendron," which literally means tree. Dendrimers are versatile, derivatisable, well-defined, compartmentalized chemical polymers with sizes and physicochemical properties resembling those of biomolecules, for example, proteins (Boas and Heegaard 2004). These are made by either a divergent or convergent route of synthesis via a succession of well-directed polymerization reactions (Zeng and Zimmerman 1997), resulting in the genesis of thoroughly branched globular macromolecules comparable to a globular protein. Dendrimers are thus generated from smaller monomer units where the units polymerize from a central core (Zeng and Zimmerman 1997). Nanoscale-sized dendrimers (10–100 nm) look very similar to those of a ball where they express their capacity to entrap molecules within the spaces of the branched structure, expressing the multiple functional groups on their surface (Panda 2012). After their discovery in the last century, the importance and use of these structures found several applications. One of which was for biological applications like drug delivery systems and as vaccine adjuvant. With more research, the potential of dendrimers to serve as adjuvants for delivering antigens related to cancer cure has become quite evident. Thus, dendrimers can be used for making anticancer drugs, they can act as efficient drug delivery system and as dendritic sensors, along with their use in phototherapy for precise anticancer targeting, gene therapy and may more (Chowdhury et al. 2022). Antigens can be loaded onto dendrimers by either complexation or by the formation of chemical bonds at the end branch points. PAMAM (polyamidoamine) has been recognized as an important dendrimer which is formed by the repetitive addition of branching units to ammonia or ethylene diamine (Panda 2012). Recent research has identified specific types of dendrimers with potential use in cancer nanotheranostics. These include glycodendrimers, phosphorous dendrimers, and several different dendrimers (PAMAM being one of them) with the ability to activate the immune system (Gao et al. 2021). Mesoporous silica nanoparticle dendrimers have been studied as adjuvants for developing foot-and-mouth disease vaccines for ruminants and have been found to be highly stable and safe for use in subunit and VLP-based vaccines (Liu et al. 2019). The branched chain of dendrimers can be altered or modified, which can be helpful for making some specific adaptations towards specific uses (Panda 2012). In a similar way, immunodominant epitope of bacteria which correlate with the protection can be incorporated and delivered through such strategy.

13.9 Mesoporous Silica NPs-Based Vaccine

High-porosity of these NPs extends ample scope for improved loading capacity of the entrapped cargo. Nanovaccine containing extracellular vesicles (EVs) of *S. aureus* coated with indocyanine green (ICG)-entrapped into magnetic mesoporous silica nanoparticles (MSN) has been used for immunization in male C57BL/6 mice. During photothermal therapy, ICG molecules can absorb photon to produce heat; therefore, endolysosomal lysis is enhanced to promote proteasome-dependent antigenic processing. Upon laser irradiation, the multi-antigenic vaccine has modulated antigen presentation pathways and improved immune responses. Moreover, this vaccine also prevented superficial infection and complications associated with drug-resistant *S. aureus* invasion (Chen et al. 2020).

13.10 Silk Fibroin (SF)

Silk fibroin (SF) is a completely natural protein polymer. Disulfide-bound heavy (~370 kDa) and light chains (~25 kDa) make up SF (Inoue et al. 2000). The core structure is made up of 12 hydrophobic protein domains with a high propensity to form an antiparallel sheet (physical cross-connections) structure. These domains are rich in alanine and glycine residues. There are 11 tiny hydrophilic amorphous domains that divide these blocks. Particularly, SF materials have shown advantageous immunological characteristics, an important factor to take into account when implanting protein-based materials (Aramwit et al. 2009). SF physically forms hydrogel matrices (Matsumoto et al. 2006). Numerous techniques, including pH, temperature, shear, vortexing, electricity, and sonication, can be used to cause the creation of silk gel (Matsumoto et al. 2006; Yucel et al. 2009; Wang et al. 2008; Leisk et al. 2010). Proteins can reportedly be stabilized and released using silk-based biomaterials (Lu et al. 2009; Yucel et al. 2009). These investigations demonstrated that either trapped or untrapped forms of proteins are integrated into silk films. Reversible interactions between untrapped protein and silk allowed it to quickly escape from the matrix, whereas irreversible connections between trapped protein and silk were discovered, requiring destruction of the silk matrix to allow release. Loading sensitive therapeutic molecules such as peptides and proteins into SF needs gentle aqueous-processing conditions, which is one of the most essential features of SF for drug delivery systems (Mottaghitalab et al. 2015). SF as drug delivery system has been under study for a long time (Guziewicz et al. 2011). With the advancement of technology now, SFNPs have been exploited for clinical use due to certain positive properties such as lower levels of toxicity, better biocompatibility, and many other structural properties (Hassani Besheli et al. 2017; Hassani Besheli et al. 2018). In addition, the SF protein has amino acids with functional groups like carboxyl and amine. These functional groups can be triggered to attach to a variety of substances (Mottaghitalab et al. 2015). Introduction of SFNPs as a nano-adjuvant for multi-epitope-based vaccine delivery against urinary tract infections (UTI) (Hasanzadeh et al. 2020) demonstrated that SFNPs with the characteristics of biocompatibility, biodegradability, and the ability to control the release of the antigen could be utilized as a nano-adjuvant for developing more efficient vaccine formulations. In recent years, SFNPs have garnered increased interest due to their unique properties, such as their tunable size, high surface area to volume ratio, and high drug loading capacity (Rezaei et al. 2020). A suitable adjuvant should be risk-free and devoid of negative effects. In addition, adjuvants should be chemically neutral, biodegradable, and quickly eliminated from the body. In addition, they must be accessible and affordable to produce in order to be commercially viable (Li et al. 2011). Given the aforementioned parameters, SF is a highly biocompatible polymer with distinctive physical and functional characteristics. In addition, the techniques for SF extraction and purification are simple and cost-effective. As a natural biopolymer, SF often undergoes proteolytic breakdown, producing nontoxic by-products (Mottaghitalab et al. 2015). Consequently, the biodegradation products of SF have negligible or nonexistent toxic effects on the body (Cao and Wang 2009). Therefore, by enhancing the particle size and structure of SFNPs, these NPs can be utilized as nano-adjuvants.

13.11 Conclusion and Future Prospects

Nanotechnology has left no stone unturned in biomedical sciences. The landscape of vaccinology is not an exception too. Basically, incorporation of NPs in vaccines has worked in two major ways: firstly as efficient delivery system of antigens and secondly, as nano-adjuvants. Nano-encapsulation of antigens in suitable carriers has reduced the antigenic biomass and cold-chain requirement, imparting less toxicity, improved antigen stability, immuno-kinetic behavior and biocompatibility, besides providing sustained and targeted release system. Nano-adjuvants have the potential to elicit humoral as well as cellular immune response. For instance, inclusion of TLR-based adjuvants in nanovaccine formulations can prime innate immune system apart from adaptive immune system. Several types of NPs including aluminum hydroxide NPs, liposomes, polyanhydride NPs, PLGA or chitosan-based polymeric NPs, etc. have already been employed against various bacterial infections including *Salmonella, Brucella, Mycobacterium, Bacillus anthracis, E. coli,* etc. with encouraging outcome. Exploitation of full potential of nanotechnology in vaccine formulation can bring revolutionary impact in prophylactic management of infectious diseases.

References

Ada G (2001) Vaccines and vaccination. N Engl J Med 345:1042–1053. https://doi.org/10.1056/NEJMra011223

Alving CR, Rao M, Steers NJ, Matyas GR, Mayorov AV (2012) Liposomes containing lipid A: an effective, safe, generic adjuvant system for synthetic vaccines. Expert Rev Vaccines 11(6): 733–744

Amini Y, Moradi B, Fasihi-Ramandi M (2017) Aluminum hydroxide nanoparticles show strong activity to stimulate Th-1 immune response against tuberculosis. Artif Cells Nanomed Biotechnol 45(7):1331–1335. https://doi.org/10.1080/21691401.2016.1233111

Andre FE (2003) Vaccinology: past achievements, present roadblocks and future promises. Vaccine 21:593–595. https://doi.org/10.1016/s0264-410x(02)00702-8

Aramwit P, Kanokpanont S, De-Eknamkul W, Srichana T (2009) Monitoring of inflammatory mediators induced by silk sericin. J Biosci Bioeng 107:556–561. [PubMed: 19393558]

Barenholz YC (2012) Doxil®—the first FDA-approved nano-drug: lessons learned. J Control Release 160(2):117–134. https://doi.org/10.1016/j.jconrel.2012.03.020

Bawa R (2007) Patents and nanomedicine. Nanomedicine 2:351–374. https://doi.org/10.2217/17435889.2.3.351

Bawarski WE, Chidlowsky E, Bharali DJ, Mousa SA (2008) Emerging nanopharmaceuticals. Nanomedicine 4:273–282. https://doi.org/10.1016/j.nano.2008.06.002

Boas U, Heegaard PMH (2004) Dendrimers in drug research. Chem Soc Rev 33(1):43–63. https://doi.org/10.1039/B309043B

Calabro S, Tortoli M, Baudner BC, Pacitto A, Cortese M, O'Hagan DT, De Gregorio E, Seubert A, Wack A (2011) Vaccine adjuvants alum and MF59 induce rapid recruitment of neutrophils and monocytes that participate in antigen transport to draining lymph nodes. Vaccine 29(9): 1812–1823. https://doi.org/10.1016/j.vaccine.2010.12.090

Cao Y, Wang B (2009) Biodegradation of silk biomaterials. Int J Mol Sci 10(4):1514–1524. https://doi.org/10.3390/ijms10041514

Chen G, Bai Y, Li Z, Wang F, Fan X, Zhou X (2020) Bacterial extracellular vesicle-coated multi-antigenic nanovaccines protect against drug-resistant Staphylococcus aureus infection by modulating antigen processing and presentation pathways. Theranostics 10(16):7131–7149. https://doi.org/10.7150/thno.44564

Chowdhury S, Toth I, Stephenson RJ (2022) Dendrimers in vaccine delivery: recent progress and advances. Biomaterials 280:121303. https://doi.org/10.1016/j.biomaterials.2021.121303

Gao Y, Shen M, Shi X (2021) Interaction of dendrimers with the immune system: an insight into cancer nanotheranostics. Viewpoints 2(3):20200120. https://doi.org/10.1002/VIW.20200120

García A, De Sanctis JB (2014) An overview of adjuvant formulations and delivery systems. APMIS : acta pathologica, microbiologica, et immunologica Scandinavica 122(4):257–267. https://doi.org/10.1111/apm.12143

Gasparini R, Schioppa F, Lattanzi M, Barone M, Casula D, Pellegrini M, Veitch K, Gaitatzis N (2010) Impact of prior or concomitant seasonal influenza vaccination on MF59-adjuvanted H1N1v vaccine (Focetria™) in adult and elderly subjects. Int J Clin Pract 64(4):432–438. https://doi.org/10.1111/j.1742-1241.2009.02309.x

Guziewicz N, Best A, Perez-Ramirez B, Kaplan DL (2011) Lyophilized silk fibroin hydrogels for the sustained local delivery of therapeutic monoclonal antibodies. Biomaterials 32(10): 2642–2650. https://doi.org/10.1016/j.biomaterials.2010.12.023

Han Y, Renu S, Patil V, Schrock J, Feliciano-Ruiz N, Selvaraj R, Renukaradhya GJ (2020) Immune response to salmonella Enteritidis infection in broilers immunized orally with chitosan-based salmonella subunit nanoparticle vaccine. Front Immunol 11:935. https://doi.org/10.3389/fimmu.2020.00935

Hanna E, Abadi R, Abbas O (2016) Imiquimod in dermatology: an overview. Int J Dermatol 55(8): 831–844. https://doi.org/10.1111/ijd.13235

Hasanzadeh S, Farokhi M, Habibi M, Shokrgozar MA, Ahangari Cohan R, Rezaei F, AsadiKaram MR, Bouzari S (2020) Silk fibroin nanoadjuvant as a promising vaccine carrier to deliver the FimH-IutA antigen for urinary tract infection. ACS Biomater Sci Eng 6(8):4573–4582. https://doi.org/10.1021/acsbiomaterials.0c00736

Hassani Besheli N, Mottaghitalab F, Eslami M, Gholami M, Kundu SC, Kaplan DL, Farokhi M (2017) Sustainable release of vancomycin from silk fibroin NPs for treating severe bone infection in rat tibia osteomyelitis model. ACS Appl Mater Interfaces 9(6):5128–5138. https://doi.org/10.1021/acsami.6b14912

Hassani Besheli N, Damoogh S, Zafar B, Mottaghitalab F, Motasadizadeh H, Rezaei F, Shokrgozar MA, Farokhi M (2018) Preparation of a codelivery system based on vancomycin/silk scaffold containing silk nanoparticle loaded VEGF. ACS Biomater Sci Eng 4(8):2836–2846. https://doi.org/10.1021/acsbiomaterials.8b00149

Hem SL, Hogenesch H (2007) Relationship between physical and chemical properties of aluminium containing adjuvants and immunopotentiation. Expert Rev Vaccines 6:685–698. https://doi.org/10.1586/14760584.6.5.685

Hilleman MR (2000) Vaccines in historic evolution and perspective: a narrative of vaccine discoveries. J Hum Virol 3:63–76. PMID: 10850891

Huang JL, Yin YX, Pan ZM, Zhang G, Zhu AP, Liu XF, Jiao XA (2010) Intranasal immunization with chitosan/pCAGGS-flaA nanoparticles inhibits campylobacter jejuni in a white Leghorn model. J Biomed Biotechnol 2010:589476. https://doi.org/10.1155/2010/589476

Inoue S, Tanaka K, Arisaka F, Kimura S, Ohtomo K, Mizuno S (2000) Silk fibroin of bombyx mori is secreted, assembling a high molecular mass elementary unit consisting of h-chain, l-chain, and p25, with a 6:6:1 molar ratio. J Biol Chem 275:40517–40528. https://doi.org/10.1074/jbc.M006897200

Jain A, Rawat M, Chakaravarti S, Chaturvedi V, Chesti L, Niranjan AK, Varshney R, Kumar B (2016) Brucella phage lysate bacterin induces elevated TLRs and cytokines response in murine model. J Pure Appl Microbiol 10(3):2063–2070

Kale SN, Deore SL (2017) Emulsion micro emulsion and nano emulsion: a review. Syst Rev Pharm 8(1):39. https://doi.org/10.5530/srp.2017.1.8

Kelly SM, Larsen KR, Darling R, Petersen AC, Bellaire BH, Wannemuehler MJ, Narasimhan B (2021) Single-dose combination nanovaccine induces both rapid and durable humoral immunity and toxin neutralizing antibody responses against bacillus anthracis. Vaccine 39(29): 3862–3870. https://doi.org/10.1016/j.vaccine.2021.05.077

Kindt T, Goldsby R, Osborne B, Kuby J, Kuby J (2007) Kuby Immunology, 6th edn. W.H. Freeman & Co, New York

Lee S, Nguyen MT (2015) Recent advances of vaccine adjuvants for infectious diseases. Immune Netw 15(2):51–57. https://doi.org/10.4110/in.2015.15.2.51

Leisk GG, Lo TJ, Yucel T, Lu Q, Kaplan DL (2010) Electrogelation for protein adhesives. Adv Mater 22(6):711–715. https://doi.org/10.1002/adma.200902643

Li X, Sloat BR, Yanasarn N, Cui Z (2011) Relationship between the size of NPs and their adjuvant activity: data from a study with an improved experimental design. Eur J Pharm Biopharm 78(1): 107–116. https://doi.org/10.1016/j.ejpb.2010.12.017

Li X, Aldayel AM, Cui Z (2014) Aluminum hydroxide NPs show a stronger vaccine adjuvant activity than traditional aluminum hydroxide microparticles. J Control Release 173:148–157. https://doi.org/10.1016/j.jconrel.2013.10.032

Liu Z, Ru J, Sun S, Teng Z, Dong H, Song P, Yang Y, Guo H (2019) Uniform dendrimer-like mesoporous silica NPs as a nano-adjuvant for foot-and-mouth disease virus-like particle vaccine. J Mater Chem B 7:3446–3454. https://doi.org/10.1039/C8TB03315C

Look M, Bandyopadhyay A, Blum JS, Fahmy TM (2010) Application of nanotechnologies for improved immune response against infectious diseases in the developing world. Adv Drug Deliv Rev 62:378–393. https://doi.org/10.1016/j.addr.2009.11.011

Lu S, Wang X, Lu Q, Hu X, Uppal N, Omenetto FG, Kaplan DL (2009) Stabilization of enzymes in silk films. Biomacromolecules 10(5):1032–1042. https://doi.org/10.1021/bm800956n

Maleki M, Salouti M, Ardestani MS, Talebzadeh A (2019) Preparation of a nanovaccine against Brucella melitensis M16 based on PLGA nanoparticles and oligopolysaccharide antigen. Artif Cells Nanomed Biotechnol 47(1):4248–4256. https://doi.org/10.1080/21691401.2019.1687490

Mann JF, Shakir E, Carter KC, Mullen AB, Alexander J, Ferro VA (2009) Lipid vesicle size of an oral influenza vaccine delivery vehicle influences the Th1/Th2 bias in the immune response and protection against infection. Vaccine 27(27):3643–3649. https://doi.org/10.1016/j.vaccine.2009.03.040

Matsumoto A, Chen J, Collette AL, Kim UJ, Altman GH, Cebe P, Kaplan DL (2006) Mechanisms of silk fibroin sol– gel transitions. J Phys Chem B 110(43):21630–21638. https://doi.org/10.1021/jp056350v

Minakshi P, Ghosh M, Brar B, Kumar R, Lambe UP, Ranjan K, Manoj J, Prasad G (2019) Nano-antimicrobials: A new paradigm for combating mycobacterial resistance. Curr Pharm Des 25(13):1554–1579. https://doi.org/10.2174/1381612825666190620094041

Mohammed GM, ElZorkany HE, Farroh KY, Abd El-Aziz WR, Elshoky HA (2020) Potential improvement of the immune response of chickens against E. Coli vaccine by using two forms of chitosan nanoparticles. Int J Biol Macromol 167:395. https://doi.org/10.1016/j.ijbiomac.2020.11.200

Monaci E, Mancini F, Lofano G, Bacconi M, Tavarini S, Sammicheli C, Arcidiacono L, Giraldi M, Galletti B, Rossi Paccani S, Torre A, Fontana MR, Grandi G, de Gregorio E, Bensi G, Chiarot E, Nuti S, Bagnoli F, Soldaini E, Bertholet S (2015) MF59- and Al(OH)3-adjuvanted Staphylococcus aureus (4C-staph) vaccines induce sustained protective humoral and cellular immune responses, with a critical role for effector CD4 T cells at low antibody titers. Front Immunol 6:439. https://doi.org/10.3389/fimmu.2015.00439

Morel S, Didierlaurent A, Bourguignon P, Delhaye S, Baras B, Jacob V, Planty C, Elouahabi A, Harvengt P, Carlsen H, Kielland A, Chomez P, Garcon N, Van MM (2011) Adjuvant system AS03 containing alpha-tocopherol modulates innate immune response and leads to improved adaptive immunity. Vaccine 29:2461–2473. https://doi.org/10.1016/j.vaccine.2011.01.011

Mottaghitalab F, Farokhi M, Shokrgozar MA, Atyabi F, Hosseinkhani H (2015) Silk fibroin nanoparticle as a novel drug delivery system. J Control Release 206:161–176. https://doi.org/10.1016/j.jconrel.2015.03.020

O'Hagan DT, Ott GS, De Gregorio E, Seubert A (2012) The mechanism of action of MF59 - an innately attractive adjuvant formulation. Vaccine 30:4341–4348. https://doi.org/10.1016/j.vaccine.2011.09.061

Panda AK (2012) Nanotechnology in vaccine development. Proc Natl Acad Sci India Sect B Biol Sci 82(Suppl 1):13–27. https://doi.org/10.1007/s40011-012-0073-6

Peek LJ, Middaugh CR, Berkland C (2008) Nanotechnology in vaccine delivery. Adv Drug Deliv Rev 60:915–928. https://doi.org/10.1016/j.addr.2007.05.017

Renu S, Han Y, Dhakal S, Lakshmanappa YS, Ghimire S, Feliciano-Ruiz N, Senapati S, Narasimhan B, Selvaraj R, Renukaradhya GJ (2020) Chitosan-adjuvanted salmonella subunit nanoparticle vaccine for poultry delivered through drinking water and feed. Carbohydr Polym 243:116434. https://doi.org/10.1016/j.carbpol.2020.116434

Rezaei F, Damoogh S, Reis RL, Kundu SC, Mottaghitalab F, Farokhi M (2020) Dual drug delivery system based on pH-sensitive silk fibroin/alginate NPs entrapped in PNIPAM hydrogel for treating severe infected burn wound. Biofabrication 13(1):015005. https://doi.org/10.1088/1758-5090/abbb82

Romero Mendez IZ, Shi Y, HogenEsch H, Hem SL (2007) Potentiation of the immune response to nonadsorbed antigens by aluminum-containing adjuvants. Vaccine 25:825–833. https://doi.org/10.1016/j.vaccine.2006.09.039

Schottler S, Becker G, Winzen S, Steinbach T, Mohr K, Landfester K, Mailänder V, Wurm FR (2016) Protein adsorption is required for stealth effect of poly (ethylene glycol)-and poly (phosphoester)-coated nanocarriers. Nat Nanotechnol 11(4):372–377. https://doi.org/10.1038/nnano.2015.330

Schwendener RA (2014) Liposomes as vaccine delivery systems: a review of the recent advances. Ther Adv Vaccines 2(6):159–182. https://doi.org/10.1177/2051013614541440

Shi S, Zhu H, Xia X, Liang Z, Ma X, Sun B (2019) Vaccine adjuvants: understanding the structure and mechanism of adjuvanticity. Vaccine 37:3167–3178. https://doi.org/10.1016/j.vaccine.2019.04.055

Singh R, Sharma R, Varshney R, Mal G, Ghosh M, Singh B (2023) Evaluation of immunological adjuvant activities of saponin rich fraction from the fruits of Asparagus adscendens Roxb. with less adverse reactions. Drug Chem Toxicol 46(3):557–565. https://doi.org/10.1080/01480545.2022.2067170

Skene CD, Sutton P (2006) Saponin-adjuvanted particulate vaccines for clinical use. Methods 40: 53–59. https://doi.org/10.1016/j.ymeth.2006.05.019

Solanki KS, Varshney R, Qureshi S, Thomas P, Singh R, Agrawal A, Chaudhuri P (2021) Non-infectious outer membrane vesicles derived from Brucella abortus S19Dper as an alternative acellular vaccine protects mice against virulent challenge. Int Immunopharmacol 90: 107148

Sundquist BG, Czifra G, Stipkovits L (1996) Protective immunity induced in chicken by a single immunization with mycoplasma gallisepticum immunostimulating complexes (ISCOMS). Vaccine 14(9):892–897. https://doi.org/10.1016/0264-410x(95)00262-y

Tretiakova DS, Vodovozova EL (2022) Liposomes as adjuvants and vaccine delivery systems. Biochem (Mosc) Suppl Ser A Membr Cell Biol 16(1):1–20. https://doi.org/10.1134/S1990747822020076

Vandepapelière P, Horsmans Y, Moris P, Van Mechelen M, Janssens M, Koutsoukos M, Van Belle P, Clement F, Hanon E, Wettendorff M, Garçon N, Leroux-Roels G (2008) Vaccine adjuvant systems containing monophosphoryl lipid A and QS-21 induce strong and persistent humoral and T cell responses against hepatitis B surface antigen in healthy adult volunteers. Vaccine 26:1375–1386. https://doi.org/10.1016/j.vaccine.2007.12.038

Varshney R, Varshney R, Chaturvedi VK, Rawat M, Saminathan M, Singh V, Singh R, Sahoo M, Gupta PK (2020) Development of novel iron-regulated Pasteurella multocida B: 2 bacterin and refinement of vaccine quality in terms of minimum variation in particle size and distribution Vis-a-Vis critical level of iron in media. Microb Pathog 147:104375. https://doi.org/10.1016/j.micpath.2020.104375

Vidal D, Alomar A (2008) Mode of action and clinical use of imiquimod. Expert Rev Dermatol 3(2):151–159. https://doi.org/10.1586/17469872.3.2.151

Vyas S, Dhoble S, Ghodake V, Patravale V (2020) Xyloglucan based mucosal nanovaccine for immunological protection against brucellosis developed by supercritical fluid technology. Int J Pharm X 2:100053. https://doi.org/10.1016/j.ijpx.2020.100053

Wagner-Muñiz DA, Haughney SL, Kelly SM, Wannemuehler MJ, Narasimhan B (2018) Room temperature stable PspA-based Nanovaccine induces protective immunity. Front Immunol 9: 325. https://doi.org/10.3389/fimmu.2018.00325

Wang X, Kluge JA, Leisk GG, Kaplan DL (2008) Sonication-induced gelation of silk fibroin for cell encapsulation. Biomaterials 29:1054–1064. https://doi.org/10.1016/j.biomaterials.2007.11.003

Wang N, Chen M, Wang T (2019 Jun) Liposomes used as a vaccine adjuvant-delivery system: from basics to clinical immunization. J Control Release 10(303):130–150. https://doi.org/10.1016/j.jconrel.2019.04.025

Wang N, Wei C, Zhang Z, Liu T, Wang T (2020) Aluminum NPs acting as a pulmonary vaccine adjuvant-delivery system (VADS) able to safely elicit robust systemic and mucosal immunity. J Inorg Organomet Polym Mater 30(10):4203–4217. https://doi.org/10.1007/s10904-020-01572-z

Wilkins AL, Kazmin D, Napolitani G, Clutterbuck EA, Pulendran B, Siegrist CA, Pollard AJ (2017) AS03-and MF59-adjuvanted influenza vaccines in children. Front Immunol 8:1760. https://doi.org/10.3389/fimmu.2017.01760

Xiang SD, Scholzen A, Minigo G, David C, Apostolopoulos V, Mottram PL, Plebanski M (2006) Pathogen recognition and development of particulate vaccines: does size matter. Methods 40:1–9. https://doi.org/10.1016/j.ymeth.2006.05.016

Yucel T, Cebe P, Kaplan DL (2009) Vortex-induced injectable silk fibroin hydrogels. Biophys J 97(7):2044–2050. https://doi.org/10.1016/j.bpj.2009.07.028

Zeng F, Zimmerman SC (1997) Dendrimers in supramolecular chemistry: from molecular recognition to self-assembly. Chem Rev 97:1681–1712. https://doi.org/10.1021/cr9603892

Chapter 14
Nanotechnology in Prophylaxis of Viral Livestock Diseases

Rajesh Kumar, Alonkrita Chowdhury, Manoj Kumar Rose, Sonia Sindhu, Shafiq M. Syed, and Mayukh Ghosh

Abstract Nanoparticles have shown great potential in diverse fields, including medicine and vaccine development. When it comes to viral diseases of animals, nanoparticles can be applied in several ways to enhance the efficiency of vaccines. Adjuvant delivery is one of the key applications of nanoparticles in animal vaccine development. Nanoparticles can be used as carriers or delivery systems for vaccine adjuvants. Adjuvants are substances that enhance the immune response to a vaccine by stimulating the immune system. Nanoparticles can encapsulate adjuvants and help deliver them directly to immune cells, improving the immune response and vaccine efficacy. Nanoparticles can also be utilized to deliver viral antigens, the immunodominant components of a virus to the immune cells in a targeted and controlled manner that elicit a specific and long-lasting immune response. By encapsulating viral antigens within nanoparticles, their stability can be improved, and their controlled release can be achieved, leading to a more targeted and sustained immune response. Targeted delivery is another key aspect in customizing vaccination strategy. Nanoparticles can be functionalized with specific ligands or antibodies that recognize and bind to receptors on immune cells. This targeted delivery approach helps direct the vaccine to the antigen-presenting cells (APCs), such as dendritic cells, macrophages, Langerhans cells, and B cells, which are crucial for

R. Kumar · M. K. Rose · S. Sindhu
Department of Veterinary Physiology and Biochemistry, Lala Lajpat Rai University of Veterinary and Animal Sciences, Hisar, Haryana, India

A. Chowdhury
Department of Plant Biotechnology, Institute of Agricultural Sciences, RGSC, Banaras Hindu University, Mirzapur, Uttar Pradesh, India

S. M. Syed
Biomedical Sciences and Pharmacy, University of Newcastle Australia, Callaghan, Australia

M. Ghosh (✉)
Department of Veterinary Physiology and Biochemistry, Faculty of Veterinary and Animal Sciences, Institute of Agricultural Sciences, RGSC, Banaras Hindu University, Mirzapur, Uttar Pradesh, India

LLR University of Veterinary and Animal, ICAR-National Research Centre on Equines, *Nanotechnology Theranostics in Livestock Diseases and Management*, Livestock Diseases and Management,
https://doi.org/10.1007/978-981-16-1610-5_14

initiating an effective immune response. By specifically targeting these cells, the vaccine can elicit a stronger and more focused immune reaction. Vaccine stabilization is also very crucial to achieve the desired outcome. Vaccines often require storage and transportation under specific conditions to maintain their efficacy. Nanoparticles can provide stability to viral antigens and other vaccine components, protecting them from degradation due to environmental factors such as temperature, humidity, or light exposure. This stability can extend the shelf-life of vaccines and increase bioavailability in remote or resource-limited areas. In addition, nanoparticles can be designed to carry multiple antigens from different viruses, enabling the development of combination vaccines. This approach is particularly beneficial when animals need protection against multiple viral diseases simultaneously. By incorporating multiple antigens within nanoparticles, the immune response can be targeted against multiple pathogens, simplifying the vaccination process and reducing the number of vaccine administrations. It is important to note that while nanoparticles have shown promising results in preclinical studies, their application in animal vaccines is still an active area of research. Further studies and clinical trials are necessary to fully evaluate their safety, efficacy, and commercial viability.

Keywords Viral diseases · Nanovaccine · Nano-adjuvants · VLPs · Liposomes · Protein nanoparticles

14.1 Introduction

Viral diseases, a bad news entrapped in protein covering, pose a significant threat to animal health, affecting livestock, poultry, and companion animals worldwide. Most of the viral diseases not only reduce performances of affected animals but also spread to healthy herd due to contagious nature of transmission. Moreover, they are not only omnipresent but possess significant threat due to their ever-evolving nature by faster mutation rate that may lead to change in their morbidity and mortality pattern, and make them adaptable to changing environment. They not only possess threat to animal subjects but can also have zoonotic effect, hence, threat to human subject also like corona virus and influenza both emerged from animals and created challenges to human health system too. So, it is always better to identify these viruses under high surveillance and make methods to contain and eliminate them. Vaccination is till now the most successful strategy which has helped to contain and eliminate such threats. There are ample examples such as polio virus, SARS-CoV-2, smallpox, foot-and-mouth disease (FMD), Newcastle disease, and others in both animal and human subjects which have been confined by vaccination in a population. Conventional vaccines have played a crucial role in controlling these diseases; however, they often have limitations, such as reduced efficacy, safety concerns, and the inability to differentiate between infected and vaccinated animals. In recent years, the field of nanotechnology has emerged as a promising solution to overcome these limitations, offering innovative strategies for vaccine development.

Unique advantages which have shown promising sign of nanomaterial-based vaccines include controlled release, targeted delivery, enhanced immunogenicity, and improved stability. By reviewing the current examples of nanoparticle-based viral vaccines in animal diseases, we will showcase their practical applicability and effectiveness (Hall 1984; Meeusen et al. 2007; Carpenter et al. 2022; WHO 2022). This chapter will delve into the applications of nanoparticles in viral vaccine development for animal diseases, discussing the principles, advantages, and challenges of this approach, as well as highlighting recent advancements and future perspectives.

14.2 Impact of Viruses on Animal Health and Economy

Viral diseases have a significant impact on animal health, causing morbidity, mortality, and reduced productivity in affected animals. Nearly 23% of livestock and poultry products in the global market is only related with vaccine, mostly against viral or bacterial diseases (Meeusen et al. 2007). Moreover, apart from animal health and thus directly effecting food security and job of humans, vaccines have other implications in controlling zoonotic diseases as 60% of human infectious disease and 75% of emerging disease are zoonotic in nature with nearly 200 known types of zoonoses, where viral disease is one of the significant causes (WHO 2020; CDC 2021). Rabies virus, influenza viruses, hantavirus hemorrhagic fevers, arenavirus hemorrhagic fever, yellow fever virus, West Nile virus, Chikungunya virus, Zika virus, Sindbis virus are some of their examples. Livestock and poultry are particularly susceptible to various viral diseases, including foot-and-mouth disease (FMD), avian influenza, and porcine reproductive and respiratory syndrome (PRRS). Companion animals, such as dogs and cats, are also prone to viral infections like canine parvovirus and feline leukemia virus. There are numerous virus-causing diseases in animals, birds, and fishes such as cowpox, fowl pox, fowlpest, rotavirus, orbivirus, bluetongue, rinderpest, duck plague, orthoreovirus, Border disease, Marek's disease, ephemeral fever, bovine leukosis, West Nile fever, canine distemper, coxsackie viruses, egg drop syndrome, caprine arthritis, Rift valley fever, contagious ecthyma, African swine fever, Wesselsbron disease, vesicular stomatitis, feline panleukope nia, reticuloendotheliosis, Nairobi sheep disease, classical swine fever, equine viral arteritis, African horse sickness, foot-and-mouth disease, swine vesicular disease, viral hepatitis of geese and ducks, bat salivary gland fever, malignant catarrhal fever, equine infectious anemia, viral hepatitis of turkeys, Akabane disease of animals, peste des petits ruminants, infectious canine hepatitis, canine parvoviral infection, vesicular exanthema of swine, canine herpes viral infection, African green monkey disease, infectious laryngotracheitis, lymphoid leukosis of chickens, feline infectious peritonitis, feline immunodeficiency virus, coronaviral encephalomyelitis of swine, Hemorrhagic anemia syndrome in chickens, encephalomyocarditis virus disease of pigs, rabies, etc. The economic consequences of viral diseases in animals are substantial, resulting in direct losses due to decreased production, mortality, and increased veterinary costs. Indirect losses arise from trade restrictions, negative impacts on consumer confidence, and the costs of disease prevention and control measures. There are several reports of such outbreaks worldwide such as the foot-and-mouth disease outbreak in the United Kingdom in 2001 led to an estimated economic loss of over £8 billion, the avian influenza outbreak in Asia between 2003 and 2004 caused the culling of over 200 million birds resulting in significant economic damage to the poultry industry, 50 million birds loses mounting nearly $3.3 billion in the United States during outbreak of avian influenza (HPAI) in 2014–15, loss of 8 million animals due to porcine epidemic diarrhea virus (PEDv) in the United States during 2013 (Costa and Akdeniz 2019). Furthermore, it may lead to devasting losses if control measure is not prompt as it is estimated in one observation that hypothetical FMD outbreak in California if not controlled in 7 day and extend to 22 days, the loss may increase from expected 3 billion to $69 billion (Carpenter et al. 2011) as per 2011 data. The loss is just not only in terms of economics but can have devasting effect on psyche and environmental concern also as happened in South Korea where 4583 burial sites were made to kill nearly 20% of livestock for eradication control measure raising environmental concern (Costa and Akdeniz 2019). Ongoing outbreak of lumpy skin disease in India during 2022 estimated to have 155,724 deaths of cattle and nearly 2,952,233 were affected. The spread of the disease was pan-India (DAHD, Govt. of India 2022). This may lead to 57 million liter milk loss in a year considering 50% milk yield by cattle of 200MT produced by 273 million cattle just due to death of animals (National Dairy Development Board 2017). This itself may lead to a total loss of 228 crore rupees, along with 467 crore rupees loss as replacement value considering milk price 0.5 dollar/kg and cost of livestock 375 dollar. Apart from this 30-lakh animals affected till now which is approximately 20 times of the animals died from the disease population; hence, it easily surmounts the cost associated with direct loss. Such devastating losses not only give concern in terms of economic loss and food security but affect the moral strength of animal owners. In the era of global warming, higher human and animal trafficking and more organized animal rearing system development, such losses are expected to be more frequent and to a higher extent.

14.3 Conventional Vaccines: Advantages and Limitations

Vaccination has remained as a cornerstone in animal disease prevention and control for many decades. Conventional vaccines, such as live attenuated, inactivated, and subunit vaccines have played a crucial role in reducing the prevalence and impact of various viral diseases. For example, the eradication of rinderpest, a deadly viral disease affecting cattle, was achieved through a successful global vaccination campaign using a live attenuated vaccine. Recently, DIVA and self-assembling vaccine are also made to make vaccination program more effective and far reaching to minimize chances of epidemiological breakout of viral diseases (Murphy et al. 2016).

Despite their successes, conventional vaccines have several limitations. Live attenuated vaccines, although highly effective, can pose safety concerns due to the risk of reversion to virulence, especially in immunocompromised animals. Inactivated and subunit vaccines, while generally safer, may have reduced efficacy as they often induce weaker immune responses compared to live vaccines. Moreover, these vaccines may require the use of adjuvants to enhance their immunogenicity, which can sometimes cause adverse reactions. DNA-based vaccines are a bit safer than earlier ones but have problem with respect to weak immune response as only a specific part of the whole organism is included in the vaccine formulation; hence, adjuvant is added to enhance the immunogenicity (Petkar et al. 2021).

Another significant limitation of conventional vaccines is their inability to differentiate between infected and vaccinated animals (DIVA). This poses a challenge for disease surveillance and control, as serological tests cannot distinguish between antibodies generated by natural infection and those resulting from vaccination. Consequently, these vaccines may impede efforts to monitor and control viral diseases effectively. Considering the limitations of conventional vaccines, there is a pressing need for novel vaccine strategies to improve the prevention and control of viral diseases in animals. These new approaches should address the issues of safety, efficacy, and DIVA capability, while also considering ease of production, cost-effectiveness, and scalability.

Emerging technologies, such as nanotechnology, have shown great promise in addressing these challenges. The use of nanoparticles in vaccine development offers various advantages, including enhanced immunogenicity through nano-adjuvants, targeted delivery, controlled release, and the potential for DIVA compatibility. Furthermore, nanoparticle-based vaccines can be designed for multiple delivery routes, such as oral, intranasal, and intramuscular administration, providing flexibility and ease of use in different animal species. The following sections of this chapter will explore the applications of nanoparticles in viral vaccine development, discussing the principles, advantages, and challenges of this approach, as well as highlighting recent advancements and future perspectives.

14.4 Possible Advantages and Precaution for Nanoparticle-Based Vaccine Production

Nanoparticles having dimension of nanometer scale, exhibit various unique physical, chemical, and biological properties differing from their bulk counterparts. These properties, such as increased surface area to volume ratio, differential approach to immune system, make nanoparticles an attractive option for various applications, including drug delivery, diagnostics, and vaccine development. Nanomaterials can be customized from wide range of components, including metals, polymers, lipids, biological and inorganic compounds, etc., which can be employed in vaccine development. For example, polymeric nanoparticles made from biodegradable and biocompatible polymers, such as poly (lactic-co-glycolic acid) (PLGA) and chitosan can encapsulate antigens or adjuvants, protecting them from degradation and facilitating their controlled release to provide enhanced immunogenicity.

Similarly, liposomes are spherical vesicles composed of lipid bilayers, which can also be used for encapsulating both hydrophilic and hydrophobic molecules. They can be used to deliver antigens, adjuvants, or nucleic acids, enhancing their stability and immunogenicity.

Various inorganic nanoparticles, such as gold nanoparticles, silica nanoparticles, and mesoporous silica nanoparticles, can be used as antigen carriers or adjuvants. They can be functionalized with various ligands to improve their targeting and immune-stimulatory properties. In general, organic nanoparticle-based vaccines produce lower toxic effect, higher biodegradability and biocompatibility while enhanced stability, lower size, magnetic properties, and relatively large functionalization are advantages of inorganic and metal-based NPs. Issues like low stability, reproducibility and drug entrapment are to be taken care in organic NPs whereas greater toxic effect, poor biodegradability and biocompatibility issues have to be taken care using inorganic-based NPs in vaccine development (Poon and Patel 2020). Both organic and inorganic components, biological compounds like proteinaceous substances, virus like particles, etc. have been tried in their nanoscale to overcome current limitations in vaccine action, storage, and production steps.

These nanoparticles are synthesized through various methods, including chemical, physical, and biological processes like solvent evaporation, self-assembly method for liposomes and other self-assembling nanoparticles which can be formed by hydrating lipids or amphiphilic molecules in an aqueous solution leading to the spontaneous formation of vesicles; reduction and precipitation method, green synthesis method, etc. Apart from general characterization which includes size assay, zeta potential, shape characterization, confirmation of functionalization, etc., the safety and biocompatibility of nanoparticles are critical factors in their application for vaccine development. Several aspects need to be considered for effective nanovaccine formulation including:

- Material selection: The choice of nanoparticle material plays a crucial role in determining its biocompatibility. Biodegradable and biocompatible materials, such as PLGA and chitosan are generally preferred due to their low toxicity and ability to be metabolized by the body. Similarly, it has been observed in general that metallic NPs exert immune-modulatory effect by virtue of promoting cell recruitment, antigen-presenting cell (APC) activation, and cytokine induction (Azharuddin et al. 2022).
- Size and shape: The size and shape of nanoparticles can influence their biodistribution, cellular uptake, and potential toxicity. For instance, smaller nanoparticles may exhibit higher uptake but also greater potential for systemic dissemination and toxicity. Therefore, optimizing the size and shape of nanoparticles is essential for balancing their efficacy and safety (Azharuddin et al. 2022). For example, particles less than 500 nm in diameter have better uptake chance by dendritic cells (DCs), it may be because during evolution,

immune cells have developed strategy to identify and process nano to micro-sized particles as most of the pathogens are in this size range; hence, accumulation of NPs are better than larger particle in immune cells (Foged et al. 2005.)

- Surface properties: Surface modifications, such as the addition of targeting ligands or stealth coatings can enhance the biocompatibility and targeting abilities of nanoparticles. However, these modifications should be carefully evaluated for their potential to induce adverse immune reactions or affect the stability of the nanoparticles.
- Mechanisms by which nanomaterials improve vaccine responses: The attraction of nanomaterial for various cells need to be assessed. For instance, targeted delivery of NPs to desired cells, their interaction with various professional APCs, B cells, macrophages, neutrophils and T cells needs to be analyzed. Confirmation of toxic effects of the NPs and desired effects must be confirmed. Various factors related to nanomaterial properties such as size, charge, hydrophobicity, flexibility, trafficking of the NPs to lymph nodes, etc. need to be assayed. Similarly, for mucosal immunity through oral delivery, penetration of the mucosal barriers is required. Hydrophilic nature of the nanoformulation for entanglement with the mucin fiber and charge like positive charge can help in mucoadhesion for persistent release. Endosomal escape nature of the vaccine formulation is also needed to be assessed (Fries et al. 2021). For DC presentation, rigid particle shows more success than softer particle; whereas, half-life of NPs increases when they are softer and mimicking RBC properties (Beningo and Wang 2002; Reddy et al. 2006; Doshi et al. 2009; Hu et al. 2011).

The selection and generation of NP-based vaccine formulation also depend upon its precise uses, target site, target of inoculum, storage condition, etc. for which specific tests are needed. To ensure the safety and biocompatibility of nanoparticle-based vaccines, rigorous preclinical testing in relevant animal models is necessary. This includes the assessment of potential toxicities, immunogenicity, biodistribution, and long-term effects. In general nanovaccine can provide several advantages such as better uptake, enhanced duration of antigen presentation and DC processing, evoking both humoral and cell-mediated immune responses, extended antigen stability, enhanced immunogenicity, providing multiple binding sites for multiple antigens targeting single or multiple diseases, protecting antigens from premature degradation, adjuvantation, targeted delivery, sustained release, etc. (Azharuddin et al. 2022). Various purposes which nanoparticles can serve in vaccine formulations include but not limited to acting as carrier, as adjuvant, enhancing specific immunogenicity, stability, providing better biodistribution, targeting, controlled or prolonged release, reduction in doses or frequency, making vaccine self-replicating or DIVA based, to supply specific antigen without degradation, etc. for which discussion of various nanoparticles and their uses is given in subsequent sections, along with animal-centric applications and constrains.

14.5 Nanoparticles as Antigen Carriers

Nanoparticles can be used to deliver viral antigens more effectively by encapsulating or conjugating them to their surface which gives protection against enzymatic degradation, mimicking cell can enhance their uptake and even sometime provide initial activation of antigen. However, choosing the right NPs and their functionalization is important and here, factors like loading capacity, protection capacity, penetration ability, uptake, release efficiency, preservation of antigen properties, epitope orientation, and the overall influence on the colloidal stability of the NPs needed to be evaluated. It can be achieved by encapsulation and conjugation of antigen or vaccine by nanomaterials.

14.5.1 Encapsulation

Antigens can be physically entrapped within polymeric or lipid-based nanoparticles during their synthesis. This approach can protect the antigens from degradation and facilitate their controlled release at the site of action. For example, liposomes incorporating nickel-binding (DGS-NTA(Ni)) and thiol-reactive (MPB) lipids covalently bound to HIV-1 trimers (MD39-HHHHHHC) and ferritin nanoparticle core has been used for vaccine delivery system against HIV virus; interbilayer-cross-linked multilamellar vesicles (ICMVs) for malaria vaccine and poly(lactic-co-glycolic acid) (PLGA) nanoparticles encapsulating ovalbumin as model of antigen delivery has been tried by various workers (Moon et al. 2012; Fries et al. 2021). It has been observed that encapsulation with NPs not only help in entry and persistence of antigens to the site but also enhance immunogenicity and facilitate targeted delivery for better uptake through improved MHC II presentation (Zupančič et al. 2017). Similarly, polyanhydride nanoparticles, an FDA approved biodegradable material used as drug delivery vehicles, when added with 1,8-bis (p-carboxyphenoxy)-3,6-dioxaoctane (CPTEG), it became more amphiphilic which otherwise are hydrophobic, and have sustainable release (Senapati 2020).

14.5.2 Surface Conjugation

Antigens can be covalently or noncovalently attached to the surface of nanoparticles, such as inorganic nanoparticles or functionalized polymeric nanoparticles. This approach can enhance the presentation of antigens to immune cells and improve their immunogenicity. Conjugation of immuno-dominant peptide epitope from the circumsporozoite protein of *Plasmodium berghei*, pb9 with noninflammatory polystyrene nanoparticles in a nanovaccine formulation induced high protective CD8 T-cell response against malarial parasites in mice model (Wilson et al. 2015).

Application of PLGA NPs has found to be an effective nonviral vector for gene delivery and therefore a suitable carrier for DNA vaccines. It also observed that surface conjugation of the antigen on PLGA NPs is a better option than encapsulation for such purpose. Similarly, antigen conjugated at the surface of the multilamellar vesicles has elicited better humoral immune response than encapsulation. In another instance, it has been observed that antigen displayed on the surface of the calcium phosphate NPs has induced prominent B-cell activation in comparison to soluble antigens (Azharuddin et al. 2022). Calcium phosphate NPs has been successfully used for conjugation of synthetic peptide P1 and P2 as vaccine candidate against herpes simplex virus type 1 (Kopp et al. 2019).

14.6 Controlled Release and Targeted Delivery

Nanoparticles can be designed to control the release of encapsulated or conjugated antigens, ensuring a sustained and localized delivery. For example, APCs can be targeted through mannose receptors by conjugating mannose with NPs-based antigens. Mannose-modified multi-walled carbon nanotubes efficiently delivered ovalbumin to the macrophages promoting phagocytosis, macrophage proliferation along with secretion of IL-1β, and IL-6. The same nano-construct was also induced dendritic cell maturation along with elevated production of antigen-specific antibodies, IFN-γ, and IL-6 in mice (Feng et al. 2022). For example, dextran-based nanovaccine with lipopolysaccharide (LPS) activates murine dendritic cells, whereas liposomes conjugated with langerin ligands are specific for antigen targeting to Langerhans cells (Shen et al. 2013; Wamhoff et al. 2019). Controlled release can be achieved by tuning the nanoparticle's composition, size, and surface properties. For example, by adjusting the degradation rate of a polymeric nanoparticle or the lipid composition of a liposome, the release kinetics of the antigen can be modulated to elicit a more robust and prolonged immune response. An in vivo experimentation in mice model has demonstrated that subcutaneous delivery of 20 nm poly(lactic-co-glycolic)-b-poly(ethylene-glycol) NPs have better trafficking efficiency and higher retention time at lymph node than 40 nm size particles while it was negligible in case of 100 nm size particles. The size threshold of 30 nm has been suggested for the NPs for efficient lymph node trafficking, retention, and uptake by the APCs (Howard et al. 2019). In a similar experimentation, pluronic-stabilized polypropylene sulphide nanoparticles of 25 nm size have shown to be efficiently transported into lymphatic capillaries through interstitial flow following intradermal administration and reach upto half of the lymph node-residing dendritic cells while 100-nm particles achieved only 10% of the targeting efficiency. Further, it has also been observed that antigen conjugated 25-nm-sized polyhydroxylated NPs was found to be the most efficient to elicit humoral as well as cellular immunity in mice because the smaller size facilitates adequate lymphoid targeting while polyhydroxy surface modification helps in complement activation to stimulate DC maturation (Reddy et al. 2007). Therefore, nanoparticles can be engineered for targeted delivery of antigens to specific immune cells or tissues by modifying the nanoparticle's surface with targeting ligands, such as antibodies, peptides, carbohydrates, or specific groups that selectively bind to receptors expressed by the target cells or activate the complement cascade. Targeted delivery can enhance the efficacy of the vaccine by promoting the uptake of antigens by the desired immune cells, such as dendritic cells, leading to a more potent and specific immune response. Pullulan, linear polysaccharide of α-1, 6-linked maltotriose is an excellent carrier and an oxygen barrier material, hence can provide better stability along with adjuvantation potential. Tumor necrosis factor-α (TNF-α) encapsulated into cholesteryl pullulan NPs has depicted good storage capability and adjuvant potential to provided humoral as well as cellular immunity in mice against H1N1 influenza virus (Nagatomo et al. 2015).

14.7 Nanoparticle for Enhancement of Immunogenicity

Nanoparticles can improve the immunogenicity of viral antigens through various mechanisms. Nanoparticles with dimensions that closely mimic the size and shape of viruses can enhance the uptake of antigens by immune cells and facilitate their processing and presentation to T cells. Further, by displaying multiple copies of the antigen on their surface, nanoparticles can promote the cross-linking of B-cell receptors, leading to a stronger and more rapid antibody response. Some nanoparticles, such as inorganic nanoparticles and certain lipid-based nanoparticles, can exhibit inherent adjuvant properties, stimulating the innate immune response and enhancing the overall immunogenicity of the vaccine. Multiwalled-carbon nanotubes and silica NPs can activate the NOD-like receptor (NLR) and Superparamagnetic iron oxide NPs potentiates LPS by enhancing inflammatory action (Kodali et al. 2013). Conjugation of amantadine, an immunoregulatory molecule at the surface of PVP-PEG-coated silver NPs promotes cytotoxic lymphocytes (CTLs) for enhanced TNF-α production in HIV vaccine (Li et al. 2018).

Aluminum oxide nanoparticles has also been reported to act as an efficient carrier for targeted antigen delivery to the autophagosomes in DCs and capable of reducing antigen dose required for DC activation. The NPs has also shown to be a promising adjuvant as tumor regression was happened in mice immunized with the NP-conjugated tumor antigens (Li et al. 2011). Similarly, gold nanoparticles conjugated with the hemagglutinin antigen from avian influenza virus demonstrated enhanced immunogenicity and protection in chickens, compared to the soluble antigen alone (Wang et al. 2017, 2018). Mucosal vaccination with a PLGA-based nanoparticle vaccine carrying the FMDV DNA vaccine encoding the FMDV capsid protein and the bovine IL-6 gene as an adjuvant molecule showed improved immunogenicity and protection compared to a conventional inactivated vaccine in cattle (Wang et al. 2011). Porcine circovirus type 2 (PCV2) vaccine has been developed where chitosan nanoparticles loaded with PCV2 virus-like particles induced a strong immune response and protection in pigs, with a lower dose than the conventional subunit vaccine (Bucarey et al. 2014; Dhakal and Renukaradhya

2019). These examples showcase the potential of nanoparticle-based vaccines in enhancing the immunogenicity and protective efficacy of viral antigens, highlighting their promise as novel solutions for the prevention and control of animal diseases. By leveraging these advantages, nanoparticle-based vaccines have the potential to overcome the limitations of conventional vaccines and improve the prevention and control of viral diseases in animals. Further research and development in this field will likely lead to creation of novel and more effective vaccination strategies to address the challenges posed by viral diseases in animal populations.

14.8 Nanoparticles as Adjuvants

Adjuvants are substances that enhance the immunogenicity of antigens when administered in combination with vaccines. They can act through various mechanisms, such as:

- Prolonging antigen exposure: Adjuvants can create a depot effect, slowly releasing antigens and prolonging their exposure to the immune system.
- Targeting antigen-presenting cells: Adjuvants can increase the uptake of antigens by antigen-presenting cells, promoting their processing and presentation to T cells.
- Activation of innate immunity: Adjuvants can stimulate innate immune responses by activating pattern recognition receptors (PRRs) on immune cells, leading to the production of pro-inflammatory cytokines and chemokines.

Nanoparticle-based adjuvants offer several advantages over conventional adjuvants, such as:

- Improved safety: Many conventional adjuvants, such as Freund's adjuvant, are associated with adverse effects. In contrast, nanoparticle-based adjuvants can be designed to be biocompatible and biodegradable, reducing their potential for toxicity.
- Enhanced immunogenicity: Nanoparticle-based adjuvants can improve the immunogenicity of antigens by providing controlled release, targeted delivery, and activation of innate immunity.
- Versatility: Nanoparticles can be easily functionalized with various ligands or molecules, allowing for the development of customized adjuvants tailored to specific antigens or immune responses.

Several nanoparticle-based adjuvants have been investigated for their potential in enhancing the immunogenicity of viral vaccines in animals. Some notable examples include:

- Aluminum-based nanoparticles: Aluminum salts, such as aluminum hydroxide and aluminum phosphate, are widely used as adjuvants in human and veterinary vaccines. Recently, aluminum nanoparticles have been explored as adjuvants,

showing improved immunogenicity compared to their bulk counterparts in animal models.
- Toll-like receptor (TLR) agonist-loaded nanoparticles: Nanoparticles can be used to deliver TLR agonists, such as CpG oligodeoxynucleotides or monophosphoryl lipid A, or a mixture of such agonists which can activate innate immune responses and enhance the immunogenicity of viral antigens. For instance, mice immunized with a multi-PRR stimulant nanovaccine formulation containing influenza hemagglutinin antigen encapsulated in PLGA NPs adjuvanted with monophosphoryl lipid A, a TLR4 agonist and muramyl dipeptide, a NOD2 agonist resulted in enhanced humoral and cellular immune responses than individual pathogen-associated molecular pattern (PAMP)-based adjuvants (Tukhvatulin et al. 2020).
- Saponin-based nanoparticles: Saponins, such as Quil A and QS-21, are natural compounds derived from plants that possess adjuvant properties. Saponin-based nanoparticles, such as ISCOMs (immunostimulating complexes), have shown to enhance the immunogenicity of various viral antigens in animal models besides reducing the toxicity of saponins (Rivera-Patron et al. 2022).
- MF59-adjuvanted nano-sized seasonal influenza vaccine (Fluad®), H1N1 pandemic influenza vaccine (Focetria® and Celtura®) have reported better antibody and T-cell responses than alum but induced pain at the injection site and reactogenicity has also been observed (Seubert et al. 2008; Ko and Kang 2018).
- Murabutide (MB), a derivative of muramyl dipeptide (MDP) had also been used as adjuvant to reduce pyrogenic effect. Similarly, polyanhydride nanoparticles, apart from carrier activity when antigen-functionalized will provide sustained release. In addition to safety, it has shown immunostimulatory effect followed by prompt uptake by the APCs and facilitate expansion of CD8+ T cells. A polyanhydride nanoparticle-based influenza A vaccine (IAV-nanovax) has shown better antibody response as well as induction of lung-resident memory CD4 and CD8 T cells in intranasal vaccination (Zacharias et al. 2018).
- Calcium phosphate NPs (CPNPs) has shown better immunostimulation against HSV-2 and Epstein–Barr virus (EBV) infections, H1N1 influenza virus in different studies (He et al. 2000; Morçöl et al. 2017).

14.9 Enhancement of Vaccine Stability and Thermal Stability During Storage

The use of nanoparticles in nucleic acid-based vaccine formulations can significantly improve the stability and bioavailability of the genetic material. By encapsulating or complexing with nucleic acids, nanoparticles can protect them from degradation by nucleases and other environmental factors, such as changes in temperature and pH. Additionally, nanoparticles can facilitate the cellular uptake of nucleic acids, ensuring their successful delivery to the target cells and enhancing the overall immunogenicity of the vaccine. Similarly, shape of NPs also has importance in cellular uptake, half-life, biodistribution and stability. For example, administration of silicon NPs having different shape has resulted in significantly differential shape-dependent organ distribution in mice; discoid-shaped NPs was found to be mostly accumulated in various organs except liver where the cylindrical particles were mostly accumulated in liver (Decuzzi et al. 2010). Size, aspect ratio, surface modification, etc. all are important factors regulating antigen delivery, presentation and immune response. It has been observed that anionic hydrogel NPs made up of PEG smaller than 100 nm can efficiently target lymph nodes and deliver antigens to the B cells and APCs to induce stronger humoral and cellular responses than free antigen (Galloway et al. 2013; Mueller et al. 2015). Similarly, nature of NP can have its own effect on stability, as polymeric NPs not only help in sustained release but improve the stability of labile drugs/antigens from in vivo enzymatic degradation (Abdelghany et al. 2019). PEG NPs is not only helpful in making RNA vaccine stable but has provided long-term storage and stability of mRNA-loaded lipid nanoparticles (LNPs) through lyophilization. This allows the nanovaccine to store at 5 °C for upto 18 months as otherwise recommended storage is at −80 °C for BNT162b2 and − 20 °C for mRNA-1273 for currently approved vaccines (Karve et al. 2018). However, recently some vaccine like FDA-approved siRNA LNPs (Patisiran) can be further stored at higher temperature like 2–8 °C for 24 months and it may be performed by removing water from RNA during capsulation (Kon et al. 2022). Similarly, nanoformulations made up of bile salts such as bilosomes, probilosomes, surface-engineered bilosomes, bile salt containing micelles have found application in oral or GIT based nanovaccines as incorporation of bile salt can stabilize the liposomes in bilosomes and also contributes to better absorption and M-cell uptake (Elnaggar 2015). Similarly, use of anionic archaebacterial membrane-based archaesomes or nanoemulsion-based carriers, uses of surfactant in nanoemulsion are other strategies which can enhance vaccine stability in GIT (Marasini et al. 2014). Recently, thermally stable protein NPs are being developed for vaccine formulation like use of a fusion protein comprised of M2 protein-derived peptide from H5N1 influenza virus and superfolder green fluorescent protein (sfGFP) which led to higher thermostability in addition to stimulation of a long-lasting humoral immune response (Wong et al. 2023). The M2e conjugated gold nanoparticle influenza vaccine in lyophilized form has shown remarkable thermal stability which can be stored at 4 °C for 3 months, 37 °C for 3 months and 50 °C for 2 weeks (Ingrole et al. 2021). Surface charge modulation of viral antigen is also another strategy that has been employed to increase the thermal stability of viral vaccines. Employing the concept that the negative surface charge of the attachment glycoprotein of Newcastle disease virus (NDV) acts as a crucial determinant of viral thermostability, surface charge engineered live and inactivated vaccines for NDV has been formulated depicting storage stability for at least 10 and 60 days, respectively, at 37 °C (Shang et al. 2022). Biomineralization of viruslike particles (VLPs) by metal-organic framework nanoparticles has also led to enhancement of thermal stability of encapsulate foot-and-mouth disease virus (FMDV) vaccine rendering it suitable for storage upto 37 °C for 7 days besides enhancing the immune responses (Teng et al. 2022). For other diseases, application of pneumococcal surface protein A

(PspA) in nanovaccine formulation using biodegradable polyanhydride NPs as encapsulating agent against *Streptococcus pneumonia* not only led to antigenic dose reduction by 25-fold for similar effect but also enhanced thermal stability for 60 days at room temperature (Wagner-Muñiz et al. 2018). Finding such novel solutions can be helpful to address the issues of temperature sensitive vaccines and logistic cold-chain maintenance.

14.10 Nanoparticles for Nucleic Acid-Based Vaccines

DNA and RNA vaccines are based on the delivery of genetic material encoding specific viral antigens to the host cells. Upon uptake, the host cells produce the encoded viral proteins, which elicit immune responses against the target virus. DNA vaccines are composed of plasmids that contain the gene encoding the target viral antigen, while RNA vaccines use either messenger RNA (mRNA) or self-amplifying RNA to achieve a similar outcome (Kim et al. 2023). Both DNA and RNA vaccines offer several advantages over conventional vaccines, including the ability to induce both humoral and cellular immune responses, the absence of live viral components, and the potential for rapid and cost-effective production. Nanoparticles have emerged as promising delivery systems for nucleic acid vaccines, due to their ability to protect the genetic material from degradation, enhance cellular uptake, and promote targeted delivery to antigen-presenting cells (Sahdev et al. 2014; Ho et al. 2021).

Cationic liposomes and lipid nanoparticles can effectively bind and encapsulate nucleic acids through electrostatic interactions. These lipid-based systems have been successfully used to deliver both DNA and RNA vaccines in animal models of viral diseases. For example, cationic liposomes have been employed for the delivery of DNA vaccines against infectious hematopoietic necrosis virus (IHNV) in fish, resulting in enhanced immune responses and protection against viral challenge (Fleiderman et al. 2007). Lipid nanoparticles (LNPs) have gained significant attention for their application in mRNA vaccine delivery. A notable example is the use of LNPs in the development of mRNA vaccines against porcine epidemic diarrhea virus (PEDV) in pigs. The LNP-encapsulated modified mRNA vaccine encoding prM and E genes of deer Powassan virus (POWV)-induced robust humoral immune response in mice along with producing neutralizing antibodies against related tick-borne flaviviruses as well as distantly related Langat virus (Van Blargan et al. 2018).

Polymeric nanoparticles, such as those based on chitosan, PLGA, and polyethylenimine (PEI), have demonstrated potential for the delivery of nucleic acid vaccines. These polymers can form complexes with DNA or RNA through electrostatic interactions, protecting the genetic material from degradation and facilitating cellular uptake. Chitosan-based nanoparticles have been used to deliver DNA vaccines against porcine circovirus type 2 (PCV2) in pigs, resulting in improved immunogenicity and protection against viral challenge (Dhakal and Renukaradhya 2019). Similarly, PLGA nanoparticles have been utilized for the delivery of a DNA vaccine against infectious bursal disease virus (IBDV) in chickens, inducing strong immune responses and conferring protection against IBDV infection (Negash et al. 2013).

Inorganic nanoparticles, such as gold nanoparticles and calcium phosphate nanoparticles, have also been explored for nucleic acid vaccine delivery. Functionalization of gold nanoparticles with dendritic cell targeting ligands can facilitate the targeted delivery of DNA vaccines as evidenced by generation of long-lasting immunity against melanoma in mice immunized with thiol-modified gold NPs functionalized with DNA encoding melanoma antigen (Gulla et al. 2019). Calcium phosphate nanoparticles have been used for the delivery of a DNA vaccine against foot-and-mouth disease virus (FMDV) in mice and guinea pigs. The calcium phosphate-FMDVP1-3CD gene construct vaccine formulation induced strong humoral and cellular immune responses, providing effective protection against FMDV infection (Joyappa et al. 2009).

14.11 Protein-Based NPs

Proteinaceous NPs are self-assembled protein structures that mimic the native conformation of viruses without containing infectious genetic material; Virus-like particles (VLPs) comprised of assembled viral proteins lacking viral genetic material, NPs made up of ferritin, encapsulin, silk fibroin, gelatine, etc. are some of their examples (Sainsbury et al. 2010; Hong et al. 2020). VLPs have been used as both antigens and adjuvants in the development of viral vaccines for animal diseases. For instance, VLP-based vaccines have been successfully developed for the control of bluetongue virus (BTV) in ruminants. Studies have demonstrated that BTV-VLP vaccines can induce potent neutralizing antibody responses and confer protection against BTV challenge in sheep (Roy et al. 1994; Van Rijn 2019). These vaccines are advantageous in terms of higher biocompatibility, safety, efficacy, and DIVA compliance, but have issues in terms of displaying versatile antigens more particularly in case of glycan antigens with complex structures, cost-intensive production, low yields and slow production process, etc. (Van Rijn 2019; Hemmati et al. 2021). Several VLP-based vaccines in recent times have been approved by the FDA, like Gardasil ® and Cervarix™ against human papillomavirus infection, Engerix-B® and Recombivax® for Hepatitis B, Hecolin® against Hepatitis E and Mosquirix™ for malaria. This indicates increase in acceptance of VLP-based vaccines among scientists and producers due to the aforesaid advantages (Martins et al. 2022). Recently, a Nano-B5 platform has been introduced for self-assembling, stable protein NPs-based nanovaccine production which can bear diverse antigens including peptides and polysaccharides showing excellent lymph node targeting and immunoresponse elicitation (Pan et al. 2020). Similarly, monovalent streptavidin (mSA)-based proteinaceous NPs showing affinity to biotin has also been employed as vaccine candidate with high immunogenicity and loading capacity for both protein and polysaccharide antigens (Shi et al. 2022). Proteinaceous NP platforms improve the immune response to antigens in vaccines, can rapidly accumulate in the thyroid and spleen through free movement into circulatory and lymph vessels, have no such toxicity and non-biodegradability concern as reported in inorganic NP-based platforms, show homogenous decoration, improved APC uptake, strong B-cell activation signalling due to multiple B-cell receptor cross-linking, activate a cell-mediated immune response, therefore can be used as vaccine candidate for various viruses and other agents (López-Sagaseta et al. 2016; Nguyen and Tolia 2021; Sung et al. 2021). Recently, bacteriophage VLPs have also been tried for vaccine production because phages are nonpathogenic, and preexisting immunity against them is absent in animals and humans, therefore, they are safer and easily taken by the DCs (Khalaj-Hedayati et al. 2020).

14.12 Nanoparticle-Based Vaccines and DIVA Strategies

Differentiating infected from vaccinated animals (DIVA) is a critical strategy in the control of infectious animal diseases, as it allows identification of the infected animals within vaccinated populations. This differentiation is essential for the implementation of appropriate disease control measures, such as culling, quarantine, or further vaccination efforts. DIVA strategies typically rely on the use of vaccines that lack specific viral proteins or epitopes present in the wild-type virus, combined with diagnostic tests that can detect the presence of antibodies or other immune markers specific to these viral components. Nanoparticle-based vaccines can be designed to support DIVA strategies by selectively presenting specific viral antigens that differentiate vaccinated from naturally infected animals. The vaccinated animals produce antibodies against non-wild-type viral protein or epitopes present in the vaccine only, which can be detected using a DIVA-compatible diagnostic test, allowing the differentiation of infected and vaccinated animals. The use of nanoparticles also facilitate controlled delivery of these antigens, which can enhance their immunogenicity and assist in the development of DIVA-compatible vaccines. VLP-based vaccines usually offer excellent DIVA compliance besides providing safety and immunogenicity against viral diseases (Singh et al. 2019).

14.13 Nanoparticle in Viral Vaccines for Animal Diseases

Polymeric nanoparticles, such as those derived from polylactic-co-glycolic acid (PLGA) and chitosan, have been extensively studied as adjuvants in viral vaccines for animal diseases. For instance, PLGA nanoparticles have been used to enhance the immunogenicity of foot-and-mouth disease virus (FMDV) vaccines. It has been observed that intranasal delivery of chitosan-coated cationic PLGA-encapsulated FMDV DNA vaccine along with IL-6 gene as a molecular adjuvant elicit strong humoral and cellular immune responses in animal models, providing a higher level of protection compared to conventional vaccines (Wang et al. 2011). Intranasal administration of PLGA-based nanovaccines against swine infuenza A virus using conserved peptide or entire inactivated virus has found to enhance epitope-specific T-cell response, enhanced CD8+ T cells and IFNγ production along with reduction in virus load in lungs of the vaccinated pigs (Dhakal and Renukaradhya 2019). The potential of PLGA NPs has also been evaluated against bovine virus diarrhea virus (BVDV) infection. Nanovaccine comprised of BVDV E2 and NS3 proteins encapsulated in PLGA NPs adjuvanted with polyinosinic : polycytidylic acid (poly(I:C) has induced specific antibody as well as cellular immune response in vaccinated Holstein–Friesian calves. The efficacy profile of the nanovaccine was found to be comparable to the commercial adjuvanted inactivated BVDV vaccine ((Riitho et al. 2017).

Chitosan nanoparticles have also shown promise as adjuvants in animal viral vaccines. Utilization of chitosan nanoparticles to deliver the hemagglutinin (HA) protein from the avian influenza virus (AIV) in chickens as mucosal booster vaccine revealed that the chitosan-HA vaccine formulation induced robust immune responses and provided effective protection against AIV challenge (Sączyńska et al. 2022). Similarly, chitosan nanoparticles have also been used to deliver HA2 and M2e from H9N2 influenza virus in both mRNA and protein formats. Incorporation of mRNAs in vaccine formulation has found to induce better immune response than delivering protein antigen only (Hajam et al. 2020). Intranasal administration of chitosan-based nanovaccines against swine infuenza A virus using whole inactivated virus has found to enhance mucosal IgA and cellular immunity; reduced virus load in lungs and also diminished virus shedding from nostrils in vaccinated pigs (Dhakal et al. 2018a). The adjuvant potential of chitosan nanoparticles has also been tested against Rift Valley fever virus (RVFV). Mice immunization with killed whole RVFV antigen adjuvanted with chitosan nanoparticles elicited strong phagocytic activity by peritoneal macrophages and neutralizing antibody response as well as elevated IL-2, IFN-γ production as the indicator of cellular immune response; while using alum adjuvant instead of chitosan nanoparticles only induced humoral immune response as evidenced by elevated IL-4 expression (El-Sissi et al. 2020). Enhanced mucosal IgA response and IFNγ expression were observed against infectious bronchitis virus (IBV) in chickens oculo-nasally immunized by chitosan nanoparticles-loaded with inactivated IBV antigen (Renu and Renukaradhya 2020). Similarly, Quil-A and chitosan (QAC) has been deployed as mucosal adjuvant for intranasal delivery of nucleic acid immunogens against infectious bronchitis. The nanovaccine was found to be safe, stable, and capable of sustained release of payload and provided enhanced humoral and cellular immune responses. In addition, the immunized birds showed reduced clinical severity and viral shedding in challenge study (Chandrasekar et al. 2020). The potential of chitosan NPs as carrier in nanovaccine has been tested against Newcastle disease virus (NDV). NDV La Sota strain encapsulated in N-2-hydroxypropyl trimethyl ammonium chloride chitosan nanoparticles elicited stronger cellular, humoral as well as mucosal immune responses as compared to commercial live attenuated Newcastle disease vaccine. The nanoformulation also exhibited high stability, low toxicity, and sustained release properties (Zhao et al. 2016). Polyanhydride nanoparticles along whole killed virus antigen without or with a TLR-9 agonist CpG oligodeoxynucleotide adjuvant have also been used for immunization in pigs against swine infuenza A virus resulting in enhanced cell-mediated and mucosal antibody responses, significantly diminished virus shedding and virus titers from respiratory tract, and better protection after heterologous challenge than intranasal administration of soluble antigens (Dhakal et al. 2017; Dhakal et al. 2019a).

A chemically modified phytoglycogen, a dendrimer-like α-D-glucan nanoparticle termed as Nano-11 has also been used for vaccine formulation. Nano-11 has a positive surface charge which enables binding of negatively charged peptide antigens through electrostatic adsorption. Nano-11 adjuvanted with whole killed virus antigen has been employed for vaccination in pigs against swine infuenza A virus resulting in enhanced cross-reactive mucosal IgA antibody response in the vaccinated animals. Intranasal vaccination mitigated influenza signs and pneumonic lesions along with mild decline in virus load in the respiratory tract of the vaccinated animals (Dhakal et al. 2019b).

Lipid-based nanoparticles, including liposomes and solid lipid nanoparticles (SLNs), have been employed as adjuvants in various viral vaccines for animal diseases. Liposomes are vesicular structures composed of phospholipid bilayers, which can encapsulate antigens and immunostimulatory molecules to enhance immunogenicity. For example, liposomal formulations have been developed for porcine circovirus 2 (PCV2). It has been observed that application of liposome as an adjuvant for PCV2 DNA vaccine can induce strong cellular and humoral immune responses in pigs, resulting in improved protection against PCV2 challenge (Park et al. 2017). Similarly, liposome has been used to encapsulate multiple highly conserved B- and T-cell epitope peptide antigens of swine influenza A virus followed by adjuvantation with monosodium urate crystals. Intranasal administration of the nanovaccine has improved specific cellular and mucosal humoral immune responses with reduction in viral load and shedding as well as influenza symptoms in the vaccinated pigs during H1N1 influenza challenge (Dhakal et al. 2018b). Solid lipid nanoparticles (SLNs) are another class of lipid-based nanoparticles that have gained attention as adjuvants in animal viral vaccines. Packaging of recombinant porcine IL-2 in SLN has enhanced the immune response as evident from the immunization study in mice with FMDV antigen combined with the SLN-porcine IL-2 adjuvant (Chen et al. 2014).

The potential of protein-based nanoparticles in nanovaccines against viral livestock diseases have also been explored in recent time. Self-assembling ferritin (fe) nanocages have been customized that display E2 glycoprotein of classical swine fever virus (CSFV) on their surface. The nanoformulation induced strong ferritin-assisted humoral and cellular immunities as compared to antigen delivered without nanoparticles (Zhao et al. 2021).

Inorganic nanoparticles, such as gold nanoparticles (AuNPs) and silica nanoparticles (SiNPs), have been explored as adjuvants for animal viral vaccines. Conjugation of matrix 2 protein (M2e) of influenza A virus with 12-nm spherical gold nanoparticles followed by intranasal immunization of mice with the nanoformulation along with soluble CpG as adjuvant conferred full protection against challenge with PR8-H1N1 influenza virus (Tao et al. 2014). Intranasal vaccination in mice with dual-conjugated gold nanoparticles containing recombinant influenza hemagglutinin trimers and TLR5 agonist flagellin (FliC) elicited higher antibody responses than the soluble antigen mixture. The nanoparticles can simultaneously provide adjuvantation as well as antigen presentation with multivalent cross-linking option to elicit stronger immune response (Wang et al. 2017). Higher antibody titers had been reported for foot-and-mouth disease virus (FMDV) vaccine with the use of citrate-stabilized AuNPs conjugated with a synthetic FMDV peptide; a similar observation in terms of induced inflammatory cytokine production, including tumor necrosis factor-α (TNF-α), interleukin (IL)-6, IL-12, and granulocyte macrophage colony-stimulating factor (GM-CSF) for West Nile virus envelope protein (WNE) containing nanovaccine was reported with AuNPs of 40 nm size (Chen et al. 2010; Niikura et al. 2013).

Layered double metal hydroxide nanoparticles have shown to be having good antigen carrying potential to elicit strong immune response in nanovaccine formulations. FMDV adsorbed on layered double metal hydroxide nanoparticles with particle size of ~87.21 nm having hexagonal shape, and positive zeta charge of 32 mV has been used to immunize pigs and mice. The nanoparticles depicted promising adjuvant potential along with neutralizing antibody level persisting even after 56 day in 90% pigs indicating slow release capacity of the nanoformulation (Wu et al. 2020).

Silica nanoparticles have also been employed as adjuvants in viral vaccines for animals. Silica nanovesicles have been used as an adjuvant as well as carrier in a bovine viral diarrhea virus (BVDV) vaccine. Large cavity (~40 nm) of the vesicles allowed enhanced antigen loading capacity along with controlled release option. Mice immunization with the nanovaccine containing codon-optimized E2 (oE2) protein, an important immunogenic determinant of BVDV and Quil A elicited stronger humoral and cellular immune responses compared to soluble antigen with Quil A (Mody et al. 2014).

Porcine reproductive and respiratory virus (PRRSV) infection is one of the prominent diseases in pigs. Intranasal administration of VLP-based nanovaccines against the disease using various subunit antigens from PRRSV such as GP5, GP4, GP3, GP2a, N, M, and E proteins or whole inactivated virus with or without adjuvants resulted increased number of NK cells, γδ cells and IFNα cytokine; enhanced CD8+ T cells and IFNγ response providing improved protection against homologous and heterologous challenge (Dhakal and Renukaradhya 2019). VLP-based nanovaccine using HA, NA, M1 antigens with Emulsigen adjuvant also induced serum IgG, mucosal IgA, and virus neutralizing antibody responses against swine influenza A virus in pigs (Pyo et al. 2012). Similarly, VLP-based vaccine using various viral antigenic proteins and immunodominant peptide epitopes have also been employed against Foot and mouth disease virus, encephalomyocarditis virus (EMCV), Japanese encephalitis virus (JEV), porcine circovirus 2 (PCV2), porcine parvo virus (PPV), etc. eliciting specific antibody response as well as cellular immune responses in the vaccinated animals (Dhakal and Renukaradhya 2019).

There are ample examples which indicate potential benefits of nanomaterials in vaccine development. However, in veterinary subject the animal diversity is enormous and responses in immune system may differ subject-wise. The aspects of innate and adaptive immune functions are mostly similar in ruminants, rodents, humans, and other animals but subtle differences may be there. For example, cattle express TLR-1 to TLR-10 similar to humans, but specific differences in disease susceptibility, pathogen recognition and subsequent signalling exists between them. Similarly, predominance of γδ T cells in ruminants is there with respect to human and rodents. Such variations must be considered during customization of NP-based vaccine formulations and experiments (Werling et al. 2009; Maina et al. 2020).

14.14 Approved or Clinically Tested Nanomaterial-Based Vaccines

As discussed in the chapter, there are diverse advantages which can be availed by using nanomaterials in vaccine formulations. The diversity in materials which can be used in nanoformulations for vaccines is enormous and can further be enhanced through functionalization. Such nanoformulations are progressively approaching towards bench-to-field transition to serve both mankind and animal health sector. Various candidates nanovaccines are under clinical trial such as mRNA-1273 SARS-CoV-2 vaccine by Moderna using mRNA encapsulated in LNP, mRNA BNT162b2 SARS-CoV-2 vaccine by BioNTech and Pfizer using mRNA encapsulated in LNPs, NVX-CoV2373 SARS-CoV-2 vaccine by Novavax using recombinant SARS-CoV-2 S glycoprotein NP with saponin-based Matrix-M1 adjuvant, SARS-CoV VLP nanovaccine by Novavax, ADDomer©, a synthetic self-assembling adenovirus-derived multi-epitope-based VLP vaccine by Imophoron and Bristol University against influenza virus, Inflexal®V using virosome with influenza virus surface antigens (hemagglutinin and neuraminidase) by Crucell, Epaxal® using virosome with inactivated virus particles against Hepatitis A developed by Crucell, Gardasil®9 and Cervarix® against human papillomavirus using capsomere (major capsid protein L1) developed by Merck and GSK respectively, NanoFlu™ against influenza virus using recombinant HA protein on Tween 80 NP with Matrix-M adjuvant developed by Novavax, ResVax™ against respiratory syncytial virus using recombinant near full-length RSV F protein on Tween 80 NP with/without alum adjuvant developed by Novavax, VaxiSome™-influenza using ceramide carbamoyl-spermine/cholesterol liposomal adjuvant/delivery system combined with commercial influenza vaccine developed by NasVax, etc. (Petkar et al. 2021; Azharuddin et al. 2022).

14.15 Challenges and Future Directions

One of the most important challenges in developing NP-based veterinary vaccines is relatively lesser focus towards livestock than humans or even lab animals in spite of the fact that a considerable percentage of infectious diseases are zoonotic and animals act as carrier or reservoir hosts. Furthermore, cold-chain maintenance during transport and storage of animal vaccines is a real challenge, particularly for under-developed and developing countries with limited facility. Intense trade, transport and growing population have enhanced the pathogen exposure, thus the chance of zoonotic transmission. So, containment of the disease within animal population can play significantly to secure 'one health' perspective. Another aspect to address is the short protection time and multiple boostering requirements of conventional vaccines. Nanovaccine with targeted delivery and sustained release can provide prolonged immunity with single administration. Further, ample surface functionality and large cargo capacity of nanoparticles can facilitate multiplexing to target multiple pathogens. Such strategies can be economical and beneficial to the wild animals also. Despite the promising results achieved using nanoparticle-based adjuvants in animal viral vaccines, there are still several challenges that need to be addressed. These include issues related to biocompatibility, immunological barriers, integration with state-of-the-art technologies and platforms, manufacturing time, cost-intensiveness, regulatory considerations, safety, and scalability. However, persistent research and development efforts, alongside advances in nanotechnology and immunology are likely to enable the development of increasingly effective nano-adjuvants and vaccine carriers for a wide-range of animal diseases. It is expected that further advancements in nanoparticle synthesis, characterization, and delivery strategies will lead to even more effective and versatile vaccine platforms. The integration of nanoparticle-based vaccines with other emerging technologies, such as gene editing, multi-omics platforms, advanced diagnostics, artificial intelligence, and novel manufacturing methods, will further enhance their potential impact on animal health and disease management.

References

Abdelghany S, Parumasivam T, Pang A, Roediger B, Tang P, Jahn K, Chan HK (2019) Alginate modified-PLGA nanoparticles entrapping amikacin and moxifloxacin as a novel host-directed therapy for multidrug-resistant tuberculosis. J Drug Delivery Sci Technol 52:642–651. https://doi.org/10.1016/j.jddst.2019.05.025

Azharuddin M, Zhu GH, Sengupta A, Hinkula J, Slater NKH, Patra HK (2022) Nano toolbox in immune modulation and nanovaccines. Trends Biotechnol 40(10):1195–1212. https://doi.org/10.1016/j.tibtech.2022.03.011

Beningo KA, Wang YL (2002) Fc-receptor-mediated phagocytosis is regulated by mechanical properties of the target. J Cell Sci 115(Pt 4):849–856. https://doi.org/10.1242/jcs.115.4.849

Bucarey SA, Pujol M, Poblete J, Nuñez I, Tapia CV, Neira-Carrillo A, Bassa O (2014) Chitosan microparticles loaded with yeast-derived PCV2 virus-like particles elicit antigen-specific

cellular immune response in mice after oral administration. Virol J 11:1–12. https://doi.org/10.1186/1743-422X-11-149
Carpenter TE, O'Brien JM, Hagerman AD, McCarl BA (2011) Epidemic and economic impacts of delayed detection of foot-and-mouth disease: a case study of a simulated outbreak in California. J Vet Diagn Investig 23(1):26–33. https://doi.org/10.1177/104063871102300104
Carpenter A, Waltenburg MA, Hall A, Kile J, Killerby M, Knust B, Negron M, Nichols M, Wallace RM, Behravesh CB, McQuiston JH, The Vaccine Preventable Zoonotic Disease Working Group (2022) Vaccine preventable zoonotic diseases: challenges and opportunities for public health Progress. Vaccine 10(7):993. https://doi.org/10.3390/vaccines10070993
CDC (2021). Zoonotic Diseases. https://www.cdc.gov/onehealth/basics/zoonotic-diseases.html
Chandrasekar SS, Kingstad-Bakke B, Wu CW, Suresh M, Talaat AM (2020) A novel mucosal adjuvant system for immunization against avian coronavirus causing infectious bronchitis. J Virol 94(19):e01016–e01020. https://doi.org/10.1128/JVI.01016-20
Chen YS, Hung YC, Lin WH, Huang GS (2010) Assessment of gold nanoparticles as a size-dependent vaccine carrier for enhancing the antibody response against synthetic foot-and-mouth disease virus peptide. Nanotechnology 21(19):195101. https://doi.org/10.1088/0957-4484/21/19/195101
Chen G, Zeng S, Jia H, He X, Fang Y, Jing Z, Cai X (2014) Adjuvant effect enhancement of porcine interleukin-2 packaged into solid lipid nanoparticles. Res Vet Sci 96(1):62–68. https://doi.org/10.1016/j.rvsc.2013.11.017
Costa T, Akdeniz N (2019) A review of the animal disease outbreaks and biosecure animal mortality composting systems, vol 90. Waste management, New York, N.Y, pp 121–131. https://doi.org/10.1016/j.wasman.2019.04.047
DAHD, Govt. of India, (2022). Lumpy skin disease. https://pqals.nic.in/annex/1710/AU1090.pdf
Decuzzi P, Godin B, Tanaka T, Lee SY, Chiappini C, Liu X, Ferrari M (2010) Size and shape effects in the biodistribution of intravascularly injected particles. J Control Release 141(3): 320–327. https://doi.org/10.1016/j.jconrel.2009.10.014
Dhakal S, Renukaradhya GJ (2019) Nanoparticle-based vaccine development and evaluation against viral infections in pigs. Vet Res 50(1):1–14. https://doi.org/10.1186/s13567-019-0712-5
Dhakal S, Goodman J, Bondra K, Lakshmanappa YS, Hiremath J, Shyu DL, Ouyang K, Kang KI, Krakowka S, Wannemuehler MJ, Won Lee C, Narasimhan B, Renukaradhya GJ (2017) Polyanhydride nanovaccine against swine influenza virus in pigs. Vaccine 35(8):1124–1131. https://doi.org/10.1016/j.vaccine.2017.01.019
Dhakal S, Renu S, Ghimire S, Shaan Lakshmanappa Y, Hogshead BT, Lu F, Hogen Esch H, Krakowka S, Lee CW, Renukaradhya GJ (2018a) Mucosal immunity and protective efficacy of intranasal inactivated influenza vaccine is improved by chitosan nanoparticle delivery in pigs. Front Immunol 9:934. https://doi.org/10.3389/fimmu.2018.00934
Dhakal S, Cheng X, Salcido J, Renu S, Bondra K, Lakshmanappa YS, Misch C, Ghimire S, Feliciano-Ruiz N, Hogshead B, Krakowka S, Carson K, McDonough J, Lee CW, Renukaradhya GJ (2018b) Liposomal nanoparticle-based conserved peptide influenza vaccine and monosodium urate crystal adjuvant elicit protective immune response in pigs. Int J Nanomedicine 13:6699–6715. https://doi.org/10.2147/IJN.S178809
Dhakal S, Ghimire S, Renu S, Ross KA, Lakshmanappa YS, Hogshead BT, Bernardo P, Lee CW, Wannemuehler MJ, Narasimhan B, Renukaradhya GJ (2019a) Evaluation of CpG-ODN-adjuvanted polyanhydride-based intranasal influenza nanovaccine in pigs. Vet Microbiol 237: 108401. https://doi.org/10.1016/j.vetmic.2019.108401
Dhakal S, Lu F, Ghimire S, Renu S, Lakshmanappa YS, Hogshead BT, Ragland D, Hogen Esch H, Renukaradhya GJ (2019b) Corn-derived alpha-D-glucan nanoparticles as adjuvant for intramuscular and intranasal immunization in pigs. Nanomedicine 16:226–235. https://doi.org/10.1016/j.nano.2018.12.003
Doshi N, Zahr AS, Bhaskar S, Lahann J, Mitragotri S (2009) Red blood cell-mimicking synthetic biomaterial particles. Proc Natl Acad Sci U S A 106(51):21495–21499. https://doi.org/10.1073/pnas.0907127106

Elnaggar YS (2015) Multifaceted applications of bile salts in pharmacy: an emphasis on nanomedicine. Int J Nanomedicine 10:3955. https://doi.org/10.2147/IJN.S82558

El-Sissi AF, Mohamed FH, Danial NM, Gaballah AQ, Ali KA (2020) Chitosan and chitosan nanoparticles as adjuvant in local Rift Valley fever inactivated vaccine. 3. Biotech 10(3):88. https://doi.org/10.1007/s13205-020-2076-y

Feng H, Feng Y, Lin L, Wu D, Liu Q, Li H, Zhang L (2022) Mannose receptor-mediated carbon nanotubes as an antigen delivery system to enhance immune response both in vitro and in vivo. Int J Mol Sci 23(8):4239. https://doi.org/10.3390/ijms23084239

Fleiderman K, Hernandez R, Cayo O,Ferrari V & Cuevas Ñ (2007). *Dna vaccines for fish.* (Paris Patent No.EP2113511A1). European Patent Office. https://patents.google.com/patent/EP2113511A1

Foged C, Brodin B, Frokjaer S, Sundblad A (2005) Particle size and surface charge affect particle uptake by human dendritic cells in an in vitro model. Int J Pharm 298(2):315–322. https://doi.org/10.1016/j.ijpharm.2005.03.035

Fries CN, Curvino EJ, Chen JL, Permar SR, Fouda GG, Collier JH (2021) Advances in nanomaterial vaccine strategies to address infectious diseases impacting global health. Nat Nanotechnol 16(4):1–14. https://doi.org/10.1038/s41565-020-0739-9

Galloway AL, Murphy A, DeSimone JM, Di J, Herrmann JP, Hunter ME, Kindig JP, Malinoski FJ, Rumley MA, Stoltz DM, Templeman TS, Hubby B (2013) Development of a nanoparticle-based influenza vaccine using the PRINT technology. Nanomedicine 9(4):523–531. https://doi.org/10.1016/j.nano.2012.11.001

Gulla SK, Rao BR, Moku G, Jinka S, Nimmu NV, Khalid S, Chaudhuri A (2019) In vivo targeting of DNA vaccines to dendritic cells using functionalized gold nanoparticles. Biomater Sci 7(3): 773–788. https://doi.org/10.1039/c8bm01272e

Hajam IA, Senevirathne A, Hewawaduge C, Kim J, Lee JH (2020) Intranasally administered protein coated chitosan nanoparticles encapsulating influenza H9N2 HA2 and M2e mRNA molecules elicit protective immunity against avian influenza viruses in chickens. Vet Res 51(1): 37. https://doi.org/10.1186/s13567-020-00762-4

Hall S (1984) The narrative construction of reality: an interview with Stuart Hall. - reprinted from a transcript of an interview by John O'Hara on the ABC programme "double take", 5 may 1983. South Rev 17(1):3–17. https://doi.org/10.3316/ielapa.840706187

He Q, Mitchell AR, Johnson SL, Wagner-Bartak C, Morcol T, Bell SJ (2000) Calcium phosphate nanoparticle adjuvant. Clin Diagn Lab Immunol 7(6):899–903. https://doi.org/10.1128/CDLI.7.6.899-903.2000

Hemmati F, Hemmati-Dinarvand M, Karimzade M, Rutkowska D, Eskandari MH, Khanizadeh S, Afsharifar A (2021) Plant-derived VLP: a worthy platform to produce vaccine against SARS-CoV-2. Biotechnol Lett 1-13:45. https://doi.org/10.1007/s10529-021-03211-0

Ho W, Gao M, Li F, Li Z, Zhang XQ, Xu X (2021) Next-generation vaccines: nanoparticle-mediated dna and mrna delivery. Adv Healthc Mater 10(8):2001812. https://doi.org/10.1002/adhm.202001812

Hong S, Choi DW, Kim HN, Park CG, Lee W, Park HH (2020) Protein-based nanoparticles as drug delivery systems. Pharmaceutics 12(7):604. https://doi.org/10.3390/pharmaceutics12070604

Howard GP, Verma G, Ke X, Thayer WM, Hamerly T, Baxter VK, Lee JE, Dinglasan RR, Mao HQ (2019) Critical size limit of biodegradable nanoparticles for enhanced lymph node trafficking and Paracortex penetration. Nano Res 12(4):837–844. https://doi.org/10.1007/s12274-019-2301-3

Hu CM, Zhang L, Aryal S, Cheung C, Fang RH, Zhang L (2011) Erythrocyte membrane-camouflaged polymeric nanoparticles as a biomimetic delivery platform. Proc Natl Acad Sci U S A 108(27):10980–10985. https://doi.org/10.1073/pnas.1106634108

Ingrole RS, Tao W, Joshi G, Gill HS (2021) M2e conjugated gold nanoparticle influenza vaccine displays thermal stability at elevated temperatures and confers protection to ferrets. Vaccine 39(34):4800–4809. https://doi.org/10.1016/j.vaccine.2021.07.032

Joyappa DH, Kumar CA, Banumathi N, Reddy GR, Suryanarayana VV (2009) Calcium phosphate nanoparticle prepared with foot and mouth disease virus P 1-3CD gene construct protects mice and guinea pigs against the challenge virus. Vet Microbiol 139(1–2):58–66. https://doi.org/10.1016/j.vetmic.2009.05.004

Karve S, De Rosa F, Bhavsar Z, Heartlein M (2018) Process of Preparing mRNA - Loaded Lipid Nanoparticles, vol 1. https://patents.google.com/patent/US20180153822A1/en

Khalaj-Hedayati A, Chua CLL, Smooker P, Lee KW (2020) Nanoparticles in influenza subunit vaccine development: Immunogenicity enhancement. Influenza Other Respir Viruses 14(1): 92–101. https://doi.org/10.1111/irv.12697

Kim B, Hosn RR, Remba T, Yun D, Li N, Abraham W, Irvine DJ (2023) Optimization of storage conditions for lipid nanoparticle-formulated self-replicating RNA vaccines. J Control Release 353:241–253. https://doi.org/10.1016/j.jconrel.2022.11.022

Ko EJ, Kang SM (2018) Immunology and efficacy of MF59-adjuvanted vaccines. Hum Vaccin Immunother 14(12):3041–3045. https://doi.org/10.1080/21645515.2018.1495301

Kodali V, Littke MH, Tilton SC, Teeguarden JG, Shi L, Frevert CW, Thrall BD (2013) Dysregulation of macrophage activation profiles by engineered nanoparticles. ACS Nano 7(8):6997–7010. https://doi.org/10.1021/nn402145t

Kon E, Elia U, Peer D (2022) Principles for designing an optimal mRNA lipid nanoparticle vaccine. Curr Opin Biotechnol 73:329–336. https://doi.org/10.1016/j.copbio.2021.09.016

Kopp M, Aufderhorst UW, Alt M, Dittmer U, Eis-Hübinger AM, Giebel B, Roggendorf M, Epple M, Krawczyk A (2019) Induction of herpes simplex virus type 1 cell-to-cell spread inhibiting antibodies by a calcium phosphate nanoparticle-based vaccine. Nanomed Nanotechnol Biol Med 16:138–148. https://doi.org/10.1016/j.nano.2018.12.002

Li H, Li Y, Jiao J, Hu HM (2011) Alpha-alumina nanoparticles induce efficient autophagy-dependent cross-presentation and potent antitumour response. Nat Nanotechnol 6(10): 645–650. https://doi.org/10.1038/nnano.2011.153

Li W, Balachandran YL, Hao Y, Hao X, Li R, Nan Z, Liu Y (2018) Amantadine surface-modified silver nanorods improves immunotherapy of HIV vaccine against HIV-infected cells. ACS Appl Mater Interfaces 10(34):28494–28501. https://doi.org/10.1021/acsami.8b10948

López-Sagaseta J, Malito E, Rappuoli R, Bottomley MJ (2016) Self-assembling protein nanoparticles in the design of vaccines. Comput Struct Biotechnol J 14:58–68. https://doi.org/10.1016/j.csbj.2015.11.001

Maina TW, Grego EA, Boggiatto PM, Sacco RE, Narasimhan B, McGill JL (2020) Applications of Nanovaccines for disease prevention in cattle. Front Bioeng Biotechnol 8:608050. https://doi.org/10.3389/fbioe.2020.608050

Marasini N, Skwarczynski M, Toth I (2014) Oral delivery of nanoparticle-based vaccines. Expert Rev Vaccines 13(11):1361–1376. https://doi.org/10.1586/14760584.2014.936852

Martins SA, Santos J, Silva RD, Rosa C, Cabo Verde S, Correia JD, Melo R (2022) How promising are HIV-1-based virus-like particles for medical applications. Front Cell Infect Microbiol 12: 997875. https://doi.org/10.3389/fcimb.2022.997875

Meeusen EN, Walker J, Peters A, Pastoret PP, Jungersen G (2007) Current status of veterinary vaccines. Clin Microbiol Rev 20(3):489–510. https://doi.org/10.1128/CMR.00005-07

Mody KT, Mahony D, Zhang J, Cavallaro AS, Zhang B, Popat A, Mahony TJ, Yu C, Mitter N (2014) Silica vesicles as nanocarriers and adjuvants for generating both antibody and T-cell mediated immune resposes to bovine viral Diarrhoea virus E2 protein. Biomaterials 35(37): 9972–9983. https://doi.org/10.1016/j.biomaterials.2014.08.044

Moon JJ, Suh H, Li AV, Ockenhouse CF, Yadava A, Irvine DJ (2012) Enhancing humoral responses to a malaria antigen with nanoparticle vaccines that expand Tfh cells and promote germinal center induction. Proc Natl Acad Sci U S A 109(4):1080–1085. https://doi.org/10.1073/pnas.1112648109

Morçöl T, Hurst BL, Tarbet EB (2017) Calcium phosphate nanoparticle (CaPNP) for dose-sparing of inactivated whole virus pandemic influenza a (H1N1) 2009 vaccine in mice. Vaccine 35(35 Pt B):4569–4577. https://doi.org/10.1016/j.vaccine.2017.07.016

Mueller SN, Tian S, DeSimone JM (2015) Rapid and persistent delivery of antigen by lymph node targeting PRINT nanoparticle vaccine carrier to promote humoral immunity. Mol Pharm 12(5): 1356–1365. https://doi.org/10.1021/mp500589c

Murphy AA, Redwood AJ, Jarvis MA (2016) Self-disseminating vaccines for emerging infectious diseases. Expert Rev Vaccines 15(1):31–39. https://doi.org/10.1586/14760584.2016.1106942

Nagatomo D, Taniai M, Ariyasu H, Taniguchi M, Aga M, Ariyasu T, Ohta T, Fukuda S (2015) Cholesteryl pullulan encapsulated TNF-α nanoparticles are an effective mucosal vaccine adjuvant against influenza virus. Biomed Res Int 2015:1. https://doi.org/10.1155/2015/471468

National Dairy Development Board (2017). Milk production in India. https://www.nddb.coop/information/stats/milkprodindia

Negash T, Liman M, Rautenschlein S (2013) Mucosal application of cationic poly (D, L-lactide-co-glycolide) microparticles as carriers of DNA vaccine and adjuvants to protect chickens against infectious bursal disease. Vaccine 31(36):3656–3662. https://doi.org/10.1016/j.vaccine.2013.06.011

Nguyen B, Tolia NH (2021) Protein-based antigen presentation platforms for nanoparticle vaccines. NPJ Vaccines 6(1):70. https://doi.org/10.1038/s41541-021-00330-7

Niikura K, Matsunaga T, Suzuki T, Kobayashi S, Yamaguchi H, Orba Y, Kawaguchi A, Hasegawa H, Kajino K, Ninomiya T, Ijiro K, Sawa H (2013) Gold nanoparticles as a vaccine platform: influence of size and shape on immunological responses in vitro and in vivo. ACS Nano 7(5):3926–3938. https://doi.org/10.1021/nn3057005

Pan C, Wu J, Qing S, Zhang X, Zhang L, Yue H, Wang H (2020) Biosynthesis of self-assembled proteinaceous nanoparticles for vaccination. Adv Mater 32(42):2002940. https://doi.org/10.1002/adma.202002940

Park C, Jeong J, Choi K, Park SJ, Kang I, Chae C (2017) Development of porcine circovirus 2 (PCV2) open reading frame 2 DNA vaccine with different adjuvants and comparison with commercial PCV2 subunit vaccine in an experimental challenge. Can J Vet Res 81(3): 171–177. PMID: 28725106

Petkar KC, Patil SM, Chavhan SS, Kaneko K, Sawant KK, Kunda NK, Saleem IY (2021) An overview of Nanocarrier-based adjuvants for vaccine delivery. Pharmaceutics 13(4):455. https://doi.org/10.3390/pharmaceutics13040455

Poon C, Patel A (2020) Organic and inorganic nanoparticle vaccines for prevention of infectious diseases. Nano Express 1:012001. https://doi.org/10.1088/2632-959X/ab8075

Pyo HM, Masic A, Woldeab N, Embury-Hyatt C, Lin L, Shin YK, Song JY, Babiuk S, Zhou Y (2012) Pandemic H1N1 influenza virus-like particles are immunogenic and provide protective immunity to pigs. Vaccine 30(7):1297–1304. https://doi.org/10.1016/j.vaccine.2011.12.083

Reddy ST, Rehor A, Schmoekel HG, Hubbell JA, Swartz MA (2006) In vivo targeting of dendritic cells in lymph nodes with poly(propylene sulfide) nanoparticles. J Control Release 112(1): 26–34. https://doi.org/10.1016/j.jconrel.2006.01.006

Reddy ST, Van Der Vlies AJ, Simeoni E, Angeli V, Randolph GJ, O'Neil CP, Hubbell JA (2007) Exploiting lymphatic transport and complement activation in nanoparticle vaccines. Nat Biotechnol 25(10):1159–1164. https://doi.org/10.1038/nbt1332

Renu S, Renukaradhya GJ (2020) Chitosan nanoparticle based mucosal vaccines delivered against infectious diseases of poultry and pigs. Front Bioeng Biotechnol 8:558349. https://doi.org/10.3389/fbioe.2020.558349

Riitho V, Walters AA, Somavarapu S, Lamp B, Rumenapf T, Krey T, Graham SP (2017) Design and evaluation of the immunogenicity and efficacy of a biomimetic particulate formulation of viral antigens. Sci Rep 7(1):13743. https://doi.org/10.1038/s41598-017-13915-x

Rivera-Patron M, Cibulski SP, Miraballes I, Silveira F (2022) Formulation of IMXQB: nanoparticles nanoparticles based on QuillajabrasiliensisQuillajabrasiliensis (Q. Brasiliensis) Saponins Saponins to be used as vaccine adjuvants vaccine adjuvants. In: Plant secondary metabolism engineering: methods and protocols. Springer US, New York, NY, pp 183–191. https://doi.org/10.1007/978-1-0716-2185-1_15

Roy P, Bishop DH, LeBlois H, Erasmus BJ (1994) Long-lasting protection of sheep against bluetongue challenge after vaccination with virus-like particles: evidence for homologous and partial heterologous protection. Vaccine 12(9):805–811. https://doi.org/10.1016/0264-410x(94)90289-5

Sączyńska V, Romanik-Chruścielewska A, Florys-Jankowska K, Cecuda-Adamczewska V, Kęsik-Brodacka M (2022) Chitosan-based formulation of hemagglutinin antigens for oculo-nasal booster vaccination of chickens against influenza viruses. Vet Immunol Immunopathol 247: 110406. https://doi.org/10.1016/j.vetimm.2022.110406

Sahdev P, Ochyl LJ, Moon JJ (2014) Biomaterials for nanoparticle vaccine delivery systems. Pharm Res 31:2563–2582. https://doi.org/10.1007/s11095-014-1419-y

Sainsbury F, Cañizares MC, Lomonossoff GP (2010) Cowpea mosaic virus: the plant virus–based biotechnology workhorse. Annu Rev Phytopathol 48:437–455. https://doi.org/10.1586/erv.10.115

Senapati, S. (2020). Combination nanoadjuvant-based vaccine against influenza virus for older adults. https://dr.lib.iastate.edu/server/api/core/bitstreams/3c71529c-190b-4d67-9997-bfec8114c2b6/content

Seubert A, Monaci E, Pizza M, O'Hagan DT, Wack A (2008) The adjuvants aluminum hydroxide and MF59 induce monocyte and granulocyte chemoattractants and enhance monocyte differentiation toward dendritic cells. J Immunol 180(8):5402–5412. https://doi.org/10.4049/jimmunol.180.8.5402

Shang Y, Li L, Zhang T, Luo Q, Yu Q, Zeng Z, Wen G (2022) Quantitative regulation of the thermal stability of enveloped virus vaccines by surface charge engineering to prevent the self-aggregation of attachment glycoproteins. PLoS Pathog 18(6):e1010564. https://doi.org/10.1371/journal.ppat.1010564

Shen L, Higuchi T, Tubbe I, Voltz N, Krummen M, Pektor S, Bros M (2013) A trifunctional dextran-based nanovaccine targets and activates murine dendritic cells, and induces potent cellular and humoral immune responses in vivo. PLoS One 8(12):e80904. https://doi.org/10.1371/journal.pone.0080904

Shi Y, Pan C, Wang K, Liu Y, Sun Y, Guo Y, Wang H (2022) Construction of orthogonal modular proteinaceous Nanovaccine delivery vectors based on mSA-biotin binding. Nano 12(5):734. https://doi.org/10.3390/nano12050734

Singh RK, Sharma GK, Mahajan S, Dhama K, Basagoudanavar SH, Hosamani M (2019) Foot-and-mouth disease virus: immunobiology, advances in vaccines and vaccination strategies addressing vaccine failures—an Indian perspective. Vaccines (Basel) 7(3):90. https://doi.org/10.3390/vaccines7030090

Sung HD, Kim N, Lee Y, Lee EJ (2021) Protein-based nanoparticle vaccines for SARS-CoV-2. Int J Mol Sci 22(24):13445. https://doi.org/10.3390/ijms222413445

Tao W, Ziemer KS, Gill HS (2014) Gold nanoparticle-M2e conjugate coformulated with CpG induces protective immunity against influenza a virus. Nanomedicine (London) 9(2):237–251. https://doi.org/10.2217/nnm.13.58

Teng Z, Hou F, Bai M, Li J, Wang J, Wu J, Guo H (2022) Bio-mineralization of virus-like particles by metal–organic framework nanoparticles enhances the thermostability and immune responses of the vaccines. J Mater Chem B 10(15):2853–2864. https://doi.org/10.1039/D1TB02719K

Tukhvatulin A, Dzharullaeva A, Erokhova A, Zemskaya A, Balyasin M, Ozharovskaia T, Gintsburg A (2020) Adjuvantation of an influenza hemagglutinin antigen with TLR4 and NOD2 agonists encapsulated in poly (D, L-lactide-co-glycolide) nanoparticles enhances immunogenicity and protection against lethal influenza virus infection in mice. Vaccine 8(3):519. https://doi.org/10.3390/vaccines8030519

Van Blargan LA, Himansu S, Foreman BM, Ebel GD, Pierson TC, Diamond MS (2018) An mRNA vaccine protects mice against multiple tick-transmitted flavivirus infections. Cell Rep 25(12): 3382–3392. https://doi.org/10.1016/j.celrep.2018.11.082

Van Rijn PA (2019) Prospects of next-generation vaccines for bluetongue. Front Vet Sci 6:407. https://doi.org/10.3389/fvets.2019.00407

Wagner-Muñiz DA, Haughney SL, Kelly SM, Wannemuehler MJ, Narasimhan B (2018) Room temperature stable PspA-based nanovaccine induces protective immunity. Front Immunol 9: 325. https://doi.org/10.3389/fimmu.2018.00325

Wamhoff EC, Schulze J, Bellmann L, Rentzsch M, Bachem G, Fuchsberger FF, Rademacher C (2019) A specific, glycomimetic langerin ligand for human langerhans cell targeting. ACS Cent Sci 5(5):808–820. https://doi.org/10.1021/acscentsci.9b00093

Wang G, Pan L, Zhang Y, Wang Y, Zhang Z, Lü J, Zhou P, Fang Y, Jiang S (2011) Intranasal delivery of cationic PLGA nano/microparticles-loaded FMDV DNA vaccine encoding IL-6 elicited protective immunity against FMDV challenge. PLoS One 6(11):e27605. https://doi.org/10.1371/journal.pone.0027605

Wang C, Zhu W, Wang BZ (2017) Dual-linker gold nanoparticles as adjuvanting carriers for multivalent display of recombinant influenza hemagglutinin trimers and flagellin improve the immunological responses in vivo and in vitro. Int J Nanomedicine 12:4747–4762. https://doi.org/10.2147/IJN.S137222

Wang C, Zhu W, Luo Y, Wang BZ (2018) Gold nanoparticles conjugating recombinant influenza hemagglutinin trimers and flagellin enhanced mucosal cellular immunity. Nanomedicine 14(4): 1349–1360. https://doi.org/10.1016/j.nano.2018.03.007

Werling D, Jann OC, Offord V, Glass EJ, Coffey TJ (2009) Variation matters: TLR structure and species-specific pathogen recognition. Trends Immunol 30(3):124–130. https://doi.org/10.1016/j.it.2008.12.001

WHO, 2020. Zoonoses. https://www.who.int/news-room/fact-sheets/detail/zoonoses

WHO, 2022. Coronavirus disease (COVID-19): Vaccines. https://www.mpg.de/18106890/animal-viral-vaccines

Wilson KL, Xiang SD, Plebanski M (2015) Montanide, poly I: C and nanoparticle based vaccines promote differential suppressor and effector cell expansion: a study of induction of CD8 T cells to a minimal plasmodium berghei epitope. Front Microbiol 6:29. https://doi.org/10.3389/fmicb.2015.00029

Wong TT, Liou GG, Kan MC (2023) A thermal-stable protein nanoparticle that stimulates long lasting humoral immune response. Vaccine 11(2):426. https://doi.org/10.3390/vaccines11020426

Wu P, Zhang Y, Yin X, He Y, Zhang Q, Chen C (2020) Layered double hydroxide nanoparticles as an adjuvant for inactivated foot-and-mouth disease vaccine in pigs. BMC Vet Res 16(1):474. https://doi.org/10.1186/s12917-020-02689-6

Zacharias ZR, Ross KA, Hornick EE, Goodman JT, Narasimhan B, Waldschmidt TJ, Legge KL (2018) Polyanhydride nanovaccine induces robust pulmonary B and T cell immunity and confers protection against homologous and heterologous influenza a virus infections. Front Immunol 9:1953. https://doi.org/10.3389/fimmu.2018.01953

Zhao K, Sun Y, Chen G, Rong G, Kang H, Jin Z, Wang X (2016) Biological evaluation of N-2-hydroxypropyl trimethyl ammonium chloride chitosan as a carrier for the delivery of live Newcastle disease vaccine. Carbohydr Polym 149:28–39. https://doi.org/10.1016/j.carbpol.2016.04.085

Zhao Z, Chen X, Chen Y, Li H, Fang K, Chen H, Li X, Qian P (2021) A self-assembling ferritin Nanoplatform for designing classical swine fever vaccine: elicitation of potent neutralizing antibody. Vaccine 9(1):45. https://doi.org/10.3390/vaccines9010045

Zupančič E, Curato C, Paisana M, Rodrigues C, Porat Z, Viana AS, Florindo HF (2017) Rational design of nanoparticles towards targeting antigen-presenting cells and improved T cell priming. J Control Release 258:182–195. https://doi.org/10.1016/j.jconrel.2017.05.014

Chapter 15
Nanotechnology in Preventive Management of Miscellaneous Livestock Diseases

Harsh Kumar Rawat, Tarun Saini, Subodh Soni, Minakshi Prasad, and Hari Mohan

Abstract One of the most significant sources of revenue is thought to be livestock worldwide. The productivity and profitability of the livestock farming system is significantly impacted by the reproductive efficiency of farm animals. Applying several tactics, such as biological, hormonal, nutritional, and reproductive illness control, are the foundation of optimal reproductive management. The introduction of nanotechnology applications in cattle farming systems can offer creative and fresh approaches to resolve problems with reproductive ineffeciencies. Using nanotechnology, a variety of medications (hormones and antibiotics), biological substances, and nutrients can acquire novel physiochemical properties. The key advantages over conventional versions include enhanced bioavailability, increased cellular absorption, regulated sustained release, and reduced toxicity. Therefore, the current chapter will outline the potential nanotechnology applications towards enhancing the reproductive performances of farm animals.

Keywords Reproduction · Livestock · Disease diagnosis · Nanotechnology

15.1 Introduction

The country's socioeconomic development is strongly influenced by the livestock industry, which also provides a significant source of revenue for farmers and the rural population (Kinsella et al. 2000). A compound annual growth rate (CAGR) of 8.15% has been recorded in the livestock sector through 2020, in India (Manjunatha et al. 2022). During the previous 5 years, 30% of the world's cattle were reported in Asia, 20% in South America, and 10% in Europe (Thibier and Wagner 2002). There

H. K. Rawat · T. Saini · S. Soni · H. Mohan (✉)
Centre for Medical Biotechnology, Maharshi Dayanand University, Rohtak, Haryana, India
e-mail: harimohan.cmbt@mdurohtak.ac.in

M. Prasad
National Research Center on Equines, Hisar, Haryana, India

LLR University of Veterinary and Animal, ICAR-National Research Centre on Equines, *Nanotechnology Theranostics in Livestock Diseases and Management*, Livestock Diseases and Management,
https://doi.org/10.1007/978-981-16-1610-5_15

are currently an estimated 1.3 billion cattle in the world (Steinfeld et al. 2006). Additionally, India has a total of 535.78 million farm animals, a rise of 4.6% from the 2012 livestock census (Jaiswal and Bhattacharjee 2022). With 121.85 million tons of milk produced in 2010–11, India is the world's top producer of milk, and majority of the output come from buffalo and cow (Raju 2013). India is home to the largest population of dairy cows in the world and produces roughly 7.22% of the world's eggs and 2.55% of its meat (Landes et al. 2017). The success of the livestock sector depends on the health of animals, which helps to raise productivity. The poorest rural households in the world practice very low-intensity subsistence animal farming, which is essential to sustaining local food sources, reducing poverty through revenue production, and improving nutritional status. Recent outbreak of Lumpy Skin Disease is a fresh example of diseases caused by highly virulent virus which can spread quickly and decimate the nation's cattle herd threatening the food security. The occurrence of fungal infections is still emerging despite the availability of both conventional and the latest antifungal medications (Morace and Borghi 2010).

In this condition, customization of novel diagnostics, therapeutics and drug delivery systems based on nano-materials or nanoparticles (NPs) is a potential substitute to successfully manage diverse diseases of farm animals and overcome drug resistance to current medications (Tewabe et al. 2021). Lipids, polymers, and metals can all be used in the production of nanoparticles (Lombardo et al. 2019). According to many researches, nanoparticles are more targeted to the infection site, have fewer side effects, do not cause medication resistance, have improved antimycotic stability and solubility, and are more effective (Singh et al. 2017). These characteristics make various nano formulations a preferred choice over conventional medications for treating various diseases including the reproductive illnesses. Diagnostics applications of various nanoparticles for diverse bacterial, viral, protozoan, fungal, and metabolic disorders have been extensively discussed in specific chapters of the current book. This chapter specially focuses to deal with reproductive diseases of livestock under the broad term of miscellaneous diseases. The reproductive management in livestock is a key to success in terms of getting better return, hence the aspect has been discussed separately in this chapter.

15.2 Reproductive Diseases In livestock

15.2.1 Microbial Diseases

Brucellosis, campylobacteriosis (vibriosis), listeriosis, leptospirosis, *Haemophilus somnus* complexes, and urea plasmosis are all causes of bacterial feticides (Barr and Anderson 1993). According to the microbiological results, bacteria such as *Bacillus, chlamydia* species, *Escherichia coli, Nocardia, Pasteurella, Proteus, Pseudomonas, Staphylococcus, Salmonella,* and *Streptococcus* can result in abortion. These microorganisms, along with a few others, have all been segregated from sporadic miscarriages instances. These may result from prolonged endometritis, septicemia in the female, mount infection through the cervix and vagina, or all the three (Muller et al. 2006).

15.2.1.1 Brucellosis (Mediterranean Fever)

The most familiar and contentious microbial infection of the cow sex organs is brucellosis. Brucellosis is a disease that is typically associated with cattle, but it can also affect pigs, sheep, goats, dogs, horses, and other animals, as well as humans. The illness poses a serious occupational risk to farmers, slaughterhouse workers, and veterinarians.

The bacteria *Brucella abortus* is the major source of brucellosis. The organism particularly infects the testicles, lymph nodes, udder, accessory sex glands, uterus, and other body parts (Godfroid et al. 2013). Miscarriage is the most common symptom because of its preference for the uterus.

Since all infected cattle exhibit the disease, brucellosis is a particularly challenging condition to diagnose. Calving or abortion are the prime times when an infected cow can spread the pathogen. The environment is contaminated with virulent *Brucella* organisms by aborted fetuses, placental fluids, placental membranes, and vaginal secretions that last for many days after an infected cow has given birth or had a miscarriage (Thorne et al. 1978). Other animals that come into contact with the surrounding that has been infected by a contaminated animal's secretions may contract the bacterium. The human population and calves are easily exposed to infectious milk and colostrum from diseased cows (Cho and Yoon 2014). Brucellosis infection cycle in cattle and human being is given in Fig. 15.1.

15.2.1.2 Leptospirosis (Weil Disease)

Leptospirosis is a bacterial infection that can affect both humans and animals. It has been characterized by a wide range of ailments in cattle, horses, pigs, sheep, goats, and dogs, including fever, icterus (jaundice), hemoglobinuria (bloody urine), miscarriage, and mortality (Bedi et al. 2022). But this disease's perception has recently evolved. Previously assumed to be a disease with a high mortality rate, it is now understood to be a common, largely asymptomatic infection affecting both wild and domestic animal species (Smits et al. 2016). Leptospirosis infection cycle and spreading to livestocks and humans is given in Fig. 15.2.

Leptospirosis symptoms in cattle can range from slight, non-detectable infections to severe infections that result in mortality (Bulach et al. 2006). The final 3 months of pregnancy are when abortions are most likely to occur, and the clinical indications that lead to abortions may include Weil disease (Divers et al. 2019).

15.2.1.3 Campylobacteriosis

Cattle general illness campylobacteriosis (vibriosis) is brought on by the bacterium *Campylobacter foetus* subspecies foetus (Hoffer 1981). Campylobacteriosis, which was previously called *Vibrio foetus* subspecies *venerealis*, is characterized by

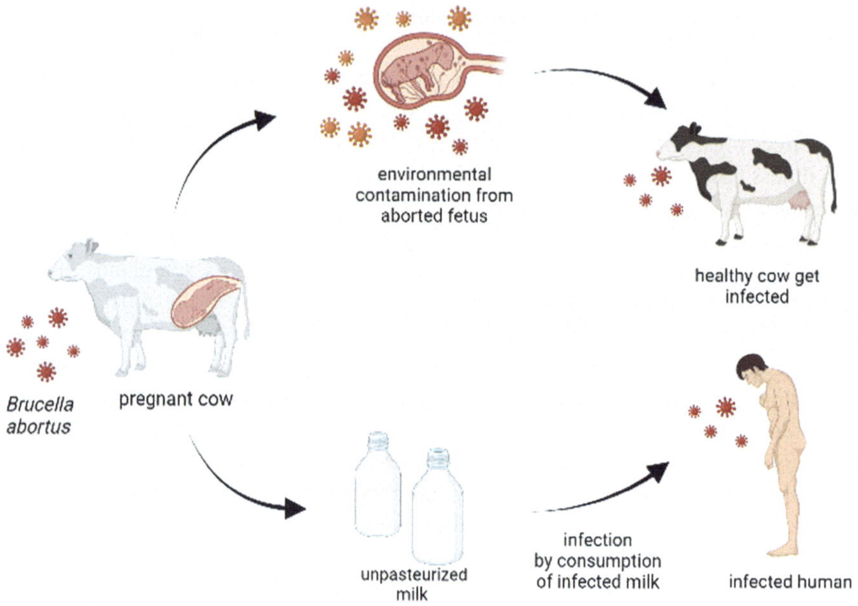

Fig. 15.1 Brucellosis infection cycle in cattle and human being

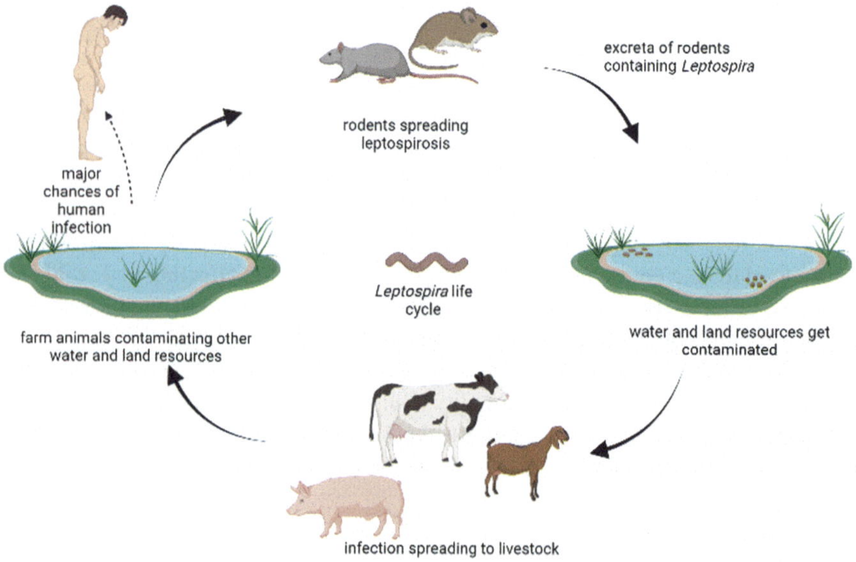

Fig. 15.2 Leptospirosis infection cycle and spreading to livestock animals and humans

sterility and a rise in the number of treatments required for conception. There are often early embryonic fatalities. In a herd that has never been exposed to the condition and in which there is no resistance, severe form of sterility problem manifests (Schaefer 1976). In this instance, endometritis-related infertility causes premature embryonic death and a lengthy interval (of 120+ days) passes before a viable pregnancy takes place. The main method of transmission of the pathogen to the male is by sexual intercourse with an infected female. It can be challenging to make a conclusive diagnosis of genital campylobacteriosis, and findings of laboratory tests are frequently unsatisfactory. Serological analyses are available; however, these are unreliable due to the condition being irregular and antibodies not frequently seen in the blood.

15.2.1.4 Listeriosis (Meningitis)

It is generally known that *Listeria monocytogenes* cause abortion, encephalitis, and septicemia in cattle. *L. monocytogenes* is more common than *L. ivanovi*; however, it has similarly been linked to the causes of abortion in cattle. Abortions and listeric infections typically start in the cold winter season or in the season of early spring (Malik et al. 2002).

15.2.1.5 Mycological Disease

The most frequent reason for periodic bovine abortion is a fungus infection of the placenta (Khodakaram-Tafti and Ikede 2005). The occurrence can be decreased by promoting excellent health (via nutrition and sound management) and averting moldy meals (Fumagalli et al. 2021). Decreasing the amount of time cows are confined, reduce cow density, and increase in ventilation when it is practical to do so, depending on supply and demand can be beneficial to manage such diseases.

15.2.2 Viral Diseases

Viral diseases pertaining to reproductive system in livestock are of concern as they not only cause health issue of animal but may also lead to abortion. For example, the infectious bovine rhinotracheitis (IBR, red nose) virus that causes bovine rhinotracheitis in cattle is contagious (Kathiriya et al. 2018). However, this virus can also result in inflammation of the vulva and vagina called vulvovaginitis, and miscarriage in cows and heifers. Typically, an abortion happens 20–45 days following infection. The usage of vaccinations can be used to control IBR infections (Parthiban et al. 2015). Animal virus infection with the diarrhea virus can result in prenatal abnormalities like miscarriage, the birth of physically weak calves, calves suffering from cerebellar hypoplasia (brain damage), and others. Fever, diarrhea, unsteady walking, and nasal discharge normally are just a few of the clinical indications that newborn calves with bovine virus diarrhea (BVD) may exhibit (ataxia). Laboratory testing of the fetus or calf is necessary for the detection of BVD virus infection. The identification of sick cattle may be facilitated by a blood test.

15.2.3 Protozoal Diseases

Sarcocystosis (also called sarcosporidiosis), neosporosis, and trichomoniasis are protozoal illnesses that cause abortion.

Trichomoniasis is an animal disease that affects livestock animals. *Tritrichomonas foetus*, a protozoan, is the culprit. These pathogens are stored in the genitalia of diseased individuals, and infected bulls can spread them from one cow to another (Givens 2018). After 2–3 months of sexual abstinence, cows will normally be cured of the sickness, but infected bulls are not able to generate immunity against the infection. The general clinical sign of the infection is sterility (Pellati et al. 2008). Miscarriage typically occurs in starting of pregnancy (first trimester). These early abortions are frequently overlooked because there is a little tissue loss. Vaccines sold commercially are now accessible. Two injections are needed for proper immunization, and they must be given 2–4 weeks apart.

15.2.4 Abortion-Related (Miscarriage) Non-Infectious Causes

15.2.4.1 Nutritional

While starvation can lead to placental insufficiency and miscarriage, it is uncommon in contemporary dairy practices (Menzies 2011). A lack of vitamin A has been linked to placental thickness, degradation, and miscarriage in late gestation (Gernand et al. 2016). Abortion has also been linked to iodine deficiency as a reason (Cabell 2007).

15.2.4.2 Chemicals, Medicines, and Toxins

Toxic substances can potentially result in early embryonic death or miscarriage. Under certain circumstances, cattle are vulnerable to plant nitrates, fertilizer nitrites, and nitrates (e.g., water scarcity) (Tarkalson et al. 2006). Abortions may happen if cattle are exposed to extended concentrations of nitrates/nitrites (-0.55% or more nitrate in feed), especially in late gestation (Hovingh 2009). It is believed that the mycotoxins produced by the fungi can result in abortion (Reddy and Raghavender 2007). Sporadic abortions are brought about by bacterial endotoxin. Prostaglandin, glucocorticoids, and estrogen are significant hormonal agents. Depending on the dosage and the gestational stage, they can trigger abortion. Abortion may also result from stress.

S. No.	Disease	Animal	Distribution	Antibiotic used	Nanoparticles (NPs)	References
1.	Brucellosis	Cows, pigs, sheep, goats, dogs, etc.	Mediterranean basin, middle, central Asia, central and south America	Doxycycline, Rifampin	ZnO-NPs	Hussain et al. (2019)
2.	Leptospirosis	Horses, pigs, sheep, goats.	Tropical and subtropical regions, Northern America, Southern Asia, etc.	Penicillin and doxycycline	Gold (Ag) based nanosensors	Barani et al. (2021)
3.	Campylobacteriosis	Poultry, cattle, sheep and ostriches, and so on	Algeria, Ethiopia, Nigeria, Cameroon	Azithromycin, Ciprofloxacin, Erythromycin	GSH-Ag NPs	Silvan et al. (2018)
4.	Listeriosis	Birds, goats, sheep, and other mammals.	Africa, South America, North America, west Asia	Ampicillin, Gentamicin	Non-UV irradiated TiO_2 NPs (Titanium dioxide), Ag-NPs	Chirathaworn et al. (2011)
5.	Bovine Virus Diarrhea	Cows, calves, buffaloes	North and South America, Russia, Australia, etc.	Gypenoside, MLV, KV (Modified Live Virus, Killed Virus) vaccines	Au NPs	Askaravi et al. (2017)
6.	Protozoal diseases (Trichomoniasis, Sarcocystis, etc)	Cattle basically in livestock farming	Middle America, Asia, China region, Africa	Metronidazole (Flagyl), Fenbendazole	ZnO NPs, (Ll-ZnO NPs)	Zoubir et al. (2021)

15.3 Nanoparticles in Prevention of Livestock Diseases

15.3.1 Metallic Nanomaterials in Reproductive Diseases

15.3.1.1 *Brucella* Diagnostics

ZnO NPs are a novel class of low-cost, low-toxicity nanomaterials that are useful in bioimaging, drug administration, and numerous therapeutic applications against cancer, microbial infections, inflammation, and diabetes (Barui et al. 2018). A ZnO-NPs-based sensor for *Brucella* identification was created by Wahab et al. They made an effort to create an accessible and low-cost biosensor that rural residents might use to identify *Brucella* in milk samples. To do this, ZnO-NPs were functionalized with the anti-*Brucella* antibody, and the resulting NPs were placed on aluminum sheets (Al-sheet). The created Al-sheets were subsequently utilized as electrode elements for voltammetric detection of the bacteria (Wahab et al. 2017).

15.3.1.2 Leptospirosis

The recently developed nanogold-based diagnostic format has yielded unsatisfactory results in resolving the aforementioned routes. *Leptospira* in urine can be found using very sensitive anti-leptospira-coated gold nanoparticles (Mohamed et al. 2021). Additionally, very sensitive markers can be discovered for an immediate and precise diagnosis employing a nanogold-based dot blot assay. The study demonstrated how to identify leptospirosis using a dot blot immunoassay based on nanogold conjugates.

15.3.1.3 Campylobacter Multidrug-Resistant Bacteria Are Susceptible to GSH-Ag NPs

Low concentrations of GSH-Ag NPs (1.23 and 4.92 g/ml) significantly (p 0.05) hindered the germination of the *Campylobacter jejuni* 11168 strains (Silvan et al. 2018; Petrus et al. 2011). These results demonstrate the potential anti-*Campylobacteria* activity of GSH-Ag NPs with particle sizes between 10 and 50 nm. Previous studies have shown that several food-borne pathogens, such as *Listeria monocytogenes, Escherichia coli, Salmonella, and Pseudomonas aeruginosa,* are more sensitive to Ag-NPs (Terzioğlu et al. 2022). The Ag ions generated from the nanoparticles react with proteins that are primarily present on cell surfaces as well as nucleic acids that contain phosphorus and sulfur (Tsolekile et al. 2017). They are known to trigger the production of ROS in the cell, which ultimately leads to cell death.

15.3.2 Quantum Dots

Due to their broad bandgap and easily controllable emission in the visible region, zinc sulfide (ZnS)-coated cadmium selenide (CdSe) nanocrystals are the most often investigated QDs for bioapplications (Philip et al. 2007). They are particularly beneficial for a variety of industrial, agricultural, and biological applications due to these unique characteristics (Murray et al. 2000). The well-known synthesis and characterization techniques used to create CdSe QDs are another factor in their appeal (Jain et al. 2018). The ability of porcine gametes to interact with self-illuminated CdSe/ZnS QDs has been demonstrated in reproductive studies (Sargazi et al. 2022). However, it is important to determine the appropriate sperm-to-QDs ratio in order to prevent or limit the QDs' toxicity to sperm function, which has been seen in previous studies using different nanoparticles. Oocytes that had grown in vitro appeared to accumulate more QDs than spermatozoa, which instead showed greater membrane staining. Although the limitations of QDs in comparison to organic dyes may not be completely ruled out, reduced QDs uptake into the spermatozoa was attributed to sperm membrane specificities (Wegner and Hildebrandt 2015). In the porcine oocyte, self-illuminated CdSe/ZnS QDs coated with anti-plasminogen antibodies displayed stronger signals than the nonconjugated QDs used for simple targeting. Ex vivo imaging of cultivated swine ovarian follicles using these CdSe/ZnS QDs was also made possible, which will soon allow for real-time monitoring of important chemicals with roles in folliculogenesis (Marvasi 2017).

In this field, the usage of nanoparticles may be clearly optimized for the administration of medicines through the umbilical cord, the amniotic sac, or the placenta. The placenta can be specifically targeted for therapeutic drug delivery using nanoparticles. In this regard, surface-functionalized nanoparticles targeted at placenta-specific receptors are showing promising results for their potential use in the targeted treatment of pregnancy problems. In this regard, we must emphasize the studies using pICSA binding peptide (pICSA-BP) nanoparticles, which target the placenta through placental chondroitin sulfate A (pICSA). With the help of these techniques, not only targeted delivery of the medication toward the fetus is possible but also boost the potency and specificity of the loaded compound while minimizing the side effects. Prospective research for potential nanoparticle therapies has primarily relied on the delivery of growth factors and nitric oxide donors (Hashem and Gonzalez-Bulnes 2020).

S. no.	Nanoparticles (NPs)	Animal species	Assisted reproductive technology	Specific Target	Results	References
1.	Fe_3O_4	Pigs, Bulls	Semen purification in fresh form	Rapid efficient and safe purification	Improve sperm quality. No negative impact on fertility. Improve sex-sorted sperm quality	(De Leeuw et al. 2009)

(continued)

S. no.	Nanoparticles (NPs)	Animal species	Assisted reproductive technology	Specific Target	Results	References
2.	CeO_2	Rams	Cooling of semen at 4 °C	Maintain semen quality and reduce the post-cooling effect	. Improved kinematic and morphologic variable. No genotoxic effect of CeO_2 NPs	(De Leeuw et al. 2009)
3.	Nano-lecithin	Goat bucks	Cryopreservation of semen	Reduce cryopreservation impacts on post-thawing sperm quality	No effects on embryo cleavage or blastocyst ratio in IVF	(De Leeuw et al. 2009)
4.	Selenium-NPs	Bulls	Cryopreservation of sperm	Reduce cryopreservation impacts on post-thawing sperm quality	Increase in vivo fertility rate. Improved seminal plasma oxidant. Decreased apoptotic and necrotic sperm cell	(De Leeuw et al. 2009)
5.	GnRH (gonadorelin)-chitosan-TPP NPs	Goat	Control menstrual cycle and egg formation	Maintain hormonal rate. Improving ovulation rate and follicle formation	. Maintained and better ovulation rate. Maintain hormonal dose	(De Leeuw et al. 2009)
6.	GnRH (buserelinacetste)-chitosan-TTP NPs	Rabbits	Induction of ovulation	The change rate of administration. Maintain hormone dose	Reduction of hormone dose. Induction of oxidation	(De Leeuw et al. 2009)

15.4 Drug-Delivery Systems During Reproductive Illness

15.4.1 Liposomes

The phospholipid vesicles known as liposomes are made up of one or more lipid bilayers suspended in water with a hydrophilic head and a lipophilic tail. They have lipophilic and hydrophilic compartments within the bilayer membranes as well as hydrophilic compartments between the membranes (Wang et al. 2017). Antibiotics delivered by a liposomal drug delivery system can fend off microbial activity against drug-resistant strains because the liposome prevents the drug from being broken down by an enzyme through an isolating process and also encourages diffusion across the bacterial membrane (El Maghraby et al. 2008). Liposomes are naturally biocompatible and biodegradable.

15.4.2 Ethosomes

Drugs are delivered transdermally using ethanolsomes, which are nanocarriers with high concentrations of ethanol, water, and phospholipids. Twenty to forty percentage of ethanol and 2% to 5% phospholipids may be present. Due to the propensity of ethanol to fluidize the membrane lipids of SC (stratum corneum), drug penetration through the skin is higher in the case of ethosomes than liposomes (Touitou et al. 2000). Ethosomes particle size is increased by the addition of ethanol, and its size was reduced by keeping the concentration of phospholipid constant (Cevc 1996).

15.4.3 Transferosomes

Transferosomes, which have a bilayer structure and are incredibly flexible, can readily pass through the SC's lipidic membrane to enter the skin. By using an osmotic gradient, it moves from a deep-sited hydrated SC layer to the dehydrated SC layer. The transferosomes surfactant aids in lipid solubilization in SC and promotes more medication penetration through vesicles (Mura et al. 2009). Due to their higher deformability, they demonstrated improved skin penetration in contrast to traditional liposomes (Zeb et al. 2016). These are easier to expel from the skin membrane because they are more hydrophilic than other traditional lipid vesicle bilayers and swell more.

15.4.4 Lipid Nanomaterials

Solid nanoparticles: The lipid core of solid lipid nanoparticles (SLNs), which are stabilized with surfactants, helps to solubilize lipophilic compounds. Monoglycerides, diglycerides, triglycerides, steroids, fatty acids, and waxes are a few examples of lipids. SLNs offer physical stability, high drug molecule tolerability, and the ability to regulate drug release (Tapeinos et al. 2017).

15.4.5 Carrier Lipids with Nanostructures

The second generation of lipid nanoparticles (lipid NPs) are called nanostructured lipid carriers, and they include both liquid and solid lipids (Beloqui et al. 2016). The greatest option for the treatment of topical skin infections, particularly for antifungals due to their lipophilic nature, is a combination of solid lipid NPs and nanostructured lipid carriers. There is little toxicity associated with these lipidic nanostructured carriers. The intimate contact of nanocarriers with SC and their small size increases the drug's cutaneous penetration (Czajka et al. 2015).

15.5 Challenges

Reproductive medicine has already benefited from nanotechnology and will likely continue to do so. Despite the fantastic solutions it can provide, there are still some difficulties with using this technology in medicine. Few researches have been done to validate all of the potential treatments or diagnostics because it is a young science. The potential for toxicity caused by the usage of nanotechnology is one of the primary issues that still need to be resolved. Although the commercialization of new diagnostic techniques is getting closer, more research needs to be done on the delivery of nanomaterials to organisms, embryos, and gametes.

(a) Some nanomaterials are harmful to living things, especially if they are ingested during pregnancy and for the development of the embryo. For instance, titanium dioxide nanoparticles can pass the placental barrier and result in anatomical abnormalities in developing fetus when exposed to pregnant mice (Ferdous and Nemmar 2020). Additionally, silver nanoparticles reduce estrogen plasma levels, increasing the number of resorbed fetuses and influencing embryonic growth (Ferdous et al. (Teng et al. 2016). According to Teng et al. (2016), carbon nanotubes also result in fewer live births per dam, fewer blood vessels in the placenta, and more abortions (Bobo et al. 2016).

(b) Ethical issues: In the future, animal health will benefit from nanotechnology. It is obvious that the potential for this technology to improve animal husbandry and food systems is very significant. As with any new technology, we have an ethical duty to use it responsibly and to be aware of any potential, unanticipated problems that might come along with its enormously great potential.

The security of animal farming and food systems could be significantly improved by the use of nanotechnology and nanomaterials. An in-depth analysis of the advantages and disadvantages of nanotechnologies that strikes a balance will be crucial to the advancement of this field. Through television, the Internet, and point-of-sale informational bulletins, the general people should be made aware of the benefits that nanotechnology applications offer, including greater safety and food security. The use of nanotechnology can boost animal husbandry and food system processes' efficiency and aid to reduce the usage of antibiotics, in addition to boosting contamination identification and the security of tracking systems. This increases consumer food safety.

15.6 Advantages

In several areas of veterinary medicine, including therapeutics, diagnostics, tissue construction, vaccine production, and disinfectants, nanotechnology has a substantial impact. The nano-applications are currently being employed in the fields of animal welfare, animal rearing, animal proliferation, and animal nourishment. Utilizing incredibly low concentrations, the treatment can be delivered directly into the target cells, reducing the need for medication buildup and withdrawal symptoms in domestic animals (Youssef et al. 2019).

15.7 Future Aspect

Research on nanotechnology will revolutionize the science and technology of animal health in the next years and will support increased livestock output. Nanotechnology has had a dramatic impact on every facet of human life, and its proponents view it as one of the most potent tools in contemporary society. One can assume that this cutting-edge technology will soon be available to us as commonplace based on the rate of advancement in nanomedicine, particularly the use of various nanoparticles in the prevention, diagnosis, and treatment of a complicated disease like cancer.

Future medical advancements may allow for the treatment of bacterial infections in patients in a matter of minutes as opposed to weeks of antibiotic therapy. With technological innovation, we can anticipate being able to perform surgery at the cellular level, removing individual sick cells and even fixing their damaged individual parts. There are already a number of nanotechnology-based products in the market, and many more are either in development or experimental stages.

Studies have proven that nanotechnology is a viable method for diagnosis, imaging, and treatments in addition to its application in nano-chemoprevention. Currently, a lot of research is being done on medicine delivery using nanoparticles. A number of nanotechnology-based constructions are currently undergoing clinical or preclinical testing, and a few of these have already received the food and drug administration (FDA) approval. Nanotechnology advancements may make it possible to obtain the larger concentrations of natural compounds required for their efficacy against a variety of ailments. It is projected that a cancer cure will be available, and nanotechnology will grow to be a US$1 trillion business with the majority of its effects concentrated on cancer treatment and health care.

Although most developing nations have very little research and development in nanotechnology, there will be more chances to import nanoproducts and processes. Other potential future applications include self-replicating nanotechnology, fabric that responds to the environment, and sensors that can find pollutants and airborne contaminants. Additionally, the movement of animals will be monitored using nanosensors surveillance equipment. Additionally, nanotechnology will be a major factor in veterinary care around the world.

15.8 Conclusion

For the poultry industry, veterinary medicine, and other sectors of animal production, nanotechnology holds great promise. Through the characterization, manufacture, and application of structures, devices, and systems by manipulating shape and size on a nanoscale scale, nanotechnologies are designed and employed for a variety of objectives. Nanotechnology (NT) can improve food qualities, make it easier for nutrients and flavors to enter the body, lengthen shelf life, and boost poultry production and health. Although there are advantages to using nanoparticles to heal illnesses and increase animal productivity, it is widely acknowledged that some nanoparticles can have hazardous and negative side effects on living things when used right away. So, a precaution is advocated to ensure nano-safety while harvesting the maximum benefits from nanotechnology interventions.

References

Askaravi M, Rezatofighi SE, Rastegarzadeh S, Shapouri SA, M.R. (2017) Development of a new method based on unmodified gold nanoparticles and peptide nucleic acids for detecting bovine viral diarrhea virus-RNA. AMB Express 7(1):1–9

Barani M, Mukhtar M, Rahdar A, Sargazi G, Thysiadou A, Kyzas GZ (2021) Progress in the Application of Nanoparticles and Graphene as Drug Carriers and on the Diagnosis of Brain Infections. Molecules 26(1):186

Barr BC, Anderson ML (1993) Infectious diseases causing bovine abortion and fetal loss. Vet Clin N Am Food Anim Pract 9(2):343–368

Barui AK, Kotcherlakota R, Patra CR (2018) Biomedical applications of zinc oxide nanoparticles. In: Inorganic frameworks as smart nanomedicines. William Andrew Publishing, pp 239–278

Bedi JS, Vijay D, Dhaka P (2022) Textbook of Zoonoses. John Wiley & Sons

Beloqui A, Solinís MÁ, Rodríguez-Gascón A, Almeida AJ, Préat V (2016) Nanostructured lipid carriers: Promising drug delivery systems for future clinics. Nanomedicine 12(1):143–161

Bobo D, Robinson KJ, Islam J, Thurecht KJ, Corrie SR (2016) Nanoparticle-based medicines: a review of FDA-approved materials and clinical trials to date. Pharm Res 33(10):2373–2387

Bulach DM, Zuerner RL, Wilson P, Seemann T, McGrath A, Cullen PA, Davis J, Johnson M, Kuczek E, Alt DP, Peterson-Burch B (2006) Genome reduction in Leptospiraborgpetersenii reflects limited transmission potential. Proc Natl Acad Sci 103(39):14560–14565

Cabell E (2007) Bovine abortion: aetiology and investigations. In Practice 29(8):455–463

Cevc G (1996) Transfersomes, liposomes and other lipid suspensions on the skin: permeation enhancement, vesicle penetration, and transdermal drug delivery. Crit Rev Ther Drug Carrier Syst 13(3-4):257–388

Chirathaworn C, Chantaramalai T, Sereemaspun A, Kongthong N, Suwancharoen D (2011) Detection of Leptospira in urine using anti-Leptospira-coated gold nanoparticles. Comp Immunol Microbiol Infect Dis 34(1):31–34

Cho YI, Yoon KJ (2014) An overview of calf diarrhea-infectious etiology, diagnosis, and intervention. J Vet Sci 15(1):1–17

Czajka M, Sawicki K, Sikorska K, Popek S, Kruszewski M, Kapka-Skrzypczak L (2015) Toxicity of titanium dioxide nanoparticles in central nervous system. Toxicol in Vitro 29(5):1042–1052

De Leeuw J, De Vijlder HC, Bjerring P, Neumann HAM (2009) Liposomes in dermatology today. J Eur Acad Dermatol Venereol 23(5):505–516
Divers TJ, Chang YF, Irby NL, Smith JL, Carter CN (2019) Leptospirosis: An important infectious disease in North American horses. Equine Vet J 51(3):287–292
El Maghraby GM, Barry BW, Williams A (2008) Liposomes and skin: from drug delivery to model membranes. Eur J Pharm Sci 34(4–5):203–222
Ferdous Z, Nemmar A (2020) Health impact of silver nanoparticles: a review of the biodistribution and toxicity following various routes of exposure. Int J Mol Sci 21(7):2375
Fumagalli F, Ottoboni M, Pinotti L, Cheli F (2021) Integrated mycotoxin management system in the feed supply chain: innovative approaches. Toxins 13(8):572
Gernand AD, Schulze KJ, Stewart CP, West KP Jr, Christian P (2016) Micronutrient deficiencies in pregnancy worldwide: health effects and prevention. Nat Rev Endocrinol 12(5):274–289
Givens MD (2018) Risks of disease transmission through semen in cattle. Animal 12(s1):s165–s171
Godfroid J, Garin-Bastuji B, Saegerman C, Blasco JM (2013) Brucellosis in terrestrial wildlife, vol 32. Office International des Epizooties, Revue Scientifiqueet Technique, p 27
Hashem NM, Gonzalez-Bulnes A (2020) State-of-the-art and prospective of nanotechnologies for smart reproductive management of farm animals. Animals 10(5):840
Hoffer MA (1981) Bovine campylobacteriosis: a review. Can Vet J 22(11):327
Hovingh, E., 2009. Abortions in dairy cattle I: Common causes of abortions
Hussain A, Singh S, Das SS, Anjireddy K, Karpagam S, Shakeel F (2019) Nanomedicines as drug delivery carriers of anti-tubercular drugs: from pathogenesis to infection control. Curr Drug Deliv 16(5):400–429
Jain S, Park SB, Pillai SR, Ryan PL, Willard ST, Feugang JM (2018) Applications of fluorescent quantum dots for reproductive medicine and disease detection. In: Unraveling the Safety Profile of Nanoscale Particles and Materials—From Biomedical to Environmental Applications
Jaiswal P, Bhattacharjee M (2022) Understanding the potential of livestock market with special reference to the export of swine meat from India: A study of time-series analysis using arima-based forecasting method. Asian J Dairy Food Res 41(3):293–297
Kathiriya J, Sindhi S, Mathapati B, Bhedi K (2018) Seroprevalence of infectious bovine rhinotracheitis (bhv-1) in dairy animals with reproductive disorders in saurashtra of Gujarat, India. Int J Curr Microbiol App Sci 7(3):1371–1376
Khodakaram-Tafti A, Ikede BO (2005) A retrospective study of sporadic bovine abortions, stillbirths, and neonatal abnormalities in Atlantic Canada, from 1990 to 2001. Can Vet J 46(7):635
Kinsella J, Wilson S, De Jong F, Renting H (2000) Pluriactivity as a livelihood strategy in Irish farm households and its role in rural development. Sociol Rural 40(4):481–496
Landes M, Cessna J, Kuberka L, Jones K (2017) India's Dairy Sector: Structure, Performance, and Prospects. United States Department of Agriculture, Washington, DC, USA
Lombardo D, Kiselev MA, Caccamo MT (2019) Smart nanoparticles for drug delivery application: development of versatile nanocarrier platforms in biotechnology and nanomedicine. J Nanomater 2019:1
Malik SVS, Barbuddhe SB, Chaudhari SP (2002) Listeric infections in humans and animals in the Indian subcontinent: a review. Trop Anim Health Prod 34(5):359–381
Manjunatha BL, Naorem A, Hajong D, Tewari P (2022) Dynamics of Household and Regional Economy in Banni Grasslands, India: A Cross-Sectional Study. Sustainability 14(18):11236
Marvasi M (2017) Potential use and perspectives of nitric oxide donors in agriculture. J Sci Food Agric 97(4):1065–1072
Menzies PI (2011) Control of important causes of infectious abortion in sheep and goats. Vet Clin North Am Food Anim Pract 27(1):81–93
Mohamed MM, Abdelaziz WR, Sayed RH, FA S, Ali AF (2021) Nanoparticles- based assay for rapid detection of caseous lymphadenitis in sheep. Adv Anim Vet Sci 9(5):709–714

Morace G, Borghi E (2010) Fungal infections in ICU patients: epidemiology and the role of diagnostics. Minerva Anestesiol 76(11):950

Muller AE, Oostvogel PM, Steegers EA, JoepDörr P (2006) Morbidity related to maternal group B streptococcal infections. Acta Obstet Gynecol Scand 85(9):1027–1037

Mura S, Manconi M, Sinico C, Valenti D, Fadda AM (2009) Penetration enhancer-containing vesicles (PEVs) as carriers for cutaneous delivery of minoxidil. Int J Pharm 380(1–2):72–79

Murray CB, Kagan CR, Bawendi MG (2000) Synthesis and characterization of monodisperse nanocrystals and close-packed nanocrystal assemblies. Annu Rev Mater Sci 30(1):545–610

Parthiban S, Malmarugan S, Murugan MS, Rajeswar J, Pothiappan P (2015) Review on emerging and reemerging microbial causes in bovine abortion. Int J Food Sci Nutr 4(4–1):1–6

Pellati D, Mylonakis I, Bertoloni G, Fiore C, Andrisani A, Ambrosini G, Armanini D (2008) Genital tract infections and infertility. Eur J Obstet Gynecol Reprod Biol 140(1):3–11

Petrus EM, Tinakumari S, Chai LC, Ubong A, Tunung R, Elexson N, Chai LF, Son R (2011) A study on the minimum inhibitory concentration and minimum bactericidal concentration of Nano Colloidal Silver on food-borne pathogens. Int Food Res J 18(1):55–66

Philip S, Keshavarz T, Roy I (2007) Polyhydroxyalkanoates: biodegradable polymers with a range of applications. J Chem Technol Biotechnol 82(3):233–247

Raju SS (2013) Assessment of feed resources and its impact on livestock output in India. Agric Situat 69(12):5–11

Reddy BN, Raghavender CR (2007) Outbreaks of aflatoxicoses in India. Afr J Food Agric Nutr Dev 7(5):01

Sargazi S, Fatima I, Kiani MH, Mohammadzadeh V, Arshad R, Bilal M, Rahdar A, Díez-Pascual AM, Behzadmehr R (2022) Fluorescent-based nanosensors for selective detection of a wide range of biological macromolecules: A comprehensive review. Int J Biol Macromol 206:115–147

Schaefer G (1976) Female genital tuberculosis. Clin Obstet Gynecol 19(1):223–239

Silvan JM, Zorraquin-Pena I, Gonzalez de Llano D, Moreno-Arribas MV, Martinez-Rodriguez AJ (2018) Antibacterial activity of glutathione-stabilized silver nanoparticles against Campylobacter multidrug-resistant strains. Front Microbiol 9:458

Singh L, Kruger HG, Maguire GE, Govender T, Parboosing R (2017) The role of nanotechnology in the treatment of viral infections. Ther Adv Infect Dis 4(4):105–131

Smits WK, Lyras D, Lacy DB, Wilcox MH, Kuijper EJ (2016) Clostridium difficile infection. Nat Rev Dis Primers 2(1):1–20

Steinfeld H, Gerber P, Wassenaar TD, Castel V, Rosales M, Rosales M, de Haan C (2006) Livestock's long shadow: environmental issues and options. Food & Agriculture Org

Tapeinos C, Battaglini M, Ciofani G (2017) Advances in the design of solid lipid nanoparticles and nanostructured lipid carriers for targeting brain diseases. J Control Release 264:306–332

Tarkalson DD, Payero JO, Ensley SM, Shapiro CA (2006) Nitrate accumulation and movement under deficit irrigation in soil receiving cattle manure and commercial fertilizer. Agric Water Manag 85(1–2):201–210

Teng C, Wang Z, Yan B (2016) Fine particle-induced birth defects: Impacts of size, payload, and beyond. Birth Defects Res C Embryo Today 108(3):196–206

Terzioğlu E, Arslan M, Balaban BG, Çakar ZP (2022) Microbial silver resistance mechanisms: recent developments. World J Microbiol Biotechnol 38(9):158

Tewabe A, Marew T, Birhanu G (2021) The contribution of nano-based strategies in overcoming ceftriaxone resistance: a literature review. Pharmacol Res Perspect 9(4):e00849

Thibier M, Wagner HG (2002) World statistics for artificial insemination in cattle. Livest Prod Sci 74(2):203–212

Thorne ET, Morton JK, Blunt FM, Dawson HA (1978) Brucellosis in elk. II. Clinical effects and means of transmission as determined through artificial infections. J Wildl Dis 14(3):280–291

Touitou E, Dayan N, Bergelson L, Godin B, Eliaz M (2000) Ethosomes—novel vesicular carriers for enhanced delivery: characterization and skin penetration properties. J Control Release 65(3): 403–418

Tsolekile N, Parani S, Matoetoe MC, Songca SP, Oluwafemi OS (2017) Evolution of ternary I–III–VI QDs: Synthesis, characterization and application. Nano-Structures & Nano-Objects 12:46–56

Wahab R, Khan ST, Ahmad J, Ansari SG, Musarrat J, Al-Khedhairy AA (2017) MWCNTs functionalization and immobilization with anti-Brucella antibody; towards the development of a nanosensor. Vacuum 146:623–632

Wang L, Hu C, Shao L (2017) The antimicrobial activity of nanoparticles: present situation and prospects for the future. Int J Nanomedicine 12:1227

Wegner KD, Hildebrandt N (2015) Quantum dots: bright and versatile in vitro and in vivo fluorescence imaging biosensors. Chem Soc Rev 44(14):4792–4834

Youssef FS, El-Banna HA, Elzorba HY, Galal AM (2019) Application of some nanoparticles in the field of veterinary medicine. Int J Vet Sci Med 7(1):78–93

Zeb A, Qureshi OS, Kim HS, Cha JH, Kim HS, Kim JK (2016) Improved skin permeation of methotrexate via nanosizedultradeformable liposomes. Int J Nanomedicine 11:3813

Zoubir J, Radaa C, Bougdour N, Idlahcen A, Bakas I, Assabbane A (2021) Electro-detection of the antibacterial metronidazole using zinc oxide nanoparticles formed on graphitic carbon sheets. Analytical application: human serum and urine. Mater Sci Energy Technol 4:177–188

Chapter 16
Nanotechnology in Veterinary Hospital Management

Alonkrita Chowdhury, Dodla Mamatha, Soumendu Chakravarti, and Rajesh Kumar

Abstract The hospital management system plays a pivotal role in the all-encompassing success of health services by facilitating maintained conditions that aid in achieving optimum therapeutic outcomes as well as patient satisfaction. Maintenance of hygiene in hospitals is the foremost concern as it is prone to spreading contagious diseases, and acquisition of infection, in addition to zoonotic and reverse zoonotic transmission, especially in veterinary clinics. Nanotechnological applications in hospital management could be the troubleshooter for overcoming such challenges. It is an expeditiously emerging field of science with great potential in different sectors, including medical care. From the beginning to the end of a hospital management system, for instance, hygiene maintenance, waste disposal, manufacturing protective equipment, construction management, water decontamination, developing a smart hospital system, and so forth nanotechnology holds the potential to be utilized in almost every aspect as well as in achieving an advanced healthcare management system. Disinfectant solutions, anti-slip materials, vacuum cleaners, etc. can be nanotechnologically customized for the maintenance of cleanliness and hygiene on hospital premises. Especially, AgNPs have got the limelight as an excellent antimicrobial agent because of the optical, chemical, electrical, and catalytic properties facilitating the modification of its shape, size, surface nature, etc. Hence, these are being exploited in all the attributes of hospital management including sanitation, waste and water management, protective wear manufacturing, surgical decontamination as well as designing building materials too. The implementation of NPs in construction materials has made an immense modification in

A. Chowdhury · D. Mamatha
Department of Genetics and Plant Breeding, Institute of Agricultural Science, Rajiv Gandhi South Campus, Banaras Hindu University, Mirzapur, Uttar Pradesh, India

S. Chakravarti
The Pirbright Institute, Surrey, UK

R. Kumar (✉)
Department of Veterinary Physiology and Biochemistry, Lala Lajpat Rai University of Veterinary and Animal Sciences, Hisar, Haryana, India

LLR University of Veterinary and Animal, ICAR-National Research Centre on Equines, *Nanotechnology Theranostics in Livestock Diseases and Management*, Livestock Diseases and Management,
https://doi.org/10.1007/978-981-16-1610-5_16

their physical and chemical properties. Titanium dioxide (TiO_2), silica (SiO_2), copper, aluminum oxide (Al_2O_3), etc. are the mostly used NPs in the construction sector. Regarding waste management, proper disposal of trash requires safe as well as sustainable methods. Semiconductors functionalized with NPs such as TiO_2, zinc oxide (ZnO), alpha hematite (α- Fe_2O_3), and tungsten oxide (WO_3) are effectively utilized in water purification through photocatalysis. Recently, TiO_2 NPs have emerged as the most efficient photocatalyst to be employed. It is also being employed in the removal of heavy metals such as Hg (II), Mn (II), Ti (I), etc. An efficacious application of nano-sensors has been found in animal data management and the determination of effective treatment steps; at the same time, it also possesses a critical role in waste management. The most widely used NPs for wastewater treatment include metal and carbon-based NPs, whereas zero-valent iron is the most effective element in the degradation of chlorinated hydrocarbons. Owing to its high surface area, it also serves as an adsorbate for removing heavy metals from water. Applications of NPs in surgical management emphasize the prevention of infection, accelerating the recovery process, and pain management together with reducing corrosion of surgical equipments and incorporating antimicrobial properties to it. NPs-aided fabrication of medical scrubs provides reusability, water-repellant capacity, self-cleaning as well as sterilized properties. In addition, TiO_2, ZnO, SiO_2, and Al_2O_3 NPs are being employed in manufacturing UV-protective suits. Hereby, this chapter enlightens the nanotechnological approaches in all aspects of hospital management systems.

Keywords Nanoscale zero-valent iron (nZVI) · Nanoelectromechanical systems (NEMS) · Microelectromechanical systems (MEMS) · Nano-sensor · Carbon nanotubes · Nanocomposite · Nano-adsorbents · Photocatalyst · Nanoboat · Nano-clay · Personal protective equipment (PPE) · Surgical management · Antimicrobial agent

16.1 Introduction

Since the ancient age, a nursing tradition has been developed in the form of charitable guesthouses in both western and eastern societies. However, a religious ethos of charity was mainly focused in such traditions. Later, upon exposure to science, the idea of recovering someone from a disease got developed and consequently, a few institutes were constructed for the educational purpose of medical practitioners. In the mid-1700s, isolation hospitals were established in the United States. According to American Hospital Association, since 1960, the number of hospitals in the United States was increased from 247 to 7123 in the next 10 years (https://www.nursing.upenn.edu/nhhc/nurses-institutions-caring/history-of-hospitals). Simultaneously, veterinary research comes of age in the late nineteenth century owing to the contribution of Sir John McFadyean. Subsequently, several veterinary hospitals become established that provide facilities to mainly livestock and pets regarding the diagnosis and treatment of their sickness, as well as immunization, vaccination, artificial insemination, surgical intervention, etc. (veterinary.assam.gov.in/portlet-innerpage/veterinary-hospitals-and-dispensaries). Along with the evolution of hospital facilities, several difficulties regarding the spread of diseases, contamination, and ineptitude of various clinical aspects arose. Hence, management of such complications was urgently needed. However, it is a challenging task as it requires concernment of multiple dimensions. That is, it includes proper care of the sufferer, at the same time, the protection of healthcare providers also has to be ensured; so that contagious diseases should not be passed from affected to healthy ones. As most veterinary diseases are zoonotic, chances of reverse zoonotic transmission are a big concern to be minimized (Rahman et al. 2020; Minakshi et al. 2020; Prasad et al. 2021). Further, in veterinary science, there is ample diversity of subject species coming for treatment. They have their own anatomy, physiology, and disease susceptibility that may require a different set of microenvironments to maintain their comfort zone. Hence, it is also a matter of concern for the maintenance of such specialized microenvironments. Besides, proper management is required to ensure the cleanliness of hospital premises, as well as proper waste disposal, sustaining decontamination of water, and reducing microbial loads. Carrying protective wear, such as a mask, apron, gloves, etc. is also aided to maintain hygiene. Considering the needs and variety of species, better instrumentation with novelties is required to enhance the efficiency of diagnostics and treatment. Hence, nanotechnology (NT)-based approaches are mostly desired to overcome the array of challenges in hospital management. Nanotechnology is the incipient area in modern sciences that deals with the enhanced efficiency of any particle in its nano-range. The discovery of molecules at the nanoscale level as nanoparticles (NPs), in the twentieth century, imparted one of the biggest impacts on current science and society regarding its ample diversity of applications along with the scope of further diversifications. In accordance with Allied Market Research, it has been predicted that the market size of NP-based applications will be expected to be reached from 16.3 billion, in 2015 to 62.8 billion by 2022 indicating the potential momentum gained in a varied field of applications (Nanomaterials Market Research 2031; Nanomaterials Market-Allied Market Research. Available online: https://www.alliedmarketresearch.com/nano-materials-market). According to further reports, 6222 companies in 32 countries are producing 1814 consumer products using NPs. Along with the applications in all the branches of existing science, nanotechnology has also found niche implementation in hospital management. The major domains where NT has found applicability in hospital management is delineated hereunder (Fig. 16.1).

16.2 Hygiene Management on Hospital Premises

Hospital cleanliness is of utmost importance for a safe and healthy world as it not only reduces the spreading of diseases but hastens the recovery of patients also. Hence, maintaining hygienic surroundings is the foremost requirement in the

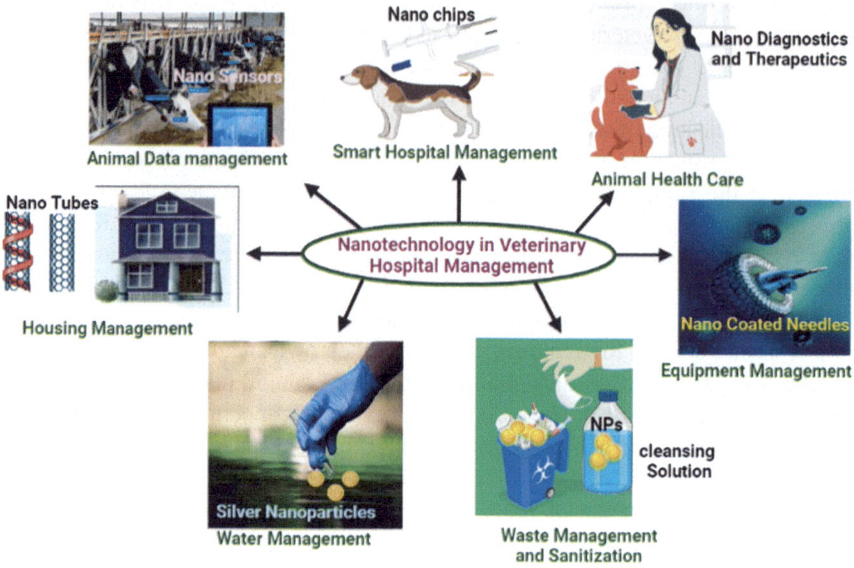

Fig. 16.1 Nanotechnology in diverse aspects of veterinary hospital management

hospital management system. Regarding the efficiency enhancement of currently available cleansing solutions and disinfectants, several NP-based formulations have been industrialized. For example, silver NPs (AgNPs) are being employed as an antimicrobial agent in cleaners with the aid of removing microorganisms. AgNPs functionalized photothermal devices or ultraviolet (UV) generators are also in use for maintaining a hygienic atmosphere (Nakamura et al. 2020). Composite AgNPs are being fabricated containing other NPs for cleansing purposes. For instance, $AgTiO_2$ is employed as a self-cleaning agent for bacterial inactivation (Saraswati et al. 2019). Similarly, owing to the photo-oxidative properties, AgZnO is being utilized as bactericides. AgCo-NPs, comprising of an antibacterial Ag-shell with a superparamagnetic Co-core are applied in the water purification system as a disinfectant (Alonso et al. 2013). Antiviral filters are also constructed employing AgNPs as an antiviral agent against aerosolized bacteriophages (Deshmukh et al. 2019). Owing to self-cleaning properties, silica-based NPs can be exploited as a cleanser in glasses and all other instruments having glass components (https://www.diva-portal.org/smash/get/diva2:665233/FULLTEXT02.pdf). Thus, several products have been developed using nano-formulations that keep glasses cleaned for a longer time by removing stains and minerals from them (https://www.cleanlink.com/casestudieswhitepapers/details.aspx?id=23741). Moreover, exploiting NPs, anti-slip materials have been developed that can be exploited to prevent skidding on hospital floors (https://www.technanoindia.com/antislip.html). Proper cleaning by using these materials also facilitates deodorizing and maintains sanitation as well as ambiance of hospital premises. Additionally, various nanomaterial-based sensors or nano-sensors have been developed to estimate the microbial loads and other contaminants. It also ensures effective cleaning along with remedial measures taken for proper cleanliness. Restoration of the microenvironment, needed for faster recovery is also facilitated by nano-sensors. It cut down drudgery as well, hence economical too. Currently, NP-incorporated thousands of commercial products are available in the market, such as several disinfectants, cleansing solutions for floors, wood, glasses, and instruments, and so on owing to its antimicrobial properties and other advantages such as water-repellent efficiency, UV resistance, scratch resistance, chemical resistance, superhydrophobicity, oil or grease removal property, durability, high performance, antiviral properties, etc. Nanomaterials are advantageous in terms of efficiency, durability, sustainability, etc. Hence, it can be employed as sealing materials or in the manufacture of shoes or foot rubbers, and other such purposes (https://product.statnano.com/search?). Nanomaterials such as graphene are being implemented as a source of energy reservoir in multipurpose vacuum cleaners through long-lasting, lightweight batteries with high storage capacity. Due to the better energy efficiency and supply, such nano-based appliances are mostly advantageous in the hospital management systems in remote areas worldwide where uninterrupted energy supply is a concern. Moreover, the utilization of such storage devices aids in reducing toil with enhanced efficiency. In the same context, the employment of various nano-sensors in automated cleaning equipment can also be beneficial for the disposal of animal excreta and body fluids or tissues from the ground. AgNP-loaded hydroxyl zirconium sodium phosphate (Ag-HZDP) is utilized as an antibacterial agent in disinfectant solutions (Jia et al. 2019). Various engineered nanoparticles (ENPs) have been found to have antiviral applications. Similarly, transparent hydrophobic and oleophobic coatings have been found to have lotus leaflike self-cleaning properties (https://patents.google.com/patent/EP0772514A1/en). Hence, NPs have the potential to be applied for the sanitation of hospital premises signifying its role in hospital hygiene management as a cleaning material in place of conventional approaches in near future.

16.3 Smart Hospital System

Environmental factors such as light (luminescence, wavelength, light period), temperature, humidity, sound, etc. provide different impacts on diverse species of animals for the maintenance of their comfort zone or homeostasis. Smart hospital system includes all these factors to develop automation devices for providing optimum relaxation. Contrastingly, several nano-devices have been designed using nano-sensors to access as well as modify the setting of a hospital ward. For instance, the temperature of different body parts of an animal can be monitored using a thermal imaging camera. Accordingly, the temperature of the microenvironment of the respective animal can be regulated to maintain internal equilibrium. Similarly, depending on operating hours or outside climate, the smart lighting system can be customized to provide proper illumination. Nowadays, smart-dust, a system of many tiny microelectromechanical systems (MEMS) such as sensors, robots, or other devices, has been developed to manage the microenvironments as it provides the detection of light, temperature, sound, vibration, magnetism, or presence of toxins. The wireless networks of submillimeter-scale autonomous computing and sensing system senses and records information about its surroundings and transmit the data wirelessly to larger computer systems generating the effect on a macroscale. With the advancement of fabrication technologies, a shrinkage in the size of smart-dust, evolving nanoelectromechanical systems or NEMS has been proposed (https://www.nanowerk.com/smartdust.php). Incorporating such devices in hospital management facilitates the reduction of contaminants and microbial loads from the surrounding atmosphere along with the maintenance of the internal environs of the ward as per patients' requirements. Furthermore, corresponding systems automated with regulatory devices such as exhaust fan, lighting sources, or other equipment, control the surrounding atmosphere as well as aid in management of time, energy, and manpower. Hence, the employment of such smart systems as the biosecurity component of hospitals is much expected in the future.

16.4 Hospital Housing Management

Damages in hospital buildings can also provide a direct or indirect impact on the health of subjects as well as other living beings on hospital premises. For instance, the presence of a crack in a hospital can be a source of various parasitic diseases. Damp areas of walls are similarly favorable for fungal and parasitic growth. The corrosive nature of confining cages is another concern for physical entities in the veterinary hospital. However, the application of NPs can be beneficial in construction purposes owing to its lightweight property, strength, and higher energy efficiency. According to AECOM Technology Corporation, an American infrastructure consulting firm, more than 50% of the building materials will contain NPs by 2025 (Jones et al. 2019). For example, nano-carbon such as carbon nanotubes (CNTs)-based building resources (cement, concrete, etc.) possess anti-crack properties along with strength and lightweight as well as desirable thermal properties in ceramics and walls (Rashad 2017). Hence, such NP-fabricated construction materials can reduce the risks of microbial growth together with less chances of building deterioration. Additionally, NEMS-based devices designed with nanocarbon can be deployed to get a real-time condition of materials and buildings. Hence, it provides prevention of fungal and parasitic growth coupled with the facilitation of lowering zoonotic and parasitic disease dispersal chances. Several NP-based materials, for instance, iron oxide (Fe_2O_3), and copper NPs (Cu-NPs) containing building materials have been found to possess higher strength and corrosion resistibility. Exploiting the biocidal property of silver NPs (AgNPs), coating and painting materials have been industrialized to resist harmful microorganisms. Similarly, titanium oxide NPs (TiO_2-NPs) can also be employed in varnishing cages due to its self-cleaning properties along with super-hydrophilicity, antifogging, and antimicrobial characteristics (Verma et al. 2022). In the management of hospital housing systems, silica (SiO_2) also possesses a pivotal role by providing improved fire resistance, light transmission, reduced water leaching from wall and cooling properties in ceramics. Hence, it can be employed in glasses for better fireproofing and anti-glare coating. Since nano-porous silica aerogel materials possess thermal insulation attributes, it is advantageous for the construction of ambient temperature-proof rooms (Mohajerani et al. 2019). Other beneficial attributes of such NP-based construction material include scratch-resistance, anti-graffiti, water-repellent, easy-to-clean or self-cleaning properties, in addition to UV-protection and bactericidal attributes. Hence, nanotechnology-aided housing construct enables improved durability and microenvironment management with high relevancy in hospitals.

16.5 Waste Management

Another critical issue in effective hospital management is related to handling organic and inorganic wastes. Since the disposed stuff possess a high risk of spreading diseases, hospital waste management is a major object for maintaining a healthy environment. Further complications can also be generated via the development of multi-drug-resistant (MDR) or extensively drug-resistant (XDR) variants of microbes. Frequent exposure of bacteria and other microorganisms with antibiotics in such waste results in evolution of such variants. Hence, proper disposal of hospital waste is required otherwise, in long run, it can create unhealthy conditions in society. It is a matter of concern in case of both human and animal hospitals. Clinical wastes may include infectious objects or chemical substances, trashes from a pathology lab or of pharmaceutical origin, genotoxic waste, radioactive elements and heavy metals originating from one-time used equipments and laboratory wastes, in addition to body fluids and tissues, and cotton swabs from isolation wards. It also adds in hardware such as unused batteries, inoperative apparatus, parts of machinery, and so on. There are a number of conventional approaches to remove waste materials by autoclaving, hydroclaving, incineration and microwave treatment, ultraviolet (UV) exposure, etc. Along with energy-based systems, management of waste is also performed by means of chemical procedures such as applications of sodium hypochlorite, chlorine dioxide, calcium hydroxide, glutaraldehyde, and peracetic acid are applied for disinfection, sterilization, and encapsulation purposes. In this context, NP-based solutions also possess the efficiency and desired potential to be applied in different forms of waste management. Hence, NPs are being effectively exploited in the resolution of hospital waste management in comparison to conventional systems. For example, polypyrrole doped with iron oxide (PPy-Fe_3O_4) magnetic nanocomposite is employed for the removal of highly toxic hexavalent chromium (Cr (VI)) from aqueous solution (Bhaumik et al. 2011). Similarly, nano-photo catalysts, widely used in environmental and ecological waste treatment, can also be employed in hospital waste management systems. Corrosion-resistant NPs functionalized with metal oxide semiconductors such as TiO_2, ZnO, Fe_2O_3, and WO_3 are generally used for cleaning waste through the photocatalyst method holding TiO_2 as the most efficient photocatalyst (Nunes et al. 2021). Oxygen is required for the effective oxidation of organic materials as well as the photoreduction of nitrogen and carbon dioxide. Additionally, bacteria along with malignant cells also get destroyed upon oxidation, leading to the reduction in odor emission and volatile organic compounds (VOCs) released from cancerous cells. Indirect UV-irradiation or destruction of the steric acid layer of the bacterial cell wall has made TiO_2 the most competent against various bacteria, as it facilitates the generation of oxygen radicals, hydrogen peroxide, or metal as by-products. Titanium-based NPs are more effective than traditional strategies to remove dyes, pharmaceuticals, pesticides, chemicals such as chlorinated phenol, nitrobenzene, cyanides, and several heavy metal pollutants like Hg (II), Mn (II), Ti (I), etc. CuO-doped titanium photocatalysts are also used in the destruction of bacteriophage T4. Apart from titanium-based photocatalysts, other NP-based materials such as ZnO can be used for the complete discoloration of acridine orange, methyl orange, and methylene blue. Fe_2O_3 is employed for the oxidation of aniline to azobenzene. Disposal of 4-chlorophenol and chloramphenicol drug is accomplished using gadolinium oxide (Gd_2O_3). Another binary metal oxide nanocomposite, CeO_2-CdO has been synthesized as an antimicrobial agent for the hospital waste management process (Hooshmand et al. 2020). Similarly, magnetic Fe_2O_3 NPs functionalized with activated carbon are employed to remove copper from waste. Furthermore, surface-engineered magnetic NPs with carboxyl, amine, and thio-group along with Fe_3O_4 NPs are engaged in the removal of heavy metal ions such as Cr^{+3}, Co^{+2}, Ni^{+2}, Cu^{+2}, Cd^{+2}, Pb^{+2}, and As^{+3}. Whereas, polyvinylpyrrolidone (PVP)-coated magnetic NPs is very effective to eradicate carcinogenic substances. NP-aided waste management system is advantageous regarding time, space, specificity, and economy over traditionally used methods (Patil and Bohara 2020). Additionally, nano-sensor can also be deployed to estimate the category of waste through continuous tracking and quantification of different components in waste matter. This facilitates data management of various wastes along with the determination of the efficiency of treatment procedures. Thus, it is beneficial for judging what is needed and in what extent for proper waste disposal leading to the development of a healthy, self-automated ecosystem based on artificial intelligence (AI).

16.6 Water Management

An ample supply of water with little or no contaminants can be obtained by removing impurities from the waste generated during laboratory tests, washing, animal preparation, etc. The use of various kits based on radioimmunoassay as well as the application of radioactive materials for diagnosis can contaminate the water. Here, titanate nanofibers can be employed as absorbents for the removal of radioactive ions from the water through filtration (Yang et al. 2008; Yang et al. 2013). Similarly, MgO functionalized carbon NP (MgO-NC), carbon nanotubes, graphene oxide (GO) or reduced graphene oxide (rGO)-based nanocomposite, hollow silica nanosphere, and nanoscale zero-valent iron (nZVI) materials are also being effectively employed in filtration and other methods for removing heavy metal ions and radioactive contaminants from water. Hence it facilitates the supply of pollutant-free water for hospital applications (Yang et al. 2019; Kumar et al. 2020). Surface-immobilized silver NPs containing desalination systems have been applied for the removal of biological as well as radioactive waste from water (Shim et al. 2018). Due to small size, large surface area, and high reactivity, zero-valent iron-based nanomaterials are being employed to remove contaminates such as U(VI), Tc (VII), Se (IV), or Se (VI), Cr (VI), As (III) or As (V), Hg (II), Cu (II), Ni (II), and other toxic materials (Wen et al. 2022). Similarly, graphene-like nanoflakes are more beneficial as electrodes for capacitive deionization in comparison to conventionally used activated carbon materials. NMs can also be deployed to remove water-borne pathogens. Such approaches are very much essential in hospital management to secure the "One Health" perspective as it facilitates the reduction of microbial load from the discharge of the hospital. A number of pathogens such as *Shigella* spp., *Clostridium botulinum*, *Mycobacterium marinum*, *Leptospira* spp., Adenoviridae virus, Enteroviruses, Astroviruses, Noroviruses, Sapoviruses, Rotavirus, Hepatitis virus A and E, *Entamoeba histolytica*, *Giardia lamblia*, *Toxoplasma gondii*, *Schistosoma* spp., *Ascaris lumbricoides*, etc. can be disseminated through the water. Hence, several NP-based membranes containing AgNP, CNTs, nano-TiO_2, etc. have been developed for the decontamination of the water from microbes. Utilizing the bactericidal property of AgNPs as well as the virus adsorption capability of CNTs, filtration membranes have been developed to prevent parasites, oocytes of *Cryptosporidium*, and so on. Similarly, TiO_2 is also being employed as a nano-thorn membrane for filtration purposes (Weir et al. 2008). Moreover, bioactive prodigiosin-impregnated cellulose matrix and fullerol NPs have been designed for the destruction of pathogenic bacteria as well as the inactivation of bacteriophages by photosensitization. ZnO nanorods are also equally employed in water purification procedures using the visible light-active photocatalytic method. Surface-modified cobalt ferrite NPs are also utilized as a membrane for rapid capture, detection, and removal of pathogens from aqueous solution. TiO_2-based decontamination is performed through the formation of ROS between 300 and 390 nm wavelength of UV irradiation. Contrastingly, graphene oxide is also employed in water treatments owing to its broad-spectrum antimicrobial activity against bacterial and fungal conidia by intertwining and membrane perturbation (Arivizhivendhan et al. 2015, 2015; Bohara et al. 2017, 2017). NMs such as photocatalytic compounds and their hybrids comprised of metal oxides such as iron oxide, hybrid colloidal nanostructures, tungstate, molybdate, vanadate-based ternary hybrids, ferrite-based or chalcogenide-based hybrid nanomaterials, etc. or as membrane support materials that include organic-inorganic HNMs, graphene-supported HNMs, CNTs, graphene oxide (GO), graphitic carbon nitride (GCN), CoFe2O4/rGO, sulfophthalocyanine/ TiO2/ FAC, etc. can be used in membrane effectively to remove water contaminants of specific type, hence indicative that in future NPs based water purification with large-scale water treatment capacity is a viable option in hospital management. Similarly, NP-aided adsorption systems such as carbon-based nano-adsorbents

(CNTs), metal-based nano-adsorbents, polymeric nano-adsorbents, zeolites, etc. have been developed for water purification purposes. Therefore, the comprehensive use of membrane filters, photocatalyst systems, adsorption systems, etc. can act as a very efficient water management approach in hospitals since these are advantageous in terms of their point-of-use, high efficiency, easiness, and space-saving property (Zhao et al. 2019).

16.7 Surgical Management

A number of NP-based formulations have been developed to be employed for surgical management. These formulations facilitate improved surgical procedures, better utility of surgical equipments as well as enhanced recovery of animals in comparison to existing approaches in different aspects of animal treatments. For instance, the nanomaterial-based coating of surgical equipment is relevant to corrosion reduction along with providing antibacterial activity and easy cleaning properties. Recently, nano-needles have been developed for pain management during surgical procedures. NPs provide lots of positive impacts on available suture materials. NP-aided suture materials possess better potential in several aspects compared to the conventional ones. For example, propolis and biogenic AgNPs-coated silk sutures possess reasonably better potential regarding antibacterial activity and fibroblast cell proliferation (Baygar 2020). Correspondingly, silver-loaded poly (lactic-co-glycolic acid) (PLGA) sutures have been reported to promote better cell migration and proliferation of 3T3 murine fibroblast cells in the wounded area. Moreover, scar-free surgery can be obtained by employing ultrathin poly-L-lactic acid (PLLA) nanosheet with excellent sealing efficiency for repairing the incision completely without scars and tissue adhesion. Curcumin-loaded gold suture material enables much faster repair with lesser inflammation contrary to conventional sutures (Sunitha et al. 2017). General nylon surgical sutures coated with biogenic AgNPs have been demonstrated to possess stable antibacterial activity up to 12 weeks along with better wound healing property (Syukri et al. 2021). Furthermore, nanoboat application has facilitated minimally invasive surgery up to the cellular level with miniaturized opening for entry. Similar benefits can be obtained from nanocoated surgical blades in complex surgery such as neurological surgery. For instance, diamond nanolayer coating on scalpel provides advantages of physical adhesion. Similarly, nanocarbon-coated surgical knives also possess higher stability (Implants, Surgery and Coatings. http://www.observatorynano.eu/project/filesystem/files/Surgery, %20implants%20and%20 coatings-April%2009.pdf). Additionally, several nano-liposomal analgesic formulations have been industrialized exploiting the intrinsic analgesic properties of zinc oxide (ZnO) and magnesium oxide (MgO) NPs. Moreover, NPs are also being employed in diagnostic modalities and guided surgical procedures. NP-aided conventional diagnostic approaches such as X-ray, MRI, PET, etc. has obtained resolution enhancement, noise reduction, diminishing adverse reactions along with cost-effectiveness. It leads to the enhancement of diagnostic value as point-of-care (POC) application with better analytical scope. Nowadays, nano-sensors attached instruments have been employed to obtain real-time data along with the post-operative inspection of internal organs (Ferreira et al. 2022a, 2022b). Gold nanorod fabricated cardiac patches are very much beneficial regarding the reduction of cardiological stress due to suturing. Using near IR radiation as a source of thermal energy, suture-free engraftment can be carried out for repairing the infarcted heart via changing the molecular structure of the fibrous scaffold for strong, and safer attachment to the myocardium (Malki et al. 2018). Besides, CNTs and AgNPs-based catheter provides reduced invasion, enhanced antibacterial property, higher strength and flexibility, and low thrombogenic effect in comparison to conventional catheters (Mali 2013). Recently, laser-responsive NM-aided tissue repairing is getting advantageous as it could lead to rapid tissue sealing and repair. Hence, owing to the smaller size, NPs are gaining momentum in hospital applications regarding pain management, better treatment, and post-operative managements opening new avenues in veterinary treatment.

16.8 Management of Protective Equipment

Medical protective coverings include aprons, surgical gloves, masks, and surgical drapes in addition to radiation shielding equipments and normal curtains. The application of nanotechnology in manufacturing such textiles or materials can be advantageous regarding comfort, durability, efficient antimicrobial barrier, easy maintenance, and self-cleaning properties. Moreover, it can also provide quantification upon exposure through nano-sensors. Nanotechnology has gained momentum in textile production for formulating smarter or better clothing materials. Furthermore, the employment of nanomaterial in smart and interactive textiles (SMIT), sensing electrical, thermal, chemical, magnetic, or other stimuli, is also beneficial for protective covering, and monitoring patients and hospital load. Carbon-based nanomaterials (graphene, carbon nanofibers, and CNTs), inorganic nanoparticles (metal oxide, metal, and nano-clay), core-shell nanoparticles, polymeric nanofibers, and others are used in manufacturing fabrics via nano-finishing, nano-coating or nanofiber applications. Hence, NM-incorporated cloths are advantageous in terms of excellent mechanical properties, water-repellant ability, and fire resistance along with porous textile structure leading to breathability, water vapor penetration, and high moisture absorbance. Owing to self-cleaning and antibacterial properties, AgNPs, TiO_2, ZnO, TiO_2-Ag and ZnO-Ag NPs can be used in the fabrication of aprons and surgical drapes. Textiles made of chitosan fibers have moisture-controlling, antibacterial, and fungicidal properties, while the application of other materials like zinc oxide and Ag NPs can also yield a similar desirable result. TiO_2, ZnO, SiO_2, and Al_2O_3 can be employed for UV protective cloth production needed during scanning diagnostic performances. A surgical mask is one of the most important personal protective equipment (PPE) to prevent communicable disease transmission. AgNP-functionalized surgical masks have been demonstrated to hold a broad-spectrum antibacterial activity. The layering of cellulose acetate (CA) and polyvinylidene fluoride (PVDF) based polymeric nanofiber in N_{95} respirator has demonstrated advanced results in many aspects relative to normal counterparts. Similarly, enhanced air permeability can be achieved by employing nanofibers in masks. Implementation of novel PVDF nanofiber filters and multilayer nanofiber filters can prevent viral entry through masks in addition to better breathability which is another advantage of NM-based masks. It is also highly preferable owing to its reusability attribute. Hence, even after washing or cleaning, it retains the uniformity in porous size assisting filtration efficiency and breathability (Damokhi et al. 2022). Due to the antimicrobial activity of NPs, surgical masks doped with MgO or CuO can destroy microorganisms upon exposure along with filtration capability (Pullangott et al. 2021, 2021; Ferreira et al. 2022a, 2022b). Therefore, many nanomaterial-based fabrications have been evolved due to water-repellant, antibacterial, anti-odor, and self-cleaning properties accompanied by protection from radioactive emission (Saleem and Zaidi 2020).

16.9 Animal Data Management

Not all individuals are the same within species and apart from the care taken in the hospital, environmental factors also influence the outcome of treatment. However, the combined effect of such studies and data influenced by these factors are still very nascent in veterinary subjects. In this context, various nano-sensors and wearable devices have been evolved for keeping records of data regarding vital parameters as well as the movements of animals to areas and different environmental conditions surrounding them. It also facilitates the transmission of data about post-care on a real-time basis which is extremely beneficial for the assessment of disease progression and treatment outcome. Additionally, epidemiological data of diseases, even timely intervention of specialized personnel on a real-time basis is very much essential in animal health care for developing personalized medicine. There are a variety of nano-sensors already available and are under development that aid in the proper identification, recording and storing, and transmission of data. Implantation of nanomaterials enhances their specificity in action, energy efficiency, etc. Hence, nanotechnology is gaining importance progressively in animal data management and may be implemented in a larger extent imminently.

16.10 Conclusion

Nanotechnology is one of the most prolific fields of modern science. Due to its vast cognitive properties, it possesses a huge number of novel applications with a better scope. There is no sector left out from the array of applicability that nanomaterial offers. Hence, NP-based applications and the advancement of products are expanding on a continuous basis. There is no denying that the nanotechnological domain can have a very positive effect on life also and have numerous potential and proven applications in various fields including hospital management. It facilitates effective hospital management practices in multiple ways regarding diagnosis, treatment, care, ambiance along with environmental security. However, the vast application of NPs also adversely impacts the environment as a form of pollutant. Furthermore, several negative effects of NPs on life had already been reported. Therefore, it should be a concern also to keep these NP-based applications under stringent regulations. These tiny particles can hold a larger threat to the ecosystem and the biomass of the earth if not used in judicial ways with their proper disposal.

References

Alonso A, Muñoz-Berbel X, Vigués N, Rodríguez-Rodríguez R, Macanás J, Muñoz M, Muraviev DN (2013) Superparamagnetic ag@ co-nanocomposites on granulated cation exchange polymeric matrices with enhanced antibacterial activity for the environmentally safe purification of water. Adv Funct Mater 23(19):2450–2458

Arivizhivendhan KV, Boopathy R, Maharaja P, Mary RR, Sekaran G (2015) Bioactive prodigiosin-impregnated cellulose matrix for the removal of pathogenic bacteria from aqueous solution. RSC Adv 5(84):68621–68631

Baygar T (2020) Characterization of silk sutures coated with propolis and biogenic silver nanoparticles (AgNPs); an eco-friendly solution with wound healing potential against surgical site infections (SSIs). Turkish Journal of Medical Sciences 50(1):258–266

Bhaumik M, Maity A, Srinivasu VV, Onyango MS (2011) Enhanced removal of Cr (VI) from aqueous solution using polypyrrole/Fe3O4 magnetic nanocomposite. J Hazard Mater 190(1–3): 381–390

Bohara RA, Throat ND, Mulla NA, Pawar SH (2017) Surface-modified cobalt ferrite nanoparticles for rapid capture, detection, and removal of pathogens: a potential material for water purification. Appl Biochem Biotechnol 182(2):598–608

Damokhi A, Yousefinejad S, Fakherpour A, Jahangiri M (2022) Improvement of performance and function in respiratory protection equipment using nanomaterials. J Nanopart Res 24(4):76

Deshmukh SP, Patil SM, Mullani SB, Delekar SD (2019) Silver nanoparticles as an effective disinfectant: a review. Mater Sci Eng C 97:954–965

Ferreira CA, Guerreiro SF, Valente JF, Patrício TM, Alves N, Mateus A, Dias JR (2022a) Advanced face mask filters based on PCL electrospun meshes Dopped with antimicrobial MgO and CuO nanoparticles. Polymers 14(16):3329.www.alliedmarketresearch.com/nano-materials-market

Ferreira CA, Guerreiro SF, Valente JF, Patrício TM, Alves N, Mateus A, Dias JR (2022b) Advanced face mask filters based on PCL electrospun meshes Dopped with antimicrobial MgO and CuO nanoparticles. Polymers 14(16):3329

Hooshmand S, Kargozar S, Ghorbani A, Darroudi M, Keshavarz M, Baino F, Kim HW (2020) Biomedical waste management by using nanophotocatalysts: the need for new options. Materials 13(16):3511

Jia C, Zhang Y, Cui J, Gan L (2019) The antibacterial properties and safety of a nanoparticle-coated parquet floor. Coatings 9(6):403

Jones W, Gibb A, Goodier C, Bust P, Song M, Jin J (2019). Implants, Surgery and Coatings. http://www.observatorynano.eu/project/filesystem/files/Surgery, %20implants%20and%20 coatings-April%2009.pdf) Nanomaterials in construction–what is being used, and where? Proc. Inst. Civ. Eng.: Constr. Mater 172(2):49–62

Kumar V, Katyal D, Nayak S (2020) Removal of heavy metals and radionuclides from water using nanomaterials: current scenario and future prospects. Environ Sci Pollut Res 27:41199–41224

Mali S (2013) Nanotechnology for surgeons. Indian Journal of Surgery 75(6):485–492

Malki M, Fleischer S, Shapira A, Dvir T (2018) Gold nanorod-based engineered cardiac patch for suture-free engraftment by near IR. Nano Lett 18(7):4069–4073

Minakshi P, Ghosh M, Kumar R, Brar B, Lambe UP, Banerjee S, Ranjan K, Kumar B, Goel P, Malik YS, Prasad G (2020) An insight into Nanomedicinal approaches to combat viral Zoonoses. Curr Top Med Chem 20(11):915–962. https://doi.org/10.2174/1568026620666200325114400

Mohajerani A, Burnett L, Smith JV, Kurmus H, Milas J, Arulrajah A, Abdul Kadir A (2019) Nanoparticles in construction materials and other applications, and implications of nanoparticle use. Materials 12(19):3052

Nakamura S, Ando N, Sato M, Ishihara M (2020) Ultraviolet irradiation enhances the Microbicidal activity of silver nanoparticles by hydroxyl radicals. Int J Mol Sci 21(9):3204

Nanomaterials Market Research (2031) Nanomaterials Market—Allied Market Research. https://www.alliedmarketresearch.com/nano-materials-market

Nunes D, Pimentel A, Branquinho R, Fortunato E, Martins R (2021) Metal oxide-based photocatalytic paper: a green alternative for environmental remediation. Catalysts 11(4):504

Patil PM, Bohara RA (2020) Nanoparticles impact in biomedical waste management. Waste Manag Res 38(11):1189–1203

Prasad M, Ghosh M, Kumar R, Brar B, Surjith KP, Lambe UP, Ranjan K, Banerjee S, Prasad G, Kumar Khurana S, Kharb P (2021) The importance of nanomedicine in prophylactic and Theranostic intervention of bacterial Zoonoses and reverse Zoonoses in the era of microbial resistance. J Nanosci Nanotechnol 21(6):3404–3452. https://doi.org/10.1166/jnn.2021.18999

Pullangott G, Kannan U, Gayathri S, Kiran DV, Maliyekkal SM (2021) A comprehensive review on antimicrobial face masks: an emerging weapon in fighting pandemics. RSC Adv 11(12): 6544–6576

Rahman MT, Sobur MA, Islam MS, Ievy S, Hossain MJ, El Zowalaty ME, Rahman AT, Ashour HM (2020) Zoonotic diseases: etiology, impact, and control. Microorganisms 8(9):1405

Rashad AM (2017) Effect of carbon nanotubes (CNTs) on the properties of traditional cementitious materials. Constr Build Mater 153:81–101

Saleem H, Zaidi SJ (2020) Sustainable use of nanomaterials in textiles and their environmental impact. Materials 13(22):5134

Saraswati, M., Permadani, R. L., & Slamet, A. (2019, April). The innovation of antimicrobial and self-cleaning using ag/TiO2 nanocomposite coated on cotton fabric for footwear application. In IOP conference series: materials science and engineering (Vol. 509, No. 1, p. 012091). IOP Publishing

Shim HE, Yang JE, Jeong SW, Lee CH, Song L, Mushtaq S, Jeon J (2018) Silver nanomaterial-immobilized desalination systems for efficient removal of radioactive iodine species in water. Nano 8(9):660

Sunitha S, Adinarayana K, Sravanthi RP, Sonia G, Nagarjun R, Pankaj T, Sujatha D (2017) Fabrication of surgical sutures coated with curcumin loaded gold nanoparticles. Pharm Anal Acta 8(1):2

Syukri DM, Nwabor OF, Singh S, Voravuthikunchai SP (2021) Antibacterial functionalization of nylon monofilament surgical sutures through in situ deposition of biogenic silver nanoparticles. Surf Coat Technol 413:127090

Verma V, Al-Dossari M, Singh J, Rawat M, Kordy MG, Shaban M (2022) A review on green synthesis of TiO2 NPs: photocatalysis and antimicrobial applications. Polymers 14(7):1444

veterinary.assam.gov.in/portlet-innerpage/veterinary-hospitals-and-dispensaries

Weir E, Lawlor A, Whelan A, Regan F (2008) The use of nanoparticles in anti-microbial materials and their characterization. Analyst 133(7):835–845

Wen T, Ma R, Liu X, Song S, Wu B, Jiang Z, Wang X (2022) Adsorptive and reductive removal of toxic and radioactive metal ions by nanoscale zero-valent iron–based nanomaterials from wastewater. In: Emerging nanomaterials for recovery of toxic and radioactive metal ions from environmental media. Elsevier, pp 195–228

www.diva-portal.org/smash/get/diva2:665233/FULLTEXT02.pdf

www.observatorynano.eu/project/filesystem/files/Surgery,%20implants%20and%20 coatings-April%2009.pdf

www.technanoindia.com/antislip.html

Yang DJ, Zheng ZF, Zhu HY, Liu HW, Gao XP (2008) Titanate nanofibers as intelligent absorbents for the removal of radioactive ions from water. Adv Mater 20(14):2777–2781

Yang D, Liu H, Zheng Z, Sarina S, Zhu H (2013) Titanate-based adsorbents for radioactive ions entrapment from water. Nanoscale 5(6):2232–2242

Yang J, Hou B, Wang J, Tian B, Bi J, Wang N, Huang X (2019) Nanomaterials for the removal of heavy metals from wastewater. Nano 9(3):424

Zhao W, Chen IW, Huang F (2019) Toward large-scale water treatment using nanomaterials. Nano Today 27:11–27

Chapter 17
Nanotechnology in Animal and Microenvironment Management

Annu Yadav, Tarun Kumar, Neelesh Sindhu, Ankit Kumar, Nikhil Mudgil, and Anil Kumar

Abstract Farm management, particularly the microenvironment around animals is a key to harvest the optimum production potential and to maintain the welfare also. The management cost in any livestock farm is one of the major recurring costs which needs to be controlled for sustainable farming system. Nanotechnology-based applications can be used for monitoring the animals, maintaining biosecurity, providing comfortable microenvironment, curtailing the cost by automation and reducing drudgery. Application of AI-based solutions using various nanosensors and nanomaterial-based housing systems can help in maintaining the microenvironment of an animal farm, similarly nanoparticle-based products facilitate enhanced biosecurity and productivity of animals. Nanosensor-based machines can also detect pollutants/contaminants and remove the biological wastes from farm in a timely manner that will reduce the human interventions and cost associated with it besides minimizing the chance of infections. All these inputs will ultimately keep the animals in a comfortable condition and within its physiological limit, hence enhance the production and sustainability of the farm. The present chapter will discuss all these

A. Yadav · N. Mudgil
Department of Veterinary Medicine, Lala Lajpat Rai University of Veterinary and Animal Sciences, Hisar, Haryana, India

T. Kumar · N. Sindhu
Department of Veterinary Clinical Complex, Lala Lajpat Rai University of Veterinary and Animal Sciences, Hisar, Haryana, India

A. Kumar
Haryana Pashu Vigyan Kendra, Lala Lajpat Rai University of Veterinary and Animal Sciences, Hisar, Haryana, India

A. Kumar (✉)
Department of Agricultural Meteorology, Chaudhary Charan Singh Haryana Agricultural University, Hisar, Haryana, India

LLR University of Veterinary and Animal, ICAR-National Research Centre on Equines, *Nanotechnology Theranostics in Livestock Diseases and Management*, Livestock Diseases and Management,
https://doi.org/10.1007/978-981-16-1610-5_17

potential applications of nanomaterials in the effective management of microenvironment in animal farms.

Keywords Microenvironment · Livestock farm · Nanotechnology · Biosensors

17.1 Introduction

Nanotechnology is an innovative discipline of science that has paved the foundation for novel and ingenious approaches in numerous areas of medical science. In the USA (2004), National Science and Technology Council described nanotechnology as a field that relates with seeing, assessing, and manipulating matter at atomic, molecular and supramolecular levels (Muktar et al. 2015). Currently, a wide variety of nanomaterials are being researched for their credible use in several applications, for example, nanoshells (for the destruction of cancerous cells by infrared radiations (IR)), Bucky balls (catching free radicals produced through allergic responses), carbon nanotubes (for sensors and drugs delivery), aluminosilicate nanoparticles (to decrease bleeding), nanocrystalline silver (as antimicrobial agent), gold nanoparticles (for diagnosis and as labeling agents), iron oxide nanoparticles (for the perfection of magnetic resonance imaging (MRI)), and nanorobots (for distinct cell repair) (Chakravarthi and Balaji 2010). Various tools of nanotechnology, such as nanomaterials, microfluidics, and biodiagnostic nanosensors, have been speculated to resolve several mysteries pertaining to animal well-being, disease prevention, reproduction, and production, as well as treatment of infections (Youssef et al. 2019). In forthcoming years, it is sensible to believe that nanotechnology would reform the discipline of science and technology for the animal strength and will aid to enhance the production of livestock (Wilczewska et al. 2012). Veterinary nanotechnology will enhance the systems of diagnosis and treatment delivery, and provide new gears for molecular and cellular breeding, animal history from birth to consumer table, and animal nutrition scenarios ranging from nutrient uptake to livestock waste management (Meena et al. 2018). This chapter aims to explore the diverse applications of nanotechnology in farm animals and their microenvironment management.

17.1.1 Nanotechnology Approaches for Developing Male Effect (Biological Stimulation Management) Procedure

17.1.1.1 Challenges of Male Effect Applications

Pheromones are volatile chemical substances secreted by specific organs and scent glands of different animals. In mammals, including livestock species, pheromones play crucial roles in the communication between animals, signaling many behavioral and physiological processes. The role of pheromones in regulating several reproductive events in farm animals is well documented and can be indicated through observing changes in animals' activity following direct (physical) and/or indirect (spatial) exposure to specific sexual pheromones (Heredia and García 2018). The most known pheromones-based phenomenon that can prime and regulate many reproductive events in farm animals is known as the "Male Effect." Pheromones in the wool, wax, and urine originating from sexually active males have a significant influence on the onset of puberty, the resumption of sexual activity of seasonal anestrous females, and the reduction in postpartum anestrus in many farm animals, specifically sheep, goats, and swine (Wang et al. 2017). Each of these effects is mediated by olfactory chemical signals, pheromones, released from the male and effecting the hypothalamic system of the stimulated female to generate pulses of gonadotropin-releasing hormone (GnRH) modulating sexual activity (Duffy et al. 2018). The male effect is well-recognized and applied in reproductive management (Ajitha et al. 2016). In small ruminants, for example, many breeds exert different degrees of seasonality, which constrains the progression of the breeding program of commercial flocks, being restricted within a definite period of the year (Darvishi et al. 2019). The resumption of sexual and ovulation activities of the seasonal anestrous females depends on the activation of the neuroendocrine system by male pheromone signals and the subsequent modulation in LH pulse frequency and amplitude after sudden exposure to males. Usually, ovulation occurs within 2–3 days after male introduction to a large number of females (Mohd Yusof et al. 2020a). From a practical point of view, the male effect not only has the advantage of advancing the breeding season by about 4–6 weeks or more, but it can also provide an acceptable degree of female synchrony at the time of breeding and thus the subsequent lambing/kidding time. In context, the male effect is successfully used to induce ovulation in prepubertal ewe lambs and lactational anestrous ewes. The male effect can be used with 35 to 40 days postpartum ewes to decrease the time between deliveries (King et al. 2018). In pigs, the presence of boars reduces the postpartum period in lactating sows and the age of onset of puberty in gilts by about 30 days. These effects were early linked tool factory cues because a pen previously occupied by a boar and presumably pervaded with a boar's odor is effective in inducing early puberty (Thames and Theradiyil Sukumaran 2020). Apparently, priming pheromones remaining in the boar's pen after his removal are sufficient to induce early puberty (Duffy et al. 2018). The same was observed for sheep and goats, as pheromones in the wool, wax, and urine are sufficient to stimulate females to ovulate. The male effect, being a sexual biostimulation method, confers an opportunity to diminish, or even eliminate in the future, the intensive use of hormones in reproductive management and promote the production of hormone residues-free animal products. However, the outputs of this method, when applied at a farm scale, are challenged by several limiting factors. In brief, the main such factors are latitude, breed, sexual activity, hormonal status, age, and experience of the male (Zhao et al. 2014), depth of female anestrous status, male to female ratio, and the length of isolation period (Abedini et al. 2017). In fact, the application of such a method for induction of estrus needs the isolation of males from females for a certain period and reintroduction thereafter to stimulate females' sexual activity (estrus and ovulation) and/or to introduce new males that have not previously been kept in the herd (Mohd Yusof et al. 2020b). In many cases, the male effect is applied during out of season, during which males also may undergo weak sexual activity. For this reason, many studies have referred to the importance of priming males with exogenous stimuli, such as hormonal treatments mainly melatonin (Haque et al. 2017) and testosterone (NRC 1994) and photostimuli (Ashengroph et al. 2020), to speed up the sexual activity of males during the natural anestrous season. Overall, these challenges can negatively impact the application of the male effect at a large/field scale as a biological reproductive management protocol (Li et al. 2010).

17.1.1.2 Nanotechnology Approaches for Developing Male Effect Procedure

Male effect as previously described depends on the effect of olfactory signals generated by male pheromones on the hypothalamic–pituitary–gonadal axis of the females stimulating release of reproductive hormones (Knight and Lynch 1980). An aerosol nano-drug delivery system may facilitate the delivery of different drugs, including pheromones, through the olfactory pathways, nose-to-brain delivery and/or through the systematic circulation into target sites, passing the brain–blood and other biological barriers (Dugasa and Williams 2016). Despite few studies performed on applications of an aerosol nano-drug delivery system in the field of livestock production and veterinary medicine. Pamungkas et al. (2016) proved the efficiency of chitosan-TPP nanoparticles for nasal human chorionic gonadotropins (hCG) delivery to induce ovulation in cows. Moreover, many vaccines have been developed using nanomaterials to be delivered through the nasal pathway (Cerbu et al. 2021).

In this regard, different nanomaterials can be used as a novel delivery system to implement sexual priming pheromones ready for female's treatment through nasal spray. Specifically, the chemical compounds that act as pheromones are identified and can be isolated from different sites of the animal body including, wool wax, saliva, urine, and feces, of sexually active males. Fabrication of pheromones in an aerosol nanoformula may aid in overcoming some obstacles facing the male effect application at a farm scale. In this way, such therapies can be used at any season of the year regardless of the sexual activity of the males and the need for isolating or priming males' sexual activity before introduction into the group of females. Actually, the emergence of aerosol nano-drug delivery technology opens the way for developing numerous pheromones-based protocols for biological control of reproductive events, such as the male effect, estrous detection, postpartum anestrous, puberty, pregnancy diagnosis, sexual activity of males, and mother offspring relationship (Kekan et al. 2017). These could be achieved through producing commercial biostimulating pheromones and/or pheromones-based bioassay kits. As an example, urinary pheromones 4-methyl phenol (4-mp, p-Cresol), 9-octadecenoic acid (oleic acid), and luteinizing hormone (LH) are suggested to develop nanoparticle-based bioassay kits for detection of estrus or ovulation in buffaloes (Archunan 2020).

17.1.2 Application of Nanotechnology in Dairy Cattle Welfare Improvement

1. Animal welfare in dairy farms should be possible in shelters with a comfortable rest area (Gill 2012). Specific areas in this aspect include thermal comfort, comfort around resting, and ease of movement of all the cows (EFSA 2009). Some areas in the dairy farms can be improved with the nanotechnology approach. But in general, it is connected with construction and facilities in the barn, for example, cow bedding, milking grease, milking machine, feeder, cubicles, and the other facilities. However, the discomfort aspect is also related with pollutants in the barn, and the nanotechnology is also able to solve the problems in this field. Mat-stall (bedding) improves the productive and reproductive performance (Singh et al. 2020). Mat-stall (bedding) materials are related with odds ratio of lameness. If the bedding is made from hard materials, it will improve the odds ratio of lameness. The bedding is also affected by the overall gaseous emissions (Hartung and Phillips 1994). Using mat-stall for the bedding for the cows will improve comfort levels in the barn. Nanotechnology can be applied for mat-stall materials. SeNPs is proven to increase the activity of anti-oxidative enzymes in serum, also relieves and improves oxidative damage in dairy cows caused by heat stress (Weixing et al. 2009). To decrease the heat stress effect in the tropic and hot climates, construction manipulation is one of the solutions. Construction manipulation is always correlated with materials used to building the barn, especially roof materials to decrease the heat radiation. Nanoparticle TiO_2 has the capability of photocatalytic activities, so this material is good for heat insulation. Furthermore, using TiO_2 nanoparticle in the feeder can improve the comfort level of the cattle because this material has self-cleaning coating which will reduce the contaminant agents in the feed. Another discomfort factor is gas emissions in the barn that causes odors. There are two approaches to reduce the emission of gas in dairy farms: reduced gas emissions inside of the body metabolic system and reduce the emissions in the slurry degradation processes. Applicable nanotechnology approach can be used to solve gas emission problems with some nanomaterials. For instance, using SiO_2 as a feed additive can reduce ammonia emission in the body metabolic system (Pieszka et al. 2018). Another example is an approach to animal waste management by adding nanoparticles to manure to enhance biogas production from anaerobic digesters or to reduce odors (Scott 2007).
2. Some food and nutrition products containing nanoscale additives are already commercially available and nanotechnology has begun to be widely used in different fields (Brunori et al. 2009). For example, various compounds in their

nanoforms are used as a supplemental source of trace minerals (MgO, Na_2O, SiO_2, CaO, K_2O, TiO_2, Al_2O_3, and Fe_2O_3) in diets (Meena et al. 2018). Each nanomaterial has a specific function in the metabolism system. For example, ZnO nanoparticles as feed additives can improve the growth of rumen microorganisms and ruminal microbial protein synthesis (Huang et al. 2015) and raise the energy utilization efficiency (Riazi et al. 2019). Then, SeNPs can stimulate rumen microbial activity, digestive microorganisms, or enzyme activity (Shi et al. 2011) besides exhibiting lower toxicity than selenite (Ciçek et al. 2021). Moreover, $CuSO_4$ can significantly improve the digestibility of crude fat and energy, levels of IgG, γ-globulin and total globulin protein, and superoxide dismutase (SOD) activity (Gonzales-Eguia et al. 2009). Further, SiO_2 can significantly increase the number of lactic acid bacteria and decrease the number of bacterial pathogens (Pieszka et al. 2018). Technique nanoencapsulation can be applied for protection of some amino acids from rumen activity. Because some researches have reported that rumen protected methionine (RPM) plus rumen protected lysine (RPL) can improve milk yield and protein contents of dairy cows (Awawdeh 2016). From the examples that are mentioned above, we can conclude that nanoparticles can become an alternative solution to improve the dairy cattle welfare quality level. Because the data show usage of nanoparticles in the feed can improve digestibility, enzyme activity in the digestive system and inhibit pathogenic bacterial growth (Huang et al. 2015; Riazi et al. 2019). As a result, the evaluation of dairy cattle welfare improvement in this aspect can be done with body condition score (BCS) method (Spigarelli et al. 2020). After nanotechnology application, the BCS of cows can be tested to see if they are in the normal range and are not overconditioned or underconditioned.

3. Dystocia and the need of human assistance might cause trauma to the cattle. The nanoparticle MgO has preventive capability in the memory impairment induced by postpartum depression (Zadedarvish et al. 2020). Furthermore, the milking process also has the potency to inflict trauma if the process is not properly done. Therefore, nanotechnology can be applied in this area. To improve the comfort levels in the milking process, generally some dairy farms use milk grease. Nanoparticles can be applied in this area, especially nano-lubricants fields. Thin and short multi-walled carbon nanotubes could improve the friction-reduction and anti-wear properties of vegetable based oil more effectively than thick and long multi-walled carbon nanotubes (Su et al. 2018). Furthermore, some part in the milking machine uses rubber, so nanotechnology can also take a part in this area to improve the comfort quality while the milking process is carried out. For example, the tensile strength, tensile stress, and elongation at break of rubber nanocomposites were found to increase with addition of cellulose nanocrystals (CNC) regardless of cellulose sources (Ditanet et al. 2019).

17.1.3 Nanotechnology Application in Farm Power and Machinery

17.1.3.1 Application of Nanotechnology in the Chassis and Body of Farm Machinery and Equipment

Farm machinery and equipments, especially tractors, combines and other machines specify the need for utilizing more durable parts. Nanocomposites with improved mechanical properties compared to conventional composites can be a new approach to lightening and even replacing metal parts in machinery (Zein El-Din et al. 2021). Theoretically, these materials can be easily extruded or molded in the final shape, while they have the same strength of the metal and are lighter (Dehkordi and Keivani 2017). This not only helps increasing the lifetime of these equipments, but also lightening them, reducing fuel consumption and, consequently, reducing environmental pollution. These composites also have two other advantages. First, nanoparticles significantly provide a more beauty by creating a smooth and uniform network structure, and thus nanocomposites have a more acceptable appearance surface and more transparent colors. Furthermore, nanocomposites are lighter than common composites up to around 22% due to the need for lesser reinforcing materials, and therefore the use of nanocomposites in the production and construction of farm equipments and machinery is very important (Vatan-Dolat-Xah 2006).

17.1.3.2 Effect of Nano-Structured Coating Material on Reducing Abrasive Wear of Tillage Tools

Wear is the most important limiting factor of life and performance of cutting tools and other tillage tools. Wear of tillage tools is a kind of roaring wear. Roaring wear mechanism is created in contact with each other in the presence of hard particles on one or both tool surfaces. The hard roughness of the surface can also be the cause of this type of wear (Rabinowicz and Tanner 1966). The wear depends on many parameters such as geometry, strength, and surface properties. Moisture content can be an important factor in this regard. Changes in moisture can affect the wear resistance of the nanostructured materials in different orders (Czichos and Misra 1978). Since the use of sunlight-resistant and wear-resistant paint in the machinery is essential, nanometric coatings for various purposes such as increased strength, improved bending and sheeting properties, protective coatings resistant to corrosion, scratches, wear and nanocomposites have been used due to high strength and durability against environmental factors. One of the most important applications of this technology in the machinery industry is the production of materials such as Mica and Talc which in addition to the apparent beauty has increased the strength and, in part, the weight loss of agricultural equipment, and finally, has caused to reduce the compaction of agricultural soils and be lacking the formation of hard layers. Furthermore, the application of nanotechnology in the production of agricultural machinery and equipments will cause reduction in fuel consumption and increase the life of parts and, ultimately, enhance the economic efficiency. Ceramic coatings of nanoparticles contribute to thermal stability and erosion resistance in engine parts (Asadi-Fard 2017). Hardness and wear resistance are two important properties of objects that are used in a wear environment. By definition, hardness of a material and its resistance to plastic deformation is localized and has a direct relationship with wear resistance. Nickel electroplating is based on using a nickel anode and applying a direct current to the anode and cathode in an aqueous environment containing nickel ions. Passing a direct current causes anode electrode dissolution in the electrolyte and nickel is coated on the cathode. It has been proved that stiffness can be increased by fine grain size materials, without reducing the flexibility of the material. In nickel nanostructured coatings which have an average grain size of about 100 nm and produced by electroplating method, wear resistance is about 100–170 times than nickel coatings with the average grain size of about 100 mm. The coefficient of friction of these coatings is about 40–50% less than conventional nickel coatings (Ebrahimi et al. 1999). Size and shape of grains in a coating have large effects on the mechanical properties such as hardness. The smaller the grain size of a coating, the less the friction coefficient will be. These differences are due to the changing in wear mechanism by alterations in the grains size. Wear resistance of a conventional nickel coating is low. However, nickel nanostructured coatings have high resistance against wear. There is a relationship between the shape, grain size of coating, and mechanical and physical properties of coating with the conditions of plating that awareness of this relationship makes nano-nickel coating replacement with chromium coating affordable. Abed et al. (2019) studied the influence of three methods of coating, including DC electroplating, pulse electroplating and the electro plasma methods for coating of nickel compounds on pieces carved from high carbon steel material (Ck60). Resistance of the coatings to abrasion was assessed with the criterion of proportion of weight reduction due to abrasion in the farm of operations. They found that coating made by DC electroplating, improved abrasion resistance in the samples by 30.41%. The coatings made by pulse electroplating on the pieces could improve abrasion resistance relative to raw samples by 55.83%. In the electro plasma method, by formation of coatings that had complex compositions, resistance to abrasion increased up to 90% relative to raw samples. It is noteworthy that such resistance to abrasion was in samples with the 480 HV and average grain size of 38 nanometers. The less the average sizes of grains of nickel nano-coatings of the structure, the more the hardness and, consequently, the more wear resistance. Zain El-Din et al. (2021) studied the reducing wear behavior of soil working tools by coating and treating processes of the tool surfaces. Three different carbon steel materials: API 52, M238 and K110 were used. Some of the previous specimens were treated with two different coating materials: Hard chrome, HCR, and carbon nanotubes-hard chrome composite, CNTs + HCR. The other specimens were treated by conventional heat treatment, HT, and nitrocarburizing treatment, NC. Both the coated and treated tool specimens were mounted on a circular soil bin containing sandy clay loam soil at 14% moisture content. The working speed was adjusted at 5 km/h (1.39 m/s). The coating thickness and the mass of tools were measured before and after each test. The test duration of tillage operation was conducted for 200 h for both coated and treated tools. Results obtained from all tests showed that the wearing rate of coated specimens by carbon nanotubes-hard chrome composite CNTs + HCR were the minimum compared with other treatments. Results also showed that the surface roughness was the minimum (Jain et al. 2021).

17.1.4 Impact of Nanotechnology for Animal Health and Nutrition

17.1.4.1 Nanofeed

In an animal production system, approximately 40–50% of the operating costs are due to animal feed (Wanapat et al. 2015). Feed industry's aim is to increase the efficiency of feed and its additives. Nutritional deficiency significantly decreases the production potential of animals, and nutrient-deficient animals are more prone to disease (Wanapat et al. 2015). The animal feed industry should formulate animal feed with NPs with the following aims: (1) improvement of overall feed efficiency; (2) increased production levels and quality of animal products (milk, eggs, and meat); (3) use of NPs having antioxidant and immune-modulatory properties to improve health; (4) reducing the requirement of antibiotics as growth promoters, as these may have negative effects on human health; and (5) removing unpleasant smells (boar taint) of animals products, particularly in meat of pigs (Hill and Li 2017). Several studies suggested that nanoform of feed can be used to enhance its nutritive value and to promote animal growth (Hill and Li 2017). Some of the examples of nanofeed additives are summarized in Table 17.1.

1. Nano-minerals are inexpensive, utilized in lower quantities, and act as growth-promoting and immune-stimulating agents, so produce many benefits for animal feed production. Similarly, they also can help to manipulate feed pathogens and improve fermentation process in the rumen. Nano zinc oxide is considered one of the promising nano-minerals that has been used to improve the growth rate, immune response, and reproductive performance of livestocks. Further, diarrhea in young piglets can be reduced by nano zinc (Yang and Sun 2006). Nano zinc decreases the somatic cell count in dairy cows suffered from subclinical mastitis (Rajendran 2013).
2. Similarly, microencapsulation of feed components is done in order to protect them from oxidation and break down by light and oxidation, further it keeps them away from the lysis by enzymes of the digestive system such as proteases, also promote stability at various values of pH, with better dispersion and good mixing of lipophilic additives, and will prolong their action. A big problem affecting animals and human is the mycotoxicosis. Mycotoxins may be detected in about 25% of the animal feeds. Nano silica and nano magnesium oxide are considered

Table 17.1 Commonly used nanoparticle additives in farm animal feed (Younis et al. 2022)

Animal species	Nanomaterial	Application
Sheep, goat, and poultry	Selenium	Stimulation of rumen microbial and enzyme activity Enhancing semen quality and immune response
Ruminants, pig, and poultry	Zinc	Improving feed conversion ratio that promotes growth Enhancing the immune response
Pig	Chromium	Anti-diarrheal Improving carcass quality, particularly lean meat production Enhancing the immune response
Poultry	Copper	Stimulating the metabolic rate in broilers during embryonic development Strengthening immunological biocompatibility
Poultry	Montmorillonite–composite	Reducing aflatoxin toxicity
Farm animals, including poultry	Nano-polystyrene with polyethylene glycol linkers and mannose targeting biomolecules	Binding and removal of food-borne pathogens in animal feed

Table 17.2 Characteristics and expected advances by the implementation of nanotechnologies in nutritional management of farm animals

Formula	Technique	Expected advances
Zinc oxide NPs; Shahin et al. (2020)	Commercial product	Increased bioavailability and reduce negative effects of toxic concentrations in in vitro assisted reproductive techniques
Selenium oxide NPs; Kachuee et al. (2019)	Commercial product	Increased absorption and bioavailability and reduce negative effects of toxic concentrations during late pregnancy
Fish oil or soy oil in water; El-Sherbiny et al. (2016)	Nanoemulsion	Improving post ruminal supply of PUFA
Solid lipid lysine; Albuquerque et al. (2020)	Ultrasonic processor	Improving post ruminal supply of lysine
Alginate chitosan glycerol NPs; Gawad and Fellner (2019)	Ionic gelatin	Encapsulating glycerol to bypass rumen fermentation

as powerful nano antimycotoxin agents that bind successfully to aflatoxins and inactivate them (Moghaddam et al. 2010).

3. Specific engineered nanomaterials can be implemented to provide various solutions to nutritional manipulation challenges in animals (Table 17.2). In the field of animal nutrition, nano-encapsulation of active feed ingredients and processing of nano-minerals, in particular, trace minerals of poor bioavailability, are rapidly

growing technologies that are used to protect targeted feed ingredients from degradation (during processing, storage, and in the rumen) and to ensure effective and sustained delivery of the valuable feed ingredients to target sites (Gawad and Fellner 2019). In addition, feed ingredients in nano-forms show novel physico-chemical properties (mainly small size and high surface area) that improve their absorption, bioavailability, and utilization by animal tissues. However, up to date, studies showing the effect of nanofeed ingredients on animal health and reproduction are limited (Kachuee et al. 2019).

In a study by El-sherbiny et al. (2016), nanoemulsification of soyabean and fish oil aimed to post rumen delivery of long chain polyunsaturated fatty acids, significantly increased proportion of long chain polyunsaturated fatty acids in rumen fluid without negative effects on rumen fermentation efficiency. Similarly, Albuquerque et al. (2020) used solid lipid nanoparticles to protect lysine from enzymatic hydrolysis by ruminal microbiota and to increase rumen bypass of lysine to small intestine. In this study, the lipid lysine nanoformulation showed stability against microbial hydrolysis in rumen for upto 24 h. In addition, nanoencapsulation extends the opportunity to use lower quantities of nanofeed ingredients to get the same effects as ordinary forms which may decrease the feeding costs. Further, it restricts the release of some materials, mainly minerals to the environment decreasing pollution (Gawad and Fellner 2019).

17.1.5 Farm Breeding and Reproduction

Management of breeding is an expensive and time-consuming problem for dairy and swine farmers. Estrus detection using nanomaterials have provided a great solution in this regard. The nanotubes (O'Connell et al. 2002) are currently being used as a means of tracking oestrus in animals because these tubes have the capacity to bind and detect the estradiol antibody at the time of oestrus by near-infrared fluorescence. The signal from this sensor will be incorporated as a part of a central monitoring and control system to actuate breeding. Microfluidics is currently used in animal science to significantly simplify traditional in vitro fertilization procedures in animal breeding. It is being used in livestock breeding to physically sort sperm and eggs (Kumar et al. 2023). Microfluidic and nanofluidic are the systems which analyze the gamets by controlling the flow of liquids or gases through a series of tiny channels and valves, thereby sorting them, much as a computer circuit sorts data through wires and logic gates. With the mapping of the human genome behind them, geneticists are now rapidly sequencing the genomes of cattle, sheep, poultry, pig, and other livestock hoping to identify gene sequences that relate to commercially valuable traits such as disease resistance and leanness of meat. By including probes for these traits on biochips, breeders will be able to speedily identify champion breeders and screen out genetic diseases (Patil et al. 2009).

17.1.5.1 Estrus Detection

Nanosensors are nanoscale devices that are highly sensitive to mobile sensing biomolecules. The application of nanosensors in livestock is to diagnose reproductive tract infections, metabolic and endocrine disorders as well as in estrus detection (Saragusty and Arav 2011). Nanotubes are placed under the skin during the estrus of animals to measure the estradiol concentration in the blood at a particular time. The nanotubes have a binding affinity towards the estradiol antibody during the estrus phase and transmit the fluorescence signal toward the signal receiver central computer. Recently, a new approach based on a colorimetric assay has been employed to determine odorant chemical compounds (pheromones) for the diagnosis of estrus in bovine by using capped L-tyrosine silver nanoparticles (L-TyrAgNPs). Sex pheromones like acetic acid or propionic acid change color from yellow to reddish-brown which might have a great potential to apply in colorimetric sensors for the identification of the accurate time of estrus (Manikkaraja et al. 2020). One of the main constraints in animal breeding is the accurate time of heat identification. Recently, Bhatia et al. (2021) developed a noninvasive method for oestrus identification in buffaloes based on seed germination inhibition test.

17.1.5.2 Nanobiosensors for the Detection of Reproductive Status

Nanobiosensors are being validated and used for disease detection and estrus management, as well as to detect the hormonal levels, and considered as a promising tool in the reproductive management (Monerris et al. 2012). Nanostructured delivery systems have shown numerous advantages and nanosized delivery systems have displayed better therapeutic efficacy in the delivery of bioactive molecules, such as reproductive hormones by the improvement of pharmacodynamic properties. These nanobiosensors can deliver a wide range of molecules which are sensitive to light, oxidation, or hydrolysis and improve their sustained release, lowering systemic toxicity and preventing impulsive degradation and inactivation (Meena et al. 2018).

17.1.5.3 Nanotechnology Approaches for Developing Hormonal Based-Treatments

Nano-delivery of hormones has emerged as a new pharmacological approach. Several engineered nanoparticles have been proposed as novel platforms for the protection and controlled release of reproductive hormones, including gonadotropin and steroid hormones (Hashem and Gonzalez-Bulnes 2021). There are, to date, several studies reporting the use of nano-hormone delivery systems in the field of livestock reproductive management. These studies showed an enhancement of the pharmacokinetics and pharmacodynamics of the hormonal treatments, specifically, those of low molecular weight and short lifespan, such as GnRH (Hassanein et al.

2021). In addition, the opportunity to use biodegradable materials as a matrix for progesterone delivery instead of silicon-based materials was shown (Fogolari et al. 2017) to sustain the release of some hormones, such as melatonin in *in vitro* production media (Noori Siahdasht et al., 2020) and to change the route of hormone administration by enhancing mucosal absorption even if these hormones have high molecular weight (Rathbone and Burke 2013). Overall, these studies show that the use of nano-hormone delivery systems provide many advantages to hormonal-based treatments, such as decreasing hormone dosage, changing the route of administration, increasing animal welfare, and decreasing the risk of exposition to different hormones by workers and technicians.

Among the different gonadotropins used for ovarian stimulation, eCG has a unique feature as it displays pronounced FSH-like activity in addition to its LH-like activity when used in species other than equine (Combarnous et al. 1981). Thus, the shortage/disappearance of this hormone will negatively impact many assisted reproductive techniques, specifically those depending on the stimulation of follicular growth. In these terms, Santos-Jimenez et al. (2020) showed the possibility of the use GnRH as an eCG alternative if GnRH release behavior is slowed and sustained for a longer time, avoiding the occurrence of early LH-surge. In this study, diluting GnRH in propyleneglycol was effective in achieving this purpose and has been used as an eCG alternative in a CIDR-based oestrous synchronization protocol tailored for sheep. Accordingly, nano-hormone delivery systems may aid in changing the biological behavior of GnRH.

17.2 Detection of Ovulation

17.2.1 Progesterone

Breeding is an integral part of livestock farming. Detection of the ovulation period in cattle is important in order to determine the time window for artificial insemination. Conventional oestrus detection involves ocular inspection of cattle by skilled labor, which is expensive as well as inefficient. Biosensors for ovulation detection have been researched for a long time. Pemberton and colleagues reported a device able to determine ovulation using a disposable screen-printed amperometric progesterone biosensor, operated in a competitive immunoassay. The biosensor included a monoclonal anti-progesterone antibody (mAb) immobilized on a screen-printed carbon electrode (SPCE). It was later incorporated into a thin-layer flow cell offering advantages such as on-line analysis and improved fluid handling with the possibility of future automation (Pemberton et al. 2001). The Herd Navigation® system was developed in 2008 for commercial use and combines five sensing systems, including progesterone in the milk. Herd Navigator™ (Durkin and DeLaval 2010) measures the level of progesterone in milk and the software suggests the insemination time, lists animals for final pregnancy confirmation, indicates early abortion and lists the cows at risk for cysts and prolonged anoestrus (Mazeris 2010). Oestrus detection rates of 95–97% have been reported in the farms in Denmark, with significantly higher pregnancy rates (up to 42–50%) than the conventional techniques (Blom and Ridder 2010). There are also reports indicating cost savings of €250 and €350/cow per year, as farmers do not have to spend money on expensive pregnancy tests (Mazeris 2010; Leonardi et al. 2013).

(a) SPR-based biosensors for progesterone

More recently, Zeidan and colleagues have reported the development of a progesterone sensor by integrating novel aptamer development with a nano-enhanced Surface Plasmon Resonance imaging sensor (SPRi). The authors first developed X-aptamers and selected them for binding to progesterone. Then, the multi-array feature of SPRi was used to develop an optimized biosensor capable of simultaneously screening the 9 X-aptamers for binding affinities. The sensor surface was further optimized in a sandwich assay, where nanoEnhancers (NIR-streptavidin-coated quantum dots) were used for ultrasensitive detection of progesterone molecules (Zeidan et al. 2016).

(b) Herd Navigator ™ for monitoring ovulation

Herd Navigator has excellent oestrus detection rates, but the system is too expensive. Aiming at improving detection reliability using low-cost sensor data, Jónsson et al. reported the combination of information from step count and leg tilt sensors. The authors developed a change-detection algorithm that can analyze cow-specific data in real time. The system has shown an increase in the successful alerts and significantly reduced false positives (Jónsson et al. 2011).

(c) Intravaginal probes

Andersson et al. (2016) tested a wireless intravaginal probe, with a possibility for automation of the process. The probe is based on the measurements of conductivity and temperature, and also senses the movement of the animal. These parameters can all be used independently to detect oestrus. Although still in the testing phase, the device has been shown to have a higher reliability and to be more resistant to external disturbances as compared to existing alternatives (Andersson et al. 2016; Andersson et al. 2015).

17.2.2 Importance of Biosensors for Livestock Health and Management

The term biosensor encompasses devices that have the potential to quantify physiological, immunological, and behavioral responses of livestock and multiple animal species. Novel biosensing methodologies offer highly specialized monitoring devices for the specific measurement of individual and multiple parameters covering an animal's physiology as well as monitoring of an animal's environment. These devices are not only highly specific and sensitive for the parameters being analyzed, but they are also reliable and easy to use, and can accelerate the monitoring process. Novel biosensors in livestock management provide significant benefits and applications in disease detection and isolation, health monitoring and detection of reproductive cycles, as well as monitoring physiological wellbeing of the animal via analysis of the animal's environment. Biosensors will contribute to the fourth revolution in agriculture by incorporating innovative technologies into cost-effective diagnostic methods that can mitigate the potentially catastrophic effects of infectious outbreaks in farmed animals. One of salient features of biological, agricultural and environmental applications of nanotechnology is that the nanoscale devices and systems are of the same size-scale as biomolecules. While conventional sensors have been used in livestock monitoring, and as tools to assist in health monitoring and disease diagnosis, nanobiosensors have the ability to multiplex the bioassays on site, thereby eliminating the need for the transportation of biological samples to centralized laboratories for analysis. Integration of these sensors for wireless data transfer via a server or through cloud-based systems would enable access to the analyzed data for any Internet-enabled device. Nanobiosensor applications will not only reduce the incumbent costs for reagents, sample handling, analysis times, and transportation costs, but will also help in adapting and promoting sustainable agricultural techniques and ethical handling of livestock.

Also researchers have developed nanobiosensors for the diagnosis of pathogens, diseases, oestrus, hormone level, and physiological changes in the reproductive tract of animals (Monerris et al. 2012). Wearable biosensors are innovative technologies nowadays becoming increasingly important for animal health and management. Biosensors and wearable technology can be deployed on animals to detect their sweat composition and analyze the sodium content present at a particular time (Glennon et al. 2016). Respiratory disease in livestock is quite common and may cause infantile death. The efficient way to early diagnose respiratory infections is to measure the rectal temperature of a calf by using a thermometer. However, this method is time consuming and required lots of effort from farmers during group feeding. The author has developed a wearable wireless biosensor device to record the body temperature of the calf automatically and save time and effort of the farmers.

These technologies shall focus on the noninvasive methodologies to assess animal welfare by quantifying the stress and metabolic disease biomarkers, welfare assessment based on activity of the animals (monitoring oestrus and lameness detection to maximize animal production) and sensors for temperature and pH sensing (to determine calving alert and rumen function). Furthermore, various noninvasive sensing technologies for early disease detection shall help in saving animals' lives as well as reducing expenses of the farmers. A combination of these technologies and the use of "smart" husbandry support systems will ensure maximum productivity while improving the wellbeing of farm animals (Caja et al. 2016). Noninvasive technologies for the chemical and biological analysis of samples from livestock, food, and feed can rapidly provide detailed information to evaluate the safety of various biological samples. Biosensors equipped with robust data collection and integration infrastructure will be able to realise this potential and shall become vital elements in real-time analysis of industrial agriculture. Not only will such systems help in maximizing the utilization of resources for farming; they will also allow for an evaluation of individual and group behavior of animals.

Development of on-site biosensing technologies will enable rapid, cost-effective and meaningful monitoring of dietary inputs, environmental conditions, genetic makeup, performance, metabolism, welfare and physiological state of animals (Nadin et al. 2012).

17.3 Biosensors as Indispensable Tools in Animal Husbandry

Future developments in biosensors are expected to result in the development of new methodological and technological approaches to measuring dynamic changes in real time, with respect to the changes in physiological state and metabolism (e.g., gastrointestinal flora, circulating levels of anabolic and catabolic hormones, immune function, gene expression). This is to better understand the factors influencing animals' responses, and to develop solutions (e.g., husbandry practices, technology and associated decision support system) that improve productivity and/or well-being of these animals.

17.3.1 Monitoring Jaw Movement of Cattle to Know the Grazing Efficiency

Cattle grazing behavior requires individual monitoring of cattle based on three important parameters, including location, posture, and animal movement, especially movements such as walking and jaw movement (Herinaina et al. 2016). Jaw movements define the grazing behavior of the cattle, and there are three different classes of biosensors that can be used to identify such movements. These include:

17.3.1.1 Mechanical Sensors (Pressure Sensors), Acoustic Sensors (Microphone) and Electromyography Sensors

Early systems developed to quantify the feeding behavior of cattle were created exclusively for research purposes. The IGER Behaviour Recorder (Institute of Grassland and Environmental Research, Okehampton, UK) (Rutter et al. 1997) and the ART-MSR pressure sensor (Agroscope Reckenholz-Tänikon ART Research Institute, Modular Signal Recorder MSR145, MSR 342 Electronics GmbH, Switzerland) (Nydegger et al. 2010) were designed specifically to be used in pastures and stables, respectively. The IGER Behaviour Recorder comprises a noseband and an electronic interface connected to a computer to record, analyze, and store data at 20 Hz (Rutter 2000). Jaw movement is identified as a pressure peak through the transmission of the movement to the halter and the change in the tube pressure. The installed software can distinguish between bites and chews (Rutter 2000). Peaks are considered to be bites when they are a combination of a major long peak followed by a smaller sub-peak, or a nonsymmetrical peak in the absence of the sub-peak (Nadin et al. 2012). The IGER Behaviour Recorder can also be used to estimate the feed intake with reasonable accuracy, using data on the number of chews and eating duration (using correlation coefficients) (Pahl et al. 2016). Miscalculations in the pressure sensing arise from practical considerations, that is, due to variations in the tightening of the halter on individual animals, which can result in different pressure values. Another practical issue relates to the measurement of the output wave signal, which can be altered if the halter is mounted too tightly or too loosely. Eating and rumination behaviors in cattle have also been reported with a noseband pressure sensor, consisting of a data logger, incorporated in the noseband to record the jaw movements via a pressure sensor (Braun et al. 2013).

17.3.1.2 Monitoring Jaw Movement Through Acoustic Sensing

Acoustic analysis of grazing behavior has been shown to accurately identify chewing and biting, and therefore can be used to estimate the feed intake of cattle (Laca and De Vries 2000). Acoustic analysis allows differentiation of three types of jaw movements: chew, bite and chew-bite, and microphones can be used to record the jaw sounds of a grazing animal (Ungar and Rutter 2006). The data can be used to classify ruminating behavior (Benvenutti et al. 2016) and are especially helpful in monitoring animal wellbeing.

Acoustic sensing of jaw movements can be classified based on the microphone location and acoustic system classification. Some systems are simple and detect jaw movements based on 10-min recordings of grazing sessions on a camera, with an accuracy of 94% (Herinaina et al. 2016). Signal patterns are analyzed by the machine-learning algorithms to determine intervals between jaw movements, intensity of each jaw movement (observed as a peak in the time domain), their duration, and their integration in a sequence of behaviors (Navon et al. 2013). The discrimination is based on the signal patterns produced during biting and chewing in a 1-kHz sound window: peak frequency, peak intensity, average intensity, and their duration (Laca and De Vries 2000). Clapham et al. (2011) have demonstrated a fully automatic Chew-Bite Real-Time algorithm for the detection and classification of ingestive events during cattle grazing. The system consists of a directional wide-frequency microphone facing inward on the forehead of the animal, and coupled with the signal analysis and decision logic algorithm, it can detect and analyze bites and chews with an accuracy of 94% (Clapham et al. 2011). Miloneet al. (2012) have reported the detection and classification of bites, chews and chew-bites with the help of the Hidden Markov model, which estimates the sequences of bites, chews and chew-bites using acoustic spectrum characteristics like decibels, by each sound. The successful classification is reported to range between 61% and 99% (Milone et al. 2012).

Microphone-based methods have demonstrated good accuracy for the detection of jaw movements and can differentiate between three different kinds of jaw movements. However, one of the key disadvantages of acoustic systems in outdoor environments is their susceptibility to environmental noises. Acoustic interpretation techniques that can overcome these disturbances still need to be advanced before these systems are deployed on the farm.

17.3.1.3 Acceleration Sensors for Jaw Movement and Feeding Behavior

Accelerometer sensors convert physical acceleration recorded from motion or gravity into a voltage output. Accelerometers can be used to measure static acceleration due to gravity, the low-frequency component of the acceleration, and the dynamic acceleration due to animal movement (Herinaina et al. 2016). Several researchers have demonstrated the use of accelerometers for analyzing the grazing behavior of animals (Mattachini et al. 2016). Andriamandroso et al. (2015) used smartphone inertial measurement units (IMUs) to count the number of bites through a frequency pattern of single-axis acceleration data.

Oudshoorn et al. (2013) were the first to use a 3-axis accelerometer to quantify cow bites. The method involved visualization of the recorded signals from the three individual orthogonal axes to determine the signal that best matched the recorded bites. The method, however, indicated an average correlation coefficient of 0.65. Umemura et al. (2009) were able to monitor the jaw movements by modifying a pedometer into a pendulum, attached to the lower jaw. The data from the device could be downloaded wirelessly and the system showed 90% accuracy in measuring jaw movements when compared to manual counts over 10-min segments (Umemura et al. 2009).

One key issue associated with the use of accelerometers is the sensitivity of the signals recorded. Undesirable signals due to rapid head movement or the movement of ears can interfere with data interpretation, and the system would require a preprocessing of the signal relative to the jaw movements in order to be useful for precision livestock farming (Tani et al. 2013). Interestingly, the IMUs, which combine several sensors such as the accelerometers, gyroscope, magnetometer, and GPS, can offer a real advantage in terms of multiple-parameter measurements related to animal feeding behavior and animal position (active or not). These multiparameter sensors collect and integrate the data from individual sensors to provide a comprehensive health picture of individual animals, as well as herd behavior (Giovanetti et al. 2017). Such integrated sensors will also enable predictions on animal health events and disease. In the event of a disease outbreak, multiparameter sensors can assist in identification and isolation of affected livestock before the spread of the outbreak and potentially prevent unnecessary culling of uninfected animals, as is the current practice in animal husbandry.

17.3.2 Biosensors for Breath Analysis

Disease diagnosis by identification of volatile organic compounds (VOCs) has long been of interest to researchers, as it offers a noninvasive methodology. VOCs can be found in the breath, blood, feces, skin, urine, and vaginal fluids of animals as well as humans (Burciaga-Robles et al. 2009; Garner et al. 2009; Spinhirne et al. 2004). These compounds are produced by a number of biochemical reactions, pathogens, and host pathogen interactions and are effected by a number of biological variables such as age, actions, and biochemical pathways (Sethi et al. 2013). Breath monitoring provides a noninvasive and easy approach to determine the physiological and general health status of animals. Advances in sampling methods, like solid phase and needle trap micro-extraction, and developments in techniques for representative breath sampling (Turner et al. 2012) can be coupled with modern analytical technologies (spectroscopy techniques and electronic noses) to allow for precise analysis of breath composition at an unprecedented level (Pereira et al. 2015). One of the key challenges that require important attention is the statistical analysis and data interpretation of large and potentially heterogeneous datasets collected from research on the exhaled breath composition from animals.

Metabolites in the breath include gasses like hydrogen and methane and volatile organic compounds such as fatty acids, which can act as biomarkers for metabolic and pathologic processes. Usually, the glucose level in blood is associated with VOCs like ketone bodies, ethanol, methanol, and exogenous compounds (Leopold et al. 2014). In cattle, analysis of VOCs has been explored to diagnose bovine respiratory disease (Burciaga-Robles et al. 2009), brucellosis (Knobloch et al. 2009), bovine tuberculosis (Peled et al. 2012), Johne's disease, ketoacidosis (Kumanan et al. 2009), and normal rumen physiology. A rapid, noninvasive identification of foot-and-mouth disease has been performed using air samples collected with a handheld prototype device equipped with electrostatic particle capture in a microchip chamber of 10–15 μL (Christensen et al. 2011).

17.3.3 Sensors Analyzing Metabolites in Perspiration

Most biosensors developed for analyzing metabolites in sweat were developed with the purpose of human health monitoring. These have been used to analyze sodium concentration (Schazmann et al. 2010) and lactate levels, and converted to portable formats (belt form) to analyze sweat. The electrochemical sensor for lactate levels includes a flexible printed tattoo that can detect lactate levels with linearity up to 20 mM. The sensor has been shown to be resilient against mechanical deformation. This sensor can also be adapted for use in animal sweat monitoring, especially as a sign of physical stress in animals (Jia et al. 2013). Others have developed an adhesive radiofrequency identification (RFID) sensor patch, which allows for potentiometric sensing of solutes and surface temperature that can be read on a smartphone application (Rose et al. 2015). New research in this area has led to the development of a fully integrated wearable sensor array for multiplexed analysis of perspiration. The system includes a mechanically flexible sensor array that measures metabolites like glucose and lactate, and electrolyte composition such as sodium and potassium ions. The sensor also integrates temperature measurement for comprehensive analysis (Gao et al. 2016).

17.3.4 Analysis of Tears for Continuous Glucose Monitoring

Metabolites in tears can provide information about the concentration of these metabolites in blood and provide a noninvasive continuous monitoring technique. Iguchi et al. have reported the development of a flexible, wearable amperometric glucose sensor using immobilized glucose oxidase on a flexible oxygen electrode (Pt working electrode and Ag/AgCl counter/reference electrode). The biosensor is fabricated using Soft-MEMS techniques onto a functional polymer membrane (Yonemori et al. 2009). Others are working toward the development of a biosensor for self-monitoring of tear glucose and are currently in the animal testing stages (La Belle et al. 2014).

17.4 Livestock Monitoring Systems for Observing Physiological Parameters and Health of Cattle

Jegadeesan and Venkatesan (2016) have proposed a two-component system. The first component is the monitoring and collection of data on the health parameters of animals in the field, and the second component is the monitoring and acquisition of data on animals from the farms. Animals are subjected to a variety of stress factors during their lives on farms. These include stressors due to changes in temperature, transport across farms, physiological stress due to ill health or improper feed intake as well as stress due to restraint. This information on the external environment and animal health can be collected and analyzed in real time using closed-circuit cameras. The complete system is expected to work independently, making necessary changes in response to the real-time data inputs. Human intervention is only expected in the event of an emergency (Jegadeesan and Venkatesan 2016). Although such a model system is expected to be a norm in the near future, steps are underway to implement these integrated systems, at least partly, in livestock and agriculture.

Contagious livestock diseases can result in economic losses and decreased productivity of cattle farms. Park et al. have developed a livestock monitoring system (LMS) integrated with a wireless sensor network to collect data on heart rate, breathing rate and cattle movement. The authors report an average correlation coefficient of 0.97 for the collected data. Such systems have the ability to change the landscape of animal farming by providing comprehensive information on animal well-being. These systems are also able to rapidly identify livestock diseases and prevent economic losses stemming from loss of productivity (Park and Ha 2015).

17.5 Monitoring CO_2 Ventilation in Farms

Carbon dioxide (CO_2) can be used as a tracer gas to measure ventilation, as well as its emission rates (Persily 2016). Currently, CO_2 concentrations from agricultural facilities are measured using a photo acoustic spectroscope (PAS) gas analyzer (Chepete et al. 2012; Hassouna et al. 2013; Zhao et al. 2012) and the open-path laser (OPlaser) (Frish 2014; He et al. 2009). Although these are appropriate for use in mechanically ventilated farms, the measurements in naturally ventilated (NV) farms show considerable variability. Moreover, these systems are cost-intensive when multiple samples are required. Nondispersive infrared (NDIR)-based sensors are suggested as an alternative for PAS and OP-laser to measure CO_2 concentrations in NV buildings. Experimental evaluation of NDIR sensors to measure CO_2 levels has been demonstrated (Calvet et al. 2014; Yasuda et al. 2012) along with the miniaturization of the method, facilitating in-field usage (Hodgkinson et al. 2013). Mendes et al. (2015) have recently demonstrated the use of NDIR CO_2 sensors for monitoring CO_2 levels in a naturally ventilated dairy cow barn, comparing them to other commercially available NDIR CO_2 sensors. The authors conclude that the CO_2 concentrations were in agreement with the platforms tested, and the number of NDIR sensors required to represent the overall CO_2 concentration of the dairy cow barn must be calculated based on the barn length and occupied barn area. The NDIR CO_2 sensors were found suitable to be used as a multipoint monitoring system of CO_2 concentrations in NV buildings as a feasible alternative to PAS and OP laser methods (Mendes et al. 2015).

17.6 Monitoring Animal Movement and Behavior

Monitoring movement and behavior can provide information on an animal's activity and well-being. A top-view camera can provide vital information if the animal is low weight. Motion detection technology and video recording coupled with the Gaussian mixture model (GMM) can be used to gather information on animal size and identify low-weight animals. MooMonitor integrates information on cow oestrus, as well as data on rumination, feeding and levels of activity. It makes use of wireless sensors for the two-way transmission of data. Other technologies, such as HerdNavigator™ and the Afimilk Silent Herdsman also serve the same purpose. The Silent Herdsman is a wearable technology and monitors all activities of cattle to analyze their behavior. Any changes in an animal's behavior pattern can be used to identify the oestrus cycles and onset of disease/sickness. Honeybees produce a variety of different sounds as a means to communicate with the colony. The sounds have characteristic low fundamental frequencies between 300 and 600 Hz (Barth et al. 2005). Honeybees' sounds have specific frequencies within this range for a number of reasons. Both the range of sound frequency and the acoustic signal pattern determine the meaning of the sound. An accurate quantification of these signal patterns can give valuable information on the hive health (Qandour et al. 2014). Electronic systems for management of beehives have been developed lately and combine hive acoustics monitoring with measurement of parameters like brood temperature (Kridi et al. 2016), humidity, hive weight, and the weather conditions of the apiary. Dietlein (1985) have developed a system for automated continuous recording of sound emission by honeybees as a measure of their activity. The specific sounds from the hives are useful in providing information on colony behavior, strength and health, and the data from the monitoring device can be accessed remotely from any Internet-enabled device (Bromenshenk et al. 2015; Evans 2015; Meikle and Holst 2015). Other systems are being developed that integrate visual, acoustic, and beehive monitoring systems and share them with the environmental monitoring platform These systems can help in collecting and analyzing the data on bee behavior for biologists (Chiron et al. 2013).

17.6.1 Nanotechnology in Disease Diagnostics

(i) Rapid, accurate, and sensitive detection of disease-causing pathogens is the basis for an effective treatment and eradication of a disease. Recently, advancements in nanotechnology have revolutionized the veterinary diagnostics. In medicine, various applications of nanotechnology proved their potential to revolutionize medical diagnosis, immunization, treatment, and even healthcare products (El-Sayed and Kamel 2021). For example, iron oxide-based NPs can be used to determine the distribution of a drug in the body using magnetic resonance imaging (MRI) (Soenen et al. 2010). Also, a specific drug can be tagged with fluorescent NPs to detect its localization inside the targeted cells (Ajmal et al. 2015). Furthermore, NP-based diagnostic chips are made available to quickly analyze a large number of samples (Craighead 2006). NP-based diagnostic chips take less time, require a small volume of starting material (blood or serum), fewer consumables, and give more accurate results compared to classical laboratory tests, such as polymerase chain reactions and enzyme-linked immunosorbent assays (Bai et al. 2018). A few examples of NP-based veterinary diagnostics are summarized in Table 17.3.

(ii) Mastitis treatment
Mastitis is a disease that is commonly observed among dairy cows with a collection of cooperative variables, normally bacterial, which entail the exploitation of some agents that are antibacterial to clear (Jamali et al. 2018). Researchers have investigated novel therapeutic approaches to replace the use

Table 17.3 Some innovative applications of NP-based veterinary diagnostics

Animal species	Nanoparticle-based diagnostic tool	References
Poultry	Quantum dot-based 3D-printed immunoassay to detect avian influenza (H7N9) virus	Xiao et al. (2019)
Pig	Nano-polymerase chain reaction assay to detect the epidemic porcine diarrhea virus	Wanzhe et al. (2015)
Poultry	Gold nanoparticle-coated polyvinylidene difluoride membrane to detect avian flu antibodies in serum	Emami et al. (2012)
Cattle	Nanoparticle-based array of sensors (NA-NOSE) to detect volatile organic compounds linked to bovine tuberculosis in breath	Peled et al. (2012)
Domestic and wild ruminants	Nanoparticle-based biosensor assay to detect the *Mycobacterium avium* subsp. paratuberculosis (MAP), the causative agent of Johne's disease (JD), in fecal samples	Kumanan et al. (2009)
Chicken and goat	Quantum dot-based fluoroimmunoassay to detect antibodies of chicken Newcastle Disease and goat pox virus in serum	Yuan et al. (2009)

of antibiotics in mastitis control (El-Sayed and Kamel 2021). In the treatment of bovine mastitis, nanoparticles show greater effects on bacteria than their micro-counterparts and may serve as potential delivery systems because they may be ingested by phagocytes (Neculai-Valeanu et al. 2021). During mastitis infection, a drug named "tilmicosin" is utilized but its high concentration has deleterious side effects whenever given to the infected animals (Hill and Li 2017). Concerning this, Han et al. (2009) observed controlling the discharge of tilmicosin by means of hydrogenated castor oil-solid lipid nanoparticle carriers. Attentive manipulation of curative nanocarriers to determine coordination among prescribed amount and half-life might assist profit makers by restraining milk disposed in periods and the quantity of milk unexploited (Wang et al. 2012). Nanotechnology can equally support assuring that the quality of milk is nontoxic for human consumption through unique foodborne pathogens/micro-organisms-revealing techniques (Yang and Duncan 2021).

17.7 Conclusion

Nanotechnology has great potential to improve farm animal health and production. Existing research studies have clearly demonstrated the practicality of nanotechnology in accurate diagnostic tools, targeted drug delivery, enhanced vaccine response, improved feed efficiency, and augmentation of reproduction in farm animals. Scientists, engineers, and biologists working in the field of nanotechnology need to continue efforts to revolutionize animal health and production, and also require significant budgetary allocation in research and development.

References

Abed N, Bahrololoom ME, Kasraei M (2019) The effect of nano-structured nickel coating on reducing abrasive wear of tillage tine. J Nanotech Res 1(2):59–74

Abedini M, Shariatmadari F, Torshizi MAK, Ahmadi H (2017) Effects of a dietary supplementation with zinc oxide nanoparticles, compared to zinc oxide and zinc methionine, on performance, egg quality, and zinc status of laying hens. Livest Sci 203:30–36

Ajitha B, Kumar Reddy YA, Reddy PS, Jeon HJ, Ahn CW (2016) Role of capping agents in controlling silver nanoparticles size, antibacterial activity and potential application as optical hydrogen peroxide sensor. RSC Adv 6:36171–36179

Ajmal M, Yunus U, Matin A, Haq NU (2015) Synthesis, characterization and in vitro evaluation of methotrexate conjugated fluorescent carbon nanoparticles as drug delivery system for human lung cancer targeting. J Photochem Photobiol B 153:111–120

Albuquerque J, Casal S, de Jorge Páscoa RNM, Van Dorpe I, Fonseca AJM, Cabrita ARJ, Neves AR, Reis S (2020) Applying nanotechnology to increase the rumen protection of amino acids in dairy cows. Sci Rep 10:1–12

Andersson LM, Okada H, Zhang Y, Itoh T, Miura R, Yoshioka K (2015) Wearable wireless sensor for estrus detection in cows by conductivity and temperature measurements. IEEE SENSORS, pp 1–4

Andersson LM, Okada H, Miura R, Zhang Y, Yoshioka K, Aso H, Itoh T (2016) Wearable wireless estrus detection sensor for cows. Comput Electron Agri 127:101–108

Andriamandroso A, Lebeau F, Bindelle J (2015) Changes in biting characteristics recorded using the inertial measurement unit of a smartphone reflect differences in sward attributes. In: Guarino M, Berckmans D (eds) Proceedings of the 7th European conference on precision livestock farming, precision livestock farming, Milan, Italy, pp 283–289

Archunan G (2020) Reproductive enhancement in buffalo: looking at urinary pheromones and hormones. Iran J Vet Res 21:163

Asadi-Fard R (2017) A brief glance at some applications of nanotechnology 2004, pp 102–106

Ashengroph M, Khaledi A, Bolbanabad EM (2020) Extracellular biosynthesis of cadmium sulphide quantum dot using cell-free extract of pseudomonas chlororaphis CHR05 and its antibacterial activity. Process Biochem 89:63–70

Awawdeh MS (2016) Rumen-protected methionine and lysine: effects on milk production and plasma amino acids of dairy cows with reference to metabolisable protein status. J Dairy Res 83: 15

Bai DP, Lin XY, Huang YF, Zhang XF (2018) Theranostics aspects of various nanoparticles in veterinary medicine. Int J Mol Sci 19(11):3299

Barth FG, Hrncir M, Tautz J (2005) Vibratory and airborne-sound signals in bee communication (hymenoptera). Insect sounds and communication: physiology, behaviour, ecology, and evolution. CRC press, pp 421–436

Benvenutti MA, Pavetti DR, Poppi DP, Gordon IJ, Cangiano CA (2016) Defoliation patterns and their implications for the management of vegetative tropical pastures to control intake and diet quality by cattle. Grass Forage Sci 71(3):424–436

Bhatia T, Nayan V, Singh R, Singh C, Bhardwaj A, Kumar S, Mohanty AK (2021) An alternative buffalo urine- based non-invasive early estrus test using wheat and mung bean seed germination. Ind J Ani Res 55(11):1279–1285

Blom, J. and Ridder, C. (2010). Reproductive management and performance can be improved by use of DeLaval herd navigator®. The First North American Conference on Precision Dairy Management

Braun U, Trösch L, Nydegger F, Hässig M (2013) Evaluation of eating and rumination behaviour in cows using a noseband pressure sensor. BMC Vet Res 9(1):164

Bromenshenk JJ, Henderson CB, Seccomb RA, Welch PM, Debnam SE, Firth DR (2015) Bees as biosensors: chemosensory ability, honey bee monitoring systems, and emergent sensor technologies derived from the pollinator syndrome. Biosensors 5(4):678–711

Brunori G, Jiggens L, Gallardo R, Schmidt O (2009) New challenges for agricultural research: climate change, food security, rural development, Agricultural Knowledge Systems. Second SCAR Foresight Report. EU Commission, SCAR, Brussels

Burciaga-Robles LO, Holland BP, Step DL, Krehbiel CR, McMillen GL, Richards CJ, Sims LE, Jeffers JD, Namjou K, McCann PJ (2009) Evaluation of breath biomarkers and serum haptoglobin concentration for diagnosis of bovine respiratory disease in heifers newly arrived at a feedlot. Amer J Vet Res 70(10):1291–1298

Caja G, Castro-Costa A, Knight CH (2016) Engineering to support wellbeing of dairy animals. *J Dairy Res* 83(2):136–147

Calvet S, Campelo JC, Estellés F, Perles A, Mercado R, Serrano JJ (2014) Suitability evaluation of multipoint simultaneous CO2 sampling wireless sensors for livestock buildings. *Sensors* 14(6): 10479–10496

Cerbu C, Kah M, White JC, Astete CE, Sabliov CM (2021) Fate of biodegradable engineered nanoparticles used in veterinary medicine as delivery systems from a one health perspective. Molecules 26:523

Chakravarthi VP, Balaji N (2010) Applications of nanotechnology in veterinary medicine. Vet World 3:477

Chepete H, Xin H, Mendes L, Li H, Bailey T (2012) Ammonia emission and performance of laying hens as affected by different dosages of Yucca schidigera in the diet. *J Appl Poult Res* 21(3): 522–530

Chiron G, Gomez-Krämer P, Ménard M (2013) Outdoor 3D acquisition system for small and fast targets. In: Application to honeybee monitoring at the beehive entrance.GEODIFF, Barcelona, France, pp 10–19

Christensen LS, Brehm KE, Skov J, Harlow KW, Christensen J, Haas B (2011) Detection of foot-and-mouth disease virus in the breath of infected cattle using a hand-held device to collect aerosols. *J Virol Methods* 177(1):44–48

Ciçek S, Turhan S, Işık S (2021) Application of selenium nanoparticles diets in ruminants. Atatürk Uni Agri Fac 52:98–107

Clapham WM, Fedders JM, Beeman K, Neel JP (2011) Acoustic monitoring system to quantify ingestive behavior of free-grazing cattle. Comput Electron Agric 76(1):96–104

Combarnous Y, Salesse R, Garnier J (1981) Physico-chemical properties of pregnant mare serum gonadotropin. Biochim Biophys Acta BBA Protein Struct 667:267–276

Craighead H (2006) Future lab-on-a-chip technologies for interrogating individual molecules. Nature 442:387–393

Czichos H, Misra A (1978) Tribology: A Systems Approach to the Science and Technology of Friction, Lubrication and Wear (Tribology Series, 1), vol 100, p 336

Darvishi E, Kahrizi D, Arkan E (2019) Comparison of different properties of zinc oxide nanoparticles synthesized by the green (using Juglans regia L. leaf extract) and chemical methods. J Mol Liq 286:110831

Dehkordi AL, Keivani F (2017) Applications of nanotechnology for improving production methods and performance of agricultural equipment. Biological Forum–An Int J 9(2):102–106

Dietlein DG (1985) A method for remote monitoring of activity of honeybee colonies by sound Analysis. J Apic Res 24(3):176–183

Ditanet, P., Somphol, W., Lampang, N., Prapainainar, P. and Loykulnan, S. (2019). Natural rubber reinforced by nanocellulose extracted from dried rubber leaves. Seventh International Conference on Nano and Materials Science (ICNMS 2019) AIP Conf Proc; 2083 (030008): 1–5

Duffy LL, Osmond-McLeod MJ, Judy J, King T (2018) Investigation into the antibacterial activity of silver, zinc oxide and copper oxide nanoparticles against poultry-relevant isolates of salmonella and campylobacter. Food Control 92:293–300

Dugasa HL, Williams RO (2016) Nanotechnology for pulmonary and nasal drug delivery. In: Nanoengineered Biomaterials for Advanced Drug Delivery, p 102

Durkin J, DeLaval BW (2010) Heat detection: trends and opportunities. In: Proceedings of the Second North American Conference Precision Dairy Management, Toronto, Canada, pp 1–10

Ebrahimi F, Bourne GR, Sl Kelly M, Matthews TE (1999) Mechanical properties of Nanocrystalline nickel produced by electrodeposition. Nanostruct Mater 11(3):343–350

EFSA (2009) Scientific report of EFSA prepared by the animal health and animal welfare unit on the effects of farming systems on dairy cow welfare and disease. Annex to the EFSA 1143:1–38

El-Sayed A, Kamel M (2021) Bovine mastitis prevention and control in the post-antibiotic era. Trop Anim Health Prod 53:1–16

El-Sherbiny M, Cieslak A, Pers-Kamczyc E, Szczechowiak J, Kowalczyk D, Szumacher-Strabel M (2016) A nanoemulsified form of oil blends positively affects the fatty acid proportion in ruminal batch cultures. J Dairy Sci 99:399–407

Emami T, Madani R, Rezayat SM, Golchinfar F, Sarkar S (2012) Applying of gold nanoparticle to avoid diffusion of the conserved peptide of avian influenza nonstructural protein from membrane in Western blot. J Appl Poult Res 21:563–566

Evans SK (2015) Electronic beehive monitoring–applications to research. *Julius-Kühn-Archiv* 0(450):121

Fogolari O, Felippe AC, Leimann FV, Gonçalves OH, Sayer C, Araújo PHHD (2017) Method validation for progesterone determination in poly (methyl methacrylate) nanoparticles synthesized via miniemulsion polymerization. Int. J. Polym Sci *2017:1*

Frish, M.B. (2014). Current and emerging laser sensors for greenhouse gas sensing and leak 1138 detection. SPIE Sensing Technology+ Applications, pp. 91010H–91012. International Society 1139 for Optics and Photonics

Gao W, Emaminejad S, Nyein HYY, Challa S, Chen K, Peck A, Fahad HM, Ota H, Shiraki H, Kiriya D, Lien D-H, Brooks GA, Davis RW, Javey A (2016) Fully integrated wearable sensor arrays for multiplexed in situ perspiration analysis. *Nature* 529:509–514

Garner CE, Smith S, Bardhan PK, Ratcliffe NM, Probert CS (2009) A pilot study of faecal volatile organic compounds in faeces from cholera patients in Bangladesh to determine their utility in disease diagnosis. Trans R Soc Trop Med Hyg 2q+ 103(11):1171–1173

Gawad R, Fellner V (2019) Evaluation of glycerol encapsulated with alginate and alginate-chitosan polymers in gut environment and its resistance to rumen microbial degradation. Asian Australas J Anim Sci 32:72

Gill R (2012).Animal welfare and the "five freedoms". https://animalscience.tamu.edu/wp-content/uploads/sites/14/2012/04/five-freedoms-Longvers

Giovanetti V, Decandia M, Molle G, Acciaro M, Mameli M, Cabiddu A, Cossu R, Serra MG, Manca C, Rassu SPG, Dimauro C (2017) Automatic classification system for grazing, ruminating and resting behaviour of dairy sheep using a tri-axial accelerometer. Livest Sci 196:42–48

Glennon T, O'Quigley C, McCaul M, Matzeu G, Beirne S, Wallace GG, Diamond D (2016) 'SWEATCH': A wearable platform for harvesting and analysing sweat sodium content. Electroanalysis 28(6):1283–1289

Gonzales-Eguia A, Fu CM, Lu FY, Lien TF (2009) Effects of nanocopper on copper availability and nutrients digestibility, growth performance and serum traits of piglets. Livest Sci 126(1–3): 122–129

Han C, Qi CM, Zhao BK, Cao J, Xie SY, Wang SL, Zhou WZ (2009) Hydrogenated castor oil nanoparticles as carriers for the subcutaneous administration of tilmicosin: in vitro and in vivo studies. J Vet Pharmacol Ther 32:116–123

Haque MA, Imamura R, Brown GA, Krishnamurthi VR, Niyonshuti II, Marcelle T, Mathurin LE, Chen J, Wang Y (2017) An Experiment-based model quantifying antimicrobial activity of silver nanoparticles on Escherichia coli. RSC Adv 7:56173–56182

Hartung J, Phillips VR (1994) Control of gaseous emissions from livestock buildings and manure stores. *Agri Eng Res* 57:173–189

Hashem NM, Gonzalez-Bulnes A (2021) Nanotechnology and reproductive Management of Farm Animals: challenges and advances. Animals 11:1932

Hassanein E, Hashem N, El-Azrak K, Gonzalez-Bulnes A, Hassan G, Salem M (2021) Efficiency of GnRH–loaded chitosan nanoparticles for inducing LH secretion and fertile ovulations in protocols for artificial insemination in rabbit does. *Animals* 11:440

Hassouna M, Robin P, Charpiot A, Edouard N, Méda B (2013) Infrared photoacoustic spectroscopy in animal houses: effect of non-compensated interferences on ammonia, nitrous oxide and methane air concentrations. Biosyst Eng 114(3):318–326

He Y, Zhang Y-J, Xia H, Geng H, Ruan J, Wang M (2009) Open-path online monitoring of ambient atmospheric co2 based on laser absorption spectrum. Spectrosc Spectr Anal 29(1):10–13

Heredia N, García S (2018) Animals as sources of food-borne pathogens: A review. Anim Nutr 4: 250–255

Herinaina AL, Bindelle J, Mercatoris B, Lebeau F (2016) A review on the use of sensors to monitor cattle jaw movements and behavior when grazing. Biotechnol Agron Soc Environ 23(S1): 273–286

Hill EK, Li J (2017) Current and future prospects for nanotechnology in animal production.*J Anim.* Sci Biotech 8:26

Hodgkinson J, Smith R, Ho WO, Saffell JR, Tatam RP (2013) Non-dispersive infra-red (NDIR) measurement of carbon dioxide at 4.2 μm in a compact and optically efficient sensor. Sensors Actuators B Chem 186:580–588

Huang S, Wang L, Liu L, Hou Y, Li L (2015) Nanotechnology in agriculture, livestock, and aquaculture in China: A review. Agron Sustain Dev 35:369–400

Jain M, Choudhary S, Kumar V (2021) Application of nanotechnology in farm power, machinery and operations: A review. Agric Eng Today 45(4):1–9

Jamali H, Barkema HW, Jacques M, Lavallée-Bourget E-M, Malouin F, Saini V, Stryhn H, Dufour S (2018) Invited review: incidence, risk factors, and effects of clinical mastitis recurrence in dairy cows. J Dairy Sci 101:4729–4746

Jegadeesan S, Venkatesan GP (2016) Smart cow health monitoring, farm environmental monitoring and control system using wireless sensor networks. Int J Adv Eng Tech 7(1):334–339

Jia W, Bandodkar AJ, Valdés-Ramírez G, Windmiller JR, Yang Z, Ramírez J, Chan G, Wang J (2013) Electrochemical tattoo biosensors for real-time noninvasive lactate monitoring in human perspiration. Anal Chem 85(14):6553–6560

Jónsson R, Blanke M, Poulsen NK, Caponetti F, Højsgaard S (2011) Oestrus detection in dairy cows from activity and lying data using on-line individual models. Comput Electron Agric 76(1):6–15

Kachuee R, Abdi-Benemar H, Mansoori Y, Sánchez-Aparicio P, Seifdavati J, Elghandour MM, Guillén RJ, Salem AZ (2019) Effects of sodium selenite, L-selenomethionine, and selenium nanoparticles during late pregnancy on selenium, zinc, copper, and iron concentrations in Khalkhali goats and their kids. Biol Trace Elem Res 191:389–402

Kekan PM, Ingole SD, Sirsat SD, Bharucha SV, Kharde SD, Nagvekar AS (2017) The role of pheromones in farm animals—A review. Agric Rev 38(2):83–93

King T, Osmond-McLeod MJ, Du LL (2018) Trends in Food Science & Technology Nanotechnology in the food sector and potential applications for the poultry industry. Trends Food Sci Technol 72:62–73

Knight T, Lynch P (1980) Source of ram pheromones that stimulate ovulation in the ewe. *Anim Reprod Sci* 3:133–136

Knobloch H, Köhler H, Commander N, Reinhold P, Turner C, Chambers M, Pardo M, Sberveglieri G (2009) Volatile organic compound (VOC) analysis for disease detection: proof of principle for field studies detecting paratuberculosis and brucellosis. In: AIP conference proceedings. AIP, pp 195–197

Kridi DS, Carvalho CGNDD, Gomes DG (2016) Application of wireless sensor networks for beehive monitoring and in-hive thermal patterns detection. Comput Electron Agric 127:221–235

Kumanan V, Nugen SR, Baeumner AJ, Chang YF (2009) A biosensor assay for the detection of Mycobacterium avium subsp paratuberculosis in fecal samples.*J.* Vet Sci 10(1):35–42

Kumar, P., Singh, P., Chauhan, S., Swaroop, M. N., Bhardwaj, A., Datta, T. K., & Nayan, V. (2023). Nanotechnology for animal sciences-new insights and pitfalls: A review

La Belle JT, Engelschall E, Lan K, Shah P, Saez N, Maxwell S, Adamson T, Abou-Eid M, McAferty K, Patel DR, Cook CB (2014) A disposable tear glucose biosensor-part 4: preliminary animal model study assessing efficacy, safety, and feasibility. J Diabetes Sci Technol 8(1): 109–116

Laca, De Vries W (2000) Acoustic measurement of intake and grazing behaviour of cattle. Grass Forage Sci 55(2):97–104

Leonardi S, Marchesi G, Tangorra FM, Lazzari M (2013) Use of a proactive herd management system in a dairy farm of northern Italy: technical and economic results. J Agri Eng 44(s2)

Leopold JH, Van Hooijdonk RT, Sterk PJ, Abu-Hanna A, Schultz MJ, Bos LD (2014) Glucose prediction by analysis of exhaled metabolites - a systematic review. BMC Anesthesiol 14:46

Li WR, Xie XB, Shi QS, Zeng HY, Ou-Yang YS, Chen Y (2010) Ben antibacterial activity and mechanism of silver nanoparticles on Escherichia coli. Appl Microbiol Biotechnol 85:1115–1122

Manikkaraja C, Mahboob S, Al-Ghanim KA, Rajesh D, Selvaraj K, Sivakumar M, Archunan G (2020) A novel method to detect bovine sex pheromones using l-tyrosine-capped silver nanoparticles: special reference to nanosensor based estrus detection. J Photochem Photobiol Biol 203:111747

Mattachini G, Riva E, Perazzolo F, Naldi E, Provolo G (2016) Monitoring feeding 1275 behaviour of dairy cows using accelerometers.*J*. Agric Eng 47(1):54–58

Mazeris F (2010) DeLaval herd navigator: proactive herd management. In: Proceedings of First North American Conference on Precision Dairy Management, pp 26–27

Meena N, Sahni Y, Thakur D, Singh R (2018) Applications of nanotechnology in veterinary. Vet World 3:477–480

Meikle W, Holst N (2015) Application of continuous monitoring of honeybee colonies. Apidologie 46(1):10–22

Mendes L, Ogink N, Edouard N, Van Dooren H, Tinôco I, Mosquera J (2015) NDIR gas sensor for spatial monitoring of carbon dioxide concentrations in naturally ventilated livestock buildings. Sensors 15(5):11239

Milone DH, Galli JR, Cangiano CA, Rufiner HL, Laca EA (2012) Automatic recognition of ingestive sounds of cattle based on hidden Markov models. Comput Electron Agric 87:51–55

Moghaddam SH, Jebali A, Daliri K (2010) The use of MgO-Sio2 nano-composite for adsorption of aflatoxin in wheat flour samples. Nano 10:12–14

Mohd Yusof H, Abdul Rahman N, Mohamad R, Zaidan UH, Samsudin AA (2020a) Biosynthesis of zinc oxide nanoparticles by cell-biomass and supernatant of lactobacillus plantarum TA4 and its antibacterial and biocompatibility properties. Sci Rep 10:19996

Mohd Yusof H, Mohamad R, Zaidan UH, Rahman NA (2020b) Sustainable microbial cell nanofactory for zinc oxide nanoparticles production by zinc-tolerant probiotic lactobacillus plantarum strain TA4. Microb Cell Factories 19:10

Monerris MJ, Arévalo FJ, Fernández H, Zon MA, Molina PG (2012) Integrated electrochemical immunosensor with gold nanoparticles for the determination of progesterone. Sensors Actuators B Chem 166:586–592

Muktar Y, Bikila T, Kefale M (2015) Application of nanotechnology for animal health and production improvement: a review. World Appl Sci J 33:1588–1596

Nadin LB, Chopa FS, Gibb MJ, Trindade JKd, Amaral GAd, de Faccio Carvalho PC, Gonda HL (2012) Comparison of methods to quantify the number of bites in calves grazing winter oats with different sward heights. Appl Anim Behav Sci 139(1–2):50–57

National Research Council (NRC) (1994) Nutrient requirements of poultry, 9th edn. National Academy Press, Washington, DC, USA

Navon S, Mizrach A, Hetzroni A, Ungar ED (2013) Automatic recognition of jaw movements in free-ranging cattle, goats and sheep, using acoustic monitoring. Biosyst Eng 114(4):474–483

Neculai-Valeanu AS, Ariton AM, Mădescu BM, Rîmbu CM, Creangă Ş (2021) Nanomaterials and essential oils as candidates for developing novel treatment options for bovine mastitis. Animals 11(6):1625

Noori Siahdasht F, Farhadian N, Karimi M, Hafizi L (2020) Enhanced delivery of melatonin loaded nanostructured lipid carriers during in vitro fertilization: NLC formulation, optimization and IVF efficacy. RSC Adv 10(16):9462–9475. https://doi.org/10.1039/c9ra10867j
Nydegger F, Gyga L, Egli W (2010) Automatic measurement of rumination and feeding activity using a pressure sensor, Cemagref, Aubiere, p 027
O'Connell MJ et al (2002) Band gap fluorescence from individual single-walled carbon nanotubes. Science 297:593–596
Oudshoorn FW, Cornou C, Hellwing ALF, Hansen HH, Munksgaard L, Lund P, Kristensen T (2013) Estimation of grass intake on pasture for dairy cows using tightly and loosely mounted di- and tri-axial accelerometers combined with bite count. Comput Electron Agric 99:227–235
Pahl C, Hartung E, Grothmann A, Mahlkow-Nerge K, Haeussermann A (2016) Suitability of feeding and chewing time for estimation of feed intake in dairy cows. Ani: Intern J Anim Bio 10(9):1507–1512
Pamungkas FA, Sianturi RSG, Wina E, Kusumaningrum DA (2016) Chitosan nanoparticle of hCG (human chorionic gonadotrophin) hormone in increasing induction of dairy cattle ovulation. J Anim Vet Sci 21:34–40
Park MC, Ha OK (2015) Development of effective cattle health monitoring system based on 1342 biosensors. Adv Sci Tech 117:180–185
Patil SS, Kore KB, Kumar P (2009) Nanotechnology and its applications in veterinary and animal science. Vet World 2(12):475–477
Peled N, Ionescu R, Nol P, Barash O, Mccollum M, Vercauteren K, Koslow M, Stahl R, Rhyan J, Haick H (2012) Detection of volatile organic compounds in cattle naturally infected with Mycobacterium bovis. Sensors Actuators B Chem 171–172:588–594
Pemberton RM, Hart JP, Mottram TT (2001) An electrochemical immunosensor for milk progesterone using a continuous flow system. Biosens Bioelectron 16(9–12):715–723
Pereira J, Porto-Figueira P, Cavaco C, Taunk K, Rapole S, Dhakne R, Nagarajaram H, Camara JS (2015) Breath analysis as a potential and non-invasive frontier in disease diagnosis: an overview. Meta 5(1):3–55
Persily AK (2016) Field measurement of ventilation rates. Indoor Air 26(1):97–111
Pieszka M, Szczurek P, Pietras M (2018) SiO2 nanostructures as a feed additive to prevent bacterial infections in piglets. J Ann Warsaw Uni life Sci SGGW Anim Sci 57
Qandour, A., Ahmad, I., Habibi, D. and Leppard, M. (2014).Remote beehive monitoring using acoustic signals
Rabinowicz E, Tanner RI (1966) Friction and wear of materials. J Appl Mech 33(2):479
Rajendran D (2013) Application of nano minerals in animal production system. Res J Biotechnol 8: 1–3
Rathbone MJ, Burke CR (2013) Controlled release intravaginal veterinary drug delivery. In: Long acting animal health drug products. Springer, New York, NY, USA, pp 247–270
Riazi H, Rezaei J, Rouzbehan Y (2019) Effects of supplementary nano-ZnO on in vitro ruminal fermentation, methane release, antioxidants, and microbial biomass. Turk Vet J Anim Sci 43: 737–734
Rose DP, Ratterman ME, Griffin DK, Hou L, Kelley-Loughnane N, Naik RR, Hagen JA, Papautsky I, Heikenfeld JC (2015) Adhesive RFID sensor patch for monitoring of sweat electrolytes. IEEE Trans Biomed Eng 62(6):1457–1465
Rutter SM (2000) Graze: A program to analyze recordings of the jaw movements of ruminants. 1377. Behav Res Methods Instrum Comput 32(1):86–92
Rutter S, Champion R, Penning P (1997) An automatic system to record foraging behaviour in free-ranging ruminants. Appl Anim Behav Sci 54(2–3):185–195
Santos-Jimenez Z, Guillen-Gargallo S, Encinas T, Berlinguer F, Veliz-Deras FG, Martinez-Ros P, Gonzalez-Bulnes A (2020) Use of propylene-glycol as a cosolvent for gnrh in synchronization of estrus and ovulation in sheep. Animals 10:897
Saragusty J, Arav A (2011) Current progress in oocyte and embryo cryopreservation by slow freezing and vitrification. Reproduction 141(1):1–19. https://doi.org/10.1530/REP-10-0236

Schazmann B, Morris D, Slater C, Beirne S, Fay C, Reuveny R, Moyna N, Diamond D (2010) A wearable electrochemical sensor for the real-time measurement of sweat sodium concentration. Anal Methods 2(4):342–348

Scott NR (2007) Impact of nanoscale technologies in animal management. Anim pro Anim Sci Worldwide 1:283–291

Sethi S, Nanda R, Chakraborty T (2013) Clinical application of volatile organic compound 1398 analysis for detecting infectious diseases. Clin Microbiol Rev 26(3):462–475

Shahin MA, Khalil WA, Saadeldin IM, Swelum AAA, El-Harairy MA (2020) Comparison between the effects of adding vitamins, trace elements, and nanoparticles to shorter extender on the cryopreservation of dromedary camel epididymal spermatozoa. Animals 10:78

Shi L, Xun W, Yue W, Zhang C, Ren Y, Liu Q, Wang Q, Shi L (2011) Effect of elemental nano-selenium on feed digestibility, rumen fermentation, and purine derivatives in sheep. Anim Feed Sci Technol 163:136–142

Singh AK, Kumari T, Rajput MS, Baishya A, Bhatt N, Roy S (2020) A review: effect of bedding material on production, reproduction and health and behavior of dairy animals. Int J Livest Res 10(7):11–20

Soenen SJH, Himmelreich U, Nuytten N, Pisanic TR, Ferrari A, De Cuyper M (2010) Intracellular nanoparticle coating stability determines nanoparticle diagnostics efficacy and cell functionality. Small 6:2136–2214

Spigarelli C, Zuliani A, Battini M, Mattiello S, Bovolenta S (2020) Welfare assessment on pasture: A review on animal-based measures for ruminants. J Anim 10(4):609

Spinhirne JP, Koziel JA, Chirase NK (2004) Sampling and analysis of volatile organic compounds in bovine breath by solid-phase microextraction and gas chromatography-mass spectrometry. J Chromatogr 1:63–69

Su Y, Tang Z, Wang G, Wan R (2018) Influence of carbon nanotube on the tribological properties of vegetable -based oil. Adv Mech Eng 10(5):1–11

Tani Y, Yokota Y, Yayota M, Ohtani S (2013) Automatic recognition and classification of cattle chewing activity by an acoustic monitoring method with a single-axis acceleration sensor. Comput Electron Agric 92:54–65

Thames HT, Theradiyil Sukumaran A (2020) A review of salmonella and campylobacter in broiler meat: emerging challenges and food safety measures. Foods 9:776

Turner C, Knobloch H, Richards J, Richards P, Mottram TTF, Marlin D, Chambers MA (2012) Development of a device for sampling cattle breath. Biosyst Eng 112(2):75–81

Umemura K, Wanaka T, Ueno T (2009) Technical note: estimation of feed intake while grazing using a wireless system requiring no halter. J Dairy Sci 92(3):996–1000

Ungar ED, Rutter SM (2006) Classifying cattle jaw movements: comparing IGER behaviour 1437 recorder and acoustic techniques. Appl Anim Behav Sci 98(1):11–27

Vatan-Dolat-Xah J (2006) Nanotechnology, Basic Science, and New Technology, pp 102–106

Wanapat M, Cherdthong A, Phesatcha K, Kang S (2015) Dietary sources and their effects on animal production and environmental sustainability. Anim Nutr 1(3):96–103

Wang XF, Zhang SL, Zhu LY, Xie SY, Dong Z, Wang Y, Zhou WZ (2012) Enhancement of antibacterial activity of tilmicosin against Staphylococcus aureus by solid lipid nanoparticles in vitro and in vivo. *Vet J* 191:115–120

Wang C, Zhang L, Su W, Ying Z, He J, Zhang L, Zhong X, Wang T (2017) Zinc oxide nanoparticles as a substitute for zinc oxide or colistin sulfate: effects on growth, serum enzymes, zinc deposition, intestinal morphology and epithelial barrier in weaned piglets. PLoS One 12: e0181136

Wanzhe Y, Yanan L, Peng L, Qinye S, Limin L, Jiguo S (2015) Development of a nanoparticle assisted PCR assay for detection of porcine epidemic diarrhoea virus. J Virol Methods 220:18–20

Weixing D, Dongmei W, Zheng L, Cheng M, Depo Y, Chaoliang L, Shaobao L (2009) Effects of nano-Se and vitamin E on anti-oxidative capability of dairy cows in heat stress. China Dairy Cattle 9:21–24

Wilczewska AZ, Niemirowicz K, Markiewicz KH, Car H (2012) Nanoparticles as drug delivery systems. Pharmacol Rep 64(5):1020–1037

Xiao M, Huang L, Dong X, Xie K, Shen H, Huang C, Xiao W, Jin M, Tang Y (2019) Integration of a 3D-printed read-out platform with a quantum dot-based immunoassay for detection of the avian influenza A (H7N9) virus. Analyst 144(8):2594–2603

Yang T, Duncan TV (2021) Challenges and potential solutions for nanosensors intended for use with foods. Nat Nanotechnol 16:251–265

Yang Z, Sun L (2006) Effects of nanometer ZnO on growth performance of early weaned piglets. J Shanxi Agric Sci 3:577–588

Yasuda T, Yonemura S, Tani A (2012) Comparison of the characteristics of small commercial NDIR CO2 sensor models and development of a portable CO2 measurement device. Sensors 12(3):3641–3655

Yonemori Y, Takahashi E, Ren H, Hayashi T, Endo H (2009) Biosensor system for continuous glucose monitoring in fish. Anal Chim Acta 633(1):90–96

Younis S, Zia R, Tahir N, Bukhari SZ, Khan WS, Bajwa SZ (2022) Nanosensors for animal health monitoring.In *nanosensors for smart agriculture*. Elsevier, pp 509–529

Youssef FS, El-Banna HA, Elzorba HY, Galal AM (2019) Application of some nanoparticles in the field of veterinary medicine. Int J Vet Sci Med 7:78–93

Yuan P, Ma Q, Meng R, Wang C, Dou W, Wang G, Su X (2009) Multicolor quantum dot-encoded microspheres for the fluoroimmunoassays of chicken Newcastle disease and goat pox virus. J Nanosci Nanotechnol 9(5):3092–3098

Zadedarvish F, Kesmati M, Khajepour L, Torabi M (2020) Effects of magnesium oxide nanoparticles on memory impairment induced by postpartum depression model. *Physiol Pharmacol* 24(1):63–73

Zeidan E, Shivaji R, Henrich VC, Sandros MG (2016) Nano-SPRi aptasensor for the detection of progesterone in buffer. Sci Rep 6(1):26714

Zein El-Din AM, Youssef Taha RM, Abdel Hamied RG (2021) Nanotechnology sector applications in power and farm machinery: a review. J Adv Agric Res 26(2):38–47. https://doi.org/10.21608/jalexu.2021.170677

Zhao Y, Pan Y, Rutherford J, Mitloehner FM (2012) Estimation of the interference in 1498 multi-gas measurements using infrared photoacoustic analyzers. Atmosphere 3(2):246–265

Zhao CY, Tan SX, Xiao XY, Qiu XS, Pan JQ, Tang ZX (2014) Effects of dietary zinc oxide nanoparticles on growth performance and antioxidative status in broilers. Biol Trace Elem Res 160:361–367

Chapter 18
Nanotechnology in Wildlife Management

Shobha Jawre, Kajal Kumar Jadav, Deepak Kumar Tiwari, Neeraj Arora, Rajesh Kumar, Dodla Mamatha, Alonkrita Chowdhury, Akshita, and Kratika Maurya

Abstract Wildlife plays a key role in sustaining the equilibrium of the ecosystem. It is helpful in several ways ranging from preserving the biological diversities to maintaining social ethos. However, an emergent transformation of the world and globalization leads to a climate crisis and habitat loss. Some of the major threats to wildlife include unlawful wildlife trade, invasive species, pollution, and climate change resulting in the deterioration of wildlife day by day. Nanotechnology offers different ways in overcoming these challenges. It is a fast-emerging field with great potential in different sectors, including animal healthcare and nutrition. It also facilitates disease diagnosis and treatment, studying the behavior of animals and prevention of poaching, etc. In therapeutic approaches, nanotechnology is employed in different ways such as the delivery of vaccines, therapeutic agents, wound and fracture management, devices for surgical interventions, etc. Various nanoparticles used in diagnosis are nano-shells, quantum dots, magnetic NPs, etc. Surveillance of the animals using NPs aids in protecting the animals from poaching as well as facilitates the study of animal behavior and maintenance of epidemiological data.

S. Jawre
School of Wildlife Forensic and Health, Nanaji Deshmukh Veterinary Sciences University, Jabalpur, Madhya Pradesh, India

K. K. Jadav
Animal Biotechnology Centre, Nanaji Deshmukh Veterinary Sciences University, Jabalpur, Madhya Pradesh, India

D. K. Tiwari · N. Arora
Department of Veterinary Surgery and Radiology, College of Veterinary Sciences, Lala Lajpat Rai University of Veterinary and Animal Sciences, Hisar, Haryana, India

R. Kumar (✉)
Department of Veterinary Physiology and Biochemistry, College of Veterinary Sciences, Lala Lajpat Rai University of Veterinary and Animal Sciences, Hisar, Haryana, India

D. Mamatha · A. Chowdhury · Akshita · K. Maurya
Department of Plant Biotechnology, Institute of Agricultural Sciences, RGSC, Banaras Hindu University, Varanasi, Uttar Pradesh, India

LLR University of Veterinary and Animal, ICAR-National Research Centre on Equines, *Nanotechnology Theranostics in Livestock Diseases and Management*, Livestock Diseases and Management,
https://doi.org/10.1007/978-981-16-1610-5_18

Some NPs also help in regulating the homeostasis of animals such as tin-doped indium oxide. It also plays a major role in enhancing the survivability of animals by providing a high interface of food particles, enhancing systemic circulation, and efficient uptake of nutrients. Some NPs act as efficient cargo materials for the delivery of vitamins, minerals, coenzymes, antioxidants and so on which will increase the availability of nutrients. The reproductive capacity of animals is also enhanced by increasing sperm motility using NPs. Here, the current chapter delineates the implementation of nanotechnology in wildlife welfare through disease prevention, diagnosis, and treatment, surveillance of animals, animal health and nutrition, enhancing reproductive efficiencies, etc.

Keywords Wildlife surveillance · Nutritional management · Metallic nanoparticle · Nano-delivery · Nano-minerals · Nano-diagnostics · Assisted reproductive techniques (ART) · Cryopreservation · Nano-biosensors · Nanotoxicity

18.1 Introduction

One of the most eye-captivating planets in the solar system is the earth. It is distinctive in terms of its chromatic appearance due to the biotic and abiotic diversities it carries. This diversity of flora and fauna makes the earth distinct in habitation in comparison to other planets. Maintaining ecosystems and diversities in their natural forms is not only essential for the amazing attractiveness but also for the existence of the human race itself. Destruction of the niche of wild animals leads to their extinction thus, unbalancing the ecosystem which is a threat to mankind as well as all the life forms on the earth. Wildlife conservation and its prosperity is the key to the sustainability of life on earth. In this regard, various exertions have been made throughout time for the maintenance of wildlife. However, the latest exploration of the area of novel scientific technologies and facts has actualized in vitro conservation of wild lives by preserving their gene pool. Corresponding approaches include stem cell conservation, embryo conservation, cloning, and semen conservation of animals living in both wild conditions and in artificial habitats (e.g., zoos). In such indispensable perspectives regarding diversity conservations, nanotechnology-based applications play a pivotal role.

Nanotechnology can be defined as the implementation of materials on an atomic, molecular, and supramolecular level, that is, on a nanometric scale (1–100 nm) (Teli et al. 2010). Applications of the novel nanoparticles have generated another research arena of nanobiotechnology, which plays a fundamental role in disease diagnosis, and drug designing along with targeted drug delivery and sustained release system (Ramos et al. 2017). Novel nanoparticles (NPs) offer revolutionary advances in comparison to conventional diagnostic and therapeutic approaches. In wildlife conservation and welfare, personnel involved, face various challenges regarding surveillance, proper and timely identification of diseased animals in herds or forests, administration of effective treatment to wild lives, vaccination, enhancement of animal reproductive fitness, restricting poaching, and identification of poachers, self-protection during surveillance and treatment, the pattern of disease epidemiology, and difficulties in studying both inter and intra species-specific interrelation and behavior under changing stimuli, etc. Hence, nanotechnology possesses implementations in these respects as well. The potential application of NT in wildlife conservation has been delineated subsequently.

18.2 Applications of Nanotechnology in Wildlife Conservation

The potential applications of nanotechnology in wildlife management have been elaborated as follows:

(a) Prophylactic measurements, diagnosis, and treatment
(b) Surveillance of animals
(c) Study of interaction or behavior
(d) Management of forest for better health and inhabitation
(e) Maintenance of environmental homeostasis during captivity
(f) Surgical management
(g) Prevention of poaching
(h) Animal health and nutritional management
(i) Enhancement of reproductive efficiencies

18.2.1 Application of Nanotechnology in Prophylactic, Diagnostic, and Therapeutic Aspect

In wild animals, NP-based approaches can be advantageous in the diagnosis, and treatment of diseases and to study the epidemiological pattern of diseases. These tiny molecule-based systems can be formulated with conventional therapies to enhance their effectiveness, and prolong the retention time of drugs in circulations. Alternatively, it can be exploited as a targeted delivery system with sustained and controlled release of available drugs to achieve improved results. Examples of such nano-delivery systems include nano-capsules, nano-co-chelates, nanoballs, nanodevices, nanomachines, and nanorobots. Multiple therapeutic agents or factors can be delivered using such devices. It also provides a new perspective to the administration of desired doses of medications masking taste as well as facilitating the reduction of stress and toxicity. In addition, it can be employed as a therapeutic agent such as silver NPs (AgNPs), which is abundantly exploited due to their antimicrobial properties. Various nano-sensors are utilized for diagnostic purposes as well as the collection of data regarding disease transmission so that the epidemiological behavior of diseases can be studied well on a real-time basis (Xu et al. 2022). Similar to drug delivery, nano-carriers, and nano-adjuvants are also beneficial for the targeted delivery of vaccines besides improving immunogenicity.

Nanotechnology also aids way out to negotiate with serious problems faced by veterinarians to combat deadly intracellular or blood pathogens such as tuberculosis (TB), brucellosis, methicillin-resistant *Staphylococcus aureus* (MRSA) infection, foot-and-mouth disease (FMD), and other infections. Instances of diseases in wildlife include Ebola virus disease (EVD) in monkeys; classical swine fever (CSF), also known as hog cholera, in domestic and wild swine; African swine fever (ASF) in domestic and wild pigs; brucellosis, commonly found in sheep, cattle, goats, pigs, and dogs; the fungal diseases such as chytridiomycosis and ranavirus disease in amphibians; rabies in mammals including raccoons, skunks, bats, and foxes as the most common wild reservoirs; anthrax in cattle, sheep, goats, antelope, and deer; salmonellosis in poultry, swine, and reptiles; West Nile fever reported in sheep, cattle, and pigs in Africa and Eurasia; Hendra virus (HeV) and Nipah virus (NiV) infection in large fruit bats (flying foxes) along with subsequent transmission to horses and pigs. (https://www.vet.upenn.edu/research/centers-laboratories/research-initiatives/wildlife-futures-program/resources/fact-sheets/fact-sheet-detail/ranavirus#:~:text=Since%20there%20is%20no%20treatment,prevent%20infection%20of%20other%20animals; https://www.cdc.gov/anthrax/basics/index.html#:~:text=Domestic%20and%20wild%20animals%20can,vaccination%20can%20help%20prevent%20outbreaks; https://www.cdc.gov/healthypets/diseases/salmonella.html#:~:text=Animals%20get%20Salmonella%20from%20their,still%20appear%20healthy%20and%20clean; https://www.cdc.gov/rabies/index.html; https://portal.ct.gov/DOAG/Regulatory/Regulatory/West-Nile-Virus-In-Domestic-Animals-and-Birds#:~:text=Does%20WNV%20infect%20other%20animals,exhibited%20fever%2C%20abortion%20and%20encephalitis; https://www.who.int/health-topics/hendra-virus-disease#tab=tab_1; Spiropoulou 2019). Furthermore, treatment or eradication of diseases is not only crucial for wildlife welfare but is significant to mankind as well, since, over 60% of pathogens responsible for human diseases are of animal origin, and wildlife acts as the major reservoir of them. Nano-drugs are advantageous over conventional medicines in many aspects. For example, binding gentamicin by a peptide linker to a hydrogel keeps the gentamicin medically inactive as long as the linker is intact. The peptide linker can only be broken down and activated by the protease enzyme produced by *Pseudomonas aeruginosa* (Soppimath et al. 2002, 2002). NPs targeting bacterial toxins and receptors were also developed in order to bind the pathogenic microflora in the gut before being excreted outside the body (Kim et al. 2010; Underwood and Van Eps 2012).

The functionalization of NPs with antibodies or nucleic acid enables the development of rapid, sensitive, specific, and portable diagnostic assays. NPs also act as a potent tool for the reduction of the risk associated with rising instances of microbial resistance to traditional drugs, avoiding some of the resistance mechanisms, increasing drug bioavailability, and potentiating the target of the treatment. Novel antiparasitic drugs can also be developed employing nanotechnology such as titanium dioxide (TiO_2)-functionalized AgNPs, an eco-friendly and inexpensive approach for use in veterinary medicine (Allahverdiyev et al. 2013). Pegylated or polyplex NPs provide advanced approaches to enhance the biodistribution of bioactive compounds. NP-aided formulations also progress veterinary medicine in mass treatments of ectoparasites, facilitating the drug-parasite interaction. The effectiveness of phytotherapeutic products can also be enhanced using NPs, for instance, synthesis of AgNPs using *Cissus quadrangularis* has shown antiparasitic activities against adult hematophagous fly *Hippobosca aculata* and the larvae of cattle tick *Rhipicephalus* (*Boophilus*) *microplus* (Santhoshkumar et al. 2012). The development of a carrier system for drug delivery in the form of single injection and multiple burst release to the target site is a major boon and requirement for wildlife. Here, a nano-delivery system with nanobots plays a pivotal role. This kind of system of real-time drug trafficking in the body and release in a pulsatile way will immensely help in the well-being of animals and hence, in their conservation. Such a system will not only help deliver drugs efficiently but can be used for vaccination also. Vaccination with a better response to immunization and targeted delivery is another arena where nanomedicine has proven to be an advanced way of achieving the desired immunity. Nanobots can further facilitate real-time imaging of animals having sickness and hence provide data for various vital symptoms to help in monitoring. Furthermore, it can be employed in drug delivery in the epidemic area through aerosol-based dispersion. Since drugs are nano-size, mass drug delivery in wilds can also be achieved from a drone in times of epidemiological need and can be remained in the environment for a longer period. Once it enters the body, it can be released in a controlled manner (Zhang et al. 2019; Snow and Giordano 2019; Bromberg et al. 2009). As repeated drug administration is extremely difficult in wild animals, so, nano-conjugated long-acting drugs can be a very useful solution. Several nanoparticle-based therapeutic and diagnostic devices or systems have been evolved which can be used for disease tracking and monitoring as well (Zavaleta et al. 2018).

18.2.2 Surveillance of Animals

Effective surveillance of wild animals is vital to protect them from poachers along with identifying their location. It also aids in studying animal behavior and interaction for their proper welfare. Such a system is highly essential for endangered animals. The nanotechnology-based surveillance system can be an effective model to achieve it owing to its long life. The application of nanobots or drones fitted with thermal imaging cameras is an approach to getting the desired work done. The application of nanoparticles assists in developing smaller devices for effective surveillance with longer working life. Advantages of infrared thermography (IRT) include the detection of minute changes in surface temperature up to 0.2°C. It extends several benefits in terms of requiring no assistance of workers, no need for sedation, anesthesia, or radiations, while providing acquisition of real-time data and finally, it can be employed to work at night irrespective of climate. It is a non-invasive technique that is handy and small, and can be easily mounted on drones. These cameras are also used for stress evaluation, studies of pollution, wildlife behavior, epidemiological data, poaching control, etc. (Cilulko et al. 2013; Smith et al. 2015; Tsai et al. 2020). The use of nanomaterial in thermal imaging cameras has enhanced its potency in wildlife maintenance. For example, upconversion NPs (UCNPs) aid in enhancing temporal and thermal resolution as well as sensitivity in temperature measurement (Carter et al. 2014). Similarly, iron oxide NPs also enhance the resolution of thermal imaging (Kirubha and Rajput 2019). Plasmonic metal NPs deposited on a metal surface as optical components for infrared imaging and sensing has been developed to reduce their cost and for better applicability (Mandal et al. 2021). These are a few of the examples where NPs can be employed for the effective monitoring of wild animals.

18.2.3 Maintenance of Environmental Homeostasis During Captivity

NPs are also employed for maintaining environmental homeostasis in the artificial habitat of wild animals, such as zoos and other scientific captivity to enhance their propagation and conservation. Moreover, providing a specific ecological environment for sustainability is one of the major factors for upholding their physiological performances including reproductive efficiencies. Nanomaterial-based devices and other modalities facilitate achieving such an aim. For example, the development of nanoparticle-impregnated self-heating fabrics for moisture management under ambient conditions helps maintain the natural environment of animals, aids in lowering the ectoparasite load, and fabricate smart clothing for wildlife people and sensors to regulate different environmental factors, all these aspects are important to maintain homeostasis. Self-heating (IRANISH) fabric impregnated with tin-doped indium oxide (ITO) nanoparticles can be used for moisture management (Wijenayaka et al. 2021). Nanorobots also permit to obtain real-time data within the animal body with respect to changes in temperature, pressure, etc. Chip-based nanodevices monitor animal health on a continuous basis (Pareek et al. 2021). In the same way, various nanodevices are also being used nowadays to assess environmental pollutant data with limited consumption of power. It can be utilized in both wild and captive conditions. For example, the development of small-sized nanoparticle monitoring instruments, Nanoscan Scanning Mobility Particle Sampler (SMPS) can replace larger size instruments with more power consumption like conventional SMPS and Fast Mobility Particle Sizer (FMPS). Moreover, it can be fitted in a drone or any place which is inaccessible to the wild in captivity to monitor hygiene studies (Jørgensen 2019). Similarly, for the reduction of the chance of viral or bacterial load in captivity, several NP-coatings such as CuNPs and CuONPs have been formulated to reduce influenza A, *Escherichia*, bacteriophages (Qβ), etc. Different NPs can be utilized in materials of cages to keep animals safer by reducing the chance of infections. Examples of NPs used for such purposes include AgNPs as a broad-spectrum antibacterial agent, graphene oxide (GO) as an antiviral agent,

AuNPs against bacteriophages and other viruses, ZnONPs and zirconium (Zo) for the prevention of herpes virus, colloidal NPs against the para-myxo virus, etc. (Imani et al. 2020). Further application of nano-SiO_2 will provide better strength than conventional portland cement and can be used for large animal captivity. Application of the coating of TiO_2NPs provides better self-cleaning properties than conventional metal oxide due to its antimicrobial effect and photocatalytic properties (Dahlan 2019). Similarly, nanomaterials can also be applied as insulating material to maintain the ambiance of captive houses. It can provide 10× more insulation than conventional materials (Pacheco-Torgal and Jalali 2011). Correspondingly, nanosensors can be used in captivity for the detection of various chemicals, toxins, or microorganisms. Recently, silica NPs have been reported to be used in sensors to detect 13 types of chemicals (Nichols 2015).

18.2.4 Pollution Control

As a result of urbanization and growing industrialization, the level of pollution is increasing day by day impacting the global biome. Even though everyone on earth is surviving apparently in a stable manner but the resources are very limited for the increasing greed. This conclusion can be summarized by the fact that the top ten countries of the world produce more pollutants than the rest. This is making the earth a threatening place to live with maximum impact over the endangered wild lives. Finding a suitable solution to control pollution is necessary for sustaining life on earth. As pollutants affect every vital function of organisms, these can limit their survivability, growth, and reproductive potential, including wildlife which is probably the most susceptible to environmental aberrations. However, nanotechnological applications can facilitate the remediation of pollutants in several ways. Different remedial measures include water purification, reduction of CO_2 emissions, and recycling approaches promoting material circulation based on need. In case of any outbreak, wildfire, or industrial leakage accidents, applications of photocatalyst nanoparticles or AlGaN-based deep UV emitting diodes and various NP-formulated membranes facilitates air purification, virus capturing, and other pollutant removal activities (Pokrajac et al. 2021; Beni and Jabbari 2022; Roy et al. 2021; Ibrahim et al. 2016). Various NP-based technologies have been developed to reduce pollution, such as nanostructured membranes that have pores small enough to separate methane or carbon dioxide from the exhaust. It also includes separation, purification, and decontamination of polluted water using NP-based ion exchange resins, the application of nano-wires composed of NPs using potassium manganese oxide for cleaning up oil spills, and other organic pollutants (Khin et al. 2012).

18.2.5 Animal Health and Nutritional Management

Another area in wildlife management is the application of nanomaterials in animal feed both in captive and natural conditions for better sustainability. It can be achieved by determining soil quality in wild and providing a mineral mixture either by spray or by brick form. In captivity it can be used for providing better nutrient solutions throughout the food chain by procuring, processing, balancing, and preserving the feed as per the requirement of the captive animals. As a result, enhanced survivability and reproductive activities can be observed. The nano-formulation provides various additional benefits such as higher surface area of nutrients for better interaction, prolonged systemic circulation due to smaller size, deeper penetration into tissues, cross epithelial lining fenestration making efficient uptake and higher bioavailability, etc. The production of nano-minerals is also cost-effective, needed in lower concentrations, and possesses growth-promoting and immune-stimulating effects. Nano-minerals also facilitate the regulation of pathogens present in the feed and the rumen fermentation process as well. Finally, they can be utilized to overcome many reproductive complications in herds or flocks in captivity (Swain et al. 2015). Many nano-minerals are now available for commercial utility such as nano-ZnO which improves the growth rate, and immune, and reproductive health of farm animals and birds. It also decreases the prevalence of diarrhea among young piglets (Mishra et al. 2014). Similarly, the administration of vitamin D for animals and neonates in captivity can be achieved by adding it to casein micelles, a major micronutrient for better bioavailability (Niemiec et al. 2021). Various essential micronutrients can be delivered by cargo systems to enhance animal productivity, reproductive fitness, and growth in captivity. The application of chitosan NP supplements loaded with chromium is another example that can lead to lowering cholesterol levels, better immunity, and egg formation. Similarly, animals in captivity undergo higher stress, particularly, in the early days. Here, nano-formulation can be applied in the form of nano-Selenium which exerts a positive effect against peroxidative damage in blood components hence, improves testicular microstructure, the ultramicroscopic structure of spermatozoa, and testicular glutathione peroxides activity than its normal counterpart. The application of Cr-nano aids in improving the immune system by enhancing immunoglobulins such as IgM and IgG in plasma. Similar effects were also observed in nano-Cu supplementation in terms of elevation in IgG, γ-globulin, and total globulin protein levels and in the superoxide dismutase (SOD) activity which will help to mitigate the stress to adapt in adversity. Several nano-formulations can act as effective cargo materials to enhance animal growth and provide specific target molecules like micelles. For example, nano-capsules provide an efficient cargo system for the targeted delivery of essential oils, flavor, antioxidants, coenzyme Q_{10}, vitamins, minerals, and phytochemicals, hence enhancing their bioavailability (Marappan et al. 2017).

18.2.6 Enhancement of Reproductive Efficiencies

Assisted reproductive techniques (ART) are used in domestic animals with increasing applications in wildlife conservation. Currently, several techniques are available for semen preparation protocol for several species. Nano-purification technique is a rapid, straightforward, effortless, and noninvasive technique for the evaluation of spermatozoa in wildlife, thus, offering an excellent effect on ART outcomes (Durfey et al. 2018). Successful cryopreservation of an embryo requires cooling of an embryo at a cryogenically stable state followed by warming at a faster rate than the rate of cooling. The aforesaid procedure involves the exploitation of an antifreezing agent or cryoprotectant to prevent the formation of ice crystals which might damage the embryo. Due to the larger embryo size in fish and other nonmammalian animals, it is difficult to thaw quickly for preventing ice crystals formation. Furthermore, the embryonic membranes of aquatic animals are mostly impenetrable, blocking the cryoprotectants. Thus, NP-based cryoprotectants can have a paradigm shift in conservation. Similarly, real-time imaging of gametes can be achieved by nano-robot and self-illuminating, inorganic nanoparticles. Bioluminescent resonance energy transfer-conjugated quantum dots (BRET-QD) NPs facilitate visualization of the molecular and cellular events during fertilization, which is helpful in wildlife research in addition to the management of pregnancy during captivity which is a matter of big concern. These modalities also help in enhancing the birth rate and survival of young ones to preserve the species. The implementations of NP-based approaches in males for improving reproductive efficiency include the application of antibodies-functionalized magnetic NPs against ubiquitin for defective sperm, magnetic NPs-coated with lectins for acrosomal damage, nano-protectant additives for better preservation, improvement in sperm motility as well as fertility, etc. Similarly, egg production, particularly, in wild birds is another type of conservation where NPs play a significant role. Application of diclofenac results in the reduction of shell membrane thickness leading to lower hatchability in eagles. However, ZnONP supplementation enhanced the thickness as well as strength of the eggshell (Abedini et al. 2017). Similarly, chromium picolinate NP supplementation improves egg quality in terms of egg weight, yolk mass, dense protein height, protein mass, and shell thickness by increasing calcium content in the eggshell (Konkol and Wojnarowski 2018). Another application includes the usage of nano-biosensors for the detection of physiological status such as pathogens, environmental conditions, metabolite profiles, etc. Similarly, biosensors are also used to detect hormones and the estrous cycle which is helpful in pregnancy diagnosis, oestrus detection as well in the determination of hormonal levels in the body to improve the reproductive potential both in wild and captivity. Some recently identified urine-based biomarkers for the diagnosis of pregnancy in ruminants can be explored to develop biosensors that can be used for the effective reproductive management of wild animals (Sarangi et al. 2022). To assist in improving the animal production system, the sustainable release of nano-systems for the delivery of reproductive hormones can be used as ART. The application of nanomaterials in the cryopreservation of gametes or stem cells or sperm is another aspect that can obtain great value in the future for wildlife conservation. For instance, genipin-cross-linked pluronic F127-chitosan nanoparticles-mediated intracellular delivery of trehalose is an attractive method for cryopreservation of adipose-derived stem cells because transport across the cell membrane is specific to trehalose encapsulated in nanoparticles (Wei et al. 2015). Nano-sensors are also beneficial in the development of artificial wombs which is perceived as the future of medicine where these devices can be utilized for determination of the condition inside the womb. The development of self-expandable materials based on smart nanotechnology can also be achieved (Partridge et al. 2017). Successful implantation of an artificial placenta to support an extremely preterm ovine fetus is at the border of viability and it seeks hope for wildlife conservation in the future to enhance the reproduction rate of extinct animals (Fallon and Mychaliska 2021). Nanotechnology also offers numerous assistances to establish such state-of-the-art technology.

18.3 Nanotoxicity and Environmental Impact Ringing the Alarm

Several instances of nanoparticle-induced toxicity, particularly from metallic and metal oxide NPs have already been perceived (Manuja et al. 2021). The toxicity mechanisms include oxidative stress and genotoxicity, disruption of the actin cytoskeleton, activation of lysosomal AcP activity, stimulation of phagocytosis in blood cells and increase of MXR transport activity, and inhibition of Na-K-ATPase in gill cells of fish (Katsumiti et al. 2015). Lead NPs can also be found in game meat if they get shot with lead-containing bullets (Kollander et al. 2017). These materials may enter into the food chain either through soil, water, or air contamination that can impart toxicological signs and deteriorate the health of animals in the wild both terrestrial and aquatic on chronic exposure. Even apart from direct aerosol exposure to animals, recently it has been demonstrated that AgNPs can be directly absorbed by leaf and transfered to stem much faster than root which is a sign of bioaccumulation arising from nonjudicial uses of nanoparticles, particularly for engineered nanoparticles (Cocozza et al. 2019). Zinc oxide and titanium oxide nanoparticles are examples that reduce in-vitro sperm viability in a dose- and time-dependent manner by membrane weakening and DNA fragmentation of sperm (Barkhordari et al. 2013; Santonastaso et al. 2020). So, biocompatibility and biodegradability are the most essential parameters that need to be critically verified before their commercial implementation. Chronic and acute toxicity both are equally important, but most of the current reports did not vividly depict both aspects, rather only focused on the efficacy. Instead of a boon, nano-intervention can be a boomerang without such critical introspection.

18.4 Conclusion

Wildlife plays a major role in balancing the environment. However, due to climate change, habitat loss, and illegal trade, the population of wild animals is facing extinction and has either attained the status of rare, endangered, or threatened species or has become a smaller isolated population. The emergence of technological solutions using NPs in wildlife research, and conservation efforts would yield positive results in various aspects such as disease diagnosis, prevention, and treatment, surveillance, prevention of poaching, health, and nutritional management, etc. However, NPs also possess various negative effects on animals such as oxidative stress, genotoxicity, disruption of the actin cytoskeleton, etc. The future aspects are positive enough as the disadvantages are getting overwhelmed progressively through the expeditious development of nanotechnology that has paved the way for designing new strategies.

References

Abedini M, Shariatmadari F, Torshizi MK, Ahmadi H (2017) Effects of a dietary supplementation with zinc oxide nanoparticles, compared to zinc oxide and zinc methionine, on performance, egg quality, and zinc status of laying hens. Livest Sci 203:30–36

Allahverdiyev AM, Abamor ES, Bagirova M, Baydar S, Ates S, Kaya F (2013) Investigation of antileishmanial activities of TiO2@Ag nanoparticles on biological properties of L. tropica and L. infantum parasites in vitro. Exp Parasitol 135:55–63

Barkhordari A, Hekmatimoghaddam S, Jebali A, Khalili MA, Talebi A, Noorani M (2013) Effect of zinc oxide nanoparticles on viability of human spermatozoa. Iran J Reprod Med 11(9):767

Beni AA, Jabbari H (2022) Nanomaterials for environmental applications. Results Eng 15:100467

Bromberg L, Raduyk S, Hatton TA (2009) Functional magnetic nanoparticles for biodefense and biological threat monitoring and surveillance. Anal Chem 81(14):5637–5645

Carter KP, Young AM, Palmer AE (2014) Fluorescent sensors for measuring metal ions in living systems. Chem Rev 114(8):4564–4601

Cilulko J, Janiszewski P, Bogdaszewski M, Szczygielska E (2013) Infrared thermal imaging in studies of wild animals. Eur J Wildl Res 59:17–23

Cocozza C, Perone A, Giordano C, Salvatici MC, Pignattelli S, Raio A, Cherubini P (2019) Silver nanoparticles enter the tree stem faster through leaves than through roots. Tree Physiol 39(7): 1251–1261

Dahlan AS (2019) Smart and functional materials based nanomaterials in construction styles in Nano-architecture. SILICON 11:1949–1953

Durfey CL, Rowlison T, Lagu CU, Sente C, Khaitsa ML, Clemente HJ, Ryan PL, Willard ST, Feugang JM (2018) Improvement of cat and bull sperm quality using nanotechnology as a model for wild species. Reprod Fertil Dev 31(1):181–182

Fallon BP, Mychaliska GB (2021) Development of an artificial placenta for support of premature infants: narrative review of the history, recent milestones, and future innovation. Transl Pediatr 10(5):1470

https://portal.ct.gov/DOAG/Regulatory/Regulatory/West-Nile-Virus-In-Domestic-Animals-and-Birds#:~:text=Does%20WNV%20infect%20other%20animals,exhibited%20fever%2C%20abortion%20and%20encephalitis https://www.cdc.gov/anthrax/basics/index.html#:~:text=Domestic%20and%20wild%20animals%20can,vaccination%20can%20help%20prevent%20outbreaks

https://www.cdc.gov/healthypets/diseases/salmonella.html#:~:text=Animals%20get%20Salmonella%20from%20their,still%20appear%20healthy%20and%20clean

https://www.cdc.gov/rabies/index.html

https://www.vet.upenn.edu/research/centers-laboratories/research-initiatives/wildlife-futures-program/resources/fact-sheets/fact-sheet-detail/ranavirus#:~:text=Since%20there%20is%20no%20treatment,prevent%20infection%20of%20other%20animals

https://www.who.int/health-topics/hendra-virus-disease#tab=tab_1

Ibrahim RK, Hayyan M, AlSaadi MA, Hayyan A, Ibrahim S (2016) Environmental application of nanotechnology: air, soil, and water. Environ Sci Pollut Res 23:13754–13788

Imani SM, Ladouceur L, Marshall T, Maclachlan R, Soleymani L, Didar TF (2020) Antimicrobial nanomaterials and coatings: current mechanisms and future perspectives to control the spread of viruses including SARS-CoV-2. ACS Nano 14(10):12341–12369

Jørgensen RB (2019) Comparison of four nanoparticle monitoring instruments relevant for occupational hygiene applications. J Occup Med Toxicol 14(1):1–14

Katsumiti A, Gilliland D, Arostegui I, Cajaraville MP (2015) Mechanisms of toxicity of ag nanoparticles in comparison to bulk and ionic ag on mussel Hemocytes and gill cells. PLoS One 10:e0129039

Khin MM, Nair AS, Babu VJ, Murugan R, Ramakrishna S (2012) A review on nanomaterials for environmental remediation. Energy Environ Sci 5(8):8075–8109

Kim BY, Rutka JT, Chan WC (2010) Nanomedicine. N Engl J Med 363:2434–2443

Kirubha SA, Rajput A (2019) Enhancement of thermal imaging by iron oxide nanoparticle–preliminary study. Biocatal Agric Biotechnol 17:352–360

Kollander B, Widemo F, Ågren E, Larsen EH, Loeschner K (2017 Mar) Detection of lead nanoparticles in game meat by single particle ICP-MS following use of lead-containing bullets. Anal Bioanal Chem 409(7):1877–1885

Konkol D, Wojnarowski K (2018) The use of nanominerals in animal nutrition as a way to improve the composition and quality of animal products. J Chem 2018:1

Mandal, J., Brewer, J., Mandal, S., & Raman, A.P. (2021). Nanostructured Plasmonic metal surfaces as optical components for infrared imaging and sensing

Manuja A, Kumar B, Kumar R, Chhabra D, Ghosh M, Manuja M, Brar B, Pal Y, Tripathi BN, Prasad M (2021) Metal/metal oxide nanoparticles: toxicity concerns associated with their physical state and remediation for biomedical applications. Toxicol Rep 8:1970–1978

Marappan G, Beulah P, Kumar RD, Muthuvel S, Govindasamy P (2017) Role of nanoparticles in animal and poultry nutrition: modes of action and applications in formulating feed additives and food processing. Int J Pharmacol 13(7):724–731

Mishra A, Swain R, Mishra S, Panda N, Sethy K (2014) Growth performance and serum biochemical parameters as affected by nano zinc supplementation in layer chicks. Indian J AnimNutr 31: 384–388

Nichols G (2015) The potential for nanotechnology to improve community resilience through better building materials, sensors, and medical applications. Nano Res Appl 1

Niemiec T, Łozicki A, Pietrasik R, Pawęta S, Rygało-Galewska A, Matusiewicz M, Zglińska K (2021) Impact of ag nanoparticles (AgNPs) and multimicrobial preparation (EM) on the carcass, mineral, and fatty acid composition of Cornuaspersumaspersum snails. Animals 11(7):1926

Pacheco-Torgal F, Jalali S (2011) Nanotechnology: advantages and drawbacks in the field of construction and building materials. Constr Build Mater 25(2):582–590

Pareek A, Zafar M, Lakshminarayanan R, Joshi SJ (2021) Nanotechnology for green applications: how far on the anvil of machine learning. In: Biobased Nanotechnology for Green Applications, pp 1–38

Partridge EA, Davey MG, Hornick MA, McGovern PE, Mejaddam AY, Vrecenak JD, Flake AW (2017) An extra-uterine system to physiologically support the extreme premature lamb. Nat Commun 8(1):15112

Pokrajac L, Abbas A, Chrzanowski W, Dias GM, Eggleton BJ, Maguire S, Mitra S (2021) Nanotechnology for a sustainable future: addressing global challenges with the international network4sustainable nanotechnology, vol 15, p 18608

Ramos AP, Cruz MAE, Tovani CB, Ciancaglini P (2017) Biomedical applications of nanotechnology. Biophys Rev 9:79–89

Roy A, Sharma A, Yadav S, Jule LT, Krishnaraj R (2021) Nanomaterials for remediation of environmental pollutants. Bioinorg Chem Appl 2021:1

Santhoshkumar T, Rahuman AA, Bagavan A, Marimuthu S, Jayaseelan C, Kirthi AV (2012) Evaluation of stem aqueous extract and synthesized silver nanoparticles using Cissus quadrangularis against Hippobosca maculata and Rhipicephalus (Boophilus) microplus. Exp Parasitol 132:156–165

Santonastaso M, Mottola F, Iovine C, Cesaroni F, Colacurci N, Rocco L (2020) In vitro effects of titanium dioxide nanoparticles (TiO2NPs) on cadmium chloride (CdCl2) genotoxicity in human sperm cells. Nano 10(6):1118

Sarangi A, Ghosh M, Sangwan S et al (2022) Exploration of urinary metabolite dynamicity for early detection of pregnancy in water buffaloes. Sci Rep 12:16295

Smith AC, Hudson MAR, Downes CM, Francis CM (2015) Change points in the population trends of aerial-insectivorous birds in North America: synchronized in time across species and regions. PLoS One 10(7):e0130768

Snow J, Giordano J (2019) Aerosolized Nanobots: parsing fact from fiction for health security—a dialectical view. Health Secur 17(1):77–79

Soppimath KS, Aminabhavi TM, Dave AM, Kumbar SG, Rudzinski W (2002) Stimulus-responsive Bsmart^ hydrogels as novel drug delivery systems. Drug Dev Ind Pharm 28:957–974

Spiropoulou CF (2019) Nipah virus outbreaks: still small but extremely lethal. J Infect Dis 219(12): 1855–1857

Swain PS, Rajendran D, Rao SB, Dominic G (2015) Preparation and effects of nano mineral particle feeding in livestock: a review. Vet World 8:888–891

Teli MK, Mutalik S, Rajanikant GK (2010) Nanotechnology and nanomedicine: going small means aiming big. Curr Pharm Des 16:1882–1892

Tsai CC, Childers RA, Nan Shi N, Ren C, Pelaez JN, Bernard GD, Yu N (2020) Physical and behavioral adaptations to prevent overheating of the living wings of butterflies. Nat Commun 11(1):551

Underwood C, Van Eps A (2012) Nanomedicine and veterinary science: the reality and the practicality. Vet J 193:12–23

Wei R, Huang H, Wang H, Zhao S, Dumbleton J, Zhao G, He X (2015) ACS Appl Mater Interfaces 7(8):5017–5028

Wijenayaka LA, Wijesena RN, Tissera ND, Nisansala Bandara WRL, Amaratunga GJ, Nalin De Silva KM (2021) Infrared absorbing nanoparticle impregnated self-heating fabrics for significantly improved moisture management under ambient conditions. R Soc Open Sci 8(5):202222

Xu C, Zhang YS, Begin S, Thanh NTK (2022) Introduction to advanced functional nanomaterials for biomedical applications. Nanoscale 14(20):7441–7443

Zavaleta C, Ho D, Chung EJ (2018) Theranostic nanoparticles for tracking and monitoring disease state. SLAS technology 23(3):281–293

Zhang Y, Li M, Gao X, Chen Y, Liu T (2019) Nanotechnology in cancer diagnosis: progress, challenges, and opportunities. J Hematol Oncol 12(1):1–13

Chapter 19
Nanotherapeutics for Bacterial Livestock Diseases

Rajesh Kumar, Somesh Banerjee, Akshita, Kratika Maurya, Komal Kumari, Minakshi Prasad, and Mayukh Ghosh

Abstract Microbial diversity and abundance is so vast that almost no single corner on this earth is unoccupied by pathogenic bacteria. Animals and human population frequently encounter arrays of microbial strains. The first line of combat strategy involves prophylaxis; however, the availability of vaccines against these diverse pathogenic genera and strains never meets the adequacy. Consequently, therapeutic intervention most conveniently with different antibiotics stands as the next predominant approach. Nonjudicious use of these antibiotics has resulted emergence of antibiotic resistance among various microbial populations. Multidrug resistant (MDR), extensively drug resistant (XDR), and pandrug-resistant (PDR) bacteria are such classes that urge critical worry. Moreover, commercialization of newer antibiotics is time- as well as recourse-intensive, urging for effective alternative therapeutics to prevent billions of losses in animal husbandry and agriculture. Nanomedicines have extended multi-way benefits in this perspective. Nanoparticles as bactericidal drugs and nano-carriers to potentiate the efficacy of conventional antimicrobial therapeutics are the major approaches for therapeutic nanointervention. Extracting synergistic

R. Kumar
Department of Veterinary Physiology and Biochemistry, Lala Lajpat Rai University of Veterinary and Animal Sciences, Hisar, Haryana, India

S. Banerjee
Department of Veterinary Microbiology, Immunology Section, Lala Lajpat Rai University of Veterinary and Animal Sciences, Hisar, Haryana, India

Akshita · K. Maurya
Department of Plant Biotechnology, Institute of Agricultural Sciences, Rajiv Gandhi South Campus, Banaras Hindu University, Mirzapur, Uttar Pradesh, India

K. Kumari · M. Ghosh (✉)
Department of Veterinary Physiology and Biochemistry, Banaras Hindu University, Mirzapur, India

M. Prasad
Department of Animal Biotechnology, Lala Lajpat Rai University of Veterinary and Animal Sciences, Hisar, Haryana, India

LLR University of Veterinary and Animal, ICAR-National Research Centre on Equines, *Nanotechnology Theranostics in Livestock Diseases and Management*, Livestock Diseases and Management,
https://doi.org/10.1007/978-981-16-1610-5_19

effect by amalgamating conventional and nano-drugs are also being explored extensively. Further, reports of arising resistance against these nano-drugs are also very limited to render this means a sustainable alternative imminently. In this chapter, a brief account of various microbial diseases with special emphasis towards the zoonotic ones has been elaborated. Various means of nanointervention towards therapeutic management of these diseases along with the challenges for clinical transition of nanomedicines and future prospects have been depicted in respective sections of the current elaboration.

Keywords Antibiotics · Resistance · Nano-drugs · Nano-carriers · Zoonotic bacterial diseases

19.1 Introduction

Antibiotics are employed as the most commonplace therapeutic intervention for treating bacterial diseases which have significantly improved the animal productivity and human life expectancy worldwide (from 59.7 years in 1930 to 78.7 years in 2010 in the USA) (Maxson and Mitchell 2016; Van et al. 2020). However, the WHO lists the spread of multidrug-resistant (MDR) and extensively drug-resistant (XDR) bacteria as one of the three major public health hazards of the twenty-first century (WHO 2014). According to their acquired non-susceptibility to various antibiotic groups, the resistant microbial populations are classified and designated as MDR, XDR, and PDR (pandrug-resistant) by the European Centre for Disease Control (ECDC) and CDC, Atlanta, which become particularly difficult to treat (Magiorakos et al. 2012). Antibiotic resistance claims numerous lives across the world every year, including alarming numbers of infant mortality (CDC 2018a,b,c,d,e,f,g,h,i,j,k,l). Antibiotic resistance also inflicts loss of billions of dollars of global economy (Munita and Arias 2016). Moreover, it has been reported that from 2005 to 2015, the Organization for Economic Co-operation and Development (OECD) countries' resistance climbed from 14 to 17% (OECD 2018a,b). Furthermore, this level is more concerning in low- and middle-income nations like Brazil, Indonesia, and Russia, where it can reach up to 60%, while in Turkey, Korea, and Greece, it is only reported to be about 35%, indicating a general concern over the rise of antibiotic resistance in the modern era. Evolutionary mutations in various microbial genes and their swapping among diverse microbial populations are important underlying factors for the development of resistance. These mutations are primarily caused by unsystematic and inappropriate use of antibiotics in human and animal subjects, which necessitates the use of newer antibiotics for therapeutic management of microbial infections (WHO 2018a,b,c,d,e,f,g,h,i,j,k,l). However, the development of novel antibiotics is a time- and resource-intensive process; for instance, after the introduction of cephalosporins and quinolones in 1960s, it took almost 40 years to get the FDA approval for newer antibiotics linezolid and daptomycin followed by the latest introduction of malacidins in 2018 (Clatworthy et al. 2007; Gould 2016; Hover et al. 2018). Moreover, the effective life of an antibiotic has been shortened in recent years due to the evolution of resistance. For example, earlier antibiotics streptomycin and chloramphenicol were introduced in 1943 and 1947, respectively, and resistance was reported against them in 1959; while resistance to the most recent antibiotics such as linezolid and doxycycline was reported within a few years of their introduction (Palumbi 2001; Tsiodras et al. 2001). The inappropriate therapeutic use of antibiotics in human or animal subjects is not the sole factor leading to emergence of resistance; the unsystematic use of antibiotics as growth promoters in animal feed is also a major concern. A significant portion of total antibiotic consumption in feed animals is used as growth promoters, oftentimes in excess of recommended dose (Cantas and Suer 2014). This could facilitate acquiring resistance among the commensal microflora in animal gut, acting as a reservoir of resistance genes for (possibly) pathogenic bacteria. Alternatively, these antibiotic residues could enter the fecal flora or food chain, which may further increase the probability of resistance development in human subjects too.

Therefore, the transmissible nature of the majority of infectious diseases and the rapid emergence of resistance may spell disaster for global health, especially for the approximately 700 million rural poor people who depend heavily on animal husbandry. In this regard, nanomedicine can be a useful substitute while developing effective preventive, diagnostic, and therapeutic strategies to manage the communicable diseases. The effectiveness of numerous targeted drug delivery systems using nanoparticles (NPs) has been demonstrated, and the usual physicochemical characteristics of several nano-formulations have shown excellent bactericidal effects. In addition to their therapeutic potential, nano-vaccines and theranostic uses of nano-formulations have drawn a lot of interest as alternatives to traditional antibiotics.

19.2 Bacterial Diseases of Livestock Animals with Zoonotic and Reverse-Zoonotic Importance

The epidemiological outbreaks from 1986 to 2007 inflicted by various bacterial, viral, parasitic, and fungal diseases have resulted in direct financial loss of over $20 billion, along with indirect losses over $200 billion with passing of plentiful human lives each year (World Bank 2010; Belay et al. 2017). Among the top transmittable diseases, major numbers are of bacterial origin and are having zoonotic and reverse-zoonotic importance; these include but not limited to Tuberculosis, Salmonellosis, Shigellosis, Pneumococcal disease, Listeriosis, Campylobacteriosis, and *E. coli* infection. They pose substantial threat to the human and animal populations (Cassini et al. 2018). The appearance and recurrence of diseases is not the only alarm, but the antibiotic resistance is also developing as an emerging threat. According to the WHO, Enterobacteriaceae group bacteria has been recognized as critical, while *Staphylococcus aureus* and *Salmonellae* infections have been stated as high priority bacterial diseases in terms of resistance (WHO 2017). Antibiotic resistance has also been observed in *Mycobacteria, Campylobacter, Brucella, Pasteurella, Clostridium,*

Acinetobacter, Pseudomonas, etc. (Kim et al. 2017; Meade et al. 2017; WHO 2018a, b,c,d,e,f,g,h,i,j,k,l). Thus, the quest for better and new antimicrobials is evident.

Anthrax: The etiological agent *Bacillus anthracis* is a spore-forming, extremely virulent gram-positive bacteria that is present throughout the world and can be employed as a probable war agent. Cutaneous, pulmonary or respiratory, and gastrointestinal forms of the disease are observed in human, while apoplectic or peracute, acute and subacute, and chronic forms are developed in animals. The apoplectic form is mostly seen in the start of an outbreak in animals. Several outbreaks in human population have been stated along the southern peninsula from 1973 to 1977, ensuing approximately 500,000 deaths (Acha and Szyfres 2001). In addition to the usual zoonotic causes of mortality, industrial releases and aerosol transmission have also been reported (Marshall 1988). The drug of choice for treatment is penicillin G in human, whereas sodium benzylpenicillin should be administered intravenously right away, followed by either long-acting benethamine penicillin or amoxicillin given intramuscularly in animals (WHO 2008). However, penicillin resistance in anthrax has been reported (Braderic and Punda-Polic 1992; Hendricks et al. 2014). Alternate therapy includes hyperimmune serum or monoclonal antibody therapy. Vaccination in the endemic area and proper disposal are recommended for the management of the disease.

Arcobacter infections: Broadly dispersed Campylobacter-like organism *Arcobacter butzleri* is one of the main enteropathogens in this group along with others like *A. cryaerophilus* and *A. skirrowii*, which causes diarrheal illness in people, pigs, and cattle and is spread through the consumption of contaminated water and food such as raw meat (Riesenberg et al. 2017; Ogan et al. 1982). It also causes abortion in pigs and mastitis in cows. Tetracyclines, aminoglycosides, or fluoroquinolones (ciprofloxacin) are suggested for treatment. However, a gyrA gene mutation has been linked to fluoroquinolone resistance (Abdelbaqi et al. 2007); however, the fact that the organism is susceptible to heavy metals (copper and zinc) indicates NPs of these metals may be able to treat the infection (Šilha et al. 2014).

Bordetellosis: The disease "kennel cough" (Bordetellosis), which primarily affects immunocompromised people, is caused by the bacterium *Bordetella bronchiseptica*, a zoonotic respiratory pathogen. It is spread by farm and companion animals, primarily dogs and cats. The major route of transmission is through saliva or sputum or aerosol infection. The presentations include sinusitis, bronchitis, pneumonia, and disseminated disease (Register et al. 2012; Chomel 2014; Veterinary Manual 2018). The organism is found to be susceptible to ciprofloxacin and tobramycin, amoxicillin–clavulanate, and trimethoprim–sulfamethoxazole (Wernli et al. 2011). However, some strains of the pathogen have been reported to be resistant to erythromycin, streptomycin, trimethoprim/sulfamethoxazole, penicillin, and ampicillin (Shin et al. 2011; Zhao et al. 2011; Dewan et al. 2018).

Borreliosis: The condition, also referred as "Lyme disease," is brought on by the spirochete group pathogen *Borrelia burgdorferi* sensu lato species complex and spread to people by ticks of the *Ixodes* genus (Stanek et al. 2012). Wild rodents, insectivores, hedgehogs, hares, and other mammals as well as birds serve as the reservoirs (Veterinary Manual 2018). In temperate regions of Europe, North America, and Asia, the illness is common. General symptoms include fever, rash, facial paralysis, and arthritis. Currently vaccine against Borreliosis is not available. Doxycycline, cefuroxime axetil, and amoxicillin are among the antibiotics used in treatment; patients who are intolerant to these medications can also use azithromycin, clarithromycin, or erythromycin (CDC 2018e,l). However, in vitro sensitivity testing has revealed resistance to a number of antibiotics, including aminoglycosides, quinolones, fusidic acid, erythromycin, aztreonam, and several members of cephalosporins, (Terekhova et al. 2002; Hunfeld and Brade 2006). The other zoonotic diseases linked to *Borrelia* pathogens include tick-borne relapsing fever, which is primarily spread by soft-tick *Ornithodoros* spp., and southern tick-associated rash illness, which is spread by *Amblyomma americanum* tick bites (Veterinary Manual 2018).

Brucellosis: The disease is caused by six facultative intracellular, gram-negative coccobacillus species, including *Brucella abortus*, biovars 1–6, 9, *B. melitensis*, biovars 1–3, *B. suis*, biovars 1, 3, and 4, *B. canis*, *B. Suis* biovar 2, and *B. maris*. *B. melitensis* is the most relevant among them for human infection (WHO 2018f,g; Mantur et al. 2007). Nearly 500,000 cases of human disease are reported annually worldwide (Acha and Szyfres 2001), with 1000 cases reported in Peru each year (Maves et al. 2011). These widely dispersed species can be passed from animals, including cattle, sheep, goats, pigs, and camels, to humans through ingestion, direct contact, airborne infection, or contaminated food (WHO 2018f,g). Bacteria initially localize in the regional lymph nodes and then disseminate to the organs of the reticuloendothelial system to multiply within phagocytic cells (Alizadeh et al. 2013). In general, nonspecific clinical features of brucellosis include undulating fever, fatigue, malaise, chills, sweats, headaches, myalgia, arthralgia, anorexia, weight loss, and hepatosplenomegaly with rare death cases (Franco et al. 2007; Gwida et al. 2010). Safe and effective human brucellosis vaccines are currently not widely used in the global environment. It is common practice to administer attenuated vaccines using *B. melitensis* strain Rev.1 for sheep and goats and *B. abortus* strain 19 for all other animal species. The best way to manage animal disease is to isolate/kill the positive animals. For human treatment, it is advised that patients take tetracycline or doxycycline for 6 weeks along with aminoglycosides for 2–3 weeks; alternative treatments include rifampicin and fluoroquinolones (Corbel 2006). However, alarmingly high MIC values of ceftriaxone, streptomycin, and rifampicin have already been reported against cattle isolates, necessitating the possibility of an alternative therapeutic approach in the future (Barbosa Pauletti et al. 2015).

Campylobacteriosis: Campylobacter, a member of the Enterobacteriaceae family with 17 species and six subspecies, is the etiological agent underlining the illness. *C. jejuni, C. fetus,* and *C. coli* are the most common zoonotic and emerging foodborne human pathogens (WHO 2018b,h; Iovine 2013). It produces enteritis, i.e., diarrheal and systemic illnesses in human. In addition to humans, farm animals and birds are also seriously impacted by the organisms, and reports of reverse-zoonotic transmission among wild animals have also been made (Nizeyi et al. 2001). A reactive arthritis and the life-threatening neuropathy, Guillain–Barre syndrome may develop as a result of inadequate treatment for enteric manifestations (de Zoete et al. 2007; Huang et al. 2010; Silvan et al. 2018). Although the U.S. FDA has authorized a protective vaccine against *C. jejuni* for human trials, no recommendation for human use is currently accessible (WHO 2018b,h). Supportive fluid and electrolyte therapy is the primary form of treatment; antibiotics are only used when necessary or in invasive cases (WHO 2018b,c,h; Peterson 1994). However, there has recently been widespread news of drug resistance to ciprofloxacin, azithromycin, other fluoroquinolones, aminoglycosides, beta-lactam and macrolide antibiotics, and tetracycline (Iovine 2013; Altekruse et al. 1999; Murphy Jr. et al. 1996; Piddock 1995; Jacobs-Reitsma et al. 1996; Smith et al. 1998).

Cat scratch disease: Intracellular, gram-negative bacterium *Bartonella henselae* is the primary cause of the disease and cat is the natural reservoir. Other *Bartonella* species that are specific to various animal hosts such as canids, rodents, and rabbits can also be the mediating factor in zoonoses. The most common method of transmission is saliva-mediated transmission through bites or scratches, but fleas, flies, lice, or ticks can also spread the illness. In people with a functioning immune system, the illness is curable and the common signs include lymphadenopathy, fever, malaise, and skin lesions at inoculation site. However, in immunocompromised cases, bacillary angiomatosis or bacillary peliosis or other complications may arise (Veterinary Manual 2018). Several antibiotics, including fluoroquinolones, gentamicin, erythromycin, azithromycin, and rifampicin, have been reported to face resistance against *Bartonella* infection, whereas azithromycin, penicillin, tetracyclines, cephalosporins, and aminoglycosides have been found to be effective.

Clostridial food poisoning: *Clostridium perfringens* (*C. welchii*) is a gram-positive, spore-forming, anaerobic bacillus that secretes four different extracellular toxins which have five strains, from A to E. Among these strains, *C. perfringens* type A is the most contagious and widely dispersed strain, while the others are only found in animal GIT; *C. perfringens* types B and E only have localized distribution and cause enterotoxemia (Acha and Szyfres 2001). Its symptoms include abdominal cramps, diarrhea, and necrotic enteritis. It is also responsible for gas gangrene in humans with the other clostridial species like *C. novyi*, *C. septicum*, *C. sordellii*, *C. histolyticum*, and *C. fallax*, whereas *C. septicum* infection is prominent in animals (Acha and Szyfres 2001; Ghosh et al. 2013). Another zoonotic species in this genus that causes diarrhea in humans is *Clostridium difficile*, which is spread by pets and other companion animals (Rabold et al. 2018). Currently, no human vaccine is available for the illness. Antibiotics are not advised in cases of clostridium food poisoning; however, oral rehydration or intravenous fluids and electrolyte therapy can be used in severe cases (CDC 2018b,l). Extensive surgical debridement and removal of all devitalized tissue along with penicillin/tazobactam treatment is generally given for clostridial gas gangrene (Stevens et al. 2014). Clindamycin, vancomycin, carbapenem, and cephalosporins are used in other antibiotic therapies, but sporadic resistance to clindamycin, cephalosporins, erythromycin, and fluoroquinolones has recently been reported (Peng et al. 2017; Spigaglia 2015; Han et al. 2018).

Corynebacteriosis: The genus *Corynebacterium* consists of a large number of lipophilic and non-lipophilic organisms and are pleomorphic, nonspore-forming, catalase positive, aerobic or facultative anaerobic, gram-positive, and fermentive or non-fermentive in nature (Funke et al. 1997). This genus consists of probable zoonotic pathogens including *C. amycolatum*, *C. bovis*, *C. diphtheriae*, *C. jeikeium*, *C. pseudotuberculosis*, *C. ulcerans*, *C. urealyticum* and *C. xerosis* with consistent reports and *C. canis*, *C. falsenii*, *C. freiburgense*, *C. glucuronolyticum*, *C. kutscheri*, and *C. mastitidis*-like organisms with rare reports (Bernard 2012). The majority of *Corynebacterium* persists in the host as commensal, opportunistic pathogens and infects immunocompromised individuals (Funke et al. 1997). The diphtheria toxin (DT), which is produced by the toxicogenic *C. diphtheriae*, *C. ulcerans*, and sporadically by *C. pseudotuberculosis*, is a potent exotoxin that is encoded by a beta-phage-carried tox gene (Meinel et al. 2015). Previously, it was believed that *C. ulcerans* spreads among people through tainted milk or farm animals, but more recently, zoonotic transmission between companion animals (dogs and cats) and their owners suggested their global spread (Dias et al. 2011). Small ruminants (sheep and goats) are primarily affected by *C. pseudotuberculosis*, which causes caseous lymphadenitis and is spread by direct contact (Baird and Fontaine 2007; Peel et al. 1997). Even though the diphtheria toxoid vaccine is a primary preventive measure that has been approved for use worldwide, diphtheria continues to be a major cause of illness and death in both younger (under 5 years) and older (over 40 years) people (CDC 2018c,l). Although *C. bovis* causes CNS impairment and *C. renale* has been linked to chest and rectal abscesses, only a small number of human cases have been linked to these zoonotic organisms to date (Acha and Szyfres 2001). It has been suggested to use erythromycin and procaine penicillin G as a direct treatment for human diphtheria (CDC 2018c,l). Antibiotic resistance is found in several species of *Corynebacterium* species against erythromycin, clindamycin, tetracycline, (Berger et al. 2018), trimethoprim, sulfamethoxazole, co-trimoxazole (Barraud et al. 2011), ciprofloxacin (Hahn et al. 2016), azithromycin, and gentamicin (Soriano et al. 1995).

Cholera/vibriosis: The gram-negative bacteria *Vibrio cholerae*, which has a worldwide distribution, causes this acute enteric diarrheal disease and is transmitted to people or animals through contaminated food and water without the use of an insect vector or an animal reservoir. Two pathological serogroups of *V. cholera*, O1 and O139, based on their somatic O antigen, cause pandemic cholera, mostly in low-income and developing countries (Dutta et al. 2013). Since 2013, mass vaccination campaigns in several countries with oral cholera vaccine have provided solution for future outbreak prevention and control. Most cases can be successfully treated with oral rehydration solution (ORS), but severe cases should be treated with intravenous fluids and antibiotics (WHO 2018j). The other zoonotic pathogens in this genus that cause gastroenteritis, sepsis, and wound infection are *Vibrio parahaemolyticus* and *V. vulnificus*. There have been reports of sulfamethoxazole, trimethoprim, chloramphenicol, and streptomycin resistance from various parts of the world, necessitating the development of future alternative strategies (Hochhut et al. 2001; Kitaoka et al. 2011; Miwanda et al. 2015). Here, NP-based methods can be helpful in battling the disease in all aspects, especially in diagnosis as current tests like alkaline peptone water (APW), isolation by culture on thiosulfate citrate bile salt sucrose agar (TCBS), and identification by slide agglutination test with specific antisera are resource- and time-intensive (Yu et al. 2011; Chatterjee et al. 2017).

Erysipeloid: *Erysipelothrix rhusiopathiae* is a gram-negative bacillus except *E. insidiosa* which is gram-positive, having worldwide distribution with 256 strains causing erysipeloid and rose disease in human, swine erysipelas in swine, and polyarthritis in sheep. Few outbreaks of the human infection affect people who work in slaughterhouses, whereas several outbreaks in swine have been reported due to serotype 15 (Gerber et al. 2018). Skin lesions, arthritis in the finger joints, lymphadenitis, endocarditis, urticarial lesions, etc., are all common symptoms in humans (Takahashi et al. 1984). According to reports, penicillin and ceftriaxone are still effective against *E. rhusiopathiae* (Fidalgo et al. 2002). However, nearly 43 strains had been found to be resistant to erythromycin, oleandomycin, oxytetracycline, or dihydrostreptomycin (Takahashi et al. 1984). Vaccination using bacterin adsorbed on aluminum hydroxide and live avirulent vaccine is found to be ineffective against chronic erysipelas but successful in inhibiting acute form (Timoney et al. 1988).

Intestinal pathogenic *Escherichia coli* infections: *E. coli,* a gram-negative Enterobacteriaceae microbe with a global distribution, is a part of the flora in the large intestine of all warm animals but acts as a major contributor to diarrhea-related deaths in people. Rooted on thermostable somatic O polysaccharide antigens, there are 170 reported *E. coli* serogroups, and each serogroup is further divided into various serotypes based on heat-labile flagellar H antigen (Chin 2000). The most hazardous serogroups of *E. coli* are O_{157}:H_7, O_{26}:H_{11}, and O_{45}:H_2, along with the serotypes of serogroups O_4, O_{103}, O_{104}, O_{111}, and O_{145}. They also infect various animals, causing mastitis in cattle, white diarrhea in sheep and calves, and neonatal enteritis in pigs and horses (Timoney et al. 1988; Gillespie and Timoney 1981). The pathogenic strains which cause enteric disease are grouped into six categories: (a) enterohemorrhagic (EHEC), (b) enterotoxigenic (ETEC), (c) enteroinvasive (EIEC), (d) enteropathogenic (EPEC), (e) enteroaggregative (EAggEC), and (f) diffuse-adherent (DAEC) (Acha and Szyfres 2001). Among these groups, EPEC and ETEC are mainly endemic in developing countries, ETEC and EAEC are associated with traveller diarrhea in endemic areas, while EHEC is mostly responsible for large outbreaks (Rojas-Lopez et al. 2018). Besides diarrhea, it also causes hemorrhagic colitis and hemolytic uremic syndrome (HUS). The spread is not only limited to humans from animals (CDC 2018i); it is also reportedly transmitted via reverse zoonosis to domesticated and wild animals (Rwego et al. 2008; Ewers et al. 2010). Although fimbria (pili) antigens have served as the basis for the development of vaccines for cattle and swine, there is currently no recommended vaccine for use in humans (Rojas-Lopez et al. 2018). Antibiotic therapies are not suggested in EHEC infections (Goldwater and Bettelheim 2012; Wong et al. 2000) as antibiotic use is still essential to control other enteric *E. coli* infections; there is a potential risk of the emergence of resistance (Torres 2017). Numerous reports of multidrug resistance by *E. coli* have been stated against some antibiotics such as fluoroquinolones, β-lactams, (Collignon 2009; Alhashash et al. 2013), ampicillin, rifampin, chloramphenicol, sulfamethoxazole, streptomycin, tetracycline (Saenz et al. 2018), and cephalosporins (Ramírez-Castillo et al. 2018). Therefore, the search for alternative therapeutics, such as monoclonal and single-chain antibodies for toxin neutralization, is ongoing, and nanomedicine can support this determined effort for successful intrusion (Bitzan et al. 2009; Mejías et al. 2016).

Glanders and melioidosis: *Burkholderia mallei* is a gram-negative zoonotic pathogen responsible for the highly contagious, widely dispersed, and frequently fatal disease known as glanders. It primarily affects members of the Equidae family of animals, but it can also infect humans and other mammals (Van Zandt et al. 2013). Most cases of transmission to humans involve contact with lacerated or rubbed skin or an aerosolized infection. The presence of nodular lesions in the lungs and other organs, ulceration of the mucous membranes of the respiratory tract, pleural effusion, and death from septicemia are the most common symptoms. It is thought to be a potential biological weapon (Woods 2005). Occasional epidemics have been observed in Central and South America, India, the Middle East, Africa, and Asia (Malik et al. 2012). Currently, there is no vaccine that has received FDA approval (CDC 2018h,f). For initial intensive therapy, ceftazidime or meropenem can be used, and for eradication therapy, trimethoprim/sulfamethoxazole, or amoxicillin/clavulanic acid are recommended medications and trimethoprim/sulfamethoxazole for postexposure prophylaxis (Lipsitz et al. 2012). However, it has been noted that this pathogen population is resistant to a variety of antibiotics, including amikacin, amoxicillin, ampicillin, cefazolin, cefoxitin, and clindamycin (Van Zandt et al. 2013). *Burkholderia pseudomallei*, a closely related species that causes melioidosis, is more common in South Asia and Northern Australia. Direct contact with the contaminated soil and water results in its transmission. The symptoms of various melioidosis infections may vary. There is currently no approved vaccine for the disease. Antibiotic treatment options include oral administration of trimethoprim/sulfamethoxazole or amoxicillin/clavulanic acid as well as intravenous administration of ceftazidime or meropenem (CDC 2018h,f). Resistance against a number of antibiotics including chloramphenicol (Dance et al. 1989), beta-lactam antibiotics (Godfrey et al. 1991; Livermore et al. 1987), ceftazidime, amoxicillin/clavulanic acid, and penicillin (Schweizer 2012) has been cited among these microbes (Chugh 2008).

***Helicobacter* infection:** More than 35 species of gram-negative, spiral-shaped, microaerophilic bacteria make up the genus, with *Helicobacter pylori* being the most significant as a human pathogen causing gastritis, peptic ulcers, and gastric malignancies. In order to cause disease in a variety of animals, other *Helicobacter* species, such as *H. heilmannii, H. felis, H. mustelae, H. acinonychis, H. bizzozeronii, H. bilis, H. rodentium, H. suis, and H. pullorum,* also colonize primarily in the stomach. Although the precise mode of transmission is still unknown, several of them, including *H. bizzozeronii*, *H. suis*, *H. pullorum*, and *H. pylori*, have been reported to be important zoonotic agents. However, more evidence is required to confirm this last claim (Mladenova-Hristova et al. 2017; Kim 2012). Australian captive marsupials have also been reported to have contracted *H. pylori* from their keepers in a reverse-zoonotic transmission (Every et al. 2011). Antibiotics like amoxicillin, tetracycline, metronidazole, or clarithromycin are used in combination with ranitidine bismuth citrate, bismuth subsalicylate, a proton pump inhibitor, or an H2 blocker as treatment for *H. pylori* infection (CDC 1998). These organisms have been found to exhibit varying degrees of resistance to a number of antibiotics including metronidazole, clarithromycin, tetracycline, and amoxicillin (Boltin et al. 2014; Alba et al. 2017).

Leptospirosis: The *Leptospira* genus of spirochete bacteria, which can be seen under a dark-field microscope, cause more than one million human infections worldwide each year, including nearly 60,000 fatalities. This is one of the most significant neglected emerging zoonotic diseases (CDC 2018g). The genus is divided into 23 species and grouped into saprophytic, intermediate, and pathogenic groups based on molecular taxonomy. The final group consists of ten pathogenic species, the most significant one in terms of zoonotic aspects being *L. interrogans*. The other categorization is based on the microscopic agglutination test (MAT) and the cross-agglutination absorption test (CAAT), which separated leptospires into two serological species. The "*Leptospira interrogans* sensu lato complex" harbors the majority of known or suspected pathogenic leptospires, while saprophytic strains are primarily grouped within the "*Leptospira biflexa* sensu lato complex" (Bharti et al. 2003; Guernier et al. 2018). It is more dominant in tropical regions with alkaline soil and rainfall (Acha and Szyfres 2001). The reservoir hosts are wild and domesticated animals, and the disease is spread by coming into contact with urine- or reproductive fluid-contaminated water, wet soil, or mucous membranes, conjunctiva, and skin cuts or abrasions. Aseptic meningitis, renal failure, hemorrhage (particularly pulmonary), aseptic meningitis, cardiac arrhythmias, pulmonary insufficiency, and hemodynamic collapse are the main symptoms in severe cases. General symptoms may include fever, headache, myalgia, conjunctival suffusion, nausea, vomiting, diarrhea, abdominal pain, cough, and skin rash. Weil's disease is a combined renal and liver failure caused by leptospirosis (CDC 2018g; WHO 2018a). SPIROLET and vax-Spiral, two commercial human vaccines, were developed in France and Cuba, but neither could produce cross-protective or long-lasting immunity (Bahaman and Bashiru 2018; SEARO.WHO 2018). For mild symptoms, doxycycline is the preferred medication; however, azithromycin, ampicillin, and amoxicillin can also be used. In contrast, intravenous penicillin G is the preferred medication for severe cases, though intravenous ceftriaxone is also equally effective (CDC 2018g,l). However, leptospirosis cases have been linked to resistance to amphotericin B, 5-fluorouracil, fosfomycin, trimethoprim, sulfamethoxazole, streptomycin, neomycin, spectinomycin, and vancomycin (Chakraborty et al. 2010; Poggi et al. 2010).

Listeriosis: The zoonotic pathogen *Listeria monocytogenes* is primarily responsible for the food-borne, fatal infection known as listeriosis (Dhama et al. 2015). *L. ivanovii*, another member of this genus with zoonotic potential, is a bacterium that primarily affects ruminants but can also infect humans and cause gastroenteritis and bacteremia making it an opportunistic enteric pathogen (Guillet et al. 2010). Pregnant women, immunocompromised individuals due to protracted immunosuppressive therapy, HIV-positive individuals, and elders are always at high risk for infection because of the patient's immunological status (World Health Organization

(2018)). It is one of the most dangerous food-borne pathogens due to its high pathogenicity and cell-to-cell invasiveness which can lead to serious infections like meningitis or septicemia, especially in newborns (Havelaar et al. 2012; Belluco et al. 2016; Liu et al. 2015; Barton Behravesh et al. 2011). Although early diagnosed cases are advised to be treated with antibiotics (either alone or in combination), detailed prevention techniques are available to stop the cycle of animal to human disease transmission (CDC 2018d). Traditionally, gentamicin, an aminoglycoside, in combination with ampicillin or penicillin G is the first medication of choice (Charpentier and Courvalin 1999), although resistance has been observed in various strains of *L. monocytogenes* against tetracyclines and fluoroquinolone (Morvan et al. 2010), erythromycin, minocycline, trimethoprim (Granier et al. 2011), and ciprofloxacin (Escolar et al. 2017).

Pasteurellosis: This is a zoonotic disease affecting both humans and animals and is present throughout the world. A gram-negative, pleomorphic, catalase and oxidase positive, non-motile, facultatively anaerobic bacteria called *Pasteurella* spp. most frequently causes the disease (Acha and Szyfres 2001; Abreu et al. 2018). The additional species of *Pasteurella* include *P. dagmatis* sp. nov., *P. canis* sp. nov., *P. stomatis* sp. nov., and *P. caballi* (Schlater et al. 1989). Capsular (A, B1, D, and E) and somatic antigens of *P. multocida* are used to classify its serotypes, whereas *P. haemolytica* has 15 serotypes that are further divided into two biotypes (A and T). The pathogen is a common commensal of upper respiratory tracts of both wild and domestic animals. *Pasteurella multocida* causes hemorrhagic septicemia, predominantly affecting cattle and water buffalo to inflict significant mortality in tropical regions of Asia, Africa, and the Middle East. Human transmission is frequently brought on by animal bites or scratches (Abreu et al. 2018). Soft tissue infection is the most frequent disease manifestation of human pasteurellosis, followed by more complicated conditions like abscesses, septic arthritis, and osteomyelitis (GOV.UK 2018). Tetracyclines, erythromycin, and some beta-lactam antibiotics (such as dicloxacillin or cephalexin) are examples of broad-spectrum antibiotics that are frequently linked to resistance and are therefore not advised for treating animal bites or scratches alone (Wilson and Ho 2013). Antimicrobial resistance against quinolones and aminoglycosides has been stated in pig and poultry (Sellyei et al. 2009). Penicillin G and streptomycin in bovine pasteurellosis (Jamali et al. 2014), while clindamycin along with a fluoroquinolone (ciprofloxacin) or trimethoprim/sulfamethoxazole combination for children or ceftriaxone for pregnant women are the suggested treatment for human pasteurellosis (Oehler et al. 2009; Fleisher 1999; Talan et al. 1999).

Plague: *Yersinia pestis,* a deadly gram-negative organism responsible for the zoonotic disease, transmitted to human beings mainly through rodents and lagomorphs. It is also a potential warfare agent. Febrile flu-like syndrome with swollen and very painful draining lymph node(s), pneumonia, and sepsis are the most common presentations (Veterinary Manual 2018). Various outbreaks have been documented in the past (Acha and Szyfres 2001), and the incidences of cases are even increasing recently (Galimand et al. 2006). For the treatment, a recommended vaccine is available along with sulfonamide and streptomycin therapy, and fortunately no significant resistance has yet emerged among these organisms (Galimand et al. 2006; Urich et al. 2011; Wagner et al. 2010).

Rat bite fever: It is brought on by gram-negative *Streptobacillus moniliformis*, universally dispersed, and manifests as a systemic sickness with fever, rigors, and polyarthralgias. If left untreated, the mortality rate can be approximately 10% (Elliott 2007). Along with dogs, cats, ferrets, and other animals, rodents are the main carrier of the disease. The most effective treatment is penicillin (Glaser et al. 2000).

Salmonellosis: *Salmonella* are the main source of food-borne zoonotic and reverse-zoonotic diseases (non-typhoidal) (WHO 2018k; Nizeyi et al. 2001). Salmonellosis in humans is a serious diarrheal disease that is responsible for 60–80% of all instances of food-borne infections (Park 2017). *Salmonella* spp. is a gram-negative, facultative, rod-shaped anaerobic bacteria of the Enterobacteriaceae family. Based on their O antigens (LPS), H or flagellar antigens, and Vi antigens, more than 2500 serovars of *Salmonella* spp. have been identified (Leedom Larson and Spickler 2013). Most serotypes of *Salmonella* can cause infections in humans and have a wide range of potential hosts (Hendriksen et al. 2004). Certain serotypes are particular to their host, such as *Salmonella enterica* Dublin in cattle. *Salmonella enterica* serovars Typhimurium and *Salmonella enterica* serovars Enteritidis are two of the most significant and widely spread zoonotic or potentially zoonotic *Salmonella* species (Leedom Larson and Spickler 2013). Due to the advent of antibiotic-resistant strains, WHO did not recommend antimicrobial therapy for the treatment of mild to moderate infections (WHO 2018k). Electrolyte replenishment and rehydration are the sole suggested treatments. As a preventative step against the spread of the disease through eggs and meat, the US Department of Agriculture (USDA) has licensed the commercial use of a specific chicken vaccination in the poultry business in the United States and other nations (PoultryWorld 2018; ScienceDaily 2018b). Resistance to up to six commonly used antimicrobials (Threlfall 2002) including sulfamethoxazole/trimethoprim, ampicillin, amoxicillin/clavulanic acid (Reshetneva et al. 2015), ampicillin, chloramphenicol, co-trimoxazole, and fluoroquinolones has been reported (Raveendran et al. 2010).

Shigellosis: The gram-negative, facultative aerobic, nonspore-forming, non-motile, rod-shaped bacteria of the genus *Shigella* of the family Enterobacteriaceae are the causal organisms of the severe diarrheal illness known as shigellosis in humans (WHO 2018i). Young children under the age of five are particularly affected by the disease, which accounts for 61% of the one million deaths caused by diarrheal diseases each year worldwide (Park 2017). *Shigella dysenteriae* (serogroup A), *S. flexneri* (serogroup B), *S. boydii* (serogroup C), and *S. sonnei* (serogroup D) are the four species that make up the genus *Shigella*, and the first two of them are the ones that cause serious illnesses. Shiga toxin, a powerful enterotoxin produced by *S. dysenteriae* is a recognized virulence factor of the bacterium and is responsible for the potentially fatal hemolytic uremic syndrome in humans, especially in impoverished nations. It has also been observed that these microbes can cause disease in wild animals by reverse-zoonotic transmission (Nizeyi et al. 2001). Ceftriaxone, pivmecillinam, and azithromycin are the antibiotics that are recommended by WHO (Taneja and Mewara

2016), while resistance has been reported against sulfonamides, tetracycline, chloramphenicol, ampicillin, co-trimoxazole, and nalidixic acid (Puzari et al. 2018).

Staphylococcal food poisoning: *Staphylococcus aureus* is a gram-positive organism which is the main causative agent of staphylococcal food poisoning present globally along with occasionally occurring *S. intermedius* and *S. hyicus*. Five enterotoxins (A, B, C, D, and E) are produced by them, out of which enterotoxin A is linked with most of the outbreaks (Acha and Szyfres 2001). The major way of zoonotic transmission is by consumption of infected milk and milk products from animals with staphylococcal mastitis. In humans, nausea, vomiting, stomach discomfort, and diarrhea are the main symptoms. Recently methicillin-resistant *S. aureus* (MRSA) is causing threat to health during epidemics (Chambers and DeLeo 2009). The major signs include bacteremia, endocarditis, empyema, osteomyelitis, pneumonia, sepsis, scalded-skin syndrome, and toxic shock syndrome (Bhatia and Zahoor 2007; Yeom et al. 2016) in humans and indicate resistance against penicillin and other β-lactam antibiotics, tetracycline, and quinolones together with newer antibiotics like linezolid and daptomycin (Chambers and DeLeo 2009; David and Daum 2010; Pantosti et al. 2007). MRSA and oxacillin-resistant *S. aureus* (ORSA) reverse-zoonotic transmission from human to livestock and companion animals has been widely observed in recent years (Seguin et al. 1999; Weese et al. 2006; Morris et al. 2006; Kwon et al. 2006; Hsieh et al. 2008; Messenger et al. 2014).

Streptococcosis: *Streptococcus suis* from pigs, *S. equi subsp. zooepidemicus* from horses, *S. canis* from dogs and cats, and *S. iniae* from fish are the main species involved in streptococcal zoonoses along with other species as minor contributors (Fulde and Valentin-Weigand 2012; Fong 2017). Despite the rarity of zoonotic streptococcosis epidemics, pandemics and some outbreaks from different parts of the world have been documented (Fulde and Valentin-Weigand 2012). Transmission is facilitated by direct contact with skin that has disintegrated or by consuming infected food, such as milk and its derivatives (Veterinary Manual 2018). Meningitis caused by *S. suis* infection, endocarditis and bacteremia in humans caused by *S. bovis* infection, and scarlet fever (streptococcal tonsillitis and pharyngitis) caused by *S. pyogenes* are some of the clinical manifestations. Other symptoms include different suppurative processes, streptococcal toxic shock syndrome, septicemias, puerperal sepsis, and ulcerative endocarditis (Acha and Szyfres 2001; Fulde and Valentin-Weigand 2012). These pathogens have depicted antibiotic resistance against tetracycline, erythromycin, clindamycin, and several other antibiotics (Arvand et al. 2000; Camara et al. 2013).

Tuberculosis: This is brought on by a bacterial group known as the *Mycobacterium tuberculosis* complex (MTBC), of which *M. tuberculosis*, *M. africanum*, *M. bovis*, *M. canetti*, *M. microti*, *M. caprae*, and *M. pinnipedii* are of prime importance (Riojas et al. 2018). It is most likely the second most lethal infectious lung illness that occurs globally. According to the WHO report, the worst-affected regions for TB are in Southeast Asia and the Western Pacific. Eight countries (India, China, Indonesia, Philippines, Pakistan, Nigeria, Bangladesh, and South Africa) accounted for two-thirds of the new TB cases in 2017, and latent TB cases affect nearly one-fourth of the world's population. Overall, 87% of new TB cases were reported in the 30 high TB burden countries in 2017 (WHO 2018d,l). The main entrance point is through inhalation, and raw milk or goods made from it can also spread the disease. Clinical signs include cough occasionally producing bloody sputum, chest aches, weakness, weight loss, fever, and night sweats. The MTBC group of diseases, particularly *M. tuberculosis* and *M. bovis*, are of enormous reverse-zoonotic importance and must be taken into account when designing an efficient eradication program (Messenger et al. 2014). One of the most commonly used vaccines, the bacille Calmette–Guérin (BCG) vaccine is only marginally effective at protecting newborns and young children but poorly effective against pulmonary disease in adolescents and adults (WHO 2018e). Isoniazid, rifampin, pyrazinamide, and either ethambutol or streptomycin therapy with a typical 6-month duration make up the initial empiric TB treatment. Ethambutol (or streptomycin, if it is used as a fourth medication) can be stopped once it is established that the TB isolate is fully susceptible. In *M. bovis* infection, pyrazinamide is ineffective against the organism (CDC 2003). New drug combinations are needed as a result of the recent reports of the global emergence of the MDR and XDR populations of mycobacteria, which are frequently resistant to regular antibiotic therapy. The traditional course of treatment is also time-consuming and expensive, and the patients frequently experience harmful side effects. Hence, nanomedicine has shown to be a promising alternative for illness prevention and theranostic intervention.

Leprosy: *Mycobacterium leprae* is the causal organism of this worldwide distributed disease and armadillos are naturally afflicted with this organism from which zoonotic transmission to humans can happen. According to a 2015 WHO report, the greatest numbers of new cases have been reported from Brazil, India, and Southeast Asia. The five drugs dapsone, rifampicin, clofazimine, ofloxacin, and minocycline make up an efficient multidrug treatment (MDT). Streptomycin, clarithromycin, fluroquinolones, and ethionamide are among the other beneficial medications (Matsuoka 2010). There have been reports of the emergence of resistant strains from around the world, which raises serious concern about the future viability of this successful treatment regimen (Matsuoka 2010; Guerrero et al. 2014; da Silva Rocha et al. 2012; Maeda et al. 2001; Kai et al. 2011; Williams et al. 2013).

Diseases caused by nontuberculous mycobacteria: Buruli ulcer, originated from Uganda and ranks third among human mycobacteriosis worldwide after tuberculosis and leprosy, is caused by nontuberculous *M. ulcerans*. Clinical manifestation includes non-ulcerated and ulcerated lesions on skin and subcutaneous tissues (Silva et al. 2009). The standard recommended antibiotic treatment guideline by WHO consists of oral rifampicin at 10 mg/kg body weight daily for 8 weeks and intramuscular streptomycin injection at 15 mg/kg body weight by daily for 8 weeks. Four weeks of rifampicin–streptomycin followed by 4 weeks of rifampicin–clarithromycin is also acceptable for recurrence-free healing (WHO 2012). Development of drug resistance is neither different in case of *M. ulcerans* as reported from different foci of the endemic areas (Owusu et al. 2018; Gupta et al. 2017; Marsollier et al. 2003; Jansson et al. 2014). Moreover, Buruli ulcer and HIV co-infection is an emerging area of concern and requires critical monitoring along with effective future control strategies (WHO 2012).

Mycobacterium avium complex (MAC) is comprised of two closely related intracellular bacteria *M. avium* and *M. intracellulare* that survive and multiply inside macrophages producing nontuberculous mycobacteria (NTM)-based disease. The pulmonary system including the lungs is the primary site of infection to produce inflammation in respiratory organs along with either lung nodules or cavities in the lobes. Until the 1980s, MAC infections were uncommon in humans mostly reported in elderly patients with chronic pulmonary disorders like silicosis, emphysema, cystic fibrosis, or even previous TB and immune-compromised leukemic patients. However, significant morbidity and mortality has been reported by MAC in human immunodeficiency virus (HIV)-positive patients during the subsequent period (Inderlied et al. 1993; Karakousis et al. 2004; Ishikane and Tanuma 2014; Scholar 2007; Hadad et al. 1995; Reed et al. 2006; Hayashi et al. 2012; Kingshuk et al. 2016; Collins et al. 2017). Recommended antibiotic therapy of MAC lung disease comprises of macrolide together with ethambutol and rifamycin (Griffith et al. 2007). However, MAC is blamed to be notoriously drug resistant mainly due to innate or "natural" drug resistance and acquired mutational resistance which render them extremely difficult to treat with unpredictable therapeutic response (Griffith 2018; Doucet-Populaire et al. 1995; Griffith 2016; Bose et al. 2007; Morimoto et al. 2016; Field et al. 2004; Lam et al. 2006; Kobashi et al. 2007; Oh et al. 2012). Although treatment with some advanced antibiotics like moxifloxacin has claimed to improve the outcome against MAC (Koh et al. 2013), newer therapeutic options with distinct mechanism of action may provide long-term solution.

The other nontuberculous mycobacterial species include *M. abscessus, M. chelonae, M. fortuitum, M. neoaurum, M. bacteremicum, M. kansasii, M. marinum, M. simiae, M. xenopi, M. parascrofulaceum, M. massiliense,* and *M. gordonae*, and most of these infections are associated with chronic respiratory manifestation or wound infection or cutaneous lesion. Effective treatment with different antibiotics mostly targeting cell wall components, protein, or nucleic acid biosynthesis can produce improvement (Cohen et al. 2012). However, the innate and acquired mechanism similar to MAC enables them to achieve resistance, and several evidences of different degree of resistance against most of the commonly used antimycobacterial agents like isoniazid, rifampin, ethambutol, streptomycin, capreomycin, amikacin, protionamide, para-aminosalicylic acid, ofloxacin, levofloxacin, ciprofloxacin, imipenem, clarithromycin, and linezolid has been depicted over the recent past necessitating the quest for newer effective future control strategies (Cohen et al. 2012; Wang et al. 2014; Jayasingam et al. 2017; Li et al. 2017; Litvinov et al. 2018).

Tularemia: The gram-negative pleomorphic coccobacillus bacterium *Francisella tularensis* is the cause of this adaptable zoonotic disease, which has the potential to be utilized as a weapon of bioterrorism. Although wild lagomorphs and rodents are natural subject, it can also harm more than 200 species of domestic and wild mammals including humans (Stidham et al. 2018). Although the disease is mainly found in North America, infections caused by some subspecies have also been documented in Europe, Asia, and Australia. Tick and deer fly bites, skin contact with infected animals, consumption of contaminated water, aerosol infection, etc., are all ways that the disease can be transmitted to humans. High fever, lymphadenitis, ulcerative skin lesions, exudative pharyngitis and stomatitis, conjunctivitis, gastroenteritis, respiratory indications or pneumonia, sepsis, etc., are some of the symptoms depending on the different forms (Veterinary Manual 2018). There is presently no advised vaccine available against the illness. Streptomycin, gentamicin, doxycycline, and ciprofloxacin are some of the antibiotics used in treatment of the disease (CDC 2018j). The organism has also been shown to be susceptible to the majority of other antibiotics, and no substantial reports of antibiotic resistance in the microorganism have been made so far (Caspar and Maurin 2017; Tomaso et al. 2017). Since antibiotic therapy has not been changed, nanomedicine can be used in the other elements of illness management.

Yersiniosis: It is a contagious illness brought on by the gram-negative bacillus *Yersinia enterocolitica* (the most prevalent strain) and the *Y. pseudotuberculosis* (less often occurring). Pigs are the main reservoir; however, *Y. enterocolitica* pathogenic strains can also be found in rodents, rabbits, sheep, cattle, horses, dogs, and cats that can affect humans (CDC 2018a,k). Young children affected with *Y. enterocolitica* typically experience fever, diarrhea with blood and mucus, and pseudoappendicitis syndrome with fever and leukocytosis in adults. In contrast, *Y. pseudotuberculosis* symptoms include fever, a scarlatiniform rash, abdominal pain, and acute pseudoappendicitis. Serotypes 8, 13, 18, 20, and 21 of *Y. enterocolitica* are the most prevalent in the USA, while serotypes 3, 5, 9, and 27 have been documented in Europe. Several pandemic outbreaks have been brought about by serotype 8 (Carniel and Mollaret 1990; Rahman et al. 2011). It is also the fourth most frequent cause of acute bacterial enteritis in Norway (MacDonald et al. 2012). No preventive vaccine for the illness is currently available, and only sanitary practices may help to avoid the disease. Most cases of the illness are self-limiting, but severe cases may require antibiotic therapy with trimethoprim–sulfamethoxazole, aminoglycosides, third-generation cephalosporins, fluoroquinolones, and tetracyclines (CDC 2018a,k). Nevertheless, resistance to first-generation cephalosporins and the majority of penicillins as well as other antibiotic classes has been documented (Fàbrega et al. 2015; Grosdent et al. 2002).

19.3 Nanotherapeutics

Several nanoformulations have already been recognized for their bactericidal action that can be used alone or in conjunction with conventional antibiotics to treat microbiological infections. Numerous nano-scale materials have customizable size, shape, and surface chemistry, chemical stability, high surface to mass ratios, enhanced reactivity, biosafety, and biocompatibility, which provide them inimitable mechanisms of cidal actions that have been skillfully exploited in nanotherapeutics to elude drug resistance and biofilm-producing microbes. These nano-materials possess a variety of antimicrobial mechanisms, including direct damage to bacterial cell membranes, discharge of toxic metal ions from metallic or metal oxide nanoparticles to stop bacterial protein and enzyme complexes, disruption of bacterial electron transport, protein oxidation, and membrane collapse through the surface positive charge of the NPs, production of reactive oxygen species (ROS) for photothermal damage, oxidation of bacterial lipids and proteins, and repression of bacterial enzymes (Razei et al. 2017). In order to create nanotherapeutic approaches against microbial superbugs, a variety of metallic or metal oxide nanomaterials, including iron oxide (Fe_3O_4), zinc oxide (ZnO), titanium oxide (TiO), and gold and silver nanoparticles, as well as organic nanoparticles or their composites are commonly used (Kulshreshtha et al. 2017a,b; Baptista et al. 2018). In addition to direct bactericidal activity, nanocarriers are endowed with an admirable therapeutic control via specific targeting of existing antibiotics to the infection sites and sustained release of the drug to keep the minimum inhibitory concentration (MIC) for extended period of time along with avoiding the potent side effects (Bermúdez et al. 2017). This can also abet to restrict the extended course of antibiotic therapy for certain intracellular and resistant microbes like *Mycobacterium* (Xu et al. 2017). Moreover, resistance because of poor penetration-mediated underdosing of free drug therapy can also be addressed via NP-based drug delivery system (Bermúdez et al. 2017). The diverse range of nanocarriers includes but not limited to liposome, niosome, micelle, nano-emulsions, solid lipid nanoparticles and nanoconstructed lipid carriers, natural and synthetic polymer-based systems, dendrimers, and different nano-structures made up of organic and inorganic materials. These nanomaterials can suit a variety of demands for enhanced drug delivery systems because of their extensive drug encapsulating capacity and adaptable surface conjugation capabilities (Bermúdez et al. 2017). To create a customized "active" delivery method that balances in vivo stability and controlled release of the therapeutic drug at the target region, the intensity of interaction between the nano-carrier and the conjugated drug moieties must be carefully considered (Burygin et al. 2009; Pissuwan et al. 2011). Table 19.1 enlisted the most recent developments about nanotherapeutic applications against various infectious and communicable microbial diseases. Thus, it is apparent from the previous discussion that nanomedicine carries ample potential to address the complementary issues of diagnosis and therapy in a comprehensive way to actuate stringent nanotheranostic approach for skirmishing contagious diseases.

19.4 Nano-Antimicrobials Approved or Under Development and Future Prospect

By 2024, the global market for nanomedicine is expected to be worth over US$ 200 billion, with several nanoparticle drugs already available on the market and more than 250 formulations either in various stages of testing or on the verge of commercialization (Prasad et al. 2021). The majority of the newly developed

Table 19.1 Nano-antibiotics with drugs delivery systems and its effects against zoonotic and reverse-zoonotic bacterial infections in different models

Nano-drug	Vehicle/Action	Species	Outcome	Reference
Glutathione-stabilized silver nanoparticles (GSH-Ag NPs)	Antimicrobial	Multidrug-resistant (MDR) *Campylobacter* strains	Exposure of GSH-Ag NPs to intestinal cells showed a dose-dependent cytotoxic effect in all cell lines between 9.85 and 39.4 μg/mL	Silvan et al. (2018)
Gamma interferon (IFN-γ)	Albumin nanoparticles	*Brucella abortus*	NP-adsorbed IFN-γ shows additional activity than the free cytokine against brucella in vitro and in vivo	Segura et al. (2007)
Ulva fasciata and *Caulerpa taxifolia* with Ag NP	–	*C. perfringens, L. monocytogenes, Mycobacterium* spp., *S. aureus, Vibrio vulnificus, Plesiomonas shigelloides, Edwardsiella tarda, Delftia acidovorans*	The silver nanoparticles of *Ulothrix flacca*, *Ulva fasciata*, and *Caulerpa taxifolia* showed the zone of inhibition ranges from 0.7 ± 0.6 to 26.3 ± 0.6 mm	Rao and Boominathan (2015)
ZnO NP	Bactericidal	*C. jejuni*	Potent bactericidal agent, MIC level 0.05 to 0.025 mg/mL	Xie et al. (2011)
MgO nanoparticle	Bactericidal	*E. coli*	Considerable bactericidal activity at 0.1 g/mL	Tabassum et al. (2014)
Cu/Zn mixed oxide	Antimicrobial	*E. coli*	Mixed oxide of Cu and Zn [73:27] exhibits the highest antimicrobial activity, with 90% growth inhibition after 8 h for a dose of 200 μg/mL, whereas CuO reaches the same performance only after 24 h	Carbone et al. (2017)
TiO2 NP	Antibacterial	*E. coli* O111	TiO_2 NPs with anatase crystal structure and smaller particle size generates more intracellular reactive oxygen species and malondialdehyde with greater antibacterial activity	Lin et al. (2014)

Iron oxide NP	Bactericidal	*E. coli* O111	Bacterial cell death caused by intracellular vacuole formation and significantly elevated ROS levels in bacteria.	Li et al. (2018)
Au/Pt bimetallic NPs	Antibacterial	MDR *E. coli*	AuPt NPs are potent antibacterial agents and harmless to human cells. By contrast, pure Au NPs or pure Pt NPs are not antibiotic at all	Zhao et al. (2014)
Fe_3O_4 NP		*E. coli, P. vulgaris. S. aureus, Xanthomonas* sp.	Antibacterial action	Prabhu et al. (2015)
Ag NP with either cefazolin, mupirocin, gentamycin, neomycin, tetracycline, or vancomycin	Antimicrobial	*E. coli., P. aeruginosa, and S. aureus*	Microbial membrane damage to the pathogen by pharmacodynamic interaction of Ag NP with topical antibiotics	Thirumurugan et al. (2016)
Al_2O_3	Bactericidal	*E. coli*	Serve as broad-spectrum bactericidal agent	Ansari et al. (2014)
Au/Ag bimetallic NPs		*E. coli, S. aureus, E. faecalis, P. aeruginosa*	Cell wall disruption, inactivate the proteins and enzymes for ATP production, more effective in inhibiting bacterial biofilm formation	Ramasamy et al. (2016)
ZnO NPs	Antimicrobial	Beta lactamases (ESBL)-positive *E. coli, Pseudomonas aeruginosa,* and methicillin-resistant *S. aureus* (MRSA)	Significant antibacterial activity	Ali et al. (2016)
Au NP combined with ampicillin	Antimicrobial	*E. coli*, *S. aureus*	Au NPs conjugated antibiotics are more efficient and might have significant therapeutic implications	Saha et al. (2007)
Cinnamon bark extract	Poly(DL-lactide-co-glycolide) (PLGA) nanoparticles	*Listeria monocytogenes* and *Salmonella typhimurium*	Effective inhibitors, better retention time up to 24 and 72 h at concentrations ranging from 224.42 to 549.23 μg/mL	Hill et al. (2013)

(continued)

Table 19.1 (continued)

Nano-drug	Vehicle/Action	Species	Outcome	Reference
Quercetin	PLGA	*L. monocytogenes, S. aureus*	Effective against gram-positive strains, bacteria cells lost their integrity and became pale	Arasoğlu et al. (2017)
Curcumin NP	Control of foodborne pathogens	*L. monocytogenes*	Enhanced antimicrobial action mediated by surface charge	No et al. (2017)
Gold nanocomposite, functionalized with antimicrobial peptide, pediocin AcH, and Listeria adhesion protein (LAP)	Antimicrobial	*L. monocytogenes*	11.1% higher zone of inhibition, and higher reduction in biofilm production	Singh et al. (2018)
TiO2 NPs	Antibacterial	*L. monocytogenes*	2.4 mg of nanoparticle-bound ampicillin (three doses of 0.8 mg each) had a greater therapeutic effect than 48 mg of free ampicillin	Youssef et al. (1988)
Amikacin	Liposome	*M. avium*	Better clearance as compared to free drug without any toxic effect	Ehlers et al. (1996)
Kanamycin	Liposome	*M. intracellulare*	Enhanced antimicrobial activity	Tomioka et al. (1991)
Rifampicin	Mannosylated dendrites	*Mycobacteria*	Higher viability, retention time, and negligible cytotoxicity	Kumar et al. (2007)
Fe_3O_4/Au bimetallic NP	Antibacterial	MDR *P. aeruginosa*	Fe_2O_3 nanoparticles functionalized with imipenem can make the imipenem resistant bacterium more susceptible to weaker concentrations of antibiotic	Khataminejad et al. (2015)
Green Fe_2O_3 NP	Antibacterial	*P. aeruginosa*	Strong antibacterial activity	Irshad et al. (2017)

Gallium NP with bioactive hydroxyapatite nanorods	Antibacterial	*P. aeruginosa*	Superior antibacterial activity with less cell toxicity	Kurtjak et al. (2016)
Au NP coated onto poly (ethyleneimine) (PEI) modified glass surfaces	Antibacterial	*S. aureus* and *E. coli*	Cefaclor reduced gold nanoparticles have potent antimicrobial activity against both gram-positive (*Staphylococcus aureus*) and gram-negative (*Escherichia coli*) bacteria as compared to cefaclor or gold nanoparticles alone	Rai et al. (2010)
Platinum NP	Broad-spectrum antibacterial activity	*E. coli*, *Enterobacter cloacae*, *K.. pneumoniae*, *Proteus vulgaris*, *S. aureus*, *Pseudomonas aeruginosa*	Biogenic platinum nanoparticles exhibited broad-spectrum antibacterial activity against coagulase-negative *Staphylococcus* sp. (20.33 ± 1.52), followed by *Proteus vulgaris* (17.66 ± 2.08), *Staphylococcus aureus* (16.33 ± 1.52), *E. cloacae* (14.00 ± 2.64), and *E. coli* (12.33 ± 1.52), while poor antibacterial activity against *P. aeruginosa* (08.33 ± 0.57) and *K. pneumoniae* (07.66 ± 1.52)	Sharma (2017)
NO precursor isosorbide mononitrate (ISMN)	PLGA	*S. aureus*	Targeted biofilm delivery, considerable antibacterial action	Hasan et al. (2017)
Cadmium oxide NP	Bactericidal	*S. aureus*	Inhibit bacterial growth in a dose-dependent manner	Salehi et al. (2014)
RNAi	Fusogenic lipid-coated pSiNPs	*S. aureus*	Enhancement of the clearance capability of macrophages via bypassing endocytosis and achieving potent gene knockdown efficacy	Kim et al. (2018)

(continued)

Table 19.1 (continued)

Nano-drug	Vehicle/Action	Species	Outcome	Reference
Al_2O_3 NPs	Bactericidal	*S. aureus*	Effective bactericidal agents producing adverse effect including cell death	Ansari et al. (2013)
Cu/Ni bimetallic NPs	Antimicrobial	*S. aureus, E. coli, S. mutans*	Adsorption of ions to the bacteria cells. Active against the microorganisms with potential use in dental case	Argueta-Figueroaa et al. (2014)
SeNPs	Antibacterial peptide UBI29–41 (PEP) linked to NP	*S. aureus, E. coli*	Theranostic application for selective imaging and targeted therapy as it disrupts microbial cell wall	Huang et al. (2017)
V. cholerae LPS	Lipopolysaccharide-loaded chitosan NPs	*V. cholerae*	Induction of higher level of antibodies via humoral immune response	Fasihi-Ramandi et al. (2018)
ZnO NPs		*V. cholerae*	10 μg/mL of ZnO NP did not affect the growth rates but block its interaction with the GM1 ganglioside receptor hence reduce uptake without measurable toxic effect	Sarwar et al. (2017)
Monosialotetrahexosylganglioside (GM1), coated onto polymeric NP	PLGA	*V. cholerae*	GM1-polymer hybrid nanoparticles were shown to act as toxin decoys by selectively and stably binding cholera toxin, and neutralizing its activity on epithelial cells in vitro and in vivo	Das et al. (2018)

Ag NP	Citrate and chitosan silver nanoparticle	MDR oxacillin-resistant *Staphylococcus aureus*, vancomycin-resistant *Enterococcus* spp., carbapenem- and polymyxin B-resistant *A. baumannii*, carbapenem-resistant *P. aeruginosa*, and carbapenem-resistant Enterobacteriaceae	In vitro inhibitory effect was stronger against gram-negative than gram-positive bacteria, and similar against multidrug-resistant and susceptible bacteria, with best result achieved using citrate and chitosan silver nanoparticles	Cavassin et al. (2015)
Chlorhexidine	Mesoporous silica nanoparticle	*Streptococcus mutans, Streptococcus sobrinus, Fusobacterium nucleatum, Aggregatibacter actinomycetemcomitans*, and *Enterococcus faecalis*	Decreased biofilm formation and strong antibacterial effects can prevent dental caries	Seneviratne et al. (2014)

Notes: Besides zoonotic importance, different strains of Mycobacterial species, methicillin-resistant *Staphylococcus aureus* (MRSA), *Streptococcus pneumonia, Campylobacter* spp., *Salmonella* spp., *Shigella* spp., *Escherichia coli,* and *Helicobacter pylori* are of reverse-zoonotic importance also nanodrugs or delivery systems are currently focused on oncotherapeutic applications, along with some antiviral applications against influenza, hepatitis B, hepatitis C, Kaposi's sarcoma virus, etc., or antifungal therapeutics using amphotericin b and nystatin, and against chronic inflammatory or immune-mediated diseases like psoriatic arthritis, Crohn's disease, and rheumatoid arthritis (Mukhopadhyay 2018; Ventola 2017). The promising hotspots where extensive research is being conducted are the use of antibacterial nanoparticles in medical equipment, implants, wound dressings, etc. (Hemeg 2017; Wang et al. 2017). With the aid of oral amikacin administration using lipid-crystal NP delivery method, Matinas BioPharma (MTNB) has disclosed positive preclinical efficacy of MAT2501 in the in vitro model of *Mycobacterium abscessus* infection (Matinasbiopharma.com 2018). A different nano-formulation called Rifanano which contains NP-coated versions of the four primary first-line antituberculosis medications has demonstrated enhanced intestinal absorption properties, and human trials for the formulation are also planned (Prasad et al. 2021). In order to treat tuberculosis, researchers from the Council for Scientific and Industrial Research (CSIR) in South Africa created a nano-formulation using biodegradable polymer encapsulated drugs that allows targeted delivery and controlled release of the antituberculosis drugs at the target site while reducing the frequency of dosing of the conventional therapy (Nanotechnology Public Engagement Programme Newsletter, October 2016). For the treatment of cystic fibrosis, liposomal administration of ciprofloxacin by inhalation has been shown to be superior in comparison to free drug therapy (Pita et al. 2018). A patent has been received for a new class of synthetic antibiotics that uses beta-lactams and nano-sized spherical vehicles. These antibiotics have a novel mechanism for treating drug-resistant microbes like MRSA and may be useful for preventing the spread of zoonotic diseases caused by these resistant organisms (ScienceDaily 2018a,b). Metal oxide nanoparticles including zinc and titanium oxide nanoparticles are present in a number of sun creams; their antibacterial properties may prevent the spread of pathogens through contact (Khan et al. 2016). Because of its antimicrobial activities, colloidal silver NPs are also applied commercially to treat skin wounds and dermatitis (Keck and Schwabe 2009). A broad-spectrum tetracycline that comprises of nano-emulsion foam of minocycline, FMX101 is now undergoing a phase III trial for topical use (Nanalyze 2018). Since *S. aureus* and *K. pneumoniae* have been shown to be effectively controlled by the antibiotic and deodorant agent cryson nano-silver, it may be beneficial to manage bacterial infections. A copper oxide nanorod-based artificial enzyme, "nanozymes" has been acknowledged to limit *E. coli* and golden staph infections through ROS generation and thus can be helpful to control nosocomial and aerosol infections (Karim et al. 2018). It has been demonstrated that titanium dioxide powder-coated food packaging film can reduce *E. coli* contamination in packaged foods. Thus, coating of such NPs can help reduce the spread of food-borne bacterial infections (Chawengkijwanich and Hayata 2008).

19.5 Challenges to Clinical Translation of the Nano-Drugs and Conclusion

Despite ample instances of nanoparticles-mediated antibacterial actions explored through academic research, clinical transition rate of the novel nano-drugs is very low as they face a number of challenges (Hua et al. 2018).

Oncotherapeutics development has always remained at the focus, while many metabolic disorders or infectious diseases of bacterial origin are mostly ignored. Uncertain financial prospect may be one of the reasons, as customization of nano-drug is a resource-intensive means (Paliwal et al. 2014; Hare et al. 2017). The cumbersome regulatory approval process for clinical use of the nano-formulations is another major constrain. Moreover, dearth of experimental data regarding the long-term toxicity profile of the nanomaterials is another crucial impediment (Xia et al. 2014; Kulshreshtha et al. 2017a,b). Determination of in vivo dose requirements and safety margins of nanodrugs is frequently ill-defined which may result in improper dosing-mediated resistance against nanoparticles (Larimer et al. 2014; Zhu et al. 2014; Qiu et al. 2012; Hasman and Aarestrup 2004; Wang et al. 2018). Overlapping and inadequately documented clinical applications along with subsequent threat to intellectual property rights is also a serious concern for the nano-pharmaceutical industry (Pelaz et al. 2017). Despite these bottlenecks, nanomedicine has progressed steadily under recent times which definitely shows good prospect in this arena.

References

Abdelbaqi K, Ménard A, Prouzet-Mauleon V, Bringaud F, Lehours P, Mégraud F (2007) Nucleotide sequence of the gyrA gene of Arcobacter species and characterization of human ciprofloxacin-resistant clinical isolates. FEMS Immunol Med Microbiol 49(3):337–345. https://doi.org/10.1111/j.1574-695X.2006.00208.x

Abreu F, Rodríguez-Lucas C, Rodicio M, Vela A, Fernández-Garayzábal J, Leiva P, Cuesta F, Cid D, Fernández J (2018) Human Pasteurella multocida infection with likely zoonotic transmission from a pet dog, Spain. Emerg Infect Dis 24(6):1145–1146. https://doi.org/10.3201/eid2406.171998

Acha P, Szyfres B (2001) Zoonoses and communicable diseases common to man and animals. Pan American Health Organization, Washington, DC. ISBN 92 75 11580 X

Alba C, Blanco A, Alarcón T (2017) Antibiotic resistance in helicobacter pylori. Curr Opin Infect Dis 30(5):489–497. https://doi.org/10.1097/QCO.0000000000000396

Alhashash F, Weston V, Diggle M, McNally A (2013) Multidrug-resistant Escherichia coli bacteremia. Emerg Infect Dis J 19(10):1699–1701. https://doi.org/10.3201/eid1910.130309

Ali K, Dwivedi S, Azam A, Saquib Q, Al-Said MS, Alkhedhairy AA, Musarrat J (2016) Aloe vera extract functionalized zinc oxide nanoparticles as nanoantibiotics against multi-drug resistant clinical bacterial isolates. J Colloid Interface Sci 472:145–156. https://doi.org/10.1016/j.jcis.2016.03.021

Alizadeh H, Salouti M, Shapouri R (2013) Intramacrophage antimicrobial effect of silver nanoparticles against Brucella melitensis 16M. Sci Iranica 20(3):1035–1038. https://doi.org/10.1016/j.scient.2013.02.018

Altekruse S, Stern N, Fields P, Swerdlow D (1999) Campylobacter jejuni—an emerging foodborne pathogen. Emerg Infect Dis 5(1):28–35. https://wwwnc.cdc.gov/eid/article/5/1/99-0104_article

Ansari MA, Khan HM, Khan AA, Pal R, Cameotra SS (2013) Antibacterial potential of Al2O3 nanoparticles against multidrug resistance strains of Staphylococcus aureus isolated from skin exudates. J Nanopart Res 15:1970. https://doi.org/10.1007/s11051-013-1970-1

Ansari MA, Khan HM, Khan AA, Cameotra SS, Saquib Q, Musarrat SJ (2014) Interaction of Al2O3 nanoparticles with Escherichia coli and their cell envelope biomolecules. J Appl Microbiol 116(4):772–783

Arasoğlu T, Derman S, Mansuroğlu B, Uzunoğlu D, Koçyiğit B, Gümüş B, Acar T, Tuncer B (2017) Preparation, characterization, and enhanced antimicrobial activity: quercetin-loaded PLGA nanoparticles against foodborne pathogens. Turk J Biol 41:127–140. https://doi.org/10.3906/biy-1604-80

Argueta-Figueroaa L, Morales-Luckie RA, Scougall-Vilchis RJ, Olea-Mejía OF (2014) Synthesis, characterization and antibacterial activity of copper, nickel and bimetallic Cu–Ni nanoparticles for potential use in dental materials. Prog Nat Sci Mater Int 24(4):321–328

Arvand M, Hoeck M, Hahn H, Wagner J (2000) Antimicrobial resistance in streptococcus pyogenes isolates in Berlin. J Antimicrob Chemother 46(4):621–624. https://doi.org/10.1093/jac/46.4.621

Bahaman A, Bashiru G (2018) Advances and challenges in leptospiral vaccine development. Indian J Med Res 147(1):15. https://doi.org/10.4103/ijmr.IJMR_1022_16

Baird GJ, Fontaine MC (2007) Corynebacterium pseudotuberculosis and its role in ovine caseous lymphadenitis. J Comp Pathol 137:179–210

Baptista P, McCusker M, Carvalho A, Ferreira D, Mohan N, Martins M, Fernandes A (2018) Nano-strategies to fight multidrug resistant bacteria—"A battle of the titans". Front Microbiol 9:1441. https://doi.org/10.3389/fmicb.2018.01441

Barbosa Pauletti R, Reinato Stynen A, da Silva P, Mol J, Seles Dorneles E, Alves T, de Sousa Moura Souto M, Minharro S, Heinemann M, Lage A (2015) Reduced susceptibility to rifampicin and resistance to multiple antimicrobial agents among brucella abortus isolates from cattle in Brazil. PLoS One 10(7):e0132532. https://doi.org/10.1371/journal.pone.0132532

Barraud O, Badell E, Denis F, Guiso N, Ploy M (2011) Antimicrobial drug resistance in corynebacterium diphtheriae mitis. Emerg Infect Dis 17(11):2078–2080. https://doi.org/10.3201/eid1711.110282

Barton Behravesh C, Jones T, Vugia D, Long C, Marcus R, Smith K, Thomas S, Zansky S, Fullerton K, Henao O, Scallan E (2011) Deaths associated with bacterial pathogens transmitted commonly through food: foodborne diseases active surveillance network (FoodNet), 1996–2005. J Infect Dis 204(2):263–267. https://doi.org/10.1093/infdis/jir263

Belay E, Kile J, Hall A, Barton-Behravesh C, Parsons M, Salyer S, Walke H (2017) Zoonotic disease programs for enhancing global health security. Emerg Infect Dis 23(13):S65–S70. https://doi.org/10.3201/eid2313.170544

Belluco S, Losasso C, Patuzzi I, Rigo L, Conficoni D, Gallocchio F, Cibin V, Catellani P, Segato S, Ricci A (2016) Silver as antibacterial toward listeria monocytogenes. Front Microbiol 7:307. https://doi.org/10.3389/fmicb.2016.00307

Berger A, Trcek J, Hogardt M, Hörmansdorfer S, Rau J, Sing A (2018) P678 Abstract: antimicrobial resistance of Corynebacterium ulcerans and C. pseudotuberculosis. Escmid.org. https://www.escmid.org/escmid_publications/escmid_elibrary/material/?mid=2698. Accessed 7 April 2023

Bermúdez J, Cid A, Romero A, Villegas M, Villegas N, Palma S (2017) New trends in the antimicrobial agents delivery using nanoparticles. Antimicrobial Nanoarchitectonics:1–28. https://doi.org/10.1016/B978-0-323-52733-0.00001-X

Bernard K (2012) The genus corynebacterium and other medically relevant coryneform-like bacteria. J Clin Microbiol 50(10):3152–3158. https://doi.org/10.1128/JCM.00796-12

Bharti A, Nally J, Ricaldi J, Matthias M, Diaz M, Lovett M, Levett P, Gilman R, Willig M, Gotuzzo E, Vinetz J (2003) Leptospirosis: a zoonotic disease of global importance. Lancet Infect Dis 3(12):757–771. https://doi.org/10.1016/S1473-3099(03)00830-2

Bhatia A, Zahoor S (2007) Staphylococcus aureus enterotoxins: A review. J Clin Diagn Res 3:188–197

Bitzan M, Poole R, Mehran M, Sicard E, Brockus C, Thuning-Roberson C, Riviere M (2009) Safety and pharmacokinetics of chimeric anti-shiga toxin 1 and anti-shiga toxin 2 monoclonal antibodies in healthy volunteers. Antimicrob Agents Chemother 53(7):3081–3087. https://doi.org/10.1128/AAC.01661-08

Boltin D, Ben-Zvi H, Perets T, Kamenetsky Z, Samra Z, Dickman R, Niv Y (2014) Trends in secondary antibiotic resistance of helicobacter pylori from 2007 to 2014: has the tide turned? J Clin Microbiol 53(2):522–527. https://doi.org/10.1128/JCM.03001-14

Bose M, Venugopal D, Kumar S, Isa M (2007) Drug resistance profile of human Mycobacterium avium complex strains from India. Indian J Med Microbiol 25(2):115. https://doi.org/10.4103/0255-0857.32716

Braderic N, Punda-Polic V (1992) Cutaneous anthrax due to penicillin-resistant bacillus anthracis transmitted by an insect bite. Lancet 340:306–307

Burygin G, Khlebtsov B, Shantrokha A, Dykman L, Bogatyrev V, Khlebtsov N (2009) On the enhanced antibacterial activity of antibiotics mixed with gold nanoparticles. Nanoscale Res Lett 4(8):794–801. https://doi.org/10.1007/s11671-009-9316-8

Camara M, Dieng A, Boye C (2013) Antibiotic susceptibility of streptococcus pyogenes isolated from respiratory tract infections in Dakar, Senegal. Microbiol Insights 6:71–75. https://doi.org/10.4137/MBI.S12996

Cantas L, Suer K (2014) Review: the important bacterial zoonoses in "one health" concept. Front Public Health 2:144. https://doi.org/10.3389/fpubh.2014.00144

Carbone M, Briancesco R, Bonadonna L (2017) Antimicrobial power of Cu/Zn mixed oxide nanoparticles to Escherichia coli. Environ Nanotechnol Monitor Manag 7:97–102. https://doi.org/10.1016/j.enmm.2017.01.005

Carniel E, Mollaret HH (1990) Yersiniosis. Comparat Immunol Microbiol Infect Dis 13(2):51–58

Caspar, Y. and Maurin, M. (2017). Francisella tularensis susceptibility to antibiotics: A comprehensive review of the data obtained in vitro and in animal models. Front Cell Infect Microbiol, 7: 122. Doi:https://doi.org/10.3389/fcimb.2017.00122

Cassini A, Colzani E, Pini A, Mangen M, Plass D, McDonald S, Maringhini G, van Lier A, Haagsma J, Havelaar A, Kramarz P, Kretzschmar M (2018) Impact of infectious diseases on population health using incidence-based disability-adjusted life years (DALYs): results from the Burden of communicable diseases in Europe study, European Union and European economic area countries, 2009 to 2013. Eur Secur 23(16):17-00454. https://doi.org/10.2807/1560-7917.ES.2018.23.16.17-00454

Cavassin ED, de Figueiredo LF, Otoch JP, Seckler MM, de Oliveira RA, Franco FF, Marangoni VS, Zucolotto V, Levin AS, Costa SF (2015) Comparison of methods to detect the in vitro activity of silver nanoparticles (AgNP) against multidrug resistant bacteria. J Nanobiotechnol 13:64. https://doi.org/10.1186/s12951-015-0120-6

Center for Disease Control and Prevention (CDC) (1998) Helicobacter pylori: Fact sheet for health care providers. https://www.cdc.gov/ulcer/files/hpfacts.pdf. Accessed 3 April 2023

Center for Disease Control and Prevention (CDC) (2018a) Yersinia enterocolitica (Yersiniosis). https://www.cdc.gov/yersinia/index.html. Accessed 4 April 2023

Center for Disease Control and Prevention (CDC) (2018b) Clostridium perfringens food poisoning. https://www.cdc.gov/foodsafety/diseases/clostridium-perfringens.html. Accessed 1 April 2023

Center for Disease Control and Prevention (CDC) (2018c) Diphtheria. https://www.cdc.gov/diphtheria/clinicians.html. Accessed 5 April 2023

Center for Disease Control and Prevention (CDC) (2018d) Prevent listeria. https://www.cdc.gov/listeria/prevention.html. Accessed 4 April 2023

Center for Disease Control and Prevention (CDC) (2018e) Educational materials: lyme disease. https://www.cdc.gov/lyme/toolkit/index.html. Accessed 5 April 2023

Center for Disease Control and Prevention (CDC) (2018f) Glanders prevention. https://www.cdc.gov/glanders/prevention/index.html. Accessed 6 April 2023

Center for Disease Control and Prevention (CDC) (2018g) Leptospirosis: fact sheet for clinicians. https://www.cdc.gov/leptospirosis/pdf/fs-leptospirosis-clinicians-eng-508.pdf. Accessed 7 April 2023

Center for Disease Control and Prevention (CDC) (2018h) Melioidosis treatment. https://www.cdc.gov/melioidosis/treatment/index.html. Accessed 7 April 2023

Center for Disease Control and Prevention (CDC) (2018i) National enteric disease surveillance: Shiga toxin-producing Escherichia coli (STEC) annual report, 2015. https://www.cdc.gov/nationalsurveillance/pdfs/STEC_Annual_Summary_2015-508c.pdf. Accessed 7 April 2023

Center for Disease Control and Prevention (CDC) (2018j) Tularemia. https://www.cdc.gov/tularemia/index.html. Accessed 8 April 2023

Center for Disease Control and Prevention (CDC) (2018k) Yersinia enterocolitica (Yersiniosis): Prevention. https://www.cdc.gov/yersinia/prevention.html. Accessed 6 April 2023

Center for Disease Control and Prevention (CDC) (2018l) CDC Global health—infographics: antibiotic resistance the global threat. https://www.cdc.gov/globalhealth/infographics/antibiotic-resistance/antibiotic_resistance_global_threat.htm. Accessed 3 April 2023

Chakraborty A, Miyahara S, Villanueva S, Gloriani N, Yoshida S (2010) In vitro sensitivity and resistance of 46 Leptospira strains isolated from rats in The Philippines to 14 antimicrobial agents. Antimicrob Agents Chemother 54(12):5403–5405. https://doi.org/10.1128/AAC.00973-10

Chambers H, DeLeo F (2009) Waves of resistance: Staphylococcus aureus in the antibiotic era. Nat Rev Microbiol 7(9):629–641. https://doi.org/10.1038/nrmicro2200

Charpentier E, Courvalin P (1999) Antibiotic resistance in Listeria spp. Antimicrob Agents Chemother 43(9):2103–2108

Chatterjee T, Chatterjee B, Chakrabarti P (2017) Modelling of growth kinetics of vibrio cholerae in presence of gold nanoparticles: effect of size and morphology. Sci Rep 7(1):9671. https://doi.org/10.1038/s41598-017-09357-0

Chawengkijwanich C, Hayata Y (2008) Development of TiO2 powder-coated food packaging film and its ability to inactivate Escherichia coli in vitro and in actual tests. Int J Food Microbiol 123(3):288–292. https://doi.org/10.1016/j.ijfoodmicro.2007.12.017

Chin J (ed) (2000) Control of communicable diseases manual, 17th edn. Washington, DC, American Public Health Association, p 624

Chomel B (2014) Emerging and re-emerging zoonoses of dogs and cats. Animals 4(3):434–445. https://doi.org/10.3390/ani4030434

Chugh TD (2008) Emerging and re-emerging bacterial diseases in India. J Biosci 33:549–555. https://doi.org/10.1007/s12038-008-0073-0

Clatworthy A, Pierson E, Hung D (2007) Targeting virulence: A new paradigm for antimicrobial therapy. Nat Chem Biol 3(9):541–548

Cohen T, van Helden P, Wilson D, Colijn C, McLaughlin M, Abubakar I, Warren R (2012) Mixed-strain mycobacterium tuberculosis infections and the implications for tuberculosis treatment and control. Clin Microbiol Rev 25(4):708–719. https://doi.org/10.1128/CMR.00021-12

Collignon P (2009) Resistant Escherichia coli—we are what we eat. Clin Infect Dis 49(2):202–204. https://doi.org/10.1086/599831

Collins LF, Clement ME, Stout JE (2017) Incidence, long-term outcomes, and healthcare utilization of patients with human immunodeficiency virus/acquired immune deficiency syndrome and disseminated Mycobacterium avium complex from 1992-2015. Open Forum Infect Dis 4(3): ofx120. https://doi.org/10.1093/ofid/ofx120

Corbel M (2006) Brucellosis in humans and animals. World Health Organization, Geneva. http://www.who.int/csr/resources/publications/Brucellosis.pdf

da Silva Rocha A, Cunha M, Diniz LM, Salgado C, Aires MA, Nery JA, Gallo EN, Miranda A, Magnanini MM, Matsuoka M, Sarno EN, Suffys PN, de Oliveira ML (2012) Drug and multidrug resistance among mycobacterium leprae isolates from Brazilian relapsed leprosy patients. J Clin Microbiol 50(6):1912–1927. https://doi.org/10.1128/JCM.06561-11. https://www.ncbi.nlm.nih.gov/pmc/articles/PMC3372169/

Dance D, Wuthiekanun V, Chaowagul W, White N (1989) The antimicrobial susceptibility of pseudomonas pseudomallei. Emergence of resistance in vitro and during treatment. J Antimicrob Chemother 24(3):295–309. https://doi.org/10.1093/jac/24.3.295

Das S, Angsantikul P, Le C, Bao D, Miyamoto Y, Gao W, Zhang L, Eckmann L (2018) Neutralization of cholera toxin with nanoparticle decoys for treatment of cholera. PLoS Negl Trop Dis 12(2):e0006266. https://doi.org/10.1371/journal.pntd.0006266

David M, Daum R (2010) Community-associated methicillin-resistant Staphylococcus aureus: epidemiology and clinical consequences of an emerging epidemic. Clin Microbiol Rev 23(3): 616–687. https://doi.org/10.1128/CMR.00081-09

de Zoete MR, van Putten JPM, Wagenaar JA (2007) Vaccination of chickens against campylobacter. Vaccine 25(30):5548–5557

Dewan K, Skarlupka A, Rivera I, Cuff L, Gestal M, Taylor-Mulneix D, Wagner S, Ryman V, Rodriguez C, Hamidou Soumana I, Levin B, Harvill E (2018) Development of macrolide resistance in Bordetella bronchiseptica is associated with the loss of virulence. J Antimicrob Chemother 73(10):2797–2805. https://doi.org/10.1093/jac/dky264

Dhama K, Karthik K, Tiwari R, Shabbir M, Barbuddhe S, Malik S, Singh R (2015) Listeriosis in animals, its public health significance (food-borne zoonosis) and advances in diagnosis and control: a comprehensive review. Vet Q 35(4):211–235. https://doi.org/10.1080/01652176.2015.1063023

Dias AA, Santos LS, Sabbadini PS, Santos CS, Silva Junior FC, Napoleão F, Nagao PE, Villas-Bôas MH, Hirata Junior R, Guaraldi AL (2011) Corynebacterium ulcerans diphtheria: an emerging zoonosis in Brazil and worldwide. Rev Saude Publica 45(6):1176–1191. http://www.scielo.br/pdf/rsp/v45n6/en_2848.pdf

Doucet-Populaire F, Truffot-Pernot C, Grosset J, Jarlier V (1995) Acquired resistance in Mycobacterium avium complex strains isolated from AIDS patients and beige mice during treatment with clarithromycin. J Antimicrob Chemother 36(1):129–136

Dutta D, Chowdhury G, Pazhani G, Guin S, Dutta S, Ghosh S, Rajendran K, Nandy R, Mukhopadhyay A, Bhattacharya M, Mitra U, Takeda Y, Nair G, Ramamurthy T (2013) Vibrio choleraeNon-O1, non-O139 serogroups and cholera-like diarrhea, Kolkata, India. Emerg Infect Dis 19(3):464–467. https://doi.org/10.3201/eid1903.121156

Ehlers S, Bucke W, Leitzke S, Fortmann L, Smith D, Hansch H, Hahn H, Bancroff G, Müller R (1996) Liposomal amikacin for treatment of M. Avium infections in clinically relevant experimental settings. Zentralbl Bakteriol 284:218–231

Elliott S (2007) Rat bite fever and streptobacillus moniliformis. Clin Microbiol Rev 20(1):13–22. https://doi.org/10.1128/CMR.00016-06

Escolar C, Gómez D, del Carmen Rota García M, Conchello P, Herrera A (2017) Antimicrobial resistance profiles of listeria monocytogenes and listeria innocua isolated from ready-to-eat products of animal origin in Spain. Foodborne Pathog Dis 14(6):357–363. https://doi.org/10.1089/fpd.2016.2248

Every A, Selwood L, Castano-Rodriguez N, Lu W, Windsor H, Wee J, Swierczak A, Marshall B, Kaakoush N, Mitchell H, Sutton P (2011) Did transmission of helicobacter pylori from humans cause a disease outbreak in a colony of stripe-faced dunnarts (Sminthopsis macroura)? Vet Res 42(1):26. https://doi.org/10.1186/1297-9716-42-26

Ewers C, Grobbel M, Stamm I, Kopp P, Diehl I, Semmler T, Fruth A, Beutlich J, Guerra B, Wieler L, Guenther S (2010) Emergence of human pandemic O25:H4-ST131 CTX-M-15 extended-spectrum-β-lactamase-producing Escherichia coli among companion animals. J Antimicrob Chemother 65(4):651–660. https://doi.org/10.1093/jac/dkq004

Fàbrega A, Ballesté-Delpierre C, Vila J (2015) Antimicrobial Resistance in Yersinia enterocolitica. Antimicrobial Resistance and Food Safety, pp. 77–104. https://doi.org/10.1016/B978-0-12-801214-7.00005-3

Fasihi-Ramandi M, Ghobadi-Ghadikolaee H, Ahmadi-Renani S, Taheri RA, Ahmadi K (2018) Vibrio cholerae lipopolysaccharide loaded chitosan nanoparticle could save life by induction of specific immunoglobulin isotype. Artif Cells Nanomed Biotechnol 46(1):56–61

Fidalgo SG, Longbottom CJ, Rjley TV (2002) Susceptibility of Erysipelothrix rhusiopathiae to antimicrobial agents and home disinfectants. Pathology 34(5):462–465

Field SK, Fisher D, Cowie RL (2004) Mycobacterium avium complex pulmonary disease in patients without HIV infection. Chest 126:566–581

Fleisher GR (1999) The management of bite wounds. N Engl J Med 340:138–140

Fong I (2017) Zoonotic streptococci: A focus on Streptococcus suis. Emerging Zoonoses, pp.189–210. https://doi.org/10.1007/978-3-319-50890-0_10

Franco MP, Mulder M, Gilman RH, Smits HL (2007) Human brucellosis. Lancet Infect Dis 7:775–786

Fulde M, Valentin-Weigand P (2012) Epidemiology and pathogenicity of zoonotic streptococci. Host-Pathogen Interactions in Streptococcal Diseases, pp. 49–81. https://doi.org/10.1007/82_2012_277

Funke G, von Graevenitz A, Clarridge JIII, Bernard KA (1997) Clinical microbiology of coryneform bacteria. Clin Microbiol Rev 10:125–159

Galimand M, Carniel E, Courvalin P (2006) Resistance of Yersinia pestis to antimicrobial agents. Antimicrob Agents Chemother 50(10):3233–3236. https://doi.org/10.1128/AAC.00306-06

Gerber P, MacLeod A, Opriessnig T (2018) Erysipelothrix rhusiopathiae serotype 15 associated with recurring pig erysipelas outbreaks. Vet Rec 182(22):635–635. https://doi.org/10.1136/vr.104421

Ghosh IN, Patil SD, Sharma TK, Srivastava SK, Pathania R, Navani NK (2013) Synergistic action of cinnamaldehyde with silver nanoparticles against spore-forming bacteria: a case for judicious use of silver nanoparticles for antibacterial applications. Int J Nanomedicine 8:4721–4731. https://doi.org/10.2147/IJN.S49649

Gillespie JH, Timoney JF (1981) Hagan and Bruner's infectious diseases of domestic animals, 7th edn. Comstock, Ithaca

Glaser C, Lewis P, Wong S (2000) Pet-, animal-, and vector-borne infections. Pediatr Rev 21(7): 219–232. https://doi.org/10.1542/pir.21-7-219

Godfrey AJ, Wong S, Dance DA, Chaowagul W, Bryan LE (1991) Pseudomonas pseudomallei resistance to beta-lactam antibiotics due to alterations in the chromosomally encoded beta-lactamase. Antimicrob Agents Chemother 35(8):1635–1640

Goldwater P, Bettelheim K (2012) Treatment of enterohemorrhagic Escherichia coli (EHEC) infection and hemolytic uremic syndrome (HUS). BMC Med 10(1). https://doi.org/10.1186/1741-7015-10-12

Gould K (2016) Antibiotics: from prehistory to the present day. J Antimicrob Chemother 71(3): 572–575

GOV.UK (2018) Pasteurellosis: characteristics, diagnosis and management. https://www.gov.uk/guidance/pasteurellosis. Accessed 5 April 2023

Granier S, Moubareck C, Colaneri C, Lemire A, Roussel S, Dao T, Courvalin P, Brisabois A (2011) Antimicrobial resistance of listeria monocytogenes isolates from food and the environment in France over a 10-year period. Appl Environ Microbiol 77(8):2788–2790. https://doi.org/10.1128/AEM.01381-10

Griffith DE (2016) Macrolide-resistant Mycobacterium avium complex: "I feel like I've been here before". Ann Am Thorac Soc 13(11):1881–1882. https://doi.org/10.1513/AnnalsATS.201609-666ED

Griffith DE (2018) Treatment of mycobacterium avium complex (MAC). Semin Respir Crit Care Med 39(3):351–361. https://doi.org/10.1055/s-0038-1660472

Griffith DE, Aksamit T, Brown-Elliott BA, Catanzaro A, Daley C, Gordin F, Holland SM, Horsburgh R, Huitt G, Iademarco MF, Iseman M, Olivier K, Ruoss S, von Reyn CF, Wallace RJ Jr, Winthrop K, ATS Mycobacterial Diseases Subcommittee; American Thoracic Society; Infectious Disease Society of America (2007) An official ATS/IDSA statement: diagnosis, treatment, and prevention of nontuberculous mycobacterial diseases. Am J Respir Crit Care Med 175(4):367–416

Grosdent N, Maridonneau-Parini I, Sory MP, Cornelis GR (2002) Role of Yops and adhesins in resistance of yersinia enterocolitica to phagocytosis. Infect Immun 70:4165–4176. https://doi.org/10.1128/IAI.70.8.4165-4176.2002

Guernier V, Goarant C, Benschop J, Lau C (2018) A systematic review of human and animal leptospirosis in the Pacific Islands reveals pathogen and reservoir diversity. PLoS Negl Trop Dis 12(5):e0006503. https://doi.org/10.1371/journal.pntd.0006503

Guerrero MI, Colorado CL, Torres JF, León CI (2014) Is drug-resistant mycobacterium leprae a real cause for concern?: First approach to molecular monitoring of multibacillary Colombian patients with and without previous leprosy treatment. Biomedica 34(Suppl 1):137–147. https://doi.org/10.1590/S0120-41572014000500016

Guillet C, Join-Lambert O, Le Monnier A, Leclercq A, Mechaï F, Mamzer-Bruneel M, Bielecka M, Scortti M, Disson O, Berche P, Vazquez-Boland J, Lortholary O, Lecuit M (2010) Human listeriosis caused by Listeria ivanovii. Emerg Infect Dis 16(1):136–138. https://doi.org/10.3201/eid1601.091155

Gupta SK, Drancourt M, Rolain JM (2017) In silico prediction of antibiotic resistance in mycobacterium ulcerans Agy99 through whole genome sequence analysis. Am J Trop Med Hyg 97(3):810–814. https://doi.org/10.4269/ajtmh.16-0478

Gwida M, Al Dahouk S, Melzer F, Rosler U, Neubauer H, Tomaso H (2010) Brucellosis: regionally emerging zoonotic disease? Croat Med J 51:289–295

Hadad D, Palhares M, Placco A, Domingues C, Castelo Filho A, Ferrazoli L, Ueki S, Telles M, Martins M, Palaci M (1995) Mycobacterium avium complex (MAC) isolated from AIDS patients and the criteria required for its implication in disease. Rev Inst Med Trop Sao Paulo 37(5):375–383. https://doi.org/10.1590/S0036-46651995000500001

Hahn WO, Werth BJ, Butler-Wu SM, Rakita RM (2016) Multidrug-resistant Corynebacterium striatum associated with increased use of parenteral antimicrobial drugs. Emerg Infect Dis 22(11):1908–1914. https://doi.org/10.3201/eid2211.160141

Han Y, King J, Janes M (2018) Detection of antibiotic resistance toxigenic Clostridium difficile in processed retail lettuce. Food Qual Saf 2(1):37–41. https://doi.org/10.1093/fqsafe/fyx032

Hare J, Lammers T, Ashford M, Puri S, Storm G, Barry S (2017) Challenges and strategies in anticancer nanomedicine development: an industry perspective. Adv Drug Deliv Rev 108:25–38. https://doi.org/10.1016/j.addr.2016.04.025

Hasan S, Thomas N, Thierry B, Prestidge CA (2017) Biodegradable nitric oxide precursor-loaded micro- and nanoparticles for the treatment of Staphylococcus aureus biofilms. J Mater Chem B 5:1005–1014. https://doi.org/10.1039/c6tb03290g

Hasman H, Aarestrup F (2004) Relationship between copper, glycopeptide, and macrolide resistance among enterococcus faecium strains isolated from pigs in Denmark between 1997 and 2003. Antimicrob Agents Chemother 49(1):454–456. https://doi.org/10.1128/AAC.49.1.454-456.2005

Havelaar A, Haagsma J, Mangen M, Kemmeren J, Verhoef L, Vijgen S, Wilson M, Friesema I, Kortbeek L, van Duynhoven Y, van Pelt W (2012) Disease burden of foodborne pathogens in The Netherlands, 2009. Int J Food Microbiol 156(3):231–238. https://doi.org/10.1016/j.ijfoodmicro.2012.03.029

Hayashi M, Takayanagi N, Kanauchi T, Miyahara Y, Yanagisawa T, Sugita Y (2012) Prognostic factors of 634 HIV-negative patients with Mycobacterium avium complex lung disease. Am J Respir Crit Care Med 185(5):575–583. https://doi.org/10.1164/rccm.201107-1203OC

Hemeg H (2017) Nanomaterials for alternative antibacterial therapy. Int J Nanomedicine 12:8211–8225. https://doi.org/10.2147/IJN.S132163

Hendricks K, Wright M, Shadomy S, Bradley J, Morrow M, Pavia A, Rubinstein E, Holty J, Messonnier N, Smith T, Pesik N, Treadwell T, Bower W (2014) Centers for disease control and prevention expert panel meetings on prevention and treatment of anthrax in adults. Emerg Infect Dis 20(2):e130687. https://doi.org/10.3201/eid2002.130687

Hendriksen S, Orsel K, Wagenaar J, Miko A, van Duijkeren E (2004) Animal-to-human transmission of salmonella typhimurium DT104A variant. Emerg Infect Dis 10(12):2225–2227. https://doi.org/10.3201/eid1012.040286

Hill LE, Taylor TM, Gomes C (2013) Antimicrobial efficacy of poly (DL-lactide-co-glycolide) (PLGA) nanoparticles with entrapped cinnamon bark extract against Listeria monocytogenes and Salmonella typhimurium. Food Sci Nutr 78(4):N626–N632

Hochhut B, Lotfi Y, Mazel D, Faruque S, Woodgate R, Waldor M (2001) Molecular analysis of antibiotic resistance gene clusters in vibrio cholerae O139 and O1 SXT constins. Antimicrob Agents Chemother 45(11):2991–3000. https://doi.org/10.1128/AAC.45.11.2991-3000.2001

Hover B, Kim S, Katz M, Charlop-Powers Z, Owen J, Ternei M, Maniko J, Estrela A, Molina H, Park S, Perlin D, Brady S (2018) Culture-independent discovery of the malacidins as calcium-dependent antibiotics with activity against multidrug-resistant gram-positive pathogens. Nat Microbiol 3(4):415–422

Hsieh J, Chen R, Tsai T, Pan T, Chou C (2008) Phylogenetic analysis of livestock oxacillin-resistant Staphylococcus aureus. Vet Microbiol 126(1–3):234–242. https://doi.org/10.1016/j.vetmic.2007.07.013

Hua S, de Matos M, Metselaar J, Storm G (2018) Current trends and challenges in the clinical translation of nanoparticulate nanomedicines: pathways for translational development and commercialization. Front Pharmacol 9. https://doi.org/10.3389/fphar.2018.00790

Huang J, Yin Y, Pan Z, Zhang G, Zhu A, Liu X, Jiao X (2010) Intranasal immunization with chitosan/pCAGGS-flaA nanoparticles inhibits Campylobacter jejuniin a White Leghorn model. J Biomed Biotechnol 2010:1–8. https://doi.org/10.1155/2010/589476

Huang N, Chen X, Zhu X, Xu M, Liu J (2017) Ruthenium complexes/polypeptide self-assembled nanoparticles for identification of bacterial infection and targeted antibacterial research. Biomaterials 141:296–313. https://doi.org/10.1016/j.biomaterials.2017.07.005

Hunfeld K, Brade V (2006) Antimicrobial susceptibility of borrelia burgdorferi sensu lato: what we know, what we don't know, and what we need to know. Wien Klin Wochenschr 118(21–22):659–668. https://doi.org/10.1007/s00508-006-0693-z

Inderlied CB, Kemper CA, Bermudez LE (1993) The Mycobacterium avium complex. Clin Microbiol Rev 6(3):266–310. https://www.ncbi.nlm.nih.gov/pmc/articles/PMC358286

Iovine N (2013) Resistance mechanisms in Campylobacter jejuni. Virulence 4(3):230–240. https://doi.org/10.4161/viru.23753

Irshad R, Tahir K, Li B, Ahmad A, Siddiqui AR, Nazir S (2017) Antibacterial activity of biochemically capped iron oxide nanoparticles: A view towards green chemistry. J Photochem Photobiol B Biol 170:241–246

Ishikane M, Tanuma J (2014) Mycobacterium avium complex enteritis in HIV-infected patient. IDCases 1(2):22–23. https://doi.org/10.1016/j.idcr.2014.02.004

Jacobs-Reitsma WF, Kan CA, Bolder NM (1996) The induction of quinolone resistance in campylobacter bacteria in broilers by quinolone treatment. In: Newell DG, Ketley JM, Feldman RA (eds) Campylobacters, helicobacters, and related organisms. Plenum Press, New York, NY, pp 307–311

Jamali H, Rezagholipour M, Fallah S, Dadrasnia A, Chelliah S, Velappan R, Wei K, Ismail S (2014) Prevalence, characterization and antibiotic resistance of Pasteurella multocida isolated from bovine respiratory infection. Vet J 202(2):381–383. https://doi.org/10.1016/j.tvjl.2014.07.024

Jansson M, Beissner M, Phillips RO, Badziklou K, Piten E, Maman I, Sarfo FS, Huber KL, Rhomberg A, Symank D, Wagner M, Wiedemann F, Nitschke J, Banla Kere A, Herbinger KH, Adjei O, Löscher T, Bretzel G (2014) Comparison of two assays for molecular determination of rifampin resistance in clinical samples from patients with Buruli ulcer disease. J Clin Microbiol 52(4):1246–1249. https://doi.org/10.1128/JCM.03119-13

Jayasingam SD, Zin T, Ngeow YF (2017) Antibiotic resistance in Mycobacterium Abscessus and Mycobacterium Fortuitum isolates from Malaysian patients. Int J Mycobacteriol 6(4):387–390. https://doi.org/10.4103/ijmy.ijmy_152_17

Kai M, Nguyen Phuc N, Nguyen H, Pham T, Nguyen K, Miyamoto Y, Maeda Y, Fukutomi Y, Nakata N, Matsuoka M, Makino M, Nguyen T (2011) Analysis of drug-resistant strains of mycobacterium leprae in an endemic area of Vietnam. Clin Infect Dis 52(5):e127–e132. https://doi.org/10.1093/cid/ciq217

Karakousis P, Moore R, Chaisson R (2004) Mycobacterium avium complex in patients with HIV infection in the era of highly active antiretroviral therapy. Lancet Infect Dis 4(9):557–565. https://doi.org/10.1016/S1473-3099(04)01130-2

Karim M, Singh M, Weerathunge P, Bian P, Zheng R, Dekiwadia C, Ahmed T, Walia S, Della Gaspera E, Singh S, Ramanathan R, Bansal V (2018) Visible-light-triggered reactive-oxygen-species-mediated antibacterial activity of peroxidase-mimic CuO nanorods. ACS Appl Nano Mater 1(4):1694–1704. https://doi.org/10.1021/acsanm.8b00153

Keck C, Schwabe K (2009) Silver-nanolipid complex for application to atopic dermatitis skin: rheological characterization, in vivo efficiency and theory of action. J Biomed Nanotechnol 5(4):428–436. https://doi.org/10.1166/jbn.2009.1053

Khan S, Musarrat J, Al-Khedhairy A (2016) Countering drug resistance, infectious diseases, and sepsis using metal and metal oxides nanoparticles: current status. Colloids Surf B: Biointerfaces 146:70–83. https://doi.org/10.1016/j.colsurfb.2016.05.046

Khataminejad MR, Mirnejad R, Sharif M, Hashemi M, Sajadi N, Piranfar V (2015) Antimicrobial effect of imipenem functionalized Fe2O3 nanoparticles on pseudomonas aeruginosa producing metallo β-lactamases. Iran J Biotechnol 13(4):43–47

Kim O (2012) Helicobacter - an emerging new zoonotic pathogen. In: Lorenzo-Morales J (ed) Zoonosis. IntechOpen, Rijeka, Croatia. https://www.intechopen.com/books/zoonosis/helicobacter-an-emerging-new-zoonotic-pathogen. Accessed 7 April 2023

Kim J, Coble D, Salyards GW, Bower JK, Rinaldi WJ, Plauche GB, Habing GG (2017) Antimicrobial use and resistance in zoonotic bacteria recovered from nonhuman primates. Comparat Med 67:79–86

Kim, B., Pang, H.B., Kang, J., Park, J.H., Ruoslahti, E. And Sailor, M.J., 2018. Immunogene therapy with fusogenic nanoparticles modulates macrophage response to Staphylococcus aureus. Nat Commun, 9(1), p.1969, DOI: https://doi.org/10.1038/s41467-018-04390-7

Kingshuk D, Prasad SV, Shashidhar V, Mukhyaprana P (2016) Disseminated Mycobacterium avium intracellulare complex (MAC) disease in a retropositive patient caused by noncompliance of HAART. Ann Trop Med Public Health 9(3):194–196. http://www.atmph.org/text.asp?2016/9/3/194/179100

Kitaoka M, Miyata S, Unterweger D, Pukatzki S (2011) Antibiotic resistance mechanisms of vibrio cholerae. J Med Microbiol 60(4):397–407. https://doi.org/10.1099/jmm.0.023051-0

Kobashi Y, Matsushima T, Oka M (2007) A double-blind randomized study of aminoglycoside infusion with combined therapy for pulmonary Mycobacterium avium complex disease. Respir Med 101:130–138

Koh WJ, Hong G, Kim SY, Jeong BH, Park HY, Jeon K, Kwon OJ, Lee SH, Kim CK, Shin SJ (2013) Treatment of refractory Mycobacterium avium complex lung disease with a moxifloxacin-containing regimen. Antimicrob Agents Chemother 57(5):2281–2285. https://doi.org/10.1128/AAC.02281-12

Kulshreshtha N, Jadhav I, Dixit M, Sinha N, Shrivastava D, Bisen P (2017a) Nanostructures as antimicrobial therapeutics. In: Grumezescu A (ed) Antimicrobial nanoarchitectonics: from synthesis to applications. Elsevier, pp 29–59

Kulshreshtha N, Jadhav I, Dixit M, Sinha N, Shrivastava D, Bisen P (2017b) Nanostructures as antimicrobial therapeutics. Antimicrobial Nanoarchitectonics, pp. 29–59. https://doi.org/10.1016/B978-0-323-52733-0.00002-1

Kumar PV, Agashe H, Dutta T, Jain NK (2007) PEGylated dendritic architecture for development of a prolonged drug delivery system for antitubercular drug. Curr Drug Deliv 4:11–19. https://doi.org/10.2174/156720107779314794

Kurtjak M, Vukomanović M, Kramer L, Suvorov D (2016) Biocompatible nano-gallium/hydroxyapatite nanocomposite with antimicrobial activity. J Mater Sci Mater Med 27:170

Kwon N, Park K, Jung W, Youn H, Lee Y, Kim S, Bae W, Lim J, Kim J, Kim J, Hong S, Park Y (2006) Characteristics of methicillin resistant Staphylococcus aureus isolated from chicken meat and hospitalized dogs in Korea and their epidemiological relatedness. Vet Microbiol 117(2–4): 304–312. https://doi.org/10.1016/j.vetmic.2006.05.006

Lam PK, Griffith DE, Aksamit TR, Ruoss SJ, Garay SM, Daley CL, Catanzaro A (2006) Factors related to response to intermittent treatment of Mycobacterium avium complex lung disease. Am J Respir Crit Care Med 173:1283–1289

Larimer C, Islam M, Ojha A, Nettleship I (2014) Mutation of environmental mycobacteria to resist silver nanoparticles also confers resistance to a common antibiotic. Biometals 27(4):695–702

Leedom Larson KR, Spickler AR (2013) Salmonellosis. http://www.cfsph.iastate.edu/DiseaseInfo/factsheets.php. [Accessed 10 April 2023

Li B, Yang S, Chu H, Zhang Z, Liu W, Luo L, Ma W, Xu X (2017) Relationship between antibiotic susceptibility and genotype in mycobacterium abscessus clinical isolates. Front Microbiol 8: 1739. https://doi.org/10.3389/fmicb.2017.01739

Li Y, Yang D, Wang S, Li C, Xue B, Yang L, Shen Z, Jin M, Wang J, Qiu Z (2018) The detailed bactericidal process of ferric oxide nanoparticles on E. coli. Molecules 23(3):E606. https://doi.org/10.3390/molecules23030606

Lin X, Li J, Ma S, Liu G, Yang K, Tong M, Lin D (2014) Toxicity of TiO2 nanoparticles to Escherichia coli: effects of particle size, crystal phase and water chemistry. PLoS One 9(10): e110247. https://doi.org/10.1371/journal.pone.0110247

Lipsitz R, Garges S, Aurigemma R, Baccam P, Blaney DD, Cheng AC, Currie BJ, Dance D, Gee JE, Larsen J, Limmathurotsakul D, Morrow MG, Norton R, O'Mara E, Peacock SJ, Pesik N, Rogers LP, Schweizer HP, Steinmetz I, Tan G, Tan P, Wiersinga WJ, Wuthiekanun V, Smith T (2012) Workshop on treatment of and postexposure prophylaxis for Burkholderia pseudomallei and B. mallei Infection, 2010. Emerg Infect Dis J 18(12):e2. https://doi.org/10.3201/eid1812.120638. https://wwwnc.cdc.gov/eid/article/18/12/12-0638_article

Litvinov V, Makarova M, Galkina K, Khachaturiants E, Krasnova M, Guntupova L, Safonova S (2018) Drug susceptibility testing of slowly growing non-tuberculous mycobacteria using slomyco test-system. PLoS One 13(9):e0203108. https://doi.org/10.1371/journal.pone.0203108

Liu Z, Zhu J, Yang C, Li X (2015) Visual detection of listeria monocytogenes using unmodified gold nanoparticles based on a novel marker. Anal Methods 7(19):8159–8164. https://doi.org/10.1039/C5AY01367D

Livermore DM, Chau PY, Wong AI, Leung YK (1987) Beta-lactamase of Pseudomonas pseudomallei and its contribution to antibiotic resistance. J Antimicrob Chemother 20(3): 313–321. https://doi.org/10.1093/jac/20.3.313

MacDonald E, Heier B, Nygard K, Stalheim T, Cudjoe KS, Skjerdal T, Wester AL, Lindstedt BA, Stavnes TL, Vold L (2012) Yersinia enterocolitica outbreak associated with ready-to-eat salad mix, Norway, 2011. Emerg Infect Dis 18(9):1496–1499. https://doi.org/10.3201/eid1809.120087

Maeda S, Matsuoka M, Nakata N, Kai M, Maeda Y, Hashimoto K, Kimura H, Kobayashi K, Kashiwabara Y (2001) Multidrug resistant mycobacterium leprae from patients with leprosy. Antimicrob Agents Chemother 45(12):3635–3639. https://doi.org/10.1128/AAC.45.12.3635-3639.2001

Magiorakos A, Srinivasan A, Carey R, Carmeli Y, Falagas M, Giske C, Harbarth S, Hindler J, Kahlmeter G, OlssonLiljequist B, Paterson D, Rice L, Stelling J, Struelens M, Vatopoulos A, Weber J, Monnet D (2012) Multidrug resistant, extensively drug-resistant and pandrug-resistant bacteria: an international expert proposal for interim standard definitions for acquired resistance. Clin Microbiol Infect 18(3):268–281

Malik P, Singha H, Khurana SK, Kumar R, Kumar S, Raut AA, Riyesh T, Vaid RK, Virmani N, Singh BK, Pathak SV, Parkale DD, Singh B, Pandey SB, Sharma TR, Chauhan BC, Awasthi V, Jain S, Singh RK (2012) Emergence and re-emergence of glanders in India: a description of outbreaks from 2006 to 2011. Vet Ital 48(2):167–178

Mantur BG, Amarnath SK, Shinde RS (2007) Review of clinical and laboratory features of human brucellosis. Indian J Med Microbiol 25:188–202

Marshall E (1988) Sverdlovsk: anthrax capital? Science 240:383–385

Marsollier L, Honore N, Legras P, Manceau A, Kouakou H, Carbonnelle B, Cole S (2003) Isolation of three mycobacterium ulcerans strains resistant to rifampin after experimental chemotherapy of mice. Antimicrob Agents Chemother 47(4):1228–1232. https://doi.org/10.1128/AAC.47.4.1228-1232.2003

Matinasbiopharma.com (2018) Matinas BioPharma Holdings, Inc. https://www.matinasbiopharma.com. Accessed 8 April 2023

Matsuoka M (2010) Drug resistance in leprosy. Jpn J Infect Dis 63(1):1–7. https://www0.niid.go.jp/JJID/63/1.pdf

Maves R, Castillo R, Guillen A, Espinosa B, Meza R, Espinoza N, Núñez G, Sánchez L, Chacaltana J, Cepeda D, González S, Hall E (2011) Antimicrobial susceptibility of Brucella melitensis Isolates in Peru. Antimicrob Agents Chemother 55(3):1279–1281. https://doi.org/10.1128/AAC.00979-10

Maxson T, Mitchell D (2016) Targeted treatment for bacterial infections: prospects for pathogen-specific antibiotics coupled with rapid diagnostics. Tetrahedron 72(25):3609–3624

Meade E, Slattery MA, Garvey M (2017) Antimicrobial resistance: an agent in zoonotic disease and increased morbidity. J Clin Exp 1(1):30–37

Meinel D, Konrad R, Berger A, König C, Schmidt-Wieland T, Hogardt M, Bischoff H, Ackermann N, Hörmansdorfer S, Krebs S, Blum H, Margos G, Sing A (2015) Zoonotic transmission of toxigenic corynebacterium ulcerans strain, Germany, 2012. Emerg Infect Dis 21(2):356–358. https://wwwnc.cdc.gov/eid/article/21/2/pdfs/14-1160.pdf

Mejías M, Hiriart Y, Lauché C, Fernández-Brando R, Pardo R, Bruballa A, Ramos M, Goldbaum F, Palermo M, Zylberman V (2016) Development of camelid single chain antibodies against Shiga toxin type 2 (Stx2) with therapeutic potential against hemolytic uremic syndrome (HUS). Sci Rep 6(1):24913. https://doi.org/10.1038/srep24913

Messenger A, Barnes A, Gray G (2014) Reverse zoonotic disease transmission (zooanthroponosis): A systematic review of seldom-documented human biological threats to animals. PLoS One 9(2):e89055. https://doi.org/10.1371/journal.pone.0089055

Miwanda B, Moore S, Muyembe J, Nguefack-Tsague G, Kabangwa I, Ndjakani D, Mutreja A, Thomson N, Thefenne H, Garnotel E, Tshapenda G, Kakongo D, Kalambayi G, Piarroux R (2015) Antimicrobial drug resistance of Vibrio cholerae, Democratic Republic of the Congo. Emerg Infect Dis 21(5):847–851. https://doi.org/10.3201/eid2105.141233

Mladenova-Hristova I, Grekova O, Patel A (2017) Zoonotic potential of Helicobacter spp. J Microbiol Immunol Infect 50(3):265–269. https://doi.org/10.1016/j.jmii.2016.11.003

Morimoto K, Namkoong H, Hasegawa N, Nakagawa T, Morino E, Shiraishi Y, Ogawa K, Izumi K, Takasaki J, Yoshiyama T, Hoshino Y, Matsuda S, Hayashi Y, Sasaki Y, Ishii M, Kurashima A, Nishimura T, Betsuyaku T, Goto H (2016) Macrolide-resistant Mycobacterium avium complex lung disease: analysis of 102 consecutive cases. Ann Am Thorac Soc 13(11):1904–1911. https://doi.org/10.1513/AnnalsATS.201604-246OC

Morris D, Mauldin E, O'Shea K, Shofer F, Rankin S (2006) Clinical, microbiological, and molecular characterization of methicillin-resistant Staphylococcus aureus infections of cats. Am J Vet Res 67(8):1421–1425. https://doi.org/10.2460/ajvr.67.8.1421

Morvan A, Moubareck C, Leclercq A, Herve-Bazin M, Bremont S, Lecuit M, Courvalin P, Le Monnier A (2010) Antimicrobial resistance of listeria monocytogenes strains isolated from humans in France. Antimicrob Agents Chemother 54(6):2728–2731. https://doi.org/10.1128/AAC.01557-09

Mukhopadhyay S (2018) Nano drugs. Characterization and biology of nanomaterials for drug delivery, pp. 523–551. https://doi.org/10.1016/B978-0-12-814031-4.00018-0
Munita J, Arias C (2016) Mechanisms of antibiotic resistance. Virulence Mech Bact Pathog 4(2). https://doi.org/10.1128/microbiolspec.VMBF-0016-2015
Murphy GS Jr, Echeverria P, Jackson LR, Arness MK, LeBron C, Pitarangsi C (1996) Ciprofloxacin- and azithromycin-resistant campylobacter causing traveler's diarrhea in U.S. troops deployed to Thailand in 1994. Clin Infect Dis 22:868–869
Nanalyze (2018) Foamix's nano-emulsion foam fights acne—nanalyze. https://www.nanalyze.com/2014/09/foamixs-nano-emulsion-foam-fights-acne/. Accessed 8 April 2023
Nanotechnology Public Engagement Programme Newsletter (October 2016). The nano drug delivery system in South Africa—NPEP. https://www.npep.co.za/newsletter/issues/october-2016/the-nano-drug-delivery-system-in-south-africa. Accessed 3 April 2023
Nizeyi J, Innocent R, Erume J, Kalema G, Cranfield M, Graczyk T (2001) Campylobacteriosis, salmonellosis, and shigellosis in free-ranging human-habituated mountain gorillas of Uganda. J Wildl Dis 37(2):239–244. https://doi.org/10.7589/0090-3558-37.2.239
No DA, Algburi A, Huynh P, Moret A, Ringard M, Comito N, Drider D, Takhistov P, Chikindas ML (2017) Antimicrobial efficacy of curcumin nanoparticles against listeria monocytogenes is mediated by surface charge. J Food Saf 37:e12353. https://doi.org/10.1111/jfs.12353
OECD (2018a) Stemming the superbug tide: just a few dollars more. OECD Health Policy Studies. OECD Publishing, Paris. https://doi.org/10.1787/9789264307599-en
OECD (2018b) Stopping antimicrobial resistance would cost just USD 2 per person a year. http://www.oecd.org/newsroom/stopping-antimicrobial-resistance-would-cost-just-usd-2-per-persona-year.htm
Oehler RL, Velez AP, Mizrachi M, Lamarche J, Gompf S (2009) Bite-related and septic syndromes caused by cats and dogs. Lancet Infect Dis 9:439–447
Ogan EF, Neill SD, Mackie DP (1982) Mastitis in dairy cows associated with an aerotolerant campylobacter. Vet Rec 110:229–230
Oh WJ, Jeong BH, Jeon K, Lee NY, Lee KS, Woo SY, Shin SJ, Kwon OJ (2012) Clinical significance of the differentiation between Mycobacterium avium and Mycobacterium intracellulare in M. avium complex lung disease. Chest 142:1482–1488
Owusu E, Newman M, Kotey N, Akumwena A, Bannerman E (2018) Susceptibility profiles of mycobacterium ulcerans isolates to streptomycin and rifampicin in two districts of the eastern region of ghana. Int J Microbiol 2016:8304524. https://doi.org/10.1155/2016/8304524
Paliwal R, Babu R, Palakurthi S (2014) Nanomedicine scale-up technologies: feasibilities and challenges. AAPS PharmSciTech 15(6):1527–1534. https://doi.org/10.1208/s12249-014-0177-9
Palumbi SR (2001) Humans as the world's greatest evolutionary force. Science 293:1786–1790
Pantosti A, Sanchini A, Monaco M (2007) Mechanisms of antibiotic resistance in Staphylococcus aureus. Future Microbiol 2(3):323–334. https://doi.org/10.2217/17460913.2.3.323
Park K (2017) Parks text book of preventive and social medicine. Banarasidas Bhanot, Jabalpur
Peel MM, Palmer GG, Stacpoole AM, Kerr TG (1997) Human lymphadenitis due to corynebacterium pseudotuberculosis: report of ten cases from Australia and review. Clin Infect Dis 24:185–191
Pelaz B, Alexiou C, Alvarez-Puebla R, Alves F, Andrews A, Ashraf S, Balogh L, Ballerini L, Bestetti A, Brendel C, Bosi S, Carril M, Chan W, Chen C, Chen X, Chen X, Cheng Z, Cui D, Du J, Dullin C, Escudero A, Feliu N, Gao M, George M, Gogotsi Y, Grünweller A, Gu Z, Halas N, Hampp N, Hartmann R, Hersam M, Hunziker P, Jian J, Jiang X, Jungebluth P, Kadhiresan P, Kataoka K, Khademhosseini A, Kopeček J, Kotov N, Krug H, Lee D, Lehr C, Leong K, Liang X, Ling Lim M, Liz-Marzán L, Ma X, Macchiarini P, Meng H, Möhwald H, Mulvaney P, Nel A, Nie S, Nordlander P, Okano T, Oliveira J, Park T, Penner R, Prato M, Puntes V, Rotello V, Samarakoon A, Schaak R, Shen Y, Sjöqvist S, Skirtach A, Soliman M, Stevens M, Sung H, Tang B, Tietze R, Udugama B, VanEpps J, Weil T, Weiss P, Willner I, Wu Y, Yang L, Yue Z, Zhang Q, Zhang Q, Zhang X, Zhao Y, Zhou X, Parak W (2017) Diverse

applications of nanomedicine. ACS Nano 11(3):2313–2381. https://doi.org/10.1021/acsnano.6b06040
Peng Z, Jin D, Kim H, Stratton C, Wu B, Tang Y, Sun X (2017) Update on antimicrobial resistance in Clostridium difficile: resistance mechanisms and antimicrobial susceptibility testing. J Clin Microbiol 55(7):1998–2008. https://doi.org/10.1128/JCM.02250-16
Peterson MC (1994) Clinical aspects of campylobacter jejuni infections in adults. West J Med 161: 148–152
Piddock LJV (1995) Quinolone resistance and campylobacter spp. Antimicrob Agents Chemother 36:891–898
Pissuwan D, Niidome T, Cortie M (2011) The forthcoming applications of gold nanoparticles in drug and gene delivery systems. J Control Release 149(1):65–71. https://doi.org/10.1016/j.jconrel.2009.12.006
Pita R, Ehmann F, Thürmer R (2018) Regulation of biomedical applications of functionalized nanomaterials in the European Union. Biomedical Applications of Functionalized Nanomaterials, pp. 653–679. https://doi.org/10.1016/B978-0-323-50878-0.00022-7
Poggi D, Oliveira de Giuseppe P, Picardeau M (2010) Antibiotic resistance markers for genetic manipulations of Leptospira spp. Appl Environ Microbiol 76(14):4882–4885. https://doi.org/10.1128/AEM.00775-10
PoultryWorld (2018) Israel: New Salmonella live vaccine approved. https://www.poultryworld.net/Broilers/Health/2010/7/Israel-New-Salmonella-live-vaccine-approved-WP007703W/. Accessed 4 April 2023
Prabhu YT, Rao KV, Kumari BS, Kumar VS, Pavani T (2015) Synthesis of Fe3O4 nanoparticles and its antibacterial application. International. Nano Lett 5:85–92
Prasad M, Ghosh M, Kumar R, Brar B, Surjith KP, Lambe UP, Ranjan K, Banerjee S, Prasad G, Kumar Khurana S, Kharb P (2021) The importance of nanomedicine in prophylactic and theranostic intervention of bacterial zoonoses and reverse zoonoses in the era of microbial resistance. J Nanosci Nanotechnol 21(6):3404–3452. https://doi.org/10.1166/jnn.2021.18999
Puzari M, Sharma M, Chetia P (2018) Emergence of antibiotic resistant shigella species: A matter of concern. J Infect Public Health 11(4):451–454. https://doi.org/10.1016/j.jiph.2017.09.025
Qiu Z, Yu Y, Chen Z, Jin M, Yang D, Zhao Z, Wang J, Shen Z, Wang X, Qian D, Huang A, Zhang B, Li JW (2012) Nanoalumina promotes the horizontal transfer of multiresistance genes mediated by plasmids across genera. Proc Natl Acad Sc USA 109(13):4944–4949
Rabold D, Espelage W, Abu Sin M, Eckmanns T, Schneeberg A, Neubauer H, Möbius N, Hille K, Wieler LH, Seyboldt C, Lübke-Becker A (2018) The zoonotic potential of Clostridium difficile from small companion animals and their owners. PLoS One 13(2):e0193411. https://doi.org/10.1371/journal.pone.0193411
Rahman A, Bonny T, Stonsaovapak S, Ananchaipattana C (2011) Yersinia enterocolitica: epidemiological studies and outbreaks. J Pathog 2011:1–11. https://doi.org/10.4061/2011/239391
Rai A, Prabhune A, Perry CC (2010) Antibiotic mediated synthesis of gold nanoparticles with potent antimicrobial activity and their application in antimicrobial coatings. J Mater Chem 20: 6789–6798
Ramasamy M, Lee JH, Lee J (2016) Potent antimicrobial and antibiofilm activities of bacteriogenically synthesized gold-silver nanoparticles against pathogenic bacteria and their physiochemical characterizations. J Biomater Appl 31(3):366–378
Ramírez-Castillo F, Moreno-Flores A, Avelar-González F, Márquez-Díaz F, Harel J, Guerrero-Barrera A (2018) An evaluation of multidrug-resistant Escherichia coli isolates in urinary tract infections from Aguascalientes, Mexico: cross-sectional study. Ann Clin Microbiol Antimicrob 17(1):34. https://doi.org/10.1186/s12941-018-0286-5
Rao BV, Boominathan M (2015) Antibacterial activity of silver nanoparticles of seaweeds. Am J Adv Drug Deliv 3(6):296–307
Raveendran R, Datta S, Wattal C (2010) Drug resistance in salmonella enterica serotype typhi and paratyphi A. Int Med Sci Acad 23(1):21

Razei A, Cheraghalii AM, Saadati M, Ramandi MF, Panahi Y (2017) Application of nanoparticles drug delivery systems in the treatment of intracellular bacterial infections. Minerva Biotecnol 29(3):156–165. https://doi.org/10.23736/S1120-4826.17.02221-2

Reed C, von Reyn C, Chamblee S, Ellerbrock T, Johnson J, Marsh B, Johnson L, Trenschel R, Horsburgh C (2006) Environmental risk factors for infection with Mycobacterium avium complex. Am J Epidemiol 164(1):32–40. https://doi.org/10.1093/aje/kwj159

Register K, Sukumar N, Palavecino E, Rubin B, Deora R (2012) Bordetella bronchiseptica in a paediatric cystic fibrosis patient: possible transmission from a household cat. Zoonoses Public Health 59(4):246–250. https://doi.org/10.1111/j.1863-2378.2011.01446.x

Reshetneva IT, Per'ianova OV, Dmitrieva GM, Ostapova TS (2015) Antibiotic resistance of salmonella spp. isolated in the territory of the krasnoyarsk region. Gig Sanit 94(2):35–38

Riesenberg A, Frömke C, Stingl K, Feßler A, Gölz G, Glocker E, Kreienbrock L, Klarmann D, Werckenthin C, Schwarz S (2017) Antimicrobial susceptibility testing of arcobacter butzleri: development and application of a new protocol for broth microdilution. J Antimicrob Chemother 72(10):2769–2774. https://doi.org/10.1093/jac/dkx211

Riojas MA, McGough KJ, Rider-Riojas CJ, Rastogi N, Hazbón MH (2018) Phylogenomic analysis of the species of the mycobacterium tuberculosis complex demonstrates that Mycobacterium africanum, Mycobacterium bovis, Mycobacterium caprae, Mycobacterium microti and mycobacterium pinnipedii are later heterotypic synonyms of mycobacterium tuberculosis. Int J Syst Evol Microbiol 68(1):324–332. https://doi.org/10.1099/ijsem.0.002507

Rojas-Lopez M, Monterio R, Pizza M, Desvaux M, Rosini R (2018) Intestinal pathogenic Escherichia coli: insights for vaccine development. Front Microbiol 9:440. https://doi.org/10.3389/fmicb.2018.00440

Rwego I, Isabirye-Basuta G, Gillespie T, Goldberg T (2008) Gastrointestinal bacterial transmission among humans, mountain gorillas, and livestock in Bwindi impenetrable National Park, Uganda. Conserv Biol 22(6):1600–1607. https://doi.org/10.1111/j.1523-1739.2008.01018.x

Saenz Y, Brinas L, Dominguez E, Ruiz J, Zarazaga M, Vila J, Torres C (2018) Mechanisms of resistance in multiple-antibiotic-resistant Escherichia coli strains of human, animal, and food origins. Antimicrob Agents Chemother 48(10):3996–4001. https://doi.org/10.1128/AAC.48.10.3996-4001.2004

Saha B, Bhattacharya J, Mukherjee A, Ghosh A, Santra C, Dasgupta AK, Karmakar P (2007) In vitro structural and functional evaluation of gold nanoparticles conjugated antibiotics. Nanoscale Res Lett 2:614

Salehi B, Mehrabian S, Ahmadi M (2014) Investigation of antibacterial effect of cadmium oxide nanoparticles on Staphylococcus aureus bacteria. J Nanobiotechnol 12:26. https://doi.org/10.1186/s12951-014-0026-8

Sarwar S, Ali A, Pal M, Chakrabarti P (2017) Zinc oxide nanoparticles provide anti-cholera activity by disrupting the interaction of cholera toxin with the human GM1 receptor. J Biol Chem 292(44):18303–18311

Schlater LK, Brenner DJ, Steigerwalt AG, Moss CW, Lambert MA, Packer RA (1989) Pasteurella caballi, a new species from equine clinical specimens. J Clin Microbiol 27(10):2169–2174

Scholar E (2007) Mycobacterium avium-Intracellulare infections. xPharm: the comprehensive pharmacology reference, pp.1-5. https://doi.org/10.1016/B978-008055232-3.60891-8

Schweizer HP (2012) Mechanisms of antibiotic resistance in Burkholderia pseudomallei: implications for treatment of melioidosis. Future Microbiol 7(12):1389–1399. https://doi.org/10.2217/fmb.12.116

ScienceDaily (2018a) Chemists pioneer new antibiotics and nano-sized delivery vehicles. https://www.sciencedaily.com/releases/2003/09/030924055708.htm. Accessed 5 April 2023

ScienceDaily (2018b) Salmonella vaccine now available for poultry. https://www.sciencedaily.com/releases/1998/12/981204074551.htm. Accessed 10 April 2023

SEARO.WHO (2018) Leptospirosis: fact sheet. http://www.searo.who.int/about/administration_structure/cds/CDS_leptospirosis-Fact_Sheet.pdf. Accessed 8 April 2023

Seguin JC, Walker RD, Caron JP, Kloos WE, George CG, Hollis RJ, Jones RN, Pfaller. (1999) Methicillin-resistant Staphylococcus aureus outbreak in a veterinary teaching hospital: potential human-to-animal transmission. J Clin Microbiol 37(5):1459–1463

Segura S, Gamazo C, Irache JM, Espuelas S (2007) Gamma interferon loaded onto albumin nanoparticles: in vitro and in vivo activities against brucella abortus. Antimicrob Agents Chemother 51(4):1310–1314

Sellyei B, Varga Z, Szentesi-Samu K, Kaszanyitzky É, Magyar T (2009) Antimicrobial susceptibility of pasteurella multocida isolated from swine and poultry. Acta Vet Hung 57(3):357–367. https://doi.org/10.1556/AVet.57.2009.3.2

Seneviratne CJ, Leung KC, Wong CH, Lee SF, Li X, Leung PC, Lau CB, Wat E, Jin L (2014) Nanoparticle-encapsulated chlorhexidine against oral bacterial biofilms. PLoS One 9(8): e103234. https://doi.org/10.1371/journal.pone

Sharma KD (2017) Antibacterial activity of biogenic platinum nanoparticles: an in vitro study. Int J Curr Microbiol App Sci 6(2):801–808. https://doi.org/10.20546/ijcmas.2017.602.089

Shin E, Barate AK, Lam TQ, Hong CH, Hahn T (2011) Antimicrobial susceptibility of Bordetella bronchiseptica isolates from slaughtered pigs. Korean J Veterin Public Health 35(3):147–153

Šilha D, Hrušková L, Brožková I, Moťková P, Vytřasová J (2014) Survival of selected bacteria from the genus arcobacter on various metallic surfaces. J Food Nutr Res 53(3):217–223

Silva MT, Portaels F, Pedrosa J (2009) Pathogenetic mechanisms of the intracellular parasite mycobacterium ulcerans leading to Buruli ulcer. Lancet Infect Dis 9:699–710

Silvan J, Zorraquin-Peña I, Gonzalez de Llano D, Moreno-Arribas M, Martinez-Rodriguez A (2018) Antibacterial activity of glutathione-stabilized silver nanoparticles against campylobacter multidrug-resistant strains. Front Microbiol 9. https://doi.org/10.3389/fmicb.2018.00458

Singh AK, Bai X, Amalaradjou MAR, Bhunia AK (2018) Antilisterial and antibiofilm activities of pediocin and LAP functionalized gold nanoparticles. Front Sustain Food Syst 2:74. https://doi.org/10.3389/fsufs.2018.00074

Smith KE, Besser JM, Leano F, Bender J, Wicklund J, Johnson B (1998) Fluoroquinolone-resistant campylobacter isolated from humans and poultry in Minnesota [abstract]. Program of the 1st international conference on emerging infectious diseases. Centers for Disease Control and Prevention, Atlanta, GA

Soriano F, Zapardiel J, Nieto E (1995) Antimicrobial susceptibilities of corynebacterium species and other non-spore-forming gram-positive bacilli to 18 antimicrobial agents. Antimicrobial Agents Chemother 39(1):208–214

Spigaglia P (2015) Recent advances in the understanding of antibiotic resistance in Clostridium difficile infection. Therap Adv Infect Dis 3(1):23–42. https://doi.org/10.1177/2049936115622891

Stanek G, Wormser G, Gray J, Strle F (2012) Lyme borreliosis. Lancet 379(9814):461–473. https://doi.org/10.1016/S0140-6736(11)60103-7

Stevens D, Bisno A, Chambers H, Dellinger E, Goldstein E, Gorbach S, Hirschmann J, Kaplan S, Montoya J, Wade J (2014) Practice guidelines for the diagnosis and management of skin and soft tissue infections: 2014 update by the Infectious Diseases Society of America. Clin Infect Dis 59(2):e10–e52. https://doi.org/10.1093/cid/ciu296

Stidham R, Freeman D, von Tersch R, Sullivan P, Tostenson S (2018) Epidemiological review of Francisella tularensis: A case study in the complications of dual diagnoses. PLoS Curr. https://doi.org/10.1371/currents.outbreaks.8eb0b55f377abc2d25031 4bbb8fc9d6d

Tabassum A, Sunita D, Raj SM (2014) Antibacterial effect of magnesium oxide nanoparticle on water contaminated with E. coli. Res Rev J Microbiol Biotechnol 3(3):P10–P13

Takahashi T, Sawada T, Ohmae K, Terakado N, Muramatsu M, Seto K, Maruyama T, Kanzak M (1984) Antibiotic resistance of Erysipelothrix rhusiopathiae isolated from pigs with chronic swine erysipelas. Antimicrob Agents Chemother 25(3):385–386

Talan DA, Citron DM, Abrahamian FM, Moran GJ, Goldstein EJ (1999) Bacteriologic analysis of infected dog and cat bites. Emergency medicine animal bite infection study group. N Engl J Med 340:85–92

Taneja N, Mewara A (2016) Shigellosis epidemiology in India. Indian J Med Res 143:565–576
Terekhova D, Sartakova M, Wormser G, Schwartz I, Cabello F (2002) Erythromycin resistance in borrelia burgdorferi. Antimicrob Agents Chemother 46(11):3637–3640. https://doi.org/10.1128/AAC.46.11.3637-3640.2002
Thirumurugan G, Rao JS, Dhanaraju MD (2016) Elucidating pharmacodynamic interaction of silver nanoparticle-topical deliverable antibiotics. Sci Report 6:29982. https://doi.org/10.1038/srep29982.74
Threlfall EJ (2002) Antimicrobial drug resistance in salmonella: problems and perspectives in food- and water-borne infections. FEMS Microbiol Rev 26(2):141–148
Timoney J, Gillespie J, Scott F, Barlough J (1988) Hagan and Bruner's microbiology and infectious diseases of domestic animals, 8th edn. Cornell University Press, Ithaca, NY, pp 1–912
Tomaso H, Hotzel H, Otto P, Myrtennäs K, Forsman M (2017) Antibiotic susceptibility in vitro of Francisella tularensis subsp. holarctica isolates from Germany. J Antimicrob Chemother 72(9): 2539–2543. https://doi.org/10.1093/jac/dkx182
Tomioka H, Saito H, Sato K, Yoneyama T (1991) Therapeutic efficacy of liposome encapsulated kanamycin against mycobacterium intracellulare infection induced in mice. Am Rev Respir Dis 145:575–579
Torres A (2017) Escherichia coli diseases in Latin America—a 'one health' multidisciplinary approach. Pathog Dis 75(2). https://doi.org/10.1093/femspd/ftx012
Tsiodras S, Gold H, Sakoulas G, Eliopoulos G, Wennersten C, Venkataraman L, Moellering R, Ferraro M (2001) Linezolid resistance in a clinical isolate of Staphylococcus aureus. Lancet 358(9277):207–208
Urich S, Chalcraft L, Schriefer M, Yockey B, Petersen J (2011) Lack of antimicrobial resistance in Yersinia pestis isolates from 17 countries in the Americas, Africa, and Asia. Antimicrob Agents Chemother 56(1):555–558. https://doi.org/10.1128/AAC.05043-11
Van Zandt KE, Greer MT, Gelhaus HC (2013) Glanders: an overview of infection in humans. Orphanet J Rare Dis 8:131. https://doi.org/10.1186/1750-1172-8-131
Van TTH, Yidana Z, Smooker PM, Coloe PJ (2020) Antibiotic use in food animals worldwide, with a focus on Africa: pluses and minuses. J Glob Antimicrob Resist 20:170–177. https://doi.org/10.1016/j.jgar.2019.07.031
Ventola CL (2017) Progress in nanomedicine: approved and investigational nanodrugs. Pharm Therap 42(12):742–755
Veterinary Manual (2018) Zoonotic diseases—public health—veterinary manual. https://www.merckvetmanual.com/public-health/zoonoses/zoonotic-diseases. Accessed 6 April 2023
Wagner D, Runberg J, Vogler A, Lee J, Driebe E, Price L, Engelthaler D, Fricke W, Ravel J, Keim P (2010) No resistance plasmid in Yersinia pestis, North America. Emerg Infect Dis 16(5): 885–887. https://doi.org/10.3201/eid1605.090892
Wang X, Li H, Jiang G, Zhao L, Ma Y, Javid B, Huang H (2014) Prevalence and drug resistance of nontuberculous mycobacteria, Northern China, 2008–2011. Emerg Infect Dis 20(7):1252–1253. https://doi.org/10.3201/eid2007.131801
Wang L, Hu C, Shao L (2017) The antimicrobial activity of nanoparticles: present situation and prospects for the future. Int J Nanomedicine 12:1227–1249. https://doi.org/10.2147/IJN.S121956
Wang X, Yang F, Zhao J, Xu Y, Mao D, Zhu X, Luo Y, Alvarez P (2018) Bacterial exposure to ZnO nanoparticles facilitates horizontal transfer of antibiotic resistance genes. NanoImpact 10:61–67. https://doi.org/10.1016/j.impact.2017.11.006
Weese J, Dick H, Willey B, Mcgeer A, Kreiswirth B, Innis B, Low D (2006) Suspected transmission of methicillin-resistant Staphylococcus aureus between domestic pets and humans in veterinary clinics and in the household. Vet Microbiol 115(1–3):148–155. https://doi.org/10.1016/j.vetmic.2006.01.004
Wernli D, Emonet S, Schrenzel J, Harbarth S (2011) Evaluation of eight cases of confirmed Bordetella bronchiseptica infection and colonization over a 15-year period. Clin Microbiol Infect 17(2):201–203. https://doi.org/10.1111/j.1469-0691.2010.03258.x]

Williams D, Lewis C, Sandoval F, Robbins N, Keas S, Gillis T, Scollard D (2013) Drug resistance in patients with leprosy in the United States. Clin Infect Dis 58(1):72–73. https://doi.org/10.1093/cid/cit628

Wilson B, Ho M (2013) Pasteurella multocida: from zoonosis to cellular microbiology. Clin Microbiol Rev 26(3):631–655. https://doi.org/10.1128/CMR.00024-13

Wong CS, Jelacic S, Habeeb RL, Watkins SL, Tarr PI (2000) The risk of the hemolytic–uremic syndrome after antibiotic treatment of Escherichia coli O157:H7 infections. N Engl J Med 342: 1930–1936

Woods LCJB (2005) USAMRIID's medical management of biological casualties handbook (PDF), 6th edn. U.S. Army Medical Institute of Infectious Diseases, Fort

World Bank (2010) People, pathogens, and our planet: volume one: towards a one health approach for controlling zoonotic diseases (English). Washington, DC: World Bank. Documents. worldbank.org. http://documents.worldbank.org/curated/en/214701468338937565/Volume-one-towards-a-onehealth-approach-for-controlling-zoonotic-diseases. Accessed 3 April 2023

World Health Organization (2018) Listeriosis. http://www.who.int/mediacentre/factsheets/listeriosis/en/. Accessed 1 April 2023

World Health Organization (WHO) (2008) Anthrax in humans and animals, Geneva: World Organization for Animal Health. http://www.who.int/csr/resources/publications/anthrax_web.pdf. Accessed 6 April 2023

World Health Organization (WHO) (2012) Treatment of Mycobacterium Ulcerans disease (Buruli Ulcer), ISBN 9789241503402. www.who.int/iris/bitstream/10665/77771/1/9789241503402_eng.pdf

World health Organization (WHO) (2014) Antimicrobial resistance: Global report on surveillance. World Health Organization, pp. 1–256. http://apps.who.int/iris/bitstream/handle/10665/112642/9789241564748_eng.pdf?sequence=1. Accessed 3 April 2023

World Health Organization (WHO) (2017) WHO publishes list of bacteria for which new antibiotics are urgently needed. http://www.who.int/news-room/detail/27-02-2017-who-publishes-list-of-bacteriafor-which-new-antibiotics-are-urgently-needed. Accessed 7 April 2023

World Health Organization (WHO) (2018a) Leptospirosis. http://www.who.int/zoonoses/diseases/Leptospirosissurveillance.pdf. Accessed 1 April 2023

World Health Organization (WHO) (2018b) Status of Vaccine Research and Development for Campylobacter Prepared for WHO PD-VAC. http://www.who.int/immunization/research/meetings_workshops/Campylobacter_VaccineRD_Sept2014.pdf. Accessed 4 April 2023

World Health Organization (WHO) (2018c) Antibiotic resistance. [online] Available at: http://www.who.int/news-room/factsheets/detail/antibiotic-resistance. Accessed 3 April 2023

World Health Organization (WHO) (2018d) Drug-resistant TB:Global situation. http://www.who.int/tb/areasof-work/drug-resistant-tb/global-situation/en/. Accessed 5 April 2023

World Health Organization (WHO) (2018e) BCG vaccine. http://www.who.int/biologicals/areas/vaccines/bcg/en. Accessed 8 April 2023

World Health Organization (WHO) (2018f) Brucellosis. http://www.who.int/zoonoses/diseases/Brucellosissurveillance.pdf. Accessed 10 April 2023

World Health Organization (WHO) (2018g) Brucellosis. http://www.who.int/zoonoses/diseases/brucellosis/en/. Accessed 8 April 2023

World Health Organization (WHO) (2018h) Campylobacter. http://www.who.int/news-room/fact-sheets/detail/campylobacter. Accessed 5 April 2023

World Health Organization (WHO) (2018i) Dysentery (shigellosis): current who guidelines and the WHO essential medicine list for children. http://www.who.int/selection_medicines/committees/expert/21/applications/s6_paed_antibiotics_appendix5_dysentery.pdf. Accessed 5 April 2023

World Health Organization (WHO) (2018j) Oral cholera vaccines. http://www.who.int/cholera/vaccines/en/. Accessed 3 April 2023

World Health Organization (WHO) (2018k) Salmonella (non-typhoidal). [online] Available at: http://www.who.int/news-room/fact-sheets/detail/salmonella-(non-typhoidal). Accessed 6 April 2023

World Health Organization (WHO) (2018l) Tuberculosis (TB). http://www.who.int/news-room/fact-sheets/detail/tuberculosis. Accessed 10 April 2023

Xia DL, Wang YF, Bao N, He H, Li XD, Chen YP, Gu HY (2014) Influence of reducing agents on biosafety and biocompatibility of gold nanoparticles. Appl Biochem Biotechnol 174(7): 2458–2470

Xie Y, He Y, Irwin PL, Jin T, Shi X (2011) Antibacterial activity and mechanism of action of zinc oxide nanoparticles against Campylobacter jejuni. Appl Environ Microbiol 77(7):2325–2331

Xu K, Liang Z, Ding X, Hu H, Liu S, Nurmik M, Bi S, Hu F, Ji Z, Ren J, Yang S, Yang Y, Li L (2017) Nanomaterials in the prevention, diagnosis, and treatment of Mycobacterium tuberculosis infections. Adv Healthc Mater 7(1):1700509. https://doi.org/10.1002/adhm.201700509

Yeom J, Lee B, Kim D, Lee J, Kim S, Bae J, Park Y, Lee K (2016) Gold nanoparticle-DNA aptamer conjugate-assisted delivery of antimicrobial peptide effectively eliminates intracellular salmonella enterica serovar typhimurium. Biomaterials 104:43–51. https://doi.org/10.1016/j.biomaterials.2016.07.009

Youssef, M., Fattal, E., Alonso, M., Roblot-Treupel, L., Sauzieres, J., Tancrede, C., Omnes, A., Couvreur, P. And Andremont, A., 1988. Effectiveness of nanoparticle-bound ampicillin in the treatment of listeria monocytogenes infection in athymic nude mice. Antimicrob Agents Chemother, 32(8), pp.1204–1207

Yu C, Ang G, Chua A, Tan E, Lee S, Falero-Diaz G, Otero O, Rodríguez I, Reyes F, Acosta A, Sarmiento M, Ghosh S, Ramamurthy T, Yean Yean C, Lalitha P, Ravichandran M (2011) Dry-reagent gold nanoparticle-based lateral flow biosensor for the simultaneous detection of vibrio cholerae serogroups O1 and O139. J Microbiol Methods 86(3):277–282. https://doi.org/10.1016/j.mimet.2011.04.020

Zhao Z, Xue Y, Wang C, Ding K, Wu B, He Q, Cheng X, Chen H (2011) Antimicrobial susceptibility of Bordetella bronchiseptica isolates from pigs with respiratory diseases on farms in China. J Vet Med Sci 73(1):103–106

Zhao Y, Ye C, Liu W, Chen R, Jiang X (2014) Tuning the composition of AuPt bimetallic nanoparticles for antibacterial application. Angew Chem Int Ed 53(31):8127–8131

Zhu X, Radovic-Moreno AF, Wu J, Langer R, Shi J (2014) Nanomedicine in the management of microbial infection: overview and perspectives. Nano Today 9(4):478–498

Chapter 20
Nanotherapeutics for Viral Diseases of Animals

Lukumoni Buragohain, Bhabesh Mili, Arpita Bharali, and Anamika Mazumdar

Abstract Prevention is no doubt the best strategy to combat viral diseases as the options of antiviral therapeutics are limited. The shortage is even more acute in case of viral diseases of animals. Considering the diversity of livestock species and their viral disease, only a few among them are having vaccines developed against them. So, a mere supportive therapy or limiting secondary infections often remain as the only therapeutic option to the clinician. In addition, limitations of conventional antiviral drugs such as drug resistance, toxicity effect, short half-life, patient noncompliance, and issue of bioavailability pose daunting task to treat viral diseases effectively. At this juncture, nanotechnology interventions have paved the way to surmount certain limitations of conventional antiviral therapeutics. The predominant role of nanoparticles in antiviral therapy is to enhance the pharmacokinetic activity of the existing antiviral drugs. Nanoencapsulation of antiviral drugs potentiates the therapeutics by slow and sustained release as well as protecting them from degradation, thus prolonging the half-life. Nanointervention can improve the solubility and bioavailability of certain drugs which cannot be administered in free form. The smaller size and favorable attributes of nanoparticles often help them to cross certain biological barriers that an antiviral drug cannot penetrate. Moreover, intelligent surface functionalization of nanoparticles can promote molecular interaction at cellular and subcellular levels. The predominant mechanisms for inhibiting viral infection include but not limited to suppressing the replication of viral genome, inhibiting the attachment of virus particles to the target cells, inhibiting the functions of viral proteins, and deactivating viruses. The potential of nanomedicine has progressively getting reflected as newer nanoformulations are now hitting the market regularly. So, it is expected that the economically important and zoonotic viral

L. Buragohain (✉) · A. Bharali · A. Mazumdar
Department of Animal Biotechnology, College of Veterinary Science, Assam Agricultural University, Khanapara, Guwahati, Assam, India

B. Mili
Department of Veterinary Physiology and Biochemistry, College of Veterinary Sciences and Animal Husbandry, CAU (I), Jalukie, Peren, Nagaland, India

LLR University of Veterinary and Animal, ICAR-National Research Centre on Equines, *Nanotechnology Theranostics in Livestock Diseases and Management*, Livestock Diseases and Management,
https://doi.org/10.1007/978-981-16-1610-5_20

diseases of animals will certainly gain more attention in upcoming days in the context of nanotechnology boom.

Keywords Polymeric nanoparticles · Liposomes · Fullerenes · Solid lipid NPs · Polymeric micelles · Dendrimers · Metallic nanomaterials · Quantum dots · Antiviral nanotherapeutics · Livestock

20.1 Introduction

Nanotechnology has been used by craftsmen around the world approximately since 2600 BC, even though it came under the limelight fairly in the year 1959 by a renowned physicist and Nobel laureate Richard P. Feynman. His visionary and historic talk at the American physical society in 1959 underlining "*There is plenty of room at the bottom*" explained about smart tiny devices that can be swallowed, swim through blood vessels, and remove damaged parts, and then the movie presents a submarine that is shrunk to the size of a microbe to fix the clot on a mission to rescue a scientist (Feynmann 2011). Later, the word "nanotechnology" was coined by Norio Taniguchi, a Japanese scientist in 1974 who defines nanotechnology as "*nanotechnology mainly consists of the processing of separation, consolidation, and deformation of materials by one atom or one molecule*" (Taniguchi et al. 1974). The word "nano" is derived from the Greek word "nanos" meaning dwarf, which generally refers to a size spanning between 1 and 100 nm in the length scale. A nanometer is one thousandth of a micron or one millionth a millimeter or 10^{-9} of meter and is roughly the length by five to ten atoms stacked in straight line. Nanoparticles (NPs) exhibit unique physical, chemical, and biological properties compared to their respective bulk material counterparts (Dolez 2015). NPs have very high specific surface area (area per unit volume), which facilitates the electronic activity and their interaction with outside (Sharma et al. 2003). It has high mobility not only in free state but also in porous media and exhibit quantum effect due to their comparable dimension with the wavelength of the electron wave.

Well-defined sizes of NPs are used for therapeutic purposes in both medical and veterinary medicine. The use of the large and very small sizes of NPs is usually avoided in medical/veterinary fields. The large NPs induce embolism and rapidly get cleared from the circulation. Likewise, very small size NPs are cleared promptly by the kidneys. They also lead to an increase in production of free radicals and reactive oxygen species in the body (Buzea et al. 2007; Underwood and van Eps 2012; Hauser et al. 2021). Therefore, selection of suitable size of NPs for therapeutic purposes is based on the nature of target cells, type/amount of payload agents, preference between tissue internalization and longer half-life time in the circulation, need for rapid excretion, way of clearance (renal/reticuloendothelial system), the intensity of the contrast image, biodistribution, and desired degree of immune system stimulation and type of immune response (Cormode et al. 2008).

Different nanotechnology-based tools and various types of NPs are being employed to deliver drugs, diagnose diseases, understand of various physiological and pathological processes, and customize new strategies for treating various diseases including viruses, vaccine production, and cancer across the world. NPs are very useful in drug delivery. NPs help to mask the active ingredients and protect them from the attacks of the immune system, degradation by catalysts, or extreme pH conditions and promote the controlled release of bound drug at the target site for a prolonged period (Mili et al. 2018). The therapeutic/diagnostic agents are stuck to the surface of the NPs or get encapsulated/trapped inside them (Sailor and Park 2012; Mili et al. 2018). This technology is employed in the treatment of African trypanosomiasis (Howsam et al. 2010). Various NPs are being used for the production of nano-vaccines. Nano-vaccines may replace the conventional adjuvants and prolong the vaccine protection period (Moyer et al. 2016). The applications of nanotechnology-based tools may provide advanced and effective therapy for viral diseases of animals.

20.2 Classification of Nanoparticles

There is no specific criteria for the classification of NPs. Many classifications of the NPs exist based on (1) origin (e.g., natural, incidental, and engineered NPs), (2) dimension (e.g., zero-dimensional, one-dimensional, two-dimensional, three-dimensional), (3) chemical composition (e.g., metallic, semiconductor, carbon-based, ceramic, oxide, lipid, polymeric NPs), (4) morphology (e.g., spherical, triangular, cubic, hexagonal, oval-shaped, prism-shaped, rod-like, tube-like, helical), and so on (Dolez 2015). NPs are also classified based on (1) origin as organic NPs (e.g., proteins, peptides, or lipids), inorganic NPs (such as gold, silver, silica, iron, magnesium, or grapheme), and hybrid NPs (Riley and Vermerris 2017); (2) shape (e.g., spheres, tubes, or liquid drops); and (3) application and type of payload (e.g., therapeutic, diagnostic, vaccine administration, nutritional) (Thulasi et al. 2013).

Broadly, NPs can be classified as (i) polymeric nanoparticles, (ii) liposomes, (iii) fullerenes and buckytubes, (iv) nanoshells, (v) solid lipid NPs, (vi) polymeric micelles, (vii) dendrimers, (viii) metallic nanomaterials, (ix) ceramic nanomaterials, (x) quantum dots, (xi) nanoemulsion, (xii) nanobubbles, (xiii) respirocytes, (xiv) microbivores, and (xv) nano/microrobots (El-Sayed and Kamel 2020).

1. *Polymeric nanoparticles*: Polymeric nanoparticles (PNPs) are prepared via adsorption of the active molecules on the surface of NP polymers to form complex spherical structures. Two types, namely, (a) natural polymers (e.g., chitosan) and (b) synthetic polymers (e.g., PEG, poly(ethylene glycol); PLA, poly(lactic acid); PLGA, poly(lactide-co-glycolide)) are the most commonplace (Torres-Sangiao et al. 2016). PNPs are easy to synthesize, biocompatible, biodegradable, non-immunogenic, and mostly nontoxic. They also promote effective permeation through cell membranes and maintain stability in the bloodstream. Advantages of these particles as drug carriers include their potential use for stimuli responsive and controlled release, their ability to protect drug

and other molecules with biological activity against the harsh environment, and their capacity to improve the bioavailability of the attached active ingredients and therapeutic index (Cano et al. 2019, 2020).

2. *Liposomes*: Liposomes are small spherical-shaped, nontoxic PEGylated nanoparticles. The particle is composed of cholesterol and natural nontoxic phospholipids, such as phosphatidylcholine. A double-layered phospholipid shell surrounds the liposomes. These particles are biodegradable, biocompatible, and load both hydrophilic and hydrophobic preparations. This feature helps to make them a potential delivery vehicle for therapeutic drugs (Elgqvist 2017), including the efficient delivery of the mRNA and DNA. However, a major disadvantage is their poor stability, short storage capacity, easy oxidation of the capsule, and rapid release of the contents (Jurj et al. 2017).
3. *Fullerenes and buckytubes (nanotubes)*: Fullerene is an allotrope of pure carbon molecules with at least 60 carbon atoms, whose molecules connect by single and double bonds to form a closed or partially closed mesh. Fullerenes may be hollow sphere, ellipsoid, tube, or other shape. The particles are inert and nontoxic and can interact easily with cells, proteins, and viruses due to their nano-scale size (El-Sayed and Kamel 2020). They can also be internalized and translocated in the cytoplasm of the cells very efficiently (Jurj et al. 2017).

   Nanotubes are made up of a sequence of at least 60 numbers of carbon atoms arranged in a long thin cylindrical structure that is stable, strong, and flexible. They are related to two other carbon crystal forms, graphite, and diamonds. Nanotubes are hollow inside, making it easy to put pharmacological agents inside them. Besides their use in cancer treatment, the nanotubes are also interesting for gene therapy and DNA delivery due to their small size and chemical inertness (Reilly 2007).
4. *Nanoshells*: Nanoshells are spherical nanoparticles that have a dielectric core covered by a thin metallic shell, e.g., gold (Loo et al. 2004). The size of a nanoshell ranges from 1 to 20 nm. The particle is biocompatible, inert, and safe for the body (Krishnan and George 2014). These nanoparticles have been often used in cancer diagnosis (Mohantya et al. 2014) and detection of immunoglobulins in the blood sample (Manuja et al. 2012).
5. *Solid lipid NPs*: Solid lipid NPs (sLNPs) are composed of lipids. They are solid at room temperature. The average diameter of sLNPs ranges between 10 and 1000 nm. The hydrophilic drug agents or antibodies can be conjugated easily in the hydrophilic outer shell coats of the sLNPs. These particles can adhere easily to mucous membranes and easily absorb via the mucosa as well as possess the ability to cross the blood–brain barrier of the central nervous system (Mohantya et al. 2014). Additionally, cationic solid lipid NPs can directly bind with DNA/RNA fragments by electrostatic interaction which enables them to be used in gene therapy (Elgqvist 2017). This property of sLNPs makes them a novel pharmaceutical drug delivery system (Saupe and Rades 2006).
6. *Polymeric micelles*: These are nanostructures formed by the self-assembly of amphiphiles in water. The size of polymeric micelles ranges from 0 to 100 nm. The micelles are characterized by a hydrophobic core in contrast to sNLPs.

These also have low toxicity and are very effective in the delivery of therapeutic agents including water-insoluble preparations. Polymeric micelles can be divided into four types: (1) phospholipid shell, (2) poly(amino acid) shell, (3) biocompatible polyester shell, and (4) pluronic micelles with a block copolymer shell made of both hydrophobic and hydrophilic substances.

7. *Dendrimers*: Dendrimers are highly defined artificial macromolecules whose size ranges between 5 and 20 nm. Dendrimers possess a high number of functional groups either inside the sphere or on the surfaces. The functional groups of the dendrimers are bound with different hydrophobic/hydrophilic therapeutic substances. These are highly water-soluble, biocompatible, and polyvalent NPs (Jurj et al. 2017). Their well-defined structure, size, surface functionalization capability, and stability make these NPs a promising vehicle system for therapeutic agents, genes, and anticancer drugs (Baker et al. 2001).
8. *Metallic nanomaterials*: Metallic nanomaterials are prepared from highly conductive metal and metal oxides by chemical, electrochemical, or photochemical synthesis. Examples of metallic nanoparticles include gold nanoparticles, silver nanoparticles, copper nanoparticles, aluminum nanoparticles, iron nanoparticles, iron oxide nanoparticles, platinum nanoparticles, lead nanoparticles, cobalt nanoparticles, and copper oxide nanoparticles. The size of a stable metallic nanoproduct usually ranges from 1 to 150 nm. They often possess superparamagnetism, catalytic behavior, superior mechanical strength, and optical properties (Li et al. 1996). These properties favor the metal nanoparticles for diverse applications such as in bio-sensing, bio-imaging, cancer thermotherapy, and targeted drug delivery by conjugating with various functional groups (Greenwood et al. 2008).
9. *Ceramic nanomaterials*: Ceramic nanoparticles are made up of oxides, carbides, phosphates, and carbonates of metals and metalloids. They are easy to design in different forms, porosity, and sizes. The particles are biocompatible and highly heat resistant and possess chemical inertness such as extreme pH values (Mohantya et al. 2014).
10. *Quantum dots*: Quantum dots are colloidal fluorescent semiconductor nanocrystals of about 2–10 nm. The shell consists of inorganic material in addition to the aqueous coating, which can conjugate to various biomolecules. Also, they possess unique characteristics such as optical, electronic, and photophysical properties that make them appealing in promising applications in biological labeling, imaging, and screening of blood samples for the presence of proteins, pathogens, tumor markers, etc.
11. *Nanoemulsion*: Nanoemulsions are stabilized nanosized oil drops in water. The outer surface is coated with a thin film to stabilize them physically by using surfactant and co-surfactant. These particles are thermodynamically stable isotropic. This feature made them an improved delivery vehicle of active pharmaceutical ingredients.
12. *Nanobubbles*: Nanobubbles are formed using any gas and injected into any liquid. The bubbles remain stable at room temperature but aggregate together to form microbubbles if slightly heated when exposed to ultrasonic waves. The

size of the nanobubbles ranges from 70 to 120 nm. Nanobubbles exhibit unique properties that improve numerous physical, chemical, and biological processes. These NPs are mostly used for selective drug targeting at tumor site (Rapoport et al. 2007), gene therapy (Mohantya et al. 2014), etc.

13. *Respirocytes*: Respirocytes are artificial nanomedical erythrocytes. The primary uses of artificial respirocytes include transfusable blood substitution, treatment for anemia, perinatal and neonatal disorders, and a variety of lung diseases (Robert 2009).
14. *Microbivores*: Microbivores are nanorobotic phagocytes. It is used to patrol the bloodstream, seeking out and digesting unwanted pathogens including bacteria, viruses, or fungi (Kang et al. 2004).
15. *Nano/microrobots*: Nano/microrobots are tiny devices that can convert energy into motion and operate at nano/microscales (Dekanovský et al. 2022). These tiny robots are designed in such a way that they propel in their surroundings at high speed, load drugs efficiently, transport them safely, and release them upon request successfully. They are also used for screening the body for the presence of malignant cells or to introduce nanocameras to allow monitoring of surgeries.

20.3 Application of Nanotherapeutics to Combat Viral Diseases of Animals

Significant advances in nanotechnology have been witnessed in the last two decades. A huge number of nanoparticle-based products have been approved for clinical diagnosis and therapies, and still many are in developmental phase. Viruses of different nature (DNA and RNA) infect livestock and poultry causing huge economic loss. Viral diseases are best prevented by vaccines, but vaccines are not available for each and every viral disease of animals. Again, available antiviral drugs are not very effective against viral infections. Thus, the uses of nanoparticles alone or in combination with other substances should be given priority because several literatures have reported the effectiveness of different nanoparticles against many viral diseases of animals (Minakshi et al. 2020). The use of nanoparticles (either in in-vitro or in-vivo conditions) in treatment of viral diseases of animals is represented in Fig. 20.1.

Nanoparticles of different types were found to exhibit virucidal and antiviral activity against diverse viruses that infect animals and humans. The basic role of nanoparticles in treatment of viral diseases is to enhance the pharmacokinetic activity of existing antiviral agents (Zhou et al. 2021). Encapsulation of antiviral drugs with nanomaterial increases the therapeutic window by slowing release of the drug as well as protecting from degradation (Wang et al. 2012; Zhou et al. 2021). Intervention of nanoparticles in antiviral therapy makes bioavailability of certain drugs for long duration in the circulation which cannot be administered in free form to the body system (Pollock et al. 2008; Zhou et al. 2021). Moreover, nanoparticles can be used to cross certain biological barriers that an antiviral drug cannot cross

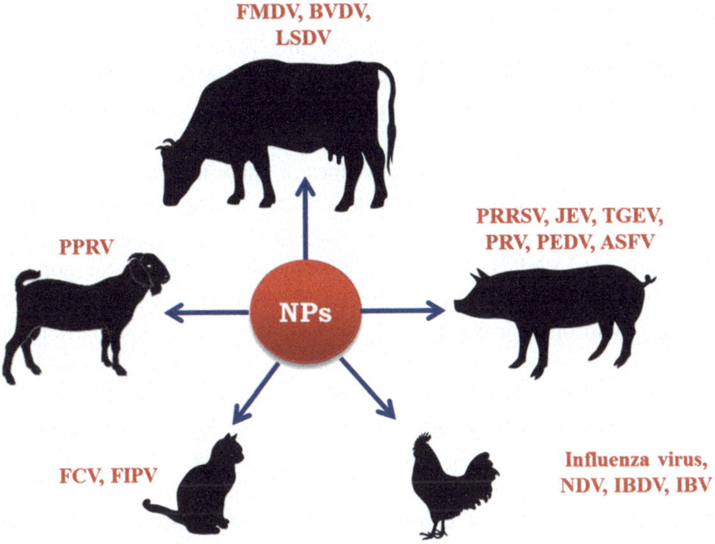

Fig. 20.1 Use of nanoparticles as antiviral agent against animal viruses; *FMDV* foot-and-mouth disease virus, *BVDV* bovine viral diarrhea virus, *LSDV* lumpy skin disease virus, *PRRSV* porcine reproductive and respiratory syndrome virus, *JEV* Japanese encephalitis virus, *TGEV* transmissible gastroenteritis virus, *PRV* pseudorabies virus, *PEDV* porcine epidemic diarrhea virus, *ASFV* African swine fever virus, *NDV* Newcastle disease virus, *IBDV* infectious bursal disease virus, *IBV* infectious bronchitis virus, *FCV* feline calicivirus, *FIPV* feline infectious peritonitis virus, *PPRV* peste des petits ruminants virus

(Cojocaru et al. 2020; Zhou et al. 2021). Limitations of conventional antiviral drugs, such as drug resistance, toxicity effect, short half-life, patient noncompliance, and issue of bioavailability, can be overcome by intervention of nanotechnology in antiviral therapy (Chakravarty and Vora 2021).

Nanoparticles can prevent viral infection in animals by various mechanisms at various stages of infection. The mechanisms include but not limited to suppressing the replication of viral genome, inhibiting the attachment of virus particles to the target cells (by binding either to the viral proteins or to the receptors of targeted host cells), inhibiting the functions of viral proteins, and deactivating viruses (Jindal and Packirisamy 2020; Sarkar et al. 2022; Yasamineh et al. 2022). However, the working principle of nanoparticles against viruses depends on its form and type (Chen and Liang 2020). Thus, the selection of right nanoparticle is an essential criterion to effectively treat a viral disease.

20.3.1 Nanotherapeutics for Viral Disease of Swine

Pigs are infected with several DNA and RNA viruses such as the African swine fever virus (ASFV), classical swine fever virus (CSFV), porcine circoviruses (PCV), porcine parvovirus (PPV), Japanese encephalitis virus (JEV), porcine reproductive and respiratory syndrome virus (PRRSV), porcine epidemic diarrhea virus (PEDV), swine influenza (SI) virus, swine pox (SP) virus, and many others. Effective vaccines are not available for many of the viral diseases of pigs. Therefore, the use of nanoparticles in treatment regimen may be helpful to cure such viral diseases that lack effective drugs or vaccines. Different types of nanomaterials are described as an antiviral agent against viral diseases of pigs.

Graham et al. (2021) reported that three nanoparticles, viz., antiviral nanoparticle second-generation (AVNP2), copper nanoparticles (CuNPs), and copper oxide nanoparticles (CuONPs), have strong antiviral activity against porcine reproductive and respiratory syndrome virus (PRRSV). Among these three nanoparticles, CuNPs exhibited the highest antiviral activity at lower concentrations. The proliferation of pseudorabies virus (PRV), the causative agent of Aujeszky's disease in swine, can be inhibited by histidine-stabilized gold nanoclusters (His-Au NCs) (Feng et al. 2018). Silver nanoparticles have antiviral activity against ASFV and can completely inhibit ASFV at the titer of 10^3 HAD_{50} (Dung et al. 2020). GSH-modified zinc sulfide nanoparticles (GSH-ZnS NPs) exhibit broad-spectrum antiviral activity against several viruses that infect pigs which include PRRSV, PRV, and porcine epidemic diarrhea virus (Zhou et al. 2020a,b). The GSH-ZnS NPs inhibit viral infection by multiple mechanisms at different stages of infection. Curcumin carbon quantum dots (Cur-CQDs) synthesized by Chen et al. (2022) can inhibit Japanese encephalitis virus infection by suppressing virus entry to target cells through binding to E protein of JEV.

Cho et al. (2014) patented an admixture of silver colloid and titanium dioxide (TiO2) NPs along with a stabilizer, a binder, and water that showed strong antiviral activity against transmissible gastroenteritis virus (TGEV) and porcine epidemic diarrhea virus (PEDV) (Nikaeen et al. 2020). Lv et al. (2014) reported that silver nanoparticles and silver nanowires can be used as antiviral therapy against TGEV of pigs. In their experiment, it was observed that both silver nanoparticles and silver nanowires have virucidal action against TGEV and also restrict the entry of the virus into host cells. Curcumin-based cationic carbon dots (CCM-CDs) were synthesized by Ting et al. (2018) and demonstrated antiviral activity against PEDV. There are several inhibitory properties of CCM-CDs against PEDV such as inhibiting viral replication by stimulating the production of pro-inflammatory cytokines and interferon-stimulating genes, suppressing viral entry, inhibiting RNA synthesis, and also restricting the accumulation of reactive oxygen species in cells caused by PEDV infection. Du et al. (2018) showed that glutathione-capped Ag2S nanoclusters are highly efficient against PEDV. The Ag2S nanoclusters can prevent the synthesis of viral RNA as well as upregulate the expression of interferon-stimulating genes and pro-inflammation cytokines which can further prevent the infection of PEDV. Du et al. (2020) synthesized hybrid gold/silver nanorods (Au/AgNRs) that exhibited strong antiviral activity against PEDV. The Au/AgNRs can suppress PEDV replication and entry of the virus into host cells.

Zhou et al. (2020a,b) generated unique triangular star-shaped bovine serum albumin (BSA)-coated tellurium nanoparticles (Te/BSA nanostars) which were found to have antiviral activity against PRRSV and PEDV. The Te/BSA nanostars mainly inhibit the internalization process of PRRSV. Tong et al. (2020) synthesized glycyrrhizic-acid-based carbon dots (Gly-CDs) that have several inhibitory properties against PRRSV. The Gly-CDs not only inhibit the replication and invasion properties of PRRSV but also it can elicit the antiviral innate immune response as well as inhibit the accumulation of intracellular reactive oxygen species in PRRSV-infected cells. Moreover, Gly-CDs can also inhibit the propagation of other swine viruses which include pseudorabies virus and porcine epidemic diarrhea virus. Recently, Tong et al. (2022) reported the antiviral and virucidal properties of gelatin-stabilized ferrous sulfide nanoparticles (Gel-FeS NPs) against PRRSV. The in vitro experiment conducted in MARC-145 cells exhibited that Gel-FeS NPs could inhibit the replication, invasion, and adsorption of PRRSV.

20.3.2 Nanotherapeutics Against Viral Disease of Sheep and Goat

Sheep and goats are infected with many viral diseases and the most commonly encountered are peste des petits ruminants, bluetongue, contagious ecthyma, and pox (sheep pox and goat pox). Peste des petits ruminants virus (PPRV) is one of the major economically important diseases of sheep and goats. PPRV causes peste des petits ruminants (PPR), which is a highly contagious and fatal viral disease of sheep and goat. Khandelwal et al. (2014) produced silver nanoparticles that could inhibit replication of PPRV in vero cells as well as it could restrict the entry of virus to the target cell. Zainab Mazhari (2022) reported that gold nanoparticles have antiviral property against blue tongue virus.

20.3.3 Nanotherapeutics Against Viral Disease of Cattle and Buffalo

The common viral diseases of bovines are foot-and-mouth disease, bovine viral diarrhea, enzootic bovine leucosis, infectious bovine rhinotracheitis, lumpy skin disease, etc.

El-Gaffary et al. (2019) evaluated the antiviral activity of PEG dithiol functionalized gold nanoparticles against bovine viral diarrhea virus (BVDV). In their experiment, it was observed that functionalized gold nanoparticles strongly bind to BVDV which induces attachment inhibitory effect of the virus. In an in vitro experiment, it was observed that silver nanoparticles at a concentration of 24 μg/mL have the capacity to inhibit the replication of bovine herpesvirus-1 (BoHV-1) (El-Mohamady et al. 2018). Evaluation of green zinc oxide and silver nanoparticles showed inhibitory effect to BoHV-1 both in in vitro and in vivo conditions conducted in MDCK cell line and rabbits, respectively (Zeedan et al. 2020). Farag et al. (2020) produced propolis–alginate nanoparticles (propolis–ALg NPs) that were safe and could be used to treat lumpy skin disease. The propolis–ALg NPs did not show any toxicity and there was 100% recovery of cattle infected with the lumpy skin disease virus; however, only 13.3% of the animals recovered when treated with tetracycline.

Rafiei et al. (2016) reported that gold nanoparticles (AuNPs) could inhibit replication of foot-and-mouth disease virus (FMDV) at post-entry stages of the virus to the target cells. Magnesium oxide nanoparticles exhibit antiviral and virucidal activities and can prevent FMDV at an early stage of infection by inhibiting virus attachment and penetration. However, magnesium oxide nanoparticles cannot restrict FMDV infection post-penetration to target cells.

20.3.4 Nanotherapeutics for Avian Viral Disease

Poultry industry is affected by many viral diseases such as avian influenza, Newcastle disease, Marek's disease, avian infectious bronchitis and laryngotracheitis, Gumboro disease, duck hepatitis, and duck plague. Moreover, certain strains of influenza virus not only hamper the poultry sector but also human health to a great extent.

AbouAitah et al. (2020) reported that nanoformulations of natural prodrugs loaded in mesoporous silica nanoparticles are safe and effective antiviral agents to fight pathogenic avian influenza H5N1 virus. Several authors have reported that silver nanoparticles in combination with other substances (epigallocatechin gallate and zinc sulfate) exhibit strong antiviral activity against various subtypes (H5N1, H9N2) of avian influenza (Saadh and Aldalaen 2021; Saadh et al. 2021).

Green synthesized silver nanoparticles are strong antivirals that inhibit entry of Newcastle disease virus (NDV) and reduce viral replication (Saadh 2022). Propolis, chitosan, and mixture of both produce antiviral activity against NDV and demonstrated to be safe and effective antiviral agent against Newcastle disease (Alkhalefa et al. 2022). Graphene oxide–silver nanocomposites have antiviral activity against infectious bursal disease virus (IBDV); however, only graphene oxide did not show antiviral activity against IBDV (Chen et al. 2016). Pangestika and Ernawati (2017) have shown that silver nanoparticles are effective against the growth of IBDV at the concentration of 20 ppm. Chou et al. (2021) reported that carbonized nanogels (CNGs) at the concentration of 30 μg/mL can be used to inhibit infectious bronchitis virus (IBV).

20.3.5 *Nanotherapeutics for Other Viral Diseases*

Besides common livestock and poultry, other animals also get infected with different types of viruses. There are also many zoonotic viruses that infect animals and human. Many nanoparticle-based agents were also tested against such viruses.

Copper iodide (CuI) nanoparticles exhibit extremely high antiviral activity against feline calicivirus (FCV) (Shionoiri et al. 2012). Graphene oxide–silver nanocomposites can inhibit feline coronavirus (FCoV) infection in higher percentage than graphene oxide alone (Chen et al. 2016). Ng et al. (2020) showed the therapeutic application of curcumin-encapsulated chitosan nanoparticles in cats against feline infectious peritonitis virus (FIPV). Diphyllin-loaded polymeric nanoparticles have strong antiviral activity against FIPV (Hu et al. 2017).

Antiviral activity of several nanoparticles has been reported against pathogenic H1N1 influenza virus. Fujimori et al. (2012) showed that nanosized copper iodide (CuI) particles can inhibit the activity of swine-origin 2009 pandemic H1N1 influenza virus. Mori et al. (2013) demonstrated that silver nanoparticle/chitosan composites have antiviral activity against H1N1 influenza A virus. The antiviral activity of iron oxide nanoparticles (IONPs) against H1N1 influenza A virus was reported by Kumar et al. (2019). PEGylated zinc oxide nanoparticles were found to be effective against H1N1 influenza virus infection (Ghaffari et al. 2019).

20.4 Conclusion and Future Prospective

Animals are infected with several viral diseases but the vaccines or effective treatments are limited. However, the development in the field of nanotechnology has revolutionized the veterinary and human medicine. It has already been discussed that several types of nanoparticles exhibit strong antiviral activity against animal or zoonotic viruses. Thus, the horizon of nanoparticles as antiviral agent should be explored further against other economically important disease of livestock and poultry so that losses to the farmers could be minimized. Moreover, the fatal diseases of animals like ASFV that do not have vaccine or effective treatment should be given priority, and possible use of nanoparticles must be tested as early as possible. The applications of multifunctional nanoparticles and other forms of nanoparticles such as nanotraps, nanobubbles, nanofibers, nanorobots, and nanodiamonds also need to be tested against viruses so that strong, durable, and effective antiviral therapy could be formulated.

References

AbouAitah K, Swiderska-Sroda A, Kandeil A, Salman A, Wojnarowicz J, Ali MA, Opalinska A, Gierlotka S, Ciach T, Lojkowski W (2020) Virucidal action against avian influenza H5N1 virus and immunomodulatory effects of nanoformulations consisting of mesoporous silica nanoparticles loaded with natural prodrugs. Int J Nanomedicine 15:5181–5202. https://doi.org/10.2147/IJN.S247692

Alkhalefa N, Khaliel S, Tahoon A, Shaban H, Magouz A, Ghabban H, Lokman MS, Elmahallawy EK (2022) Vitro investigation of the antiviral activity of propolis and chitosan nanoparticles against the genotype VII Newcastle disease virus. Front Vet Sci 9:947641. https://doi.org/10.3389/fvets.2022.947641

Baker JR, Quintana A, Piehler L, Banazak-Holl TD, Raczka E (2001) The synthesis and testing of anti-cancer therapeutic nanodevices. Biomed Microdevices 3:61–69

Buzea C, Pacheco II, Robbie K (2007) Nanomaterials and nanoparticles: sources and toxicity. Biointerphases 2(4):MR17–MR71. https://doi.org/10.1116/1.2815690

Cano A, Ettcheto M, Chang JH, Barroso E, Espina M, Kuhne BA, Barenys M, Auladell C, Folch J, Souto EB et al (2019) Dual-drug loaded nanoparticles of Epigallocatechin-3-gallate (EGCG)/ascorbic acid enhance therapeutic efficacy of EGCG in a APPswe/PS1dE9 Alzheimer's disease mice model. J Control Release 301:62–75

Cano A, Sánchez-López E, Ettcheto M, López-Machado A, Espina M, Souto EB, Galindo R, Camins A, García ML, Turowski P (2020) Current advances in the development of novel polymeric nanoparticles for the treatment of neurodegenerative diseases. Nanomed 15:1239–1261

Chakravarty M, Vora A (2021) Nanotechnology-based antiviral therapeutics. Drug Deliv Transl Res 11(3):748–787. https://doi.org/10.1007/s13346-020-00818-0

Chen L, Liang J (2020) An overview of functional nanoparticles as novel emerging antiviral therapeutic agents. Mater Sci Eng C Mater Biol Appl 112:110924. https://doi.org/10.1016/j.msec.2020.110924

Chen YN, Hsueh YH, Hsieh CT, Tzou DY, Chang PL (2016) Antiviral activity of graphene-silver nanocomposites against non-enveloped and enveloped viruses. Int J Environ Res Public Health 13(4):430. https://doi.org/10.3390/ijerph13040430

Chen HH, Lin CJ, Anand A, Lin HJ, Lin HY, Mao JY, Wang PH, Tseng YJ, Tzou WS, Huang CC, Wang R (2022) Development of antiviral carbon quantum dots that target the Japanese encephalitis virus envelope protein. J Biol Chem 298(6):101957. https://doi.org/10.1016/j.jbc.2022.101957

Chou DL, Mao JY, Anand A, Lin HJ, Lin JH, Tseng CP, Huang CC, Wang HY (2021) Carbonized lysine-nanogels protect against infectious bronchitis virus. Int J Mol Sci 22(11):5415. https://doi.org/10.3390/ijms22115415

Cojocaru FD, Botezat D, Gardikiotis I, Uritu CM, Dodi G, Trandafir L, Rezus C, Rezus E, Tamba BI, Mihai CT (2020) Nanomaterials designed for antiviral drug delivery transport across biological barriers. Pharmaceutics 12(2):171. https://doi.org/10.3390/pharmaceutics12020171

Cormode DP, Skajaa T, van Schooneveld MM, Koole R, Jarzyna P, Lobatto ME, Calcagno C, Barazza A, Gordon RE, Zanzonico P, Fisher EA, Fayad ZA, Mulder WJ (2008) Nanocrystal core high-density lipoproteins: a multimodality contrast agent platform. Nano Lett 8(11):3715–3723. https://doi.org/10.1021/nl801958b

Dekanovský L, Li J, Zhou H, Sofer Z, Khezri B (2022) Nano/microrobots line up for gastrointestinal tract diseases: targeted delivery, therapy, and prevention. Energies 15:426

Dolez PI (2015) Nanomaterials definitions, classifications, and applications. Nanoengineering. Elsevier, pp 3–40

Du T, Liang J, Dong N, Lu J, Fu Y, Fang L, Xiao S, Han H (2018) Glutathione-capped Ag2S nanoclusters inhibit coronavirus proliferation through blockage of viral RNA synthesis and budding. ACS Appl Mater Interfaces 10(5):4369–4378. https://doi.org/10.1021/acsami.7b13811

Du T, Zhang J, Li C, Song T, Li P, Liu J, Du X, Wang S (2020) Gold/silver hybrid nanoparticles with enduring inhibition of coronavirus multiplication through multisite mechanisms. Bioconjug Chem 31(11):2553–2563. https://doi.org/10.1021/acs.bioconjchem.0c00506

Dung TTN, Nam VN, Nhan TT, Ngoc TTB, Minh LQ, Nga BTT et al (2020) Silver nanoparticles as potential antiviral agents against African swine fever virus. Mater Res Express 6:1250g9. https://doi.org/10.1088/2053-1591/ab6ad8

El-Gaffary M, Bashandy MM, Ahmed AR, El-Borady OM (2019) Self-assembled gold nanoparticles for in-vitro inhibition of bovine viral diarrhea virus as surrogate model for HCV. Mater Res Express 6(7):075075. https://doi.org/10.1088/2053-1591/ab18b6

Elgqvist J (2017) Nanoparticles as theranostic vehicles in experimental and clinical applications—focus on prostate and breast cancer. Int J Mol Sci 18:1102

El-Mohamady RS, Ghattas TA, Zawrah MF, Abd El-Hafeiz Y (2018) Inhibitory effect of silver nanoparticles on bovine herpesvirus-1. Int J Vet Sci Med 6(2):296–300. https://doi.org/10.1016/j.ijvsm.2018.09.002

El-Sayed A, Kamel M (2020) Advanced applications of nanotechnology in veterinary medicine. Environ Sci Pollut Res Int 27(16):19073–19086. https://doi.org/10.1007/s11356-018-3913-y

Farag TK, El-Houssiny AS, Abdel-Rahman EH, Hegazi AG (2020) A new approach to the treatment of lumpy skin disease infection in cattle by using propolis encapsulated within Algnps. Adv Anim Vet Sci 8(12):1346–1355. https://doi.org/10.17582/journal.aavs/2020/8.12.1346.1355

Feng C, Fang P, Zhou Y, Liu L, Fang L, Xiao S, Liang J (2018) Different effects of His-Au NCs and MES-Au NCs on the propagation of pseudorabies virus. Glob Chall 2(8):1800030. https://doi.org/10.1002/gch2.201800030

Feynman RP (2011) There's plenty of room at the bottom. Resonance 16:890–905. https://doi.org/10.1007/s12045-011-0109-x

Fujimori Y, Sato T, Hayata T, Nagao T, Nakayama M, Nakayama T, Sugamata R, Suzuki K (2012) Novel antiviral characteristics of nanosizedcopper(I) iodide particles showing inactivation activity against 2009 pandemic H1N1 influenza virus. Appl Environ Microbiol 78(4): 951–955. https://doi.org/10.1128/AEM.06284-11

Ghaffari H, Tavakoli A, Moradi A et al (2019) Inhibition of H1N1 influenza virus infection by zinc oxide nanoparticles: another emerging application of nanomedicine. J Biomed Sci 26:70. https://doi.org/10.1186/s12929-019-0563-4

Graham SP, Cheong YK, Furniss S, Nixon E, Smith JA, Yang X, Fruengel R, Hussain S, Tchorzewska MA, La Ragione RM, Ren G (2021) Antiviral efficacy of metal and metal oxide nanoparticles against the porcine reproductive and respiratory syndrome virus. Nanomaterials (Basel, Switzerland) 11(8):2120. https://doi.org/10.3390/nano11082120

Greenwood DLV, Dynonc K, Kalkanidis M, Sue X, Plebanskid M, Scheerlinck JY (2008) Vaccination against footand-mouth disease virus using peptides conjugated to nano-beads. Vaccine 26(22):2706–2713

Hauser PV, Chang H-M, Yanagawa N, Hamon M (2021) Nanotechnology, nanomedicine, and the kidney. Appl Sci 11(16):7187. https://doi.org/10.3390/app11167187

Howsam M, Mossalayi D, Vincendeau P, Betbeder D (2010) Development of a nanoparticulate formulation of diminazene to treat African trypanosomiasis. Nanotechnology 21(50):1–8

Hu CJ, Chang WS, Fang ZS, Chen YT, Wang WL, Tsai HH, Chueh LL, Takano T, Hohdatsu T, Chen HW (2017) Nanoparticulate vacuolar ATPase blocker exhibits potent host-targeted antiviral activity against feline coronavirus. Sci Rep 7(1):13043. https://doi.org/10.1038/s41598-017-13316-0

Jindal S, Packirisamy G (2020) Nanotechnology based approaches for combatting COVID-19 viral infection. Nano Express 1(2):022003. https://doi.org/10.1088/2632-959X/abb714

Jurj A, Braicu C, Pop L-A, Tomuleasa C, Gherman CD, BerindanNeagoe I (2017) The new era of nanotechnology, an alternative to change cancer treatment. Drug Des Dev Ther 11:2871–2890

Kang BK, Chon SK, Kim SH, Jeong SY, Kim MS, Cho SH et al (2004) Controlled release of paclitaxel from microemulsion containing PLGA and evaluation of antitumor activity in vitro and in vivo. Int J Pharm 286:147–156

Khandelwal N, Kaur G, Chaubey KK, Singh P, Sharma S, Tiwari A, Singh SV, Kumar N (2014) Silver nanoparticles impair Peste des petitsruminants virus replication. Virus Res 190:1–7. https://doi.org/10.1016/j.virusres.2014.06.011

Krishnan SR, George SK (2014) Nanotherapeutics in cancer prevention, diagnosis and treatment. Pharmacology and therapeutics, Sivakumar Joghi Thatha Gowder, IntechOpen

Kumar R, Nayak M, Sahoo GC, Pandey K, Sarkar MC, Ansari Y, Das VNR, Topno RK, Bhawna, Madhukar M, Das P (2019) Iron oxide nanoparticles based antiviral activity of H1N1 influenza a virus. J Infect Chemother 25(5):325–329. https://doi.org/10.1016/j.jiac.2018.12.006

Li S, Silvers SJ, Samy El-Shall M (1996) Preparation, characterization and optical properties of zinc oxide nanoparticles. MRS Proc 452:389

Loo C, Lin A, Hirsch L, Lee M, Barton J, Halas N, West J, Drezek R (2004) Nanoshell-enabled photonics-based imaging and therapy of cancer. Technol Cancer Res Treat 3(1):33–40

Lv X, Wang P, Bai R, Cong Y, Suo S, Ren X, Chen C (2014) Inhibitory effect of silver nanomaterials on transmissible virus-induced host cell infections. Biomaterials 35(13): 4195–4203. https://doi.org/10.1016/j.biomaterials.2014.01.054

Manuja A, Kumar B, Singh RK (2012) Nanotechnology developments: opportunities for animal health and production. Nanotechnol Dev 2:4

Mili B, Das K, Kumar A, Saxena AC, Singh P, Ghosh S, Bag S (2018) Preparation of NGF encapsulated chitosan nanoparticles and its evaluation on neuronal differentiation potentiality of canine mesenchymal stem cells. J Mater Sci 29(4):1–13

Minakshi P, Ghosh M, Kumar R, Brar B, Lambe UP, Banerjee S, Ranjan K, Kumar B, Goel P, Malik YS, Prasad G (2020) An insight into nanomedicinal approaches to combat viral zoonoses. Curr Top Med Chem 20(11):915–962. https://doi.org/10.2174/1568026620666200325114400

Mohantya N, Palaib T, Prustyc B, Mohapatrad J (2014) An overview of nanomedicine in veterinary science. Vet Res 2:90–95

Mori Y, Ono T, Miyahira Y et al (2013) Antiviral activity of silver nanoparticle/chitosan composites against H1N1 influenza A virus. Nanoscale Res Lett 8:93. https://doi.org/10.1186/1556-276X-8-93

Moyer TJ, Zmolek AC, Irvine DJ (2016) Beyond antigens and adjuvants: formulating future vaccines. J Clin Invest 126(3):799–808. https://doi.org/10.1172/JCI81083

Ng SW, Selvarajah GT, Hussein MZ, Yeap SK, Omar AR (2020) In vitro evaluation of curcumin-encapsulated chitosan nanoparticles against feline infectious peritonitis virus and pharmacokinetics study in cats. BioMed Res Int 2020:3012198. https://doi.org/10.1155/2020/3012198

Nikaeen G, Abbaszadeh S, Yousefinejad S (2020) Application of nanomaterials in treatment, anti-infection and detection of coronaviruses. Nanomedicine (Lond) 15(15):1501–1512. https://doi.org/10.2217/nnm-2020-0117

Pangestika R, Ernawati R (2017) Antiviral activity effect of silver nanoparticles (Agnps) solution against the growth of infectious bursal disease virus on embryonated chicken eggs with elisa test. KnE Life Sci 3:536–548. https://doi.org/10.18502/kls.v3i6.1181

Pollock S, Dwek RA, Burton DR, Zitzmann N (2008) N-butyldeoxynojirimycin is a broadly effective anti-HIV therapy significantly enhanced by targeted liposome delivery. AIDS (London, England) 22(15):1961–1969. https://doi.org/10.1097/QAD.0b013e32830efd96

Rafiei S, Rezatofighi SE, RoayaeiArdakani M, Rastegarzadeh S (2016) Gold nanoparticles impair foot-and-mouth disease virus replication. IEEE Trans Nanobioscience 15(1):34–40. https://doi.org/10.1109/TNB.2015.2508718

Rapoport N, Gao Z, Kennedy A (2007) Multifunctional nanoparticles for combining ultrasonic tumor imaging and targeted chemotherapy. J Natl Cancer Inst 99:1095–1106

Reilly RM (2007) Carbon nanotubes: potential benefits and risks of nanotechnology in nuclear medicine. J Nucl Med 48:1039–1042

Riley MK, Vermerris W (2017) Recent advances in nanomaterials for gene delivery—a review. Nanomaterials 7:94

Robert AFJ (2009) Medical nanorobotics: the long-term goal for nanomedicine. In: Schulz MJ, Shanov VN (eds) Nanomedicine design of particles, sensors, motors, implants, robots, and devices. Norwood, MA, Artech House, pp 367–392

Saadh M (2022) Potent antiviral effect of green synthesis silver nanoparticles on Newcastle disease virus. Arab J Chem 15:103899. https://doi.org/10.1016/j.arabjc.2022.103899

Saadh MJ, Aldalaen SM (2021) Inhibitory effects of epigallocatechingallate (EGCG) combined with zinc sulfate and silver nanoparticles on avian influenza a virus subtype H5N1. Eur Rev Med Pharmacol Sci 25(6):2630–2636. https://doi.org/10.26355/eurrev_202103_25427

Saadh MJ, Aggag MM, Alboghdadly A, Kharshid AM, Aldalaen SM, Abdelrazek MA (2021) Silver nanoparticles with epigallocatechingallate and zinc sulphate significantly inhibits avian influenza a virus H9N2. Microb Pathog 158:105071. https://doi.org/10.1016/j.micpath.2021.105071

Sailor MJ, Park J-H (2012) Hybrid nanoparticles for detection and treatment of cancer. Adv Mater 24:3779–3802. https://doi.org/10.1002/adma.201200653

Sarkar J, Das S, Aich S, Bhattacharyya P, Acharya K (2022) Antiviral potential of nanoparticles for the treatment of coronavirus infections. J Trace Elem Med Biol 72:126977. https://doi.org/10.1016/j.jtemb.2022.126977

Saupe A, Rades T (2006) Solid lipid nanoparticles. Nanocarrier Technologies, p 41

Sharma P, Ganti S, Bhate N (2003) Effect of surfaces on the size-dependent elastic state of nano-inhomogeneities. Appl Phys Lett 82(4):535–537. https://doi.org/10.1063/1.1539929

Shionoiri N, Sato T, Fujimori Y, Nakayama T, Nemoto M, Matsunaga T, Tanaka T (2012) Investigation of the antiviral properties of copper iodide nanoparticles against feline calicivirus. J Biosci Bioeng 113(5):580–586. https://doi.org/10.1016/j.jbiosc.2011.12.006

Taniguchi N, Arakawa C, Kobayashi T. On the basic concept of nano-technology. In Proceedings of the International Conference on Production Engineering, Tokyo, Japan, 26–29 August 1974

Thulasi A, Rajendran D, Jash S, Selvaraju S, Jose VL, Velusamy S, Mathivanan S (2013) Nanobiotechnology in animal nutrition. Satish Serial Publishing House, New Delhi

Ting D, Dong N, Fang L, Lu J, Bi J, Xiao S, Han H (2018) Multisite inhibitors for enteric coronavirus: antiviral cationic carbon dots based on curcumin. ACS Appl Nano Mater 1: 5451–5459. https://doi.org/10.1021/acsanm.8b00779

Tong T, Hu H, Zhou J, Deng S, Zhang X, Tang W, Fang L, Xiao S, Liang J (2020) Glycyrrhizic-acid-based carbon dots with high antiviral activity by multisite inhibition mechanisms. Small 16(13):e1906206. https://doi.org/10.1002/smll.201906206

Tong T, Deng S, Zhang X et al (2022) Inhibitory effect and mechanism of gelatin stabilized ferrous sulfide nanoparticles on porcine reproductive and respiratory syndrome virus. J Nanobiotechnol 20:70. https://doi.org/10.1186/s12951-022-01281-4

Torres-Sangiao E, Holban AM, Gestal MC (2016) Advanced nanobiomaterials: vaccines, diagnosis and treatment of infectious diseases. Molecules 21:867

Underwood C, van Eps AW (2012) Nanomedicine and veterinary science: the reality and the practicality. Vet J (London, England: 1997) 193(1):12–23. https://doi.org/10.1016/j.tvjl.2012.01.002

Wang L, Sassi AB, Patton D, Isaacs C, Moncla BJ, Gupta P, Rohan LC (2012) Development of a liposome microbicide formulation for vaginal delivery of octylglycerol for HIV prevention. Drug Dev Ind Pharm 38(8):995–1007. https://doi.org/10.3109/03639045.2011.637048

Yasamineh S, Kalajahi HG, Yasamineh P et al (2022) An overview on nanoparticle-based strategies to fight viral infections with a focus on COVID-19. J Nanobiotechnol 20:440. https://doi.org/10.1186/s12951-022-01625-0

Zainab Mazhari BB (2022) Antiviral properties of Streptomyces tuirus DBZ39 mediated gold nanoparticles against bluetongue virus. Pak J Biol Sci 25(1):90–99. https://doi.org/10.3923/pjbs.2022.90.99

Zeedan GSG, EL-Razik KAA, Allam AM, Abdalhamed AM, Zeina HAA (2020) Evaluations of potential antiviral effects of green zinc oxide and silver nanoparticles against bovine herpesvirus-1. Adv Anim Vet Sci 8(4):433–443. https://doi.org/10.17582/journal.aavs/2020/8.4.433.443

Zhou Y, Jiang X, Tong T, Fang L, Wu Y, Liang J, Xiao S (2020a) High antiviral activity of mercaptoethanesulfonate functionalized Te/BSA nanostars against arterivirus and coronavirus. RSC Adv 10(24):14161–14169. https://doi.org/10.1039/d0ra01387k

Zhou Y, Tong T, Jiang X, Fang L, Wu Y, Liang J, Xiao S (2020b) GSH-ZnS nanoparticles exhibit high-efficiency and broad-Spectrum antiviral activities via multistep inhibition mechanisms. ACS Appl Bio Mater 3(8):4809–4819. https://doi.org/10.1021/acsabm.0c00332

Zhou J, Krishnan N, Jiang Y, Fang RH, Zhang L (2021) Nanotechnology for virus treatment. Nano Today 36:101031. https://doi.org/10.1016/j.nantod.2020.101031

Chapter 21
Nanointerventions in the Treatment of Protozoan Livestock Diseases

Rajesh Kumar, H. N. Malik, S. K. Mohapatra, Komal Kumari, Alonkrita Chowdhury, and Mayukh Ghosh

Abstract Protozoan diseases of veterinary importance such as trypanosomiasis, theileriosis, babesiosis, anaplasmosis, toxoplasmosis, and coccidiosis pose a major threat to the livestock and poultry population in terms of mortality, decreased production, and draught strength, which has prompted the customization of effective control strategies. Conventional synthetic antiprotozoal drugs often fail to target intracellular parasites enhancing drug-dose requirement along with poor biodistribution and pharmacokinetic properties. Off-target cytotoxicity, residual effect, and emergence of drug resistance are the outcomes. Although ethnobotanicals have brought some promise as an alternative strategy, but variability in their bioactive constituents imposing difficulty to dose determination and lack of interest of the stakeholders towards their application in parasitic diseases has resulted underutilization of the strategy. Another aspect, control of arthropod vectors of these protozoan diseases using synthetic acaricides has threatened the environmental safety along with evolution of acaricide resistance among the vectors. Therefore, nanointervention has become inevitable to address the multi-directional issues of controlling protozoan diseases. In terms of enhancing the efficacy of conventional drugs, nano-carrier based targeting and controlled release system lowers the drug-dose requirement, improves pharmacokinetics, as well as curtails the toxicity. Polymeric nanoparticles have found niche application to formulate various nano-

R. Kumar
Department of Veterinary Physiology and Biochemistry, COVS, Lala Rajpat Rai University of Veterinary and Animal Sciences, Hisar, Haryana, India

H. N. Malik · S. K. Mohapatra
Odisha University of Agriculture and Technology, Odisha, India

K. Kumari · M. Ghosh (✉)
Department of Veterinary Physiology and Biochemistry, Faculty of Veterinary and Animal Sciences, RGSC, Banaras Hindu University, Mirzapur, Uttar Pradesh, India

A. Chowdhury
Department of Plant Biotechnology, Institute of Agricultural Sciences, RGSC, Banaras Hindu University, Mirzapur, Uttar Pradesh, India

LLR University of Veterinary and Animal, ICAR-National Research Centre on Equines, *Nanotechnology Theranostics in Livestock Diseases and Management*, Livestock Diseases and Management,
https://doi.org/10.1007/978-981-16-1610-5_21

delivery systems. Various metallic and metal oxide nanoparticles have shown inherent biocidal actions. Therefore, green synthesis of these materials has drawn significant attention as an eco-friendly strategy of protozoan control. Further, several plant-derived essential oils in their nanoformulation or under nano-encapsulation have found to be effective in the control of protozoan parasites and their arthropod vectors too, strengthening their application as eco-friendly biocompatible pesticides. The report of emerging resistance against these novel nanoformulations is also negligible, thus extending another advantage. So, the potential of nanotherapeutics in protozoal control is ample, but the execution adhering to the standards and regulations needs substantial stringency to convert this promise to a sustainable strategy for the control of protozoan diseases.

Keywords Protozoan diseases · Antiparasitic · Chemotherapeutic agents · Ethnobotanicals · Nanotherapeutics · Metallic nanoparticles · Vector control · Nano-vaccines

21.1 Introduction

Parasitic diseases are associated with major concern for animal and human health worldwide. Surra, trichomoniasis, babesiosis, anaplasmosis, fasciolosis, trypanosomiasis, theileriosis, leucocytozoonosis, coccidiosis, toxoplasmosis, histomonosis, sarcocystis, giardiasis, and plasmodium infection are some of examples of parasitic diseases of animals and poultry (Maharana et al. 2016; Mofokeng et al. 2020). Among parasitic diseases, protozoa are one of the most important groups of parasites causing animal diseases leading to loss of production and death. Among protozoan diseases, there are ample species diversity which includes Telosporidia, Piroplasma, Leishmania, Neosporidia, Sarcosporidia, Trypanosoma, Cryptobia, Cercomonadida, Herpetomonas, Orithidia, Bodo, Costia, Monocercomonas, Trichomonas, Hexamitus, and Lamblia of veterinary importance (Crawley 1912). Protozoans are ubiquitous and control of these diseases is a very cumbersome job as pathogens not only present in plenty but have multiple host system distributed in diverse geography. However, effective control and treatment is required as they have great potential to inflict production losses in farm animals and additionally these are considered as significant health problem also, as several are of zoonotic and reverse-zoonotic importance (Esch and Petersen 2013). Most of these diseases are considered to be neglected tropical diseases. These diseases have been treated customarily by chemotherapeutic agents. The chemotherapeutics are antiprotozoal drugs like antimalarials, antiamoebic, antigiardial, trypanocidal, antileishmanial, and anti-toxoplasmic that are used to treat various protozoan diseases (Kaiser et al. 2015). These agents act by disrupting the parasite DNA or inhibiting the protein synthesis, damaging the parasite membrane, and killing the protozoa (Horn and Duraisingh 2014). However, due to the development of resistance by these protozoa, some of these treatments are no longer effective (Capela et al. 2019). Furthermore, uses of different herbal preparations against different parasites have shown good efficacy

(Wink 2012). There is limited research in this field with concrete outcome, mostly due to a lack of interest from the commercial stakeholders. A prominent disadvantage of ethnobotanicals is that the chemical composition of the plants varies from plant to plant due to differences in soils and ecological areas. The dosage of active ingredients is also not controlled, therefore limiting the efficacy of this strategy (Sahoo and Manchikanti 2013). The parasites are also equipped with various mechanisms that help them to evade the host immune system. Again, this area is also understudied. So, the quest for efficient antiprotozoal therapeutics still persists. Advancement of new biotechnological tools has upgraded the specificity of diagnostic tests and prospered the path for customizing novel therapeutic strategies to combat these protozoan diseases.

Nanotechnology is an emerging field of science which deals with synthesis, strategy, and manipulation of particle's size and structure for a variety of applications. The use of nanotechnology in parasitic diseases is expected to open new opportunities by overcoming the increasing problem of drug resistance in several parasites by providing more effective and safer alternatives to traditional therapy. Therapeutic applications of nanomedicine against protozoan diseases mostly employ various types of drug-carrier nano-devices for targeted delivery of conventional or nano-drugs, customization of vector control strategies using nanoformulations, and development of prophylactic measures through nano-vaccines (Duncan 2003; Minakshi et al. 2019; Ramos Campos et al. 2020). Therapeutic utility of nanoparticles (NPs) and nano-carriers against bacterial, viral, fungal, and parasitic diseases has already been explored with promising outcomes. The current chapter is focused to discuss the nanotherapeutics employed for controlling various protozoan diseases with special emphasis on livestock and pet animals.

21.2 Parasites and Parasitic Diseases

A parasite is a living organism that harbors on another and takes its nourishment from that organism or host. Parasites are of many forms, i.e., helminths (worms), arthropods (lice, ticks, mosquitoes, etc.), and protozoa. There are around 1000 species of parasites affecting livestock throughout the world. They can be classified as external or internal, depending on where they live on their host. External parasites often annoy their hosts by biting, embedding, or otherwise irritating the skin. They can cause serious diseases, such as mange and scabies, which affect animals' health and growth. Internal parasites live in the blood or tissues inside an animal's body. Some organisms enter into an animal when it swallows contaminated feed or water. Others burrow through the skin, reach the blood stream, and settle in a preferred location to mature and reproduce. Internal parasites often interfere with digestion and assimilation of feed, causing poor growth, temporary or permanent injuries, or death (https://www.nal.usda.gov/exhibits/speccoll/exhibits/show/parasitic-diseases-with-econom/parasitic-diseases-with-econom). Hemoprotozoan disease of veterinary importance such as trypanosomiasis, theileriosis, babesiosis, and anaplasmosis poses a major threat to the cattle population in terms of mortality, decreased milk production, and decreased draught strength. One of the most significant hemoprotozoan illnesses affecting human and animal health in the tropics, particularly in Southeast Asia, is trypanosomiasis, often known as surra (Maharana et al. 2016). In addition to other contributing factors such as concurrent sickness, inadequate nutrition, innate and acquired resistance, parasite pathogenicity, and parasite strain, the prevalence of surra rises around the monsoon when the animals are under the most work stress due to intense agricultural activities. Lack of sensitive diagnostic tests and data on *Trypanosoma evansi* distribution in India limit the effectiveness of its' surveillance. Although veterinary professionals and diagnostic labs have almost adequate knowledge of *T. evansi*, however, lack of highly sensitive field diagnostic methods makes it difficult for quick identification as microscopy remains as the sole pen-side diagnostic method oftentimes.

Babesiosis has long been recognized as a significant disease of cattle, horses, and dogs, but the infection is now receiving more attention as a re-emerging zoonotic disease. Erythrocytes of both domestic and wild animals become paralyzed by babesiosis. Red water fever, often known as bovine babesiosis, affects cattle throughout the world. Indian cattle are mostly infected by *Babesia bigemina*. Bovine infection is characterized by increased body temperature, rapid heartbeat, and respiration as well as a sharp decline in appetite and reluctance to move. In severe infection, hemoglobinuria and hemolytic anemia are signs of a catastrophic outcome when chemotherapy is not used (Narurkar et al. 2017). The main flaw of available field-level microscopic diagnostic technique is low sensitivity, which makes it challenging to identify low parasitemia throughout the chronic stage of infection which further serves as carrier for rest of the herd (Almeria et al. 2001).

Bovine tropical theileriosis is a tick-borne infection caused by *Theileria annulata*, an intracellular protozoan parasite. It is a lymphoproliferative disease with high mortality and morbidity in cattle. *Theileria sergenti*, *T. buffeli*, and *T. orientalis* cause mild or asymptomatic disease in cattle, well-known as bovine benign theileriosis. Certain *Ixodid* ticks, such as *Hyalomma anatolicum anatolicum*, *H. marginatum*, and *H. excavatum* known to transmit *T. annulata*, are found in the Mediterranean region, especially in semiarid areas (Tsamesidis et al. 2021).

Anaplasmosis or gall sickness is an infectious, noncontagious intra-erythrocytic rickettsial disease. Bovine anaplasmosis is primarily inflicted by *Anaplasma marginale* and *Anaplasma centrale* in tropical and subtropical areas. In a manner like that used to diagnose babesiosis, the diagnosis of bovine anaplasmosis can be performed by microscopic detection of *Anaplasma* parasites in a well-prepared thin smear free of foreign materials. This approach is appropriate for clinically infected animals, but it is unreliable for identifying infection in presymptomatic or carrier animals (Aubry and Geale 2011).

Globally prevalent toxoplasmosis is caused by *Toxoplasma gondii*. Most often, eating raw or undercooked meat or other animal products that contain infectious parasitic stages of *T. gondii* is linked to a significant portion of human illness. This parasite is of tremendous zoonotic importance and affects animal reproduction with output losses. Toxoplasmosis in livestock animals consequently causes monetary losses due to death, abortion, and neonatal mortality (Mammari et al. 2019). Molecular, serological, histological, or combinations of these approaches are used to diagnose toxoplasmosis. Toxoplasmosis symptoms are nonspecific and insufficiently distinctive to make a conclusive diagnosis. Indirect hemagglutination assay, indirect fluorescent antibody assay, latex agglutination test, and ELISA are the major options for antemortem diagnosis. Pyrimethamine is considered to be the most effective drug against toxoplasmosis; while treating acute toxoplasmosis, toltrazuril, ponazuril, and diclazuril can be used to prevent cats from shedding oocysts.

Protozoan parasites of the genus *Eimeria*, a class of monoxenous obligate intracellular parasites, are responsible for coccidiosis. Chickens (*Gallus gallus*) are very susceptible to several species of this genus; each of the species has unique pathogenic traits and predilection sites at digestive tract. *Eimeria* inflicts aberration in the usual intestinal functions, causing inadequacies in nutrient absorption, reduced productivity, and even mortality resulting significant revenue losses.

The most crucial step in controlling hemoprotozoan infections is early and precise diagnosis of the etiological agent on which the therapeutic regimen is directed (Maharana et al. 2016). Microscopy-based detection is still the most affordable and efficient approach for diagnosis, despite the techniques' poor sensitivity and specificity, particularly for subclinical and chronic infections. Besides serological tests, specific nucleic acid–based diagnostic techniques offer faster throughput (Nadeem et al. 2011). Several such molecular diagnostic tools, including Reverse Line Blot (RLB), PCR, PCR-ELISA semi-nested PCR, and real-time PCR, are currently in trend; however, these techniques are meagerly accessed by the terminal farmers due to lack of sophisticated laboratories. Thus, the quest for specific as well as sensitive animal-side diagnostic tool is going on besides developing novel therapeutics as the reports of drug resistance is nowadays arising very frequently. Currently, the nanotechnology pursuit for customizing effective antiprotozoal therapeutics has been intensified which has been delineated in a much-needed and timely manner under the subsequent sections.

21.3 Nanotechnology and Nanoparticles

Nanotechnological sciences (nanoparticle-based science) means technology usage of physical, chemical, or biological systems at scales starting from individual atoms or molecules to submicron dimensions and has diverse prophylactic, diagnostic, monitoring, and therapeutic applications constituting the arena of nanomedicine. Nanotechnology has advanced the development of many fields because of the unique properties of nanomaterials in biomedical applications including tissue engineering, drug delivery, bioimaging, and nano-diagnostics which have been promoted by nanotechnology significantly. Nano-sensors are evolving at a rapid pace to accomplish noninvasive diagnosis of various diseases with high sensitivity and precision.

As the bulk material is reduced to nanosize, there are drastic changes in their physicochemical properties which render them competent for diverse nanomedicinal applications. The nanoparticle repository is extensive and diverse, comprised of several metallic, organic, and composite materials. The use of noble metal nanoparticles, particularly gold and silver, in biomolecular applications has surged, ranging from innovative strategies for molecular diagnostics to radically new ways of treatment. Taking advantage of the optical–chemical characteristics of these metal NPs, every year new methods of molecular diagnostics of infectious diseases are reported to provide higher analytical capability, sensitivity, and throughput at lower costs and with the possibility to be used where needed. Gold NPs (AuNPs) and silver NPs (AgNPs), or a combination of both, possess amazing optical/spectral properties. AuNPs are noncytotoxic in nature with an additional advantage of a huge surface area, which makes their surfaces accessible for modification with targeting molecules. This renders them advantageous over other nanoparticles for various biomedical applications. Targeted drug delivery is the most efficient therapy since it is possible to target only the affected cells or part. This minimizes the side effects of drugs. This is useful in treating cancer where the medicines can be delivered directly to the affected cells without damaging healthier cells in any way. Metal NPs are a promising anti-biofilm agent. They are already used as additives, mainly to medical devices and food packaging and preservation. Among metal NPs, nano-silver is the most active and commonly used in medicine and hygiene. Similarly, various non-metallic and carbon-based NPs along with biomolecules in their nanoforms have been deployed to counter various diseases. The roles of nano-carriers such as liposomes, polymeric nanoparticles, solid lipid NPs, and dendrimers in improving the delivery of antibiotics, reducing the minimum inhibitory concentration (MIC), and minimizing side effects have been demonstrated. The therapeutic implications of these NPs against protozoan diseases have been delineated hereunder.

21.4 Conventional Nanoparticles as Antiprotozoal Agents

21.4.1 Gold Nanoparticles for Therapeutic Control of Protozoan Diseases

AuNPs has depicted potential in the treatment of various diseases (Sultana et al. 2020). It occurs in different forms, particularly two oxidized forms (Au^{+1}, Au^{+3}) and one nonoxidized form (Au^{0}) (Moradi et al. 2016). Oxidized forms are reduced to nonoxidized forms through the use of chloroauric acid as a precursor. AuNPs are prepared by the use of either in vitro techniques or conventional methods. After preparation, the fresh batches are stored in the dark to avoid their oxidation by light (Baek et al. 2020). The effectiveness of AuNPs has already been shown against various parasites of public health importance including the protozoa *Toxoplasma gondii*, *Trypanosoma* spp., *Leishmania* spp. (*L. tropica* and *L. donovani*), and

Cryptosporidium spp. (*C. parvum*); helminths including trematodes (*Schistosoma* spp.) and cestodes (*Raillietina* spp.); vectors including mosquito, genus *Aedes* (spp. *A. aegypti L.*), *Anopheles* (spp. *A. stephensi*), and Culex (spp. *C. quinquefasciatus*); and dipteran flies. Currently, AuNPs are considered as novel products in the field of drug discovery (Norgan and Pritt 2018). Protozoan parasites are notoriously difficult to control. Success has been achieved with the use of AuNPs against these parasites. Some of these protozoan parasites and their control using AuNPs are described in the succeeding texts.

21.4.1.1 Control of Plasmodium Using AuNPs

Plasmodium is an intracellular parasite that is transmitted through the *Anopheles* mosquito and causes malaria in infected individuals. Partial success has been achieved in the fight against malaria by controlling the vector through the application of insecticides, but the development of resistance in the parasite to insecticides has worsened the situation. Treatment of infected individuals with AuNPs yielded good results (Bajwa et al. 2022). In one study, patients were treated with AuNPs that had been synthesized from *C. guianensis* flower extract. Tremendous antimalarial properties were observed. Similarly, in another study, mice were experimentally infected with *Plasmodium berghei* and treated with *Streptomyces LK-3*-mediated AuNPs (Varela-Aramburu et al. 2020). The findings suggested that these NPs could be considered as potential antimalarial drug candidates. AuNPs have also shown good results in the diagnosis of malaria. Recently, a cost-effective diagnostic technique was developed that was inspired by the "coffee-ring" phenomenon—the occurrence of histidine-enriched protein-II, which is a biomarker of *Plasmodium falciparum* (Prakas et al. 2020). This protein was detected by the use of a surface-coupled ring of nitrilotriacetic acid aurum-plated polystyrene microparticles in infected individuals. This kit, based on AuNPs, indicates the presence of the protein by the formation of a ring that can be analyzed visually. Hence, this kit can be used to detect malarial parasites (Gruessner and Weathers 2021).

21.4.1.2 Application of AuNPs in Leishmaniasis Control

Leishmaniasis is a vector-borne protozoan disease inflicting fatality worldwide and is highly prevalent in tropical areas. Transmission occurs through the sandfly vector. About seven million cases are reported annually (Abdellahi et al. 2022). The prevalence of *Leishmania* is high because there is a lack of suitable treatment options. Antileishmanial pharmaceutical drugs that are used against infection are expensive and not widely available (Boukthir et al. 2020). However, these drugs have certain limitations: (i) severe side effects in parenteral use; (ii) treatment failure and drug resistance; (iii) unequal distribution of drug at target site; (iv) lack of research in herbal medicine (Uliana et al. 2018). In later studies, it was revealed that *Leishmania* sp. have developed resistance against these chemotherapeutic agents.

Further studies were conducted in which genomic studies of *Leishmania* for drug development were carried out (Bruni et al. 2017). It was revealed that *Leishmania* sp. modified its genes that were targeted by these drugs. The major issue was unequal distribution/reach of the drug at the target site because of the intracellular nature of the parasite. Then, another milestone was achieved by the use of nanomaterials as drug delivery vehicles to target *Leishmania* sp. In some studies, these nanomaterials have shown greater efficacy and controlled the release of the drug at the target site along with imparting less toxicity (Yasinzai et al. 2013). Currently, this area of drug development is still understudied. It will take more time to establish more suitable nanomaterial-based chemotherapeutic agents that may have no resistance (Karimkhani et al. 2016). In recent years, nanointervention against leishmaniasis has shown successful results in the diagnosis and treatment of the infection. In a study, promastigotes of *Leishmania braziliensis* were targeted in skin lesions with AuNPs that were incorporated into membranes of natural rubber (Sazgarnia et al. 2013). The results showed a decreased growth rate of promastigotes, a change in the physiological behavior of the promastigotes, and decreased lifespan. These particles also aided in the angiogenesis of infected areas of skin. There was no issue of resistance. Treatment with AuNPs coupled with microwave radiation at a frequency of 2450 MHz showed greater efficacy of the particles. Various NP-based diagnostic techniques are also used, including nano-sensors to detect *Leishmania* infection (Karimkhani et al. 2016). Furthermore, a diagnostic kit has been developed in which AuNPs are conjugated with four oligonucleotide probes that target *Leishmania* DNA. A color change from red (the medium) to purple indicates the presence of leishmanial DNA. The minimum detection limit is 11.5 ng/μL of the sample (Sazgarnia et al. 2013). This technique has shown good results in the detection of canine leishmaniasis. It is suggested that AuNP-based treatments and diagnostic tools have provided fresh impetus to drug development.

21.4.1.3 Treatment of Toxoplasmosis Using AuNPs

Toxoplasmosis is caused by the intracellular protozoan parasite *Toxoplasma gondii*. The infection is usually asymptomatic but can be life-threatening in immunocompromised individuals (Bastien 2002). Recently, *Toxoplasma gondii* was treated with NPs (Fig. 21.1). In vitro studies have targeted *Toxoplasma gondii* tachyzoites with different concentrations of AuNPs: 100–1000 ppm for exposure time periods between 30 and 180 min (Adeyemi et al. 2017). The concentration of AuNPs and period of treatment were found to be directly proportional to the number of *Toxoplasma gondii* tachyzoites killed. The results revealed that AuNPs have offered potential as drugs for the treatment of toxoplasmosis (Vazini and Br 2018).

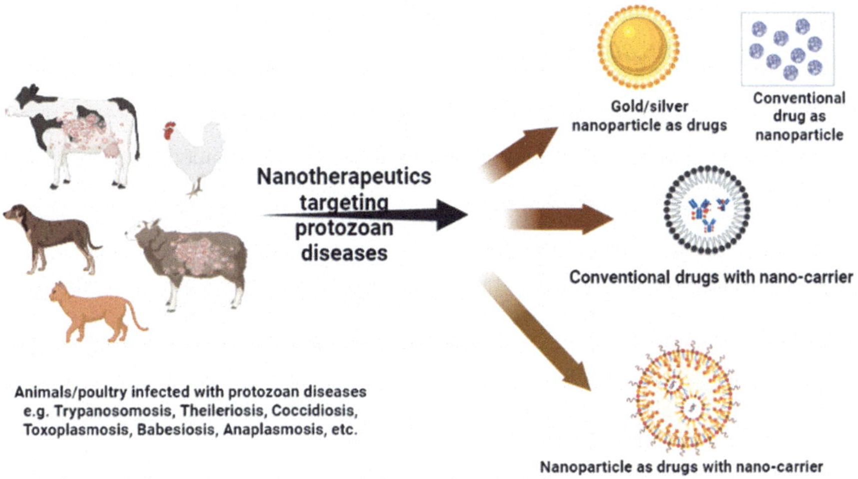

Fig. 21.1 Nanointervention in the treatment of protozoan livestock diseases

21.4.1.4 Application of AuNPs in Trypanosomiasis

African trypanosomiasis is a neglected tropical disease caused by *Trypanosoma brucei gambiense* or *T. brucei rhodesiense*. The tsetse fly transmits the infectious agent. In Latin America, the infection is caused by *T. cruzi* and *Triatoma* bugs that transmit the parasite which causes a disease commonly known as Chagas disease. Currently, there are no reliable treatment options for the control of this disease. Recently, treatment of trypanosomiasis with NPs has shown promising results (Fig. 21.1). Various *Trypanosoma* spp. were targeted by inorganic metal NPs. AuNPs and AgNPs were used against *T. brucei gambiense*, *T. cruzi*, *T. evansi*, and *T. congolense* (Eger et al. 2012; Rehman et al. 2021). These NPs showed great efficacy against *T. brucei gambiense* and *T. cruzi*, and protozoal growth was arrested by 50% when targeted *T. evansi* and *T. congolense*. Similarly, arginine kinase was targeted through treatment with AuNPs and AgNPs. Arginine kinase (phosphotransferase) is essential for *Trypanosoma* energy metabolism (Adeyemi and Whiteley 2013). These NPs also showed effective arrest of parasite growth. AuNPs were also used in the diagnosis of *Trypanosoma* with the use of confocal length microscopy. Endocytosis was observed in epimastigotes of *T. cruzi*. Epimastigotes ingested the AuNPs and the formation of gold ring complexes was observed. It was concluded that, in theory, this could be a reliable technique for the diagnosis of trypanosomiasis (Eger et al. 2012).

21.4.2 Silver Nanoparticles for Therapeutic Control of Protozoan Diseases

Silver NPs have many applications in the field of biomedical sciences. They have been used extensively in pharmaceuticals, bioimaging, and biomedical devices and as antipathogenic agents. Recently, their benefits have also been extended in the control of parasites. AgNPs are synthesized through the application of various in vitro techniques, such as electrochemical reduction, the sol–gel technique, and green synthesis. Green synthesis is favored because chemical methods are expensive and the reagents used for electrochemical reduction are toxic. Plant extracts, such as *Solanum trilobatum*, *Sechium edule*, and the fruit of the *Ficus* tree are used in green synthesis (Roy et al. 2019). These show promise in the treatment of parasitic infections, but the NPs are produced in uneven sizes of 10–250 nm. The antiparasitic impact of AgNPs is indicated in the following.

21.4.2.1 Silver Nanoparticles for Vector Control

Mosquitoes serve as vectors for various protozoan parasites. Various chemical and biological methods have been applied for their control. However, due to immense genetic variation and development of resistance in mosquitoes, these techniques are no longer effective adequately. In the last decade, excellent results have been obtained in mosquito control through the use of green synthesized AgNPs. In one study, AgNPs synthesized from *Mimosa pudica* leaf extract were used against *Anopheles subpictus* and *C. quinquefasciatus* to good effect (Marimuthu et al. 2011). When AgNPs were synthesized through the use of methanol with *Mimosa pudica* leaf extract, the efficacy of the AgNPs was increased (Priyadarshini et al. 2012). AgNPs synthesized from the fungal species *Chrysosporium keratinophilum* and *Verticillium lecanii* showed good activity against the pupae and larvae of *A. stephensi*, *C. quinquefasciatus*, and *A. aegypti* (Soni and Prakash 2013). These reports clearly depicted that green synthesized AgNPs can be used as an alternative to other control strategies for mosquitoes.

21.4.2.2 Silver Nanoparticles for Controlling Plasmodium Infection

Similar to gold-based NPs, AgNPs are also effective in the treatment and control of malaria caused by plasmodium. Green synthesized AgNPs using *Azadirachta indica* and *Ocimum sanctum* leaf extracts have shown good antiplasmodial activity with smaller particles being more effective (Soni and Prakash 2013). Similarly, green synthesized AgNPs using leaf extracts of Ashoka has also depicted antiplasmodial activity with IC_{50} of 8 μg/mL, while the same from the aqueous leaf extract of *Salvia officinalis* yielded IC_{50} of 3.6 μg/mL (Mishra et al. 2013; Okaiyeto et al. 2021).

Similarly, *S. officinalis* leaf extract synthesized AgNPs depicted antiplasmodial and hepatoprotective effects against *Plasmodium chabaudi* (Metwally et al. 2021).

21.4.2.3 Application of Silver Nanoparticles for the Treatment of Leishmaniasis

In vitro studies of AgNP treatment against *Leishmania tropica* reported the formation of reactive oxygen species (ROS), to which *Leishmania* is known to be highly sensitive (Chakravarty and Sundar 2019). The morphological features and proliferation, infectivity, and survival rates of *Leishmania* were studied in the presence of AgNPs. These NPs were found to impair the metabolism, decrease the survival rates, and inhibit the proliferation of *Leishmania tropica* promastigotes (Abdellahi et al. 2022). Moreover, the reactivity of AgNPs was enhanced by application of UV light, which increased their impact by 2- to 6.5-fold (Dos Santos et al. 2014; Ali et al. 2021). Moreover, the effects of AgNPs were enhanced through the use of both infrared and UV light. ROS were formed, which led to the development of toxicity in *Leishmania* and resulted in the killing of the parasite (Saleem et al. 2019). In another study, AgNPs, prepared from *Anethum graveolens* leaf extract, were structurally analyzed through the use of X-ray diffraction and their size (35 nm) was estimated through the use of transmission electron microscopy. The combination of AgNP with miltefosine (12 μM) resulted in the death of amastigotes and promastigotes of *Leishmania* (Morones et al. 2005). The multiple ways in which NPs act mean that they can be used alone or in combination with other agents. This may be a reliable form of treatment soon.

21.4.2.4 Application of Silver Nanoparticles Against *Toxoplasma gondii*

Toxoplasma gondii is a coccidian protozoan parasite that is prevalent across the globe. Various regimes can be used for the treatment and control. However, due to vaccine failure and resistance to chemotherapeutic agents, most of these options are not reliable. Therefore, the utility of NPs against *Toxoplasma gondii* has been explored by several research groups. The application of silver, gold, and platinum NPs against *Toxoplasma gondii* resulted in damage to the mitochondrial membrane of the parasite and 90% inhibition of the infection (Bertin et al. 2019). There was no adverse effect on host cells. In another study, AgNPs were combined with chitosan and changes that were observed in the *Toxoplasma gondii* parasite included tachyzoite shape deformity, movement arrest, and a decrease in the parasite burden (Chikkanna et al. 2019). The gamma interferon was also increased in treatment groups. Moreover, in in vitro studies, biogenic AgNPs were tested against *Toxoplasma gondii* in HeLA cell lines at concentrations ranged from 0.25 μM to 12 μM. Parasite proliferation rates were significantly reduced (Cheraghipour et al. 2020). Furthermore, AgNPs that were prepared from *Phoenix dactylifera* and *Ziziphus christi* extract were used against *Toxoplasma gondii*, and they showed improvement in serum profile and reduced immunoreactivity of infected mice. These studies show that AgNPs can offer a reliable treatment regime for the control of toxoplasmosis (Fig. 21.1) (El-Kosary et al. 2021).

21.4.3 Application of Other Metallic NPs in Protozoan Disease Control

Other metallic NPs used in protozoan disease control includes selenium nanoparticles (SeNPs) which inhibit the proliferation of the amastigote and promastigote forms of *Leishmania major* with inhibitory concentrations (IC_{50}) of 1.62 ± 0.6 and 4.4 ± 0.6 μg mL^{-1} against the promastigote and amastigote (Beheshti et al. 2013). These NP treatments were further able to abolish the localized lesions after 14 days in mice. Similar results were obtained using Mn_2O_3 NPs with IC_{50} values of 15 and 40 μg mL^{-1}, respectively, against *Leishmania major* promastigotes and macrophages (Tavakoli et al. 2019). Similarly, *Toxoplasma gondii*-infected mice treated with various oral doses of SeNP for 14 days once daily led to a dose-dependent significant reduction of tissue cysts (Keyhani et al. 2020). There was significant reduction in tachyzoites load and mortality rate at the doses of 5 and 10 mg/kg body weight once daily (Shakibaie et al. 2020). Similarly, SeNPs generated using *Azadirachta indica* leaf extract administered in mice infected with coccidian parasite *Eimeria papillata* at 0.5 mg/kg of body weight dose for 5 days inhibited 97.21% oocyst output which was comparable to standard coccidiostat drug amprolium. In addition, it provided protection to jejunum indicated by drastically lower oxidative stress, inflammation, and apoptotic indicators (Abdel-Gaber et al. 2023). Various other metallic NPs such as calcium oxide (CaO) nanoparticles, magnesium oxide (MgO) NPs, and tellurium oxide (TeO_2) nanorods had also yielded antiprotozoal effects under in vitro experiment (Dalir Ghaffari et al. 2022; Ghaffari et al. 2023; Tavakoli et al. 2022). Cobalt NPs had been used against sight threatening protozoa *Acanthamoeba keratitis* and resulted consistent inhibition at 100 μg/mL indicating potent amoebicidal activity of the NPs without causing any significant damage to host cells (Anwar et al. 2019). Silver NPs had shown antiplasmodial activity against *P. falciparum*, and effective killing had also been observed by gold NPs at a concentration of 0.3 mg mL^{-1} for *Giardia* cysts; NPs of copper and silver had shown effectiveness against *E. histolytica* (Saad et al. 2015). Against *Eimeria papillata*, zinc oxide nanoparticles reduced the oocyte load post five day treatment to $12.5 \times 10^3 \pm 1000$ oocysts from $29.7 \times 10^3 \pm 1500$ oocysts/g feces when treated with 10 mg/Kg dose of NPs orally once daily (Dkhil et al. 2015).

21.4.4 Application of Nonmetallic NPs in Protozoa Control

Nonmetallic NP-based drugs like silica NPs have shown efficacy in controlling malaria- and leishmaniasis-causing protozoan parasites. It can further be used with antimalarial drug artemisinin to facilitate sustainable release of the drug so that dose can be maintained for longer duration (Tsamesidis et al. 2021). Nano-SiO_2 has not only been helpful in reducing parasitic load but also found its application in vector control against *Hyalomma* spp. of ticks acting as vector of theileriosis. The dose-dependent acaricidal action was found where maximum activity up to 97% was reported at the dose of 250 mg/mL of SiO_2-NPs with low toxicity assayed by MTT toxicity test in the initial studies (Norouzi et al. 2022). Micro- and nanoparticles generated from ellagic acid, a plant-derived polyphenol or its derivative were found to produce antiprotozoal activity against different species of *Babesia* such as *B. bovis*, *B. bigemina*, *B. divergens*, *B. caballi*, and *Theileria equi* under in vitro condition with IC_{50} in the range of about 3–20 μM, while against *B. microti* in mice model infection at a dose rate of 70 mg/kg (Beshbishy et al. 2019). NPs of natural compounds such as curcumin and berberine (CurNPs and BerNPs) have found to exert considerable antiprotozoan activity against *Entamoeba histolytica* and *Trichomonas vaginalis*. The IC_{50} values of CurNPs against *E. histolytica* and *T. vaginalis* were found to be 9.48 and 4.25 μg/mL, respectively, while for BerNPs, the same were 0.24 and 0.71 μg/mL (Pacheco-Ordaz et al. 2022). Self-assembled nanoparticles customized from ester-linked dihydroartemisinin trimer (DHA3 NPs) depicted superior antimalarial effects as compared to antimalarial drug dihydroartemisinin in a murine experimental model, with diminished recurrence rate and improved cure rate and survival in the *Plasmodium yoelii* BY265-infected animals (Guo et al. 2021). So, it is evident that conventional standard antiprotozoal drugs can also be potentiated by administering them in their nanoforms.

21.5 Nano-Drugs as Antiprotozoal Therapeutics

Even, efficacy of various drugs in their nanoforms had been studied against protozoan diseases. Nano-nitazoxanide (NTZ) has shown better result than its free form in combating *Cryptosporidium* (Sedighi et al. 2016). Amphotericin B-based NPs have shown similar effect on cure of leishmaniasis as of free drug but in lower dose (Sharma et al. 2021). Nano-zinc oxide and nano-propolis combination has been used to provide protection against *Eimeria* infestation in rabbits (Abd El Megid et al. 2018). Ivermectin or avermectin has achieved an unmatched success against parasitic disease like onchocerciasis, lymphatic filariasis, and strongyloidiasis, scabies, and lice with an extended application potential as antiviral serving human society since 1987, but nowadays resistance against these drugs have been reported in animals (van Wyk and Malan 1988). Application of nanoformulation can potentiate their applicability. Ivermectin nano-emulsion can be used as an antiparasitic in treating broad-spectrum organism including nematode and arthropod, vectors of various protozoan diseases, and protozoan itself like *Trypanosoma evansi* and *Leishmania*. The application can help in providing multifaceted gain like improved solubility, bioavailability, and lesser toxicity as well as side effects (Crump 2017; Ahmadpour et al. 2019). Similarly, tafenoquine is usually used to treat protozoan disease like *Babesia* and *Plasmodium* species (Mordue and Wormser 2019), but when the drug is used as nanoformulation, it has led to enhancement of bioavailability and hence can provide better treatment efficiency against *Plasmodium* and reduce the blood loss as depicted in mice model (Melariri et al. 2015). Silver NPs-loaded metronidazole has found to be favorable to combat *Giardia lamblia* than their single counterpart; in addition, it has extended desirable effects such as lower toxic effects along with significant enhancement in the levels of antioxidant reduced glutathione (GSH) and decline in the level of lipid peroxidation product malondialdehyde (MDA) (Hamad 2020). A similar report regarding enhancement of the potency of the conventional drug has been found when low doses of metronidazole were encapsulated of in poly(D, L-lactide-co-glycolide) polymeric vehicle and administered to treat *Giardia intestinalis* (Madbouly et al. 2020). Combination of AuNPs and *Citrullus colocynthis* L. NPs yielded better result to treat *Giardia lamblia* infection in mice than single components without any detrimental effect on hematological and serum biochemical parameters (Al-Ardi 2020). Similarly, enhanced therapeutic effects were observed when chitosan NPs and metronidazole were used in combination against *Giardia lamblia* with more than 90% inhibition of *Giardia* cyst and trophozoite (El-Gendy et al. 2021). Diminazene, an anti-trypanocidal drug effective against various *Trypanosoma* species, was formulated in nanoform as porous cationic nanoparticle with an oily core ($_{70}DGNP^{+}$) leading to increase in efficacy of diminazene as nano-drug was more active than the free form in cidal action (Kroubi et al. 2010). Nano-extracts of *Citrullus colocynthis* and *Capparis spinosa* have shown comparable result to metronidazole against *Trichomonas vaginalis* without any hemolysis during an in vitro trials (Al-Ardi 2021). Application of nanoform of chitosan extracted from *Penicillium* fungi has yielded inhibitory activity of 59.5%, 99.4%, and 31.3% against *P. falciparum*, *T. vaginalis*, and *G. lamblia*, respectively, under in vitro experiment at the dose rate of 50 μg/mL without exerting significant toxicity and side effects (Elmi et al. 2021). Metronidazole (MTDZ), a conventional drug to treat trichomoniasis, when administered in a thermo-reversible gel containing chitosan-ensembled nanoform (CHS-MTDZ NPs) was able to cure trichomoniasis in BALB/c mice within three days, while regular gel and pure MTDZ failed to cure the infection (Fang et al. 2022).

Various herbal drugs in their nanoformulation have shown augmented potential against various protozoan diseases. Curcumin-based NPs had shown better effect than its free counterpart against *Giardia lamblia* and *Plasmodium* (Said et al. 2012; Nayak et al. 2010). Nano-curcumin has also shown its effectiveness against coccidiosis in broiler chicks with additional benefits of resulting in better feed conversion ratio (FCR) and growth (Gogoi et al. 2019). Nanoformulation of eugenol, a clove oil constituent, limits the deleterious effects of coccidiosis and provides additional benefit of antioxidant and anti-inflammatory properties (Youssefi et al. 2023).

Nano-emulsion of *Schinus molle* essential oil showed better cidal activity against *Trypanosoma evansi*, while *Lavandula officinalis* essential oil was found to be more effective against *Trichomonas vaginalis*. These evidences are ample enough to decipher the antiprotozoal activity of plant/herb-derived essential oils in their nanoform besides showing better biocompatibility and less toxicity. Similarly, nano-emulsionated essential oil of *Lavandula angustifolia* and *Rosmarinus officinalis* has depicted anti-leishmanial effect, and *Zataria multiflora* showed cidal effect against *Echinococcus granulosus* and hydatid cysts. Such nanoformulations of plant-derived components have either shown better potency in comparison to their free counterpart or needed in lower doses to achieve therapeutic efficacy besides exerting less systemic toxicity with respect to their standard compounds (Echeverría and Duarte Galhardo de Albuquerque 2019). Presence of various constituents in herbal nanoformulations like monoterpenes, sesquiterpenes, and phenylpropanoids can act against protozoa through various mechanisms, and nanoformulation can enhance their potency due to better pharmacokinetic properties, enhanced surface area for functionalization, and other advantageous properties.

Nanoformulation of conventional chemotherapeutic agents and herbal drugs can impart several positive attributes to already available drugs; hence, this aspect needs in-depth introspection in order to establish such nano-drugs for future uses. This will also reduce the dose requirement of conventional drugs; hence, the risk of residual effects and resistance development can be mitigated. This segment illuminates the preparations of nanoformulation of existing conventional drugs that have been used for enhancing efficacy and reducing doses or time lag to get the desired response. Similarly, effective nano-carriers are used to improve the pharmacokinetic properties of conventional drugs and facilitate targeted delivery and sustained release in order to enhance the efficacy of the therapeutic agents. In addition, nanoformulation employing different functionalized groups or types amalgamated with drug component can also produce improved outcome. Another possible way of application is using various kinds of NPs possessing inherent therapeutic properties along with nano-carriers for the treatment of protozoan diseases which has been elaborated subsequently.

21.6 Nano-Delivery Systems for Antiprotozoal Drug Administration

Nano-encapsulation of various conventional antiprotozoal drugs has shown promising results in terms of enhanced effectiveness. For example, lipid matrix nanoparticle delivery systems have been used for carrying various drugs like paromomycin, trifluralin, amphotericin B, oryzalin, buparvaquone, and miltefosine in solo or in combination for treating leishmaniasis. In such applications, nano-carriers have been formulated in the form of liposome, nano-structured lipid carriers, solid lipid nanoparticles, nano-emulsions using diverse lipids or derivatives such as dipalmitoyl phosphatidylglycerol (DPPG), dipalmitoyl phosphatidylcholine (DPPC), dimyristoylphosphatidylcholine (DMPC), dimyristoylphosphatidylglycerol (DMPG), dioleoylphosphatidylcholine (DOPC), dioleoylphosphatidylglycerol (DOPG), 1,2-dipalmitoyl-sn-glycero-3-phosphocholine (DPPC), cholesterol (CHOL), 1,2-dipalmitoyl-sn-glycero-3-phosphate (monosodium salt) (DPPA), dioleoylphosphatidylserine (DOPS), 1,2-distearoyl-sn-glycero-3-phosphoethanolamine-N-[methoxy(polyethylene glycol)-2000] (ammonium salt) (DSPE), and chalcones and their derivatives. Similarly, various other drugs including amphotericin B, amphotericin B with magnetic citrate-coated maghemite, and hydroxymethylnitrofurazone have been deployed using various polymeric nanoparticle delivery systems like dimercaptosuccinic acid, poly(glycolic acid), and poly (n-butyl cyanoacrylate) (de Souza et al. 2018). Buparvaquone loaded in solid lipid nanoparticles (BPQ SLN) has shown higher uptake by macrophage cell line, good stability, and low hemolytic potential during in vitro trial for targeted theileriosis treatment. The evidences indicate that various antiprotozoal drugs can be destined for targeted delivery employing nano-carrier-based delivery system (Soni et al. 2014). Similarly, chitosan–tripolyphosphate conjugated chloroquine has produced better result at a dose rate of 250 mg kg^{-1} body weight for 15 days of treatment against *Plasmodium berghei*. Encapsulation of trifluralin (TFL) in lipid nanoparticles for targeted drug delivery has shown superior anti-leishmanial activity in mice (Lopes et al. 2016). Similarly, nerolidol loaded in carrier nanospheres has reported to percolate through blood–brain barrier resulting into 100% cure against *T. evansi* in the brain tissue which was revealed by PCR-based assay (Baldissera et al. 2017). Piperine-loaded nano-capsules comprised of gum rosin as polymer and oleic acid as oil core were also evaluated against *Trypanosoma evansi* yielding times lesser IC_{50} in comparison to free form of the drug as generation of reactive oxygen species (ROS) by the encapsulated drug was found to be higher than the free form. Lesser toxicity of the nano-capsule on horse peripheral blood mononuclear cells was also evidenced as an additional benefit (Rani et al. 2020). Targeted delivery of drug to RBC is very useful against the intra-erythrocytic hemoprotozoan diseases. Application of dendronized hyperbranched polymers (DHPs) has noted to be very effective as a carrier to provide drug to cells infected with *Plasmodium falciparum*. The intrinsic antimalarial activity of heparin has been enhanced by coating it with DHP. Rhodamine B labeling with the polymer along with heparin coating has further reduced the IC_{50} (San Anselmo et al. 2022). Such nano-delivery systems for other protozoan diseases can also be very helpful in target-based elimination of protozoa, reducing toxicity and doses, and enhancing bioavailability in feature. Gum-based nano-capsules loaded with naphthoquinones, a plant-based phenolic compound, exhibited more ROS generation and required in 1.5–4 times lesser dose to exert inhibitory effect on the growth of *T. evansi* than its free form. Further, the encapsulated form was found to be less cytotoxic on horse peripheral blood mononuclear cells than its free counterpart (Rani et al. 2022). Various drugs employing nano-delivery systems are under preclinical trial such as amphotericin B and stearylamine with liposomes carrier, ursolic acid, bis-triazole D0870, lychnopholide, nifurtimox, allopurinol, benznidazole using polymeric nanoparticles, S-benzyldithiocarbazate with solid lipid nanoparticles, benznidazole using nano-emulsion, and CdTe entrapped in quantum dots, indicating better promise for therapeutic control of protozoan diseases (Choudhury 2021).

21.7 Nano-Drug with Nano-Delivery System Targeting Protozoan Parasites

Targeted delivery of nano-drugs with proper nano-carrier system often enhances the inherent therapeutic potential of the drug component(s) accompanied by reduction in dose requirement and cytotoxicity. For example, o-naphthoquinone category drug β-lapachone, naturally yielded from bark of the Lapacho tree, when loaded in its nanoform within lecithin-chitosan nanoparticles, reduced the inflammatory response without affecting the parasite killing potential during topical treatment of cutaneous leishmaniasis (Espuelas et al. 2015). Although the nanoformulation was unable to diminish parasitic load in mice, but topical application improved the drug penetration through skin and reduced the progression of ulcerative skin lesions by anti-inflammatory action. Another mode of delivery of the biocidal agent can be encapsulating the agent in a multilayered hydrophobic nano-sized vehicle (liposome consisting of lecithin, gum, gel, and polymers) which sustains integrity in aqueous medium. These particles mimicking the size of bacteria can be engulfed by the protozoa due to their natural grazing behavior. For instance, *Acanthamoeba* and *Tetrahymena* are two protozoan species that engulf the bacteria *Legionella pneumophila* from the surface. After internalization, the protozoa tries to digest the bacteria but without any success, converting it into cysts. The cyst containing bacteria become inaccessible to chemotherapeutic agents. *Legionella pneumophila* is a pathogenic one causing Legionnaires' disease and Pontiac fever in humans. But the delivery of biocidal agent(s) employing this hydrophobic delivery system has found to be effective in controlling trophozoites of the protozoa and cysts containing microbial pathogens (Whitekettle et al. 2011). Another example involves niosomal delivery of the combination of selenium NPs and Glucantime for the treatment of leishmaniasis. Glucantime is a drug for first-line treatment of cutaneous leishmaniasis, while selenium NPs has found to be effective against *L. major*, *L. tropica*, and *L. infantum*. Niosomal delivery of the combination has found to potentiate inhibitory action of the individual components significantly against *L. tropica* besides depicting least cytotoxicity (Mostafavi et al. 2019). Similarly, toltrazuril, a standard anticoccidial agent encapsulated in a nano-capsule, comprised of poly-ε-caprolactone and administered through drinking water in broiler chicken to control coccidiosis. The nanoformulation reduced the lesion scores and oocysts excretion along with providing higher stability in gastrointestinal fluid (Baron et al. 2022). Application of chitosan nanoparticles (CsNPs) as nano-carrier for ethanolic extract of *Rosmarinus officinalis* (UEERO) against *Eimeria tenella* mediated coccidiosis in poultry stopped the sporulation after 72 h, and dose-dependent action was found in terms of inhibiting sporulation, distortion in wall, and creating sporocyst abnormality (Kasem et al. 2022). The aforesaid examples vividly justify about augmentation of antiprotozoal activity of the bioactive components when delivered through suitable nano-carriers along with improving the safety profiles of the cidal agents.

21.8 Nanoformulations for Vector Control

Various nano-emulsions of essential oil have also shown effectiveness against control of vector-borne diseases than their free counterparts. Such natural compounds can be targeted to reduce the use of pesticides and overcome the rising concern of acaricide resistance in combating vector-borne protozoan diseases. These nanoformulations of eco-friendly natural compounds can also be used as low-risk insecticides, antifungals, antifeedants, anti-vectors, growth regulators, oviposition deterrents, and repellents which can facilitate in vector control to reduce parasitic load for farm animal management. There is no dearth of plant-derived aromatic compounds of diverse groups as there are approximately 17,500 plant species distributed worldwide that can provide essential oils. Plant-derived bioactive constituents include but not limited to camphor, carvacrol, eugenol, terpinen-4-ol, or thymol that has reported to exert antiprotozoal action. Therefore, producing their nano-emulsion can provide additional benefits in terms of ease of preparation, scalability option, simple composition, higher thermodynamic stability, less residual damage, and low-cost production. Larvicidal action against arthropod carriers or vectors of various parasitic infestations including the protozoan diseases has been demonstrated by various nanoformulations such as nano-emulsion-based preparation of eucalyptus essential oil (EO) against *Culex quinquefasciatus* larvae, EO of *Ocimum basilicum L.* against third instar larvae from *Aedes aegypti*, *Rosmarinus officinalis* EO against fourth instar larvae of *A. aegypti*, *Cinnamomum zelanycum* EO against L8 stage larvae of mealworm, *Baccharis reticulate* EO against *A. aegypti*, *Ocimum basilicum* EO on second and third instar larvae of *C. quinquefasciatus*, *Artemisia dracunculus* (Asteraceae) EO against *Anopheles stephensi*, etc. The positive outcome is imparted by either curtailing the dose requirements or enhancing larvicidal activity or both (Echeverría and Duarte Galhardo de Albuquerque 2019). Apart from restricting larval activity, nano-emulsions have also depicted potent insecticidal action and repellent activity. Nanoformulation of *Ocimum sanctum* EO has found to be effective in knocking down *A. aegypti* and *C. quinquefasciatus*, whereas repellent effect of nanoformulation has been observed for EO of citronella grass *Cymbopogon nardus* (citronella oil), *Ocimum americanum* (hairy basil), and *Vetiveria zizanioides* (vetiver) against *A. aegypti* (Sritabutra and Soonwera 2013; Nuchuchua et al. 2009). Acaricidal activity against *Rhipicephalus microplus* tick has been observed in nano-emulsion of *Cinnamomum verum* EO, where inhibition of oviposition was observed in both in vitro and in vivo by Santos et al. (2017), while Monteiro et al. (2017) reported that the nano-emulsioned EO did not affect oviposition rate but only produce larvicidal action (Echeverría and Duarte Galhardo de Albuquerque 2019). Similarly, acaricidal activity of nano-emulsion of *Eucalyptus globulus* EO was observed against *Rhipicephalus microplus*, and *Ocimum sanctum* EO against *C. quinquefasciatus* and *A. aegypti*. Application of solid lipid nanoparticles (SLN) or nano-structured lipid carriers (NLC) to deliver synthetic acaricide drug cypermethrin (cip) or chlorpyrifos (chlo) along with vegetable compounds (citral, menthol, or limonene) has been reported to reduce the lethal concentrations of the synthetic drugs against *R. microplus* ticks; further, the nanoformulations assisted to stabilize the drugs leading to lower degradation in solution against the tick larvae (Figueiredo et al. 2022). Similarly, AgNPs synthesized by laser ablation using resins of myrrh (*Commiphora molmol*) and ginger (*Zingiber officinale*) depicted enhanced acaricidal activity against cattle tick *Hyalomma dromedary* that have shown resistance to Butox®. Further, the nanoformulations exerted effect without resulting in any skin irritation or abnormal health observation that is usually encountered in case of using synthetic acaricides (Nabil et al. 2023). Nanoformulations of imidacloprid were also found to be more effective against fourth instar larvae of *Culex pipiens* than the conventional form of the drug (Ahmed and Othman 2020). Likewise, *Cymbopogon citratus* leaf extract mediated green synthesized AuNPs have shown excellent efficacy against different species of mosquito, including *Aedes* (*A. aegypti*) and *Anopheles* (*A. stephensi*), as per the measured LC_{50} (Bajwa et al. 2022). The aforesaid discussion clearly advocates for the extensive use of eco-friendly nanoformulations in vector control imminently.

21.9 Nanoparticles in Prophylactic Control of Protozoan Diseases

Animals are tended in form of livestock or pets for either materialistic purposes or spiritual satisfaction of humans. However, pathogenic attack hampers the normal physiological behavior of animals, which sometimes turns out to be lethal. Hence, accountable otherwise unaccountable loss of animals results in. Prophylactic measures relying upon vaccination modalities are one of the most effective approaches to prevent such fatalities. Billions of dollars are regularly exhausted from global livestock earnings due to flies, biting midges Culicoides, ticks, and other arthropod vector-borne diseases. Furthermore, human transmission of diseases should not be overlooked as several diseases possess zoonotic potential. Many biopharmaceutical corporations and governmental organizations provide vaccination against coccidiosis of chickens and turkeys, East Coast fever of cattle, toxoplasmosis abortion in sheep, bovine tropical theileriosis, bovine babesiosis, and other such diseases (Fig. 21.1) (Cornelissen and Schetters 1996; Mcallister 2014). Hence, advancement in vaccinations is the most desirable aspect to reduce the pathogenic load as well as acaricide hazard. In this regard, nanoformulations can be employed to device cutting-edge vaccines with advantages including prolonged protection, lower doses, higher efficacy, better biodistribution, sustained release, faster action, an minor side effects. The drawbacks of some latest subunit-based vaccines such as lower retention time, enzyme degradation, poor solubility, and low immunogenicity in addition to reduced uptake by antigen-presenting cells can also be overcome by employing nanomaterials. For example, protein NPs aid in targeted delivery by providing endosomal osmotic swelling followed by rupturing at target site leading to exposure of the antigen, along with facilitation to the membrane fusion mechanisms for better cellular immune responses (Butkovich et al. 2021; Zottig et al. 2020). Manganese, when applied as nano-adjuvant (MnARK), is found to produce stronger cellular immune responses even at low doses (Garçon et al. 2007). Therefore, more than a few nanoformulation-aided vaccines have been developed against microorganisms to achieve such gains. Two leishmaniasis vaccines using LEISH-F3 (recombinant protein antigen) + GLA-SE (adjuvant) and LEISH-F3 (recombinant protein antigen) + MPL-SE (adjuvant) nanoemulsions are undergoing phase 1 clinical trial (Coler et al. 2015). Advanced control in the spread of diseases and replication of pathogen as well can be obtained by formulation of nano-vaccine subcloning *Trypanosoma cruzi antigens, TcG_2* and TcG, in a nanoplasmid to ensure prophylactic protection against *Trypanosoma infection.* PLGA-based nano-encapsulations of antigen, fragments of proteins or polysaccharides with pathogenic origin, and recombinant proteins are employed to provide vaccination against leishmaniosis and *Haemonchus contortus* with higher efficacy and immune response (Celis-Giraldo et al. 2021). Nano-silica vesicles functionalized with Bm86 antigen (SV-140-C_{18} vesicles) were employed as vaccine against cattle tick *Rhipicephalus (Boophilus) microplus* for better bioavailability. In this context, efficient uptake of the vaccine by the host cells has been reported in mice models under both in vitro and in vivo trial provoking immunological pathways to induce essential immune responses (Mody et al. 2021). Similarly, against toxoplasmosis, self-assembling protein NPs (SAPNs) were applied as scaffolds or platforms for vaccine delivery with high synergistic effect (El Bissati et al. 2020). Such platforms tempt to develop upgraded vaccines against various protozoa and their associated vectors for farm animal welfare. Protection against *Eimeria* infection in chickens and rabbits can be achieved by immunization with whole sporozoites vaccine of *E. tenella* emulsified in the nanoparticle adjuvant IMS 1313 N VG Montanide™ (EtSz-IMS1313). It facilitates identification of B-cell epitopes to build up immune protection against *E. tenella* (Juárez-Estrada et al. 2021). Similarly, vaccination against *Plasmodium* infection employing *P. chabaudi* recombinant chabaupain-1 (PcrCP-1) alone or in combination with chitosan NPs or hydroxide aluminum has been reported in mice. A strong humoral immune response was accomplished with high titers of IgG1 and IgG2a antibodies indicating potent immunogenic potential of chabaupain-1, but employment of chitosan NPs did not enhance protection (Armada et al. 2013). For immunization against coccidial agent in chick, recombinant TA4 protein of *E. tenella* (rEtTA4) was employed along with PLGA nano-carrier. Intramuscular administration of the preparation in chicks at the age of two weeks followed by *E. tenella* challenge resulted in a significant enhancement of immune protection due to enhancement of EtTA4-specific IgY antibody, CD81 T cells, and cytokines such as IL-17 in EtTA4-PLGA formulation. Furthermore, PLGA was not only advantageous as a carrier but it also significantly enhanced the levels of cytokines including interferon gamma, IL-4, IL-6, IL-17, and anti-coccidial index (ACI), indicative of better immune responses and promotion of the protective immunity (Huang et al. 2021). Nano-adjuvants are capable of eliciting better immune response than conventional system and nano-carriers can facilitate efficient delivery as well as sustained release of the antigen to decipher improved protection. Oral administration of recombinant *Eimeria maxima* protein functionalized with polystyrene NPs (NP-EMaxIMP1) in chicken resulted in improved complete or partial protection against *E. maxima* challenge (Jenkins et al. 2018). Synthesis of such nano-vaccines is also advantageous as they are reported to impart lower side effects. Chickens vaccinated with ginsenoside-based NPs show adjuvant effects including better weight gain, significantly enhanced lymphocyte proliferation, and IL-1 secretion along with lower lesion score when challenged with *Eimeria tenella* (Zhang et al. 2012). Overall, nano-adjuvanted vaccine has exhibited superior efficacy and immunity against protozoan diseases in comparison to their conventional counterpart. However, implementations of nano-vaccine in terms of nanoDNA-vaccine or nano-adjuvanted subunit vaccine are fewer with respect to the efforts extended towards conventional vaccines. Additionally, multiplexing of vaccine candidates in a single formulation can also be a further scope in this regimen.

21.10 Nanotoxicity and Safety

Parasitic diseases are more often chronic in nature and therefore can inflict greater economic loss in production herd that frequently remains overlooked. They may survive for years in the environment as well as host organisms due to their complex life cycle stages. Remarkably, each maturation phase produces a distinct sensitivity against the same drug. Moreover, the bioavailability of antiparasitic drugs is often restricted at the site of infection due to poor targeting and shorter half-life. For instance, some prominent antiparasitic drugs such as ivermectin and praziquantel are prone to enzymatic degradation along with the incompetency of penetration across the biological membranes. Consequently, a reduction in the bioavailability of the drug as well as their expected therapeutic effect occurs. Therefore, an immense problem is encountered by the clinicians while treating such parasitic infections (Parish 2019). Drug resistance is another critical problem increasing steadily due to nonjudicial usage of some chemotherapeutic agents including antibiotics, antifungals, antiprotozoal, and anthelmintics formulations (Yang et al. 2018). In addition, excessive application of such chemotherapeutic agents may result in the modification of pathogen response mechanisms leading to the insensitivity of older medicine regimes. Therefore, in such tricky juncture, nanomedicine has provided multifaceted edges to enforce effective control of parasitic diseases (Li et al. 2018). Targeted delivery of antiparasitic drugs is a major concern which is facilitated by nano-carriers

(Sun et al. 2019). Different nano-carriers in the form of nanocrystals, solid lipid nanoparticles (SLNs), and liposomes have been successfully introduced that can be administered through oral, intestinal, intravenous, and pulmonary routes. These nano-carriers provide physical stability as well as targeted and controlled release of drugs (Kashyap et al. 2018; Jalil et al. 2021; Aziz et al. 2021; Xie et al. 2010; Rathore et al. 2011; Souris et al. 2010). Furthermore, they prevent premature enzymatic degradation of the loaded drug (Xie et al. 2010). For example, the bioavailability and mean retention time of praziquantel was increased by 5.67- and 4.94-folds subsequently when formulated with hydrogenated castor oil SLNs. NPs are metabolized easily into the liver modifying their physicochemical properties through enzymatic degradation; otherwise, it can also be engulfed and broken down by endocytosis process facilitated by the reticuloendothelial system (Rathore et al. 2011). Another example in this context, the amphotericin B liposome can fuse with the cell to counter the intracellular parasites. In the cell, NPs can be hydrolyzed with the help of lysosomes resulting in the release of the loaded drug to effectively inhibit intracellular parasites. Besides, NPs can be excreted by the glomerular excretion system of the kidneys as well. However, the excretion process mainly depends on their size, shape, and charge. It has been observed that NPs with smaller sizes (80–120 nm) can be excreted by the kidneys easily, while larger particles undergo hepatobiliary excretion. Whereas, regarding the charges, NPs with higher charge (+34.4 mV) were excreted through feces, bypassing the gastrointestinal tract in contrast to the particles with lesser charge (−17.6 mV) that remained isolated in the liver (Souris et al. 2010). Thus, the sustainability and efficacy of NPs greatly rely on their physicochemical properties (Shin et al. 2015). Therefore, to achieve the implementations of NPs for satisfactory therapeutic outcomes, not only the NPs with higher rate of absorption (Bilia et al. 2017), sustained release, or effective intracellular delivery with proper concentration and expected retention time is required but the NPs should also be safe, cheaper, and easily reproducible at larger scales (Rai et al. 2019). However, there is a lack of analysis regarding the safety of NPs due to less systemic study and research (Dobrovolskaia and McNeil 2007). The core concept of nanomedicine is based on the precise formulation of NPs that facilitates fewer applications of drugs with lower concentrations to achieve more effective results. Although most of NPs are safe to be directed and may be able reduce the adverse effects of various drugs, a glitch still remains regarding its toxicity issues. Therefore, it needs to keep an eye out on the discovery and synthesis of novel biofriendly nanomaterials with no toxicity in addition to excellent physiological and biological compatibility (Najahi-Missaoui et al. 2020). It must be nontoxic, non-immunotoxic, and capable of getting eliminated from the body within a reasonable time (Suresh et al. 2013). Moreover, the safety of NPs should be evaluated using proper biosafety evaluation systems at every level, including molecule, cell, tissue, organ, and organism levels after its preparation (Tsuji et al. 2006).

21.11 Conclusions and Future Prospects

Neglected tropical parasites have resulted several major disease outbreaks around the world in recent time. Parasites have been treated through a variety of methods during previous several years but quite a lot of these methods have become blunt in efficacy. Overuse, rather to say, misuse of drugs, and some contraindications to chemotherapeutic agents have limited their use. Secondary infections can sometimes damage host physiology and parasites have developed evading mechanisms, thus indicating a need for other options. Ethnobotanicals have shown promise but variability in bioactive ingredients along with lack of adequate knowledge regarding dosage information restricts their applications. Further, lack of funding and limited interest of the investors has resulted in very few clinical transitions of the herbal drugs against parasitic diseases. The alternative option for ethnobotanicals is vaccine development; however, due to homogeneous antigenic variation, this task is challenging in parasitology. Parasites develop mechanisms to escape the host immune system. Hence, the application of nanomedicine to control parasites has become indispensable. The unique properties of NPs offer advantages over other treatment options, along with minimum probability of resistance from parasites. Nanomodification of curative agents or as carrier has shown benefits either in term of potency or efficacy, or safety and effectiveness as a pesticide or antiparasitic agents. Nanoformulation usually requires a smaller dose and herbal-based formulation of nano-pesticides are conducive and eco-friendly to help in developing "greener" strategy for the control of vectors of pathogenic diseases. Substantial work has been carried out regarding parasite control by NPs and protozoa such as *Plasmodium*, *Trypanosoma*, and *Leishmania* has already captured the focus. NPs prepared through chemical, mechanical, or green synthesis methods have shown cidal action against protozoan parasites or arrest their growth besides significant application in nano-diagnosis of parasitic diseases. Gold and silver NPs have been explored extensively, but other metallic NPs such as iron, zinc, nickel, platinum, copper, and their oxides or composites have remained understudied in the current context. Among nonmetallic ones, polymeric NPs have been extensively employed, particularly for drug encapsulation and targeting. The results clearly justify the introduction of nanointervention in therapeutic control of protozoan diseases. However, in-depth introspection regarding the mechanism of action of the nanotherapeutics along with their short- as well as long-term toxicity analysis and most importantly following the international standards and regulations are the need of the hour for smooth lab-to-market transition of the potential nanotherapeutics. A plethora of incomplete efficacy studies without toxicity analysis following non-standard protocols and guidelines with contraindicating results are overcasting the future of nanomedicine towards which every stakeholders need to be extremely careful so that nanotechnology can offer a sustainable solution in prophylactic and therapeutic control of parasitic diseases in near future.

References

Abd El Megid AD, Khaled M, Emam MA, Adel A (2018) Biochemical role of zinc oxide and propolis nanoparticles in protection rabbits against coccidiosis. Benha Vet Med J 34:314–328. https://doi.org/10.21608/bvmj.2018.54256

Abdel-Gaber R, Hawsah MA, Al-Otaibi T, Alojayri G, Al-Shaebi EM, Mohammed OB et al (2023) Biosynthesized selenium nanoparticles to rescue coccidiosis-mediated oxidative stress, apoptosis and inflammation in the jejunum of mice. Front Immunol 14:1139899

Abdellahi L, Iraji F, Mahmoudabadi A, Hejazi SH (2022) Vaccination in leishmaniasis: a review article. Iran Biomed J 26(1):1–35. https://doi.org/10.52547/ibj.26.1.35

Adeyemi OS, Whiteley CG (2013) Interaction of nanoparticles with arginine kinase from Trypanosoma brucei: kinetic and mechanistic evaluation. Int J Biol Macromol 62:450–456. https://doi.org/10.1016/j.ijbiomac.2013.09.008

Adeyemi OS, Murata Y, Sugi T, Kato K (2017) Inorganic nanoparticles kill toxoplasma gondii via changes in redox status and mitochondrial membrane potential. Int J Nanomedicine 12:1647–1661. https://doi.org/10.2147/IJN.S122178

Ahmadpour E, Godrati-Azar Z, Spotin A, Norouzi R, Hamishehkar H, Nami S et al (2019) Nanostructured lipid carriers of ivermectin as a novel drug delivery system in hydatidosis. Parasit Vectors 12(1):1–9

Ahmed MAI, Othman AAE (2020) Piperonyl butoxide enhances the insecticidal toxicity of nanoformulation of imidacloprid on Culex pipiens (Diptera: Culicidae) mosquito. Vector Borne Zoonotic Dis 20(2):134–142

Al-Ardi MH (2020) The uses of gold nanoparticles and Citrullus colocynthis L. nanoparticles against Giardia lamblia in vivo. Clin Epidemiol Global Health 8(4):1282–1286

Al-Ardi MH (2021) Anti-parasitic activity of nano Citrullus colocynthis and nano Capparis spinose against Trichomonas vaginalis in vitro. J Parasit Dis 45:845–850. https://doi.org/10.1007/s12639-021-01371-4

Ali KM, Hassan EA, Abuowarda MM, Mahmoud MA, Torad F (2021) Bilateral panophthalmia as a late sequel of leishmaniasis in dogs. Pak Vet J 41(1):13–18. https://doi.org/10.29261/pakvetj/2021.006

Almeria S, Castellà J, Ferrer D, Ortuño A, Estrada-Peña A, Gutiérrez JF (2001) Bovine piroplasms in Minorca (Balearic Islands, Spain): a comparison of PCR-based and light microscopy detection. Vet Parasitol 99(3):249–259. https://doi.org/10.1016/s0304-4017(01)00464-2

Anwar A, Numan A, Siddiqui R, Khalid M, Khan NA (2019) Cobalt nanoparticles as novel nanotherapeutics against Acanthamoeba castellanii. Parasit Vectors 12(1):280. https://doi.org/10.1186/s13071-019-3528-2

Armada A, Gazarini ML, Gonçalves LM, Antunes S, Custódio A, Rodrigues A et al (2013) Generation of an antibody that recognizes Plasmodium chabaudi cysteine protease (chabaupain-1) in both sexual and asexual parasite life cycle and evaluation of chabaupain-1 vaccine potential. Exp Parasitol 135(1):166–174

Aubry P, Geale DW (2011) A review of bovine anaplasmosis. Transbound Emerg Dis 58(1):1–30. https://doi.org/10.1111/j.1865-1682.2010.01173.x

Aziz S, Abdullah S, Anwar H, Latif F, Mustfa W (2021) Effect of engineered nickel oxide nanoparticles on antioxidant enzymes in freshwater fish, Labeo rohita. Pak Vet J 41:424–428. https://doi.org/10.29261/pakvetj/2021.044

Baek SH, Roh J, Park CY, Kim MW, Shi R, Kailasa SK, Park TJ (2020) Cu-nanoflower decorated gold nanoparticles-graphene oxide nanofiber as electrochemical biosensor for glucose detection. Mater Sci Eng C 107:110273

Bajwa HUR, Khan MK, Abbas Z, Riaz R, Rehman TU, Abbas RZ et al (2022) Nanoparticles: synthesis and their role as potential drug candidates for the treatment of parasitic diseases. Life 12(5):750

Baldissera MD, Souza CF, Boligon AA, Grando TH, De Sa MF, Da Silva AS et al (2017) Solving the challenge of the blood–brain barrier to treat infections caused by Trypanosoma evansi: evaluation of nerolidol-loaded nanospheres in mice. Parasitology 144(11):1543–1550

Baron LF, Fonseca FND, Maciag SS, Bellaver FAV, Ibeli AMG, Mores MAZ et al (2022) Toltrazuril-loaded polymeric nanocapsules as a promising approach for the preventive control of coccidiosis in poultry. Pharmaceutics 14(2):392

Bastien P (2002) Molecular diagnosis of toxoplasmosis. Trans R Soc Trop Med Hyg 96(Suppl 1): S205–S215. https://doi.org/10.1016/s0035-9203(02)90078-7

Beheshti N, Soflaei S, Shakibaie M, Yazdi MH, Ghaffarifar F, Dalimi A, Shahverdi AR (2013) Efficacy of biogenic selenium nanoparticles against Leishmania major: in vitro and in vivo studies. J Trace Elem Med Biol 27(3):203–207. https://doi.org/10.1016/j.jtemb.2012.11.002

Bertin E, Münzer A, Reichenberger S, Streubel R, Vinnay T, Wiggers H, Schulz C, Barcikowski S, Marzun G (2019) Durability study of platinum nanoparticles supported on gas-phase synthesized graphene in oxygen reduction reaction conditions. Appl Surf Sci 467:1181–1186. https://doi.org/10.1016/J.APSUSC.2018.10.061

Beshbishy AM, Batiha GES, Yokoyama N et al (2019) Ellagic acid microspheres restrict the growth of Babesia and Theileria in vitro and Babesia microti in vivo. Parasit Vectors 12:269. https://doi.org/10.1186/s13071-019-3520-x

Bilia AR, Piazzini V, Guccione C, Risaliti L, Asprea M, Capecchi G, Bergonzi MC (2017) Improving on nature: the role of nanomedicine in the development of clinical natural drugs. Planta Med 83:366–381. https://doi.org/10.1055/s-0043-102949

Boukthir A, Bettaieb J, Erber AC, Bouguerra H, Mallekh R, Naouar I et al (2020) Psycho-social impacts, experiences and perspectives of patients with cutaneous Leishmaniasis regarding treatment options and case management: an exploratory qualitative study in Tunisia. PLoS One 15(12):e0242494

Bruni N, Stella B, Giraudo L, Della Pepa C, Gastaldi D, Dosio F (2017) Nanostructured delivery systems with improved leishmanicidal activity: a critical review. Int J Nanomedicine 12:5289

Butkovich N, Li E, Ramirez A, Burkhardt AM, Wang SW (2021) Advancements in protein nanoparticle vaccine platforms to combat infectious disease. Wiley Interdiscip Rev Nanomed Nanobiotechnol 13(3):e1681

Capela R, Moreira R, Lopes F (2019) An overview of drug resistance in protozoal diseases. Int J Mol Sci 20(22):5748. https://doi.org/10.3390/ijms20225748

Celis-Giraldo CT, López-Abán J, Muro A, Patarroyo MA, Manzano-Román R (2021) Nanovaccines against animal pathogens: the latest findings. Vaccine 9(9):988

Chakravarty J, Sundar S (2019) Current and emerging medications for the treatment of leishmaniasis. Expert Opin Pharmacother 20(10):1251–1265. https://doi.org/10.1080/14656566.2019.1609940

Cheraghipour K, Masoori L, Ezzatkhah F, Salimikia I, Amiri S, Makenali AS, Taherpour F, Mahmoudvand H (2020) Effect of chitosan on toxoplasma gondii infection: a systematic review. Parasite Epidemiol Control 11:e00189. https://doi.org/10.1016/j.parepi.2020.e00189

Chikkanna MM, Neelagund SE, Rajashekarappa KK (2019) Green synthesis of zinc oxide nanoparticles (ZnO NPs) and their biological activity. SN Appl Sci 1:117. https://doi.org/10.1007/s42452-018-0095-7

Choudhury SD (2021) Nano-medicines a hope for Chagas disease! Front Mol Biosci 8:655435. https://doi.org/10.3389/fmolb.2021.655435

Coler RN, Duthie MS, Hofmeyer KA, Guderian J, Jayashankar L, Vergara J, Rolf T, Misquith A, Laurance JD, Raman VS, Bailor HR, Cauwelaert ND, Reed SJ, Vallur A, Favila M, Orr MT, Ashman J, Ghosh P, Mondal D et al (2015) From mouse to man: safety, immunogenicity and efficacy of a candidate leishmaniasis vaccine LEISH-F3+GLA-SE. Clin Transl Immunol 4(4): e35. https://doi.org/10.1038/cti.2015.6

Cornelissen AWCA, Schetters TP (1996) Vaccines against protozoal diseases of veterinary importance. FEMS Immunol Med Microbiol 15(2–3):61–72

Crawley H (1912) The protozoan parasites of domesticated animals. In: Circular (U.S. Department of Agriculture. Bureau of Animal Industry), vol 194. U.S. Department of Agriculture, Bureau of Animal Industry, Washington, DC, pp 465–498. https://naldc.nal.usda.gov/download/5420799/PDF

Crump A (2017) Ivermectin: enigmatic multifaceted 'wonder' drug continues to surprise and exceed expectations. J Antibiot 70(5):495–505. https://doi.org/10.1038/ja.2017.11

Dalir Ghaffari A, Barati M, Ghaffarifar F, Pirestani M, Ebrahimi M, KarimiPourSaryazdi A (2022) Investigation of antileishmanial activities of CaO nanoparticles on L. tropica and L. infantum parasites, in vitro. J Parasit Dis 47(1):73–81

de Souza A, Marins DSS, Mathias SL, Monteiro LM, Yukuyama MN, Scarim CB, Löbenberg R, Bou-Chacra NA (2018) Promising nanotherapy in treating leishmaniasis. Int J Pharm 547(1–2): 421–431. https://doi.org/10.1016/j.ijpharm.2018.06.018

Dkhil MA, Al-Quraishy S, Wahab R (2015) Anticoccidial and antioxidant activities of zinc oxide nanoparticles on Eimeria papillata-induced infection in the jejunum. Int J Nanomedicine 10: 1961–1968. https://doi.org/10.2147/IJN.S79944

Dobrovolskaia MA, McNeil SE (2007) Immunological properties of engineered nanomaterials. Nanosci Technol 2:469–478. https://doi.org/10.1038/nnano.2007.223

Dos Santos CA, Seckler MM, Ingle AP, Gupta I, Galdiero S, Galdiero M, Gade A, Rai M (2014) Silver nanoparticles: therapeutical uses, toxicity, and safety issues. J Pharm Sci 103(7): 1931–1944. https://doi.org/10.1002/jps.24001

Duncan R (2003) The dawning era of polymer therapeutics. Nat Rev Drug Discov 2:347–360. https://doi.org/10.1038/nrd1088

Echeverría J, Duarte Galhardo de Albuquerque RD (2019) Nanoemulsions of essential oils: new tool for control of vector-borne diseases and in vitro effects on some parasitic agents. Medicines 6(2):42. https://doi.org/10.3390/medicines6020042

Eger I, Soares MJ (2012) Endocytosis in Trypanosoma cruzi (Euglenozoa: Kinetoplastea) epimastigotes: visualization of ingested transferrin–gold nanoparticle complexes by confocal laser microscopy. J Microbiol Methods 91:101–105. https://doi.org/10.1016/j.mimet.2012.07.013

El Bissati K, Zhou Y, Paulillo SM, Raman SK, Karch CP, Reed S et al (2020) Engineering and characterization of a novel self assembling protein for toxoplasma peptide vaccine in HLA-A* 11: 01, HLA-A* 02: 01 and HLA-B* 07: 02 transgenic mice. Sci Rep 10(1):1–13

El-Gendy AML, Mohammed MAA, Ghallab MMI, Aziz MOA, Ibrahim SM (2021) Therapeutic effect of chitosan nanoparticles and metronidazole in treatment of experimentally giardiasis infected hamsters. Iran J Parasitol 16(1):32

El-Kosary S, Allatif AM, Stino RG, Hassan MM, Kinawy AA (2021) Effect of silver nanoparticles on micropropagation of date palm (Phoenix Dactylifera L, CV. Sewi and Medjool). Plant Arch 20(2):9701–9706

Elmi T, Rahimi Esboei B, Sadeghi F et al (2021) In vitro antiprotozoal effects of Nano-chitosan on plasmodium falciparum, Giardia lamblia and trichomonas vaginalis. Acta Parasit 66:39–52. https://doi.org/10.1007/s11686-020-00255-6

Esch KJ, Petersen CA (2013) Transmission and epidemiology of zoonotic protozoal diseases of companion animals. Clin Microbiol Rev 26(1):58–85. https://doi.org/10.1128/CMR.00067-12

Espuelas S, Irache JM, Sanmartin-Grijalba C, Font M, Conde I, Larrea E et al (2015) Assesment of β-lapachone loaded in lecithin-chitosan nanoparticles for the topical treatment of cutaneous leishmaniasis in L. major infected BALB/c mice. Nanomedicine 11(8):2003–2012

Fang L, Lu X, Cui C, Shi Q, Wang H (2022) Metronidazole-loaded nanoparticulate thermoreversible gel for gynecologic infection of trichomonas vaginalis. Am J Transl Res 14(6):4015–4023

Figueiredo A, Anholeto LA, Cola DF, Fantatto RR, Santos IB, Gainza YA et al (2022) Nanoformulations with synthetic and plant-derived compounds for cattle tick control. Vet Parasitol 309:109756

Garçon N, Chomez P, Van Mechelen M (2007) GlaxoSmithKline adjuvant systems in vaccines: concepts, achievements and perspectives. Expert Rev Vaccines 6(5):723–739

Ghaffari AD, Barati M, KarimiPourSaryazdi A, Ghaffarifar F, Pirestani M, Ebrahimi M (2023) In vitro and in vivo study on antiprotozoal activity of calcium oxide (CaO) and magnesium oxide (MgO) nanoparticles on promastigote and amastigote forms of Leishmania major. Acta Trop 238:106788

Gogoi C, Sarma J, Barua CC, Tamuly S, Upadhyaya T, Islam S, Sonowal J, Borthakur U, Banerjee D, Niyamath B (2019) Evaluation of nano-curcumin on experimentally induced coccidiosis in broiler chicks. IJCS 7(3):4514–4520

Gruessner BM, Weathers PJ (2021) In vitro analyses of Artemisia extracts on Plasmodium falciparum suggest a complex antimalarial effect. PLoS One 16(3):e0240874

Guo W, Li N, Ren G, Wang R, Chai L, Li Y et al (2021) Murine pharmacokinetics and antimalarial pharmacodynamics of dihydroartemisinin trimer self-assembled nanoparticles. Parasitol Res 120:2827–2837

Hamad HK (2020) Estimation of activity and toxicity of silver nanoparticles loaded metronidazole against giardia lamblia. Syst Rev Pharmacy 11(10):1129–1133. https://doi.org/10.31838/srp.2020.10.163

Horn D, Duraisingh MT (2014) Antiparasitic chemotherapy: from genomes to mechanisms. Annu Rev Pharmacol Toxicol 54:71–94. https://doi.org/10.1146/annurev-pharmtox-011613-135915

Huang J, Chen S, Zhou Z, Sun X, Haseeb M, Lakho SA, Zhang Y, Liu J, Shah MAA, Song X, Xu L, Yan R, Li X (2021) Poly (D, L-lactide-co-glycolide) delivery system improve the protective efficacy of recombinant antigen TA4 against Eimeria tenella infection. Poult Sci 100(5):101083. https://doi.org/10.1016/j.psj.2021.101083

Jalil PJ, Shnawa BH, Hamad SM (2021) Silver nanoparticles: green synthesis, characterization, blood compatibility and protoscolicidal efficacy against echinococcus granulosus. Pak Vet J 41: 393–399. https://doi.org/10.29261/pakvetj/2021.039

Jenkins MC, Stevens L, O'Brien C, Parker C, Miska K, Konjufca V (2018) Incorporation of a recombinant Eimeria maxima IMP1 antigen into nanoparticles confers protective immunity against E. Maxima challenge infection. Vaccine 36(8):1126–1131. https://doi.org/10.1016/j.vaccine.2017.11.014

Juárez-Estrada MA, Gayosso-Vázquez A, Tellez-Isaias G, Alonso-Morales RA (2021) Protective immunity induced by an Eimeria tenella whole Sporozoite vaccine elicits specific B-cell antigens. Animals 11(5):1344. https://doi.org/10.3390/ani11051344

Kaiser M, Mäser P, Tadoori LP, Ioset J-R, Brun R (2015) Antiprotozoal activity profiling of approved drugs: a starting point toward drug repositioning. PLoS One 10(8):e0135556. https://doi.org/10.1371/journal.pone.0135556

Karimkhani C, Wanga V, Coffeng LE, Naghavi P, Dellavalle RP, Naghavi M (2016) Global burden of cutaneous leishmaniasis: a cross-sectional analysis from the global burden of disease study 2013. Lancet Infect Dis 16(5):584–591

Kasem SM, Mira NM, Mahfouz ME, Helal IB (2022) In vitro study to evaluate the efficacy of ultrasonicated ethanolic extract of rosmarinus officinalis and its chitosan-based nanoparticles against Eimeria tenella oocysts of chickens. AAPS PharmSciTech 23(8):295

Kashyap A, Kaur R, Baldi A, Jain UK, Chandra R, Madan J (2018) Chloroquine diphosphate bearing dextran nanoparticles augmented drug delivery and overwhelmed drug resistance in Plasmodium falciparum parasites. Int J Biol Macromol 114:161–168. https://doi.org/10.1016/j.ijbiomac.2018.03.102

Keyhani A, Ziaali N, Shakibaie M, Kareshk AT, Shojaee S, Asadi-Shekaari M et al (2020) Biogenic selenium nanoparticles target chronic toxoplasmosis with minimal cytotoxicity in a mouse model. J Med Microbiol 69(1):104–110

Kroubi M, Daulouede S, Karembe H, Jallouli Y, Howsam M, Mossalayi D, Vincendeau P, Betbeder D (2010) Development of a nanoparticulate formulation of diminazene to treat African trypanosomiasis. Nanotechnology 21(50):505102. https://doi.org/10.1088/0957-4484/21/50/505102

Li P, Rios Coronado PE, Longstaff XRR, Tarashansky AJ, Wang B (2018) Nanomedicine approaches against parasitic worm infections. Adv Healthc Mater 7:1701494. https://doi.org/10.1002/adhm.201701494

Lopes RM, Pereira J, Esteves MA, Gaspar MM, Carvalheiro M, Eleutério CV, Gonçalves L, Jiménez-Ruiz A, Almeida AJ, Cruz ME (2016) Lipid-based nanoformulations of trifluralin analogs in the management of Leishmania infantum infections. Nanomedicine (Lond) 11(2): 153–170. https://doi.org/10.2217/nnm.15.190

Madbouly NA, Nashee H, Elgendy AA, Rabee I, El Amir A (2020) Encapsulation of low metronidazole dose in poly (D, L-lactide-co-glycolide)(PLGA) nanoparticles improves giardia intestinalis treatment. Infect Chemother 52(4):550

Maharana BR, Tewari AK, Saravanan BC, Sudhakar NR (2016) Important hemoprotozoan diseases of livestock: challenges in current diagnostics and therapeutics: an update. Vet World 9(5): 487–495. https://doi.org/10.14202/vetworld.2016.487-495

Mammari N, Halabi MA, Yaacoub S, Chlala H, Dardé L, Courtioux B (2019) Toxoplasma gondii modulates the host cell responses: an overview of apoptosis pathways. Biomed Res Int 2019: 6152489. https://doi.org/10.1155/2019/6152489

Marimuthu S, Rahuman AA, Rajakumar G, Santhoshkumar T, Kirthi AV, Jayaseelan C, Bagavan A, Zahir AA, Elango G, Kamaraj C (2011) Evaluation of green synthesized silver nanoparticles against parasites. Parasitol Res 108(6):1541–1549. https://doi.org/10.1007/s00436-010-2212-4

Mcallister MM (2014) Successful vaccines for naturally occurring protozoal diseases of animals should guide human vaccine research. A review of protozoal vaccines and their designs. Parasitology 141(5):624–640

Melariri P, Kalombo L, Nkuna P, Dube A, Hayeshi R, Ogutu B, Gibhard L, deKock C, Smith P, Wiesner L, Swai H (2015) Oral lipid-based nanoformulation of tafenoquine enhanced bioavailability and blood stage antimalarial efficacy and led to a reduction in human red blood cell loss in mice. Int J Nanomedicine 10:1493–1503. https://doi.org/10.2147/IJN.S76317

Metwally DM, Alajmi RA, El-Khadragy MF, Al-Quraishy S (2021) Silver nanoparticles biosynthesized with Salvia officinalis leaf exert protective effect on hepatic tissue injury induced by plasmodium chabaudi. Front Vet Sci 7:620665. https://doi.org/10.3389/fvets.2020.620665

Minakshi P, Ghosh M, Brar B, Kumar R, Lambe UP, Ranjan K, Manoj J, Prasad G (2019) Nano-antimicrobials: a new paradigm for combating mycobacterial resistance. Curr Pharm Des 25(13):1554–1579. https://doi.org/10.2174/1381612825666190620094041

Mishra A, Kaushik NK, Sardar M, Sahal D (2013) Evaluation of antiplasmodial activity of green synthesized silver nanoparticles. Colloids Surf B Biointerfaces 111:713–718. https://doi.org/10.1016/j.colsurfb.2013.06.036

Mody KT, Zhang B, Li X, Fletcher NL, Akhter DT, Jarrett S et al (2021) Characterization of the biodistribution of a silica vesicle nanovaccine carrying a Rhipicephalus (Boophilus) microplus protective antigen with in vivo live animal imaging. Front Bioeng Biotechnol 8:606652

Mofokeng LS, Taioe OM, Smit NJ, Thekisoe OMM (2020) Parasites of veterinary importance from domestic animals in uMkhanyakude district of KwaZulu-Natal province. J S Afr Vet Assoc 91(0):e1–e11. https://doi.org/10.4102/jsava.v91i0.2023

Monteiro IN, dos Santos Monteiro O, Costa-Junior LM, da Silva Lima A, de Aguiar Andrade EH, Maia JGS, Mouchrek Filho VE (2017) Chemical composition and acaricide activity of an essential oil from a rare chemotype of Cinnamomum verum Presl on Rhipicephalus microplus (Acari: Ixodidae). Vet Parasitol 238:54–57

Moradi M, Sattarahmady N, Hatam GR, Heli H (2016) Electrochemical genosensing of Leishmania major using gold hierarchical nanoleaflets. J Biol Today's World 5:128–136

Mordue DG, Wormser GP (2019) Could the drug tafenoquine revolutionize treatment of Babesia microti infection? J Infect Dis 220(3):442–447

Morones JR, Elechiguerra JL, Camacho A, Holt K, Kouri JB, Ramírez JT, Yacaman MJ (2005) The bactericidal effect of silver nanoparticles. Nanotechnology 16(10):2346–2353. https://doi.org/10.1088/0957-4484/16/10/059

Mostafavi M, Khazaeli P, Sharifi I, Farajzadeh S, Sharifi H, Keyhani A et al (2019) A novel niosomal combination of selenium coupled with glucantime against Leishmania tropica. Korean J Parasitol 57(1):1

Nabil M, Khater HF, Selim A, Baz MM, Govindarajan M, Taie HA, Negm S (2023) Acaricidal efficacy of silver nanoformulations of Commiphora molmol and Zingiber officinale against the camel tick, Hyalomma dromedarii (Ixodida: Ixodidae). Inorg Chem Commun 147:110229. https://doi.org/10.1016/j.inoche.2022.110229

Nadeem A, Aslam A, Chaudhary Z, Ashraf K, Saeed K, Ahmad N, Ahmed I (2011) Indirect fluorescent antibody technique based prevalence of surra in equines. Pak Vet J 31

Najahi-Missaoui W, Arnold RD, Cummings BS (2020) Safe nanoparticles: are we there yet? Int J Mol Sci 22:385. https://doi.org/10.3390/ijms22010385

Narurkar R, Mamorska-Dyga A, Nelson JC, Liu D (2017) Autoimmune hemolytic anemia associated with babesiosis. Biomark Res 5:14. https://doi.org/10.1186/s40364-017-0095-6

Nayak AP, Tiyaboonchai W, Patankar S, Madhusudhan B, Souto EB (2010) Curcuminoids-loaded lipid nanoparticles: novel approach towards malaria treatment. Colloids Surf B Biointerfaces 81(1):263–273. https://doi.org/10.1016/j.colsurfb.2010.07.020

Norgan AP, Pritt BS (2018) Parasitic infections of the skin and subcutaneous tissues. Adv Anat Pathol 25(2):106–123

Norouzi R, Irani S, Katiraee F, Hejazy M (2022) In vitro acaricidal activity of nano-SiO2 against Hyalomma spp. ticks and its toxicity. Nanomed Res J 7(1):93–98

Nuchuchua O, Sakulku U, Uawongyart N, Puttipipatkhachorn S, Soottitantawat A, Ruktanonchai U (2009) In vitro characterization and mosquito (Aedes aegypti) repellent activity of essential-oils-loaded nanoemulsions. AAPS PharmSciTech 10:1234–1242

Okaiyeto K, Hoppe H, Okoh AI (2021) Plant-based synthesis of silver nanoparticles using aqueous leaf extract of Salvia officinalis: characterization and its antiplasmodial activity. J Clust Sci 32: 101–109. https://doi.org/10.1007/s10876-020-01766-y

Pacheco-Ordaz A, Sánchez-García E, Quintanilla-Licea R, Bazaldúa-Rodríguez AF, Pérez-Hernández RA, Hernández-García ME et al (2022) Amoebicidal and trichomonicidal capacity of polymeric nanoparticles loaded with extracts of the plants Curcuma longa (Zingiberaceae) and Berberis vulgaris (Berberidaceae). Rev Biol Trop 70(1):319–331

Parish T (2019) Steps to address anti-microbial drug resistance in today's drug discovery. Expert Opin Drug Discov 14:91–94. https://doi.org/10.1080/17460441.2019.1550481

Prakas P, Strazdaitė-Žielienė Ž, Januškevičius V, Chiesa F, Baranauskaitė A, Rudaitytė-Lukošienė E et al (2020) Molecular identification of four Sarcocystis species in cattle from Lithuania, including S. Hominis, and development of a rapid molecular detection method. Parasit Vectors 13(1):1–9

Priyadarshini KA, Murugan K, Panneerselvam C, Ponarulselvam S, Hwang JS, Nicoletti M (2012) Biolarvicidal and pupicidal potential of silver nanoparticles synthesized using Euphorbia hirta against Anopheles stephensi Liston (Diptera: Culicidae). Parasitol Res 111(3):997–1006. https://doi.org/10.1007/s00436-012-2924-8

Rai A, Comune M, Ferreira L (2019) Nanoparticle-based drug delivery systems: promising approaches against bacterial infections. In: Antibacterial drug discovery to combat MDR. Springer, Berlin/Heidelberg, pp 605–633. https://doi.org/10.1590/S1516-89132013005000011

Ramos Campos EV, Abrantes DC, Rogério CB, Bueno C, Miranda VR, Monteiro RA, Fraceto LF (2020) Recent developments in nanotechnology for detection and control of Aedes aegypti-borne diseases. Front Bioeng Biotechnol 8:102. https://doi.org/10.3389/fbioe.2020.00102

Rani R, Kumar S, Dilbaghi N, Kumar R (2020) Nanotechnology enabled the enhancement of antitrypanosomal activity of piperine against Trypanosoma evansi. Exp Parasitol 219:108018. https://doi.org/10.1016/j.exppara.2020.108018

Rani R, Narsiman B, Varma RS, Kumar R (2022) Gum-based nanocapsules comprising naphthoquinones enhance the apoptotic and trypanocidal activity against Trypanosoma evansi. Eur J Pharm Sci 171:106118. https://doi.org/10.1016/j.ejps.2022.106118

Rathore A, Jain A, Gulbake A, Shilpi S, Khare P, Jain A, Jain SK (2011) Mannosylated liposomes bearing Amphotericin B for effective management of visceral Leishmaniasis. J Liposome Res 21:333–340. https://doi.org/10.3109/08982104.2011.575381

Rehman TU, Saeed Z, Zaman MA, Sikandar A, Ali HM (2021) Factors influencing the incidence of Eimeria leuckarti in horses. Agrobiol Rec 6:13–17. https://doi.org/10.47278/journal.abr/2021.004

Roy A, Bulut O, Some S, Mandal AK, Yilmaz MD (2019) Green synthesis of silver nanoparticles: biomolecule-nanoparticle organizations targeting antimicrobial activity. RSC Adv 9(5): 2673–2702. https://doi.org/10.1039/c8ra08982e

Saad H, Soliman MI, Azzam AM, Mostafa B (2015) Antiparasitic activity of silver and copper oxide nanoparticles against Entamoeba histolytica and Cryptosporidium parvum cysts. J Egypt Soc Parasitol 45(3):593–602. (10.15412/J.JBTW.01061003)

Sahoo N, Manchikanti P (2013) Herbal drug regulation and commercialization: an Indian industry perspective. J Altern Complement Med (New York, N.Y.) 19(12):957–963. https://doi.org/10.1089/acm.2012.0275

Said DE, Elsamad LM, Gohar YM (2012) Validity of silver, chitosan, and curcumin nanoparticles as anti-giardia agents. Parasitol Res 111:545–554

Saleem K, Khursheed Z, Hano C, Anjum I, Anjum S (2019) Applications of nanomaterials in Leishmaniasis: a focus on recent advances and challenges. Nanomaterials (Basel, Switzerland) 9(12):1749. https://doi.org/10.3390/nano9121749

San Anselmo M, Lantero E, Avalos-Padilla Y, Bouzón-Arnáiz I, Ramírez M, Postigo A et al (2022) Heparin-coated dendronized hyperbranched polymers for antimalarial targeted delivery. ACS Appl Polymer Mater 5(1):381–390. https://doi.org/10.1021/acsapm.2c01553

Santos DS, Boito JP, Santos RCV, Quatrin PM, Ourique AF, dos Reis JH, Gebert RR, Glombowsky P, Klauck V, Boligon AA (2017) Nanostructured cinnamon oil has the potential to control Rhipicephalus microplus ticks on cattle. Exp Appl Acarol 73:129–138. https://doi.org/10.1007/s10493-017-0171-5

Sazgarnia A, Taheri AR, Soudmand S, Parizi AJ, Rajabi O, Darbandi MS (2013) Antiparasitic effects of gold nanoparticles with microwave radiation on promastigots and amastigotes of Leishmania major. Int J Hyperthermia 29(1):79–86. https://doi.org/10.3109/02656736.2012.758875

Sedighi F, Abbasali PR, Maghsood A, Fallah M (2016) Comparison of therapeutic effect of anti-cryptosporidium nano-nitazoxanide (ntz) with free form of this drug in neonatal rat. Sci J Hamadan Univ Med Sci 23:134–140

Shakibaie M, Ezzatkhah F, Gabal E, Badparva E, Jahanbakhsh S, Mahmoudvand H (2020) Prophylactic effects of biogenic selenium nanoparticles on acute toxoplasmosis: an in vivo study. Ann Med Surg 54:85–88

Sharma L, Dhiman M, Singh A, Sharma MM (2021) Green approach: "a forwarding step for curing Leishmaniasis—a neglected tropical disease". Front Mol Biosci 8:655584

Shin SW, Song IH, Um SH (2015) Role of physicochemical properties in nanoparticle toxicity. Nano 5:1351–1365. https://doi.org/10.3390/nano5031351

Soni N, Prakash S (2013) Possible mosquito control by silver nanoparticles synthesized by soil fungus (Aspergillus niger 2587). Adv Nanopart 2(02):125–132. https://doi.org/10.4236/anp.2013.22021

Soni MP, Shelkar N, Gaikwad RV, Vanage GR, Samad A, Devarajan PV (2014) Buparvaquone loaded solid lipid nanoparticles for targeted delivery in theleriosis. J Pharm Bioallied Sci 6(1): 22–30. https://doi.org/10.4103/0975-7406.124309

Souris JS, Lee C-H, Cheng S-H, Chen C-T, Yang C-S, Ja-an AH, Mou C-Y, Lo L-W (2010) Surface charge-mediated rapid hepatobiliary excretion of mesoporous silica nanoparticles. Biomaterials 31:5564–5574. https://doi.org/10.1016/j.biomaterials.2010.03.048

Sritabutra D, Soonwera M (2013) Repellent activity of herbal essential oils against Aedes aegypti (Linn.) and Culex quinquefasciatus (Say.). Asian Pac J Trop Dis 3(4):271–276. https://doi.org/10.1016/S2222-1808(13)60069-9

Sultana J, Mazzaglia G, Luxi N, Cancellieri A, Capuano A, Ferrajolo C, de Waure C, Ferlazzo G, Trifirò G (2020) Potential effects of vaccinations on the prevention of COVID-19: rationale, clinical evidence, risks, and public health considerations. Expert Rev Vaccines 19(10):919–936. https://doi.org/10.1080/14760584.2020.1825951

Sun Y, Chen D, Pan Y, Qu W, Hao H, Wang X, Liu Z, Xie S (2019) Nanoparticles for antiparasitic drug delivery. Drug Deliv 26:1206–1221. https://doi.org/10.1080/10717544.2019.1692968

Suresh AK, Pelletier DA, Doktycz MJ (2013) Relating nanomaterial properties and microbial toxicity. Nanoscale 5:463–474. https://doi.org/10.1039/C2NR32447D

Tavakoli P, Ghaffarifar F, Delavari H, Shahpari N (2019) Efficacy of manganese oxide (Mn_2O_3) nanoparticles against Leishmania major in vitro and in vivo. J Trace Elem Med Biol 56:162–168

Tavakoli P, Ghaffarifar F, Delavari H, KarimiPourSaryazdi A, Dayer MS, Nasiri V, Ahmadi S (2022) Synthesis of tellurium oxide (TeO2) nanorods and nanoflakes and evaluation of its efficacy against Leishmania major in vitro and in vivo. Acta Parasitol 67(1):143–152. https://doi.org/10.1007/s11686-021-00445-w

Tsamesidis I, Lymperaki E, Egwu CO, Pouroutzidou GK, Kazeli K, Reybier K, Bourgeade-Delmas S, Valentin A, Kontonasaki E (2021) Effect of silica based nanoparticles against Plasmodium falciparum and Leishmania infantum parasites. J Xenobiot 11(4):155–162. https://doi.org/10.3390/jox11040011

Tsuji JS, Maynard AD, Howard PC, James JT, Lam C, Warheit DB, Santamaria AB (2006) Research strategies for safety evaluation of nanomaterials, part IV: risk assessment of nanoparticles. Toxicol Sci 89:42–50. https://doi.org/10.1093/toxsci/kfi339

Uliana SR, Trinconi CT, Coelho AC (2018) Chemotherapy of leishmaniasis: present challenges. Parasitology 145(4):464–480

van Wyk JA, Malan FS (1988) Resistance of field strains of Haemonchus contortus to ivermectin, closantel, rafoxanide and the benzimidazoles in South Africa. Vet Rec 123(9):226–228. https://doi.org/10.1136/vr.123.9.226

Varela-Aramburu S, Ghosh C, Goerdeler F, Priegue P, Moscovitz O, Seeberger PH (2020) Targeting and inhibiting plasmodium falciparum using ultra-small gold nanoparticles. ACS Appl Mater Interfaces 12(39):43380–43387

Vazini H, Br E (2018) The investigation of nanoparticles of gold's fatality effect on toxoplasma gondii parasite in an in vitro study. https://doi.org/10.4172/2472-1212.1000172

Whitekettle WK, Melzer JI, Hunter MM, Wang L, Pierce CC, Zhang G, Reynolds DG (2011) WO. Patent No.US2011/020432

Wink M (2012) Medicinal plants: a source of anti-parasitic secondary metabolites. Molecules 17(11):12771–12791. https://doi.org/10.3390/molecules171112771

Xie S, Pan B, Wang M, Zhu L, Wang F, Dong Z, Wang X, Zhou W (2010) Formulation, characterization and pharmacokinetics of praziquantel-loaded hydrogenated castor oil solid lipid nanoparticles. Nanomedicine 5:693–701. https://doi.org/10.2217/nnm.10.42

Yang Y-T, Zhu J-F, Liao G, Xu H-J, Yu B (2018) The development of biologically important spirooxindoles as new antimicrobial agents. Curr Med Chem 25:2233–2244. https://doi.org/10.2174/0929867325666171129131311

Yasinzai M, Khan M, Nadhman A, Shahnaz G (2013) Drug resistance in leishmaniasis: current drug-delivery systems and future perspectives. Future Med Chem 5(15):1877–1888

Youssefi MR, Alipour R, Fakouri Z, Shahavi MH, Nasrabadi NT, Tabari MA et al (2023) Dietary supplementation with eugenol nanoemulsion alleviates the negative effects of experimental coccidiosis on broiler chicken's health and growth performance. Molecules 28(5):2200

Zhang DF, Xu H, Sun BB, Li JQ, Zhou QJ, Zhang HL, Du AF (2012) Adjuvant effect of ginsenoside-based nanoparticles (ginsomes) on the recombinant vaccine against Eimeria tenella in chickens. Parasitol Res 110:2445–2453

Zottig X, Côté-Cyr M, Arpin D, Archambault D, Bourgault S (2020) Protein supramolecular structures: from self-assembly to nanovaccine design. Nano 10(5):1008

Chapter 22
Nanomedicinal Approaches to Treat Fungal Livestock Diseases

Mayukh Ghosh and Alonkrita Chowdhury

Abstract Fungal diseases, such as candidiasis, dermatophytosis, mycotoxicosis, aspergillosis, zygomycosis, histoplasmosis, cryptococcosis, sporotrichosis, and lymphangitis, can affect various livestock animals, including cattle, sheep, goats, pigs, horses, and poultry. These diseases in farm animals can result in fatality, otherwise diminish their yields resulting in significant economic losses. Conventionally, antifungal drugs, such as azoles, polyenes, allylamines, and griseofulvin, along with topical antifungal agents, such as miconazole, clotrimazole, and chlorhexidine, are employed as therapeutic agents against fungal diseases in livestock. However, they possess limitations including poor drug solubility, limited drug penetration, drug resistance, and adverse side effects. Along with overcoming such confinements, nanotechnology-aided therapeutic approaches have emerged as one of the most promising fields in the management of fungal disease. Several metal NPs, such as silver nanoparticles, copper nanoparticles, zinc oxide nanoparticles, titanium oxide nanoparticles, and gold nanoparticles, have been employed as antifungal agents owing to their antimicrobial properties. Nanoparticles-based drug delivery systems or controlled drug release systems against fungal infections have been constructed incorporating organic nanoparticles, such as lipid nanoparticles, polymeric nanoparticles, protein-based nanoparticles, and carbon-based nanoparticles. Nanoemulsion is another formulation for delivering antifungal agents with enhanced stability, improved solubility, and increased efficacy. It is also exploited in combination therapy against fungal infections. Nanomaterials-coated textiles, infrastructures, animal housing, medical devices, etc., are also industrialized to prevent the spread of fungal infections within animals. Hence, diverse implementations of nanoformulations have revolutionized veterinary therapeutics for controlling fungal

M. Ghosh (✉)
Department of Veterinary Physiology and Biochemistry, RGSC, Banaras Hindu University, Mirzapur, Uttar Pradesh, India

A. Chowdhury
Department of Plant Biotechnology, Institute of Agricultural Sciences, RGSC, Banaras Hindu University, Mirzapur, Uttar Pradesh, India

LLR University of Veterinary and Animal, ICAR-National Research Centre on Equines, *Nanotechnology Theranostics in Livestock Diseases and Management*, Livestock Diseases and Management,
https://doi.org/10.1007/978-981-16-1610-5_22

infections. However, there are still risks of nanotoxicity, and therefore further research and advancement is required for safe and effective applications of nanomaterial-based therapies in the veterinary field.

Keywords Candidiasis · Dermatophytosis · Mycotoxicosis · Aspergillosis · Antifungal drugs · Nanotherapeutics

22.1 Introduction

Nanomedicine holds great promise in the field of veterinary medicine for the treatment of various diseases, including fungal infections in livestock. Fungal diseases in livestock, such as candidiasis, dermatophytosis, mycotoxicosis, aspergillosis, zygomycosis, histoplasmosis, cryptococcosis, sporotrichosis, and lymphangitis, can cause significant economic losses and pose a threat to animal health. Several of them are of zoonotic importance also (Adebiyi and Oluwayelu 2018; Dworecka-Kaszak et al. 2020; Seyedmousavi et al. 2018). There are several conventional approaches for diagnosis and treatment of such diseases. Mostly clinical diagnosis, microscopic examinations, laboratory trials, histological observation, and serological tests are performed for the diagnosis of fungal diseases. The traditional practices for the treatment of fungal diseases in animals include applications of antifungal medications with supportive care including nutritional support, oxygen therapy, management of secondary complications, minimizing stress, and hygiene maintenance. Topical treatment with antifungal shampoo or cream is also performed for localized infection (Seyedmousavi et al. 2018).

Although such conventional therapeutics are effective in controlling fungal infections to some extent, they possess certain limitations. Application of antifungal drug for a longer period can give rise to drug-resistant fungal strain (Kanafani and Perfect 2008). Narrow-spectrum drugs are also confined to act against limited fungal species only. In addition, several antifungal drugs are unable to penetrate deep into tissues. Hence, fungal infection in visceral organs is difficult to be cured (Lewis 2011). Furthermore, systemic administration of antifungal drugs can result in toxicity, allergic reactions, or other side effects (Deamer and Weber 2010). However, nanomedicinal approaches offer several advantages in tackling these diseases, including targeted drug delivery, enhanced therapeutic efficacy, and reduced side effects. Nanoparticles can be designed to encapsulate antifungal drugs, such as azoles or polyenes, and improve their stability, bioavailability, and efficacy. Nanoparticles can protect the drugs from degradation, enhance their penetration into the affected tissues, and provide sustained release profiles, resulting in prolonged drug action (Souza and Amaral 2017). Combination therapy employing nanoemulsion has emerged providing synergistic effect, better than individual components. Hence, it reduces the risk of drug resistance. Furthermore, it enhances drug penetration to deeper tissues due to small droplet size; thus, it is also employed as effective drug carrier (Huang et al. 2023). Lipid-based nanoparticles such as liposomes or solid lipid nanoparticles (SLN) and nanostructured lipid carriers (NLC) are utilized as carrier for hydrophobic antifungal drug delivery to the target site (Jain et al. 2010). Antifungal drugs incorporated nanostructured hydrogels have been developed that can enhance drug stability, control drug release, and provide prolonged drug action at the site of infection (Abdellatif et al. 2019). Therefore, these hydrogels can be applied topically as dressings or coatings to treat fungal skin infections in livestock. Nanoscale vaccines can be employed to deliver antigens from fungal pathogens and stimulate a targeted immune response in livestock. It can enhance the immunogenicity of antigens, improve their stability, and provide controlled release profiles (Kischkel et al. 2020). This approach can help prevent or reduce the severity of fungal infections in livestock. It is important to note that while nanomedicinal approaches show great potential, further research and development are needed to optimize their efficacy, safety, and scalability for widespread use in veterinary medicine. Additionally, regulatory considerations and cost-effectiveness should be taken into account when translating these approaches from the laboratory to the field.

22.2 Prominent Fungal Diseases of Livestock and Poultry

Fungal diseases can affect various livestock animals, including cattle, sheep, goats, pigs, horses, and poultry. It causes several health issues as well as reduce production resulting in significant economic losses. Some common fungal diseases affecting livestock animals and poultry birds have been delineated hereunder.

22.2.1 Ringworm (Dermatophytosis)

Ringworm, also known as dermatophytosis, is a common fungal infection that frequently occurs in pet animals and livestock, sometimes in wildlife also. It is caused by different species of dermatophyte fungi, such as *Microsporum* and *Trichophyton*. Despite its name, ringworm is not caused by a worm but by these fungi that invade the skin, hair, and occasionally the nails of affected animals. Ringworm in animals is highly contagious and can spread through direct contact with an infected animal or via contaminated fomites. The infection typically occurs in animals with compromised immune systems, young animals with developing immune systems, or those living in crowded and unsanitary conditions (Chermette et al. 2008).

The clinical signs of ringworm in animals can vary, but common symptoms include skin lesions, changes in hair and coat color and appearance, and deformation of nails and claw.

Diagnosing ringworm in animals typically involves a combination of clinical examination, fungal culture, and microscopic examination of skin, hair, or nail samples. Fungal cultures are performed to isolate and identify the causative dermatophyte fungi (Watkins 2014).

The treatment of ringworm in animals involves a multifaceted approach, including antifungal medications, topical treatments, environmental decontamination, and supportive care. Antifungal medications, such as griseofulvin, terbinafine, or itraconazole, are commonly prescribed to eliminate the infection (Moriello 2014; Rengasamy et al. 2020). Supportive care measures, such as boosting the animal's immune system, providing a nutritious diet, and minimizing stress, can also aid in the recovery process.

Prevention of ringworm in animals involves maintaining good hygiene, regular grooming, and minimizing exposure to infected animals or contaminated environments. Prompt isolation and treatment of infected animals can help prevent the spread of the infection. Regular veterinary checkups and routine screening for ringworm can also aid in early detection and treatment (Moriello and Coyner 2021).

22.2.2 Aspergillosis

Aspergillosis is a fungal infection caused by fungi belonging to the genus *Aspergillus*. It can affect a wide range of animal species, including mammals, birds, reptiles, and amphibians. Aspergillosis is caused primarily by *Aspergillus fumigatus*, although other species of *Aspergillus* can also be involved. The spores of *Aspergillus* fungi are commonly found in the environment, particularly in soil, decaying vegetation, and organic matter. Animals can become infected with *Aspergillus* through inhalation of the spores, ingestion of contaminated food or bedding, or through the skin in cases of wounds or burns (Desoubeaux and Cray 2018).

The clinical signs and severity of aspergillosis in animals can vary depending on the species, route of infection, and extent of the infection. Common signs may include respiratory distress, coughing, sneezing, bloody or pus-like nasal discharge, reduced appetite, weight loss, lethargy, and neurological symptoms in advanced cases. In birds, aspergillosis can cause respiratory signs, such as difficult breathing, open-mouth breathing, and voice changes. In reptiles, skin lesions and abscesses may be observed.

The diagnosis of aspergillosis in animals typically involves a combination of clinical evaluation, laboratory tests, and imaging studies. Fungal cultures, microscopic examination of tissues or fluids, and serological tests can help identify the presence of *Aspergillus* and confirm the diagnosis. Radiographs, computed tomography (CT), or other imaging modalities may be used to assess the extent of the infection and identify lesions in affected organs (Tell 2005).

The treatment of aspergillosis in animals usually involves a combination of antifungal medications, supportive care, and management of underlying conditions. Antifungal drugs, such as itraconazole or voriconazole, are commonly used to combat the fungal infection. In severe or systemic cases, intravenous administration of antifungal medications may be necessary. Supportive care measures, including nutritional support, oxygen therapy, and management of secondary complications, are essential for the overall management of the affected animal (Jenks and Hoenigl 2018).

Prevention of aspergillosis in animals and poultry involves maintaining proper hygiene and reducing exposure to contaminated environments. This includes regular cleaning and disinfection of living spaces, proper ventilation, and minimizing the accumulation of damp or decaying organic materials. Good nutrition, stress reduction, and appropriate housing conditions are also important factors in preventing aspergillosis (Arné et al. 2011).

22.2.3 Candidiasis

Candidiasis, also known as yeast infection, is an opportunistic fungal infection caused by the *Candida* spp., particularly *Candida albicans*. While candidiasis is mostly associated with infections in humans, it can also affect animals, including domesticated pets and livestock. Candidiasis in animals can occur when there is an overgrowth of *Candida* fungi in specific areas of the body. Factors that can contribute to the development of candidiasis in animals include a weakened immune system, underlying diseases, prolonged antibiotic use, hormonal imbalances, poor hygiene, and stress.

The symptoms of candidiasis in animals can vary depending on the affected area. Common signs include skin and ear infections causing redness, itching, inflammation, hair loss, pustules, crusts, or waxy discharge from skinfolds, ears, paws, or other moist areas; gastrointestinal issues such as diarrhea, vomiting, abdominal discomfort, and poor appetite; respiratory infections involving nasal discharge, coughing, sneezing, and breathing difficulties; and genital or urinary tract infections causing inflammation, discharge, discomfort, and difficulty in urination or defecation (Seyedmousavi et al. 2018).

The diagnosis of candidiasis in animals is typically based on clinical signs, physical examination, and laboratory tests. Fungal cultures, microscopic examination of affected tissues or fluids, and molecular evaluations such as polymerase chain reaction (PCR) can help identify the presence of *Candida* and confirm the diagnosis.

Treatment of candidiasis in animals usually involves a combination of antifungal medications and supportive care. Antifungal drugs, such as fluconazole or itraconazole, may be prescribed to combat the fungal infection. Topical treatments, such as antifungal shampoos or creams, may be used for localized infections. In severe or systemic cases, oral or injectable antifungal medications may be necessary. Supportive care measures, such as maintaining good hygiene, managing underlying health conditions, and addressing any contributing factors, are also important for successful treatment (Pappas et al. 2004).

Prevention of candidiasis in animals involves maintaining proper hygiene, providing a balanced diet, minimizing stress, and promptly treating any underlying health conditions. Regular grooming, cleaning of ears and skinfolds, and proper sanitation in kennels or living spaces can help reduce the risk of candidiasis.

22.2.4 Mycotoxicosis

Mycotoxicosis refers to the toxic effects caused by ingestion of fungal toxins, known as mycotoxins, present in contaminated feed or forage. These mycotoxins are produced by various fungi, including *Aspergillus*, *Fusarium*, and *Penicillium* species. Several types of mycotoxins can affect animals and poultry, including aflatoxins, trichothecenes, zearalenone, ochratoxins, and fumonisins (El-Sayed et al. 2022; Awuchi et al. 2021). Each mycotoxin has its own specific toxic effects and can target different organs and systems in animals.

The clinical signs and symptoms of mycotoxicosis can vary depending on the mycotoxin involved, the level and duration of exposure, and the species of animal. Common symptoms of mycotoxicosis in animals and poultry may include reduced feed intake, weight loss, poor growth, impaired immune function, reproductive issues, liver damage, gastrointestinal disturbances, neurologic signs, and even increased susceptibility to other diseases (Zain 2011).

The diagnosis of mycotoxicosis can be challenging as the symptoms can be non-specific and resemble other diseases or nutritional deficiencies. However, laboratory analysis of feed or animal samples can help identify the presence and levels of mycotoxins. This may involve techniques such as high-performance liquid chromatography (HPLC), enzyme-linked immunosorbent assays (ELISA), or mass spectrometry (Abidin et al. 2011).

Strategies involved in prevention and management of mycotoxicosis in animals and poultry include good agricultural practices such as implementing proper storage, drying, and handling practices for feed and forage crops to minimize fungal growth and mycotoxin contamination (Zaki et al. 2012). Regularly testing feed and forage samples for mycotoxin levels can help in identifying potential contamination issues and allow for timely intervention. Application of specific additives, such as activated charcoal, clay minerals, yeast cell walls, or other mycotoxin binders, can help mitigate the effects of mycotoxins by binding and reducing their absorption in the digestive tract (Patil et al. 2014). Managing feed stuff by segregating contaminated feed, diluting contaminated batches, or replacing contaminated feed with uncontaminated alternatives can also be effective strategies to reduce mycotoxin exposure. Furthermore, providing nutritional support with a balanced and nutrient-dense diet can improve the overall health and immune function of animals, minimizing the impact of mycotoxicosis (Kępińska-Pacelik and Biel 2021).

22.2.5 *Zygomycosis*

Mucormycosis, also known as zygomycosis, is a rare but serious fungal infection caused by fungi of the order Mucorales. While it is predominantly seen in humans, mucormycosis can also affect animals and poultry. The infection occurs when the spores of Mucorales fungi enter the body through inhalation, ingestion, or direct contact with open wounds. In animals, mucormycosis typically affects species with compromised immune systems, such as those with underlying diseases, extensive injuries, or undergoing immunosuppressive therapies. The infection often starts at the site of a wound or through the nasal passages and can spread to involve nearby tissues and organs.

The clinical signs of mucormycosis in animals can vary depending on the affected area and the species involved. Common signs include facial swelling, nasal discharge, ulceration or necrosis of tissues, difficult breathing, loss of appetite, weight loss, and general signs of illness. In severe cases, the infection can progress to involve the central nervous system, resulting in neurological signs (de Almeida Jr et al. 2018).

The diagnosis of mucormycosis in animals typically involves a combination of clinical evaluation, laboratory tests, and imaging studies. Fungal cultures, histopathology, and polymerase chain reaction (PCR) testing can be used to identify the causative fungus and confirm the presence of mucormycosis. Radiographs, computed tomography (CT), or magnetic resonance imaging (MRI) may be performed to assess the extent of the infection and the involvement of surrounding tissues.

The treatment of mucormycosis in animals generally involves a combination of antifungal medications, surgical debridement of affected tissues, and supportive care. Antifungal drugs such as amphotericin B or posaconazole are commonly used, but the choice of medication depends on the species, severity of the infection, and available treatment options. Surgical intervention may be necessary to remove necrotic tissues and improve the chances of successful treatment. Supportive care measures, including fluid therapy, pain management, and nutritional support, are important for the overall management of the affected animal (Skiada et al. 2018; Spellberg and Ibrahim 2010).

Prevention of mucormycosis in animals and poultry involves maintaining good hygiene, preventing injuries or wounds, and minimizing exposure to contaminated environments. Regular cleaning and disinfection of living spaces, proper wound management, and appropriate handling of immune compromised animals are essential. Mucormycosis can have severe consequences in animals and poultry. Hence, timely intervention is crucial for successful outcomes.

22.2.6 *Histoplasmosis*

Histoplasmosis can affect both domesticated and wild animals. Animals can become infected by inhaling fungal spores or through ingestion of contaminated soil, bird droppings, or other organic matter containing *Histoplasma*. The severity of histoplasmosis in animals can vary depending on the species, immune status, and extent of the infection.

In dogs, histoplasmosis is one of the most common systemic fungal infections. It primarily affects the respiratory system, but the fungus can also disseminate to other organs. Dogs with histoplasmosis may show symptoms such as cough, difficulty breathing, weight loss, lethargy, loss of appetite, and fever. Gastrointestinal involvement can lead to vomiting, diarrhea, and abdominal pain. Skin lesions, lymph node enlargement, and eye problems may also occur. Histoplasmosis can be severe and life-threatening in dogs, particularly if left untreated, or in animals with compromised immune systems. Cats can also develop histoplasmosis, but the disease is generally less common and often milder compared to dogs. Respiratory signs, weight loss, poor appetite, and fever are commonly observed. Cats may also develop eye problems, skin lesions, and gastrointestinal symptoms. Histoplasmosis in cats can be more challenging to diagnose since the signs can be non-specific, resembling other diseases. Feline histoplasmosis can have a chronic and relapsing course if not effectively treated (Brömel and Sykes 2005). Other animals, including horses, cattle, birds, and exotic species, can also be affected by histoplasmosis. Horses may develop respiratory signs, weight loss, and eye problems. In birds, particularly poultry, the disease can lead to respiratory distress, reduced egg production, and increased mortality. However, the clinical signs and severity of histoplasmosis can vary from species to species.

The diagnosis of histoplasmosis in animals typically involves a combination of clinical evaluation, laboratory tests, and imaging studies. Fungal cultures, histopathology, serology, or antigen testing can be used to confirm the presence of *Histoplasma* infection. Radiographs or other imaging modalities may be performed to assess organ involvement (Rezabek et al. 1993).

The treatment of histoplasmosis in animals often involves the use of antifungal medications, such as itraconazole or amphotericin B, depending on the severity of the infection and the species involved. Treatment may need to be prolonged and require close monitoring of the animal's response to therapy. Supportive care, including nutritional support and management of secondary complications, may also be necessary (Mocherla and Wheat 2001).

Prevention of histoplasmosis in animals involves minimizing exposure to environments where the fungus is present, such as avoiding contact with contaminated soil or organic matter. Maintaining good hygiene, providing proper ventilation, and reducing exposure to bird or bat droppings can help reduce the risk of infection in susceptible animals (de Perio et al. 2021). Effective management and prevention strategies for fungal diseases in livestock animals include proper hygiene, isolation of affected animals, disinfection of premises, appropriate antifungal treatment, and the use of clean and uncontaminated feed. In severe cases or outbreaks, it is important to consult a veterinarian for diagnosis, treatment, and guidance on disease control measures.

22.2.7 Bovine Cryptococcosis

Bovine cryptococcosis, also known as cryptococcal meningoencephalitis or torulosis, is a rare fungal disease that affects cattle. It is caused by the *Cryptococcus neoformans* or *Cryptococcus gattii* fungi, which are environmental pathogens commonly found in soil and bird droppings (Riet-Correa et al. 2011).

Bovine cryptococcosis usually occurs when cattle inhale the fungal spores, which can lead to an infection in the respiratory tract. From there, the infection can spread to other organs, including the central nervous system, causing meningoencephalitis.

The clinical signs of bovine cryptococcosis can vary depending on the organs affected and the stage of the disease. Common symptoms may include coughing, nasal discharge, difficult breathing, and pneumonia-like symptoms if the respiratory tract is affected. As the infection progresses to the central nervous system, cattle may exhibit neurological symptoms such as head pressing, circling, aimless wandering, depression, ataxia (loss of coordination), blindness, and seizures. Besides, few generalized signs such as weight loss, decreased appetite, lethargy, and fever may be observed as the infection progresses and affects the overall health of the animal (Carmo et al. 2020).

Diagnosing bovine cryptococcosis can be challenging, as the symptoms can mimic other diseases. A definitive diagnosis is typically made through laboratory tests including fungal culture, microscopic examination of affected tissues or fluids, and molecular techniques like polymerase chain reaction (PCR) to detect the presence of *Cryptococcus* spp.

Unfortunately, treatment options for bovine cryptococcosis are limited, and the disease often has a poor prognosis. Antifungal medications, such as fluconazole or amphotericin B, may be prescribed to manage the infection, but they may not completely eliminate the fungus. Supportive care, including the administration of anti-inflammatory drugs and management of neurological symptoms, may be employed to improve the animal's comfort and quality of life (Spadari et al. 2020).

Preventing bovine cryptococcosis involves minimizing exposure to the fungi and their spores. This can be achieved by maintaining good hygiene, including proper sanitation of animal housing and feeding areas, limiting access to contaminated environments (such as bird roosting sites), and avoiding overcrowding. Additionally, controlling bird populations near cattle facilities and minimizing contact between cattle and bird droppings can reduce the risk of infection.

22.2.8 Mycotic Mastitis

Mycotic mastitis, also known as fungal mastitis, is a type of mastitis that is caused by fungal pathogens including *Candida* spp., *Aspergillus* spp., and *Cryptococcus* spp. (Bakr et al. 2015). Mycotic mastitis can occur due to various factors, including poor hygiene, environmental contamination, immune suppression, or the presence of skin injuries or teat lesions that allow fungi to enter the mammary gland. Fungal organisms can colonize the teat canal and invade the mammary tissues, leading to an inflammatory response and subsequent mastitis.

The clinical signs of mycotic mastitis may include abnormal milk production from the affected quarter. The milk may appear watery and clumpy or have visible clots with changes in color, consistency, or odor. There may be visible enlargement of the affected quarter in comparison to other. The udder may exhibit swelling, heat, and pain. Sometimes teat appearance get changed with signs of inflammation, such as redness, swelling, or discharge. Moreover, the cow may exhibit reduced milk production from the affected quarter (Dworecka-Kaszak et al. 2012).

Diagnosing mycotic mastitis involves a combination of clinical examination, milk analysis, and laboratory tests. Milk samples can be collected and sent for culture and identification of the fungal pathogen. Microscopic examination of milk smears or histopathological evaluation of mammary tissue samples may also be performed to confirm the presence of fungi.

The treatment of mycotic mastitis typically involves a combination of antifungal therapy, supportive care (anti-inflammatory drug), and management practices to prevent further spread of the infection. Antifungal medications, such as fluconazole or intramammary nystatin infusions or systemic therapy, may be prescribed to target the specific fungal pathogen. Supportive care measures, such as frequent milking, proper hygiene, and the use of teat dips or disinfectants, can help reduce the spread of the infection and promote healing (Mahendra 1991).

Preventing mycotic mastitis involves maintaining good hygiene practices, including regular cleaning and disinfection of milking equipment and proper udder preparation before milking. Minimizing skin injuries or teat lesions and ensuring a clean and dry environment for the cows can also help reduce the risk of fungal infections (Zigo et al. 2021).

22.2.9 Epizootic Lymphangitis

Epizootic lymphangitis, also known as epizootic equine lymphangitis or African horse sickness, is a chronic infectious disease that primarily affects horses, donkeys, and mules. It is caused by the fungus *Histoplasma capsulatum var. farciminosum.* The disease is characterized by the development of firm, nodular lesions in the subcutaneous tissues, lymphatic vessels, and lymph nodes. These nodules can ulcerate and discharge pus-like material, leading to the formation of draining tracts and chronic inflammation. Epizootic lymphangitis can affect multiple body parts, including the limbs, abdomen, chest, and neck. The exact mode of transmission of epizootic lymphangitis is not fully understood, but it is believed to occur through direct contact with infected animals, contaminated fomites, or through insect vectors, such as flies and mosquitoes (al-Ani 1999).

The clinical signs of epizootic lymphangitis can vary depending on the stage and severity of the disease. Firm, painless nodules with varying size may be present in multiple locations on the body. The nodules may ulcerate, leading to the formation of draining tracts which can discharge a thick, pus-like material. Lymph nodes near the affected areas may become enlarged and firm. If the limbs are affected, lameness and swelling may be observed.

Diagnosing epizootic lymphangitis involves a combination of clinical examination, laboratory tests, and imaging studies. A definitive diagnosis is typically made by isolating and identifying the fungus from the affected tissues or discharge. This can be done through fungal culture, microscopic examination, or molecular techniques such as PCR (Seid et al. 2019; Hadush et al. 2020).

The treatment of epizootic lymphangitis is often unsatisfactory and requires long-term therapy and management. Antifungal medications, such as potassium iodide or itraconazole, are commonly used to combat the fungal infection. Supportive care, including wound management, bandaging, and pain relief, is crucial for the comfort and healing of the affected animals. In severe cases, surgical intervention combined with antifungal drugs (amphotericin B) may be necessary (Scantlebury and Reed 2009).

Preventing the spread of epizootic lymphangitis involves strict quarantine and isolation measures for infected animals. Fly and mosquito control can also help reduce the transmission of the disease. Good hygiene practices, such as regular cleaning and disinfection of stables and equipments, are important for preventing the spread of the fungus (Seid et al. 2019).

22.2.10 Sporotrichosis

Sporotrichosis is a fungal infection that can affect animals, including both domesticated and wild animals. The most commonly affected animals include cats, horses, dogs, and occasionally livestock such as cows and goats. Sporotrichosis in animals is caused by the fungus *Sporothrix schenckii*, which is found in soil, plants, and organic matter. In animals, sporotrichosis typically occurs as a result of traumatic inoculation of the fungus into the skin through wounds, scratches, or puncture injuries. Cats are particularly susceptible to sporotrichosis and can acquire the infection through bites from infected animals or by coming into contact with contaminated materials or soil.

The clinical signs of sporotrichosis in animals can vary depending on the species, route of infection, and the individual's immune response. The most common form of the infection is cutaneous or subcutaneous, characterized by the development of nodules or ulcers at the site of entry. These lesions may be slow-growing and non-healing, and may progress along the lymphatic vessels, causing a sporotrichoid lymphocutaneous infection characterized by a chain of nodules or ulcers (Pereira et al. 2015; Schubach et al. 2004). Other manifestations of sporotrichosis in animals can include respiratory involvement, osteomyelitis (infection of the bone), and dissemination to internal organs in severe cases or animals with compromised immune systems (Queiroz-Telles et al. 2019).

Diagnosing sporotrichosis in animals involves a combination of clinical examination, cytology or histopathological examination of affected tissues, and fungal culture or molecular tests such as PCR and ELISA to identify the presence of *Sporothrix schenckii*. It is important to differentiate sporotrichosis from other skin diseases or infections with similar clinical signs (Pereira et al. 2015; Fernandes et al. 2011; Rodrigues et al. 2013).

The treatment of sporotrichosis in animals generally involves systemic antifungal therapy. The most used antifungal medications for sporotrichosis include itraconazole and potassium iodide. The duration of treatment can vary depending on the severity and response to therapy. In some cases, surgical intervention may be necessary to remove large or non-responsive lesions (Mahajan 2014).

Preventing sporotrichosis in animals involves minimizing their exposure to contaminated materials or environments. This includes avoiding contact with infected animals, using proper wound care techniques, and practicing good hygiene. Regular veterinary checkups and prompt treatment of any skin lesions or signs of illness are also important for early detection and management of sporotrichosis.

22.2.11 Fungal Abortions

Fungal abortions, also known as mycotic abortions, can occur in cattle when pregnant cows are exposed to certain fungal organisms that can cause infections in the uterus, leading to abortion. Several fungal species have been implicated in causing fungal abortions in cattle, including *Aspergillus* spp., *Mucor* spp., *Rhizopus* spp., and others (Ali and Khan 2006). Fungal abortions in cattle can occur through the ingestion of contaminated feed or forage, inhalation of fungal spores, or through direct contact with contaminated materials. The fungi can colonize in the uterus and cause inflammation and damage to the fetal membranes, resulting in abortion. The exact mechanisms of infection and pathogenesis may vary depending on the specific fungal species involved.

The most prominent sign of fungal abortions in cattle is the spontaneous loss of pregnancy, typically occurring during the second or third trimester. The aborted fetus may exhibit signs of decomposition, and placental membranes may appear thickened and necrotic or have visible fungal growth. In addition, the cow may exhibit a foul-smelling vaginal discharge, which can be accompanied by signs of uterine inflammation (Prameela 2019; Anderson 2007).

Diagnosing fungal abortions in cattle involves a combination of clinical examination, examination of the aborted fetus and placenta, and laboratory testing. Veterinarians may perform a postmortem examination of the fetus and placenta to assess their condition and collect samples for fungal culture and identification. Histopathological examination of the placenta and fetal tissues may also be conducted to confirm the presence of fungal organisms (Anderson 2007).

Treatment options for fungal abortions in cattle are limited. In most cases, prevention is key to minimizing the occurrence of fungal abortions. This includes proper management of feed and forage to prevent fungal contamination, maintaining good hygiene practices in the barn or housing facilities, and promptly removing and disposing of aborted fetuses and placental material to reduce the risk of environmental contamination.

22.3 Conventional Therapeutics Against Fungal Diseases of Livestock and Poultry

Conventional therapeutics for fungal diseases in livestock and poultry primarily involve the use of antifungal drugs. These drugs are administered orally, topically, or via injection, depending on the specific disease and the affected animal (Foy and Trepanier 2010). Here are some commonly used conventional therapeutics against fungal diseases in livestock and poultry:

Azoles: Azole antifungal drugs, such as fluconazole, ketoconazole, and itraconazole, are widely used for the treatment of fungal infections in animals. Azoles inhibit the synthesis of ergosterol, an essential component of fungal cell membranes, leading to cell death (Ghannoum and Rice 1999). They can be administered orally or topically, depending on the type and location of the infection.

Polyenes: Polyene antifungal drugs, such as amphotericin B and nystatin, are effective against a broad spectrum of fungal infections. They work by binding to ergosterol in fungal cell membranes, causing disruption and leakage of cellular contents (Serhan et al. 2014). Polyenes are often used topically or as injections for systemic infections.

Allylamines: Allylamines, such as terbinafine, are antifungal drugs that inhibit the enzyme squalene epoxidase, an important step in the synthesis of fungal cell membranes (Ghannoum and Rice 1999). Allylamines are commonly used topically for the treatment of dermatophytosis (ringworm) in livestock animals.

Griseofulvin: Griseofulvin is an antifungal drug used primarily for the treatment of dermatophytosis in animals. It works by inhibiting fungal mitosis, preventing the growth and spread of fungal cells (Borgers 1980). Griseofulvin is usually administered orally and is effective against certain species of dermatophytes.

Topical antifungal agents: In addition to systemic antifungal drugs, various topical antifungal agents, such as creams, ointments, shampoos, and sprays, are used for localized fungal infections in livestock and poultry. These formulations may contain antifungal drugs, such as miconazole, clotrimazole, or chlorhexidine, and are applied directly to the affected area (Rochette et al. 2003).

It is important to note that the use of antifungal drugs in livestock and poultry should be done under the guidance of a veterinarian. The dosage, duration of treatment, and withdrawal periods should be strictly followed to ensure proper efficacy and minimize the risk of drug residues in animal products. Additionally, as with any medication, the development of antifungal drug resistance is a concern (Ghannoum and Rice 1999); proper management practices, including good hygiene and prevention strategies, should be implemented to reduce the incidence of fungal infections in livestock and poultry (Rochette et al. 2003).

22.4 Limitations of Conventional Therapeutics Against Fungal Diseases

While conventional therapeutics, such as antifungal drugs, have been effective in treating fungal diseases in livestock and poultry, they do have certain limitations. These limitations include the following:

Development of drug resistance: Fungal pathogens have the potential to develop resistance to antifungal drugs over time. This can occur due to the genetic variability of the fungi and their ability to adapt to the drugs. The emergence of drug-resistant strains can reduce the effectiveness of conventional therapeutics and complicate treatment options (Fisher et al. 2022).

Limited spectrum of activity: Different antifungal drugs have varying spectra of activity against different fungal species. Some drugs may be effective against a wide range of fungi, while others have a more limited spectrum. This limitation can pose challenges when dealing with fungal infections caused by less common or emerging fungal species (Ashley et al. 2006).

Systemic side effects: Systemic administration of antifungal drugs can sometimes lead to adverse effects in animals. These side effects can include gastrointestinal disturbances, liver and kidney toxicity, allergic reactions, and drug interactions. Careful monitoring and dosage adjustments may be necessary to minimize these side effects (Scorzoni et al. 2017).

Limited tissue penetration: Certain antifungal drugs may have limited ability to penetrate deep into tissues, especially in cases of fungal infections affecting internal organs or deep-seated tissues or targeting biofilm-producing fungi. This limited tissue penetration can hinder the effectiveness of treatment and require alternative therapeutic strategies (Scorzoni et al. 2017).

Withdrawal periods and residue concerns: In food-producing animals, the use of antifungal drugs is regulated to ensure that residues do not exceed acceptable levels in animal products, such as meat, milk, and eggs (Beigi et al. 2011). This leads to the establishment of withdrawal periods, during which treated animals should not be slaughtered or their products should not be used for human consumption. These withdrawal periods can impact the timeline for effective treatment and the management of disease outbreaks.

Lack of options for certain infections: Some fungal infections, particularly those caused by highly resistant or uncommon fungal species, may not respond well to conventional antifungal therapies. In such cases, alternative treatment approaches, such as combination therapies, novel drugs, or adjuvant therapies, may be necessary (Deamer and Weber 2010).

To overcome these limitations, ongoing research is focused on developing new antifungal agents, combination therapies, and alternative treatment strategies. These efforts aim to enhance the efficacy, broaden the spectrum of activity, and minimize the development of drug resistance in the treatment of fungal diseases in livestock and poultry.

22.5 Nanotherapeutics Against Fungal Diseases of Livestock and Poultry

Nanotherapeutics, which involve the application of nanotechnology in the development of therapeutic agents, show promise in addressing the limitations of conventional therapeutics for fungal diseases in livestock and poultry (Alghuthaymi et al. 2021). Nanoparticle-based approaches offer several advantages, including improved drug delivery, enhanced efficacy, reduced toxicity, and targeted action (Kischkel et al. 2020). Nanotherapeutics as antifungal agent can exert functions either by their inherent fungicidal action or as delivery vehicle of different antifungal agents. Nanoparticle-based antifungal agents can be designed to encapsulate or conjugate antifungal drugs with nanoparticles, enhancing their stability, solubility, and bioavailability. These nanoparticles can improve drug delivery to the target site and provide sustained-release profiles, leading to prolonged therapeutic action. Additionally, the small size of nanoparticles enables enhanced penetration into tissues, including deep-seated infections (Kischkel et al. 2020). Further, nanoformulations can induce synergistic action or facilitate the option for combinatorial therapy by incorporating multiple antifungal agents to increase the spectrum of fungicidal action. Different prominent nanoparticles that have been extensively explored to combat fungal infections have been delineated hereunder.

22.5.1 Metal Nanoparticles as Antifungal Agents

Metal nanoparticles have gained significant attention as potential antifungal agents due to their unique properties and antimicrobial activity. Metal nanoparticles can be synthesized by different chemical, physical, or biogenic methods such as chemical reduction, laser ablation, and green synthesis. Green-synthesized nanoparticles derived from plant extracts and bacterial or fungal filtrate have gained attention as potential antifungal agents due to several advantages such as their eco-friendly character, low toxicity, renewable and sustainable nature, cost-effectiveness, potential for large-scale production, and inherent antimicrobial properties. Even the plant extracts used for the green synthesis of nanoparticles often contain bioactive compounds, such as polyphenols, flavonoids, or terpenoids, which contribute to the antimicrobial activity of the resulting nanoparticles. These nanoparticles are synthesized by reducing metal salts using biogenic extracts or phytochemicals, resulting in the formation of nanoparticles with antimicrobial activity. The prominent metal and metal oxide NPs depicting antifungal action have been delineated hereunder.

22.5.1.1 Silver Nanoparticles

Silver nanoparticles (AgNPs) are one of the most extensively studied metal nanoparticles for their antifungal properties. AgNPs exhibit broad-spectrum antifungal activity against various fungi such as *Aspergillus brasiliensis*, *Aspergillus fumigatus*, *Aspergillus oryzae*, *Bipolaris sorokiniana*, *Candida albicans*, *Candida auris*, *Candida glabrata*, *Chaetomium globosum*, *Cladosporium cladosporioides*, *Cryptococcus* sp., *Magnaporthe grisea*, *Mortierella alpine*, *Penicillium brevicompactum*, *Penicillium chrysogenum*, *Rhizoctonia solani*, and *Stachybotrys chartarum* (Huang et al. 2023). Green-synthesized silver nanoparticles have been extensively studied for their antifungal properties. Plant extracts, such as those from neem (*Azadirachta indica*), turmeric (*Curcuma longa*), or green tea (*Camellia sinensis*), have been used to synthesize AgNPs. These nanoparticles have shown efficacy against various fungal pathogens, including *Candida* and *Aspergillus* sp. (Mansoor et al. 2021). Similarly, AgNPs synthesized by using *Achillea santolina* extract depicted anti-dermatophytic potential in an in vitro and in vivo rat model against dermal *Trichophyton rubrum* infection (Abdallah et al. 2023). AgNPs have also shown antifungal activity against *Microsporum canis*, *Trichophyton mentagrophytes*, and *Microsporum gypseum* although it was found to be less effective than standard antifungal griseofulvin according to the yielded minimum inhibitory concentration (MIC) values (Ayatollahi Mousavi et al. 2015). Antimicrobial action of AgNPs mostly exerted through the attachment of Ag + ions and/or AgNPs to the cell membrane by electrostatic interaction and/or penetration inside the microorganism. The outcome may lead to alteration in cellular architecture, disruption of the cell wall and membranes resulting into leakage of intracellular components, damage to the lipids and macromolecules, and enhancement of reactive oxygen species (ROS) production mediating mitochondrial dysfunctional, fungal growth inhibition, and cell death. The phenotypic antifungal manifestations include inhibition of the conidial germination, restriction in the growth of sclerotia and mycelia, and damage to fungal spores (conidia) and hyphae (Huang et al. 2023; Mansoor et al. 2021). Further, AgNPs can be conjugated with different standard antifungal drugs to potentiate the fungicidal action. For instance, conjugation of the antifungal drug miconazole to AgNPs has enhanced the fungicidal action than either of the antifungal agents individually. Inhibition of ergosterol biosynthesis and biofilm formation by increasing ROS levels has been reported as the major antifungal action of the nanoformulation. Therefore, AgNPs can act as inherent antifungal agent as well as delivery vehicle for conventional fungicidals (Ayeshamariam et al. 2014). In a different approach, monodispersed amphotericin B-coated silver nanoparticles (AmB-AgNPs) have been synthesized employing standard antifungal drug amphotericin B as a reducing and stabilizing/capping agent. The formulation has reported to depict synergistic antifungal activity against *Candida albicans*, *Aspergillus niger*, and *Fusarium culmorum* (Tutaj et al. 2016). AgNPs have also been used as antifungal coating to restrict fungal growth which has been discussed elsewhere.

22.5.1.2 Copper Nanoparticles

Copper nanoparticles (CuNPs) and copper oxide nanoparticles have been reported to exert antifungal action against diverse fungi, including *Alternaria alternate*, *Aspergillus flavus*, *A. niger*, *C. albicans*, *C. tropicalis*, *Corticium salmonicolor*, *Curvularia lunata*, *Fusarium equiseti*, *F. oxysporum*, *F. culmorum*, *F. kuroshium*, *Fusarium solani*, *Neofusicoccum* sp., *Phoma destructive*, *Stachybotrys chartarum*, *Penicillium italicum*, and *Phytophthora capsici* (Huang et al. 2023; Pariona et al. 2019; Osonga et al. 2022). Green synthesis of CuNPs using different plant extracts has also gained popularity in recent times and has shown antifungal activity against various fungal organisms such as *Fusarium solani*, *Neofusicoccum* sp., *Fusarium oxysporum*, *Aspergillus niger*, *Fusarium moniliforme*, *F. culmorum*, *F. tricinctum*, and *Candida* sp. (Bhavyasree and Xavier 2022; Ingle and Rai 2017; Pariona et al. 2019). Antifungal mechanism of action of CuNPs include induction of morphological changes in the mycelium, damage to the hyphae and alteration in the characteristic form of macroconidia, cell membrane damage, changes in colony morphology, and growth inhibition due to diminished ergosterol biosynthesis induced by free cytosolic copper ions. The nanoparticles size, shape, and concentration may be important variables to determine the antifungal activity (Pariona et al. 2019; Ibarra-Laclette et al. 2022; Pham et al. 2019). Copper nanoparticles are also known for their potential to inhibit biofilm formation, which is an important aspect in fungal infections (Mudiar and Kelkar-Mane 2020).

22.5.1.3 Zinc Oxide Nanoparticles

Zinc oxide nanoparticles (ZnO NPs) possess antifungal activity against wide range of fungal species, including *Alternaria alternata*, *A. solani*, *Aspergillus aculeatus*, *A. brasiliensis*, *A. flavus*, *A. fumigatus*, *A. niger*, *A. nidulans*, *Botrytis cinereal*, *Candida albicans*, *Erythricium salmonicolor*, *Fusarium oxysporum*, *F. graminearum*, *Microsporum canis*, *Mucor plumbeus*, *Penicillium expansum*, *Pythium debaryanum*, *Rhizopus stolonifer*, *Sclerotium rolfsii*, *Trichoderma harzianum*, and *Trichophyton mentagrophytes* (Huang et al. 2023). ZnO NPs can inhibit the growth of dermatophytes and reduce fungal burden. They can be formulated into creams, ointments, or powders for topical application, providing improved drug delivery to the site of infection. In addition, nanoparticles of griseofulvin with ZnO (GF-ZnO NPs) have yielded stronger inhibitory action against dermatophytes *Trichophyton mentagrophytes* and *Trichophyton verrucosum* than conventional griseofulvin according to the MIC values (Al-Janabi and Bashi 2022). Green-synthesized zinc oxide nanoparticles using plant extracts, such as leaf extract of *Cinnamomum camphora* (L.) Presl, have exhibited antifungal properties against *Alternaria alternate*. The lowest MIC value (20 mg/L) was obtained when the NPs were synthesized at pH 7. The NPs restricted spore germination and germ tube elongation as well as induced oxidative stress resulting in cell membrane damage causing leakage of protein and nucleic acid from fungal cells (Zhu et al. 2021). ZnO NPs derived from pomegranate peel extract have shown efficacy against several multidrug-resistant candidal strains such as *C. albicans*, *C. glabrata*, and *C. tropicalis*. Further, the NPs exhibited synergistic effect against the fungal pathogens with different standard antifungal drugs such as fluconazole, nystatin, terbinafine, itraconazole, and clotrimazole (Yassin et al. 2022). Further, ZnO NPs synthesized using leaf extract of *Crinum latifolium* have also been reported to significantly reduce biofilm formation by *C. albicans*. The NPs penetrated into the fungal cells led to extensive damage to the cell wall and cell membrane followed by inhibition of fungal growth and various virulent factors (Jalal et al. 2018). The antifungal actions of ZnO NPs include inducing ROS production, damage to the fungal cell membrane, reducing fungal enzymatic activity, obstruction to conidial development and distorting the conidiophores, damage to fungal hyphae and conidia, diminishing the spore germination capabilities, altering nucleic acids and carbohydrate content, and inducing protein leakage from the fungal cells. Moreover, light illumination has been reported to further enhance the antifungal action of ZnO NPs by additional enhancement in ROS production in fungal cells (Huang et al. 2023).

22.5.1.4 Gold Nanoparticles

Green synthesized gold nanoparticles (AuNPs) using plant extracts and bacterial or fungal filtrate have also depicted antifungal activity against diverse fungal species such as *Candida albicans*, *Puccinia graminis tritici*, *Aspergillus flavus*, *Aspergillus niger*, *Aspergillus terreus*, *Fusarium Oxysporum*, *Microsporum gypseum*, and *Trichophyton rubrum* (Huang et al. 2023; Jayaseelan et al. 2013). AuNPs can attach to the surface of microorganisms to inflict physical damage to the cells including destruction of the flagella. Further, AuNPs can inhibit the transmembrane proton efflux in fungal cells and hamper the H^+-ATPase activity resulting into acidification and death of the fungal cells (Huang et al. 2023; Jebali et al. 2014).

22.5.1.5 Titanium Dioxide Nanoparticles

Titanium dioxide nanoparticles (TiO_2 NPs) also exhibited antifungal action against various fungal species such as *Candida albicans* and *Fusarium oxysporum* with significant inhibition of biofilm formation by fluconazole-sensitive as well as fluconazole-resistant *C. albicans* (Huang et al. 2023). Green-synthesized TiO_2 NPs using ethanolic leaf extract of *Lawsonia inermis* was reported to show anti-dermatophytic activity against *Trichophyton mentagrophytes* and *Microsporum gypseum* (Lutfana Banu and Soundhari 2019). Green-synthesized TiO_2 NPs using Caricaceae (papaya) shell extracts have shown antifungal activity against different fungal plant pathogens such as *Sclerotinia sclerotiorum*, *Rosellinia necatrix*, and *Fusarium* spp. (Saka et al. 2022). TiO_2 NPs prepared by pulsed laser ablation have also shown inhibitory action against *Trichophyton rubrum* (Kadhim et al. 2016). Comparison of the antifungal properties of TiO_2 and ZnO versus ZnO-TiO_2 nanocomposites against *Aspergillus flavus* has yielded that ZnO-TiO_2 can exert maximum fungicidal activity at low concentration due to higher oxidative stress induction through ROS production. TiO_2 NPs (crystallite size 17.68 nm) showed a higher fungicidal action than ZnO NPs (33.21 nm) as smaller particle size facilitates better cell internalization and thereby produces more ROS. The MIC value of the ZnO-TiO_2 was the lowest (39 μg/mL) with respect to TiO_2 NPs (78 μg/mL) and ZnO NPs (156 μg/mL). The higher fungicidal action of ZnO-TiO_2 nanocomposites can be underlined by its high specific surface area facilitating chemical reactions and ROS production besides the sharp structure and smaller size of the nanocomposite promotes cellular internalization by the fungi leading to cell wall and membrane damage (NajibiIlkhechi et al. 2021). Inhibition of fungal mycelium development in *Sclerotinias sclerotium*, *Rosellinia necatrix*, and *Fusarium* spp. inflicted by green-synthesized TiO_2 NPs using Caricaceae (papaya) shell extracts has also been reported (Saka et al. 2022; AhmadpourKermani et al. 2021).

22.5.1.6 Iron Oxide Nanoparticles

Maghemite (γ-Fe_2O_3) and magnetite (Fe_3O_4) NPs have also been reported to show antifungal activity against several fungi, such as *Alternaria alternata*, *Aspergillus niger*, *Cladosporium herbarum*, *Penicillium chrysogenum*, and *Trichothecium roseum*. A concentration-dependent inhibition of spore germination was observed by the Fe_2O_3 NPs to restrict the fungal growth (Parveen et al. 2018). Fe_3O_4 NPs have also been reported to inhibit biofilm production by *Candida albicans* (Huang et al. 2023). Interaction of Fe_2O_3 NPs with the fungal cell surface is one of the probable mechanisms of its cidal action (Parveen et al. 2018). Green-synthesized Fe_3O_4 NPs using black coffee have been reported to inhibit the mycelial growth of *F. oxysporum* in in vitro culture (Ashraf et al. 2022). Induction of ROS production, DNA and protein damage, membrane damage and disruption have already been documented as the antimicrobial mechanism of iron oxide NPs; however, further exploration is required to elucidate the antifungal mechanisms of these nanomaterials. Different properties of NPs such as size, shape, surface functionalization, and concentration effect need to be considered to achieve its biomedical applications. Cytotoxicity is another aspect which needs special attention despite the report of apparently good safety profile of iron oxide NPs documented in several literatures.

22.5.1.7 Selenium Nanoparticles

Selenium nanoparticles (SeNPs) have also exhibited antifungal activities against *Aspergillus flavus*, *Aspergillus fumigatus*, *Aspergillus niger*, *Candida albicans*, *Candida glabrata*, *Candida krusei*, *Candida tropicalis*, *Trichophyton rubrum*, *Fusarium oxysporum*, *Colletotrichum gloeosporioides*, *Pyricularia grisea*, *Colletotrichum capsici*, and *Alternaria solani*. Biogenic SeNPs and chitosan-capped SeNPs have also generated anti-biofilm formation activity against *Candida albicans*. The antifungal mechanisms of SeNPs include membrane destruction, inhibiting spore germination, morphological changes in the fungal colonies, retarding the growth of hyphae, and alteration in expression *CYP51A* and *HSP90* resistance genes in *A. fumigatus* and *A. flavus* (Huang et al. 2023; Lazcano-Ramírez et al. 2023; Bafghi et al. 2022; Pescuma et al. 2023). Synthesis of SeNPs using chitosan as capping agent has been reported to yield synergistic fungicidal action. Adhesive and permeabilizing properties of chitosan have propelled SeNPs to penetrate the fungal cell wall causing distortion and disruption of the fungal cell (Lara et al. 2018). Green-synthesized SeNPs using *Melia azedarach* leaves extract have also been reported to inhibit conidial germination of *Fusarium mangiferae* at 300 μg/mL concentration (Shahbaz et al. 2023). Even silk-based films containing SeNPs has also been reported to generate antifungal action against *Candida parapsilosis* (Toprakcioglu et al. 2023).

It is important to note that while metal nanoparticles have demonstrated antifungal activity in vitro and in some animal models, their translation into clinical use requires further research and evaluation. Factors such as nanoparticle size, concentration, surface coating, and delivery methods need to be carefully optimized for safety and effectiveness. Regulatory considerations and potential toxicity issues also need to be addressed before the widespread use of metal nanoparticles as antifungal agents in clinical settings.

22.5.2 Nanoemulsions as Antifungal Agents

Nanoemulsions have emerged as a promising approach for delivering antifungal agents, offering enhanced stability, improved solubility, and increased efficacy against fungal infections such as *Candida albicans*, *Cryptococcus* spp., *Aspergillus fumigatus*, and Mucorales (Shahid et al. 2022; Garcia et al. 2019). Nanoemulsions are thermodynamically stable systems consisting of oil droplets dispersed in an aqueous phase, stabilized by surfactants or co-surfactants. Several plant essential oil-based emulsions have been explored to depict antifungal activity against *Candida albicans*, *Mucor ramosissimus*, *Penicillium verrucosum*, *Fusarium* spp., etc. (Wu et al. 2019; Krishnamoorthy et al. 2021; Mosa et al. 2023).The important implications regarding the use of nanoemulsions as antifungal agents are as follows.

Enhanced drug solubility and bioavailability: Nanoemulsions can solubilize hydrophobic antifungal drugs, improving their aqueous solubility and bioavailability. This allows for increased drug concentrations at the site of infection, leading to improved antifungal activity. The unique properties of nanoemulsions, such as high surface area and increased drug contact with fungal cells, can enhance the antifungal activity of encapsulated drugs. Nanoemulsions can interact with the fungal cell membrane, leading to membrane disruption, leakage of intracellular components, and inhibition of fungal growth (Sosa et al. 2017; Shahid et al. 2022;Garg et al. 2020).

Increased drug penetration and Potential for topical application: The small droplet size of nanoemulsions facilitates their penetration into the deeper layers of the skin or mucosal membranes, enhancing drug delivery to the site of infection. This improved penetration can be advantageous for treating superficial fungal infections or targeting fungi residing in difficult-to-reach areas (Shahid et al. 2022; Soriano-Ruiz et al. 2019).

Sustained drug release: Nanoemulsions can provide controlled and sustained release of antifungal drugs, ensuring a prolonged therapeutic effect. This sustained release can help maintain effective drug concentrations over an extended period, reducing the frequency of administration and improving patient compliance (Soriano-Ruiz et al. 2019).

Combination therapy: Nanoemulsions can be formulated to incorporate multiple antifungal agents, allowing for combination therapy. Combination therapy can have synergistic effects, targeting different aspects of fungal growth and reducing the risk of drug resistance (do Carmo Silva et al. 2020; Garcia et al. 2019).

It is important to note that the development and optimization of nanoemulsions as antifungal agents require careful consideration of factors such as droplet size, stability, surfactant selection, and compatibility with antifungal drugs. Additionally, further research and clinical studies are needed to evaluate the safety, efficacy, and long-term effects of nanoemulsion-based antifungal therapies in target subject-specific manner.

22.5.3 Organic Nanoparticles as Antifungal Agents

Organic nanoparticles have gained attention as potential antifungal agents due to their unique properties and the ability to deliver antifungal drugs effectively. Nanoparticles-based drug delivery systems have been investigated as promising approaches for the treatment of fungal infections. These systems offer advantages such as improved drug stability, enhanced bioavailability, targeted drug delivery, and controlled drug release. Some of the prominent examples of organic nanoparticles used as antifungal agents have been discussed hereunder.

22.5.3.1 Polymeric Nanoparticles

Polymeric nanoparticles, such as polymeric micelles or dendrimers, have been investigated for their antifungal properties. These nanoparticles can encapsulate antifungal drugs, improving their stability, solubility, and targeted delivery to the site of infection. Polymeric nanoparticles have shown efficacy against various fungal species, including *Candida* and *Aspergillus* (Alakkad et al. 2022; Krawczyk-Santos et al. 2021). Polymeric nanoparticles, such as chitosan nanoparticles or poly(lactic-co-glycolic acid) (PLGA) nanoparticles, have been explored for the delivery of antifungal agents against ringworm infections. These nanoparticles can encapsulate antifungal drugs, enhancing their stability, controlled release, and targeting to the affected skin. They can improve drug penetration into the deeper layers of the skin, where dermatophytes reside. For instance, entrapment of hydrophobic antifungal itraconazole in PLGA nanoparticles has found to improve antifungal action of the drug by increasing the bioavailability and penetration through improving aqueous dispersibility, thereby resulting inhibition of *Aspergillus flavus* for 11 days while itraconazole in water restricted fungal growth only for two days (Patel et al. 2010). Similarly, amphotericin B-loaded PLGA nanoparticles and nanosuspensions have found to enhance the antifungal efficacy of the active drug against *Candida albicans*, *Aspergillus fumigatus*, and *Trichophyton rubrum*; the nanoformulations have shown even better efficacy than AmBisome (amphotericin B with liposomal delivery system) and Fungizone (Ven 1988). Different other polymers and functionalization types have also been explored to encapsulate various standard antifungal agents depicting differential drug loading and release properties influencing the antifungal action of the nanoformulations on diverse fungal types (Souza and Amaral 2017). Biodegradable chitosan polymeric NPs have also been frequently used to encapsulate antifungal drugs like amphotericin B to treat candidiasis, aspergillosis, etc. Such encapsulation has extended multifaceted benefits such as facilitating oral delivery of the drug by preventing gastrointestinal degradation, enhancing antifungal action in topical delivery, and making it suitable for prolonged ocular application during fungal keratitis. (Souza and Amaral 2017).

22.5.3.2 Lipid-Based Nanoparticles

Lipid-based nanoparticles, such as liposomes or solid lipid nanoparticles (SLNs), have been utilized as carriers for antifungal drugs. Lipid nanoparticles can encapsulate hydrophobic antifungal agents, enhancing their solubility and stability. They can deliver the drugs to the target site, increasing their bioavailability, penetration, and efficacy against fungal infections (Füredi et al. 2017; Faustino and Pinheiro 2020). These nanoparticles can encapsulate antifungal drugs, protecting them from degradation and facilitating their targeted delivery to the site of infection. Commercially available liposomal amphotericin B sold under the brand name AmBisome® is a prominent example of such nanoformulation which has been extensively used to control candidiasis and aspergillosis. The formulation has gained popularity due to its capability of site-specific visceral delivery of hydrophobic as well as hydrophilic antifungal drugs, even in a pH-responsive manner, therefore reducing the drug dose and toxicity (Souza and Amaral 2017). Oral and topical terbinafine therapy is a conventional strategy to control dermatophytic infections. Several novel nanoformulations incorporating terbinafine with liposome have been developed such as terbinafine-loaded liposome film, terbinafine-loaded liposome, ethosome and transfersome, liposome poloxamer gel, and microemulsion-based gel, which may enable enhanced drug penetration to increase the efficacy (Sahni et al. 2018). Terbinafine hydrochloride-loaded liposome film has been found to be very effective in the treatment of onychomycosis with an improved drug release and site-specific accumulation potential (Tanrıverdi et al. 2016). Application of griseofulvin-entrapped SLNs gel twice daily for eight days has resulted complete recovery of *Microsporum canis* infection in guinea pig model. Besides providing good skin permeation, the nanoformulation has found to be nonsensitizing, histopathologically safe, and stable at a temperature range of 5–40 °C for a period of six months (Aggarwal and Goindi 2013). Similarly, fluconazole-loaded SLNs topical gels have resulted prolonged in vitro release and superior significant fast therapeutic index in treatment of pityriasis versicolor, a common fungal skin infection, over commercially available Candistan® cream (clotrimazole) (El-Housiny et al. 2018).

22.5.3.3 Protein-Based Nanoparticles

Protein-based nanoparticles, such as albumin nanoparticles or silk fibroin nanoparticles, have shown promise in delivering antifungal drugs (Füredi et al. 2016). These nanoparticles can encapsulate the drugs and offer controlled release properties. Protein-based nanoparticles can be biocompatible and biodegradable, making them suitable for biomedical applications (Kianfar 2021).

22.5.3.4 Cyclodextrin-Based Nanoparticles

Cyclodextrin-based nanoparticles, such as hydroxypropyl-β-cyclodextrin nanoparticles, have been explored as antifungal carriers. Cyclodextrins can form inclusion complexes with antifungal drugs, improving their solubility and stability. These nanoparticles can enhance drug delivery and increase antifungal efficacy (Deshkar and Palve 2019; Li et al. 2020).

22.5.4 *Other Nanoparticles*

22.5.4.1 Carbon-Based Nanoparticles

Carbon-based nanomaterials, such as carbon nanotubes and graphene derivatives, have shown antifungal properties against different fungal species such as *Fusarium graminearum*, *Fusarium poae*, and *Botrytis cinerea*. These nanoparticles conjugated or loaded with antifungal drugs can physically interact with the fungal cell membrane, leading to membrane disruption and inhibition of fungal growth (Benincasa et al. 2011; Huang et al. 2016). They can also alter oxidative enzyme activities, reduce mycelial biomass and branching, as well as alter metabolites involved in carbon and nitrogen flows (Huang et al. 2023). Further research is ongoing to explore their potential for the treatment of ringworm infections.

22.5.4.2 Mesoporous Silica Nanoparticles

Mesoporous silica nanoparticles (MSNs) have attracted attention as carriers for antifungal drugs. These nanoparticles possess a large surface area and pore volume, allowing for high drug-loading capacity. MSNs can be functionalized to provide controlled drug release and targeted delivery to the site of infection. They have shown promise in enhancing the therapeutic efficacy of antifungal agents. For instance, MSNs incorporated with zinc oxide (MSNs-ZnO) has been depicted to act as a potential antifungal agent against *Fusarium graminearum* and *Aspergillus flavus* strains (Savi et al. 2013). Negatively as well as positively charged surface functionalized MSNs have been found to enhance solubility and release of antifungal drug griseofulvin, but hydrophobic modification of MSNs affected the drug release capacity (Jambhrunkar et al. 2014).

22.5.4.3 Nanofibers and Nanocomposites

Nanofibers and nanocomposites, such as electrospun fibers or composite nanoparticles, have been explored for localized drug delivery against fungal infections. These systems can provide a high surface area-to-volume ratio and sustained drug release properties. They can be incorporated with antifungal agents to create dressings or scaffolds for topical or localized treatment of fungal infections. For instance, multilayer electrospun poly(vinyl alcohol)/gelatin nanofibers have been employed to load antifungal agents fluconazole and cinnamaldehyde, separately as well as in combination to treat fungal keratitis. The formulations promisingly inhibited fungal growth and biofilm formation by *Candida albicans* without producing any cytotoxicity (Ilhan et al. 2022).

22.5.5 Nanomaterials in Antifungal Coatings

Diverse nanomaterials have been explored for their potential in developing antifungal coatings to prevent or inhibit fungal growth on various surfaces. These coatings can be applied to medical devices, building materials, textiles, and other surfaces to provide long-lasting protection against fungal infections.

Silver nanoparticles are widely studied for their antimicrobial properties, including antifungal activity. They can be incorporated into coatings to provide effective protection against fungal colonization. AgNPs can disrupt fungal cell membranes, inhibit enzyme activity, and interfere with cellular processes, leading to fungal growth inhibition. Silver nanoparticle-based coatings have shown efficacy against various fungal species (Boivin et al. 2023; Bellotti et al. 2015). Similarly, other metal and metal oxides NPs such as copper NPs, selenium NPs, ZnO NPs, and TiO_2 NPs have also been employed into coatings to prevent fungal growth. These NPs has found to inhibit fungal enzymes, generate reactive oxygen species (ROS), induce oxidative stress, and damage the fungal cell membrane, resulting in antifungal effects (Sohail et al. 2022; Ilkhechi et al. 2021; Salem et al. 2022). Chitosan nanoparticles derived from chitin, a natural polysaccharide found in crustacean shells, have exhibited antifungal properties. Chitosan nanoparticles can disrupt fungal cell membranes, inhibit fungal enzyme activity, and interfere with fungal growth. Coatings containing chitosan nanoparticles or composites have demonstrated efficacy against various fungal pathogens (Xing et al. 2021).

These nanomaterials can be incorporated into coatings through various methods such as surface modification, blending, or layer-by-layer deposition. The nanomaterials provide durable and long-lasting antifungal properties to the coated surfaces, making them suitable for applications in healthcare, construction, textiles, and agriculture. It is important to consider factors such as the stability of nanomaterials in the coating matrix, controlled release of the active components, and compatibility with the surface to be coated. The safety, regulatory compliance, and potential environmental impact of these nanomaterial-based coatings should also be thoroughly evaluated before their widespread use.

22.6 Conclusion and Future Outlook

In conclusion, nanomaterials hold great promise for the treatment of fungal diseases in livestock and poultry. Their unique properties, such as small size, large surface area, and tunable surface chemistry, enable targeted drug delivery, enhanced drug stability, improved bioavailability, and controlled release. Nanomaterials can overcome the limitations of conventional therapeutics, including poor drug solubility, limited drug penetration, drug resistance, and adverse side effects.

Currently, there are several nanomaterial-based approaches being explored for the treatment of fungal diseases in livestock and poultry. Nanoparticles, such as metal nanoparticles, polymer nanoparticles, lipid-based nanoparticles, and green-synthesized nanoparticles, offer advantages in terms of improved drug delivery, sustained release, and enhanced antifungal activity. Nanoemulsions and nanofibers have also shown promise as antifungal agents in veterinary applications.

However, there are still challenges and future considerations that need to be addressed. Further research is needed to optimize the formulation and manufacturing processes of nanomaterial-based antifungal agents to ensure their safety, efficacy, and scalability. In-depth studies on the pharmacokinetics, biodistribution, and long-term effects of nanomaterials in animals are crucial.

Regulatory approvals and guidelines specific to nanomaterial-based veterinary therapeutics are also necessary. Standardization of testing methods and quality control measures will help ensure the consistent performance and safety of these nanomaterial-based products.

Furthermore, it is important to assess the environmental impact and potential development of resistance associated with the use of nanomaterials in veterinary medicine. Studies on the fate and behavior of nanomaterials in the environment, as well as their impact on non-target organisms are required.

Overall, the use of nanomaterials for the treatment of fungal diseases in livestock and poultry shows great promise and can revolutionize veterinary therapeutics. Continued research and development efforts, along with collaboration between researchers, veterinarians, regulatory authorities, and industry stakeholders, will pave the way for the safe and effective implementation of nanomaterial-based therapies in the veterinary field.

References

Abdallah BM, Rajendran P, Ali EM (2023) Potential treatment of dermatophyte trichophyton rubrum in rat model using topical green biosynthesized silver nanoparticles with Achillea santolina extract. Molecules 28(4):1536

Abdellatif AA, El-Telbany DFA, Zayed G, Al-Sawahli MM (2019) Hydrogel containing PEG-coated fluconazole nanoparticles with enhanced solubility and antifungal activity. J Pharm Innov 14:112–122

Abidin Z, Khatoon A, Numan M (2011) Mycotoxins in broilers: pathological alterations induced by aflatoxins and ochratoxins, diagnosis and determination, treatment and control of mycotoxicosis. Worlds Poult Sci J 67(3):485–496

Adebiyi AI, Oluwayelu DO (2018) Zoonotic fungal diseases and animal ownership in Nigeria. Alexandria J Med 54(4):397–402

Aggarwal N, Goindi S (2013) Preparation and in vivo evaluation of solid lipid nanoparticles of griseofulvin for dermal use. J Biomed Nanotechnol 9(4):564–576. https://doi.org/10.1166/jbn.2013.1569

AhmadpourKermani S, Salari S, GhasemiNejadAlmani P (2021) Comparison of antifungal and cytotoxicity activities of titanium dioxide and zinc oxide nanoparticles with amphotericin B against different Candida species: in vitro evaluation. J Clin Lab Anal 35(1):e23577

Alakkad A, Stapleton P, Schlosser C, Murdan S, Odunze U, Schatzlein A, Uchegbu IF (2022) Amphotericin b polymer nanoparticles show efficacy against candida species biofilms. Pathogens 11(1):73

al-Ani FK (1999) Epizootic lymphangitis in horses: a review of the literature. Rev Sci Tech 18(3): 691–699

Alghuthaymi MA, Hassan AA, Kalia A, Kalia A, Sayed El Ahl RM, El Hamaky AA, Oleksak P, Kuca K, Abd-Elsalam KA (2021) Antifungal nano-therapy in veterinary medicine: current status and future prospects. J Fungi 7(7):494. https://doi.org/10.3390/jof7070494

Ali R, Khan IH (2006) Mycotic abortion in cattle. Pak Vet J 26(1):44

Al-Janabi AAHS, Bashi AM (2022) Synthesis and antifungal activity of novel griseofulvin nanoparticles with zinc oxide against dermatophytic fungi: Trichophyton mentagrophytes and Trichophyton verrucosum: A primary study. Curr Med Mycol 8(2):40

Anderson ML (2007) Infectious causes of bovine abortion during mid-to late-gestation. Theriogenology 68(3):474–486

Arné P, Thierry S, Wang D, Deville M, Le Loch G, Desoutter A, Guillot J (2011) Aspergillus fumigatus in poultry. Int J Microbiol 2011

Ashley ES, Lewis R, Lewis JS, Martin C, Andes D (2006) Pharmacology of systemic antifungal agents. Clin Infect Dis 43(Supplement_1):S28–S39. https://doi.org/10.1086/504492

Ashraf H, Batool T, Anjum T, Illyas A, Li G, Naseem S, Riaz S (2022) Antifungal potential of green synthesized magnetite nanoparticles black coffee–magnetite nanoparticles against wilt infection by ameliorating enzymatic activity and gene expression in Solanum lycopersicum L. Front Microbiol 13:754292. https://doi.org/10.3389/fmicb.2022.754292

Awuchi CG, Ondari EN, Ogbonna CU, Upadhyay AK, Baran K, Okpala COR, Guiné RP (2021) Mycotoxins affecting animals, foods, humans, and plants: types, occurrence, toxicities, action mechanisms, prevention, and detoxification strategies—A revisit. Foods 10(6):1279

Ayatollahi Mousavi SA, Salari S, Hadizadeh S (2015) Evaluation of antifungal effect of silver nanoparticles against microsporumcanis, trichophyton mentagrophytes and microsporumgypseum. Iran J Biotechnol 13(4):38–42. https://doi.org/10.15171/ijb.1302

Ayeshamariam A, Kashif M, Vidhya VS, Sankaracharyulu MGV, Swaminathan V, Bououdina M, Jayachandran M (2014) Biosynthesis of (ZnO–Aloe Vera) nanocomposites and antibacterial/ antifungal studies. J Opto Bio Mat 6(3):85–99

Bafghi MH, Nazari R, Darroudi M, Zargar M, Zarrinfar H (2022) The effect of biosynthesized selenium nanoparticles on the expression of CYP51A and HSP90 antifungal resistance genes in aspergillus fumigatus and aspergillus flavus. Biotechnol Prog 38(1):e3206

Bakr EM, El-Tawab AEKMA, Elshemey TM, Abd-Elrhman AH (2015) Diagnostic and therapeutic studies on mycotic mastitis in cattle. Alexandria J Vet Sci 46(1):138

Beigi FAK, Imani M, Payehghadr M, Hosseini H (2011) SPE-HPLC method for determination of ketoconazole and clotrimazole residues in cow's milk. J Braz Chem Soc 22(9):1679–1685. https://doi.org/10.1590/S0103-50532011000900009

Bellotti N, Romagnoli R, Quintero C, Domínguez-Wong C, Ruiz F, Deyá C (2015) Nanoparticles as antifungal additives for indoor water borne paints. Prog Org Coat 86:33–40

Benincasa M, Pacor S, Wu W, Prato M, Bianco A, Gennaro R (2011) Antifungal activity of amphotericin B conjugated to carbon nanotubes. ACS Nano 5(1):199–208

Bhavyasree P, Xavier T (2022) Green synthesised copper and copper oxide based nanomaterials using plant extracts and their application in antimicrobial activity: review. Curr Res Green Sustain Chem 5:100249. https://doi.org/10.1016/j.crgsc.2021.100249

Boivin G, Ritcey AM, Landry V (2023) Silver nanoparticles as antifungal agents in acrylic latexes: influence of the initiator type on nanoparticle incorporation and Aureobasidium pullulans resistance. Polymers 15(6):1586

Borgers M (1980) Mechanism of action of antifungal drugs, with special reference to the imidazole derivatives. Rev Infect Dis 2(4):520–534

Brömel C, Sykes JE (2005) Histoplasmosis in dogs and cats. Clin Tech Small Anim Pract 20(4): 227–232

Carmo PMSD, Uzal FA, Pedroso PM, Riet-Correa F (2020) Conidiobolomycosis, cryptococcosis, and aspergillosis in sheep and goats: a review. J Vet Diagn Investig 32(6):826–834

Chermette R, Ferreiro L, Guillot J (2008) Dermatophytoses in animals. Mycopathologia 166:385–405

de Almeida Jr JN, Peçanha-Pietrobom PM, Colombo AL (2018) Paracoccidioidomycosis in immunocompromised patients: a literature review. J Fungi 5(1):2

de Perio MA, Benedict K, Williams SL, Niemeier-Walsh C, Green BJ, Coffey C, Nett RJ (2021) Occupational histoplasmosis: epidemiology and prevention measures. J Fungi 7(7):510

Deamer D, Weber AL (2010) Bioenergetics and life's origins. Cold Spring Harb Perspect Biol 2: a004929

Deshkar SS, Palve VK (2019) Formulation and development of thermosensitive cyclodextrin-based in situ gel of voriconazole for vaginal delivery. J Drug Deliv Sci Technol 49:277–285

Desoubeaux G, Cray C (2018) Animal Models of Aspergillosis. Comparat Med 68(2):109–123

do Carmo Silva L, Miranda MA, de Freitas JV, Ferreira SF, de Oliveira Lima EC, de Oliveira CM, Kato L, Terezan AP, Rodriguez AF, Faria FS, de Almeida Soares CM (2020) Antifungal activity of Copaíba resin oil in solution and nanoemulsion against Paracoccidioides spp. Braz J Microbiol 51(1):125–134. https://doi.org/10.1007/s42770-019-00201-3

Dworecka-Kaszak B, Biegańska MJ, Dąbrowska I (2020) Occurrence of various pathogenic and opportunistic fungi in skin diseases of domestic animals: a retrospective study. BMC Vet Res 16(1):1–8

Dworecka-Kaszak B, Krutkiewicz A, Szopa D, Kleczkowski M, Biegańska M (2012) High prevalence of Candida yeast in milk samples from cows suffering from mastitis in Poland. Sci World J 2012:1

El-Housiny S, Shams Eldeen MA, El-Attar YA, Salem HA, Attia D, Bendas ER, El-Nabarawi MA (2018) Fluconazole-loaded solid lipid nanoparticles topical gel for treatment of pityriasis versicolor: formulation and clinical study. Drug Deliv 25(1):78–90. https://doi.org/10.1080/10717544.2017.1413444

El-Sayed RA, Jebur AB, Kang W, El-Demerdash FM (2022) An overview on the major mycotoxins in food products: characteristics, toxicity, and analysis. J Fut Foods 2(2):91–102

Faustino C, Pinheiro L (2020) Lipid systems for the delivery of amphotericin B in antifungal therapy. Pharmaceutics 12(1):29

Fernandes GF, Lopes-Bezerra LM, Bernardes-Engemann AR, Schubach TMP, Dias MAG, Pereira SA, de Camargo ZP (2011) Serodiagnosis of sporotrichosis infection in cats by enzyme-linked immunosorbent assay using a specific antigen, SsCBF, and crude exoantigens. Vet Microbiol 147(3–4):445–449

Fisher MC, Berman J, Bicanic T, Bignell EM, Bowyer P, Bromley M, Brüggemann R, Garber G, Cornely OA, Gurr SJ, Harrison TS, Kuijper E, Rhodes J, Sheppard DC, Warris A, White PL, Xu J, Zwaan B, Verweij PE (2022) Tackling the emerging threat of antifungal resistance to human health. Nat Rev Microbiol 20(9):557–571. https://doi.org/10.1038/s41579-022-00720-1

Foy DS, Trepanier LA (2010) Antifungal treatment of small animal veterinary patients. Vet Clin 40(6):1171–1188

Füredi P, Kovács K, Ludányi K, Antal I, Klebovich I (2016) Development and characterization of voriconazole loaded nanoparticles for parenteral delivery. Int J Pharm 510(1):159–163

Füredi P, Pápay ZE, Kovács K, Kiss BD, Ludányi K, Antal I, Klebovich I (2017) Development and characterization of the voriconazole loaded lipid-based nanoparticles. J Pharm Biomed Anal 132:184–189

Garcia A, Fan YY, Vellanki S, Huh EY, Vanegas D, Wang SH, Lee SC (2019) Nanoemulsion as an effective treatment against human-pathogenic fungi. MSphere 4(6):e00729–e00719

Garg A, Sharma GS, Goyal AK, Ghosh G, Si SC, Rath G (2020) Recent advances in topical carriers of anti-fungal agents. Heliyon 6(8):e04663

Lazcano-Ramírez HG, Garza-García JJ, Hernández-Díaz JA, León-Morales JM, Macías-Sandoval AS, García-Morales S (2023) Antifungal activity of selenium nanoparticles obtained by plant-mediated synthesis. Antibiotics 12(1):115. https://doi.org/10.3390/antibiotics12010115

Ghannoum MA, Rice LB (1999) Antifungal agents: mode of action, mechanisms of resistance, and correlation of these mechanisms with bacterial resistance. Clin Microbiol Rev 12(4):501–517

Hadush B, Michaelay M, Menghistu HT, Abebe N, Genzebu AT, Bitsue HK, Ameni G (2020) Epidemiology of epizootic lymphangitis of carthorses in northern Ethiopia using conventional diagnostic methods and nested polymerase chain reaction. BMC Vet Res 16(1):1–7

Huang JF, Zhong J, Chen GP, Lin ZT, Deng Y, Liu YL, Jiang GB (2016) A hydrogel-based hybrid theranostic contact lens for fungal keratitis. ACS Nano 10(7):6464–6473

Huang T, Li X, Maier M, O'Brien-Simpson NM, Heath DE, O'Connor AJ (2023) Using inorganic nanoparticles to fight fungal infections in the antimicrobial resistant era. Acta Biomater 158:56–79. https://doi.org/10.1016/j.actbio.2023.01.019

Ibarra-Laclette E, Blaz J, Pérez-Torres CA, Villafán E, Lamelas A, Rosas-Saito G, Pariona N (2022) Antifungal effect of copper nanoparticles against fusarium kuroshium, an obligate symbiont of Euwallaceakuroshio ambrosia beetle. J Fungi 8(4):347

Ilhan E, Cesur S, Sulutas RB, Pilavci E, Dalbayrak B, Kaya E, Gunduz O (2022) The role of multilayer electrospun poly (vinyl alcohol)/gelatin nanofibers loaded with fluconazole and cinnamaldehyde in the potential treatment of fungal keratitis. Eur Polym J 176:111390

Ilkhechi NN, Mozammel M, Khosroushahi AY (2021) Antifungal effects of ZnO, TiO2 and ZnO-TiO2 nanostructures on aspergillus flavus. Pestic Biochem Physiol 176:104869

Ingle AP, Rai M (2017) Copper nanoflowers as effective antifungal agents for plant pathogenic fungi. IET Nanobiotechnol 11(5):546–551

Jain S, Jain S, Khare P, Gulbake A, Bansal D, Jain SK (2010) Design and development of solid lipid nanoparticles for topical delivery of an anti-fungal agent. Drug Deliv 17(6):443–451

Jalal M, Ansari MA, Ali SG, Khan HM, Rehman S (2018) Anticandidal activity of bioinspired ZnO NPs: effect on growth, cell morphology and key virulence attributes of Candida species. Artif Cells Nanomed Biotechnol 46(sup1):912–925

Jambhrunkar S, Qu Z, Popat A, Karmakar S, Xu C, Yu C (2014) Modulating in vitro release and solubility of griseofulvin using functionalized mesoporous silica nanoparticles. J Colloid Interface Sci 434:218–225

Jayaseelan C, Ramkumar R, Rahuman AA, Perumal P (2013) Green synthesis of gold nanoparticles using seed aqueous extract of Abelmoschus esculentus and its antifungal activity. Ind Crop Prod 45:423–429. https://doi.org/10.1016/j.indcrop.2012.12.019

Jebali A, Hajjar FH, Pourdanesh F, Hekmatimoghaddam S, Kazemi B, Masoudi A, Daliri K, Sedighi N (2014) Silver and gold nanostructures: antifungal property of different shapes of these nanostructures on Candida species. Med Mycol 52(1):65–72. https://doi.org/10.3109/13693786.2013.822996

Jenks JD, Hoenigl M (2018) Treatment of aspergillosis. J Fungi 4(3):98

Kadhim AA, Haleem A, Abass RH (2016) Anti-dermatophyte activity of TiO2 NPs colloidal prepared by pulsed laser ablation in liquid environment. Adv Environ Biol 10(12):43–54

Kanafani ZA, Perfect JR (2008) Resistance to antifungal agents: mechanisms and clinical infectious diseases impact. Clin Infect Dis 46:120–128

Kępińska-Pacelik J, Biel W (2021) Mycotoxins-prevention, detection, impact on animal health. PRO 9(11):2035

Kianfar E (2021) Protein nanoparticles in drug delivery: animal protein, plant proteins and protein cages, albumin nanoparticles. J Nanobiotechnol 19:159. https://doi.org/10.1186/s12951-021-00896-3

Kischkel B, Rossi SA, Santos SR, Nosanchuk JD, Travassos LR, Taborda CP (2020) Therapies and vaccines based on nanoparticles for the treatment of systemic fungal infections. Front Cell Infect Microbiol 10:575904. https://doi.org/10.3389/fcimb.2020.00463

Krawczyk-Santos AP, da Rocha PB, Kloppel LL, Souza BDS, Anjos JLV, Alonso A, Taveira SF (2021) Enhanced nail delivery of voriconazole-loaded nanomicelles by thioglycolic acid pretreatment: A study of protein dynamics and disulfide bond rupture. Int J Pharm 602:120597

Krishnamoorthy R, Gassem MA, Athinarayanan J, Periyasamy VS, Prasad S, Alshatwi AA (2021) Antifungal activity of nanoemulsion from Cleome viscosa essential oil against food-borne pathogenic Candida albicans. Saudi J Biol Sci 28(1):286–293. https://doi.org/10.1016/j.sjbs.2020.10.001

Lara HH, Guisbiers G, Mendoza J, Mimun LC, Vincent BA, Lopez-Ribot JL, Nash KL (2018) Synergistic antifungal effect of chitosan-stabilized selenium nanoparticles synthesized by pulsed laser ablation in liquids against Candida albicans biofilms. Int J Nanomedicine 13: 2697–2708

Lewis RE (2011) Current concepts in antifungal pharmacology. Mayo Clin Proc 86(8):805–817

Li Y, Zhu C, Wu H, Pan H, Liu H (2020) Kolliphor® HS 15-cyclodextrin complex for the delivery of voriconazole: preparation, characterization, and antifungal activity. Curr Drug Metab 21(5): 379–389

Lutfana Banu A, Soundhari C (2019) Anti-dermatophytic activity of titanium dioxide nanoparticle synthesized using LawsoniaInermis. Ind J Appl Microbiol 22(3):38–46

Mahajan VK (2014) Sporotrichosis: an overview and therapeutic options. Dermatol Res Pract 2014: 272376. This review paper gives an overview of the disease. PubMed PubMed Central

Mahendra P (1991) Diagnosis and chemotherapy of mycotic mastitis. J Vet Clin 8(1):53–57

Mansoor S, Zahoor I, Baba TR, Padder SA, Bhat ZA, Koul AM, Jiang L (2021) Fabrication of silver nanoparticles against fungal pathogens. Front Nanotechnol 3:679358. https://doi.org/10.3389/fnano.2021.679358

Mocherla S, Wheat LJ (2001, June) Treatment of histoplasmosis. Sem Respir Infect 16(2):141–148

Moriello K (2014) Feline dermatophytosis: aspects pertinent to disease management in single and multiple cat situations. J Feline Med Surg 16(5):419–431

Moriello KA, Coyner K (2021) Dermatophytosis. In Greene's infectious diseases of the dog and cat. WB Saunders, pp 961–977

Mosa MA, Youssef K, Hamed SF, Hashim AF (2023) Antifungal activity of eco-safe nanoemulsions based on Nigella sativa oil against Penicillium verrucosum infecting maize seeds: biochemical and physiological traits. Front Microbiol 13:1108733. https://doi.org/10.3389/fmicb.2022.1108733

Mudiar R, Kelkar-Mane V (2020) Original research article (experimental): targeting fungal menace through copper nanoparticles and Tamrajal. J Ayurveda Integr Med 11(3):316–321. https://doi.org/10.1016/j.jaim.2018.02.134

Osonga FJ, Eshun G, Kalra S, Yazgan I, Sakhaee L, Ontman R, Sadik OA (2022) Influence of particle size and shapes on the antifungal activities of greener nanostructured copper against Penicillium italicum. ACS Agric Sci Technol 2(1):42–56

Pappas PG, Rex JH, Sobel JD, Filler SG, Dismukes WE, Walsh TJ, Edwards JE (2004) Guidelines for treatment of candidiasis. Clin Infect Dis 38(2):161–189

Pariona N, Mtz-Enriquez AI, Sánchez-Rangel D, Carrión G, Paraguay-Delgado F, Rosas-Saito G (2019) Green-synthesized copper nanoparticles as a potential antifungal against plant pathogens. RSC Adv 9(33):18835–18843

Parveen S, Wani AH, Shah MA, Devi HS, Bhat MY, Koka JA (2018) Preparation, characterization and antifungal activity of iron oxide nanoparticles. Microb Pathog 115:287–292

Patel NR, Damann K, Leonardi C, Sabliov CM (2010) Itraconazole-loaded poly (lactic-co-glycolic) acid nanoparticles for improved antifungal activity. Nanomedicine 5(7):1037–1050

Patil RD, Sharma R, Asrani RK, Negi G (2014) Mycotoxicosis and its control in poultry: A review. J Poultry Sci Technol 2(1):1

Pereira SA, Gremião IDF, Menezes RC (2015) Sporotrichosis in animals: zoonotic transmission. Sporotrichosis: new developments and future prospects:83–102

Pescuma M, Aparicio F, Zysler RD, Lima E, Zapata C, Marfetán JA, Vélez M, Ordoñez OF (2023) Biogenic selenium nanoparticles with antifungal activity against the wood-rotting fungus Oligoporuspelliculosus. Biotechnol Rep 37:e00787. https://doi.org/10.1016/j.btre.2023.e00787

Pham N, Duong M, Le M, Hoang HA, Pham L (2019) Preparation and characterization of antifungal colloidal copper nanoparticles and their antifungal activity against fusarium oxysporum and Phytophthora capsici. Comptes Rendus Chimie 22(11–12):786–793. https://doi.org/10.1016/j.crci.2019.10.007

Prameela DR (2019) Mycotic abortions in bovine: A review. J Anim Feed Sci Technol 7(2)

Queiroz-Telles F, Buccheri R, Benard G (2019) Sporotrichosis in immunocompromised hosts. J Fungi 5(1):8

Rengasamy M, Shenoy MM, Dogra S, Asokan N, Khurana A, Poojary S, Kaushal V (2020) Indian association of dermatologists, venereologists and leprologists (IADVL) task force against recalcitrant tinea (ITART) consensus on the management of glabrous tinea (INTACT). Indian Dermatol Online J 11(4):502

Rezabek GB, Donahue JM, Giles RC, Petrites-Murphy MB, Poonacha KB, Rooney JR, Tramontin RR (1993) Histoplasmosis in horses. J Comp Pathol 109(1):47–55

Riet-Correa F, Krockenberger M, Dantas AF, Oliveira DM (2011) Bovine cryptococcal meningo-encephalitis. J Vet Diagn Investig 23(5):1056–1060

Rochette F, Engelen M, Vanden Bossche H (2003) Antifungal agents of use in animal health–practical applications. J Vet Pharmacol Ther 26(1):31–53

Rodrigues AM, de Melo Teixeira M, de Hoog GS, Schubach TMP, Pereira SA, Fernandes GF, de Camargo ZP (2013) Phylogenetic analysis reveals a high prevalence of Sporothrixbrasiliensis in feline sporotrichosis outbreaks. PLoS Negl Trop Dis 7(6):e2281

Sahni K, Singh S, Dogra S (2018) Newer topical treatments in skin and nail dermatophyte infections. Indian Dermatol Online J 9(3):149

Saka A, Shifera Y, Jule LT, Badassa B, Nagaprasad N, Shanmugam R, Priyanka Dwarampudi L, Seenivasan V, Ramaswamy K (2022) Biosynthesis of TiO2 nanoparticles by Caricaceae (Papaya) shell extracts for antifungal application. Sci Rep 12(1):1–10. https://doi.org/10.1038/s41598-022-19440-w

Salem MF, Abd-Elraoof WA, Tayel AA, Alzuaibr FM, Abonama OM (2022) Antifungal application of biosynthesized selenium nanoparticles with pomegranate peels and nanochitosan as edible coatings for citrus green mold protection. J Nanobiotechnol 20(1):182

Savi GD, Bortoluzzi AJ, Scussel VM (2013) Antifungal properties of zinc-compounds against toxigenic fungi and mycotoxin. Int J Food Sci Technol 48(9):1834–1840

Scantlebury C, Reed K (2009) Epizootic lymphangitis. Infectious diseases of the horse. EVJ Ltd, Ely, Cambridgeshire, p 390

Schubach TM, Schubach A, Okamoto T, Barros MB, Figueiredo FB, Cuzzi T, Wanke B (2004) Evaluation of an epidemic of sporotrichosis in cats: 347 cases (1998–2001). J Am Vet Med Assoc 224(10):1623–1629

Scorzoni L, de Paula e Silva AC, Marcos CM, Assato PA, de Melo WC, de Oliveira HC, Costa-Orlandi CB, Mendes-Giannini MJ, Fusco-Almeida AM (2017) Antifungal therapy: new advances in the understanding and treatment of mycosis. Front Microbiol 8:242257. https://doi.org/10.3389/fmicb.2017.00036

Seid A, Fedlu M, Mama A (2019) Review on epizootic lymphangitis: epidemiology and its diagnosis. J Dairy Vet Sci 12:1–8

Serhan G, Stack CM, Perrone GG, Morton CO (2014) The polyene antifungals, amphotericin B and nystatin, cause cell death in Saccharomyces cerevisiae by a distinct mechanism to amphibian-derived antimicrobial peptides. Ann Clin Microbiol Antimicrob 13(1):1–4

Seyedmousavi S, Bosco SDM, De Hoog S, Ebel F, Elad D, Gomes RR, Guillot J (2018) Fungal infections in animals: a patchwork of different situations. Med Mycol 56(suppl_1):S165–S187

Shahbaz M, Akram A, Raja NI, Mukhtar T, Mehak A, Fatima N, Abasi F (2023) Antifungal activity of green synthesized selenium nanoparticles and their effect on physiological, biochemical, and antioxidant defense system of mango under mango malformation disease. PLoS One 18(2): e0274679

Shahid M, Hussain A, Khan AA, Alanazi AM, Alaofi AL, Alam M, Ramzan M (2022) Antifungal cationic nanoemulsion ferrying miconazole nitrate with synergism to control fungal infections: in vitro, ex vivo, and in vivo evaluations. ACS Omega 7(15):13343–13353

Skiada A, Lass-Floerl C, Klimko N, Ibrahim A, Roilides E, Petrikkos G (2018) Challenges in the diagnosis and treatment of mucormycosis. Med Mycol 56(suppl_1):S93–S101

Sohail Y, Raza N, Shakeel N, Raza H, Manzoor S, Yasmin G, Mohammad Wabaidur S (2022) Polyaniline-coated nanoparticles of zinc oxide and copper oxide as antifungal agents against Aspergillus parasiticus. Front Plant Sci 13:925451

Soriano-Ruiz JL, Calpena-Capmany AC, Canadas-Enrich C, Bozal-de Febrer N, Suñer-Carbó J, Souto EB, Clares-Naveros B (2019) Biopharmaceutical profile of a clotrimazole nanoemulsion: evaluation on skin and mucosae as anticandidal agent. Int J Pharm 554:105–115

Sosa L, Clares B, Alvarado HL, Bozal N, Domenech O, Calpena AC (2017) Amphotericin B releasing topical nanoemulsion for the treatment of candidiasis and aspergillosis. Nanomedicine 13(7):2303–2312. https://doi.org/10.1016/j.nano.2017.06.021

Souza AC, Amaral AC (2017) Antifungal therapy for systemic mycosis and the nanobiotechnology era: improving efficacy, biodistribution and toxicity. Front Microbiol 8:336

Spadari CDC, Wirth F, Lopes LB, Ishida K (2020) New approaches for cryptococcosis treatment. Microorganisms 8(4):613

Spellberg B, Ibrahim AS (2010) Recent advances in the treatment of mucormycosis. Curr Infect Dis Rep 12:423–429

Tanrıverdi ST, Hilmioğlu Polat S, Yeşim Metin D, Kandiloğlu G, Özer Ö (2016) Terbinafine hydrochloride loaded liposome film formulation for treatment of onychomycosis: in vitro and in vivo evaluation. J Liposome Res 26(2):163–173. https://doi.org/10.3109/08982104.2015.1067892

Tell LA (2005) Aspergillosis in mammals and birds: impact on veterinary medicine. Med Mycol 43 (Supplement_1):S71–S73

Toprakcioglu Z, Wiita EG, Jayaram AK, Gregory RC, Knowles TP (2023) Selenium silk nano-structured films with antifungal and antibacterial activity. ACS Appl Mater Interfaces 15(8): 10452–10463

Tutaj K, Szlazak R, Szalapata K, Starzyk J, Luchowski R, Grudzinski W, Gruszecki WI (2016) Amphotericin B-silver hybrid nanoparticles: synthesis, propertiesand antifungal activity. Nanomedicine 12(4):1095–1103

Ven JA (1988) Practical theology: from applied to empirical theology. J Empir Theol 1(1):7–7

Watkins J (2014) Symptoms and treatment of ringworm infections. Br J School Nurs 9(3):76

Wu D, Lu J, Zhong S, Schwarz P, Chen B, Rao J (2019) Influence of nonionic and ionic surfactants on the antifungal and mycotoxin inhibitory efficacy of cinnamon oil nanoemulsions. Food Funct 10(5):2817–2827

Xing Y, Yi R, Yang H, Xu Q, Huang R, Tang J, Yu J (2021) Antifungal effect of chitosan/nano-TiO2 composite coatings against Colletotrichum gloeosporioides, Cladosporium oxysporum and Penicillium steckii. Molecules 26(15):4401

Yassin MT, Mostafa AA, Abdulrahman A, Al-Otibi FO (2022) Facile green synthesis of zinc oxide nanoparticles with potential synergistic activity with common antifungal agents against multidrug-resistant candidal strains. Crystals 12(6):774. https://doi.org/10.3390/cryst12060774

Zain ME (2011) Impact of mycotoxins on humans and animals. J Saudi Chem Soc 15(2):129–144

Zaki MM, El-Midany SA, Shaheen HM, Rizzi L (2012) Mycotoxins in animals: occurrence, effects, prevention and management. J Toxicol Environ Health Sci 4(1):13–28

Zhu W, Hu C, Ren Y, Lu Y, Song Y, Ji Y, Han C, He J (2021) Green synthesis of zinc oxide nanoparticles using Cinnamomum camphora (L.) Presl leaf extracts and its antifungal activity. J Environ Chem Eng 9(6):106659. https://doi.org/10.1016/j.jece.2021.106659

Zigo F, Vasil' M, Ondrašovičová S, Výrostková J, Bujok J, Pecka-Kielb E (2021) Maintaining optimal mammary gland health and prevention of mastitis. Front Vet Sci 8:607311

Chapter 23
Nanomedicinal Approaches for Metabolic Disorders of Livestock Animals

Alonkrita Chowdhury and Mayukh Ghosh

Abstract Metabolic disorders in livestock usually appear due to deficiency or malfunctioning of certain enzyme(s) leading to derangement of normal metabolic pathway(s). The metabolic syndrome may also arise from disbalanced hormonal level predisposing the subject towards cardiovascular and neurodegenerative diseases as well as reproductive illness. Consequently, morbidity or mortality of animals and hindrance to optimum productivity and profitability from livestock farming systems arise. Therefore, suitable diagnostic and therapeutic interventions of metabolism related disorders is of utmost significance that conforms the current discussion. Syndromes related to carbohydrate metabolism predominantly include but not limited to hyperinsulinism, hypoglycemia, and galactosemia, resulting into diabetes mellitus (DM) and several glycogen storage disorders. Similarly, anomaly in lipid and amino acid metabolism causes fatty liver syndrome, ketosis, gangliosidoses, Krabbe's disease, Niemann-Pick disease, Faber's disease, hyperhomocysteinemia, phenylketonuria, cystinuria, alkaptonuria, albinism, and several other diseases. Conventional treatment of these diseases often poses limitations due to instability, poor bioavailability, limited solubility, constrained absorption, and adverse side effects of drugs. Also, traditional therapeutics have restrictions in permeability through biological barriers such as the blood–brain barrier (BBB) of the central nervous system (CNS). Owing to the novel physiochemical properties of NPs, nanotechnology-aided therapeutic approaches possess the superiority in terms of drug absorption, therapeutic concentration, stability, permeability, effective drug targeting, reproducibility, as well as prolonged release of drug within target tissue. Hence, nanotherapeutics may be preferred for treatment of the metabolic disorders as exploitation of NPs enhances the effectiveness of treatment, decreasing the side

A. Chowdhury
Department of Plant Biotechnology, Institute of Agricultural Sciences, RGSC, Banaras Hindu University, Mirzapur, Uttar Pradesh, India

M. Ghosh (✉)
Department of Veterinary Physiology and Biochemistry, RGSC, Banaras Hindu University, Mirzapur, Uttar Pradesh, India

LLR University of Veterinary and Animal, ICAR-National Research Centre on Equines, *Nanotechnology Theranostics in Livestock Diseases and Management*, Livestock Diseases and Management,
https://doi.org/10.1007/978-981-16-1610-5_23

effects and the amount of drug usage. Therefore, the current chapter will illuminate about different state-of-the art nano-interventions in therapeutic regimen of the metabolic disorders focusing towards the livestock and pet animals.

Keywords Nanotherapeutics · Diabetes mellitus · Glycogen storage diseases (GSD) · Lipid storage disease · Ketosis · Cardiovascular diseases · Hormonal delivery · Nano-minerals

23.1 Introduction

Dietary nutrients undergo complex metabolic pathways to release energy via biological oxidation. Aberration in the normal metabolic processes causes deficiency or excess of certain metabolites leading to deranged metabolic homeostasis. Metabolic disorders predominantly arise due to lack/inadequacy of enzymes, coenzymes or cofactors, hormones, second messengers, or compromised cognate receptors resulting in multi-organ dysfunctions, loss of optimum productive or reproductive performances, and even mortality in livestock. Metabolic disorders can be acquired or inherited. In cattle, it often arises at the time of high physiological stress such as during late pregnancy and early lactation period when there is a high demand of nutrients (Bezerra et al. 2014). These metabolic syndromes often manifest in terms of cardiovascular diseases, gastrointestinal disorders, neurodegenerative diseases and also reproductive illness. These aberrations can manifest as herd disease or individual basis. Management issue mostly leads to herd disease, while inherited disorders mostly affect individuals. Accordingly, the conventional therapeutic approaches have been developed that include dietary management, nutritional enhancement, oral otherwise intravenous or intramuscular administration of drugs, etc. Yet, there are many hurdles in classical systemic drug administration, mostly related to unsatisfactory pharmacokinetic properties that include but not limited to drug instability in systemic circulation, poor bioavailability, compromised solubility, absorption and ununiform bioavailability, drug accumulation, and adverse side effects. Similarly, corresponding to neurodegenerative disorders, blood–brain barrier (BBB) creates a limitation in transporting therapeutics to the central nervous system (CNS) for maintaining homeostasis and shielding the CNS from pharmacological substances so that an optimal environment is retained for neuronal signaling (Banks 2016). However, owing to the novel physical and chemical properties, NPs have improved the barrier permeability and transportation, including paracellular transport, carrier-mediated transport, and adsorptive- and receptor-mediated transcytosis. Nanotherapeutics is an emerging application of nanotechnology in diverse disciplines of medical science (Osama et al. 2020). Nano-based drug delivery system is the most broadly exploitable approach in various aspects of medicine administration. The first and foremost step towards nano-scale drug delivery systems is the functionalization of NPs to enhance the absorptivity, stability, effective drug targeting, as well as reproducibility and sustained release of the drug within the target tissue. Subsequently, several NP-based therapeutic formulations are constructed such as polymeric NPs-based preparations (e.g., Decapeptyl®, Gonapeptyl Depot®), nano gel-based preparations (e.g., Sane Care Nanogel, ZyflexNanogel, Augen Nanogel Eye-care Gel), nanoemulsion formulations (e.g., Norvir, Restasis, Gengraf, Etomidate-Lipuro, Ropion, Diprivan, Troypofol), dendrimers (poly(propylene imine) (PPI, AstromolR, DAB) and poly(amidoamine) (PAMAM; Starburst)), nanocapsule-based preparation (Paclitaxel lipid nanocapsules), nanosponge (e.g., Glymasason, Prostavastin, Brexin, and Mena-Gargle), and so forth against diverse diseases (Prasad et al. 2018). The scope of nanotherapeutics against different metabolic diseases of livestock and pet has been delineated hereunder.

23.2 Metabolic Disorders in Livestock

23.2.1 Disorders Related to Carbohydrate Metabolism

23.2.1.1 Diabetes Mellitus

Diabetes mellitus (DM) signifies a diverse group of disorders including hyperglycemia, polydipsia, polyuria, and weight loss that may lead to lethal diabetic ketoacidosis and hyperosmolar non-ketotic syndrome. The prolonged diabetic condition also induces cardiovascular diseases, diabetic retinopathy, deterioration of kidney function (nephropathy), and damage to the peripheral nervous system. Inadequate insulin secretion or low tissue reactivity, otherwise acquisition of both conditions, generates the hyperglycemic state in DM (Wolf et al. 2014). Type 1 DM develops due to a deficiency in insulin production as a result of the destruction of pancreatic beta cells. Thus, it is also known as insulin-dependent diabetes mellitus (IDDM). Whereas, as a consequence of the insulin resistance by tissues, type 2 diabetes mellitus or non-insulin-dependent diabetes mellitus (NIDDM) emerges (Rogers 2016). DM may also be induced by several bovine diseases such as fatty liver syndrome, fat cow syndrome, parturition, and chronic insulitis (Tajima et al. 1999). Bovine viral diarrhea virus (BVDV) is also suspected to be one of the possible reasons behind DM in certain virus-infected cattle (Tajima et al. 1992). As a result of overfeeding and physical inactivity, pet animals such as cats and dogs are prone to develop a higher amount of glucose accumulation, in comparison to dairy animals.

23.2.1.2 Hyperinsulinism

It refers to the enhanced insulin level in the blood. As a result of persistent hyperactivity of the insulin-secreting islet cell in the pancreas, or tumor in the pancreatic islets, insulin content increases extremely resulting in hyperinsulinism. Dogs are mostly susceptible to this disorder (Kaneko 1997).

23.2.1.3 Hypoglycemia and Pregnancy Toxemia

Deficiency in glucose ingestion or reduction in its endogenous synthesis in comparison to higher utilization of glucose develops a hypoglycemic condition. In ruminants, deficiency in Vit B12 prevents the conversion of propionate to glucose. Consequently, hypoglycemia arises. Baby pigs are prone to develop this disorder due to underdeveloped gluconeogenesis pathway that could be fatal in neonates. Pregnancy toxemia is another disorder that arises from high metabolic demand of fetus leading to hypoglycemia and ketosis to the dam. Ovines are mostly susceptible to this disorder.

23.2.1.4 Glycogen Storage Diseases

Excessive accumulation of glycogen in the liver or skeletal muscle results in glycogen storage diseases (GSD) or glycogenosis. Consequently, this irregular accumulation triggers the enlargement and dysfunction of many organs, including the liver, heart, and kidneys. Depending on the enzyme(s) responsible for unusual glycogen metabolism and accumulation site of glycogen, a number of GSDs have been described. For example, glycogen buildup in hepatocytes and renal cells due to aberrations in the glucose-6-phosphatase enzyme results in von Gierke's disease. In Pompe's disease, glycogen gets accumulated in the lysosomes of most of the organ's cells, especially in the heart due to a mutation in the lysosomal α-1,4 glucosidase enzyme coding gene. Another congenital GSD is Cori's disease that is caused by the buildup of branched-chain glycogen in the liver, muscle, heart, and leucocytes. A deficiency of amylo-α-1,6-glucosidase enzyme, responsible for the hydrolysis of 1–6-glycosidic bonds in glycogen, leads to this disorder. Besides, there are some other GSDs, such as Anderson's disease, McArdle's disease, Her's disease, and Tarui's disease developed as a result of inoperative glucosyl 4–6 transferase (branching enzyme), muscle glycogen phosphorylase, liver glycogen phosphorylase, and phosphofructokinase enzymes, respectively (Kaneko 2008).

23.2.1.5 Galactosemia

It is a condition when the body becomes unable to metabolize galactose, a component of lactose. Generally, an inherited deficiency of key enzymes (e.g., galactokinase or galactose-1-phosphate uridylyltransferase) for galactose degradation results in the accumulation of galactose in the blood. It encompasses a number of symptoms including hepatomegaly, cirrhosis of the liver, renal failure, cataracts, vomiting, seizure, and brain injury (McAuley et al. 2016).

23.2.2 Disorders Related to Lipid Metabolism

23.2.2.1 Fatty Liver Syndrome

Fatty liver syndrome, also known as hepatic lipidosis, is one of the foremost metabolic disorders in animals, mostly affecting dairy cows. Higher uptake of lipids relative to the oxidation and secretion by the liver or impaired reverse cholesterol transport results in the accumulation of lipids, mostly in the form of triacylglycerol that consequences in decreased metabolic functions of the liver (Grummer 1993). It is also associated with frailness, diminished reproductive performances, extended calving intervals, reduced milk production, and lastly, declination in the average lifespan of cows (Bobe et al. 2004).

23.2.2.2 Ketosis

Ketosis is a metabolic disorder that mostly arises in high-yielding dairy cows. Bovine ketosis mostly manifests as herd disease comprised of three syndromes underfeeding ketosis, alimentary ketosis, and spontaneous ketosis. Poor quality fodder or deficient feeding leads to underfeeding ketosis; spoiled silage containing high butyrate leads to alimentary ketosis; and spontaneous ketosis, the most common one, arises in the high yielders having accessibility to abundant high-quality feed. Ketosis often develops during the early lactation period (Guliński 2021). After calving, there is a depression in the appetite of cows; hence, dietary energy intake cannot meet the increased metabolic demands for milk production. Therefore, as a result, the excessive oxidation of fatty acids to acetyl-CoA occurs comparative to the oxidation of acetyl-CoA into CO_2 and water, resulting in accumulation of ketone bodies (acetoacetate, β-hydroxybutyrate (BHBA), and acetone) in blood (Gordon et al. 2013). Subsequent elimination of high amount of ketones through urine can also be observed resulting ketonuria. Estimation of BHBA through ELISA from blood can provide diagnostic indication of negative energy balance and ketosis.

23.2.2.3 Gangliosidoses

It is an autosomal recessive disorder characterized by the accumulation of a complex glycosphingolipids, which are gangliosides, in central and peripheral nervous systems. Deficiency or improper catalytic activity of lysosomal enzymes brings about the aggregation of ganglioside–monosialic acids (GM) in the lysosome leading to the destruction of the nerve (Lawson and Martin 2016). GM1 gangliosides (monosialotetrahexosylganglioside) and GM2 (Tay–Sachs disease) gangliosides are affiliated with the insufficiency of β-galactosidase and β-hexosaminidase enzyme-inducing accumulation of GM1 and GM2 gangliosidoses, respectively (Patterson 2013).

23.2.2.4 Gaucher's Disease (Glucocerebrosidosis or Glucosylceramide Lipidosis)

Another hereditary lysosomal storage disease found in dogs, pigs, and sheep is glucocerebrosidosis, also known as Gaucher's disease. Inadequate hydrolysis of glucocerebroside (gangliosides derivative) due to the deficiency of β-glucocerebrosidase enzyme is the underlying reason behind this disease. Consequently, the accretion of lipids in certain organs, such as the spleen and liver, takes place which leads to the enlargement and dysfunction of those organs (Jolly and Walkley 1997).

23.2.2.5 Krabbe's Disease

Krabbe's disease or globoid cell leukodystrophy comprises of a family of lysosomal storage disease (LSD) characterized by progressive and profound demyelination in the nerve tissues leading to fatality in early life. Reduced enzymatic activity of β-galactosidase results in accumulation of galactocerebrosides. Demyelination of nerve tissues eventuates due to a deficiency of the acid hydrolase galactosyl ceramidase (GALC) enzyme responsible for the degradation of galactosyl ceramides and sphingolipids, abundant in myelin sheaths. Mental retardation, convulsions, blindness, deafness, etc., are the prevalent manifestation of this disorder (Bradbury et al. 2021).

23.2.2.6 Niemann–Pick Disease

It refers to a group of inherited metabolic disorders that induces excessive lipid buildup predominantly in the liver and spleen. Lack of sphingomyelinase enzyme results in accumulation of respective substrate, i.e., sphingomyelin causing organomegaly.

23.2.2.7 Faber's Disease

It is another recessively inherited autosomal lipid storage disease characterized by excessive lipid buildup in the central nervous system, joints, and tissues. Deficiency in ceramidase enzyme consequences in the accumulation of ceramide leading to skeletal deformities, subcutaneous nodulation, dermatitis, and mental retardation (Haskins and Giger 2008).

23.2.3 Disorders Related to Amino Acids

23.2.3.1 Hyperhomocysteinemia

A remarkably high concentration of homocysteine, a naturally occurring amino acid, in the blood implies hyperhomocysteinemia disorder. Homocysteine is a by-product of methionine metabolism (Azad et al. 2018). Accumulation of this amino acid in plasma reduces blood flow inducing cardiovascular dysfunction. Insufficient occurrence of cystathionine β-synthase or *N*-methylenetetrahydrofolate reductase enzyme results in an increase in homocysteine concentration in plasma (Smith and Refsum 2016).

23.2.3.2 Phenylalanine and Tyrosine Metabolism-Related Disorders

Mutation in the phenylalanine hydroxylase gene causes deficiency in the phenylalanine hydroxylase enzyme, the first enzyme in the catabolic pathway for phenylalanine metabolism. Consequently, a buildup in phenylalanine amino acid takes place, resulting in phenylketonuria disorder. Malfunctioning of some other enzymes in phenylalanine and tyrosine metabolic pathways, such as tyrosine transaminase, p-hydroxy phenyl pyruvate dioxygenase, homogentisate oxidase, maleylacetoacetate isomerase or fumarylacetoacetate hydrolase, and tyrosinase, also develops several disorders, namely, tyrosinemia type II, neonatal tyrosinemia, alkaptonuria, tyrosinosis or tyrosinemia type I, and albinism, respectively (Kretchmer and Etzwiler 1958).

23.2.3.3 Sulfur-Containing Amino Acid-Related Disorder

Methionine and cysteine are the two sulfur-containing amino acids. Accumulation of these amino acids due to certain inborn metabolic aberrations generates respective diseases. For example, cystinuria disorder is concerned with an inordinate buildup of undissolved cystine in the urine due to improper renal absorption. Accretion of such amino acid inside the cells due to improper functioning of lysosome damages the cell leading to crystal formation. Consequently, respective tissues or organs get affected. This condition is termed as cystinosis. Correspondingly, homocystinuria type I, homocystinuria type II, homocystinuria type III, and cystathioninuria are inflicted as a reason for the malfunctioning of relevant enzymes (Kožich and Stabler 2020).

23.2.3.4 Diseases due to Dysmetabolism of Branched-Chain Amino Acid

Branched-chain amino acids possess an aliphatic side chain with branching. Valine, leucine, and isoleucine are three respective branched-chain amino acids.

Dysmetabolism of such amino acids due to improper enzymatic activity of α-keto acid dehydrogenase, valine transaminase, and isovaleryl-CoA dehydrogenase enzymes gives rise to maple syrup urine disease, hypervalinemia, and isovaleric acidemia disorder, respectively.

Besides, glycinuria, primary hyperoxaluria, Hartnup disease, histidinemia, and hyperprolinemia type I are some other metabolic disorders that occur as a result of an inoperative synthesis of respective amino acids, such as glycine, tryptophan, histidine, and proline (Ogier de Baulny et al. 1995).

23.3 Conventional Therapeutic Approaches for the Management of Metabolic Disorders

Diabetes is the most common metabolic disorder in companion animals, such as cats and dogs, because of overfeeding and physical inactivity, relative to food-producing animals. Therefore, controlled dietary intake of fat and starch can cause a reduction in the excess accumulation of fats in tissues. For instance, in dogs, meals with starch blends, carboxymethyl cellulose, and fermentable fiber blends help in managing DM. Exercise and dietary supplementation with 1% arginine or oral interferon-τ also enhance tissue sensitivity towards insulin that results in a decrease in the concentration of glucose in the blood of animals with type 2 DM. Similarly, type 1 DM may be prevented or deferred by the consumption of casein-based diets (Wu 1995). Oral administration of hypoglycemic agents such as sulfonylureas (glipizide) and alpha-glucosidase inhibitors (acarbose) in cats provides attenuation to the physiological abnormalities of type 2 DM. For the treatment of DM, insulin therapy is the most followed approach where canines are injected with human recombinant neutral protamine Hagedorn (NPH), purified pork-source Lente, beef/pork-source protamine zinc insulin (PZI), and the human synthetic insulin analogue glargine (Norsworthy et al. 2009).

Treatment of fatty liver disorder and fat cow syndrome comprises prolonged active therapy for 7–10 days or more employing a large amount of glucose as the crucial agent in this therapy. Intravenous administration of 10% or 50% dextrose is provided at a rate of 60 grams per hour. An equivalent dose can also be directed two to three times daily. A recommended dosage of protamine zinc insulin (150–200 units) is proposed to be subcutaneously injected every 36–48 h. Moreover, supplementation with insulin and potassium chloride is also advised for cellular glucose intolerance in cattle (Morrow 1976). In addition, endogenous corticosteroid secretion gets stimulated by adrenocorticotropic hormone (ACTH). Hence, intramuscular administration of ACTH along with an oral dosage of propylene glycol as a glucose precursor is suggested (Haass and Eness 1984).

Hepatic ketogenesis can be prevented by providing a higher concentration of glucose. Glucose metabolism produces oxaloacetate which in turn facilitate acetyl-CoA to enter the Krebs cycle for oxidation purpose. Consequently, there is a reduction in the conversion of the same into ketone bodies. Oral administration of propylene glycol or intravenous administration of glucose provides restoration of physiological concentrations of glucose in the blood (Gordon et al. 2013).

In hypoglycemia, hyperinsulinism, galactosemia, or glycogenesis, regulation in dietary intake or supplementation is the major strategy to prevent these disorders. Similarly, reducing dietary intake of lipids ameliorates glucocerebrosidosis and gangliosidoses in animals. For the impediment of nutritional homocysteinemia, excess intake of methionine should be avoided along with enhancement of water-soluble vitamin consumption. As well, dietary supplementation with methyl group donors or with serine facilitates conversion of homocysteine into methionine or cystathionine, respectively, leading to the alleviation of the syndromes (Wu 2020).

23.4 Nanotherapeutic Approaches for Metabolic Disorders

23.4.1 Nanotherapeutics for Diabetes Management and Insulin Delivery

Nanotechnology-aided drug delivery system has a significant impact on the treatment of DM with the objective of insulin delivery. Using polymeric NPs, nanocapsules have been constructed that encapsulate the drug in a cavity surrounded by a single polymeric membrane. After releasing the drug in the target tissue, such membranes get decomposed into biocompatible components upon hydrolysis (Bahadori et al. 2019). Due to temperature, pH, solvent, and some other variables inside gut environment, stability of protein gets hindered. Also, proteolytic enzyme activity may cause denaturation of protein. Additionally, epithelial lining in the GI tract acts as a physical barrier to large molecules hindering their absorption. As a consequence of such large number of interferences, insulin cannot be consumed intactly through digestive tract. Moreover, in diabetic condition, the gut motility is also impaired adding delay in absorption and randomness to the problems (Carino and Mathiowitz 1999). Here, nanotechnology paves the path for effective delivery of insulin with encapsulation efficiency. Hence, the oral administration of insulin loaded (vitamin B12) VB12-coated dextran NPs has considerable potential in the therapy of DM. It can prevent denaturation of insulin by intestinal proteases along with enhancing permeability via paracellular transport (Chalasani et al. 2007). Another such example, a biodegradable poly(vinyl alcohol) (PVA) and sodium alginate (NaAlg) electrospun composite nanofiber-based transmucosal patch, loaded with insulin, has been developed. Administration of the anti-diabetic drugs (proteins and peptides) via the oral route using electrospun composite nanofiber transmucosal patches is provided with a sustained- and controlled-release pattern of drug from the subsequent patch (Sharma et al. 2013).

Orally introduced hydrophilic drugs also get obturated in the intestinal epithelium as they cannot diffuse through lipid-bilayer membranes of epithelial cells to the bloodstream. Therefore, enhancement in paracellular transport of hydrophilic drugs has been emphasized. However, it also has restriction owing to the presence of tight junctions which are located at the luminal aspect of adjacent epithelial cells. Hence, a nano-system is proposed for oral administration of macromolecules that will facilitate its paracellular absorption. Since chitosan (CS), a nontoxic polycationic polysaccharide, contains the potential of adhering to the mucosal surface and transiently disrupting the tight junctions between gut epithelial cells, a novel NP system, comprising hydrophilic CS and poly-γ-glutamic acid (γ-PGA), was formulated for oral insulin delivery via the paracellular pathway (Lin et al. 2007).

23.4.2 Nanotherapeutics to Combat Neurodegenerative Disorders

Nanotechnologically formulated drugs are also employed in overcoming blood–brain barrier in central nervous system (CNS)-related disorders. Alzheimer's disease (AD) and Parkinson's disease (PD) are the two most common neurodegenerative diseases that are related to aging and dysmetabolism. Two forms of these diseases, genetic and sporadic forms, have been reported with a vast majority of sporadic types of incidents that are mostly associated with metabolic disorders (De Bem et al. 2021). Although the detailed molecular mechanisms of dysmetabolism is controversial, it is still considered as a consequence of insulin resistance. The metabolic conflicts identified that type 2 DM is strongly correlated with AD and PD. As per an early sign of AD, anomaly in brain insulin sensitivity is also captured. Hence, AD is often suggested as "type 3 diabetes" due to its resemblance towards type 2 DM (Suzanne 2014). Correspondingly, insulin resistance and impaired glucose tolerance also come across to be exhibited by patients with PD (Aviles-Olmos et al. 2013). Hence, clinical trials concerning anti-diabetic or insulin sensitizing agents have yielded positive results. However, the therapeutic effects of the conventional drug molecules have limitations in crossing the blood–brain barrier. In this regard, novel nanotherapies are receiving significant consideration in drug delivery systems and often employed for stimuli-responsive targeting. Owing to their nano-scale size, these medicinal approaches are advantageous for overcoming the blood–brain barrier. These are more beneficial over conventional therapy by offering sustained release of drug over time, which can minimize the risk of dose dumping. Hydromorphone hydrochloride prolonged-release capsules, marketed as Palladone™, have been developed earlier that also employs sustained-release system. It also aids in reducing the frequency of drug administration. NPs provide passive targeting, by making it easier for the drug to cross the blood–brain barrier. It can also facilitate active targeting by functionalization with blood–brain barrier cell-specific ligands (e.g., transferrin receptor antibodies) that enhance the passage of NPs to the brain parenchyma, improving the overall systemic effects (Torres-Ortega et al. 2019). For example, curcumin and piperine co-loaded glyceryl monooleate

(GMO) NPs coated with various surfactants, such as pluronic F-68 and vitamin E D-α-tocopherol polyethylene glycol 1000 succinate (vitamin E-TPGS), have been formulated to enhance the bioavailability of curcumin and penetration of both drugs to the brain tissue crossing the blood–brain barrier, hence delivering combinational anti-parkinsonism effect of both drugs in a single platform (Kundu et al. 2016). Rivastigmine tartrate (RT)-loaded NPs, customized by using biodegradable polymers, poly(lactide-co-glycolide) (PLGA) and poly(*N*-butyl cyanoacrylate) (PBCA) coated with polysorbate 80 as carriers, have also been designed that provide effective brain targeting and sustainable RT release for a prolonged period. Hence, it is assumed to extend a significant improvement for the treatment of AD (Joshi et al. 2010).

23.4.3 Nanotherapeutics for the Management of Cardiovascular Disorders

Targeting ligand containing NPs has been formulated to fight against cardiovascular disorders (CVD) via heart-targeted drug delivery. Accordingly, these NPs generate nanocarriers that deliver a drug to a particular therapeutic site without significant off-target distribution. Recently, cardiovascular nanomedicines targeting specific body parts have been developed for improved diagnosis and treatment. It is also exploited for a relief from postoperative pain. Liposomes are the mostly studied nanocarrier which are nanometric vesicles encompassed with concentric lipid bilayers. Atherosclerotic lesions in injured vessels produce collagen, receptor proteins, and cell adhesion molecules (CAMs), such as intercellular adhesion molecule-1 (I-CAM1), vascular cell adhesion molecule-1 (VCAM-1), platelet endothelial cell adhesion molecule (PECAM-1), and endothelial leucocyte adhesion molecule-1 (ELAM-1) on the surfaces of endothelial and smooth muscle cells (Wang et al. 2021). From the expressions of these molecules, liposomes can be modified with antibody for vascular targeting. Liposomes are also advantageous for their low immunogenicity and ability to diffuse through the blood–brain barrier, thereby employing as drug delivery agent for the treatment of cerebrovascular occlusion and ischemia. Copolymers, such as poly(lactic-co-glycolic acid) (PLGA), poly (cyclohexane-1,4-diylacetone ketone) (PCADK), and poly(ethyleneimine) (PEI) loaded with stem cells, proteins, and siRNAs, are also employed for the treatment of cardiovascular inflammatory diseases. A clot-binding peptide cysteine–arginine–glutamic acid–lysine–alanine (CREKA) conjugated as a targeting moiety to poly (ethylene glycol)–poly(lactic acid) (PEG–PLA) nanoparticles containing thymosin beta 4 (Tβ4) protein has been formulated as a cardiovascular nanomedicine (Huang et al. 2017). Fibrin, a fibrous protein involved in blood clot contraction, is absent in healthy myocardium. Contrastingly, it becomes abundant after cardiac myocyte necrosis due to myocardial injury. Hence, another nano-delivery system containing liposome along with CREKA has been designed to generate CREKA-modified mesenchymal stem cells. It provides cardiac tissue restoration and fibrin targeting. With the aid of PLA and PLGA polymeric NPs, interleukin 10 (IL-10), a therapeutic biological, can be incorporated via leaky endothelial junctions to atherosclerotic plaques which provides controlled release of IL-10 diminishing necrotic core size (Hu et al. 2022). Another widely employed regime in CVD is implementation of streptokinase as thrombolytic agent. Conflictingly, due to its prokaryotic source, it is immunogenic in nature leading to shorter half-life. However, prolonged circulation time can be obtained by encapsulating the enzyme with NPs (Hasanpour et al. 2021). For instance, PEG-grafted chitosan NPs-encapsulated streptokinase has been used in the treatment of cardiovascular disorders and deep vein thrombosis (Baharifar et al. 2020).

23.4.4 Nano-Delivery of Hormones to Combat Reproductive Disorders

Reproductive illness is another major consequence of metabolic derangement. Chronic hyperglycemia with oxidative stress is responsible for reproductive system deficiencies like testicular dysfunctions. Diabetes can also induce structural changes in the testis and spermatozoa and promote germ cell apoptosis, impairment of sperm parameters, and hormonal changes leading to infertility. Reproductive efficiency of farm animals is very much essential as it reflects on the productivity, profitability, and sustainability of farms. Therefore, various reproductive management strategies are widely and predominantly followed in different livestock production systems, such as one or more bio-stimulation tools (e.g., male effect), assisted reproductive techniques (mainly, estrous synchronization and artificial insemination), nutritional management, and prevention/treatment of reproductive diseases. However, efficiency of these approaches has been challenged by several practical and ethical aspects. Nowadays, hormones-based treatments are profoundly exploited tools in fertility improvement and farm business profitability. Mainly, gonadotrophins, progesterone, estradiol, testosterone, melatonin, and prostaglandins are used for managing fertility in animals. Their effectiveness is highly dependent on their pharmacokinetics and pharmacodynamics (Hashem and Gonzalez-Bulnes 2020). However, conventional hormonal delivery systems also have many limitations, such as restriction in sustained delivery of gonadotropin-releasing hormone (GnRH) and prostaglandin $F_{2\alpha}$ ($PGF_{2\alpha}$) to the target sites due to their low molecular weights and a shorter lifespan. Some other hormones, such as glycoprotein gonadotropins follicle-stimulating hormones (FSH), luteinizing hormone (LH), human chorionic gonadotropin (hCG), and equine chorionic gonadotropin (eCG), result in formation of specific antibodies, by stimulating the immune system, leading to a refractoriness to repeated gonadotropic treatments. This causes low fertility and other reproductive outcomes in animals. Additionally, the traditional hormone delivery strategy also has many adverse effects on the environment by spreading hormonal residues and carrier materials into the surroundings. For instance, progesterone-impregnated intravaginal devices formulated with silicon polymers are mainly employed in controlling the estrous cycle in different farm animals. These devices are loaded with high progesterone concentrations increasing the risk of hormone emissions to the environment and the transmission of the hormone directly to workers along with the probability of consuming via animal products. However, a new nanotechnology-aided pharmacological approach has been developed for the nano-delivery of hormones allowing protection and controlled release of reproductive hormones, including gonadotropin and steroid hormones. It is also advantageous by means of decreasing hormone dosage, altering the way of administration, increasing animal welfare, and diminishing the risk of exposition to different hormones by workers and technicians (De Graaff and Grimard 2018). For example, GnRH-loaded chitosan NPs (GnRH-ChNPs) have been evaluated for inducing ovulation in rabbits (Hassanein et al. 2021).

23.4.5 Application of Nano-Minerals for Productive and Reproductive Management

At different developmental stages of reproductive organs such as embryonic stage, puberty, and fertile stage, nutrition plays a vital role in influencing the reproductive efficiency. According to functional effect, which is the role of particular nutrients against specific reproductive events, amino acids are the major nutritional element for normal embryonic and fetal development. However, imbalanced nutrition results in a measurable decrease in body mass, unhealthy body condition, delayed puberty, ovarian dysfunction, hormonal imbalances, and lastly, infertility. Therefore, to obtain the optimal reproductive fitness, nutritional needs should be effectively supplemented. As feed ingredients show novel physicochemical properties in Nano-minerals forms, nano-encapsulation of active feed ingredients and processing of nano-minerals are the emerging nanotechnological strategies for improving nutritional management outputs. This manipulation improves absorption, bioavailability, and utilization of feed by animals due to small size and high surface area (Hashem and Gonzalez-Bulnes 2021). Some examples of nano-minerals include nano-selenium (Nano-Se), nano-zinc oxide (Nano-ZnO), copper oxide nanoparticles (CuONPs), selenium nanoparticles (SeNPs), and zinc nanoparticles (ZnNPs). Nano-zinc oxide (Nano-ZnO) supplementation has been evidenced to enhance milk production in dairy cows, whereas improved feed digestibility has been found in sheep treated with nano-selenium (Nano-Se). In cashmere goat, the same nano-mineral is found to be beneficial for promoting better-quality hair follicles and advanced fetal growth (Abdelnour et al. 2021).

23.5 Conclusion

Due to excessive consumption of energy, idleness, nutritional deficiency, and food-chain contamination, metabolic syndromes are getting more frequent day by day. Since metabolic disorders are the inducer of several secondary complications including diabetes, cardiovascular diseases, reproductive illness, gastrointestinal diseases, and so forth, looking for advanced therapies for healthy metabolism is crucial. Nano-platform provides an opportunity for the therapeutic intervention of these diseases. Consequently, NP-based formulations have been developed providing advantages over traditional therapies. However, it also possesses many challenges in the direction of nanoformation, such as particle size, scalability, cost-effectiveness, repeatability, stability, toxicity, and regulation that need to be overcome. Progressive development in nanotechnology arena generates ample optimism to surmount the challenges and ensure a bright future prospect.

References

Abdelnour SA, Alagawany M, Hashem NM, Farag MR, Alghamdi ES, Hassan FU, Bilal RM, Elnesr SS, Dawood MAO, Nagadi SA, Elwan HAM, ALmasoudi AG, Attia YA (2021) Nanominerals: fabrication methods, benefits and hazards, and their applications in ruminants with special reference to selenium and zinc nanoparticles. Animals 11(7):1916

Aviles-Olmos I, Limousin P, Lees A, Foltynie T (2013) Parkinson's disease, insulin resistance and novel agents of neuroprotection. Brain 136(2):374–384

Azad MAK, Huang P, Liu G, Ren W, Teklebrh T, Yan W, Yin Y (2018) Hyperhomocysteinemia and cardiovascular disease in animal model. Amino Acids 50(1):3–9

Bahadori E, Farjami Z, Rezayi M, Lngari H, Darroudi M, Avan A, Ghayour-Mobarhan M (2019) Recent advances in nanotechnology for the treatment of metabolic syndrome. Diabetes Metab Syndr Clin Res Rev 13(2):1561–1568

Baharifar H, Khoobi M, Bidgoli SA, Amani A (2020) Preparation of PEG-grafted chitosan/streptokinase nanoparticles to improve biological half-life and reduce immunogenicity of the enzyme. Int J Biol Macromol 143:181–189

Banks WA (2016) From blood–brain barrier to blood–brain interface: new opportunities for CNS drug delivery. Nat Rev Drug Discov 15(4):275–292

Bezerra LR, de Oliveira Neto CB, de Araújo MJ, Edvan RL, de Oliveira WD, Pereira FB (2014) Major metabolic diseases affecting cows in transition period. Int J Biol 6(3):85–94

Bobe G, Young JW, Beitz DC (2004) Invited review: pathology, etiology, prevention, and treatment of fatty liver in dairy cows. J Dairy Sci 4:3105–3124

Bradbury AM, Bongarzone ER, Sands MS (2021) Krabbe disease: new hope for an old disease. Neurosci Lett 752:135841

Carino GP, Mathiowitz E (1999) Oral insulin delivery. Adv Drug Deliv Rev 35(2–3):249–257

Chalasani KB, Russell-Jones GJ, Yandrapu SK, Diwan PV, Jain SK (2007) A novel vitamin B12-nanosphere conjugate carrier system for peroral delivery of insulin. J Control Release 117(3):421–429

De Bem AF, Krolow R, Farias HR, de Rezende VL, Gelain DP, Moreira JCF, de Oliveira J (2021) Animal models of metabolic disorders in the study of neurodegenerative diseases: an overview. Front Neurosci 1457

De Graaff W, Grimard B (2018) Progesterone-releasing devices for cattle estrus induction and synchronization: device optimization to anticipate shorter treatment durations and new device developments. Theriogenology 112:34–43
Gordon JL, LeBlanc SJ, Duffield TF (2013) Ketosis treatment in lactating dairy cattle. Vete Clin Food Anim Pract 29(2):433–445
Grummer RR (1993) Etiology of lipid-related metabolic disorders in periparturient dairy cows. J Dairy Sci 76(12):3882–3896
Guliński P (2021) Ketone bodies–causes and effects of their increased presence in cows' body fluids: a review. Vet World 14(6):1492
Haass CL, Eness PG (1984) Bovine fatty liver syndrome. Iowa State Univ Vet 46(108):111
Hasanpour A, Esmaeili F, Hosseini H, Amani A (2021) Use of mPEG-PLGA nanoparticles to improve bioactivity and hemocompatibility of streptokinase: in-vitro and in-vivo studies. Mater Sci Eng C 118:111427
Hashem NM, Gonzalez-Bulnes A (2020) State-of-the-art and prospective of nanotechnologies for smart reproductive management of farm animals. Animals 10(5):840
Hashem NM, Gonzalez-Bulnes A (2021) Nanotechnology and reproductive management of farm animals: challenges and advances. Animals 11(7):1932
Haskins M, Giger U (2008) Lysosomal storage diseases. In: Jerry Kaneko J, Harvey JW, Bruss ML (eds) Clinical biochemistry of domestic animals, 6th edn. Academic Press, San Diego, CA, pp 731–749
Hassanein EM, Hashem NM, El-Azrak KEDM, Gonzalez-Bulnes A, Hassan GA, Salem MH (2021) Efficiency of GnRH–loaded chitosan nanoparticles for inducing LH secretion and fertile ovulations in protocols for artificial insemination in rabbit does. Animals 11(2):440
Hu Q, Fang Z, Ge J, Li H (2022) Nanotechnology for cardiovascular diseases. Innovation 3(2): 100214
Huang Z, Song Y, Pang Z, Zhang B, Yang H, Shi H et al (2017) Targeted delivery of thymosin beta 4 to the injured myocardium using CREKA-conjugated nanoparticles. Int J Nanomedicine 12: 3023
Jolly RD, Walkley SU (1997) Lysosomal storage diseases of animals: an essay in comparative pathology. Vet Pathol 34(6):527–548
Joshi SA, Chavhan SS, Sawant KK (2010) Rivastigmine-loaded PLGA and PBCA nanoparticles: preparation, optimization, characterization, in vitro and pharmacodynamic studies. Eur J Pharm Biopharm 76(2):189–199
Kaneko JJ (1997) Carbohydrate metabolism and its diseases. In: Jerry Kaneko J, Harvey JW, Bruss ML (eds) Clinical biochemistry of domestic animals, 5th edn. Academic Press, San Diego, CA, pp 45–81
Kaneko JJ (2008) Carbohydrate metabolism and its diseases. In: Jerry Kaneko J, Harvey JW, Bruss ML (eds) Clinical biochemistry of domestic animals, 6th edn. Academic Press, San Diego, CA, pp 45–80
Kožich V, Stabler S (2020) Lessons learned from inherited metabolic disorders of sulfur-containing amino acids metabolism. J Nutr 150(Supplement_1):2506S–2517S
Kretchmer N, Etzwiler DD (1958) Disorders associated with the metabolism of phenylalanine and tyrosine. Pediatrics 21(3):445–475
Kundu P, Das M, Tripathy K, Sahoo SK (2016) Delivery of dual drug loaded lipid-based nanoparticles across the blood–brain barrier impart enhanced neuroprotection in a rotenone induced mouse model of Parkinson's disease. ACS Chem Neurosci 7(12):1658–1670
Lawson CA, Martin DR (2016) Animal models of GM2 gangliosidosis: utility and limitations. Appl Clin Genet 9:111–120
Lin YH, Chen CT, Liang HF, Kulkarni AR, Lee PW, Chen CH, Sung HW (2007) Novel nanoparticles for oral insulin delivery via the paracellular pathway. Nanotechnology 18(10): 105102
McAuley M, Kristiansson H, Huang M, Pey AL, Timson DJ (2016) Galactokinase promiscuity: a question of flexibility. Biochem Soc Trans 44(1):116–122

Morrow DA (1976) Fat cow syndrome. J Dairy Sci 59(9):1625–1629
Norsworthy GD, Lynn R, Cole C (2009) Preliminary study of protamine zinc recombinant insulin for the treatment of diabetes mellitus in cats. Vet Ther 10(1):24
Ogier de Baulny H, Wendel U, Saudubray JM (1995) Branched-chain organic acidurias. Inborn metabolic diseases: diagnosis and treatment, pp. 207–221
Osama E, El-Sheikh SM, Khairy MH, Galal AA (2020) Nanoparticles and their potential applications in veterinary medicine. J Adv Vet Res 10(4):268–273
Patterson MC (2013) Gangliosidoses. Handb Clin Neurol 113:1707–1708
Prasad M, Lambe UP, Brar B, Shah I, Manimegalai J, Ranjan K, Prasad G (2018) Nanotherapeutics: an insight into healthcare and multi-dimensional applications in medical sector of the modern world. Biomed Pharmacother 97:1521–1537
Rogers CS (2016) Genetically engineered livestock for biomedical models. Transgenic Res 25(3): 345–359
Sharma A, Gupta A, Rath G, Goyal A, Mathur RB, Dhakate SR (2013) Electrospun composite nanofiber-based transmucosal patch for anti-diabetic drug delivery. J Mater Chem B 1(27): 3410–3418
Smith AD, Refsum H (2016) Homocysteine, B vitamins, and cognitive impairment. Annu Rev Nutr 36:211–239
Suzanne M (2014) Type 3 diabetes is sporadic Alzheimer's disease: mini-review. Eur Neuropsychopharmacol 24(12):1954–1960
Tajima M, Yazawa T, Hagiwara K, Kurosawa T, Takahashi K (1992) Diabetes mellitus in cattle infected with bovine viral diarrhea mucosal disease virus. J Veterinary Med Ser A 39(1–10): 616–620
Tajima M, Yuasa M, Kawanabe M, Taniyama H, Yamato O, Maede Y (1999) Possible causes of diabetes mellitus in cattle infected with bovine viral diarrhoea virus. J Veterinary Med Ser B 46(3):207–216
Torres-Ortega PV, Saludas L, Hanafy AS, Garbayo E, Blanco-Prieto MJ (2019) Micro-and nanotechnology approaches to improve Parkinson's disease therapy. J Control Release 295: 201–213
Wang DK, Rahimi M, Filgueira CS (2021) Nanotechnology applications for cardiovascular disease treatment: current and future perspectives. Nanomedicine 34:102387
Wolf E, Braun-Reichhart C, Streckel E, Renner S (2014) Genetically engineered pig models for diabetes research. Transgenic Res 23:27–38
Wu G (1995) Nitric oxide synthesis and the effect of aminoguanidine and N G-monomethyl-L-arginine on the onset of diabetes in the spontaneously diabetic BB rat. Diabetes 44(3):360–364
Wu G (2020) Management of metabolic disorders (including metabolic diseases) in ruminant and nonruminant animals. In Animal agriculture. Academic Press, pp 471–491

Chapter 24
Cancer Nanotherapeutics

S. Suvetha, Alla Yaswanth Naveen Kumar, Mayukh Ghosh, Ravindra Prasad, and Pavan Kumar Yadav

Abstract Diverse nanomodules, engineered nanomaterials, and methods have paved the way for state-of-the-art cancer chemotherapies and immunotherapies selectively targeting the cancerous cells. The conventional treatment methods often lead to paucity in drug delivery at the desired location and off-target side-effects due to improper targeting. Nano-intervention in cancer treatment assists to eliminate several issues of conventional therapeutics such as drug insolubility, early drug release and removal from the system, drug-efflux and resistance issues, etc. through precise site-specific drug delivery and controlled release system. The strategies that are predominantly used for nanoparticles-assisted targeted drug delivery are passive, active, and stimuli-responsive drug release. Besides the strategies, the modules in which the drugs are loaded play a major role in targeted delivery. It is widely depending on the properties of the modules such as size, shape, pH, and thermal and acoustic properties. For instance, the smaller the size, the more effective penetration will occur. Nano-modules help in temperature, pH, enzyme-sensitive drug release; it also reduces the amount of free drugs within the system which in turn decreases the side effects of drugs. There are several organic as well as inorganic nanoparticles that are used for coating and targeted delivery of drugs; in addition, other nanoparticles such as virus-like particles, a viral based nano-formulation that are processed in non-infectious way has also been used for the purpose. The prominent mode of nano-interventions in cancer therapy and nano-modalities used for the treatment of diverse types of cancer has been delineated hereunder along with the riveting outline of therapeutic nano-vaccines.

S. Suvetha · A. Y. N. Kumar · R. Prasad
Department of Genetics and Plant Breeding, Institute of Agricultural Science, Rajiv Gandhi South Campus, Banaras Hindu University, Mirzapur, Uttar Pradesh, India

M. Ghosh · P. K. Yadav (✉)
Department of Veterinary Physiology and Biochemistry, Faculty of Veterinary and Animal Sciences, Rajiv Gandhi South Campus, Banaras Hindu University, Mirzapur, Uttar Pradesh, India
e-mail: pavan.yadav@bhu.ac.in

LLR University of Veterinary and Animal, ICAR National Research Centre on Equines, *Nanotechnology Theranostics in Livestock Diseases and Management*, Livestock Diseases and Management,
https://doi.org/10.1007/978-981-16-1610-5_24

Keywords Tumor · Chemotherapy · Immunotherapy · Nano-formulation · Nano-vaccines · Passive targeting · Active targeting · Stimuli-responsive drug delivery

24.1 Introduction

Cancer is attributed with the abnormal and uncontrolled division of the cells which further invade and destroy adjacent or distant healthy tissues through metastasis. It spreads to other organs or parts of the body through blood or lymphatic system of the organisms. The term cancer is medically denoted as "malignant neoplasm." More than 90% of the cancers are due to environmental factors and lifestyle while remaining 10% occurs as a result of genetic changes that derange the cell growth and apoptosis (Anand et al. 2008). As per the World Health Organization (WHO), more than 70% of the deaths due to cancer are observed in low- and middle-income nations. Cancer is not entirely a hereditary disease; it will not pass from the parent to the offspring until the defective gene is inherited and the chances of it is very low. These genes become defective due to mutations which transform normal genes into oncogenes. The cells with defective gene proliferate unlimitedly without undergoing apoptosis which is otherwise stringently regulated as the part of normal cell cycle. Therefore cancer cells grow as a mass which sometimes get detached from the site of origin and cause metastatic cancer. In this case, the integrity of extracellular matrices around the cancer cells gets deranged, frequently by the over-expression of matrix metalloproteases (MMPs) and these altered cells enter into the blood or lymphatic stream (Yadav et al. 2020, 2021, 2022). After reaching the distant locations, the cancer cells start growing at the secondary sites to be termed as metastatic cancer and this process is referred as metastasis. It is of the following types: local seeding, blood-borne metastasis, and lymphatic spread.

The most common factors inducing cancerous mutations in organisms are carcinogens. Carcinogen is a substance which causes cancer. It usually falls under three categories namely (i) Physical carcinogen—electromagnetic radiation such as ultraviolet rays and ionizing radiation. (ii) Chemical carcinogen—Benzene, Cadmium, Arsenic, Chromium, Poly vinyl chloride, Poly cyclic hydrocarbons, aflatoxin (a liver carcinogen) and tobacco (Smoke from the tobacco is the cause of 80–90% of the lung cancer. It affects oral cavity, pharynx, larynx, esophagus and causes cancer). (iii) Biological carcinogen–Certain viruses that have oncogenic properties, infect and make the host cancerous. Viruses such as hepatitis B and C are associated with hepatocellular cancer; Epstein Barr Viruses (EBV) is associated with Burkitt's lymphoma, gastric cancers, and nasopharyngeal cancer whereas Kaposi sarcoma—associated herpesvirus (KSHV) from the same family, Herpes virus, causes Kaposi's sarcoma, Castleman diseases and pulmonary effusion lymphoma; Papilloma viruses cause anogenital cancer; Human T-cell lymphotropic I virus causes adult T cell leukemia.

Besides these there are some dietary factors such as smoked fish, beef, and high fat diet which carry high risk of gastric cancer, bowel cancer, breast cancer, pancreatic cancer, prostate cancer, etc. (Norat et al. 2005; Han et al. 2015). The level of hormones and their receptors also play crucial role in inducing cancer and deciding the prognosis (Travis and Key 2003). For instance, estrogens and their receptors such as ERα and ERβ enhance the incidence of lung cancer in women, as well as the mortality rate (Chakraborty et al. 2010). Cancer is an umbrella term which encompasses several types of aberrations and frequently classified based on the location in the host. Some of the most common types are lung cancer, breast cancer, cervical cancer, bladder cancer, endometrial cancer, thyroid cancer, prostate cancer, kidney cancer, pancreatic cancer, colorectal cancer, etc. Among the aforesaid cancers, in 2020 the incidence rate of breast (2.26 million cases); lung (2.21 million cases); colon and rectum (1.93 million cases); prostate (1.41 million cases); skin (non-melanoma) (1.20 million cases); and stomach (1.09 million cases) cancer is observed to be significantly higher among the diverse cancer types. Further, in 2020, lung cancer (1.80 million deaths); colon and rectum cancer (935,000 deaths); liver cancer (830,000 deaths); stomach cancer (769,000 deaths); and breast cancer (685,000 deaths) have topped with the most significant mortality rate (www.who.int). Over the last few decennary, it has been observed that cancer is not same in all the individuals, since it is mostly by reason of single nucleotide polymorphism (SNP) and heterogenous nature. Thence, WHO recommended diverse mode of therapeutic interventions which involve removal of the mass from the affected part (surgery), killing of the cancerous cells using synthetics or biologicals (chemotherapy) and rays of energy (radiotherapy).

The most common methods of treatment of cancer are chemotherapy, surgery, radiation therapy, targeted therapy, and immunotherapy. Curative surgery is used to remove the affected tissues from the host body. Diagnostic surgery is performed as biopsy to diagnose the suspected tissue for cancer. Staging surgery helps to determine the size of the cancerous mass and extent of spread by examining the lymph nodes in surrounding area. Debulking surgery removes certain part of the cancerous mass but not the entire cancerous growth and performed when removal of the whole cancerous part can hamper the organ function. Palliative surgery does not cure but reduces the discomfort such as pain and helps in welfare of cancer patients (https://stanfordhealthcare.org/medical-treatments/c/cancer-surgery/types.html). The different types of surgery followed are laparoscopic, cryosurgery (uses liquid nitrogen or CO_2 or argon spray to freeze cells), laser surgery, and Mohs micrographic surgery for certain type of skin cancers and malignant cutaneous neoplasia. Surgery is often used in combination with radiotherapy and chemotherapy. Radiotherapy is a technique in which high energy beams such as X-rays, gamma rays, or electron or proton beams are applied to destroy the cancer cells by damaging the DNA of the cells. It is used to control the tumor growth after surgery, reduce the size of the cancerous cells. It involves short treatment span and it helps to reduce the discomfort undergone by the patient alike pain. The three-dimensional (3D) conformal radiotherapy, image-guided radiotherapy, intensity-modulated radiotherapy, volumetric-modulated arc radiotherapy, and stereotactic ablative radiotherapy are the types of radiotherapy (Sapkaroski et al. 2015; Grégoire et al. 2020; Gupta et al. 2020). Radiotherapy can be used effectively only in the case where the cancer is localized while chemotherapy is suitable for systemic way of treatment. In chemotherapy the chemicals used for treating cancer mainly aims at activation or deactivation of the target which further disturbs the cancer cell proliferation or metabolic activity, mobility of the cell; sometimes they act directly on DNA to create double-strand break. Besides these treatments, there is also an emerging treatment strategy, immunotherapy. The aim of immunotherapy is to use the natural immune response of the organisms in an effective way by educating the immune system to recognize and kill the cancerous cells. Immunotherapy is categorized into three, namely adoptive cell transfer therapy, cellular therapy, and genetic modifications of the immune cells. On advancement of cancer, malignant cells adopt different mechanisms to escape from the host immune response by downregulating the tumor antigen expression and blocking the human leukocyte antigen (HLA) expression, reducing the counts of CD4 and CD25 cells (regulatory T cells) etc. Thus, immunotherapy aims at increasing the number of the regulatory T cells, tumor associated macrophages, and myeloid-derived suppressor cells (Yildizhan et al. 2018). Targeted therapy is a method of treating the cancer with earmark on the proteins that controls the growth and spread of the cancer juxtaposing with immunotherapy. Targeted therapy is performed using either the small molecule drugs that are usually targeted inside the cell as it can set foot in easily because of its micro size or monoclonal antibodies directed to attach to the proteins in the cancer cell. Monoclonal antibodies (MAbs) are frequently used to mark the cancer cells with an aim of enhancing the recognition by immune system. A few are directly involved in arresting the cancer cell growth and leads to apoptosis. Some of the MAbs act as angiogenesis inhibitors. For example, Axitinib, Ziv-aflibercept block the vascular endothelial growth factor or its receptors that are on the surface of endothelial cell. These are considered as immune modulators. Hormonal therapies also come under targeted therapy which follows two strategies. It stops the mechanism of body that is involved in making specific hormones and another strategy is to protect cells from the hormones (cancer.gov/about-cancer/treatment/types/targeted-therapies). Howbeit, these commonly used therapies cause side effects in which only some are reversible.

The most curative treatment in many cases, surgery, has a problem of recurrence of cancer, damages to nearby tissues, causes blood clots, hemorrhage, pain, and infection. Radiotherapy leads to mild faint, fatigue, skin irritation, etc. whereas the chemotherapy is associated with mild to severe systemic side-effects and causes cell damage in organs such as heart, kidney, lungs, liver, bladder, nervous system, and even hair follicles. The modern treatments, immune therapy and targeted therapy cause allergies in skin, intense itching, breathing trouble, dizziness, etc. (Yildizhan et al. 2018; Brar et al. 2021). Even the angiogenesis inhibitors result in hemorrhage clots in arteries which lead to cardiac arrest, impaired wound healing, and reversible posterior leukocephalopathy: a brain disorder, hand foot syndrome (Moore-Maxwell et al. 2004; Kamba and McDonald 2007). Most of the adverse events are inflicted through off-target drug actions such as effect of drugs on the healthy cells as they have the similar surface proteins that are found over the cancer cells and inappropriate drug delivery. Nanoparticles (NPs) can be exploited for the biodistribution of medicines ideally at the target site. It enhances the therapeutic index and also potentially modulate the pharmacokinetics and pharmacodynamic profile of drugs.

Nanoparticles have idiosyncratic properties such as nanoscale size, high surface-to-volume ratio, and favorable physio-chemical characteristics which make them noteworthy (Brar et al. 2021). Moreover, using NPs in drug delivery increases the in vivo stability of the drug; drugs like radiosensitizer wortmannin and PI3K inhibitor have the problem of poor stability which is then sorted using a nanolipid-based nano-carrier by increasing its solubility from 4 mg/L to 20 g/L (Karve et al. 2012). Also, the blood circulation time of the drug or compound to be delivered can be increased using nano-carriers and it assists the controlled delivery of drug in the system. Encapsulation of the drugs by NPs or biodegradable compounds protects them from enzymatic cleavage; it also helps in effective penetration of the drug and converge the effect of chemotherapy towards the cancerous tissue. It can also be achieved by either active or passive targeting. Nano-carriers can be designed in such a way where it will release the loaded drugs in response to the stimuli such as micro-environmental pH, thermo-sensitivity, reactive oxygen species (ROS), etc. (Brar et al. 2021). The best instance to explain the case is the delivery of anti-cancer drug doxorubicin which is not conventionally relying upon pH dependent mechanism but when it is conjugated with pH-sensitive liposomal NPs (Doxil® has been approved by FDA), it enhances cellular drug uptake and intracellular drug release along with reduced systemic toxicity (Dos Reis et al. 2021). This specific uptake of the drugs reduces the drug resistance of the cancer cells; thereby decreasing the probability of indeterminate MDR/ATP efflux pump-driven elimination. Moreover, the nano-carriers have several modes of action in transporting the drug to the targeted site in the body (Wei et al. 2021). Similarly, several other anti-cancer drugs such as Docetaxel, Paclitaxel, Daunorubicin, etc. have been efficiently delivered to the target site through stimuli-responsive nano-carriers based upon diverse stimuli such as pH, temperature, electric and magnetic field, enzyme activity, light, etc. (Brar et al. 2021; Rahim et al. 2021).

24.2 Mode of Nanointerventions in Cancer Therapy

The development of the nanotechnology has extended the opportunity for cancer therapy to facilitate the specific drug delivery which inevitably benefits the patients by overcoming the obstacle of conventional therapeutics. There are three main strategies of nano-intervention for delivering drugs at the site of action. They are passive targeting, active targeting, and stimuli response release.

24.2.1 Passive Targeting

Passive targeting exploits different properties of the normal and tumor cells in order to deliver drugs at the site of action (Fig. 24.1). Neovascularization which is induced by the increased proliferation of cancer cells and massive vent in the walls of

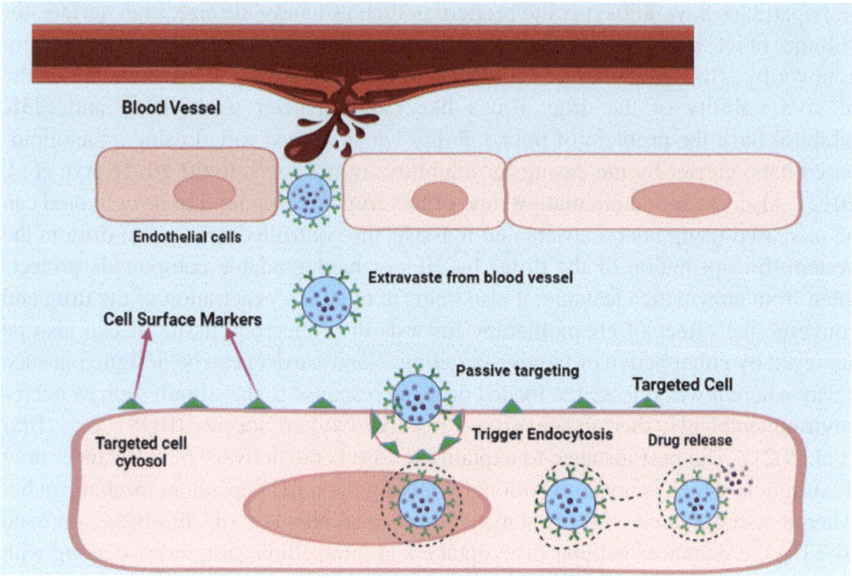

Fig. 24.1 Nanoparticle-assisted passive targeting of cancer therapeutics

vascular tissues unlike normal vessels results in the poor perm-selectivity of the tumor vessels. Due to the defective angiogenesis, nano-conjugated drugs leak from the vessels and cumulate at the cancerous tissue. Retention of the NPs is increased in the meantime because of the defective lymphatic drainage system in cancer tissue which leads to the release of contents from the nano-carriers in the cancer cells. This process is responsible for the propulsion factor of the passive targeting, the enhanced permeability, and retention (EPR) effect. The EPR effect and the lack of lymphatics around the cancer tissues facilitate the passive targeting. Size of the NPs also plays a major role in the penetrability into the site of delivery. The smaller the NPs, the greater the efficacy because they do not leak through the normal vessels whereas the large particles are cleared by the immune system. EPR effect depends on the leaky blood vessels that are intra-tumoral, which have gaps between 100 nm and 780 nm of size in its endothelium. Kaposi sarcoma is such a case which has fenestrated vasculature. Moreover, microenvironment of the tumor is considered as another important factor in passive delivery mechanism. This factor can be explained with one of the metabolic characteristics that acts as the major source of energy for cancer cells, glycolysis. The pH of the tumor micro-environment becomes acidic as a result of enhanced glycolysis (Warburg Effect). This pH change at the tumor micro-environment sets off pH sensitive NPs to release the drugs into the cancer cells. Almost all the pharmaceutical nanomedicine drugs are passive target mediated first generation drugs. They mainly focus on modifying the physio-chemical properties in order to control the distribution of the compounds and the pharmacokinetics. Some of the passively targeted first-generation drugs are nab-paclitaxel (Abraxane®) and pegylated liposomal doxorubicin (Doxil®). (Wicki et al. 2015). The side effects caused by the cytotoxic drugs cannot be controlled only by the EPR effect though it is the most common method of passive targeting. Tumor heterogeneity such as hypoxic gradient has an effect on the efficiency of drug delivery in the host. Increased interstitial fluid pressure also decreases the efficiency of the nanomedicine delivery, in which the drugs get cumulated at organ instead of tumor. Thus, the development of the second-generation nanomedicine drug is forsooth of enhanced targeting with increase in efficacy. The compounds in second generation are based on the active targeting technologies and smart nano-carriers with stimuli responsive release property (Yao et al. 2020).

24.2.2 Active Targeting

In active targeting, direct interaction between the ligands on the surface of NPs and the receptors that are expressed over the cancer cells takes place (Yao et al. 2020). These receptors are specific to the cancer cells as they either express or overexpress only in the cancer cells and not in normal cells. The drugs encapsulated in the NPs are released at the site by receptor mediated endocytosis, induced by the ligand-receptor interaction of NPs and cancer cell surface. Thus, active targeting is appropriate for delivery of proteins, siRNAs, MAbs and other biological, etc. i.e., macromolecular drug delivery. The most ubiquitous targeting components used are carbohydrates, peptides, amino acids, vitamins, and MAbs (Bazak et al. 2015). These ligands recognize and bind to the specific receptors which are expressed over the cancer cells (Fig. 24.2). Often targeted receptors are transferrin receptor, the epidermal growth factor receptor (EGFR), folate receptor, and glycoproteins. Transferrin is a blood-serum glycoprotein that helps in iron transport through blood to various organ tissues such as the liver. Most of the solid tumor cells expresses the receptor of transferrin, comparatively high in number than the healthy cells. On juxtaposing, the NPs conjugated with the transferrin show effective uptake into the cells than the non-conjugated NPs. These transferrin conjugated NPs show increased intracellular drug delivery and succours to manage the drug resistance chemotherapy. Folate receptors are for folic acid, which are essential vitamins of nucleotide synthesis. Almost in 40% of the human cancer folate receptor of alpha isoform is over-expressed whereas in case of hematopoietic cancers folate receptor of beta isoform is expressed on the cell surface. As a result, folate conjugated NPs are used in drug delivery targeting the folate receptors that are over expressed in the cancer cells. Nanoparticles conjugated with lectin targets carbohydrates over the cancer cell surface and initiate the lectin targeting pathway; there is a reverse lectin targeting pathway in which intracellular lectins of cancer cells are targeted by the carbohydrate conjugated NPs. Another promising method of drug delivery is targeting the epidermal growth factor receptor which is found in most of the cancer cases involving in the process of tumor growth. This has been exploited since a long time as a target for cancer therapy. Targeting the human epidermal growth factor

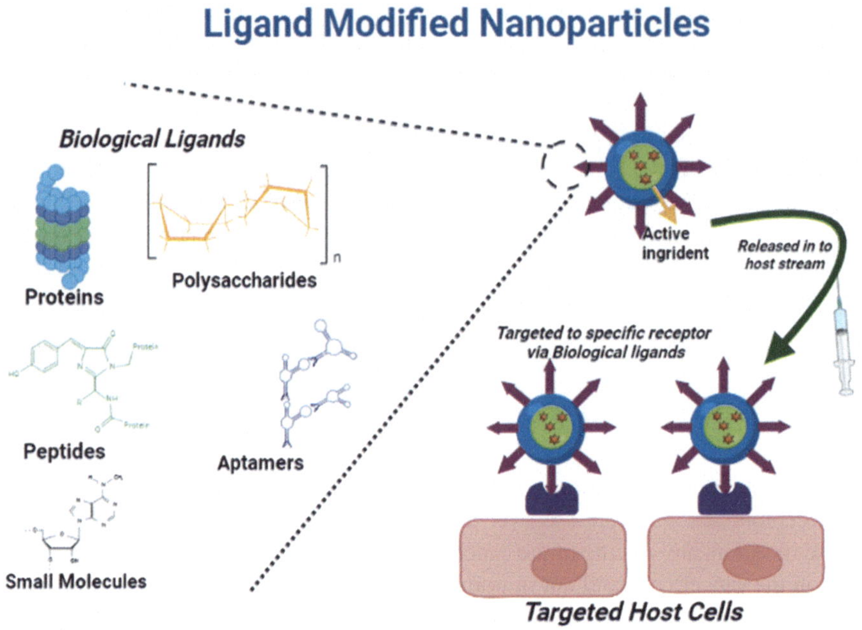

Fig. 24.2 Active targeting of cancer nanotherapeutics

receptor-2 (HER-2) is the most commonly used therapy for breast and gastric cancer which are positive for HER-2. The ligands of the NPs are modified in such a way that facilitate binding with the EGFR. Nowadays the specificity of the nanomedicine is increased by conjugating two or more cancer specific ligands over a single NP, while targeting the endothelial growth factor, not all the formulations directly target cancer cells. Some end up affecting angiogenesis. Interaction of vascular endothelial growth factor receptors (VEGFRs) and VEGF is important for vascularization. Targeting the major VEGFRs, namely VEGFR-2 and VEGFR-3, concurrently results in improved efficiency in treating cancer. A cell surface receptor, integrin, which has important role in calcium-dependent pathway results in initiating the endothelial migration, and helps in tumor cell invasion. Neovascular endothelial cells have large amount of αvβ3 integrin than resting endothelial cells. Thus, in association with VEGFR-2 signalling, blocking αvβ3 integrin binding site can lead to a reduction in VEGF signalling, which shows that designing the NPs targeting αvβ3 integrin increases the efficacy of anti-VEGFR therapy. The main focus of the nano-drug formulation is attaining the balance between low opsonization of the nano-carrier and high avidity of the ligand. Nanoparticles that are actively targeted improve the drug retention in the cancer cells due to cellular binding, minimized non-specific uptake, and mechanisms that is used to overcome resistance (Yao et al. 2020).

24.2.3 Stimuli Responsive Drug Release

Stimuli responsive drug formulations act according to the physical, biological, and chemical triggers that interferes with the conformation of the NPs and pave way for the release of conjugated drugs at the targeted site. The nano-carriers used in this case functions similar to the feedback mechanism exerted by the biological systems in which the physio-chemical properties decide the biochemical signalling of the system. The main advantage of this system is the systemic exposure of the compound is minimized as the drugs are released from nano-carriers only due to the stimuli from neoplastic tissue. Stimuli are categorized into external (physical stimuli such as temperature, light, ultrasound, magnetic force, and electric fields) and internal (patho-physiological/patho-chemical condition) stimuli (Rahim et al. 2021). On comparing, it is considered that external stimuli provide better spatially and temporally controllable features for activation and controlled release of the loaded drugs from the nano-carriers at the target sites than internal stimuli. The change in pH of the tumor microenvironment is exploited for internal stimuli-responsive NPs as the pH of cancerous tissue (less than 6.5) is different from the healthy tissue. When paclitaxel is loaded in pH-responsive cationic polymer, poly (β-amino ester) (PbAE), its bioavailability is found to be increased than obtained when paclitaxel is loaded in non-pH-sensitive polymer (Shenoy et al. 2005; Ma and Mumper 2013). There are other notable anti-cancer drugs that are loaded in pH-responsive NPs such as liposomes, polymeric micelles, and dendrimers which also results in increased efficiency of targeted drug delivery in tumor tissue (AlSawaftah et al. 2022). Oxygen-dearth hypoxic tumor microenvironment has reducing agents such as glutathione acting as internal stimuli which cleaves the disulfide bonds and releases the payloads from the redox-responsive nano-carriers at the target sites. Doxorubicin packed in micellar nano-carrier made of block polymer of poly(ε-caprolactone) and poly (ethyl ethylene phosphate) with disulfide bridges improves the targeted drug release (Ulbrich et al. 2016).

Hyperthermia is used as one of the strategies for cancer therapy because it shows increased tumor destroying efficiency when it is given in combination with radiation or chemotherapy. Also, cancer cells are more susceptible to heat-induced damage than normal cells. Hyperthermia helps in mediating the passive targeting of drug by increasing the tumor vasculature permeability. Temperature-responsive polymers such as liposomes, polymeric micelles, nanocomposites, nano-capsules, nanogels, and vesicles have amphiphilic property containing temperature-sensitive hydrophilic portions and hydrophobic core in their structure. Yatvin and his colleagues described the temperature-responsive drug release for the first-time using the thermo-sensitive liposomes in 1978 (Yatvin et al. 1978). ThermoDox® is one such notable example of temperature-responsive liposome; it is explained using the murine tumor model in which it shows prominent anti-tumor efficacy at 42 °C (Borys and Dewhirst 2021). Another most widely known copolymer is N-isopropylacrylamide (NIPAM); it is used to obtain liposomes of temperature-sensitivity that has been extensively used for spatiotemporal delivery of anticancer drugs in solid tumors (Amin et al. 2020).

Drug delivery system directed using ultrasound has many applications in nanomedicine-mediated cancer treatment. The main advantages of using ultrasound technology are non-invasiveness, non-exposure or exposure to the ionizing radiation is minimized, penetration depth in the tissue can be adjusted by modifying the frequency and increased uptake of the drug in cancer cells. The drugs confined in the polymeric micelle are commonly used along with ultrasound technology as it facilitates targeted drug delivery by reducing the amount of free drugs within the system which in turn decreases the side effects of drugs. The high-intensity focused ultrasound (HIFU) results in cancer cell apoptosis and cell necrosis by increasing the temperature of the tissue to 70–80 °C (Alkhorayef et al. 2015). When the HIFU is applied to the cells with phase change material (PCM) which changes into liquid at high temperature, it results in drug release from the nanocage in the cancer tissue. Low frequency ultrasound (LFUS) is the sonic variation, it is an unfocused ultrasonic wave. Schroeder and co-workers have done study on this in 2009, which shows prominently that more than 70% anti-cancer drug (cisplatin) encapsulated in liposome NPs that are nano-sterically stabilized could be released at the target site when exposed to the LFUS (Schroeder et al. 2009). Similarly, Lee et al. in 2016 has proved that the tumor size can be reduced when the gas generating nanoparticles that produces carbon dioxide under acidic condition is injected into the host and exposed to the ultrasound periodically (Lee et al. 2016). As gas generating nanoparticles, phase change material, pH-responsive cationic polymer, etc., there are different types of nano-carriers with unique properties that have been exploited in nano-carrier-mediated stimuli-responsive anti-cancer drug delivery systems (Rahim et al. 2021).

24.3 Nanomodalities in Oncotherapy

There are various types of nano-carriers that include but not limited to polymeric NPs, solid lipid NPs, polymeric micelles, nano-suspension, nano-shells, dendrimers, liposomes, ceramic nanoparticles, magnetic nanoparticles, ferrofluids, quantum dots, nanowires, nanopores, etc. All of the above can be broadly categorized as organic nano-carriers and inorganic nano-carriers. Some of the nano-formulations are already approved by U.S. FDA and several are in the pipeline (Table 24.1).

24.3.1 Organic Nano-carriers

Organic nano-carriers are made up of compounds mainly polysaccharides, lipids, and proteins. These types of nano-carriers are designed to perform targeted and non-targeted drug delivery.

Table 24.1 FDA approved anti-cancer nano-formulations

Name of the formulation	Nano-component Nanovehicle/ nanodrug	Active—ingredients	Types of cancer
VYXEOS—CPX-351	Liposomal formulation	Cytarabine: Daunorubicin	Acute myeloid leukemia
Doxil/Caelyx (Janssen)	PEGylated liposome	Doxorubicin	Ovarian cancer Kaposi's sarcoma Multiple myeloma
DaunoXome (Galen)	Non-PEGylated liposome	Daunorubicin	HIV-associated Kaposi's sarcoma
Myocet (Teva UK)	Non-PEGylated liposome	Doxorubicin	Metastatic breast cancer
Abraxane (Celgene)	Albumin-particle	Paclitaxel	Advanced non-small cell lung cancer Metastatic breast cancer Metastatic pancreatic cancer
Marqibo (Spectrum)	Non-PEGylated liposome	Vincristine	Acute lymphoblastic leukemia (tertiary)
MEPACT (Millennium)	Non-PEGylated liposome	Mifamurtide	Osteosarcoma
Onivyde MM-398 (Merrimack)	PEGylated liposome	Irinotecan	Metastatic pancreatic cancer (secondary)
Eligard (Tolmar)	Polymer	Leuprolide acetate	Prostate cancer
Ontak (Eisai)	Protein nanoparticles	Denileukin diftitox	Cutaneous T-cell lymphoma

24.3.1.1 Liposome-Based Nano-carriers

Liposomes are the small spherical vesicles made of naturally occurring phospholipids and cholesterol. It was the first nanoscale drug to be approved for therapeutic purpose as it provides the effective channel for the in-vivo drug delivery with enhanced therapeutic efficiency. Liposomes has a core that entrap drugs of either hydrophilic or hydrophobic nature, with lipid outer layer. The functions such as mobility and deformation of the liposomes are modified by altering the structure of the outer lipid layer. Poly ethylene glycol (PEG) is often attached to the lipid layer in order to increase the systemic time of the compounds in blood. Thus, the drugs encapsulated with liposomes exhibit very low systemic toxicity. The liposomes are further classified based on size and the number of bilayers present in it. They are small unilamellar vesicles (SUV) which is of size 25–50 nm in diameter, large unilamellar vesicles (LUV), multilamellar vesicles (MLV), etc. Liposome based anti-cancer drugs have become the most common in case of prostrate and breast cancer. Han et al. in 2020 has observed that drug delivery system employing the paclitaxel liposomes has increased bioavailability and anti-cancerous efficiency than free paclitaxel supplied into the system (Han et al. 2020). Similarly, liposomal doxorubicin is found to have reduced cardiotoxicity and also relatively increased efficacy in case of breast cancer (O'Brien et al. 2004). In recent days there are a greater number of liposome-mediated anti-cancer nano-drugs with improved properties such as multiple drug combinations, better therapeutic efficiency with reduced toxicity, surmounting drug resistance, etc. that are mostly exploited for cancer therapy (Yao et al. 2020). Vyxeos™/daunorubicin (Daunorubicin and Cytarabine), Doxil® and Lipo-Dox® (Doxorubicin), Onivyde® (Irinotecan), Marqibo® (Vincristine sulfate), DaunoXome (Daunorubicin), DepoCyte (Cytarabine), Myocet (Doxorubicin), etc. are some of the FDA and EMA approved liposomal anti-cancer nanoformulations which are already in use while several more are in the clinical trial or pipeline.

24.3.1.2 Polymer-Based Nano-carriers

Polymeric NPs are solid colloidal particles of size 1–1000 nm. It is of different forms: nano-capsules in which the drugs are encapsulated by the unique polymer membrane; nano-spheres are the drug system in which the drugs are dispersed. It is used to mediate the controlled delivery of drugs at the targeted site. This kind of NPs penetrates easily through the capillaries because of its size. They have structural arrangement of monomers that facilitates drug delivery. One of the most common polymeric NP for drug delivery is polylactic-co-glycolic acid (PLGA) which encompasses glycolic acid and lactic acid co-polymerization. PLGA is used because of its improved EPR effect, biodegradation, and biocompatibility. Cyclodextrin and PEG co-polymerize to form CALAA-01, a nanoparticle of 50–70 nm, developed for si-RNA delivery. Human transferrin receptors are found to be over expressed in the cancer cells. CALAA-01 targets human transferrin receptors in order to silence the M2 subunit of the ribonucleotide reductase. The release of siRNA within the cell is mediated by the acidic endosomal pH (Wicki et al. 2015).

Polymeric micelle is the system used for systemic delivery of water insoluble drugs. Amphiphilic block copolymers self-assembles to form micelles with a hydrophilic corona and a hydrophobic core having small size less than 100 nm helps in avoiding renal excretion and it also increases the permeability of endothelial cell. All these factors together facilitate in greater accumulation of drugs at the cancer cells than at normal cell. The absorption and smooth delivery of the drugs are enabled by the hydrophobic core of the micelle, whereas the stability of the molecule is increased by hydrophilic segment. As a result, there is a prominent reduction in the drug uptake by the reticuloendothelial system and the systemic circulatory time is increased (Yao et al. 2020). The stimuli-responsive micelles have been used for targeted drug delivery using pH, temperature alteration, enzymes, ultrasound, and ROS as external or internal stimuli. Micelles surface-functionalized with aptamers, si-RNAs, peptides, MAbs, carbohydrates, folic acid, etc. have also been used for active targeting. Genexol-PM (PEG-PLA micelle containing Paclitaxel) has already been approved by FDA for the treatment of breast, ovarian, and lung cancer while several others such as NK105 (PEG-PAA micelle containing Paclitaxel), NK911 (PEG-PAA micelle containing Doxorubicin), NC-6004 (PEG-Polyglutamic acid micelle containing Cisplatin), etc. are some of the prominent examples that are currently undergoing through different phases of clinical trial (Kenchegowda et al. 2021).

24.3.1.3 Dendrimers

Dendrimers are synthetic polymers with branches and layered architectures which consist of a central core, an internal region along with numerous terminal groups. These terminal groups help in determining the characteristics of dendrimer. It has intrinsic drug properties thus used to repair tissue scaffolds. They have multiple functional groups over the surface which increase the drug loading capacity and delivery at targeted sites. PLL dendrimer Gly-Lys63 (NH2)64 with doxorubicin is found to trigger the anti-angiogenic activity in a systemic manner in cancer-affected mice (Al-Jamal et al. 2013), whereas Wang et al. in 2017 found that bortezomib and doxorubicin loaded in telodendrimers exhibit increased anti-tumor effect in case of ovarian cancer (Wang et al. 2017). Similarly, PEGylated polyamidoamine (PAMAM) dendrimer conjugated with doxorubicin and tumor-penetrating peptide iRGD has found to enhance the tumor vascular permeability, accumulation, and anti-angiogenic effect in brain tumor (Wang et al. 2014). In addition, diaminobutyric polypropylenimine (DAB), poly (amidoamine-organosilicon) (PAMAMOS), poly (Lysine), and poly (propylene imine) (PPI) are the other different dendrimers that have been frequently exploited as delivery system for diverse anti-cancer drugs and biologicals (Mandal 2021). PAMAM dendrimer loaded with Cisplatin and si-RNA, PAMAM-PEG dendrimer loaded with Doxorubicin or 5-Flouro uracil, Folic acid-PAMAM dendrimer conjugated Methotrexate, etc. are some of the prominent dendrimer-based anti-cancer nano-formulations which are currently undergoing different phases of clinical trials with promising outcome (Kenchegowda et al. 2021).

24.3.2 Inorganic Nano-carriers

In the present-day material science, inorganic NPs are emerging as novel drug delivery systems because of their unique physical properties such as size-dependent magnetic, optical, catalytic, and electronic properties. Such nano-carriers exhibit large surface area, high stability, physicochemical multifunctionality, and some particular biological behavior. Inorganic material-based biocompatible nano-systems provide an unprecedented choice to effectively prevail over the stumbling block of traditional organic materials in therapeutic applications, particularly to overcome the multidrug resistance. Some of the most common inorganic NPs and their applications in cancer therapy have been enumerated hereunder.

24.3.2.1 Gold Nanoparticles

One of the most widely recognized inorganic NPs in cancer therapy is gold (Au) NPs because of its unique properties such as electronic, optical, biochemical, and sensing attributes. AuNPs are fine particles of gold which has its diameters between 1 and 100 nm. The color of AuNPs varies from red to purple and the color change depends on the size of the particle. AuNPs of desired size and shape is synthesized by various biological, physical, and chemical methods. These special characteristics of the gold NPs have rendered them useful in therapeutic approaches such as biosensing, diagnosing, and targeted drug delivery (Li et al. 2019). Its main advantage is that it protects the drugs from degrading and getting leaked into physiological environment. Besides being used in treating cancer with its photothermal ablation properties, AuNPs is used along with anticancer drugs. Using drugs and AuNPs together makes prominent changes in surface, size, shape, and pharmacokinetic properties of the drugs which further enhances its systemic consumption. In case of cancer therapy, AuNPs is exploited as drug carriers because of its unique properties. It is principally used in three different morphologies, namely hollow nanosphere, hollow gold nano-shells, and gold nanocages. Thus, the drugs are loaded in the carriers via the pores over them and packed in it. The development of AuNPs drug carrier system was pioneered by Khutale and Casey that facilitates intracellular delivery of the doxorubicin (Khutale and Casey 2017). The doxorubicin conjugated with AuNPs stabilized by coupling with PEG-PAMAMG4 dendrimer depicted enhanced drug release at acidic microenvironment in a pH-responsive manner, thus acting as a promising intracellular drug delivery vehicle for anticancer drugs. Several other AuNPs-based drug carrier systems have been recently employed with promising results. For instance, Folate-AuNP containing Cyclophosphamide has yielded promising outcome against breast cancer. MTX-AuNP containing Methotrexate has depicted higher cytotoxicity, greater accumulation, and better tumor inhibition than the free drug. Similarly, VCR-AuNP containing Vincristine, 6MP-AuNP containing 6-mercaptopurine and 5-FU-Glutathione-AuNP containing 5-Flourouracil have also depicted better anti-tumor activity as compared to the respective free drugs (Kenchegowda et al. 2021).

24.3.2.2 Carbon Nanotubes

Carbon nanotubes (CNTs) are made of carbon in which the atoms are arranged in hexagonal form similar to the graphene. CNTs are obtained by rolling of graphene sheets to generate single-walled carbon nanotube (SWCNTs) or multi-walled carbon nanotubes (MWCNTs). CNTs have ambient optical, thermal, and electrical properties along with high intracellular bioavailability, pay-load capacity, and ultra-high aspect ratio which are exploited in biomedical stream for detection and drug delivery system. These properties of CNTs have enabled them to be chosen for delivering the anticancer drugs such as methotrexate, paclitaxel, doxorubicin, and biological like MAbs, siRNA, etc. which have been used for different types of cancers. It has been observed that the CNTs release heat energy under the exposure of near-infrared radiation, which can kill the cancer cells and thus used for thermal ablation in cancer therapy (Luo et al. 2013; Singh and Torti 2013). The CNTs also accommodate diverse functionalization to act as adept carrier of several biologicals and drugs. The inner hollow space of CNTs is modified in such a way that it encapsulates the chosen drug to be dispersed in targeted manner and also it protects the drug material from getting oxidized whereas the external walls are modified using organic materials either peptide or substrates. For instance, Zhao et al. in 2018 have designed PEGylated multi-walled CNTs of 300 nm which possess a high doxorubicin-loading capacity along with enhanced cumulative release of drug within 24 h and a low premature leakage which is of only 10% (Zhao et al. 2018). Further, CNTs have been explored as promising delivery agents for gene therapy. Aptamers, micro-RNA (miRNAs) and siRNAs, double-stranded DNA (dsDNA), etc. which cannot enter into the cell like the small molecule drugs and gets easily degraded by the action of nucleases can conveniently delivered as attached at the surface of CNTs, thus enhancing the therapeutic efficiency (Augustine et al. 2017; Kenchegowda et al. 2021). Alike CNTs, functionalized carbon nanohorns are also being successfully used as delivery vehicle of several anti-cancer drugs, MAbs, and agent for photothermal and photodynamic therapy (Curcio et al. 2021). Nanomodalities based on the other carbon allotropes such as graphene, fullerenes, and nanodiamonds are also accommodate diverse functionalization attributes and found to be of enormous utility in multiple facets of cancer management such as in cancer imaging, targeted drug delivery, acoustic wave-assisted theranostics, photothermal and photodynamic therapy, etc. (Augustine et al. 2017).

24.3.2.3 Mesoporous Silica Nanoparticles

Mesoporous silica NPs are synthesized with pore channels of size ranging between 2 and 50 nm. Such MSNs are exploited as drug carrier and for biomedical applications. Its high pore volume capacity and large surface area has made it to be used as drug carrier, its controlled structure and adjustable pore size helps in drug dissolution as well as prevention of crystallization of the molecule to be delivered. Howbeit, the density of the silanol is high, it facilitates the interior and exterior surface modification of the MSNs which helps in drug loading and unloading kinetics. The surface modifications of MSNs also result in the hybrid silica nanoparticles with multifunctional characteristics. The stability of the water insoluble drugs is found to be enhanced when the MSNs are attached to the hydrophobic materials. Further, such modification limits the premature drug release which ultimately results in increased concentration of the drug at the target site. This kind of targeted transport helps to minimize the undesired effects of the drug in host system and also improves therapeutic efficiency. Thus, because of the increased mechanical, chemical, and thermal stabilities over wide range of temperature and pH, MSNs are used as an ideal nano-vehicle for anti-cancer drugs. MCM-41, MCM-48, SBA-15, etc. are some of the common ordered MSNs that have been extensively introspected for biomedical applications (Yu et al. 2022). Also, Fontana et al. in 2017 have found that the porous silica NPs show greater efficiency in case of immunotherapy as it has the properties of immune adjuvants such as increased antigen cross presentation, interferon-γ secretion, and lymphocyte polarization (Fontana et al. 2017; Yu et al. 2022). Surface functionalized MSNs have been extensively used for active targeting using diverse ligands such as transferrin, mannose, folic acid, etc. Anti-cancer drugs like Doxorubicin, Sorafenib, Axitinib, si-RNA, etc. have been delivered using functionalized MSNs. MSNs are also being used as nano-carriers for photodynamic and photothermal therapy of diverse cancer types (Kenchegowda et al. 2021).

24.3.2.4 Quantum Dots

Quantum dots (QDs) are light absorbing luminescent semiconductor particles of nano-size (2–8 nm) which frequently consist of an organic coating and heavy metal core-shell. QDs are used as bioimaging agents in order to monitor certain substances, such as prostate protein antigen, glutathione (GSH), ATP, Cu^{2+}. They are mostly exploited for bioimaging because of unique optical properties and also possess some potential toxicity which limits their role as drug delivery tool. Surface functionalization with biocompatible polymers or biological targeting agents can improve the performance of quantum dots. In relation to semiconductor QDs, graphene quantum dots (GQDs) are having low toxicity which encourages its application as tool for drug delivery and also depicted high stability. For instance, GQDs conjugated with MSNs and doxorubicin has facilitated chemo-photothermal therapeutic potential. Similarly, GQD-Biotin-Doxorubicin nanoconjugate or nitrogen functionalized GQD-methotrexate has also been successfully exploited for cancer therapy (Kenchegowda et al. 2021). Currently, carbon QDs are extensively employed for targeted gene delivery, photodynamic therapy, carriers for drug delivery in various types of cancer (Devi et al. 2022). QDs are also having extended applications in various fields such as biosensors, bioimaging, in photo voltaic devices, etc.

24.3.2.5 Magnetic Nanoparticles

Magnetic nanoparticles (MNPs) are the materials of nano-size with magnetic properties. It includes elements such as nickel, iron, chromium, cobalt, gadolinium, manganese, and their compounds. This kind of NPs has been exploited for diverse biomedical purposes including targeted drug delivery. It is modified in such a way that facilitates localized drug delivery in case of cancer therapy. The drug is infused in the patients bloodstream and once it reaches the targeted site, tissues are exposed to the magnetic field in order to retain the particles at the site of the tumor. The main principle behind this technology is that the MNPs infused in the host are attracted with the external magnetic field and directed towards the cancer cell. Thus, the drugs are delivered at targeted site. Super paramagnetic iron oxide nanoparticle (SPiON) is a type of MNP which has a wide application because of its size and super paramagnetic property (net moment of the particle is zero in absence of external magnetic field whereas it shows the property of paramagnet in presence of external magnetic field). The SPIONs with modified surface (with targeting molecules/ligandin their surfaces) are used along with an external magnetic field is considered as a modern technology to target particles where the drug is to be released locally. This system has an advantage of minimized side effects lowering the required dose of the drug for the treatment. However, burst effect is observed in SPIONs. In order to overcome the burst effect, Mahmoudi and co-workers modified the surface of the NPs with iron oxide particles and crosslinked PEG cofumarate (PEGF). This NPs with surface modification are loaded with Tamoxifen, an anti-estrogen drug which is used for treating the breast cancer, to check the efficiency of the modifications done. It is found that the burst release of the payload, i.e., Tamoxifen is reduced by 21% in case of NPs surfaces are modified with iron oxide particles and crosslinked PEGF than the unmodified NPs (Mahmoudi et al. 2011). Besides drug delivery, MNPs are extensively used in magnetic resonance based bioimaging; Gastromark® containing an aqueous formulation of silicone coated SPIONs, Feridex® (dextran coated SPIONs), and Feraheme® (polyglucose sorbitol carboxymethylether coated SPIONs) are some of the FDA approved MNP-based nano-formulations used in bioimaging and several more are in clinical trial (Kenchegowda et al. 2021).

24.3.3 *Virus-Like Nanoparticles (VLP)*

Viruses-like nanoparticles (VLPs) are obtained from bacteriophages, plant, or animal viruses. The scope of VLPs is increased in the field of cancer therapy. VLPs are virus-based nano-formulations that are mostly designed in non-infectious way. The unique characters of VLP-based nano-carriers that make them conspicuous are morphological uniformity, biocompatibility, biodegradable nature, water solubility, high uptake efficiency, and easy functionalization. Plant viruses that are used for developing the VLPs are Brome mosaic virus (BMV), Cowpea chlorotic mottle virus (CCMV), Cowpea mosaic virus (CPMV), Hibiscus chlorotic ringspot virus (HCRSV), Red clover necrotic mottle virus (RCNMV), physalis mottle virus (PhMV), and potato virus X (PVX). MS2, P22, Qβ, and the filamentous phage M13 are the most commonly used bacteriophage formulation of VLPs. Cai et al. in 2020 encapsulated the CpG oligodeoxynucleotides (ODNs) within CCMV and employed in cancer immunotherapy with a payload of Doxorubicin. This has shown a significant cytotoxicity in in vitro cancer cell line model (Cai et al. 2020). It has been observed that when the platinum-based drugs namely phenanthriplatin and cisplatin are delivered using VLPs such as TMV, malignant cells uptake those complexes effectively and exhibit superior cytotoxicity. VLPs are conjugated to contrasting and targeting substances to be used for cancer cell imaging and targeted delivery at the desired site. The VLPs are also used as vehicle for nano-vaccines to treat cancer targeting the tumor associated antigens (TAAs) such as human epidermal growth factor receptor 2 (HER2) epitope, MUC1 glycopeptide antigens, Tn antigen, etc. (Chung et al. 2020).

24.4 Therapeutic Nano-vaccines

Conventionally vaccine is the product obtained from the part of causative agent of the disease or the part itself; used to the trigger the host immune system to produce the antibodies in order to either prevent the disease occurrence or to fight against the antigens of the disease occurred. In order to enhance the efficiency of vaccines, they are conjugated to adjuvants of nano-size; this helps in improving the immune response. Adjuvants are the substances that are conjugated with the antigens in order to enhance the immune response by increasing the immunogenicity of a weak antigen, modifying the subclass distribution, avidity, specificity of the antibody. Some of the adjuvants that are commonly used are MF59, monophosphoryl lipid, Montanid, calcium phosphate nanoparticles, immuno-stimulating complexes, and cholesterol-bearing hydrophobized pullulan nanoparticles. MF59 is an approved nano-vaccine adjuvant, it is an oil in water emulsion which increases the efficiency of glycoprotein D several times against herpes simplex virus (HSV) as studied in Guinea pigs (Byars et al. 1994). Such amalgamation of nanotechnology with immunotherapy results in the development of nano-vaccines. Nano-vaccines are emerging rapidly; they can target the cancer cell sites specifically because of self-adjuvating effects, release kinetics, and delivery routes of nanoparticles. Though the traditional vaccines have all these formulations, they have limitations in terms of poor storage property and cold-chain requirement. The soluble proteins used in case of subunit vaccines have short life span which causes the requirement of multiple booster doses in order to elicit the immunity for longer period. The particle size of the nano-vaccine facilitates direct entry of the antigen into the lymphatic system and it is considered as an important factor for biodistribution and successful internalization by the antigen presenting cells (APCs). The shape of the nano-vaccine product used also plays a role in internalization and retention period. The rod-shaped particles are processed effectively by the dendritic cell than spherical shaped particles. Similarly, it has found to have a longer circulatory period in gastro intestinal tract. In case of Bovine Respiratory Disease that are caused by Bovine Respiratory Syncytial Virus (BRSV) in cattle, McGill et al. has developed a CPTEG: CPH (50: 50) a polyanhydride based nano-vaccine encapsulating the proteins of BRSV namely post-fusion F and G (BRSV-F/G). The 3-week-old calves were immunized with BRSV-F/G nano-vaccine via intranasal route. It was observed that it induced a viral specific response in lower respiratory tract and peripheral blood; also the bronchoalveolar lavage fluid and nasal secretions were found have viral specific IgA. Moreover the encapsulation enhanced antigen release kinetics over 30 days. Thus, BRSV nano-vaccine was found to be effective as they responded even in the presence of maternal antibodies; also, it reduced the virus burden of the host and shedding (McGill et al. 2018). Kavanagh and co-workers delivered two synthetic peptides of BRSV F and G proteins encapsulated in Poly lactic-co-glycolic acid (PLGA) NPs, a biodegradable polymer. As a result, mucosal IgA response was observed whereas the serum IgG or cellular immune responses were not observed (Kavanagh et al. 2013). Thukral et al. in 2020 have used CPTEG: CPH (20:80) nanoparticles containing cell lysate and filtrate proteins of *M. paratuberculosis* culture as a vaccine in mice to study its effect against Johne's disease. Primarily, response was found to be prominent by observing the reduction in bacterial load in different organs. The vaccine was found to induce polyfunctional CD8+ T cells that were characterized by IFN-γ, IL-2, and TNF-α production in mice and also it did not show any inflammatory reaction at the site of introduction (Thukral et al. 2020; Maina et al. 2020). Recently Ma et al. have found that encapsulation of water-soluble as well as water-insoluble components from the tumor cells into nano-vaccine has promoted phagocytosis by APCs yielding better therapeutic response in melanoma and breast cancer than using only water-soluble components. Further, loading the nano-vaccine with the anti PD-1 (anti Programmed Cell Death protein) antibody and metformin enhances the immune response and prevents the occurrence of the lung cancer completely and almost 70% occurrence of the melanoma in mice. It has also activated innate as well as adaptive immune system of the host (Ma et al. 2021). Thus, it is becoming evident that nano-vaccines are playing major role in case of cancer immunotherapy.

Nano-vaccines possess several advantages as compared to the subunit vaccines. For instance, encapsulation of tumor associated antigens (TAAs, shared antigens that are expressed by cancer cells as well as healthy cells) or neoantigens (exclusively expressed by cancer cells) in nano-carriers can avert early degradation and improve antigen stability. Nano-vaccines containing antigen as well as adjuvant facilitate co-delivery enhancing the immunogenicity and potentiating the therapeutic efficacy. Nano-size of these vaccines promotes phagocytosis and antigen processing by the APCs. Intra-cellular delivery of antigens through nano-vaccines facilitates major histocompatibility complex I (MHC-I) presentation, thereby eliciting adaptive immune responses including cytotoxic T-cell lymphocyte (CTL) responses which are pivotal for successful cancer immunotherapy. Surface functionalization of the nano-vaccine with targeting ligands facilitates site-specific targeting of the antigens to the lymphoid tissues and APCs for precise immunomodulation. Cross-linking of B-cell receptor can be taken place due to the polyvalent presentation of antigens at the nano-vaccine surface that also potentiates humoral immune responses. Several biogenic nano-carriers such as exosomes and outer membrane vesicles (OMVs); semi-biogenic/semi-synthetic nano-carriers such as VLPs and cancer cell membrane-coated nanoparticle (CCNP); synthetic nano-carriers such as PLGA, PLA, PEG-based polymeric NPs, liposomes, AuNPs or inorganic NPs have been used as delivery vehicle in anti-cancer nano-vaccines (Zhang et al. 2019).

In conclusion, nanotechnology has certainly provided several smart weapons in the armory against cancer. Engineered nanomaterials have provided the opportunity for targeted delivery of diverse anti-cancer therapeutics through active or passive targeting and stimuli-responsive intelligent delivery systems which enabled surmount certain limitations of conventional therapeutics such as water-insolubility, premature delivery, degradation and early elimination of drugs from the system, drug-efflux and resistance, off-target side-effects, etc. Nanoparticle-assisted photothermal and photodynamic therapy has certainly provided unique advantages in cancer treatment. Therapeutic nano-vaccines and adjuvants have also paved the path for cancer immunotherapy through specific targeting and antigen delivery. The nanomodality-assisted concerted effort will certainly make cancer therapy a realistic and viable mean of cancer management imminently.

References

Al-Jamal KT, Al-Jamal WT, Wang JT, Rubio N, Buddle J, Gathercole D, Zloh M, Kostarelos K (2013) Cationic poly-L-lysine dendrimer complexes doxorubicin and delays tumor growth in vitro and in vivo. ACS Nano 7(3):1905–1917. https://doi.org/10.1021/nn305860k

Alkhorayef M, Mahmoud MZ, Alzimami KS, Sulieman A, Fagiri MA (2015) High-intensity focused ultrasound (HIFU) in localized prostate cancer treatment. Pol J Radiol 80:131–141. https://doi.org/10.12659/PJR.892341

AlSawaftah NM, Awad NS, Pitt WG, Husseini GA (2022) pH-responsive nanocarriers in cancer therapy. Polymers 14(5):936. https://doi.org/10.3390/polym14050936

Amin M, Huang W, Seynhaeve A, Ten Hagen T (2020) Hyperthermia and temperature-sensitive nanomaterials for spatiotemporal drug delivery to solid tumors. Pharmaceutics 12(11):1007. https://doi.org/10.3390/pharmaceutics12111007

Anand P, Kunnumakkara AB, Sundaram C, Harikumar KB, Tharakan ST, Lai OS, Sung B, Aggarwal BB (2008) Cancer is a preventable disease that requires major lifestyle changes. Pharm Res 25(9):2097–2116. https://doi.org/10.1007/s11095-008-9661-9

Augustine S, Singh J, Srivastava M, Sharma M, Das A, Malhotra BD (2017) Recent advances in carbon based nanosystems for cancer theranostics. Biomater Sci 5(5):901–952. https://doi.org/10.1039/c7bm00008a

Bazak R, Houri M, El Achy S, Kamel S, Refaat T (2015) Cancer active targeting by nanoparticles: a comprehensive review of literature. J Cancer Res Clin Oncol 141(5):769–784. https://doi.org/10.1007/s00432-014-1767-3

Borys N, Dewhirst MW (2021) Drug development of lyso-thermosensitive liposomal doxorubicin: combining hyperthermia and thermosensitive drug delivery. Adv Drug Deliv Rev 178:113985. https://doi.org/10.1016/j.addr.2021.113985

Brar B, Ranjan K, Palria A, Kumar R, Ghosh M, Sihag S, Minakshi P (2021) Nanotechnology in colorectal cancer for precision diagnosis and therapy. Front Nanotechnol 3:699266. https://doi.org/10.3389/fnano.2021.699266

Byars NE, Fraser-Smith EB, Pecyk RA, Welch M, Nakano G, Burke RL, Hayward AR, Allison AC (1994) Vaccinating Guinea pigs with recombinant glycoprotein D of herpes simplex virus in an efficacious adjuvant formulation elicits protection against vaginal infection. Vaccine 12(3): 200–209. https://doi.org/10.1016/0264-410x(94)90195-3

Cai H, Shukla S, Steinmetz NF (2020) The antitumor efficacy of CpG oligonucleotides is improved by encapsulation in plant virus-like particles. Adv Funct Mater 30:1908743. https://doi.org/10.1002/adfm.201908743

Chakraborty S, Ganti AK, Marr A, Batra SK (2010) Lung cancer in women: role of estrogens. Expert Rev Respir Med 4(4):509–518. https://doi.org/10.1586/ers.10.50

Chung YH, Cai H, Steinmetz NF (2020) Viral nanoparticles for drug delivery, imaging, immunotherapy, and theranostic applications. Adv Drug Deliv Rev 156:214–235. https://doi.org/10.1016/j.addr.2020.06.024

Curcio M, Cirillo G, Saletta F, Michniewicz F, Nicoletta FP, Vittorio O, Hampel S, Iemma F (2021) Carbon nanohorns as effective nanotherapeutics in cancer therapy. C 7(1):3. https://doi.org/10.3390/c7010003

Devi S, Kumar M, Tiwari A, Tiwari V, Kaushik D, Verma R, Bhatt S, Sahoo BM, Bhattacharya T, Alshehri S, Ghoneim MM, Babalghith AO, Batiha GE-S (2022) Quantum dots: an emerging approach for cancer therapy. Front Mater 8:798440. https://doi.org/10.3389/fmats.2021.798440

Dos Reis SB, de Oliveira Silva J, Garcia-Fossa F, Leite EA, Malachias A, Pound-Lana G, Mosqueira V, Oliveira MC, de Barros A, de Jesus MB (2021) Mechanistic insights into the intracellular release of doxorubicin from pH-sensitive liposomes. Biomed Pharmacother 134:110952. https://doi.org/10.1016/j.biopha.2020.110952

Fontana F, Shahbazi M-A, Liu D, Zhang H, Mäkilä E, Salonen J, Hirvonen JT, Santos HA (2017) Nanovaccines: multistaged nanovaccines based on porous silicon@acetalated dextran@cancer cell membrane for cancer immunotherapy. Adv Mater 29(7):1603239. https://doi.org/10.1002/adma.201770049

Grégoire V, Guckenberger M, Haustermans K, Lagendijk JJW, Ménard C, Pötter R, Slotman BJ, Tanderup K, Thorwarth D, van Herk M, Zips D (2020) Image guidance in radiation therapy for better cure of cancer. Mol Oncol 14:1470–1491. https://doi.org/10.1002/1878-0261.12751

Gupta T, Sinha S, Ghosh-Laskar S, Budrukkar A, Mummudi N, Swain M, Phurailatpam R, Prabhash K, Agarwal JP (2020) Intensity-modulated radiation therapy versus three-dimensional conformal radiotherapy in head and neck squamous cell carcinoma: long-term and mature outcomes of a prospective randomized trial. Radiat Oncol (London, England) 15(1):218. https://doi.org/10.1186/s13014-020-01666-5

Han J, Jiang Y, Liu X, Meng Q, Xi Q, Zhuang Q, Han Y, Gao Y, Ding Q, Wu G (2015) Dietary fat intake and risk of gastric cancer: a meta-analysis of observational studies. PLoS One 10(9):e0138580. https://doi.org/10.1371/journal.pone.0138580

Han B, Yang Y, Chen J, Tang H, Sun Y, Zhang Z, Wang Z, Li Y, Li Y, Luan X, Li Q, Ren Z, Zhou X, Cong D, Liu Z, Meng Q, Sun F, Pei J (2020) Preparation, characterization, and pharmacokinetic study of a novel long-acting targeted paclitaxel liposome with antitumor activity. Int J Nanomedicine 15:553–571. https://doi.org/10.2147/IJN.S228715

Kamba T, McDonald DM (2007) Mechanisms of adverse effects of anti-VEGF therapy for cancer. Br J Cancer 96(12):1788–1795. https://doi.org/10.1038/sj.bjc.6603813

Karve S, Werner ME, Sukumar R, Cummings ND, Copp JA, Wang EC, Li C, Sethi M, Chen RC, Pacold ME, Wang AZ (2012) Revival of the abandoned therapeutic wortmannin by nanoparticle drug delivery. Proc Natl Acad Sci U S A 109(21):8230–8235. https://doi.org/10.1073/pnas.1120508109

Kavanagh OV, Adair BM, Welsh M, Earley B (2013) Immunogenetic responses in calves to intranasal delivery of bovine respiratory syncytial virus (BRSV) epitopes encapsulated in poly (DL-lactide-co-glycolide) microparticles. Res Vet Sci 95(2):786–793. https://doi.org/10.1016/j.rvsc.2013.06.023

Kenchegowda M, Rahamathulla M, Hani U, Begum MY, Guruswamy S, Osmani R, Gowrav MP, Alshehri S, Ghoneim MM, Alshlowi A, Gowda DV (2021) Smart nanocarriers as an emerging platform for cancer therapy: a review. Molecules (Basel, Switzerland) 27(1):146. https://doi.org/10.3390/molecules27010146

Khutale GV, Casey A (2017) Synthesis and characterization of a multifunctional gold-doxorubicin nanoparticle system for pH triggered intracellular anticancer drug release. Eur J Pharm Biopharm 119:372–380. https://doi.org/10.1016/j.ejpb.2017.07.009

Lee J, Min HS, You DG, Kim K, Kwon IC, Rhim T, Lee KY (2016) Theranostic gas-generating nanoparticles for targeted ultrasound imaging and treatment of neuroblastoma. J Controll Rel 223:197–206. https://doi.org/10.1016/j.jconrel.2015.12.051

Li W, Cao Z, Liu R, Liu L, Li H, Li X, Chen Y, Lu C, Liu Y (2019) AuNPs as an important inorganic nanoparticle applied in drug carrier systems. Artif Cells Nanomed Biotechnol 47(1):4222–4233. https://doi.org/10.1080/21691401.2019.1687501

Luo E, Song G, Li Y, Shi P, Hu J, Lin Y (2013) The toxicity and pharmacokinetics of carbon nanotubes as an effective drug carrier. Curr Drug Metab 14(8):879–890. https://doi.org/10.2174/1389200211314000110

Ma P, Mumper RJ (2013) Paclitaxel nano-delivery systems: a comprehensive review. J Nanomed Nanotechnol 4(2):1000164. https://doi.org/10.4172/2157-7439.1000164

Ma L, Diao L, Peng Z, Jia Y, Xie H, Li B, Ma J, Zhang M, Cheng L, Ding D, Zhang X, Chen H, Mo F, Jiang H, Xu G, Meng F, Zhong Z, Liu M (2021) Immunotherapy and prevention of cancer by nanovaccines loaded with whole-cell components of tumor tissues or cells. Adv Mater (Deerfield Beach, Fla.) 33(43):e2104849. https://doi.org/10.1002/adma.202104849

Mahmoudi M, Sant S, Wang B, Laurent S, Sen T (2011) Superparamagnetic iron oxide nanoparticles (SPIONs): development, surface modification and applications in chemotherapy. Adv Drug Deliv Rev 63(1–2):24–46. https://doi.org/10.1016/j.addr.2010.05.006

Maina TW, Grego EA, Boggiatto PM, Sacco RE, Narasimhan B, McGill JL (2020) Applications of nanovaccines for disease prevention in cattle. Front Bioeng Biotechnol 8:608050. https://doi.org/10.3389/fbioe.2020.608050

Mandal AK (2021) Dendrimers in targeted drug delivery applications: a review of diseases and cancer. Int J Polym Mater Polym Biomater 70(4):287–297. https://doi.org/10.1080/00914037.2020.1713780

McGill JL, Kelly SM, Kumar P, Speckhart S, Haughney SL, Henningson J, Narasimhan B, Sacco RE (2018) Efficacy of mucosal polyanhydride nanovaccine against respiratory syncytial virus infection in the neonatal calf. Sci Rep 8(1):3021. https://doi.org/10.1038/s41598-018-21292-2

Moore-Maxwell CA, Datto MB, Hulette CM (2004) Chemotherapy-induced toxic leukoencephalopathy causes a wide range of symptoms: a series of four autopsies. Mod Pathol 17(2):241–247. https://doi.org/10.1038/modpathol.3800049

Norat T, Bingham S, Ferrari P, Slimani N, Jenab M, Mazuir M, Overvad K, Olsen A, Tjønneland A, Clavel F, Boutron-Ruault MC, Kesse E, Boeing H, Bergmann MM, Nieters A, Linseisen J, Trichopoulou A, Trichopoulos D, Tountas Y, Berrino F et al (2005) Meat, fish, and colorectal cancer risk: the European prospective investigation into cancer and nutrition. J Natl Cancer Inst 97(12):906–916. https://doi.org/10.1093/jnci/dji164

O'Brien ME, Wigler N, Inbar M, Rosso R, Grischke E, Santoro A, Catane R, Kieback DG, Tomczak P, Ackland SP, Orlandi F, Mellars L, Alland L, Tendler C, CAELYX Breast Cancer Study Group (2004) Reduced cardiotoxicity and comparable efficacy in a phase III trial of pegylated liposomal doxorubicin HCl (CAELYX/Doxil) versus conventional doxorubicin for first-line treatment of metastatic breast cancer. Ann Oncol 15(3):440–449. https://doi.org/10.1093/annonc/mdh097

Rahim MA, Jan N, Khan S, Shah H, Madni A, Khan A, Jabar A, Khan S, Elhissi A, Hussain Z, Aziz HC, Sohail M, Khan M, Thu HE (2021) Recent advancements in stimuli responsive drug delivery platforms for active and passive cancer targeting. Cancers 13(4):670. https://doi.org/10.3390/cancers13040670

Sapkaroski D, Osborne C, Knight KA (2015) A review of stereotactic body radiotherapy—is volumetric modulated arc therapy the answer? J Med Radiat Sci 62(2):142–151. https://doi.org/10.1002/jmrs.108

Schroeder A, Honen R, Turjeman K, Gabizon A, Kost J, Barenholz Y (2009) Ultrasound triggered release of cisplatin from liposomes in murine tumors. J Controll Rel 137(1):63–68. https://doi.org/10.1016/j.jconrel.2009.03.007

Shenoy D, Little S, Langer R, Amiji M (2005) Poly(ethylene oxide)-modified poly(beta-amino ester) nanoparticles as a pH-sensitive system for tumor-targeted delivery of hydrophobic drugs: part 2. In vivo distribution and tumor localization studies. Pharm Res 22(12):2107–2114. https://doi.org/10.1007/s11095-005-8343-0

Singh R, Torti SV (2013) Carbon nanotubes in hyperthermia therapy. Adv Drug Deliv Rev 65(15):2045–2060. https://doi.org/10.1016/j.addr.2013.08.001

Thukral A, Ross K, Hansen C, Phanse Y, Narasimhan B, Steinberg H, Talaat AM (2020) A single dose polyanhydride-based nanovaccine against paratuberculosis infection. npj Vacc 5:15. https://doi.org/10.1038/s41541-020-0164-y

Travis RC, Key TJ (2003) Oestrogen exposure and breast cancer risk. Breast Cancer Res 5(5): 239–247. https://doi.org/10.1186/bcr628

Ulbrich K, Holá K, Šubr V, Bakandritsos A, Tuček J, Zbořil R (2016) Targeted drug delivery with polymers and magnetic nanoparticles: covalent and noncovalent approaches, release control, and clinical studies. Chem Rev 116(9):5338–5431. https://doi.org/10.1021/acs.chemrev.5b00589

Wang K, Zhang X, Liu Y, Liu C, Jiang B, Jiang Y (2014) Tumor penetrability and anti-angiogenesis using iRGD-mediated delivery of doxorubicin-polymer conjugates. Biomaterials 35(30):8735–8747. https://doi.org/10.1016/j.biomaterials.2014.06.042

Wang L, Shi C, Wright FA, Guo D, Wang X, Wang D, Wojcikiewicz R, Luo J (2017) Multifunctional Telodendrimer nanocarriers restore synergy of Bortezomib and Doxorubicin in ovarian cancer treatment. Cancer Res 77(12):3293–3305. https://doi.org/10.1158/0008-5472.CAN-16-3119

Wei X, Song M, Li W, Huang J, Yang G, Wang Y (2021) Multifunctional nanoplatforms co-delivering combinatorial dual-drug for eliminating cancer multidrug resistance. Theranostics 11(13):6334–6354. https://doi.org/10.7150/thno.59342

Wicki A, Witzigmann D, Balasubramanian V, Huwyler J (2015) Nanomedicine in cancer therapy: challenges, opportunities, and clinical applications. J Controll Rel 200:138–157. https://doi.org/10.1016/j.jconrel.2014.12.030

Yadav PK, Gupta SK, Kumar S, Ghosh M, Yadav BS, Kumar D, Kumar A, Saini M, Kataria M (2020) IL-18 immunoadjuvanted xenogeneic canine MMP-7 DNA vaccine overcomes immune tolerance and supresses the growth of murine mammary tumor. Int Immunopharmacol 82: 106370. Advance online publication. https://doi.org/10.1016/j.intimp.2020.106370

Yadav PK, Gupta SK, Kumar S, Ghosh M, Yadav BS, Kumar D, Kumar A, Saini M, Kataria M (2021) MMP-7 derived peptides with MHC class-I binding motifs from canine mammary tumor tissue elicit strong antigen-specific T-cell responses in BALB/c mice. Mol Cell Biochem 476: 311–320. https://doi.org/10.1007/s11010-020-03908-2

Yadav PK, Ghosh M, Kataria M (2022) Matrix metalloproteinases (MMPs) in cancer immunotherapy. In: Chakraborti S (ed) Handbook of oxidative stress in cancer: therapeutic aspects. Springer Nature Singapore Pte Ltd, pp 1–26. https://doi.org/10.1007/978-981-16-1247-3_182-1

Yao Y, Zhou Y, Liu L, Xu Y, Chen Q, Wang Y, Wu S, Deng Y, Zhang J, Shao A (2020) Nanoparticle-based drug delivery in cancer therapy and its role in overcoming drug resistance. Front Mol Biosci 7:193. https://doi.org/10.3389/fmolb.2020.00193

Yatvin MB, Weinstein JN, Dennis WH, Blumenthal R (1978) Design of liposomes for enhanced local release of drugs by hyperthermia. Science (New York, NY) 202(4374):1290–1293. https://doi.org/10.1126/science.364652

Yildizhan H, Barkan NP, Karahisar Turan S, Demiralp Ö, Özel Demiralp FD, Uslu B, Ōzkan SA (2018) Chapter 1—Treatment strategies in cancer from past to present. In: Grumezescu AM (ed) Drug targeting and stimuli sensitive drug delivery systems. William Andrew Publishing, Norwich, NY, pp 1–37. https://doi.org/10.1016/B978-0-12-813689-8.00001-X

Yu A, Dai X, Wang Z, Chen H, Guo B, Huang L (2022) Recent advances of mesoporous silica as a platform for cancer immunotherapy. Biosensors 12(2):109. https://doi.org/10.3390/bios12020109

Zhang Y, Lin S, Wang XY, Zhu G (2019) Nanovaccines for cancer immunotherapy. Wiley Interdiscip Rev Nanomed Nanobiotechnol 11(5):e1559. https://doi.org/10.1002/wnan.1559

Zhao X, Tian K, Zhou T, Jia X, Li J, Liu P (2018) PEGylated multi-walled carbon nanotubes as versatile vector for tumor-specific intracellular triggered release with enhanced anti-cancer efficiency: optimization of length and PEGylation degree. Colloids Surf B Biointerfaces 168: 43–49. https://doi.org/10.1016/j.colsurfb.2018.02.041

Chapter 25
Nanotechnology in Wound Healing

Neelesh Sindhu, Sandeep Kumar, and Rajesh Kumar

Abstract Wound care is one of the critical managements in healthcare services that urge well-being of the patient. The concerned USD 20.8 billion (in 2022) market has several components which include dressing materials such as foam, film, hydrocolloid, collagen, etc., devices for negative pressure wound therapy (NPWT), assessment, debridement, and graft, matrix, suture, stapler, etc. for various kind of wounds including normal cut to diabetic foot ulcers (DFUs), trauma, surgical wound, burn, etc. Faster and better wound healing is desired for favorable outcome of the treatment, hence require improved wound healing materials for repair of the disintegrated tissue, maintaining proper homeostasis, inflammation, proliferation and remodeling, removing any sources of infection and other secondary complications along with case-specific carrier system to deliver specific drugs or factors at the wound site in both animal and human subjects. The urge is even more prominent in cases like diabetic patients or animals in production that need little residual effect. Hence, there is a continuous quest for finding such formulations to achieve aforesaid targets, more importantly, in the era of antibiotic resistance. Recent enrichment in knowledge about nanomaterials and their properties has enhanced the hope for formulating dressing materials with all ideal properties required on cases to case basis. Various nanomaterials have shown positive attributes in healing of different wounds due to their specific inherited advantages. Present chapter has emphasized on applications of various nanomaterials as dressing materials, antimicrobial agents, anti-

N. Sindhu
Department of Veterinary Clinical Complex, Lala Lajpat Rai University of Veterinary and Animal Sciences, Hisar, Haryana, India

S. Kumar
Department of Veterinary Surgery and Radiology, Lala Lajpat Rai University of Veterinary and Animal Sciences, Hisar, Haryana, India

R. Kumar (✉)
Department of Veterinary Physiology and Biochemistry, Lala Lajpat Rai University of Veterinary and Animal Sciences, Hisar, Haryana, India

LLR University of Veterinary and Animal, ICAR-National Research Centre on Equines, *Nanotechnology Theranostics in Livestock Diseases and Management*, Livestock Diseases and Management,
https://doi.org/10.1007/978-981-16-1610-5_25

inflammatory action, delivery system, angiogenic materials, etc., for effective and faster wound healing, along with their advantages at current perspective.

Keywords Wound dressing · Antimicrobial agents · Angiogenic potential · Anti-inflammatory action · Chronic wound · si-RNA · Stem cell delivery · Herbal nanomaterials

25.1 Introduction

The animals of veterinary importance may be broadly categorized into three domain applications which includes animal for economic gain, animals in wildlife, and animals as pets. The animals maintained for economic gains are basically for ensuring food security along with harvesting of other valuable resources and serve as a crucial pillar in the economy of various countries. Animals in wildlife have a significant influence in term of providing diversity to the ecosystem. Delineating the third domain applications, the value of pets, it is important to note that, in the era of global connectivity, wherein our lives are hyperlinked to the virtual world, we have now started to realize the poignant emptiness it has brought upon in our existence. Today, we human beings, as self-aware sentient, are frenetically seeking relief from this state of mind all over the globe and towards this end—nothing has brought solace to fill up this void other than the companionship and confidante that we find in our pets, especially dogs. Their working roles might have reduced over the last few decades, but their importance has increased enormously from pets to sporting companions and now they are becoming part and parcel of the civilized establishments they share their lives with. Hence, wellbeing of all these three classes of animals is of paramount importance to the veterinarians.

Veterinary subjects including the wild animals are more prone to wound because of several factors including but not limited to in-fighting, biting, improper management such as slippery ground, barbed wired fencing, environmental factors like high humidity will facilitate maggot wound, skin burn due to sun, etc., post-disease effects like ulcers in Foot-and-Mouth Disease, fractures, behavior-based fighting for space and partners are few examples which create both open and closed wounds in animals. The wounds may be acute or chronic in nature, can be personalized due to metabolic disorders like diabetes or heard-related that often give tough times to the clinicians both in term of animal welfare and economical losses. The reason for wound formation may be verse but broadly the successful management of wound have generalized steps in term of efficient closing of wound followed by managing the wound to reduce the chance of infection, maggot infestation and bleeding, providing the environment to stimulate healing process by enhancing growth factors, extracellular matrix features, angiogenesis, and proper tissue regeneration at the site in shortest possible time with restoration of proper physiology and reducing the scar formation. All these processes need to be completed as soon as possible without much reduction in animal productivity which is the major demand of the animal owners to the veterinarian. Multiple set of challenges have to be overcome to meet the demands such as type of dressing materials best suited for wound closing, type of antibiotics and anti-maggot agents required to prevent the infection for faster healing, considering the residual effects of such agents vis-à-vis ensuring the efficacy in the era of antibiotic resistance, supplying additional factors like growth factors, stem cells, antibiotic carriers for prolonged supply (very important in wild animals), angiogenic factors, etc. during the treatment of wounds. The type and size of the wound, its place, and species involved are certain important factors need to be considered for choosing these treatment materials or strategies for successful wound healing at the earliest.

Nanotechnology has provided wonderful solutions to these multi-faceted challenges as the engineered nanomaterials possess certain advantages in terms of their unique physicochemical attributes unlike to the conventional materials. For instance, smaller size of the NPs facilitating penetrability and targeted drug delivery, higher surface-to-volume ratio enhancing the scope for functionalization and molecular interaction, inherent antimicrobial action of the NPs, significant drug loading capacity and stimuli-responsive release system to promote sustained drug release as well as enhanced circulatory time, etc. are certain prominent features of the NMs advocating their application in wound management (Table 25.1). Nano-intervention in wound management is predominantly oriented towards either facilitating wound closure by the intrinsic therapeutic properties of the NMs or acting as delivery vehicle of therapeutic agents that promote the wound healing. Several organic and inorganic NMs, nanocomposites, hydrogels, nanofibers, films and coatings, or scaffolds have been reported to be employed for wound healing by either of the two aforesaid means (Hamdan et al. 2017; Mihai et al. 2019; Wang et al. 2021).

25.2 Wound Healing Process

During wound healing, loss of tissue needed to be repaired in order to maintain anatomical and physiological integrity. The mechanism of wound healing is a complex molecular phenomenon which includes phase of maintaining hemostasis, inflammation, proliferation, and remodeling but in overlapping manner (Hamdan et al. 2017). The first action in wound healing process is stoppage of bleeding, removal of necrosed tissues and exudates in order to maintain hemostasis. So, all the elements of clotting mechanism are recruited for effective clot formation. Subsequently, inflammatory process is activated to control bleeding, prevent infection, and to remove damaged cells, pathogens, and bacteria, etc. It is also supported by the supply of blood cells including macrophages, release of inflammatory mediators, activation of complement, growth factors, nutrients like excess lactate, and enzymes in order to hasten the process of healing. Filling the wound, contraction of margins and epithelialization of the wound with proliferation of vascular, epithelial and connective tissue take place for angiogenesis, stimulation of epithelialization and fibrosis in proliferation phase. These require a number of factors to accomplish the processes at the earliest. Any defect in these processes or enhancement in the rate of

Table 25.1 Application of nanomaterials in wound healing

Nanomaterial	Model	Application	Remarks	References
Ag/AgCl/PDDA-functionalized rGO nanosheets and solution	Burn wound in mice	Dressing material	Antibacterial, better adsorption	Zhou et al. (2016)
Chitosan-Ag/ZnO	Mice model	Dressing material	Antibacterial, better porosity, adsorption	Lu et al. (2017)
AgNPs	Excision wound	Ointment	Antibacterial, better wound healing post day 14	Pourali and Yahyaei (2016)
Hydrogel composed of natural polymers sodium alginate and gelatin along with AgNPs	Biopsy punch wound in Wister rats	Hydrogel	Bactericidal activity against *P. aeruginosa*, and *S. aureus*; non-cytotoxic to the fibroblast cells; reduce the wound size	Diniz et al. (2020)
Cotton fabrics coated with green synthesis of AgNPs from *Curcuma longa* L.	Scratch assay on fibroblast (L929) cells	Prevention of skin infection and wound healing	Antimicrobial activity against the *S. aureus*, *P. aeruginosa*, *S. pyogenes*, and *C. albicans*; enhanced cell proliferation and migration to promote wound healing	Maghimaa and Alharbi (2020)
AgNPs	Scratch assay on human dermal fibroblasts (NHDFs) and normal human epidermal keratinocytes (NHEKs)	More effective in epidermal keratinocytes	Decreased inflammatory cytokines TNF-α and IL-12, growth factors (VEGF), and expression of COX-2	Franková et al. (2016)
XKP nanocomposite hydrogel	Skin wound in rat-tail amputation model	Dressing material, antibacterial	Adaptable to wound, kill bacteria by heat, Cytocompatibility	Zeng et al. (2021)
AuNPs with light excitation	Rat model	Photobiomodulation therapy	Re-epithelialization, collagen deposition and triggered inflammatory response	Lau et al. (2017)
AuNPs functionalized with keratinocyte growth factor (KGF)	HDFs and HEKs; full-thickness dorsal excisional wound in rats	As a promising wound healing drug	Enhanced wound healing by promoting the proliferation of keratinocytes and wound re-epithelialization rather than granulation	Pan et al. (2018)

Gallic acid-isoflavone-covered AuNPs and protocatechuic acid-isoflavone-covered AuNPs	Surgical and burned wound skin in Sprague-Dawley (SD) rats	Transdermal treatment	Suppressed metalloproteinase-1 (MMP-1), higher superoxide dismutase (SOD) activity and promoted VEGF, angiopoietin-2 and collagen, enhanced skin regeneration	Lee et al. (2015)
Retinoic acid-loaded solid lipid nanoparticles surrounded by chitosan film	Diabetic mice	Increase the collagen deposition	High drug entrapment efficiency, accelerated the closure of wounds of diabetic mice	Arantes et al. (2020)
Fe_3O_4 doped Polyhydroxybutyrate-co-hydroxyvalerate copolymer functionalized graphite oxide nanocomposite	Scratch assay on mouse fibroblast cells	3D-scaffold for tissue engineering	Antibacterial action against gram-positive and negative bacteria such as *E. coli*, *P. aeruginosa*, *S. aureus*, and *B. subtilis*; facilitate cell adhesion, proliferation and accelerated wound contraction	Pramanik et al. (2019)
Interleukin (IL)-4 loaded TiO_2 nanotubes	RAW 264.7 murine macrophage cells	Biomedical implants having controlled release system of the loaded cargo	Facilitated controlled release of IL-4 and polarization of macrophages from the M1 to M2 phenotype	Li et al. (2018)
Nanobioglass incorporated into chitosan hydrogel	Liver injury with biopsy punch; femoral artery injury in rats	As injectable	Rapid blood clotting for effective bleeding control	Sundaram et al. (2019)
Curcumin encapsulated silica nanoparticles	Scratch assay on human dermal fibroblast (HDF) cells	Antimicrobial photodynamic therapy	Reduction in the number of bacteria and bacterial biofilm production, accelerated wound healing	Mirzahosseinipour et al. (2020)
ZnO NPs	Full-thickness excised skin wound in rabbit	Topical application	Depicted biocompatibility, anti-inflammatory action, facilitated cell adhesion and proliferation without any scar formation	Manuja et al. (2020)
Cuprous oxide nanoparticles	Rabbit ear scar model after full thickness excisional wound	Intralesional injection	Improve the scar appearance and collagen arrangement, and	Wang et al. (2017)

(continued)

Table 25.1 (continued)

Nanomaterial	Model	Application	Remarks	References
			decreased scar elevation index by inhibiting hypertrophic scar fibroblasts proliferation and inducing apoptosis	
Chitosan-PVA (polyvinyl alcohol) trilayer nanofibrous membrane	Rat full-thickness skin defects model and mice diabetic chronic wound model	As functional wound dressings	Depicted excellent biocompatibility, antibacterial activity, promotion of re-epithelialization, improved collagen alignment and formation of skin appendages; upregulated growth factors including VEGF and TGF-β, downregulated inflammatory cytokines such as TNF-α and IL-1β	Chen et al. (2019)
Single-wall or multi-wall carbon nanotubes complexed with chitosan hydrogel	Mouse model with full-thickness wounds	Topical application	Improved the re-epithelialization of wounds, but increased fibrosis	Kittana et al. (2018)
CNTs/PVA/EGF composite	In vitro in L929 fibroblasts and in vivo in rat skin wound model	Biological dressing	Facilitated sustained release of EGF and hastened wound healing	Liao et al. (2017)
Nanodiamond-based hydrogel	Human umbilical vein endothelial cells (HUVECs)	Injectable	Facilitated sustained release of exogenous VEGF for accelerated wound healing	Pacelli et al. (2017)
Nanofiber scaffolds (NFSs) functionalized with bone-marrow-derived mesenchymal stem cells (BM-MSCs)	Acute full-thickness skin wounds in rat model	Implant	Promoted epithelial edge in growth and collagen synthesis, epidermal differentiation at early and intermediate stages, facilitated complete wound closure	Ma et al. (2011)

Aloe vera-polycaprolactone nanoscaffold containing human umbilical cord Wharton's jelly stem cells (hWJSCs)	Excisional and diabetic wounds in murine model	Wound dressing	Rapid wound closure, reepithelialization, and increased numbers of sebaceous glands and hair follicles, antibacterial action of Aloe vera, nanoscaffolds along with hWJSCs yielded synergistic benefits	Tam et al. (2014)

these processes warrant medical support. The healing process gets accomplished with restoration of all mechanical strength, physiological, and anatomical structures during the final phase of maturation (Martin and Nunan 2015). Delay in the healing process may either result in scar formation or healing will become chronic leading to morbidity or mortality of the animal. Various factors critically influence the wound healing that include but not limited to type and size of the wound, age and physiological status of the animal, availability of systemic mediators of wound healing and environmental factors including infection, etc. Animal suffering from compromised disease like diabetes mellitus exhibits slower rate of wound healing; similarly, animals in high temperature and humid condition with badly managed environment are more susceptible to infection or maggot in wound as complication. Systemic mediators including immune cells, availability of growth factors, etc. also influence wound healing. Defective stem cells, higher oxidant production, lower growth factors and extra-cellular matrix (ECM) production or their higher degradation, impaired microvasculature and cell function, infection and maggots formation, enhanced inflammatory cytokines, or clotting disorder are the major challenges which can impair the wound healing process (Nunan et al. 2014). Evidently, providing additional assistance to the animals that are compromised in these factors can improve the efficiency of wound healing; the assistance may include special type dressing materials, supply of growth factors and drugs through targeted delivery systems, providing scaffolds for enhanced molecular interaction, etc. However, there are several other additional factors that may affect the wound healing process such as contamination, hematoma, foreign body, immunosuppression, necrotic tissue, malignant cells, poor perfusion, hypoxia, anemia, vitamin and mineral deficiency, hypothermia, hypoproteinemia, post-operative hypermetabolism, obesity, diabetes mellitus, cardiovascular disease, pulmonary disease, upper respiratory tract obstruction, uremia, pain/stress, dehydration, hepatic disease, paraneoplastic syndromes, viral infection (FIV/FeLV), adjacent skin tension, chemotherapeutic agents, immunosuppressive drugs NSAIDs (early in healing), exogenous steroids, and radiation therapy. In terms of providing additional assistance to the animals, application of nano-formulations can help in various ways to hasten the wound recovery and reduce the complications. This can be achieved by targeted drug, growth factors or stem cell delivery, silencing of gene through targeted nucleic acid delivery, gene therapy, providing various other factors that may be deficient to accelerate the wound healing process, act as an antimicrobial to limit the infection, reducing the dose requirement, enhancing the activity of conventional drugs by controlled release system, providing microenvironment for better healing in the form of bandage materials, etc. Evidently, the most common utility of nanomaterials in wound healing has been experienced in the following ways: (a) As/In wound bandage (b) Antimicrobials either as topical colloidal agents in the form of antiseptics and antibacterials or as internal drugs to prevent infections (c) As other matrix agents like growth promoters, gene therapeutics, etc., (d) As carrier system for efficient and targeted delivery of drugs or other biologicals required for healing.

25.3 Nanomaterials in Wound Dressing

The wound whatever the type is first dressed in order to facilitate prompt healing, remove unwanted products like exudates and cell debris, reduce the chance of infection and maggot formation, and provide the scaffold and environment for faster regeneration. An ideal material for faster wound healing must have properties such as it should not allow growth of bacteria, reduces the pain, have infection prevention capacity, should be biocompatible to tissue and biodegradable, provide essential environment for growth, facilitate proper gas exchange, reduce exudates and better moisture restoration capacity without attaching to the tissue, should also have enough mechanical strength to be durable at animal wound site, without any toxicity issue and must be economical in application. Finding such ideal candidate with additional property of less frequent requirement to change and long-term drug release system are additional criteria required for animals in wild to enhance their recovery. Various dressing materials in the form of hydrogel, film, foam, or as nanofiber have been employed depending upon the nature and size of wounds. Commonly used polymeric nanoparticle-based such dressing materials include synthetic polymers such as PLGA, PLA, PLC, PVA, etc. as well as natural polymers such as cellulose, chitosan, hyaluronic acid, etc. (Table 25.1). For instance, chitosan (CS) is having antimicrobial activity, better drainage, ability to form gel, good swelling behavior and healing stimulation, therefore, it is one of the most extensively studied polymer-based nanomaterial for wound healing but has poor mechanical properties and unsatisfactory anti-bacterial performance (Okur et al. 2020). Purity, high porosity, biocompatibility, and availability as the most abundant carbohydrate on the earth are some of the advantages of cellulose as dressing material apart from better drug loading capacity for local delivery; whereas enhanced cellular migration through stimulate and recruitment of cells like macrophages and fibroblasts and component to new tissue development are the prominent advantageous features of collagen-based dressing materials. Maintaining the tissue integrity, facilitating adhesion and differentiation of cells during wound repair and inflammation, being a natural component of soft connective tissue and hygroscopicity are the prominent expedient features of hyaluronic acid (HA). Alginate, due to high water absorptivity, keeps the wound moist for longer period. Nanomaterial-based dressing materials provide additional benefits in terms of increased mechanical strength, increased adhesion, antibacterial effect, sustained drug release, and as nanoenzymes. For example, metals such as silver, gold, zinc, etc. when incorporated in dressing materials in their nano-forms have extended several beneficial effects towards stimulation of wound healing and antibacterial activity. Nanoparticles of gold, chitosan, alginate, cellulose, hyaluronic acid, and zinc oxide (ZnO) have been reported to have antibacterial effect; in addition, they also enhanced keratinocyte growth and differentiation to improve the re-epithelialization at the wound site (Barroso et al. 2020). Wound healing material using silver/silver chloride NPs anchored on reduced graphene oxide (Ag/AgCl/rGO) showed bactericidal activity and stability due to higher production of oxidative radicals against *Escherichia coli* and *Staphylococcus aureus*. The property of the material was further magnified by using (diallyldimethyl ammonium chloride) (PDDA) that led higher adsorptive capacity and regeneration of epidermis layer of burn wound (Zhou et al. 2016). The enhanced wound healing and antibacterial activity of NP-based dressing materials has further been documented by the application of chitosan-Ag/ZnO composite dressing in spongy microporous form with enhanced blood clotting, porosity and swelling as additional advantages (Lu et al. 2017). The antibacterial activity was found against *E. coli*, *S. aureus,* and *P. aeruginosa*, and the clotting time was also shortened due to efficient activation of coagulation cascade. Another challenge in wound dressing material is its inability to adapt as per the shape of the wound which also hampers the healing process. Application of hydrogel using polydopamine nanoparticles (PDA NPs) in matrix of FDA-approved food thickening agent xanthan gum and konjac glucomannan has demonstrated rapid shape adaptability. The material is very helpful in bacteria-infected wound healing as it also acts as photothermal transduction agent. This material helps in conversion of near-infrared laser radiation to heat, hence acting as bactericidal agent and applicable to skin wound. In addition, the material also helps to reduce the inflammatory response and promote vascular reconstruction (Zeng et al. 2021). Recent application of mechanoactive materials for tissue regeneration has shown various advantages in terms of thermoresponsiveness, tough adhesiveness, high stretchability, toughness, tissue adhesion, and antimicrobial function in wound healing. They have shown the property of good adhesion to the skin and with respect to change in temperature of the skin, the materials actively contract the wounds (Blacklow et al. 2019). These active adhesive dressing materials mimic the contraction ability of the embryonic wounds, comprised of certain materials like poly(N-isopropyl acrylamide) (PNIPAm) for thermoresponsiveness, chitosan, and carbodiimide for tissue adhesion and strength, AgNPs, and alginate hydrogels (e.g., ALGICELL) as antibacterial and tissue dressing material. This is probably the first step to develop smart wound healing material which responds according to the tissue status and even it may release materials required in the future steps of healing process according to the need.

25.4 Nanomaterials as Antimicrobial Agents in Wound Healing

Antibacterial activity of AgNPs against *S. aureus* and methicillin-resistant *S. aureus* has been reported and when doped with enoxaparin conjugate in gel natrosol (hydroxyethylcellulose) ointments, showed excellent wound healing up to 100% post 28 days compared to 55% in control and 89% in enoxaparin gel in Wistar rats (Barabadi et al. 2022). Similarly, in excision wound, the healing was higher using AgNPs against control neomycin (Sundaramoorthi et al. 2009). Biosynthesized CuNPs has also shown antimicrobial activity against *E. coli* (NCIM 5011), *Bacillus subtilis* (NCIM 2063), *P. aeruginosa* (NCIM 2036), and *S. aureus*, and anti-inflammatory activities with respect to native copper but predominantly against gram positive bacteria. Evidently, the gel-containing CuNPs has found to enhance wound healing activity (Tiwari et al. 2014). Various metal oxide NPs have also depicted strong antibacterial properties. For instance, CdONPs inflicted antibacterial action through ROS production, affecting active transport, membrane permeability, and enzyme inactivation after binding on outer membrane of bacteria. Similarly, ZnONPs has done the same by generating more ROS due to Zn ion release. TiO_2NPs, iron oxide NPs, SeNPs, and CoNPs have also shown antibacterial activity and can be used as topical gel in solo or in combination for better healing.

Various NP-based carriers have been used for targeted delivery of conventional drugs at the wound site for efficient healing (Table 25.1). For example, liposomes (cationic), a lipid bilayer nano-vehicle with low immunogenicity, acts as carrier for vancomycin to be more effective against methicillin-resistant *S. aureus* or *S. epidermidis*. Other carriers like lipid NPs are propitious vehicles for various agents required for wound healing such as drugs, growth factors, and small interfering RNA (siRNA). An endogenous host defense peptide LL37 along with Serpin A1 (A1), an elastase inhibitor having wound healing property, was co-delivered by solid-lipid nanoparticles (SLNs) that facilitated prolonged release of the bioactive agents at a precise ratio to elicit synergistic antibacterial action against *S. aureus* and *E. coli*, for accelerated wound healing (Fumakia and Ho 2016). Various drugs that have been used in wound healing like Madecassoside, Usnic acid, Baicalin, Norfloxacin, and Amphotericin B depicted better efficacy when delivered with nanocarriers such as liposomes, deformable liposomes, SLNs, hydrogel, nanohydrogel, etc. (Wang et al. 2019).

25.5 Nanomaterials for Anti-inflammatory Action in Wound Healing

Besides the antibacterial action, CuNPs has also been reported to have anti-inflammatory activity leading to better wound healing capacity (Tiwari et al. 2014). In excisional wounds, phenytoin-loaded copper nanoparticles (PHT-loaded CuNPs) have reduced the expression of inflammatory *JAK3* to decrease the inflammation along with facilitation of epidermal regeneration, granulation, and tissue formation leading to an effective and rapid healing (Saddik et al. 2020). Similarly, topical application of AgNPs in Wistar albino rats showed anti-inflammatory activities, inhibition of edema, increased collagen deposition, lesser macrophages, and more fibroblasts to promote better wound healing (Selvam and Joicesky 2021). Cotton fabrics containing AgNPs have also been reported to show anti-bacterial and anti-inflammatory action in rat using paw oedema method. The dressing material showed a dose-dependent reduction in inflammation and was comparable to the standard anti-inflammatory drug indomethacin (Hebeish et al. 2014). Topically applied AgNPs has shown positive attributes in wound healing such as anti-microbial action, controlling inflammation, modulating the expression of fibrogenic cytokines, etc. that resulted decreased neutrophil counts, lower IL-6, and increased level of IL-10, TGF-β, vascular endothelial growth factor (VEGF), interferon (IFN)-γ facilitating faster healing in mice with burn wounds (Tian et al. 2006). Similarly, application of Mupirocin nanoparticle-loaded hydrogel in rat burn wound showed lower lipopolysaccharide (LPS) activity in macrophage-like cells, reduced nitric oxide and inflammatory cytokine IL-1β production indicating anti-inflammatory action in wound healing apart from other attributes like enhanced granulation tissue proliferation, angiogenesis, generation of hair follicle cells, reduced secretion, and lesser necrosis (Kamlungmak et al. 2021). Better anti-inflammatory action was also observed in case of topical application of zinc oxide and estrogen nanoparticles in oophorectomized rats indicating the NP-based formulation have a better capacity to modulate the inflammation phase for efficient wound healing (Koga et al. 2021). Similarly, application of α-gal (Galα1-3Galβ1- (3)4GlcNAc-R) nanoparticle helps in overcoming aberrant inflammatory responses in wound created due to irradiation in tissue and lead to faster healing. It is also effective in both acute as well as chronic wound (Samadi et al. 2021).

25.6 Nanomaterials in Angiogenesis for Wound Healing

The breakdown of tissue integrity may lead to loss of blood vessels in the wounded area. These vessels are necessary to maintain the homeostasis as they not only provide oxygen and nutrients along with removal of waste materials, but are also responsible for supply of various growth factors, matrix components, bioactive molecules, and cells required for faster wound healing, that too in a proper way. So, angiogenesis is a key process to achieve success in wound healing. Adequate supplies of the materials which hasten the process are necessary for proper wound healing, more importantly for chronic wound. Application of copper NPs has found to increase the production of interleukin-2 leading to faster repair of the damaged tissue in excision wound model. In addition to facilitating angiogenesis, copper NPs are also having anti-bacterial and anti-inflammatory properties that also assist in rapid wound healing (Tiwari et al. 2014). Topical application of copper has found to regulate various pathways associated with the healing process such as remodeling of extra-cellular matrix, angiogenesis, modulating the expression of vascular endothelial growth factor (VEGF) resulting in rapid contraction and closure of dermal wound (Sen et al. 2002). Various nano-formulations are helpful in mimicking the natural extracellular matrix (ECM) of blood vessels and thus provide environment for faster healing through enhanced adhesion, proliferation, and migration of endothelial cells and vascular endothelialization. Nanofibrin, micro/nano-structured surfaces of Cu_x-HA, nano-sized bioactive glass, Poly-l-Lactide (PLLA) electrospun fibrous membranes, reduced graphene oxide (rGO), etc. have also been reported to have angiogenic effects in various models (Liu et al. 2020). Nanomaterials are not only involved in angiogenic action but also reported to be helpful in nerve damage repair at wound sites. Application of poly (lactic-co-glycolic acid) NPs encapsulating $VEGF_{165}$ promoted both angiogenesis and myelination in a rat spinal cord hemisection model. Similar dual action has also been reported for reduced graphene oxide (GO). Pro-angiogenic activity of AuNPs has found to increase the expression of VEGF, Angiopoietin-1 (ANG-1), and Angiopoietin-2 (ANG-2) that promote the wound healing process. The CuSNPs of 200-600 nm size has facilitated the generation of vascular networks and CD31 positive vessels by controlled and sustained release of copper ions from the NPs. VEGF and CD31 are reliable markers of angiogenesis, and prominently expressed on the surface of endothelial cells. Similarly, hydroxyapatite (HA) NPs has found to induce the expression of different angiogenic markers such as VEGF, CD31, hypoxia-inducible factor (HIF)-1, Von Willebrand Factor (vWF), vascular endothelial growth factor receptor (VEGFR)-2, fibroblast growth factor (FGF), and ANG-1, along with facilitating the tube formation, and improve the viability, adhesion, and proliferation of endothelial cells. The cell culture tube formation assay is one of the most commonly employed in vitro assays for angiogenesis. Tricalcium phosphate (TCP) NPs of 50 nm size has found to enhance the proliferation of Human umbilical vein endothelial cells (HUVECs), and upregulate the expression of VEGF, VEGFR2, and HIF-1α. In addition, other inorganic NPs such as zinc oxide NPs, Terbium hydroxide and Europium hydroxide nano-rods or spheres, Neodymium NPs, TiO_2, and Cerium oxide NPs, etc. are also found to impart angiogenic action to promote the wound healing process. Even, the triangular tetrahedral DNA organic NPs generated by assembling four ss-DNAs fragments has found to facilitate angiogenesis by promoting proliferation, migration, and tube formation of endothelial cells accompanied by inducing the expression of VEGF-A and VEGFR2 (Liu et al. 2020).

25.7 Nanoformulation-Mediated siRNA and Stem Cell Delivery for Wound Management

In complex wound healing cases, the application of gene silencing or targeted stem cell delivery is warranted. Apart from delivery of drugs and growth factors in sustained and desirable way, NPs have been used for siRNA and stem cell delivery along with different therapeutics (Table 25.1). Application of bone marrow-derived mesenchymal stem cells (BMSCs) has shown better wound healing potential, and when doped in the silk nanofiber hydrogels, it resulted better maintenance of stemness character. This application in wound has led to better angiogenesis, inflammation, collagen deposition, and hair follicle regeneration in rat model (Li et al. 2021). Similarly, delivery of human induced pluripotent stem cells (iPSC)-derived endothelial cells with NANOG and LIN28 transcription factors has led to better reprogramming in cell epithelialization at the wounded area. Enhanced granulation tissue formation, angiogenesis, ECM deposition, and pro-regenerative response to accelerate wound healing have been observed when BMSCs are delivered with nanofibrous scaffold (Chen et al. 2020). Similarly, the delivery of human adipose stem cells (hASC) in PLGA scaffold has resulted better wound healing due to enhanced cellular activity, cell adhesion, proliferation, and survival (Tang et al. 2019). Apart from these, cell-based therapy and gene-based therapy has also been used as alternate methods in effective wound healing. Utilization of polyplex nanoparticles for carrying stromal-derived factor-1 alpha (SDF-1α) in encoded gene in collagen–chondroitin sulfate (coll–CS) scaffold has yielded better regenerative functionality of human diabetic adipose-derived stem cells (ADSC) (Aragona et al. 2017). Similarly, siRNA-based spherical nucleic acids with gold nanoparticle conjugates have found to reverse impaired wound healing in diabetic mice by knocking down ganglioside GM3 synthase (Randeria et al. 2015). Likewise, siRNA has been employed for macrophage reprogramming from the M1 to M2 phenotype through knockdown of interferon regulatory factor 5 (irf5) to achieve better wound healing (Sharifiaghdam et al. 2021). Selenium-based layer-by-layer (LBL) nano-complexes (NCs) carrier has been used for the delivery of siRNA at the desired site. In another model, siRNA-loaded lipid nanoparticles (LNPs) have been employed to knockout tumor necrosis factor α (TNFα) in order to accelerate wound healing in chronic wound (Kasiewicz and Whitehead 2018). All these aforesaid nanointerventions have prospered the way for management of more complex wounds that had been otherwise difficult to treat. Here nano-carriers have been found to be effective and even application of cell markers (epitope-based) with such carriers can further be used for the delivery of therapeutic agents to more prone site in internal wounds also.

25.8 Miscellaneous Other Applications of Nanomaterials in Wound Management

Diverse NPs have also depicted synergistic action for wound healing in combination with different conventional therapeutics or bioactive agents, such as acting as antibacterial agents, removing other infections, providing inputs for healing like growth factors, angiogenic activity and anti-inflammatory action (Table 25.1). For instance, proliferation and migration of keratinocytes and dermal fibroblasts were enhanced by AuNPs conjugated with α-lipoic acid and epigallocatechin gallate on topical applications. AuNPs has also found to exert stimulatory action for accelerated expression of the endothelial growth factor, angiopoietin 1, and angiopoietin 2 (Leu et al. 2012; Kim et al. 2015). Various forms and functionalization of AuNPs have resulted diverse actions in wound healing process, for example hydrogels of poly ethylene glycol (PEG)-gold nanorods (AuNRs) and cationic poly allyl amine hydrochloride (PAH)-AuNRs have found to hasten skin re-epithelization and collagen deposition, expression of various inflammatory and anti-inflammatory mediators in addition to antibacterial action against organisms like *S. aureus* and *P. aeruginosa*

(Mahmoud et al. 2019). It indicates both functionalization and shape in addition to concentration are some of the key factors influencing wound healing capacity of candidate NPs. Besides growth factors, cytokines and other immune-mediators have found to be effective agents controlling the healing process. Epigallocatechin-3-O-gallate (EGCG)-based NMs have found to regulate the secretion of cytokines and thus influence the wound healing process as evidenced by the increase in immunoreactivity of Ki-67 (re-epithelialization at the wound site) and CD31 (formation of blood vessels) at the area of wound in mice. The healing in full thickness wound has been achieved due to enhanced cell infiltration, re-epithelialization, and angiogenesis process in these animals (Kim et al. 2013). Another method of NP-mediated healing process is by manipulating the expression of various receptors in and around the wound site. For instance, application of NPs-based formulation containing Au, epigallocatechin gallate (EGCG), and α-lipoic acid (ALA) has led to downregulation of the receptor for advanced glycation end-products (RAGE) resulting in accelerated healing of cutaneous diabetic wound (Chen et al. 2012). Mesenchymal stem cell (MSC)-based therapies in addition to epigallocatechin-3-gallate (EGCG) have enhanced the epidermal thickness, reduced tumor necrosis factor-α (TNF-α), interleukin-1β (IL-1β) and IL-6 expression. Thus better healing was achieved due to angiogenesis, inflammation, cell migration, proliferation, and epidermal terminal differentiation (Li et al. 2016). Application of α-lipoic acid has found to be associated with better promotion of repair in intestinal epithelial injury by regulating MAPK signaling pathways (Yang et al. 2022). NPs can also play crucial role in providing matrix environment for better healing. For instance, peptide nanoparticles (PNPs) frameworks closely impersonate the natural ECM, hence may lead to upsurge their contact with other cells displaying a growing cell-responsive environment in addition to showing high biocompatibility and cytocompatibility (Kushwaha et al. 2022). Similarly, natural material, fibrin, when used in wound healing, reduces inflammation, increases immunological response, and improves cell adhesion. Application of siRNA targeted against Keap1 in cationic lipid nanoparticle (CLN) composed of 1,2-dioleoyl-3-trimethylammonium-propane (DOTAP) and the edge activator sodium cholate (NaChol), in a 6:1 ratio has restored Nrf2 antioxidant function that led to accelerated diabetic tissue regeneration, and augmented reduction-oxidation homeostasis in diabetic wounds. This is indicative of the clinical potential of the NM in skin-related wound healing (Rabbani et al. 2017). Delivery of epidermal growth factor (EGF), insulin-like growth factor-I (IGF-I), and platelet-derived growth factor-A (PDGF-A) for wound healing has been executed through a NM-based carrier by conjugating a low-molecular-weight protamine (LMWP) cationic elastic liposomes to the N-termini of these growth factors that resulted better growth of various cells and thus enhanced the wound healing process (Choi et al. 2017a,b). Various growth factors such as EGF, PDGF, CSF, TGF-β, basic fibroblast growth factor (bFGF), acidic fibroblast growth factor (aFGF), etc. have often been used topically to induce the process of healing in more effective ways. Various nanocarriers such as SLN, liposome, nanohydrogels, nanosccafold, nanofibers, niosomes, etc. have been employed to deliver the substance of interest in the area of wound in an effective way for healing process (Wang et al. 2019). Application of graft is another method in wound healing where immunosuppressive is added to avoid graft rejection. Application of cyclosporine has found to be assistive in wound healing without inhibiting the inflammatory reaction of granulation tissue formation or the regenerative process. Nano-based bioadhesive gels containing cyclosporine A in solid lipid nanoparticle have been used in the treatment of aphthous stomatitis in rabbits with enhanced rate of healing and mucosal repair with respect to control. Similarly, application of nitric oxide nanoparticles (NO-NP) has depicted to stimulate the migration of fibroblasts and collagen leading to better healing. Application of calcium-based NPs has also found to enhance the rate of proliferation of fibroblasts to promote the healing process (Silva et al. 2018).

25.9 Nanomaterials in the Management of Chronic Wound

Apart from wounds of acute nature, there are wounds of chronic nature also which can be associated with various metabolic malfunctioning like diabetes. NPs have shown promising result in management of such chronic wounds (Table 25.1). Application of chitosan-based hydrogel containing AgNPs and extracts of calendula as topical formulation has shown curative result in cutaneous wound in human subjects with complication of aging and diabetes. It has also depicted antibacterial effects with considerable reduction in pain and inflammation (Rodríguez-Acosta et al. 2022). Curcumin loaded chitosan NPs impregnated into collagen-alginate scaffolds has resulted efficient healing of diabetic wound where curcumin acted as anti-inflammatory and antioxidant agent, chitosan acted as carrier and helped in sustained drug release for wound healing, and collagen functioned as scaffold (Karri et al. 2016). The efficiency of the nano-formulation in treatment of adult male Wistar rats post-diabetic induction and wound creation was evident by complete epithelialization with thick granulation tissue formation, better water uptake, biocompatibility, sustained drug availability and significant wounds contraction. Similarly, application of hydrogel-loaded curcumin NPs (Cur-NP/HG) in diabetic wounds revealed complete re-epithelization, intact dermo-epidermal junction, reorganization of the dermis with significantly increased collagen deposition as evidenced in the histological and immuno-histochemical studies along with VEGF and AQP3 expression as signs of improved healing potential of Cur-NP/HG (Kamar et al. 2018). Various pathways have been identified to be altered in chronic wound; for example receptor for advanced glycation end-products (RAGE) expression may have critical role in the diabetic cutaneous wound healing. Application of AuNPs, epigallocatechingallate (EGCG), and α-lipoic acid (ALA) has led to downregulation of RAGE resulting in faster wound healing through angiogenesis regulation and anti-inflammatory effects (Chen et al. 2012). Chronic wound may need additional supply of various factors which can be delivered in an orderly manner using nanocarriers. For instance, liposome-mediated delivery of epidermal growth factor (EGF), insulin-like growth factor-I (IGF-I), and platelet-derived growth factor-A

(PDGF-A) has found to enhance the chronic wound healing process (Choi et al. 2017a,b).

25.10 Plant/Herbal Nanomaterials

Different stakeholders in biomedical science are in continuous search for improved wound care products, either natural or synthetic. Newer methods that not only improve patient compliance but also facilitate wound healing are being evaluated with a great effort. Natural medicinal products and plant extracts have been widely used as topical applications for wound healing (Hsu et al. 2003) (Table 25.1). It has been estimated by World Health Organization (WHO) that more than 80% of the world population, mainly in the developing countries, dependent on herbal products for their primary health care needs. Emergence of novel wound healing agents using medicinal plants with nanotechnology has the potential to develop efficacious wound healing therapeutics with enhanced wound healing potential. Several plant and herb extracts and their phyto-constituents are known as promising alternatives for conventional synthetic wound healing agents due to the presence of diverse active components, ease of access, and their limited side effects. Plant-based wound healing products not only accelerate healing but also maintain the aesthetics in a natural way (Chitra and Patil 2009). Developments in nanotechnology perspective can help to improve the efficacy of different herbal-based therapeutics. Due to their small size and high surface-to-volume ratio, NPs can pass through the skin barrier and intracellularly (Flora et al. 2013), thereby, presenting themselves as an ideal candidate for topical drug delivery.

Medicinal plants have been mingled with the field of nanotechnology to enhance the activity of the phyto-constituents (Table 25.1). Different metallic NPs including silver (Lakkim et al. 2020), gold (Ananth et al. 2018), titanium dioxide (Sivaranjani and Philominathan 2016) and selenium (Menon et al. 2019) have been successfully synthesized using plants. Moreover, nanostructure of plant products and their phyto-constituents can enhance their bioavailability, release in the form of sustained delivery at the wound site and uplift the permeability of these phyto-constituents to the underlying skin layers, which are all necessary for the efficient wound healing process. In drug delivery systems, control over the release of drug at the target site is critically important to limit drug toxicity. Nanoparticles, nanoemulsion, nanohydrogels, nanofilms, and nanoliposomes are frequently used in controlled drug delivery systems of these medicinal plant extracts. These NPs are diminishing the side effects and increasing the efficacy of different phyto-chemicals by providing a sustained and controlled drug delivery system, and they increase the dissolution rate of the drugs based on their surface characteristics (El-Refaie et al. 2015). In addition, slow and sustained release of encapsulated NPs are further an added advantage as it limits toxicity because the theoretical maximum amount of medicine is never in contact with wound site at one time. Leaf extracts of *Pelargonium graveolens*, *Azadirachta indica*, *Cymbopogon flexuosus*, *Tamarindus indica*, Aloe

Vera, *Coriandrum sativum*, *Cinnamomum camphora*, *Capsicum annuum*, *Gliricidia sepium*, *Pongamia pinnata*, *Datura metel*, latex of *Jatropha curcas*, *Zingiber officinale*, *Cinnamon zeylanicum* bark extract, Eclipta leaf, Cycas Leaf, *Hibiscus rosasinensis*, *Terminalia chebula*, *Camellia sinensis*, and Orange peel extract have been successfully used for an efficient and rapid extracellular synthesis of silver nanoparticles (Raut et al. 2014).

Bromelain, a proteinase found in the stem and fruit of pineapple plant, *Ananas comosus*, has been reported to possess several therapeutic properties. It has been reported to be helpful in healing minor injuries, especially strains and sprains, tendonitis, muscle injuries, bruising by reducing swelling, healing time, and pain (Aiyegbusi et al. 2011). Bromelain has been shown to exert beneficial effects including platelet aggregation inhibition, anti-edematous, anti-inflammatory, anti-thrombotic, modulation of cytokines and immunity, skin debridement and fibrinolytic activity, promote wound healing, and reduce post-surgical discomfort (Manzoor et al. 2016). Further, encapsulation of bromelain in NPs has been reported to show enhanced therapeutic potential. Bromelain nano-formulations have been earlier evaluated for preclinical studies on cancer, inflammation, and other conditions, which demonstrate enhanced therapeutic efficacy with nano- versus non-encapsulated bromelain (Bhatnagar et al. 2014; Bernela et al. 2016; Rachmawati et al. 2016; Xia et al. 2018; Sharma and Sharma 2018).

Neem (*Azadirachta indica*) is a very well-known medicinal plant, having a wide spectrum of therapeutic activity in ayurveda and homoeopathic medicine. Ingredients such as nimbidin, nimbin, and nimbidolinneem have anti-inflammatory, antibacterial, antifungal, and antiviral properties that help in the wound healing process. In addition, neem contains an excellent amount of vitamin, amino acid, and mineral which are very important for wound healing processes in proliferation phase (Subapriya and Nagini 2005). In diabetic rats, the silver nanoparticles of *Azadirachta indica* show increased wound contraction and accelerated wound healing compared to the control and the drug alone (Bhagavathy and Kancharla 2016). Synthesized silver nanoparticles of *Azadirachta indica* displayed enhanced antibacterial and antioxidant effects when compared to the leaf extracts alone (Chinnasamy et al. 2021). The wound beds where the NPs of *Azadirachta indica* were topically applied showed no microbial growth, hemorrhage, or formation of pus throughout treatment in female BALB/c mice when compared to control group (Lakkim et al. 2020).

Indian dietary spice, turmeric (the common name for *Curcuma longa*, *Zingiberaceae* family), has been widely used as home remedy from centuries to treat a variety of inflammatory conditions and other diseases. Curcumin a component of turmeric powder possesses anti-inflammatory, anti-proliferative, and antioxidant properties that are mediated through the regulation of several inflammatory cytokines, protein kinases, growth factors, transcription factors and other enzymes (Thangapazham et al. 2013). Curcumin a component of turmeric powder reduces not only the wound-healing time but also increases fibroblast number, improves collagen deposition and vascular density (Gopinath et al. 2004). Curcumin is well known potent antioxidant compound, and nano-sizing this herbal-based compound can be achieved without any significant alteration in its intrinsic antioxidant activity. Synthesized NPs of *Curcuma longa* increased fibroblast cell proliferation and migration indicating an effective wound healing (Maghimaa and Alharbi 2020).

Calendula officinalis L. (pot marigold) is one of the commonly used herbal plants in India, China, Europe, and the United States (Muley et al. 2009). Calendula flower have a high degree of activity against 18 different strains of anaerobic and facultative aerobic periodontal bacteria in vitro (Iauk et al. 2003), and against four different types of fungi, with the inhibitory effect comparable to antifungal agents Nystatin and Amphotericin B (Kasiram et al. 2000). *Calendula officinalis* possesses a number of properties that make this herbal plant ideal for wound management in animals (Kumar et al. 2019). *Calendula officinalis* with *Curcuma longa* in glycerine base is economical, easy to apply, and can be prepared at home or field level demanding more use of these herbal agents for large open wound management in ruminants (Kumar et al. 2020) and equine (Kumar et al. 2022). Silver nanoparticle of Calendula extract could be used as a good natural source of antioxidants due to the high antioxidant activity and can be used in the development of valuable products for wound healing and biomedical purposes (Fierascu et al. 2014).

Falcaria vulgaris, a medicinal plant has been used for its antioxidant, anti-inflammatory, antibacterial, antifungal, antiviral, and bleeding inhibitory activities (Jivad and Bahmani 2006; Jaberian et al. 2013). The silver nanoparticles of *Falcaria vulgaris* promoted the healing of cutaneous wound on male rats by decreasing the wound area and increasing wound contracture, antimicrobial and antioxidant activity when compared to the activity of *Falcaria vulgaris* extracts alone and tetracycline (Zangeneh 2019).

Moringa oliefera has exhibited anti-hyperthyroidism, antispasmodic, antitumor, antioxidant, hepato-protective, and antimicrobial activities (Dillard and German 2000). The synthesized NPs of *Moringa oliefera* promoted significant wound healing in Albino rats when compared to standard commercial wound therapeutic agents (Majoumouo et al. 2019).

Aloe vera (AV) is used in diverse biomedical applications as anticancer, antimicrobial, antiviral as well as wound healing agent due to its numerous phyto-constituents (Azevedo et al. 2019). The biological activities of aloe vera include analgesic, antibacterial, antioxidant, antifungal, antiviral, wound healing, anti-inflammatory, cleansing, antiseptic, etc. (Hashemi et al. 2015). *Aloe vera* with chitosan NPs thin membranes resulted in significant improvement of full thickness skin wound healing in rats (Ranjbar and Yousefi 2018).

Flowers of *Woodfordia fruticosa* are in high demand for preparation of herbal medicines and other parts of the plant also have wide variety of medicinal properties (Oudhia 2003). In tribal areas of Chhattisgarh state and some districts of Central India, fresh flowers of this plant are used for healing purpose of cuts and wounds, whereas powder of dried flowers help in rapid wound healing. Most part of the plant is important due to its usage in ayurvedic formulations (Chopra et al. 1956). Dried flowers of *W. fruticosa* are reported to treat wounds and peptic ulcers (Khorya and Katrak 1984). Topical application of *W. fruticosa* with synthesized silver NPs on excised wounds in Wistar albino rat model resulted in effective prevention of microbial adhesion and accelerated wound closure in comparison to the activity of 5% Povidone-iodine (Raghuwanshi et al. 2017).

Delonix elata copper oxide NPs applied on human wounds after anal rectum surgery caused rapid wound epithelialization (Chai et al. 2018). Plant-based NPs in wound healing showed a weaker inflammatory response may be resulting from cytokine modulation. An association between wound healing and the anti-inflammatory or antioxidant activities of therapeutics has been established. Anti-inflammatory agents are important in controlling inflammation while antioxidants address oxidative stress at the site of the wound thereby accelerates wound healing mechanism (Fitzmaurice et al. 2011). Both these properties have also been reported to decrease the exudation at the wound site. Numerous NPs synthesized using herbal plants have been shown to exhibit either antioxidant or anti-inflammatory activity by various mechanisms.

25.11 Conclusion

Nanotechnology has extended multi-dimensional utility in wound care and management that include but not limited to advanced dressing materials, antimicrobial agents, anti-inflammatory action, angiogenic facilitator, and delivery systems of diverse therapeutic agents such as drugs, growth factors, si-RNAs, and stem cells. Even the chronic wounds that are difficult to treat through conventional modalities can be taken care of efficiently by nanoformulations. Although toxicity concern always looms ahead when using nanomaterials, but several of them have been depicted to be biocompatible and non-toxic on different in vitro and in vivo models. Persistent growth in nanomedicinal market clearly suggests the effectiveness and popularity of nanoformulations in biomedical applications including wound care; however, the regulatory aspects should be followed stringently to avoid any nanotoxicological consequences. Currently, nanoformulations for wound management are mostly going through experiments on laboratory animals or directed towards human application, while the veterinary application is yet to get enough attention. However, animals are usually more prone to wound injury than human subjects, so, the benefits of nanointervention in wound care for the well-being of animals will certainly be exploited more frequently in upcoming days.

References

Aiyegbusi AI, Duru FIO, Anunobi CC, Noronha CC, Okanlawon AO (2011) Bromelain in the early phase of healing in acute crush Achilles tendon injury. Phytother Res 25(1):49–52

Ananth S, Induja M, Thangamathi P, Prabha D, Vinotha K (2018) In vitro antibacterial activity of biogenic gold nanoparticles from Murraya koenigii seed extract against pathogens associated with traumatic wound infections. Int J Fauna Biol Stud 5:137–144

Aragona M, Dekoninck S, Rulands S, Lenglez S, Mascre G, Simons BD, Blanpain C (2017) Defining stem cell dynamics and migration during wound healing in mouse skin epidermis. Nat Commun 8:14684. https://doi.org/10.1038/ncomms14684

Arantes VT, Faraco AAG, Ferreira FB, Oliveira CA, Martins-Santos E, Cassini-Vieira P, Barcelos LS, Ferreira LAM, Goulart GAC (2020) Retinoic acid-loaded solid lipid nanoparticles surrounded by chitosan film support diabetic wound healing in in vivo study. Colloids Surf B Biointerfaces 188:110749. https://doi.org/10.1016/j.colsurfb.2019.110749

Azevedo J, Juliao E, Dantas J, Reis J (2019) Is Aloe vera effective for wound healing? The state of the art abstract. J Oral Diagn 04:e20190005

Barabadi H, Mostafavi E, Truong LB, Cruz DM, Vahidi H, Mahjoub MA, Hosseini O, Saravanan M (2022) Microbial nanotechnology–based approaches for wound healing and infection control. Handbook of microbial nanotechnology. Academic Press, pp 1–15. https://doi.org/10.1016/B978-0-12-823426-6.00009-7

Barroso A, Mestre H, Ascenso A, Simões S, Reis C (2020) Nanomaterials in wound healing: from material sciences to wound healing applications. Nano Select 1:443. https://doi.org/10.1002/nano.202000055

Bernela M, Ahuja M, Thakur R (2016) Enhancement of anti-inflammatory activity of bromelain by its encapsulation in katira gum nanoparticles. Carbohydr Polym 143:18–24

Bhagavathy S, Kancharla S (2016) Wound healing and angiogenesis of silver nanoparticles from Azadirachta indica in diabetes induced mice. Int J Herb Med 4:24–29

Bhatnagar P, Patnaik S, Srivastava AK, Mudiam MK, Shukla Y, Panda AK, Gupta KC (2014) Anti-cancer activity of bromelain nanoparticles by oral administration. J Biomed Nanotechnol 10(12):3558–3575

Blacklow SO, Li J, Freedman BR, Zeidi M, Chen C, Mooney DJ (2019) Bioinspired mechanically active adhesive dressings to accelerate wound closure. Sci Adv 5(7):eaaw3963. https://doi.org/10.1126/sciadv.aaw3963

Chai S-H, Wang Y, Qiao Y, Wang P, Li Q, Xia C, Ju M (2018) Bio fabrication of silver nanoparticles as an effective wound healing agent in the wound care after anorectal surgery. J Photochem Photobiol B Biol 178:457–462

Chen S-A, Chen H-M, Yao Y-D, Hung C-F, Tu C-S, Liang Y-J (2012) Topical treatment with anti-oxidants and au nanoparticles promote healing of diabetic wound through receptor for advance glycation end-products. Eur J Pharm Sci 47(5):875–883. https://doi.org/10.1016/j.ejps.2012.08.018

Chen Q, Wu J, Liu Y, Li Y, Zhang C, Qi W, Yeung KWK, Wong TM, Zhao X, Pan H (2019) Electrospun chitosan/PVA/bioglass Nanofibrous membrane with spatially designed structure for accelerating chronic wound healing. Mater Sci Eng C Mater Biol Appl 105:110083. https://doi.org/10.1016/j.msec.2019.110083

Chen S, Wang H, Su Y, John JV, McCarthy A, Wong SL, Xie J (2020) Mesenchymal stem cell-laden, personalized 3D scaffolds with controlled structure and fiber alignment promote diabetic wound healing. Acta Biomater 108:153–167. https://doi.org/10.1016/j.actbio.2020.03.035

Chinnasamy G, Chandrasekharan S, Koh TW, Bhatnagar S (2021) Synthesis, characterization, antibacterial and wound healing efficacy of silver nanoparticles from Azadirachta indica. Front Microbiol 12:204

Chitra MB, Patil R (2009) Preliminary phytochemical investigation and wound healing activity of allium. Cepalin (Liliaceae) Int J Pharm Pharm Sci 2:7–9

Choi JU, Lee SW, Pangeni R, Byun Y, Yoon I-S, Park JW (2017a) Preparation and in vivo evaluation of cationic elastic liposomes comprising highly skin-permeable growth factors combined with hyaluronic acid for enhanced diabetic wound-healing therapy. Acta Biomater 57:197–215. https://doi.org/10.1016/j.actbio.2017.04.03

Choi JU, Lee SW, Pangeni R, Byun Y, Yoon I-S, Park JW (2017b) Preparation and in vivo evaluation of cationic elastic liposomes comprising highly skin-permeable growth factors combined with hyaluronic acid for enhanced diabetic wound-healing therapy. Acta Biomater 57:197–215. https://doi.org/10.1016/j.actbio.2017.04.034

Chopra RN, Nayar SL, Chopra IC (1956) Glossary of Indian medicinal plants. Delhi, CSIR
Dillard CJ, German JB (2000) Phytochemicals: nutraceuticals and human health. J Sci Food Agric 80:1744–1756
Diniz FR, Maia RCAP, Rannier L, Andrade LN, Chaud MV, da Silva CF, Corrêa CB, de Albuquerque Junior RLC, da Costa LP, Shin SR, Hassan S, Sanchez-Lopez E, Souto EB, Severino P (2020) Silver nanoparticles-composing alginate/gelatine hydrogel improves wound healing in vivo. Nanomaterials (Basel) 10(2):390. https://doi.org/10.3390/nano10020390
El-Refaie WM, Elnaggar YSR, El-Massik MA, Abdallah OY (2015) Novel curcumin-loaded gel-core hyaluosomes with promising burn-wound healing potential: development, in-vitro appraisal and in-vivo studies. Int J Pharm 486(1–2):88–98
Fierascu I, Bunghez I-R, Raduclaudiu F, Ion R-M, Cristina ED-P, Nuţă D (2014) Characterization and antioxidant activity of phytosynthesised silver nanoparticles using calendula officinalis extract. Farmacia 62(1):129–136
Fitzmaurice S, Sivamani R, Isseroff R (2011) Antioxidant therapies for wound healing: A clinical guide to currently commercially available products. Skin Pharmacol Physiol 24:113–126
Flora G, Gupta D, Tiwari A (2013) Nanocurcumin: a promising therapeutic advancement over native curcumin. Crit Rev Ther Drug Carrier Syst 30(4):331
Franková J, Pivodová V, Vágnerová H, Juráňová J, Ulrichová J (2016) Effects of silver nanoparticles on primary cell cultures of fibroblasts and keratinocytes in a wound-healing model. J Appl Biomater Funct Mater 14(2):e137–e142. https://doi.org/10.5301/jabfm.5000268
Fumakia M, Ho EA (2016) Nanoparticles encapsulated with LL37 and serpin A1 promotes wound healing and synergistically enhances antibacterial activity. Mol Pharm 13(7):2318–2331. https://doi.org/10.1021/acs.molpharmaceut.6b
Gopinath D, Ahmed MR, Gomathi K, Chitra K, Sehgal PK, Jayakumar R (2004) Dermal wound healing processes with curcumin incorporated collagen films. Biomaterials 25:1911
Hamdan S, Pastar I, Drakulich S, Dikici E, Tomic-Canic M, Deo S, Daunert S (2017) Nanotechnology-driven therapeutic interventions in wound healing: potential uses and applications. ACS Cent Sci 3(3):163–175. https://doi.org/10.1021/acscentsci.6b00371
Hashemi SA, Madani SA, Abediankenari S (2015) The review on properties of Aloe vera in healing of cutaneous wounds. Rev Prop Aloe Vera Heal Cutan Wounds 2015:714216. https://doi.org/10.1155/2015/714216
Hebeish A, El-Rafie M, El-Sheikh M, Seleem AA, El-Naggar ME (2014) Antimicrobial wound dressing and anti-inflammatory efficacy of silver nanoparticles. Int J Biol Macromol 65:509–515. https://doi.org/10.1016/j.ijbiomac.2014.01.071
Hsu S, Bollag WB, Lewis J et al (2003) Green tea polyphenols induce differentiation and proliferation in epidermal keratinocytes. J Pharmacol Exp Ther 306(1):29–34. https://doi.org/10.1124/jpet.103.049734
Iauk L, Lo Bue A, Milazzo I (2003) Antibacterial activity of medicinal plant extracts against periodontopathic bacteria. Phytother Res 17:599–604
Jaberian H, Piri K, Nazari J (2013) Phytochemical composition and in vitro antimicrobial and antioxidant activities of some medicinal plants. Food Chem 136(1):237–244
Jivad N, Bahmani M (2006) A review of important medicinal plants native to Iran effective on recovery from peptic ulcer. Der Pharm Lett 8(2):347–352
Kamar SS, Abdel-Kader DH, Rashed LA (2018) Beneficial effect of curcumin nanoparticles-hydrogel on excisional skin wound healing in type-I diabetic rat: histological and immunohistochemical studies. Ann Anat 222:94. https://doi.org/10.1016/j.aanat.2018.11.005
Kamlungmak S, Nakpheng T, Kaewpaiboon S, Mudhar Bintang MAK, Prom-In S, Chunhachaichana C, Suwandecha T, Srichana T (2021) Safety and biocompatibility of mupirocin nanoparticle-loaded hydrogel on burn wound in rat model. Biol Pharm Bull 44(11):1707–1716. https://doi.org/10.1248/bpb.b21-00397
Karri VVSR, Kuppusamy G, Talluri SV, Mannemala SS, Kollipara R, Wadhwani AD et al (2016) Curcumin loaded chitosan nanoparticles impregnated into collagen-alginate scaffolds for

diabetic wound healing. Int J Biol Macromol 93:1519–1529. https://doi.org/10.1016/j.ijbiomac.2016.05

Kasiewicz LN, Whitehead KA (2018) Lipid nanoparticles silence tumor necrosis factor α to improve wound healing in diabetic mice. Bioeng Transl Med 4(1):75–82. https://doi.org/10.1002/btm2.10123

Kasiram K, Sakharkar P, Patil A (2000) Antifungal activity of Calendula officinalis. Indian J Pharma Sci 62(6):464–466

Khorya RN, Katrak NN (1984) Materia Medica of India and their therapeutics, vol 1984. Neeraj Publishing House, Ashok Vihar, Delhi, pp 278–279

Kim H-L, Lee J-H, Kwon BJ, Lee MH, Han D-W, Hyon S-H, Park J-C (2013) Promotion of full-thickness wound healing using epigallocatechin-3-O-gallate/poly (lactic-co-glycolic acid) membrane as temporary wound dressing. Artif Organs 38(5):411–417. https://doi.org/10.1111/aor.12190

Kim JE, Lee J, Jang M, Kwak MH, Go J, Kho EK et al (2015) Accelerated healing of cutaneous wounds using phytochemically stabilized gold nanoparticle deposited hydrocolloid membranes. Biomater Sci 3(3):509–519. https://doi.org/10.1039/c4bm00390j

Kittana N, Assali M, Abu-Rass H, Lutz S, Hindawi R, Ghannam L, Zakarneh M, Mousa A (2018) Enhancement of wound healing by single-wall/multi-wall carbon nanotubes complexed with chitosan. Int J Nanomedicine 13:7195–7206. https://doi.org/10.2147/IJN.S183342

Koga AY, Carletto B, Farago PV, Lipinski LC (2021) Evaluation of cutaneous wound healing effect of associated zinc oxide (ZnO) and estrogen (E2) nanoparticles. Brazilian J Dev Curitiba 7(3):32382–32393. https://doi.org/10.34117/bjdv7n3-790

Kumar S, Kumar A, Chaudhary RN, Niwas R, Kumar A, Dishant (2019) Efficacy of Calendula officinalis for large open wound in animals. Pharma Innovat J 8(12):257–260

Kumar S, Kumar A, Dangi A, Chaudhary RN, Niwas R, Dinesh (2020) Efficacy of Calendula officinalis with Curcuma longa for open wound in large ruminants. Ruminant Sci 9(2):395–398

Kumar S, Kumar A, Singh B, Chaudhary RN, Niwas R, Priyanka. (2022) Use of calendula officinalis with curcuma longa in glycerine base for management of large wound in equines-a clinical study. Haryana Vet 61:105–107

Kushwaha A, Goswami L, Kim BS (2022) Nanomaterial-based therapy for wound healing. Nanomaterials (Basel) 12(4):618. https://doi.org/10.3390/nano12040618

Lakkim V, Reddy MC, Pallavali RR, Reddy KR, Reddy CV, Bilgrami AL, Lomada D (2020) Green synthesis of silver nanoparticles and evaluation of their antibacterial activity against multidrug-resistant bacteria and wound healing efficacy using a murine model. Antibiotics 9:902

Lau P, Bidin N, Islam S, Shukri W, Zakaria N, Musa N, Krishnan G (2017) Influence of gold nanoparticles on wound healing treatment in rat model: Photobiomodulation therapy. Lasers Surg Med 49:380–386

Lee J, Kim J, Go J, Lee JH, Han DW, Hwang D, Lee J (2015) Transdermal treatment of the surgical and burned wound skin via phytochemical-capped gold nanoparticles. Colloids Surf B Biointerfaces 135:166–174. https://doi.org/10.1016/j.colsurfb.2015.07.058

Leu J-G, Chen S-A, Chen H-M, Wu W-M, Hung C-F, Yao Y-D et al (2012) The effects of gold nanoparticles in wound healing with antioxidant epigallocatechin gallate and α-lipoic acid. Nanomedicine 8(5):767–775. https://doi.org/10.1016/j.nano.2011.08.013

Li M, Xu J, Shi T, Yu H, Bi J, Chen G (2016) Epigallocatechin-3-gallate augments therapeutic effects of mesenchymal stem cells in skin wound healing. Clin Exp Pharmacol Physiol 43(11):1115–1124. https://doi.org/10.1111/1440-1681.12652

Li M, Gao L, Chen J, Zhang Y, Wang J, Lu X, Duan K, Weng J, Feng B (2018) Controllable release of interleukin-4 in double-layer sol-gel coatings on TiO2 nanotubes for modulating macrophage polarization. Biomed Mater 13(4):045008. https://doi.org/10.1088/1748-605X/aa9526

Li J, Ding Z, Zheng X, Lu G, Lu Q, Kaplan DL (2021) Injectable silk nanofiber hydrogels as stem cell carriers to accelerate wound healing. J Mater Chem B 9:7771–7781. https://doi.org/10.1039/D1TB01320C

Liao JL, Zhong S, Wang SH, Liu JY, Chen J, He G, He B, Xu JQ, Liang ZH, Mei T, Wu S, Cao K, Zhou JD (2017) Preparation and properties of a novel carbon nanotubes/poly(vinyl alcohol)/ epidermal growth factor composite biological dressing. Exp Ther Med 14(3):2341–2348. https://doi.org/10.3892/etm.2017.4752

Liu W, Zhang G, Wu J, Zhang Y, Liu J, Luo H, Shao L (2020) Insights into the angiogenic effects of nanomaterials: mechanisms involved and potential applications. J Nanobiotechnol 18(1):9. https://doi.org/10.1186/s12951-019-0570-3

Lu Z, Gao J, He Q, Wu J, Liang D, Yang H, Chen R (2017) Enhanced antibacterial and wound healing activities of microporous chitosan-Ag/ZnO composite dressing. Carbohydr Polym 156: 460–469. https://doi.org/10.1016/j.carbpol.2016.09.051

Ma K, Liao S, He L, Lu J, Ramakrishna S, Chan CK (2011) Effects of nanofiber/stem cell composite on wound healing in acute full-thickness skin wounds. Tissue Eng Part A 17(9–10):1413–1424. https://doi.org/10.1089/ten.TEA.2010.0373

Maghimaa M, Alharbi SA (2020) Green synthesis of silver nanoparticles from Curcuma longa L. and coating on the cotton fabrics for antimicrobial applications and wound healing activity. J Photochem Photobiol B Biol 204:111806

Mahmoud NN, Hikmat S, Abu Ghith D, Hajeer M, Hamadneh L, Qattan D, Khalil EA (2019) Gold nanoparticles loaded into polymeric hydrogel for wound healing in rats: effect of nanoparticles' shape and surface modification. Int J Pharm 565:174–186. https://doi.org/10.1016/j.ijpharm. 2019.04.07

Majoumouo MS, Sibuyi NRS, Tincho MB, Mbekou M, Boyom FF, Meyer M (2019) Enhanced anti-bacterial activity of biogenic silver nanoparticles synthesized from Terminalia mantaly extracts. Int J Nanomedicine 14:9031–9046

Manuja A, Raguvaran R, Kumar B, Kalia A, Tripathi BN (2020) Accelerated healing of full thickness excised skin wound in rabbits using single application of alginate/acacia based nanocomposites of ZnO nanoparticles. Int J Biol Macromol 155:823–833. https://doi.org/10. 1016/j.ijbiomac.2020.03.221

Manzoor Z, Nawaz A, Mukhtar H, Haq I (2016) Bromelain: methods of extraction, purification and therapeutic applications. Braz Arch Biol Technol 59

Martin P, Nunan R (2015) Cellular and molecular mechanisms of repair in acute and chronic wound healing. Br J Dermatol 173(2):370–378. https://doi.org/10.1111/bjd.13954

Menon S, Devi KSS, Agarwal H, Shanmugam VK (2019) Efficacy of biogenic selenium nanoparticles from an extract of ginger towards evaluation on anti-microbial and anti-oxidant activities. Colloid Interfac Sci Commun 29:1–8

Mihai MM, Dima MB, Dima B, Holban AM (2019) Nanomaterials for wound healing and infection control. Materials 12(13):2176. https://doi.org/10.3390/ma12132176

Mirzahosseinipour M, Khorsandi K, Hosseinzadeh R, Ghazaeian M, Shahidi FK (2020) Antimicrobial photodynamic and wound healing activity of curcumin encapsulated in silica nanoparticles. Photodiagn Photodyn Ther 29:101639. https://doi.org/10.1016/j.pdpdt.2019. 101639

Muley BP, Khadabadi SS, Banarase NB (2009) Phytochemical constituents and pharmacological activities of Calendula officinalis L. (Asteraceae): A review. Trop J Pharma Res 8:455–465

Nunan R, Harding KG, Martin P (2014) Clinical challenges of chronic wounds: searching for an optimal animal model to recapitulate their complexity. Dis Model Mech 7(11):1205–1213. https://doi.org/10.1242/dmm.01678

Okur ME, Karantas ID, Şenyiğit Z, Üstündağ Okur N, Siafaka PI (2020) Recent trends on wound management: new therapeutic choices based on polymeric carriers. Asian J Pharm Sci 15(6): 661–684. https://doi.org/10.1016/j.ajps.2019.11.008

Oudhia P (2003) Interaction with the herb collectors of Gandai region, Chhatisgarh, MP, India. www.botanical.com

Pacelli S, Acosta F, Chakravarti AR, Samanta SG, Whitlow J, Modaresi S, Ahmed RPH, Rajasingh J, Paul A (2017) Nanodiamond-based injectable hydrogel for sustained growth factor release: preparation, characterization and in vitro analysis. Acta Biomater 58:479–491. https://doi.org/10.1016/j.actbio.2017.05.026

Pan A, Zhong M, Wu H, Peng Y, Xia H, Tang Q, Huang Q, Wei L, Xiao L, Peng C (2018) Topical application of keratinocyte growth factor conjugated gold nanoparticles accelerate wound healing. Nanomedicine 14(5):1619–1628. https://doi.org/10.1016/j.nano.2018.04.007

Pourali P, Yahyaei B (2016) Biological production of silver nanoparticles by soil isolated bacteria and preliminary study of their cytotoxicity and cutaneous wound healing efficiency in rat. J Trace Elem Med Biol 34:22–31. https://doi.org/10.1016/j.jtemb.2015.11.004

Pramanik N, Bhattacharya S, Rath T, De J, Adhikary A, Basu RK, Kundu PP (2019) Polyhydroxybutyrate-co-hydroxyvalerate copolymer modified graphite oxide based 3D scaffold for tissue engineering application. Mater Sci Eng C Mater Biol Appl 94:534–546. https://doi.org/10.1016/j.msec.2018.10.009

Rabbani PS, Zhou A, Borab ZM, Frezzo JA, Srivastava N, More HT et al (2017) Novel lipoproteoplex delivers Keap1 siRNA based gene therapy to accelerate diabetic wound healing. Biomaterials 132:1–15. https://doi.org/10.1016/j.biomaterials.2017.0

Rachmawati H, Sulastri E, Immaculata Iwo M, Safitri D, Rahma A (2016) Bromelain encapsulated in self assembly nanoemulsion exhibits better debridement effect in animal model of burned skin. J Nano Res 40:158–166. Trans Tech Publications, Ltd. https://doi.org/10.4028/www.scientific.net/jnanor.40.158

Raghuwanshi N, Kumari P, Srivastava AK, Vashisth P, Yadav TC, Prasad R, Pruthi V (2017) Synergistic effects of Woodfordia fruticosa gold nanoparticles in preventing microbial adhesion and accelerating wound healing in Wistar albino rats in vivo. Mater Sci Eng C 80:252–262

Randeria PS, Seeger MA, Wang X-Q, Wilson H, Shipp D, Mirkin CA, Paller AS (2015) siRNA-based spherical nucleic acids reverse impaired wound healing in diabetic mice by ganglioside GM3 synthase knockdown. Proc Natl Acad Sci 112(18):5573–5578. https://doi.org/10.1073/pnas.1505951112

Ranjbar R, Yousefi AR (2018) Effects of Aloe Vera and Chitosan nanoparticle thin-film membranes on wound healing in full thickness infected wounds with methicillin resistant staphylococcus aureus. Bull Emerg Trauma 6(1):8–15. https://doi.org/10.29252/beat-060102

Raut RW, Mendhulkar VD, Kashid SB (2014) Photosensitized synthesis of silver nanoparticles using Withania somnifera leaf powder and silver nitrate. J Photochem Photobiol B 132:45–55

Rodríguez-Acosta H, Tapia-Rivera JM, Guerrero-Guzmán A, Hernández-Elizarraráz E, Hernández-Díaz JA, Garza-García JJO, Pérez-Ramírez PE, Velasco-Ramírez SF, Ramírez-Anguiano AC, Velázquez-Juárez G, Velázquez-López JM, Sánchez-Toscano YG, García-Morales S, Flores-Fonseca MM, García-Bustos DE, Sánchez-Chiprés DR, Zamudio-Ojeda A (2022) Chronic wound healing by controlled release of chitosan hydrogels loaded with silver nanoparticles and calendula extract. J Tissue Viability 31(1):173–179. https://doi.org/10.1016/j.jtv.2021.10.004

Saddik MS, Alsharif FM, El-Mokhtar MA, Al-Hakkani MF, El-Mahdy MM, Farghaly HS, Abou-Taleb HA (2020) Biosynthesis, characterization, and wound-healing activity of phenytoin-loaded copper nanoparticles. AAPS PharmSciTech 21(5):175. https://doi.org/10.1208/s12249-020-01700-5

Samadi A, Buro J, Dong X, Weinstein A, Lara DO, Celie K-B et al (2021) Topical α-gal nanoparticles enhance wound healing in radiated skin. Skin Pharmacol Physiol 35:31–40. https://doi.org/10.1159/000518015

Selvam SI, Joicesky SMB, Dashli AA (2021) Assessment of anti bacterial, anti inflammation and wound healing activity in wistar albino rats using green silver nanoparticles synthesized from Tagetes erecta leaves. J Appl Nat Sci 13(1):343–351. https://doi.org/10.31018/jans.v13i1.2519

Sen CK, Khanna S, Venojarvi M, Trikha P, Ellison EC, Hunt TK, Roy S (2002) Copper-induced vascular endothelial growth factor expression and wound healing. Am J Phys Heart Circ Phys 282(5):H1821–H1827. https://doi.org/10.1152/ajpheart.01015.2001

Sharifiaghdam M, Shaabani E, Sharifiaghdam Z, De Keersmaecker H, Lucas B, Lammens J, Ghanbari H, Teimoori-Toolabi L, Vervaet C, De Beer T, Faridi-Majidi R, De Smedt SC,

Braeckmans K, Fraire JC (2021) Macrophage reprogramming into a pro-healing phenotype by siRNA delivered with LBL assembled nanocomplexes for wound healing applications. Nanoscale 13(36):15445–15463. https://doi.org/10.1039/d1nr03830c

Sharma M, Sharma R (2018) Implications of designing a bromelain loaded enteric nanoformulation on its stability and anti-inflammatory potential upon oral administration. RSC Adv 8(5): 2541–2551

Silva MMP, Aguiar MI, Rodrigues AB, Miranda MDC, Araújo MÂM, Rolim ILTP, Souza AMA (2018) Utilização de nanopartículas no tratamento de feridas: revisão sistemática. Rev Esc Enferm USP 51. https://doi.org/10.1590/s1980-220x20160435032

Sivaranjani V, Philominathan P (2016) Synthesize of titanium dioxide nanoparticles using Moringa oleifera leaves and evaluation of wound healing activity. Wound Med 12:1–5

Subapriya R, Nagini S (2005) Medicinal properties of neem leaves: a review. Curr Med Chem Anticancer Agents 5(2):149–146

Sundaram MN, Amirthalingam S, Mony U, Varma PK, Jayakumar R (2019) Injectable chitosan-nano bioglass composite hemostatic hydrogel for effective bleeding control. Int J Biol Macromol 129:936–943. https://doi.org/10.1016/j.ijbiomac.2019.01.220

Sundaramoorthi C, Kalaivani M, Mathews DM, Palanisamy S, Kalaiselvan V, Rajasekaran A (2009) Biosynthesis of silver nanoparticles from aspergillus Niger and evaluation of its wound healing activity in experimental rat model. Int J PharmTech Res 1:15231529

Tam K, Cheyyatraviendran S, Venugopal J, Biswas A, Choolani M, Ramakrishna S, Bongso A, Fong C-Y (2014) A nanoscaffold impregnated with human Wharton's Jelly stem cells or its secretions improves healing of wounds. J Cell Biochem 115:794–803. https://doi.org/10.1002/jcb.24723

Tang KC, Yang KC, Lin CW, Chen YK, Lu TY, Chen HY, Cheng NC, Yu J (2019) Human adipose-derived stem cell secreted extracellular matrix incorporated into electrospunpoly(lactic-co-glycolic acid) nano-fibrous dressing for enhancing wound healing. Polymers 11:1609. https://doi.org/10.3390/polym11101609

Thangapazham RL, Sharad S, Maheshwari RK (2013) Skin regenerative potentials of curcumin. Biofactors 39:141

Tian J, Wong KKY, Ho C-M, Lok C-N, Yu W-Y, Che CM, Chiu J-F, Tam PKH (2006) Topical delivery of silver nanoparticles promotes wound healing. ChemMedChem 2:129–136. https://doi.org/10.1002/cmdc.200600171

Tiwari M, Thakar MB, Jagani HV, Narayanan K, Venkata Rao J (2014) Biosynthesis and wound healing activity of copper nanoparticles. IET Nanobiotechnol 8(4):230–237. https://doi.org/10.1049/iet-nbt.2013.0052

Wang L, Hu C, Shao L (2017) The antimicrobial activity of nanoparticles: present situation and prospects for the future. Int J Nanomedicine 12:1227–1249. https://doi.org/10.2147/IJN.S121956

Wang W, Lu K, Yu C, Huang Q, Du Y-Z (2019) Nano-drug delivery systems in wound treatment and skin regeneration. J Nanobiotechnol 17(1):82. https://doi.org/10.1186/s12951-019-0514-y

Wang M, Huang X, Zheng H, Tang Y, Zeng K, Shao L, Li L (2021) Nanomaterials applied in wound healing: mechanisms, limitations and perspectives. J Control Release 337:236–247. https://doi.org/10.1016/j.jconrel.2021.07.017

Xia J, Asiri AM, Alamry KA (2018) Nanogels for bromelain stabilization. Nano LIFE 8(01): 1850002

Yang Y, Xiao Y, Jiang Y, Luo J, Yuan J, Yan J, Tong Q (2022) Alpha-lipoic acid promotes intestinal epithelial injury repair by regulating MAPK signaling pathways. Mediat Inflamm 2022:1894379. https://doi.org/10.1155/2022/1894379

Zangeneh MM (2019) Green synthesis and chemical characterization of silver nanoparticles from aqueous extract of Falcaria vulgaris leaves and assessment of their cytotoxicity and antioxidant, antibacterial, antifungal and cutaneous wound healing properties. Appl Organomet Chem 33: e4963

Zeng Q, Qian Y, Huang Y, Ding F, Qi X, Shen J (2021) Polydopamine nanoparticle-dotted food gum hydrogel with excellent antibacterial activity and rapid shape adaptability for accelerated bacteria-infected wound healing. Bioact Mater 6(9):2647–2657. https://doi.org/10.1016/j.bioactmat.2021.01

Zhou Y, Chen R, He T, Xu K, Du D, Zhao N et al (2016) Biomedical potential of ultrafine ag/AgCl nanoparticles coated on graphene with special reference to antimicrobial performances and burn wound healing. ACS Appl Mater Interfaces 8(24):15067–15075. https://doi.org/10.1021/acsami.6b03021

Chapter 26
Nanotechnology in Fracture Management

Neeraj Arora, Sandeep Saharan, and Rajesh Kumar

Abstract Nanomaterials have extended numerous advantages in fracture management in terms of providing mechanical strength, biocompatibillity, biointegration, customizable porosity, recruitment of cells and growth factors for faster healing, antimicrobial action to prevent infection, etc. Nanomaterials can be employed for orthopedic intervention in different forms such as membrane, rods, plates, screw, wire, etc. Nanomaterial-based implants, apart from enhancing fracture healing efficiency than their conventional counterpart, also facilitate targeted delivery as well as sustained release of the loaded drugs and biologicals which is certainly an important attribute not only for the well-being of companion and farm animals but also has immense potential in fracture management of old or compromised animals or animals of wild origin where the option of day to day care is limited. Various nanomaterials in solo or in combinations as composites such as metal-based NPs, polymeric NPs, carbon-based NPs, ceramic, and silicon-based other NPs have shown ample potential for such cases. However, before commercialization, extensive study is required to eliminate the toxicity issues and establish proper safety profile in organ to organ basis. Despite stringent regulation for nanointervention in biomedical application, recently various nanoparticle-based formulations/implants have obtained FDA approval for fracture treatment that enhanced the hope regarding success of the strategy. Combination of emerging 3D fabrication technology along with the advent of novel nanomaterials can effectively resolve several age-old issues of conventional therapies and can be a boon for developing devoted membranes or implants for veterinary application that battles the major challenges of enormous

N. Arora · S. Saharan
Department of Veterinary Surgery and Radiology, Lala Lajpat Rai University of Veterinary and Animal Sciences, Hisar, Haryana, India

R. Kumar (✉)
Department of Veterinary Physiology and Biochemistry, Lala Lajpat Rai University of Veterinary and Animal Sciences, Hisar, Haryana, India

LLR University of Veterinary and Animal, ICAR-National Research Centre on Equines, *Nanotechnology Theranostics in Livestock Diseases and Management*, Livestock Diseases and Management,
https://doi.org/10.1007/978-981-16-1610-5_26

diversity of animal species as well as wide array of bone size and shape requiring diverse and compatible specialized implant materials.

Keywords Nanomaterials · Implants · 3D printing · Bone healing · Drug delivery system · Antimicrobials

26.1 Introduction

Fracture in food and pet animals is one of the common pathological representations in veterinary clinics. Faulty housing and nutrition, selection of genetically fast growing animals, infighting, trauma including bending, compression, tension, and rotation during accident, dystocia and handling are the most important predisposing conditions to bone fracture. Among all displaced, non-displaced, open, and closed fractures, limb fracture is the most common in animals. Proper management of fracture is very important for animal welfare and higher productivity. For effective and faster managements of fracture, various points need to be considered such as fracture type, location, various adjacent anatomical structures, and their damage, major and supportive functioning of concerned bone, nutritional, physiological, behavioral and immunological status along with infection managements. For instance, bone can be viscoelastic and anisotropic and fracture can be incomplete, complete, comminuted and segmented; accordingly the treatment needs to be applied. Major regimen of therapy is to complete all four steps of fracture healing that includes hematoma formation (Days 1–5), fibrocartilaginous callus formation (Days 5–11), bony callus formation (Days 11–28), and bone remodelling (Day 18 onwards, lasting months to years) in effective and earliest manner. For this, effective immobilization, generation of broken tissue (gap filling), pain, inflammation, and infection management need to be carried out. Surgical intervention is required to remove undesired materials, fixing, and filling for the proper alignment. Various support materials and fillers like splints, plate, interlocking nail, wire, pin, external fixators, screw, cerclage, stem cell, 3D-graft, etc. are used for proper managements of bone deformities including fractures. Such surgical interventions are required in congenital or acquired pathologies in the form of trauma, infection, neoplasm, and failed arthroplasty where either there is a need of faster healing or the self-healing is not possible because of defects beyond a critical-size. Apart from immediate bone healing, the root cause of the bone deformities must also be addressed which may lead to reduction in bone density causing osteopenia and osteoporosis. Hence a comprehensive bone management that too in economical way frequently pose therapeutic and reconstructive challenges in veterinary science as the species and breed variations are numerous.

The management of fracture includes an effective delivery of material to reduce the cause of infection, inflammation, and pain to the animal. Further, it is also required to supply nutrition, hormone, or growth factors at the site of injury for faster and effective healing. Another primary activity in bone healing is the application of material or implant to reduce the gap, fixation of the bone to its mimetic anatomical structure, grow the broken part and fill the native tissues and living cells, in an effective manner. All these require multi-step process where various materials had been tried in the past through surgical as well as non-surgical interventions. Irrespective of the cause of fracture, whenever the size of loss of the tissue is beyond a critical point, it urges support for healing through grafting and fixing. Bone transplant is the second most used transplant of tissue after skin but possess the risk of infection and immune reaction in allograph whereas in autographs, a gold standard, the size of graft restricts its application. In general, metalwork and bone substitute are used for the surgical interventions but issues like stress shielding, wear, and tear along with risk of revision surgery are the major concern. Ultrasound as non-invasive technique has also shown promising results but mainly applicable in case of stable and reduced fractures. So the quest for novel materials persists that have advantages over the pre-existing conventional materials in terms of mechanical properties, biocompatibility, scaffold architecture, bioactivity, reactivity, and durability. The emergence of nanomaterials (NMs) and 3D-scaffolds employing Poly (lactic acid) (PLA), poly(glycolic acid) (PGA), poly(caprolactone) (PCL), and poly (ethylene glycol) (PEG) have yielded ample promise in terms of the aforesaid attributes. For instance, co-polymers of poly(lactic-co-glycolic acid) (PLGA) inherit advantages of desired and tailored porosity for nutrient supply, adjustable degradation rate and mechanical strength; however, hydrophobicity and lack of biological recognition sites may pose limitation to these synthetic polymers. These synthetic polymer-based NMs can also be modified through functionalization to provide growth factors, promote cell adhesion and proliferation in a better way (Turnbull et al. 2017).

Apart from the application in reducing gap for fracture management, NMs have also depicted other applications in bone healing. As a composite of organic (protein) and inorganic (mineral) phase, bone requires all the nourishing materials for healing along with their stem cells. Hence, maintaining osteoconductive, osteoinductive, and osteogenic properties through engineered materials is the key for proper bone healing. Spatial, physical, and biochemical stimuli are provided by the osteoconductive scaffolds which help in formation of cell lineages from their respective stem or progenitor cells along with proper attachment, survival, cell migration, and distribution across the gap area. Here NMs-based 3D scaffolds provide biomimetic environment not only for proliferation of osteoblastic cells but also help in vascularization, repeating the hierarchical organization through providing the stimulus like bone morphogenetic proteins (BMPs), growth factors, etc. It has been reported in various in vitro osteogenesis studies that nano-surface due to higher surface area increase the wettability, protein adsorption, cell adhesion, and mineralization, hence have better osteoblastic and osteoclastic responses in comparison to conventional alumina, titanium, and hydroxyapatite (HA)-based implants. Similarly, calcium phosphate-based NMs apart from acting as carriers for antibiotics, anti-inflammatory agents, analgesics, anticancer drugs, growth factors, proteins, and genes, also provide 3D scaffold for mechanical support and leading to better cell growth or tissue generation. However, result may vary with respect to generation of 3D scaffolds which includes methods such as thermal-induced phase separation, co-precipitation, freeze-drying, self-assembly, selective laser sintering, electro-spinning, salt-leaching, solvent-casting, sol-gel method, thermal pressing, etc. in solo or combinations, as the generation method influences size and level of doping of NMs, their shape, and properties. Diverse NMs have been used both as scaffolds or membranes which inherit various assistive properties for faster healing along with facilitating stem cell seeding and providing drug or bone matrix materials for proper healing or acting as acellular implants to reduce the gap.

26.2 Nanomaterials in Bone Defect and Fracture Management

Various NMs have been used so far for treating bone defects and fracture in order to extend multiple benefits such as tailor-made nano-porous 3D scaffolds provide better mechanical support and growth promotion, providing structure for better nutrient and growth promoter adsorption through enhanced surface area and functionalization and also act as carrier of various drugs for slow and sustainable release at precise site where it is required (Table 26.1). Scaffolds are exogenous material structures that once implanted allow to encourage cells to grow and proliferate to form the desired tissue structure. In addition, it has other benefits in terms of eliciting less immunoreaction, reduced toxicity, etc. Due to the scope of functionalization, their size, higher surface area to volume ratio, they in general provide scaffold of desired porosity and proper pore size (for exchange of materials like nutrients, growth factors, facilitating cell seeding, migration, growth, mass transport, etc.), biodegradability with proper degradation rate (when mixed with biodegradable material), higher contact and adsorption area (for attachment and proliferation), biocompatibility, sufficient mechanical support, predictable and reproducible mechanical properties, and microstructure along with matrix environment, and are capable of sterilization without loss of the desired properties (Fig. 26.1).

26.2.1 Polymeric Nanomaterials

Traditional metallic devices or implants to fix bone fracture requires second surgery to remove it. It has other complications too such as mismatch of the mechanical properties which advocates the application of polymeric NMs as implant materials. These NMs have additional attributes of biocompatibility and biodegradability. These polymers fit in other properties required for bone healing like providing scaffold to promote bone apposition, containing or promoting osteoprogenitor cells and providing the signals to induce osteogenic differentiation of local stem cells, i.e., osteoconductivity, osteogenicity, and osteoinductivity. Various

Table 26.1 Different nanomaterials depicting osteogenic activity for potential application in bone repair

Nanomaterials	Implants	Subject	Bone	Key observation	Reference
Self-assembled helical rosette nanotubes (HRNs)	Hydrogel scaffold	In vitro	Biomimics natural nanostructured components of bone (such as collagen)	Enhanced adhesion	Zhang et al. (2008)
Poly (lactic-co-glycolic acid) (PLGA)/multi-walled carbon nanotubes (MWCNTs)	Scaffold	In vitro	Rat bone marrow-derived mesenchymal stem cells (BMSCs)	Increased the thermal stability, improved mechanical properties, enhanced attachment and proliferation of BMSCs	Zhang et al. (2011)
Collagen fibers and synthetic apatite nanocrystals	Foam-like scaffold	Rat and pig	Trabecular bone, femur, tibia	Successfully healed a non-union fracture	Pek et al. (2008)
Collagen-apatite (col-Ap) nanocomposite	Scaffolds	Rodent	Mandibular defect	Stimulated bone-forming cells and vascular tissue growth	Liu et al. (2016)
Polylactic acid (PLA) with antibiotics and carbon nanotubes	Scaffolds and membrane	In vitro	Rat bone marrow-derived mesenchymal stem cells (BMSCs)	Antibacterial function, increased biological activity	de Moura et al. (2020)
Silver nanoparticles	Scaffolds	In vitro	Human bone marrow derived mesenchymal stem cells (hMSCs)	High concentration cytotoxic (50 μg/mL) and low concentration longer exposure help in differentiation without effecting cell viability (10 μg/mL)	He et al. (2016a)
Mesoporous bioactive glass [$80SiO_2$.(15 - x)CaO.$5P_2O_5$.xAg_2O	Scaffolds	In vitro	Human hepatocellular carcinoma cell line HepG2	Antibacterial action against *S. aureus* and *E. coli*, high porosity and had relatively high surface area and lower cancerous cell count	Phetnin and Rattanachan (2015)
3D printed dicalcium silicate with aluminum ions solid solution (Al-C2S) scaffolds	3D scaffold of β-Ca_2SiO_4 and $Ca_2Al_2SiO_7$	In vitro	Rat bone mesenchymal stem cells (rBMSCs)	Better proliferation and osteogenic genetic expression of osteocalcin, osteopontin, collagen type I, and Runx2	Yu et al. (2020)

(continued)

Table 26.1 (continued)

Nanomaterials	Implants	Subject	Bone	Key observation	Reference
Ni^{2+} – doped nano-hydroxyapatite	Scaffold of TiNi-based shape memory alloy (SMA)	In vitro and in vivo	Bone marrow mesenchymal cells (in vitro), hepatocytes, and pancreatic islet cells (in vivo)	Proangiogenic–osteogenic, unique biocompatible incubator	Kokorev et al. (2016)
Reduced-graphene oxide/polypyrrole/hydroxyapatite composite	Scaffold	In vitro	MC3T3-E1 cells	Very high proliferation rate, excellent osteoblastic performance, good biocompatibility, high specific surface area and large surface roughness	Song et al. (2016)
Composite of polycaprolactone and copper decorated reduced graphene oxide	Scaffold	In vitro	MC3T3-E1cells	Enhanced expression of angiogenic markers VEGF and ANG-2, elevated osteogenic activity with 87% more mineralization, and prominent antibacterial action by reducing the viability of *E. coli*	Jaidev et al. (2017)
Bio-composite of chitosan (CS) polymer, nano-hydroxyapatite (nHAp), copper (Cu), and zinc (Zn) nanoparticles	Scaffold	In vitro	Rat osteoprogenitor cells	Increased protein adsorption, and increased antibacterial activity of the scaffold without any toxicity.	Tripathi et al. (2012)
Copper-containing bioactive Glass (Cu-BG)	Polyethylene terephthalate (PET) synthetic fiber ligaments	Anterior cruciate ligament injury	PET synthetic ligaments	Non-degradable, better mechanical properties	Hamido et al. (2011), Parchi et al. (2013)
TiO(2) nanotubes	Implant	Rabbit	Tibias	Significantly improved bone bonding strength upto nine folds, greater contact area, better bone formation, mineralization indicted by higher calcium and phosphorus levels	Bjursten et al. (2010)

Silver–copper–boron composite nanoparticles	Nanoparticles administered through i.v. or i.m. route	Mice	Treatment of osteomyelitis	Eradicate *S. aureus* bone infection@1 mg/kg dose	Qadri et al. (2017)
Titanium implant with nano-structured zirconia surface	Implant	1-year-old rats	Intramedullary implants	Can be used in elderly patients as it increased expression of genes involved in bone maturation and adhesion	Dusad et al. (2013)
Titanium (Ti) implants comprised of titania nanotubes (TiO2-NTs) incorporated with silver nanoparticles		In vitro	Primary rat osteoblasts	Killed all the planktonic bacteria, prevented their adhesion, satisfactory osteoconductivity indicating possible application in orthopedics, dentistry, and biomedical devices	Zhao et al. (2011)
HA deposited Ti nano-tubes	CAD assisted 3D printing scaffolds	In vitro		Better porosity, interconnecting bimodal pore size distribution	Maleksaeedi et al. (2013)
Magnesium nanoparticles		In vitro	Human osteoblasts	Increase in cell viability, alkaline phosphate activity and osteocalcin levels, significantly promoted gap junction intercellular communication in osteoblasts	He et al. (2016b)
Nanocellulose-collagen-apatite		In vitro	Osteoblastic cells	Associated with osteogenic growth peptide for bone regeneration with no cytotoxic, genotoxic or mutagenic effects.	Marestoni et al. (2020)
Polycaprolactone (PCL), collagen I, and hydroxyapatite (HA) nanoparticles	Scaffold	Rat	Tibias	Favorable cell responses, increased phosphorylated FAK, biodegradable	Phipps et al. (2011)

(continued)

Table 26.1 (continued)

Nanomaterials	Implants	Subject	Bone	Key observation	Reference
Magnetic NP-incorporated PCL nanofibers	Scaffold	In vitro and rat	Radius segmental defect	Better apatite formation, improved initial cell adhesion and subsequent penetration, more expression of collagen I, osteopontin and bone sialoprotein, substantial neoblood vessel formation and lesser toxicity	Singh et al. (2014)
Copper loaded chitosan scaffold	Chitosan matrix	Rat	Calvarial defects	Promotes bone regeneration	Dong and Virtanen (2022)
Superparamagnetic iron oxide nanoparticles	PLGA based matrix	Murine bone BMSCs, rat	Femoral bone	Promoted reconstruction of bone defects by increasing bone mineral density, trabecular thickness, and bone tissue volume/total tissue volume were increased significantly at injury site	Zhao et al. (2021)
CNT–HA composites	Implant	Rabbit	Cortical bone	Improved lexural strength and impact strength, better implant-bone interface compatibility	Mukherjee et al. (2016)
CNT–gelatin–HA core–shell nanoparticles	Sheet	In vitro	Primary mesenchymal stem cells (MSCs)	Higher biocompatibility, cell viability, elastic modulus, tensile strength, and elongation rate	Yoon et al. (2014)
Graphene nanoribbons	Ribbon	Tibia	Osteopenic rats	Improved biocompatibility and bone mineralization activity	de Carvalho et al. (2018)
Graphene/multi-walled carbon nanotube oxides (MWCNTO-GO, 50% w/w)	With combining poly (d, l-lactic acid) (PDLLA) scaffold	In vitro and adult male rats	Tibias	Better tensile strength tests, cell adhesion and cell viability, alkaline phosphatase and mineralization	Silva et al. (2017)

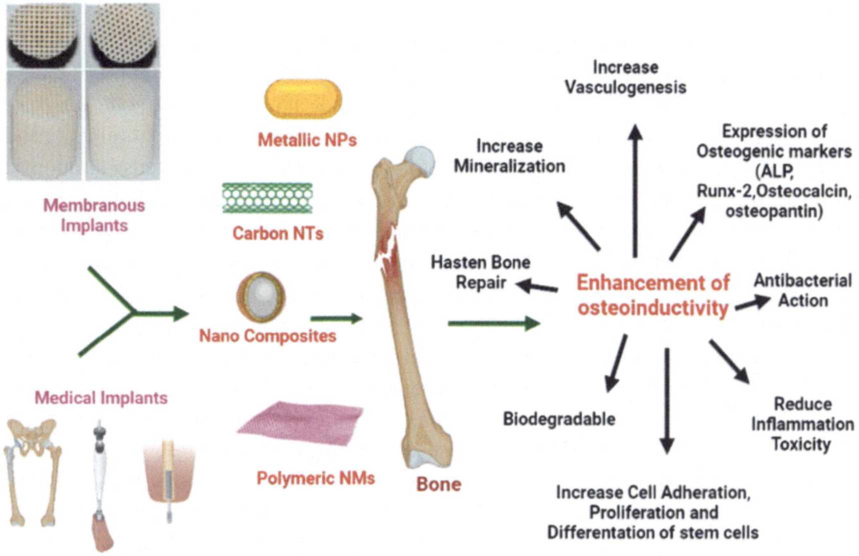

Fig. 26.1 Nanomaterials in fracture management

biodegradable polymers have been used in biomedical interventions such as starch, cellulose, agar, alginates, collagen, keratin, elastin, heparin, fibrin, cellulose, dextran, and Poly(hydroxyalkanoates) (PHA), as biodegradable fibers sourced from plants, animals, or microorganisms, while grafted chitosan, dextran-coated polystyrene latex (PSL), glycosaminoglycans, Poly(butylene adipate-co-terephthalate) (PBAT), Methyl-cellulose(MC), Ethyl-cellulose (EC), Poly(butylene succinate) (PBS), Poly(lactic acid)(PLA), Poly(vinyl alcohol) (PVA), Poly(glycolic acid) (PGA), Poly(butylene succinate-co-adipate) (PBSA), Poly(hydroxylketones) (PHK), Poly-ε-caprolactone (PCL), Poly(methyl methacrylate) (PMMA), Poly (hydroxybutyrates) (PHB), Poly(ethylene terephthalate) (PET), Poly (hydroxybutyrate-co-hydroxyvalerate) (PHVB), Poly(methylene adipate/terephthalate) (PTMAT), etc. are some of the prominent examples of synthetic fibers used in biomedical application. Employing different functionalization, their number and properties can be changed enormously, hence found a lot of application as naoformulates. Among them, various polymer-based grafting has been employed in repairing the defect of the bone, teeth, and other hard tissues using collagen and chitosan as natural biodegradable polymers and Poly (α-hydroxy acids) class synthetic polymers like PGA, PLA, and their copolymer poly (lactic-co-glycolide) (PLGA). The polymeric NMs with diverse functionalization potential have found to be even more effective than metallic class NMs with further addition of another property that it can provide electrical stimulation for faster healing. Electrical stimulation has observed to be an important stimulant for cells regeneration (Das et al. 2020). Application of these polymers in nano-form as solo or in combination has further potentiated their healing utility of fracture. For instance, poly (lactic-co-glycolic acid) (PLGA)/multi-walled carbon nanotubes (MWNTs) as composite scaffolds lead to enhanced attachment and proliferation of BMSCs (bone marrow-derived mesenchymal stem cells) than their normal counterparts (Zhang 2011). Similarly, nano-scaffold made from natural nanofibers through polymer keratin has led to change in mechanical strength along with compatibility, good spinnability, and increased antithrombogenicity. Application of nano-structured poly-ε-caprolactone/$CaCO_3$ as composite nano-fibers to form guided bone regeneration (GBR) membranes depicted high tensile strength and better cell attachment manner (Fujihara et al. 2005). Likewise, nanocomposites of poly(ε-caprolactone) and silica have better Young's modulus, no cytotoxicity, better cell adhesion, and proliferation potential than its normal counterparts which advocates its applicability in bone graft or fracture as scaffold or membrane material (Calandrelli et al. 2010). Collagen-apatite is a having a similar structure of hard tissue when used in nanoformulation for scaffold and membrane preparation, it demonstrated better mechanical strength and chemical stability (Song et al. 2007). Polymeric NMs possess advantages as scaffold and membrane in fracture management due to the possibility of loading with various cells, enrichment with various biofactors and their combinations in order to enhance mechanical, physical, chemical, and biological properties. Porous bioresorbable bone scaffold made from collagen fibers and synthetic apatite nano-crystals have similar mechanical strength, chemical composition, excellent bioactivity, crystalline phase, and grain size that lead to good osteoconductivity and osteoblastic action due to enhanced attachment and proliferation; hence, rapid bone healing is possible (Pek et al. 2008). Application of collagen-apatite (Col-Ap) nano-composite to formulate biosimilar structure as bone matching the structural properties and functionality has observed to be advantageous in terms of greater growth factor accumulation, induced extracellular matrix secretion and mineralization leading to more vascularity and better healing in mandibular defect (Liu et al. 2016). The biodegradable NMs like PLA, PGA, and their composites are becoming choice materials and even some of them have obtained FDA approval for various applications due to the aforesaid properties. The degradation is mostly through hydrolysis of ester bond followed by monomeric conversion. For example, PLGA form lactic and glycolic acids which are metabolized by the cells through their natural pathways hence fully cleared. PLA forms lactic acid monomer which enters into the metabolic cycle of cells to be converted into pyruvic acid and ultimately through TCA cycle gets cleared in the form of carbon dioxide and water. Similarly, PGA is converted to glycolic acid monomers and via glycine-serine to pyruvic acid enters into TCA cycle. Some is also cleared by urine. In general, the degradation rate of the polymers is as follows: PGA > PDLLA > PLLA > PCL ~ PPF (Okamoto and John 2013). Overall, these nano-composite polymers have advantageous properties such as suitable surface area, better mechanical capacity, stability, enhanced cell adhesion, proliferation, and differentiation along with better adsorption, degradation rate, osteogenesis, clinical operability, relatively inexpensive, and non-cytotoxic. For instance, nano-composite rods of 15 mm long and 1×2 mm in cross section area containing poly(butylene succinate-butylene dilinoleate) (PBS-DLA) copolymer along with nano-sized hydroxyapatite (HAP) particles as polymer/ceramic composite implanted in tibial fracture in male Wistar rats has depicted well-preserved elastomeric properties, stimulated bone regeneration, and remodelling, thus indicative of high potential in bone tissue engineering and repair (Prowans et al. 2019). Functionalization of these polymers and their amalgamation with other composites pave the way for customizing wide ranges of NM-based implants having more distinct and desirable properties suitable for fracture healing and transplant. For example, surface modification of nano-HA crystals by grafting poly(L-lactide) (PLLA) using stannous octanoate ($Sn(Oct)_2$) catalyst has yielded distinct dispersion, increased Young's modulus according to filler content, increased tensile strength, enhanced bending strength and modulus, and ductility (Hong et al. 2004). Similar effect was obtained in PCL-grafted HA nanoparticles and by grafting with L-lactic acid (Wang et al. 2010). These grafting not only increased the material strength but also yielded higher levels of chondrocyte attachment and proliferation in comparison to their non-NM-based resources as evidenced in both in vitro as well as in vivo studies. In two different studies, when PLGA and hydroxyapatite surface-grafted with L-lactic acid oligomer-based nano-composites scaffolds was implanted in dorsal muscles, and radius with critical size defect in rabbit, better mineralization and osteogenesis was observed (Zhang et al. 2009; Cui et al. 2009).

26.2.2 Metallic Nanomaterials

Various metals have been used in bone and dentistry-related surgery in order to restore, maintain, reconstruct, and sometimes to improve the normal functions primarily because of high tensile strength, non-reactivity, and non-toxicity of the materials along with their rough surface for cell adhesion. Now a days, metal-based NMs has gained significance in biological applications due to providing desired porosity, higher surface area for interactions, inherent antibacterial effects, shape memory phenomenon, lower cytotoxicity, enhanced proliferation process; thus extending good osteogenic potential, acceptable biocompatibility, and ability to regulate cell growth pathways, etc. as additional features. For instance, NMs of zirconia, titanium, silver, and zinc possess antimicrobial activity, particularly bactericidal action, thus, are vastly used in tissue engineering. Nano-sized silver (Ag) along with poly(L-lactide) NMs have shown prolonged antibacterial action even up to 20 days. In addition, metallic NMs also provide thermal conductivity, surface wettability and roughness, hence enhance the mechanical property and adherence of cells conducive of fracture healing (Okamoto and John 2013). Zirconium oxide (ZrO_2)-based NMs are also very commonly used in dental implants, coatings, and fracture management due to versatility, extensive chemical stability, biocompatibility, suitable fracture resistance due to phase transformation toughening, etc. (Hu et al. 2019). Various other metallic and metal oxide NMs have also been employed for scaffold that include but not limited to Au, Fe, Al, Ni, Cu, and magnetic nanoparticles, etc. Calcium and phosphorus-based NPs have also been used but discussed elsewhere in subsequent sections.

Among metals, gold (Au) NMs is one of the extensively studied NMs in various fields including diagnostics, theranostics, and tissue engineering. Due to inertness, low toxicity and high biocompatibility in addition to antibacterial effects, it has been used for generation of scaffold microenvironment. In addition to the general benefits extended by the metallic scaffold, gold-based NMs has shown positive attributes in osteogenic differentiation of stem cells along with cell proliferation rates, upregulation of different growth factors, enzymes, stimulation of various essential gene expression pathways like Runx2, OCN, Wnt/β-catenin pathway, ERK/MAPK pathway and p38/MAPK pathway, etc. (Heo et al. 2014; Eivazzadeh-Keihan et al. 2020). Furthermore, enhanced bone tissue regeneration was achieved in doses dependent manner when AuNPs was incorporated into gelatine-based biodegradable hydrogels (Heo et al. 2014). Similarly AuNanoRods using gelan gum (GG) polyelectrolytes, poly (acrylic acid) (PAA) and poly (allylaminehydrochloride) (PAH) enhanced the mineralization of osteoblast-like cells without any cytotoxic effect. However, the reports of AuNPs-induced cytotoxic effect inflicted mainly by increased ROS production are also available but it generally occurs in dose-dependent manner. Ultra-small AuNPs of <10 nm has been used in the delivery of miRNA to enhance osteoblastic differentiation of the bone marrow stem cells through targeted delivery. In this way, miRNAs having the potential of epigenetic regulation can be used for faster healing. The miRNAs induce desired modulation of the expression of target genes at post-transcriptional stage by binding at the untranslated region of mRNA where Au@PEI@Lipo nanoparticles work as an efficient carrier of the miRNAs and save them from the hostile environment of nucleases to facilitate their targeted delivery (Yu et al. 2017). Likewise, the positive effects of miRNAs in bone tissue regeneration have also been successfully tested (Oliveira et al. 2011; Li et al. 2013). Another commonly employed metallic NPs is AgNPs that demonstrated antimicrobial and antiviral properties. In addition, due to rough surface and other unique properties, AgNPs demonstrated good adhesion, spreading, and growth rate of osteoblast cells, high proliferation activity, enhanced osteogenicity, osteoconductivity, cytocompatibility, acceptable mechanical strength that rendered it suitable to be used as potent scaffold material. However, silver-based NMs need proper doping in scaffold to lower the toxicity as AgNPs can be cytotoxic at higher dose. Similarly hydroxyapatite/chitosan porous scaffolds along with Ag and strontium (Sr) NPs separately reduced bacterial activity upto 98.1% and 98.9% respectively along with significant enhancement of osteoinductivity (Xu et al. 2016). The AgNPs has been extensively used as antimicrobial and various commercial products are already available both as topical formulations and also in various instrumentation of surgical applicability. In addition, the role of AgNPs in wound healing has already been reported due to various other positive attributes such as enhancement in the proliferation and migration of keratinocyte leading to wound closure and wound contraction by differentiation of fibroblasts into myofibroblasts, enhancement of cell proliferation by regulating the expression of genes like HIF-1α and IL-8, etc. (Tian et al. 2007; Liu et al. 2010). Its osteogenic property was exerted through cbfa1 expression which leads to effective mMSCs' differentiation towards osteoblast with little effect on adipocyte or chondrocyte differentiation (Zhang et al.

2015). Antibacterial effect of Ag-based scaffolds was also demonstrated using Gt/Ag-NPs/BG nanocomposite membranes, gelatin (Gt), alginate (Alg), and poly (vinyl alcohol) (PVA) silver hydroxyapatite (Ag-Hap) scaffold, chitosan, carboxymethyl cellulose and silver nanoparticle modified cellulose nano-whiskers (CS/CMC/CCNWs-AgNPs) scaffold, and borosilicate bioactive glass scaffolds (BG) against *E. coli, B. subtilis, S. aureu,* and *E. hirae* in more efficient way than non-silver scaffold without compromising on mechanical strength (Wang et al. 2015; Hassan et al. 2018; Kumar Saini et al. 2019; Akturk et al. 2020). Similarly iron-doped HAp (FeHAp) along with poly(ε-caprolactone) (PCL) matrix has been used as scaffold that lead to better loading and growth of BMSCs but may have issue of biodegradability which can be overcome by using tungsten NP (Eivazzadeh-Keihan et al. 2020). Aluminum oxide NP-based scaffold is another example depicting low toxicity and better proliferation of osteoclastic cells along with downregulation of genes of inflammatory cytokines and reactive oxygen species (ROS). Further, osteopontin (OPN) and collagen 1 (Col1) gene were also upregulated but size of NPs influenced these properties. Expression of osteocalcin, osteopontin, Collagen type I, and Runx2 was upregulated when 3D scaffold was used employing Al_2O_3 NPs (Yu et al. 2020). Copper nanoparticles (CuNPs) are another known antimicrobial agent that has been used to treat osteoporosis (Hejazy et al. 2018). Bone mineralization, adhesion, and proliferation of osteoblast cells are the important properties of CuNPs-based scaffolds useful for fracture management but also associated with the issues of strong inflammatory response and cytotoxicity. To overcome these limitations and utilize the advantageous properties like antibacterial activity, anti-cancerous properties etc., it has been dopped with other compounds. For instance, CuNPs dopped reduced graphene oxide sheets (RGO-Cu) in polycaprolactone (PCL) matrix (PCL/RGO-Cu) enhanced angiogenic activity, bone mineralization, and osteogenic activity. Cu-based PET ligament has been used in both in vivo and in vitro trials for ligament injury along with fracture and has shown better osteogenic, angiogenic differentiation from stem cells and graft-bone integration (Eivazzadeh-Keihan et al. 2020). Antibacterial properties of titanium sheet (Ti-6Al-4V alloy sheets) have been added with HAp-copper (HAp-Cu) nanocomposite that reduced the growth of *E. coli* and *S. aureus* (Eivazzadeh-Keihan et al. 2020). Similarly zirconium NPs inherits the desired properties of low cytotoxicity, ideal mechanical strength, corrosion resistance, biocompatibility, osteoinductivity, enhanced porosity, etc. Thus it has been extensively explored as good candidate for scaffolds to hasten bone healing process. Chitosan (CS), hydroxyapatite (HA), wollastonite (WS), and zirconium nanoparticles (Zr NPs) based nanocomposites have shown all above discussed properties along with formation of apatite structures (Roohani-Esfahani et al. 2010; Altuna et al. 2016). It has been evidenced that 15 wt.% HA-ZrO_2 with CS-PEG (55:30 wt.%) have similar mechanical strengths and porosity as of cancellous or spongy bone with desirable quality (Bhowmick et al. 2017).

26.2.3 Nano-ceramics and Other Composite Materials

Bioactive ceramics such as tricalcium phosphate (TCP), silicate and phosphate bioglasses, hydroxyapatite (HA), and calcium phosphates are the major class of biomaterials used for bone repair with appropriate osteoconductivity and biocompatibility as the bone contains nearly 65% of these hydroxyapatite ($Ca_{10}(PO_4)_6(OH)_2$) crystals with a Ca/P ratio of 1.50–1.67. They have the advantage of similar chemical and structural construct as the mineral phase to native bone. However, brittleness and the difficulty in getting desired shape from them are the major concerns for using in fracture healing. To avoid such limitations, nano-polymer/bioactive ceramic composite scaffolds have been introduced which provide additional benefit of larger surface area to enhance protein adsorption and osteoblast adhesion, good osteoconductivity, osteoinductivity, biodegradability, and high mechanical strengths to facilitate bone tissue engineering (Okamoto and John 2013). Various nanocomposites based on HA have been tested in both in vitro and in vivo condition to have desirable properties like ceramic composites, such as collagen/HA composite, polyamide 6/HA composite, and PLGA/TCP composite (McMahon et al. 2013) at in vivo level, while at in vitro level, nano-hydroxyapatite (n-HA), nano-zirconia (nZrO; ZrO_2), nano-calcium zirconate (nCZ; $CaZrO_3$) used with chitosan (CS) scaffolds showed better or enhanced mechanical strength, absorption capacity, cell proliferation, ideal porosity, and cell spreading (Gaihre et al. 2018). Among them CS-nCZ has facilitated higher proliferation of pre-osteoblasts in relation to CS-nZrO. Due to the advantages of better bone growth through increased adhesion, proliferation, and mineralization, no toxicity, immune reaction, or carcinogenicity Zirconia (ZrO_2) ceramics has widely been applied in dental, orthopedic, and bone repair. Synthesis of nHA using poly(L-lactic acid) (PLLA)-based nanocomposites scaffold has found to have regular internal ladder-like pore structure with better compressive strength. Similarly, nano-carbonated HA/collagen/PLGA composite membrane has been developed which possess three layers of these materials in different composition, depicted flexibility, enough mechanical strength, biodegradability, high biocompatibility, and osteoconductivity (Liao et al. 2005). Various combinations of nano-apatite (nAp) and poly(-ε-caprolactone) membrane have also been developed for bone and hard tissue-related applications. During evaluation of structural, mechanical, chemical, and biological properties, the highest strength and toughness was found in 25% nAp along with enhanced bioactive behavior, early cell differentiation, augmented osteoblast-like cell proliferation, and differentiation (Yang et al. 2009). Membrane-based scaffold using electrospun nanofibers supports cell survival, attachment, protein absorption, cellular reactions, and promotes activation of specific gene expression and intracellular signalling, reorganization ability due to higher surface area and emulates natural extracellular matrix (ECM) (Zhuang et al. 2019). Nanoparticulate hydroxyapatite with polycaprolactone and collagen I in the ration of 2:5:3 was electrospun and made scaffold that facilitated prompt cell spreading and proliferation of MSCs with much alike response in in vitro as well as in vivo condition (Phipps et al. 2011). The collagen used in the scaffold has depicted high biocompatibility, bio-affinity, and resorbability but is highly degradable and provides less mechanical strength, thus can also be replaced by chitosan. Nano-formulation of chitosan/bioactive glass-based scaffold has induced the precipitation of bone-like apatite, cell metabolic activity, and mineralization with good shape memory (Mota et al. 2012; Correia et al. 2015). Nano-macro porous Na_2O-CaO-SiO_2-P_2O_5 ceramic glasses is one of the bioactive glass, subgroup of ceramic materials has been employed in hard tissue fixation. Usually they are comprised of Ca-Si based inorganic materials: a network of SiO_2-CaO-P_2O_5. This has been established as an excellent material for repairing the bone defects due to an excellent osteoconductivity and good osteoinductivity. Recently these bioactive glasses have been studied with various blending like PCL/chitosan, PCL/gelatin, and PVA polymers. Application of nanoparticles of mesoporous bioactive glass (mBGn) has been further demonstrated to be beneficial in bone tissue regeneration, adhesion, and proliferation as they have long-lasting action, enhanced the differentiation of stem cells or progenitors into osteoblasts and even drug can be loaded on them. Similarly 3D-based scaffold has been developed using $CaCO_3$ aragonite and implanted in rabbit; it withstanded cortical bone strength, promoted bone growth, showed ideal porosity, and improved integration (Mahmood et al. 2017).

Self-assembled nanostructure helical rosette nanotubes (HRNs) functionalized with lysine amino acids were embedded in poly(2-hydroxyethyl methacrylate) or pHEMA to form hydrogel was used as scaffold that showed faster healing due to significantly enhanced osteoblast adhesion (Zhang et al. 2008). Similarly, nanocellulose-collagen-apatite composite along with osteogenic growth peptide (OGP) or its C-terminal pentapeptide [OGP(10–14)]-based nanocomposite has also been evaluated for bone healing properties and it showed enhanced cell growth, early osteoblastic development without any kind of toxicity (Saska et al. 2017).

26.2.4 Carbon-Based Nanocomposites

Carbon due to its specific atomic configuration can form various kinds of bonds, hence has number of compounds and isomorphs which makes it very versatile. When it is used as nano form, the versatility further increases to leap and bound due to addition of nano features including the scope of functionalization. Carbon has been used in various nano forms which include but not limited to single and multi-walled carbon nanotubes, nanodiamonds, graphene, etc. in various biological and non-biological applications. Light-weight, higher density, specific quasi-static strength, stiffness, fatigue loads, and fatigue life are the major attributes which make nano-carbon-based fibers attractive in fracture healing. One-dimensional hollow tube is formed by using single and multi-walled carbon nanotubes (SWCNT and MWCNT) which can be generated by various methods like electrical arc discharge, laser ablation, or chemical vapor deposition having the length in ranges from nano to millimeter. CNT coatings have shown resemblance to bone collagen matrix, hence positively induce the orientation and growth of new bones when used in fracture healing. Type I collagen can be incorporated easily when SWCNT scaffolds are used without causing toxicity, hence does not affect the viability of cells so has been reported in MWCNT as rhBMP-2/collagen deposition is ectopic. Apart from better cell recruitment and adhesion due to high specific surface area, it is also light-weight and provides desired mechanical strength. Through controlling the diameter and length of the nano-tubes, the electrical conductivity can also be regulated which can be useful to modulate the behavior of cells with desirable events. It has also been used with hydroxyapatite leading to better nucleation/precipitation and self-assembly of HA molecules along with significant induction of osteoblast differentiation and matrix mineralization. However in certain cases such as in MWCNT-chitosan based scaffold, dispersion of MWCNT has been observed away from the implanted scaffold which may cause opsonization, blood coagulation, and activation of the complement system, hence may have adverse effects. SWCNTs has also produced toxic effect in free form on various cell lines and through respiratory entry, hence needed to be dealt with caution when using as candidate for implant materials. However the advantages of CNTs as scaffold cannot be overlooked like nano-tube mimicking collagen role. Hence applying CNTs with other materials as composite and using different functionalization and treatments to reduce the toxic effect is a key to enhance their utilities. Application of CNTs with HA increases the mechanical strength, tensile strength, and fracture toughness which otherwise is limiting factor for HA due to its brittle nature. CNT composites have also shown good biocompatibility and significantly improved flexural and impact strength, proliferation of fibroblasts and osteoblasts in both in vitro and in vivo studies. Similarly, beta tricalcium phosphate (β-TCP)–1% CNT nanocomposite has induced apatite formation due to negative charge (Tanahashi and Matsuda 1997). Likewise, MWCNT–HA composites coated titanium has depicted better compatibility, mineralization, and bone growth so was with CNT–collagen–HA which was having 10x more strength than collagen along with better proliferation and growth (Jing et al. 2017; Park et al. 2019). MWCNT–conductive polymer composites (CPC) composite has also shown 10x more strength than pure cementing material β-TCP and di-calcium phosphate anhydrous in addition to better nucleation, growth, and crystal formation (Chew et al. 2011). In bone implants, 3D collagen scaffold coated with MWCNT has found to induce early osteoblast differentiation leading to faster healing as compared to uncoated collagen scaffolds (White et al. 2007; Hirata et al. 2011). Similarly 0.5% MWCNT–collagen–HA scaffold implantation in rat calvarial defect has led to healing within 12 week without sign of toxicity, inflammation or inflammatory cell infiltration (Jing et al. 2017; Pei et al. 2019). Nano-diamond (ND) based multifunctional bone scaffold enhances bone energy upto 310% when used as PLLA and octadecylamine-functionalized ND composite (10 wt.% ND-ODA) along with early apatite nucleation and growth at faster rate than PLLA alone. Such type of functionalization with ND significantly increases the Young's modulus and hardness of the nanocomposite, almost matching to the properties of human cortical bone (Zhang et al. 2011). Graphene NP-based nano-hydroxyapatite (nHAp)/CNT with graphene oxides termed as graphene nano-ribbon composite were also found to be effective as scaffolds when employed in rat tibial fracture management (Oliveira et al. 2019). Overall, various carbon-based nanocomposites have depicted better applicability in bone fracture healing due to conducive surface morphology, biocompatibility, mechanical strength, drug release, and osteogenic potential.

26.3 Nanomaterials in Drug Delivery and Antimicrobial Action

Various NMs due to their inherent antibacterial effect have extended additional benefits in scaffold like membranes and implant materials apart from the usual attributes of other conventional orthopedic materials. By virtue of the prosperity in nanotechnology, currently a bulk of literature is available depicting the antimicrobial effect of various metallic NPs against diverse infectious agents like bacteria, virus, fungi, etc. (Minakshi et al. 2020; Prasad et al. 2021). Even in occasions, these metallic NPs have found to be effective against resistant species or strains also. Moreover, due to the favorable porosity and shape effect, these NMs-based implants also act as carrier for slow and targeted release of various drugs and biologicals such as mRNA, DNA, antibacterial ingredients, or anti-inflammatory drugs or even supply of the additional growth factors for rapid healing of bones. For instance, copper NPs have been considered as potential antimicrobial agents and also provide large surface area to volume ratio (Ingle et al. 2014). The potential of this NPs has also been explored as a drug for treating the osteoporosis along with pro-angiogenic properties (Ewald et al. 2012; Hejazy et al. 2018; Saeed et al. 2019; Pinna et al. 2021; Thurner et al. 2021) as it increases the release of hypoxia-inducible factor (HIF-1α) hence stimulates the secretion of VEGF for the development of new vascular network promoting blood supply (Rigiracciolo et al. 2015). Similarly assays on Gram-negative and Gram-positive microorganisms have demonstrated that ZrO_2 NPs have notable antibacterial properties (Gouda 2012; Jangra et al. 2012). Silver-Copper-Boron (AgCuB) nanoparticle at 1 mg/kg body weight dose has shown significant antibacterial activity against *Staphylococcus aureus*, a key pathogen for bone infection. Further AgCuB NPs has also been found to be effective as an alternative therapeutic strategy for osteomyelitis in rabbit and mouse as well as in targeted delivery of biologicals (Qadri et al. 2018). Orthopedic titanium implants fabricated with titanium oxide nanotubes through electrochemical anodization containing AgNPs has demonstrated long-lasting antibacterial effect against methicillin-resistant *Staphylococcus aureus* in both in vivo and in vitro for even upto 30 days along with excellent bio-integration properties, indicating enormous potential of the NM in orthopedic applications (Cheng et al. 2014). The coating implant was introduced in rat fractured tibial plateau, from the proximal end to the distal end along with administration of 2 μL bacterial suspension containing 1.5×10^8 cells/mL and exhibited a long-lasting antibacterial effects during antibacterial and anti-adhesive assay which was similar in in vitro observation where the bacterial load was 10^5 CFU/mL (Cheng et al. 2014). A Mg-Zn-Ca alloy implants coated with chitosan nanofibers containing 0.5 wt.% silver sulfadiazine-MWCNTs depicted satisfactory cellular compatibility and significant antibacterial activity against *Staphylococcus aureus* and *Escherichia coli* to prevent bone infection in orthopedic intervention (Bakhsheshi-Rad et al. 2019). As a potential medium for the targeted delivery of drugs and biologicals, chitosan-based micro-porous titanium implant was used with lyophilized miR-148b MicroRNAs that depicted accelerated osseointegration along with delayed release of the payload and the transfection was also prolonged with enhanced osteogenic differentiation of the primary rat bone marrow mesenchymal stem cells (MSCs) indicating better regeneration capacity (Song et al. 2013). Similarly, the 3D pattern fabricated by multi-layer coatings of strontium-doped hydroxyapatite nano-rods with regulated inter-rod lateral spacing on microporous titanium oxide has induced the proliferation and differentiation of osteoblasts at below 96 nm lateral-spacing while the effect decreased with the increase in inter-rod spacing, thereby elucidating the potential of the nano-composite as orthopedic implant material (Zhou et al. 2013).

Magnetic nanoparticles have also been used to load drugs which release the payload under stimulation through certain magnetic field. Application of magnetite with HA scaffolds not only provides better porosity, biocompatibility, and osteointegration for bone regeneration as observed when applied in the management of critical size lesion in rabbit condyle but also can provide additional medium for targeted drug delivery (Panseri et al. 2012). Similarly ceramic nano-scaffolds in addition to directing cell growth or tissue generation, and providing mechanical support, has also been found as an excellent drug delivery and release system, so has been the calcium phosphate scaffolds (Anil et al. 2015). Porous hollow silica NP-based scaffold having the diameter of 60–70 nm and around 10 nm wall-thickness can be loaded with several times more drug with time-delayed three-stage release from the surface, pore channels of the wall, and inner core of the NPs, hence better drug targeting and bioavailability can be achieved as observed in case of drug cefradine (Chen et al. 2004). Similarly penicillin-based antibiotics loaded nanotube scaffold made from titania has also shown promising result as drug delivery platform. Slow and sustainable release of dexamethasone has been achieved through biopolymer/mesoporous bioactive glass scaffolds which efficiently carried the drug and provided long term release via targeted delivery and controlled release system apart from delivering the other conventional attributes of scaffold for better bone healing (El-Fiqi et al. 2015). Ceramics comprised of calcium phosphate nano-formulates (HAp, beta-tricalcium phosphate, alpha-tricalcium phosphate, tetracalcium phosphate, dibasic calcium phosphate, amorphous calcium phosphate, etc.) has also been used for the delivery of drug, gene, protein, etc. where the delivery efficiency depends upon solubility. Liposome coated calcium phosphate NPs based construct has been used for gene delivery; for instance, the delivery of small interfering RNA (siRNA) to treat xenograft in tumor model where pH-dependent release of drug has been achieved. Range of anticancer drugs such as ocetaxel, doxorubicin hydrochloride, methotrexate, and 5-fluorouracil have also been effectively loaded and delivered through these NP-based targeted delivery and sustained release system under in vitro or in vivo trials. Similarly nano-structured calcium phosphates has also been found as an excellent delivery medium for plasmid DNA which is free from conventional viral vector system indicating the potential of these carrier systems for gene delivery (Choi et al. 2017). Silica-based mesophores has been found to be an excellent carrier for drug delivery to the target tissues but bone scaffold made solely from it may have lower strength, lesser apatite formation and bone regeneration, hence are usually doped with other materials to get desirable properties in addition to act as delivery system. The scope of supplying various growth and differentiation factors through these formulates also exists which can hasten the process of bone healing, more importantly in elderly and genetic or immuno-compromised subjects. For example, supply or regulation of bone morphogenetic protein (BMP) for mesenchymal cells differentiation into osteogen and cartilage tissue, transforming growth factor-b (TGF-b) for induction of stem cell, fibroblast growth factor (FGF) for stimulation, proliferation and differentiation of different cells including vascular endothelial cells, platelet-derived growth factor (PDGF) to promote mature osteoblast synthesis, interleukin-1 (IL-1) for type I collagen and activity of alkaline phosphatase inhibition etc. can be carried out using the scaffold-based delivery systems as per need and demand of subjects. CNT-mesoporous silica composites not only enhance adhesion, proliferation, and differentiation through electric stimulation but also help load more drugs as demonstrated by Vila et al. (2013) for zoledronic acid (Zol) and prolong drug action time (Jeong et al. 2019). Gold-based NPs has been found to generate immunomodulatory effect when loaded with mesoporous silica NPs where they stimulate secretion of osteogenic cytokines from macrophages in addition to producing anti-inflammatory response when studied on critical-sized cranial defect in rats (Liang et al. 2019). Various other NPs, particularly nano-composites, have been used for targeted delivery like liposomes for targeted release of p-coumaric acid to reduce inflammation, polymeric micelles for docetaxel and doxorubicin delivery to counter metastatic bone tumors, BMSC-exosome for osteoporosis, STExos for faster healing, exosome for dexona delivery and magnetic NPs for the delivery of diclofenac sodium are few examples in order to get additional advantage for particular drug supply to targeted locations. Similarly, bone morphogenetic proteins (BMPs) which is required for proper healing of bone as they have various activities like, BMP-1 induces cartilage formation, BMP-2, BMP-7 (OP1), BMP-9 (GDF2) and BMP-4 (BMP-2b) for skeletal repair and regeneration, BMP-3 (osteogenin) a negative regulator of bone morphogenesis, BMP-5 for the development of limb, morphogenesis of bone and cartilage and connecting soft tissues, BMP-6 (Vrg1, Dvr6) for morphogenesis of bone and cartilage hypertrophy and BMP-12 (GDF7/CDMP2) for ligament and tendon development, etc. can also be precisely supplied on case to case basis for faster healing of fractures as various clinical studies have established the role of such BMPs in bone healing. For instance BMP-2 has found utility for the management of mandibular fracture, open tibial shaft fractures, high risk ankle and hind foot fusion, BMP-7 in long bone non-unions, persistent non-union involving different bones and non-union of sacral fracture, etc. (Sheikh et al. 2015). Various researchers have used biomaterial-based nano-formulations for targeted delivery of these factors like BMP-2 loaded in PGA in dog periodontal repair, cervical spinal fusion in goats; in PLGA for orbital floor reconstruction in sheep, rabbit femoral head necrosis, rat calvaria, repair of canine mandibular defects; in PLGA-gelatine for the management of rabbit ulnar defects, rabbit knee cartilage/bone interface, dog tooth, and tibial defect and in PLGA-fibrin for the repairing of rabbit radial bone defect (Sheikh et al. 2015). Carbon nanotubes have also been used successfully as intracellular protein transporters indicating its potential as additional drug delivery system along with the conventional function of implant materials (Kam and Dai 2005). Continuous progress in nano-fabrication techniques and material science will certainly enhance the utility of these systems as an alternative delivery vehicle for diverse drugs and biologicals along with performing the usual role of orthopedic implants in order to hasten the healing process with lesser complications in animal subjects.

26.4 Future Prospects

Several NP-based formulations, implants, and devices have reached the market in recent times after regulatory approvals from different regulatory agencies for various diseases indicating the potential of nanotechnology for bench to field transition. In recent times, 3D printed HA-nano surface modified porous titanium implant has got regulatory approval for spinal injury correction (First 3D Printed HAnano Surface Modified Implant Gets FDA Approval, https://www.odtmag.com/contents/view_breaking-news/2020-08-25/first-3d-printed-hanano-surface-modified-implant-gets-fda-approval/). Various other bone substitutes have also received FDA clearance such as Vitoss® (Stryker): a nanoformulation of calcium phosphate, EquivaBone® (Zimmer Biomet), Ostim® (HeraseusKulzer), OsSatura® (IsoTisOrthobiologics), NanOss® (Rti surgical) and EquivaBone® (Zimmer Biomet): a Hydroxyapatite based formulation which mimics bone structure to help in adhesion and growth of cell for fracture healing. To overcome the challenges faced by veterinarians in fracture fixation in different animals, NP-based formulations and implants can have various positive implications not only in terms of increasing cell viability, adhesion, propagation, homing, osteogenic isolation and differentiation, vascularization, host integration, and load bearing but also supplying drugs for sustained release which can bring additional benefits to the table of fracture management in animals. Due to an enormous diversity of animal species, the range of size and composition of the bone varies dramatically that poses the major constrain for veterinarians; hence, fracture management in animals sometimes urges specialized solutions. For example, restraining animals and restricting movement is certainly a key challenge, so the post-operative care needs drugs having longer duration impact, can be implanted for patch burst release to control the infection and hasten the recovery process. This is even more useful in wild animals where post-operative care is very challenging. Application of 3D and 4D printing along with nano-formulations can provide implant materials of desirable quality and morphology that can be more suitable for the animal subjects. So, nanointervention in fracture management carries bright prospect but more focus towards animal subjects and the ample diversity of animal species needs to be taken care of adequately by capitalizing on the state-of-the art nano-fabrication techniques and advancements in material science.

References

Akturk A, Erol Taygun M, Goller G (2020) Optimization of the electrospinning process variables for gelatin/silver nanoparticles/bioactive glass nanocomposites for bone tissue engineering. Polym Compos 41:2411–2425. https://doi.org/10.1002/pc.25545

Altuna P, Lucas-Taulé E, Gargallo-Albiol J, Figueras-Álvarez O, Hernández-Alfaro F, Nart J (2016) Clinical evidence on titanium-zirconium dental implants: a systematic review and meta-analysis. Int J Oral Maxillofac Surg 45(7):842–850. https://doi.org/10.1016/j.ijom.2016.01.004

Anil S, Al-Sulaimani AF, Beeran AE, Chalisserry EP, Varma HP, Amri MDA (2015) Drug delivery systems in bone regeneration and implant dentistry. In: Turkyilmaz I (ed) Current concepts in dental implantology. IntechOpen, Rijeka, Croatia. https://doi.org/10.5772/60003

Bakhsheshi-Rad HR, Chen XB, Ismail AF, Aziz M, Abdolahi E, Mahmoodiyan F (2019) Improved antibacterial properties of an Mg-Zn-Ca alloy coated with chitosan nanofibers incorporating silver sulfadiazine multiwall carbon nanotubes for bone implants. Polym Adv Technol 30:1333–1339. https://doi.org/10.1002/pat.4563

Bhowmick A, Pramanik N, Jana P, Mitra T, Gnanamani A, Das M, Kundu PP (2017) Development of bone-like zirconium oxide nanoceramic modified chitosan based porous nanocomposites for biomedical application. Int J Biol Macromol 95:348–356. https://doi.org/10.1016/j.ijbiomac.2016.11.052

Bjursten LM, Rasmusson L, Oh S, Smith GC, Brammer KS, Jin S (2010) Titanium dioxide nanotubes enhance bone bonding in vivo. J Biomed Mater Res A 92(3):1218–1224. https://doi.org/10.1002/jbm.a.32463

Calandrelli L, Annunziata M, Della Ragione F, Laurienzo P, Malinconico M, Oliva A (2010) Development and performance analysis of PCL/silica nanocomposites for bone regeneration. J Mater Sci Mater Med 21(11):2923–2936. https://doi.org/10.1007/s10856-010-4156-8

Chen JF, Ding HM, Wang JX, Shao L (2004) Preparation and characterization of porous hollow silica nanoparticles for drug delivery application. Biomaterials 25(4):723–727. https://doi.org/10.1016/s0142-9612(03)00566-0

Cheng H, Li Y, Huo K, Gao B, Xiong W (2014) Long-lasting in vivo and in vitro antibacterial ability of nanostructured titania coating incorporated with silver nanoparticles. J Biomed Mater Res Part A 2014(102A):3488–3499. https://doi.org/10.1002/jbm.a.35019

Chew KK, Low KL, Sharif Zein SH, McPhail DS, Gerhardt LC, Roether JA, Boccaccini AR (2011) Reinforcement of calcium phosphate cement with multi-walled carbon nanotubes and bovine serum albumin for injectable bone substitute applications. J Mech Behav Biomed Mater 4(3):331–339. https://doi.org/10.1016/j.jmbbm.2010.10.013

Choi AH, Macha IJ, Akyol S, Cazalbou S, Ben-Nissan B (2017) Nanostructured calcium phosphates for drug, gene, DNA and protein delivery and as anticancer chemotherapeutic devices. In: Kaur G (ed) Clinical applications of biomaterials. Springer, Cham. https://doi.org/10.1007/978-3-319-56059-5_6

Correia CO, Leite ÁJ, Mano JF (2015) Chitosan/bioactive glass nanoparticles scaffolds with shape memory properties. Carbohydr Polym 123:39–45. https://doi.org/10.1016/j.carbpol.2014.12.076

Cui Y, Liu Y, Cui Y, Jing X, Zhang P, Chen X (2009) The nanocomposite scaffold of poly(lactide-co-glycolide) and hydroxyapatite surface-grafted with L-lactic acid oligomer for bone repair. Acta Biomater 5(7):2680–2692. https://doi.org/10.1016/j.actbio.2009.03.024

Das R, Curry EJ, Le T, Awale G, Liu Y, Li S, Contreras J, Bednarz C, Millender J, Xin X, Rowe DW, Emadi S, Lo KW, Nguyen TD (2020) Biodegradable nanofiber bone-tissue scaffold as remotely-controlled and self-powering electrical stimulator. Nano Energy 76:105028

de Carvalho JO, de Carvalho OF, Freitas SAP, Soares LM, de Cássia Barros Lima R, de Sousa Gonçalves L, Webster TJ, Marciano FR, Lobo AO (2018) Carbon nanomaterials for treating osteoporotic vertebral fractures. Curr Osteoporos Rep 16(5):626–634. https://doi.org/10.1007/s11914-018-0476-2

de Moura NK, Martins EF, Oliveira RLMS, de Brito Siqueira IAW, Machado JPB, Esposito E, Amaral SS, de Vasconcellos LMR, Passador FR, de Sousa TE (2020) Synergistic effect of adding bioglass and carbon nanotubes on poly (lactic acid) porous membranes for guided bone regeneration. Mater Sci Eng C Mater Biol Appl 117:111327. https://doi.org/10.1016/j.msec.2020.111327

Dong H, Virtanen S (2022) Influence of bovine serum albumin on biodegradation behavior of pure Zn. J Biomed Mater Res B Appl Biomater 110(1):185–194. https://doi.org/10.1002/jbm.b.34901

Dusad A, Chakkalakal DA, Namavar F, Haider H, Hanisch B, Duryee MJ, Diaz A, Rensch A, Zhang Y, Hess R, Thiele GM, Fehringer EV (2013) Titanium implant with nanostructured zirconia surface promotes maturation of peri-implant bone in osseointegration. Proc Inst Mech Eng H 227(5):510–522. https://doi.org/10.1177/0954411913479300

Eivazzadeh-Keihan R, Bahojb Noruzi E, Khanmohammadi Chenab K, Jafari A, Radinekiyan F, Hashemi SM, Ahmadpour F, Behboudi A, Mosafer J, Mokhtarzadeh A, Maleki A, Hamblin MR (2020) Metal-based nanoparticles for bone tissue engineering. J Tissue Eng Regen Med 14(12): 1687–1714. https://doi.org/10.1002/term.3131

El-Fiqi A, Kim JH, Kim HW (2015) Osteoinductive fibrous scaffolds of biopolymer/mesoporous bioactive glass nanocarriers with excellent bioactivity and long-term delivery of osteogenic drug. ACS Appl Mater Interfaces 7(2):1140–1152. https://doi.org/10.1021/am5077759

Ewald A, Käppel C, Vorndran E, Moseke C, Gelinsky M, Gbureck U (2012) The effect of Cu(II)-loaded brushite scaffolds on growth and activity of osteoblastic cells. J Biomed Mater Res A 100(9):2392–2400. https://doi.org/10.1002/jbm.a.34184

Fujihara K, Kotaki M, Ramakrishna S (2005) Guided bone regeneration membrane made of polycaprolactone/calcium carbonate composite nano-fibers. Biomaterials 26(19):4139–4147. https://doi.org/10.1016/j.biomaterials.2004.09.014

Gaihre B, Uswatta S, Jayasuriya AC (2018) Nano-scale characterization of nano-hydroxyapatite incorporated chitosan particles for bone repair. Colloids Surf B Biointerfaces 165:158–164. https://doi.org/10.1016/j.colsurfb.2018.02.034

Gouda M (2012) Nano-zirconium oxide and nano-silver oxide/cotton gauze fabrics for antimicrobial and wound healing acceleration. J Ind Text 41(3):222–240. https://doi.org/10.1177/1528083711414960

Hamido F, Misfer AK, Al Harran H, Khadrawe TA, Soliman A, Talaat A, Awad A, Khairat S (2011) The use of the LARS artificial ligament to augment a short or undersized ACL hamstrings tendon graft. Knee 18(6):373–378. https://doi.org/10.1016/j.knee.2010.09.003

Hassan M, Fetecau C, Majeed A, Zeeshan A (2018) Effects of iron nanoparticles' shape on convective flow of ferrofluid under highly oscillating magnetic field over stretchable rotating disk. J Magn Magn Mater 465:531–539. https://doi.org/10.1016/j.jmmm.2018.06.019

He W, Kienzle A, Liu X, Müller WE, Elkhooly TA, Feng Q (2016a) In vitro effect of 30 nm silver nanoparticles on adipogenic differentiation of human mesenchymal stem cells. J Biomed Nanotechnol 12(3):525–535. https://doi.org/10.1166/jbn.2016.2182

He LY, Zhang XM, Liu B, Tian Y, Ma WH (2016b) Effect of magnesium ion on human osteoblast activity. Braz J Med Biol Res 49(7):e5257. https://doi.org/10.1590/1414-431X20165257

Hejazy M, Koohi M, Bassiri Mohamad Pour A, Najafi D (2018) Toxicity of manufactured copper nanoparticles—A review. Nanomed Res J 3(1):1–9. https://doi.org/10.22034/nmrj.2018.01.001

Heo DN, Ko WK, Bae MS, Lee JB, Lee W, Byun W, Lee CH, Kim EC, Jung BY, Kwon IK (2014) Enhanced bone regeneration with a gold nanoparticle-hydrogel complex. J Mater Chem B 2(11):1584–1593. https://doi.org/10.1039/c3tb21246g

Hirata E, Uo M, Takita H, Akasaka T, Watari F, Yokoyama A (2011) Multiwalled carbon nanotube-coating of 3D collagen scaffolds for bone tissue engineering. Carbon 49(10):3284–3291. https://doi.org/10.1016/j.carbon.2011.04.002

Hong Z, Qiu X, Sun J, Deng M, Chen X, Jing X (2004) Grafting polymerization of L-lactide on the surface of hydroxyapatite nano-crystals. Polymer 45(19):6699–6706. https://doi.org/10.1016/j.biomaterials.2005.04.018

Hu C, Sun J, Long C, Wu L, Zhou C, Zhang X (2019) Synthesis of nano zirconium oxide and its application in dentistry. Nanotechnol Rev 8(1):396–404. https://doi.org/10.1515/ntrev-2019-0035

Ingle AP, Duran N, Rai M (2014) Bioactivity, mechanism of action, and cytotoxicity of copper-based nanoparticles: a review. Appl Microbiol Biotechnol 98(3):1001–1009. https://doi.org/10.1007/s00253-013-5422-8

Jaidev LR, Kumar S, Chatterjee K (2017) Multi-biofunctional polymer graphene composite for bone tissue regeneration that elutes copper ions to impart angiogenic, osteogenic and bactericidal properties. Colloids Surf B Biointerfaces 159:293–302. https://doi.org/10.1016/j.colsurfb.2017.07.083

Jangra SL, Stalin K, Dilbaghi N, Kumar S, Tawale J, Singh SP, Pasricha R (2012) Antimicrobial activity of zirconia (ZrO2) nanoparticles and zirconium complexes. J Nanosci Nanotechnol 12(9):7105–7112. https://doi.org/10.1166/jnn.2012.6574

Jeong JH, Back SK, An JH, Lee N-S, Kim D-K, Na CS, Jeong Y-G, Han SY (2019) Topical film prepared with Rhus verniciflua extract-loaded pullulan hydrogel for atopic dermatitis treatment. J Biomed Mater Res Part B 2019(107B):2325–2334. https://doi.org/10.1002/jbm.b.34325

Jing Z, Wu Y, Su W, Tian M, Jiang W, Cao L, Zhao L, Zhao Z (2017) Carbon nanotube reinforced collagen/hydroxyapatite scaffolds improve bone tissue formation in vitro and in vivo. Ann Biomed Eng 45(9):2075–2087. https://doi.org/10.1007/s10439-017-1866-9

Kam NW, Dai H (2005) Carbon nanotubes as intracellular protein transporters: generality and biological functionality. J Am Chem Soc 127(16):6021–6026. https://doi.org/10.1021/ja050062v

Kokorev OV, Hodorenko VN, Chekalkin TL, Kim JS, Kang SB, Dambaev GTs, Gunther VE. (2016) In vitro and in vivo evaluation of porous TiNi-based alloy as a scaffold for cell tissue engineering. Artif cells Nanomed Biotechnol 44(2):704–709. https://doi.org/10.3109/21691401.2014.982799

Kumar Saini R, Prasad Bagri L, Bajpai AK (2019) Nano-silver hydroxyapatite based antibacterial 3D scaffolds of gelatin/alginate/poly (vinyl alcohol) for bone tissue engineering applications. Colloids Surf B Biointerfaces 177:211–218. https://doi.org/10.1016/j.colsurfb.2019.01.064

Li Y, Fan L, Liu S, Liu W, Zhang H, Zhou T, Wu D, Yang P, Shen L, Chen J, Jin Y (2013) The promotion of bone regeneration through positive regulation of angiogenic-osteogenic coupling using microRNA-26a. Biomaterials 34(21):5048–5058. https://doi.org/10.1016/j.biomaterials.2013.03.052

Liang H, Jin C, Ma L, Feng X, Deng X, Wu S, Liu X, Yang C (2019) Accelerated bone regeneration by gold-nanoparticle-loaded mesoporous silica through stimulating immunomodulation. ACS Appl Mater Interfaces 11(44):41758–41769. https://doi.org/10.1021/acsami.9b16848

Liao S, Wang W, Uo M, Ohkawa S, Akasaka T, Tamura K, Cui F, Watari F (2005) A three-layered nano-carbonated hydroxyapatite/collagen/PLGA composite membrane for guided tissue regeneration. Biomaterials 26(36):7564–7571. https://doi.org/10.1016/j.biomaterials.2005.05.050

Liu X, Lee P-Y, Ho C-M, Lui V, Chen Y, Che C-M, Tam P, Wong K (2010) Silver nanoparticles mediate differential responses in keratinocytes and fibroblasts during skin wound healing. ChemMedChem 5:468–475. https://doi.org/10.1002/cmdc.200900502

Liu S, Sun Y, Fu Y, Chang D, Fu C, Wang G, Liu Y, Tay FR, Zhou Y (2016) Bioinspired collagen-apatite nanocomposites for bone regeneration. J Endod 42(8):1226–1232. https://doi.org/10.1016/j.joen.2016.04.027

Mahmood SK, Razak IA, Ghaji MS, Yusof LM, Mahmood ZK, Rameli MABP, Zakaria ZAB (2017) In vivo evaluation of a novel nanocomposite porous 3D scaffold in a rabbit model: histological analysis. Int J Nanomedicine 12:8587–8598. https://doi.org/10.2147/IJN.S145663

Maleksaeedi S, Wang J, El-Hajje A, Harb L, Guneta V, He Z, Wiria F, Choong C, Ruys A (2013) Toward 3D printed bioactive titanium scaffolds with bimodal pore size distribution for bone ingrowth. Procedia CIRP 5:158–163. https://doi.org/10.1016/j.procir.2013.01.032

Marestoni LD, Barud HS, Gomes RJ, Catarino RPF, Hata NNY, Ressutte JB, Spinosa WA (2020) Commercial and potential applications of bacterial cellulose in Brazil: ten years review. Polímeros Ciência Tecnol 30(4):e2020047. https://doi.org/10.1590/0104-1428.09420

McMahon RE, Wang L, Skoracki R, Mathur AB (2013) Development of nanomaterials for bone repair and regeneration. J Biomed Mater Res 101B:387–397. https://doi.org/10.1002/jbm.b.32823

Minakshi P, Ghosh M, Kumar R, Brar B, Lambe UP, Banerjee S, Ranjan K, Kumar B, Goel P, Malik YS, Prasad G (2020) An insight into nanomedicinal approaches to combat viral zoonoses. Curr Top Med Chem 20(11):915–962. https://doi.org/10.2174/1568026620666200325114400

Mota J, Yu N, Caridade SG, Luz GM, Gomes ME, Reis RL, Jansen JA, Walboomers XF, Mano JF (2012) Chitosan/bioactive glass nanoparticle composite membranes for periodontal regeneration. Acta Biomater 8(11):4173–4180. https://doi.org/10.1016/j.actbio.2012.06.040

Mukherjee S, Nandi SK, Kundu B, Chanda A, Sen S, Das PK (2016) Enhanced bone regeneration with carbon nanotube reinforced hydroxyapatite in animal model. J Mech Behav Biomed Mater 60:243–255. https://doi.org/10.1016/j.jmbbm.2016.02.005

Okamoto M, John B (2013) Synthetic biopolymer nanocomposites for tissue engineering scaffolds. Prog Polym Sci 38(10–11):1487–1503. https://doi.org/10.1016/j.progpolymsci.2013.06.001

Oliveira JM, Sousa RA, Malafaya PB, Silva SS, Kotobuki N, Hirose M, Ohgushi H, Mano JF, Reis RL (2011) In vivo study of dendronlike nanoparticles for stem cells "tune-up": from nano to tissues. Nanomedicine 7(6):914–924. https://doi.org/10.1016/j.nano.2011.03.002

Oliveira FC, Carvalho JO, Gusmão SBS, Gonçalves LS, Soares Mendes LM, Freitas SAP, Gusmão GOM, Viana BC, Marciano FR, Lobo AO (2019) High loads of nano-hydroxyapatite/graphene nanoribbon composites guided bone regeneration using an osteoporotic animal model. Int J Nanomedicine 14:865–874. https://doi.org/10.2147/IJN.S192456

Panseri S, Cunha C, D'Alessandro T, Sandri M, Russo A, Giavaresi G, Marcacci M, Hung CT, Tampieri A (2012) Magnetic hydroxyapatite bone substitutes to enhance tissue regeneration: evaluation in vitro using osteoblast-like cells and in vivo in a bone defect. PLoS One 7(6): e38710. https://doi.org/10.1371/journal.pone.0038710

Parchi PD, Gianluca C, Dolfi L, Baluganti A, Nicola P, Chiellini F, Lisanti M (2013) Anterior cruciate ligament reconstruction with LARS™ artificial ligament results at a mean follow-up of eight years. Int Orthop 37(8):1567–1574. https://doi.org/10.1007/s00264-013-1917-2

Park JE, Jang YS, Bae TS, Lee MH (2019) Biocompatibility characteristics of titanium coated with multi walled carbon nanotubes-hydroxyapatite nanocomposites. Materials (Basel) 12(2):224. https://doi.org/10.3390/ma12020224

Pei B, Wang W, Dunne N, Li X (2019) Applications of carbon nanotubes in bone tissue regeneration and engineering: superiority, concerns, current advancements, and prospects. Nanomaterials (Basel) 9(10):1501. https://doi.org/10.3390/nano9101501

Pek YS, Gao S, Arshad MS, Leck KJ, Ying JY (2008) Porous collagen-apatite nanocomposite foams as bone regeneration scaffolds. Biomaterials 29(32):4300–4305. https://doi.org/10.1016/j.biomaterials.2008.07.030

Phetnin R, Rattanachan ST (2015) Preparation and antibacterial property on silver incorporated mesoporous bioactive glass microspheres. J Sol-Gel Sci Technol 75:279–290. https://doi.org/10.1007/s10971-015-3697-1

Phipps MC, Clem WC, Catledge SA, Xu Y, Hennessy KM, Thomas V, Jablonsky MJ, Chowdhury S, Stanishevsky AV, Vohra YK, Bellis SL (2011) Mesenchymal stem cell responses to bone-mimetic electrospun matrices composed of polycaprolactone, collagen I and nanoparticulate hydroxyapatite. PLoS One 6(2):e16813. https://doi.org/10.1371/journal.pone.0016813

Pinna A, Torki Baghbaderani M, Vigil Hernández V, Naruphontjirakul P, Li S, McFarlane T, Hachim D, Stevens MM, Porter AE, Jones JR (2021) Nanoceria provides antioxidant and osteogenic properties to mesoporous silica nanoparticles for osteoporosis treatment. Acta Biomater 122:365–376. https://doi.org/10.1016/j.actbio.2020.12.029

Prasad M, Ghosh M, Kumar R, Brar B, Surjith KP, Lambe UP, Ranjan K, Banerjee S, Prasad G, Kumar Khurana S, Kharb P (2021) The importance of nanomedicine in prophylactic and theranostic intervention of bacterial zoonoses and reverse zoonoses in the era of microbial resistance. J Nanosci Nanotechnol 21(6):3404–3452. https://doi.org/10.1166/jnn.2021.18999

Prowans P, Kowalczyk R, Wiszniewska B, Czapla N, Bargiel P, El Fray M (2019) Bone healing in the presence of a biodegradable PBS-DLA copolyester and its composite containing hydroxyapatite. ACS Omega 4(22):19765–19771. https://doi.org/10.1021/acsomega.9b02539

Qadri S, Haik Y, Mensah-Brown E, Bashir G, Fernandez-Cabezudo MJ, Al-Ramadi BK (2017) Metallic nanoparticles to eradicate bacterial bone infection. Nanomedicine 13(7):2241–2250. https://doi.org/10.1016/j.nano.2017.05.013

Qadri SM, Abdulrehman T, Haik Y (2018) Targeted osteomyelitis management using metallic nanoparticles. Proceedings of the 2nd International Conference of Theoretical and Applied Nanoscience and Nanotechnology (TANN'18), Niagara Falls, Canada – June 10–12, 2018, Paper No. 123. https://doi.org/10.11159/tann18.123

Rigiracciolo DC, Scarpelli A, Lappano R, Pisano A, Santolla MF, De Marco P, Cirillo F, Cappello AR, Dolce V, Belfiore A, Maggiolini M, De Francesco EM (2015) Copper activates HIF-1α/GPER/VEGF signalling in cancer cells. Oncotarget 6(33):34158–34177. https://doi.org/10.18632/oncotarget.5779

Roohani-Esfahani SI, Nouri-Khorasani S, Lu Z, Appleyard R, Zreiqat H (2010) The influence hydroxyapatite nanoparticle shape and size on the properties of biphasic calcium phosphate scaffolds coated with hydroxyapatite-PCL composites. Biomaterials 31(21):5498–5509. https://doi.org/10.1016/j.biomaterials.2010.03.058

Saeed BA, Lim V, Yusof NA, Khor KZ, Rahman HS, Abdul Samad N (2019) Antiangiogenic properties of nanoparticles: a systematic review. Int J Nanomedicine 14:5135–5146. https://doi.org/10.2147/IJN.S199974

Saska S, Teixeira LN, de Castro Raucci LMS, Scarel-Caminaga RM, Franchi LP, Dos Santos RA, Santagneli SH, Capela MV, de Oliveira PT, Takahashi CS, Gaspar AMM, Messaddeq Y, Ribeiro SJL, Marchetto R (2017) Nanocellulose-collagen-apatite composite associated with osteogenic growth peptide for bone regeneration. Int J Biol Macromol 103:467–476. https://doi.org/10.1016/j.ijbiomac.2017.05.086

Sheikh Z, Javaid MA, Hamdan N, Hashmi R (2015) Bone regeneration using bone morphogenetic proteins and various biomaterial carriers. Materials (Basel). 8(4):1778–1816. https://doi.org/10.3390/ma8041778

Silva E, Vasconcellos LMR, Rodrigues BVM, Dos Santos DM, Campana-Filho SP, Marciano FR, Webster TJ, Lobo AO (2017) PDLLA honeycomb-like scaffolds with a high loading of superhydrophilic graphene/multi-walled carbon nanotubes promote osteoblast in vitro functions and guided in vivo bone regeneration. Mater Sci Eng C Mater Biol Appl 73:31–39. https://doi.org/10.1016/j.msec.2016.11.075

Singh RK, Patel KD, Lee JH, Lee EJ, Kim JH, Kim TH, Kim HW (2014) Potential of magnetic nanofiber scaffolds with mechanical and biological properties applicable for bone regeneration. PLoS One 9(4):e91584. https://doi.org/10.1371/journal.pone.0091584

Song J-H, Kim H-E, Kim H-W (2007) Collagen-apatite nanocomposite membranes for guided bone regeneration. J Biomed Mater Res 83B:248–257. https://doi.org/10.1002/jbm.b.30790

Song W, Kaimin W, Yan J, Zhang Y, Zhao L (2013) MiR-148b laden titanium implant promoting osteogenic differentiation of rat bone marrow mesenchymal stem cells. RSC Adv 3(28): 11292–11300. https://doi.org/10.1039/C3RA00007A

Song F, Jie W, Zhang T, Li W, Jiang Y, Wan L, Li X, Liu B, Liu W (2016) Room-temperature fabrication of a three-dimensional reduced-graphene oxide/polypyrrole/hydroxyapatite composite scaffold for bone tissue engineering. RSC Adv 6(95):92804–92812. https://doi.org/10.1039/C6RA15267H

Tanahashi M, Matsuda T (1997) Surface functional group dependence on apatite formation on self-assembled monolayers in a simulated body fluid. J Biomed Mater Res 34(3):305–315. https://doi.org/10.1002/(sici)1097-4636(19970305)34:3<305::aid-jbm5>3.0.co;2-o

Thurner GC, Haybaeck J, Debbage P (2021) Targeting drug delivery in the elderly: are nanoparticles an option for treating osteoporosis? Int J Mol Sci 22(16):8932. https://doi.org/10.3390/ijms22168932

Tian J, Wong KK, Ho CM, Lok CN, Yu WY, Che CM, Chiu JF, Tam PK (2007) Topical delivery of silver nanoparticles promotes wound healing. ChemMedChem 2(1):129–136. https://doi.org/10.1002/cmdc.200600171

Tripathi A, Saravanan S, Pattnaik S, Moorthi A, Partridge NC, Selvamurugan N (2012) Bio-composite scaffolds containing chitosan/nano-hydroxyapatite/nano-copper-zinc for bone tissue engineering. Int J Biol Macromol 50(1):294–299. https://doi.org/10.1016/j.ijbiomac.2011.11.013

Turnbull G, Clarke J, Picard F, Riches P, Jia L, Han F, Li B, Shu W (2017) 3D bioactive composite scaffolds for bone tissue engineering. Bioact Mater 3(3):278–314. https://doi.org/10.1016/j.bioactmat.2017.10.001

Vila M, Cicuéndez M, Sánchez-Marcos J, Fal-Miyar V, Manzano M, Prieto C, Vallet-Regi M (2013) Electrical stimuli to increase cell proliferation on carbon nanotubes/mesoporous silica composites for drug delivery. J Biomed Mater Res 101A:213–221. https://doi.org/10.1002/jbm.a.34325

Wang Y, Dai J, Zhang Q, Xiao Y, Lang M (2010) Improved mechanical properties of hydroxyapatite/poly(ε-caprolactone) scaffolds by surface modification of hydroxyapatite. Appl Surfac Sci 256(20):6107–6112. https://doi.org/10.1016/j.apsusc.2010.03.127

Wang H, Zhao S, Cui X, Pan Y, Huang W, Ye S, Luo S, Rahaman MN, Zhang C, Wang D (2015) Evaluation of three-dimensional silver-doped borate bioactive glass scaffolds for bone repair: biodegradability, biocompatibility, and antibacterial activity. J Mater Res 30(18):2722–2735. https://doi.org/10.1557/jmr.2015.243

White AA, Best SM, Kinloch IA (2007) Hydroxyapatite–carbon nanotube composites for biomedical applications: a review. Int J Appl Ceram Technol 4:1–13. https://doi.org/10.1111/j.1744-7402.2007.02113.x

Xu ZL, Lei Y, Yin WJ, Chen YX, Ke QF, Guo YP, Zhang CQ (2016) Enhanced antibacterial activity and osteoinductivity of Ag-loaded strontium hydroxyapatite/chitosan porous scaffolds for bone tissue engineering. J Mater Chem B 4(48):7919–7928. https://doi.org/10.1039/c6tb01282e

Yang F, Both SK, Yang X, Walboomers XF, Jansen JA (2009) Development of an electrospun nano-apatite/PCL composite membrane for GTR/GBR application. Acta Biomater 5(9): 3295–3304. https://doi.org/10.1016/j.actbio.2009.05.023

Yoon IK, Hwang JY, Seo JW, Jang WC, Kim HW, Shin US (2014) Carbon nanotube-gelatin-hydroxyapatite nanohybrids with multilayer core–shell structure for mimicking natural bone. Carbon 77:379–389. https://doi.org/10.1016/j.carbon.2014.05.041

Yu M, Lei B, Gao C, Yan J, Ma PX (2017) Optimizing surface-engineered ultra-small gold nanoparticles for highly efficient miRNA delivery to enhance osteogenic differentiation of bone mesenchymal stromal cells. Nano Res 10:49–63. https://doi.org/10.1007/s12274-016-1265-9

Yu B, Fu S, Kang Z, Zhu M, Ding H, Luo T, Zhu Y, Zhang Y (2020) Enhanced bone regeneration of 3D printed β-Ca2SiO4 scaffolds by aluminum ions solid solution. Ceram Int 46:7783–7791. https://doi.org/10.1016/j.ceramint.2019.11.282

Zhang H (2011) Electrospun poly (lactic-co-glycolic acid)/multiwalled carbon nanotubes composite scaffolds for guided bone tissue regeneration. J Bioact Compat Polym 26(4):347–362. https://doi.org/10.1177/0883911511413450

Zhang L, Ramsaywack S, Fenniri H, Webster TJ (2008) Enhanced osteoblast adhesion on self-assembled nanostructured hydrogel scaffolds. Tissue Eng Part A 14(8):1353–1364. https://doi.org/10.1089/ten.tea.2006.0436

Zhang P, Hong Z, Yu T, Chen X, Jing X (2009) In vivo mineralization and osteogenesis of nanocomposite scaffold of poly(lactide-co-glycolide) and hydroxyapatite surface-grafted with poly(L-lactide). Biomaterials 30(1):58–70. https://doi.org/10.1016/j.biomaterials.2008.08.041

Zhang Q, Mochalin VN, Neitzel I, Knoke IY, Han J, Klug CA, Zhou JG, Lelkes PI, Gogotsi Y (2011) Fluorescent PLLA-nanodiamond composites for bone tissue engineering. Biomaterials 32(1):87–94. https://doi.org/10.1016/j.biomaterials.2010.08.090

Zhang R, Lee P, Lui VC, Chen Y, Liu X, Lok CN, To M, Yeung KW, Wong KK (2015) Silver nanoparticles promote osteogenesis of mesenchymal stem cells and improve bone fracture healing in osteogenesis mechanism mouse model. Nanomedicine 11(8):1949–1959. https://doi.org/10.1016/j.nano.2015.07.016

Zhao L, Wang H, Huo K, Cui L, Zhang W, Ni H, Zhang Y, Wu Z, Chu PK (2011) Antibacterial nano-structured titania coating incorporated with silver nanoparticles. Biomaterials 32(24): 5706–5716. https://doi.org/10.1016/j.biomaterials.2011.04.040

Zhao YZ, Chen R, Xue PP, Luo LZ, Zhong B, Tong MQ, Chen B, Yao Q, Yuan JD, Xu HL (2021) Magnetic PLGA microspheres loaded with SPIONs promoted the reconstruction of bone defects through regulating the bone mesenchymal stem cells under an external magnetic field. Mater Sci Eng C Mater Biol Appl 122:111877. https://doi.org/10.1016/j.msec.2021.111877

Zhou J, Li B, Lu S, Zhang L, Han Y (2013) Regulation of osteoblast proliferation and differentiation by interrod spacing of Sr-HA nanorods on microporous titania coatings. ACS Appl Mater Interfaces 5(11):5358–5365. https://doi.org/10.1021/am401339n

Zhuang Y, Lin K, Yu H (2019) Advance of nano-composite electrospun fibers in periodontal regeneration. Front Chem 7:495. https://doi.org/10.3389/fchem.2019.00495

Chapter 27
Nanotechnology Based Strategies for the Improvement of Conventional Diagnostics and Therapeutics in Veterinary Medicine

Mayukh Ghosh, Alonkrita Chowdhury, and Harshad Sudhir Patki

Abstract Nanotechnology offers promising strategies to enhance conventional diagnostics and therapeutics in veterinary medicine. The unique properties of materials at the nanoscale enable precise targeting, controlled drug delivery, improved imaging, and innovative diagnostic techniques. This abstract provides an overview of nanotechnology-based approaches for the improvement of diagnostics and therapeutics in veterinary medicine. In diagnostics, nanotechnology enables the development of sensitive and specific detection methods. Nanosensors and nanoparticles can be utilized as diagnostic tools to detect biomarkers, pathogens, or abnormal cellular processes in animals. Nanomaterials act as contrast agents in imaging techniques, providing enhanced visualization of tissues and structures, aiding in accurate diagnosis and treatment planning. Nanotechnology may also extensively transform therapeutic interventions in veterinary medicine. In addition to their inherent therapeutic potential, nanoparticles serve as carriers for drugs, allowing for targeted and controlled drug delivery to specific tissues or cells. This enhances therapeutic efficacy while minimizing the drug-dose requirement and systemic side effects. Nanomaterials can be engineered to encapsulate therapeutic agents, enabling sustained release and improved bioavailability. The unique antimicrobial potential of novel nanomaterials also helps to counter the emerging issue of drug resistance. Moreover, nanotechnology offers opportunities for personalized medicine by

M. Ghosh (✉)
Department of Veterinary Physiology and Biochemistry, RGSC, Banaras Hindu University, Mirzapur, Uttar Pradesh, India

A. Chowdhury
Department of Plant Biotechnology, Institute of Agricultural Sciences, RGSC, Banaras Hindu University, Mirzapur, Uttar Pradesh, India

H. S. Patki
Department of Veterinary Anatomy and Histology, College of Veterinary and Animal Sciences, Kerala Veterinary and Animal Science University, Pookode, Wayanad, Kerala, India

LLR University of Veterinary and Animal, ICAR-National Research Centre on Equines, *Nanotechnology Theranostics in Livestock Diseases and Management*, Livestock Diseases and Management,
https://doi.org/10.1007/978-981-16-1610-5_27

tailoring treatments to individual animals' needs. The integration of nanotechnology with other emerging fields, such as genomics and artificial intelligence, opens doors to precision medicine in animals. Nanotechnology can facilitate the convergence of technologies, enabling precise disease characterization, targeted treatment interventions, and optimized therapeutic regimens. However, the translation of nanotechnology into routine veterinary practice requires careful consideration of safety, regulatory frameworks, and education. Persistent research and development are necessary to refine nanomaterial design, improve biocompatibility, and assess potential risks associated with their use. Overall, nanotechnology-based strategies hold tremendous potential to revolutionize conventional diagnostics and therapeutics in veterinary medicine. Further advancements in the field, coupled with regulatory guidance and increased awareness, will pave the way for improved patient outcomes and personalized animal healthcare.

Keywords Nanodiagnostics · Nanosensors · Lab-on-a-chip devices · Diagnostic imaging · Nanocarriers · Nanovaccines · Nanotherapeutics · Combination therapies · Livestock

27.1 Introduction

Nanotechnology, the science and engineering of manipulating matter at the nanoscale, has revolutionized various fields including medicine and healthcare. Its application in veterinary medicine is gaining significant attention due to its potential to improve conventional diagnostics and therapeutics. Nanotechnology offers unique advantages such as increased surface area, enhanced drug delivery, and improved targeting capabilities, which can address challenges faced in veterinary diagnostics and therapeutics. This paper aims to explore the potential of nanotechnology-based strategies in veterinary medicine, highlighting their role in improving the accuracy and efficiency of diagnostics and enhancing the efficacy of therapeutic interventions.

Conventional diagnostics and therapeutics in veterinary medicine often rely on traditional methods that may have limitations in terms of sensitivity, specificity, effectiveness, time and resource requirement, and point-of-care option. Nanotechnology-based approaches can overcome these limitations by introducing innovative techniques and materials at the nanoscale. Nanoparticles, nanosensors, nanocarriers, and nanodevices are some of the nanotechnology tools that can be harnessed to enhance veterinary diagnostics and therapeutics.

In the field of diagnostics, nanotechnology offers opportunities for improved disease detection and monitoring. Nanoparticles functionalized with specific ligands can be used as diagnostic agents to detect biomarkers associated with various veterinary diseases. These nanoparticles can be engineered to enhance signal amplification, enabling sensitive and accurate detection even at low biomarker concentrations. Additionally, nanosensors can be developed to provide real-time monitoring of physiological parameters in animals, enabling early detection of diseases and facilitating timely intervention (Minakshi et al. 2020a; Prasad et al. 2021).

Nanotechnology also holds promise in the field of therapeutics in veterinary medicine. Several nanoparticles have exhibited inherent therapeutic action against diverse microbial pathogens of veterinary importance. Even, nanomedicine has found utility in managing different metabolic and reproductive disorders. In addition, nanocarriers, such as liposomes and polymeric nanoparticles, can be utilized to improve drug delivery systems. These nanocarriers can protect drugs from degradation, enhance their stability, and provide controlled release, leading to improved drug efficacy and reduced side effects. Furthermore, the surface of nanocarriers can be functionalized with targeting ligands, allowing site-specific delivery of therapeutic agents to the desired tissues or cells, thereby increasing the treatment efficiency (Minakshi et al. 2019, 2020a,b; Prasad et al. 2021).

Moreover, nanotechnology-based strategies can aid in the development of novel veterinary vaccines. Nanoparticle-based vaccine formulations have the potential to enhance immune responses by improving antigen presentation and targeting specific immune cells. These formulations can also facilitate the delivery of multiple antigens simultaneously, leading to broader immune protection against various pathogens. Additionally, nanotechnology can play a crucial role in developing new generation veterinary antimicrobials, tackling the challenge of antimicrobial resistance by enabling targeted and controlled drug delivery to combat infectious diseases in animals (Dhakal and Renukaradhya 2019).

Overall, nanotechnology-based strategies offer immense potential for the improvement of conventional diagnostics and therapeutics in veterinary medicine. By harnessing the unique properties of nanomaterials and nanoscale devices, veterinary diagnostics can be made more sensitive, accurate, and efficient. Therapeutics can be enhanced through improved drug delivery systems, targeted therapies, and the development of innovative vaccines and antimicrobials. The integration of nanotechnology into veterinary medicine holds promise for advancing animal healthcare and welfare, ultimately benefiting both animals and their human companions.

27.2 Conventional Diagnostics in Veterinary Medicine

Conventional diagnostics in veterinary medicine refer to the traditional methods and techniques used by veterinarians to diagnose diseases and conditions in animals. These methods have been widely used for many years and form the foundation of veterinary diagnostics. Some of the commonly employed conventional diagnostic techniques in veterinary medicine include the following:

27.2.1 Physical Examination

A thorough physical examination is often the first step in diagnosing a veterinary patient. The veterinarian visually inspects the animal, palpates different body parts, listens to heart and lung sounds, and checks for any abnormalities. This initial assessment helps in identifying visible signs of illness or injury.

27.2.2 Medical History

Obtaining a detailed medical history from the animal's owner or caretaker is crucial for diagnostic purposes. Information about the animal's symptoms, duration of illness, previous medical conditions, and medications can provide valuable insights and guide further diagnostic procedures (Rich 1991; Tully 1998).

27.2.3 Hematobiochemical Analyses

Blood tests are commonly used in veterinary medicine to assess the overall health status of an animal. Complete blood count (CBC) provides information about red and white blood cells, hemoglobin levels, and platelet counts. Biochemical profiles measure various parameters such as liver and kidney function, electrolyte levels, and blood glucose.

27.2.4 Urinalysis

Urinalysis involves the examination of urine to evaluate kidney function, detect urinary tract infections, and identify abnormalities such as the presence of blood, crystals, or abnormal cells. It provides valuable information about the animal's overall health and can aid in diagnosing urinary system disorders (Das et al. 2017; Sarangi et al. 2022).

27.2.5 Imaging Techniques

Radiography (X-rays), ultrasonography, and computed tomography (CT) scans are commonly used imaging techniques in veterinary medicine. They help visualize the internal structures of the body, detect fractures, identify tumors or abnormal growths, and assess organ size and shape (Prasad et al. 2021b).

27.2.6 Microscopic Examination

Microscopic examination involves the analysis of samples such as blood smears, fecal samples, or tissue biopsies under a microscope. It helps in identifying parasites, abnormal cells, bacteria, or fungi, aiding in the diagnosis of various infections and diseases (Kadam et al. 2022).

27.2.7 Culture and Sensitivity Testing

Culture and sensitivity testing involves collecting samples, such as blood, urine, or swabs, and culturing them in a laboratory to identify the specific microorganisms causing an infection. Additionally, sensitivity testing determines the most effective antimicrobial treatment against the identified pathogens.

27.2.8 Histopathology

Histopathology involves the examination of tissues collected through biopsy or during surgery. The tissues are processed, sectioned, stained, and analyzed under a microscope to identify abnormalities, such as tumors or inflammatory conditions, and provide a definitive diagnosis (Kadam et al. 2022).

27.2.9 Serological Diagnostics

Serological diagnostics in veterinary medicine involve the detection and measurement of antibodies or antigens in an animal's blood serum to diagnose specific diseases or assess the immune response (Vainionpää and Leinikki 2008). Serological tests play a crucial role in identifying infectious agents, monitoring vaccination status, and evaluating the immune status of animals. Here are some commonly used serological diagnostic techniques in veterinary medicine:

27.2.9.1 Enzyme-Linked Immunosorbent Assay (ELISA)

ELISA is a widely used serological test that detects and quantifies either antigens or antibodies in the animal's serum. It involves coating a microplate with the target antigen or antibody, adding the serum sample, and then detecting the presence of specific antibodies or antigens using an enzyme-linked detection system. ELISA can be used to diagnose various infectious diseases, including viral, bacterial, and parasitic infections.

27.2.9.2 Indirect Fluorescent Antibody Test (IFA)

IFA is a serological technique that utilizes fluorescently labeled antibodies to detect specific antigens in the animal's serum. The test involves incubating the serum with a known antigen, followed by the addition of fluorescently labeled antibodies that bind to the antigen-antibody complex. If the animal has specific antibodies against the target antigen, fluorescence will be observed under a fluorescence microscope. IFA is commonly used for the diagnosis of diseases such as leptospirosis and certain viral infections.

27.2.9.3 Rapid Diagnostic Tests

Rapid diagnostic tests, also known as lateral flow assays, are simple and rapid serological tests that provide quick results. These tests often use immunochromatographic techniques to detect specific antibodies or antigens in the animal's serum. They are convenient for on-site or point-of-care testing and are commonly used for diseases such as heartworm infection, parvovirus, and certain bacterial and viral infections.

27.2.9.4 Hemagglutination Inhibition (HI) Test

The HI test is a serological technique used primarily for the diagnosis of viral diseases. It involves incubating the animal's serum with a viral antigen and then adding red blood cells (usually from specific animal species). If antibodies against the virus are present in the serum, they will inhibit the agglutination (clumping) of the red blood cells. The HI test is commonly used for diseases such as influenza, Newcastle disease, and equine infectious anemia.

27.2.9.5 Neutralization Assay

Neutralization assays are used to detect neutralizing antibodies in the animal's serum, which are antibodies that can block the infectivity of viruses or toxins. The assay involves incubating the serum with a standardized amount of virus or toxin and then assessing the ability of the antibodies to neutralize the infectivity or toxicity. Neutralization assays are essential for evaluating the effectiveness of vaccines and diagnosing diseases such as rabies and botulism.

27.2.10 *Molecular Diagnostics*

Molecular diagnostics in veterinary medicine involve the use of advanced molecular techniques to detect and analyze nucleic acids (DNA or RNA) from animals for diagnostic purposes. These techniques allow for highly sensitive and specific detection of pathogens, genetic mutations, and other molecular markers associated with diseases (Vainionpää and Leinikki 2008). Molecular diagnostics have revolutionized veterinary medicine by enabling rapid and accurate diagnosis, monitoring of infectious diseases, and identification of genetic disorders. Here are some commonly used molecular diagnostic techniques in veterinary medicine:

27.2.10.1 Polymerase Chain Reaction (PCR)

PCR is a widely used molecular technique that amplifies specific regions of DNA or RNA in a sample. It involves a series of heating and cooling cycles to replicate the target nucleic acid sequences. PCR can detect and identify pathogens such as viruses, bacteria, and parasites by amplifying their genetic material. It is also used for genotyping, identification of genetic mutations, and detection of inherited diseases in animals.

27.2.10.2 Different Polymerase Chain Reaction (PCR) Types in Diagnostics

Real-time PCR (qPCR): Real-time PCR, also known as quantitative PCR, is a variation of PCR that allows for the quantification of DNA or RNA in real-time during the amplification process. It provides accurate and quantitative information about the initial amount of the target nucleic acid present in the sample. qPCR is commonly used in veterinary medicine for the quantification of pathogens, monitoring of viral load, and detection of gene expression levels.

Reverse Transcription PCR (RT-PCR): RT-PCR is a technique used to detect and quantify RNA molecules. It involves the reverse transcription of RNA into complementary DNA (cDNA), followed by PCR amplification of the cDNA. RT-PCR is particularly useful for detecting RNA viruses and assessing gene expression levels in animals.

Nested PCR: Nested PCR is a two-step PCR technique that involves two rounds of amplification. In the first round, external primers amplify a specific DNA target. Then, a portion of the first PCR product is used as a template in the second round with internal primers, resulting in higher specificity. Nested PCR is particularly useful for detecting low-abundance targets or when contamination is a concern.

Multiplex PCR: Multiplex PCR allows simultaneous amplification and detection of multiple target sequences within a single PCR reaction. It involves the use of multiple primer sets, each specific to a different target sequence. By incorporating differently labeled probes or primers, multiplex PCR enables the identification and differentiation of multiple pathogens or genetic markers in a single reaction. Multiplex PCR is efficient for diagnosing co-infections and detecting multiple mutations or genetic variations.

Digital PCR (dPCR): Digital PCR is a newer technique that provides absolute quantification of target nucleic acids in a sample. It partitions the PCR reaction into thousands of individual reaction chambers to distribute the target molecules. Each chamber is assessed for amplification, allowing for the precise determination of target concentration. dPCR is highly sensitive and useful for rare mutation detection, absolute quantification, and copy number variation analysis.

Hot Start PCR: Hot Start PCR incorporates modified DNA polymerases or inhibitory molecules to prevent non-specific amplification during the initial PCR setup. The technique helps eliminate primer-dimer formation and non-specific amplification, leading to increased specificity and sensitivity. Hot Start PCR is commonly used in diagnostic applications to improve specificity and minimize false-positive results.

Polymerase Spiral Reaction (PSR): PSR utilizes a DNA polymerase with strand-displacement activity and a set of specially designed primers to amplify a target DNA sequence at a constant temperature. The primers contain a spiral-forming sequence that enables the formation of a DNA helical structure during the amplification process. This spiral structure promotes efficient strand displacement and continuous DNA synthesis, leading to exponential amplification of the target sequence (Tomar et al. 2020).

These different types of PCR techniques offer flexibility and specificity for various diagnostic purposes. The choice of PCR type depends on the specific diagnostic requirements, target nucleic acid, desired sensitivity, and quantification needs in a given diagnostic application.

27.2.10.3 Nucleic Acid Sequencing

Nucleic acid sequencing techniques, such as Sanger sequencing and next-generation sequencing (NGS), are used to determine the exact sequence of nucleotides in a DNA or RNA sample. Sequencing allows for the identification of specific genetic variations, mutations, or pathogens present in the sample. It is used in veterinary medicine for genotyping, identification of drug resistance mutations, and characterization of infectious agents.

27.2.10.4 Fluorescent In Situ Hybridization (FISH)

FISH is a molecular technique used to detect and visualize specific nucleic acid sequences within cells or tissues. It involves the hybridization of fluorescently labeled DNA or RNA probes to complementary target sequences. FISH can be used in veterinary diagnostics to identify pathogens, detect chromosomal abnormalities, and study gene expression patterns.

27.2.10.5 Microarray Technology

Microarrays are powerful tools used to analyze the expression of thousands of genes simultaneously. They contain thousands of immobilized DNA or RNA probes that hybridize with nucleic acids from the sample. Microarrays can provide information about gene expression profiles, identify genetic mutations, and detect pathogens in veterinary medicine.

27.2.10.6 Loop-Mediated Isothermal Amplification (LAMP)

LAMP is an isothermal amplification technique that allows for the rapid and sensitive detection of specific DNA or RNA sequences. It operates at a constant temperature and amplifies the target sequences using multiple primers and DNA polymerases. LAMP is used in veterinary medicine for the detection of pathogens, including viral, bacterial, and parasitic infections (Kashir and Yaqinuddin 2020).

These conventional diagnostic techniques serve as the basis for identifying diseases and conditions in veterinary medicine. They are often complemented by advanced diagnostic modalities and molecular techniques as the field continues to evolve.

27.3 Nanotechnology to Improve Convention Diagnostics in Veterinary Medicine

Nanotechnology offers promising opportunities to improve conventional diagnostics in veterinary medicine by enhancing sensitivity, accuracy, and efficiency. Here are some ways in which nanotechnology can be utilized to improve conventional diagnostics in veterinary medicine:

27.3.1 Biosensors

Nanotechnology-based biosensors can be developed for the detection of specific biomarkers associated with veterinary diseases. These biosensors incorporate nanomaterials such as nanoparticles or nanowires, which provide a large surface area for enhanced detection sensitivity. Functionalization of these nanomaterials with specific ligands or antibodies enables selective binding to target biomarkers, resulting in highly sensitive and specific detection. Biosensors can be used for the rapid diagnosis of diseases such as infections, cancer, and metabolic disorders in animals (Prasad et al. 2021).

27.3.2 Point-of-Care Diagnostics

Nanotechnology can facilitate the development of portable and easy-to-use point-of-care diagnostic devices for veterinary medicine. These devices can be based on nanomaterials and nanoscale structures, enabling rapid and on-site testing of animals. Point-of-care devices can include nanosensors, lab-on-a-chip systems, or handheld devices integrated with nanomaterial-based assays. They can provide quick results, allowing for timely diagnosis and treatment decisions in veterinary practice, particularly in remote or resource-limited settings (Minakshi et al. 2020a).

27.3.3 Enhanced Imaging

Nanotechnology can improve conventional imaging techniques used in veterinary diagnostics. Nanoparticles can be engineered to act as contrast agents for various imaging modalities such as magnetic resonance imaging (MRI), computed tomography (CT), or ultrasound. These nanoparticles, when targeted to specific tissues or cells, can enhance the visualization of anatomical structures, detect lesions, or identify early stages of diseases in animals (Prasad et al. 2021b).

27.3.4 Nanocarriers in Diagnosis

Nanocarriers can be designed to carry imaging contrast agents, such as fluorescent dyes, magnetic nanoparticles, or radioactive labels. These nanocarriers improve the detection sensitivity and specificity of imaging techniques, such as MRI, CT, positron emission tomography (PET), or optical imaging. By targeting specific tissues or cells, nanocarriers can accumulate at the disease site, providing enhanced contrast and enabling the visualization of pathological changes. Nanocarriers can be functionalized with targeting ligands, such as antibodies or peptides, to specifically deliver diagnostic agents to diseased tissues or cells (Prasad et al. 2021b).

27.3.5 Amplification Techniques

Nanotechnology-based amplification techniques, such as rolling circle amplification or nanoparticle-based amplification, can enhance the sensitivity of diagnostic assays. These techniques allow for the amplification of target nucleic acids, enabling the detection of low-abundance biomarkers. Amplification methods based on nanotechnology can improve the accuracy and reliability of diagnostic tests in veterinary medicine.

27.3.6 Lab-on-a-Chip Devices

Lab-on-a-chip devices, incorporating nanotechnology, can integrate multiple diagnostic functions onto a single chip. These devices can perform sample preparation, detection, and analysis in a miniaturized and automated format. Lab-on-a-chip platforms offer advantages such as reduced sample and reagent requirements, faster analysis times, and portability. They can be utilized for various veterinary diagnostic applications, including infectious disease testing, genetic screening, or monitoring of animal health parameters (Minakshi et al. 2020a; Prasad et al. 2021).

By leveraging the unique properties of nanomaterials and nanoscale devices, nanotechnology-based strategies can enhance conventional diagnostics in veterinary medicine. The integration of nanotechnology into diagnostics has the potential to improve disease detection, monitoring, and treatment decisions, ultimately enhancing animal healthcare and welfare. However, further research and development are necessary to translate these advancements into practical applications in veterinary medicine.

27.4 Conventional Therapeutics in Veterinary Medicine

Conventional therapeutics in veterinary medicine refer to the standard and widely accepted treatments used to manage various health conditions in animals. These therapies may include medications, surgeries, and other interventions that have been proven effective through scientific research and clinical practice. Here are some examples of conventional therapeutics used in veterinary medicine:

27.4.1 Pharmaceuticals

Veterinarians commonly prescribe medications to treat specific diseases and symptoms in animals. These can include antibiotics, antiparasitics, pain relievers, anti-inflammatory drugs, antifungals, and more. Pharmaceutical treatments are often tailored to the specific needs of the individual animal, taking into account factors such as species, breed, age, weight, and overall health.

27.4.2 Surgeries

Surgical interventions are routinely performed in veterinary medicine to address various conditions. Common surgeries include spaying and neutering, tumor removal, fracture repairs, joint surgeries (such as cruciate ligament repair), exploratory surgeries, and many more. Surgical procedures aim to correct anatomical abnormalities, remove diseased tissue, or alleviate symptoms caused by certain conditions (Bain 2020).

27.4.3 Vaccinations

Therapeutic vaccines are a type of vaccine that are designed to treat or manage specific diseases or conditions in animals. Unlike preventive vaccines, which are administered to prevent the onset of a disease, therapeutic vaccines are aimed at stimulating the immune system to target and eliminate existing diseases. Therapeutic vaccines play a crucial role in veterinary medicine and can have several important benefits including disease treatment, reduced disease severity, extended treatment options, personalized medicine application, prevention of disease spread, potential for long-term protection, etc. (Meeusen et al. 2007).

27.4.4 Prescription Diets

Specialized diets can play a significant role in managing certain medical conditions in animals. Prescription diets are formulated to meet the nutritional requirements of animals with specific health concerns such as kidney disease, urinary tract issues, allergies, gastrointestinal disorders, and obesity. These diets often contain specific ingredients or nutrient compositions to support the animal's overall health and address their specific condition.

27.4.5 Physical Therapy and Rehabilitation

Physical therapy techniques, including therapeutic exercises, massage, hydrotherapy, and electrotherapy, can be employed to aid in the recovery and rehabilitation of animals after surgery, injury, or illness. These treatments help improve mobility, strength, and function while reducing pain and promoting healing (Levine et al. 2005).

It's important to note that veterinary medicine is a rapidly evolving field, and new therapies and treatment modalities continue to emerge. Veterinarians rely on their clinical experience, scientific research, and continuing education to stay up-to-date with the latest advances in conventional therapeutics and provide the best care for their animal patients.

27.5 Nanotechnology to Improve Conventional Therapeutics in Veterinary Medicine

Nanotechnology holds great potential to enhance conventional therapeutics in veterinary medicine. By utilizing nanoscale materials and techniques, researchers and veterinarians can develop more targeted and efficient treatments.

27.5.1 Nanotechnology to Improve the Efficacy of Veterinary Pharmaceuticals

Nanotechnology can indeed enhance the efficacy of veterinary pharmaceuticals in several ways. By utilizing nanoscale materials and techniques, it becomes possible to improve drug delivery, increase drug stability, and enhance therapeutic outcomes (Youssef et al. 2019). Here are some specific applications of nanotechnology to improve the efficacy of veterinary pharmaceuticals:

27.5.1.1 Inherent Therapeutic Potential of Nanomaterials

Nanomaterials possess inherent antimicrobial potential due to their unique physical and chemical properties at the nanoscale. These properties enable nanomaterials to interact with microorganisms and inhibit their growth, making them promising candidates for antimicrobial applications. Smaller size-dependent effects, broad-spectrum activity, inflicting physical damage to microbial membranes and cellular ultra-structures, induction of oxidative stress and reactive oxygen species (ROS)-mediated damage are the prominent attributes of inherent antimicrobial potential of common therapeutic nanomaterials. The unique antimicrobial mechanisms also help to counter the emerging issue of microbial resistance to conventional therapeutics.

27.5.1.2 Targeted Drug Delivery

Nanoparticles can be engineered to encapsulate veterinary drugs and deliver them specifically to the desired site of action. These nanoparticles can be designed to release the drug gradually or in response to specific stimuli, such as pH or enzymes present at the target site. By achieving targeted drug delivery, nanotechnology enables higher drug concentrations at the intended site while minimizing systemic side effects.

27.5.1.3 Improved Drug Solubility and Stability

Many veterinary drugs face challenges related to their poor solubility or susceptibility to degradation. Nanotechnology can address these issues by formulating drugs into nanoparticles or nanoemulsions, enhancing their solubility and stability. This improved formulation allows for better drug absorption, distribution, and bioavailability.

27.5.1.4 Extended Drug Release

Nanoparticles can provide controlled and sustained drug release over an extended period. By encapsulating drugs within biodegradable or stimuli-responsive nanoparticles, veterinarians can achieve a slow and steady release of the drug. This controlled release can enhance the drug's therapeutic effect, reduce the frequency of administration, and improve patient compliance.

27.5.1.5 Enhanced Cellular Uptake

Nanoparticles can facilitate the cellular uptake of veterinary drugs, particularly in cases where the drugs have limited access to target cells. By attaching drugs to nanoparticles or modifying the surface properties of nanoparticles, it becomes possible to improve their interaction with cell membranes and enhance drug uptake into cells. This can lead to increased drug efficacy and better treatment outcomes.

27.5.1.6 Combination Therapy

Nanotechnology enables the development of combination therapies where multiple drugs or therapeutic agents are co-encapsulated or conjugated within nanoparticles. This approach allows for synergistic effects and improved therapeutic outcomes by targeting multiple aspects of a disease simultaneously. For example, nanoparticles can carry both an antimicrobial drug and an anti-inflammatory agent to address both infection and inflammation in veterinary medicine (VanDyke et al. 2016).

27.5.1.7 Photothermal and Photodynamic Therapies

Certain nanomaterials possess unique optical properties that allow them to convert light energy into heat or generate reactive oxygen species. This enables photothermal therapy (PTT) or photodynamic therapy (PDT) for targeted destruction of diseased cells, such as cancer cells, while sparing healthy tissues (Hou et al. 2018).

Nanotechnology has the potential to revolutionize veterinary pharmaceuticals by addressing challenges related to drug delivery, solubility, stability, and targeted therapy. However, it is important to conduct thorough research and ensure the safety and efficacy of nanotechnology-based veterinary pharmaceuticals before their widespread implementation in clinical practice.

27.5.2 Nanotechnology to Improve Surgical Interventions

Nanotechnology can play a significant role in improving surgical interventions in veterinary medicine by providing innovative tools and materials that enhance precision, promote healing, and reduce post-operative complications (Mohantya et al. 2014). Here are some ways nanotechnology can improve surgical interventions:

27.5.2.1 Nanoscale Surgical Tools

Nanotechnology enables the development of surgical tools and instruments at the nanoscale. These tools can be designed to have higher precision and maneuverability, allowing veterinarians to perform delicate and minimally invasive procedures. For example, nanosurgical blades and probes can be used to make precise incisions and manipulate tissues with greater accuracy.

27.5.2.2 Tissue Regeneration, Wound Healing, and Fracture Management

Nanomaterials can be utilized to promote tissue regeneration and wound healing after surgery. Scaffolds made of nanofibers or nanocomposites can provide a structural framework for cells to grow and regenerate damaged tissues. Additionally, nanoparticles can be incorporated into surgical dressings or implanted materials to deliver growth factors, antimicrobial agents, or other therapeutics that enhance healing and reduce the risk of infection.

27.5.2.3 Nanoengineered Implants

Nanotechnology allows for the development of advanced implants with improved properties and functionalities. For instance, nanostructured coatings on implants can enhance biocompatibility, reduce the risk of rejection, and promote tissue integration. Nanoscale modifications can also be applied to implant surfaces to enhance their mechanical stability and prevent bacterial colonization, reducing the risk of implant-associated infections.

27.5.2.4 Drug Delivery Systems

Nanotechnology-based drug delivery systems can be employed to deliver therapeutic agents directly to the surgical site. By encapsulating drugs within nanoparticles, the release kinetics can be controlled, and higher concentrations of the drug can be maintained locally. This targeted drug delivery approach can reduce systemic side effects while maximizing the therapeutic efficacy of the drugs.

27.5.2.5 Imaging and Navigation

Nanoparticles can serve as contrast agents for imaging techniques, allowing veterinarians to visualize the surgical site with improved accuracy. Nanoparticle-based imaging agents can enhance the visibility of specific tissues or structures, aiding in surgical planning and precise navigation during procedures. Real-time imaging can help surgeons avoid critical structures and optimize surgical outcomes.

27.5.2.6 Biosensors and Diagnostics

Nanotechnology-based biosensors can be used intraoperatively to detect specific molecules or biomarkers in real-time. These biosensors can provide valuable information about the surgical site, such as tissue oxygenation, pH levels, or the presence of infections. Real-time monitoring can guide surgical decision-making and help prevent complications (Ramesh et al. 2022).

27.5.3 Nanotechnology to Improve Therapeutic Vaccines in Veterinary Science

Nanotechnology has the potential to enhance therapeutic vaccines in veterinary science by improving their effectiveness, targeted delivery, and immune response modulation (Maina et al. 2020; Celis-Giraldo et al. 2021). Here are some ways nanotechnology can be utilized to improve therapeutic vaccines in veterinary medicine:

27.5.3.1 Targeted Delivery

Nanoparticles can be used as carriers for therapeutic vaccine components, such as antigens or immunomodulators. By encapsulating these components within nanoparticles, their targeted delivery to specific cells or tissues can be achieved. This targeted delivery enhances the uptake of antigens by immune cells, increasing the vaccine's efficacy.

27.5.3.2 Controlled Release

Nanoparticles can be engineered to release vaccine components in a controlled manner, ensuring a sustained and prolonged immune response. This sustained release can lead to prolonged antigen presentation to the immune system, improving the vaccine's ability to induce a robust and long-lasting immune response.

27.5.3.3 Adjuvant Delivery

Adjuvants are substances that enhance the immune response to vaccines. Nanoparticles can serve as effective adjuvants it-self or carriers for adjuvants, allowing for controlled and targeted delivery to immune cells. This targeted delivery enhances the adjuvant's interaction with immune cells, resulting in a more potent and specific immune response.

27.5.3.4 Immune Response Modulation

Nanoparticles can be designed to modulate the immune response and direct it towards the desired outcome. For example, nanoparticles can be functionalized with specific ligands that selectively target immune cells involved in the desired immune response. Additionally, nanoparticles can be engineered to release immune-stimulating molecules or immunomodulators that promote the desired immune response profile.

27.5.3.5 Combination Therapies

Nanotechnology enables the development of combination therapies, where multiple therapeutic components can be co-delivered using nanoparticles. This approach allows for synergistic effects and a more comprehensive immune response. For instance, nanoparticles can carry both antigens and immunomodulators, enhancing the immune response and promoting specific therapeutic outcomes.

27.5.3.6 Improved Stability and Shelf Life

Nanoparticles can protect vaccine components from degradation, improving their stability and extending their shelf life. This is particularly beneficial for therapeutic vaccines that require long-term storage or transportation to remote areas. By encapsulating antigens or other vaccine components within nanoparticles, their susceptibility to environmental factors can be reduced.

27.5.4 Nanotechnology in Personalized Nutritional Therapy in Animals

Nanotechnology has the potential to revolutionize personalized nutritional therapy in animals by offering precise delivery systems, enhanced bioavailability, and targeted nutrient delivery. Here's how nanotechnology can be utilized in personalized nutritional therapy for animals:

27.5.4.1 Nutrient Delivery Systems

Nanoparticles can serve as carriers for essential nutrients, vitamins, minerals, and other bioactive compounds (Ban et al. 2015). These nanoparticles can protect the nutrients from degradation in the digestive system, improve their solubility, and enhance their absorption in the body. By encapsulating nutrients within nanoparticles, personalized nutritional therapies can be tailored to meet the specific dietary needs of individual animals (Abdelnour et al. 2021).

27.5.4.2 Improved Bioavailability

Many nutrients exhibit poor bioavailability, meaning that they are not effectively absorbed or utilized by the body. Nanotechnology can enhance the bioavailability of nutrients by improving their stability, solubility, and transport across biological barriers. Nanoemulsions or nanoscale delivery systems can enhance the absorption and utilization of nutrients, ensuring that animals receive the maximum benefit from their personalized nutritional therapy (Haham et al. 2012; Hill and Li 2017).

27.5.4.3 Targeted Nutrient Delivery

Nanoparticles can be designed to selectively deliver nutrients to specific tissues or cells in the body. By functionalizing nanoparticles with specific ligands or receptors, they can be targeted to specific sites of action. This targeted delivery ensures that nutrients reach the intended tissues or cells, enhancing the therapeutic effect and minimizing potential side effects (Arshad et al. 2021).

27.5.4.4 Controlled Release

Nanotechnology enables the controlled release of nutrients over time. By encapsulating nutrients within nanoparticles with specific release properties, the nutrients can be delivered gradually, mimicking natural nutrient absorption patterns or addressing specific metabolic needs. This controlled release can optimize nutrient uptake and utilization, leading to improved therapeutic outcomes (McClements 2015).

27.5.4.5 Nutraceutical Formulations

Nutraceuticals are bioactive compounds that provide health benefits beyond basic nutrition. Nanotechnology can be used to develop nanoscale formulations of nutraceuticals, improving their stability and bioavailability (Manocha et al. 2022). These formulations can deliver therapeutic doses of nutraceuticals, such as antioxidants, probiotics, or omega-3 fatty acids, to support specific health conditions or promote overall well-being in animals.

27.5.4.6 Diagnostic and Monitoring Tools

Nanotechnology-based sensors and diagnostics can be employed to monitor the nutritional status and metabolic parameters of animals. Nanosensors can detect and quantify specific biomarkers related to nutritional deficiencies or imbalances, allowing for personalized adjustments to the nutritional therapy. Real-time monitoring can provide valuable insights into the animal's response to the personalized nutritional therapy and guide adjustments as needed.

It's important to note that while nanotechnology holds tremendous potential in personalized nutritional therapy for animals, further research is necessary to ensure its safety, efficacy, and long-term effects. Regulatory considerations and extensive testing are vital before widespread implementation of nanotechnology-based personalized nutritional therapies in veterinary practice.

27.5.5 Nanointervention to Improve Therapeutic Management of Reproductive Disorders

Nanotechnology interventions hold promise in the therapeutic management of reproductive disorders in animals by offering targeted drug delivery, improved diagnostics, and regenerative approaches (Hashem and Gonzalez-Bulnes 2021; Wang et al. 2022). Here are some potential applications of nanotechnology in the reproductive health of animals:

27.5.5.1 Targeted Drug Delivery

Nanoparticles can be used to deliver therapeutic agents directly to reproductive tissues or cells. By encapsulating drugs within nanoparticles, their targeted delivery to the reproductive organs can be achieved, improving drug efficacy while minimizing systemic side effects. This approach can be beneficial in treating conditions such as uterine infections, hormonal imbalances, or reproductive organ tumors.

27.5.5.2 Hormonal Regulation

Nanoparticles can be designed to deliver hormones or hormone-modulating agents to regulate reproductive processes in animals. These nanoparticles can mimic the natural hormone secretion patterns or provide sustained hormone release, helping to manage conditions like infertility, estrous cycle disorders, or reproductive hormonal imbalances.

27.5.5.3 Diagnostic Imaging

Nanoparticles can act as contrast agents or imaging probes to enhance diagnostics in reproductive disorders. They can be used in imaging techniques such as ultrasound, MRI, or CT to improve visualization and detection of reproductive abnormalities or tumors. Nanoparticles can help in early diagnosis, guiding appropriate therapeutic interventions (Prasad et al. 2021b).

27.5.5.4 Regenerative Medicine

Nanotechnology-based approaches can promote tissue regeneration and repair in reproductive disorders. Nanomaterials, such as scaffolds or hydrogels, can provide support and structure for tissue regeneration in cases of damaged reproductive organs, such as the uterus or ovaries. Nanoparticles can also deliver growth factors or stem cells to promote tissue regeneration and restore reproductive functions.

27.5.5.5 Assisted Reproductive Technologies (ART)

Nanotechnology can be employed in ART techniques to enhance their success rates. For example, nanoscale carriers can protect fragile gametes, such as sperm or oocytes, during cryopreservation and improve their viability upon thawing. Nanotechnology-based approaches can also improve sperm selection techniques, enhance fertilization rates, or aid in controlled release of hormones during assisted reproduction procedures (Hashem and Gonzalez-Bulnes 2021).

27.5.5.6 Contraception and Sterilization

Nanoparticles can be utilized for contraception or sterilization purposes in animal populations. They can deliver contraceptive agents to target reproductive tissues, providing temporary or long-term contraception. Nanomaterials can also be used to deliver sterilizing agents directly to reproductive organs, offering a non-surgical alternative for population control.

It's important to note that while nanotechnology shows promise in reproductive health management, further research and development are necessary to ensure safety, efficacy, and long-term effects.

27.6 Conclusion and Future Prospect

Nanotechnology holds immense potential for improving conventional diagnostics and therapeutics in veterinary medicine. The unique properties and characteristics of nanomaterials enable precise targeting, enhanced drug delivery, improved imaging, and innovative diagnostic techniques. By leveraging nanotechnology, veterinarians can achieve more effective and personalized treatments for animals, leading to better health outcomes.

Nanotechnology offers the opportunity to develop targeted drug delivery systems that enhance the efficacy of therapeutics while minimizing side effects. Nanoparticles can encapsulate drugs, nutrients, or bioactive compounds, allowing for controlled release and improved bioavailability. This targeted delivery approach can significantly enhance the treatment of various diseases, including cancer, infectious diseases, and autoimmune disorders.

Moreover, nanotechnology-based diagnostic tools provide more accurate and sensitive detection of diseases in animals. Nanosensors and contrast agents enable early disease diagnosis and real-time monitoring, enabling veterinarians to make timely and informed treatment decisions. Nanotechnology also enhances imaging techniques, allowing for precise visualization of tissues and structures, aiding in diagnosis and treatment planning.

Looking ahead, the future prospects of nanotechnology in veterinary medicine are promising. Continued research and development will refine nanomaterial design and synthesis, ensuring their safety, biocompatibility, and cost-effectiveness. Additionally, advances in nanotoxicology studies will provide a better understanding of the potential risks associated with nanomaterials.

Furthermore, the integration of nanotechnology with other emerging fields, such as genomics, proteomics, and artificial intelligence, can unlock new possibilities for personalized medicine in animals. This convergence of technologies can enable precise disease characterization, targeted treatment interventions, and optimized therapeutic regimens.

As with any emerging field, the translation of nanotechnology into routine veterinary practice will require collaboration among researchers, veterinarians, and regulatory authorities. Regulatory frameworks must be established to ensure the safe and ethical use of nanotechnology in veterinary medicine. Additionally, efforts should be made to increase awareness and education about nanotechnology among veterinary professionals.

In conclusion, the integration of nanotechnology into veterinary diagnostics and therapeutics has the potential to revolutionize the field, leading to improved patient outcomes and a more personalized approach to animal healthcare. With ongoing research and development, nanotechnology is poised to play a transformative role in advancing veterinary medicine in the years to come.

References

Abdelnour SA, Alagawany M, Hashem NM, Farag MR, Alghamdi ES, Hassan FU, Bilal RM, Elnesr SS, Dawood MAO, Nagadi SA, Elwan HAM, ALmasoudi AG, Attia YA (2021) Nanominerals: fabrication methods, benefits and hazards, and their applications in ruminants with special reference to selenium and zinc nanoparticles. Animals 11(7):1916. https://doi.org/10.3390/ani11071916

Arshad R, Gulshad L, Haq I-U et al (2021) Nanotechnology: a novel tool to enhance the bioavailability of micronutrients. Food Sci Nutr 9:3354–3361. https://doi.org/10.1002/fsn3.2311

Bain M (2020) Surgical and behavioral relationships with welfare. Front Vet Sci 7:519. https://doi.org/10.3389/fvets.2020.00519

Ban C, Park SJ, Lim S, Choi SJ, Choi YJ (2015) Improving flavonoid bioaccessibility using an edible oil-based lipid nanoparticle for oral delivery. J Agric Food Chem 63(21):5266–5272. https://doi.org/10.1021/acs.jafc.5b01495

Celis-Giraldo CT, López-Abán J, Muro A, Patarroyo MA, Manzano-Román R (2021) Nanovaccines against animal pathogens: the latest findings. Vaccine 9(9):988. https://doi.org/10.3390/vaccines9090988

Das MR, Patra RC, Das RK, Rath PK, Mishra BP (2017) Hemato-biochemical alterations and urinalysis in dogs suffering from benign prostatic hyperplasia. Vet World 10(3):331–335. https://doi.org/10.14202/vetworld.2017.331-335

Dhakal S, Renukaradhya GJ (2019) Nanoparticle-based vaccine development and evaluation against viral infections in pigs. Vet Res 50:90. https://doi.org/10.1186/s13567-019-0712-5

Haham M, Ish-Shalom S, Nodelman M, Duek I, Segal E, Kustanovich M, Livney YD (2012) Stability and bioavailability of vitamin D nanoencapsulated in casein micelles. Food Funct 3: 737–744. https://doi.org/10.1039/C2FO10249H

Hashem NM, Gonzalez-Bulnes A (2021) Nanotechnology and reproductive management of farm animals: challenges and advances. Animals 11(7):1932. https://doi.org/10.3390/ani11071932

Hill EK, Li J (2017) Current and future prospects for nanotechnology in animal production. J Anim Sci Biotechnol 8:26. https://doi.org/10.1186/s40104-017-0157-5

Hou X, Tao Y, Pang Y, Li X, Jiang G, Liu Y (2018) Nanoparticle-based photothermal and photodynamic immunotherapy for tumor treatment. Int J Cancer 143(12):3050–3060. https://doi.org/10.1002/ijc.31717

Kadam RG, Karikalan M, Siddappa CM, Mahendran K, Srivastava G, Rajak KK, Bhardwaj Y, Varshney R, War ZA, Singh R, Ghosh M, Beena V, Pawde AM, Singh KP, Sharma AK (2022) Molecular and pathological screening of canine distemper virus in Asiatic lions, tigers, leopards, snow leopards, clouded leopards, leopard cats, jungle cats, civet cats, fishing cat, and jaguar of different states, India. Infect Gen Evol 98:105211. https://doi.org/10.1016/j.meegid.2022.105211

Kashir J, Yaqinuddin A (2020) Loop mediated isothermal amplification (LAMP) assays as a rapid diagnostic for COVID-19. Med Hypotheses 141:109786. https://doi.org/10.1016/j.mehy.2020.109786

Levine D, Millis DL, Marcellin-Little DJ (2005) Introduction to veterinary physical rehabilitation. Vet Clin North Am Small Anim Pract 35(6):1247–1vii. https://doi.org/10.1016/j.cvsm.2005.07.002

Maina TW, Grego EA, Boggiatto PM, Sacco RE, Narasimhan B, McGill JL (2020) Applications of nanovaccines for disease prevention in cattle. Front Bioeng Biotechnol 8:608050. https://doi.org/10.3389/fbioe.2020.608050

Manocha S, Dhiman S, Grewal AS, Guarve K (2022) Nanotechnology: an approach to overcome bioavailability challenges of nutraceuticals. J Drug Deliv Sci Technol 72:103418. https://doi.org/10.1016/j.jddst.2022.103418

McClements DJ (2015) Nanoscale nutrient delivery systems for food applications: improving bioactive dispersibility, stability, and bioavailability. J Food Sci 80:N1602–N1611. https://doi.org/10.1111/1750-3841.12919

Meeusen EN, Walker J, Peters A, Pastoret PP, Jungersen G (2007) Current status of veterinary vaccines. Clin Microbiol Rev 20(3):489–510. https://doi.org/10.1128/CMR.00005-07

Minakshi P, Ghosh M, Brar B, Kumar R, Lambe UP, Ranjan K, Manoj J, Prasad G (2019) Nano-antimicrobials: a new paradigm for combating mycobacterial resistance. Curr Pharm Des 25(13):1554–1579. https://doi.org/10.2174/1381612825666190620094041

Minakshi P, Ghosh M, Kumar R, Brar B, Lambe UP, Banerjee S, Ranjan K, Kumar B, Goel P, Malik YS, Prasad G (2020a) An insight into nanomedicinal approaches to combat viral zoonoses. Curr Top Med Chem 20(11):915–962. https://doi.org/10.2174/1568026620666200325114400

Minakshi P, Kumar R, Ghosh M, Brar B, Barnela M, Lakhani P (2020b) Application of polymeric nano-materials in management of inflammatory bowel disease. Curr Top Med Chem 20(11): 982–1008. https://doi.org/10.2174/1568026620666200320113322

Mohantya N, Palaib T, Prustyc B, Mohapatrad J (2014) An overview of nanomedicine in veterinary science. Vet Res 2(4):90-5. https://doi.org/10.1007/s12262-012-0726-y

Prasad M, Ghosh M, Kumar R, Brar B, Surjith KP, Lambe UP, Ranjan K, Banerjee S, Prasad G, Kumar Khurana S, Kharb P (2021) The importance of nanomedicine in prophylactic and theranostic intervention of bacterial zoonoses and reverse zoonoses in the era of microbial resistance. J Nanosci Nanotechnol 21(6):3404–3452. https://doi.org/10.1166/jnn.2021.18999

Prasad M, Ghosh M, Suman, Patki HS, Kumar S, Brar B, Sindhu N, Goel P, Kaushik S, Mohan H, Shafiq S, Kumar R (2021b) Imaging techniques in veterinary disease diagnosis. In: Gahlawat SK, Maan S (eds) Advances in animal disease diagnosis. CRC Press, p 43. https://doi.org/10.1201/9781003080282-8

Ramesh M, Janani R, Deepa C, Rajeshkumar L (2022) Nanotechnology-enabled biosensors: a review of fundamentals, design principles, materials, and applications. Biosensors 13(1):40. https://doi.org/10.3390/bios13010040

Rich GA (1991) Basic history taking and the avian physical examination. Vet Clin North Am Small Anim Pract 21(6):1135–1145. https://doi.org/10.1016/s0195-5616(91)50128-5

Sarangi A, Ghosh M, Sangwan S, Kumar R, Balhara S, Phulia SK, Sharma RK, Sahu S, Kumar S, Mohanty AK, Balhara AK (2022) Exploration of urinary metabolite dynamicity for early detection of pregnancy in water buffaloes. Sci Rep 12(1):16295. https://doi.org/10.1038/s41598-022-20298-1

Tomar PS, Kumar JS, Patel S, Sharma S (2020) Polymerase spiral reaction assay for rapid and real time detection of West Nile virus from clinical samples. Front Cell Infect Microbiol 10:426. https://doi.org/10.3389/fcimb.2020.00426

Tully TN Jr (1998) Health examinations and clinical diagnostic procedures of ratites. Vet Clin North Am Food Anim Pract 14(3):401–420. https://doi.org/10.1016/s0749-0720(15)30228-0

Vainionpää R, Leinikki P (2008) Diagnostic techniques: serological and molecular approaches. Encyclop Virol:29–37. https://doi.org/10.1016/B978-012374410-4.00585-9

VanDyke D, Kyriacopulos P, Yassini B, Wright A, Burkhart E, Jacek S, Pratt M, Peterson CR, Rai P (2016) Nanoparticle based combination treatments for targeting multiple hallmarks of cancer. Int J Nano Stud& Technol Suppl 4:1–18. https://doi.org/10.19070/2167-8685-SI04001

Wang K, Lu X, Lu Y, Wang J, Lu Q, Cao X, Yang Y, Yang Z (2022) Nanomaterials in animal husbandry: research and prospects. Front Genet 13:915911. https://doi.org/10.3389/fgene.2022.915911

Youssef FS, El-Banna HA, Elzorba HY, Galal AM (2019) Application of some nanoparticles in the field of veterinary medicine. Int J Vet Sci Med 7(1):78–93. https://doi.org/10.1080/23144599.2019.1691379

Chapter 28
Nanotheranostics in Personalized Veterinary Medicine

Mayukh Ghosh, Alonkrita Chowdhury, and Rajesh Kumar

Abstract Nanotheranostics, a rapidly advancing field at the intersection of nanotechnology and theranostics, holds great promise for revolutionizing personalized veterinary medicine. The integration of nanoscale materials, imaging modalities, and therapeutic agents offers new opportunities for precise diagnosis, targeted treatment, and individualized care in animals. This abstract provides an overview of the applications, benefits, and challenges of nanotheranostics in personalized veterinary medicine. Nanotheranostic platforms can extend multi-arrayed benefits that include but not limited to improve animal production, reproductive performances, facilitates in stem cell therapy and regenerative medicine, veterinary transplant medicine, wildlife conservation, etc. In animal production, nanotheranostic platforms can facilitate comprehensive disease management by enhancing prophylactic, diagnostic and therapeutic interventions, nutritional enhancement, environmental monitoring, and animal welfare monitoring to optimize the output from animal husbandry. Towards improving reproductive performances of animals, nanotechnology can help in reproductive health evaluations, hormonal monitoring and management, genetic evaluation and selection, assisted reproductive technologies such as artificial insemination (AI), embryo transfer (ET), in vitro fertilization (IVF), or sex-sorted semen production, gamete quality improvement, etc. Nanotechnology in personalized veterinary medicine also facilitates in multiple aspects of stem cell therapy and regenerative medicine that include isolation and purification of stem cells, stem cell delivery and tracking, scaffold and matrix engineering, release of growth factors to

---

M. Ghosh (✉)
Department of Veterinary Physiology and Biochemistry, Banaras Hindu University, Mirzapur, Uttar Pradesh, India

A. Chowdhury
Department of Plant Biotechnology, Institute of Agricultural Sciences, RGSC, Banaras Hindu University, Mirzapur, Uttar Pradesh, India

R. Kumar
Department of Veterinary Physiology and Biochemistry, Lala Lajpat Rai University of Veterinary and Animal, Hisar, Haryana, India

LLR University of Veterinary and Animal, ICAR-National Research Centre on Equines, *Nanotechnology Theranostics in Livestock Diseases and Management*, Livestock Diseases and Management,
https://doi.org/10.1007/978-981-16-1610-5_28

maturation media, monitoring stem-cell therapy progress by theranostic imaging, etc. Transplant medicine is also an important domain where nanotechnology can deliver in terms of organ preservation, targeted drug delivery and immunomodulation, detection of rejection and complications, tissue engineering and regeneration, etc. Wildlife management and conservation is another area where nanotechnology can help in wildlife disease management, analysis of genetic diversity, physiological monitoring, animal movement monitoring and surveillance, nutritional management, monitoring environmental pollutants, and securing "One Health" perspective. Overall, nanotheranostics hold immense potential for personalized veterinary medicine, enabling precise diagnostics and targeted therapies in animals. By harnessing the power of nanotechnology, veterinarians can advance individualized care, improve treatment outcomes, and enhance the overall well-being of veterinary patients. Further research and collaborations are needed to overcome existing challenges and translate nanotheranostics into routine clinical practice in veterinary medicine.

Keywords Nanotechnology · Animal production · Assisted reproductive technologies · Sex-sorting · Genetic selection · Stem cell therapy · Veterinary transplant medicine · Wildlife conservation

28.1 Introduction

Personalized veterinary medicine, also known as precision veterinary medicine or individualized veterinary medicine, refers to an approach that tailors medical care and treatments to meet the unique needs of individual animals. It takes into account factors such as an animal's genetics, environment, lifestyle, and specific health conditions to provide targeted and effective veterinary care. The key aspects of personalized veterinary medicine include but not limited to genomic medicine, precision diagnostics, targeted therapies, nutritional and environmental considerations, systematic monitoring and follow-up care, etc. (Lloyd et al. 2016; Eduardo et al. 2021).

Advances in genomic technologies have enabled veterinarians to better understand the genetic makeup of animals. Genetic testing can identify inherited diseases, assess an animal's susceptibility to certain conditions, and help determine the most appropriate treatments or preventive measures based on their genetic profile (O'Connor and Crystal 2006; Zakirova et al. 2022). This allows for personalized recommendations and interventions. Precision diagnostics is another key aspect of personalized veterinary medicine that emphasizes the use of advanced diagnostic techniques to obtain a detailed understanding of an animal's health. This may include imaging technologies, molecular diagnostics, and specialized laboratory tests that provide insights into the animal's specific condition, helping veterinarians make accurate diagnoses and tailored treatment plans (Huang and Qian 2020). Personalized veterinary medicine focuses on using treatments that are specifically tailored to an animal's individual needs. This may involve selecting medications or therapies based on the animal's specific disease characteristics, genetic predispositions, or even their response to previous treatments. The goal is to optimize treatment outcomes while minimizing side effects (Lloyd et al. 2016; Fosse et al. 2023). Nutritional requirements and factors regarding the micro-environment of individual animal are considered in personalized veterinary medicine. Animal's nutritional needs and environmental factors play a significant role in their overall health and production. Veterinarians may develop personalized nutrition plans, considering an animal's breed, age, activity level, and underlying health conditions. Environmental factors, such as living conditions and exposure to toxins or allergens, are also taken into account to provide individualized recommendations (Avci and Baran 2014; Wang et al. 2018; Hogan et al. 2017). Personalized veterinary medicine also emphasizes the importance of ongoing monitoring and follow-up care. Regular check-ups, diagnostic screenings, and evaluations of treatment responses are conducted to assess the effectiveness of interventions and make any necessary adjustments. This allows veterinarians to continuously optimize the care provided based on an animal's changing needs (Goetz and Schork 2018). Conclusively, personalized veterinary medicine aims to enhance the quality of care, improve treatment outcomes, and minimize adverse effects by tailoring veterinary services to each animal's unique characteristics and circumstances. It requires a multidisciplinary approach, integrating various technologies, diagnostic tools, and expertise to provide the most effective and individualized care for animals.

Nanotheranostics is an emerging field that combines nanotechnology, therapeutics, and diagnostics to customize personalized medicine approaches (Mura and Couvreur 2012). While the application of nanotheranostics in veterinary medicine is still in its early stages, it holds great promise for advancing personalized veterinary care. In personalized veterinary medicine, the goal is to tailor diagnostics and treatment to individual animals based on their specific needs, characteristics, and medical conditions. Nanotheranostics can play a significant role in achieving this goal by offering precise diagnostic tools and targeted therapeutic interventions.

In diagnostics, nanoparticles can be designed to carry imaging agents, such as fluorescent dyes or magnetic nanoparticles, enabling enhanced imaging modalities like MRI, CT scans, or optical imaging. These nanoparticles can help visualize diseases, detect specific biomarkers, or identify tumor margins more accurately, aiding in the early diagnosis and monitoring of veterinary conditions. In targeted drug delivery, nanoparticles can be engineered to encapsulate and deliver therapeutic agents to specific sites in the body, such as tumors or infected tissues. By incorporating targeting ligands on the nanoparticle surface, they can selectively bind to receptors on diseased cells, increasing drug concentration at the desired location while minimizing side effects on healthy tissues. Nanoparticles can also be designed as drug delivery systems with controlled release properties. This allows for sustained drug release over an extended period, ensuring therapeutic levels are maintained and reducing the frequency of drug administration. Such systems can enhance treatment efficacy and patient compliance. Nanotheranostic platforms can combine imaging capabilities for real-time monitoring of treatment response with targeted drug delivery. This enables veterinarians to evaluate the effectiveness of a treatment while simultaneously delivering therapy, facilitating treatment adjustments when necessary. By utilizing nanotheranostics, veterinarians can gain insights into an animal's individual response to therapy. By monitoring treatment response in real-time, they can tailor the therapeutic regimen and adjust dosages based on the specific needs of the animal, optimizing treatment outcomes. Overall, nanotheranostics in personalized veterinary medicine has the potential to revolutionize diagnostics, treatment, and monitoring of various diseases in animals (Mura and Couvreur 2012; Kim et al. 2013; Youssef et al. 2019). However, it's important to note that this field is still in the early stages of development, and more research is needed to fully explore and harness the potential of nanotheranostics in the veterinary setting.

28.2 Nanotheranostics in Personalized Veterinary Medicine to Improve Animal Production

Personalized veterinary medicine can greatly contribute to improving animal production in several ways. By considering the unique needs and characteristics of individual animals, personalized veterinary medicine can optimize animal health, enhance productivity, and improve overall production outcomes.

Personalized veterinary medicine emphasizes proactive and preventive healthcare measures. Veterinarians can conduct individual health assessments, including physical examinations, diagnostic screenings, and genetic testing, to identify potential health risks and intervene early. By implementing tailored vaccination protocols, parasite control programs, and biosecurity measures, the overall health and disease resistance of animals can be enhanced, leading to improved productivity. Personalized veterinary medicine takes into account the specific nutritional needs of animals based on their age, breed, reproductive status, and production goals. Veterinarians can develop customized nutrition plans that optimize nutrient intake, leading to improved growth, reproduction, milk production, or weight gain. Personalized nutrition also helps prevent nutritional deficiencies or excesses, reducing the risk of metabolic disorders and improving overall performance. Personalized veterinary medicine plays a vital role in maximizing reproductive efficiency in animals. By monitoring reproductive health parameters, conducting fertility evaluations, and utilizing reproductive technologies, veterinarians can identify and address individual factors that may affect fertility, such as hormonal imbalances, infections, or genetic predispositions. This leads to better reproductive outcomes, including increased conception rates, reduced embryonic loss, and improved genetic selection. Personalized veterinary medicine incorporates genetic information to optimize animal production. Genetic testing can identify desirable traits, such as disease resistance, growth rate, milk production, or meat quality, allowing producers to make informed breeding decisions. By selecting animals with favorable genetic profiles, personalized veterinary medicine helps improve the quality and productivity of future generations, leading to more efficient and profitable animal production. Personalized veterinary medicine utilizes individual animal monitoring and data analysis to assess performance and identify areas for improvement. By tracking parameters such as weight gain, feed conversion efficiency, milk production, or egg quality on an individual basis, veterinarians can identify outliers, detect health issues, and implement interventions to optimize production. This data-driven approach enables more precise management decisions and enhances productivity. Personalized veterinary medicine recognizes the importance of animal welfare in optimizing production outcomes. By addressing individual animal needs, providing appropriate housing conditions, managing pain or discomfort, and implementing stress-reduction strategies, veterinarians can ensure animals are in optimal physiological and psychological states. Improved animal welfare leads to reduced stress-related production losses, better immune function, and overall enhanced productivity. In summary, personalized veterinary medicine offers significant benefits for animal production by tailoring healthcare, nutrition, reproductive management, genetic selection, and welfare practices to meet the specific needs of individual animals. By implementing personalized approaches, producers can optimize production outcomes, improve efficiency, enhance animal welfare, and contribute to sustainable and profitable animal production systems (Berckmans 2017; Lloyd et al. 2016; Schillings et al. 2021; Campion and Dowell 2019; Aquilani et al. 2022).

Nanotheranostics has the potential to significantly impact animal production by improving various aspects of livestock health, productivity, and welfare.

1. **Disease Management:** Nanoparticles can be used to deliver therapeutic agents directly to diseased tissues or organs in animals. This targeted drug delivery system can help treat infectious diseases, such as bacterial or viral infections, in livestock. It can enhance the efficacy of treatments, reduce the required dosage, and minimize the development of drug resistance. Additionally, nanoparticles can carry diagnostic agents that enable early detection of diseases, allowing for prompt interventions and preventing disease outbreaks in animal production systems (Patra et al. 2018).
2. **Nutritional Enhancement:** Nanotechnology-based formulations can improve the delivery and absorption of nutrients in animals. For instance, encapsulating vitamins, minerals, or feed additives within nanoparticles can protect them from degradation in the digestive tract and enhance their bioavailability. This can lead to better nutrient utilization, improved growth, and overall health of livestock (Arshad et al. 2021; Albuquerque et al. 2020).
3. **Reproductive Technologies:** Nanotheranostics can enhance assisted reproductive technologies (ART) in animals. Nanoparticles can serve as carriers for reproductive hormones, growth factors, or genetic materials, facilitating targeted delivery to reproductive organs. This targeted delivery can improve the success rates of artificial insemination, embryo transfer, and other ART procedures, thus enhancing breeding programs and genetic progress in animal production (Hashem and Gonzalez-Bulnes 2020).
4. **Environmental Monitoring:** Nanosensors can be employed to monitor environmental parameters, such as temperature, humidity, and air quality, in animal production facilities. This can help optimize environmental conditions for

livestock, ensuring their well-being and productivity. Nanosensors can also detect and monitor pollutants, pathogens, or contaminants in the environment, enabling early detection and mitigation of potential threats to animal health (McConnell et al. 2020; Brijs et al. 2021).

5. **Animal Welfare Monitoring:** Nanotechnology-based sensors can be utilized to monitor the behavior and well-being of individual animals in real-time (Handcock et al. 2009; Sharma and Koundal 2018). For example, wearable nanosensors can track vital parameters, such as heart rate, body temperature, or activity levels, providing valuable insights into the health and welfare status of livestock. This information can aid in early disease detection, prompt intervention, and optimization of animal management practices (Tyler et al. 2020; Younis et al. 2022).

It is important to note that the field of nanotheranostics in animal production is still in its early stages, and more research is needed to develop and validate these applications. Additionally, the safety, ethical considerations, and regulatory aspects of nanomaterial use in animal production should be carefully evaluated to ensure the well-being of both animals and the environment.

28.3 Nanotheranostics in Personalized Veterinary Medicine to Improve Reproductive Performance of Animals

Personalized veterinary medicine can play a crucial role in improving the reproductive performance of animals by focusing on individualized assessment, management, and interventions. Here are some ways in which personalized veterinary medicine can be applied to enhance reproductive performance:

1. **Reproductive Health Evaluations:** Personalized veterinary medicine begins with comprehensive reproductive health evaluations of individual animals. This includes assessing overall health, conducting breeding soundness examinations, evaluating reproductive anatomy, and performing diagnostic tests to identify any underlying reproductive issues or abnormalities. By understanding the specific reproductive challenges of each animal, veterinarians can develop targeted management plans (Giordano et al. 2022; Manríquez et al. 2021; Madureira et al. 2015; Pinedo et al. 2020).
2. **Hormonal Monitoring and Management:** Hormonal monitoring is an essential aspect of personalized veterinary medicine for optimizing reproductive performance. Veterinarians can assess hormonal profiles through blood tests or hormone assays to determine the timing of breeding, detect reproductive disorders, and evaluate reproductive functions. Based on the results, appropriate hormonal interventions, such as synchronization protocols or hormone supplementation, can be implemented to improve breeding success (Bisinotto et al. 2015; Wijma et al. 2018; Burnett et al. 2022; Holman et al. 2011; Peralta et al. 2005).

3. **Genetic Evaluation and Selection:** Personalized veterinary medicine integrates genetic evaluation and selection techniques to enhance reproductive performance. Genetic testing can identify animals with desirable reproductive traits, such as fertility, litter size, or calving ease. By selectively breeding animals with favorable genetic profiles, producers can improve reproductive efficiency and offspring quality, ultimately leading to better reproductive performance in subsequent generations (Berry et al. 2014; Lima et al. 2020; Chebel and Veronese 2020).
4. **Reproductive Technologies:** Personalized veterinary medicine incorporates the use of advanced reproductive technologies to overcome reproductive challenges and improve performance. Techniques such as artificial insemination (AI), embryo transfer (ET), in vitro fertilization (IVF), or sex-sorted semen can be utilized based on individual animals' requirements. These technologies offer increased breeding efficiency, genetic selection opportunities, and the ability to overcome infertility issues (Morotti et al. 2014; Giordano et al. 2015; Cardoso Consentini et al. 2021; Rial et al. 2022).
5. **Nutritional Management:** Proper nutrition is critical for reproductive success. Personalized veterinary medicine takes into account the specific nutritional needs of individual animals based on their reproductive stage and requirements. Veterinarians can develop customized nutrition plans that optimize fertility, promote healthy gestation, and support lactation. Adequate nutrient intake and balanced diets are essential for optimal reproductive performance (Tamminga et al. 1997; Barton et al. 1996; Cools et al. 2014; Silvestre et al. 2011; Pontes et al. 2015).
6. **Disease Prevention and Management:** Disease can significantly impact reproductive performance. Personalized veterinary medicine emphasizes preventive measures to minimize the risk of reproductive diseases and infections. This includes implementing vaccination programs, managing bio-security practices, and addressing specific diseases or conditions that can affect fertility and reproductive health. Early detection and appropriate treatment of reproductive infections or disorders can help improve overall reproductive performance (LeBlanc et al. 2002; Donofrio et al. 2007; Sheldon et al. 2009).
7. **Reproductive Monitoring and Data Analysis:** Personalized veterinary medicine emphasizes the importance of monitoring and analyzing reproductive data on an individual animal basis. This includes tracking reproductive parameters, such as heat cycles, conception rates, pregnancy rates, and calving/lambing/kidding outcomes. By analyzing this data, veterinarians can identify trends, assess performance, and make informed decisions to optimize reproductive management practices (LeBlanc 2010; Mulligan et al. 2006; Aungier et al. 2012, 2015).

By adopting personalized veterinary medicine approaches, producers can optimize reproductive performance in animals, resulting in improved fertility rates, increased conception rates, reduced embryonic loss, and better overall reproductive efficiency. This not only enhances productivity but also contributes to genetic advancement, profitability, and sustainable breeding programs (Crowe et al. 2018;

Giordano et al. 2022). Nanotheranostics have a significant potential to contribute in all of the aforesaid avenues of reproductive management in livestock animals (Hashem and Gonzalez-Bulnes 2020; Hashem 2021). A few potential applications of nanotheranostics in enhancing animal reproductive performance have been depicted hereunder.

28.3.1 Nano-delivery Systems to Improve Reproductive Performance in Animals

Nano-delivery systems have shown promising potential in improving reproductive performance in animals. These systems involve the use of nanoparticles to deliver drugs or bioactive compounds directly to the reproductive organs, allowing for targeted and controlled release. Here are a few examples of how nano-delivery systems can enhance reproductive performance in animals:

Controlled hormone delivery: Hormones play a crucial role in reproductive processes, such as estrus synchronization, ovulation induction, and fertility enhancement. Nano-delivery systems can encapsulate hormones within nanoparticles and release them in a controlled manner, mimicking the natural hormone secretion patterns. This approach helps optimize hormonal profiles, leading to improved reproductive performance (Hashem 2021; Oliveira et al. 2013; Hashem and Sallam 2020). For instance, controlled release systems can deliver hormones like gonadotropin-releasing hormone (GnRH), follicle-stimulating hormone (FSH) or luteinizing hormone (LH) to regulate ovarian function, ovulation, or estrus cycle in animals (Hassanein et al. 2021; Deguettes et al. 2020).

Gamete Quality Improvement: Gamete quality is vital for successful breeding in animals. Nanotechnology can help improve the quality and viability of sperm and ova. Nano-delivery systems can be used to encapsulate antioxidants, vitamins, or other bioactive compounds that promote sperm and ova viability during cryopreservation, sperm motility, and DNA integrity or assist in overcoming oxidative stress, thereby improving fertilization rates and reproductive outcomes (Saadeldin et al. 2020; Tar et al. 2021; Najafi et al. 2022; Remião et al. 2016). These nanoparticles can protect the active ingredients from degradation, deliver them directly to the reproductive tract, and improve gamete quality.

Targeted drug delivery: In cases of reproductive disorders or infections, nano-drug delivery systems can target the affected reproductive organs. For example, nanoparticles can be functionalized with specific ligands that recognize and bind to receptors present in the reproductive tract. This targeted approach ensures higher drug concentrations at the desired site, improving the effectiveness of the treatment while minimizing systemic side effects. As a consequence, different nano-delivery systems have been successfully employed for the treatment of endometritis (Singh et al. 2015; Garzon et al. 2021). In this context, different nano-delivery systems have also been used for the treatment of mastitis to increase the efficacy of the conventional antibiotics by their targeted delivery and prolonged action (Wang et al. 2012). This also reduces the drug dose requirement and frequency along with curtailing the risk of antibiotic resistance (Algharib et al. 2020).

Controlled release of growth factors: Growth factors are essential for follicular development, oocyte maturation, and embryo implantation. Nano-delivery systems can be designed to encapsulate growth factors and release them in a sustained and controlled manner. This approach provides a prolonged and optimal environment for reproductive processes, leading to improved fertility rates (Viana Silva et al. 2021). In this context, engineered biomaterial-based macroporous scaffold mimicking ovarian microstructure has found application in female fertility management. This scaffold can also be used to deliver growth factors to promote follicular maturation (Felder et al. 2019).

Vaccine delivery: Vaccination plays a crucial role in preventing reproductive diseases in animals. Nano-delivery systems can enhance the effectiveness of vaccines by improving antigen stability, facilitating antigen uptake by immune cells, and promoting sustained immune responses. This approach can lead to better protection against reproductive pathogens, reducing the incidence of diseases that can negatively impact the reproductive performance. Mannosylated Chitosan Nanoparticles-assisted delivery of immunogenic FliC protein has found to confer significant protection in mice against *Brucella melitensis* and *Brucella abortus* infection (Sadeghi et al. 2020a,b). Not only for antigen delivery, nanomaterials can also act as adjuvants to provide humoral as well as cellular immunity against different pathogens associated with reproductive diseases. For instance, calcium phosphate nanoparticles have found to act as adjuvants as well as delivery system for different *Brucella* antigens and epitopes to elicit protection against *B. melitensis* and *B. abortus* in *BALB/c* mice (Sadeghi et al. 2020a,b).

Overall, nano-delivery systems offer exciting opportunities to enhance reproductive performance in animals. However, it is important to note that further research and developments are needed to optimize these systems, evaluate their safety, and ensure their effectiveness in different animal species.

28.3.2 Nanotheranostics in Diagnosis, Imaging, and Therapy of Animal Reproductive Diseases

Nanotheranostics can provide real-time monitoring and diagnosis of reproductive disorders. Nanosensors and imaging agents can be designed to detect biomarkers associated with reproductive health, such as hormone levels, uterine conditions, or genetic abnormalities, allowing early detection and targeted treatment. Nanotheranostics can also be used to treat reproductive disorders, such as uterine infections or sexually transmitted diseases. Nanoparticles carrying antimicrobial agents or gene-editing tools can target infected cells or pathogens, improving reproductive health and fertility. For example, a gold nanoparticle (AuNP)-based biosensor has found to detect genomic DNA of *Brucella* spp. This method was capable of detecting up to 1.09 pg μL^{-1} of unamplified *Brucella* genomic DNA (Sattarahmady et al. 2015). Similarly, leptospirosis is another important reproductive disease of farm animals with zoonotic importance. Most of the available diagnostic tools for leptospirosis oftentimes lack accuracy, sensitivity, portability, and repeatability. A disposable immunosensor based on AuNP-modified paper electrodes with carbon nanotube-labeled monoclonal anti-LipL32 antibody has enabled simple, rapid, and sensitive determination of *Leptospira* in clinical samples with a very low detection limit of 348 fg/mL (Sapna et al. 2022). Campylobacteriosis is another zoonotic infectious disease affecting reproductive performances of farm animals. Gold nanorods functionalized with specific ssDNA probes of *cadF* gene has enabled detection of the organism from clinical samples yielding detection limit of 10^2 copy number/mL (Shams et al. 2019). For diagnostic imaging, ultra-small superparamagnetic iron oxide NPs have been used as contrast agent in MRI to detect endometriosis, particularly non-hemorrhagic ectopic endometrial lesions. Even several nano-carriers like lipid-grafted chitosan micelles have shown potential for gene therapy-based management of endometriosis. Nano-delivery systems have also shown promise for therapeutic management of uterine fibroids, ectopic pregnancy and trophoblastic diseases (Barkalina et al. 2014). In addition, nanotechnology can facilitate gene therapy for managing reproductive diseases, sperm-mediated transfer of genes and biological compounds, selection of gametes and embryos, etc. which can be useful in Assisted Reproductive Technologies (Barkalina et al. 2014). Similarly, sperm targeting using mesoporous silica nanoparticles conjugated with a cell-penetrating peptide (C105Y) can address issues of sperm quality and male infertility (Barkalina et al. 2015). Recently, silver nanoparticles (AgNPs) of 10 nm size have found to induce metabolic toxicity and oxidative stress, thereby, inhibiting cell viability and biofilm production by multi-drug resistant *Prevotella melaninogenica* and *Arcanobacterium pyogenes*, two most common bacteria causing bovine clinical endometritis (Gurunathan et al. 2018). Nanosensors have also been developed for monitoring reproductive hormone level as a diagnostic indicator of reproductive disorder. For example, a Robotic APTamer-enabled Electrochemical Reader (RAPTER) has been developed to monitor LH pulsatility. The DNA aptamer was selected through selective evolution of ligands by exponential enrichment (SELEX) method which extended specific binding with LH and sparing the other related hormones. Such system facilitates real-time in vivo monitoring of hormonal levels to indicate reproductive aberrations (Liang et al. 2019).

28.3.3 Nanotheranostics in Assisted Reproductive Technologies (ART)

Nanotechnology can enhance the efficiency of ART procedures like in vitro fertilization (IVF) or embryo transfer (Siahdasht et al. 2020; Helbling et al. 2018).

Nanoscale tools can improve embryo selection, increase embryo implantation rates, or enhance the viability of cultured embryos, ultimately leading to higher success rates in reproductive technologies.

Functionalized nanoparticles can bind with defective spermatozoa and remove them to improve the semen quality and fertility rate. Iron oxide (Fe_3O_4) magnetic NPs functionalized with anti-ubiquitin monoclonal antibody or lectin PNA/PSA (peanut and *pisum sativum* agglutinin) bind with spermatozoa having acrosomal defect. Subsequently, these defective spermatozoa can be removed using magnetic field (MACS, magnetic activated cells sorting) (Odhiambo et al. 2014). Another important application of nanotechnology in sperm sorting is using the Cell-SELEX technique. Biotin-labeled single strand DNA aptamers identified by SELEX technique that specifically bind to the sperms having membrane damage has been used to separate the defective sperms using avidin-coated super-paramagnetic iron oxide nanoparticles (SPION) using magnetic field (Farini et al. 2016). The enhanced optical properties of NPs have also been explored towards labeling the sperms for sorting as an alternative to fluorochromes and flow cytometry. Oligonucleotide/DNA tags conjugated gold NPs has been employed to label specific sites of the Y chromosome, thereby, paved the path for sex sorted semen production to improve the reproductive performance (Gamrad et al. 2017).

The antibacterial property of nanoparticles has also been explored for sperm storage. The bactericidal action of silver nanoparticles (AgNPs) of 10–20 nm size against *Staphylococcus aureus* in storage of swine sperm has been explored along with its potential toxic effects. The NPs depicted concentration-dependent bacterial inhibition in a concentration range of 0.4–10 mM. Regarding toxicity, sperm cell morphology, viability, sperm capacitation, and acrosome reactions have been analyzed and at below 2 mM concentration, none of the parameter got affected. This indicates the use of NPs as potential alternative to the conventional antibiotics in semen extender for sperm storage (Pérez-Duran et al. 2020). Application of zinc nanoparticles has also been suggested increasing mitochondrial activity, improving membrane integrity, enhancing total antioxidant capacity, and reducing lipid peroxidation in bovine sperm, therefore may extend significant benefits if added to semen extender (Jahanbin et al. 2021). Cerium oxide (CeO_2), zinc oxide (ZnO), and selenium nanoparticles have also been depicted to restore sperm viability and motility in cryopreserved semen by similar mechanisms. Despite the potential benefits of nanoparticles in sperm sorting, sexing, and cryopreservation, evidences of in vivo study in animals is limited as metallic and metal oxide NPs may also induce toxic effects in a concentration-dependent manner along with marked species-wise variations (Viana Silva et al. 2021).

Nanoparticles comprising gold, iron oxide, and poly-L-lysine have facilitated introduction of three-dimensional culture for in vitro growth of bovine secondary follicles using magnetic levitation method. The 3-D cultured secondary follicles at 200-μl/ml treatment with magnetic NPs depicted higher oocyte viability, antrum formation, and lower degeneration rates than two-dimensionally cultured follicles (de Cássia Antonino et al. 2019). Green synthesized gold NPs using the extract of *Achillea millefolium* has also been depicted to facilitate the maturation of preantral follicles in vitro by increasing estradiol production from granulosa cells (Abdollahi et al. 2020). Treatment of bovine spermatozoa with zinc NPs has found to improve semen quality without affecting embryo development rate; however, supplementation of zinc NPs in in vitro maturation (IVM) media has found to enhance the embryo development rate. Addition of the NPs in IVM media enhanced superoxide dismutase activity in cumulus cells, reduced DNA damage, and apoptosis in cumulus-oocyte complexes to facilitate the embryo development (Jahanbin et al. 2021). Similarly, supplementation of nano-selenium and nano-zinc oxide particles to IVM media has found to reduce DNA damage of cumulus cells, and increase glutathione (GSH) concentration in bovine oocytes (Abdel-Halim and Helmy 2017). Supplementation of polymeric and lipid-core nanocapsules to IVM media has reported to increase embryo cell number, decreased cell apoptosis and ROS level. The NPs has found to upregulate catalase (CAT) and superoxide dismutase (SOD)-2 genes to promote antioxidant function, while downregulated CASP3, SHC1, and proapoptotic BAX genes, thereby augmenting the protective effects of melatonin against oxidative stress and apoptosis in bovine embryos (Komninou et al. 2016).

It's worth noting that while nanotheranostics show great promise, further research is needed to optimize their design, safety, and efficacy in animal reproductive medicine. Additionally, ethical considerations and regulatory frameworks should be taken into account when translating these technologies into practical applications.

28.3.4 Nanotheranostics for the Management of Breeding Bulls

Nanotheranostics offer potential applications in the management of breeding bulls, focusing on enhancing reproductive performance and addressing fertility-related issues. Semen quality improvement, reproductive tract imaging and diagnostics, targeted drug delivery, nutritional and hormonal management are some ways where nanotheranostics have already shown potential as mentioned in aforesaid sections (Falchi et al. 2018). Genetic modifications in different productive, reproductive, or disease-associated genes of individual breeding bull is another potential area where nanotechnology can deliver significantly (Gaj et al. 2016).

Nanotechnology can facilitate genetic modifications in breeding bulls. Nanoparticles integrated with gene-editing tools, such as CRISPR-Cas9, can be employed to introduce specific genetic modifications or edits to the germ cells (Duan et al. 2021). This may aid in addressing genetic defects, improving desirable traits, or enhancing fertility potential of bulls in the breeding population (Young et al. 2020). However, it is important to note that ethical considerations and regulatory frameworks must be considered when translating these technologies into practical applications in veterinary medicine.

28.4 Nanotheranostics in Stem Cell Therapy and Regenerative Medicine in Veterinary Science

Nanotheranostics have significant potential in the field of stem cell therapy and regenerative medicine in veterinary science. These advanced nanoscale systems can be utilized to isolate and purify stem cells, enhance the delivery, tracking, and therapeutic effects of stem cells, thereby improving their efficacy in veterinary applications. Here are some ways nanotheranostics can be employed in this context:

Isolation and purification of Stem Cells: Magnetic nanoparticles labeled with monoclonal antibody against surface antigens of stem cells are frequently used for stem cell isolation and purification. For instance, super-paramagnetic magnetite (Fe_3O_4) nanoparticles conjugated with anti-CD105 and anti-CD73 monoclonal antibodies have been successfully used for separation of mesenchymal stem cells from peripheral blood, human bone marrow cells, and fibroblastic cells (Odabaş et al. 2008). Similarly, anti-$CD34^+$ monoclonal antibody-conjugated magnetic nanoparticles have been frequently used for stem cell separation from apheresis samples or whole blood (Krasteva et al. 2020; Jing et al. 2007).

Stem Cell Delivery and Tracking: Nanoparticles can serve as carriers for targeted delivery of stem cells to specific tissues or organs (Sun et al. 2020). They can be functionalized with ligands or surface modifications that facilitate cell attachment, migration, and engraftment. Nanomaterials such as QD, magnetic nanoparticles, gold nanorods, and polymeric nanoparticles are commonly employed for cell tracking. Additionally, nanoparticles can be engineered with imaging agents, such as fluorescent dyes or magnetic nanoparticles, to track and monitor the distribution and fate of transplanted stem cells in real-time (Huang et al. 2021). In this context, the most common imaging methods for cell imaging include MRI, photoacoustic imaging, fluorescent imaging, and radioactive cell imaging (Dong et al. 2021).

Scaffold and Matrix Engineering: Nanotechnology can be employed to design and fabricate nanoscale scaffolds or matrices that mimic the natural extracellular matrix (ECM). These scaffolds provide structural support for stem cells and can be functionalized with nanomaterials to enhance cellular adhesion, proliferation, and differentiation. They can also be loaded with growth factors or signaling molecules to guide stem cell behavior and tissue regeneration. Graphene and graphene oxide (GO) nanomaterials are probably the most widely used scaffold material to support stem cell growth and proliferation. For instance, graphene-based scaffold increased osteogenic differentiation of mesenchymal stem cells, supported neurospheres shaping and facilitated cell migration of neural stem cells, improved integrin clustering, focal adhesion, and neuronal differentiation of human neural stem cells, etc. (Dong et al. 2021). Similarly, as laminins are vital glycoprotein component of extracellular matrices, therefore, laminin-coated graphene scaffold has facilitated differentiation of human neural stem cells into neurons (Park et al. 2011). Among other nanomaterials, gold NP-coated collagen-nanofiber has promoted neuronal differentiation and proliferation of placental derived mesenchymal stem cells; polymeric

NPs such as gelatin-immobilized poly(L-lactide-co-caprolactone) or PCL/PLA-3D scaffold has enhanced osteogenic differentiation of human mesenchymal stem cells (Dong et al. 2021). Nanopore patterned polystyrene surfaces have also significantly enhanced the pancreatic differentiation of human embryonic and induced pluripotent stem cells in a nanotopography-dependent manner with 200 nm nanopore-patterned surface has proved to be the most conducive for stem cell differentiation (Kim et al. 2016).

Controlled Release of Bioactive Molecules: Nanotheranostics can be designed to release bioactive molecules, such as growth factors, cytokines, or immune modulatory agents, in a controlled and sustained manner. These molecules can enhance the regenerative capacity of stem cells and promote tissue repair and healing. Nanoparticles can act as reservoirs, protecting and preserving the bioactivity of therapeutic agents until they are released at the target site. Hyaluronan based porous microparticles have been employed for targeted delivery and sustained release of Platelet-derived growth factor (PDGF) and Transforming Growth Factor-β (TGF-β) to promote stem cell proliferation and differentiation at full-thickness wound site on Wistar rats (Zavan et al. 2009). Similarly, various scaffold materials, hydrogels, and nano-composites have been employed for the delivery of different bone morphogenetic proteins (BMPs), TGF-β, PDGF, fibroblast growth factors (FGFs), tumor necrosis factors (TNFs), vascular endothelial growth factor (VEGF), interleukins (ILs), etc. for stem cell growth and differentiation to promote bone repair and angiogenesis (Li et al. 2023).

Tissue Regeneration: Nanotheranostics can be utilized to promote tissue regeneration and repair in veterinary medicine. Nanoparticles can facilitate the differentiation of stem cells into specific cell types required for tissue reconstruction. They can also be used to deliver genetic materials or potentiate gene-editing tools for targeted modification of stem cells, promoting their therapeutic potential. Various types of hydrogels such as citrate-based polyester hydrogel, hyaluronic acid-based hydrogel, glutathione-modified collagen hydrogel, etc., different polymeric NPs have been used to deliver various growth factors for cardiac tissue regeneration. Nanomaterials loaded with stem cells are also being employed for cardiac tissue repair. Such type of therapy can be useful to manage myocardial infarction (Smagul et al. 2020).

Nanopatterns have also been reported to have crucial role to direct stem cell differentiation towards a specific cell lineage which is a key aspect of stem cell therapy. Various extracellular matrix proteins interact with diverse cell adhesion molecules through arginine-glycine-aspartate (RGD) sequence, therefore, facilitate stem cell growth. Quartz substrates surface modified with RGD coupled with negatively or positively charged or neutral oligopeptide motifs have depicted strikingly different osteogenic differentiation of mesenchymal stem cells. The negatively charged and neutral oligopeptide motifs have found to promote osteogenic differentiation in presence of RGD, while positively charged oligopeptide motif inhibited osteogenic differentiation. So, nanopatterns surface functionalized with different charged moieties may behave differentially in stem cell differentiation (Cao et al. 2015). In another experiment, self-assembled layers of vertically oriented TiO_2

nanotubes of diameters between 15 and 100 nm were generated on titanium surfaces to facilitate adhesion, growth, and differentiation of mesenchymal stem cells. It has been observed that tube diameter larger than 50 nm severely restricted cell adhesion and spreading with significantly reduced cellular activity and enhanced apoptosis as raising the tube diameter reduced the focal contact for the stem cells. The lateral spacing geometry of the nanotube surfaces strongly regulated the integrin clustering/focal contact formation and thus, the cellular activities (Park et al. 2007). In addition, several biomimetic 3D scaffolds are currently being explored that facilitated stem cell growth and differentiation towards a specific lineage (Gandhimathi et al. 2015).

Immunomodulation and Anti-inflammatory Effects: Nanotheranostics can be utilized to modulate the immune response and mitigate inflammation in the context of stem cell therapy. Nanoparticles can carry immunomodulatory agents or anti-inflammatory drugs, promoting stem cell survival and reducing adverse immune reactions that can hinder the therapeutic effects of stem cell treatments. Nano-delivery system can facilitate immunomodulation through gene editing. In vivo delivery of large CRISPR-Cas system or si-RNA or microRNA is possible through nanocarriers (Chuang et al. 2022). Local delivery of immunomodulators can be mediated through nanohydrogels or scaffolds to modulate the microenvironment of the transplant site (Chua et al. 2023). For instance, activation of Fas receptor/Fas ligand (FasL) pathway is a predominant immune mechanism of islet allograft rejection; co-transplantation of allogeneic islets with FasL-modified microgel induced tolerance, therefore, resulted long-term islet engraftment and normoglycemia in diabetic mice (Headen et al. 2018).

Monitoring Stem-Cell Therapy Progress by Theranostic Imaging: Nanoparticles can act as contrast agents for various imaging modalities, such as magnetic resonance imaging (MRI), computed tomography (CT), or ultrasound. By labeling stem cells with these nanoparticles, researchers and veterinarians can non-invasively visualize and monitor the distribution, migration, and integration of transplanted cells, as well as assess the progress of tissue regeneration (Masoudi Asil et al. 2020). For instance, the superparamagnetic iron oxide nanoparticles (SPIONs) can improve the sensitivity of MRI, therefore, acting as in vivo or in vitro contrast agents (Unterweger et al. 2023). Further, it has been depicted that uptake of these particles by human mesenchymal stem cells does not alter proliferation and differentiation of the stem cells but still generates strong contrast MRI signals, therefore, can be employed for stem cell tracking after transplant (Rosenberg et al. 2016). The surface plasmon resonance (SPR) property of gold NPs can be exploited to use them as CT contrast agents (Xi et al. 2012). Human mesenchymal stem cells labeled with gold NPs coupled poly-l-lysine (PLL) and rhodamine B isothiocyanate facilitated in vivo cell tracking in rat brain through Micro-CT imaging (Kim et al. 2017). Quantum Dots (QDs) such as carbon quantum dots (CQDs) and graphene quantum dots (GQDs) have also been used for labeling stem cells for in vitro and in vivo imaging (Muller-Borer et al. 2007; Majood et al. 2022).

These applications of nanotheranostics in stem cell therapy and regenerative medicine have the potential to revolutionize veterinary medicine by improving the treatment outcomes for various conditions, including musculoskeletal disorders, wound healing, neurological disorders, and organ regeneration. However, it's important to note that further research is needed to optimize nanotheranostic systems, understand their safety profiles, and evaluate their long-term effects in veterinary patients.

28.5 Nanotheranostics in Veterinary Transplant Medicine

Nanotheranostics can play a significant role in veterinary transplant medicine by improving the outcomes of organ transplantation and reducing complications. Here are some ways nanotheranostics can be applied in this field:

28.5.1 Organ Preservation

Nanoparticles can be utilized to enhance the preservation of organs before transplantation. By encapsulating cryoprotective agents within nanoparticles, it is possible to improve the efficiency of cryopreservation and reduce the damage caused by ice crystal formation. These nanoparticles can help protect organ tissues during the freezing and thawing processes, preserving organ viability and functionality. It has been depicted that mesoporous and microporous silica-coated iron oxide nanoparticles can improve cryopreservation of large samples or volumes like porcine arterial tissue and rat kidneys by exploiting nanowarming that rely upon local heating from the magnetic nanoparticles under an alternating magnetic field, facilitating homogeneous and rapid rewarming (Murray and Gibson 2022). Several synthetic polymers such as poly(vinyl alcohol)-co-poly(vinyl acetate) (PVA-co-PVAc) copolymers, carboxylated ε-poly-L-lysine, etc. have been employed to inhibit devitrification preventing ice nucleation and growth during warming (Wowk et al. 2000; Matsumura et al. 2011). Nanoparticles derived by polymerization-induced self-assembly of different polymers such as poly(ethylene glycol), poly (dimethylacrylamide), and poly(vinylpyrrolidone) depicted ice recrystallization inhibition (IRI) while the polymers in non-assembled form yielded no such activity (Georgiou et al. 2021). Similarly, cellulose nanocrystals and 2,2,6,6-tetramethylpiperidine-1-oxyl oxidized cellulose nanofibrils have also produced IRI activity at different concentrations in 0.01 M NaCl, phosphate-buffered saline and sucrose solutions (Li et al. 2019).

Trehalose is a non-reducing disaccharide that protects mammalian cells during cryopreservation. The presence of trehalose, both extracellularly as well as intracellularly, assists in optimum cryoprotection. Several NPs have found to be efficient for intracellular delivery of trehalose to facilitate cryopreservation. For instance, a pH responsive genipin-cross-linked Pluronic F127-chitosan nanoparticles has found to efficiently encapsulate trehalose and deliver to the cryopreserved primary human adipose-derived stem cells (hADSCs). The nanoparticles can be used as sole non-toxic cryoprotectant without the requirement of using any DMSO which is toxic for mammalian cells at physiological temperature (Rao et al. 2015). A cold-responsive nanoparticle of polymer (poly(N-isopropylacrylamide-co-butyl acrylate)) has reported to encapsulate and deliver trehalose to MDA-MB-231 cells and hADSCs efficiently. The NPs carrying trehalose cargo can be taken up by the cells easily followed by rapid cold-triggered disassembly of the NPs and intracellular release of trehalose. This also eliminates the necessity of using DMSO as cryoprotectant which requires a hectic and cell-offensive washing step for in vivo application of the cryopreserved cell (Zhang et al. 2019). Colloidal apatite nanoparticles have also been successfully employed for intracellular delivery of trehalose to RBCs, although the NPs do not cross the membrane but it modulates the permeability of RBC membrane to deliver trehalose and facilitate cryopreservation. The approach can substitute the requirement of toxic glycerol used for cell cryopreservation (Stefanic et al. 2017).

28.5.2 Targeted Drug Delivery and Immunomodulation

Nanoparticles can serve as carriers for targeted drug delivery to the transplanted organ or the recipient. They can be designed to release immunosuppressive agents or anti-inflammatory drugs at the site of transplantation, reducing the risk of rejection and minimizing the need for systemic immunosuppression. This targeted approach can enhance therapeutic efficacy while minimizing side effects.

Nanotheranostics can be used to modulate the immune response in organ transplantation. Nanoparticles can be loaded with immunomodulatory agents, such as regulatory T cells (Tregs) or tolerogenic molecules, which can help create an immunotolerant environment and reduce the risk of rejection. Additionally, nanoparticles can be engineered to selectively interact with immune cells and regulate their function, promoting immune tolerance towards the transplanted organ.

Rapamycin is a macrolide exhibiting potent immunosuppressive activity, prevents rejection in organ transplantation. Grafts pre-treated with 100 ng mL^{-1} of targeted rapamycin micelles have enhanced preservation with a significant inhibition of chronic rejection post-transplantation as compared to tenfold higher dose of free rapamycin (Zhu et al. 2018). Pretreatment of human vessels with Poly(amine-co-ester) NPs loaded siRNA targeting class II transactivator silenced major histocompatibility complex class II expression on endothelial cells that protect organs from acute transplant rejection (Cui et al. 2017). PLGA-based nano and microparticles have been used extensively to deliver immunomodulators and avert allograft rejection. Being biodegradable, PLGA-based carrier system has found to be safe with large loading capacity and sustained release system to deliver antigenic peptides, growth factors, monoclonal antibodies, and diverse immunomodulators through different routes (Shahzad et al. 2020).

In transplant medicine, cryopreservation of cells or organs induces apoptotic pathways to affect the outcome. Activation of caspase-mediated pathways is a predominant underlying factor in this process. Therefore, nanoparticle-based delivery of caspase inhibitors in cryopreservation media or cells can improve the outcome to facilitate transplantation. Pretreatment of human mesenchymal stem cells (MSCs) with a broad-spectrum caspase inhibitor Z-VAD-FMK prior cryopreservation has found to improve preservation efficacy (Bissoyi and Pramanik 2014). Delivery of such type of caspase inhibitor with nano-carriers may further improve the outcome.

Intra-lymph node delivery of tolerogenic NPs customized by conjugation of antigen with biodegradable polymers can induce antigen-specific regulatory T cells (Tregs) to provide a systemic and antigen specific immunotolerance by reprogramming the local lymph node, therefore can prevent graft rejection in transplanted subject. Such strategy has already been explored against autoimmune disorders such as autoimmune encephalomyelitis in mice model (Tostanoski et al. 2016; Pearson et al. 2017; LaMothe et al. 2018).

28.5.3 Detection of Rejection and Complications

Nanoparticles can be functionalized with imaging agents, such as fluorescent dyes or magnetic nanoparticles, to detect early signs of rejection or complications after transplantation. By tracking the distribution and behavior of these nanoparticles, veterinarians can monitor the viability and functionality of the transplanted organ. This real-time monitoring can facilitate early intervention and appropriate management of post-transplant complications.

A nanosensor comprised of a fluorescent reporter-labeled peptide which is a substrate for serine protease granzyme B can help to detect early rejection with high sensitivity and specificity. Granzyme B is produced by the recipient T cells at the early phase of acute cellular rejection. Systemic administration of the nanosensor accumulate in the allograft tissue followed by cleavage of the labeled peptide by granzyme B and release of the fluorescent reporter which comes in the urine to be detected before graft failure and complication. The modality has been successfully tested in mouse models of skin-graft rejection (Mac et al. 2019). Dextran NPs derivatized with the PET isotope copper-64 has been used for the detection of cardiac allograft rejection and immunomodulation in C57BL/6 recipients from BALB/c donor using positron emission tomography-computed tomography (PET-CT) imaging (Ueno et al. 2013). Similarly, dextranated and diethylenetriamine pentaacetic acid (DTPA)-modified magnetofluorescent nanoparticle labeled with copper-64 isotope, ^{89}Zr-labeled high-density lipoprotein nanoparticles, ^{89}Zr-labeled dextran NPs have been used for high-resolution PET-CT imaging of macrophages. Superparamagnetic or ultrasmall superparmagnetic iron oxide NPs as superior MRI contrast agents, PEGylated gold nanoparticles for X-ray CT can be the other potential options (Ochando and Braza 2017).

28.5.4 *Tissue Engineering and Regeneration*

Nanotheranostics can be employed in tissue engineering approaches to create customized grafts or scaffolds for transplantation. Nanoparticles can be incorporated into scaffolds to enhance cellular adhesion, proliferation, and differentiation. They can also be loaded with growth factors or signaling molecules to guide tissue regeneration and facilitate the integration of the transplanted tissue with the host. Tacrolimus, a typical immunosuppressant encapsulated in PLGA nanoparticles-based hydrogel scaffold significantly reduced immune responses and increased survival of endothelial progenitor cells (EPCs) both in vitro and in vivo as compared to its free form. The drug encapsulated scaffold laden with EPCs has been successfully employed for the treatment of hind limb ischemia in mice facilitating tissue regeneration by transplanted cells (Li et al. 2018).

28.5.5 *Antibacterial Coatings*

Nanotheranostics can be utilized to prevent infections and improve the survival of transplanted organs. Nanoparticles with antimicrobial properties can be applied as coatings on the surface of the medical devices or the implants to inhibit bacterial growth and reduce the risk of post-transplant infections (Furno et al. 2004). Nanomaterials of silver, copper, titanium, etc. have been frequently used in surface coating of various biomedical devices (Liu et al. 2020).

By leveraging nanotheranostics in veterinary transplant medicine, it is possible to enhance organ preservation, improve therapeutic outcomes, and minimize complications associated with organ transplantation. However, it's important to note that further research is necessary to optimize the design, safety, and efficacy of nanotheranostic systems specifically in the context of veterinary transplant medicine.

28.6 Nanotheranostics in Personalized Veterinary Medicine for Wildlife Conservation

Personalized veterinary medicine plays a crucial role in wildlife conservation efforts by helping to ensure the health and well-being of individual animals and populations. Here are some key reasons why personalized veterinary medicine is important in wildlife conservation:

Disease management in wildlife is a key aspect of personalized veterinary medicine. Wildlife populations are susceptible to various diseases, including infectious diseases, parasites, and chronic conditions. Personalized veterinary medicine allows for the identification and monitoring of specific diseases or health conditions within wildlife populations (Kadam et al. 2022). By understanding the unique health needs of individual animals and populations, veterinarians can develop targeted disease management strategies, including vaccination programs, treatment protocols, and preventive measures. Further generation of medical data, multi-omics data, and environmental data along with their integration can be of great help for wildlife disease management (Whilde et al. 2017; Reading et al. 2013).

Genetic diversity is essential for the long-term survival of wildlife populations. Personalized veterinary medicine can provide valuable genetic information through techniques such as genetic testing and monitoring. This information helps identify and manage genetic diseases or disorders within a population, develop breeding programs to maintain genetic diversity, and make informed decisions regarding translocation or reintroduction efforts (Brown et al. 2009; Pérez-González et al. 2021; Hohenlohe et al. 2021; Oldenbroek 2021).

Personalized veterinary medicine is critical in the rehabilitation and release of injured or orphaned wildlife. Individualized care and treatment plans can be developed based on the specific needs and conditions of each animal. This includes medical interventions, nutritional support, behavioral rehabilitation, and monitoring to ensure successful recovery and appropriate release back into the wild (Whilde et al. 2017; Karesh and Cook 1995).

Personalized veterinary medicine contributes to scientific research and understanding of wildlife health and conservation. By studying individual animals, researchers can gain insights into disease prevalence, impacts of environmental factors, and responses to conservation interventions. This knowledge informs conservation strategies, enables early detection of emerging diseases, and helps mitigate threats to wildlife populations (Karesh and Cook 1995; Reading et al. 2013).

Personalized veterinary medicine in wildlife conservation aligns with the One Health concept, which recognizes the interconnectedness of human health, animal health, and ecosystem health. Understanding and addressing the health needs of wildlife populations can have positive impacts on human and domestic animal health, as wildlife can serve as reservoirs for zoonotic diseases that can affect other species, including humans (Buttke et al. 2015; Cunningham et al. 2017; Mavingui et al. 2018).

Overall, personalized veterinary medicine in wildlife conservation is essential for promoting the health and sustainability of wildlife populations. By tailoring veterinary care and interventions to individual animals and populations, we can mitigate disease risks, manage genetic diversity, facilitate successful rehabilitation and release, contribute to scientific knowledge, and support the overall conservation of wildlife and their ecosystems.

In the field of wildlife conservation, nanotheranostics holds promise for various applications, including disease detection, drug delivery, tracking and monitoring of wild animals, environmental monitoring, genetic analysis and management, etc.

Nanoparticles can be designed to target specific pathogens or disease biomarkers in wildlife. These nanoparticles can carry diagnostic agents that help in the early detection of diseases in animals, allowing for timely interventions and preventing the spread of infectious diseases within wildlife populations (Tallury et al. 2010; Minakshi et al. 2020, 2019).

Nanoparticles can serve as carriers for drugs or therapeutic agents, enabling targeted delivery to specific tissues or organs in animals. This targeted drug delivery and controlled release system can enhance the effectiveness of treatments while minimizing side effects and reducing the required dosage as well as frequency. It can be particularly useful in treating wildlife populations affected by diseases, parasites, or other health issues where repeated drug administration is extremely difficult (Prasad et al. 2021, 2018; Yetisgin et al. 2020).

Nanotechnology-based tracking devices such as Radio Frequency ID (RFid) Tags, "Senser" tags (Silicon Enhanced Nanoparticles for Surface Enhanced Raman Scattering), etc. can be developed to monitor the movements, behavior, and vital parameters of wildlife. Implantable nanosensors can provide real-time data on factors like body temperature, heart rate, and environmental exposure. Such monitoring tools can aid researchers in understanding the behavior, migration patterns, and overall health status of the endangered or vulnerable species (AZoNano.com 2005; Bridge et al. 2019; Ruckh and Clark 2014; Lard et al. 2010).

Nanosensors can be employed to detect and monitor environmental pollutants, food chain or water contaminants that affect wildlife habitats. These sensors can detect specific pollutants at low concentrations, providing early warning systems to identify and mitigate potential threats to ecosystems (Willner and Vikesland 2018; Mahmoud and Fawzy 2021; Pandey and Mishra 2014; Nikoleli et al. 2018; Satish et al. 2022; Ibrahim et al. 2016).

Nanotechnology-based platforms such as nanopore-based single-molecule DNA/RNA sequencing can facilitate genetic analysis and sequencing of wildlife species (Ying et al. 2022; Ding et al. 2020). This can aid in population genetics studies, conservation genetics, and the identification of specific genes associated with adaptations or vulnerabilities. Nanotheranostics can help accelerate and improve the accuracy of genetic research in wildlife conservation.

It is important to note that nanotheranostics in wildlife conservation is still an emerging field, and much research and development are needed to fully explore its potential applications and address any ethical or environmental concerns. However, the integration of nanotechnology and theranostics has the potential to revolutionize wildlife conservation efforts by providing new tools for disease management, monitoring, and genetic analysis in wildlife populations.

28.7 Conclusion and Future Outlook

Nanotheranostics, the integration of nanotechnology with therapeutics and diagnostics, holds great promise in the field of personalized veterinary medicine. The ability to deliver targeted therapies and simultaneously monitor treatment responses through nanoscale imaging and sensing techniques has the potential to revolutionize veterinary healthcare. Through the use of nanotheranostics, veterinarians can tailor treatments to individual animals based on their specific disease characteristics, genetic makeup, and responses to therapy. This personalized approach can enhance treatment outcomes, minimize side effects, and optimize the overall well-being of animals.

The application of nanotheranostics in veterinary medicine has already shown promising results in various areas, including cancer therapy, infectious disease treatment, and regenerative medicine. Nanoparticles can be designed to selectively target cancer cells, deliver chemotherapeutic agents directly to the tumor site, and enable real-time monitoring of treatment efficacy. Additionally, nanotheranostics can aid in the diagnosis and treatment of infectious diseases by enabling rapid and sensitive detection of pathogens, targeted delivery of antimicrobial agents, and monitoring of treatment response.

The future of nanotheranostics in personalized veterinary medicine is bright, with several exciting prospects on the horizon. The potential areas of development and research include but not limited to advancements in nanoparticle design, integration of multimodal imaging and sensing techniques, expansion of veterinary nanomedicine, translation to clinical practice, and collaborative approach between veterinarians, nanotechnologists, engineers, and pharmaceutical companies.

In conclusion, nanotheranostics holds immense potential in personalized veterinary medicine. By harnessing the power of nanotechnology, veterinarians can provide tailored treatments, improve patient outcomes, and advance our understanding of veterinary diseases. Continued research, development, and collaboration are crucial to unlock the full potential of nanotheranostics and shape the future of personalized veterinary medicine.

References

Abdel-Halim BR, Helmy NA (2017) Effect of nano-selenium and nano-zinc particles during in vitro maturation on the developmental competence of bovine oocytes. Anim Prod Sci 58(11): 2021–2028

Abdollahi M, Baharara J, Shahrokhabadi K, Bamoharram FF, Saadatfar Z (2020) The effect of green gold nanoparticles synthesized with the Achillea millefolium extract on the morphometric and developmental indices of preantral follicles in immature NMRI mice. Thrita 9(2)

Albuquerque J, Casal S, Páscoa RN, Van Dorpe I, Fonseca AJ, Cabrita AR, Neves AR, Reis S (2020) Applying nanotechnology to increase the rumen protection of amino acids in dairy cows. Sci Rep 10(1):1–12. https://doi.org/10.1038/s41598-020-63793-z

Algharib SA, Dawood A, Xie S (2020) Nanoparticles for treatment of bovine Staphylococcus aureus mastitis. Drug Deliv 27(1):292–308. https://doi.org/10.1080/10717544.2020.1724209

Aquilani C, Confessore A, Bozzi R, Sirtori F, Pugliese C (2022) Review: precision livestock farming technologies in pasture-based livestock systems. Animal 16(1):100429. https://doi.org/10.1016/j.animal.2021.100429

Arshad R, Gulshad L, Haq IU, Farooq MA, Al-Farga A, Siddique R, Karrar E (2021) Nanotechnology: a novel tool to enhance the bioavailability of micronutrients. Food Sci Nutr 9(6): 3354–3361. https://doi.org/10.1002/fsn3.2311

Aungier SP, Roche JF, Sheehy M, Crowe MA (2012) Effects of management and health on the use of activity monitoring for estrus detection in dairy cows. J Dairy Sci 95(5):2452–2466. https://doi.org/10.3168/jds.2011-4653

Aungier SP, Roche JF, Duffy P, Scully S, Crowe MA (2015) The relationship between activity clusters detected by an automatic activity monitor and endocrine changes during the periestrous period in lactating dairy cows. J Dairy Sci 98(3):1666–1684. https://doi.org/10.3168/jds.2013-7405

Avci ÇB, Baran Y (2014) Use of microRNAs in personalized medicine. miRNomics: MicroRNA biology and computational. Analysis:311–325. https://doi.org/10.1007/978-1-62703-748-8_19

AZoNano.com (2005) Tagging, monitoring and tracking using nanotechnology methods and devices. https://www.azonano.com/article.aspx?ArticleID=1344

Barkalina N, Charalambous C, Jones C, Coward K (2014) Nanotechnology in reproductive medicine: emerging applications of nanomaterials. Nanomedicine 10(5):921–938. https://doi.org/10.1016/j.nano.2014.01.001

Barkalina N, Jones C, Townley H, Coward K (2015) Functionalization of mesoporous silica nanoparticles with a cell-penetrating peptide to target mammalian sperm in vitro. Nanomedicine (Lond) 10(10):1539–1553. https://doi.org/10.2217/nnm.14.235

Barton B, Rosario H, Anderson G, Grindle B, Carroll D (1996) Effects of dietary crude protein, breed, parity, and health status on the fertility of dairy cows. J Dairy Sci 79(12):2225–2236. https://doi.org/10.3168/jds.S0022-0302(96)76599-2

Berckmans D (2017) General introduction to precision livestock farming. Anim Front 7(1):6–11. https://doi.org/10.2527/af.2017.0102

Berry DP, Wall E, Pryce JE (2014) Genetics and genomics of reproductive performance in dairy and beef cattle. Animal 8(s1):105–121. https://doi.org/10.1017/S1751731114000743

Bisinotto R, Pansani M, Castro L, Narciso C, Sinedino L, Martinez N, Carneiro P, Thatcher W, Santos J (2015) Effect of progesterone supplementation on fertility responses of lactating dairy cows with corpus luteum at the initiation of the Ovsynch protocol. Theriogenology 83(2):257–265. https://doi.org/10.1016/j.theriogenology.2014.09.021

Bissoyi A, Pramanik K (2014) Role of the apoptosis pathway in cryopreservation-induced cell death in mesenchymal stem cells derived from umbilical cord blood. Biopreserv Biobank 12(4):246–254. https://doi.org/10.1089/bio.2014.0005

Bridge ES, Wilhelm J, Pandit MM, Moreno A, Curry CM, Pearson TD, Proppe DS, Holwerda C, Eadie JM, Stair TF, Olson AC, Lyon BE, Branch CL, Pitera AM, Kozlovsky D, Sonnenberg BR, Pravosudov VV, Ruyle JE (2019) An Arduino-based RFID platform for animal research. Front Ecol Evol 7:458932. https://doi.org/10.3389/fevo.2019.00257

Brijs J, Føre M, Gräns A, Clark TD, Axelsson M, Johansen JL (2021) Bio-sensing technologies in aquaculture: how remote monitoring can bring us closer to our farm animals. Philos Trans R Soc B 376(1830):20200218. https://doi.org/10.1098/rstb.2020.0218

Brown AR, Hosken DJ, Balloux F, Bickley LK, LePage G, Owen SF, Tyler CR (2009) Genetic variation, inbreeding and chemical exposure-combined effects in wildlife and critical considerations for ecotoxicology. Philos Trans R Soc B Biol Sci 364(1534):3377–3390

Burnett TA, Madureira AM, Bauer JW, Cerri RL (2022) Impact of gonadotropin-releasing hormone administration at the time of artificial insemination on conception risk and its association with estrous expression. J Dairy Sci 105(2):1743–1753. https://doi.org/10.3168/jds.2021-20156

Buttke DE, Decker DJ, Wild MA (2015) The role of one health in wildlife conservation: a challenge and opportunity. J Wildl Dis 51(1):1–8

Campion DP, Dowell FJ (2019) Translating pharmacogenetics and pharmacogenomics to the clinic: progress in human and veterinary medicine. Front Vet Sci 6:429286. https://doi.org/10.3389/fvets.2019.00022

Cao FY, Yin WN, Fan JX, Tao L, Qin SY, Zhuo RX, Zhang XZ (2015) Evaluating the effects of charged oligopeptide motifs coupled with RGD on osteogenic differentiation of mesenchymal stem cells. ACS Appl Mater Interfaces 7(12):6698–6705

Cardoso Consentini CE, Wiltbank MC, Sartori R (2021) Factors that optimize reproductive efficiency in dairy herds with an emphasis on timed artificial insemination programs. Animals 11(2):301. https://doi.org/10.3390/ani11020301

Chebel RC, Veronese A (2020) Associations between genomic merit for daughter pregnancy rate of Holstein cows and metabolites postpartum and estrus characteristics. J Dairy Sci 103(11): 10754–10768. https://doi.org/10.3168/jds.2020-18207

Chua CYX, Jiang AY, Eufrásio-da-Silva T, Dolatshahi-Pirouz A, Langer R, Orive G, Grattoni A (2023) Emerging immunomodulatory strategies for cell therapeutics. Trends Biotechnol 41(3): 358–373

Chuang ST, Conklin B, Stein JB, Pan G, Lee KB (2022) Nanotechnology-enabled immunoengineering approaches to advance therapeutic applications. Nano Converg 9(1):19

Cools S, Van den Broeck W, Vanhaecke L, Heyerick A, Bossaert P, Hostens M, Opsomer G (2014) Feeding soybean meal increases the blood level of isoflavones and reduces the steroidogenic capacity in bovine corpora lutea, without affecting peripheral progesterone concentrations. Anim Reprod Sci 144(3–4):79–89. https://doi.org/10.1016/j.anireprosci.2013.12.008

Crowe MA, Hostens M, Opsomer G (2018) Reproductive management in dairy cows - the future. Ir Vet J 71:1. https://doi.org/10.1186/s13620-017-0112-y

Cui J, Qin L, Zhang J, Abrahimi P, Li H, Li G, Tietjen GT, Tellides G, Pober JS, Mark Saltzman W (2017) Ex vivo pretreatment of human vessels with siRNA nanoparticles provides protein silencing in endothelial cells. Nat Commun 8(1):1–11. https://doi.org/10.1038/s41467-017-00297-x

Cunningham AA, Daszak P, Wood JL (2017) One health, emerging infectious diseases and wildlife: two decades of progress? Philos Trans R Soc B Biol Sci 372(1725):20160167

de Cássia Antonino D, Soares MM, de Melo Júnior J, de Alvarenga PB, Mohallem RDFF, Rocha CD, Alves KA (2019) Three-dimensional levitation culture improves in-vitro growth of secondary follicles in bovine model. Reprod Biomed Online 38(3):300–311. https://doi.org/10.1016/j.rbmo.2018.11.013

Deguettes Q, Fattal E, Moreau M, Lego E, Bochot A (2020) Controlled delivery of follicle-stimulating hormone in cattle. Int J Pharm 590:119904. https://doi.org/10.1016/j.ijpharm.2020.119904

Ding T, Yang J, Pan V, Zhao N, Lu Z, Ke Y, Zhang C (2020) DNA nanotechnology assisted nanopore-based analysis. Nucleic Acids Res 48(6):2791–2806. https://doi.org/10.1093/nar/gkaa095

Dong Y, Wu X, Chen X, Zhou P, Xu F, Liang W (2021) Nanotechnology shaping stem cell therapy: recent advances, application, challenges, and future outlook. Biomed Pharmacother 137:111236

Donofrio G, Herath S, Sartori C, Cavirani S, Flammini CF, Sheldon IM (2007) Bovine herpesvirus 4 is tropic for bovine endometrial cells and modulates endocrine function. Reproduction (Cambridge, England) 134(1):183–197. https://doi.org/10.1530/REP-07-0065

Duan L, Ouyang K, Xu X, Xu L, Wen C, Zhou X, Qin Z, Xu Z, Sun W, Liang Y (2021) Nanoparticle delivery of CRISPR/Cas9 for genome editing. Front Genet 12:673286. https://doi.org/10.3389/fgene.2021.673286

Eduardo C, Palmieri C, Dagli ML (2021) Editorial: precision medicine in veterinary oncology. Front Vet Sci 8:718891. https://doi.org/10.3389/fvets.2021.718891

Falchi L, Khalil WA, Hassan M, Marei WF (2018) Perspectives of nanotechnology in male fertility and sperm function. Int J Vet Sci Med 6(2):265–269

Farini VL, Camaño CV, Ybarra G, Viale DL, Vichera G, Yakisich JS, Radrizzani M (2016) Improvement of bovine semen quality by removal of membrane-damaged sperm cells with DNA aptamers and magnetic nanoparticles. J Biotechnol 229:33–41. https://doi.org/10.1016/j.jbiotec.2016.05.008

Felder S, Masasa H, Orenbuch A, Levaot N, Shachar Goldenberg M, Cohen S (2019) Reconstruction of the ovary microenvironment utilizing macroporous scaffold with affinity-bound growth factors. Biomaterials 205:11–22. https://doi.org/10.1016/j.biomaterials.2019.03.013

Fosse V, Oldoni E, Bietrix F, Budillon A, Daskalopoulos EP, Fratelli M, McCormack E (2023) Recommendations for robust and reproducible preclinical research in personalised medicine. BMC Med 21(1):14. https://doi.org/10.1186/s12916-022-02719-0

Furno F, Morley KS, Wong B, Sharp BL, Arnold PL, Howdle SM, Bayston R, Brown PD, Winship PD, Reid HJ (2004) Silver nanoparticles and polymeric medical devices: A new approach to prevention of infection? J Antimicrob Chemother 54(6):1019–1024. https://doi.org/10.1093/jac/dkh478

Gaj T, Sirk SJ, Shui SL, Liu J (2016) Genome-editing technologies: principles and applications. Cold Spring Harb Perspect Biol 8(12):a023754

Gamrad L, Mancini R, Werner D, Tiedemann D, Taylor U, Ziefuß A, Rath D (2017) Triplex-hybridizing bioconjugated gold nanoparticles for specific Y-chromosome sequence targeting of bull spermatozoa. Analyst 142(11):2020–2028. https://doi.org/10.1039/C6AN02461K

Gandhimathi C, Venugopal JR, Tham AY, Ramakrishna S, Kumar SD (2015) Biomimetic hybrid nanofibrous substrates for mesenchymal stem cells differentiation into osteogenic cells. Mater Sci Eng C 49:776–785

Garzon S, Laganà AS, Barra F, Casarin J, Cromi A, Raffaelli R, Ferrero S (2021) Novel drug delivery methods for improving efficacy of endometriosis treatments. Expert Opin Drug Deliv 18(3):355–367. https://doi.org/10.1080/17425247.2021.1829589

Georgiou PG, Marton HL, Baker AN, Congdon TR, Whale TF, Gibson MI (2021) Polymer self-assembly induced enhancement of ice recrystallization inhibition. J Am Chem Soc 143(19): 7449–7461. https://doi.org/10.1021/jacs.1c01963

Giordano JO, Stangaferro ML, Wijma R, Chandler WC, Watters RD (2015) Reproductive performance of dairy cows managed with a program aimed at increasing insemination of cows in estrus based on increased physical activity and fertility of timed artificial inseminations. J Dairy Sci 98(4):2488–2501. https://doi.org/10.3168/jds.2014-8961

Giordano JO, Sitko EM, Rial C, Pérez MM, Granados GE (2022) Symposium review: use of multiple biological, management, and performance data for the design of targeted reproductive management strategies for dairy cows. J Dairy Sci 105:4669–4678. https://doi.org/10.3168/jds.2021-21476

Goetz LH, Schork NJ (2018) Personalized medicine: motivation, challenges, and progress. Fertil Steril 109(6):952–963. https://doi.org/10.1016/j.fertnstert.2018.05.006

Gurunathan S, Choi YJ, Kim JH (2018) Antibacterial efficacy of silver nanoparticles on endometritis caused by prevotellamelaninogenica and arcanobacterum pyogenes in dairy cattle. Int J Mol Sci 19(4):1210. https://doi.org/10.3390/ijms19041210

Handcock RN, Swain DL, Bishop-Hurley GJ, Patison KP, Wark T, Valencia P, Corke P, O'Neill CJ (2009) Monitoring animal behaviour and environmental interactions using wireless sensor networks, GPS collars and satellite remote sensing. Sensors 9(5):3586–3603. https://doi.org/10.3390/s90503586

Hashem NM (2021) Nanotechnology and reproductive management of farm animals: challenges and advances. Animals 11(7):1932. https://doi.org/10.3390/ani11071932

Hashem NM, Gonzalez-Bulnes A (2020) State-of-the-art and prospective of nanotechnologies for smart reproductive management of farm animals. Animals 10(5):840. https://doi.org/10.3390/ani10050840

Hashem NM, Sallam SM (2020) Reproductive performance of goats treated with free gonadorelin or nanoconjugated gonadorelin at estrus. Domest Anim Endocrinol 71:106390. https://doi.org/10.1016/j.domaniend.2019.106390

Hassanein EM, Hashem NM, El-Azrak KE, Gonzalez-Bulnes A, Hassan GA, Salem MH (2021) Efficiency of GnRH–loaded chitosan nanoparticles for inducing LH secretion and fertile ovulations in protocols for artificial insemination in rabbit does. Animals 11(2):440. https://doi.org/10.3390/ani11020440

Headen DM, Woodward KB, Coronel MM, Shrestha P, Weaver JD, Zhao H, Shirwan H (2018) Local immunomodulation with Fas ligand-engineered biomaterials achieves allogeneic islet graft acceptance. Nat Mater 17(8):732–739

Helbling IM, Busatto CA, Fioramonti SA, Pesoa JI, Santiago L, Estenoz DA, Luna JA (2018) Preparation of TPP-crosslinked chitosan microparticles by spray drying for the controlled delivery of progesterone intended for estrus synchronization in cattle. Pharm Res 35(3):66. https://doi.org/10.1007/s11095-018-2363-z

Hogan MC, Norton JN, Reynolds RP (2017) Environmental factors: macroenvironment versus microenvironment. Management of Animal Care and Use Programs in Research, Education, and Testing, pp. 461–477. https://doi.org/10.1201/9781315152189-20

Hohenlohe PA, Funk WC, Rajora OP (2021) Population genomics for wildlife conservation and management. Mol Ecol 30(1):62–82

Holman A, Thompson J, Routly JE, Cameron J, Jones DN, Grove-White D, Dobson H (2011) Comparison of oestrus detection methods in dairy cattle. Vet Rec 169(2):47–47. https://doi.org/10.1136/vr.d2344

Huang L, Qian K (2020) Precision diagnosis for the era of personalized medicine. Small Methods 4(4):1900717. https://doi.org/10.1002/smtd.201900717

Huang H, Du X, He Z, Yan Z, Han W (2021) Nanoparticles for stem cell tracking and the potential treatment of cardiovascular diseases. Front Cell Dev Biol 9:662406. https://doi.org/10.3389/fcell.2021.662406

Ibrahim RK, Hayyan M, AlSaadi MA, Hayyan A, Ibrahim S (2016) Environmental application of nanotechnology: air, soil, and water. Environ Sci Pollut Res 23:13754–13788

Jahanbin R, Yazdanshenas P, Rahimi M, Hajarizadeh A, Tvrda E, Nazari SA, Ghanem N (2021) In vivo and in vitro evaluation of bull semen processed with zinc (Zn) nanoparticles. Biol Trace Elem Res 199:126–135. https://doi.org/10.1007/s12011-020-02153-4

Jing Y, Moore LR, Williams PS, Chalmers JJ, Farag SS, Bolwell B, Zborowski M (2007) Blood progenitor cell separation from clinical leukapheresis product by magnetic nanoparticle binding and magnetophoresis. Biotechnol Bioeng 96(6):1139–1154

Kadam RG, Karikalan M, Siddappa CM, Mahendran K, Srivastava G, Rajak KK, Sharma AK (2022) Molecular and pathological screening of canine distemper virus in Asiatic lions, tigers, leopards, snow leopards, clouded leopards, leopard cats, jungle cats, civet cats, fishing cat, and jaguar of different states, India. Infect Genet Evol 98:105211. https://doi.org/10.1016/j.meegid.2022.105211

Karesh WB, Cook RA (1995) Applications of veterinary medicine to in situ conservation efforts. Oryx 29(4):244–252

Kim TH, Lee S, Chen X (2013) Nanotheranostics for personalized medicine. Expert Rev Mol Diagn 13(3):257–269. https://doi.org/10.1586/erm.13.15

Kim JH, Kim HW, Cha KJ, Han J, Jang YJ, Kim DS, Kim JH (2016) Nanotopography promotes pancreatic differentiation of human embryonic stem cells and induced pluripotent stem cells. ACS Nano 10(3):3342–3355

Kim T, Lee N, Arifin DR, Shats I, Janowski M, Walczak P, Bulte JW (2017) In vivo micro-CT imaging of human mesenchymal stem cells labeled with gold-poly-L-lysine nanocomplexes. Adv Funct Mater 27(3):1604213

Komninou ER, Remião MH, Lucas CG, Domingues WB, Basso AC, Jornada DS, Collares T (2016) Effects of two types of melatonin-loaded nanocapsules with distinct supramolecular structures: polymeric (NC) and lipid-core nanocapsules (LNC) on bovine embryo culture model. PLoS One 11(6):e0157561

Krasteva DR, Ivanov YL, Chervenkov TG, Gabrovska KI, Godjevargova TI (2020) CD34+ stem cell counting using labeled immobilized anti-CD34 antibody onto magnetic nanoparticles and EasyCounter BC image cytometer. Anal Biochem 610:113929

LaMothe RA, Kolte PN, Vo T, Ferrari JD, Gelsinger TC, Wong J, Kishimoto TK (2018) Tolerogenic nanoparticles induce antigen-specific regulatory T cells and provide therapeutic efficacy and transferrable tolerance against experimental autoimmune encephalomyelitis. Front Immunol 9:281. https://doi.org/10.3389/fimmu.2018.00281

Lard M, Bäckman J, Yakovleva M, Danielsson B, Hansson A (2010) Tracking the small with the smallest – using nanotechnology in tracking zooplankton. PLoS One 5(10):e13516. https://doi.org/10.1371/journal.pone.0013516

LeBlanc S (2010) Monitoring metabolic health of dairy cattle in the transition period. J Reprod Dev 56(Suppl):S29–S35. https://doi.org/10.1262/jrd.1056s29

LeBlanc SJ, Duffield TF, Leslie KE, Bateman KG, Keefe GP, Walton JS, Johnson WH (2002) Defining and diagnosing postpartum clinical endometritis and its impact on reproductive performance in dairy cows. J Dairy Sci 85(9):2223–2236. https://doi.org/10.3168/jds.S0022-0302(02)74302-6

Li R, Liang J, He Y, Qin J, He H, Lee S, Pang Z, Wang J (2018) Sustained release of immunosuppressant by nanoparticle-anchoring hydrogel scaffold improved the survival of transplanted stem cells and tissue regeneration. Theranostics 8(4):878–893. https://doi.org/10.7150/thno.22072

Li T, Zhao Y, Zhong Q, Wu T (2019) Inhibiting ice recrystallization by nanocelluloses. Biomacromolecules 20(4):1667–1674. https://doi.org/10.1021/acs.biomac.9b00027

Li Y, Xu C, Lei C (2023) The delivery and activation of growth factors using nanomaterials for bone repair. Pharmaceutics 15(3):1017

Liang S, Kinghorn AB, Voliotis M, Prague JK, Veldhuis JD, Tsaneva-Atanasova K, McArdle CA, Li RHW, Cass AEG, Dhillo WS, Tanner JA (2019) Measuring luteinising hormone pulsatility with a robotic aptamer-enabled electrochemical reader. Nat Commun 10(1):852. https://doi.org/10.1038/s41467-019-08799-6

Lima F, Silvestre F, Peñagaricano F, Thatcher W (2020) Early genomic prediction of daughter pregnancy rate is associated with improved reproductive performance in Holstein dairy cows. J Dairy Sci 103(4):3312–3324. https://doi.org/10.3168/jds.2019-17488

Liu J, Liu J, Attarilar S, Wang C, Tamaddon M, Yang C, Xie K, Yao J, Wang L, Liu C, Tang Y (2020) Nano-modified titanium implant materials: a way toward improved antibacterial properties. Front Bioeng Biotechnol 8:576969. https://doi.org/10.3389/fbioe.2020.576969

Lloyd KK, Khanna C, Hendricks W, Trent J, Kotlikoff M (2016) Precision medicine: an opportunity for a paradigm shift in veterinary medicine. J Am Vet Med Assoc 248(1):45–48. https://doi.org/10.2460/javma.248.1.45

Mac QD, Mathews DV, Kahla JA, Stoffers CM, Delmas OM, Holt BA, Kwong GA (2019) Non-invasive early detection of acute transplant rejection via nanosensors of granzyme B activity. Natu Biomed Eng 3(4):281–291. https://doi.org/10.1038/s41551-019-0358-7

Madureira AML, Silper BF, Burnett TA, Polsky LB, Drago Filho EL, Soriano S, Cerri RLA (2015) Effects of expression of estrus measured by activity monitors on ovarian dynamics and conception risk in Holstein cows. J Dairy Sci 98(Suppl 1):875. https://doi.org/10.3168/jds.2015-9672

Mahmoud A, Fawzy M (2021) Nanosensors and nanobiosensors for monitoring the environmental pollutants. In: Makhlouf ASH, Ali GAM (eds) Waste recycling technologies for nanomaterials manufacturing. Topics in mining, metallurgy and materials engineering. Springer, Cham. https://doi.org/10.1007/978-3-030-68031-2_9

Majood M, Garg P, Chaurasia R, Agarwal A, Mohanty S, Mukherjee M (2022) Carbon quantum dots for stem cell imaging and deciding the fate of stem cell differentiation. ACS Omega 7(33):28685–28693

Manríquez D, Thatcher WW, Santos JEP, Chebel RC, Galvão KN, Schuenemann GM, Pinedo P (2021) Effect of body condition change and health status during early lactation on performance and survival of Holstein cows. J Dairy Sci 104(12):12785–12799. https://doi.org/10.3168/jds.2020-20091

Masoudi Asil S, Ahlawat J, Guillama Barroso G, Narayan M (2020) Application of nanotechnology in stem-cell-based therapy of neurodegenerative diseases. Appl Sci 10(14):4852

Matsumura K, Bae JY, Kim HH, Hyon SH (2011) Effective vitrification of human induced pluripotent stem cells using carboxylated ε-poly-l-lysine. Cryobiology 63(2):76–83. https://doi.org/10.1016/j.cryobiol.2011.05.003

Mavingui P, Boetsch G, Boissier J, Darriet F, Duboz P, Fritsch C, Giraudoux P, Le Roux F, Morand S, Paillard C, Pontier D, Sueur C, Voituron Y (2018) The one health concept: 10 years old and a long road ahead. Front Vet Sci 5:306018. https://doi.org/10.3389/fvets.2018.00014

McConnell EM, Nguyen J, Li Y (2020) Aptamer-based biosensors for environmental monitoring. Front Chem 8:539439. https://doi.org/10.3389/fchem.2020.00434

Minakshi P, Ghosh M, Brar B, Kumar R, Lambe UP, Ranjan K, Prasad G (2019) Nano-antimicrobials: a new paradigm for combating mycobacterial resistance. Curr Pharm Des 25(13):1554–1579

Minakshi P, Ghosh M, Kumar R, Brar B, Lambe UP, Banerjee S, Prasad G (2020) An insight into nanomedicinal approaches to combat viral zoonoses. Curr Top Med Chem 20(11):915–962

Morotti F, Sanches BV, Pontes JH, Basso AC, Siqueira ER, Lisboa LA, Seneda MM (2014) Pregnancy rate and birth rate of calves from a large-scale IVF program using reverse-sorted semen in Bos indicus. Bos indicus 81:696–701. https://doi.org/10.1016/j.theriogenology.2013.12.002

Muller-Borer BJ, Collins MC, Gunst PR, Cascio WE, Kypson AP (2007) Quantum dot labeling of mesenchymal stem cells. J Nanobiotechnol 5:1–9

Mulligan FJ, O'Grady L, Rice DA, Doherty ML (2006) A herd health approach to dairy cow nutrition and production diseases of the transition cow. Anim Reprod Sci 96(3–4):331–353. https://doi.org/10.1016/j.anireprosci.2006.08.011

Mura S, Couvreur P (2012) Nanotheranostics for personalized medicine. Adv Drug Deliv Rev 64(13):1394–1416. https://doi.org/10.1016/j.addr.2012.06.006

Murray KA, Gibson MI (2022) Chemical approaches to cryopreservation. Nat Rev Chem 6(8):579–593. https://doi.org/10.1038/s41570-022-00407-4

Najafi A, Daghigh-Kia H, Mehdipour M, Mohammadi H, Hamishehkar H (2022) Comparing the effect of rooster semen extender supplemented with gamma-oryzanol and its nano form on post-thaw sperm quality and fertility. Poult Sci 101(3):101637. https://doi.org/10.1016/j.psj.2021.101637

Nikoleli G, Nikolelis D, Siontorou CG, Karapetis S (2018) Lipid membrane nanosensors for environmental monitoring: the art, the opportunities, and the challenges. Sensors 18(1):284. https://doi.org/10.3390/s18010284

O'Connor TP, Crystal RG (2006) Genetic medicines: treatment strategies for hereditary disorders. Nat Rev Genet 7(4):261–276. https://doi.org/10.1038/nrg1829

Ochando J, Braza MS (2017) Nanoparticle-based modulation and monitoring of antigen-presenting cells in organ transplantation. Front Immunol 8:315611. https://doi.org/10.3389/fimmu.2017.01888

Odabaş S, Sayar F, Güven G, Yanıkkaya-Demirel G, Pişkin E (2008) Separation of mesenchymal stem cells with magnetic nanosorbents carrying CD105 and CD73 antibodies in flow-through and batch systems. J Chromatogr B 861(1):74–80

Odhiambo JF, DeJarnette JM, Geary TW, Kennedy CE, Suarez SS, Sutovsky M, Sutovsky P (2014) Increased conception rates in beef cattle inseminated with nanopurified bull semen. Biol Reprod 91(4):97–91. https://doi.org/10.1095/biolreprod.114.121897

Oldenbroek JK (2021) The use of genomic information for the conservation of animal genetic diversity. Animals 11(11):3208. https://doi.org/10.3390/ani11113208

Oliveira JE, Medeiros ES, Cardozo L, Voll F, Madureira EH, Mattoso LHC, Assis OBG (2013) Development of poly(lactic acid) nanostructured membranes for the controlled delivery of progesterone to livestock animals. Mater Sci Eng C 33(2):844–849. https://doi.org/10.1016/j.msec.2012.10.032

Pandey S, Mishra SB (2014) Chemical nanosensors for monitoring environmental pollution. Appl Nanotechnol Water Res:309–332. https://doi.org/10.1002/9781118939314.ch12

Park J, Bauer S, von der Mark K, Schmuki P (2007) Nanosize and vitality: TiO2 nanotube diameter directs cell fate. Nano Lett 7(6):1686–1691

Park SY, Park J, Sim SH, Sung MG, Kim KS, Hong BH, Hong S (2011) Enhanced differentiation of human neural stem cells into neurons on graphene. Adv Mater 23(36):H263–H267

Patra JK, Das G, Fraceto LF et al (2018) Nano based drug delivery systems: recent developments and future prospects. J Nanobiotechnol 16:71. https://doi.org/10.1186/s12951-018-0392-8

Pearson RM, Casey LM, Hughes KR, Wang LZ, North MG, Getts DR, Shea LD (2017) Controlled delivery of single or multiple antigens in tolerogenic nanoparticles using peptide-polymer bioconjugates. Mol Ther 25(7):1655–1664. https://doi.org/10.1016/j.ymthe.2017.04.015

Peralta OA, Pearson RE, Nebel RL (2005) Comparison of three estrus detection systems during summer in a large commercial dairy herd. Anim Reprod Sci 87(1–2):59–72. https://doi.org/10.1016/j.anireprosci.2004.10.003

Pérez-Duran F, Acosta-Torres LS, Serrano-Díaz PN, Toscano-Torres IA, Olivo-Zepeda IB, García-Caxin E, Nuñez-Anita RE (2020) Toxicity and antimicrobial effect of silver nanoparticles in swine sperms. Syst Biol Reprod Med 66(4):281–289. https://doi.org/10.1080/19396368.2020.1754962

Pérez-González J, Carranza J, Martínez R, Benítez-Medina JM (2021) Host genetic diversity and infectious diseases. Focus on wild boar, red deer and tuberculosis. Animals 11(6):1630

Pinedo P, Santos J, Chebel R, Galvão K, Schuenemann G, Bicalho R, Gilbert R, Rodriguez-Zas S, Seabury C, Rosa G, Thatcher W (2020) Associations of reproductive indices with fertility outcomes, milk yield, and survival in Holstein cows. J Dairy Sci 103(7):6647–6660. https://doi.org/10.3168/jds.2019-17867

Pontes GC, Monteiro PL Jr, Prata AB, Guardieiro MM, Pinto DA, Fernandes GO, Wiltbank MC, Santos JE, Sartori R (2015) Effect of injectable vitamin E on incidence of retained fetal membranes and reproductive performance of dairy cows. J Dairy Sci 98(4):2437–2449. https://doi.org/10.3168/jds.2014-8886

Prasad M, Lambe UP, Brar B, Shah I, Manimegalai J, Ranjan K, Rao R, Kumar S, Mahant S, Khurana SK, Iqbal HM (2018) Nanotherapeutics: an insight into healthcare and multi-dimensional applications in medical sector of the modern world. Biomed Pharmacother 97:1521–1537. https://doi.org/10.1016/j.biopha.2017.11.026

Prasad M, Ghosh M, Kumar R, Brar B, Surjith KP, Lambe UP, Kharb P (2021) The importance of nanomedicine in prophylactic and theranostic intervention of bacterial zoonoses and reverse zoonoses in the era of microbial resistance. J Nanosci Nanotechnol 21(6):3404–3452

Rao W, Huang H, Wang H, Zhao S, Dumbleton J, Zhao G, He X (2015) Nanoparticle-mediated intracellular delivery enables cryopreservation of human adipose-derived stem cells using trehalose as the sole cryoprotectant. ACS Appl Mater Interfaces 7(8):5017–5028. https://doi.org/10.1021/acsami.5b00655

Reading RP, Kenny DE, Fitzgerald KT (2013) The crucial contribution of veterinarians to conservation biology. Top Companion Anim Med 28(4):131–134

Remião MH, Lucas CG, Domingues WB, Silveira T, Barther NN, Komninou ER, Basso AC, Jornada DS, Beck RC, Pohlmann AR, Junior AS, Seixas FK, Campos VF, Guterres SS, Collares T (2016) Melatonin delivery by nanocapsules during in vitro bovine oocyte maturation decreased the reactive oxygen species of oocytes and embryos. Reprod Toxicol (Elmsford, N.Y.) 63:70–81. https://doi.org/10.1016/j.reprotox.2016.05.016

Rial C, Laplacette A, Giordano JO (2022) Effect of a targeted reproductive management program designed to prioritize insemination at detected estrus and optimize time to insemination on the reproductive performance of lactating dairy cows. J Dairy Sci 105(10):8411–8425. https://doi.org/10.3168/jds.2022-22082

Rosenberg JT, Yuan X, Grant S, Ma T (2016) Tracking mesenchymal stem cells using magnetic resonance imaging. Brain Circ 2(3):108

Ruckh TT, Clark HA (2014) Implantable nanosensors: toward continuous physiologic monitoring. Anal Chem 86(3):1314–1323

Saadeldin IM, Khalil WA, Alharbi MG, Lee SH (2020) The current trends in using nanoparticles, liposomes, and exosomes for semen cryopreservation. Animals 10(12):2281. https://doi.org/10.3390/ani10122281

Sadeghi Z, Fasihi-Ramandi M, Bouzari S (2020a) Nanoparticle-based vaccines for brucellosis: calcium phosphate nanoparticles-adsorbed antigens induce cross protective response in mice. Int J Nanomedicine 15:3877–3886. https://doi.org/10.2147/IJN.S249942

Sadeghi Z, Fasihi-Ramandi M, Azizi M, Bouzari S (2020b) Mannosylated chitosan nanoparticles loaded with FliC antigen as a novel vaccine candidate against Brucella melitensis and Brucella abortus infection. J Biotechnol 310:89–96. https://doi.org/10.1016/j.jbiotec.2020.01.016

Sapna K, Sonia J, Shim YB, Arun AB, Prasad KS (2022) Au nanoparticle-based disposable electrochemical sensor for detection of leptospirosis in clinical samples. ACS Appl Nano Mater 5(9):12454–12463. https://doi.org/10.1021/acsanm.2c01978

Satish S, Dey A, Tharmavaram M, Khatri N, Rawtani D (2022) Risk assessment of selected pharmaceuticals on wildlife with nanomaterials based aptasensors. Sci Total Environ 836: 155622. https://doi.org/10.1016/j.scitotenv.2022.155622

Sattarahmady N, Tondro G, Gholchin M, Heli H (2015) Gold nanoparticles biosensor of Brucella spp. genomic DNA: visual and spectrophotometric detections. Biochem Eng J 97:1–7. https://doi.org/10.1016/j.bej.2015.01.010

Schillings J, Bennett R, Rose DC (2021) Exploring the potential of precision livestock farming technologies to help address farm animal welfare. Front Anim Sci 2:639678. https://doi.org/10.3389/fanim.2021.639678

Shahzad KA, Naeem M, Zhang L, Wan X, Song S, Pei W, Zhao C, Jin X, Shen C (2020) Design and optimization of PLGA particles to deliver immunomodulatory drugs for the prevention of skin allograft rejection. Immunol Investig 49(7):840–857. https://doi.org/10.1080/08820139.2019.1695134

Shams S, Bakhshi B, Tohidi Moghadam T et al (2019) A sensitive gold-nanorods-based nanobiosensor for specific detection of Campylobacter jejuni and campylobacter coli. J Nanobiotechnol 17:43. https://doi.org/10.1186/s12951-019-0476-0

Sharma B, Koundal D (2018) Cattle health monitoring system using wireless sensor network: a survey from innovation perspective. IET Wireless Sensor Syst 8(4):143–151. https://doi.org/10.1049/iet-wss.2017.0060

Sheldon IM, Cronin J, Goetze L, Donofrio G, Schuberth HJ (2009) Defining postpartum uterine disease and the mechanisms of infection and immunity in the female reproductive tract in cattle. Biol Reprod 81(6):1025–1032. https://doi.org/10.1095/biolreprod.109.077370

Siahdasht FN, Farhadian N, Karimi M, Hafizi L (2020) Enhanced delivery of melatonin loaded nanostructured lipid carriers during in vitro fertilization: NLC formulation, optimization and IVF efficacy. RSC Adv 10(16):9462–9475. https://doi.org/10.1039/C9RA10867J

Silvestre F, Carvalho T, Francisco N, Santos J, Staples C, Jenkins T, Thatcher W (2011) Effects of differential supplementation of fatty acids during the peripartum and breeding periods of Holstein cows: I. Uterine and metabolic responses, reproduction, and lactation. J Dairy Sci 94(1):189–204. https://doi.org/10.3168/jds.2010-3370

Singh AK, Chakravarty B, Chaudhury K (2015) Nanoparticle-assisted combinatorial therapy for effective treatment of endometriosis. J Biomed Nanotechnol 11(5):789–804. https://doi.org/10.1166/jbn.2015.2020

Smagul S, Kim Y, Smagulova A, Raziyeva K, Nurkesh A, Saparov A (2020) Biomaterials loaded with growth factors/cytokines and stem cells for cardiac tissue regeneration. Int J Mol Sci 21(17):5952

Stefanic M, Ward K, Tawfik H, Seemann R, Baulin V, Guo Y, Drouet C (2017) Apatite nanoparticles strongly improve red blood cell cryopreservation by mediating trehalose delivery via enhanced membrane permeation. Biomaterials 140:138–149. https://doi.org/10.1016/j.biomaterials.2017.06.018

Sun Y, Lu Y, Yin L, Liu Z (2020) The roles of nanoparticles in stem cell-based therapy for cardiovascular disease. Front Bioeng Biotechnol 8:947

Tallury P, Malhotra A, Byrne LM, Santra S (2010) Nanobioimaging and sensing of infectious diseases. Adv Drug Deliv Rev 62(4–5):424–437

Tamminga S, Luteijn PA, Meijer RGM (1997) Changes in composition and energy content of liveweight loss in dairy cows with time after parturition. Livest Prod Sci 52(1):31–38. https://doi.org/10.1016/S0301-6226(97)00115-2

Tar M, Towhidi A, Zeinoaldini S, Zhandi M, Mohammadi-Sangcheshmeh A, Moazeni Zadeh MH (2021) Effects of different ultrastructures of lecithin on cryosurvival of goat spermatozoa. Andrologia 53(10):e14183. https://doi.org/10.1111/and.14183

Tostanoski LH, Chiu YC, Gammon JM, Simon T, Andorko JI, Bromberg JS, Jewell CM (2016) Reprogramming the local lymph node microenvironment promotes tolerance that is systemic and antigen specific. Cell Rep 16(11):2940–2952. https://doi.org/10.1016/j.celrep.2016.08.033

Tyler J, Choi SW, Tewari M (2020) Real-time, personalized medicine through wearable sensors and dynamic predictive modeling: A new paradigm for clinical medicine. Curr Opin Syst Biol 20: 17–25. https://doi.org/10.1016/j.coisb.2020.07.001

Ueno T, Dutta P, Keliher E, Leuschner F, Majmudar M, Marinelli B, Nahrendorf M (2013) Nanoparticle PET-CT detects rejection and immunomodulation in cardiac allografts. Circulation 6(4):568–573. https://doi.org/10.1161/CIRCIMAGING.113.000481

Unterweger H, Janko C, Folk T, Cicha I, Kovács N, Gyebnár G, Lyer S (2023) Comparative in vitro and in vivo evaluation of different iron oxide-based contrast agents to promote clinical translation in compliance with patient safety. Int J Nanomedicine Volume 18:2071–2086

Viana Silva JR, Aguiar Barroso PA, Nascimento DR, Figueira CS, Nunes Azevedo VA, Silva BR (2021) Benefits and challenges of nanomaterials in assisted reproductive technologies. Mol Reprod Dev 88(11):707–717. https://doi.org/10.1002/mrd.23536

Wang XF, Zhang SL, Zhu LY, Xie SY, Dong Z, Wang Y, Zhou WZ (2012) Enhancement of antibacterial activity of tilmicosin against Staphylococcus aureus by solid lipid nanoparticles in vitro and in vivo. Vet J 191(1):115–120. https://doi.org/10.1016/j.tvjl.2010.11.019

Wang AYM, Kalantar-Zadeh K, Fouque D, Wee PT, Kovesdy CP, Price SR, Kopple JD (2018) Precision medicine for nutritional management in end-stage kidney disease and transition to dialysis. Sem Nephrol 38(4):383–396. https://doi.org/10.1016/j.semnephrol.2018.05.008

Whilde J, Martindale MQ, Duffy DJ (2017) Precision wildlife medicine: applications of the human-centred precision medicine revolution to species conservation. Glob Chang Biol 23(5): 1792–1805. https://doi.org/10.1111/gcb.13548

Wijma R, Pérez M, Masello M, Stangaferro M, Giordano J (2018) A resynchronization of ovulation program based on ovarian structures present at nonpregnancy diagnosis reduced time to pregnancy in lactating dairy cows. J Dairy Sci 101(2):1697–1707. https://doi.org/10.3168/jds.2017-13489

Willner MR, Vikesland PJ (2018) Nanomaterial enabled sensors for environmental contaminants. J Nanobiotechnol 16:95. https://doi.org/10.1186/s12951-018-0419-1

Wowk B, Leitl E, Rasch CM, Mesbah-Karimi N, Harris SB, Fahy GM (2000) Vitrification enhancement by synthetic ice blocking agents. Cryobiology 40(3):228–236. https://doi.org/10.1006/cryo.2000.2243

Xi D, Dong S, Meng X, Lu Q, Meng L, Ye J (2012) Gold nanoparticles as computerized tomography (CT) contrast agents. RSC Adv 2(33):12515–12524

Yetisgin AA, Cetinel S, Zuvin M, Kosar A, Kutlu O (2020) Therapeutic nanoparticles and their targeted delivery applications. Molecules 25(9):2193

Ying Y, Hu Z, Zhang S, Qing Y, Fragasso A, Maglia G, Meller A, Bayley H, Dekker C, Long Y (2022) Nanopore-based technologies beyond DNA sequencing. Nat Nanotechnol 17(11): 1136–1146. https://doi.org/10.1038/s41565-022-01193-2

Young AE, Mansour TA, McNabb BR, Owen JR, Trott JF, Brown CT, Van Eenennaam AL (2020) Genomic and phenotypic analyses of six offspring of a genome-edited hornless bull. Nat Biotechnol 38(2):225–232

Younis S, Zia R, Tahir N, Bukhari SZ, Khan WS, Bajwa SZ (2022). Nanosensors for animal health monitoring. Nanosensors for Smart Agriculture, pp. 509–529. https://doi.org/10.1016/B978-0-12-824554-5.00026-4

Youssef FS, El-Banna HA, Elzorba HY, Galal AM (2019) Application of some nanoparticles in the field of veterinary medicine. Int J Vet Sci Med 7(1):78–93. https://doi.org/10.1080/23144599.2019.1691379

Zakirova EY, Malanyeva AG, Rutland CS, Aimaletdinov AM (2022) Genetic therapy in veterinary medicine. BioNano Sci 12(4):1397–1403. https://doi.org/10.1007/s12668-022-00986-y

Zavan B, Vindigni V, Vezzù K, Zorzato G, Luni C, Abatangelo G, Cortivo R (2009) Hyaluronan based porous nano-particles enriched with growth factors for the treatment of ulcers: a placebo-controlled study. J Mater Sci Mater Med 20:235–247

Zhang Y, Wang H, Stewart S, Jiang B, Ou W, Zhao, G, He X (2019) Cold-responsive nanoparticle enables intracellular delivery and rapid release of trehalose for organic-solvent-free cryopreservation. Nano Lett 19(12):9051–9061. https://doi.org/10.1021/acs.nanolett.9b04109

Zhu P, Atkinson C, Dixit S, Cheng Q, Tran D, Patel K, Nadig SN (2018) Organ preservation with targeted rapamycin nanoparticles: a pre-treatment strategy preventing chronic rejection in vivo. RSC Adv 8(46):25909–25919. https://doi.org/10.1039/c8ra01555d

Chapter 29
Nanotoxicity and Mechanisms

Rajesh Kumar, Preeti Lakhani, Alla Yaswanth Naveen Kumar, and Mayukh Ghosh

Abstract Nanotechnology has already contributed prominently in diverse scientific avenues and the scope is expanding progressively. The nanomaterial (NM) repository is also prospering rapidly, propelled by the extensive demand for nanomedicine to combat emerging situations such as viral outbreaks or antimicrobial resistance and to provide solution to decades old problems like cancer along with other potential applications in wound healing, fracture management, nano-vaccines, advanced imaging and diagnostics, etc. Consequently, the load of nanoparticles (NPs) on the ecosystem is increasing rapidly, enhancing the probability of NM exposure to all the living organisms via diverse routes. Several NMs such as metallic and metal oxide NPs, carbon-based NMs, silica-based NMs, polymeric or composite NMs, quantum dots, etc. have found to inflict toxicity by diverse mechanisms that include but not limited to generation of oxidative stress by overproduction of free radicals and depletion of cellular antioxidants, inflammation by upregulating the secretion of pro-inflammatory cytokines, cell cycle alteration, affecting cell viability, inducing the formation of autophagosome and apoptosis by activating caspase-mediated pathways, alteration in transcriptional regulations, causing DNA damage and genotoxicity, etc. The toxicity signatures can either be manifested at molecular level or by phenotypic changes to the cellular architectures. Enumeration of different tissue- or organ-specific toxicity markers, cell viability assays and morphological examinations, quantifying different enzymatic and non-enzymatic antioxidants, measuring lipid peroxidation, evaluating the expression of different cytokines and transcription factors, monitoring DNA damage and oxidation products along with

R. Kumar · P. Lakhani
Department of Veterinary Physiology and Biochemistry, Lala Lajpat Rai University of Veterinary and Animal Sciences, Hisar, Haryana, India

Department of Genetics and Plant Breeding, Mirzapur, Uttar Pradesh, India

A. Y. N. Kumar · M. Ghosh (✉)
Department of Veterinary Physiology and Biochemistry, RGSC, Banaras Hindu University, Mirzapur, Uttar Pradesh, India

LLR University of Veterinary and Animal, ICAR-National Research Centre on Equines, *Nanotechnology Theranostics in Livestock Diseases and Management*, Livestock Diseases and Management,
https://doi.org/10.1007/978-981-16-1610-5_29

histological examination are usually performed to elucidate the nanotoxicological implications. Two-dimensional monolayer cell culture is the most commonly employed in-vitro model for nanotoxicity evaluation, however, it lacks the barrier function, thus, often overestimates the toxic effects. So, three-dimensional cultures, organoids, spheroids, organ-on-chip models are gaining in popularity as alternative in-vitro models to provide findings close to the *in vivo* models. Different laboratory animals, particularly rats and mice are the most extensively used in vivo models, but again the species-specific anatomical and physiological variations can influence the ultimate nanotoxicity outcome. Although the significance of a nanotoxicity study is being realized progressively, the stakeholders are still predominantly focused on customizing novel engineered NMs rather than on toxicity analysis. In addition, existing nanotoxicity studies are mostly on short-term basis and on a single testing model, however, the long-term effects has remained neglected. No international standards for nanotoxicity study are currently available and the reproducibility is also a major issue. So, formation of a consortium to study toxicological effects of different NPs is urgently needed along with the establishment of a central databank for NM study that may help to develop predictive models for enumerating toxicological consequences of present and emerging nanomaterials.

Keywords Nanotoxicity · Oxidative stress · Lipid peroxidation · Pro-inflammatory cytokines · Genotoxicity · Apoptosis · Cell viability assays

29.1 Introduction

Probably, no branch in the present scientific avenues has been left untouched by nano-interventions. Recently, applications of nanomaterials (NMs) have seen an explosion and been commercialized for various applications in nearly every field of science including chemical, materials engineering, biomedical, luminescent, fuel, catalysts, coating, and pharmaceutical industries in the last few years. It is being used in versatile applications including environmental monitoring, remediation, engineering and medical applications panning over rocket science to nano-vaccines, drugs, bioimaging, cosmetics, sensor, nano-diagnostics, and theranostics (Minakshi et al. 2019, 2020a,b; Prasad et al. 2021a). The pace at which it has been commercialized can be promoted by the fact that demand of nano-interventions is growing at a rate of 27% CAGR, which is phenomenal as no such technology or product or industry with such a wide range of applications is growing at a comparable pace in the present era. It can be proven by the fact that the nanomedicine market worldwide has been predicted to attain the valuation of over US$ 200 billion by 2024 at a CAGR of 10% (Prasad et al. 2021b). The class of nanoparticles (NPs) is as diverse as materials available in the world that include but are not limited to metals/metal oxide NPs, carbon-based NPs, biological material-based NPs (siRNA, protein, etc.), non-carbon non-metallic NPs (ex. Silica, ceramic), organic and inorganic polymer-based NPs, nano-composites, metal-organic framework NPs, etc. Increase in applications of such materials has certainly elevated the load of such particles in the biosphere.

These NPs due to diverse possible shapes, size, ionic strength, functionalization, and bioaccumulation properties have the potential to interact adversely with the flora and fauna in a diverse ways (Manuja et al. 2021). Enrichment of these NPs on the earth environment can contaminate air, water and, soil and thus enter into the flora and fauna indirectly apart from the direct exposure. This can impart a deteriorating effect on their survival and, furthermore, it can easily increase the load of NPs in animals through the food chain. The indiscriminate use of NPs and improper disposal of NP-containing wastes without any proper guidelines can aggravate the issue imminently as many NPs have been reported to be toxic to various organisms at organ, tissue, or cellular levels. Any negligence in the study of toxicity aspects of the commercialized NPs in the context of their increased accumulation in the local environment, along with a non-stringent protocol for post-use disposal, may create havoc just like the plastic industry has already created to sea fauna. Further, direct uses of NPs in feed and pharmaceutical applications in overdoses can also create resistance in various micro-organisms as reported frequently in the case of antibiotics. It has already been reported that even the United States Food and Drug Administration (FDA) approved NP-based drugs have imparted both serious and non-serious toxicological effects (Prasad et al. 2021b). Thus, short- as well as long-term toxicity evaluation of nano-formulation prior to commercialization is necessary to warrant the nano-safety perspective (Gatoo et al. 2014; Garduño-Balderas et al. 2015; Ji et al. 2021). Because the realistic possibility of NPs accumulation in higher organisms is certainly very high through various routes such as oral ingestion through food chain and water contamination, inhalation through air contaminants, drug administration, and topical applications. Therefore, understanding the mechanisms of toxicity of various NP classes is the first step to formulate the strategy for future nano-safety. Few groups, such as the diseased, neonate, pregnant, and aged populations, can be easily afflicted from the exposure to NMs due to their compromised immunity, abridged protection mechanism, and diminished self-repair ability, hence, more emphasis must be given to nanotoxicological evaluation of these vulnerable groups. The elevated accumulation of NPs is already perceived, as various reports have depicted, to be increasing NPs in the environment at different corners of the globe. For example, nano TiO_2, a major product in the paint industry, has been found to be increasing in Switzerland, reaching a concentration of 0.021 mg L^{-1} in surface water, and an annual increase of 1.203 mg kg^1 $year^{-1}$ for sediment and 11.2 ng kg^1 $year^{-1}$ for soil, whereas in the United States and Europe, the estimates are 0.015 and 0.002 mg L^{-1} in surface water, with an upward trend of 358 and 53 mg kg^1 $year^{-1}$ in sediment and 1.28 and 0.53 mg kg^1 $year^{-1}$ in soil on an annual basis (Gottschalk et al. 2009; Farré et al. 2009; Schultz et al. 2014). Similarly, the burning of petroleum products due to automobile and other fuel applications is another quantum source of carbon-based NPs released in large amounts on an annual basis in the atmosphere (Dobbins 2007). In spite of poignant regulations in nano-intervention for biomedical purposes, nano-science has currently achieved considerable amelioration in the 'bench to bedside' transition, building up the possibility of serving the livestock sector as well as humans with numerous impending nano-formulations.

Diversification of inherited properties of NPs can have a differential outcome for the same NPs in terms of nanotoxicity due to the distinction of properties they possess such as size, surface-attributes, shape, charges, functional groups, chemical compositions, opto-electronic, and magnetic properties leading to differential accumulation, aggregation, dissolution, and their interaction with different cells and organs (Manuja et al. 2021). The mode of toxicity may range from specific to general ROS production or genotoxicity effect, which can vary with respect to cell type and organs; hence a concrete and broad study is required before the release of the NPs for commonplace applications. Altered cell morphology, viability and growth, mitochondrial dysfunction, apoptosis, aberrant cell signaling, necrosis, oxidative stress, generation of non-specific immune response, inflammation, fibrosis, and carcinogenicity are other prominent mechanisms of nanotoxicity, apart from the major action of reactive oxygen species (ROS) production and genotoxicity (Prasad et al. 2021b). Various attributes of NPs have a role in manifestations of toxicity (Manuja et al. 2021); for instance, size of NP is one of the major determinants as it not only determines the efficiency of uptake by clathrin- or caveolin-mediated endocytosis, pinocytosis or macropinocytosis but also influences aggregation in a time-dependent manner. Similarly, the shape of NP has also its contributory effect as it has been observed that in the case of gold NPs (AuNPs) the intake is five-fold higher for spherical-shaped NPs than for rod-shaped ones. Charge is another factor affecting the outcome, for example, the negatively charged property of the membrane enhances the uptake of positively charged NPs, and oppositely positive charged NPs have more toxicological implications on a membrane than their negatively charged counterparts. The NPs with an inner hydrophobic core have more internalization within the cell due to easy passage through the membrane. Functionalization is another determinant as positive amine (–NH2) group and –COOH functional groups have been found to have a higher intake or uptake within the cell (Foroozandeh and Aziz 2018). Another attribute determining the internalization is elasticity; stiffer NPs have been depicted to be internalized more efficiently within the cell as compared to softer NPs. This chapter will provide an insight on various NPs and their toxicity effects reported so far to perceive the urge of nanotoxicological introspection prior to the commercialization of such novel nano-formulations.

29.2 Overview of Nanotoxicity

In general, nanotoxicity is imparted due to the biodistribution of NPs in the body fluids followed by their penetration into the cellular as well as sub-cellular compartments and interaction with various macromolecules, leading to alteration in cellular architecture along with hindrance to the cellular processes that ultimately manifests in the aberration of normal organ functions. The physico-chemical attributes of the NPs such as morphology, crystallinity, electro-magnetic features, thermal and optical properties, dissolution, agglomeration, chemical compositions,

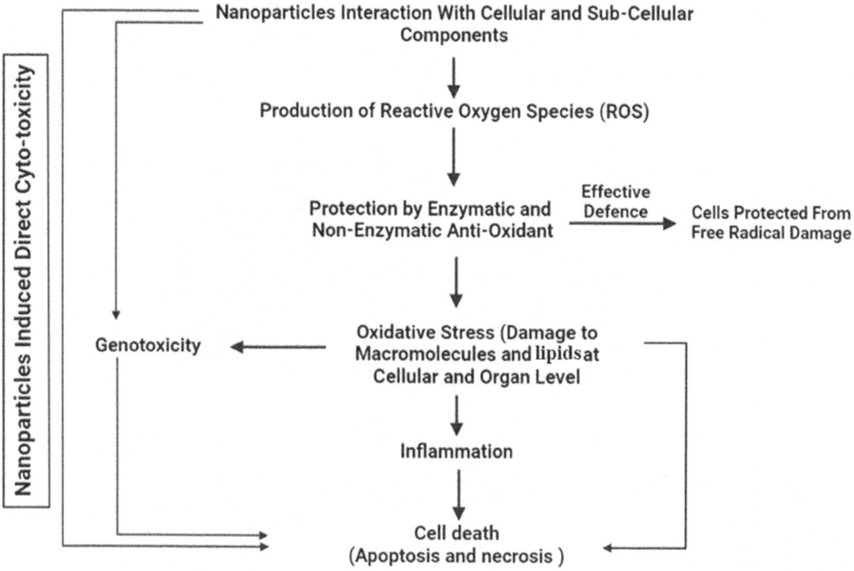

Fig. 29.1 General mechanisms of nanotoxicity

etc. are the major determinants of nanotoxicity. NPs that remain unrecognized by the protective systems may degrade slowly resulting in their accumulation or high cellular concentrations and toxicity. Most of the NPs exert toxicity in a concentration-dependent manner accompanied by the duration of exposure to the cellular components. As morphology and the surface area of NPs critically determine the pattern of their interaction with the living components, these properties of the NPs also influence their recognition by the host defense system and their pharmacokinetics patterns such as adsorption, distribution, metabolism, and clearance, or accumulation in the biological compartments. For instance, NPs of less than 5 nm size usually percolate cellular barriers nonspecifically, whereas bigger particles are taken up by the cells through phagocytosis, macropinocytosis, and other specific or nonspecific transporter systems. In-vivo elucidation of the biodistribution pattern of AuNPs has revealed that intravenous administration of NPs less than 10 nm in size yielded extensive organ distribution, while larger NPs sized between 50 to 250 nm mostly accumulated in the liver, spleen, and blood (De Jong et al. 2008).

The most common ways of inflicting cytotoxicity and genotoxicity include but are not limited to generation of ROS and other free radicals having a high oxidation potential, DNA damage through chromosomal fragmentation, mutagenesis and strand breakage, altered transcription, metabolism and cell signaling processes, induction of inflammatory mediators and apoptosis, mitochondrial damage, intrusion in lysosomal processes and release of hydrolytic enzymes, impaired cell-division by cytoskeletal mutilation, cell membrane dysfunction resulting alteration in cellular transport and communication, etc. (Figs. 29.1 and 29.2). For instance, zinc oxide (ZnO) NPs exert oxidative stress through ROS production and inflict DNA

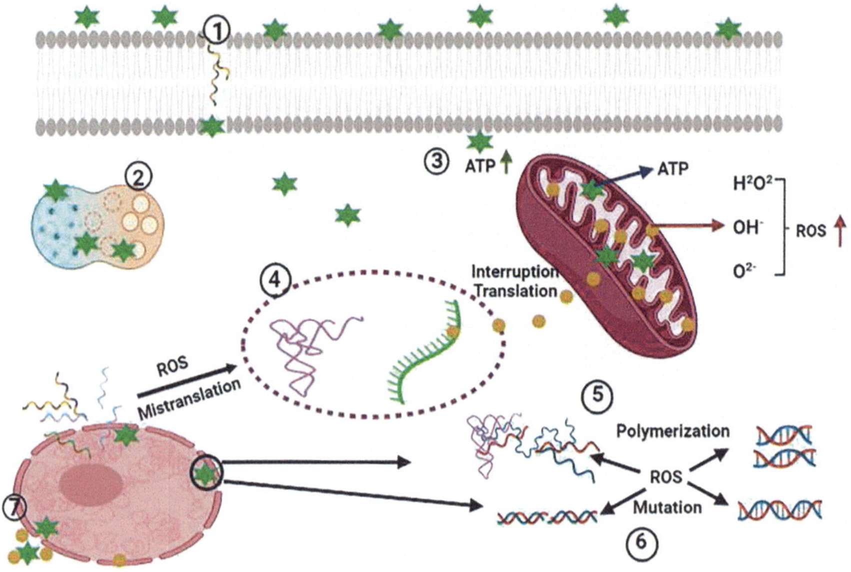

Fig. 29.2 Nanoparticle-induced cytotoxicity: (1) Nanoparticle-mediated cell membrane damage and lipid peroxidation, (2) Lysosomal membrane damage and release of hydrolytic enzymes (3) Mitochondrial membrane damage and release of contents, reduction of ATP generation and elevated ROS production, (4) Mis-translation of RNA by ROS and Hindrance to tRNA binding to ribosomes, (5) ROS induced protein and nucleic acid, (6) ROS induced DNA mutation and strand break, (7) Destruction of the nuclear membrane and release of contents

damage causing cytotoxicity and genotoxicity (Ng et al. 2017). Calcium is one of the major ions involved in various signaling processes and metabolic reactions liable to homeostasis. Alteration in Ca^{2+} (intracellular calcium) trafficking leading to imbalance in cellular function both structurally and metabolically is another major general toxicological implication of various NPs like ZnO, FeO, SiO_2, TiO_2, and Bi_2O_3. (Egbuna et al. 2021).

29.2.1 *Cellular and Subcellular Toxicity*

29.2.1.1 Overproduction of Reactive Oxygen Species (ROS)

Reactive oxygen species such as superoxide radicals (O^{2-}), hydroxyl radicals (•OH), and hydrogen peroxide (H_2O_2) are normally produced through oxidative phosphorylation and electron transport for ATP synthesis in a measured way under the physiological context. These ROS are neutralized or scavenged by different enzymatic antioxidants like superoxide dismutase (SOD), catalase, glutathione peroxidase (GPx), etc. in association with various non-enzymatic antioxidants like vitamin C, vitamin E, carotenoids, glutathione, uric acid, bilirubin, melatonin, polyamines, etc. However, due to strong oxidative potential, overproduction of ROS beyond the physiological scavenging potential results in the mutilation of different biomolecules and organelle organizations, induction of altered signaling cascades and redox-sensitive transcription factors, inflammatory response, protein oxidative carboxylation and generation of protein radicals, lipid peroxidation, mutagenesis, DNA/RNA strand breakage, and membrane structure destruction, which eventually lead to autophagy, necrosis, and apoptosis (Fu et al. 2014; Yu et al. 2020). Diverse NPs having smaller sizes, a high surface-to-volume ratio, and high local charge densities carry enhanced potential to penetrate and interact with the surrounding biomolecules. Such interactions and ionic dissolution of NPs often inflict toxicity by overproduction of ROS and dysregulating the stringent redox balance of various biological reaction milieu. Metal ions released from the NPs interfere with the redox cycling and mediate chemocatalysis through the Fenton reaction or Fenton-like reaction generating free radicals. In addition, the surface charges on NPs or ions released from them interact with the biological components, leading to enzyme deactivation, membrane disruption, hindrance to electron-shuttling process, diminished mitochondrial membrane potentials, redox imbalance, impaired electron transport, and ATP synthesis. All such interference synergistically merge to increase the intracellular ROS production; besides, direct interaction of NP-derived electrons with oxygen may also promote extracellular ROS generation (Yu et al. 2020). Several transition metals such as iron, copper, silica, chromium, etc. are found to produce ROS through the Fenton reaction and Haber-Weiss reaction. For instance, iron and copper ions frequently generate highly reactive hydroxyl radicals by reducing H_2O_2 along with the production of oxidized metal ions, while other transition metals such as vanadium, chromium, or cobalt can catalyze Haber-Weiss as well as Fenton reaction by the interaction of oxidized metal ions with H_2O_2 (Manke et al. 2013).

Several reports of IONP-induced ROS-mediated toxicity on diverse cell types have already been documented. For instance, spherical iron oxide NPs (IONPs) of 15–30 nm size has inflicted concentration as well as exposure time-dependent damage to the murine macrophage (J774) cells through the overproduction of ROS. A considerable extent of apoptosis and cell membrane injury has occurred at SPION concentration over 300–500 μg/mL with an exposure time of 6 h (Naqvi et al. 2010). Similarly, zero-valent IONPs have also been found to exert a cytotoxic effect on human bronchial epithelial cells (cell line 16HBE14o) through ROS production when zero-valent iron and ferrous ions are oxidized to ferric ions accompanied by oxidative burst. Therefore, the oxidation state of iron in nanoformulations is a crucial determinant to induce such types of toxic effects where the ferric form is more stable thermodynamically and can be favored over the ferrous form (Keenan et al. 2009). Iron oxide (Fe_2O_3) NPs have been found to increase the permeability of human microvascular endothelial cells by ROS-mediated modulation of Akt/GSK-3β signaling pathways and microtubular remodeling (Apopa et al. 2009). Dextran-coated SPION has also been found to exert oxidative stress and induce inflammatory response in human epidermal keratinocytes (HEK) and murine epidermal cells (JB6 P(+)) through activation of

AP-1 and NF-κB (Murray et al. 2013). Further, the size of NPs is another crucial determinant to induce cytotoxicity as iron oxide (Fe_3O_4) NPs smaller than 5 nm in size have been found to inflict ROS-mediated injury to several organs in mice by generating highly reactive hydroxyl radicals. Oxidative stress was also induced by silica and gold NPs of similar ultra-small size, although the production of hydroxyl radicals was observed only in IONPs, whereas larger-sized NPs (>9 nm) were found to be safer than the smaller ones (Wu et al. 2022). The concentration-dependent cytotoxicity and genotoxicity of IONPs have also been elucidated through in-vitro and in-vivo studies on rat lymphocytes and Wistar rats, respectively. Interaction with DNA and intercalation along with increased ROS production, diminished antioxidant enzymes, reduced glutathione levels, and elevated lipid peroxidation by NP exposure have been observed in a concentration-dependent manner (Ansari et al. 2019).

Besides inducing genotoxic effects, copper oxide NPs (CuONPs) have also been found to exert oxidative stress and cytotoxicity in human pulmonary epithelial cells (A549) in a concentration-dependent manner. Elevation in lipid peroxidation, along with modulation of cellular antioxidant components like catalase, superoxide dismutase, and glutathione has been observed as a result of oxidative stress. Further induction of Hsp70, cell cycle regulating protein p53, and DNA damage repair proteins Rad51 and MSH2 also points towards oxidative stress-mediated genotoxic effects of CuONPs (Ahamed et al. 2010). Size-dependent toxicity of CuONPs has also been elucidated in human origin cell line A549. The nano-sized particles exerted significant oxidative DNA damage and mitochondrial impairment as compared to the microparticles of the same composition (Karlsson et al. 2009). In addition, CuONPs have been found to be the most cytoxic to the respiratory airway epithelial (HEp-2) cells as compared to silicon oxide and ferric oxide NPs. Generation of ROS and surmounting the host antioxidant defense was depicted as the principal mode of cytotoxicity by the CuONPs in a dose-dependent manner (Fahmy and Cormier 2009). Further, CuONPs have been shown to produce oxidative damage to the human umbilical vein endothelial cells (HUVECs) through activation of the p38 mitogen-activated protein kinase (MAPK) signaling pathway resulting in DNA damage and cell death. Enhancement of superoxide ions, heme oxygenase-1 (HO-1) protein level, MDA production along with a decline in reduced oxidized glutathione ratio has been evidenced in the treated cells as a vivid indication of CuONP-induced oxidative stress. Antioxidant *N*-acetyl-L-cysteine (NAC) and copper ion chelator tetrathiomolybdate have been found to ameliorate the oxidative stress and cytotoxicity produced by the CuONPs (He et al. 2020). Similar toxic effects and oxidative stress induced by CuONPs have also been observed in human ovarian granulosa cells (COV434) through activation of MAPK14-Nrf2 signaling pathway, resulting in induction of HO-1 expression and reduction in antioxidant components like reduced glutathione, catalase, superoxide dismutase, etc. (Zou et al. 2021). NMR-based metabolomics analysis of body fluids and tissue extracts of rats exposed to CuONPs at 200 mg/kg/d dose for 5 days inflicted severe hepatotoxicity and nephrotoxicity as evidenced by mitochondrial dysfunction, elevated ketogenesis, fatty acid oxidation, and glycolysis (Lei et al. 2008). Not only in animals, the toxicity of copper NPs was also elucidated in plants. Bioaccumulation of the copper NPs of 10–30 nm size was found in the roots of treated cucumber plants that induced electrolyte leakage, lipid peroxidation by elevation in H_2O_2 and malondialdehyde (MDA) production, as well as upregulation in copper-zinc superoxide dismutase expression, which press the alarm for proper disposal of NPs which could otherwise result in food-chain contamination (Mosa et al. 2018).

Comparative toxicity analyses of ZnO, TiO_2, SiO_2, and Al_2O_3 NPs of around 20 nm size on human fetal lung fibroblast (HFL1) cells have revealed that these metal oxide NPs inflicted nanotoxicity in a concentration-dependent manner and the effects were pronounced at the concentration range of 0.25–1.50 mg/mL. Mitochondrial dysfunction, morphological aberrations, and apoptosis have been observed as toxic effects and the order of nanotoxicity potential was $ZnO > TiO_2 > SiO_2 > Al_2O_3$ (Zhang et al. 2011). ZnO NPs have found to produce cytotoxicity and oxidative stress on RAW 264.7 and BEAS-2B cells. Enhanced ROS production and induction of cytoprotective enzymes such as HO-1, NAD(P)H:quinone oxidoreductase 1 (NQO1), SOD, catalase, etc. was observed in response to ZnO NPs exposure. Further, activation of pro-inflammatory signaling pathways like JNK activation, Nrf2 upregulation was also found as a response to ROS-mediated oxidative stress. ZnO NPs are also capable of intracellular calcium ion release as a major indicator of oxidative stress, which may lead to mitochondrial damage and apoptosis (Xia et al. 2008). Similarly, toxic effects of ZnO NPs at a concentration of 14–20 μg/mL for 12 h have also been depicted in human liver cells (HepG2). A decrease in cell viability, DNA damage due to oxidative stress, and depletion in mitochondria membrane potential inducing apoptosis through elevation in the ratio of apoptosis regulator Bax/Bcl2 were the underlying causes of the toxicity. Further, ZnO NPs also induced JNK, p38, and p53(Ser15) phosphorylation to activate the relevant pathways that may also be involved in inflicting toxicity (Sharma et al. 2012). ZnO NPs also inflicted toxic effects on rat retinal ganglion cells (RGC-5) by depleting mitochondrial membrane potential, producing ROS to activate caspase-12 that led to endoplasmic reticulum stress, apoptosis, or necrosis (Guo et al. 2013). ZnO NPs can exert toxic effects not only on normal cells of different organ systems but also on different cancer cells too. The exposure of ZnO NPs for 24 h on human colon carcinoma (LoVo) cells has been found to increase the production of H_2O_2, hydroxyl and superoxide radicals, diminish GSH level, alter mitochondrial membrane potential, release pro-inflammatory cytokine IL-8, and induce apoptosis (De Berardis et al. 2010). Similarly, ZnO NPs at 5–20 μg/mL concentration have been observed to alter cell morphology and viability, generate free radicals, deplete cellular antioxidant levels, and induce oxidative DNA damage in human skin melanoma (A375) cells in short-term exposure (24–48 h) toxicity study (Alarifi et al. 2013). The toxic effects of ZnO nanorods on human alveolar adenocarcinoma (A549) cells were found to be exerted in a concentration and time-dependent manner. Induction of cytotoxicity, ROS, activation of caspase-3 & caspase-9, up-regulation of pro-apoptotic factor Bax and down-regulation of anti-apoptotic factors survivin, bcl-2 along with activation of associated oxidative stress pathways such as p53 and caspase pathways have been found to be a major cause of nanotoxicity causing cellular damage (Ahamed et al.

2011). A similar type of ZnO NPs mediated oxidative stress-induced toxic effects were also observed in zebrafish embryos (Zhao et al. 2016). ZnO NPs induced cytotoxicity and genotoxicity was also observed in human lung fibroblasts (MRC5) and *Drosophila melanogaster*. Over-expression of different endoplasmic reticulum stress-related genes and elevated levels of DNA oxidation products, 8-hydroxydeoxyguanosine (8-OHdG) have been evidenced as prominent indications of ROS-induced toxicity imparted by the ZnO NPs (Ng et al. 2017). Animal cells are not the only ones to become prey to ZnO-mediated nanotoxicity, radiation-resistant extremophilic bacteria *Deinococcus radiodurans* has also suffered from similar toxic effects. Significant elevation in ROS production, protein oxidation, and DNA damage has been observed in this organism in the presence of ZnO NPs (1–80 μg/mL) in a concentration-dependent manner (Singh et al. 2020). Several studies have depicted the anti-bacterial action of ZnO NPs, but it is of immense concern that NPs have also been found to be toxic to human cells. In a single study, it has been depicted that ZnO NPs exert oxidative stress to the microorganism *Pseudomonas* sp. in a concentration-dependent manner, while at the same time, the NPs induced cytotoxic and genotoxic effects on human promyelocytic leukemia (HL-60) cells and peripheral blood mononuclear cells (PBMC) in a dose-dependent manner (Soni et al. 2017). In addition to the cytotoxic effects, ZnO NPs have also been found to produce epigenetic modifications by global DNA hypomethylation via upregulation of Ten-Eleven Translocation (TET)-methylcytosine dioxygenase enzyme. The cytotoxic effects were imposed on human embryonic kidney (HEK-293) cells by inducing ROS-responsive genes such as AOX1, HMOX1, NCF2, SOD3, etc. along with diverse sub-cellular modifications like actin depolymerization, increased release of mitochondrial cytochrome C, nuclear enlargement, etc. The epigenomic toxicity imparted by the ZnO NPs in terms of global demethylation is a prominent indication of diverse cancer types that can be propagated generation-wise (Choudhury et al. 2017). This can be a dangerous event in case of long-term toxicity of such metal oxide NPs and warrants immediate attention for indiscriminate use and improper disposal of the nano-formulations and wastes.

Silver nanoparticles (AgNPs) are one of the most frequently employed NPs in diverse nano-formulations. However, it also induces oxidative stress to diverse normal as well as cancerous cell types. AgNPs have been found to induce ROS and reactive nitrogen species (RNS)-mediated oxidative damage to pancreatic ductal adenocarcinoma cells (PANC-1) as well as pancreatic non-cancer cells (hTERT-HPNE); although the extent of oxidative stress was higher in cancerous cells. Depletion in mitochondrial membrane potential, alteration in cell cycle events, alteration in cellular antioxidant enzymes like SOD1, SOD2, and SOD3, glutathione peroxidase and catalase, changes in mitochondrial ultrastructure along with cell death mediated by oxidative damage have been evidenced clearly in presence of AgNPs (Barcińska et al. 2018). Sublethal exposure by dietary AgNPs has been reported to cause tissue-wide accumulation of ROS and activation of the Nrf2-dependent antioxidant pathway in *Drosophila melanogaster* that ultimately leads to DNA damage, apoptosis, and autophagy (Mao et al. 2018). It was observed that citrate-coated AgNPs (20 nm) interacted with catalase and SOD antioxidant enzymes to form surface complexes. The AgNP-Catalase complex was found to promote dissolution of the NPs and brought conformational changes to the catalase enzyme resulting in impairment of the enzyme activity while AgNP-SOD complex was having a marginal effect on the conformational alteration of the enzyme (Liu et al. 2020a). Size-dependent toxic effects of AgNPs through ROS production have been observed in BALB/3T3 A31-1-1 cells. Spatio-temporal evaluation of ROS generation through live-cell imaging has depicted that exposure at 5 μg/mL concentration, AgNPs of 1 nm size rapidly produced ROS within 5 min in the cytoplasm and mitochondria while AgNPs of 70 nm size generated ROS after 60 min of exposure only in the mitochondria, while $AgNO_3$ was found to exert minimum or no toxicity. Further, AgNPs of 1 nm size were far more toxic and extensively reduced the cell viability as compared to AgNPs of 70 nm size (Onodera et al. 2015). Not only on animal cells, the size-dependent toxicity of AgNPs was also evidenced in nitrogen-fixing bacteria *Azotobacter vinelandii*. Exposure at 10 mg/L concentration for 12 h, AgNPs of 10 nm size imparted more apoptotic effect than AgNPs of 50 nm size. Cell membrane damage, ROS generation, and enzyme inactivation, alteration in gene expression, and induction of cell death were found to be the cardinal effects of AgNPs-induced toxicity (Zhang et al. 2018). AgNPs-induced oxidative stress and toxicity were also evidenced in *Staphylococcus aureus*, *Escherichia coli*, and *Pseudomonas aeruginosa*; in addition, the elevation in ROS generation was found to be correlated with the anti-microbial activity. The affinity of AgNPs with the membranes causing impairment in the electron transport chain, ROS, and reactive nitrogen intermediates inflicted macromolecular damage were the major toxic effects of the NPs leading to bacterial death (Quinteros et al. 2016). The toxic effects of AgNPs have also been observed in plants. Concentration-dependent toxic effects were imparted in *Solanum tuberosum* L. by AgNPs as well as Ag ions; however, the toxicity was higher in the case of AgNPs as evidenced by significant depletion in the level of antioxidants like glutathione, ascorbate, etc. (Bagherzadeh Homaee and Ehsanpour 2016). The surface coatings or functionalization may significantly influence the effects of AgNPs on the antioxidant/pro-oxidant systems of the host. For instance, ethylene glycol-coated AgNPs (EG-Ag NPs) have been found to elicit a protective antioxidant effect on Wistar rats in a dose-dependent manner as evidenced by the increase in the total antioxidant capacity and the catalase activity along with a decrease in thiobarbituric reactive species (TBARS) and protein carbonyl levels, while Polyvinylpyrrolidone (PVP) functionalized AgNPs (PVP-EG-AgNPs) have been found to elicit pro-oxidant activity as evidenced by the elevated protein carbonyl level along with a reduced glutathione (GSH) level and a total antioxidant capacity under similar dose regimen of (0–4 mg/kg body weight/day, intraperitoneally) (Docea et al. 2020).

Gold NPs (AuNPs) are other potential metallic NPs that have received significant attention in biomedical applications. However, the safety profile of AuNPs on different organ systems for short as well as long-term exposure needs keen introspection. The effects of AuNPs having different sizes (13 nm and 60 nm), shapes (spheres and stars), and functionalizations (coated with 11-mercaptoundecanoic acid (MUA) or sodium citrate) on proximal tubular cells of normal human kidney (HK-2) have been evaluated. The smaller 13 nm spherical AuNPs coated with MUA and citrate were proved to be the most toxic with elevation in ROS production and diminishing mitochondrial membrane potential. Modulation in the level of GSH, ATP production, and activation of caspase 8 and 9-mediated apoptotic pathways have also been observed as toxicity signatures. Further, the larger 60 nm spherical as well as star-shaped AuNPs depicted a better safety profile as compared to the smaller ones (Enea et al. 2020). Cold Atmospheric Plasma (CAP) therapy has recently gained significant attention for wide-biomedical applications such as in cancer therapy, wound healing, etc. CAP is an ionized gas produced by irradiation or by applying an electric field on atmospheric air, argon, helium, etc. and operates near room temperature or below 40 °C as cold plasma. So, the CAP acts as a well-regulated source of reactive species, neutral particles, electromagnetic field, and ionizing radiation (Yan et al. 2018; Dubey et al. 2022). The size-dependent apoptotic action and ROS generation by the variable size of AuNPs with helium-based CAP therapy have depicted that only 2 nm-sized smaller AuNPs can impart a synergistic effect and induce significant apoptosis in human myelomonocytic lymphoma U937 cells while larger AuNPs of 40 nm and 100 nm size failed to do so. In addition, helium-based CAP alone was unable to induce intracellular ROS formation, but in combination with 2 nm-sized AuNPs, significant GSH depletion and superoxide and hydroxyl radical generation were observed (Jawaid et al. 2020). Surface coating of AuNPs with different polymers such as polyethylene glycol (PEG) and polyethyleneimine (PEI) may also alter the toxicity profile of the NPs as evidenced by the induction of oxidative stress on HepG2 cells by high molecular weight PEI-coated AuNPs through elevation in lipid peroxidation and protein oxidation while low molecular weight PEI and PEG-coated AuNPs did not induce any oxidative stress to the cells (Sen et al. 2020). Similarly, pegylated AuNPs (Au-PEG) of 4.5 nm have been found to induce neurotoxic effects on human neuroblastoma SHSY5Y cells as the NPs have induced apoptosis through enhanced ROS generation and have altered mitochondrial membrane potential. However, surface functionalization of Au-PEG with antioxidant Trolox was found to improve oxidative neurotoxicity significantly in SHSY5Y cells. Under in-vivo condition, Au-PEG was also found to induce neurotoxicity in mice as evidenced by the reduction in antioxidant enzymes such as glutathione peroxidase, superoxide dismutase, and catalase activity along with neuronal apoptosis and activation of astrocytes. A similar protective effect of Trolox-functionalized Au-PEG was also evidenced in mice under in-vivo study (Zhang et al. 2020). Green synthesized AuNPs using leaf extracts of different plants depicted differences in toxicity profiles. The AuNPs synthesized using leaf extract of *Pterygota alata* (Pa–AuNPs) depicted significant dose-dependent anti-cancer efficacy on MCF-7 breast cancer cells as it induced ROS generation and apoptosis through mitochondrial dysfunction as compared to other green synthesized AuNPs (Kowsalya et al. 2021). Charge densities and hydrophobicity of AuNPs have been depicted as crucial attributes to induce oxidative stress by the NPs. Hydrophobic AuNPs having positive charge density have been found to induce cellular oxidative stress on different human cell lines. Hydrophobicity-mediated oxidative stress has been observed to be inflicted through

NADPH oxidase activation whereas positive charge density-associated effects are imparted through mitochondrial dysfunction and regulation of intracellular calcium ion concentration (Sun et al. 2018).

Single-walled and multi-walled carbon nanotubes (SWCNTs and MWCNTs) have found wide applications in various biomedical modalities. These CNTs are also capable of producing oxidative stress and DNA damage to diverse cell types. MWCNTs have imparted a dose and exposure time-dependent toxicity on mouse fibroblast (L929) cells. Enhancement of lipid production through elevated MDA level, SOD activity, diminished GSH level, induction of caspase 3-mediated apoptotic pathways, and DNA fragmentation as detected by comet assay were the pronounced toxic effects produced by the MWCNTs (Alarifi and Ali 2015). Similarly, SWCNTs and MWCNTs have been found to generate oxidative stress in rat brain mitochondria as evidenced by impairment in mitochondrial membranes, elevated MDA production, and depleted GSH level through ROS generation (Rasras et al. 2019). The influence of various functional groups such as pristine, carboxyl and hydroxyl group, etc., as well as the length of SWCNTs (1–3 μm and 5–30 μm) on their cytotoxic potential has been evaluated on HepG2 cells. Reduction in cell viability, enhancement in intracellular ROS generation, and impairment of plasma membrane were evidenced in all types of SWCNTs tested in a concentration-dependent manner. In addition, longer SWCNTs yielded a greater cytotoxic effect than the shorter ones. Despite all the six SWCNTs being toxic, as evidenced by their effects towards alteration in membrane fluidity, inhibition of membrane efflux pump, and producing a chemosensitive effect, the hydroxyl group functionalized short SWCNTs depicted a better safety profile as compared to other types (Shen et al. 2019). Similarly, long and rigid type MWCNTs have been found to cause acute inflammation and enhanced ROS production even after 90 days of intrapleural injection in C57BL/6 mice, whereas short and tangled type MWCNTs, as well as SWCNTs, were found to be safer as they did not induce marked pleural inflammation or ROS production. However, SWCNT-induced DNA strand breaks at 90 days post-exposure (Wils et al. 2021a). Further, SWCNTs have been found to impact the telomere length in a chronic exposure scenario (Wils et al. 2021b). Intra-tracheal instillation of MWCNTs in rat lungs at different dose rates (0.2, 1, and 5 mg/kg body weight) has been found to generate oxidative stress in a dose-dependent manner in terms of decline in blood total antioxidant capacity, GSH, SOD, catalase activity, and enhancement in MDA level through lipid peroxidation (Reddy et al. 2011). Elucidation of the effects of oxidized MWCNTs on human articular chondrocytes (NHAC-kn cell line) has revealed a concentration-dependent reduction in cell viability, however, CNTs were found to interfere with the luminescence-based measurement of ROS generation (Szymański et al. 2020).

It is evident from the aforesaid discussion that several of the common types of NPs have significant toxic potential under short as well as long-term exposure in terms of producing oxidative stress, impairing membrane integrity, mitochondrial function, DNA damage, altering cell viability through activation of apoptotic pathways and inflammatory mediators, etc. Size and morphology, charge density, hydrophobicity, functionalization, method of synthesis, dose, exposure time, etc. are found to be the crucial factors influencing the toxicity of different NPs under in-vitro or in-vivo conditions. In-depth introspection of the mechanistic processes and factors influencing the toxicity of the NPs on short as well as long-term basis is the need of the hour as the nano-exposure in global biome is increasing rapidly and steadily, which may render the nano-safety aspect quite vulnerable during the coming years.

29.2.1.2 Inflammation-Mediated Nanotoxicity

Whatever the root of entry of NPs in the body, it has to encounter both active as well as passive defense systems of the body. Hence, the first direct effect of NPs may occur due to the interaction of these NPs with the immune cells. The inflammatory response is one of the manifestations of nanotoxicity that is influenced by different modes of action of NPs having diverse attributes. The mechanism of inflammatory reactions can be induced either by an increase in pro-inflammatory molecule secretion like cytokines, such as TNF-α, IL-1, and IL-8, or hypersensitivity reactions and anaphylaxis as shown in some lipid-based NPs. AuNPs have also shown inflammatory effects but it is mostly transitory and size-dependent (Noël et al. 2016). In general, smaller particles have more inflammatory action than the larger ones. Neutrophils and macrophages are the major inflammatory phagocytes that induce oxidative outbursts as an immune defense mechanism as the body encounters environmental pollutants, tumor cells, and pathogens. Activation of macrophages and neutrophils by various NPs is the predominant inflammatory mechanism that results in oxidative burst through increased production of ROS. In addition, modulation of intracellular calcium ion concentrations, activation of various transcription factors, induction of pro-inflammatory cytokines, and activation of downstream inflammatory pathways are the modes of actions adopted by NPs to produce inflammation. For instance IONPs, AgNPs, and their combinations have been found to produce oxidative stress and increase the expression of TNF-α, IL-6, and p53 in rat brain. The other neurotoxic effects include DNA fragmentation, decrease in neurotransmitters, alteration in the expression of transcription factors like mitochondrial transcription factor A (mtTFA) and peroxisome proliferator activator receptor gamma-coactivator 1α (PGC-1α), etc. (Yousef et al. 2019a). Inhalation of metal oxide NPs such as ZnO, TiO_2, Al_2O_3, or CeO_2 has been found to produce acute airway irritation, pulmonary inflammation, DNA damage, and toxicity in the murine model. Significant elevation in infiltration of neutrophils, lymphocytes, and macrophages in bronchoalveolar lavage fluid indicated acute inflammatory response induced by the exposure of NPs. The lung inflammatory potential of the metallic NPs has been ranked as $Al_2O_3 = TiO_2 < CeO_2 \ll ZnO$ (Larsen et al. 2016). The influence of the size of the NPs to induce inflammation has been depicted by treating RAW 264.7 macrophages with AgNPs of 20–113 nm size. Analysis of inflammatory markers such as IL-1α, IL-1β, IL6, IL10, MIP1α, MIP1β, MIP2, G-CSF, and TNF-α in the culture supernatant of RAW 264.7 cells 24 h post-exposure revealed a many-fold increase in most of the cytokines, while the small 20 nm AgNPs were found to be the most potent inflammatory and cytotoxic agent, although the extent of toxicity depends on both the size of the NPs as well as the cell types used (Park et al. 2011). Although 2D in-vitro cell culture is the most popular toxicity testing model, however, it often overestimates the toxic effects of NPs. Under in-vivo conditions, the cellular architecture is far more complex and the NPs have to face several barriers as well as many-fold greater interactions to impart a toxicological impression on the cells. So, the use of 3D cell culture and organoids as a toxicity testing model can provide a much more realistic interpretation (Prasad et al. 2021b). Treatment of 2D and 3D culture of Caco-2 cells with various concentrations (10 to 100 μg mL^{-1}) of ZnONPs having different sizes (24 nm, 56 nm, and 87 nm) for 12 and 24 h markedly elevated the levels of IL-8 and IL-1β in 2D as well as 3D culture supernatant. The post-exposure pro-inflammatory cytokine response varied significantly between 2D and 3D cultures. Particle size and exposure time were also found to be crucial regulators of inflammatory cytokine response. The 3D culture system was also depicted to be a more realistic model than the 2D culture system (Wu et al. 2020). Application of precision-cut murine liver slices as a representative 3D model depicted that amino-modified polystyrene NPs can induce a strong inflammatory response in the liver slice under serum-free condition, while no inflammatory response was elicited by the same NPs when the slices were maintained under serum-supplemented medium, providing an opportunity for corona formation (Bartucci et al. 2021). CNTs are also recognized to produce pulmonary inflammation and fibrosis in different experimental animals (Dong and Ma 2016, 2018, 2019; Dong 2020a). During the early acute phase, neutrophils and M1 macrophages are dominant and infiltrate into lung tissues from blood vessels to produce pro-inflammatory cytokines, chemokines, and growth factors, causing acute inflammation. The late acute phase response is driven by the immune cells eliciting type 2 immune response such as mast cells, eosinophils, Th2 cells, and M2 macrophages that produce type 2 mediators leading to transition from acute to chronic inflammation and fibrosis (Dong 2020b). Low-toxicity NPs such as carbon black and polystyrene also stimulate macrophages via ROS generation, activation of transcription factor AP-1, and calcium signaling to induce the secretion of pro-inflammatory cytokines like TNF-α (Brown et al. 2004).

29.2.1.3 Effects on Mitochondria

Mitochondria plays a central role in cellular ATP production through oxidative phosphorylation involving the electron transport chain where maintaining the mitochondrial membrane potential and integrity is crucial for ATP synthesis. During the course of ATP synthesis, the reduction of molecular oxygen to water via electron transfer reactions leads to the formation of superoxide anion radicals which may subsequently generate hydroxyl radicals and H_2O_2. Thus, mitochondria are the major organelle to produce ROS as the byproducts of cellular oxidative metabolism. Smaller size and favorable properties of NPs promote their access into the mitochondria where these engineered particles induce ROS production via impaired electron transport chain, impairment of structural integrity, activating NADPH-like enzyme system, altering the redox ratio (NADP/NADPH), and depolarization of the mitochondrial membrane. Oxidative stress by ROS generation, depletion in cellular antioxidants, mt DNA damage, impaired ATP synthesis, increased outer mitochondrial membrane permeability, loss of inner mitochondrial membrane potential, mitochondrial swelling, disruption of structural integrity, release of pro-apoptotic BCl-2 family of proteins, apoptosis, and necrosis are the cardinal nanotoxicological implications on mitochondria (Manke et al. 2013; Rim et al. 2013). As the interplay of ROS generation and mitochondrial dysfunction goes on concurrently, so, most of the metallic and metal oxide NPs that induce oxidative stress also frequently lead to mitochondrial toxicity. Various metal-based NPs such as silver, gold, iron oxide, zinc oxide, aluminum oxide, molybdenum oxide, titanium dioxide, copper oxide, cerium oxide, silica, etc. depicted oxidative stress and mitochondrial dysfunction under different in-vitro as well as in-vivo set-ups. Modulations in several signaling pathways such as HIF-1, NF-κB, PI3 K, and MAPK also take place concurrently as a result of redox imbalance, transcriptional activation of several bioactive factors, inflammatory response, and mitochondrial dysfunction induced by different NPs that ultimately lead to toxicological manifestation (Manke et al. 2013). Similar to metal-based NPs, a study on the mechanism of toxicity of SWCNT showed a concentration-dependent toxicity on mitochondria due to inhibition of F0F1-ATPase, which has been found to be similar to oligomycin A inhibition by blocking of the c-ring of the F0 subunit (González-Durruthy et al. 2021).

29.2.1.4 Effects on Cell Morphology and Cytoskeletal Integrity

The NPs that alter the cell viability and induce apoptosis are supposed to also affect the cell morphology as a nanotoxicological consequence as depicted in the aforesaid discussion. Alteration in cell morphology oftentimes arises from cytoskeletal defects imparted by diverse nanomaterials. For instance, ultrafine AgNPs of 20 nm size produced cytoskeletal defects such as loss of β-tubulin and filamentous actin (F-actin) in primary rat cortical cells at 1–50 μg/mL concentrations (Xu et al. 2013). Similarly, 20 nm size AuNPs have been found to be internalized in the cytosolic vacuoles of human neutrophils and induced degradation of different cytoskeletal proteins such as vimentin, lamin B1, and gelsolin. However, AuNPs did not induce cell surface expression of these cytoskeletal proteins; therefore, the toxicity from autoantibody generation did not take place (Noël et al. 2016). Degradation of tubulin monomers through 26S proteasomal activity and elimination of microtubules have been observed in *Arabidopsis thaliana* under exposure to ultrasmall TiO_2 NPs of about 2.8 nm size (Wang et al. 2011). The size is one of the major regulators by which the NPs interact with cytoskeleton components. Usually, the larger particles tend to interact with the cytoskeletal mesh while the smaller ones generally penetrate deep and distribute internally following the Van Hove correlation function $\Delta = x(t + \tau) - x(t)$ and $y(t + \tau) - y(t)$ for steps in the x and y directions, respectively, where Δ is the probability of a particle moving a distance along a single axis within a specific time interval τ. (Grady et al. 2017). Apart from that electrophysiological property like *N*-methyl-D-aspartate receptor function, ion channels, transient receptor potential channels, and the Rho pathway also influence this interaction. Various reports have suggested that exposure to inorganic nanomaterials like Ag, Au, Si, Ti, SiO_2, TiO, Fe_2O_3, etc. affect the morphology of cells by a loss of actin-tubulin network, dissolution of synaptic protein, cell cycle arrest, shrinkage or selling of cell, etc. Severing of F-actin, oxidation of the thiol group of tubulin monomers, and the disturbance of microtubule polymerization are major cytoskeletal insults that have been reported due to increased ROS and RNS formation by inorganic NPs (Kang et al. 2016). Alteration of cellular morphology, cytoskeleton remodeling, damage to different organelles, activation of the endocytic pathway, ubiquitination, and proteasomal degradation have been observed in Chinese Hamster Ovary (CHO-K1) cells by exposure to mesoporous silica NPs (MSN), MWCNT, and ZnO NPs (Yadav et al. 2021). The eukaryotic cells are not only susceptible to nanotoxicological effects but prokaryotic cells also suffer. Destruction of cell topography, morphology, membrane structure, inhibition of respiration, and biofilm formation are some of the toxicological effects observed in bacteria inflicted by Ag, ZnO, CuO, TiO_2, and Al_2O_3 NPs (Ahmed et al. 2020). Further, nickel oxide NPs (NiO NPs) altering G0/G1 transition in different cell lines, CuO and ZnO NPs causing G2/M phase arrest, TiO_2 resulting S phase arrest, and Al_2O_3 inducing enhancement in sub-G0 are major toxicological implications found to be affecting the cell morphology by altering the cell cycle progression (Egbuna et al. 2021)

29.2.1.5 Effects on Nucleus

Various NPs have been reported to assault the nucleus by various mechanisms such as gene mutation, DNA adduct formation, DNA strand break, and genotoxicity, which may lead to diverse pathological consequences including cancer. Enhanced ROS generation by the NPs is one of the predominant reasons leading to DNA toxicity as evident from enhanced 8-hydroxy-20-deoxyguanosine (8-OHdG) formation by CuONPs (Sarma et al. 2021). Not only the nuclear DNA but ROS may also damage mitochondrial DNA. Both direct contact with the macromolecules and indirect effect through ROS generation are possible toxic effects of NPs, apart from impairment in membrane structure and function. The toxicity mechanism and effects vary differentially with respect to the type and nature of NPs. For instance, induction of oxidative stress and DNA-adduct formation are the major toxicological effects of TiO_2NPs in lung cells without much DNA breakage, while the predominant mode of toxicity of Fe_2O_3-NPs has been imparted through DNA strand break (Bhattacharya et al. 2009). Similarly, nano-CuO induces DNA damage and forms a considerable extent of 8-OHdG, whereas nano-TiO_2 has little toxicity effect with minimum 8-OHdG formation (Sarma et al. 2021). Platinum-DNA adduct formation was found to be a major effect in cisplatin cross-linked platinum-gadolinium hybrid block copolymer in glioblastoma cells (Lajous et al. 2018). Another common mode of DNA-related damage is due to enhanced lipid peroxidation related to oxidative stress, which has been reported in NPs of silicate, TiO_2, buckminster fullerene (C_{60}), and CNTs. Sometimes metallic NPs themselves are not so much genotoxic but generation of their ionic counterparts has a more damaging effect on DNA (Rim et al. 2013). AuNPs due to ROS production lead to the generation of 8-OHdG, caspase-3 activation, and heat shock protein 70, sequentially leading to DNA damage and cell death (Rim et al. 2013). Concentration-dependent DNA strand breaks have been observed in AgNPs in testicular embryonic carcinoma cells, lymphoma cells, and G1 hypo-diploid cells. Nanoparticulate CarbonBlack (NPCB) has also been reported to alter the cell cycle through single-strand DNA break, activating p53, and proteins related to DNA repair, mimicking irradiation-related carcinogenesis pathways (Mroz et al. 2008). Oxidation stress generation leading to DNA degradation is not the only the effect of NPs but also affects the repair process of damaged DNA, which is slower in TiO_2-NP than in ZnO and potassium bromate ($KBrO_3$) (Carriere et al. 2017).

29.2.1.6 Effects on Plasma Membrane

Disrupting the integrity of the plasma membrane is the predominant toxicological implication of several NPs. Membrane integrity is the utmost important factor for barrier function, which otherwise leads to intracellular leakage and cell death. For instance, metallic copper and copper-zinc alloy NPs have induced strong and rapid membrane-damaging action at 20 μg/mL particle mass dose, while metallic CuNPs and CuoNPs did not produce such an effect. The membrane damaging effect has been suggested due to metal release process at the membrane surface and the difference in reactivity among the metallic (Cu & Cu-Zn alloy) and metal oxide (CuO) NPs (Karlsson et al. 2013). The interaction between NPs and lipid bilayer membrane can be influenced by the asymmetry in the lipid component of membranes and the functionalization of NPs. Disruption of the erythrocyte membrane and hemolysis have been reported by non-functionalized 30 nm silica NPs, while amine-modified and carboxyl-modified silica NPs did not induce hemolysis (Bigdelou et al. 2020). The toxicity to the plasma membrane is also dependent upon the particle size of the NPs; smaller particles are found to be more potent for membrane damage than larger particles. For instance, size-dependent toxic effects of AgNPs have been elucidated on four different cell lines, viz., A549, SGC-7901, HepG2, and MCF-7. Significant adverse effects on cellular morphology (microscopy and Hoechst 33342 staining) and membrane integrity (lactate dehydrogenase assay) have been observed by exposure to 5 nm and 20 nm size particles as compared to 50 nm particles. The EC_{50} of 5 nm size AgNPs for the four different cells ranged between 0.51 and 1.02 μg/ml and for 20 nm size particles ranged between 9.96 and 50.94 μg/ml, while 50 nm particles were found to be less toxic with EC_{50} ranging between 14.31 to 112.03 μg/ml. The cell type is also a crucial factor determining the extent of the toxic effect (Liu et al. 2010). Similar types of size-dependent toxic effects of amorphous silica NPs of 30 nm, 70 nm, and 300 nm size have been observed on lung epithelial (NCI H441) cells and endothelial (ISO-HAS-1) cells, where the smaller particles produced more toxic effects as compared to larger particles (Kasper et al. 2013).

The shape of the NPs is another crucial attribute that may influence the extent of cellular uptake of NPs as well as the cell viability and membrane integrity. The toxicological effects of AuNPs having different shapes and sizes on cancerous HeLa and normal HEK293T cell lines depicted that nanospheres and nanorods have higher toxicological potential than star, flower, and prism-shaped AuNPs (Woźniak et al. 2017). However, in case of PLGA-PEG NPs, the spherical particles have been depicted to be safer for HepG2 cells while the needle-shaped particles elicited cytotoxicity through lysosomal membrane disruption along with activation of apoptotic pathways (Zhang et al. 2017). So, the shape effect over toxicity is not uniform, rather it is dependent upon the type of NPs also. The concentration- and time-dependent toxic potential of Tween 80-coated superparamagnetic IONPs on membrane integrity has been evidenced in murine macrophage (J774) cells through lactate dehydrogenase assay (Naqvi et al. 2010). In addition to the oxidative stress associated with chemical damage, NPs also induce membrane damage through physical mechanisms irrespective of chemical oxidation as evidenced by the extent of membrane integrity loss inflicted by graphene oxide nanosheet which was found to be dependent on the total surface area of the nanosheet (Zucker et al. 2017).

29.2.1.7 Effects on Endoplasmic Reticulum

The Endoplasmic reticulum (ER) is a crucial organelle that carries out several important functions such as protein and lipid biosynthesis, post-translational modifications, ion homeostasis, and signaling. Evidently, ER stress and dysfunction may lead to the accumulation of unfolded or misfolded proteins, activation of autophagy, and apoptotic pathways (Ghosh et al. 2019). A dose-dependent toxicity on *Saccharomyces cerevisiae* has been reported by 10–25 nm sized Mn_3O_4 NPs through ER stress as evident from significant upregulation of the unfolded protein response (UPR) genes (Yi et al. 2017). Similarly, ER stress induced by AgNPs has been observed in human bronchial epithelial cells (16HBE), hepatocellular carcinoma cells (HepG2), and umbilical vein endothelial cells (HUVECs) under in-vitro and in mice under in-vivo condition. Upregulations of different ER stress response genes or proteins such as spliced X-box binding protein 1 (xbp-1 s), chop/DDIT3, TRIB3, ADM2, BIP, Caspase-12, Asparagine synthetase (ASNS), and HERP cells indicated the activation of ER stress response pathways in 16HBE following the exposure to AgNPs, but such response was missing in HepG2 or HUVECs cells. AgNPs also increased ER stress in the liver, lung, and kidney in mice which was evident from elevation in α-subunit of eukaryotic translation initiation factor 2 (eIF2α), caspase-12 and C/BEP homologous protein (CHOP) protein expression in ER followed by apoptosis in lung and kidney. The observation indicated that the time- and concentration-dependent ER toxicity of AgNPs is neither universal for all kinds of cells nor for all organs (Huo et al. 2015). Similar dose-time dependent ER stress was also observed to be induced by ZnO NPs as evident from increased IRE1α, XBP1s, BIP, and CHOP expression in male reproductive organs. Further, elevated expression of caspase-3 was also observed as an indication of apoptosis (Tang et al. 2019).

Activation of apoptotic and inflammatory pathways may be the predominant pathway modulations induced by ER stress apart from the impairment in ER morphology by different NPs. ER toxicity was also observed in human umbilical vein endothelial cells due to the localization of CdTe QDs along with elevation in ER stress markers. Orally administered ZnO NPs induced liver injury through ER stress in mice. Similarly, inhalation of TiO_2NPs induced ER swelling and expression of ER stress markers in mice in a dose-dependent manner (Cao et al. 2017a).

29.2.1.8 Effects on Lysosome

The intake of NPs by cells is carried out through the active mechanisms of internalization, also known as the Trojan horse effect that facilitates the release of high levels of toxic ions from internalized NPs, causing cytotoxicity. The acidic environment of lysosomes promotes the release of toxic ions like Ag^+, Cd^{2+}, $Fe^{2+/3+}$, $Au^{1+/3+}$, etc., and thus this mechanism is termed as the "lysosome-enhanced Trojan horse effect" (Sabella et al. 2014). Various NPs inflict toxicity to lysosomes through various mechanisms depending on their differential physico-chemical attributes. It includes but is not limited to disruption of lysosomal membrane, autophagy and lysosomal dysfunctions, inhibition of lysosomal enzymes, lysosomal perturbation, etc. Rupture of the lysosomal membrane is associated with the release of hydrolases and inflammation, followed by activation of caspase-1-mediated apoptotic pathways (Lima Jr et al. 2013). Further, NPs can accumulate in lysosomes, which may lead to alteration in the lysosomal degradative pathway and autophagy dysfunction. The induction of autophagy to degrade the NPs or the blockade of autophagic flow by the persistent NPs can result in cytotoxicity (Stern et al. 2012; Wang et al. 2018a). Size- and time-dependent intracellular localization of AgNPs has been reported in B16 cells. The 5 nm sized AgNPs were found to be uniformly distributed in the cytoplasm and nucleus, whereas, 20 nm, 50 nm, and 100 nm sized AgNPs were mainly localized in the early endosomes and lysosomes. In addition, the process of cellular uptake of the NPs may also vary depending upon the size of NPs, and clathrin-mediated endocytosis was found to be the predominant mechanism for the internalization of smaller particles (Wu et al. 2019). Lysosomal impairment and autophagy dysfunction were reported in human brain-derived endothelial cells due to exposure to uncoated Fe_3O_4 NPs and TiO_2 NPs (Halamoda Kenzaoui et al. 2012). The AuNPs can also decrease the lysosomal functions by alkalization of the lysosomal lumen, leading to the accrual of autophagosomes, which in turn reduces cellular degradative capability (Manshian et al. 2018). Physical damage to lysosomes is one of the major assaults by the NPs, which occurs due to direct intra-luminal accumulation of NPs. Autophagy dysfunction leading to diminished lysosomal activity and resulting in altered cytoskeletal organization and cell cycle progression is another mode of nanotoxicity imparted through lysosomes. Toxicity due to accumulation of autophagosomes has also been reported in the case of diverse metal-based NPs such as gold (Au), ferric oxide (Fe_2O_3 and Fe_3O_4), manganese oxide (MnO), copper oxide (CuO), neodymium oxide (Nd_2O_3), titanium dioxide (TiO_2), and silica oxide (SiO_2) NPs (Zhang et al. 2016), as well as by amine-modified polystyrene nanoparticles (Wang et al. 2018a). Similarly, lysosomal perturbation has been observed in case of several types of NPs such as G5-PAMAM dendrimer, Fullerene (C_{60}) NPs, MWCNTs, mercaptopropanoic acid-coated gold NPs, TNF-bp20-K PEG monomer (38 kDa), Polyalkyl-sulfonated C_{60}, Polystyrene NPs, along with TiO_2, SiO_2, ZnO, and Ag NPs (Stern et al. 2012).

29.2.1.9 Effects on Golgi Body

Golgi body is another vital organelle that performs several key roles such as downstream processing of proteins received from the ER followed by their packaging into membrane-bound vesicles and sorting to targeted sub-cellular locations or secretion, synthesis of different lipid components including glycolipids and sphingomyelin, etc. Small 4 nm silicon NPs have been found to be accumulated mainly in lysosomes after a short presence in early endosomes along with minor accretion in the Golgi apparatus after 24 h (Cao et al. 2017b). Retinoic acid-conjugated chondroitin sulfate NPs have been found to get accumulated in the Golgi apparatus of hepatic stellate cells that helped to target anti-cancer drugs and prevent metastatic signaling process (Li et al. 2019). The important role of the Golgi apparatus for intra-cellular trafficking of 250 nm sized witepsol/miglyol nanostructured lipid carriers has also been depicted, however, no toxicological consequences have been mentioned (Silva et al. 2017). Variable-sized AuNPs (10 nm, 25 nm, 50 nm, 75 nm, and 100 nm) have been found to enhance cytoplasmic calcium ion release and Golgi fragmentation in a size-dependent manner, resulting in impairment in normal Golgi functions, abnormal protein processing, aberrant integrin distribution, and a decrease in cellular adhesion (Ma et al. 2019). Photothermal ablation of hepatocellular carcinoma (HepG2) cells following exposure to albumin-conjugated AuNPs has been found to induce Golgi apparatus-ER dysfunction along with activation of caspase-3 mediated apoptotic pathway (Mocan et al. 2015).

29.2.2 Toxicity of Organs

Nanomaterials can enter the human body via different routes: intravenous, dermal, subcutaneous, inhalation, intraperitoneal, and oral followed by their cellular internalization through diverse processes such as clathrin-mediated endocytosis, caveolae-mediated endocytosis, macropinocytosis, and clathrin- and caveolae-independent endocytosis. Subsequently, biodistribution of NPs through intracellular trafficking and cellular processing may lead to diverse undesirable effects at a cellular and sub-cellular level that are phenotypically manifested as structural and functional aberrations of organs and nanotoxicological consequences. These nanomaterials in the bloodstream can move through the body and are taken up by different organs and tissues, including the brain, heart, liver, kidneys, spleen, bone marrow, and nervous system. Not all the NPs have similar biodistribution and effects on different tissues; the route of administration and organ-wise differences are found to be crucial regulators of nanotoxicity. For example, intravenous administration of CdTe/ZnS aqQDs in male ICR mice at 0.2 μmol/kg dose yielded differential blood kinetics and organ distribution of cadmium and tellurium ions as the former metal ions mostly accumulated in the liver, kidneys, and spleen while the later predominantly accumulated in kidneys (Liu et al. 2013). Oral administration of PEGylated GO did not show a significant tissue uptake in mice, while intraperitoneal administration facilitated considerable uptake of the NPs by the reticuloendothelial system and accumulation in the liver and spleen (Yang et al. 2013). So the general nanotoxicological effects of diverse NPs on various vital organs have been discussed hereunder.

29.2.2.1 Brain Toxicity and Neurodegenerative Effects

Various NPs can access the brain as many of them can cross the blood-brain barrier (BBB) when applied for imaging, diagnosis, and treatment of brain diseases such as tumors, neurodegenerative disorders, and stroke. The brain is one of the critical organs coordinating numerous functions in higher organisms. Therefore, NPs passing through the BBB must be used judicially after confirming their pharmacokinetic properties and toxicological effects. Diverse NPs such as neurons, astrocytes, and microglial cells have been found to accumulate in various parts of the brain and produce ROS and inflammation. Apart from that synaptic plasticity, loss of cellular functions, apoptosis, impairment of cellular components, signaling pathways, cognitive dysfunctions, and genotoxicity are the other detrimental effects that can be produced by various NPs. Several reports of neurotoxicity are available for TiO_2 NPs, IONPs, AgNPs, AuNPs, silica NPs, and carbon-based NPs. For instance, TiO_2 NPs have been depicted to show nearly all the aforesaid toxicological effects on the brain. Aberration in intra-cellular calcium level related to transporter/receptor mechanisms, mitochondrial damage, and enhanced ROS production have been reported from exposure to AgNPs. Death of neuronal cells, increased number of astrocytes, reticence in axon regeneration, increased seizure due to alteration in the electrical activity of the brain, and cognition defects such as impairments in attention, memory, and judgment have been reported in AuNP-induced toxicity. Inflammatory response by the activating microglial cells, depleted dopamine in the striatum, and down-regulation of tyrosine hydroxylase activity have been depicted to be the predominant toxicological implications of silica NPs (Teleanu et al. 2018; Wu et al. 2011). Intra-nasal exposure to nano-particulate materials often jeopardizes the development of the brain and is empirical to underline neurodegenerative diseases involving the deposition of misfolded proteins responsible for maternal and fetal neuroinflammation. In mice, maternal exposure to low doses of carbon black nanoparticles (CB-NPs) has been found to induce ER stress in the perivascular macrophages and reactive astrocytes only around brain blood vessels along with accumulation of β-sheet-rich misfolded proteins in the offsprings. Upregulation of

ER stress markers such as activating transcription factor 6 (ATF6) and CHOP has been observed only in the astrocytes and macrophages that are located in the proximity of the brain blood vessels but not in the distant ones (Onoda et al. 2020). TiO_2 NPs in diameter < 100 nm are one of the frequently used components of several paints, coatings, and cosmetics. Entry of these NPs into the circulatory system may facilitate their distribution in the brain, leading to ROS production by the microglia phagocytic cells in response to the NP-induced oxidative stress along with neurodegeneration and cognitive dysfunctions (Zeman et al. 2018).

29.2.2.2 Liver Toxicity

The liver is the most crucial innate immune organ that carries out the metabolism and detoxification of drugs as well as the toxicants. Obviously, it is very susceptible to toxicity-associated damage at the cellular and molecular level which subsequently gets manifested in terms of organ damage and dysfunction. Several metallic NPs have been reported to exert profound hepatotoxic effects when administered through different routes under in-vivo setup. Oral administration of 80 nm sized CuNPs at 100, 200, and 400 mg/kg BW dose for 7 days in male SD rats has been found to induce liver injury by elevation in the levels of inflammatory cytokines such as IL-2, IL-6, IFN-γ, etc., along with producing oxidative stress. The CYP450 activity involved in the detoxification process was also affected by the CuNPs along with the suppression of some nuclear receptors *via* the NF-κB signaling pathway (Tang et al. 2018). Intravenous injection of 10 nm sized CuNPs at 10 mg/kg BW dose in Male CD-1 (ICR) mice produced extensive hepatocyte necrosis along with numerous bleeding and hemorrhage around the biliary tract and gallbladder wall as well as injury to the portal vein endothelium (Recordati et al. 2016). Oral administration of 20–30 nm sized PVP-Ag NPs induced hepatic damage by pro-inflammatory activation of Kupffer cells and inflammation along with inhibition of fatty acid oxidation in mice (Jia et al. 2017); whereas in rats, the PVP-Ag NPs initially induced SOD and catalase activity as protective means which subsequently decreased following exposure to high amounts of the NPs. Similarly, at lower doses (50 and 100 mg/kg BW), IRS-1, AKT, mTOR, p53, p21, and caspase 3 activity was found to be increased as a protective response to the oxidative assault, which diminished accordingly with an increase in the dose rate to 200 mg/kg BW due to cellular damage by excessive ROS production and autophagy (Blanco et al. 2018). Intravenous injection of bioactive peptide-functionalized gold nanorods did not induce significant hepatotoxicity in healthy C57BL/6 mice; however, animals having immune-mediated hepatitis may suffer from liver damage due to AuNP-mediated preactivation of hepatic macrophages. Therefore, the use of such NPs in clinical settings warrants ample precaution (Bartneck et al. 2012). Intravenous injection of 50 nm Ni NPs also induced liver and spleen injury in SD rats as evidenced by the serum biochemical changes and the histopathological findings (Magaye et al. 2014). However, no prominent NM-induced hepatic damage was observed when PEGylated Pd nanosheets were introduced orally or intraperitoneally at 200 μg/animal dose for up to 8 days in ICR mice (Chen et al. 2017). Several other functionalized metallic NPs have also been reported to induce liver damage in different experimental animals, which include but are not limited to silicon-coated Ag, Citrate-modified Ag, superparamagnetic zinc-cobalt ferrite, Au/FeO_4 NMs coated with amphiphilic polymer or PEG, Au prisms, Chitosan-capped star Au, PEGylated Au, Dextran coated Au, etc. (Yao et al. 2019; Kermanizadeh et al. 2020). Various metal oxide NPs are also found to be potent hepatotoxic agents and the dose, route of administration, as well as exposure time, influence the extent of hepatic damage. For instance, oral, intravenous, or intraperitoneal administration of TiO_2, ZnO, Al_2O_3, NiO, and SiO_2 NPs in rats and mice have been found to elicit liver damage by inducing oxidative stress and altering the level of SOD, Catalase, Glutathione peroxidase, MDA, etc., inflammation through elevation in the level of pro-inflammatory cytokines IL-1α., IL-1β, IL-6, IL-10 and TNFα, proliferation of Kupffer cells, venous congestion and dilatation of hepatic sinus, fibrosis, DNA damage, focal hemorrhage, swollen mitochondria and vacuoles, damaged nucleolus, necrosis, and apoptotic pathway activation, along with an elevation in the circulatory levels of different liver- and gall bladder- or bile duct-origin plasma non-functional enzymes such AST, ALT, LDH, ALP, γGT, etc. (Yao et al. 2019; Kermanizadeh et al. 2020). Carbon-based NMs such as CNTs, Graphene oxide, Graphite nanoplates, Fullerene C_{60}, etc. have also been reported to induce hepatotoxicity in experimental animals when administered through various routes, and the toxicological manifestations are similar to the aforesaid ones along with altering the expression of different genes related to the stress response. Similarly, intravenous administration of functionalized QDs of different materials such as PEGylated-ZnO QDs, CdSe/CdS/ZnS-QDs, MEA and MPA coated Ag_2Se QDs, Oleic acid PEGylated CdSeS/ZnS QDs, CdTe-QDs, Carboxyl coated CdSe/ZnS QDs, etc. in mice and rat have also found to inflict hepatic injury, although the extent depends on diverse attributes of NPs, dose, exposure time, route, etc. (Kermanizadeh et al. 2020).

29.2.2.3 Kidney Damage

Renal excretion is one of the prominent routes of elimination of the NPs and the system is also very sensitive to xenobiotics. So, nanotoxicity in the kidney is obvious and has been reported for diverse NPs under in-vivo conditions. Oral administration of graphene oxide (40 nm diameter) at 10–40 mg/kg BW for 5 days in SD rats showed a dose-dependent elevation in the level of antioxidant enzymes such as SOD, catalase, and glutathione peroxidase as a response to the oxidative stress imposed by the NPs. In addition, serum creatinine and blood urea nitrogen levels also increased due to nephrotoxic effect. Dilation of tubules, tubular necrosis, disintegration of tubules, and degeneration of hematopoietic tissue and tubular lumen have also been observed as nanotoxicity consequences in histopathological investigation (Patlolla et al. 2016). Intraperitoneal administration of carboxylic group functionalized MWCNTs of <8 nm diameter in Wistar rats at 2.5–20 mg/kg bw dose has led to kidney toxicity as evidenced by alteration in the level of urea, uric acid, and creatinine in a dose-dependent manner along with different histopathological alterations such as inflammatory changes in the cortex and medulla, amassing of hyaline-like substances, presence of inflammatory cells, glomerular degeneration, and proximal tubular necrosis, etc. (Poormohammad Matouri and Noori 2018). Intra-tracheal installation of pristine and amylated MWCNTs in albino rats at 1 mg/kg BW dose for 30 days produced diverse toxicological consequences in the kidney. Oxidative stress has been induced by the NPs as evidenced by alteration in antioxidant enzymes such as SOD, glutathione peroxidase, elevation in MDA level, activation of caspase 3 mediated apoptotic activity, and elevation in pro-inflammatory cytokines like IL1β, etc. Major histological alterations included thin collapsed and edematous hemorrhagic glomeruli, packed mesangial and endothelial cells, along with apoptotic imprints as indicative of renal damage that subsequently led to renal dysfunction as evidenced by high serum levels of urea and creatinine in the affected rats. However, polyethylene glycol-coated MWCNTs did not induce any toxicological evidence. So, the functionalization of NPs can be critically introspected to reduce the toxicity of various NPs (Abu Gazia and El-Magd 2019). Several in-vivo experiments in rats or mice have identified diverse toxicological implications of copper NPs on the renal system through single or repeated oral exposure. The general histopathological signs include swollen glomeruli and alteration in the structural integrity, tubular necrosis, presence of proteinaceous liquid and cellular fragments in renal tubules, degeneration of epithelial cells of renal tubules, etc. Impairment in renal functions has also been observed as compromised renal clearance and elevation in serum BUN and creatinine levels as the consequence. Induction of oxidative stress by increased ROS and RNS production and the resulting alteration in the level of antioxidant enzymes, inducing inflammatory response through enhancement of pro-inflammatory cytokines, alteration in the level of different renal injury marker genes, activation of apoptotic pathways, etc. are the other common signatures of nanotoxicity by copper NPs (Iavicoli et al. 2016). Different other metal-based NPs such as AuNPs, BSA and GSH-conjugated Au nanocluster, AgNPs, TiO_2NPs, ZnO NPs, Oleic acid or Amino poly vinylalcohol coated or uncoated Fe_3O_4 NPs, PtNPs, Mesoporous SiNPs, Cd-SiNP, SiO_2 NPs, etc. have also been found to induce similar types of toxicological consequences on the kidney in different laboratory animals and the extent of toxicity often corresponds to the dose of NPs, exposure time, and route of administration. In addition to the in-vivo studies, several in-vitro experiments on the kidney cells and renal organoid model have also vividly depicted diverse nephrotoxic effects of various NPs (Iavicoli et al. 2016; Prasad et al. 2021b).

29.2.2.4 Pulmonary Effect

Nanoparticles can be accumulated in the lungs through inhalation or systemic delivery from either incidental/accidental exposure or intentional pulmonary targeting. Different attributes of NPs such as size, shape, aspect ratio, surface charge, functionalization, solubility, agglomeration, chemical properties, etc. determine their potential and probability of instigating pulmonary impairment. Nanotoxicity can be imparted through various mechanisms that include but are not limited to induction of oxidative stress, inflammation, and genotoxicity leading to pulmonary fibrosis, pneumoconiosis, exacerbation of asthma, etc. (Li et al. 2010). Intratracheal administration of polyvinylpyrrolidone (PVP) and citrate-capped AgNPs of 20 nm and 110 nm size at 0.1 mg/Kg BW dose has been found to inflict pulmonary inflammation by eliciting eosinophilic and neutrophilic response as well as bronchial hyperresponsiveness in rats as prominent asthmatic features. The smaller particles have been reported to be more pro-inflammatory than the larger ones, however, variation in the capping agent did not influence the toxicity too much. Disruption in the permeability barrier of the capillary or alveolar epithelial lining, induction of oxidative stress and eosinophilic response as evidenced by enhancement in the level of bronchoalveolar lavage (BAL) protein, MDA level, Keratinocyte Chemoattractant, IL-13, CCL11, and IgE have also been observed as the toxic effects of AgNPs in Brown-Norway (BN) and Sprague-Dawley (SD) rats, although both types of rats showed marked variations in toxicity response (Seiffert et al. 2015). Similarly, intratracheal instillation of 11 and 137 nm ZnO NPs in rats induced eosinophilia, proliferation of airway epithelial cells, and goblet cell hyperplasia within 24 h of exposure along with pulmonary fibrosis 4-weeks post-exposure; further, smaller particles depicted stronger effect than the larger ones (Cho et al. 2011). Concentration-dependent increases in inflammatory parameters of BAL fluid and granulocyte infiltration in the lung and increases in alveolar macrophages are the other histopathological changes observed as acute toxicity of ZnO NPs in rats (Landsiedel et al. 2014). The nanotoxicological response may vary profoundly between short-term and long-term effects. For instance, intratracheal instillation of 25 nm size TiO_2 NPs in Wistar rats at 0.66 mg/kg and 3.3 mg/kg dose has been found to elicit pulmonary inflammation, but the response varied considerably during the examination period from 3 days to 24 months post-exposure. A transient dose-dependent increase in inflammatory cells, heme oxygenase-1 activity, and elevation in the level of neutrophil chemoattractants in the lung tissue and BAL fluid were observed for 1-month post-exposure that slowly got normalized during the recovery period without producing any tumors, fibrosis, or irreversible lung disorder in the long run (Yoshiura et al. 2015). The other toxicological effects of intratracheal instillation or inhalation of TiO_2 NPs in rats can be summarized as enhancement of inflammatory parameters in BAL fluid, neutrophilic inflammation and macrophage activation, bronchoalveolar hyperplasia, epithelial and fibroproliferative changes, impaired clearance, inflammation-mediated mutagenesis, etc., often in a dose-dependent manner, while some changes may be transient acute response and a few are even irreversible (Landsiedel et al. 2014). Similarly, cerium dioxide (CeO_2), SiO_2 NPs, MWCNTs, etc. have also been found to induce similar type of oxidative stress, inflammatory changes and histopathological impressions in BAL fluid and lung parenchyma such as immune cell infiltration, fibroproliferative changes, focal necrosis, activation of caspase 3 pathways, production of inflammatory cytokines like IL-1α, IL-1β, TNF-α, IL-6, CINC1 (IL-8), MCP-1, M-CSF, OPN, etc. (Landsiedel et al. 2014).

29.2.2.5 Effects on Cardiovascular System

Various cardiotoxic effects have been reported by diverse NPs under in-vitro as well as in-vivo conditions. The cationic NPs including gold and polystyrene have been shown to cause hemolysis and blood clotting. CuO NPs are found to be highly toxic to the heart by inducing electrical, troponin T and contractile dysfunction, enhanced B-type natriuretic peptide, *N*-terminated pro-hormone BNP, and Troponin I production. Similarly, SiO_2 also induced cardiotoxicity but by another method through modulating the intracellular Ca^{2+} handling and pro-inflammatory cytokines secretion (Lu et al. 2021). ZnO NPs have also been found to increase the heart rate, troponin T, CRP, and secretion of IL-6, etc. The secretion of these stress inducers and cardiac degenerative markers was also enhanced by TiO_2 NPs. However, IO NPs exhibit toxicological effects by decreasing the heart rate and cell proliferation along with enhancement in vascular permeability. Carbon-based NPs have also increased the production of IL-6, IL-10, IL-12, MCP-1, and ACE as the toxicological effects (Bostan et al. 2016). Enhanced ROS production and redox imbalance have been reported to be the major cardiotoxic manifestation of mesoporous silica NPs in Wistar rats (Hozayen et al. 2019). PEG-coated AuNPs have been demonstrated to induce reversible cardiac hypertrophy in mice (Yang et al. 2016). Histological examinations of cardiac tissues of rats exposed to IONPs (5 mg/kg) and AgNPs (50 mg/kg) orally for 79 days have been found to have fragmentation of sarcoplasm and myocardial degeneration (Yousef et al. 2019b). Intraperitoneal administration of cadmium oxide NPs thrice a week for 6 weeks have been found to alter hemodynamic parameters and myocardial contractility in rats as the subchronic toxicological effects (Klinova et al. 2022). Cardiotoxicity of TiO_2 NPs and AgNPs has also been reported in planktonic crustacean *Daphnia magna* in terms of reducing heart rate and alteration in behavioral parameters (Park et al. 2022). So, diverse NPs have been found to induce cardiotoxicity in terms of molecular changes such as alteration in α-actinin, cardiac troponin or cytokine level, and/or phenotypic changes such as beat rates, Ca^{2+} flux, myocardial degeneration, etc. Thus, any novel nanoformulation must be checked for acute or chronic cardiotoxic implications to ensure nano-safety.

29.2.2.6 Dermatological Effects

Particles with a size of approximately 50–500 nm are widely used in cosmetic products. Because of their scattering properties, NPs increase the optical pathway of UV photons entering the upper part of the horny layer and provide enhanced protection from UV rays. However, several NPs have been reported to inflict dermatological adverse effects. In-vitro studies have shown that MWCNTs are capable of localizing within and initiating an irritation response in human epidermal keratinocytes, which is a primary route of occupational nano-exposure. TiO_2 NPs have generated toxic effects on human fibroblast skin cells in terms of producing cytotoxicity and oxidative damage, while collagen deformation and inflammation have been reported in mouse fibroblasts. Fe_3O_4 NPs have also been found to produce different toxic effects on human skin epithelial A431 cells such as elevated ROS production, lipid peroxidation and depleted glutathione level, up-regulated caspase 3 expression, etc. Similarly, ZnO NPs have also been found to produce oxidative stress, mitochondrial damage, and cytotoxicity in mouse epidermal skin cells. Platinum NPs inflicted DNA damage, inhibited replication, altered cell cycle through p53 mediated pathways and induced apoptosis in human foreskin fibroblasts (Wang et al. 2018b). The use of Ag NPs in enhancing several topical applications including X-ray films and cosmetics is of serious concern as it has toxic effects on several organs including the skin in the reducing order kidney>muscle>bone>skin>liver>heart>spleen (Korani et al. 2013). A comparative toxicity study of AgNPs on a 3D epidermal model and 2D keratinocytes monolayer culture model has revealed that marked oxidative damage and inflammation were inflicted in the 2D model with elevated production of ROS, MDA pro-inflammatory cytokines such as IL-1α, IL-6, and IL-8 along with reduced SOD activity, while no such toxic effects were observed in the 3D model. Improved barrier function and lesser penetration of NPs in the 3D model are probably the underlying cause for such differential toxic effects (DOI Chen et al. 2019). Elucidation of toxic effects of washed/unwashed and uncoated/carbon-coated AgNPs under in-vitro and in-vivo setup yielded differential results. Under in-vitro testing on human epidermal keratinocytes, unwashed AgNPs have been found to accumulate in cytoplasmic vacuoles and reduce the cell viability in a dose-dependent manner along with significant elevation in the level of pro-inflammatory cytokines such as IL-1β, IL-6, IL-8, and TNF-α, while washed or carbon-coated AgNPs are safer. Under in-vivo condition, macroscopically no gross irritation was observed on porcine skin but focal inflammation and localization of NPs in skin layers were found under microscopic examination (Samberg et al. 2010). So, it is also advocated that in-vivo studies or at least 3D culture, or the organoid model are necessary to get a realistic toxicity profile of diverse nanomaterials, choosing only in-vitro monolayer culture may yield elusive findings.

29.2.2.7 Toxicity on the Reproductive System

Intravenous administration of silicon NPs can cross the placental barrier and get distributed in different organs, particularly, if the size is about 500 nm or smaller (Refuerzo et al. 2011). Trans-placental translocation of 50 nm sized polystyrene NPs through passive diffusion has been reported in the in-vitro BeWo b30 placental model, while 20- and 50-nm Au NPs were found to enter fetal circulation by crossing the placental barrier by clathrin- and caveolin-mediated endocytosis in mice (Rattanapinyopituk et al. 2014; Kloet et al. 2015). PEG-functionalized or non-functionalized liposomes of 70 to 160 nm size loaded with various drugs or bioactive molecules have been found to penetrate the placental barrier efficiently in mice as well as in different placental models such as Villous placental explants, human placenta perfusion, BeWo cells, etc. for targeted delivery of the cargo (Shojaei et al. 2021). Similarly, AgNPs (2–35 nm), AuNPs (3–30 nm), TiO_2 NPs (6–25 nm), ZnO NPs (<100 nm), ZrO_2 NPs (54 nm), CeO_2 NPs (3–5 nm), Cu and CuO NPs (10–20 nm), Silica NPs (25–322 nm), IONPs (5–13 nm), different polymeric NPs (104–298 nm), dendrimers (5–7 nm), carbon-based NPs like SWCNTs and Fullerene, Quantum dots (3–20 nm), etc. in their coated/ functionalized form or unmodified form can pass through the placental barrier as evidenced in rat or mice or different in-vitro models (Shojaei et al. 2021). So, the probability of reproductive toxicity often looms when nanoformulation, particularly of smaller size, is used in pregnant subjects. In-vitro treatment of mouse Sertoli and ovarian follicle granulosa cells with AgNPs of 20 nm diameter resulted in increased cytotoxicity, autophagosome and autolysosome formation in Sertoli cells, and enhanced mitochondrial-mediated apoptosis. While under the in-vivo condition, the expression of pro-inflammatory cytokines such as TNF-α, IFN-γ, IL-6, IL-1β, and monocyte chemoattractant protein-1 was found to be markedly increased along with a significant loss of germ cells in male as well as female mice in histopathological examination (Han et al. 2016). Thus, it seems that exposure to such metallic NPs may have catastrophic consequences leading to the induction of apoptosis and/or acute inflammatory responses in the ovaries. Similarly, subcutaneous administration of TiO_2 NPs has been found to be aggregated in spermatids, Leydig, and Sertoli cells showing effects of disorganized and disrupted tubule lumens along with reduction in mature and overall seminal count in daily production and epididymal motility. An increase in testosterone, accumulation and testicular damage, a decrease in thickness of the seminiferous epithelium along with enhanced vacuolation were the toxic effects observed in various forms of carbon-based NPs. Similarly, AuNPs of 5 nm size were found to be internalized into the Leydig cells, resulting in formation of autophagosomes, enhanced ROS production, cell cycle disruption, cytotoxicity and DNA damage, reduced testosterone production, and malformation of sperm in BALB/c mice in a dose-dependent manner (Liu et al. 2020b). Unlike the aforesaid, intravenous and/or intra-abdominal injection of silicon NPs of 10–100 nm size (coated with either sodium chloride or sodium iodide) at a dose of 0–225 mg/kg of body weight in mice resulted in NP accumulation in the testicles, among other organs, including in the glands and interstitial cells, 96 h after administration without producing any pronounced toxic effect (Chen et al. 2003). So, a plethora of parameters such as nanomaterial properties, experimental conditions, subject species, dose, route, etc. need to be carefully considered to come up with a concrete and reproducible pattern of toxicological consequences of certain types of NPs.

29.2.2.8 Effects on Bone and Bone Marrow Cells

The skeletal system is responsible for attaining the morphology of an organism, providing the capability of movement and strength. In addition, it is the organ system responsible for the generation of connective tissue of blood through its marrow. This is one of the most sensitive parts where nanotoxicty has been studied in detail.

Various groups of NPs have shown toxicity to the bone or its marrow. For instance, intravenous injection of doxorubicin-bound polyisohexylcyanoacrylate (doxo-PIHCA) NPs in mice at the dose of 11 mg/kg body weight produced bone marrow toxicity, induced the release of colony-stimulating factors, and greater cytopenia. Further, besides strong myelosuppressive effects and reduced bone marrow granulocyte progenitors (CFU-GM), the nanoformulation also led to a faster decrease in spleen WBC count that was evident post 2-days of treatment (Gibaud et al. 1994). Furthermore, AgNPs showed concentration-dependent toxic effects on bone marrow-derived dendritic cells (BMDC) reducing the cell viability just after 12 h post-exposure at a concentration of 1.0 μg AgNP/ml. The BMDC are important constituents of the immunological function as they are important antigen-presenting cells. However, no significant damage to the metabolic status or cell membrane integrity, genotoxic effects, and morphological disorientation was observed at a lower dose (Castro-Gamboa et al. 2019). Considering the ample applications of Al_2O_3 NPs in coatings, catalysts, drug delivery, medical imaging and optoelectronic devices, the toxic effects of Al_2O_3 NPs have also been observed in bone composition as these NPs at higher doses @ 250, 500, and 1000 μg/mL significantly reduced the density of bone probably due to reduction in trabeculae and empty lacunae formation (Soltaninejad et al. 2020). Intravenous administration of Tween® 20-conjugated gold NPs in mice for 21 days has been depicted to accrual of gold particles in the hematopoietic bone tissue producing toxic effects such as leucopoiesis and megakaryopoiesis (Berce et al. 2016). Similarly, IONPs, a magnetic NP have recently gained considerable attention in various applications including orthopedic implants and devices but have also been found to have detoriating effects on bone sometimes. Concentrations up to 1% were not found to be toxic to bone marrow cells but their internalization within the cell was significantly increased (Paik et al. 2015). But intravenous administration of IONPs in rats at the dose rate of 7.5, 15, and 30 mg /kg body weight once in a week for 4 weeks significantly diminished bone marrow-mononuclear cell proliferation, elevated oxidative stress, and DNA damage that led to growth arrest and apoptosis, even in undifferentiated hematopoietic stem cells at higher doses. Elevation in lipid peroxidation, alteration in antioxidant profile, upregulation in pro-inflammatory cytokine secretion induced by splenic M1 macrophages, and alteration in immunological response were also found to be the major toxicological consequences of the IONPs (Gaharwar et al. 2020).

29.2.2.9 Developmental Toxicity

Nanotoxicity can hamper the developmental process in two major ways: one by producing genotoxicity in germ cell lines and second by producing development-related issues through transplacental crossing and imparting toxicological implications in a developing fetus. Reproduction is one of the prime aims of every organism and any NPs showing such detrimental effects must be handled with utmost care as it not only hampers the reproduction of food animals but also can lead to the population decline of various wild species just like the population decline of wild fultures from visceral gout caused by diclofenac toxicity (Becker 2016). Several NPs have shown such toxic effects in different forms and characters. For instance, CNTs directly interfere in various cell division cycles; silica NPs of 70 nm size have increased the infiltration of inflammatory cells at the placenta; SWCNTs can impart occasional teratogenic effects at higher doses; MWCNTs have been reported to induce dose-dependent fetal abnormalities, increase abortion rate, and cause placental vascular alterations; cadmium oxide NPs have a negative effect on fetal development and growth; TiO_2 NPs have been found to reduce sperm count; Carbon black has been found to have a deteriorating effect on daily sperm production, seminiferous tubules, and serum testosterone. Further, placental dysfunctions, increased fetal resorption, and reduced fetal size are the toxic effects of SiO_2 NPs observed in mice. In rats, ZnO NPs having a size of <100 nm showed a reduced number of born/live pups, decreased body weights of pups, and increased fetal resorption. (Hougaard et al. 2015). Similarly, CdSe quantum dots have been demonstrated to produce cytotoxic effects on the maturation of mice oocytes, fertilization, and fetal development. Maternal exposure to some NPs has multiple organ-specific toxic effects on the fetus such as AgNPs translocate the fetus, causing resorption, and growth retardation along with various histological and functional abnormalities to different fetal organs that include the central nervous system (CNS), the reproductive system, the immune system, etc. (Zhang et al. 2021). IONPs of any charge in lower doses are non-toxic but at higher doses, negatively charged poly(acrylic acid)-Fe_2O_3-NPs (PAA-NPs) induced fetal loss of 22% while positively charged polyethyleneimine-Fe_2O_3-NPs (PEI-NPs) resulted in the fetal loss of up to 42% (Di Bona et al. 2015). Maternal exposure to NPs can take place through various routes; the time and dose may vary, the type of NPs and their properties as well as the subject species are also variable factors that regulate the developmental nanotoxicity, which is a serious concern and needs to be introspected diligently before lab-to-market transition of any nanoformulation.

29.3 Conclusions

Proper application of any new invention can help society prosper while misuse of the same can lead to disaster. For instance, there was a time when the advent of antibiotics saved millions of lives while currently, the indiscriminate use of antibiotics has resulted in scary effects on human and animal populations by promoting the evolution of numerous antibiotic-resistant microbial strains throughout the world. Similarly, nanotechnology has arrived as a boon to society with multi-arrayed utilities in diverse scientific arenas. The nanomedicine market is growing steadily, propelled by enormous demand and a plethora of novel nanoformulations are hitting the market on a regular basis despite the stringent regulatory issues with respect to nano-intervention in biomedical applications. While most of the stakeholders are currently focused on the development of novel engineered NPs with enhanced utility, we are certainly in a critical temporal juncture where the aspect of nanotoxicity needs diligent introspection. Otherwise, similar to antibiotics, nanotechnology will meet the fate of irreversible damage to the health arena and the advantages of nanotechnology may be negated by the limitation of nanotoxicity. No form of life and no component of life has been left unaffected by the toxicological consequences of different NPs. Hence, a comprehensive and meticulous nanotoxicity analysis, particularly on a short as well as long-term basis on different models and species, study regarding bioaccumulation of NPs, and disposal of nanowastes are essentially required before giving a no-objection to any nanoformulation. We should not be lured by the potential benefits of nano-intervention to discount the regulatory stringency; the load of NPs on earth and their toxicological consequences on the global biome should be considered as the top priority. We are already perceiving a faster rate of extinction of species in recent times due to global warming and environmental insult, hence, perseverance may not allow any further nanotoxicological catastrophe. Nanotoxicity can be inflicted slowly but can be extremely lethal once the NPs reach a critical concentration in the earth's ecosystem. Furthermore, there is so much diversity in NPs that studying all forms of toxicological effects on vast flora and fauna in a comprehensive way is a daunting task. Therefore, only certain biodegradable and biocompatible non-toxic NPs may be allowed to be commercialized post long and short-term toxicity studies in suitable subjects with uniform and reproducible results. There is an urgent need to constitute a consortium to study the toxicological effects of different NPs in order to pave its safe and proper utilization. Further, a central databank for nanomaterial study should be created and applications of artificial intelligence (AI) must be explored to develop some formula or predictive models for enumerating the toxicological effects of present and emerging nanomaterials.

References

Abu Gazia M, El-Magd MA (2019) Effect of pristine and functionalized multiwalled carbon nanotubes on rat renal cortex. Acta Histochem 121(2):207–217. https://doi.org/10.1016/j.acthis.2018.12.005

Ahamed M, Siddiqui MA, Akhtar MJ, Ahmad I, Pant AB, Alhadlaq HA (2010) Genotoxic potential of copper oxide nanoparticles in human lung epithelial cells. Biochem Biophys Res Commun 396(2):578–583. https://doi.org/10.1016/j.bbrc.2010.04.156

Ahamed M, Akhtar MJ, Raja M, Ahmad I, Siddiqui MK, AlSalhi MS, Alrokayan SA (2011) ZnO nanorod-induced apoptosis in human alveolar adenocarcinoma cells via p53, survivin and bax/bcl-2 pathways: role of oxidative stress. Nanomedicine 7(6):904–913. https://doi.org/10.1016/j.nano.2011.04.011

Ahmed B, Ameen F, Rizvi A, Ali K, Sonbol H, Zaidi A, Khan MS, Musarrat J (2020) Destruction of cell topography, morphology, membrane, inhibition of respiration, biofilm formation, and bioactive molecule production by nanoparticles of Ag, ZnO, CuO, TiO2, and Al2O3 toward beneficial soil bacteria. ACS Omega 5(14):7861–7876. https://doi.org/10.1021/acsomega.9b04084

Alarifi S, Ali D (2015) Mechanisms of multi-walled carbon nanotubes-induced oxidative stress and genotoxicity in mouse fibroblast cells. Int J Toxicol 34(3):258–265. https://doi.org/10.1177/1091581815584799

Alarifi S, Ali D, Alkahtani S, Verma A, Ahamed M, Ahmed M, Alhadlaq HA (2013) Induction of oxidative stress, DNA damage, and apoptosis in a malignant human skin melanoma cell line after exposure to zinc oxide nanoparticles. Int J Nanomedicine 8:983–993. https://doi.org/10.2147/IJN.S42028

Ansari MO, Parveen N, Ahmad MF, Wani AL, Afrin S, Rahman Y, Jameel S, Khan YA, Siddique HR, Tabish M, Shadab GGHA (2019) Evaluation of DNA interaction, genotoxicity and oxidative stress induced by iron oxide nanoparticles both in vitro and in vivo: attenuation by thymoquinone. Sci Rep 9(1):6912. https://doi.org/10.1038/s41598-019-43188-5

Apopa PL, Qian Y, Shao R, Guo NL, Schwegler-Berry D, Pacurari M, Porter D, Shi X, Vallyathan V, Castranova V, Flynn DC (2009) Iron oxide nanoparticles induce human microvascular endothelial cell permeability through reactive oxygen species production and microtubule remodeling. Part Fibre Toxicol 6:1. https://doi.org/10.1186/1743-8977-6-1

Bagherzadeh Homaee M, Ehsanpour AA (2016) Silver nanoparticles and silver ions: oxidative stress responses and toxicity in potato (Solanum tuberosum L) grown in vitro. Hortic Environ Biotechnol 57:544–553. https://doi.org/10.1007/s13580-016-0083-z

Barcińska E, Wierzbicka J, Zauszkiewicz-Pawlak A, Jacewicz D, Dabrowska A, Inkielewicz-Stepniak I (2018) Role of oxidative and nitro-oxidative damage in silver nanoparticles cytotoxic effect against human pancreatic ductal adenocarcinoma cells. Oxidative Med Cell Longev 2018: 8251961. https://doi.org/10.1155/2018/8251961

Bartneck M, Ritz T, Keul HA, Wambach M, Bornemann J, Gbureck U, Ehling J, Lammers T, Heymann F, Gassler N, Lüdde T, Trautwein C, Groll J, Tacke F (2012) Peptide-functionalized gold nanorods increase liver injury in hepatitis. ACS Nano 6(10):8767–8777. https://doi.org/10.1021/nn302502u

Bartucci R, van der Meer AZ, Boersma YL, Olinga P, Salvati A (2021) Nanoparticle-induced inflammation and fibrosis in ex vivo murine precision-cut liver slices and effects of nanoparticle exposure conditions. Arch Toxicol 95(4):1267–1285. https://doi.org/10.1007/s00204-021-02992-7

Becker R (2016) Cattle drug threatens thousands of vultures. Nature 29:1038. https://doi.org/10.1038/nature.2016.19839

Berce C, Lucan C, Petrushev B, Boca S, Miclean M, Sarpataki O, Astilean S, Buzoianu A, Tomuleasa C, Bojan A (2016) In vivo assessment of bone marrow toxicity by gold nanoparticle-based bioconjugates in Crl:CD1(ICR) mice. Int J Nanomedicine 11:4261–4273. https://doi.org/10.2147/IJN.S108711

Bhattacharya K, Davoren M, Boertz J, Schins RP, Hoffmann E, Dopp E (2009) Titanium dioxide nanoparticles induce oxidative stress and DNA-adduct formation but not DNA-breakage in human lung cells. Part Fibre Toxicol 6:17. https://doi.org/10.1186/1743-8977-6-17

Bigdelou P, Vahedi A, Kiosidou E, Farnoud AM (2020) Loss of membrane asymmetry alters the interactions of erythrocytes with engineered silica nanoparticles. Biointerphases 15(4):041001. https://doi.org/10.1116/6.0000246

Blanco J, Tomás-Hernández S, García T, Mulero M, Gómez M, Domingo JL, Sánchez DJ (2018) Oral exposure to silver nanoparticles increases oxidative stress markers in the liver of male rats and deregulates the insulin signalling pathway and p53 and cleaved caspase 3 protein expression. Food Chem Toxicol 115:398–404. https://doi.org/10.1016/j.fct.2018.03.039

Bostan HB, Rezaee R, Valokala MG, Tsarouhas K, Golokhvast K, Tsatsakis AM, Karimi G (2016) Cardiotoxicity of nano-particles. Life Sci 165:91–99. https://doi.org/10.1016/j.lfs.2016.09.017

Brown DM, Donaldson K, Borm PJ, Schins RP, Dehnhardt M, Gilmour P, Jimenez LA, Stone V (2004) Calcium and ROS-mediated activation of transcription factors and TNF-alpha cytokine gene expression in macrophages exposed to ultrafine particles. Am J Physiol Lung Cell Mol Physiol 286(2):L344–L353. https://doi.org/10.1152/ajplung.00139.2003

Cao Y, Long J, Liu L, He T, Jiang L, Zhao C, Li Z (2017a) A review of endoplasmic reticulum (ER) stress and nanoparticle (NP) exposure. Life Sci 186:33–42. https://doi.org/10.1016/j.lfs.2017.08.003

Cao Z, Peng F, Hu Z, Chu B, Zhong Y, Su Y, He S, He Y (2017b) In vitro cellular behaviors and toxicity assays of small-sized fluorescent silicon nanoparticles. Nanoscale 9(22):7602–7611. https://doi.org/10.1039/c7nr00530j

Carriere M, Sauvaigo S, Douki T, Ravanat JL (2017) Impact of nanoparticles on DNA repair processes: current knowledge and working hypotheses. Mutagenesis 32(1):203–213. https://doi.org/10.1093/mutage/gew052

Castro-Gamboa S, Garcia-Garcia MR, Piñon-Zarate G, Rojas-Lemus M, Jarquin-Yañez K, Angel Herrera-Enriquez M, Fortoul TI, Toledano-Magaña Y, Garcia-Iglesias T, Pestryakov A, Eliu Castell-Rodriguez A, Bogdanchikova N (2019) Toxicity of silver nanoparticles in mouse bone marrow-derived dendritic cells: implications for phenotype. J Immunotoxicol 16(1):54–62. https://doi.org/10.1080/1547691X.2019.1584652

Chen Y, Xue Z, Zheng D, Xia K, Zhao Y, Liu T, Long Z, Xia J (2003) Sodium chloride modified silica nanoparticles as a non-viral vector with a high efficiency of DNA transfer into cells. Curr Gene Ther 3(3):273–279. https://doi.org/10.2174/1566523034578339

Chen X, Li J, Huang Y, Wei J, Sun D, Zheng N (2017) The biodistribution, excretion and potential toxicity of different-sized Pd nanosheets in mice following oral and intraperitoneal administration. Biomater Sci 5(12):2448–2455. https://doi.org/10.1039/c7bm00769h

Chen L, Wu M, Jiang S, Zhang Y, Li R, Lu Y, Liu L, Wu G, Liu Y, Xie L, Xu L (2019) Skin toxicity assessment of silver nanoparticles in a 3D epidermal model compared to 2D keratinocytes. Int J Nanomedicine 14:9707–9719. https://doi.org/10.2147/IJN.S225451

Cho WS, Duffin R, Howie SE, Scotton CJ, Wallace WA, Macnee W, Bradley M, Megson IL, Donaldson K (2011) Progressive severe lung injury by zinc oxide nanoparticles; the role of Zn2+ dissolution inside lysosomes. Part Fibre Toxicol 8:27. https://doi.org/10.1186/1743-8977-8-27

Choudhury SR, Ordaz J, Lo CL, Damayanti NP, Zhou F, Irudayaraj J (2017) From the cover: zinc oxide nanoparticles-induced reactive oxygen species promotes multimodal cyto- and epigenetic toxicity. Toxicol Sci 156(1):261–274. https://doi.org/10.1093/toxsci/kfw252

De Berardis B, Civitelli G, Condello M, Lista P, Pozzi R, Arancia G, Meschini S (2010) Exposure to ZnO nanoparticles induces oxidative stress and cytotoxicity in human colon carcinoma cells. Toxicol Appl Pharmacol 246(3):116–127. https://doi.org/10.1016/j.taap.2010.04.012

De Jong WH, Hagens WI, Krystek P, Burger MC, Sips AJ, Geertsma RE (2008) Particle size-dependent organ distribution of gold nanoparticles after intravenous administration. Biomaterials 29(12):1912–1919. https://doi.org/10.1016/j.biomaterials.2007.12.037

Di Bona KR, Xu Y, Gray M, Fair D, Hayles H, Milad L, Montes A, Sherwood J, Bao Y, Rasco JF (2015) Short- and long-term effects of prenatal exposure to iron oxide nanoparticles: influence of surface charge and dose on developmental and reproductive toxicity. Int J Mol Sci 16(12): 30251–30268. https://doi.org/10.3390/ijms161226231

Dobbins RA (2007) Hydrocarbon nanoparticles formed in flames and diesel engines. Aerosol Sci Technol 41(5):485–496. https://doi.org/10.1080/02786820701225820

Docea AO, Calina D, Buga AM, Zlatian O, Paoliello MMB, Mogosanu GD, Streba CT, Popescu EL, Stoica AE, Bîrcă AC, Vasile BŞ, Grumezescu AM, Mogoanta L (2020) The effect of silver nanoparticles on antioxidant/pro-oxidant balance in a murine model. Int J Mol Sci 21(4):1233. https://doi.org/10.3390/ijms21041233

Dong J (2020a) Microenvironmental alterations in carbon nanotube-induced lung inflammation and fibrosis. Front Cell Dev Biol 8:126. https://doi.org/10.3389/fcell.2020.00126

Dong J (2020b) Signaling pathways implicated in carbon nanotube-induced lung inflammation. Front Immunol 11:552613. https://doi.org/10.3389/fimmu.2020.552613

Dong J, Ma Q (2016) Myofibroblasts and lung fibrosis induced by carbon nanotube exposure. Part Fibre Toxicol 13(1):60. https://doi.org/10.1186/s12989-016-0172-2

Dong J, Ma Q (2018) Type 2 immune mechanisms in carbon nanotube-induced lung fibrosis. Front Immunol 9:1120. https://doi.org/10.3389/fimmu.2018.01120

Dong J, Ma Q (2019) Integration of inflammation, fibrosis, and cancer induced by carbon nanotubes. Nanotoxicology 13(9):1244–1274. https://doi.org/10.1080/17435390.2019.1651920

Dubey SK, Parab S, Amit A, Agrawal M, Achalla VPK, Pal UN, Pandey MM, Kesharwani P (2022) Cold atmospheric plasma therapy in wound healing. Process Biochem 112:112–123. https://doi.org/10.1016/j.procbio.2021.11.017

Egbuna C, Parmar VK, Jeevanandam J, Ezzat SM, Patrick-Iwuanyanwu KC, Adetunji CO, Khan J, Onyeike EN, Uche CZ, Akram M, Ibrahim MS, El Mahdy NM, Awuchi CG, Saravanan K, Tijjani H, Odoh UE, Messaoudi M, Ifemeje JC, Olisah MC, Ezeofor NJ, Chikwendu CJ, Ibeabuchi CG (2021) Toxicity of nanoparticles in biomedical application: nanotoxicology. J Toxicol. 2021:9954443. https://doi.org/10.1155/2021/9954443

Enea M, Pereira E, de Almeida MP, Araújo AM, de Lourdes BM, Carmo H (2020) Gold nanoparticles induce oxidative stress and apoptosis in human kidney cells. Nanomaterials (Basel) 10(5):995. https://doi.org/10.3390/nano10050995

Fahmy B, Cormier SA (2009) Copper oxide nanoparticles induce oxidative stress and cytotoxicity in airway epithelial cells. Toxicol In Vitro 23(7):1365–1371. https://doi.org/10.1016/j.tiv.2009.08.005

Farré M, Gajda-Schrantz K, Kantiani L, Barceló D (2009) Ecotoxicity and analysis of nanomaterials in the aquatic environment. Anal Bioanal Chem 393(1):81–95. https://doi.org/10.1007/s00216-008-2458-1

Foroozandeh P, Aziz AA (2018) Insight into cellular uptake and intracellular trafficking of nanoparticles. Nanoscale Res Lett 13(1):339. https://doi.org/10.1186/s11671-018-2728-6

Fu PP, Xia Q, Hwang HM, Ray PC, Yu H (2014) Mechanisms of nanotoxicity: generation of reactive oxygen species. J Food Drug Anal 22(1):64–75. https://doi.org/10.1016/j.jfda.2014.01.005

Gaharwar US, Kumar S, Rajamani P (2020) Iron oxide nanoparticle-induced hematopoietic and immunological response in rats. RSC Adv 10(59):35753–35764. https://doi.org/10.1039/d0ra05901c

Garduño-Balderas LG, Urrutia-Ortega IM, Medina-Reyes EI, Chirino YI (2015) Difficulties in establishing regulations for engineered nanomaterials and considerations for policy makers: avoiding an unbalance between benefits and risks. J Appl Toxicol 35(10):1073–1085. https://doi.org/10.1002/jat.3180

Gatoo MA, Naseem S, Arfat MY, Dar AM, Qasim K, Zubair S (2014) Physicochemical properties of nanomaterials: implication in associated toxic manifestations. Biomed Res Int 2014:498420. https://doi.org/10.1155/2014/498420

Ghosh C, Nandi A, Basu S (2019) Lipid nanoparticle-mediated induction of endoplasmic reticulum stress in cancer cells. ACS Appl Bio Mater 2(9):3992–4001. https://doi.org/10.1021/acsabm.9b00532

Gibaud S, Andreux JP, Weingarten C, Renard M, Couvreur P (1994) Increased bone marrow toxicity of doxorubicin bound to nanoparticles. Eur J Cancer 30A(6):820–826. https://doi.org/10.1016/0959-8049(94)90299-2

González-Durruthy M, Concu R, Ruso JM, Cordeiro MNDS (2021) New mechanistic insights on carbon nanotubes' nanotoxicity using isolated submitochondrial particles, molecular docking, and nano-QSTR approaches. Biology (Basel) 10(3):171. https://doi.org/10.3390/biology10030171

Gottschalk F, Sonderer T, Scholz RW, Nowack B (2009) Modeled environmental concentrations of engineered nanomaterials (TiO(2), ZnO, Ag, CNT, Fullerenes) for different regions. Environ Sci Technol 43(24):9216–9222. https://doi.org/10.1021/es9015553

Grady ME, Parrish E, Caporizzo MA, Seeger SC, Composto RJ, Eckmann DM (2017) Intracellular nanoparticle dynamics affected by cytoskeletal integrity. Soft Matter 13(9):1873–1880. https://doi.org/10.1039/c6sm02464e

Guo D, Bi H, Liu B, Wu Q, Wang D, Cui Y (2013) Reactive oxygen species-induced cytotoxic effects of zinc oxide nanoparticles in rat retinal ganglion cells. Toxicol In Vitro 27(2):731–738. https://doi.org/10.1016/j.tiv.2012.12.001

Halamoda Kenzaoui B, Chapuis Bernasconi C, Guney-Ayra S, Juillerat-Jeanneret L (2012) Induction of oxidative stress, lysosome activation and autophagy by nanoparticles in human brain-derived endothelial cells. Biochem J 441(3):813–821. https://doi.org/10.1042/BJ20111252

Han JW, Jeong JK, Gurunathan S, Choi YJ, Das J, Kwon DN, Cho SG, Park C, Seo HG, Park JK, Kim JH (2016) Male- and female-derived somatic and germ cell-specific toxicity of silver nanoparticles in mouse. Nanotoxicology 10(3):361–373. https://doi.org/10.3109/17435390.2015.1073396

He H, Zou Z, Wang B, Xu G, Chen C, Qin X, Yu C, Zhang J (2020) Copper oxide nanoparticles induce oxidative DNA damage and cell death via copper ion-mediated P38 MAPK activation in vascular endothelial cells. Int J Nanomedicine 15:3291–3302. https://doi.org/10.2147/IJN.S241157

Hougaard KS, Campagnolo L, Chavatte-Palmer P, Tarrade A, Rousseau-Ralliard D, Valentino S, Park MVDZ, de Jong WH, Wolterink G, Piersma AH, Ross BL, Hutchison GR, Hansen JS, Vogel UB, Jackson P, Slama R, Pietroiusti A, Cassee FR (2015) A perspective on the developmental toxicity of inhaled nanoparticles. Reprod Toxicol 56(SI):118–140. https://doi.org/10.1016/j.reprotox.2015.05.015

Hozayen WG, Mahmoud AM, Desouky EM, El-Nahass ES, Soliman HA, Farghali AA (2019) Cardiac and pulmonary toxicity of mesoporous silica nanoparticles is associated with excessive ROS production and redox imbalance in Wistar rats. Biomed Pharmacother 109:2527–2538. https://doi.org/10.1016/j.biopha.2018.11.093

Huo L, Chen R, Zhao L, Shi X, Bai R, Long D, Chen F, Zhao Y, Chang YZ, Chen C (2015) Silver nanoparticles activate endoplasmic reticulum stress signaling pathway in cell and mouse models: The role in toxicity evaluation. Biomaterials 61:307–315. https://doi.org/10.1016/j.biomaterials.2015.05.029

Iavicoli I, Fontana L, Nordberg G (2016) The effects of nanoparticles on the renal system. Crit Rev Toxicol 46(6):490–560. https://doi.org/10.1080/10408444.2016.1181047

Jawaid P, Rehman MU, Zhao QL, Misawa M, Ishikawa K, Hori M, Shimizu T, Saitoh JI, Noguchi K, Kondo T (2020) Small size gold nanoparticles enhance apoptosis-induced by cold atmospheric plasma via depletion of intracellular GSH and modification of oxidative stress. Cell Death Discov 6:83. https://doi.org/10.1038/s41420-020-00314-x

Ji Z, Guo W, Sakkiah S, Liu J, Patterson TA, Hong H (2021) Nanomaterial databases: data sources for promoting design and risk assessment of nanomaterials. Nano 11:1599. https://doi.org/10.3390/nano11061599

Jia J, Li F, Zhou H, Bai Y, Liu S, Jiang Y, Jiang G, Yan B (2017) Oral exposure to silver nanoparticles or silver ions may aggravate fatty liver disease in overweight mice. Environ Sci Technol 51:9334–9343. https://doi.org/10.1021/acs.est.7b02752

Kang Y, Liu J, Song B, Feng X, Ou L, Wei L, Lai X, Shao L (2016) Potential links between cytoskeletal disturbances and electroneurophysiological dysfunctions induced in the central nervous system by inorganic nanoparticles. Cell Physiol Biochem 40(6):1487–1505. https://doi.org/10.1159/000453200

Karlsson HL, Gustafsson J, Cronholm P, Möller L (2009) Size-dependent toxicity of metal oxide particles—a comparison between nano- and micrometer size. Toxicol Lett 188(2):112–118. https://doi.org/10.1016/j.toxlet.2009.03.014

Karlsson HL, Cronholm P, Hedberg Y, Tornberg M, De Battice L, Svedhem S, Wallinder IO (2013) Cell membrane damage and protein interaction induced by copper containing nanoparticles—importance of the metal release process. Toxicology 313(1):59–69. https://doi.org/10.1016/j.tox.2013.07.012

Kasper J, Hermanns MI, Bantz C, Koshkina O, Lang T, Maskos M, Pohl C, Unger RE, Kirkpatrick CJ (2013) Interactions of silica nanoparticles with lung epithelial cells and the association to flotillins. Arch Toxicol 87(6):1053–1065. https://doi.org/10.1007/s00204-012-0876-5

Keenan CR, Goth-Goldstein R, Lucas D, Sedlak DL (2009) Oxidative stress induced by zero-valent iron nanoparticles and Fe(II) in human bronchial epithelial cells. Environ Sci Technol 43(12): 4555–4560. https://doi.org/10.1021/es9006383

Kermanizadeh A, Powell LG, Stone V (2020) A review of hepatic nanotoxicology—summation of recent findings and considerations for the next generation of study designs. J Toxicol Environ Health B Crit Rev 23(4):137–176. https://doi.org/10.1080/10937404.2020.1751756

Klinova SV, Protsenko YL, Minigalieva IA, Sutunkova MP, Gerzen OP, Nabiev SR, Balakin AA, Lisin RV, Lookin ON, Panov VG, Katsnelson BA, Nikitina LV (2022) Cardiotoxicity in rats following subchronic exposure to cadmium chloride solution and cadmium oxide nanoparticles. Public Health Toxicol 2(2):11. https://doi.org/10.18332/pht/150787

Kloet SK, Walczak AP, Louisse J, van den Berg HH, Bouwmeester H, Tromp P, Fokkink RG, Rietjens IM (2015) Translocation of positively and negatively charged polystyrene nanoparticles in an in vitro placental model. Toxicol In Vitro 29(7):1701–1710. https://doi.org/10.1016/j.tiv.2015.07.003

Korani M, Rezayat SM, Arbabi BS (2013) Sub-chronic dermal toxicity of silver nanoparticles in guinea pig: special emphasis to heart, bone and kidney toxicities. Iran J Pharm Res 12(3): 511–519

Kowsalya E, MosaChristas K, Jaquline CRI, Balashanmugam P, Devasena T (2021) Gold nanoparticles induced apoptosis via oxidative stress and mitochondrial dysfunctions in MCF-7 breast cancer cells. Appl Organomet Chem 35:e6071. https://doi.org/10.1002/aoc.6071

Lajous H, Riva R, Lelièvre B, Tétaud C, Avril S, Hindré F, Boury F, Jérôme C, Lecomte P, Garcion E (2018) Hybrid Gd3+/cisplatin cross-linked polymer nanoparticles enhance platinum accumulation and formation of DNA adducts in glioblastoma cell lines. Biomater Sci 6(9):2386–2409. https://doi.org/10.1039/c8bm00346g

Landsiedel R, Sauer UG, Ma-Hock L, Schnekenburger J, Wiemann M (2014) Pulmonary toxicity of nanomaterials: a critical comparison of published in vitro assays and in vivo inhalation or instillation studies. Nanomedicine (Lond) 9(16):2557–2585. https://doi.org/10.2217/nnm.14.149

Larsen ST, Jackson P, Poulsen SS, Levin M, Jensen KA, Wallin H, Nielsen GD, Koponen IK (2016) Airway irritation, inflammation, and toxicity in mice following inhalation of metal oxide nanoparticles. Nanotoxicology 10(9):1254–1262. https://doi.org/10.1080/17435390.2016.1202350

Lei R, Wu C, Yang B, Ma H, Shi C, Wang Q, Wang Q, Yuan Y, Liao M (2008) Integrated metabolomic analysis of the nano-sized copper particle-induced hepatotoxicity and nephrotoxicity in rats: a rapid in vivo screening method for nanotoxicity. Toxicol Appl Pharmacol 232(2): 292–301. https://doi.org/10.1016/j.taap.2008.06.026

Li JJ, Muralikrishnan S, Ng CT, Yung LY, Bay BH (2010) Nanoparticle-induced pulmonary toxicity. Exp Biol Med (Maywood) 235(9):1025–1033. https://doi.org/10.1258/ebm.2010.010021

Li H, Zhang P, Luo J, Hu D, Huang Y, Zhang ZR, Fu Y, Gong T (2019) Chondroitin sulfate-linked prodrug nanoparticles target the golgi apparatus for cancer metastasis treatment. ACS Nano 13(8):9386–9396. https://doi.org/10.1021/acsnano.9b04166

Lima H Jr, Jacobson LS, Goldberg MF, Chandran K, Diaz-Griffero F, Lisanti MP, Brojatsch J (2013) Role of lysosome rupture in controlling Nlrp3 signaling and necrotic cell death. Cell Cycle 12(12):1868–1878. https://doi.org/10.4161/cc.24903

Liu W, Wu Y, Wang C, Li HC, Wang T, Liao CY, Cui L, Zhou QF, Yan B, Jiang GB (2010) Impact of silver nanoparticles on human cells: effect of particle size. Nanotoxicology 4(3):319–330. https://doi.org/10.3109/17435390.2010.483745

Liu N, Mu Y, Chen Y, Sun H, Han S, Wang M, Wang H, Li Y, Xu Q, Huang P, Sun Z (2013) Degradation of aqueous synthesized CdTe/ZnS quantum dots in mice: differential blood kinetics and biodistribution of cadmium and tellurium. Part Fibre Toxicol 10:37. https://doi.org/10.1186/1743-8977-10-37

Liu W, Worms I, Slaveykova VI (2020a) Interaction of silver nanoparticles with antioxidant enzymes. Environ Sci Nano 7:1507–1517. https://doi.org/10.1039/C9EN01284B

Liu Y, Li X, Xiao S, Liu X, Chen X, Xia Q, Lei S, Li H, Zhong Z, Xiao K (2020b) The effects of gold nanoparticles on leydig cells and male reproductive function in mice. Int J Nanomedicine 15:9499–9514. https://doi.org/10.2147/IJN.S276606

Lu RXZ, Lai BFL, Benge T, Wang EY, Davenport L, Rafatian N, Radisic M (2021) Heart-on-a-chip platform for assessing toxicity of air pollution related nanoparticles. Adv Mater Technol 6: 2000726. https://doi.org/10.1002/admt.202000726

Ma X, Sun J, Zhong L, Wang Y, Huang Q, Liu X, Jin S, Zhang J, Liang XJ (2019) Evaluation of turning-sized gold nanoparticles on cellular adhesion by golgi disruption in vitro and in vivo. Nano Lett 19(12):8476–8487. https://doi.org/10.1021/acs.nanolett.9b02826

Magaye RR, Yue X, Zou B, Shi H, Yu H, Liu K, Lin X, Xu J, Yang C, Wu A, Zhao J (2014) Acute toxicity of nickel nanoparticles in rats after intravenous injection. Int J Nanomedicine 9:1393–1402. https://doi.org/10.2147/IJN.S56212

Manke A, Wang L, Rojanasakul Y (2013) Mechanisms of nanoparticle-induced oxidative stress and toxicity. Biomed Res Int 2013:942916. https://doi.org/10.1155/2013/942916

Manshian BB, Pokhrel S, Mädler L, Soenen SJ (2018) The impact of nanoparticle-driven lysosomal alkalinization on cellular functionality. J Nanobiotechnology. 16(1):85. https://doi.org/10.1186/s12951-018-0413-7

Manuja A, Kumar B, Kumar R, Chhabra D, Ghosh M, Manuja M, Brar B, Pal Y, Tripathi BN, Prasad M (2021) Metal/metal oxide nanoparticles: toxicity concerns associated with their physical state and remediation for biomedical applications. Toxicol Rep 8:1970–1978. https://doi.org/10.1016/j.toxrep.2021.11.020

Mao BH, Chen ZY, Wang YJ, Yan SJ (2018) Silver nanoparticles have lethal and sublethal adverse effects on development and longevity by inducing ROS-mediated stress responses. Sci Rep 8(1):2445. https://doi.org/10.1038/s41598-018-20728-z

Minakshi P, Ghosh M, Brar B, Kumar R, Lambe UP, Ranjan K, Manoj J, Prasad G (2019) Nano-antimicrobials: a new paradigm for combating mycobacterial resistance. Curr Pharm Des 25(13):1554–1579. https://doi.org/10.2174/1381612825666190620094041

Minakshi P, Kumar R, Ghosh M, Brar B, Barnela M, Lakhani P (2020a) Application of polymeric nano-materials in management of inflammatory bowel disease. Curr Top Med Chem 20(11): 982–1008. https://doi.org/10.2174/1568026620666200320113322

Minakshi P, Ghosh M, Kumar R, Brar B, Lambe UP, Banerjee S, Ranjan K, Kumar B, Goel P, Malik YS, Prasad G (2020b) An insight into nanomedicinal approaches to combat viral zoonoses. Curr Top Med Chem 20(11):915–962. https://doi.org/10.2174/1568026620666200325114400

Mocan L, Matea C, Tabaran FA, Mosteanu O, Pop T, Mocan T, Iancu C (2015) Photothermal treatment of liver cancer with albumin-conjugated gold nanoparticles initiates Golgi Apparatus-ER dysfunction and caspase-3 apoptotic pathway activation by selective targeting of Gp60 receptor. Int J Nanomedicine 10:5435–5445. https://doi.org/10.2147/IJN.S86495

Mosa KA, El-Naggar M, Ramamoorthy K, Alawadhi H, Elnaggar A, Wartanian S, Ibrahim E, Hani H (2018) Copper nanoparticles induced genotoxicty, oxidative stress, and changes in superoxide dismutase (SOD) gene expression in cucumber (Cucumis sativus) plants. Front Plant Sci 9:872. https://doi.org/10.3389/fpls.2018.00872

Mroz RM, Schins RP, Li H, Jimenez LA, Drost EM, Holownia A, MacNee W, Donaldson K (2008) Nanoparticle-driven DNA damage mimics irradiation-related carcinogenesis pathways. Eur Respir J 31(2):241–251. https://doi.org/10.1183/09031936.00006707

Murray AR, Kisin E, Inman A, Young SH, Muhammed M, Burks T, Uheida A, Tkach A, Waltz M, Castranova V, Fadeel B, Kagan VE, Riviere JE, Monteiro-Riviere N, Shvedova AA (2013) Oxidative stress and dermal toxicity of iron oxide nanoparticles in vitro. Cell Biochem Biophys 67(2):461–476. https://doi.org/10.1007/s12013-012-9367-9

Naqvi S, Samim M, Abdin M, Ahmed FJ, Maitra A, Prashant C, Dinda AK (2010) Concentration-dependent toxicity of iron oxide nanoparticles mediated by increased oxidative stress. Int J Nanomedicine 5:983–989. https://doi.org/10.2147/IJN.S13244

Ng CT, Yong LQ, Hande MP, Ong CN, Yu LE, Bay BH, Baeg GH (2017) Zinc oxide nanoparticles exhibit cytotoxicity and genotoxicity through oxidative stress responses in human lung fibroblasts and Drosophila melanogaster. Int J Nanomedicine 12:1621–1637. https://doi.org/10.2147/IJN.S124403

Noël C, Simard JC, Girard D (2016) Gold nanoparticles induce apoptosis, endoplasmic reticulum stress events and cleavage of cytoskeletal proteins in human neutrophils. Toxicol In Vitro 31:12–22. https://doi.org/10.1016/j.tiv.2015.11.003

Onoda A, Kawasaki T, Tsukiyama K, Takeda K, Umezawa M (2020) Carbon nanoparticles induce endoplasmic reticulum stress around blood vessels with accumulation of misfolded proteins in the developing brain of offspring. Sci Rep 10(1):10028. https://doi.org/10.1038/s41598-020-66744-w

Onodera A, Nishiumi F, Kakiguchi K, Tanaka A, Tanabe N, Honma A, Yayama K, Yoshioka Y, Nakahira K, Yonemura S, Yanagihara I, Tsutsumi Y, Kawai Y (2015) Short-term changes in intracellular ROS localisation after the silver nanoparticles exposure depending on particle size. Toxicol Rep 2:574–579. https://doi.org/10.1016/j.toxrep.2015.03.004

Paik SY, Kim JS, Shin SJ, Ko S (2015) Characterization, quantification, and determination of the toxicity of iron oxide nanoparticles to the bone marrow cells. Int J Mol Sci 16(9):22243–22257. https://doi.org/10.3390/ijms160922243

Park MV, Neigh AM, Vermeulen JP, de la Fonteyne LJ, Verharen HW, Briedé JJ, van Loveren H, de Jong WH (2011) The effect of particle size on the cytotoxicity, inflammation, developmental toxicity and genotoxicity of silver nanoparticles. Biomaterials 32(36):9810–9817. https://doi.org/10.1016/j.biomaterials.2011.08.085

Park J, Park C, Lee Y, Ryu C, Park J, Kim Y (2022) Acute adverse effects of metallic nanomaterials on cardiac and behavioral changes in daphnia magna. Environments 9(2):26. https://doi.org/10.3390/environments9020026

Patlolla AK, Randolph J, Kumari SA, Tchounwou PB (2016) Toxicity evaluation of graphene oxide in kidneys of sprague-dawley rats. Int J Environ Res Public Health 13(4):380. https://doi.org/10.3390/ijerph13040380

Poormohammad Matouri Z, Noori A (2018) Effect of multi-wall carbon nanotubes toxicity on kidney function and tissue in rats. J Gorgan Univ Med Sci 20(1):22–28. http://goums.ac.ir/journal/article-1-3273-en.html

Prasad M, Ghosh M, Kumar R, Brar B, Surjith KP, Lambe UP, Ranjan K, Banerjee S, Prasad G, Kumar Khurana S, Kharb P (2021a) The importance of nanomedicine in prophylactic and theranostic intervention of bacterial zoonoses and reverse zoonoses in the era of microbial resistance. J Nanosci Nanotechnol 21(6):3404–3452. https://doi.org/10.1166/jnn.2021.18999

Prasad M, Kumar R, Buragohain L, Kumari A, Ghosh M (2021b) Organoid technology: a reliable developmental biology tool for organ-specific nanotoxicity evaluation. Front Cell Dev Biol 9:696668. https://doi.org/10.3389/fcell.2021.696668

Quinteros MA, Cano Aristizábal V, Dalmasso PR, Paraje MG, Páez PL (2016) Oxidative stress generation of silver nanoparticles in three bacterial genera and its relationship with the antimicrobial activity. Toxicol In Vitro 36:216–223. https://doi.org/10.1016/j.tiv.2016.08.007

Rasras S, Kalantari H, Rezaei M, Dehghani MA, Zeidooni L, Alikarami K, Dehghani F, Alboghobeish S (2019) Single-walled and multiwalled carbon nanotubes induce oxidative stress in isolated rat brain mitochondria. Toxicol Ind Health 35(7):497–506. https://doi.org/10.1177/0748233719856983

Rattanapinyopituk K, Shimada A, Morita T, Sakurai M, Asano A, Hasegawa T, Inoue K, Takano H (2014) Demonstration of the clathrin- and caveolin-mediated endocytosis at the maternal-fetal barrier in mouse placenta after intravenous administration of gold nanoparticles. J Vet Med Sci 76(3):377–387. https://doi.org/10.1292/jvms.13-0512

Recordati C, De Maglie M, Bianchessi S, Argentiere S, Cella C, Mattiello S, Cubadda F, Aureli F, D'Amato M, Raggi A, Lenardi C, Milani P, Scanziani E (2016) Tissue distribution and acute toxicity of silver after single intravenous administration in mice: nano-specific and size-dependent effects. Part Fibre Toxicol 13:12. https://doi.org/10.1186/s12989-016-0124-x

Reddy AR, Rao MV, Krishna DR, Himabindu V, Reddy YN (2011) Evaluation of oxidative stress and anti-oxidant status in rat serum following exposure of carbon nanotubes. Regul Toxicol Pharmacol 59(2):251–257. https://doi.org/10.1016/j.yrtph.2010.10.007

Refuerzo JS, Godin B, Bishop K, Srinivasan S, Shah SK, Amra S, Ramin SM, Ferrari M (2011) Size of the nanovectors determines the transplacental passage in pregnancy: study in rats. Am J Obstet Gynecol 204(6):546.e5–546.e9. https://doi.org/10.1016/j.ajog.2011.02.033

Rim KT, Song SW, Kim HY (2013) Oxidative DNA damage from nanoparticle exposure and its application to workers' health: a literature review. Saf Health Work 4(4):177–186. https://doi.org/10.1016/j.shaw.2013.07.006

Sabella S, Carney RP, Brunetti V, Malvindi MA, Al-Juffali N, Vecchio G, Janes SM, Bakr OM, Cingolani R, Stellacci F, Pompa PP (2014) A general mechanism for intracellular toxicity of metal-containing nanoparticles. Nanoscale 6:7052–7061. https://doi.org/10.1039/C4NR01234H

Samberg ME, Oldenburg SJ, Monteiro-Riviere NA (2010) Evaluation of silver nanoparticle toxicity in skin in vivo and keratinocytes in vitro. Environ Health Perspect 118(3):407–413. https://doi.org/10.1289/ehp.0901398

Sarma A, Bania R, Devi JR, Deka S (2021) Therapeutic nanostructures and nanotoxicity. J Appl Toxicol 41(10):1494–1517. https://doi.org/10.1002/jat.4157

Schultz AG, Boyle D, Chamot D, Ong KJ, Wilkinson KJ, McGeer JC, Sunahara G, Goss GG (2014) Aquatic toxicity of manufactured nanomaterials: challenges and recommendations for future toxicity testing. Environ Chem 11:207–226. https://doi.org/10.1071/EN13221

Seiffert J, Hussain F, Wiegman C, Li F, Bey L, Baker W, Porter A, Ryan MP, Chang Y, Gow A, Zhang J, Zhu J, Tetley TD, Chung KF (2015) Pulmonary toxicity of instilled silver nanoparticles: influence of size, coating and rat strain. PLoS One 10(3):e0119726. https://doi.org/10.1371/journal.pone.0119726

Sen GT, Ozkemahli G, Shahbazi R, Erkekoglu P, Ulubayram K, Kocer-Gumusel B (2020) The effects of polymer coating of gold nanoparticles on oxidative stress and DNA damage. Int J Toxicol 39(4):328–340. https://doi.org/10.1177/1091581820927646

Sharma V, Anderson D, Dhawan A (2012) Zinc oxide nanoparticles induce oxidative DNA damage and ROS-triggered mitochondria mediated apoptosis in human liver cells (HepG2). Apoptosis 17(8):852–870. https://doi.org/10.1007/s10495-012-0705-6

Shen Z, Wu J, Yu Y, Liu S, Jiang W, Nurmamat H, Wu B (2019) Comparison of cytotoxicity and membrane efflux pump inhibition in HepG2 cells induced by single-walled carbon nanotubes with different length and functional groups. Sci Rep 9(1):7557. https://doi.org/10.1038/s41598-019-43900-5

Shojaei S, Ali MS, Suresh M, Upreti T, Mogourian V, Helewa M, Labouta HI (2021) Dynamic placenta-on-a-chip model for fetal risk assessment of nanoparticles intended to treat pregnancy-associated diseases. Biochim Biophys Acta Mol basis Dis 1867(7):166131. https://doi.org/10.1016/j.bbadis.2021.166131

Silva E, Barreiros L, Segundo MA, Costa Lima SA, Reis S (2017) Cellular interactions of a lipid-based nanocarrier model with human keratinocytes: Unravelling transport mechanisms. Acta Biomater 53:439–449. https://doi.org/10.1016/j.actbio.2017.01.057

Singh R, Cheng S, Singh S (2020) Oxidative stress-mediated genotoxic effect of zinc oxide nanoparticles on Deinococcus radiodurans. 3 Biotech 10(2):66. https://doi.org/10.1007/s13205-020-2054-4

Soltaninejad H, Zare-Zardini H, Hamidieh AA, Sobhan MR, Saeed-Banadaky SH, Amirkhani MA, Tolueinia B, Mehregan M, Mirakhor M, Eshaghi FS (2020) Evaluating the toxicity and histological effects of Al2O3 nanoparticles on bone tissue in animal model: a case-control study. J Toxicol 2020:8870530. https://doi.org/10.1155/2020/8870530

Soni D, Gandhi D, Tarale P, Bafana A, Pandey RA, Sivanesan S (2017) Oxidative stress and genotoxicity of zinc oxide nanoparticles to pseudomonas species, human promyelocytic leukemic (HL-60), and blood cells. Biol Trace Elem Res 178(2):218–227. https://doi.org/10.1007/s12011-016-0921-y

Stern ST, Adiseshaiah PP, Crist RM (2012) Autophagy and lysosomal dysfunction as emerging mechanisms of nanomaterial toxicity. Part Fibre Toxicol 9:20. https://doi.org/10.1186/1743-8977-9-20

Sun H, Liu Y, Bai X, Zhou X, Zhou H, Liu S, Yan B (2018) Induction of oxidative stress and sensitization of cancer cells to paclitaxel by gold nanoparticles with different charge densities and hydrophobicities. J Mater Chem B 6(11):1633–1639. https://doi.org/10.1039/c7tb03153j

Szymański T, Kempa M, Giersig M, Dalibor RJ (2020) Carbon nanotubes interference with luminescence-based assays. Materials (Basel) 13(19):4270. https://doi.org/10.3390/ma13194270

Tang H, Xu M, Shi F, Ye G, Lv C, Luo J, Zhao L, Li Y (2018) Effects and mechanism of nano-copper exposure on hepatic cytochrome P450 enzymes in rats. Int J Mol Sci 19(7):2140. https://doi.org/10.3390/ijms19072140

Tang Y, Chen B, Hong W, Chen L, Yao L, Zhao Y, Aguilar ZP, Xu H (2019) ZnO nanoparticles induced male reproductive toxicity based on the effects on the endoplasmic reticulum stress signaling pathway. Int J Nanomedicine 14:9563–9576. https://doi.org/10.2147/IJN.S223318

Teleanu DM, Chircov C, Grumezescu AM, Volceanov A, Teleanu RI (2018) Impact of nanoparticles on brain health: an up to date overview. J Clin Med 7(12):490. https://doi.org/10.3390/jcm7120490

Wang S, Kurepa J, Smalle JA (2011) Ultra-small TiO2 nanoparticles disrupt microtubular networks in Arabidopsis thaliana. Plant Cell Environ 34:811–820. https://doi.org/10.1111/j.1365-3040.2011.02284.x

Wang F, Salvati A, Boya P (2018a) Lysosome-dependent cell death and deregulated autophagy induced by amine-modified polystyrene nanoparticles. Open Biol 8(4):170271. https://doi.org/10.1098/rsob.170271

Wang M, Lai X, Shao L, Li L (2018b) Evaluation of immunoresponses and cytotoxicity from skin exposure to metallic nanoparticles. Int J Nanomedicine 13:4445–4459. https://doi.org/10.2147/IJN.S170745

Wils RS, Jacobsen NR, Vogel U, Roursgaard M, Møller P (2021a) Inflammatory response, reactive oxygen species production and DNA damage in mice after intrapleural exposure to carbon nanotubes. Toxicol Sci 183(1):184–194. https://doi.org/10.1093/toxsci/kfab070

Wils RS, Jacobsen NR, Di Ianni E, Roursgaard M, Møller P (2021b) Reactive oxygen species production, genotoxicity and telomere length in FE1-Muta™Mouse lung epithelial cells exposed to carbon nanotubes. Nanotoxicology 15(5):661–672. https://doi.org/10.1080/17435390.2021.1910359

Woźniak A, Malankowska A, Nowaczyk G, Grześkowiak BF, Tuśnio K, Słomski R, Zaleska-Medynska A, Jurga S (2017) Size and shape-dependent cytotoxicity profile of gold nanoparticles for biomedical applications. J Mater Sci Mater Med 28(6):92. https://doi.org/10.1007/s10856-017-5902-y

Wu J, Wang C, Sun J, Xue Y (2011) Neurotoxicity of silica nanoparticles: brain localization and dopaminergic neurons damage pathways. ACS Nano 5(6):4476–4489. https://doi.org/10.1021/nn103530b

Wu M, Guo H, Liu L, Liu Y, Xie L (2019) Size-dependent cellular uptake and localization profiles of silver nanoparticles. Int J Nanomedicine 14:4247–4259. https://doi.org/10.2147/IJN.S201107

Wu Z, Guan R, Tao M, Lyu F, Cao G, Liu M, Gao J (2020) Correction: assessment of the toxicity and inflammatory effects of different-sized zinc oxide nanoparticles in 2D and 3D cell cultures. RSC Adv 10(72):44397. https://doi.org/10.1039/d0ra90118k. Erratum for: 10.1039/C6RA27334C

Wu L, Wen W, Wang X, Huang D, Cao J, Qi X, Shen S (2022) Ultrasmall iron oxide nanoparticles cause significant toxicity by specifically inducing acute oxidative stress to multiple organs. Part Fibre Toxicol 19(1):24. https://doi.org/10.1186/s12989-022-00465-y

Xia T, Kovochich M, Liong M, Mädler L, Gilbert B, Shi H, Yeh JI, Zink JI, Nel AE (2008) Comparison of the mechanism of toxicity of zinc oxide and cerium oxide nanoparticles based on dissolution and oxidative stress properties. ACS Nano 2(10):2121–2134. https://doi.org/10.1021/nn800511k

Xu F, Piett C, Farkas S, Qazzaz M, Syed NI (2013) Silver nanoparticles (AgNPs) cause degeneration of cytoskeleton and disrupt synaptic machinery of cultured cortical neurons. Mol Brain 6:29. https://doi.org/10.1186/1756-6606-6-29

Yadav K, Ali SA, Mohanty AK, Muthusamy E, Subaharan K, Kaul G (2021) MSN, MWCNT and ZnO nanoparticle-induced CHO-K1 cell polarisation is linked to cytoskeleton ablation. J Nanobiotechnology 19(1):45. https://doi.org/10.1186/s12951-021-00779-7

Yan D, Xu W, Yao X, Lin L, Sherman JH, Keidar M (2018) The cell activation phenomena in the cold atmospheric plasma cancer treatment. Sci Rep 8(1):15418. https://doi.org/10.1038/s41598-018-33914-w

Yang K, Gong H, Shi X, Wan J, Zhang Y, Liu Z (2013) In vivo biodistribution and toxicology of functionalized nano-graphene oxide in mice after oral and intraperitoneal administration. Biomaterials 34(11):2787–2795. https://doi.org/10.1016/j.biomaterials.2013.01.001

Yang C, Tian A, Li Z (2016) Reversible cardiac hypertrophy induced by PEG-coated gold nanoparticles in mice. Sci Rep 6:20203. https://doi.org/10.1038/srep20203

Yao Y, Zang Y, Qu J, Tang M, Zhang T (2019) The toxicity of metallic nanoparticles on liver: the subcellular damages, mechanisms, and outcomes. Int J Nanomedicine 14:8787–8804. https://doi.org/10.2147/IJN.S212907

Yi X, Zhao W, Li J, Zhang B, Yu Q, Li M (2017) Mn3O4nanoparticles cause endoplasmic reticulum stress-dependent toxicity to Saccharomyces cerevisiae. RSC Adv 7:46028–46035. https://doi.org/10.1039/C7RA07458A

Yoshiura Y, Izumi H, Oyabu T, Hashiba M, Kambara T, Mizuguchi Y, Lee BW, Okada T, Tomonaga T, Myojo T, Yamamoto K, Kitajima S, Horie M, Kuroda E, Morimoto Y (2015) Pulmonary toxicity of well-dispersed titanium dioxide nanoparticles following intratracheal instillation. J Nanopart Res 17(6):241. https://doi.org/10.1007/s11051-015-3054-x

Yousef MI, Abuzreda AA, Kamel MAE-N (2019a) Neurotoxicity and inflammation induced by individual and combined exposure to iron oxide nanoparticles and silver nanoparticles. J Taibah Univ Sci 13:570–578. https://doi.org/10.1080/16583655.2019.1602351

Yousef MI, Abuzreda AA, Kamel MA (2019b) Cardiotoxicity and lung toxicity in male rats induced by long-term exposure to iron oxide and silver nanoparticles. Exp Ther Med 18(6):4329–4339. https://doi.org/10.3892/etm.2019.8108

Yu Z, Li Q, Wang J, Yu Y, Wang Y, Zhou Q, Li P (2020) Reactive oxygen species-related nanoparticle toxicity in the biomedical field. Nanoscale Res Lett 15(1):115. https://doi.org/10.1186/s11671-020-03344-7

Zeman T, Loh EW, Čierný D, Šerý O (2018) Penetration, distribution and brain toxicity of titanium nanoparticles in rodents' body: a review. IET Nanobiotechnol 12(6):695–700. https://doi.org/10.1049/iet-nbt.2017.0109

Zhang XQ, Yin LH, Tang M, Pu YP (2011) ZnO, TiO(2), SiO(2,) and Al(2)O(3) nanoparticles-induced toxic effects on human fetal lung fibroblasts. Biomed Environ Sci 24(6):661–669. https://doi.org/10.3967/0895-3988.2011.06.011

Zhang X, Zhang H, Liang X, Zhang J, Tao W, Zhu X, Chang D, Zeng X, Liu G, Mei L (2016) Iron oxide nanoparticles induce autophagosome accumulation through multiple mechanisms: lysosome impairment, mitochondrial damage, and ER stress. Mol Pharm 13(7):2578–2587. https://doi.org/10.1021/acs.molpharmaceut.6b00405

Zhang B, Sai Lung P, Zhao S, Chu Z, Chrzanowski W, Li Q (2017) Shape dependent cytotoxicity of PLGA-PEG nanoparticles on human cells. Sci Rep 7(1):7315. https://doi.org/10.1038/s41598-017-07588-9

Zhang L, Wu L, Si Y, Shu K (2018) Size-dependent cytotoxicity of silver nanoparticles to Azotobacter vinelandii: Growth inhibition, cell injury, oxidative stress and internalization. PLoS One 13(12):e0209020. https://doi.org/10.1371/journal.pone.0209020

Zhang X, Guo X, Kang X, Yang H, Guo W, Guan L, Wu H, Du L (2020) Surface functionalization of pegylated gold nanoparticles with antioxidants suppresses nanoparticle-induced oxidative stress and neurotoxicity. Chem Res Toxicol 33(5):1195–1205. https://doi.org/10.1021/acs.chemrestox.9b00368

Zhang J, Liu S, Han J, Wang Z, Zhang S (2021) On the developmental toxicity of silver nanoparticles. Mater Des 203. https://doi.org/10.1016/j.matdes.2021.109611

Zhao X, Ren X, Zhu R, Luo Z, Ren B (2016) Zinc oxide nanoparticles induce oxidative DNA damage and ROS-triggered mitochondria-mediated apoptosis in zebrafish embryos. Aquat Toxicol 180:56–70. https://doi.org/10.1016/j.aquatox.2016.09.013

Zou L, Cheng G, Xu C, Liu H, Wang Y, Li N, Fan X, Zhu C, Xia W (2021) Copper nanoparticles induce oxidative stress via the heme oxygenase 1 signaling pathway in vitro studies. Int J Nanomedicine 16:1565–1573. https://doi.org/10.2147/IJN.S292319

Zucker I, Werber JR, Fishman ZS, Hashmi SM, Gabinet UR, Lu X, Osuji CO, Pfefferle LD, Elimelech M (2017) Loss of phospholipid membrane integrity induced by two-dimensional nanomaterials. Environ Sci Technol Lett 4:404–409. https://doi.org/10.1021/acs.estlett.7b00358

Chapter 30
Nanotoxicity Evaluation Methods

Neeti Lakhani, Alonkrita Chowdhury, Akshita, Kratika Maurya, Komal Kumari, Manoj Kumar Rose, Sonia Sindhu, and Mayukh Ghosh

Abstract Nanomaterials are utilized across a wide spectrum of scientific disciplines, encompassing biomedicine among others. Though nanomaterials hold a great impact on the advancement of conventional systems, the toxicological implications of such nano-formulations must not be overlooked. Hence, an assessment of the toxicity profile of such compounds must be carried out in addition to their application-specific evaluation for animals as well as human well-being. The extent of nanotoxicity induction depends on their size, shape, nature, area/volume ratio, functionalization, and other physico chemical properties. Animals including human exposure to nano-fabrications may give rise to cytotoxicity leading to apoptosis and necrosis, generation of Reactive Oxygen Species (ROS) and oxidative stresses, abnormality in gene expression causing genotoxicity, immunological disturbances, etc. Several in-vivo as well as in-vitro methods for assessing nanotoxicity potential are required for the fruitful application of nanomaterials with minimal or zero side- effects. For instance, colorimetric assays such as MTT assay, LDH assay, etc. can be used for the determination of cell viability. TUNEL assay, COMET assay, Caspase assay, Annexin V assay, CLSM, FACS, and such other methods are involved in detection of apoptosis and necrosis. Along with DCFH assay and plasmid assay, determination of Glutathione level using diagnostic kits or microscopic observation is also carried out for evaluating ROS production. Comet assay, Ames test, etc. are performed for determining the potential of DNA damage,

N. Lakhani · M. K. Rose · S. Sindhu
Department of Animal Nutrition, College of Veterinary and Animal Sciences, Guru Angad Dev Veterinary and Animal Sciences University, Rampuraphul, Punjab, India

A. Chowdhury · Akshita · K. Maurya
Department of Plant Biotechnology, Institute of Agricultural Sciences, Rajiv Gandhi South Campus, Banaras Hindu University, Mirzapur, Uttar Pradesh, India

K. Kumari · M. Ghosh (✉)
Department of Veterinary Physiology and Biochemistry, RGSC, Banaras Hindu University, Mirzapur, Uttar Pradesh, India

LLR University of Veterinary and Animal, ICAR-National Research Centre on Equines, *Nanotechnology Theranostics in Livestock Diseases and Management*, Livestock Diseases and Management,
https://doi.org/10.1007/978-981-16-1610-5_30

mutagenicity, and cell proliferation by NPs. Effects of nanomaterials on immunological health can be elicited through diverse in-vitro technologies including EpiScreen™, Epibase®, REVEAL®, and so on. The psychological impact of NPs can also be determined through several behavioral studies such as hole-board test (HBT), Elevated Plus Maze (EPM) test, Porsolt swim test (PST) as well as the hot plate test and the tail flick test. Toxic effects of NPs at the organ-tissue level are evaluated by histopathological tests using fluorescence microscopy. Furthermore, several Omics-based platforms such as Single-molecule real-time (SMRT) sequencing by using a droplet-based (Drop-Seq) system, DBiT-seq, single-cell genome long-read sequencing technology (SMOOTH-seq) along with, Mass Cytometry, Fluorescence Flow Cytometry, Enzyme-Linked Immunospot Assay, Single-Molecule Array (SiMoA), Microfluidic Antibody Capture Chip (MACS Chip)-based assay, Proximity Extension Assay/Abseq, are also exploited for toxicological studies. However, every method has its limitations as a single modality can depict the toxicity effect to an extent only. Hence, a comprehensive assessment including diverse models as well as modalities would be beneficial. Further, short as well as long-term toxicity impacts need to be investigated to get a wholesome depiction. Thus, the following chapter will enlighten the applications of different state-of-art modalities regarding the determination of nanotoxicity while focusing on animal health, including humans.

Keywords Genotoxicity · Apoptosis and necrosis · Oxidative stress · Histopathology · Next generation sequencing (NGS) · Mass spectrometry · NMR

30.1 Introduction

The Majority of compounds intended for human or animal applications undergo various assessments to determine the safety or toxicity profile on the concerned subjects in addition to their application-specific evaluation. The safety or toxicity evaluation include determination of lethal doses for 50% population (LD%), Lethal Concentration (LC_{50}), effects on behavioral pattern of the subject population, clearance route and rate, half-life, impacts on different tissues and most importantly, to target contacts through histopathological examination, effects on various biochemical parameters and mineral status, consequences on neurotransmitters, implications on cell cycle regulation, outcomes on regulation of gene expression at genomic, transcriptomic, and proteomic levels as well as on associated pathways, impacts on various metabolic pathways, implications on cellular and sub-cellular architecture along with cell viability, induction of free radical generation and inflammation, effect on reproductive processes, etc. (Ciappellano et al. 2016). All of these toxicological evaluations are essential for preparations such as food grade compounds, vaccines, drugs, cosmetics, pharmaceuticals, growth promoters, and substances that can enter through the food chain like pesticides, heavy metal residues from diagnostics or surgical stuffs and implants, etc. These materials can be approved for implementations only after establishment that they impart minimal or tolerable damage to the subject. Recently, massive floodgate applications of nanomaterials (NMs) along with continuous addition of novel ones in the nanomaterial repository and their recent utility in diverse arena including the biomedical science has urged robust regulatory intervention on the basis of the aforesaid toxicological evaluations to ensure the nano-safety perspective.

Nano-interventions are not only implemented as biomedical appliances in humans or livestock but in other dimensions also such as engineering, water purification, material sciences, etc. Hence, it has enhanced the load of NMs on the earth resulting in environmental pollution and bioaccumulation through food chain contamination. Due to the nano-range size, NMs gain astonishingly different physico-chemical properties unlike their bulk material or even micro-particles. Hence, the consequences of the implementations of nano-formulations on biological systems are required to be assessed distinctly (Gupta et al. 2019). The toxicological implications of diverse NMs, termed as nanotoxicity have already been documented in different in-vitro and in-vivo models (Table 30.1) (Aguilar 2013). Various indices have been used to denote the extent of nanotoxicity such as LC_{50} (Lethal Concentration 50%), LD_{50} (Lethal Dose 50%), NOAEL (No Observed Adverse Effect Level), LOAEL (Lowest Observed Adverse Effect Level), EC_{50} (median effective concentration), NOEC (No observed effect concentration), DT_{50} (Half-life), etc. (Zielińska et al. 2020). The most commonly employed model for in-vitro toxicity assessment include cell-culture based methods such as colorimetric, fluorescent and luminescent end-point observations in LDH assay for cell membrane integrity, microscopy for necrosis and live/dead ratio of cells using trypan blue or fluorescent dyes, morphological analysis, MTT assay for cell viability, proliferation and cytotoxicity, luminescent ATP assays for mitochondrial energy metabolism, glucose/galactose assay for mitochondrial toxicity, Ames test for genotoxicity assessment, Annexin-V and propidium iodide assay, Comet assay for apoptosis along with expression analysis of various toxicity markers (Riss et al. 2019). Estimation of different cellular antioxidants such as superoxide dismutase (SOD), catalase, glutathione peroxidase activity, reduced glutathione (GSH) level along with malondialdehyde (MDA) production as a product of lipid peroxidation, hydroxyl radical and nitric oxide radical scavenging activity, etc. are usually performed for the analysis of oxidative stress (Horie and Tabei 2021). Cell-culture based in-vitro assessment of toxicity is relatively inexpensive and free from ethical issues of employing live animals. In-vivo toxicity testing methods include estimation of LD_{50}, LC_{50}, growth study, behavioral and pain study, examination of histopathological aberrations in vital organs, organ function tests, evaluating reproductive parameters and embryonic development, estimation of different antioxidants and oxidative radical scavenging activity, assessment of inflammation through measurement of different cytokines, etc. in various laboratory model animals (Kumar et al. 2017). Employment of state-of-the art OMICS platforms (genomics, transcriptomics, proteomics and metabolomics) and their integration is part of the current trend to find out the early and minute toxicity signatures at various organs, even at a cellular or organelle level. This chapter will delineate the prominent nanotoxicity evaluation methods in the limited purview.

Table 30.1 Effect of nanotoxicity based on LD_{50} on different species

Nanoparticles	Species	Route of administration	Toxicity dose	References
Silver NPs (20–25 nm) Silver NPs (30–35 nm)	Mice	Intraperitoneal administration	250 mg/kg (more cytotoxic; major hepatic injury) 350 mg/kg (less cytotoxic; minor hepatotoxicity)	Hoseini-Alfatemi et al. (2020)
Citrate-functionalized AgNPs (~20 nm) (~ 50 nm)	BALB/c mice	Intraperitoneal administration	Employing Dixon's up-and-down method and AOT425statPgm method, 169 and 213.8 mg/kg, respectively (more toxic; fatality occurs before 48 h after injection; passive behavior, difficulties in breathing, impaired movement, piloerection, loss of appetite, back pain, and diarrhea) 354 and 391.5 mg/kg, respectively (less toxic, results in fatality within 24 h of injection, inactiveness and passive behavior)	Elkhawass et al. (2015)
Non-doped CQD NPs (citric acid as carbon source) Nitrogen-doped (N-doped) CQD NPs Nitrogen and sulfur co-doped carbon dots (N,S/C-dots) NPs	Zebra fish	Zebrafish bioassays	548.48 μg/mL (minimal toxicity; pericardial edema and slight bending, relatively higher growth rate and least locomotor activity) 399.95 μg/mL (moderate toxicity, severe yolk sac edema and tail bending, average growth rate and moderate locomotor activity) 149.92 μg/mL (most toxic; yolk sac edema and severe tail deformation, poor growth rate, and highest locomotor activity)	Chousidis et al. (2020)
CuONPs ZnONPs TiO_2NPs SiO_2NPs	Albino Mice	Intra-peritoneal Administration	2012 mg/kg (reduced cell viability and death) 2225 mg/kg (reduced cell viability and death) 2775 mg/kg (reduced cell viability and death)	Shaker et al. (2017)

(continued)

Table 30.1 (continued)

Nanoparticles	Species	Route of administration	Toxicity dose	References
			3000 mg/kg (reduced cell viability and death)	
CoONPs MgONPs PbONPs	Mice	Oral	7835 mg/kg (hyperactivity for 2–3 days followed by drowsiness, scratching of body, hair loss, roughness in hair texture and also skin color started changing to bluish black, finally toxicity causes death) 3954 mg/kg (hyperactivity, moving of tail but the hair remained smooth throughout the exposure; all these symptoms were followed by death) 14,677 mg/kg (hyperactivity initially for 6–7 days followed by drowsiness and the hairs were semi smooth; ultimately death due to toxicity)	Shaikh et al. (2015)
ZnONPs	ICR male mice	Intratracheally installation	493.5 µg/kg (disheveled hair, loss of appetite, defecation, dry stool, dyspneic respiration, and acrocyanosis	Wang et al. (2017)
CuNPs (single dose)	Sprague-Dawley rats	Oral	5000 mg/kg (high level of blood Cu; prolonged blood circulation time; low Cu levels in the urine; high levels of Cu in fecal matter)	Lee et al. (2016)
CuNPs	Mice	Oral	413 mg/kg (loss of appetite, diarrhea and vomiting; passive behavior, hypopnea, tremor and arching of back)	Chen et al. (2006)
Silica nanoparticles (SNPs)	Mice	Intravenously	262.45 mg/kg (Dixon's up-and-down method; the levels of serum LDH, AST and ALT were all elevated in the SNP treated groups)	Yu et al. (2013)

(continued)

Table 30.1 (continued)

Nanoparticles	Species	Route of administration	Toxicity dose	References
AuNPs	Male ICR mice	Oral, intraperitoneal and tail vein injection	2200 μg/kg (in low concentrations no significant toxicity is present however, in high concentrations orally administered AuNP resulted in decreased body weight, spleen index, and red blood cells; oral and intraperitoneal routes showed highest toxicity while tail vein injection causes least toxicity)	Zhang et al. (2010)
ZnONPs	Chick embryo	Intravascular injection on tenth day of the egg development	1.4 μg/mL (Chorioallantoic membrane assay; nanoparticles agglomerated in the heart, liver, and kidney of the embryo 7 and 24 h after treatment; increase of the apoptosis marker)	Buhr et al. (2021)
Selenium functionalized NPs SeNPs SeO_2	Male NMRI mice	Oral	(male mice body weight decreased; reduced growth hormone, somatomedin C, insulin-like growth factor-binding proteins and insulin-like growth factor-I production; mortality) 198.1 mg/kg 7.3 mg/kg	Shakibaie et al. (2013)
Tilmicosin-loadedhydrogenated castor oil-solid lipid nanoparticles (Til-HCO-SLN)	ICR mice	Subcutaneously	More than 5 g/kg body weight (surviving mice showed hypoactivity, accelerated breath, gloomy spirit and lethargy)	Xie et al. (2011)
Cobalt-zinc-ferrite nano-complex (CZF NPs) Cobalt ferrite NPs (CFNPs)	Female Swiss albino mice	IP injection	4.6 g/kg (significant change in the total body weight; increased WBCs and decreased platelets count; increase in the ALT and AST activities; disorganization of the hepatic manner, irregular and congested central veins, some hepatocytes with pyknotic nuclei, others	El-Nahass et al. (2021)

(continued)

Table 30.1 (continued)

Nanoparticles	Species	Route of administration	Toxicity dose	References
			with vacuolated cytoplasm and widening blood sinusoids with distinct phagocytic Kupffer cells; and mortality) 4.3 g/kg (no significant changes except mortality)	

30.2 In-Vitro Nanotoxicity Assessment

30.2.1 Cytotoxicity and Cell Death

Cell viability and cytotoxicity are two paramount factors used to estimate the toxicity induced by nanoparticles. Determining permeability and integrity of cells is important to judge the extent of cytotoxicity of nanomaterials. Lactate dehydrogenase (LDH), present within cell cytoplasm, is a soluble cytosolic enzyme released when the cell loses its integrity (Kendig and Tarloff 2007). Thus, LDH release serves as a good indicator of cytotoxicity following exposure to nanomaterials (Henderson et al. 1985). LDH release levels can be quantified by assessing catalytic interconversion of nicotinamide adenine nucleotide (NADH) to its oxidized form nicotinamide adenine dinucleotide (NAD^+). LDH is also assessed by colorimetric LDH assay based on the transformation of yellow tetrazolium salt oxidation to red formazan (Wroblewski and Ladue 1955). However, the level of LDH release does not sufficiently determine the cytotoxicity of a nanoparticle.

Several other colorimetric assays are also available to evaluate the potential cytotoxicity caused by cell proliferation. Tetrazolium salt assay is indicative of metabolically active mitochondrial cells. Under tetrazolium salt assay, cells are treated multiple times with nanoparticles followed by the addition of MTS or MTT salts at 37 °C for 2–4 h. Metabolically viable mitochondrial cells reduce the tetrazolium salt into insoluble purple formazan, which is then quantified on a spectrophotometer (Nagalakshmi et al. 2013). These assays enable fast, reproducible results determining cell viability and proliferation with a simple absorbance recording. Interpretation of results depends on increase or decrease of absorbance, indicating cell proliferation, and reduced cellular metabolic activity.

Above stated methods focus on cytotoxicity of nanoparticles on cell membrane and mitochondria but assessing the ability of nanoparticles to cause cell death should also be considered. Apoptosis and necrosis are two specific forms of cell death that occur due to exposure of cells to nanomaterials and are assessed using specific fluorescent solutions. Programmed cell death or apoptosis imply inspection of the morphological changes of cells. The assay involved to study apoptosis comprise of TUNEL assay, COMET assay, Caspase assay and Annexin V assay. Generally,

Annexin V assay is the most commonly used assay to determine cell death induced by nanoparticles (Koopman et al. 1994). The apoptotic cells bind to phosphotidylserine present on the intracellular side of the plasma membrane, causing its flip between the extracellular and intracellular membranes. This serves as a marker for apoptotic cells (Trotter and Voelker 1995). Necrosis, the second form of cell death, is a pathological condition occurrs in response to externally induced toxicity. The necrotic cells use the propidium iodide fluorescent dye, which identifies cell DNA and is analyzed by CLSM and FACS.

30.2.2 Cell Proliferation Assay

Cell proliferation is also a key factor in determining the genotoxic potential of nanoparticles. This can be achieved via a number of different assays, such as MTT and MTS assays. But due to the difficulties of these assays, such as lack of adequate sensitivity, chemical interference, and toxicity, an alternative test, that is, the BrdU Assay is performed for determining the cell proliferation ability. As an immunochemistry technique, BrdU Assay assesses newly synthesized DNA strands in cells undergoing active proliferation via fluorescent microscopy. This technique enables a reliable and valid testing strategy for cell proliferation.

30.2.3 Reactive Oxygen/ Nitrogen Species and Oxidative Stress

Any imbalance or disturbance in oxidative balance generated by administered nanoparticles leads to the oxidative stress paradigm. Oxidative stress is caused by an excessive accumulation or insufficient removal of reactive oxygen species produced by nanoparticle exposure. Both ROS formation and oxidative stress form the foundation for nanotoxicology research and study. It also serves to study any sublethal effects produced by nanomaterials. ROS include the free radical and non-radical species such as superoxide, hydroxyl, peroxyl, hydroxyperoxyl, and hydrogen peroxide. RNS include free radicals like nitric oxide and nitrogen dioxide. These ROS and RNS react with lipid, protein, or nucleic acids causing severe toxicological implications at the cellular level.

For ROS production measurement, a fluorogenic marker namely 2,7-dichlorodihydro fluorescein diacetate (DCFH-DA) is commonly used in 2, 7-dichlorofluorescin (DCFH) assay in which the dye is cleaved at high pH to a non-fluorescent product (Jakubowski and Bartosz 2000). However, DCFH-DA is cleaved to a fluorescent DCFH molecule during elevated production of ROS. Therefore, the increase in fluorescence indicates the level of increased ROS production. The method is extremely advantageous as it provides information on nanoparticles alone, independent of the cellular environment. Dihydroethidium (DHE) is another method similar to DCFH-DA that detects ROS production and oxidative stress within the cells. The electroparamagnetic resonance technique is also used to assess ROS generation through nanoparticles. This involves the use of spin traps and probes. A spin-trapping agent (5,5-dimethyl-lpyrroline N-oxide) for hydroxide or superoxide radicals and a radical-consuming spin probe (4-hydroxy-2,2,6,6-tetramethylpiperidine-1-oxyl) are introduced into the nanoparticle solution. The sample is then analyzed on an EPR spectrometer for the determination of ROS (Singh et al. 2007). ROS production is also estimated using the plasmid assay where relaxed and linear bacterial DNA plasmid is used to evaluate ROS release or exposure. However the technique is not very accurate and highly insensitive.

ROS production leads to a decline in the antioxidant status, particularly, superoxide dismutase or glutathione production, resulting in oxidative imbalance and ultimately oxidative stress (MacNee 2001). Upregulation or exhaustion of SOD and Glutathione store indicates imbalance due to oxidation by ROS or RNS and serves as evidence of oxidative stress. Assessment of the antioxidant status, therefore, becomes relevant for a clear and thorough understanding of the effect of nanoparticles on the redox balance. Glutathione, an essential antioxidant, presents intracellularly and is readily oxidized to form a disulfide bond between two GSH molecules yielding GSSG under oxidative stress (Hissin and Hilf 1976). The GSH and GSSG levels are determined by the colorimetric reaction of GSH with DTNB using HPLC, but the process is time consuming and sometimes leads to overestimation. Therefore, the fluorescent dye method using ophthaldehyde is generally conducted. Other alternative methods available to determine glutathione levels include diagnostic kits and microscopic assays. Superoxide dismutase activity is determined indirectly using nitro blue tetrazolium. It causes inhibition to oxidation of the colored substrate, generating superoxide through exogenous xanthine oxidase.

30.2.4 Inflammation

Inflammatory assays involve the usage of the biochemical technique, enzyme linked immunosorbent assay (ELISA), which detects the presence of antigen or antibody in a sample. ELISA diagnostic kits measure plenty of pro-inflammatory cytokines and chemokines released when cells are exposed to nanoparticles. The most commonly tested human and murine inflammatory markers are the chemokine Interleukin-8 (IL-8), followed by TNF alpha and IL-6. Cytokine/chemokine production can also be measured either by PCR or FACS analysis. However, these methods are laborious and expensive and have no additional advantages when compared to the ELISA diagnostic method.

30.2.5 Cell Signaling and Gene Expression

Nanoparticles are known to influence cell signaling, particularly Ca^{2+} signaling, which specifically contributes to oxidative stress (Donaldson et al. 2003). Ca^{2+} signaling is relatively less-studied for the evaluation of nanotoxic effects on cell as compared to ROS and inflammatory cytokine analysis. Fluorescent marker Fura-2-AM is the most commonly used indicator for measuring intracellular alterations of calcium content. A change in Ca^{2+} level during incubation of cells with nanomaterial leads to removal of the acetoxymethyl (AM) group from Fura-2-AM molecule by cellular esterases. Ca^{2+} levels are then determined by finding the ratio of free and Ca^{2+} bound Fura-2-AM (Martínez et al. 2017).

Gene profiling or gene expression assays are important tools for screening nanoparticles. Techniques for gene profiling include ribonuclease protection assays (RPA), PCR assays, and microarrays. Detection and quantification of specific RNA in a mixture of total RNA is done by the RPA method. RPA utilizes a probe complimentary to target RNA and after hybridization of the mixture, the target RNA is detected by a specific label on the probe. Use of PCR assays involves detection or expression of a focused panel of genes. It is a probe based quantitative method of analysis. However, the method requires optimum time for obtaining valid and reliable results. Gene expression by microarray allows the measurement of thousands of mRNA levels in a single RNA sample. Purified RNA is then detected by autoradiographic imaging.

30.2.6 Genotoxicity

Distinct physicochemical properties of engineered nanomaterials related to size, structure, and shape (Kabanov 2006) allow interaction of nanoparticles with different biological systems. This may alter cell signaling and lipid membrane alteration. The high reactivity of NPs can be related directly to their genotoxic potential, which includes but is not limited to general breakage, base alkylation, deamination and oxidation, single and double strand break, cross-links, pyrimidine dimers, mismatches, bulky adduct formation, etc. (Lan et al. 2014). Specific assays are used to study the ability of NPs to elicit DNA damage, mutagenicity, and cell proliferation. Comet assay or single-cell gel electrophoresis is the most common method to study DNA damage. In this technique, agarose embedded cells are lysed to release the DNA followed by electrophoresis, resulting in a comet-like electropherogram in which the head of the comet contains intact DNA, while the tail accommodates the damaged DNA. The extent of damage corresponds to the DNA migration. The NP is stained with propidium iodide fluorescent dye, which is measured in the head and tail. The ability of nanoparticles to cause mutagenicity serves as an important strategy to ascertain the genotoxic potential of engineered nanomaterials. The most appropriate testing protocol for mutagenicity is micronucleus assay. This assay tests the genotoxicity level of micro- and nano-sized particles (Singh et al. 1988). A recent testing protocol includes Ames assay based on the bacterial strain namely *Salmonella typhimurium* being deficient in mechanism of DNA repair and unable to grow in the absence of histidine (Ames et al. 1973). If the nanoparticles inserted is mutagenic, it will enable bacteria to produce histidine and cause proliferation. The replication ability is then quantified to measure the mutagenicity. The specificity of this assay is still doubtful as nanoparticles interact with bacteria rather than the host cells to demonstrate its mutagenic effect.

In addition to the micronucleus test, a recent testing protocol for NP genotoxicity is the Ames test. This test is based on the strains of bacteria that have point mutations on the histidine (His) operon. Due to the position of the mutation on the His-gene, the bacteria are incapable of producing histidine and therefore unable to proliferate unless histidine is present (Vijay et al. 2018). In presence of mutagen, additional mutations occur on the His-gene, such as base substitutions or frameshift mutations, which enable the bacteria to produce histidine and subsequently proliferate. The mutagenicity of the substance is then determined via quantification of the ability to replicate the histidine-free bacteria. Previously, this test has been used in the industry to screen chemicals; however, recently it has all been successfully used to study the mutagenicity of NPs (Yoshida et al. 2009). There is a debate, however, on the specificity of this mutagenicity test because of the interaction of NPs with bacteria and not with the body cells. Similarly, unscheduled DNA synthesis at S-Phase, somatic mutation, and recombination tests due to exposure of toxic material in *Drosophila sp.* determines the NP-induced genotoxic effects (Shukla et al. 2021).

30.2.7 Immunological Studies

Nanomaterials due to their distinct characteristics affect the immune system of an organism and the alteration is required to be studied in order to ensure the health of animals on a long-term basis. Study of Immunogenicity in terms of immune safety and biocompatibility is essential as more than 900 drugs, biologicals, and vaccines have been developed until 2013 according to Pharmaceutical Research and Manufacturers of America (PhRMA) website. $CD4^+$ T cell responses, presence and amount (titer) of ADA responses, MHC Class II-associated peptide proteomic, adverse drug reactions, inflammation, expression profiling of various immune-cells, histopathology of spleen and bone marrow, nanomedicine-mediated C activation, erythrocytes hemolysis, etc. are various immunological methods which are studied either at an in-vivo or an in-vitro level. Diverse *in-silico* predictive platforms like iTope™, TCED™, Epibase®, EpiMatrix™ may not work effectively in nanomedicine as they may not be updated for all interaction studies related to nanosize molecules, hence, either in-vitro tools like EpiScreen™, Epibase®, REVEAL® or in-vivo models such as conventional mouse models, immune-tolerant transgenic mice, or non-human primate models may be used to get more comprehensive results (Bawa et al. 2019).

30.3 In-Vivo Nanotoxicity Assessment

30.3.1 LD_{50} and Behavior Studies

LD_{50} and behavior studies involve the study of doses which leads to the death of 50% of the population. Numerous nanomaterials have been assayed for LD_{50} on various animals, for instance, Carbon quantum dots (CQDs) doped through various materials had been used to assay LD_{50} in the zebra fish during embryonic development. N-doped CQD was discovered to have a decreased LD_{50} and to have toxicological effects on embryos, including slowed growth and malfunctions even at low concentrations of >150 g/mL (Chousidis et al. 2020). In an LD study different doses are administered to animals and the effects are then examined. Here, mice, guinea pigs and rabbits are the commonly used model animals as per standard guidelines like the protocol of the Organization for Economic Cooperation and Development (OECD). The examples of LD_{50} for some NPs have been demonstrated in Table 30.1. However, it must be remembered that NP properties depend on various parameters which include but are not limited to the method of generation, size, shape, charge, functionalization, route of entry etc. Therefore, assessment of each type of NMs planned for use in livestock and humans must be carried out for measuring numerous toxicological parameters including the LD_{50}. Study of AgNPs of different sizes on mice by Dixon's up-and-down method and "Acute Oral Toxicity (Guideline 425) Statistical Program" (AOT425StatPgm) determined that lower-sized NPs have more toxicity (Elkhawass et al. 2015). Although in the case of ZnO, LD_{50} had been reported for more than 2000 mg/kg but the toxicity effect was reported in the dose range of 5 mg/kg body weight, indicating LD_{50} along with other studies must be included to get a comprehensive toxicity profile. Similar to this, several behavioral tools used for monitoring general health, sensory functions, and motor abilities are studied routinely in toxicological studies like pole test, wire hang test, beam walking, rotarod, accelerod, footprint for motor function water maze, radial maze, conditioned fear, and avoidance tasks. A number of behavioral test are prevailing for the assessment of learning and memory, such as open field test, hole-board test (HBT), Elevated Plus Maze (EPM) test, object exploration for emotional status, Porsolt swim test (PST), as well as hot plate test and tail flick test for nociception, evaluation of acoustic startle response and prepulse inhibition to determine psychiatric-like conditions, along with study of aggressive behavior by isolation-induced aggression, impulsive aggression, and territorial aggression. Fluctuation in various physical factors such as heartbeat wave, the respiratory wave, oxygen saturation (SpO_2) values, body temperature, etc. can be analyzed precisely through different vital sign monitors for determination of physiological or psychological functions as well as the anatomical conditions of body systems during imaging procedures (Oliveira et al. 2022).

30.3.2 Histopathology

One of the common tests used to look for malformations in tissue-organs during disease is histopathological analysis of organ tissue. Regular and speciality stains are generally employed while immunofluorescence stains may be employed in specific situations. For drugs and other materials intended to be used in animal bodies, toxicological examination is carried out by various histopathological techniques. At first, post-exposure tissues must be rapidly and properly fixed by immersion or other suitable fixatives for retinal or testicular histopathology, otherwise through the perfusion fixation method or other special techniques as per requirement, followed by staining. All special stains are accompanied by positive control and nomenclature is performed as per standardized nomenclature and diagnostic criteria (such as the Standardized System of Nomenclature and Diagnostic Criteria Guides for Toxicologic Pathology) by a pathologist and evaluated as per The Society of Toxicologic Pathology's Recommendations on Statistical Analysis of Rodent Carcinogenicity Studies, 2002 etc. There are various stains that have been used for targeted materials of tissues like hematoxylin and eosin stains (H&E) alone or in combination with iron, chromium copper, aluminum, etc. Usually, nuclear staining is performed to visualize general cell deformity, whereas other routine stains such as, Gram-stain or Lugol's iodine, hematoxylin and eosin stain (H&E), hematoxylin phloxine saffron (HPS), etc. are used for discrimination of various tissue components. For a targeted component study, special stains such as picric acid-methyl blue staining is used for differentiating basement membrane, reticulum, and collagen from nuclei, cytoplasm, and muscle cell. In addition to this, several other staining techniques are used for staining different cell components, for instance, Luna-ishak staining for highlighting bile and iron present in bile canaliculi; Van Gieson's staining procedure for elastic fiber, collagen, and mast cells staining; Fraser-Lendrum stain and Weigert's elastic stain are applied for identifying elastic fibers and fibrin; muscle striations are stained with Bromphenol blue; the Ayoub-Shklar stain is specifically used to stain keratin. Similarly, cell organelles such as golgi apparatus, mitochondria, reticulum, etc. are highlighted using the Defano method, Cains method, and reticulin method, respectively. Taft's stain, Giemsa stain, Feulgen stain, etc. are used for dying nucleic acid. Several other staining methods are used for identifying particular tissues, such as staining of juxtaglomerular apparatus is performed using Smith's method; Maldonado's method, Gomori's method, or Bouin's methods are used for pancreatic islet cells differentiation; for pituitary gland identification Glenner-Lillie method, Monroe-Frommer method, or Wilson-Ezrin methods, etc. are followed. Different organic molecules are also detected using staining procedures. For example, cholestrol, fat, fatty acid, lipid, frozen fat sample, phospholipids, polysaccharides such as glycogen, mucopolysacharides, etc. are highlighted by respective staining methods namely, Schultz method, Oil Red O (ORO) staining, Fischler's method, Osmium tetroxide staining, Baker's method or Menschik's nile blue sulphate method, Periodic acid Schiff (PAS) staining, and Saunder's method. Furthermore, amyloids can be stained by congo red, crystal violet, Sirius Red, or Methyl violet dyes. Urate crystals formed of abnormally accumulated uric acids are also stained through De-Galanthas dye. Histochemical identifications of several inorganic molecules are carried out by following staining procedures, viz., Dahl's method or Von Kossa staining is performed for detection of abnormal calcium depositions; Pizzolatos methods are followed for the detection of calcium oxalate; histochemical detection of copper is performed by Uzman's method; iron as well as ferrous or ferric ion from Hemosiderin can be determined through Meguro staining or Perl's Prussian blue (PPB) staining procedures. Neural cells and tissues viz., astrocytes, axon terminals, nerve fibers, etc. are stained through Cajal's staining method, Nauta-Gygax technique, Gallya's method, and other such methodologies. Likewise many different fluorescence-stains for receptor based studies and EMVs are used for a more minute structure and size details. Most of the studies in this area are governed by OECD Guidance Document on Histopathology (Luna 1968). Most of above discussed techniques can be employed in nano-toxicological studies of animals depending upon facilities of labs. In various in vivo studies, different nanoparticles show pathological lesions or toxicity effect in different organs/tissues of different animals. The toxicity of MNPs in mice has been elucidated through histopatholgy and genotoxicity assessment using the comet assay. Histopathological examination revealed inflammatory cell infiltration between the hepatocytes along with minor congestion and hepatocyte degeneration, focal gliosis in brain, inflammatory cell infiltration in splenic capsule and lymphoid depletion in the white pulp, and inflammatory cell infiltration in the renal capsule or between the renal tubules. MNPs induced significant increase in percentage of DNA in comet tail, indicating the extent of DNA damage (Bassiony et al. 2022). In another study, significant pathological changes in the liver and spleen architecture of mice was observed in AuNPs administration which was size dependent; smaller particles (5 nm) were found to be more toxic than the larger particles (20 and 50 nm) (Ibrahim et al. 2018). Single-walled carbon nanotubes (SWCNTs) and thin multi-walled carbon nanotubes (MWCNTs) have also shown to interact with microtubules, which leads to abnormalities of the mitotic spindles, while iron oxide NPs disrupt the normal structural microtubule network in microvascular endothelial cells (Hubbs et al. 2013). Intravital fluorescence microscopy can also be used to monitor dynamic changes in tissue and direct targeting of NP by visualizing fluorescent tracers and fluorescent nanoparticles. AuNPs have found to bring about widespread histological alterations, aggregation in the neurons of the cerebral cortex and hippocampus as well as in the epithelium of the choroid plexus, along with hyalinization of the wall of some blood vessels and disruption of the capillaries (El-Drieny et al. 2015).

30.4 OMICS Platforms in Toxicological Study

Recent advancement in various disciplines has resulted in the possibility of studying numerous molecular changes at one go by elucidating either the whole genomic or the transcriptomic changes or variations in total metabolites or lipid or protein profiles separately or in an integrated manner. Currently, the enhancement in sensitivity of these analytical modalities has enabled resolution even up to a single cell level on a real time basis. Such precise and high-throughput modalities can elucidate qualitative and quantitative changes in analyte profile induced by the exposure to nanomaterials or similar assailants. Such a comprehensive analysis can pave the way for elucidation of molecular, biochemical, and pathways changes in organelle, cell, tissue, organ, or at an organism level. The analytical platforms include but are not limited to next generation sequencing or Single-molecule real-time (SMRT) sequencing by using droplet-based (Drop-Seq) system, DBiT-seq, single-cell genome long-read sequencing technology (SMOOTH-seq), etc. for both genomics and transcriptomics studies; and various MS and NMR-based platform for proteomics, metabolomics and lipidomics studies (Deng et al. 2021). This platform has shown remarkable efficiency and can be used for drug safety assessment of all drugs including nanomaterial-based formulations. However, exploration through any single omics platform may not capture all the signatures of toxicological assault; therefore, an integrated omics approach is always recommended for a comprehensive analysis.

30.4.1 Genomics Approach for Toxicological Assessment

Recently, with the advent of numerous genomics study platforms, the study of whole genomics has become faster and economical, along with reliable tools for toxicological study. Whole-genome sequencing (WGS) or next generation sequencing (NGS) at pre- as well as post-exposure with nanomaterials can reveal mutation in genome, along with its rate and site. This will not only disclose the susceptibility of different cell lines and gene fragments for nanomaterial-induced genotoxicity but also elucidate the impact at downstream cellular and metabolic pathway levels. Such exploration can be beneficial to determine the doses, route, possible target of amelioration of the nano-drugs, and provide a mechanism to diagnose the toxicity effects. The toxic impact of individual NPs is regulated through its inherent physicochemical properties and exerts a variable impact on different tissues and the organ system. Comprehensive toxicity analysis through omics data can ensure nano-safety at specific uses such as systemic or topical applications, food/feed supplementations, environmental applications, etc. For example, mutagenicity of AgNPs causing transversions, transitions, and deletion of nucleotides at G:C site has been detected in mouse lymphoma cells through whole-genome sequencing (WGS) (Pan et al. 2021). Once the information on the toxicity of nanomaterials has been gathered, further methods, such as microarray-based assays, can be carried out. These methods can also be utilized to identify relevant gene mutations, as was done for the evaluation of the toxicity of MWCNTs (multi walled carbon nanotubes) (Nath 2018). A comprehensive review of the application of genomics platforms to study the effects of different NPs on different organisms has been conducted (Dwivedi et al. 2018).

30.4.2 Trancriptomics

In central dogma of life, m-RNA is the prefinal step of protein formation and has lower stability than DNA, indicating a higher susceptibility to any toxicological effect of nanomaterial and can have a more pronounced effect on cells or organs. For instance, Quantum dots (QDs), which have numerous industrial applications, have toxicological effects when exposed to cells. The toxicity resulting from both cadimium-induced alterations in transcriptomics and non-cadmium-based QDs has been noted (Nath 2018). The platform used for whole transcriptomics analyses is similar to the genomics study including NGS and microarray. These platforms help in identifying the changes in the sequence of RNA and also relative changes in the copy number of RNAs due to exposure to nanomaterials. Once established, a less expensive platform, such as one based on q-PCR, can be used to regularly confirm one or more RNA copy number changes. To ensure nano-safety, these emerging platforms need to be incorporated into the OECD guidelines for accurate toxicity analysis (Doak et al. 2012). Similarly, when toxicity testing in animal model is not feasible, different in-vitro models such as three dimensional cell cultures, organ on chip, organoids or spheroid models can be used for nanotoxicity analysis through trancriptomic profiling of the engineered nanomaterials (ENMs) (Rahman et al. 2020; Saarimäki et al. 2021; Lian et al. 2022; Prasad et al. 2021).

30.4.3 Proteomics Platform

Proteomics platforms including Mass Spectrometry-based techniques, Mass Cytometry, Fluorescence Flow Cytometry, Enzyme-Linked Immunospot Assay, Single-Molecule Array (SiMoA), Microfluidic Antibody Capture Chip (MACS Chip)-based assay, Proximity Extension Assay/Abseq, etc. and associated softwares help in generating and analyzing all the proteins in a cell, tissue, organ, organism, or biological fluid at a particular temporal framework to obtain the complete protein profile or any deviations in terms of both quality and quantity. Proteins are the work horses of the central dogma that carry out the functional and regulatory assignments in cellular and metabolic pathways necessary for survival, reproduction, production, development, etc.. Understanding the dynamicity of protein profiling within the organism in pre and post exposure to nanomaterial is one of the major evaluation steps to elucidate the time and concentration-dependent nanotoxicity impact on specific subjects (Minakshi et al. 2019a). In a nanomaterial based toxicological study, the application of proteomics approaches is progressively getting attractive. For instance, MS-based proteomics analysis of human macrophage revealed several potential protein biomarkers involved in the oxidative stress response, protein chaperone functions, and autophagy as indicators of engineered nanomaterials-induced cellular stress (Zhang et al. 2020). An integrated proteomics and metabolomics approach using LC-MS has also demonstrated that gold NPs have a significant impact on the cellular proteomic profile associated with DNA synthesis and repair, protein synthesis, and amino acid transport, along with deregulation of glutathione metabolism (Gioria et al. 2016). Proteomics exploration has also unearthed diverse information regarding NP-based effects on organisms which include accumulation of various protein precursors at certain sites, effect on membrane stability, protein digestion, alteration in protein metabolism at a cellular and organelle level, etc. Over- or under-expression of oxidative stress response and tricarboxilic acid (TCA) cycle proteins, redox-related proteins, alteration in biosynthesis and sorting of proteins having enzymatic activities, etc. can be valuable indicators of nanotoxicity at a proteomic level (Tubatsi et al. 2022).

30.4.4 Metabolomics Techniques

Metabolites are the end products which regulate the life cycle and alteration in metabolic products leads to genesis of various diseases and disorder. Understanding NP-induced toxicity through the qualitative or quantitative changes in metabolites, alteration in pathways in a comprehensive manner within the cell, tissue, organ, and at an organism level in a concentration and time-dependent manner is the key to determine the safety of nanomaterials for animal and human application of nanodrugs. This can be achieved by many platforms such as classical methods using autoanalyzer for numerous biochemical analytes, chromatography-coupled with MS-based techniques, or recently developed various sensors etc. Both LC or GC-MS and NMR-based platforms are versatile and are conveniently employed for metabolomics analyses. In addition, other platforms such as microarray, Fourier transform infrared (FTIR) spectroscopy, PCR-Based WGA Methods, fluorescence sensor-based metabolite identification are also common place in this arena which can also be relevant for nanotoxicity assessment. State-of-the-art mass spectrometry imaging (MSI) has been evolved as one of the most precise modality for single-cell analyses in real-time which also hold potential for toxicological introspection (Single-Cell Omics) (Minakshi et al. 2019b; Kumar et al. 2020). Employing GC/MS-based approach, size-and dose-dependent toxicity of SiO_2 particles has been elucidated in mice in terms of altered metabolites of Krebs cycle and redox machinery. Similarly, an NMR-based metabolomics has revealed various alterations in urinary metabolite profile induced by 5 nm titanium dioxide (TiO_2) particles in rat (Schnackenberg et al. 2012). Alike, several metabolites were significantly altered in human embryo kidney 293 cells on exposure to silica-coated magnetic nanoparticles as elucidated by GC-MS analyses (Shim et al. 2012). Furthermore, LC-HRMS/MS analysis has also revealed several altered metabolites and corresponding biological pathways in Caco-2 cells on exposure to AuNPs (Gioria et al. 2016).

30.5 Conclusions

Nanomaterials have certainly provided multi-faceted benefits in diverse applications leading to progressive enhancement in nano-interventions and nano-accumulation in the global biome. Majority of the investigations focus on highlighting the beneficial effects of all nano-formulations but the toxicity aspect often remains neglected. Some introspection has been carried out, but only to elucidate the short-term effect. Long-term toxicity aspect, bio-remediation, or environmental impacts of these formulations are mostly missing. Moreover, the cell culture model has occupied the bulk of the literature but it hardly mimics the near-physiological microenvironment. Organoids or 3D culture systems may be preferred over the monolayer culture as in-vitro model as the former mimics the in-vivo set up to a significant extent. Even in-vivo labs, animal models never perfectly enumerate the subject-wise toxicity effects due to the enormous anatomical and physiological variations among species. Despite the availability of multiple analytical modalities, most of the toxicological studies often rely upon a few of the conventional analytical methods which only depict some discrete screen-shots of toxicological impacts. A comprehensive toxicity analysis involving multiple models and integrated analytical platforms may be beneficial to elucidate a near-perfect picture. The promising effects of nanotechnology and the flourishing potential of nanomedicine are certainly propelling the flooding of markets with numerous nano-products, but without proper measures to check on the nanotoxicity of these products, it is going to be a fiasco.

References

Aguilar ZP (2013) Nanotoxicology and remediation. Nanomat Med Appl:361–408. https://doi.org/10.1016/B978-0-12-385089-8.00008-X

Ames BN, Lee FD, Durston WE (1973) An improved bacterial test system for the detection and classification of mutagens and carcinogens. Proc Natl Acad Sci U S A 70(3):782–786. https://doi.org/10.1073/pnas.70.3.782

Bassiony H, El-Ghor AA, Salaheldin TA, Sabet S, Mohamed MM (2022) Tissue distribution, histopathological and genotoxic effects of magnetite nanoparticles on Ehrlich solid carcinoma. Biol Trace Elem Res 200(12):5145–5158. https://doi.org/10.1007/s12011-022-03102-z

Bawa R, Szebeni J, Webster TJ, Audette GF (2019) Immune aspects of biopharmaceuticals and nanomedicines. Jenny Stanford Publishing, Singapore

Buhr CR, Eckrich J, Kluenker M, Bruns K, Wiesmann N, Tremel W, Brieger J (2021) Determination of the LD50 with the chick embryo chorioallantoic membrane (CAM) assay as a promising alternative in nanotoxicological evaluation. Nanotoxicology 15(5):690–705

Chen Z, Meng H, Xing G, Chen C, Zhao Y, Jia G, Wan L (2006) Acute toxicological effects of copper nanoparticles in vivo. Toxicol Lett 163(2):109–120

Chousidis I, Stalikas CD, Leonardos ID (2020) Induced toxicity in early-life stage zebrafish (Daniorerio) and its behavioral analysis after exposure to non-doped, nitrogen-doped and nitrogen, sulfur-co doped carbon quantum dots. Environ Toxicol Pharmacol 79:103426. https://doi.org/10.1016/j.etap.2020.103426

Ciappellano SG, Tedesco E, Venturini M, Benetti F (2016) In vitro toxicity assessment of oral nanocarriers. Adv Drug Deliv Rev 106:381–401. https://doi.org/10.1016/j.addr.2016.08.007

Deng Y, Zhang D, Liu Y, Su G, Enninful A, Bai Z, Fan R (2021) Spatial epigenome sequencing at tissue scale and cellular level. BioRxiv:2021–2003

Doak SH, Manshian B, Jenkins GJS, Singh N (2012) In vitro genotoxicity testing strategy for nanomaterials and the adaptation of current OECD guidelines. Mutat Res 745(1–2):104–111

Donaldson K, Stone V, Borm PJ, Jimenez LA, Gilmour PS, Schins RP, Knaapen AM, Rahman I, Faux SP, Brown DM, MacNee W (2003) Oxidative stress and calcium signaling in the adverse effects of environmental particles (PM10). Free Radic Biol Med 34(11):1369–1382. https://doi.org/10.1016/s0891-5849(03)00150-3

Dwivedi S, Saquib Q, Ahmad B, Ansari SM, Azam A, Musarrat J (2018) Toxicogenomics: a new paradigm for nanotoxicity evaluation. Cell Mol Toxicol Nanopart:143–161

El-Drieny EAEA, Sarhan NI, Bayomy NA, Elsherbeni SAE, Momtaz R, Mohamed HED (2015) Histological and immunohistochemical study of the effect of gold nanoparticles on the brain of adult male albino rat. J Microsc Ultrastruct 3(4):181–190. https://doi.org/10.1016/j.jmau.2015.05.001

Elkhawass EA, Mohallal ME, Soliman MF (2015) Acute toxicity of different sizes of silver nanoparticles intraperitonally injected in Balb/C mice using two toxicological methods. Int J Pharm PharmSci 7(2):94–99

El-Nahass EE, El-Naggar SA, Salem BI (2021) Evaluation the toxic effects of cobalt/zinc/ferrite nano-complex in experimental mice

Gioria S, Lobo Vicente J, Barboro P, La Spina R, Tomasi G, Urbán P, Chassaigne H (2016) A combined proteomics and metabolomics approach to assess the effects of gold nanoparticles in vitro. Nanotoxicology 10(6):736–748

Gupta A, Kumar S, Kumar V (2019) Challenges for assessing toxicity of nanomaterials. Biochemical toxicology-heavy metals and nanomaterials. https://doi.org/10.5772/intechopen.89601

Henderson RF, Benson JM, Hahn FF, Hobbs CH, Jones RK, Mauderly JL, McClellan RO, Pickrell JA (1985) New approaches for the evaluation of pulmonary toxicity: bronchoalveolar lavage fluid analysis. Fundam Appl Toxicol 5(3):451–458. https://doi.org/10.1016/0272-0590(85)90092-2

Hissin PJ, Hilf R (1976) A fluorometric method for determination of oxidized and reduced glutathione in tissues. Anal Biochem 74(1):214–226. https://doi.org/10.1016/0003-2697(76)90326-2

Horie M, Tabei Y (2021) Role of oxidative stress in nanoparticles toxicity. Free Radic Res 55(4):331–342. https://doi.org/10.1080/10715762.2020.1859108

Hoseini-Alfatemi SM, Fallah F, Armin S, Hafizi M, Karimi A, Kalanaky S (2020) Evaluation of blood and liver cytotoxicity and apoptosis-necrosis induced by nanochelating based silver nanoparticles in mouse model. Iran J Pharm Res 19(2):207

Hubbs AF, Sargent LM, Porter DW, Sager TM, Chen BT, Frazer DG, Mercer RR (2013) Nanotechnology: toxicologic pathology. Toxicol Pathol 41(2):395–409. https://doi.org/10.1177/0192623312467403

Ibrahim KE, Al-Mutary MG, Bakhiet AO, Khan HA (2018) Histopathology of the liver, kidney, and spleen of mice exposed to gold nanoparticles. Molecules 23(8):1848. https://doi.org/10.3390/molecules23081848

Jakubowski W, Bartosz G (2000) 2,7-dichlorofluorescin oxidation and reactive oxygen species: what does it measure? Cell Biol Int 24(10):757–760. https://doi.org/10.1006/cbir.2000.0556

Kabanov AV (2006) Polymer genomics: an insight into pharmacology and toxicology of nanomedicines. Adv Drug Deliv Rev 58(15):1597–1621

Kendig DM, Tarloff JB (2007) Inactivation of lactate dehydrogenase by several chemicals: implications for in vitro toxicology studies. Toxicol In Vitro 21(1):125–132

Koopman G, Reutelingsperger CP, Kuijten GA, Keehnen RM, Pals ST, van Oers MH (1994) Annexin V for flow cytometric detection of phosphatidylserine expression on B cells undergoing apoptosis. Blood 84(5):1415–1420

Kumar V, Sharma N, Maitra SS (2017) In vitro and in vivo toxicity assessment of nanoparticles. Int Nano Lett 7(4):243–256

Kumar R, Ghosh M, Kumar S, Prasad M (2020) Single cell metabolomics: a future tool to unmask cellular heterogeneity and virus-host interaction in context of emerging viral diseases. Front Microbiol 11:1152

Lan J, Gou N, Gao C, He M, Gu AZ (2014) Comparative and mechanistic genotoxicity assessment of nanomaterials via a quantitative toxicogenomics approach across multiple species. Environ Sci Technol 48(21):12937–12945

Lee IC, Ko JW, Park SH, Lim JO, Shin IS, Moon C et al (2016) Comparative toxicity and biodistribution of copper nanoparticles and cupric ions in rats. Int J Nanomedicine 11:2883

Lian J, Liu W, Sun Y, Men S, Wu J, Zeb A, Zhou Q (2022) Nanotoxicological effects and transcriptome mechanisms of wheat (Triticumaestivum L.) under stress of polystyrene nanoplastics. J Hazard Mater 423:127241

Luna LG (1968) Manual of histologic staining methods of the armed forces Institute of Pathology, 3rd edn. McGraw-Hill, New York

MacNee W (2001) Oxidative stress and lung inflammation in airways disease. Eur J Pharmacol 429(1–3):195–207

Martínez M, Martínez NA, Silva WI (2017) Measurement of the intracellular calcium concentration with fura-2 AM using a fluorescence plate reader. Bio-protocol 7(14):e2411–e2411

Minakshi P, Kumar R, Ghosh M, Saini HM, Ranjan K, Brar B, Prasad G (2019a) Single-cell proteomics: technology and applications. In Single-cell omics. Academic Press, Cambridge, MA, pp 283–318

Minakshi P, Ghosh M, Kumar R, Patki HS, Saini HM, Ranjan K, Brar B, Prasad G (2019b) Single-cell metabolomics: technology and applications. In Single-cell omics. Academic Press, Cambridge, MA, pp 319–353

Nagalakshmi G, Maity T, Maiti BC (2013) Synthesis, characterization and anticancer evaluation of some novel 2-(substitutedphenyl)-5-methyl-3-(phenylamino)-1,3-thiazolidin-4-ones. Pharm Lett 5:261–277

Nath D (2018) Genomic approach of nanotoxicity evaluation. In Nanotoxicology: toxicity evaluation, risk assessment and management. CRC Press/Taylor & Francis Group, Boca Raton, FL, pp 449–462

Oliveira RG, Correia PM, Silva AL, Encarnação PM, Ribeiro FM, Castro IF, Veloso JF (2022) Development of a new integrated system for vital sign monitoring in small animals. Sensors 22(11):4264. https://doi.org/10.3390/s22114264

Pan B, Kaldhone PR, Alund AW, Du H, Guo X, Yan J, Chen T (2021) Mutagenicity of silver nanoparticles evaluated using whole-genome sequencing in mouse lymphoma cells. Nanotoxicology 15(3):418–432

Prasad M, Kumar R, Buragohain L, Kumari A, Ghosh M (2021) Organoid technology: a reliable developmental biology tool for organ-specific nanotoxicity evaluation. Front Cell Dev Biol 9: 696668

Rahman L, Williams A, Gelda K, Nikota J, Wu D, Vogel U, Halappanavar S (2020) 21st century tools for nanotoxicology: transcriptomic biomarker panel and precision-cut lung slice organ mimic system for the assessment of nanomaterial-induced lung fibrosis. Small 16(36):2000272

Riss T, Niles A, Moravec R, Karassina N, Vidugiriene J (2019) Cytotoxicity assays: in vitro methods to measure dead cells. Assay guidance manual https://www.ncbi.nlm.nih.gov/books/NBK540958/

Saarimäki LA, Federico A, Lynch I, Papadiamantis AG, Tsoumanis A, Melagraki G, Greco D (2021) Manually curated transcriptomics data collection for toxicogenomic assessment of engineered nanomaterials. Sci Data 8(1):49

Schnackenberg LK, Sun J, Beger RD (2012) Metabolomics techniques in nanotoxicology studies. Nanotoxicity: Methods and Protocols, pp 141–156

Shaikh S, Shyama SK, Desai PV (2015) Absorption, LD50 and effects of CoO, MgO and PbO nanoparticles on mice "musmusculus"

Shaker AM, El-Shahawy A, Zaki AH, Abdel-Rahim EF, Khedr MH (2017) Estimation the median lethal dose and inhibitory concentration of TiO2, SiO2, ZnO and CuO nanoparticles on human hepatoma HEPG2 cells. Int J Pharm Phytopharmacol Res 7(6):18–23

Shakibaie M, Shahverdi AR, Faramarzi MA, Hassanzadeh GR, Rahimi HR, Sabzevari O (2013) Acute and subacute toxicity of novel biogenic selenium nanoparticles in mice. Pharm Biol 51(1):58–63

Shim W, Paik MJ, Nguyen DT, Lee JK, Lee Y, Kim JH et al (2012) Analysis of changes in gene expression and metabolic profiles induced by silica-coated magnetic nanoparticles. ACS Nano 6(9):7665–7680

Shukla RK, Badiye A, Vajpayee K, Kapoor N (2021) Genotoxic potential of nanoparticles: structural and functional modifications in DNA. Front Genet 12:728250

Singh NP, McCoy MT, Tice RR, Schneider EL (1988) A simple technique for quantitation of low levels of DNA damage in individual cells. Exp Cell Res 175(1):184–191

Singh S, Shi T, Duffin R, Albrecht C, van Berlo D, Höhr D et al (2007) Endocytosis, oxidative stress and IL-8 expression in human lung epithelial cells upon treatment with fine and ultrafine TiO2: role of the specific surface area and of surface methylation of the particles. Toxicol Appl Pharmacol 222(2):141–151

Trotter PJ, Voelker DR (1995) Identification of a non-mitochondrial phosphatidylserine decarboxylase activity (PSD2) in the yeast Saccharomyces cerevisiae (*). J Biol Chem 270(11): 6062–6070

Tubatsi G, Kebaabetswe LP, Musee N (2022) Proteomic evaluation of nanotoxicity in aquatic organisms: a review. Proteomics 22(21):2200008

Vijay U, Gupta S, Mathur P, Suravajhala P, Bhatnagar P (2018) Microbial mutagenicity assay: ames test. Bio-protocol 8(6):e2763–e2763. https://doi.org/10.21769/BioProtoc.2763

Wang D, Li H, Liu Z, Zhou J, Zhang T (2017) Acute toxicological effects of zinc oxide nanoparticles in mice after intratracheal instillation. Int J Occup Environ Health 23(1):11–19

Wroblewski F, Ladue JS (1955) Lactic dehydrogenase activity in blood. Proc Soc Exp Biol Med 90(1):210–213. https://doi.org/10.3181/00379727-90-21985

Xie S, Wang F, Wang Y, Zhu L, Dong Z, Wang X, Zhou W (2011) Acute toxicity study of tilmicosin-loaded hydrogenated castor oil-solid lipid nanoparticles. Part Fibre Toxicol 8:1–10

Yoshida R, Kitamura D, Maenosono S (2009) Mutagenicity of water-soluble ZnO nanoparticles in Ames test. J Toxicol Sci 34(1):119–122. https://www.jstage.jst.go.jp/article/jts/34/1/34_1_119/_pdf

Yu Y, Li Y, Wang W, Jin M, Du Z, Li Y, Sun Z (2013) Acute toxicity of amorphous silica nanoparticles in intravenously exposed ICR mice. PLoS One 8(4):e61346

Zhang XD, Wu HY, Wu D, Wang YY, Chang JH, Zhai ZB, Fan FY (2010) Toxicologic effects of gold nanoparticles in vivo by different administration routes. Int J Nanomedicine:771–781

Zhang T, Gaffrey MJ, Thomas DG, Weber TJ, Hess BM, Weitz KK, Thrall BD (2020) A proteome-wide assessment of the oxidative stress paradigm for metal and metal-oxide nanomaterials in human macrophages. NanoImpact 17:100194

Zielińska A, Costa B, Ferreira MV, Miguéis D, Louros JMS, Durazzo A, Lucarini M, Eder P, Chaud VM, Morsink M, Willemen N, Severino P, Santini A, Souto EB (2020) Nanotoxicology and nanosafety: safety-by-design and testing at a glance. Int J Environ Res Public Health 17(13): 4657. https://doi.org/10.3390/ijerph17134657

Chapter 31
Regulatory Issues in Nanotechnology

Mayukh Ghosh and Rajesh Kumar

Abstract Nanotechnology, the manipulation and application of matter at the nanoscale, presents numerous opportunities for scientific advancement and innovation. However, it also raises regulatory concerns due to its potential impact on health, safety, and the environment. The key regulatory issues associated with nanotechnology include but are not limited to risk assessment and toxicity of nanomaterials, definition and characterization, labeling of nano-products and consumer protection, environmental impact assessment, international harmonization, intellectual property and patent protection, ethical and societal considerations, etc.

Nanomaterials exhibit different properties compared to their bulk counterparts, which may introduce unique risks. Regulators face challenges in assessing the toxicity and potential hazards associated with these materials. Developing standardized testing methods and risk assessment frameworks specific to nanomaterials is crucial. Defining and characterizing nanomaterials is a complex task. Regulatory agencies often struggle to establish clear definitions and measurement techniques to identify and differentiate nanomaterials. The size, shape, and surface properties of nanoparticles can significantly influence their behavior and potential risks. Proper labeling of products containing nanomaterials is essential for consumer awareness and choice. Regulators need to determine whether specific labeling requirements are necessary for products that incorporate nanotechnology. Nanomaterials may have unforeseen environmental consequences. They can potentially accumulate in ecosystems and impact wildlife. Additionally, the release of nanoparticles during manufacturing, use, or disposal of nano- products can raise concerns. Regulatory frameworks need to address the environmental risks and implement appropriate monitoring and mitigation measures. Efforts to establish common standards and

M. Ghosh (✉)
Department of Veterinary Physiology and Biochemistry, Banaras Hindu University, Mirzapur, Uttar Pradesh, India

R. Kumar
Department of Veterinary Physiology and Biochemistry, Lala Lajpat Rai University of Veterinary and Animal Sciences, Hisar, India

LLR University of Veterinary and Animal, ICAR-National Research Centre on Equines, *Nanotechnology Theranostics in Livestock Diseases and Management*, Livestock Diseases and Management,
https://doi.org/10.1007/978-981-16-1610-5_31

guidelines across countries through international harmonization can facilitate the responsible development and commercialization of nanotechnology while ensuring global safety and regulatory consistency. Nanotechnology inventions raise unique intellectual property challenges due to their multidisciplinary nature. Determining patent eligibility, novelty, and non-obviousness criteria for nanotechnology-related inventions should also be addressed to encourage innovation while protecting intellectual property rights. The ethical implications of nanotechnology, such as privacy concerns, potential for misuse, and equitable access, should be considered. Regulatory bodies need to ensure that societal values and concerns are incorporated into the regulatory decision-making process. Regulatory agencies around the world are actively addressing these issues through collaborations, research initiatives, and the development of specific guidelines and regulations for nanotechnology. The goal is to foster the safe and responsible development, production, and use of nanomaterials and nanotechnology-enabled products.

Keywords Nanomedicine · Nanomaterials · Regulatory challenges · Risk assessment · Ecotoxicity · Nanotoxicology · Regulatory organizations

31.1 Introduction

Nanomedicine refers to the use of nanotechnology in the field of medicine, specifically for diagnosis, treatment, and prevention of diseases. It involves the manipulation and control of materials and devices at the nanoscale (typically 1–100 nanometers) to create innovative medical products and therapies. While nanomedicine holds great promise for advancements in healthcare, it also raises concerns regarding safety, efficacy, and ethical considerations. Therefore, the need for nanomedicine regulation becomes crucial for several reasons.

The first reason is "Safety and Efficacy." Nanomedicine involves the use of engineered nanomaterials that can interact with biological systems in unique ways due to their small size and properties. These interactions can have both beneficial and potentially harmful effects on the human body. Regulation helps ensure that nanomedicines undergo rigorous testing to evaluate their safety and efficacy before they are approved for clinical use. The second reason includes "Risk Assessment." Nanomaterials may have different properties compared to their bulk counterparts, which can influence their behavior in the body. It is essential to assess the potential risks associated with exposure to nanomedicines, such as toxicity, environmental impact, and long-term effects. Regulation provides guidelines and standards for risk assessment, helping to protect public health and the environment. The third aspect is "Quality Control." Nanomedicine products need to meet specific quality standards to ensure their reliability and consistency. Regulation helps establish manufacturing standards, quality control processes, and guidelines for product characterization and testing. This ensures that nanomedicines are produced consistently and meet the required quality standards for safe and effective use. The fourth reason is "Ethical Considerations." Nanomedicine raises ethical questions related to privacy, informed consent, equity of access, and potential unintended consequences. Regulation can address these concerns by providing guidelines for ethical conduct, ensuring patient rights, and promoting transparency and accountability in research and development. The fifth reason includes "International Harmonization." Nanomedicine is a global field with potential cross-border implications. Harmonized regulation and standards facilitate international collaboration, exchange of knowledge, and consistent oversight of nanomedicine products across different jurisdictions. This ensures that the benefits of nanomedicine are maximized while minimizing potential risks and inconsistencies (Foulkes et al. 2020; Đorđević et al. 2021).

Therefore, nanomedicine regulation is necessary to ensure the safety, efficacy, and ethical use of nanomedicine products. By establishing guidelines, standards, and oversight, regulation helps protect public health, ensure quality control, address ethical concerns, and facilitate international collaboration in this rapidly advancing field.

31.2 Approval Process for Small Drug Molecules

The approval process for small drug molecules typically involves several stages of rigorous testing and evaluation by regulatory authorities (Van Norman 2016; Foulkes et al. 2020). A general overview of the process has been depicted hereunder.

1. Preclinical Testing: Before a drug can be tested on humans, it undergoes extensive preclinical testing in the laboratory and on animals. This stage involves in vitro (test tube) studies and animal testing to assess the drug's safety profile, efficacy, and potential side effects.
2. Investigational New Drug (IND) Application: If the preclinical testing results are promising, the drug's sponsor (usually a pharmaceutical company) files an IND application with the regulatory authority, such as the U.S. Food and Drug Administration (FDA). The IND application contains comprehensive data from preclinical studies, as well as information about the proposed clinical trials.
3. Clinical Trials: Clinical trials are conducted in three phases, typically involving increasing numbers of human subjects:

   Phase 1: Small-scale trials conducted on a few healthy volunteers to evaluate the drug's safety, dosage, and potential side effects.

   Phase 2: Larger trials involving several hundred patients with the target condition or disease to assess the drug's efficacy and further evaluate its safety.

   Phase 3: Extensive trials with thousands of patients to confirm the drug's effectiveness, monitor side effects, and compare it to existing treatments.

   New Drug Application (NDA) or Biologics License Application (BLA): If the clinical trial results demonstrate the drug's safety and efficacy, the drug's sponsor submits an NDA (for small drug molecules) or BLA (for biologic drugs) to the regulatory authority. This application includes comprehensive data from preclinical and clinical studies, manufacturing details, proposed labeling, and information on the drug's pharmacology and toxicology.

4. Regulatory Review: The regulatory authority reviews the NDA or BLA to ensure that the drug meets the necessary standards for safety and efficacy. This review process involves multiple experts who evaluate the submitted data, conduct inspections of manufacturing facilities, and assess the drug's benefit-risk profile.
5. Approval Decision: Based on the review findings, the regulatory authority decides whether to approve the drug. If approved, the drug can be marketed for the specified indications, and the sponsor receives a marketing authorization or license.
6. Post-Marketing Surveillance: Once a drug is approved and on the market, ongoing surveillance occurs to monitor its safety and effectiveness in larger populations. Adverse events and new side effects are reported, and regulatory authorities can take action if safety concerns arise.

It's important to note that the approval process can vary between countries, and this overview represents a general framework. The process can be time-consuming, expensive, and may take several years to complete.

31.3 Regulatory Issues in Nanotechnology

Regulatory issues in nanotechnology refer to the challenges and concerns surrounding the regulation of nanomaterials, nanodevices, and nanotechnology-based products. The major regulatory issues in nanotechnology include but are not limited to definition and classification of nanomaterials, risk assessment and characterization, toxicity and health effects, environmental impact, labeling and disclosure of nanotechnology-based products, international harmonization to avoid regulatory disparities among regulatory agencies from different countries, and the inherent limitations associated with rapidly evolving materials, applications, and processes in nanotechnology (Allan et al. 2021; Sores et al. 2018). Addressing these regulatory issues requires collaboration among governments, scientific communities, industry stakeholders, and regulatory agencies. Striking a balance between promoting innovation and ensuring safety is crucial for the responsible development and commercialization of nanotechnology.

31.3.1 Challenges in Defining and Classifying Nanomaterials

Defining and classifying nanomaterials presents several challenges due to their unique properties and dimensions (Allan et al. 2021; Sores et al. 2018; Miernicki et al. 2019). The key challenges in this perspective are as follows:

1. **Size-based Definition:** Nanomaterials are typically defined based on their size, where at least one dimension falls within the nanoscale range (1–100

nanometers). However, determining the precise measurement of nanoscale dimensions can be challenging, as different measurement techniques may yield varying results. There is ongoing debate about whether to consider the external dimensions or internal structure of nanomaterials for classification purposes.

2. **Aggregation and Agglomeration:** Nanomaterials often tend to agglomerate or aggregate, forming larger clusters or particles. These agglomerates may have different properties than individual nanoparticles. Determining how to define and classify nanomaterials when they exist in various forms (individual particles, agglomerates, or aggregates) presents a challenge.
3. **Shape and Surface Area:** Nanomaterials exhibit a wide range of shapes, including spheres, rods, tubes, and more complex structures. Different shapes can result in varying properties and behaviors. Additionally, the high surface area-to-volume ratio of nanomaterials contributes to their unique characteristics. Developing a classification system that accounts for these shape-dependent properties and surface area considerations is complex.
4. **Dynamic Nature:** Nanomaterials can exhibit dynamic properties, undergoing changes in size, shape, or structure under different conditions or environments. Defining and classifying nanomaterials that display such dynamic behavior is a challenge, as they may not fit into fixed categories based on static properties.
5. **Heterogeneity:** Nanomaterials can exhibit heterogeneity within a given sample, meaning that particles of the same nominal size may have variations in size, shape, or composition. Developing classification systems that account for this heterogeneity is difficult, as it requires defining criteria that capture the range of properties present in a sample.
6. **Multi-component Nanomaterials:** Many nanomaterials are composed of multiple components, such as core-shell nanoparticles or nanoparticles functionalized with various coatings or ligands. Determining how to define and classify these multi-component nanomaterials adds complexity to the classification process.

Addressing these challenges requires a comprehensive understanding of the physicochemical properties and behavior of nanomaterials. Collaboration among scientists, regulators, and stakeholders is necessary to establish consistent definitions and classification schemes that can accommodate the diverse range of nanomaterials encountered in research, manufacturing, and product development.

31.3.2 Risk Assessment and Characterization of Nanomaterials

Risk assessment and characterization of nanomaterials are essential steps in ensuring their safe handling and use. Nanomaterials, which have unique properties due to their small size (typically less than 100 nanometers), require specific considerations for their potential hazards and risks (Allan et al. 2021; Sores et al. 2018). An overview of the risk assessment and characterization process for nanomaterials has been delineated hereunder.

31.3.2.1 Characterization of Nanomaterials

Characterization involves determining the physical, chemical, and morphological properties of nanomaterials. This step helps in understanding their behavior, potential interactions with biological systems, and environmental fate. Characterization techniques include electron microscopy, spectroscopy, surface analysis, and particle size analysis.

31.3.2.2 Hazard Identification

Hazard identification aims to assess the potential adverse effects of nanomaterials on human health and the environment. It involves studying their intrinsic properties, such as size, shape, surface area, surface charge, chemical composition, and reactivity. Existing toxicological data on similar materials and related literature are also considered. This step helps in identifying the hazards associated with nanomaterials.

31.3.2.3 Exposure Assessment

Exposure assessment evaluates the potential routes and levels of human or animal and environmental exposure to nanomaterials. It considers various scenarios, such as occupational exposure during manufacturing, consumer exposure during product use, and release into the environment. Factors like particle size distribution, concentration, duration, and frequency of exposure are taken into account. Computational models, workplace monitoring, and experimental studies contribute to exposure assessment.

31.3.2.4 Risk Characterization

Risk characterization integrates the hazard identification and exposure assessment results to estimate the potential risks posed by nanomaterials. It involves the interpretation and integration of available data to determine the likelihood and severity of adverse effects. Risk characterization also considers the population at risk, such as animals getting nano-exposure, consumers, or vulnerable groups. Uncertainties and variability in data are addressed through sensitivity analysis and modeling.

31.3.2.5 Risk Management

Risk management involves the development and implementation of strategies to minimize or eliminate the risks associated with nanomaterials. This includes regulatory measures, engineering controls, safe handling practices, personal protective equipment, and risk communication. Risk management measures aim to protect human health and the environment while enabling the safe and sustainable use of nanomaterials.

It is important to note that risk assessment and characterization of nanomaterials are ongoing processes due to the evolving nature of scientific knowledge. Continued research, monitoring, and collaboration among scientists, regulators, and industry stakeholders are necessary to enhance our understanding of nanomaterials and ensure their safe use.

31.3.3 Regulatory Issues Regarding Potential Toxicity and Health Effects of Nanomaterials

Regulatory issues regarding the potential toxicity and health effects of nanomaterials are of significant concern due to their unique properties and potential risks (Dunphy Guzmán et al. 2006; Borm et al. 2006; Oberdorster et al. 2005). Some key regulatory issues associated with nanomaterial toxicity and health effects have been described hereunder.

31.3.3.1 Risk Assessment

One of the primary challenges is conducting accurate and reliable risk assessments for nanomaterials. Traditional toxicity testing methods may not be suitable for assessing the potential risks of nanomaterials due to their small size and unique properties. Developing appropriate testing protocols and standards specific to nanomaterials is crucial for effective risk assessment.

31.3.3.2 Exposure Assessment

Assessing the exposure routes and levels of nanomaterials is essential for determining potential health risks. Nanomaterials can enter the body through inhalation, dermal contact, or ingestion, and exposure can occur during manufacturing, use, or disposal. Understanding and quantifying exposure scenarios are critical for evaluating their toxicity and establishing exposure limits.

31.3.3.3 Hazard Characterization

Characterizing the toxicity and potential health effects of nanomaterials is challenging due to their unique physicochemical properties. Nanomaterials may exhibit different behavior, reactivity, and bioavailability compared to their bulk counterparts, potentially leading to different toxicity profiles. Identifying and understanding the specific hazards associated with different types of nanomaterials are crucial for effective regulation.

31.3.3.4 Standardization and Testing Methods

Developing standardized testing methods and protocols specific to nanomaterials is essential for accurate and reproducible toxicity assessments. The existing testing methods may need to be adapted or new methods developed to account for the unique characteristics of nanomaterials. Harmonizing testing protocols across regulatory agencies can ensure consistency in the assessment of nanomaterial toxicity.

31.3.3.5 Risk Communication

Communicating the potential risks associated with nanomaterials to animal owners, consumers, and the public is vital for informed decision-making and risk management. Regulatory agencies need to establish clear guidelines for risk communication, including labeling requirements, safety data sheets, and public information campaigns to ensure that relevant stakeholders are aware of the potential hazards and appropriate safety measures.

31.3.3.6 Regulatory Frameworks

Updating and adapting existing regulatory frameworks to address nanomaterials' unique toxicity concerns are necessary. Some regulatory agencies have already started to integrate nanomaterial-specific regulations into their frameworks, while others are in the process of developing specific guidelines. Coordinating regulatory efforts at the national and international levels is crucial to ensure consistent and effective regulation of nanomaterial toxicity.

31.3.3.7 Emerging Risks and Nanomaterials

As nanotechnology advances and new nanomaterials are developed, it is essential for regulatory agencies to proactively anticipate and address potential health risks. The rapid pace of innovation in nanotechnology requires regulatory frameworks that can keep up with emerging materials and applications to ensure the safety of workers, consumers, and the environment.

Addressing these regulatory issues requires collaboration among regulators, researchers, industry stakeholders, and policymakers. The development of robust testing methods, standardized protocols, and harmonized regulations specific to nanomaterials is crucial to ensure the safe and responsible development and use of nanotechnology products.

31.3.4 Environmental Impact of Nanomaterials

The environmental impact of nanomaterials is an important consideration due to their unique properties and potential interactions with ecosystems. While nanomaterials offer various benefits, their release into the environment raises concerns about their potential effects on living organisms and ecosystems (Ch 2009; Kolesnikov et al. 2021). Some of the key considerations regarding the environmental impact of nanomaterials have been described hereunder.

31.3.4.1 Fate and Transport

Nanomaterials can behave differently from their bulk counterparts in terms of fate and transport in the environment. Their small size and high surface area-to-volume ratio can enhance their mobility, facilitating their dispersion and transport in air, water, and soil. This can lead to a wider distribution and potential exposure to organisms.

31.3.4.2 Ecotoxicity

Nanomaterials can have direct or indirect effects on organisms and ecosystems. The unique physicochemical properties of nanomaterials, such as their small size and reactivity, can result in altered biological interactions. They may have toxic effects on aquatic and terrestrial organisms, affecting their growth, reproduction, behavior, and overall ecosystem functioning. Different nanomaterials may exhibit varying degrees of ecotoxicity.

31.3.4.3 Bioaccumulation and Biomagnification

Some nanomaterials have the potential to bioaccumulate in organisms. Due to their small size, they can be taken up by organisms through various routes such as ingestion, inhalation, or dermal contact. Once inside the organisms, nanomaterials can interact with tissues and organs, potentially leading to accumulation over time. Biomagnification, which is the increased concentration of substances through the food chain, is also a concern if nanomaterials are ingested by organisms at lower trophic levels.

31.3.4.4 Interactions with Biota

Nanomaterials can interact with different components of the ecosystems, including microorganisms, plants, and animals. These interactions can have both positive and negative effects. For example, nanomaterials may have antimicrobial properties that could be beneficial in certain applications, such as water treatment. However, they can also disrupt the balance of microbial communities, affect nutrient cycling, or interfere with the growth and development of aquatic animals.

31.3.4.5 Long-Term Effects and Unknowns

The long-term effects of nanomaterials on the environment are still being investigated. The potential for nanomaterials to persist in the environment and their cumulative effects over time are areas of ongoing research. Additionally, there may be nanomaterials with unique properties or compositions that have not yet been fully characterized, posing uncertainties about their environmental impact.

To minimize the environmental impact of nanomaterials, it is important to employ responsible and sustainable practices throughout their lifecycle, from synthesis and manufacturing to use and disposal. This includes implementing proper waste management strategies, considering the potential for exposure and release, and conducting thorough risk assessments. Regulatory frameworks and guidelines are being developed to ensure the safe and environmentally sound use of nanomaterials.

31.3.5 Regulatory Issues Regarding Labeling and Disclosure of Nanotechnology-Based Products

Nanotechnology-based products pose unique challenges when it comes to labeling and disclosure due to their novel properties and potential impacts on health and the environment (Sores et al. 2018; Amenta 2015). Regulatory bodies around the world have been working to establish guidelines and frameworks to address these issues (Allan et al. 2021). While the specific regulations may vary from country to country, an overview of the general regulatory considerations for labeling and disclosure of nanotechnology-based products has been depicted hereunder.

31.3.5.1 Product Labeling

Labeling requirements aim to provide consumers, workers, and regulators with information about the presence and characteristics of nanomaterials in products. Labels may include information about the presence of nanomaterials, their properties, potential risks, safe handling instructions, and disposal guidelines. The specific labeling requirements can vary depending on the jurisdiction and the product type.

31.3.5.2 Ingredient Disclosure

Some regulatory frameworks require manufacturers and importers to disclose the presence of nanomaterials in their products, either through ingredient lists or specific disclosure mechanisms. This information allows consumers and regulators to make informed decisions about product usage and potential risks.

31.3.5.3 Health and Safety Data

Regulatory authorities often require manufacturers to provide health and safety data related to nanomaterials used in their products. This can include toxicity studies, exposure assessments, and any relevant information on potential risks or protective measures. The disclosure of such data helps in assessing the safety of nanotechnology-based products.

31.3.5.4 International Harmonization

As nanotechnology is a global field, efforts are being made to harmonize regulatory approaches internationally. Organizations like the International Organization for Standardization (ISO) and the Organisation for Economic Co-operation and Development (OECD) work towards developing common principles, guidelines, and standards for labeling and disclosure of nanotechnology-based products.

31.4 Regulatory Organizations in Nanotechnology-Based Product Development

There are several regulatory organizations involved in the development and regulation of nanotechnology-based products. These organizations play a crucial role in ensuring the safety, efficacy, and appropriate use of such products (Allan et al. 2021; Foulkes et al. 2020). Here are some notable regulatory organizations involved in nanotechnology-based product development:

The U.S. Food and Drug Administration (FDA) is responsible for regulating the safety and efficacy of medical products, including nanotechnology-based drugs, devices, and diagnostics. The FDA provides guidance and evaluates nanotechnology-based products through its various centers, such as the Center for Drug Evaluation and Research (CDER) and the Center for Devices and Radiological Health (CDRH).

The European Medicines Agency (EMA) is responsible for the evaluation and authorization of medicinal products within the European Union (EU). It assesses the safety, quality, and efficacy of nanomedicines and provides regulatory guidance and support throughout the development process.

Health Canada is the regulatory authority for health products and food in Canada. It assesses and regulates nanotechnology-based health products and food, ensuring their safety and appropriate use. Health Canada provides guidance and requirements for pre-market assessment, labeling, and safety evaluation.

The National Institute for Occupational Safety and Health (NIOSH), a part of the Centers for Disease Control and Prevention (CDC) in the United States, is responsible for research and recommendations to prevent work-related injuries and illnesses. NIOSH evaluates and provides guidance on occupational safety and health issues related to nanotechnology, including exposure assessment and risk management.

The Environmental Protection Agency (EPA) in the United States regulates the use of nanomaterials that may have environmental implications. It assesses the potential risks posed by nanomaterials and develops regulations and guidelines for their safe use and disposal.

National Nanotechnology Initiative (NNI) is a U.S. federal research and development initiative that coordinates nanotechnology-related activities across various government agencies. While not a regulatory organization itself, the NNI collaborates with regulatory agencies to develop strategies, guidelines, and frameworks for the responsible development and regulation of nanotechnology.

The International Organization for Standardization (ISO) develops international standards that provide guidelines and best practices for the development, characterization, and safety assessment of nanotechnology-based products. These standards help ensure consistency, quality, and safety in the field of nanotechnology.

It's important to note that the regulatory landscape may vary between countries and regions, and new organizations or initiatives may emerge. It's advisable to refer to the specific regulatory bodies and guidelines relevant to the jurisdiction in which the nanotechnology-based product is being developed or marketed.

31.4.1 U.S. Food and Drug Administration (FDA) Guidelines and Regulations on Nanotechnology-Based Product Development

The U.S. Food and Drug Administration (FDA) has been actively involved in evaluating and regulating nanotechnology-based products, including those intended for use in the food and drug sectors. A general overview of the FDA's approach to nanotechnology regulation includes the following (U.S. Food & Drug Administration 2022, https://www.fda.gov/).

31.4.1.1 Safety Assessments

The FDA's primary concern is the safety of nanotechnology-based products. The agency evaluates the potential risks associated with the use of nanomaterials in various applications, such as drugs, medical devices, cosmetics, and food. The FDA examines the physicochemical properties of nanomaterials, their potential toxicity, and any unique characteristics that may affect safety.

31.4.1.2 Regulatory Framework

The FDA regulates nanotechnology-based products within its existing regulatory framework. Depending on the intended use and the specific product, different regulations may apply. For example, drugs and medical devices are subject to the Federal Food, Drug, and Cosmetic Act, while food products are regulated under the Federal Food, Drug, and Cosmetic Act and the Federal Food, Drug, and Cosmetic Act.

31.4.1.3 Pre-Market Approval

In most cases, nanotechnology-based products require pre-market approval before they can be legally marketed in the United States. The FDA assesses the safety and efficacy of these products through various regulatory pathways, such as new drug applications, biologic license applications, premarket notification (510(k)) for medical devices, and generally recognized as safe (GRAS) notifications for food.

31.4.1.4 Product-Specific Guidance

The FDA has issued specific guidance documents to provide recommendations on the development and regulation of nanotechnology-based products. These documents address various aspects, including the characterization of nanomaterials, manufacturing considerations, preclinical and clinical testing, labeling requirements, and post-market surveillance.

31.4.1.5 Collaboration and Research

The FDA actively collaborates with other government agencies, industry stakeholders, and scientific experts to advance the understanding and regulation of nanotechnology. The agency also conducts its own research to evaluate the potential risks and benefits associated with nanomaterials.

It's important to note that the field of nanotechnology is rapidly evolving, and regulatory guidelines and requirements are updated time-to-time.

31.4.2 Regulatory Guidelines by the European Medicines Agency (EMA) and the European Commission for Nanomedicine Preparation Standards

The European Medicines Agency (EMA) and the European Commission have provided general guidance on the regulatory aspects of nanomedicines within the framework of existing regulations for pharmaceutical products (European Medicines Agency (2023), https://www.ema.europa.eu/en; European Commission (2023), https://commission.europa.eu/index_en). These guidelines cover various aspects of nanomedicine development, characterization, and evaluation.

31.4.2.1 European Medicines Agency (EMA)

Reflection Paper on the Regulatory Guidance for the Development of Nanomedicines: The EMA has published a reflection paper that provides an overview of regulatory considerations for nanomedicines. It addresses the quality and the non-clinical and clinical aspects of nanomedicine development, including specific challenges and considerations for these products.

Guideline on the Quality, Non-Clinical, and Clinical Requirements for Investigational Medicinal Products Containing Nanomaterials: The EMA has issued a guideline that outlines the requirements for investigational medicinal products containing nanomaterials. It provides guidance on the characterization and control of nanomaterials, as well as considerations for non-clinical and clinical development.

General Guidelines on Pharmaceutical Development: The EMA has published general guidelines on pharmaceutical development that cover various aspects applicable to nanomedicines. These guidelines address topics such as pharmaceutical development, quality control, stability testing, and manufacturing processes.

31.4.2.2 European Commission

Regulation on Medical Devices (MDR) and *In Vitro Diagnostic Medical Devices (IVDR)*: The European Commission has established regulations on medical devices (MDR) and in vitro diagnostic medical devices (IVDR), which may be applicable to certain types of nanomedicines that fall under the definition of medical devices or in vitro diagnostic devices.

Guidelines on Safety Assessment of Nanomaterials in Cosmetics: The European Commission has issued guidelines on the safety assessment of nanomaterials used in cosmetic products. While these guidelines specifically target cosmetics, they provide useful information on the safety assessment of nanomaterials that can be relevant to nanomedicines.

31.4.3 Regulation of Nanomedicines by the Nanotechnology Characterization Laboratory of the National Cancer Institute (NCL-NCI)

The Nanotechnology Characterization Laboratory (NCL) of the National Cancer Institute (NCI) is responsible for the preclinical characterization and evaluation of nanomedicines intended for cancer treatment. The NCL-NCI plays a crucial role in assessing the safety, efficacy, and quality of various nanotechnology-based products, including nanomedicines (Nanotechnology Characterization Laboratory of the National Cancer Institute (NCL-NCI) 2022, https://www.cancer.gov/nano/research/ncl/about).

The NCL-NCI works closely with researchers, developers, and regulatory agencies to facilitate the translation of nanotechnologies from the laboratory to clinical applications. It provides comprehensive preclinical evaluation services to support the development of nanomedicines, helping to bridge the gap between basic research and clinical trials. The NCL-NCI follows a rigorous evaluation process to assess nanomedicines. This process includes:

Physicochemical Characterization: The NCL-NCI examines the physicochemical properties of nanomedicines, such as size, shape, surface charge, stability, and drug release characteristics. This evaluation helps understand how these properties may affect the behavior and performance of nanomedicine in biological systems.

In Vitro Testing: The NCL-NCI conducts in vitro studies to assess the nanomedicine's interactions with biological systems, including its effects on cell viability, cellular uptake, and intracellular fate. These tests help evaluate the nanomedicine's potential toxicity and efficacy.

In Vivo Testing: The NCL-NCI performs in vivo studies to evaluate the nanomedicine's pharmacokinetics (how the body absorbs, distributes, metabolizes, and excretes the drug), biodistribution (where the drug accumulates in the body), and toxicity in animal models. These tests provide critical data on the nanomedicine's safety profile and help identify any potential side effects.

Immunological Evaluation: The NCL-NCI assesses the nanomedicine's immunological effects, including its potential to induce an immune response or trigger adverse immune reactions.

The NCL-NCI collaborates with regulatory agencies such as the U.S. Food and Drug Administration (FDA) to establish regulatory standards and guidelines for nanomedicine development. Its evaluations and findings help inform regulatory decisions and ensure that nanomedicines meet the necessary safety and efficacy requirements before advancing to clinical trials.

31.4.4 Regulation of Nanomedicines by the European Nanomedicine Characterization Laboratory (EU-NCL)

The European Nanomedicine Characterization Laboratory (EU-NCL) is a pan-European infrastructure established to provide comprehensive characterization and evaluation of nanomedicines. The EU-NCL aims to support the translation of nanomedicines from the research stage to clinical applications by ensuring their safety, quality, and efficacy (European Nanomedicine Characterization Laboratory (EU-NCL) 2022, https://www.euncl.org/).

The EU-NCL operates under the framework of the European Commission's Horizon 2020 program and collaborates with academic institutions, research organizations, and industry partners across Europe. It offers a range of services and expertise in the characterization and preclinical evaluation of nanomedicines, following standardized protocols and methodologies.

The regulation of nanomedicines in the European Union (EU) involves multiple regulatory bodies and frameworks. The EU-NCL itself does not have regulatory authority, but it plays a crucial role in providing data and characterization information that can support regulatory decision-making.

The regulatory pathway for nanomedicines in the EU generally follows the existing regulatory framework for medicinal products, which includes the European Medicines Agency (EMA) and national regulatory agencies of individual EU member states. Nanomedicines are evaluated and authorized based on their safety, quality, and efficacy through a centralized, decentralized, or mutual recognition procedure.

The EU-NCL collaborates with regulatory agencies and provides data and characterization results to support the regulatory assessment of nanomedicines. Its role includes method development and standardization, physicochemical characterization of nanomedicines, in vitro and in vivo testing, safety and toxicology assessment, etc.

The EU-NCL's characterization and evaluation data are typically included in the regulatory submission dossiers of nanomedicines, helping to support the safety and efficacy claims made by the developers. Ultimately, the regulatory decisions regarding the authorization and approval of nanomedicines in the EU are made by the relevant regulatory authorities based on their assessments and evaluations.

31.4.5 General Guidance on Nanotechnology-Based Health Products and Food Issued by Health Canada

Health Canada, the federal department responsible for public health in Canada, has provided general guidance on nanotechnology-based health products and food. Health Canada recognizes the potential benefits and risks associated with nanotechnology and has issued guidance documents to ensure the safety and appropriate regulation of nanotechnology-based products (Health Canada 2023, https://www.canada.ca/en/health-canada.html). Some of the key aspects of their guidance include the following:

Definition of Nanomaterials: Health Canada defines nanomaterials as materials that have a size range typically between 1 and 100 nanometers and exhibit unique properties or phenomena due to their small size. This definition helps in determining whether a product falls under the category of nanotechnology-based products.

Pre-market Assessment: Nanotechnology-based health products and food are subject to pre-market assessment by Health Canada. Manufacturers are required to provide information on the characterization, safety, and efficacy of the nanomaterials used in their products.

Safety Assessment: Health Canada emphasizes the importance of evaluating the safety of nanotechnology-based products. This assessment involves the examination of potential hazards associated with nanomaterials, including their toxicity, potential for bioaccumulation, and environmental impact.

Labeling Requirements: Health Canada requires clear and accurate labeling of nanotechnology-based products to provide consumers with information about the presence of nanomaterials and any specific safety precautions or usage instructions.

Novel Foods: Health Canada's guidance also covers the assessment and regulation of nanotechnology-based food products, including novel foods. Manufacturers are required to submit data on the safety of nanomaterials used in food and provide evidence that their products do not pose risks to human health.

Collaboration and Information Sharing: Health Canada encourages collaboration and information sharing among stakeholders, including industry, academia, and other government agencies, to foster the safe development and use of nanotechnology-based products.

31.4.6 International Organization for Standardization (ISO) Guidelines, and Standards for Labeling and Disclosure of Nanotechnology-Based Products

The International Organization for Standardization (ISO) has developed several guidelines and standards related to nanotechnology-based products. These standards aim to provide a framework for the labeling and disclosure of such products to ensure the safety, quality, and appropriate information for consumers and other stakeholders (International Organization for Standardization 2023, https://www.iso.org/home.html).

One relevant ISO standard is ISO/TS 80004–1:2015, "Nanotechnologies—Vocabulary—Part 1: Core terms." Although this standard primarily focuses on establishing a common vocabulary for nanotechnology, it provides the foundational terminology used in other ISO nanotechnology standards, including those related to labeling and disclosure.

ISO/TS 80004–2:2015, "Nanotechnologies—Vocabulary—Part 2: Nano-objects" is another standard that provides definitions and descriptions of different nano-objects. Understanding the nature and characteristics of nano-objects is essential for appropriate labeling and disclosure.

ISO/TR 19601:2017, "Nanotechnologies—Compilation and description of material-specific functionalization and modification methods—Surface modification of inorganic nanoparticles" offers guidance on the labeling and characterization of surface-modified inorganic nanoparticles. It outlines the information that should be provided, such as the nature of surface modifications, potential hazards, and any specific handling instructions.

ISO/TS 13830:2013, "Nanotechnologies—Guidance on voluntary labelling for consumer products containing manufactured nano-objects" provides recommendations for the voluntary labeling of consumer products that contain manufactured nano-objects. It includes suggestions for information to be included on labels, such as the presence of nano-objects, their characteristics, and potential risks.

It's important to note that while these ISO standards provide guidance on labeling and disclosure, they may not be mandatory requirements in all jurisdictions. Different countries or regions might have their own regulations and standards specific to nanotechnology-based products. Therefore, it's advisable to consult local regulations and authorities to ensure compliance with relevant requirements.

31.4.7 Organisation for Economic Co-Operation and Development (OECD) Guidelines, and Standards for Labeling and Disclosure of Nanotechnology-Based Products

The Organisation for Economic Co-operation and Development (OECD) has also developed guidelines and standards for the labeling and disclosure of nanotechnology-based products. These guidelines provide a framework for ensuring the safe and responsible development, production, and use of nanomaterials (The Organisation for Economic Co-operation and Development, U.S. Environmental Protection Agency 2023, https://www.oecd.org/). Here are some relevant OECD documents in this context:

"OECD Guidelines for the Testing of Chemicals, Section 2: Effects on Biotic Systems": This document includes guidelines for the testing and assessment of nanomaterials, including guidance on how to conduct hazard assessments and safety evaluations for nanomaterials.

"OECD Safety Testing of a Representative Nanomaterial: OECD Sponsorship Programme": This document provides a framework for the safety testing of representative nanomaterials. It outlines the testing strategies and methods to be used for evaluating the hazards and risks associated with nanomaterials.

"OECD Series on the Safety of Manufactured Nanomaterials": This series of publications includes several documents that address the safety assessment and risk management of nanomaterials. These documents provide guidance on various aspects, such as physicochemical characterization, toxicological testing, exposure assessment, and risk assessment of nanomaterials.

"OECD Working Party on Manufactured Nanomaterials Sponsorship Programme for the Testing of Manufactured Nanomaterials: Detailed Review Paper on the State of the Science on Novel In Vitro Test Methods and Technologies for Genotoxicity and Carcinogenicity Testing of Nanomaterials": This paper focuses on the state of the science regarding in vitro test methods and technologies for evaluating the genotoxicity and carcinogenicity of nanomaterials.

These OECD guidelines and documents provide valuable guidance for the labeling and disclosure of nanotechnology-based products. They emphasize the importance of characterizing and assessing the potential hazards and risks associated with nanomaterials and promoting transparency in communicating relevant information to consumers and stakeholders.

It's important to note that these guidelines are not legally binding, but they serve as a reference for member countries and stakeholders in developing their own regulations and standards related to nanotechnology-based products.

31.4.8 Regulation and Guidelines of Nanomedicines by the National Institute for Occupational Safety and Health (NIOSH)

The National Institute for Occupational Safety and Health (NIOSH), which is part of the Centers for Disease Control and Prevention (CDC) in the United States, provides guidelines and recommendations related to occupational safety and health issues associated with nanotechnology, including the handling and use of nanomedicines. While NIOSH's primary focus is on worker safety, its guidance can also inform broader safety considerations (The National Institute for Occupational Safety and Health (NIOSH) Centers for Disease Control and Prevention 2023, https://www.cdc.gov/niosh/index.htm).

NIOSH has issued several documents that address nanotechnology and nanomaterials, including nanomedicines. These documents provide guidance on risk assessment, exposure control, and recommended practices to protect workers and minimize potential health risks associated with nanomaterials. Here are some key resources and initiatives:

Current Intelligence Bulletin (CIB) 65: "Occupational Exposure to Carbon Nanotubes and Nanofibers": This NIOSH publication provides information on the potential health risks of occupational exposure to carbon nanotubes and nanofibers. It includes recommendations for exposure assessment, engineering controls, personal protective equipment (PPE), and medical surveillance.

Nanotechnology Research Center (NTRC): NIOSH established the NTRC to conduct research and provide recommendations on occupational safety and health issues related to nanotechnology. The NTRC collaborates with other organizations and conducts studies to better understand potential hazards and develop guidance on the safe handling of nanomaterials, including those used in nanomedicines.

NIOSH Recommended Exposure Limit (REL): Although NIOSH does not currently have specific recommended exposure limits for nanomaterials, it actively conducts research to assess the potential health effects of nanomaterial exposure. This research informs the development of guidelines and best practices.

NIOSH Nanotechnology Workplace Design Solutions: NIOSH offers guidance on engineering controls and workplace design to reduce nanomaterial exposure risks. This includes recommendations on ventilation systems, containment strategies, and the use of appropriate equipment to minimize exposure during the handling, production, and disposal of nanomedicines.

31.4.9 Regulation and Guidelines of Nanomedicines by the Environmental Protection Agency (EPA)

The Environmental Protection Agency (EPA) in the United States plays a role in assessing and regulating certain aspects of nanotechnology, including nanomedicines, with a particular focus on potential environmental implications (U.S. Environmental Protection Agency (EPA) 2023, https://www.epa.gov/). The EPA's regulatory activities related to nanomaterials and nanomedicines primarily fall under the Toxic Substances Control Act (TSCA) and the Federal Insecticide, Fungicide, and Rodenticide Act (FIFRA). Here are some key points regarding the regulation and guidelines of nanomedicines by the EPA:

Risk Assessment: The EPA evaluates the potential risks and impacts of nanomaterials, including nanomedicines, on human health and the environment. This assessment includes examining the fate, transport, exposure, and potential toxicity of nanomaterials when released into the environment.

Reporting and Recordkeeping: Companies that manufacture or import nanomaterials, including those used in nanomedicines, may be subject to certain reporting and recordkeeping requirements under TSCA. This helps the EPA gather information on the production, use, and potential risks associated with nanomaterials.

Pesticide Regulation: Nanotechnology-based products, such as nanomedicines with pesticidal properties, are subject to regulation under FIFRA. The EPA evaluates the efficacy and safety of nanomaterials used in pesticidal products to protect human health and the environment.

Nanoscale Materials Stewardship Program (NMSP): The EPA initiated the NMSP to encourage companies that manufacture or use nanomaterials, including nanomedicines, to voluntarily provide information on their production, use, and potential health and environmental effects. This program helps the EPA gather data to better understand nanomaterials and their potential risks.

Collaboration and Research: The EPA collaborates with other government agencies, academic institutions, industry stakeholders, and international organizations to support research, develop testing methods, and establish best practices for the safe handling and regulation of nanomaterials.

31.4.10 Regulation and Guidelines of Nanomedicines by the National Nanotechnology Initiative (NNI)

The National Nanotechnology Initiative (NNI) in the United States is not a regulatory organization but a federal research and development initiative that coordinates nanotechnology-related activities across various government agencies. While the NNI does not have direct regulatory authority, it plays a significant role in developing strategies, guidelines, and frameworks to support the responsible development and regulation of nanotechnology, including nanomedicines (National Nanotechnology Initiative (NNI) 2023, https://www.nano.gov/). Here are some key aspects of the NNI's involvement:

Interagency Coordination: The NNI facilitates coordination and collaboration among federal agencies involved in nanotechnology research and regulation. It helps align efforts, share information, and promote consistency in the understanding and evaluation of nanomedicines.

Development of Standards and Best Practices: The NNI promotes the development of standards and best practices for the characterization, safety assessment, and regulation of nanotechnology-based products, including nanomedicines. These standards help ensure quality, consistency, and safety in the field.

Responsible Development and Use: The NNI emphasizes the importance of responsible development and use of nanotechnology, taking into account potential risks and ethical considerations. It encourages the integration of responsible practices throughout the lifecycle of nanomedicines, including research, manufacturing, and disposal.

Environmental, Health, and Safety Research: The NNI supports research efforts to better understand the potential environmental, health, and safety implications of nanotechnology, including nanomedicines. This research helps inform regulatory decisions and the development of guidelines.

Public Engagement and Education: The NNI promotes public engagement and education on nanotechnology and nanomedicines. It seeks to enhance understanding among the general public, stakeholders, and policymakers about the benefits, risks, and regulations associated with nanotechnology-based products.

It's important to note that while the NNI contributes to the development of guidelines and frameworks, the primary regulatory authorities responsible for the regulation of nanomedicines are organizations such as the U.S. Food and Drug Administration (FDA) and the Environmental Protection Agency (EPA). The NNI's role is to provide coordination, support research, and promote best practices in the field of nanotechnology, including nanomedicines.

31.5 Conclusions and Future Outlook of Nanomedicine Regulations

Nanomedicine is a rapidly advancing field that combines nanotechnology with medicine to develop innovative diagnostic, therapeutic, and preventive approaches. As nanomedicine continues to evolve, it is crucial to establish appropriate regulations to ensure the safety and efficacy of these novel technologies. The salient aspects of nanomedicine regulations include safety and efficacy assessment, risk assessment, evolving regulatory frameworks to address the specific challenges posed by nanoscale materials, international cooperation, and collaborative efforts establish consistent standards, facilitate information sharing, and avoid unnecessary duplication of regulatory efforts.

The major perspectives regarding the future outlook of nanomedicine regulations will certainly revolve around the following crucial aspects:

1. *Risk-based regulations*: Nanomedicine regulations will likely continue to evolve toward a risk-based approach. This approach would involve tailoring regulatory requirements based on the specific characteristics and intended use of the nanomedicine product, considering both the benefits and potential risks associated with its use.
2. *Pre-market approval*: As the field progresses, regulatory agencies may require pre-market approval for certain nanomedicine products to ensure their safety and efficacy. This approval process would involve rigorous evaluation of preclinical and clinical data, including specific considerations for nanoscale materials.
3. *Characterization and standardization*: Developing standardized methods for characterizing nanoparticles and assessing their biological effects will be crucial. Establishing reference materials, measurement techniques, and analytical tools will enable consistent evaluation of nanomedicine products and facilitate regulatory decision-making.
4. *Post-market surveillance*: Continuous monitoring of nanomedicine products after they enter the market will become increasingly important. Post-market surveillance programs can identify any emerging safety concerns, enable prompt regulatory actions, and ensure ongoing patient safety.
5. *Ethical and societal implications*: Nanomedicine regulations should also address ethical and societal implications, such as privacy, informed consent, and equitable access to these advanced technologies. Future regulations may need to consider these broader issues to ensure responsible development and deployment of nanomedicine.

In summary, nanomedicine regulations are crucial for ensuring the safety, efficacy, and responsible development of nanomedicine products. The future will likely involve risk-based approaches, pre-market approvals, standardized characterization methods, post-market surveillance, and considerations for ethical and societal implications. Continued collaboration among regulatory agencies, researchers, and industry stakeholders will be vital to develop effective regulatory frameworks that keep pace with the advancements in nanomedicine.

References

Allan J, Belz S, Hoeveler A, Hugas M, Okuda H, Patri A, Anklam E (2021) Regulatory landscape of nanotechnology and nanoplastics from a global perspective. Regul Toxicol Pharmacol 122: 104885. https://doi.org/10.1016/j.yrtph.2021.104885

Amenta V (2015) Karin Aschberger, 18. https://doi.org/10.1016/j.yrtph.2015.06.016

Borm PJ, Robbins D, Haubold S, Kuhlbusch T, Fissan H, Donaldson K, Oberdorster E (2006) The potential risks of nanomaterials: a review carried out for ECETOC. Part Fibre Toxicol 3(11): 1–35. https://doi.org/10.1186/1743-8977-3-11

Ray PC, Yu H, Fu PP (2009) Toxicity and environmental risks of nanomaterials: challenges and future needs. J Environ Sci Health C Environ Carcinog Ecotoxicol Rev 27(1):1–35. https://doi.org/10.1080/10590500802708267

Đorđević S, Gonzalez MM, Conejos-Sánchez I, Carreira B, Pozzi S, Acúrcio RC, Vicent MJ (2021) Current hurdles to the translation of nanomedicines from bench to the clinic. Drug Deliv Transl Res 12(3):500–525. https://doi.org/10.1007/s13346-021-01024-2

Dunphy Guzmán KA, Taylor MR, Banfield JF (2006) Environmental risks of nanotechnology: national nanotechnology initiative funding, 2000–2004. Environ Sci Technol 40(5):1401–1407. https://doi.org/10.1021/es0515708

European Commission (2023). https://commission.europa.eu/index_en

European Medicines Agency (2023). https://www.ema.europa.eu/en

Foulkes R, Man E, Thind J, Yeung S, Joy A, Hoskins C (2020) The regulation of nanomaterials and nanomedicines for clinical application: current and future perspectives. Biomater Sci 8(17): 4653–4664. https://doi.org/10.1039/d0bm00558d

Health Canada (2023). https://www.canada.ca/en/health-canada.html

International Organization for Standardization (2023). https://www.iso.org/home.html

Kolesnikov S, Timoshenko A, Minnikova T, Tsepina N, Kazeev K, Akimenko Y, Singh RK (2021) Impact of metal-based nanoparticles on cambisol microbial functionality, enzyme activity, and plant growth. Plan Theory 10(10):2080. https://doi.org/10.1002/etc.4147

Miernicki M, Hofmann T, Eisenberger I, von der Kammer F, Praetorius A (2019) Legal and practical challenges in classifying nanomaterials according to regulatory definitions. Nat Nanotechnol 14(3):208–216. https://doi.org/10.1038/s41565-019-0396-z

Nanotechnology Characterization Laboratory of the National Cancer Institute (NCL-NCI) (2022). https://www.cancer.gov/nano/research/ncl/about

National Nanotechnology Initiative (NNI) (2023). https://www.nano.gov/

Oberdorster G, Maynard A, Donaldson K, Castranova V, Fitzpatrick J, Ausman K, Yang H, Group, ArftIRFRSINTSW (2005) Principles for characterizing the potential human health effects from exposure to nanomaterials: elements of a screening strategy. Part Fibre Toxicol 2(1):8. https://doi.org/10.1186/1743-8977-2-8

Sores S, Sousa J, Pais A, Vitorino C (2018) Nano medicine: principles, properties and regulatory issues. Front Chem 6:1–15. https://doi.org/10.3389/fchem.2018.00360

The National Institute for Occupational Safety and Health (NIOSH), Centers for Disease Control and Prevention (2023). https://www.cdc.gov/niosh/index.htm

The Organisation for Economic Co-operation and Development (OECD) (2023). https://www.oecd.org/

U.S. Environmental Protection Agency (2023). https://www.epa.gov/

U.S. Food & Drug Administration (2022). https://www.fda.gov/

Van Norman GA (2016) Drugs, devices, and the FDA: part 1: an overview of approval processes for drugs. JACC Basic Transl Sci 1(3):170–179. https://doi.org/10.1016/j.jacbts.2016.03.002

SPRINGER NATURE

GPSR Compliance

The European Union's (EU) General Product Safety Regulation (GPSR) is a set of rules that requires consumer products to be safe and our obligations to ensure this.

If you have any concerns about our products, you can contact us on ProductSafety@springernature.com

In case Publisher is established outside the EU, the EU authorized representative is:

Springer Nature Customer Service Center GmbH
Europaplatz 3
69115 Heidelberg, Germany

The manufacturer's authorised representative in the EU is Springer Nature Customer Service Centre GmbH, Europaplatz 3, 69115 Heidelberg, Germany. If you have any concerns regarding our products, please contact ProductSafety@springernature.com

Printed and bound by CPI Group (UK) Ltd, Croydon, CR0 4YY
01/10/2026
02225862-0017